Element	Symbol	Atomic Number	Relative Atomic Mass[a,b]
actinium	Ac	89	(227)
aluminum	Al	13	26.981538
americium	Am	95	(243)
antimony	Sb	51	121.760
argon	Ar	18	39.948
arsenic	As	33	74.92160
astatine	At	85	(210)
barium	Ba	56	137.327
berkelium	Bk	97	(247)
beryllium	Be	4	9.012182
bismuth	Bi	83	208.98038
bohrium	Bh	107	(262)
boron	B	5	10.811
bromine	Br	35	79.904
cadmium	Cd	48	112.411
calcium	Ca	20	40.078
californium	Cf	98	(251)
carbon	C	6	12.0107
cerium	Ce	58	140.116
cesium	Cs	55	132.90545
chlorine	Cl	17	35.4527
chromium	Cr	24	51.9961
cobalt	Co	27	58.933200
copper	Cu	29	63.546
curium	Cm	96	(247)
dubnium	Db	105	(262)
dysprosium	Dy	66	162.50
einsteinium	Es	99	(252)
erbium	Er	68	167.26
europium	Eu	63	151.964
fermium	Fm	100	(257)
fluorine	F	9	18.9984032
francium	Fr	87	(223)
gadolinium	Gd	64	157.25
gallium	Ga	31	69.723
germanium	Ge	32	72.61
gold	Au	79	196.96655
hafnium	Hf	72	178.49
hassium	Hs	108	(265)
helium	He	2	4.002602
holmium	Ho	67	164.93032
hydrogen	H	1	1.00794
indium	In	49	114.818
iodine	I	53	126.90447
iridium	Ir	77	192.217
iron	Fe	26	55.845
krypton	Kr	36	83.80
lanthanum	La	57	138.9055
lawrencium	Lr	103	(262)
lead	Pb	82	207.2
lithium	Li	3	6.941
lutetium	Lu	71	174.967
magnesium	Mg	12	24.3050
manganese	Mn	25	54.938049
meitnerium	Mt	109	(266)
mendelevium	Md	101	(258)
mercury	Hg	80	200.59
molybdenum	Mo	42	95.94
neodymium	Nd	60	144.24
neon	Ne	10	20.1797
neptunium	Np	93	(237)
nickel	Ni	28	58.6934
niobium	Nb	41	92.90638
nitrogen	N	7	14.00674
nobelium	No	102	(259)
osmium	Os	76	190.23
oxygen	O	8	15.9994
palladium	Pd	46	106.42
phosphorus	P	15	30.973761
platinum	Pt	78	195.078
plutonium	Pu	94	(244)
polonium	Po	84	(209)
potassium	K	19	39.0983
praseodymium	Pr	59	140.90765
promethium	Pm	61	(145)
protactinium	Pa	91	231.03588
radium	Ra	88	(226)
radon	Rn	86	(222)
rhenium	Re	75	186.207
rhodium	Rh	45	102.90550
rubidium	Rb	37	85.4678
ruthenium	Ru	44	101.07
rutherfordium	Rf	104	(261)
samarium	Sm	62	150.36
scandium	Sc	21	44.955910
seaborgium	Sg	106	(263)
selenium	Se	34	78.96
silicon	Si	14	28.0855
silver	Ag	47	107.8682
sodium	Na	11	22.989770
strontium	Sr	38	87.62
sulfur	S	16	32.066
tantalum	Ta	73	180.9479
technetium	Tc	43	(98)
tellurium	Te	52	127.60
terbium	Tb	65	158.92534
thallium	Tl	81	204.3833
thorium	Th	90	232.0381
thulium	Tm	69	168.93421
tin	Sn	50	118.710
titanium	Ti	22	47.867
tungsten	W	74	183.84
uranium	U	92	238.0289
vanadium	V	23	50.9415
xenon	Xe	54	131.29
ytterbium	Yb	70	173.04
yttrium	Y	39	88.90585
zinc	Zn	30	65.39
zirconium	Zr	40	91.224
[110]		110	(269)
[111]		111	(272)
[112]		112	(277)

[a]Source: Commission on Atomic Weights and Isotopic Abundances, International Union of Pure and Applied Chemistry, *Pure and Applied Chemistry*, Vol. 70, 237–257 (1998). © 1998 IUPAC

[b]Values in parentheses give the mass number of the radioactive isotope with the longest half-life.

P9-EDC-817

Principles
of
Chemistry

Principles
of
Chemistry

MICHAEL MUNOWITZ

W • W • Norton & Company
New York • London

Editor: Joseph Wisnovsky
Consulting Editor: Alexander Pines
Project Editor: Mary Kelly
Copy Editor: Susan Middleton
Associate Managing Editor: Jane Carter
Director of Manufacturing: Roy Tedoff
Editorial Assistants: Shon Schooler, Peter Wei
Text Design: Jack Meserole
Illustrations: George Kelvin, John McAusland
Layout: Susan Crooks

The text and display of this book are composed in Perpetua
Composition by TSI Graphics
Manufacturing by Quebecor, Hawkins

First Edition

Library of Congress Cataloging-in-Publication Data

Munowitz, M. (Michael)
    Principles of chemistry / Michael Munowitz.
        p.   cm.
    Includes index.
    ISBN 0-393-97288-7
    1. Chemistry.   I. Title.
QD33.M984   2000
540—dc21

W. W. Norton & Company, Inc., 500 Fifth Avenue, New York, N.Y. 10110
http://www.wwnorton.com

W. W. Norton & Company Ltd., 10 Coptic Street, London WC1A 1PU

3 4 5 6 7 8 9 0

The chief problem in every science is that of endeavouring to arrange and collate the numerous individual observations and details which present themselves, in order that they may become part of one comprehensive picture.

—MAX PLANCK

BRIEF CONTENTS

PREFACE   xxiii

Introduction   1

1. Fundamental Concepts   13

2. Atoms and Molecules   43

3. Prototypical Reactions   75

4. Light and Matter—Waves and/or Particles   103

5. Quantum Theory of the Hydrogen Atom   145

6. Periodic Properties of the Elements   177

7. Covalent Bonding and Molecular Orbitals   217

8. Some Organic and Biochemical Species and Reactions   261

9. States of Matter   297

10. Macroscopic to Microscopic—Gases and Kinetic
    Theory   351

11. Disorder–Order and Phase Transitions   393

12. Equilibrium—The Stable State   435

13. Energy, Heat, and Chemical Change   461

14. Free Energy and the Direction of Change   489

15. Making Accommodations—Solubility and Molecular Recognition   527

16. Acids and Bases   569

17. Chemistry and Electricity   607

18. Kinetics—The Course of Chemical Reactions   643

19. Chemistry Coordinated—The Transition Metals and Their Complexes   685

20. Spectroscopy and Analysis   721

21. Worlds Within Worlds—The Nucleus and Beyond   751

APPENDIX A. Nomenclature and Vocabulary   A1

APPENDIX B. Pertinent Mathematics   A33

APPENDIX C. Data   A59

APPENDIX D. Glossary   A103

ANSWERS TO SELECTED EXERCISES   A183

CREDITS   A199

INDEX   A201

CONTENTS

PREFACE   xxiii

Introduction   1

1. Scale and Scope   1
2. The Basics: Force and Motion   4
3. Atoms, Molecules, and Phases   6
4. Equilibrium, Thermodynamics, and Kinetics   11

1. Fundamental Concepts   13

1-1. Science and the Scientific Method   13
1-2. Space and Time   15
1-3. Motion: Changes in Space and Time   16
1-4. Force and Mass   18
1-5. Momentum   20
1-6. Fundamental Forces   21
1-7. Work and Energy   25
1-8. Measurement   33
     Units   34
     Unit Conversion   35
     Uncertainty, Accuracy, and Precision   37

Experimental Error and Significant Figures   39
Handling of Significant Figures   40
REVIEW AND GUIDE TO PROBLEMS   R1.1
EXERCISES   R1.16

2. Atoms and Molecules   43

2-1. The Existence of Atoms   43

2-2. The Elements   46

2-3. Chemical Combination   54

2-4. Molecules   60

2-5. Stoichiometry   66

REVIEW AND GUIDE TO PROBLEMS   R2.1

EXERCISES   R2.17

3. Prototypical Reactions   75

3-1. The Nature of Chemical Reaction   75

3-2. Acids, Bases, and Salts   79

3-3. Reduction and Oxidation   86

3-4. Dissolution and Precipitation   93

3-5. Radical Reactions   96

3-6. Catalysis   98

REVIEW AND GUIDE TO PROBLEMS   R3.1

EXERCISES   R3.17

4. Light and Matter—Waves and/or
   Particles   103

4-1. A New World   103

4-2. Light: Electromagnetic Waves   105
What, When, and Where   106
The Electromagnetic Spectrum   110
Diffraction and Interference   111

4-3. Matter: Particles of the Atom   119

4-4. Light as a Particle: Photons and the Photoelectric Effect   124

4-5. Matter as a Wave   129

Atomic Spectra   129

Standing Waves   130

De Broglie Wavelength   133

4-6. The Ideas of Quantum Mechanics   136

Probability   137

Indeterminacy   142

REVIEW AND GUIDE TO PROBLEMS   R4.1

EXERCISES   R4.14

5. Quantum Theory of the Hydrogen Atom   145

5-1. The Atom   145

5-2. The Wave Equation   146

5-3. Orbitals of a One-Electron Atom   148

Three Dimensions, Three Quantum Numbers   148

Sorting Out the Orbitals   152

Energy   153

The s Orbitals: Size and Shape   156

The p Orbitals: Shape   163

The p Orbitals: Orientation   165

The d Orbitals   166

5-4. Atomic Spectroscopy   168

5-5. A Fourth Quantum Number   172

REVIEW AND GUIDE TO PROBLEMS   R5.1

EXERCISES   R5.10

6. Periodic Properties of the Elements   177

6-1. Orbitals in a Many-Electron Atom   178

The Pauli Exclusion Principle   181

Penetration and Shielding   182
Electron–Electron Repulsion   186

6-2. Electron Configurations: Building Up the Atoms   191
The First Shell: Hydrogen and Helium   192
The Second Shell: From Lithium to Neon   193
The Third Shell (s and p): From Sodium to Argon   194
d Orbitals: The Transition Elements   195
f Orbitals: Lanthanides and Actinides   198
Organization of the Periodic Table   200

6-3. Valence Properties and the Periodic Law   202
Across   203
Down   205
Atomic Size   207
Ionization Energy   210
Electron Affinity   213

REVIEW AND GUIDE TO PROBLEMS   R6.1

EXERCISES   R6.12

7. Covalent Bonding and Molecular Orbitals   217

7-1. From Atoms to Molecules   217
7-2. Diatomic Molecules   225
The p Orbitals: σ and π   229
Building Up   233
Heteronuclear Diatomic Systems   236

7-3. Bonds in Molecules   241
7-4. Molecular Geometry   245
Hybridization   247

7-5. Delocalization   253

REVIEW AND GUIDE TO PROBLEMS   R7.1

EXERCISES   R7.19

8. Some Organic and Biochemical Species
   and Reactions   261

   8-1. Why Carbon?   261

   8-2. Hydrocarbon Structure and Bonding   262

      Isomerism Among the Alkanes   263

      Double and Triple Bonds: A Review   271

      Geometric Isomerism   276

   8-3. Functional Groups   276

   8-4. Biopolymers   285

   8-5. Summary of Organic Reactions   293

   REVIEW AND GUIDE TO PROBLEMS   R8.1

   EXERCISES   R8.16

9. States of Matter   297

   9-1. The World Within   297

   9-2. Traditional Classifications   298

   9-3. Noncovalent Interactions   301

   9-4. Order and Disorder in Gases, Liquids, and Solids   308

   9-5. Symmetry: Crystals and Quasicrystals   312

   9-6. The Solid State   322

      Network Crystals   322

      Ionic Crystals   324

      Molecular Crystals   326

      Metallic Crystals   326

   9-7. Clusters   330

   9-8. Liquid Crystals   334

   9-9. Polymers   336

      Biopolymers   339

      Synthetic Polymers   344

   REVIEW AND GUIDE TO PROBLEMS   R9.1

   EXERCISES   R9.17

10. Macroscopic to Microscopic— Gases and Kinetic Theory   351

10-1. The Ideal Gas   351

10-2. Equation of State   353

Measurement of Pressure: Barometer and Manometer   354

Pressure: Boyle's Law and Compressibility   355

Boyle's Law in Action   361

Temperature: Charles's Law   362

Number of Moles: Avogadro's Law   365

Summing Up: The Ideal Gas Law   366

10-3. Kinetic Theory   369

The Statistics of Pressure   370

Energy, Temperature, and Motion   374

Collisions   378

Thermal Energy   379

Speed and Temperature   384

The Boltzmann Distribution   389

REVIEW AND GUIDE TO PROBLEMS   R10.1

EXERCISES   R10.16

11. Disorder–Order and Phase Transitions   393

11-1. Intermolecular Potential   393

11-2. Real Equations of State   399

11-3. Real Isotherms: Critical Temperature and Phase Transitions   404

11-4. Phase Equilibria   409

Vapor Pressure of a Pure Liquid   410

Vapor Pressure of a Solution   412

11-5. Phase Diagrams   420

11-6. A Survey of Disorder–Order Transitions   428

REVIEW AND GUIDE TO PROBLEMS   R11.1

EXERCISES   R11.20

12. Equilibrium—The Stable State   435

12-1. The Nature of Equilibrium   435

12-2. Energy, Entropy, and the Drive to Equilibrium: A Preview   437

12-3. Approaches to Equilibrium: A Preview   442

12-4. The Equilibrium Constant and the Law of Mass Action   446

12-5. Vapor Pressure and the Equilibrium Constant   449

12-6. Stressed Equilibria   452

12-7. Solving Problems Using the Equilibrium Constant   458

REVIEW AND GUIDE TO PROBLEMS   R12.1

EXERCISES   R12.23

13. Energy, Heat, and Chemical Change   461

13-1. Work and Heat   461

13-2. The First Law of Thermodynamics   467

13-3. Energy and Enthalpy   475

13-4. Thermochemical Computations   480

13-5. Exothermic and Endothermic Processes   483

REVIEW AND GUIDE TO PROBLEMS   R13.1

EXERCISES   R13.23

14. Free Energy and the Direction of Change   489

14-1. Statistical Inevitability   489

14-2. Distributions, Microstates, and Disorder   490

Positional Disorder   491

Energetic Disorder   497

14-3. Entropy   501

14-4. The Second Law of Thermodynamics   507

14-5. Free Energy   509

14-6. Standard Free Energy and Entropy   513

14-7. The Free Energy of Reaction   516

14-8. Free Energy and the Equilibrium Constant   519

14-9. Thermal Equilibrium and the Meaning of
Temperature   522

REVIEW AND GUIDE TO PROBLEMS   R14.1

EXERCISES   R14.20

15. Making Accommodations—Solubility
and Molecular Recognition   527

15-1. Thermodynamics: Prologue to Reaction   527

15-2. Solutes and Solutions   530

Order, Disorder, and Equilibrium   531

The Solubility Product   535

Selective Precipitation   537

Space, Time, and Equilibrium   540

Enthalpy and Entropy   543

Structure and Solubility   546

Hydrophilic and Hydrophobic Effects   550

15-3. Guests and Hosts   555

Molecular Recognition   558

Thermodynamics of Binding   564

REVIEW AND GUIDE TO PROBLEMS   R15.1

EXERCISES   R15.20

16. Acids and Bases   569

16-1. A Wandering Ion   569

16-2. Conjugate Acids and Bases   570

Strength and Weakness   572

Stabilization of a Conjugate Base   574

16-3. Aqueous Equilibria   577
  Acidic Solutions: pK_a and pH   577
  Basic Solutions: pK_b and pOH   580
  Autoionization of Water   581
  K_b for a Conjugate Base   583

16-4. Neutralization   584
  Strong Acid and Strong Base   585
  Weak Acid and Strong Base   589

16-5. Weak Acids, Conjugate Bases, and Buffers   592
  Weak Acid: Mostly HA   592
  Conjugate Base: Mostly A^-   594
  Buffers   595

16-6. Titration   601

REVIEW AND GUIDE TO PROBLEMS   R16.1

EXERCISES   R16.25

17. Chemistry and Electricity   607

17-1. Putting Reactions to Work   607

17-2. Pushing Electrons   609

17-3. Redox Redux   610

17-4. From Chemical Energy to Electrical Work   612

17-5. Completing the Circuit   616

17-6. Electrochemical Life and Death   621

17-7. Half-Reactions   624
  Standard Electrode Potentials   625
  Reductants and Oxidants   630

17-8. Charge and Mass: Balancing the Redox Equation   632
  Acidic Solution   633
  Basic Solution   636

17-9. Going Uphill: Electrolysis   637

REVIEW AND GUIDE TO PROBLEMS   R17.1

EXERCISES   R17.22

18. Kinetics—The Course of Chemical Reactions   643

18-1. Of Potentiality and Actuality   643

18-2. Prerequisites   645

18-3. The Road to Equilibrium: A Macroscopic View   647

Rate of Reaction   648

Rates and Concentrations   650

Concentrations and Time   653

Rates and Temperature   656

18-4. Crisis   659

Collision Theory   661

Thermodynamics of the Transition State   665

18-5. Control: Thermodynamics, Kinetics, and Catalysis   669

18-6. Mechanism: Step by Step   674

Elementary Rate Laws   675

Rate-Determining Step   678

Pre-equilibria   680

At Equilibrium: A Detailed Balance   682

REVIEW AND GUIDE TO PROBLEMS   R18.1

EXERCISES   R18.27

19. Chemistry Coordinated—The Transition Metals and Their Complexes   685

19-1. Coordination Complexes   686

Structure and Coordination   689

The Metal   691

The Ligands: Electrons and Isomers   693

19-2. Bonding   700

Crystal Field Theory   700

Beyond Crystal Field Theory   709

19-3. Thermodynamics   713

Water: Ligand and Solvent   713

Aqueous Equilibria   714

Chelation   716

19-4. Kinetics   718

REVIEW AND GUIDE TO PROBLEMS   R19.1

EXERCISES   R19.17

20. Spectroscopy and Analysis   721

20-1. The Interaction of Light and Matter   722

Spectroscopic Vision   724

Probing Matter with Light   725

Spectroscopy: An Appreciation   740

20-2. Manipulations in Space and Time   743

Mass Spectrometry   744

Chromatography   747

REVIEW AND GUIDE TO PROBLEMS   R20.1

EXERCISES   R20.25

21. Worlds Within Worlds—The Nucleus and Beyond   751

21-1. Unfinished Business   752

21-2. Beginnings   753

21-3. Relativity and the Meaning of $E = mc^2$ 760

Space and Time   761

Energy and Mass   763

Matter and Antimatter   766

21-4. Nuclear Structure: Binding Energy and the Strong Force   767

21-5. Nuclear Reactions   773

Kinetics of Radioactive Decay   774

Mothers and Daughters   776

Building a Chemical Universe: Fusion   780

Splitting Up: Fission   782

Latter-Day Alchemy: Transmutation   785

21-6. Fields and Particles   785
    Interactions: The Medium and the Message   786
    Mesons and the Strong Force   788
    Quarks   789

21-7. Epilogue   792

REVIEW AND GUIDE TO PROBLEMS   R21.1

EXERCISES   R21.21

APPENDIX A. Nomenclature
               and Vocabulary   A1

A-1. Naming the Atoms   A2

A-2. Prefixes and Suffixes   A7
    Number of Atoms   A7
    Charge and Oxidation State   A8
    Arrangement   A10

A-3. Combining the Elements: Inorganic Systems   A11
    Ionic Compounds   A14
    Binary Molecular Compounds   A14
    Inorganic Acids   A15
    Coordination Complexes   A18

A-4. Root and Branch: Organic Molecules   A21
    Alkanes   A24
    Alkenes and Alkynes   A25
    Alcohols and Esters   A26

A-5. Last Words: Toward a Chemical Vocabulary   A27

EXERCISES   A29

APPENDIX B. Pertinent Mathematics   A33

B-1. Powers and Logarithms   A34
    Scientific Notation   A39
    Common Logarithms: Definition and Properties   A42
    Evaluation of Common Logarithms   A43
    Natural Logarithms   A45

B-2. Functions and Graphs   A47

B-3. Quadratic Equations   A50

EXERCISES   A55

APPENDIX C. Data   A59

C-1. Dimensions, Constants, and Symbols   A62

C-2. Elements   A67

C-3. Molecules and Interactions   A81

C-4. Equilibrium   A93

APPENDIX D. Glossary   A103

ANSWERS TO SELECTED EXERCISES   A183

CREDITS   A199

INDEX   A201

PREFACE

The wonder of the world is not its complexity, but its simplicity. Given enough color and canvas, anybody can make a mess; that, we do ourselves. More to admire is the artist who makes do with little, the artist whose art is to conceal an economy of form and design. That, nature does unsuspected—in a world hidden from the senses.

. . . a world (maybe a universe) teeming with life, billions of different forms, each unique. Yet all come alive using parts and pieces drawn from the same stock and built in the same style: the same molecules, the same reactions, the same laws.

. . . millions of different molecules, no two exactly alike, yet only dozens of atoms as their building blocks.

. . . dozens of different atoms, yet only a few elementary particles to make them all: the electron, proton, and neutron, most conspicuous; here and there some others, ghostly, hard to find.

. . . a few elementary particles, and just a few fundamental forces to bring them together: gravity, electricity and magnetism, nuclear binding and decay.

. . . three or four forces and, as capstone, a minimalist cosmic constitution to legislate their use: Article I, the laws of conservation and symmetry; Article II, the laws of blind chance and dumb luck.

There, I think, is the surprise and irony of the universe and also its incomparable beauty: that all complexity is an illusion; that nature's rich texture emerges from such a small bag of tricks; that things happen simply because they *can* happen and because they are statistically *most likely* to happen.

And there, too, is my view of chemistry—a happy blend of lofty principles and gritty practicality that, no matter how entangled it gets, remains awesomely simple at the core. "Seek simplicity" is the motto I urge on the reader, whether the apprehensive first-year student or the harried instructor, for the simple approach seems to me both the most pleasurable and practical.

Start small. Celebrate what molecules and reactions have in common, not what masks their essential unity. Build a foundation. Understand why a university may want to maintain separate departments of chemistry, physics, and biology, but understand also that elementary particles, atoms, molecules, forces, interactions, and the laws of nature respect no academic boundaries. Nature is one.

Go from the general to the specific, from the simple to the complex. Confident of the similarities, begin to explore the differences. Gradually deal with the exceptions that make the rules seem all the more powerful. To understand just the smallest bit is already to rejoice. Even before the scientific revolutions and technology of the 20th century, the chemist had entertainment enough for a lifetime.

To serve this union of lofty principles and gritty practicality, I offer here a similar mix of principles and practice. Each chapter in the body of the book is partitioned into a narrative core and a recapitulation/study guide in tutorial style. The narrative first part, something like a lecture, deals single-mindedly with the *ideas* of chemistry; the tutorial second part, like a college recitation or review session, goes to work on problems and solutions.

In the first part, there is the proverbial big picture: the nation's constitution, not its local ordinances. Ideas, images, analogies. Parallels, connections, appeals to reason. Readers will find scarcely any tables of data here (but, for those, look any time to the extensive collection in Appendix C). In the same way, expect to find few numerical examples or detailed procedures on how to, say, draw Lewis structures or compute formal charges or balance equations or any of the other skills necessary for a satisfactory study of general chemistry. But again, everything you want is ready and waiting in the second part of the chapter: step-by-step examples to illuminate the concepts, together with additional commentary and problems to solve.

The second part opens with an abridged yet substantial recapitulation of the first part, followed by a how-to-do-it treatment of a series of related exercises. Questions and answers in dialogue style, supplemented by brief comments and excursions, are interspersed throughout the worked examples. Treating the preceding chapter *as a whole*, this built-in study guide thus allows the reader to view the original exposition from a more informed perspective—closer to a brief exegesis than to a literal summary.

The recapitulation proceeds not bit by bit, spitting back every point in the same order as before, but rather it interprets and provides a fresh look. Some of the conceptual material is reemphasized; some is deemphasized; some is ignored. Supplementary material is sometimes added (especially for the aforementioned Lewis structures and formal charges and chemical equations), but all the technical apparatus is now presented to a more receptive reader: a reader less in the dark, a reader better able to deal with the numbers.

For the student, the intrachapter division between principles and practice evokes the traditional separation between lecture and recitation. And like most boundaries, this division is meant to be blurred and eventually erased. The opportunities to do so are many: If a nagging question arises during the lecture portion, why not skip to a relevant worked example in the study guide? The problems chosen are closely related to issues raised specifically in the first part, and the pages of the two parts are shaded differently for easy identification. Or just keep going, if you wish, and defer the question until later, giving the primary argument more time to unfold. Later, flip back to both parts of the chapter for help with the unsolved exercises. Before long, the line between principles and practice melts away on its own.

And nowhere does this line blur so completely—so critically—as in the exercises found at chapter's end. Many of them, perhaps the most valuable of the bunch, demand mostly repetition and drill; others, more challenging, try cautiously to embellish and extend the material. Either way, they should all be attempted. The exercises form a connected set, almost a second recapitulation of the themes developed in the chapter. They are intended to furnish ample opportunity for practice without bludgeoning the solver with their sheer number.

Consistent also with my stand on simplicity and the proper relationship between principles and practice, I have declined to include a rulebased explanation of chemical nomenclature anywhere in Chapters 1 through 21. Nomenclature is too important and too interesting a matter to be scattered throughout the book. The subject begs for a reasoned appreciation of what it is and what it is not. What it is: a systematic attempt to use language to mimic the regularity of nature. What it is not: a boring collection of rules that must be learned before one can have some fun.

Granted the status of a chapter, Appendix A therefore aims for a unitary treatment of the rationale behind chemical nomenclature. Readers should turn to it early and often, beginning just as the *ites* and *ates* and *whozis* and *whatzis* of chemical names start to intrude in Chapters 2 and 3. Sections and exercises dealing with inorganic compounds, coordination complexes, and organic molecules stand ready to accompany subsequent chapters.

On the subject of language, too, be sure to consult the Glossary (Appendix D) for a capsule definition of any unfamiliar term, together with a concise breakdown of the word's roots.

Aspects of mathematics, another kind of language, are reviewed in Appendix B, wherein readers will find help with those few operations needed for general chemistry: powers and logarithms, scientific notation, linear algebraic equations, quadratic equations. Calculus is used nowhere in the book, in keeping with the low level of mathematics adopted throughout. I prefer the idea to precede the equation, even in works more sophisticated than this one.

I do value data, though, and I respect the difficulties involved in measuring numbers with accuracy and precision. The more than 20 tables compiled in Appendix C can only hint at chemistry's vast experimental foundations, but at least they provide a start. Readers may scan the list of tables and consult them where relevant in a chapter. See the appendix itself for further comments.

I make these remarks for fear that readers without a map might get lost in such a large volume. Nothing I say, however, should be taken as a commandment on how best to read this book or any other. That question is for you to decide, not me. Proceed from cover to cover, in sequence, and you will follow just one of the routes possible: the one that *I* selected, out of necessity, because the chapters had to be numbered consecutively. This particular approach has its virtues, but there are other ways to do the same thing just as well. To get my personal view of the order of battle, read the Introduction and the opening sections of most chapters (particularly the early ones).

Your own approach may be better, and, if so, you should pursue it: insert, delete, go backward, go forward. Ample cross-referencing (and don't forget the index) should help readers over any rough spots that might crop up when taking chapters out of turn. For that, a few suggestions:

1. Treat the Introduction and Chapters 1 through 3 as a broad-brush sketch, useful for any subsequent pathway taken. The Lewis-style picture of atoms, molecules, bonding, and fundamental reactions remains, to this day, the picture that professional chemists conjure up most readily. The prequantum model is simple, remarkably sensible, and requires little prior background to grasp.

2. The discussions of quantum mechanics and bonding (Chapters 4 through 7) and equilibrium thermodynamics (Chapters 12 through 14) are reversible. But note that thermodynamics is best preceded by the material on gases (Chapter 10) and, to a lesser extent, phase transitions (Chapter 11). The liquid–vapor equilibrium in Chapter 11 is used to introduce the more general concept in Chapter 12.

3. The sequence dealing with states of matter and phase transitions (Chapters 9 through 11) also may be put after the overview (which ends in Chapter 3) and before the bonding module (which begins in Chapter 4).

4. Most of the material on spectroscopy (Chapter 20) is well suited to come any time after Chapter 7. Its position at the back of the book is to avoid slowing down the transition from the microscopic world of molecules (through Chapter 8) to the macroscopic world of bulk phases and energy flows (beginning in Chapter 9). That said, an awareness of thermal equilibrium (hinted at in Chapter 10 and fully developed in Chapter 14) can only enhance one's appreciation of spectroscopy.

5. The insertion of organic and biological chemistry (Chapter 8) directly after bonding is a reward to the reader: a panorama of applied quantum mechanics and chemical bonding, meant to advertise the most intricate molecular designs found anywhere in the universe. Again, as with spectroscopy, readers wishing to go from molecules to mixtures may pass over the organic chemistry in favor of Chapter 9 (states of matter).

6. Situated near the end, the chapter on coordination compounds (Chapter 19) offers a comprehensive look at not just bonding, but thermodynamics and kinetics of complex formation as well—a reward for reading the whole book. Almost as good is to read just the material on bonding, constituting the bulk of the chapter, directly after Chapter 7.

7. Solubility and molecular recognition (Chapter 15), acids and bases (Chapter 16), and electrochemistry (Chapter 17) may be taken in any order after thermodynamics. Of the three, Chapter 17 makes the heaviest use of the Gibbs free energy.

8. Those parts of kinetics (Chapter 18) dealing with rate laws, the time course of reactions, activation energies, and catalysis can be read any time—with the usual cautions. Full understanding of the difference between thermodynamics and kinetics, as well as transition-state theory, requires the Gibbs free energy.

9. The view of the nucleus in Chapter 21 reflects my wish to place a principled view of chemistry on a more general map of the world. For some readers the taste of relativity, $E = mc^2$, antimatter, fields, quarks, and color interactions may stimulate interest in a world even smaller than the atom (and to those readers, my apologies for a presentation oversimplified to the point of danger). Other readers, with other needs, will feel free to skip such material. The sections on radioactivity, kinetics and half-life, binding energy and mass defect, nuclear structure (particularly the shell model), fission, fusion, and transmutation stand on their own, independent of the less conventional topics.

ACKNOWLEDGMENTS

I would never have dreamed of writing this book were it not for Neil Patterson, who appeared out of nowhere one day and told me to get to work. For that, and for his strong advocacy and assistance during much of the project, I am profoundly grateful. To Ippy Patterson as well, my thanks for warm and sustained support.

Portions of the manuscript were read by nearly 30 reviewers, many of whom had constructive comments to offer. Particularly important were: Edward Samulski, who taught me about synthetic polymers and then shamed me into discussing them in Chapter 9; Maitland Jones, Jr., who clarified various subtle points concerning organic chemistry in Chapter 8; Henry Griffin, who convinced me to include number conservation in Chapter 21; and James Harrison, who suggested some salutary cuts in Chapter 14 and recommended also that I provide a summary of force and energy early on (Chapter 1). I should note that Professor Harrison, along with David Vezzetti, helped greatly over the years to shape my ideas about chemistry and physics.

The difficulty of turning several hundred rough sketches into polished pieces of art was a nagging concern, but my worries evaporated when the first drawings came in from George Kelvin. A scientific illustrator in the finest tradition, he combined functionality with a simple elegance to create the art we had all hoped for: art that teaches, not merely adorns. He also knows how to draw, a rare talent in today's world of computer graphics.

With a similar blend of functionality and elegance, Jack Meserole designed a book that allows a reader to cut through the clutter and get to the point. He gave us the clean lines, white space, and simplicity of layout that we felt was needed to carry the message.

Many others have contributed too, although space allows me to list only those with whom I have had direct contact. Morgan Ryan read the first 16 chapters with a developmental editor's keen eyes and helped improve the presentation in many places. Richard Prigodich, in addition to serving as a reviewer, independently solved all the end-of-chapter exercises and helped to produce the solutions manual. His input enabled me to correct a number of flaws in the exercises.

John McAusland prepared most of the charts and graphs displayed in the text, all the while having to comply with repeated changes of instructions. Susan Crooks, who laid out the book, ultimately brought the designer's principles to practical fruition, expertly (and patiently) fitting together the words and pictures on each page. Mary Kelly, project editor, and editorial assistants Peter Wei and Shon Schooler were invaluable to everyone involved, somehow managing to impose order on thousands of loose sheets, hundreds of drawings, and a seemingly unending stream of correspondence and last-minute revisions. Copyeditor Susan Middleton made many helpful observations throughout.

My special thanks and admiration go also to Marsha Courson and her colleagues at TSI Graphics, whose skill and attention to detail in composing this book cannot be overstated. And to Roy Tedoff and Jane Carter, who supervised all aspects of production at Norton, let me express my gratitude as well.

Phyllis Munowitz asked not to be acknowledged for all the work she has done, which includes typing, telephoning, photocopying, sorting, packing, shipping, receiving, proofreading, checking exercises, and personally recording gas chromatograms and mass spectra in the laboratory. I am therefore enjoined from even mentioning any of these activities.

Joseph Wisnovsky, an editor's editor, ran the show with grace, intelligence, and unfailing good taste. In all decisions he acted with a keen appreciation of both the needs of the book and the requirements of a demanding author, no mean feat. His stamp is all over the finished work.

When Joe agreed to take on the project, he told me that he wanted "to make a difference." I hope that we have.

I reserve acknowledgment of my greatest debt for last—to Alexander Pines, whose lectures at Berkeley showed me how one could bring intellectual excitement and rigor to the introductory chemistry course. Professor Pines contributed body and soul to this work, generously sharing material and ideas and, most of all, strengthening me with unwavering friendship and encouragement through hard times. Without his participation, the book would not be what it is today.

M. M.

REVIEWERS AND ADVISORS

John Alexander, University of Cincinnati
Rachel Austin, Bates College
Joseph BelBruno, Dartmouth College
Robert Cave, Harvey Mudd College
Regina Frey, Washington University (St. Louis)
Michael Fuson, Denison University
Cynthia Goh, University of Toronto
Henry Griffin, University of Michigan (Ann Arbor)
David Harris, University of California (Santa Barbara)
James Harrison, Michigan State University
Dudley Herschbach, Harvard University
Ronald Johnson, Emory University
Maitland Jones, Jr., Princeton University
Mark Jones, Vanderbilt University
Susan Kegley, University of California (Berkeley)
Colin MacKay, Haverford College
Jeffrey Mathys, Kenyon College
Cortlandt Pierpont, University of Colorado (Boulder)
Alexander Pines, University of California (Berkeley)
Richard Prigodich, Trinity College
Wallace Pringle, Wesleyan University
Edward Samulski, University of North Carolina (Chapel Hill)
James Skinner, University of Wisconsin (Madison)
Jonathan Smith, Bowdoin College
Wayne Steinmetz, Pomona College
Mark Thomson, Xavier University
James Walsh, John Carroll University
Daniel Williams, Kennesaw State College
Robert Williams, California State University (Fresno)

Principles
of
Chemistry

Introduction

1. Scale and Scope
2. The Basics: Force and Motion
3. Atoms, Molecules, and Phases
4. Equilibrium, Thermodynamics, and Kinetics

1. SCALE AND SCOPE

Central to any understanding of the physical world is one discovery of paramount importance, a truth disarmingly simple yet profound in its implications: *matter is not continuous*. Every substance will reveal a distinct and grainy structure if examined on a sufficiently fine scale. *All* material stuff is a collection of parts—some large and some small—integrated to form a coherently functioning and whole unit. Put simply, the business of science is to identify whatever building blocks are appropriate for a given system and to expose their structure and interactions. At stake is nothing less than an explanation of how the universe is put together, how it works, and where it might be going.

Start with the most parochial of viewpoints: our own physical selves, arguably the most complex assemblies of matter anywhere to be found. Advanced organisms display structure at the level of organs, tissues, cells, cellular organelles, molecules, atoms, and so on down to the most basic elementary particles. Which, we ask, is a fundamental building block and which is not? When is an electron too small and a cell too large?

Look to the complexity of the problem. If the question involves, say, a malfunctioning heart, then we might well adopt a large-scale picture of tissues and organs. But if the goal is to understand specifically how a blockage forms in a coronary artery, then we must descend ultimately to the atomic and molecular level, for it is there that such events occur.

1

Chemistry is one of the broadest and most utilitarian of the sciences, concerning itself with the structure and transformation of matter on a microscopic scale. Here the building blocks are atoms and molecules, with a special role assumed by the simplest of the subatomic particles— the electron. Here, too, is where most of the phenomena familiar to us actually play out, from biological reactions to atmospheric changes to drug metabolism to industrial processes.

Only by thinking small can we hope to grasp the large, for nature has a finer grain than one suspects. However big it seems, the world works in miniature. A bird in flight; a forest fire; the slow erosion of a cliff; the mass production of a synthetic chemical . . . each, in its own way, involves a transformation of atoms and molecules. And from small particles and small changes, little by little, nature shapes all that we perceive: from the invisible to the visible, from the **microscopic** to the **macroscopic**. Chemistry, with its disciplined way of observing and making sense of the natural world, provides the bridge between large and small.

The subject is vast, but not disjointed. There is an underlying unity and simple beauty to chemistry despite its superficial complexity and despite its wide-ranging excursions into other fields. Just a handful of physical laws seem to govern the behavior of matter, while a few basic models and questions outline the entire discipline. Aware of this unifying framework, we remain unintimidated by chemistry's ramification into dozens of areas: organic chemistry, biochemistry, inorganic chemistry, physical chemistry, analytical chemistry, theoretical chemistry, polymer chemistry, materials chemistry, organometallic chemistry, solid-state chemistry, geochemistry, medicinal chemistry, and on and on. They are all connected.

Every phenomenon need not be analyzed at the same level of detail, but everything does fall within a consistent physical picture. Ever practical, we soon learn to ask the right questions and to adopt a description suited to the size of the problem. To understand, for instance, the structure of a huge molecule such as DNA (Figure 1) surely requires a broader, more general approach than should be brought to bear on only a single atom. Yet the difference is mostly a question of focus, for it is the same universe with the same laws. At bottom, the two examples just cited are not as disparate as they first appear.

In the chapters that follow, we shall begin to discover the natural laws and investigate their application to chemical processes. Back and forth between macroscopic and microscopic we shall go, trying always to connect one to the other. Realize, right away, that it is a story without end, but also a story that can be appreciated at many different levels of interpretation—the simpler, the better. Mostly it is the still evolving picture of structure and transformation in the molecular world, where changes are small and consequences are big.

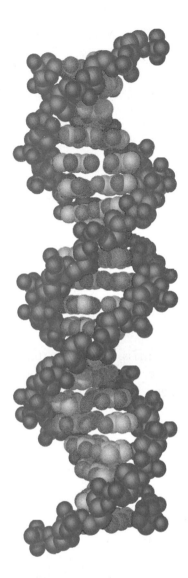

FIGURE 1. A small portion of deoxyribonucleic acid (DNA), repository of the genetic code. Millions of carbon, oxygen, hydrogen, nitrogen, and phosphorus atoms are bonded together and coiled into Watson and Crick's celebrated double helix. Hopelessly complex? Astonishingly, no: Every piece of the gigantic molecular puzzle is put together according to the same laws—every atom, every bond, every group of atoms, every twist and turn of the helix. Order and regularity prevail. (Chapters 8 and 9)

So rich indeed is this panorama of small change (even when framed by just a few fundamental laws) that the narrative cannot be told all at once. So much falls under the rubric "chemistry" that we should pause briefly to sketch out what lies ahead.

Let that be the remaining goal of this introduction: to preview, in the broadest of terms, the bare principles of chemistry before getting down to serious study. The skeletal outline to come is just the beginning, something like a road map showing the major thoroughfares. We shall return to all these ideas again and again, but in the meantime let us anticipate, quickly and informally, how the big pieces might eventually fit together.

2 . THE BASICS: FORCE AND MOTION

With what, then, do we begin? With apparently simple notions such as matter, mass, force, electric charge, and energy.

Matter is a catchall designation for physical material; matter is that which takes up space and has mass. A rock is a piece of matter, but a hope or dream is not.

Mass measures the quantity of matter, or simply how much material there is. More precisely, mass is a measure of *inertia*—the resistance of a material object to being moved. The more matter, the more effort is needed to move it or alter its speed.

Force is a push or a pull on an object, and from a force comes *motion*. Push something, and it moves. Gravity, for example, is an attractive force arising between any two objects possessing mass. When one of the objects is the earth, the force of gravity pulls the other mass toward the center of the planet and gives it a *weight*.

Electric charge is a property that, like mass, enables matter to interact by way of a corresponding force. Charge may be positive, negative, or neutral. Oppositely charged bodies attract each other, whereas like-charged bodies repel. Neutral particles are unaffected by electric forces.

Energy provides the ability to do *work*—to exert a force over some distance and thereby move a mass. (Or, if not actually to move the object right now, to store up a capability to do so at a later time.)

These ideas are both familiar and mysterious: as familiar as bouncing balls and electric appliances; as mysterious as the ultimate workings of the universe. What, exactly, is an electric charge? One can neither see nor taste it, yet charge acquires reality through the forces and motion it causes. Mass? Energy? The brief definitions given above are only hints, offered just as a start. Later we shall expand upon them as needed.

Even so, let nobody believe that such fundamental ideas can ever be wrapped up neatly with a few words. Some of the questions never go away. We can only relegate them to a deeper level and take a practical, operational approach to the subject at hand.

So where is the chemistry? It is, we shall see, bound up inextricably in force and motion, in mass and charge, in matter and energy. Tennis balls and electrons alike are tugged by the forces of nature, and the ensuing motion gives rise to the movable universe. Chemists, interested in how small things combine to produce big things, concentrate on the electrons rather than the tennis balls. The difference is in the details.

Electrons, having barely any mass, move under the influence of electric forces alone—not gravity. Tennis balls, moving as massive objects lacking an overall electric charge, are ruled by gravity. Still, the basic question is the same for both: Given the forces, what is the corresponding *law of motion*? How fast and in what direction does the object move?

Mechanical laws of motion govern the behavior of all matter. Forming the basis of **classical mechanics**, Newton's famous three laws describe the relationship between force, mass, and motion in precise mathematical and deterministic terms. By *deterministic* we mean that, starting at any instant of time, one can solve Newton's equations to determine the complete past and future path of a particle. There are only two prerequisites, both of which seem possible to satisfy: (1) knowledge of all the forces present and (2) knowledge of the initial positions and velocities of all the objects. In principle we know and can measure the forces (perhaps gravitational, perhaps electrical, perhaps something else), and we feel confident, moreover, in measuring positions and velocities to arbitrarily high accuracy. Newton's deterministic world—a world (Figure 2) that includes the solar system and ocean tides and tennis balls and figure skaters, a world where given enough information we

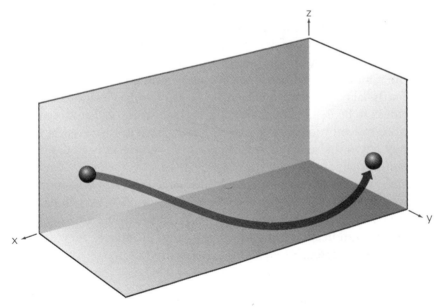

FIGURE 2. Newton's world of classical mechanics. Each particle follows a path blazed by a well-defined force acting on it at every instant. Past and future are forever determined, and there are no surprises. Specify, *now*, the whereabouts, speed, and direction of the object, together with the force it encounters, and we know all there is to know: where and when the particle once was; where and when it will someday be. True for planets. True for billiard balls. True for dust motes. Not true for electrons bound into atoms and molecules. (Chapters 1, 4 through 7)

can know all that has happened and all that will ever happen—explained all of physics for over two hundred years.

But Newton's world could not explain adequately the microscopic world of chemistry, in which subatomic particles subject to the usual electric forces do not behave according to the macroscopic laws of motion. Rather, what emerged in the first three decades of the 20th century was a revolutionary new picture called *quantum mechanics*. The new mechanics revealed the seemingly weird rules governing the microscopic world, and, at the same time, produced a model that reduced to Newton's familiar laws once objects attained a sufficient size.

If the quantum laws of motion, when we address them more fully later on, appear weird, it is only because our notion of common sense remains unrepentantly macroscopic. We take too much for granted, as with the casual statement made above concerning the exact measurement of positions and velocities. In the macroscopic world, accurate measurements are achieved by developing the right tools and using them carefully. When an object is very small, however, it happens that any measurement—no matter how gentle, no matter how finely crafted the equipment—is inevitably rocked by a disturbance. The act of measurement itself always pushes or pulls the object to some extent, frustrating any attempt to measure the position without simultaneously altering the velocity. The very notion of a well-defined path, and with it the tidy and deterministic universe of Newton, disappears with this *uncertainty principle* (or *indeterminacy principle*) of quantum mechanics.

The minimum disturbance is admittedly a small effect, barely felt by sufficiently large objects. "Sufficiently" large, though, is another one of those ideas to be defined operationally, and we shall see that electrons in atoms are indeed too insubstantial to have a path in the classical sense. They belong to the realm of the microscopic, subject fully to the laws of the quantum.

Understanding just how quantum mechanics describes the motion of electrons is one of the major unifying themes of chemistry. It explains why atoms are stable, why atoms have gaps in their energies, why atoms combine into molecules, why molecules have the shapes they do, why some substances are colored, and much more.

3. ATOMS, MOLECULES, AND PHASES

The fundamental particle of chemistry, the atom, is above all an electrical entity. At the center of an atom sits a small but heavy kernel called a *nucleus*, which carries a net positive charge and is itself an amalgam of positively charged *protons* and electrically neutral *neutrons*. Distributed around the nucleus are the lightweight *electrons*, each bearing a negative charge exactly opposite to that of a proton. In its native state, the atom contains just as many electrons as protons and therefore is electrically neutral. Addition or subtraction of any other electrons leaves a charged species known as an *ion*. Further, two atoms containing the same number of protons but different numbers of neutrons are termed *isotopes*.

Those are the parts; and from this small set of building blocks, pictured symbolically in Figure 3, nature puts together a chemical universe of inexhaustible variety. It begins with the atom. Pure substances composed of only one kind of atom (or, at most, various isotopes of the same atom) are called *elements*, examples of which include hydrogen, carbon, nitrogen, oxygen, and sodium. Each atom in an elemental substance has the same number of protons (*atomic number*), although the *mass number* (the combined total of protons and neutrons) will differ among the isotopes. Altogether, more than 100 elements are known either to exist in nature or to arise artificially through transformations in nuclear reactors and particle accelerators.

A macroscopic sample of a given element possesses measurable properties such as melting point, boiling point, and density. Certain attributes (mass and volume among them) are characteristic not of one atom but vary, instead, with the size of the sample and are said to be *extensive* properties. They derive from the "extent" of matter. *Intensive* properties, by contrast, are independent of total size and thus are specified

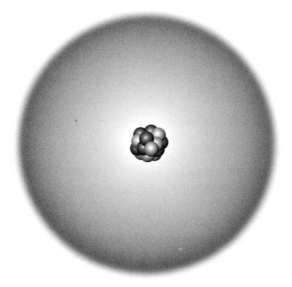

FIGURE 3. Generic nucleus and atom, not drawn to scale. Protons are distinguished from the uncharged neutrons by lighter shading. The large outer sphere (not large enough) represents the cloud of negatively charged electrons surrounding the tiny, fantastically dense, positively charged nucleus. Opposites attract. (Chapters 2 through 7, 20, 21)

from point to point. Density, the ratio of mass to volume, is such a property. A drop of water from the ocean contains the same mass as a drop from a small cup.

Much of chemistry resides in the electrons. Through the interaction of their outermost electrons, atoms come together to form ***molecules*** in a process of ***chemical bonding***. The result of the union typically is a new ***compound*** with properties different from any of its constituent elements. What we recognize as the material world is, in great measure, molecules: blood, hail, locusts, nearly everything.

Chemical bonding, important though it may be, is not a fundamental force in nature. The formation of molecules is governed entirely by ordinary electric forces under the laws of quantum mechanics. Nevertheless, bonding becomes such a key concept, so rich in variety, that it merits particular attention and helps define what we call chemical processes. Chemical interactions between atoms are recognized, according to this view, as a *sharing* of electrons by two or more nuclei. The stabilizing force that results is typically strong enough to allow molecules to exist under normal conditions of temperature and pressure.

When electrons are truly shared between nuclei, as in a molecule of water (Figure 4), the bonds are described qualitatively as ***covalent***. Another mechanism for very strong bonding, however, is through the point-to-point electric forces that bind freestanding ions into regularly spaced arrays. Such ***ionic bonds*** hold together positive sodium ions and

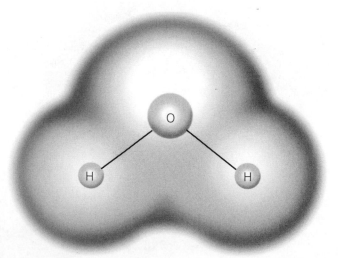

FIGURE 4. A molecule of water, H_2O. The hydrogen and oxygen nuclei, positively charged, cement themselves together through a quantum mechanical paste of negative electrons. The bonding is dubbed *covalent*, from the Latin for "shared strength." Opposites attract. (Chapters 2 through 8, 19)

negative chloride ions in crystals of sodium chloride, the primary component of ordinary table salt. See Figure 5.

There are, as well, weaker manifestations of electric forces *between* atoms and molecules (sometimes called "physical" interactions to distinguish them from the "chemical" interactions *within* a molecule). To disentangle a mixture of compounds into pure components, one then needs to address those properties that distinguish molecule from molecule. Among these attributes are size, shape, charge distribution, and so forth, from which come boiling point, melting point, density, and other macroscopic characteristics. Selective boiling allows us, for example, to refine crude oil into the various compounds found in gasoline and jet fuel, but to subdivide further we must break—with greater violence—the chemical bonds between atoms in a molecule.

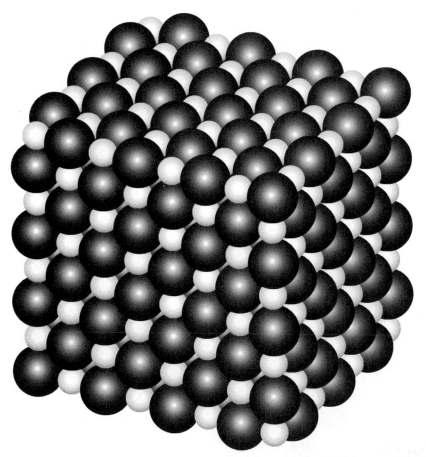

FIGURE 5. A crystal of sodium chloride, NaCl. Positive sodium ions (small spheres) and negative chlorine ions fall symmetrically into place, knitted together by a tight web of electrical interactions. Opposites attract. (Chapters 2, 3, 9, 15)

It is the weaker electrical interactions (those that do not lead to chemical bonding) that determine how atoms and molecules associate into aggregated *phases* or *states of matter*, forming solids, liquids, and gases along the lines shown in Figure 6. In the gaseous state, simplest of all, the constituent particles have little effect upon one another. Each behaves substantially independently of the rest while moving at high speed through a largely empty container. Whatever possibilities there are for interactions between molecules are minimized by rapid motion and by the infrequency of any prolonged intimate contacts. But circumstances change when the molecules are forced closer together, either by decreasing the temperature (which slows them down) or by squeezing on the gas. Allowed more opportunity to interact, each particle now must accommodate itself to its neighbors under the constraints imposed by the increasingly influential intermolecular forces.

Eventually something dramatic happens under the pressure: a change in state. A *phase transition* occurs as the gas condenses into a liquid, going from a completely disordered state to one of partial order. And there occurs yet another disorder–order transition, from liquid to

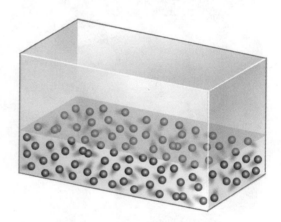

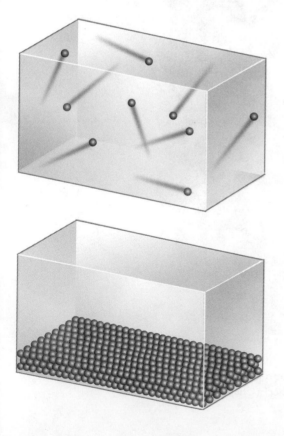

FIGURE 6. States of matter, clockwise from upper left: gas, liquid, solid. Once formed, atoms and molecules enter still further into looser associations among themselves—suggestive, perhaps, of the way individuals and families exist within a larger community. The extent of particle-to-particle association determines the ordering and properties of the aggregated state. Opposites attract. (Chapters 9 through 11, 15)

solid, as the molecule-to-molecule interactions become still more important at even higher pressures and lower temperatures.

With an appreciation of atoms, molecules, and their interactions, we shall come to view the most homely experiences of everyday life from a new perspective. Ordinary things—observations as familiar as snow melting, or steam rising, or the condensation and subsequent freezing of water vapor on a cold windshield—show themselves as simple, macroscopic manifestations of intense activity at the microscopic level.

4. EQUILIBRIUM, THERMODYNAMICS, AND KINETICS

The coexistence of any two phases immediately suggests the notion of *equilibrium*, a precise balancing of two opposing tendencies. In a liquid–gas system such as, say, water and steam (Figure 7), some molecules are always condensing into the liquid and some are evaporating into the gas. An equilibrium between condensation and evaporation is attained when the ongoing liquid-to-gas and gas-to-liquid transitions proceed at the same rate. The macroscopic behavior of the system then ceases to change. Competition at the microscopic level continues undiminished, but it remains a perfectly balanced competition and neither contender gains any additional advantage. The overall population of each phase stays constant even though individual molecules still move back and forth across the boundary.

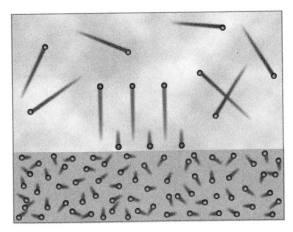

FIGURE 7. Equilibrium, a balancing of forces (from the Latin for "equal weight"): Some molecules do one thing (here, evaporate), and some molecules do precisely the opposite (condense); but at equilibrium they do so at the same rate. As many go one way as the other, and the competition ends in a draw. On the macroscopic outside, nothing changes. On the microscopic inside, activity never ceases. (Chapters 11 through 18)

So it goes for equilibria of all sorts, which may involve acids and bases, dissolved and undissolved solute, or, most generally, reactants and products in any transformation. An ability to characterize the equilibrium state quantitatively is crucial for the chemist, since the final proportions of products are absolutely fixed at a given temperature. Favorable or not, one needs to know how much product to expect and also how to obtain different amounts under different conditions.

The laws regulating equilibrium and change are provided by ***thermodynamics***, a thoroughly macroscopic and universal description of how energy converts itself into different forms. Energy powers the world. It enables muscles to contract and birds to fly; it moves mountains and motorcars, molecules and electrons. One particular measure of the energy contained within a system, a quantity we shall come to know as the ***free energy***, establishes a benchmark for all material changes. Its message: A transformation can proceed spontaneously only if the free energy is lowered, an outcome that is influenced heavily by the tendency to create greater disorder, or ***entropy***, in the universe. The change in free energy carries with it an immense quantitative significance, for it explicitly determines the final, unchanging state of the system: the state of equilibrium.

Yet if thermodynamics allows us to predict *what* will happen eventually, it is mute on the question of *when*. Given a mixture of hydrogen and oxygen, to take a well-known example, we can use thermodynamic laws to compute how much water will be present at equilibrium, along with how much hydrogen and oxygen will remain unreacted. But how long will the transformation take and by what route will it proceed? Just how, we ask, will the molecules smash together and rearrange their electrons and nuclei? Here the answers come from ***kinetics***, the study of rate and mechanism, which connects the speed and course of reaction with specific events on the microscopic level.

Standing before us, then, is the physical framework supporting all of chemistry: energy and motion, atoms and molecules, quantum mechanics, bonding, intermolecular interactions, phase transitions, thermodynamics, equilibrium, kinetics. Built on that scaffolding is an imposing edifice of chemical change, a limitless variety of reactions and transformations, of microscopic comings and goings. Some things we understand very well; others not so well. Research continues.

With this brief indication of the scope of chemistry—having deliberately suggested more questions than answers—we now begin in earnest.

1

Fundamental Concepts

1-1. Science and the Scientific Method
1-2. Space and Time
1-3. Motion: Changes in Space and Time
1-4. Force and Mass
1-5. Momentum
1-6. Fundamental Forces
1-7. Work and Energy
1-8. Measurement
   Units
   Unit Conversion
   Uncertainty, Accuracy, and Precision
   Experimental Error and Significant Figures
   Handling of Significant Figures
   REVIEW AND GUIDE TO PROBLEMS
   EXERCISES

1-1. SCIENCE AND THE SCIENTIFIC METHOD

The pursuit of science has been likened to watching a chess match without knowing the rules. Silently the observer looks on, seeking to learn how the game is played. The players know, but they aren't talking.

The game is nature, long the object of human inquiry and wonder. There is a belief, deeply rooted, that natural events conform to a pattern, and our goal has always been to decipher the code. We want to see how things are put together and thereby gain a measure of control over the world.

We look on. With disciplined **observation** we begin to recognize some of the pieces and how they move. Patterns and relationships start to appear, enabling us to guess at the rules. An educated guess, called a **hypothesis**, is tested against reality through an **experiment**: a limited, systematic manipulation of nature designed to confirm or deny a proposed explanation. Successful hypotheses might be woven into a broader picture of related occurrences, and the resulting **theory** may then be challenged further to predict new phenomena. Still there is doubt, though, even if the current challenge is met, because any theory is only as good as its last prediction. It is a temporary model of some part of nature, never the final word.

Observation, hypothesis, experiment, theory. Therein is our **scientific method**, but every child does the same thing. Small children reach out to probe their surroundings, developing personal theories through observation and experiment. Matter of a certain shape comes to be recognized as a chair, and soon the toddler learns to use such objects without detailed reconsideration at each encounter.

Science differs from child's play mostly in its more systematic approach, its discipline, and its reliance on measurement. Science is a quantitative business, concerned with numbers and exactness, and it deals only with events that can be treated accordingly. Not to be lost amidst the numbers, certainly, is a qualitative, nonnumerical understanding of nature, but ultimately something must be measured.

Everything we know comes from observation and measurement. For although the mechanism of nature is beautiful, it is not a beauty born of human design. Pursuing science is not like writing poetry; our ideas of beauty and of how things *ought* to be count for nothing. Lacking a full set of data, Aristotle mistakenly believed that the heavenly bodies move in circles. He liked the mathematical symmetry of circles, but he was wrong. Two thousand years elapsed before Kepler, Galileo, and Newton used quantitative observations first to describe and then to predict the elliptical orbits of the planets. With that giant stride, a mere 400 years ago, the modern age of science began.

Our initial task is to become better acquainted with some of the key players and rules of the game: matter and energy, mass, charge, force, momentum, and a few more. These concepts, touched upon lightly in the introductory chapter, are fundamental to nature in all her manifestations. What is the universe if not a swirl of matter and energy? All the natural sciences are studies of matter in motion, differing mainly in scale and focus. Each, in its own way, is concerned with the forces that move matter and with the special endowments (like mass and charge) that enable matter to move. We could watch a comet fly and call it astronomy. We could watch a fish swim and call it biology. We could watch photosynthesis occur in a green plant and call it chemistry. But always it is matter in motion, whether such matter be a planet or an electron. When the matter happens to be electrons and atoms and molecules—tossed about by the electric forces of the microworld—*that* we shall call chemistry.

Let us start to examine these most basic quantities, paying special attention to the demands of scientific measurement.

1-2. SPACE AND TIME

The events of chemistry unfold in space and time, prompting first the questions "where" and "when." *Where*, we ask, is the electron? How far is it from the nucleus? *When* did it appear here? When did it turn up over there?

"Where" means a position in space relative to some origin, an exact interval that can be compared to a standard **length**. A specially maintained bar of metal, for instance, once defined the **meter (m)**. The modern definition relies instead on the speed of light in a vacuum, but still the meter retains its length of approximately 3.28 feet.

Divided into 100 **centimeters** (**cm**; 2.54 cm = 1 inch, exactly), the meter marks off intervals in space: length in one dimension, **area** in two, and **volume** in three. The logical unit of area is the **square meter** (**m²**), which encompasses a square equal to 1 m on a side. Volume is

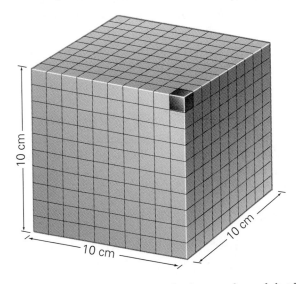

FIGURE 1-1. Space in three dimensions. The large cube, subdivided into 1000 smaller cubes, measures 10 centimeters on an edge and has a total volume of 1 liter. Each small cube, measuring 1 centimeter on an edge, has a capacity of 1 cubic centimeter; equivalently, 1 milliliter. To appreciate the scale, note that 1 centimeter extends over approximately the width of a fingernail (0.4 inch).

measured in ***cubic meters (m³)***, a unit indicating the space enclosed within a 1 m × 1 m × 1 m cubic box. Smaller units, particularly useful in chemistry, include both the ***liter (L)*** and also the ***milliliter (mL)*** or ***cubic centimeter (cm³, cc)***. One liter corresponds to a 10 cm × 10 cm × 10 cm cube, whereas one milliliter represents a 1 cm × 1 cm × 1 cm cube; 1 L = 1000 cm³. See Figure 1-1 on the preceding page.

"When" means a separation in time or, more precisely, a sequence of events by which one can distinguish "before" from "simultaneously" from "after." The standard unit is the familiar ***second (s)***, an interval approximately equal to 1/86,400 of the time required for the earth's daily rotation. A more exact definition relies on the frequency of light emitted by excited atoms of cesium.

Note that both meter and second belong to the International System of Units, abbreviated as ***SI*** from the French *Système International d'Unités*. A synopsis of SI units is provided in Appendix C.

1-3. MOTION: CHANGES IN SPACE AND TIME

Where and when; now what. *What* can be observed? What does it do? Let the "what" be a ***particle***, a generic piece of matter having unspecified characteristics beyond the basics of mass and charge. An object becomes a particle when its own extent in space has a negligible effect on the motion we perceive, making the actual size and shape of no immediate concern. Viewed from a distance, any chunk of matter becomes a particle—and all particles look the same. Sometimes an atom or molecule might behave as an undifferentiated particle (as might an electron, proton, or planet), while in other circumstances we shall need to treat every bit of matter as unique. For now, let us consider a particle that is macroscopic and easy to grasp: a small solid ball, gliding smoothly over an endless level surface with grid lines spaced one meter apart, as in Figure 1-2. We observe the motion, and keep time with a suitable clock.

Imagine the ball moving in a straight line, perpendicular to the grid, and passing one marker every second:

Time (s)	Distance Marker (m)
0	0
1	1
2	2
3	3

The particle's position is changing with time; it is in motion. The ball is advancing one meter for every second of travel, rolling along at a constant

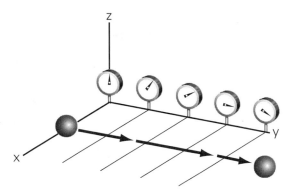

FIGURE 1-2. Space and time. The motion of a body is referred to a grid of coordinates centered about an arbitrary point of origin: $x = 0$, $y = 0$, $z = 0$. Two perpendicular dimensions (the x- and y-coordinates) cover all points in the x-y plane, while a third perpendicular dimension (z) registers the body's height above. A clock, by recording a particle's arrival at a given location, imparts a sense of "before" and "after" to any sequence of events.

speed of 1 m/s (or 1 m s^{-1}).* **Speed**, in general, measures the variation of position with time,

$$\text{Speed} = \frac{\text{distance traveled}}{\text{time elapsed}}$$

with no reference to direction. **Velocity**, a more complete description of the motion, is speed in a certain direction: 1 m s^{-1} *north*, or 1 m s^{-1} in the x direction, or 1 m s^{-1} perpendicular to the grid lines.

Velocity need not be constant. Suppose that, under some new set of conditions, we make the following measurements of time and distance:

TIME (s)	DISTANCE MARKER (m)
0	0
1	1
2	3
3	6
4	10

Now the ball covers one meter in the first second (1 m s^{-1}), two meters in the next second (2 m s^{-1}), three meters in the third second (3 m s^{-1}), and four meters in the fourth second (4 m s^{-1}), apparently *increasing* its velocity by one meter per second during each second of travel. Any

*Note that these two representations are equivalent and are used interchangeably.

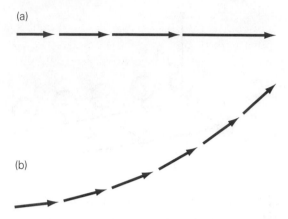

FIGURE 1-3. Acceleration, depicted as a sequence of velocities in time. The length of an arrow gives the particle's speed at a particular instant; the direction indicates its line of motion. There are two ways to accelerate: (a) Change the speed, but not the direction. (b) Change the direction, but not the speed. Either option, or both together, describes a change in velocity over time—a speeding up or slowing down, an acceleration.

change in velocity over time (either positive or negative) is called an *acceleration*, defined as

$$\text{Acceleration} = \frac{\text{change in velocity}}{\text{time elapsed}}$$

so that here we have a uniform acceleration of one meter per second *per* second: 1 m s^{-2}. The direction of travel remains the same in this example (Figure 1-3a), while the speed increases continually. Equally possible (Figure 1-3b) would be an acceleration during which a constant speed is maintained but the direction changes.

1-4. FORCE AND MASS

A particle can either sit still, move at a constant velocity, or accelerate; and from these options comes the question "why." *Why* does the ball move or not move, and why does the motion follow a particular path?

It moves because it is forced to move. The particle is pushed or pulled by *something*, and we call that agency a *force*. Left undisturbed, the ball keeps doing whatever it happens to be doing. If stationary, it remains stationary. If traveling at constant velocity, it continues along in the same direction at the same speed. It makes no changes. The ball possesses, according to *Newton's first law of motion*, the property of *inertia*—the tendency for a body to retain its current state of motion unless shoved, purposefully, this way or that. An object changes speed or direction only when prodded by a force, symbolized by the directed arrows in Figure 1-4.

Force must be applied if a particle is to overcome its inertia and thus alter its current velocity. A force (F) produces an acceleration (a) described by **Newton's second law,**

$$F = \text{mass} \times \text{acceleration} = ma$$

which he deduced through quantitative analysis of planetary motions. The force is directly proportional to the acceleration, with the constant of proportionality (m) equal to the mass. **Mass,** the quantity of matter, emerges now as a measure of inertia, or the natural tendency of a body to resist acceleration. The more matter, the more force is needed to produce a given acceleration. Someone or something must push harder to achieve the same effect.

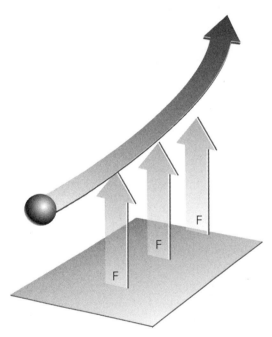

FIGURE 1-4. Newton's first two laws of motion. The first law says that motion tends to persist; or, put another way, that a body maintains its current speed and direction indefinitely unless otherwise provoked. The second law (F = ma) says that such provocation comes packaged as a *force*—an oriented push or pull, producing an acceleration in direct proportion. Inertia, resistance to change, thus becomes a consequence of mass: A large mass accelerates proportionately less than a smaller mass subjected to the same force.

The SI unit of mass is the **kilogram (kg)**, for which the smaller **gram** (**g**; 1 kg = 1000 g) is often used as a convenient substitute. A typical adult man has a mass of 70 kg (1 kg ≡ 2.2 lb); a five-cent coin has a mass of approximately 0.005 kg (5 g); a proton has a mass of 1.67×10^{-27} kg; an electron has a mass of 9.11×10^{-31} kg.

The SI unit of force is the **newton (N)**. The product of mass and acceleration, a newton has the dimensions kg m s^{-2}.

1-5. MOMENTUM

Force moves matter, but *how* might such pushes and pulls develop? What could induce either a stationary particle to move or a moving particle to change speed or direction?

The impetus can come, clearly, from some other particle already on the move. Simply by being in motion a particle acquires a **momentum**, p, taken as the product of mass and velocity ($p = mv$). This *impulse to move* can be transferred between bodies when one piece of matter collides with another, as in the encounter depicted in Figure 1-5. Two particles come together for some finite interval of time, during which there is an exchange of momentum; and from that transaction—that bump in time—comes the requisite push: a force.

See why. Inspecting the units, we find that any change in momentum over time has precisely the dimensions of a force (mass × velocity/time, or kg m s^{-2}):

$$\text{Force} = \frac{\text{change in momentum}}{\text{change in time}} = \frac{\text{change in (mass} \times \text{velocity)}}{\text{change in time}}$$

Since to be forced is, by Newton's definition, to accelerate, the motion changes when the second object acquires new momentum and a corresponding new velocity from the first. It happens all the time on a pool table.

We discover, in this relationship between force and momentum, an important lesson about the world. A force arises when there is a change in momentum over time. No change in momentum; no force. And, conversely, momentum and velocity change *only* if a force is applied. No force; no change in momentum.

The eight ball does not start moving on its own accord (just imagine such a world). Rather the cue ball hits the eight ball, and then both take off with new velocities. But the total momentum shared by the balls must be the same before and after the collision, because there are no external forces acting on the system. There is no sudden gust of wind to blow the balls away, no earthquake to shake up the table, and evidently

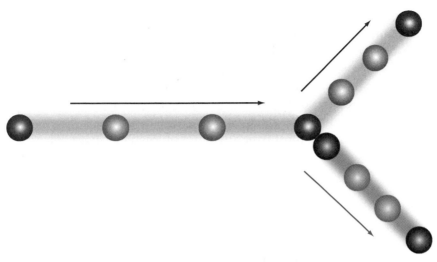

FIGURE 1-5. When particles collide: conservation of momentum. For every action, there is an equal and opposite *re*action (Newton's third law); and during every collision there is a loss-free reshuffling of momentum. Having both magnitude (mass times velocity) and direction (along the line of motion), the combined momentum remains the same before and after the glancing collision. Each particle individually, however, suffers a change, and those individual jumps in momentum send the bodies off in new directions with new speeds. Short, sharp raps generate the greatest change in momentum per unit time and hence the largest forces during the collision. Here the moving black ball comes from the left to hit the stationary colored ball.

no unbalanced force of any kind to alter the total momentum. Momentum is **conserved**, neither created nor destroyed. Momentum acquired by the eight ball comes entirely from the cue ball; it is a zero-sum game, and the rules of this game apply to all forms of matter in all manner of motion.

The conservation of momentum propels a rocket into space. Exhaust gases shoot out in one direction, and the rocket accelerates in the opposite direction. Oxygen and nitrogen molecules flying about in the air conserve momentum when they collide. Electrons in a video monitor conserve momentum when they are shot against the screen. No violation of this principle (closely related to **Newton's third law** of equal and opposite "action" and "reaction") has ever been known to occur.

1-6. FUNDAMENTAL FORCES

Matter collides with matter. Momentum is transferred. A force arises. Velocities change. It seems to occur in so many different ways, but do not think that every conceivable matter-to-matter encounter involves some uniquely specific force. There is no special cue-ball force to be distinguished from a baseball-bat force, nor an earthquake force to be differentiated from a hurricane force. Doubtless there is a transfer of momentum from, say, cue stick to cue ball to eight ball, yet we should not confuse the *effect* of the force with the cause of the force. To understand what actually makes the balls bounce, one must look deep within matter itself—into its most basic attributes of *mass* and *charge*. These are the endowments that make matter what it is; these are the properties that allow matter to interact with matter.

The truly fundamental forces of nature are few, no more than four. There is, first, **gravity**, the intrinsic attraction of mass for mass. Although the mechanism remains imperfectly understood, matter attracts matter *because* it has mass. Gravity holds together the macroscopic structures of the universe: the planets, the stars, the earth and everything in it. Gravity is the force that pulls our feet to the ground and controls the big things, but gravity is not the force that governs the small electrons, atoms, and molecules of the microscopic world within. Gravity demands sufficient mass to be effective, and the small particles just do not have it.

To sew together atoms and molecules is rather the job of the **electromagnetic force**, the interaction involving electric charge. Charged particles are either drawn to or pushed away from other charged particles. Like charges repel, and opposites attract. Electrons are attracted to protons. Electrons are repelled by electrons. Matter pushes and pulls upon itself *because* it has charge.

Sometimes one speaks of an *electrostatic* force between stationary particles, contrasting it with a *magnetic* force arising from charged particles in motion (electric currents). Both are complementary aspects of a single package, however, and together they appear as a combined *electromagnetic* force. The electromagnetic force may be manifested in different ways, but always we shall repeat the mantra: like charges repel; opposite charges attract.

From these elementary pushes and pulls are formed atoms, held together by electromagnetic interactions between positive and negative particles. From atoms come molecules, held together by essentially the same electromagnetic interactions. And from molecules are assembled the associated states of matter (solids, liquids, and gases), also knitted together by the basic electromagnetic force. It is all one. The same force pulls a lightning bolt to earth; the same force turns an electric motor; the same force makes a diamond.

Not only does the electromagnetic force bring matter together, it also pushes matter apart. Electrons repel each other. They resist crowding into the same region of space. One electron excludes another, and this powerful repulsion prevents matter from collapsing and also prevents us from walking through walls. These same electromagnetic repulsions keep the cue ball from driving right through the eight ball. There is a hard bounce instead, which is registered as a mechanical force accompanied by a change in velocity.

The electromagnetic force is fundamentally unaltered regardless of its particular environment. We routinely use the term "chemical interactions" to describe the bonding between atoms in a molecule, as if these forces differed from the "physical interactions" that bind molecules into a collective state of matter. Such distinctions often help to organize our knowledge, but they are made solely for convenience. The difference between chemical and physical interactions is a matter of scientific convention only, for nature bestows but one meaningful force on atoms and molecules: the electromagnetic force.

The electromagnetic force is not sufficient, though, to hold together the particles making up the dense kernel of an atom—its nucleus. There we find two other modes of interaction, which take us outside the realm of chemistry but which we mention in order to complete the picture (leaving the details to wait until Chapter 21). The **weak force** is responsible for beta decay, a nuclear reaction whereby the number of protons, Z, is altered. The **strong force** binds together the protons and neutrons within a nucleus, thus overpowering the repulsive electrical interactions between positively charged particles. It is the force behind the hydrogen bomb, a strong force indeed.

Nevertheless it is gravity and electromagnetism that rule the universe, shaping matter from the galaxies down to individual atoms. Gravity arises from mass and electromagnetism from charge, yet careful experimentation reveals that the two agencies have some remarkable features in common.

Consider first the mechanics of gravity. Two massive particles are attracted by a force directly proportional to the product of their masses, m_1m_2. If one mass is doubled, the gravitational force doubles as well. If both masses are doubled, the attraction increases fourfold. The force is also inversely proportional to the square of the distance between the particles, going as $1/r^2$ for a separation r. If r is doubled, the force is reduced by a factor of 4. Putting everything together, we have Newton's law of gravitation:

$$F = -\frac{Gm_1m_2}{r^2}$$

Take note of the details. The negative sign reminds us that the force is always attractive—the two masses are pulled together. G is a proportionality factor called the **gravitational constant**. Its value, determined experimentally to be 6.67×10^{-11} N m^2 kg^{-2}, gives the gravitational force its proper units of newtons.

Most of our direct experience with gravity involves the earth, a good-sized chunk of matter with a mass (m_1) of 5.98×10^{24} kg and a radius (r) of 6.378×10^6 m. The force acts as if all the earth's mass were concentrated into a point at its center, and so the factor Gm_1/r^2 becomes 9.81 m s^{-2}. Denoting this **standard acceleration due to gravity** (9.81 m s^{-2}) by the symbol g, we obtain the gravitational force for a body of mass m near the surface of the earth:

$$F = \text{mass} \times \text{acceleration due to gravity} = mg$$

The units, kg m s^{-2}, work out correctly to newtons, and this force is nothing other than the **weight** of a body. It pulls the mass down toward the center of the earth, accelerating the object by 9.81 m/s every second. Since the force is always downward, the minus sign now can be omitted without ambiguity.

Mass, we see, is a fundamental attribute of matter itself, whereas weight is a variable force arising from an attraction to other matter nearby. Although an object's mass is the same everywhere in the universe, its weight depends on the strength of the local gravitational force. A mass of 1 kg weighs 9.81 N (about 2.2 lb) on the earth, for example, but only 1.62 N on the moon. The moon is a smaller and less massive body, and its gravitational acceleration is reduced proportionately.

Every object endowed with mass is accelerated by gravity toward the center of the earth at a rate of 9.81 m s^{-2}. We do not fall through, of course, thanks in part to the electromagnetic force, which keeps our electrons from getting too close to those in the earth's surface.

And so we come again to the electrostatic interaction, which arises not from mass but from charge. Mass has nothing to do with it. Two *charged* particles interact, instead, through a force directly proportional to the product of their charges, $q_1 q_2$. The electric force also is inversely proportional to the square of the distance between the particles, from which is stated **Coulomb's law**,

$$F = \frac{1}{4\pi\epsilon_0} \frac{q_1 q_2}{r^2}$$

where $1/4\pi\epsilon_0$ is a proportionality constant. The SI unit of charge is called a **coulomb (C)**, defined so that an electron carries a charge of -1.60×10^{-19} coulombs. The factor ϵ_0, known as the permittivity constant, has the value 8.85×10^{-12} C^2 N^{-1} m^{-2}. Charges of the same sign

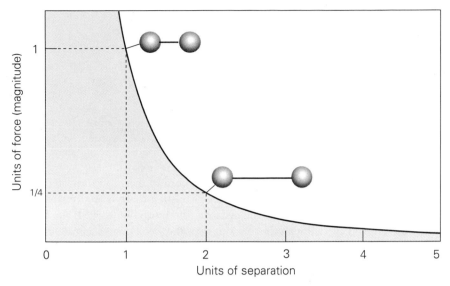

FIGURE 1-6. The inverse-square law: where the force between two particles varies inversely with the square of the distance separating them. Halve the distance; quadruple the force. Double the distance; cut the force to a fourth. Both the gravitational and electrostatic (Coulomb) interactions obey an inverse-square law.

(two positives or two negatives) react to a positive force and are pushed apart; unlike charges react to a negative force and are pulled together.

Both gravity and the electrical interaction are ***inverse-square-law forces***, sharing the same $1/r^2$ dependence illustrated in Figure 1-6. The forces decrease as the particles move apart, but still they remain effective over long distances. Each force depends only on its respective attribute—mass for one, charge for the other—and each is directly proportional to the product of the appropriate quantities. The electric force, dependent on the charges alone, is the same regardless of mass. Similarly, the gravitational force is determined only by the mass.

1-7. WORK AND ENERGY

Nothing starts or stops by itself; it must be pushed. It must be forced. With these simple words, largely unappreciated until 400 years ago, we capture the essence of matter in motion. It is the key to understanding not just the bounce of a billiard ball, but the subtleties of chemistry as well.

To move something from one place to another takes *work*. We push a package across the floor, steadily exerting a force along the way. The wind blows a cluster of dried leaves; the water carries along a piece of wood. The electrons in a wire are driven forward by the voltage in a battery.

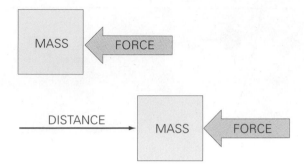

FIGURE 1-7. Work: the cost of traveling from here to there, accumulated bit by bit, step by step. To move an object against a resisting force is to do work—the stronger the force and the greater the distance, the more the work. If the force is both constant and aligned with the motion, then the work done is equal simply to force times distance.

All of it is **work**: *the application of force over a distance*, set out pictorially in Figure 1-7. Defined as

$$Work = force \times distance$$

for a constant force parallel to the motion, this physical quantity gives meaning to an intuitive sense of effort. Work is the effort expended to displace an object from one point to another. The heavier the force, the greater the effort. The greater the distance, the more work is done. We push harder and longer, grunting and groaning under the strain, and we feel that more work has been done. The equation above tells us exactly how much.

From this equation is defined the SI unit for work: the newton-meter (N m), or **joule (J)**. It is the product of force and distance, expressible in fundamental units of mass, length, and time as kg m^2 s^{-2}. Note also that work can be positive or negative, depending on how the motion is related to the applied force. If the object moves in the same direction as the force, the work is considered positive. The wind is at our backs.

Now suppose, as suggested in Figure 1-8, that the task is to move one *electrical* particle (having charge q_2) in the presence of another (with charge q_1). Particle 2 comes in bit by bit, subject at each distance to a force proportional to $q_1 q_2 / r^2$. Every step of the way, the particle responds to a continuously changing force. Every step of the way, it undergoes a small displacement. Every step of the way, there is some electrical work done. Each bit of work is the product of the local force times the small change in distance, and all these little bits add up en route.

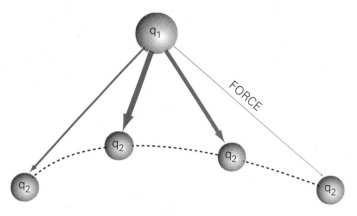

FIGURE 1-8. Electrical work, portrayed conceptually: One particle (charge = q_2) moves either with or against the force produced by another (charge = q_1), paying for its progress step by step. The force, suggested here by the thickness of the arrow, increases as q_2 draws near. Total work depends on the signs and magnitudes of the charges, together with the path that q_2 is made to follow.

It appears almost as if q_1 has been in touch with q_2 throughout, transmitting its electric force through empty space. Space itself, viewed from the perspective of particle 2, seems then to take on a special role as a messenger. Particle 2 need not "know" about or "see" particle 1 directly, but it does react to a force when it encounters one. And q_2 learns the strength of this force merely by a position in space, since at each location it retrieves an electrical message broadcast by q_1. We say, accordingly, that q_2 moves in an ***electric field***, a *field of force*.

Meaning: At every point in the field, q_2 suffers an electrical push or pull in a certain direction. Such force is plain to see, a property of the space itself, a reality made visible in the "lines of force" computed in Figure 1-9. Or, if seeing is believing, make your own (magnetic) field and your own lines of force: Hold a magnet beneath a sheet of paper, and observe the disposition of a thin layer of iron filings. The magnet stays hidden, but the force field emerges in full view on the paper.

It is a grand intellectual concept, this concept of a field, by which we assert that the very presence of a charge alters the quality of space nearby. The charge q_1 establishes a field (proportional to q_1/r^2), through which q_2 moves and through which q_2 senses a force. The original charge, q_1, stays hidden in the background while the force field does all the work on q_2. It is through the intervening field that a charge is forcibly pushed or pulled from point to point. There is a force and a displacement, and so work of some sort is surely done. Effort is required to relocate the charge in the field of force.

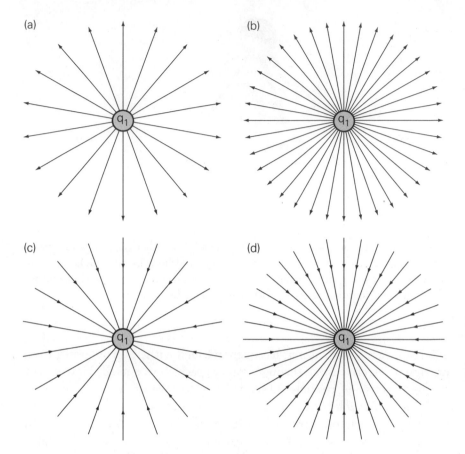

FIGURE 1-9. Electric fields generated by various source charges q_1, shown as lines of force in two dimensions. The lines represent the force F conveyed by q_1 to a (positive) test charge at each position in the field, interpreted as follows: F points in the direction of the arrow, with magnitude proportional to the spacing of the lines. Regions where the field is strong contain many lines, drawn close together; regions of weak field contain few. Each field extends into a third dimension, unseen in the diagrams. (a) $q_1 = 1$, a positive source charge expressed in arbitrary units. (b) $q_1 = 2$, a positive charge with double the magnitude. (c) $q_1 = -1$, a negative charge. (d) $q_1 = -2$, a negative charge with double the magnitude.

We now go one step further to introduce the notion of **energy**, defining this familiar word as an *ability to do work*. At point A in the electric field, let us say, the charge q_2 possesses a certain "energy" just by virtue of where it is. The particle need not literally move anywhere, but it has this energy E_A and it *could* move at any time. There is a latent, stored-up capacity to do electrical work, and we call such a capacity **potential energy**—energy dependent on location alone. *If* the charge eventually does move to point B in the force field, then it will find itself with a new energy, E_B, at the new location. What enables the particle to make the journey from A to B? *Work*. How much work? The difference in energy, $E_B - E_A$.

Work and energy. Energy (measured in joules, just like work) is the ability to do work by whatever means available, and energy is expended whenever a particle moves. The doing of any work is accompanied by a matching change in energy. In a field, the change corresponds to the difference in potential energy between positions A and B (Figure 1-10).

We already have considerable practical experience with various forms of energy. Electrical potential energy, for instance, is recognized most commonly as a *voltage*: the amount of work needed to move one coulomb of charge in a field. The volt (V), an SI unit, is equal to one joule per coulomb. Voltage therefore has nothing to do with an actual *flow* of electricity, but indicates instead the work needed to accelerate a stream of charged particles. It offers the *potential* to make electricity.

Even more familiar is gravitational potential energy. We fight the force of gravity every time we lift something, doing work in the amount mgh—force (mg) times distance (h)—to raise the mass to a height h above the earth's surface. This exertion is clearly a movement in a field, but here it is a *gravitational* field established by the mass of the earth.

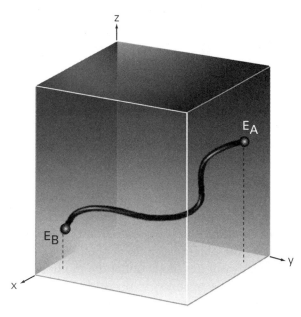

FIGURE 1-10. Potential energy in a conservative field: energy that depends on where a body sits (x, y, z), not how fast it moves. In relocating from point A to point B, the object changes its potential energy by the amount $E_B - E_A$ along whatever path is taken. Gradations in shading suggest the point-to-point variation of E.

Having been elevated, the mass now has a gravitational potential energy (*mgh*) precisely analogous to an electron's energy in an electric field.

Picture a bag of beans sitting on a high shelf, ready and able to move in the earth's gravitational field. Its energy, like money in the bank, is being saved up for later work. The elevated mass has the ability to do work; it has potential energy.

Suddenly there is a disturbance and the sack falls off the shelf, moving downward in the field of gravity. Closer to the earth it comes, its potential energy (*mgh*) decreasing in direct proportion to the height *h*. Faster and faster it goes. Work is being done. The bag is being pulled by the force of gravity and is acquiring momentum of its own. With this momentum, too, comes a corresponding ability to do work, packaged in a new form of energy—a **kinetic energy**, an energy not of position but of *motion*. Undoubtedly a falling mass can do work on another object, for we know that motion begets motion. Motion begets motion, and hence motion carries energy. Energy is the capacity to do work.

The amount of kinetic energy depends on both mass and velocity,

$$\text{Kinetic energy} = \frac{1}{2} \text{ mass} \times (\text{velocity})^2$$

going as $\frac{1}{2}mv^2$ or, equivalently, $p^2/2m$ (recall that the momentum, p, is equal to mv). This result is derived by computing the product of force and distance for each step of a particle's motion. The force is given by Newton's second law, $F = ma$, and the factor of $\frac{1}{2}$ arises when all the bits of work are properly added up along the route.

Thus the beans fall, losing potential energy but gaining kinetic energy during the descent. It is an exact trade—one for the other, energy of position for energy of motion—and nothing is lost. On the shelf, the beans in the bag have gravitational energy equal to *mgh* and kinetic energy equal to zero. Just before hitting the floor, their potential energy is zero ($h = 0$) but their combined kinetic energy ($\frac{1}{2}mv^2$) is equal to the *original* quantity *mgh*. Nothing is lost and nothing is gained.

Eventually the bag lands with a thud, unable to penetrate the electromagnetic barrier erected by the atoms and molecules in the floor. At rest now on the ground, the object has neither the kinetic energy of the fall nor the potential energy of the shelf. But at the moment before impact it does have kinetic energy—an ability to do work—and it delivers this energy to the floor during the crash, in a way suggested by Figure 1-11.

The bigger they are, the harder they fall: kinetic energy is directly proportional to mass. The *higher* they are, the harder they hit as well: more gravitational potential energy (*mgh*) is ultimately converted into kinetic energy. The bag picks up more and more speed along the way,

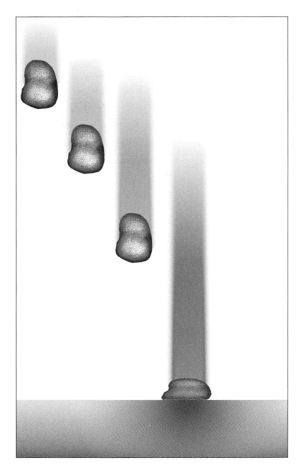

FIGURE 1-11. Conservation of energy: nothing lost, nothing gained. Energy, both kinetic and potential, passes through various modes as objects move and interact, but the total capacity to do work is fixed globally. Amounts of energy invested in the random, disordered agitation of atoms and molecules are exchanged as a flow of *heat*, an umbrella term pertaining to microscopic work and motion.

with its kinetic energy proportional to v^2 at each instant. The falling object acquires kinetic energy; it can make something move.

It makes the floor move. Some of the kinetic energy goes into the floor, stirring up the particles within. The floor's atoms and molecules move faster than they did before the crash. Work is done.

The crashing object makes *itself* move as well, but only on the inside now. Some of the kinetic energy goes back both into the beans and also into the atoms and molecules inside the beans, which jiggle around for some time after the impact. The particles move faster. Work is done.

We do not perceive the work done on atoms and molecules as visible motion simply because the microscopic particles are too small to see, yet it is work nonetheless—work invested in internal motion. Such work becomes sensible to us as *heat*, a perception of hot and cold with which we associate a *temperature*.

Shaking up the interiors of floor and beans, the crash produces heat and a corresponding rise in temperature (perhaps very slight). Temperature, as we shall appreciate later, is a measure of a material's internal energy, a reflection of the atomic and molecular motion within. The SI unit is the *kelvin (K)*, defined so that a value of 0 K corresponds to the lowest internal energy conceivable. Also useful is the *Celsius* scale ($°C = K - 273.15$), according to which water freezes at $0°C$ and boils at $100°C$. Less common among scientists is the *Fahrenheit* scale ($°F$), calibrated so that the freezing and boiling points of water are $32°F$ and $212°F$, respectively. Degrees Fahrenheit can be converted into degrees Celsius using the relationship $°C = \frac{5}{9}(°F - 32)$.

From shelf to floor, then, we see how energy is transmuted from one form into another. There is a great web of interrelated events. The sun first shines on the earth, providing energy for green plants to carry out photosynthesis. We nourish ourselves with the molecules produced by such plants, deriving our own energy through further chemical reactions. Using this energy, our muscles contract and enable us to raise the bag in the earth's gravitational field. The mass later crashes down and heats up the floor.

Yet the total of energy from all sources remains forever unchanged, however intricate the sequence of transfers and interconversions. Energy, like momentum, is conserved. Energy is neither created nor destroyed. Converted into different forms, yes. Shared, yes. Manufactured from scratch, no. It is a principle true in general, thoroughly established by innumerable experimental observations. It is one of the most solidly established rules of nature's game.

Other proposed rules have been revised in the face of improved experimental observations over the years, but not the conservation of energy and not the conservation of momentum. These principles, along with such basic concepts as mass (conserved in chemical reactions; look ahead one chapter), charge (also conserved; look ahead two chapters), and the fundamental forces, are the foundation upon which nature is built.

Reflect on what we already know, beyond the definitions and beyond the equations. We know that the universe is a universe of matter and energy, a scene of matter in motion. We know, even more, that all matter is the same—that a stone can be disassembled into its electrons, protons, and neutrons and then reassembled into a giraffe. Reassembled not by us, granted, but we understand that the same forces hold together both stone and giraffe. Matter is matter.

This search for commonality and simplicity will guide us through the chemical world of atoms and molecules. That road is just beginning, and there will be surprises along the way. Soon we shall discover that Newton's laws, stunningly accurate for macroscopic events, are insufficiently general to describe the motion of electrons in atoms. Electrons do have mass and charge. They do have energy and momentum. They do move in the presence of electric forces. But they do not move like little charged billiard balls.

Nor should they. Why should anything so small behave the same as something else so big? There are profound differences between big and small, all the more startling because (being big) we naturally tend to think big. Scientists truly came to grips with the microscopic world less than 100 years ago, and the revolution they launched enables us to understand chemistry properly as a microscopic phenomenon.

Nobody suspected that small particles behave so differently, not until our eyes were opened by more powerful techniques of observation. Take a lesson from that: always watch the game; always measure.

1-8. MEASUREMENT

Scientists do not measure the *laws* of nature. They measure, far more modestly, the numerical values of quantities such as mass, length, time, velocity, acceleration, and charge. With disciplined objectivity they monitor a portion of the world, attempting to recognize patterns in the numbers. Sometimes there is no discernible order; sometimes there is. If there is, then these relationships become the equations, theories, and laws of science.

By measuring the path of a falling particle, for example, we accumulate a set of times and positions. From that we note how the sum of two quantities, mgh and $\frac{1}{2}mv^2$, seems always to be the same, suggesting the relationship

$$mgh + \frac{1}{2}mv^2 = \text{constant}$$

sketched in Figure 1-12. Small differences might show up from one experiment to the next, but still there is this striking regularity. Intrigued, we give that constant sum a name—total energy—and break it down further into a gravitational potential energy (mgh) and a kinetic energy ($\frac{1}{2}mv^2$). The rule seems to hold wherever and whenever one looks, and with further observations there also appear other energy-like quantities

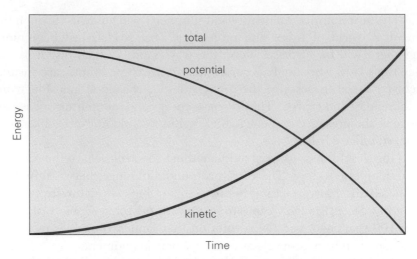

FIGURE 1-12. Kinetic energy ($\frac{1}{2}mv^2$) and potential energy (mgh) of a body falling freely in a gravitational field, plotted against the time elapsed. The sum of the two contributions—the total energy—never changes.

that fit the pattern: electrical potential energy, energy of vibration and rotation, electronic energy, elastic energy, thermal energy, chemical energy, nuclear energy . . . all kinds of energy. Always the sum of these quantities is constant, and through these diverse observations we are led to a much broader assertion. The total of energy in all its forms, we propose, is constant. Energy is conserved.

It sounds beautiful, and the whole notion appeals to our love of permanence and symmetry. But is it true? Can we trust the numbers?

For if we claim to discover principles as sweeping as the conservation of energy, then our numbers ought to be reliable. Aesthetic qualities aside, energy conservation might well be a fantasy if those values of time and position had been recorded incorrectly. Theories and interpretations, however beautiful, can come only after there have been observations. First there must be measurement.

Physical science is rooted in the numbers, not in philosophical speculation. Let us devote the rest of this chapter to developing some ground rules for scientific measurement.

Units

Associated with every quantitative observation are two pieces of information: a **unit** (or **dimension**) indicating *what it is*, and a number indicating *how much*. Mass, 3.0 g. Length, 0.153 m. Time, 8.64×10^4 s. Velocity, 2.998×10^8 m s^{-1}. Energy, 8.3145 J.

Without a number there is no measurement, but without the unit there is no meaning to the measurement. The unit tells us what we have. It can never be forgotten, never omitted.

Without the units there would be no equations—no potential energy, no kinetic energy, no conservation of energy. How could we even recognize an "energy" if given a bare number with no dimensions? We might, in complete ignorance, add apples to oranges and say that the sum is constant.

But, *with* units, the quantity *mgh* does have dimensions of mass $\times$ acceleration $\times$ length, or kg m^2 s^{-2}: *joules*. Gravitational potential energy.

And the quantity $\frac{1}{2}mv^2$ similarly has dimensions of mass $\times$ (velocity)^2, or kg m^2 s^{-2}: also *joules*. Kinetic energy.

The quantity $(1/4\pi\epsilon_0)(q_1q_2/r)$, too, has dimensions of

$$\frac{\text{N m}^2}{\text{C}^2} \times \frac{\text{C}^2}{\text{m}} = \text{N m}$$

or kg m^2 s^{-2}: *joules*. Electrical potential energy.

On and on we can go, defining in each instance some combination of variables with overall units of joules. Each would be a kind of energy, and one could be justifiably added to another.

Without units we could not tell whether a proposed equation made sense. Was gravitational potential energy equal to *mgh* or *mgh*^{2}? The specific expression is easily forgotten, but by inspecting the units one realizes immediately that *mgh*^{2} (kg m^3 s^{-2}) can never be a form of energy. Rather it is force (kg m s^{-2}) times length (m) times an *extra* dimension of length (m); it is energy times length, not energy.

Unit Conversion

More than mere labels, units are quantities in themselves: "one" of something, *one unit*. The symbol g, for instance, means "1 gram," or one multiple of some standard mass. A mass of exactly 2 g is interpreted as g + g, whereas a mass of 5.6 g means 5.6 multiples of this *unit* gram quantity. Units are thus mathematical objects and can be manipulated as such. They add, subtract, multiply, and divide just like ordinary numbers.

When analyzing relationships and solving problems, we often must convert from one set of units into another. If so, then a simple technique known as the **unit-factor method** (also called either the **factor-label method** or, more generally, **dimensional analysis**) becomes a tool of extraordinary power. It begins with the recognition that a unit is a legitimate mathematical quantity, a fact implicit in a statement like

One inch equals 2.54 centimeters (exactly).

Expressed compactly, the relationship is

$$1 \text{ in} = 2.54 \text{ cm}$$

or

$$\frac{2.54 \text{ cm}}{1 \text{ in}} = \frac{1 \text{ in}}{2.54 \text{ cm}} = 1$$

Each of these ratios has the value 1 (exactly), and hence each can be used to multiply some other expression without changing the intrinsic value. Inflicting only a multiplication by the *unit factor* 1, the operation is algebraically neutral.

Suppose we measure a length as 36.5 inches and need to convert it into centimeters. The most straightforward approach is to multiply the inches by a conversion factor with dimensions of centimeters/inches, yielding

$$36.5 \text{ in} \times \frac{2.54 \text{ cm}}{1 \text{ in}} = 92.7 \text{ cm}$$

The units of inches, functioning as mathematical objects, cancel out in the same way that any other pair of common factors would cancel. From that we have the answer, quickly and correctly, without needing to remember whether the 2.54 should be in the numerator or denominator. No special formula or trick is required. The proper procedure follows automatically from an analysis of the dimensions.

Now if it is safe to use a single unit factor, then it must also be permissible to use any number of them in succession. Each ratio has the value 1 and is thoroughly harmless. We might, as a demonstration, undertake to express the speed of light (3.00×10^8 m s^{-1}) in miles per hour, working our way systematically from left to right with a string of unit factors:

$$\frac{3.00 \times 10^8 \text{ m}}{1 \text{ s}} \times \frac{100 \text{ cm}}{1 \text{ m}} \times \frac{1 \text{ in}}{2.54 \text{ cm}} \times \frac{1 \text{ ft}}{12 \text{ in}} \times \frac{1 \text{ mi}}{5280 \text{ ft}} \times \frac{3600 \text{ s}}{1 \text{ h}}$$

$$= 6.71 \times 10^8 \text{ mi/h}$$

Step by step, the conversion goes from (1) meters to centimeters, (2) centimeters to inches, (3) inches to feet, (4) feet to miles, and (5) seconds to hours. The procedure is foolproof, provided that the units are positioned properly and the numerical factors are correct.

Uncertainty, Accuracy, and Precision

No instrument is perfect. No measurement is perfect. A number obtained by measurement is, at best, an approximation to a presumably "correct" yet unattainable value. Whether the result be close to the mark or way off, it must be treated with the skepticism accorded to any estimate.

Demons of various sorts conspire against a measuring device. In the simplest instances, there are limits to the fineness with which markings can be calibrated and read. A mercury thermometer that has gradations every degree, for example, forces an observer to guess whether the temperature is 20.2°C or 20.3°C or 20.4°C (Figure 1-13). Different observers will agree that the value is "twenty point something" without being able to fix the last digit with any certainty. Such disagreement is guaranteed by the design of the thermometer itself, and so we are forced to estimate the temperature as "20.3 plus or minus 0.1°C" or, written more concisely, 20.3 ± 0.1°C. This range of variation, different for every instrument, reflects the ***uncertainty*** of the measurement.

Observers must understand the uncertainty of their instruments and act prudently when making a measurement, displaying neither too much ambition nor too much timidity. Unable to agree on 20.2°C or 20.3°C or 20.4°C, we have no right to add another digit and call it 20.34°C. But neither should we give up entirely and report the value as 20°C, for that bit of laziness would rob us of valuable information. The best course is to report all the digits known with certainty (here, 20.) and then include

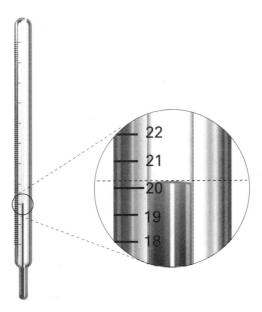

FIGURE 1-13. Uncertainty in measurement. The last significant figure, reported differently by different observers, is always in doubt. No instrument is perfect.

the first one in doubt (giving 20.3). These numbers are the **significant figures** or **significant digits** of the measurement, and they communicate roughly the uncertainty of the instrument. To assert that a temperature reading is 20.3°C is to imply that the thermometer has markings coarser than 0.1°C, putting the tenths place in dispute. To argue, though, that the reading is 20.30°C (with two digits past the decimal point) is to suggest that the measurement comes from a more finely calibrated thermometer. The argument would be about the hundredths place (20.30°C or 20.31°C?), whereas the tenths place would be solidly established.

Uncertainty is present everywhere. Any instrument is reliable only within a certain range, ready to go awry for a host of reasons: fluctuations in temperature and air currents, external vibrations, random glitches in electronic circuits, and all the other vagaries of the real world. Such **random error** (equally likely to go up or down) causes the measured value to fluctuate from one reading to the next, just as we saw for the thermometer. These errors can be minimized but not eliminated, for there is an unavoidable uncertainty associated with every sufficiently ambitious act of measurement. Some amount of plus or minus is built into all instruments.

Beyond plus or minus, however, a measurement may be subject to **systematic error**—a consistent deviation in the same direction. The thermometer might be improperly calibrated, say, so that the mercury column rises to the 1° marking when the bulb is placed in ice water. All measurements would then appear high by 1°C, meaning that the true temperature is really 19.3 ± 0.1°C. Conceivably we could locate and correct such a miscalibration, but not all systematic errors are so obvious. They can lurk unsuspected in an instrument or experimental design.

The difference between an observed quantity and its ideal value reflects the **accuracy** of the measurement—whether or not the answer is *right*. Of two pieces of data, the more accurate is the one closer to the correct value. Accuracy is determined primarily by the degree to which systematic errors affect the measurement.

Precision, by contrast, is an indication of consistency rather than rightness or wrongness. A set of measurements is considered precise if the values are clustered together, deviating only by a comparatively small random error. The sequence "20.2°C, 20.3°C, 20.4°C" is more precise than "19.3°C, 20.3°C, 21.3°C," even though both sets have the same average value: 20.3°C.

A precise measurement need not be accurate, since it can be distorted by a systematic error. An accurate measurement, too, may be imprecise if it arises as the average of a set of widely scattered numbers. Best of all are those results that are both accurate and precise, with minimal uncertainty as well: correct values, registered consistently.

Experimental Error and Significant Figures

Consider how one might determine whether the sum of kinetic and potential energy is indeed constant for a falling body. Taking a suitable particle (perhaps a small metallic ball), an observer first measures the object's mass (m) and height (h) above the surface. Initially the gravitational energy is mgh, and the kinetic energy is zero. Then the ball is released and its velocity is recorded just at the instant of impact, whereupon the kinetic energy is $\frac{1}{2}mv^2$ and the potential energy is zero. The observer, expecting energy to be conserved, is stunned to discover that the measured value of $\frac{1}{2}mv^2$ is always less than mgh. Is energy not conserved?

A principle like energy conservation is supported by too many other consistent observations to be overturned lightly. Eventually, under some special circumstances, we might find that energy is indeed not conserved—but before reaching that conclusion we should reexamine the experiment. When something is *always* wrong (as here, where $\frac{1}{2}mv^2$ is always less than mgh), a skeptical experimenter immediately looks for a systematic error. And error there is: it comes from the air. The ball is not falling through empty space, but rather is fighting its way through the nitrogen and oxygen molecules of the air. Such resistance by the atmosphere slows the ball's descent, leaving its final kinetic energy less than the initial potential energy. The "missing" energy is not missing after all; it shows up in the frictional heat produced as the air rubs against the ball.

To test this hypothesis, the experimenter takes the apparatus to a mountaintop. Here, in the rarefied atmosphere at high altitude, the final value of $\frac{1}{2}mv^2$ still proves to be less than mgh, but now the two numbers are considerably closer than before. The improvement seems reasonable because there are fewer molecules to oppose the body's fall. The particle hits the ground faster, and the systematic error is reduced.

Nevertheless, determined to eliminate *all* systematic error, our persistent observer repeats the experiment in the vacuum of space—on the moon, where the ball falls unimpeded by any atmosphere at all. The particle's mass is measured as 1.0000 ± 0.0002 kg, the same as on the earth (although its weight is cut by $\frac{1}{6}$ under the moon's weaker gravity). The initial height is determined to be 101.00 ± 0.01 m, and the final velocity is measured as 18.1 ± 0.1 m s^{-1}. Using the value $g = 1.62$ m s^{-2} for the acceleration due to lunar gravity, the experimenter then computes mgh and $\frac{1}{2}mv^2$ with the aid of a calculator (diligently writing down all the digits):

$$mgh = 1.0000 \text{ kg} \times 1.62 \text{ m s}^{-2} \times 101.00 \text{ m} = 163.6200 \text{ kg m}^2 \text{ s}^{-2}$$

$$\frac{1}{2}mv^2 = \frac{1}{2} \times 1.0000 \text{ kg} \times (18.1 \text{ m s}^{-1})^2 \quad = 163.8050 \text{ kg m}^2 \text{ s}^{-2}$$

The numbers are different once again, and now there is no air resistance to account for the discrepancy. Reluctantly we begin to suspect that the conservation of energy, a supposed "law" of nature, is about to be repealed.

But the law is really in no danger, at least not on the basis of this flawed analysis of the data. For we have forgotten that the quantities observed are all inexact numbers, reported so that their final digit is understood to be in error. Mass is known, with certainty, only through the third decimal place; its measured value is 1.0000 kg *plus or minus* 0.0002 kg. Height is 101.00 m *plus or minus* 0.01 m. Terminal velocity is 18.1 m s^{-1} *plus or minus* 0.1 m s^{-1}. The moon's g factor—1.62 m s^{-1}, also an experimental quantity—is given to only three significant figures, and consequently the 2 in the hundredths place is in doubt.

Presented with these uncertainties, we cannot claim that the two values are 163.6200 J and 163.8050 J. Who supplied all those digits? The calculator multiplies numbers without thinking, but a thoughtful observer is obliged to consider the limitations imposed on each factor. Here, with both g and v known to only three significant figures, the final answer is limited strictly to three figures as well. In view of this uncertainty we must round off the two energies to 164 J, making them equal within the overall experimental error. Energy *is* conserved.

Handling of Significant Figures

Clearly we shall have to attend carefully to significant figures, lest our numbers fail to reflect the realities of measurement. Two principal questions need to be addressed: (1) determining significance for a single quantity, and (2) determining significance for a computation.

Concerning the first issue, note the following general rules:

1. Any nonzero digit is significant. The value 7.34 cm has three significant digits.

2. Included zeros are always significant. Both 1.066 m and 1066 m have four significant digits.

3. Trailing zeros in a decimal number are significant. The value 5.1000 g has five significant figures.

4. Leading zeros are *not* significant. The value 0.0051000 kg has five significant figures. The two zeros to the left of the 5 serve only to locate the decimal point.

5. All digits are significant when a number is expressed using scientific notation. The value 8.6400 × 10^4 s has five significant figures.

6. Trailing zeros in a nondecimal number are *not* significant. There are just three significant digits in 86,400 s. Explicit use of a decimal

point, however, renders the whole number significant. The value 86,400. s has five significant figures.

7. *Exact* numbers, not subject to measurement, are considered to be infinitely significant. The factor $\frac{1}{2}$ in $\frac{1}{2}mv^2$, for example, arises purely from a mathematical operation and therefore carries no uncertainty.

Concerning the second issue (significance of a calculation), note that the result of any multiplication or division can be no more significant than the *least* significant factor. In the experiment on the moon, the kinetic energy was obtained by multiplying $\frac{1}{2}$ (an exact number) times 1.0000 kg (five significant figures) times the square of 18.1 m s^{-1} (three significant figures). The product is limited, accordingly, to only three digits and its value is rounded off to 164 J.

When measured quantities are added and subtracted, the number of decimal places is determined by the term with the fewest digits after the decimal point. Thus the sum of 101.00 m and 5 m is reported as 106 m:

$$
\begin{array}{r}
101.00 \text{ m} \\
+ \quad 5 \quad\;\; \text{m} \\
\hline
106 \quad\;\; \text{m}
\end{array}
$$

The additional information contained in the 101.00 m is lost when combined with the less exact 5 m.

We pause, then, on this practical note, underscoring that natural science stands or falls on its numbers. Remember that it was measurement—quantitative observations of planets and rolling balls—that gave us the picture of macroscopic force, motion, and energy sketched in this chapter.

Armed with such understanding, we are ready to descend into the alien world of atoms and molecules to look for the chemistry. Things will be different. Although it remains a world of force, motion, and energy, this place is no longer under the jurisdiction of Newton's laws. Some features of the landscape will be familiar and others will be strange, but there is no alternative except to push ahead. That's where the chemistry is.

We have to think small.

REVIEW AND GUIDE TO PROBLEMS

Understand this: There is no permanence in nature. Everything moves. Everywhere there is change; everywhere there is motion.

The earth orbiting the sun, a cue ball rattling around a table, a book falling from a shelf—the way of the world is to move. All around us we see large pieces of matter in motion, pushed and pulled according to the laws of macroscopic physics.

And chemistry? Chemistry, too, reduces fundamentally to motion and energy, but here it is motion and energy writ small. Chemistry unfolds inside a world in miniature, a world of small particles governed by the electromagnetic force. It is a world of electrons, nuclei, atoms, and molecules; a world where small particles move and exchange energy in small amounts.

Where there is chemistry, there is motion; and where there is motion, there is energy. *Energy* gives shape to the cloud of electrons moving about a nucleus. *Energy* drives the molecules of a gas into chaotic, mob-like action. *Energy* controls the way a molecule revises itself into a new structure, just as energy controls the way a stone falls to earth.

Gradually we shall learn how to define and manipulate the energy of chemical reactions, and we shall learn something else too: Energy is energy. Chemical energy, electrical energy, gravitational energy, nuclear energy, solar energy, kinetic energy, and all the other forms of energy are actors in the same play. One is equivalent to another; one can be transformed into another. Energy is energy.

But large is not always the same as small, where either matter or energy is concerned. The relationship between force (how hard a push? in what direction is it applied?) and motion (where does the pushed particle go? how fast does it get there?) depends on size and scale. Large, massive objects moving over long distances obey Newton's laws of classical mechanics. Small objects confined to small spaces—the electrons in an atom, for example—obey the laws of quantum mechanics. Look therefore to quantum mechanics, not classical mechanics, to explain the internal structure, motion, and energy of atoms and molecules (Chapters 2 through 8).

Yet hold on to the ideas of classical mechanics, even in the microworld, for even in the microworld we still have recourse to Newton's equations of motion. Sometimes, small as they are, atoms and molecules do masquerade as little billiard balls. We shall see (in Chapters 9 through 11) that the particles in a gas behave just so, traveling long distances and bouncing around like balls on a table. Each moves with a kinetic energy equal to $\frac{1}{2}mv^2$ and a momentum equal to mv. The fundamental things apply.

And then there are the great conservation laws, universal and inviolable: the conservation of momentum and the conservation of energy. Both the total energy and the total momentum of a system, drawn from all sources, remain constant. If momentum increases over here, then it decreases over there. If energy increases over there, then it decreases over here.

Since neither momentum nor energy can be created from scratch or utterly destroyed, the total motive power in the universe is fixed. To move something, we must *find* the (preexisting) energy somewhere and redirect it. Nobody is making any new energy; all we can do is borrow old energy and eventually give it back.

The conservation laws apply to all things, large and small. The same strictures govern the free fall of a stone as regulate the transformation of graphite into diamond. Viewed as a reshuffling of energy, small changes and large changes eventually do find common ground; and from this broad view come the laws of *thermodynamics* (Chapters 12 through 14 and beyond), the energetic rationale underlying all chemical transformations.

So there, once again, is our map for the road ahead. In this first chapter we have begun with the large, intending all the while to apprehend the small. The fundamental ideas of energy and motion will guide us through the small-scale electromagnetic interactions that shape atoms and molecules. To follow that road, however, we need classical mechanics and thermodynamics as well as quantum mechanics . . . and we need also to think quantitatively, to make measurements, to handle numbers.

Here, then, is a review of the principal definitions and formulas introduced above, condensed and collected now into one place. This summary will launch us immediately into a set of worked-out problems, enabling us further to translate ideas into mathematical quantities and relationships.

IN BRIEF: MOTION AND ENERGY

Newton's three laws of motion, determined centuries ago by experimental measurement, form the foundation of classical mechanics.

1. THE FIRST LAW. A body retains its original state of motion unless provoked by an external force. Left undisturbed, a body at rest remains at rest and a body in motion remains in motion. If already in motion, it continues in a straight line at constant velocity. We say that a massive object possesses this property of **inertia**, a resistance to any change in its speed and direction.

2. THE SECOND LAW. *Force*, a push or a pull, is needed to overcome inertia; force is needed to change an existing state of motion. We push an object and it starts to move. It *accelerates*, altering either speed or direction or both. Since a body, when pushed, changes velocity at a rate (*a*) inversely proportional to its mass, a heavy object will accelerate slowly whereas a light object will accelerate rapidly:

$$a = \frac{\Delta v}{\Delta t} = \frac{F}{m} \qquad \begin{array}{l} \Delta v = \text{change in velocity} \\[6pt] \Delta t = \text{change in time} \end{array}$$

Proportionally more force is needed to accelerate a more massive body.

Recall also that force is alternatively expressed as a change in momentum per unit time, where the *momentum* (*p*) is defined as the product of mass and velocity:

$$p = mv$$

A Newtonian (classical) force then takes the equivalent form

$$F = \frac{\Delta(mv)}{\Delta t} \equiv \frac{\Delta p}{\Delta t}$$

in which $\Delta(mv)$ symbolizes the change in momentum.

To realize this force, we must change the momentum during some interval Δt. The larger the change, the stronger the force. The shorter the interval, the stronger the force as well.

3. THE THIRD LAW. For every *action*, there is an equal and opposite *reaction*. When one body pushes against another, the second body pushes back with an opposing force of equal magnitude. A force in one direction is balanced by a counterforce in the opposite direction. Momentum is conserved.

Given these three empirical laws we thus have a relationship between force and motion, but from where do the forces come? By what means are particles actually pushed and pulled? Does matter have some intangible "hooks" with which it grabs onto some invisible force?

It does: in the form of mass, charge, and other such attributes. Any material particle will have some attribute that renders it susceptible to a corresponding force; and, with that recognition, our worldview immediately becomes broader and simpler. For we discover that nature employs just a few all-purpose tools to move matter and to exchange energy. These agencies, the *fundamental forces*, grapple with and rearrange the pieces of the material world. Particle by particle, they latch on to bits of matter and bring about all the world's changes.

4. There are four fundamental forces: the gravitational force, which acts on mass; the electromagnetic force, which acts on charge; and, inside the nucleus, the strong and weak forces. We have considered only gravitational and static electrical interactions in this chapter, leaving the nuclear forces for Chapter 21.

First, gravity. Any two pieces of matter are attracted by a *gravitational force*

$$F = -\frac{Gm_1 m_2}{r^2} \qquad G = 6.67 \times 10^{-11} \text{ N m}^2 \text{ kg}^{-2}$$

where r, the separation between the bodies, is measured in meters and the two masses, m_1 and m_2, are measured in kilograms. Like all forces, gravity is expressed in the SI units of newtons ($\text{N} = \text{kg m s}^{-2}$), a direct consequence of Newton's equation $F = ma$.

The force due to gravity is directly proportional to the mass of each body, but inversely proportional to the square of the separation between the two. Double one mass; we double the force. Double both masses; we quadruple the force. But *halve* the separation, and we quadruple rather than double the force. The interaction obeys an *inverse-square law*, going as $1/r^2$.

So does the *electrostatic force* (or *Coulomb force*), given by

$$F = \frac{1}{4\pi\epsilon_0} \frac{q_1 q_2}{r^2} \qquad \epsilon_0 = 8.85 \times 10^{-12} \text{ C}^2 \text{ N}^{-1} \text{ m}^{-2}$$

where the charges (q_1 and q_2) are measured in coulombs (C). An electron has a charge of -1.602×10^{-19} C.

Acting only on charged particles, the electric force is independent of mass. It is attractive between unlike charges and repulsive between like charges. Far stronger than the gravitational force, it is the controlling influence in the chemical microworld.

5. To move against a force is to do *work*, defined formally as

$$W = Fr$$

for instances where the force (F, in newtons) is constant and where the particle's displacement (r, in meters) is in the same direction as the applied force. The SI unit of work is the joule ($\text{J} = \text{N m}$).

Closely related to work and motion is *energy*, for to possess energy is to be *able* to do work; and, manifestly, to do work is to move. Here we find that a particle (mass m) moving at some velocity v displays a *kinetic energy* equal to

$$E = \frac{1}{2}mv^2$$

It is an energy inherent in the motion of any massive body.

But there is also energy inherent in just *being* somewhere; there is **potential energy** associated with each *position* in space. This potentiality represents stored-up energy, latent energy, a future ability to do work. If, later, the body really does move from one point to another, then the potential energy (of position) is converted into kinetic energy (of motion) and expended as work.

We imagine, accordingly, that one particle of matter creates around itself a **field** of force, both a gravitational field due to its mass and (should the particle be charged) an electric field as well. Think of the field as a *force per unit charge*, proportional to m/r^2 (gravitational force) or q/r^2 (electric force) for a single, pointlike piece of matter.

If a particle is to go from point A to point B in a field of force, it must do work. The moving body either draws energy *from* the field or donates energy *to* the field, depending on whether the displacement is with the force or against the force. In simple cases such work becomes proportional to $1/r$ (not $1/r^2$, note, but $1/r$ . . . force *times* distance).

How much work? The move will be financed by the potential energy released during the displacement, thereby preserving the total energy. All changes in energy must be paid for in full.

The field acts as a reservoir. To each point in a gravitational or electric field there is assigned a corresponding potential energy: E_A at point A, E_B at point B, E_C at point C. The work needed to go from A to B is equal to the change in potential energy ($E_B - E_A$).

In the earth's gravitational field, for example, potential energy depends on the height (h, in meters) of a mass above the surface:

$$E = mgh \qquad g = 9.81 \text{ m s}^{-2}$$

The standard acceleration due to gravity, g, combines the mass and radius of the earth with the gravitational constant, G.

SAMPLE PROBLEMS

Assume, first, that we have a rubber ball possessed of perfect *elasticity*, a ball able to bounce off another object without losing any energy during the collision. Assume further that the ball meets no resistance from any molecules in the air. See how much we can already say about this simple body's energy and motion:

EXAMPLE 1-1. How the Ball Bounces

PROBLEM: Released from a point 3.28 feet above the earth's surface, the ball (mass = 100.0 g) falls straight down and bounces straight up. (a) What is the total energy just before impact? (b) What is the kinetic energy just before impact? (c) What is the ball's velocity just before impact? (d) What is the ball's velocity at the instant it turns around? (e) How high does it travel on the way up? (f) Summarize the complete trajectory with reference to Newton's three laws.

SOLUTION: Since the ball is dropped from rest, it has only gravitational potential energy (mgh) at the start. There is no kinetic energy yet (zero velocity), and so the *total* energy remains forever equal to the original value of mgh. The ball's subsequent motion is determined entirely by its initial height above the ground.

Begin by converting all data into SI units (meters and kilograms), taking note also of the significant figures. It is a small but crucial first step, for failure to use the appropriate units will yield numerically incorrect results:

$$3.28 \text{ ft} \times \frac{12 \text{ in}}{\text{ft}} \times \frac{2.54 \text{ cm}}{\text{in}} \times \frac{1 \text{ m}}{100 \text{ cm}} = 1.00 \text{ m}$$

$$100.0 \text{ g} \times \frac{1 \text{ kg}}{1000 \text{ g}} = 0.1000 \text{ kg}$$

Next compute the gravitational potential energy present at the start, mgh, understanding this quantity to be our inviolate total energy. Call it E_{tot}. The value E_{tot} represents all the energy the ball will ever have, and this amount will remain constant as the ball bounces:

$$E_{tot} = mgh = 0.1000 \text{ kg} \times 9.81 \text{ m s}^{-2} \times 1.00 \text{ m} = 0.981 \text{ J}$$

At this stage we should recall that

$$1 \text{ joule} = 1 \text{ newton-meter} = \text{kg m}^2 \text{ s}^{-2}$$

and we should also use the number so obtained (nearly 1) to give some physical meaning to a joule: 1 J represents the energy delivered by a mass of approximately 100 g falling from a height of one meter.

Now answer the questions.

(a) *What is the total energy just before impact?* The same as it was before the fall: 0.981 J. Total energy remains constant.

(b) *What is the kinetic energy just before impact?* Also 0.981 J. At ground level (zero height), the potential energy is depleted and the ball possesses only kinetic energy.

(c) *What is the velocity just before impact?* Knowing the kinetic energy at ground level ($E_{k0} = \frac{1}{2}mv_0^2$), we simply solve for v_0, the speed at ground level:

$$E_{k0} = \frac{1}{2}mv_0^2$$

$$v_0 = \sqrt{\frac{2E_{k0}}{m}} = \sqrt{\frac{2 \times 0.981 \text{ kg m}^2 \text{ s}^{-2}}{0.1000 \text{ kg}}} = 4.43 \text{ m s}^{-1}$$

To specify the velocity fully, we say further that its direction is straight *down*. The subscript 0 reminds us that this value applies to the ball at ground level only.

(d) *What is the velocity just after the rebound?* Also 4.43 m s^{-1}. Since the ball is still on the ground (height = 0), its potential energy remains zero. The total energy is resident entirely in the body's motion.

(e) *How high does the ball travel on the way up?* If nothing is lost to the surroundings, then the total energy remains equal to the initial gravitational potential energy. The ball climbs back precisely to the point from which it was released (1.00 m); and there, drained of kinetic energy, the mass comes to rest momentarily (zero velocity) before falling once again.

(f) *What has happened?* Released from rest, the ball is pulled down to the earth by the force of gravity. The gravitational force overcomes the inertia of the object (Newton's first law) and accelerates the ball downward at a rate of 9.81 m s^{-2} (Newton's second law). Since no force acts in the horizontal direction, the ball is never deflected from its vertical path (again, Newton's first law).

Upon hitting the ground, the ball abruptly suffers a force delivered *upward*—a force supplied by the charged particles in the ground itself, a force rooted not in gravity but in the electromagnetic interaction. Pushed upward, the ball now changes its momentum; it reverses course and gains altitude while fighting against gravity.

If the momentum just before impact is positive (mv_0), then it turns negative ($-mv_0$) as the object starts up just after impact. The *ball's* change in momentum (Δp_0) therefore is

$$\Delta p_0 = p_{\text{after}} - p_{\text{before}} = -mv_0 - mv_0 = -2mv_0$$

immediately after the bounce, but the *total* change in momentum (ball plus ground) must be zero. Total momentum must always be conserved.

Indeed it is. Momentum is conserved here because the ground recoils away from the ball with a momentum equal to $+2mv_0$, a change equal in magnitude and opposite in direction. The ball, touching the ground for some time Δt, thus receives a force equal to $-2mv_0/\Delta t$ *from the ground*. The ground, in contact with the ball for the same time, simultaneously receives a force equal to $+2mv_0/\Delta t$ *from the ball*. For every action, there is an equal and opposite *re*action (Newton's third law).

The ground delivers just enough of a push to propel the ball back to its initial height. Climbing upward, the body loses speed continuously under the unremitting downward tug of gravity.

EXAMPLE 1-2. The Bigger They Are . . .

PROBLEM: Suppose some other object has a mass of 50.00 kg. Let it be dropped, again, from the same height used in Example 1-1 (1.00 m). With what velocity does the heavier mass hit the ground? Assume that the earth's atmosphere offers no frictional resistance.

SOLUTION: Absent resistance from the air, all objects fall from rest with the same acceleration due to gravity: g, a constant. On earth we sometimes forget this rule of motion, because masses falling near the earth usually are slowed by friction from the atmosphere; but not so in the vacuum of outer space, where the velocity of a freely falling body is independent of mass. A heavier piece of matter hits the moon, say, with more momentum than a lighter one—but not with more velocity. Dropped simultaneously from the same height (h), a feather and a boulder will reach the lunar surface at the same time. Note why:

$$\text{Total energy before release} = mgh$$

$$\text{Total energy just before impact} = \frac{1}{2}mv_0^2$$

Energy conservation demands that

$$mgh = \frac{1}{2}mv_0^2$$

and thence we obtain

$$v_0 = \sqrt{2gh}$$

for the final velocity. The value of v_0 is the same for all masses. The *m*'s cancel out.

What is the velocity at impact (on earth) in the absence of any air fric-tion? Just as in Example 1-1(c): 4.43 m s^{-1}. But the more massive ob-ject does deliver correspondingly more momentum (*mv*) and kinetic energy ($\frac{1}{2}mv^2$).

EXAMPLE 1-3. The Way the Ball Really Bounces

Return now to the real world, where perfectly elastic balls do not exist. With each real-world bounce, a bouncing ball loses more and more alti-tude until finally it comes to rest. Portions of the original energy are ab-sorbed by ground and ball during each impact.

PROBLEM: Repeat the experiment described in Example 1-1, but this time with an *inelastic* body of the same mass. Suppose that we observe the following behavior: Dropped from the same point ($h = 1.00$ m), the 100.0 g ball returns to a height of only 0.500 m after the first bounce. After that, it continues to bounce some 20 or more times before coming to rest. (a) How fast is the ball moving at the instant it begins to rebound from the first impact? (b) How much energy is eventually redistributed into the internal motion of ground and ball?

SOLUTION: The total energy is fixed by the initial height and mass of the object at rest, giving us the value already computed in Example 1-1(a):

$$E_{tot} = mgh = 0.981 \text{ J}$$

After one bounce, however, the potential energy evidently has been cut in half, because the ball reaches only half its original height (0.500 m rather than 1.00 m). Pausing at the top, with zero kinetic energy, the ball thus has an external energy of only 0.4905 J. Meanwhile, an equal amount (0.4905 J) has gone into the combined internal energy of ground and ball after the first impact.

Now if the ball has a potential energy of 0.4905 J at the top of its path, then it must have a kinetic energy of 0.4905 J at the bottom. Just that amount, combined with the 0.4905 J absorbed *internally* by ground and ball, preserves the total energy at 0.981 J.

(a) *How fast is the ball moving as it begins the first rebound?* Simply set $\frac{1}{2}mv_0^2$ equal to 0.4905 J and solve as in Example 1-1(c). The answer is 3.13 m s^{-1}. With less external energy available for the return trip, the ball starts out slower and rises to a lesser height than before.

(b) *How much energy eventually goes into internal motion?* With the ball at rest on the ground, the system's energy has become entirely internal. Ground and ball will have absorbed whatever potential energy the raised body started out with: here, 0.981 J. The total energy remains constant throughout all other changes. Energy is never lost, only misplaced.

EXAMPLE 1-4. **Molecular Billiard Balls**

Consider this problem an advance look at the properties of gases, a subject we shall take up extensively in Chapter 10. There we shall see more clearly how the properties of a gas arise from the billiard-ball-like behavior of the molecules.

PROBLEM: Molecular speeds vary from very slow to very fast in a gas, but under normal conditions (25°C, atmospheric pressure) a typical nitrogen molecule travels at approximately 500 m s^{-1} while suffering 5×10^9 collisions per second. If a single nitrogen molecule has a mass of 4.6517×10^{-26} kg, how much force develops if two such particles meet head on in an elastic collision? See the illustration below:

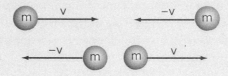

SOLUTION: The scenario is no different from a ball bouncing off the ground. Two molecules meet, interact for a time Δt, and then move off in opposite directions. We can solve the problem by requiring that momentum be conserved.

Assume that each molecule travels at a speed (v) of 500 m s^{-1} and has a mass (m) of 4.6517×10^{-26} kg. The velocities have algebraic signs corresponding to their respective directions (positive $\equiv$ right, negative $\equiv$ left):

$$\text{\textcircled{A}} \quad mv \longrightarrow \quad \longleftarrow -mv \quad \text{\textcircled{B}}$$

Molecule A originally carries a momentum of $+mv$ to the right, while molecule B carries a momentum of $-mv$ to the left.

Before the collision, the total momentum

$$p_A + p_B = mv - mv = 0$$

is *zero*; and after the collision, too, the total momentum is still zero. Each molecule reverses course and goes off in the opposite direction at the same speed:

$$\text{\textcircled{A}} \quad \longleftarrow -mv \qquad mv \longrightarrow \quad \text{\textcircled{B}}$$

Molecule A, now going to the left, has a momentum equal to $-mv$. Molecule B, rebounding to the right, has a momentum equal to $+mv$. The total momentum,

$$p_A + p_B = -mv + mv = 0$$

never changes.

But the momentum of molecule A, considered separately, does change by $-2mv$ over some interval equal to Δt (an interval we can compute). Likewise, molecule B's momentum undergoes a change of $+2mv$ during the same collision. Hence the force imparted to A is $-2mv/\Delta t$, whereas the force imparted to B (equal and opposite) is $+2mv/\Delta t$. Note the different algebraic signs.

All that remains is to compute Δt, the duration of the encounter. Now, since there are five billion (5×10^9) collisions *per second*, we estimate that *each* collision endures for 0.2 billionths of a second:

$$\Delta t = \frac{1 \text{ s}}{5 \times 10^9 \text{ collisions}} = 2 \times 10^{-10} \text{ s/collision}$$

The force delivered to, say, molecule B is then

$$F_B = \frac{2mv}{\Delta t} = \frac{2 \times 4.6517 \times 10^{-26} \text{ kg} \times 500 \text{ m s}^{-1}}{2 \times 10^{-10} \text{ s}}$$

$$= 2 \times 10^{-13} \text{ N} \qquad (1 \text{ N} = \text{kg m s}^{-2})$$

The force delivered to molecule A is equal and opposite: $F_A = -2 \times 10^{-13}$ N.

We discover, not surprisingly, a *small* force appropriate to a small particle. It amounts to only 0.2 trillionths of a newton or, equivalently, just over 0.04 trillionths of a pound (2.2 lb = 9.81 N). Let that be our invitation to begin thinking small.

The next problems highlight the electric forces and energies that govern the behavior of electrons, protons, and neutrons.

EXAMPLE 1-5. Mass versus Charge in the Microworld

Say goodbye to gravity, a force woefully insufficient to hold together atoms and molecules. The molecular world is purely an *electrical* world, and with just a few representative calculations we can see why.

Here are the parameters of that small world: The mass of a single neutron (m_n) is 1.675×10^{-27} kg, while the mass of a single proton (m_p) is only slightly smaller at 1.673×10^{-27} kg. The electron, lighter by a factor of approximately 1836, has a mass (m_e) of 9.109×10^{-31} kg. Both electron and proton carry a charge of magnitude 1.602×10^{-19} C, a quantity represented by the letter e. An electron bears a charge of $-e$; a proton, $+e$. The neutron is uncharged overall.

These elementary masses and charges

$$0.00000000000000000000000000000167 \text{ kg}$$

$$0.0000000000000000000000000000000009109 \text{ kg}$$

$$0.0000000000000000001602 \text{ C}$$

are almost incomprehensibly small, and so too are the distances between particles in molecules, atoms, and nuclei. A typical chemical bond between atoms extends over only 1×10^{-10} m, whereas the separation between protons and neutrons inside a nucleus is even less: an infinitesimal 1×10^{-15} m.

PROBLEM: For each pair of particles, compute both the Coulomb electric force and the gravitational force: (a) Two electrons separated by 1.00×10^{-10} m. (b) One proton and one electron separated by 1.00×10^{-10} m. (c) Two electrons separated by 0.500×10^{-10} m. (d) Two protons separated by 1.00×10^{-10} m. (e) Two protons separated by 1.00×10^{-15} m.

SOLUTION: For the Coulomb interaction between two electrons separated by a distance r, we have

$$F = \frac{1}{4\pi\epsilon_0} \frac{e^2}{r^2} \qquad \epsilon_0 = 8.85 \times 10^{-12} \text{ C}^2 \text{ N}^{-1} \text{ m}^{-2}$$

and thus a *repulsive* force of

$$F = \frac{1}{4\pi \times 8.85 \times 10^{-12} \text{ C}^2 \text{ N}^{-1} \text{ m}^{-2}} \frac{(1.602 \times 10^{-19} \text{ C})^2}{(1.00 \times 10^{-10} \text{ m})^2}$$

$$= 2.31 \times 10^{-8} \text{ N}$$

between the two particles at 1.00×10^{-10} m.

The corresponding gravitational force, also obeying an inverse-square law, differs by 43 powers of 10:

$$F = -\frac{Gm_e^2}{r^2} \qquad G = 6.67 \times 10^{-11} \text{ N m}^2 \text{ kg}^{-2}$$

$$F = (-6.67 \times 10^{-11} \text{ N m}^2 \text{ kg}^{-2}) \frac{(9.109 \times 10^{-31} \text{ kg})^2}{(1.00 \times 10^{-10} \text{ m})^2}$$

$$= -5.53 \times 10^{-51} \text{ N}$$

It is an enormous ratio. The repulsive electric force is nearly 10,000,000,000,000,000,000,000,000,000,000,000,000,000,000 times greater than the attractive gravitational force.

With just these two benchmark values for the forces, we can go on to answer the remaining questions with minimal computation.

(a) *Two electrons, r = 1.00 × 10⁻¹⁰ m.* Already done. Electric force $= 2.31 \times 10^{-8}$ N. Gravitational force $= -5.53 \times 10^{-51}$ N.

(b) *One electron and one proton, r = 1.00 × 10⁻¹⁰ m.* The magnitude of the electric force is the same as between two electrons at this distance, but now the interaction is *attractive*. Electric force $= -2.31 \times 10^{-8}$ N. To determine the force of gravity, replace one factor of m_e by m_p (the mass of a proton). The associated force is increased approximately 1836-fold, corresponding to the ratio m_p / m_e. Gravitational force $= -1.02 \times 10^{-47}$ N.

(c) *Two electrons, r = 0.500 × 10⁻¹⁰ m.* At a separation of 1.00×10^{-10} m, as in (a), we found the repulsive electric force to be 2.31×10^{-8} N. The inverse-square relationship guarantees that this quantity will quadruple when the distance is halved. Electric force $= 9.24 \times 10^{-8}$ N. The gravitational force quadruples as well, thereby maintaining the same ratio between the forces. Gravitational force $= -2.21 \times 10^{-50}$ N.

(d) *Two protons, r = 1.00 × 10⁻¹⁰ m.* Electric force $= 2.31 \times 10^{-8}$ N, the same as in (a). The Coulomb repulsion is unaffected by mass. To compute the gravitational force, we substitute two factors of m_p for the two factors of m_e used above in (a). Gravitational force $= -1.87 \times 10^{-44}$ N.

(e) *Two protons, r = 1.00 × 10⁻¹⁵ m.* Relative to the values in (d), these forces are enhanced by 10^{10} owing to the 100,000-fold smaller separation. Electric force $= 2.31 \times 10^{2}$ N. Gravitational force $= -1.87 \times 10^{-34}$ N.

Even at this tiny distance, typical of protons within a nucleus, we have a gravitational force still too small to matter; yet suddenly there is a huge, macro-sized repulsive electrical interaction in excess of 50 *pounds*—clearly large enough to prevent two protons from ever coming together.

But come together they do inside the nucleus, for there is an even stronger nuclear binding force present (about which, see Chapter 21).

Outside the nucleus, where chemistry actually happens, the relationships between atoms and molecules are controlled solely by the electrical interaction. And out there it is the electron, acquiring energy from other charged particles, that serves as a lightweight traveler. Our final example provides a first look at how a single electron might move through an electric field in free space.

EXAMPLE 1-6. Electrical Potential Energy

Electrical potential energy is usually related to a *voltage*, or *potential difference*, between two points in space. We say that a potential difference of 1 volt exists if 1 coulomb of charge changes energy by 1 joule upon moving from point A to point B. So defined, the voltage measures the work done when a charged particle moves through an electric field:

$$1 \text{ V} = 1 \text{ J C}^{-1}$$

PROBLEM: Let a single electron be accelerated through a potential difference of 1.000 V. (a) If the electron starts from rest, what is its final speed after undergoing this change in voltage? (b) How fast and in what direction would a proton travel if subjected to the same treatment? (c) A neutron?

SOLUTION: We have merely a disguised version of the free-falling ball, this time with an electric force substituted for gravity. The key relationship is

$$\text{Electrical work} = \text{voltage} \times \text{charge transferred}$$

which for an electron accelerated through 1.000 V becomes

$$\text{Electrical work} = 1.000 \text{ J C}^{-1} \times (-1.602 \times 10^{-19} \text{ C}) = -1.602 \times 10^{-19} \text{ J}$$

The negative sign corresponds to the electron's acceleration from a more negative region to a more positive region.

The particle thus changes energy by 1.602×10^{-19} J, an amount defined as 1 electron volt (eV). Just as a ball falls from a point of high gravitational potential down to the earth's surface, so does the electron fall from high *electrical* potential to an analogous "ground" level. At the end, all the energy imparted by the field is invested in kinetic form.

Duplicating the procedure of Example 1-1, we then obtain the speed at the electrical ground level as

$$E_{k0} = \frac{1}{2}m_e v_0^2$$

$$v_0 = \sqrt{\frac{2E_{k0}}{m_e}} = \sqrt{\frac{2 \times 1.602 \times 10^{-19} \text{ kg m}^2 \text{ s}^{-2}}{9.109 \times 10^{-31} \text{ kg}}}$$

$$= 5.931 \times 10^5 \text{ m s}^{-1}$$

(a) *What is the electron's final speed?* 5.931×10^5 m s^{-1}

(b) *What would be the final speed and direction of a proton?* The final kinetic energy is the same. Since only the mass is different, we substitute m_p for m_e in the equation above to obtain 1.384×10^4 m s^{-1}. The positive particle travels in a direction opposite to the electron (and slower as well, by a factor of $\sqrt{m_e/m_p}$).

(c) *What happens to a neutron?* Electrically, nothing. A neutral particle is unaffected by the voltage.

EXERCISES

1. Supply the missing values:

   (a) 1 day = ___ s
   (b) 1 year = ___ s
   (c) 8.1764×10^{87} s = ___ centuries
   (d) 3.45×10^{-3} mile = ___ cm
   (e) 7.5000×10^{4} m = ___ km
   (f) 32.6 cm = ___ m

2. Supply the missing values:

   (a) 1 cm^2 = ___ in^2
   (b) 1 cm^3 = ___ in^3
   (c) 1 L = ___ mL
   (d) 1 L = ___ cm^3
   (e) 1 L = ___ in^3
   (f) 1 L = ___ ft^3

3. Reduce each expression to a dimensionless number, taking care (as always) to preserve the correct number of significant digits:

   (a) $\dfrac{1.0 \text{ ft} + 3.2 \text{ ft}}{75.00 \text{ m}}$ = ___

   (b) $\dfrac{5185.7 \text{ mi} \times 31765.4 \text{ cm}^2}{1.0000 \text{ m}^3}$ = ___

   (c) $\dfrac{36.2000 \text{ ft}}{3.0007 \text{ in}}$ = ___

   (d) $\dfrac{36.2000 \text{ ft}}{3.0007 \text{ in}} - 725.2$ = ___

   Note that the relationship between centimeters and inches is expressed, *exactly*, as

   $$2.54 \text{ cm} = 1 \text{ in}$$

4. If distance is measured in centimeters (not meters) and mass is measured in grams (not kilograms), the corresponding unit of force is then called a *dyne*. (a) Use Newton's second law,

$$\text{Force} = \text{mass} \times \text{acceleration}$$

to express the dyne in fundamental units of cm, g, and s. (b) How many dynes are equal to a force of 1 newton?

5. In the centimeter-gram-second system of measurement described just above, the unit of work and energy is termed an *erg*. (a) Use the definition

$$\text{Work} = \text{force} \times \text{distance}$$

to express the erg in fundamental units of cm, g, and s. (b) How many ergs are equal to an energy of 1 joule?

6. On the basis of dimensions alone, which of the following relationships *must* be wrong?

(a) Energy = $(\text{momentum})^2 \times \text{mass}$

(b) Energy = mass + temperature

(c) Work = force $\times (\text{distance})^2/\text{volume}$

(d) Distance = momentum $\times$ time/mass

(e) Acceleration = distance $\times$ time

7. Which combinations are forbidden? Why?

(a) distance/time

(b) distance $\times$ time

(c) distance + time

(d) distance − time

(e) distance + velocity $\times$ time

8. Let x_0 denote the initial position of a particle, and let $x(t)$ denote the body's position after it moves for a time t with a constant velocity v. Which equation has the correct dimensions? Does it make sense physically?

(a) $x(t) = x_0 + v + t$

(b) $x(t) = x_0 t + v$

(c) $x(t) = x_0 + vt$

(d) $x(t) = x_0 vt$

9. Try an explicit example involving rate, time, and distance: At midnight, an automobile passes a road marker that reads "Mile 100." At what time will the car, traveling at a constant 60 miles per hour, reach the "Mile 500" marker?

10. Absent all forces other than gravity, a freely falling object covers a distance

$$d = \frac{1}{2}gt^2$$

in a time t. Suppose, then, that g and t have the values

$$g = 9.81 \text{ m s}^{-2}$$

$$t = 2.0165 \text{ s}$$

and we wish to calculate d:

$$d = \frac{9.81 \text{ m s}^{-2}}{2} (2.0165 \text{ s})^2$$

(a) How many digits are significant in each of the factors 9.81, 2, and 2.0165? (b) How many digits are significant in the final result? (c) Complete the numerical calculation, and show that d has the proper units.

11. Which of the following objects is *not* undergoing an acceleration? (a) A car rounding a curve at a constant speed of 30 miles per hour. (b) An apple falling to earth. (c) An airplane cruising at 1000 kilometers per hour on a fixed compass heading.

12. A stone at rest on the ground has zero velocity and zero acceleration. What forces act on the stone to keep it that way?

13. The familiar pound (lb) is a unit of force in the English system of measurement. Criticize the following statement: "One kilogram is *equal* to 2.2 pounds." Consider, in particular: How must we interpret the word "equal" given the relationship between force and mass? Is the statement valid at all locations throughout the universe?

14. Believe it or not, the unit of mass in the English system is called a *slug*, equivalent to 14.594 kg. Slug or kilogram, however, force is force; and the relationship

$$\text{Force} = \text{mass} \times \text{acceleration}$$

holds within any consistent system of units. (a) Expressed in SI units, the acceleration due to gravity has the value

$$g = 9.81 \text{ m s}^{-2}$$

What is the equivalent value in ft s^{-2}? (b) A force of 1 lb will accelerate a mass of 1 slug by 1 ft s^{-2}. How many pounds are equal to 1 newton?

15. A mass of 1.00 kg falls from a height of 100.0 m. With what velocity and with what momentum does it hit the ground?

16. As before, a mass of 1.00 kg falls from a height of 100.0 m. (a) By how much does the potential energy change between heights of 100.0 and 90.0 m? Between 10.0 m and 0 m? (b) By how much does the kinetic energy change between heights of 100.0 and 90.0 m? Between 10.0 m and 0 m?

17. Once more, and for the last time, a mass of 1.00 kg falls from a height of 100.0 m: Calculate the potential energy, the kinetic energy, and the total energy at 100.0 m, 90.0 m, 10.0 m, and 0 m above the ground.

18. Without doing an explicit numerical calculation, answer: Which of these two objects carries more momentum? (a) A mass of 50.0 g moving at a velocity of 100.0 m s^{-1}. (b) A mass of 100.0 g moving at a velocity of 25.0 m s^{-1}. By what factor do the two momenta differ?

19. Compute the momenta for the two bodies described in the preceding exercise: (a) A mass of 50.0 g moving at a velocity of 100.0 m s^{-1}. (b) A mass of 100.0 g moving at a velocity of 25.0 m s^{-1}.

20. Again, without doing an explicit calculation: Which object carries more kinetic energy? (a) A mass of 50.0 g moving at a velocity of 100.0 m s^{-1}. (b) A mass of 100.0 g moving at a velocity of 25.0 m s^{-1}. By what factor do the two kinetic energies differ?

21. Compute the kinetic energies of the two bodies described in the preceding exercise: (a) A mass of 50.0 g moving at a velocity of 100.0 m s^{-1}. (b) A mass of 100.0 g moving at a velocity of 25.0 m s^{-1}.

22. These same two particles

$$m_A = 50.0 \text{ g} \qquad v_A = 100.0 \text{ m s}^{-1}$$
$$m_B = 100.0 \text{ g} \qquad v_B = 25.0 \text{ m s}^{-1}$$

now bounce elastically off a wall. Which head-on collision delivers the greater force to the wall? (a) An impact of particle A that endures for 0.01 s. (b) An impact of particle B that endures for 0.05 s. Answer first without doing an explicit calculation; after that, compute the numerical values.

23. Ball A, with a mass of 25.0 g, is moving at 5.00 m s^{-1} from left to right. Ball B, with a mass of 75.0 g, is moving at 2.00 m s^{-1} from right to left. They meet head-on in an elastic collision, whereupon ball A moves away from right to left at a speed of 5.50 m s^{-1}. (a) In what direction and at what speed does ball B move away? (b) Show, explicitly, that both momentum and energy are the same before and after the collision.

24. Compute the kinetic energy (in joules) of a 4000-lb automobile traveling at 60 miles per hour. Be careful with the units.

25. Imagine that the 4000-lb car of the preceding exercise is now suspended above the earth. At what height would the potential energy of the stationary car match the kinetic energy generated at 60 miles per hour on the highway?

26. Who does more work against gravity? Weightlifter A, who raises a 50-kg barbell to a height of 2.5 m, . . . or weightlifter B, who raises a 100-kg barbell to a height of only 1.5 m. How many joules of energy does each lifter expend?

27. Energy can neither be created nor destroyed. If so, then from where do the weightlifters get their energy and to where does it go?

28. Compute the force of gravitational attraction between the earth and the moon, making the following assumptions: (1) The mass of the earth is 5.98×10^{24} kg. (2) The mass of the moon is 7.35×10^{22} kg. (3) The distance between earth and moon is 3.84×10^{5} km.

29. A 1.00-g mass of hydrogen atoms contains 6.02×10^{23} protons and 6.02×10^{23} electrons. Suppose—fantastically—that the combined electric charge of *all* these protons might be gathered into a single, supercharged, positive particle; and suppose, likewise, that the combined charge of all the electrons is gathered into a single, super-charged, negative particle. At what separation would the attractive electric force then be equal to the gravitational attractive force

between the earth and the moon? Refer to the problem immediately above for the gravitational force, and use the result to assess the strength of the electrical interaction.

30. Assume, as before, that the earth and moon behave as pointlike particles separated by 3.84×10^5 km. How much electric force would exist between them if each body had a net positive charge equivalent to 6.02×10^{23} protons? Think: Is electromagnetism ever an important consideration for determining the motion of heavenly bodies? Is gravity an important consideration for determining the motion of electrons and protons?

31. Starting from rest, an electron passes through a potential difference of 1.00×10^3 V. What final velocity does the particle attain?

32. An electron, accelerated from rest, reaches a speed of 1.00×10^6 m s^{-1} after passing through a difference in electrical potential. How many volts were applied?

2

Atoms and Molecules

2-1. The Existence of Atoms
2-2. The Elements
2-3. Chemical Combination
2-4. Molecules
2-5. Stoichiometry
   REVIEW AND GUIDE TO PROBLEMS
   EXERCISES

2-1. THE EXISTENCE OF ATOMS

Atoms and molecules are small, so we tend to overlook them. It takes a million hydrogen atoms, for example, to equal the thickness of one human hair. And in just a few grams of water there are some 10^{23} molecules—an enormous number, 1 followed by 23 zeros. Ordinary experience remains blind to the particulate nature of matter, for the building blocks are too small and too many to make any impression on everyday life.

Nevertheless it is these unseen particles, held together in various states of association, that form the very fabric of material reality. Atoms exist. Molecules exist. We can see individual atoms, remarkably, in the shadowy landscape of Figure 2-1, an image made by an instrument called a scanning tunneling microscope, or STM. Each lump in the pattern arises from an atom of silicon adhering to a metallic surface, and the total effect is of a grainy snapshot, something like the dots in a magnified newspaper photograph. What we see is the graininess of matter: atoms, one by one, each a distinct chemical entity.

Not until recently, with the invention of the scanning tunneling microscope, has it become routinely possible to visualize a single atom or

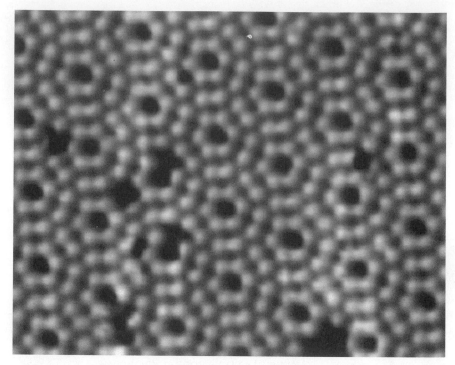

FIGURE 2-1. Matter is not continuous, as this STM image of individual silicon atoms so graphically attests. Sensing minute changes in electrical potential, the scanning tunneling microscope (STM) maps out the geography of electrons on an atomic scale. The 1986 Nobel prize for physics was awarded to E. Ruska, G. Binnig, and H. Rohrer for development of the technique.

small molecule. Yet even the first STM images, for all their power to stir the imagination, brought no surprises with their basic message that atoms exist. The reality of atoms and molecules had been thoroughly proved during the last two centuries, by means less direct although no less convincing. The newly acquired ability to resolve individual atoms was rather like the sudden illumination of a dark room, but only long after the room's contents were already mapped out by touch alone.

Scientists see and touch their atoms in various ways, usually by observing how light interacts with matter. Think of light, for now, as a vehicle by which electric and magnetic fields are conveyed through space, an agency, ghostly and intangible, that we shall explore further in Chapters 4 and 20. Picture this agency sweeping forward as a wave, bringing along an oscillating field of force capable of moving charged particles. Its target? Matter. Matter, as we now understand, is a collection of charged particles, each subject to electromagnetic forces and consequently susceptible to the push and pull of a light wave. Through such electrical interactions is the structure of matter discovered, bit by bit.

Light, understood more broadly as *electromagnetic radiation*, goes beyond the ordinary visible light perceived by our senses, the rainbow of colors. Electromagnetic disturbances travel in packets ranging from low-frequency radio waves to highly energetic X rays, with microwave, infrared, visible, and ultraviolet light falling in between. Every kind of radiation has its own energy, and each one interacts with atoms and molecules in distinctive ways. Some forms of light cause molecules to vibrate; some forms induce rotations; some forms excite the electrons.

We observe and we learn. We learn that light, in any of its forms, leaves matter temporarily disturbed in its wake, but always the disturbance is regular, discrete, and specific. There is a grittiness to the interaction, and the effect makes sense only if matter is particulate and grainy. The energy delivered by the light is accepted in well-defined, regulated doses. Matter is not continuous.

Of related interest here, while we reflect on the existence of atoms, is the method of *X-ray diffraction*, which reveals precisely where the atoms sit inside the ordered rank and file of a crystal. The electric field of an X-ray beam sets in motion the electrons contained in the material, causing them to redirect—scatter—the electromagnetic waves in transit. The waves emerging from the structure then form a distinct pattern of rings and spots from which the interatomic spacings can be inferred. Figure 2-2 shows the diffraction pattern for DNA, the molecular repository of the genetic code. The helical structure of this very large molecule, which serves as life's universal blueprint, was ascertained by X-ray diffraction measurements.

From such evidence we know that atoms and molecules exist. We can, in a way, see them and touch them, map out their structures and learn their characteristic properties. This brief sketch serves merely to suggest a few of the methods used to probe the nature of matter, not to trace the arduous path by which atoms became real to scientific observers. There were many more experimental probes, each with its own particular focus but each ultimately proving that matter is discontinuous.

We should remember throughout that the most elementary question—namely, are there indeed fundamental chemical entities in the first place?—was already answered nearly two hundred years ago, and answered even without benefit of complicated instrumentation. Careful study of chemical reactions revealed that elements combine only in certain whole-number ratios, such as when two volumes of hydrogen gas consume one volume of oxygen gas to produce two volumes of water vapor. There was always two of this and three of that, or one of this and one of that, or some other rational relationship between quantities of matter. And the ratios never changed. If it was 2:1, then so it remained; not 2:1 today and 3:2 tomorrow, but always 2:1. These were consistent, integer relationships.

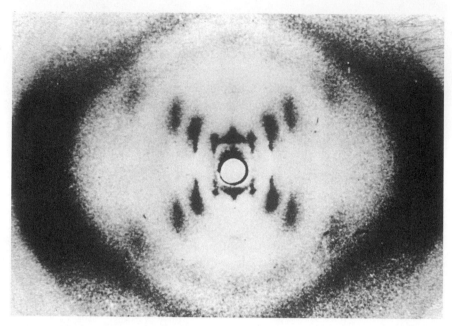

FIGURE 2-2. Matter is not continuous, as this X-ray diffraction photograph of crystalline DNA also proves. The dark spots testify to the presence of an ordered array of microscopic antennae, each antenna acting as a receiver and retransmitter of X rays—in other words, an array of separated *atoms*. Deflected by the electrons in the atoms, the electromagnetic waves scatter and eventually recombine to produce a pattern unique to the crystal (see Chapter 4). The 1962 Nobel prize for medicine was awarded to J. D. Watson, M. H. F. Wilkins, and F. H. C. Crick for elucidation of the structure of DNA.

The eventual conclusion was that chemical combinations took place between integral and indivisible units, interpreted provisionally as atoms and molecules. At the microscopic level, two *molecules* of hydrogen somehow had to interact with one *molecule* of oxygen to produce two *molecules* of water. Understanding the structure of atoms and molecules came later, but the observed laws of combination—nothing more—demanded the existence of discrete particles of matter.

2-2. THE ELEMENTS

Simplest of all the elements, an atom of **hydrogen** is built around a nucleus consisting of just one proton. A single electron, negatively charged, exactly balances the positive charge of the proton and occupies a volume of space outside the nucleus. We use the abbreviation H for hydrogen, sometimes recognizing the electron explicitly by a dot in the symbol H·. The notation ^{1}H is also employed to distinguish between isotopes, with the superscript indicating the ***mass number*** *A*: 1 in this case, originating from *one* proton.

The nucleus gives the atom its mass; the electron, its volume. Fully 99.95% of the mass of H is accounted for by the proton, concentrated into a dense sphere approximately 10^{-15} m in diameter. The electron, with a mass equal to only 1/1836 that of a proton, distributes itself over a much larger spherical region with a diameter of more than 10^{-10} m. There is a difference of five orders of magnitude, and so the hydrogen atom, like all atoms, contains mostly empty space. Were the nucleus as big as, say, a golf ball, then the electron typically would be found at a distance of 2000 m—a conflict in scale impossible to show in a drawing. Nevertheless, the stylized rendering offered in Figure 2-3 should convey a sense, at least, of the atom's basic architecture.

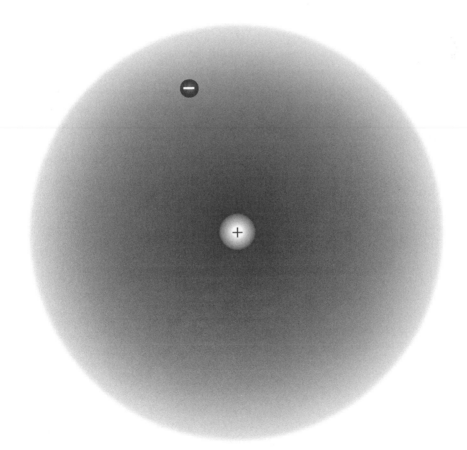

FIGURE 2-3. An atom of hydrogen: one proton, one electron. The electron roams far and wide from the nucleus, equally likely to show up in any direction. Exactly where it may be at this instant, no one can say; but, by applying the laws of quantum mechanics, we can compute the probability of the electron's appearance at any point in space (details in Chapters 4, 5, and beyond).

Hydrogen is the lightest of the elements and the most abundant, amounting to three-quarters of the known mass of the universe. Although most hydrogen nuclei exist as the isotope ^{1}H (the single proton just described), nearly 1 in 6000 contains a neutron as well. The neutron, approximately the same mass as a proton but electrically neutral, creates a heavier isotope called *deuterium* ("heavy hydrogen"), notated as ^{2}H or D. The superscript 2 records the mass number resulting here from one proton and one neutron. Another form of hydrogen, rarer still than deuterium, contains one proton and *two* neutrons. Designated *tritium* (^{3}H or T), this species is radioactive and decays over a period of years.

^{1}H, ^{2}H, and ^{3}H—three isotopes, one element. Each isotope is a form of hydrogen. Each isotope contains one proton, balanced by one electron to create a neutral atom. Each isotope has an *atomic number*, Z, of 1. Having *one proton* is the defining characteristic of a hydrogen nucleus. The three isotopes (Figure 2-4) are all legitimate forms of the same element, akin to three cousins all bearing the surname "$Z = 1$." They share much the same chemistry and differ only in mass.

Now if elements and their atoms are the building blocks of matter, then hydrogen nuclei are the source from which all heavier elements are assembled. The next element, helium (He), is formed by the *nuclear fusion* of hydrogen, a violent transformation occurring continually in the stars. Hydrogen isotopes meld together at extremely high temperature and pressure to create helium nuclei ($Z = 2$), of which the most common variety is ^{4}He with two protons and two neutrons. Heavier nuclei are built up in succession by this process of *nucleosynthesis*, and thus come into being all the atoms in the universe. Past and present, they all spring from hydrogen and helium.

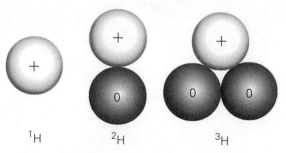

^{1}H ^{2}H ^{3}H

FIGURE 2-4. Isotopes of hydrogen: nuclei only. Each atom contains one electron outside the nucleus and one proton inside, giving it an atomic number of 1. Deuterium, twice as massive as hydrogen, has one neutron in the nucleus; tritium, with three times the mass, has two. Note that the structures pictured are purely schematic, not meant to depict the actual packing of protons (+) and neutrons (0) in the nuclei.

The series continues as

lithium	Li	$Z =$	3
beryllium	Be	$Z =$	4
boron	B	$Z =$	5
carbon	C	$Z =$	6
nitrogen	N	$Z =$	7
oxygen	O	$Z =$	8
fluorine	F	$Z =$	9
neon	Ne	$Z =$	10

all the way to uranium, $Z = 92$. Transformations in nuclear reactors and particle accelerators have produced elements beyond uranium, from neptunium (Np, $Z = 93$) through element number 112, with more expected. The atomic number for each atom is, as with hydrogen, the defining characteristic of the element, its unique identity; and the nucleus is the atom's foundation, its heart. A few selected nuclei are presented schematically in Figure 2-5.

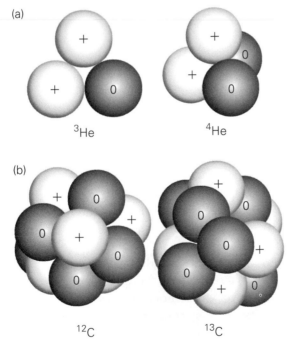

FIGURE 2-5. Examples of isotopic nuclei, grouped according to atomic number (Z, the number of protons). (a) Helium-3 and helium-4; $Z = 2$. (b) Carbon-12 and carbon-13; $Z = 6$. As part of a neutral atom, any of these nuclei would be surrounded by a much larger cloud of Z electrons.

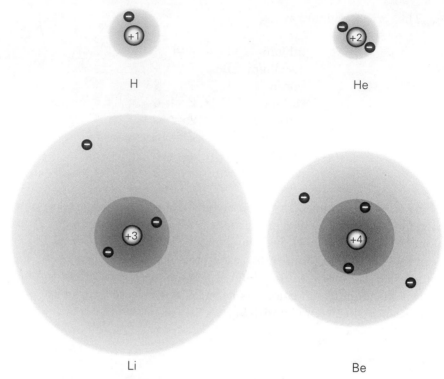

FIGURE 2-6. Atoms imagined, each built in the same way: a pinprick of a nucleus, astonishingly heavy, set amidst a spherical swarm of electronic dust; mostly empty space. Elements differ in their atomic number: Z protons inside the nucleus, Z electrons outside.

The mass of any atom, we see, is borne largely by this tiny but fantastically dense nucleus, wherein are bound Z protons and however many neutrons correspond to the particular isotope.* The volume of the atom, largely empty space, is determined by the Z electrons outside. For each element, an atomic structure then emerges according to a common blueprint.

Starting with hydrogen, one additional electron attaches itself to the structure for every increment of the atomic number, as suggested by Figure 2-6. A growing cloud of mobile electrons builds up in this way, and these lightweight particles (particularly those farthest from the nucleus) become the agents of chemical change. They forge the links between atoms and enable the formation of molecules. Chemical combination takes

*Remember that the nucleus is not held together by electrostatic forces, for these would tend to tear apart the positively charged protons. Rather the protons and neutrons are bound by the so-called **strong force**, a special interaction effective only at the very short distances found within nuclei.

several dozen elements and fashions from them a universe of millions of compounds.

Immediately with the two electrons of helium (He∶), however, we observe a striking difference compared with hydrogen: an *absence* of reactivity. Helium hardly does anything, but acts rather like a building block unable to find a place in any structure. It tends not to combine either with itself or with other atoms. It is the first of the **noble gases**, so named because these elements seem to stand aloof from the others. Helium's two electrons appear to offer a stability that makes further combination superfluous, as if a quota of two slots—this much and no more—has been filled.

A quota. Call it, instead, a filled **shell** of electrons, the term originally used to describe the apparently hierarchical organization of the atom. And treat this presumed shell, for the moment, not as an explanation but as a bare description of the facts. The hydrogen atom has one electron, and it reacts readily. The helium atom has two electrons, and it reacts not at all. Helium's first shell is complete, closed to newcomers. Hydrogen's quota is yet unfilled, still open to new combinations.

Why? We shall have to determine why, although not before all the facts are gathered. First we must examine the elements objectively, looking only to categorize and not to explain. First we must patiently count protons and electrons, seeking to uncover patterns in the structure. Only later, in the light of quantum mechanics (Chapters 4 through 7), will those patterns become understandable; only then will fact be illuminated by explanation.

Patience, then, during this brief first tour through the dozens of elements, which begins with reactive hydrogen and inert helium. The next atom, lithium, adds one proton to the nucleus and one electron *beyond* the already filled shell of two **core electrons**, as below:

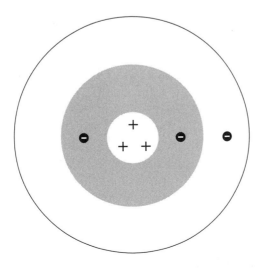

Only this outer, so-called **valence electron** now becomes available for subsequent chemical combination, whereas the two inner electrons remain close to the core and inaccessible to other atoms. It is as if lithium begins the buildup of electrons anew, having only that lone valence electron to share and therefore presenting itself as Li·.

Then comes beryllium (·Be·) with two valence electrons to offer—but which, unlike helium, is prepared to use them in the formation of molecules. Next is boron (·B·) with three valence electrons, and then carbon (·C·) with four, and so on through neon (:Ne:, $Z = 10$). Here, with neon's two electrons in the first shell and *eight* electrons beyond, comes the renewed stability of a noble gas. A second shell has been completed through the attainment of an *octet*: eight electrons.

Just past neon is sodium (Na, from the Latin *natrium*), which begins a new shell as Na·. Next is magnesium (·Mg·), followed by aluminum (·Al·). And the buildup continues until a second octet is completed with the formation of argon (Ar), where the 18 electrons occupy shells of $2 + 8 + 8$. The pattern recurs with potassium (K, $Z = 19$), which initiates another, more extensive series incorporating the first 10 *transition metals* (scandium through zinc) and ending with the noble gas krypton (Kr, $Z = 36$). All throughout the elements, points of stability reappear as shells of electrons are completed and noble-gas configurations are attained. See Figure 2-7.

There is something important going on, something still to be explained but nonetheless critical to recognize: a *periodicity* associated with the elements, a regular recurrence of electronic patterns and material properties. Lithium, sodium, and potassium, for example, all have one outer electron to share, having taken just one step past a noble gas. With that in common, they should have similar chemical properties and in fact they do. These elements, together with rubidium (Rb), cesium (Cs), and francium (Fr), are **alkali metals**, their behavior governed by a single valence electron. All the other electrons are housed in inner shells within the **core.**

The recurring properties of the elements are summarized most effectively in the **periodic table** (Figure 2-8), which lays out horizontal **periods** corresponding to the filling of electron shells. H and He make up the first row, followed by Li through Ne in the second, then Na through Ar in the third, and so forth. The columns of the table are called **groups**, or families in which the atoms display the same numbers of valence electrons. The aforementioned alkali metals, characterized by one valence electron, constitute Group I; here the members each have one electron more than the nearest noble gas. Group II consists of the **alkaline earth metals**—beryllium, magnesium, calcium, strontium, barium, and radium—elements whose chemical destiny is determined

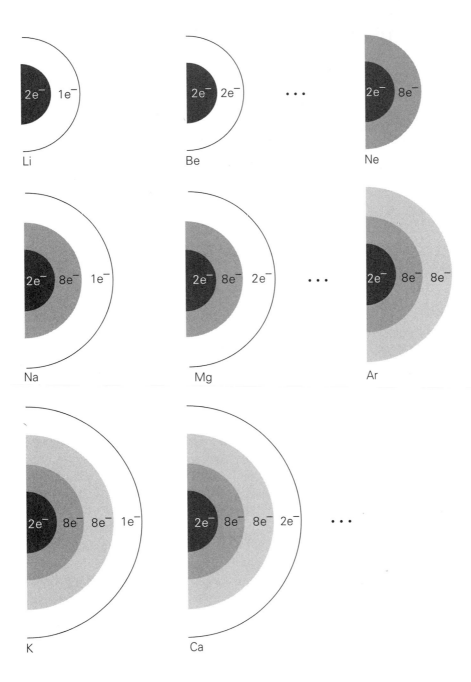

FIGURE 2-7. Core, valence, and chemical periodicity. Atoms fall naturally into groups having both similar valence shells and similar chemical properties. One implies the other: The valence electrons, farthest from the nucleus and easiest to detach, determine the reactivity of an atom. Less active chemically are the core electrons, which lie closer to the nucleus and rarely venture far afield. Core shells are represented by shaded regions; open valence shells are left white.

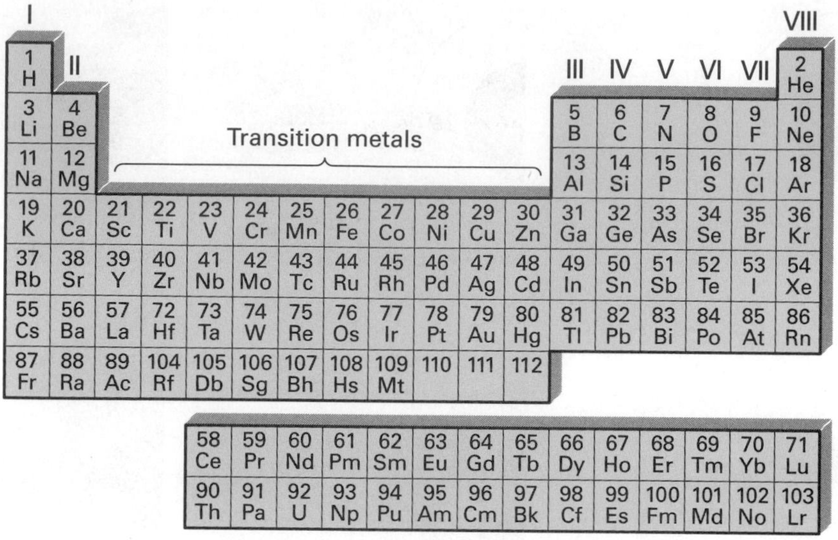

FIGURE 2-8. The periodic table, roster of the elements. The vertical columns contain *groups* of atoms with analogous valences; the horizontal rows contain *periods* of atoms set forth in order of increasing atomic number. Every row terminates with a noble gas, marking completion of the current valence shell.

by their *two* valence electrons. There are, in addition, 10 groups of **transition metals** with distinctive electronic properties, and six more groups containing various other metals, **semimetals**, and **nonmetals**. Important among the nonmetals are the **chalcogens** of Group VI, to which belong the elements oxygen, sulfur, selenium, tellurium, and polonium. Each of these atoms has six valence electrons, two short of a noble gas. Near the end of the table are the **halogens** of Group VII (fluorine, chlorine, bromine, iodine, and astatine), with seven valence electrons and hence one fewer than the next noble gas. The noble gases themselves, standing off to the right in Group VIII, complete the array.

We shall have more to say about the periodic table and the properties of the elements in subsequent chapters. To understand *why* the elements line up so neatly is to bring order and logical consistency to much of chemistry. Our needs at present, though, are satisfied by just a rough, preliminary awareness of atomic structure and particularly valence, which we now expand into a similarly qualitative picture of molecules and chemical bonds.

2-3. CHEMICAL COMBINATION

Consider that an atom is nothing more than a swarm of electrons localized around a nucleus, the whole construction a product of the electromagnetic force. Each negatively charged electron, although repelled by others of like kind, is attracted to the positively charged nucleus in such a way that everything works out right. The atom neither blows apart nor collapses, for the quantum mechanical laws of motion guarantee a measure of stability. The nucleus and its electrons can survive on their own.

But so can *two* nuclei and an appropriate number of accompanying electrons, and so can three nuclei, and four nuclei, and indeed thousands of nuclei. Composite structures—molecules—also may achieve stability if, once again, everything works out right. Why not? A molecule, after all, is nothing more than a swarm of electrons localized around two or more nuclei. It is in some ways a big atom, governed by the same fundamental electromagnetic force and the same laws of quantum mechanics.

These laws admittedly are more difficult to interpret for molecules than for atoms, and to understand the circumstances where "everything works out right" for a molecule is a major challenge. Molecular electronic structure is still an active area of chemical research, which we shall consider in due course. At this stage, however, a much simpler picture (even with its inevitable omissions, exceptions, and inaccuracies) can offer us a useful model to envision molecules and their transformations.

Our first thoughts of molecules might go as follows: Two nuclei, each bearing a positive charge, normally are pushed apart by electrostatic repulsion. Only by simultaneously attracting negatively charged electrons are they able to hold together and form a chemical bond. We expect, consequently, to find an increased concentration of electrons between two bonded nuclei, perhaps as imagined in Figure 2-9.

The most basic idea, due to G. N. Lewis in 1916, is that a chemical bond develops when two nuclei share a *pair* of electrons. A hydrogen molecule thus is formed when two atoms come together to make H_2, a process represented by **Lewis structures** or **dot diagrams** as

$$H\cdot + \cdot H \longrightarrow H\!:\!H$$

FIGURE 2-9. Sharing electrons: the covalent bond. Nuclei repel nuclei, and electrons repel electrons; but opposites attract: nuclei attract electrons. Two nuclei, each pulling on the same electron, draw together in a covalent chemical bond. An enhanced presence of negative charge between the positive particles cements the interaction.

The pair of dots between the hydrogen nuclei symbolizes the two electrons shared in the chemical bond, denoted alternatively by a dash in a structural formula such as H—H. This picture of bonding, a concept that predates quantum mechanics, has each hydrogen atom contributing its one electron to the joint enterprise of making a molecule. By doing so, the hydrogen atoms share the benefit of a helium-like configuration of two electrons and thereby gain the stability of a closed shell. They also lose their individual identities in the process, for no longer are there two hydrogen atoms. Now there is only a molecule: two nuclei and two electrons, bound together into a new structure with properties different from those of the atoms.

Such an arrangement is the classic electron-pair bond, and much of its conceptual basis will remain intact when (later) we develop a more accurate quantum mechanical interpretation. Quantum mechanics shows that electron density *is* increased between the nuclei, and shows further that the closed shell does lower the energy of the molecule. There is more to come, clearly, but still we can profit from this rudimentary view of a **covalent bond** as a shared pair of electrons.

Beyond hydrogen, the Lewis model for bonding similarly assumes that atoms tend to acquire noble-gas configurations when forming molecules. For elements in the second and third rows of the periodic table, the requirement is satisfied by an octet among the valence electrons. Hydrogen chloride (HCl),

$$\text{H} \cdot + \cdot \overset{\cdot\cdot}{\underset{\cdot\cdot}{\text{Cl}}} : \longrightarrow \text{H} : \overset{\cdot\cdot}{\underset{\cdot\cdot}{\text{Cl}}} :$$

provides an example where hydrogen gets its two electrons and chlorine gets its eight through a covalent bond. Note that only the valence electrons are represented. Chlorine's 10 core electrons do not participate in the bonding and consequently are suppressed in the diagram.

Again, the formation of HCl is a process of *sharing* in which the atoms merge their identities into a molecule. There is *not* a chlorine species with eight valence electrons and a hydrogen species with two. There is only a molecule of HCl, with two nuclei (one with $Z = 1$, the other with $Z = 17$) and a total of 18 electrons distributed between core and valence shells. The Lewis structure is just a shorthand way to show that electrons are concentrated between the hydrogen and chlorine nuclei.

Sometimes, though, the sharing becomes so unequal that one of the atoms effectively captures an electron from another. Ions are then formed. An **ion**, recall, is an atom or molecule bearing a net charge, positive or negative. In sodium chloride, for example, sodium's lone valence electron is so thoroughly detached that the charge goes over almost entirely to the chlorine. The electron transfer produces a positive sodium ion (a **cation**, Na^+) and a negative chloride ion (an **anion**, Cl^-), each with a noble-gas configuration. Sodium, by losing its one outer electron, acquires a neon-like distribution, whereas chlorine becomes similarly **isoelectronic** with argon (from the Greek *iso*, "same"). We now have two ions

$$Na\cdot + \cdot \ddot{\underset{\cdot\cdot}{Cl}}: \longrightarrow Na^+ + :\ddot{\underset{\cdot\cdot}{Cl}}:^-$$

which, in a crystal of sodium chloride (Figure 2-10), attract each other directly as two oppositely charged bodies—point to point, through empty space, governed by Coulomb's law. The attraction contributes to an **ionic bond**, a term to suggest that a covalent interaction has gone far past the point of equitable sharing.

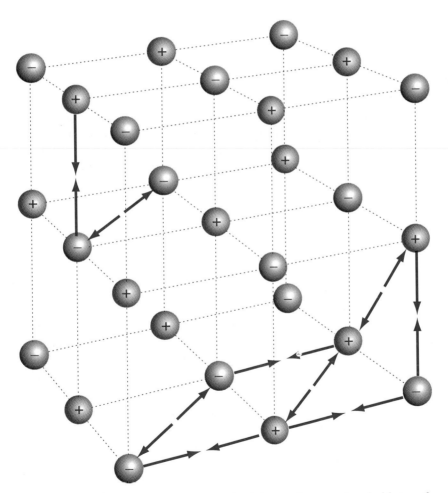

FIGURE 2-10. Ionic bonding. A tight web of electrical interactions holds together positive Na^+ and negative Cl^- ions in an ordered crystal. The attractions and repulsions weaken with distance and increase with ionic charge. For simplicity, ions are represented as points.

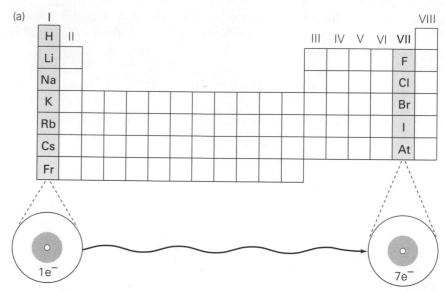

FIGURE 2-11. Valence and chemical combination. Atoms combine selectively in fixed ratios, often by giving and taking just enough electrons to mimic the filled valence of a noble gas. Two examples, of many, are shown: (a) The alkali metals (1 e⁻) combine in a 1:1 ratio with the halogens (7 e⁻). Compounds include LiF, NaF, KF, . . . , LiCl, NaCl, KCl, . . . , LiBr, NaBr, KBr, and so forth. (b) The alkali metals (1 e⁻) react in a 2:1 ratio with the chalcogens (6 e⁻). Compounds include Li_2O, Na_2S, and similar combinations.

Chlorine is said to be substantially more ***electronegative*** than sodium, indicating that it has greater capacity to draw electrons around its nucleus. It is an electron taker, only one short of an octet. Right away, too, we should expect similarly high-valence atoms also to attract electrons (nonmetals like F and O, for instance, which lie toward the right side of the periodic table). At the left of the table, however, the metallic atoms more readily *lose* their few valence electrons and drop back to the previous noble-gas configuration. These species, "electropositive" elements such as Li, Na, and Mg, are the electron givers in the give-and-take of chemical interaction.

Coming back to covalent bonds, we soon discover that more elaborate bonding arrangements are yet possible. Consider, say, the bonding of two oxygen atoms, represented as :Ö::Ö: or :Ö=Ö: or simply O=O, where simultaneous octets are achieved only through the formation of a ***double bond***. *Two* pairs of electrons are shared between the atoms, generally making for a stronger and shorter linkage than the corresponding single bond.

Triple bonds exist as well, exemplified by the nitrogen molecule :N⋮⋮N: (:N≡N: or N≡N). Here each nitrogen brings with it five valence

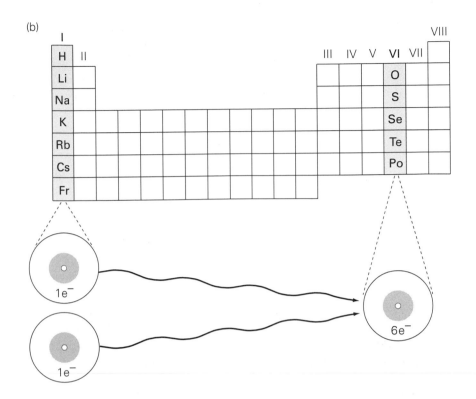

electrons, and the total pool of 10 electrons is divided into three bonding pairs and two unshared **lone pairs**. Once formed, the triple bond of nitrogen is hard to break—as demonstrated by atmospheric nitrogen, which exists as N_2 and remains unreactive under normal conditions.

With these various examples we already see hints of regularity in chemical combination. Atoms do not mix promiscuously but conform instead to well-defined requirements of *valence*. Thus (Figure 2-11): Sodium and the other alkali metals have one electron to give; chlorine and the other halogens, by contrast, need just one electron to make an octet. No surprise, then, that the alkali metals and the halogens form 1:1 compounds such as NaCl. And no surprise either that Group I elements and Group VI elements form 2:1 compounds such as Na_2O. The Group VI atoms lack two electrons to make an octet, so they grab one apiece from two Group I atoms.

Nor are two elements restricted to only one ratio, as is plain for the compounds CO and CO_2. Carbon and oxygen atoms form a triple bond in carbon monoxide ($:C:::O:$) to yield a structure isoelectronic with $:N \equiv N:$, whereas the same two atoms use double bonds to produce carbon dioxide ($\ddot{O}::C::\ddot{O}$). But not every combination is allowed: There is no neutral CO_3 molecule (although the doubly negative *carbonate* ion, CO_3^{2-}, does exist), and no CO_4 and no CO_5.

Clearly, there are fixed *combining capacities* that determine the atomic ratios and limit the variety of possible molecules. It was precisely these ratios that originally gave rise to the atomic theory in the early 19th century. The achievement of the 20th century was to explain the phenomena of valence and chemical combination through the quantum theory.

2-4. MOLECULES

Nuclei, small and dense, form the skeleton of molecules—the bricks. The electrons are the mortar holding the bricks together, the flesh on the bones. Lightweight and diffuse, the electrons determine molecular size, shape, and reactivity. What, at this point, can we guess about the electrical and geometric structure of molecules?

First there is the question of how the electrons are distributed. For molecules such as H_2, O_2, N_2, and F_2, the answer is dictated by the most basic considerations of symmetry. Each structure (the singly bonded halogen F—F, for example) remains the same whether rotated along the bond axis or flipped left and right. The two nuclei, both of the same kind, are indistinguishable and therefore incapable of attracting the electrons unequally. Since the electron distribution must respect the symmetry of the nuclear framework, the negative charge spreads out evenly around and between the nuclei. All these *diatomic* (having two nuclei), *homonuclear* (of like kind) molecules are *nonpolar* as a result, meaning that there is no skewing of the electronic charge to either end. This result we demand strictly on the basis of symmetry, requiring no specific theory of molecular structure for its justification. Go no further than the arrangement sketched in Figure 2-12(a); nothing else would make sense.

Heteronuclear diatomic molecules, containing two unlike nuclei, make for a different story. One of the nuclei always attracts the electrons more than the other, and so proportionately more negative charge accumulates on that end. Ionic bonds like those already cited in NaCl are an extreme case of unequal sharing. Where the bonding is covalent, as with HCl, the difference in electronegativity creates a skewed bond and a correspondingly *polar* molecule. The chlorine side grows richer in electrons than the hydrogen side, leaving a partially negative Cl and partially positive H. This separation of charge gives rise to a *dipole moment* (Figure 2-12b), and we use a lowercase Greek delta to represent the partial charges: $H^{\delta+}$—$Cl^{\delta-}$.

Triatomic molecules bring another element of choice. Consider water, H_2O, where the singly bonded atoms HOH form a bent structure:

$$H \overset{O}{\diagdown} H$$

(a)

(b)

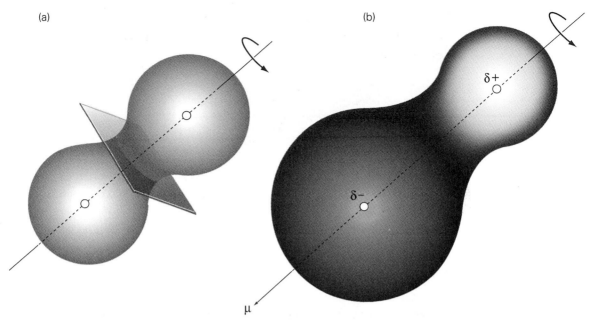

FIGURE 2-12. Symmetry and structure in diatomic molecules. (a) Homonuclear systems: Since both nuclei are the same, no distinctions exist for points either rotated around the bond axis or flipped end to end. Electrons are wrapped uniformly around the axis, and the distribution at one end is always the mirror image of the other. Neither side accumulates more electrons. The molecule is nonpolar. (b) Heteronuclear systems: With different nuclei, the molecule now possesses two clearly recognizable ends—and one of them attracts electrons at the expense of its neighbor. The structure has both an electron-rich negative site ($\delta-$) and an electron-poor positive site ($\delta+$): a *dipole moment*, μ.

Here each of the O—H bonds is polar ($O^{\delta-}$—$H^{\delta+}$) since oxygen is more electronegative than hydrogen, and the bent configuration gives the molecule a net dipole moment. The two polar bonds contribute jointly to a combined dipole along the HOH bisector:

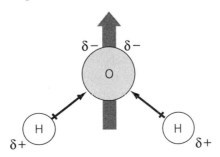

A molecule of O=C=O, however, is nonpolar despite its polar $C^{\delta+}=O^{\delta-}$ double bonds. The structure is linear, and the bond dipoles cancel out as shown below:

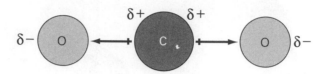

One points to the right, one points to the left, and their combined effect is nil. That does not mean the electrons are evenly distributed throughout the linear structure, which remains skewed as $O^{\delta-}=C^{2\delta+}=O^{\delta-}$. There is just no overall dipole moment for the species, even though different portions of the molecule display unequal concentrations of charge.

We should wonder why carbon dioxide is linear but water is bent, given their similarities in other ways. Both are triatomic molecules, and both contain two nuclei of one type bonded to another nucleus in the center. Where they differ, certainly, is in the particular mechanism of bonding (single versus double), so perhaps there is some relationship between electronic structure and molecular geometry. We need to uncover it.

The simplest explanation is to attribute the shape to repulsive forces suffered by the electrons around the central nucleus, an approach known as the ***valence-shell electron-pair repulsion (VSEPR)*** model. According to this picture, the negatively charged electron pairs will minimize electrical repulsions by moving away from each other—as far away as possible. Thus for O=C=O, the two sets of doubly bonded pairs on carbon lie farthest apart when the C=O bonds stand diametrically opposed to form a linear molecule. In H:Ö:H, by contrast, there are four individually acting pairs of electrons around the oxygen nucleus: two bonding pairs and two lone pairs. All four pairs move to avoid each other, and they can do so most effectively by pointing toward the corners of a ***tetrahedron*** as shown in Figure 2-13.

VSEPR's first prediction for water is almost correct. The H—O—H bond angle would be $109.5°$ in the forecasted tetrahedral geometry, a parameter reasonably close to the experimentally determined value of $104.5°$. If one assumes further that the two lone pairs interact more strongly than the bonding pairs, then VSEPR theory predicts a compression of the H—O—H bond and a smaller, more satisfactory angle. The presumed difference between lone-pair and bonding-pair repulsions is ascribed to their respective environments. The bonding electrons, shared between nuclei, take up less space and consequently are repelled less than the more diffuse lone pairs.

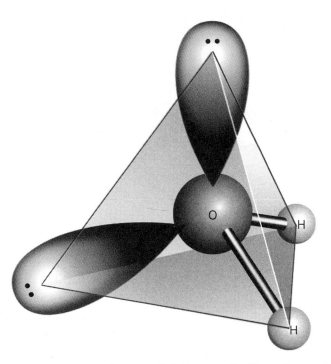

FIGURE 2-13. A geometric figure with four triangular faces, the *tetrahedron* plays a leading role in molecular structure. Here, each of the four electron pairs in H_2O points to one of the four vertices of a tetrahedron. Hydrogen atoms occupy two of the sites (with bonding pairs represented by solid rods), whereas the two other vertices (with lone pairs represented by dots) remain vacant. Such an arrangement, if realized, would produce a molecule bent at the characteristic tetrahedral angle of 109.5°—some 5° larger than the actual H—O—H bond angle. Differences in the repulsion between lone pairs and bonding pairs account partially for the distortion.

These arguments, admittedly incomplete, are consistent with the observed shapes, even if questions still remain. Taken in the whole spirit of Lewis structures and electron-pair bonds, the VSEPR model is to be appreciated as a tentative step toward a more elaborate theory, a rough guide which gives us a useful predictive capability right away. That said, we can use the method cautiously to guess the geometry of other polyatomic molecules as well.

For example: Three pairs of electrons achieve maximum separation by pointing to the vertices of an equilateral triangle. VSEPR therefore tells us that boron trifluoride, represented by the Lewis structure should be a ***trigonal planar*** molecule. It is. The four nuclei lie in one plane, and each F—B—F bond angle is 120°:

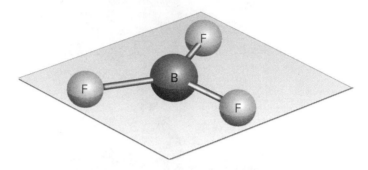

Note also that BF_3 is an apparent exception to the octet rule, since the electron configuration places only six electrons around the boron. Such deviations from the simple Lewis model are hardly surprising, and we shall resolve the difficulty eventually by using molecular quantum mechanics. Yet even in defeat the concept of an octet has a measure of truth, for the deficient structure possesses special properties precisely *because* of its "missing" electrons. The boron indeed would benefit energetically by acquiring those electrons, and hence BF_3 proves to be a strong ***Lewis acid***: a species ready to accept a pair of electrons from some other source.

Ammonia, a ***Lewis base*** (an electron-pair donor), is just such a source. With three bonding pairs and one lone pair, the molecule

$$H \overset{..}{\underset{..}{N}} H$$
$$H$$

adopts a ***trigonal pyramidal*** geometry:

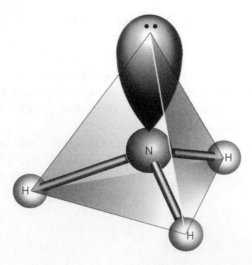

The four pairs of electrons are oriented tetrahedrally just as in water, but here only three vertices of the tetrahedron are actually occupied by nuclei. The nitrogen sits at the center; each of the three hydrogens claims a corner; and the lone pair remains unshared at the fourth and last vertex of the tetrahedron. The three hydrogens all lie in one plane, topped by a nitrogen bearing the nonbonded electron pair and ready to combine with an electron-deficient Lewis acid.

BF$_3$ and NH$_3$, each having what the other lacks, then come together to form the united compound

$$
\begin{array}{ccc}
\text{F} & \text{H} \\
| & | \\
\text{F—B—N—H} \\
| & | \\
\text{F} & \text{H}
\end{array}
$$

in which the nitrogen shares its lone pair to give boron an octet. With that fourth electron pair, moreover, the boron subsequently adopts a tetrahedral configuration just as VSEPR predicts:

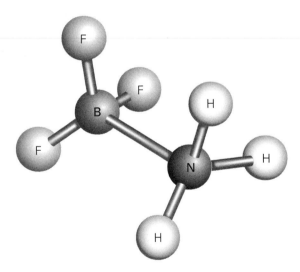

Boron and nitrogen each have four bonding pairs, and now each central atom directs its four bonds toward the corners of a tetrahedron.

Similar geometric arguments apply to central atoms with five and six electron pairs, as illustrated selectively in Figure 2-14. The basic arrangement for five electron pairs is a ***trigonal bipyramid***, in which one vertex lies above the center of an equilateral triangle and one vertex lies below. For six pairs, the shape is an ***octahedron***: a plane containing four vertices set in a square, finished off again by one additional vertex above and one below.

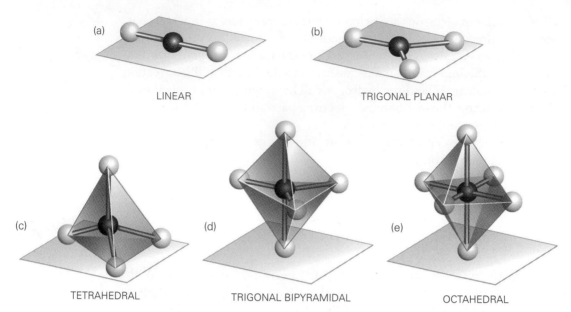

(a) LINEAR

(b) TRIGONAL PLANAR

(c) TETRAHEDRAL

(d) TRIGONAL BIPYRAMIDAL

(e) OCTAHEDRAL

FIGURE 2-14. VSEPR, a summary: orientation of electrons about a central atom. In each configuration, the pairs of electrons are farthest apart and thus suffer minimal repulsion. Bonding and lone pairs both play a role, but sites containing lone pairs are not occupied by atoms. (a) Two pairs: linear. (b) Three: trigonal planar. (c) Four: tetrahedral. (d) Five: trigonal bipyramidal. (e) Six: octahedral. To predict the molecular skeleton, place one atom at each vertex assigned to a bonding pair.

2-5. STOICHIOMETRY

Viewed at the most basic level, a chemical reaction is a microscopic encounter among individual atoms and molecules. Some integral number of particles is transformed to some other number of particles, the event summarized concisely by a balanced ***chemical equation*** such as

$$2H_2 + O_2 \longrightarrow 2H_2O$$

The process involves *whole* species, each a separate entity. Here two molecules of hydrogen and one molecule of oxygen go in; two molecules of water come out.

It is only a rearrangement, with particles neither created nor destroyed. Four hydrogen atoms and two oxygen atoms go in; four hydrogen atoms and two oxygen atoms come out. More precisely: Four hydrogen nuclei, two oxygen nuclei, and twenty electrons start out as molecular hydrogen and oxygen; the same four hydrogen nuclei, two oxygen nuclei, and twenty electrons end up as water. The numbers of electrons and nuclei are strictly conserved during any such chemical (nonnuclear) reaction, and in this recycling of building blocks is embodied the particulate, gritty nature of matter.

The apparent indestructibility of electrons and nuclei suggests a world in which matter, broken down into small bits, is permanent, a world in which mass cannot be created or destroyed. So it seems, too, until we probe into the nucleus itself (in Chapter 21) and discover there that mass is actually a form of energy. Mass and energy can be transformed one into the other. When there is a change in energy, there is a proportional change in mass. Where mass disappears, energy arises in its stead.

It comes as a shock, because neither in everyday life nor in chemical processes do we commonly see mass created or destroyed. The changes are too small, usually undetectable. To an excellent approximation, then, we observe that total mass appears to be conserved during all chemical reactions; and, with that, we derive a basic rule for the combination of atoms and molecules: the law of **conservation of mass**. The mass going in is equal to the mass going out. It is the law of the microscopic chemical realm.

Our own realm, however, is the macroscopic world of real stuff, where quantities of matter typically are large enough to see and manipulate using laboratory equipment. To do chemistry we must go from microscopic particles to macroscopic lumps and establish quantitative relationships between the amounts of matter participating in chemical reactions. The study of such relationships is called **stoichiometry**.

Usually one relies on masses—grams—to measure quantities of matter, but masses by themselves are not stoichiometrically significant. Grams do not react one by one. *Atoms* do. *Molecules* do. What we need is a way to mark off quantities of matter containing equal numbers of microscopic particles. We need a way to count atoms.

Stoichiometry begins with an understanding of **relative atomic mass** as derived from experimental measurements. The ^{12}C isotope of carbon, containing six protons, six neutrons, and six electrons, is taken as a reference and its mass is *defined* as exactly 12.0000 **atomic mass units**. This figure is consistent with the mass number 12, but nevertheless it is simply a definition. The masses of all other atoms are then measured relative to ^{12}C to give the dimensionless values displayed in the periodic table. Hence an atom of hydrogen (relative mass = 1.00794) is approximately 1/12 as massive as ^{12}C, while oxygen (15.9994) is nearly 16/12 more massive. A relative **molecular mass**, appropriate for a covalently bonded molecule, follows directly as the sum of the relevant atomic masses. H_2O therefore has a mass of 18.0153 (1.00794 + 1.00794 + 15.9994).

All these values are really weighted averages that reflect the distribution of isotopes at natural abundance. Chlorine, for instance, exists as a randomly dispersed mixture of ^{35}Cl (mass = 34.9689) and ^{37}Cl (mass = 36.9659), with isotopic populations of 75.77% and 24.23%, respectively. The average relative mass for such a distribution is computed as

$$0.7577 \times 34.9689 + 0.2423 \times 36.9659 = 35.45$$

and recorded as a single number in the periodic table. Carbon at natural abundance is itself a mixture of 98.9% ^{12}C (mass = 12.0000) and 1.1% ^{13}C (mass = 13.0034), so its average relative mass works out to 12.011.

Given this idea of relative mass, we need a stoichiometrically meaningful quantity useful for macroscopic purposes. It must be a manageable amount of material with some conveniently large but *fixed* number of particles, which we shall now take to be the number of atoms in exactly 12 grams of ^{12}C. This amount of substance, said to contain **Avogadro's number** of species (N_0), is called a **mole** and is abbreviated, just barely, as *mol*. Avogadro's number, so defined, is really an invitation and a challenge to make a measurement: to count 12 grams of isotopically pure ^{12}C atoms, using whatever techniques are suitable.

N_0 is indeed an experimentally measurable quantity, an almost unimaginably large number: 6.02×10^{23}. There are N_0 atoms of ^{12}C in 12.0000 grams of isotopically pure carbon-12, and from there the mole is generalized to mean Avogadro's number of anything. We can specify a mole of lithium atoms, a mole of oxygen molecules, a mole of glucose, a mole of bricks, even a mole of elephants. It is a collective unit, similar to "dozen" or "score" or, in days past, "myriad."

The **molar mass**, $\mathcal{M}$, of any species is taken as the mass of one mole. Carbon-12 has a molar mass of 12.0000 g mol^{-1}, by agreement, and $\mathcal{M}$ for everything else scales in direct proportion. The molar mass of oxygen, related to ^{12}C by the factor 15.9994/12.0000, is 15.9994 g mol^{-1} for the atoms and 31.9988 g mol^{-1} for the diatomic O_2 molecules. For hydrogen atoms, $\mathcal{M}$ is 1.00794 g mol^{-1}; for H_2O the value is 18.0153 g mol^{-1}.

A pattern is apparent: The molar mass is always the relative mass expressed in grams. This result is guaranteed by the contrived definition of the mole combined with the choice of ^{12}C as the standard for relative mass. And now, equipped with the molar mass, we know exactly how to choose quantities of material containing fixed numbers of atoms or molecules. By taking a mass in grams equal to the molar mass, we assure ourselves of getting 6.02×10^{23} *particles* whether the lump of stuff be one mole of molecular hydrogen (2.0159 g) or one mole of lead atoms (207.2 g). Weights and volumes may differ, but a mole is a mole. It is Avogadro's number of anything.

Just as two *molecules* of H_2 react with one molecule of O_2 to form two molecules of water, two *moles* of hydrogen react with one mole of oxygen to form two moles of water. Reactions go molecule by molecule; reactions also go mole by mole. Avogadro's number is the scale factor between microscopic and macroscopic, and the same integer stoichiometric ratios carry over from the molecule to the mole.

Imagine, as an example, carrying out the reaction above to form water. We might first confine one mole of H_2 gas in a balloon, and then initiate the process (very cautiously) by applying a spark. An explosion ensues; there is a burst of flame, and an invisible mist of water condenses out of the air. What has happened?

Interpreted stoichiometrically, the events begin when 6.02×10^{23} molecules of hydrogen (one mole) emerge from the balloon and encounter some much larger number of molecules of atmospheric oxygen. Hydrogen and oxygen species—molecules, one by one—come together as individual chemical actors, and these meetings produce two new H_2O molecules for every one O_2 molecule and every two H_2 molecules that are consumed.

So simple, yet so profound; for even the cartoonlike view of stoichiometry taken in Figure 2-15 asserts no less than this: Stoichiometry implies structure. Matter is lumpy. Atoms, rearranged in countless ways, exist as the hard kernels inside all substances.

Now at this point we still have no idea how such rearrangements take place microscopically, because the balanced equation supplies only a before-and-after accounting. To elucidate the *mechanism* of reaction step by step, one needs an entirely different set of experiments. Yet we do know that these molecular encounters, which occur at high speed and are by no means gentle, continue until all the hydrogen molecules disappear. In the end, one mole of H_2 has reacted completely with one-*half* mole of oxygen to produce one mole of water—about 18 mL of liquid, a volume much smaller than the approximately 22 L occupied by the hydrogen gas at room temperature and pressure. The surroundings remain filled with oxygen, of course, because this reactant was present in vast excess compared to the hydrogen. Oxygen is consumed only according to the stoichiometric ratio in the balanced equation ($1 O_2$ per $2 H_2$), and the reaction necessarily ends when the hydrogen is gone. The scarce reactant, hydrogen in this case, is termed the **limiting reactant** (see Figure 2-16).

Were we to use other proportions of hydrogen and oxygen, we might uncover some additional features of the process. We could, say, mix the reactants together according to their limiting stoichiometric ratios: perhaps 1 mole of H_2 and $\frac{1}{2}$ mole of O_2, or maybe 0.46 mole of H_2 and 0.23 mole of O_2, all in the same container. Remember that moles are macroscopic amounts, so there is no conceptual difficulty with the fractions; the correct molecule-to-molecule *microscopic* ratios are always

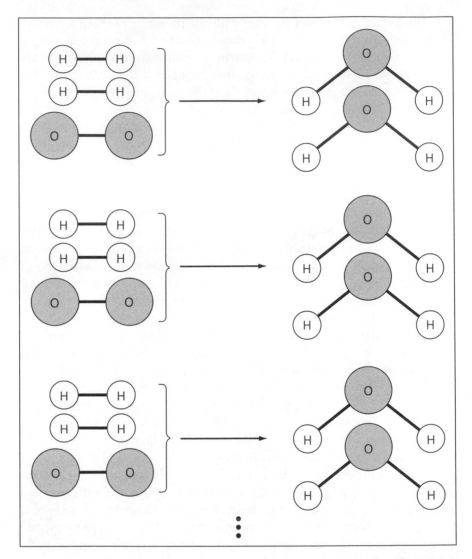

FIGURE 2-15. Molecule by molecule, mole by mole: For every two molecules of H_2 that disappear, one molecule of O_2 also vanishes whenever two molecules of H_2O are formed. For every two *moles* of H_2, so goes one *mole* of O_2 as well. Why? Because matter is not continuous; matter is palpably and countably lumpy; matter contains atoms and molecules in fixed amounts. Reactions proceed lump by lump and grain by grain.

observed. Provided that H_2 and O_2 are present in a 2:1 ratio, the transformation (activated, as before, by some heat) should proceed *stoichiometrically*. Two hydrogens will pair off with each oxygen, and no unreacted molecules will remain.

This time, though, the explosion occurs with greater force, producing more noise and a more violent, even more hazardous shock wave.

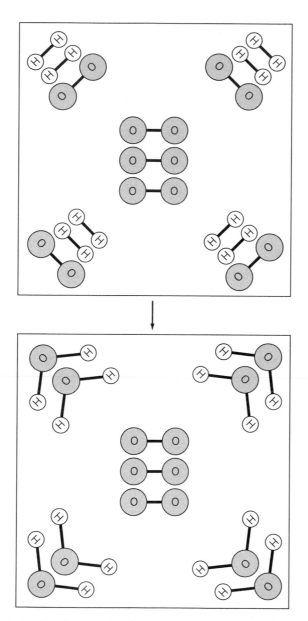

FIGURE 2-16. Molecule by molecule, the reactants react until one species disappears completely. The amount of product therefore depends on the reactant present in the smallest stoichiometric quantity. This *limiting reactant* (here, H_2) is consumed entirely while an excess of the other species (O_2) remains untouched.

With both reactants intimately mixed inside the balloon, hydrogen and oxygen come together without delay. The reaction is over that much sooner, and its outcome is that much louder.

As for the explosions themselves, we shall discover in our treatment of thermodynamics that hydrogen and oxygen together have, in some sense, a higher energy than water. This difference in energy, liberated as heat during the reaction, is a fixed quantity, and therefore exactly the same amount of heat escapes during both procedures. In each instance the heat causes a rapid expansion of the air, from which develops a shock wave and the two explosions of different intensities. But it is only the time scale that differs—not the total energy—and here we have our first example illustrating the difference between reaction *kinetics* (how fast?) and thermodynamics (how much?).

Let us briefly consider one more reaction, both to illustrate the stoichiometric determination of **empirical formula** and to appreciate the different ways in which energy can be released. The process involves the combustion of phosphorus in an atmosphere of pure oxygen, during which an intense light is emitted and a white powder is deposited. The white solid is an oxide of phosphorus in the stoichiometric ratio P_xO_y, where x and y are unknown numbers to be determined experimentally. The bright light produced by this *luminescent* reaction carries away at least part of the excess energy, with heat accounting for much of the rest. From these observations we learn that energy can be packaged in forms other than heat, and that chemical reactions might be put to various uses.

Concerning the oxide, we want to know the number of moles of P and O in any macroscopic sample. Various techniques of analytical chemistry, some of which are described in Chapter 20, enable us to establish the percent elemental composition. Suppose, then, that the unknown oxide is found to contain 43.6% P and 56.4% O by weight. We know that for every 1.00 g of P_xO_y there are 0.436 g P and 0.564 g O, from which the number of moles (n) follows directly:

$$n_P = \frac{m}{\mathcal{M}} = \frac{0.436 \text{ g}}{30.97 \text{ g/mol}} = 0.014 \text{ mol}$$

$$n_O = \frac{m}{\mathcal{M}} = \frac{0.564 \text{ g}}{16.00 \text{ g/mol}} = 0.035 \text{ mol}$$

This calculation is a simple but crucial stoichiometric conversion. A mass in grams (m) is expressed as the corresponding number of moles (n) by use of the molar mass ($\mathcal{M}$). Analyzing the dimensions, we have

$$\text{mol} = \text{g} \times \frac{\text{mol}}{\text{g}}$$

whereby $\mathcal{M}$ plays the role of a conversion factor between grams and moles.

Calculation thus shows that the oxide has an empirical formula $P_{0.014}O_{0.035}$, which must be interpreted provisionally as a *molar* rather than molecular ratio. If the compound exists as molecules, then the **molecular formula** might be P_2O_5 or P_4O_{10} or P_6O_{15} or any other structure consistent with a ratio of two phosphorus atoms to every five oxygen atoms. For the moment, however, we cannot even say that there *are* molecules; the compound might be ionic, like NaCl. Without additional information, the analysis succeeds only in determining an empirical formula of $PO_{2.5}$ or, expressed as whole numbers in lowest terms, P_2O_5.

Other analytical techniques are used to establish whether or not a material is molecular and to measure the molecular weight if appropriate. Here we would find that the oxide does form molecules, each with relative mass 283.9 and hence with molecular formula P_4O_{10}.

At this point we believe in atoms and molecules and also have some elementary notion of how they can share electrons. Stoichiometry, which establishes the mathematical rules for such combinations, emerges not merely as a kind of atomic bookkeeping but takes on far deeper meaning as a manifestation of the particulate nature of matter.

Transformation and change are at the heart of chemistry. Molecules constantly compete with other molecules for atoms, knocking about blindly as they stumble—by chance—into structures of greater stability. The possibilities for such rearrangements are manifold, as we shall begin to discover in the next chapter.

REVIEW AND GUIDE TO PROBLEMS

We start with the knowledge that matter is not continuous; that matter is discrete and grainy; that matter is particulate, lumpy, more like a bag of beans than a bowl of Jell-O.

There is a hierarchy. Protons and neutrons bind together into nuclei. Electrons and nuclei join to form atoms. Atoms combine into molecules, and molecules weave themselves into solids, liquids, and gases. Look closely and see: Everything in the world is built from small particles, each a lump of mass and charge.

From microscopic to macroscopic, then, there is a definiteness to the composition of matter—a definiteness because particles are, after all, lumpy and *countable*. Each atom has a definite structure. Each atom has its unique blueprint, its unique combination of protons, neutrons, and electrons. Each atom is what it is, always the same structure, always with the same mass, always the same number of subatomic particles.

One atom of hydrogen? One electron and one proton. Every H atom has a definite mass. Two atoms of hydrogen? Twice the mass of one. Three atoms? A million? A mole (6.02×10^{23})? The task is no different from counting beans with identical masses. If we know the number of atoms, we know the total mass. If we know the total mass, we know the number of atoms.

Molecules too. Two molecules of H_2 have twice the mass of one; four have twice the mass of two. If we know the number of molecules, we know the total mass. If we know the total mass, we know the number of molecules.

We know the number of nuclei and electrons as well, for these particles are the permanent, interchangeable, recyclable building blocks of matter in chemical reactions. Like beans in the bag, they persist. They endure. They are here to stay. To do chemistry is only to rearrange the beans, not to destroy them. From beginning to end of a chemical (non-nuclear) reaction, the total mass remains virtually the same. This ***conservation of mass***, the hallmark of all chemical reactions, derives expressly from the lumpy, particulate structure of matter.

And not only is matter particu*late*, it is also particu*lar*—particular in its choice of partners, particular in the combinations that can be formed. Some atoms react easily with others; some, like the noble gases, do not. Some atoms engage many partners; some, few. Some molecules are strong, and some are weak. Some are plentiful, and some are rare. Ample variety there is, but atoms and molecules do not combine indiscriminately.

Rather, each atom or molecule has its own structure and its own capacity to add or subtract electrons and nuclei. Carbon forms bonds to one, two, three, or four other atoms. Hydrogen normally connects itself to only one. Oxygen forms two bonds; nitrogen, usually three. All throughout the periodic table there are regular differences in *combining capacity*, or *valence*.

Then, when atoms and molecules do come together, they react one by one. Atom by atom, molecule by molecule, a chemical transformation proceeds microscopically and in fixed proportion. If one molecule of A reacts with one molecule of B, then *two* molecules of A will react with *two* molecules of B; and three with three, and a million with a million, and a mole with a mole. Neither the total mass nor the number of electrons nor the number of nuclei changes during a chemical reaction. Exact numerical relationships are maintained during each microscopic encounter, precisely because matter is particulate and particular. Because matter has a definite composition. Because matter has a structure. Because matter is not continuous. That, in short, is the real meaning of *stoichiometry*; the rest is just a matter of counting beans.

IN BRIEF: STOICHIOMETRY AND STRUCTURE

Recall the broad-brush picture of atoms and molecules sketched in this chapter.

1. LARGE AND SMALL. Atoms and molecules are small, so small that we need enormous numbers of them just to have a modest-sized macroscopic sample (a few grams or tens of grams, say, a literal "handful").

Mediating between large and small is the *mole*, defined as the quantity of matter containing *Avogadro's number* of particles: 6.02×10^{23} atoms, molecules, electrons, or whatever else they may be. This value, one mole (1 mol), is precisely the number of carbon-12 atoms in 12.0000 grams of isotopically pure ^{12}C. It is a very big number.

Microscopic or macroscopic, small world or large? In chemistry there are elements of both, for always we must deal with a very large number of very small particles. Let us build from the ground up, beginning with the atom.

2. PROTONS, NEUTRONS, AND ELECTRONS. An *atom* is an electrically neutral species in which Z electrons (total charge $= -Ze$) are associated with a nucleus containing Z protons (total charge $= +Ze$). Z denotes the element's *atomic number*.

Bound into the nucleus are N neutrons, each with zero charge and each with nearly the same mass as a proton. The sum of the protons and neutrons, $A = Z + N$, is the atom's *mass number*. Atoms with identical Z but different A are *isotopes* of the same element, differing only in the number of neutrons. They share the same electronic structure and consequently the same (or mostly the same) chemical properties.

An *ion* is an atom or molecule with a net electronic charge, negative for an *anion* and positive for a *cation*. Ions containing one nucleus are called *monatomic*; ions with two nuclei are called *diatomic*; with three, *triatomic*; more, *polyatomic*. The same terminology applies to neutral molecules as well.

3. VALENCE. With a mass of only 9.11×10^{-31} kg, the electron is nearly 2000 times lighter than either a proton or a neutron. Swarming about a very small, very dense nucleus, the electrons give an atom size, shape, and the ability to interact with other atoms. They allow atoms to combine, to make molecules.

Already we have a simple working picture of the electrons in an atom: a hierarchical *shell* structure. Electrons are organized around the nucleus according to energy and average distance, falling into groupings tentatively called shells. Successive shells contain two, eight, and eight electrons for the first 18 elements. A chlorine atom, for example, houses two electrons in its first shell, eight in its second, and seven in its third.

The outermost levels, usually not filled to capacity, function as the *valence shell*. Out here the electrons are farthest from the nucleus and most reactive, and out here is where chemical interactions usually take place. Electrons in the innermost, filled shells (the *core*) generally are unable to participate in chemical reactions.

What happens when atoms do react? They give and take electrons. Some atoms capture electrons to complete an unfilled valence shell, as when Cl goes to Cl^- to create an eight-electron *octet*. Other atoms surrender electrons to leave behind a closed shell, as when Na goes to Na^+. *Metals* (like sodium) tend to lose electrons, whereas *nonmetals* (like chlorine) tend to gain them.

Arranged in the *periodic table*, the various atoms are grouped according to the number of electrons in their valence shells. The *alkali metals* (Li, Na, . . .), with one outermost electron, all stand in Group I. The *alkaline earth metals* (Be, Mg, . . .), with their two valence electrons, are in Group II. And so on: The elements of Group III (B, Al, . . .) have three valence electrons exposed; those in Groups IV (C, Si, . . .) and V (N, P, . . .) have four and five, respectively, and the *chalcogens* of Group VI (O, S, . . .) have six. Completing these *main-group elements* (or *representative elements*) are the seven-electron *halogens* of Group VII (F, Cl, . . .) and the *noble gases* of Group VIII (He, Ne, . . .) with their filled valence shells.

There are, in addition, horizontal *periods* of *transition elements* (or *transition metals*) occupying the center of the table. We shall consider the valence properties of these atoms later, in Chapters 6 and 19.

4. CHEMICAL BONDING. Pick an atom, any atom, and you will probably find a structure deficient in some way: a structure either underbuilt or overbuilt, a structure ripe for revision. There is either too much or too little. This atom might benefit by gaining electrons; that one, by losing some. Hydrogen atoms, for instance, profit by pairing off into molecules of H_2, sharing two electrons between them. Atoms of carbon, by contrast, form continuous, interconnected sheets of graphite (the "lead" in a pencil), with electrons spread throughout the whole structure. Whatever it is, though, expect the atoms to do *something*, because they do have options. They combine with other atoms. They make bonds.

Two atoms form a *covalent bond* by sharing a pair of electrons, and for that there are two possibilities: Either (1) one nucleus attracts more of the negative charge than the other (a *polar bond*) or (2) the two nuclei share the bonding electrons equally (a *nonpolar bond*). A polar bond thus displays a negative end and a positive end, and these two electrical *poles* create a *bond dipole moment*—a two-point separation of charge.

Now if two nuclei of the same element come together in isolation (think of the molecule H_2), then the resulting *homonuclear bond* must be nonpolar. What else? One nucleus is the same as the other. Neither is special. Neither can attract more electrons.

In a *heteronuclear bond*, however, where the two nuclei originate from different elements, one of the two members will indeed prevail. A bond between heteronuclei is polar. It has a negative end and a positive end. The electrons are shared unequally.

Because: Every atom is different; every atom has its unique structure; every atom gives or takes electrons to a different extent. Atoms differ in their ability to attract electrons from other atoms. Every atom has its own *electronegativity*.

Grouped at the left of the periodic table, the metals tend to *lose* electrons. They are less electronegative than the nonmetals at the right-hand side, which tend to *gain* them. The greater the difference in electronegativity, the less equal the sharing and the more polar the bond. In the extreme where one atom (Cl) strips away and captures the electron of another (Na), an *ionic bond* is formed. Here a positive ion (Na^+) finds itself attracted to a negative ion (Cl^-) by the ordinary Coulomb interaction described in Chapter 1.

5. MOLECULES. Atoms give, take, and share valence electrons to form bonds: a *single bond* if one pair of electrons is involved, a *double bond* if there are two pairs, a *triple bond* if there are three. Thus are molecules made, atom by atom, bond by bond, molecules ranging from the simplest diatomic species to the giant structures inside living systems.

Think of the many questions. What pairs with what? At what distances? At what angles? What reacts with what? Why this and not that?

Surely there is much to understand, a little at a time, but we need at least a toehold right now. We need a quick, reasonable way to guess at the shapes of simple molecules, and for that we have the *valence-shell electron-pair repulsion (VSEPR)* model.

Intended for molecules built around a central atom, VSEPR theory argues that pairs of valence electrons will spread out optimally to minimize the electrical repulsion. It predicts: Two pairs attached to a central atom will produce a straight-line configuration. Three pairs, an equilateral triangle around the central nucleus. Four, a tetrahedron. Five, a trigonal bipyramid; six, an octahedron.

SAMPLE PROBLEMS

Small particles, but a lot of them—such are the conditions governing most chemical processes. What better way to begin than by trying to visualize one mole?

EXAMPLE 2-1. How Big Is a Mole?

PROBLEM: Proceeding at a rate of one molecule per second, how much time would Johnny need to count all the H_2 molecules present in 2.0159 g of hydrogen gas?

SOLUTION: All problems in stoichiometry connect mass (grams) with moles. Wherever some macroscopic quantity is demanded in a problem, we need *moles*. Always.

Remember, therefore, that the *molar mass* of any atom or molecule is the particle's *relative mass expressed in grams*. For the H_2 molecule, composed of two H atoms (each with relative mass = 1.00794; see periodic table), the molar mass is exactly 2.0159 g. Thus from grams to moles to molecules we compute:

$$2.0159 \text{ g} \times \frac{1 \text{ mole}}{2.0159 \text{ g}} \times \frac{6.02 \times 10^{23} \text{ molecules}}{\text{mole}}$$

$$= 6.02 \times 10^{23} \text{ molecules}$$

The numbers are simple here, by intention, but the operation

$$\text{grams} \rightleftharpoons \text{moles} \rightleftharpoons \text{particles}$$

never becomes more complex. It is the beginning and end of all stoichiometric calculations.

Now count from 1 to 6.02×10^{23} at a steady rate of one count per second,

$$6.02 \times 10^{23} \text{ molecules} \times \frac{1 \text{ s}}{\text{molecule}} \times \frac{1 \text{ h}}{3600 \text{ s}} \times \frac{1 \text{ d}}{24 \text{ h}} \times \frac{1 \text{ y}}{365 \text{ d}}$$

$$= 1.91 \times 10^{16} \text{ years}$$

and try to make sense of the result . . . nearly 20,000 trillion years to count one mole. For comparison, all of recorded history spans less than 10,000 years, humans first appeared on earth just 3,000,000 years ago, and the solar system was born only 4,500,000,000 years before that.

Too much time for Johnny alone? How about 1 *billion* seconds to count a mere billion (10^9) molecules? It can be done; with sufficient dedication it can be done. Johnny needs only to work nonstop for 31.7 years, by which time he will have counted to 1 billion. Let him now recruit 600 trillion coworkers, each counting for 31.7 years, and the task is complete: one full mole, counted by a team 100,000 times larger than the entire population of the earth.

No doubt, then, that a mole of hydrogen contains an unspeakably large crowd of tiny molecules packed into a modest space (roughly the size of a small carton; see Chapter 10). Other substances, more dense than hydrogen, use even less space to house one mole of particles.

But enough about size, for by now we accept that atoms are small. Our attention turns next to their internal structure and combining capacities.

EXAMPLE 2-2. Atoms and Ions

PROBLEM: Refer to the periodic table while answering the following questions. (a) An atomic species with 18 electrons has a net charge of +2. Name the element. (b) Another monatomic ion also has 18 electrons, but a net charge of −1. Identify it. (c) The radioactive I^- anion used to treat thyroid disease has a mass number of 131. How many neutrons are in its nucleus, and how many electrons are outside? (d) A neutral atomic species contains 54 electrons and 77 neutrons. What is it?

SOLUTION: Count protons (Z) and neutrons (N), remembering that (1) the mass number A is equal to $Z + N$, and (2) neutral atoms have just as many electrons as protons. Where needed, use the notation $_Z^A X$ to describe the nuclear structure of a given isotope.

(a) *18 electrons, ionic charge = +2.* If the nucleus contains 20 protons, then the net charge (20 − 18) will be +2 as specified. The atomic number is 20, corresponding to calcium. This doubly positive Ca^{2+} cation thus has the same arrangement of electrons as the neutral argon atom, Ar, with filled shells of 2, 8, and 8 electrons. Ca^{2+} is *isoelectronic* with Ar.

Nevertheless, ionized calcium is by no means the same as neutral argon. Electrons do not make an element; protons do. An element takes its identity not from the number of electrons, but from the number of protons in its nucleus. Massive and slow-moving, the positive nucleus controls the electrons and ultimately directs the chemistry. See how in Chapter 6.

(b) *18 electrons, ionic charge* $= -1$. Use the same reasoning. If the negative ion has 18 electrons, then the neutral atom must have 17 electrons and 17 protons. Our species is the chloride ion, Cl^-, a structure also isoelectronic with Ar.

(c) I^-, *A* $= 131$. Iodine has an atomic number of 53. Given a mass number of 131, we count 78 neutrons in each nucleus of $^{131}_{53}I$:

$$N = A - Z = 131 - 53 = 78$$

Since the charge is -1, there must be one more electron than proton; 54 electrons lie outside the nucleus.

(d) *54 electrons, 77 neutrons, neutral atom.* A structure with 54 electrons plus 54 protons corresponds to the xenon atom, a noble gas isoelectronic with the I^- anion. The specified isotope, $^{131}_{54}Xe$, has the same mass number as well ($54 + 77 = 131$).

EXAMPLE 2-3. An Ionic Compound

Do calcium and chlorine atoms form ionic or molecular compounds? Remember that these two species are widely separated in the periodic table, one an alkaline earth metal (Group II) and the other a halogen (Group VII). They differ significantly in electronegativity; and, similar to sodium and chlorine, they produce an ionic compound: calcium chloride.

PROBLEM: Calcium chloride is a granular, white solid that melts at 782°C. (a) What is its likely empirical formula? (b) If a sample of calcium chloride contains 0.5045 g of chlorine, how many Ca^{2+} ions are present to balance the charge? (c) What mass of atomic chlorine (in grams) is contained in 0.732 mol of calcium chloride? (d) What is the percentage of chlorine by weight?

SOLUTION: We know, from Example 2-2, that calcium realizes its octet by losing two electrons to become Ca^{2+}; and we also know that a chlorine atom will gain one electron to give an octet to Cl^-. Hence there must be two Cl^- ions for every Ca^{2+} if the compound is to have zero net charge.

(a) *What is calcium chloride's likely empirical formula?* $CaCl_2$. The ions are held in a rigid array, with twice as many anions as cations in the crystal. There is no $CaCl_2$ molecule.

(b) *How many calcium ions coexist with 0.5045 g of chloride ions?* From the empirical formula we determine first the moles of Cl^-, then the moles of Ca^{2+}, and finally the number of calcium ions:

$$0.5045 \text{ g Cl}^- \times \frac{1 \text{ mol Cl}^-}{35.453 \text{ g Cl}^-} \times \frac{1 \text{ mol Ca}^{2+}}{2 \text{ mol Cl}^-} \times \frac{6.022 \times 10^{23} \text{ ions}}{\text{mol}}$$

$$= 4.285 \times 10^{21} \text{ Ca}^{2+} \text{ ions}$$

In doing so, we ignore the very slight differences in mass between ions and atoms. A few more electrons, give or take, will have little effect on the calculation.

(c) *How many grams of atomic chlorine are contained in 0.732 mol of calcium chloride?* Notice by now how all stoichiometry problems are the same—grams to moles to atoms in all possible combinations:

$$0.732 \text{ mol CaCl}_2 \times \frac{2 \text{ mol Cl}}{1 \text{ mol CaCl}_2} \times \frac{35.453 \text{ g Cl}}{1 \text{ mol Cl}} = 51.9 \text{ g Cl}$$

(d) *What is the percentage chlorine by weight?* One mole of $CaCl_2$ contains one mole of calcium (40.078 g) and two moles of chlorine (70.906 g). Their combined formula weight is 110.984 g mol^{-1}:

$$\text{Weight \% Cl} = \frac{70.906 \text{ g Cl}}{110.984 \text{ g CaCl}_2} \times 100\% = 63.888\%$$

Calcium's weight percentage is 36.112%.

EXAMPLE 2-4. A Simple Molecule: Electrons, Bonds, and Geometry

Carbon and hydrogen form covalent bonds, not ionic. Compounds containing just these two elements are called *hydrocarbons*, many of which we know as common fuels: methane (70% of natural gas), ethane, ethylene, acetylene, propane, butane, pentane, hexane, heptane, octane, and so forth. They exist as independent molecules.

PROBLEM: A colorless, odorless, highly flammable gas is found to have a molecular weight of 16.04 g mol^{-1}. Analysis shows the material to be a pure hydrocarbon, 75.0% carbon by weight. (a) What is the empirical formula? (b) What is the molecular formula? (c) Draw a reasonable Lewis structure for the molecule. (d) Propose a geometric structure.

SOLUTION: Take some arbitrary amount, say 100 grams. Since every 100 grams will contain 75.0 grams of carbon and 25.0 grams of hydrogen, we can immediately compute moles of C and moles of H:

$$75.0 \text{ g C} \times \frac{1 \text{ mol C}}{12.011 \text{ g C}} = 6.24 \text{ mol C}$$

$$25.0 \text{ g H} \times \frac{1 \text{ mol H}}{1.00794 \text{ g H}} = 24.8 \text{ mol H}$$

(a) *What is the empirical formula?* The molar ratio suggests $C_{6.24}H_{24.8}$, or $CH_{3.97}$ when reduced to lowest terms. Expressed as integers, the empirical formula is CH_4.

(b) *What is the molecular formula?* The formula weight for CH_4 is 16.04, exactly equal to the molecular weight previously given. The molecular formula therefore is CH_4 (and the gas is methane).

(c) *Suggest a Lewis structure.* First, understand what a Lewis structure is and what it is not. A dot diagram is *not* a flawless, complete, or even unique description of a system's bonds. It is a sketch, a cartoon, a shorthand way to portray the gross distribution of electrons. It shows the electrons either spread between nuclei (in pair bonds) or localized on one nucleus (as lone pairs).

Exactly how many electrons are here and how many are there? Why do they fall where they do? What are their energies? Why is 8 such a special number—but why, also, are there so many exceptions to the octet rule? Expect no answers from a Lewis structure, for such diagrams serve only to summarize what we know from experiment. The answers, the reasons, and even more questions will come, soon enough, from the quantum theory of bonding (Chapters 4 through 7).

Yet after all that, the Lewis structure will still give us the bonding in a readily appreciated, almost graphical fashion. It survives as a useful, persuasive way to reduce a complex system to a simple picture. Here is what to do, step by step:

1. Ignoring the core, count the total number of valence electrons.

2. Arrange the atoms according to the connectivity of the molecule. Lay them out to show what is connected to what, who is neighbor to whom.

3. Distribute the valence electrons around the skeleton. Start by joining each of the connected atoms with a single bond, drawing either a pair of dots or, equivalently, a single dash.

4. Where necessary, add electrons to complete the valence of any atom bonded to the central atom. Hydrogen takes two electrons,

and most of the elements up through argon take eight (but be prepared for exceptions).

5. Use the remaining electrons, if any, to ensure an octet around the
   central atom. One option is to introduce lone pairs; another, to
   employ multiple bonds.

For methane, the bonding is comparatively simple. Carbon, a Group IV
element, contributes four valence electrons, and each of the four hydrogen atoms supplies one more to yield a total of eight (step 1). Then,
knowing that H forms one bond and C four, we place the carbon at the
center (step 2) and draw four single bonds (step 3):

$$
\begin{array}{c}
\text{H} \\
| \\
\text{H}-\text{C}-\text{H} \\
| \\
\text{H}
\end{array}
$$

And, by sheer good luck, we are done: Each hydrogen has its full quota
of two electrons; the carbon has its octet; there are no electrons to spare.
We have nothing more to do.

(d) *Propose a geometric structure.* A molecule of CH_4 must accommodate four bonding electron pairs about the central carbon atom. The
VSEPR model therefore predicts a tetrahedral geometry, with H—C—H
bond angles of $109.5°$ all around. Moreover, since nature cannot distinguish among the four identical hydrogen atoms, all the C—H bonds must
have the same length and polarity (skewed slightly toward the more electronegative carbon). But there is no overall *molecular* dipole moment,
because the four polar bonds (oriented tetrahedrally) oppose each other
to create a nonpolar molecule.

The related molecule CH_3Cl, also tetrahedral, does have a molecular dipole moment. Why?

EXAMPLE 2-5. More Molecules

PROBLEM: Suggest Lewis and VSEPR structures for the following
species: (a) Sulfur tetrafluoride, SF_4. (b) Acetylene, C_2H_2. (c) Nitric
oxide, NO. (d) The nitrosonium ion, NO^+.

SOLUTION: For each system, lay out the atomic skeleton and distribute the valence electrons in the most plausible way. Then: Orient the
electron pairs, nonbonding and bonding alike, in the directions specified
by the VSEPR model.

(a) *Sulfur tetrafluoride, SF_4.* Sulfur, a Group VI atom, has six valence
electrons. Fluorine has seven. With 34 electrons to distribute, we first

connect each F to the central S with a single bond:

$$\begin{array}{cc} F & F \\ \diagdown & \diagup \\ & S \\ \diagup & \diagdown \\ F & F \end{array}$$

The next step is to complete the octet on each F by adding six electrons:

There are two electrons left over from the original 34, even though both sulfur and all four fluorines already have octets. Still, octet rule notwithstanding, the 33rd and 34th electrons shall be assigned to the sulfur, where the fifth pair now contributes to an "expanded octet" of 10 electrons:

$$\begin{array}{cc} :\ddot{F}: & :\ddot{F}: \\ \diagdown & \diagup \\ & S: \\ \diagup & \diagdown \\ :\ddot{F}: & :\ddot{F}: \end{array}$$

These five pairs of electrons (four bonding, one nonbonding) we expect to point toward the vertices of a trigonal bipyramid. Four of the corners will contain fluorine nuclei, while the fifth will remain unoccupied. Sulfur sits at the center, as shown in the drawing below:

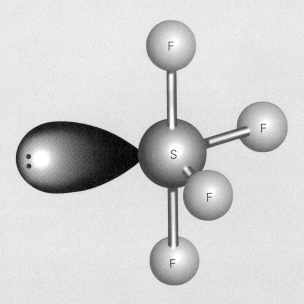

Take note as well: The nonbonding pair falls specifically in the trigonal plane between the upper and lower pyramids, taking up an *equatorial* position rather than an *axial* position on top or bottom. Nominally spaced 120° apart from the two other equatorial pairs and 90° apart from the two axial pairs, a diffuse lone pair in this configuration suffers the weakest repulsions possible. In general, a central atom surrounded by 10 electrons will preferentially house up to three nonbonding pairs in equatorial sites.

(b) *Acetylene, C_2H_2.* After drawing three single bonds to obtain the skeleton H—C—C—H, we have four valence electrons of the original 10 still to deploy. They go automatically to the carbons, since the two hydrogens are already satisfied; and there, between the carbons, the four electrons make two additional pair bonds: H—C≡C—H. The triple bond leaves each carbon with an octet.

With no lone pairs in the molecule, VSEPR predicts a linear arrangement symmetric about the HCCH axis. The two C—H bonds, indistinguishable and symmetric, must have the same length.

(c) *Nitric oxide, NO.* Pooling five valence electrons from the nitrogen and six from the oxygen, NO has a total of 11 electrons. The sum yields an odd number, a set impossible to arrange into two octets. We can give the octet either to oxygen ($:\ddot{N}=\ddot{O}$) or to nitrogen ($\ddot{N}=\ddot{O}:$), but not to both at the same time. Such are the failings of this too-simple model.

Each of these two Lewis structures offers only a partial representation of the bonding in nitrogen oxide. For a more complete interpretation, see the description of molecular orbitals in Chapter 7.

(d) *The nitrosonium ion, NO^+.* The positively charged ion, deficient by one electron, has a valence of 10. Allowed an even number of electrons, we can satisfy the octet rule with the structure $[:N≡O:]^+$.

EXAMPLE 2-6. Molecules and Mass: The Balanced Chemical Equation

From nuclei and electrons to atoms, from atoms to molecules, from molecules to reactions—the law of mass conservation is at last expressed in a balanced chemical equation. All the stoichiometric relationships are embodied within.

PROBLEM: An unidentified compound (relative molecular weight = 46.07) reacts with molecular oxygen to produce carbon dioxide, water, and molecular nitrogen. Analysis shows that the unknown material consists of 26.1% carbon, 13.1% hydrogen, and 60.8% nitrogen by mass. (a) What is its molecular formula? (b) Briefly describe a possible bonding pattern. (c) Write a balanced equation for the reaction.

SOLUTION: Given the mass percentages and molecular weight, we know how to determine both the empirical and molecular formulas.

From there we arrange the atoms into a molecule consistent with the valence properties of C, H, and N; and then, using a simple but systematic procedure, we construct a balanced equation.

(a) *What is the molecular formula?*

$$26.1 \text{ g C} \times \frac{1 \text{ mol C}}{12.011 \text{ g C}} = 2.17 \text{ mol C}$$

$$13.1 \text{ g H} \times \frac{1 \text{ mol H}}{1.00794 \text{ g H}} = 13.0 \text{ mol H}$$

$$60.8 \text{ g N} \times \frac{1 \text{ mol N}}{14.0067 \text{ g N}} = 4.34 \text{ mol N}$$

The empirical formula is $C_{2.17}H_{13.0}N_{4.34}$ ($CH_{5.99}N_{2.00}$), which is sufficiently close to CH_6N_2. Since the formula weight

$$12.011 + (6 \times 1.00794) + (2 \times 14.0067) = 46.072$$

is equal to the molecular weight, the molecular formula is also CH_6N_2.

(b) *Describe a possible bonding pattern.* There are 20 valence electrons (four from carbon, one each from six hydrogens, and five each from two nitrogens). Arranged thus,

$$
\begin{array}{c}
\quad\; \text{H} \;\; \text{H} \qquad \text{H} \\
\quad\; | \quad\; | \qquad\; / \\
\text{H} - \text{C} - \underset{\displaystyle \cdot\cdot}{\text{N}} - \ddot{\text{N}} \\
\quad\; | \qquad\qquad \backslash \\
\quad\; \text{H} \qquad\qquad \text{H}
\end{array}
$$

each atom enjoys a completed valence with all 20 electrons in place. The nitrogens form three single bonds apiece, the hydrogens one, and the carbon four. Expect tetrahedral electronic geometry about the carbon and two nitrogens, with a lone pair occupying one position on each nitrogen.

QUESTION: Are there other possibilities?

ANSWER: Yes. Rather than follow the sequence C-N-N, the molecule might equally well be connected as N-C-N:

$$
\begin{array}{c}
\text{H} \qquad\;\; \text{H} \qquad \text{H} \\
\;\backslash \qquad\;\; | \qquad\;\; / \\
\ddot{\text{N}} - \text{C} - \ddot{\text{N}} \\
\;/ \qquad\;\; | \qquad\;\; \backslash \\
\text{H} \qquad\;\; \text{H} \qquad \text{H}
\end{array}
$$

Given the scanty information available, we have no way to tell the difference. We need more experimental information about the sequence, lengths, and angles of the bonds. Nevertheless, the same chemical equation will apply to either structure.

(c) *Write a balanced equation for the reaction.* Start with the unbalanced equation

$$\_\_\ CH_3NHNH_2 + \_\_\ O_2 \longrightarrow \_\_\ CO_2 + \_\_\ H_2O + \_\_\ N_2$$

and proceed as follows:

1. Look for elements that appear in only one compound on both the left and right side of the equation, with the *same number of atoms* in each. The two substances involved must have the same coefficient, which we enter provisionally as 1.

   Carbon qualifies for such treatment here. Already in balance, C occurs only in the reactant CH_3NHNH_2 and the product CO_2: one carbon in each. So, too, is N isolated (and serendipitously balanced) in the reactant CH_3NHNH_2 and the product N_2.

2. Look now for elements that appear, again, only in one reactant and one product, but which show *different* numbers of atoms in the two species. Adjust the coefficients to bring these elements into balance.

   In our still unbalanced equation, note, the six H atoms in CH_3NHNH_2 are countered by just two H atoms in H_2O. We correct the imbalance by tripling the number of waters:

   $$CH_3NHNH_2 + \_\_\ O_2 \longrightarrow CO_2 + 3H_2O + N_2$$

3. Balance the elements that remain. On the right are five oxygens: two in CO_2, three from the three H_2O's. On the left, however, are just the two in O_2. Therefore we complete the equation by assigning oxygen a coefficient of $\frac{5}{2}$,

   $$CH_3NHNH_2 + \tfrac{5}{2}O_2 \longrightarrow CO_2 + 3H_2O + N_2$$

   after which we may (optionally) multiply through by 2 to eliminate fractions:

   $$2CH_3NHNH_2 + 5O_2 \longrightarrow 2CO_2 + 6H_2O + 2N_2$$

Either way, the stoichiometric relationships are the same. Five molecules (or moles) of O_2 react with two molecules (or moles) of CH_3NHNH_2 to produce two molecules (or moles) of CO_2, six molecules (or moles) of H_2O, and two molecules (or moles) of N_2. A fixed number of electrons and nuclei is rearranged. Mass is conserved.

EXAMPLE 2-7. Limiting Reactant

PROBLEM: A 9.21-g sample of CH_3NHNH_2 reacts with 32.0 g O_2. (a) Which reactant is limiting? (b) At most, how many grams of CO_2, H_2O, and N_2 should we expect?

SOLUTION: Take the balanced equation from above,

$$CH_3NHNH_2 + \tfrac{5}{2}O_2 \longrightarrow CO_2 + 3H_2O + N_2$$

and: (1) Convert grams of reactants into moles of reactants. (2) Identify the limiting reactant. (3) Compute moles of products. (4) Compute grams of products.

(a) *Which reactant is limiting?* Initially there are 0.200 mole of CH_3NHNH_2 and 1.00 mole of O_2:

$$9.21 \text{ g } CH_3NHNH_2 \times \frac{1 \text{ mol } CH_3NHNH_2}{46.07 \text{ g } CH_3NHNH_2} = 0.200 \text{ mol } CH_3NHNH_2$$

$$32.0 \text{ g } O_2 \times \frac{1 \text{ mol } O_2}{32.00 \text{ g } O_2} = 1.00 \text{ mol } O_2$$

Since O_2 and CH_3NHNH_2 combine in a 5:2 molar ratio, we know that 1 mole of O_2 will react completely with 0.4 mole of CH_3NHNH_2. But there is only 0.200 mole of CH_3NHNH_2 on hand, not enough to consume 1.00 mole of O_2 in a reaction with 5:2 stoichiometry. The oxygen is present in excess. Only 0.500 mole of O_2 will react with 0.200 mole of CH_3NHNH_2, the limiting reactant.

(b) *What are the expected masses of product?* Assume that the limiting reactant is wholly consumed. Then, referring to the balanced equation, convert moles of reactant into moles of product into grams of product:

$$0.200 \text{ mol } CH_3NHNH_2 \times \frac{1 \text{ mol } CO_2}{1 \text{ mol } CH_3NHNH_2} \times \frac{44.01 \text{ g } CO_2}{\text{mol } CO_2} =$$

$$8.80 \text{ g } CO_2$$

$$0.200 \text{ mol } CH_3NHNH_2 \times \frac{3 \text{ mol } H_2O}{1 \text{ mol } CH_3NHNH_2} \times \frac{18.02 \text{ g } H_2O}{\text{mol } H_2O} =$$

$$10.8 \text{ g } H_2O$$

$$0.200 \text{ mol } CH_3NHNH_2 \times \frac{1 \text{ mol } N_2}{1 \text{ mol } CH_3NHNH_2} \times \frac{28.01 \text{ g } N_2}{\text{mol } N_2} =$$

$$5.60 \text{ g } N_2$$

Observe: 9.21 g CH_3NHNH_2 (0.200 mole) react with 16.0 g O_2 (0.500 mole) to produce 8.80 g CO_2, 10.8 g H_2O, and 5.60 g N_2. 25.2 grams of reactants yield 25.2 grams of products. Mass is conserved.

And observe: 0.200 mole of C (1.20×10^{23} atoms) emerges undiminished as 0.200 mole of C (1.20×10^{23} atoms)—as do 1.20 moles of H (7.22×10^{23} atoms) and 0.400 mole of N (2.41×10^{23} atoms) and 1.00 mole of O (6.02×10^{23} atoms), along with all their component protons, neutrons, and electrons. See the reaction as a gritty, grainy coming together of small particles, a microscopic disassembly and reassembly. Mass is conserved while the electrons and nuclei are reshuffled.

EXAMPLE 2-8. Real Life: Yield

We do not always get what we want.

PROBLEM: Sally burns 9.21 g CH_3NHNH_2 in an excess of O_2, just as described in Example 2-7. But she obtains only 6.60 g CO_2 instead of the expected 8.80 g, leaving her to worry about the conservation of mass in chemical reactions. Has the law been violated?

SOLUTION: Although the stoichiometric promise of 8.80 grams was unfulfilled, the total mass *did* remain the same. No law was violated. Mass was diverted, but not destroyed; misplaced, but not lost. Look somewhere else for the 2.20 grams—in unintended products, on the walls of the reaction vessel, in the atmosphere, on the benchtop, on the spatula. Everything is there, in full measure.

The masses and moles we calculate by stoichiometry are maximum amounts, not guaranteed outcomes. *If* the limiting reactant is entirely consumed . . . *if* the reaction goes to completion . . . *if* there are no competing reactions that produce unexpected products . . . *if* there are no impurities . . . *if* every bit of product is recovered without loss, without spills, without slop . . . only if everything goes right, only then, do we obtain the ***theoretical yield*** (8.80 g in this example). If not, we settle for a smaller *actual* yield.

Sally's ***percent yield***, defined below, is 75%:

$$\text{Percent yield} = \frac{\text{actual yield}}{\text{theoretical yield}} \times 100\%$$

$$= \frac{6.60 \text{ g}}{8.80 \text{ g}} \times 100\% = 75.0\%$$

The higher the yield, the better. Unrealized products mean lost time, lost resources, lost money.

EXERCISES

1. Write the name of the element associated with each symbol:
   (a) H   (b) He   (c) Li   (d) Be   (e) C
   (f) F   (g) Ne   (h) Si

2. Write the name of the element associated with each symbol:
   (a) Ca   (b) Co   (c) Cu   (d) Ag   (e) Au
   (f) Hg   (g) At   (h) Lr

3. Write the symbol and atomic number belonging to each of the following elements: (a) Boron.   (b) Nitrogen.   (c) Aluminum. (d) Vanadium.   (e) Iron.   (f) Lanthanum.

4. Write the symbol and atomic number belonging to each of the following elements: (a) Cerium.   (b) Lutetium.   (c) Hafnium. (d) Lead.   (e) Uranium.   (f) Plutonium.

5. Specify the number of protons and neutrons in each isotope:

   (a) ^{9}Be, ^{10}Be

   (b) ^{12}C, ^{13}C

   (c) ^{14}N, ^{15}N

   (d) ^{16}O, ^{17}O

6. Specify the number of protons and neutrons in each isotope:

   (a) ^{31}P, ^{32}P

   (b) ^{35}Cl, ^{37}Cl

   (c) ^{206}Pb, ^{208}Pb

   (d) ^{235}U, ^{238}U

7. State the group (column) and period (row) assigned to each element in the periodic table:
   (a) He   (b) Li   (c) K   (d) Ar   (e) Cs   (f) Xe

8. State the group (column) and period (row) assigned to each element in the periodic table:
   (a) Mg   (b) Ne   (c) Ba   (d) O   (e) F   (f) Te

9. Specify the number of protons and electrons in each atom or ion:

   (a) Li, Li^+, He

   (b) K, K^+, Ar

   (c) Cs, Cs^+, Xe

   (d) Mg, Mg^{2+}, Ne

10. Specify the number of protons and electrons in each atom or ion:

    (a) Ba, Ba^{2+}, Xe

    (b) O, O^{2-}, Ne

    (c) F, F^-, Ne

    (d) Ti, Ti^{2+}, Ti^{4+}, Ar

11. How many valence electrons are present in each species?

    (a) O^-, Cl^-, Ar

    (b) Rb, Rb^+, Sr

    (c) O^{2-}, Ne, Mg^{2+}

12. Represent each of the nine species in the preceding exercise by a Lewis diagram:

    (a) O^-, Cl^-, Ar

    (b) Rb, Rb^+, Sr

    (c) O^{2-}, Ne, Mg^{2+}

    In which set are all the members isoelectronic?

13. Classify each element as a metal or nonmetal: (a) Beryllium. (b) Nitrogen. (c) Fluorine. (d) Potassium. (e) Calcium.

14. Classify each element as a metal or nonmetal: (a) Iron. (b) Rubidium. (c) Yttrium. (d) Iodine. (e) Xenon.

15. Which pair or pairs of elements below will most likely form an ionic compound?

    (a) cesium and chlorine

    (b) sodium and calcium

    (c) nitrogen and oxygen

    (d) carbon and hydrogen

Which pair or pairs will react to produce covalent molecules? Which are not likely to react together at all?

16. Predict a likely empirical formula for ionic compounds composed of:

    (a) rubidium and fluorine

    (b) magnesium and oxygen

    (c) aluminum and oxygen

    (d) calcium and iodine

17. For each of the ionic compounds given, state the charge on both cation and anion:

    (a) $CaCl_2$ (calcium chloride)

    (b) KBr (potassium bromide)

    (c) CsCl (cesium chloride)

    (d) $AlCl_3$ (aluminum chloride)

    (e) $Mg(CN)_2$ (magnesium cyanide)

18. Which of the following compounds are ionic, and which are molecular?
    (a) Cl_2O (b) $LiNO_3$ (c) Na_2CrO_4 (d) Si_2Br_6
    (e) $ZnCO_3$

19. More of the same: which of the following compounds are ionic, and which are molecular?
    (a) AsF_5 (b) $RbMnO_4$ (c) N_2O (d) $BaBr_2$
    (e) AgCl

20. Compute the percentage chlorine by *mass* in NaCl, KCl, and RbCl. Does it change? How about the percentage by *moles*?

21. Compute the percentage chlorine by mass and by moles in NaCl, $MgCl_2$, and $AlCl_3$.

22. A pale-yellow ionic solid is found to contain only tin and bromine, distributed by mass as

$$Sn: \quad 42.6\%$$

$$Br: \quad 57.4\%$$

What is its empirical formula?

23. Analysis shows that a certain ionic solid contains potassium plus one other element, a halogen X. Determine the identity of X from the mass percentages given below:

$$K: \quad 52.4\%$$

$$X: \quad 47.6\%$$

What is the empirical formula?

24. How many grams of sodium are contained in 0.500 mole of NaOH (sodium hydroxide)?

25. How many grams, atoms, and moles of sodium are contained in 28.62 grams of NaOH?

26. How many electrons are contained in each of the following molecules?

   (a) NH_3 (ammonia)
   (b) SO_2 (sulfur dioxide)
   (c) HI (hydrogen iodide)
   (d) C_2H_5OH (ethanol)
   (e) CH_3COOH (acetic acid)

   What are the molar masses?

27. Determine the percentage of carbon and hydrogen by mass in CH_4, C_2H_6, C_3H_8, C_4H_{10}, and C_5H_{12}.

28. Compute the number of moles and the number of molecules contained in 1.00 g each of CH_4, C_2H_6, C_3H_8, C_4H_{10}, and C_5H_{12}.

29. Compute the mass of 1.732 mol each of CH_4, C_2H_6, C_3H_8, C_4H_{10}, and C_5H_{12}.

30. A compound contains carbon, hydrogen, and oxygen in the mass percentages given below:

$$C: \quad 40.0\%$$

$$H: \quad 6.7\%$$

$$O: \quad 53.3\%$$

What is the empirical formula?

31. Further analysis shows that the compound described in the preceding exercise has a molar mass of 30.0 g mol^{-1}. What is its molecular formula?

32. Suggest Lewis structures for C_2H_6 (ethane) and C_2H_4 (ethylene). Note that ethylene contains a double bond.

33. Use the VSEPR model to predict the arrangement of hydrogen atoms around each carbon in: (a) C_2H_6 (b) C_2H_4

34. Draw Lewis structures for HF, HCl, HBr, and HI. What do you notice?

35. Show the Lewis structure for CH_3Cl. What shape does the VSEPR theory predict for the molecule?

36. Draw Lewis and VSEPR structures for PCl_3 (phosphorus trichloride).

37. Draw Lewis and VSEPR structures for SF_6 (sulfur hexafluoride).

38. A molecule of ozone, O_3, contains three oxygen atoms in a bent configuration:

    Knowing that both O—O bonds are of equal length, draw two equivalent Lewis structures to represent the molecule.

39. All three N—O bond lengths in the nitrate ion, NO_3^-, are equal. Draw three equivalent Lewis structures consistent with this observation.

40. The carbonate ion, CO_3^{2-}, similar to the nitrate ion, contains three oxygens bound to a central atom, again with equal bond lengths. Draw three equivalent Lewis structures to represent the molecular carbonate ion.

41. Balance the following equations:

    (a) __ N_2 + __ H_2 ⟶ __ NH_3
    (b) __ H_2 + __ I_2 ⟶ __ HI
    (c) __ CH_4 + __ O_2 ⟶ __ CO_2 + __ H_2O
    (d) __ C_6H_{14} + __ O_2 ⟶ __ CO_2 + __ H_2O

42. Balance the following equations:

    (a) ___ Se + ___ BrF_5 ⟶ ___ SeF_6 + ___ BrF_3
    (b) ___ HBr + ___ F_2 ⟶ ___ HF + ___ Br_2
    (c) ___ CH_3OH + ___ O_2 ⟶ ___ CO_2 + ___ H_2O
    (d) ___ $C_6H_{12}O_6$ + ___ O_2 ⟶ ___ CO_2 + ___ H_2O

43. Supply the value of *n* and *m* needed to balance each equation:

    (a) $B_nH_m + 3O_2$ ⟶ $B_2O_3 + 3H_2O$
    (b) $H_2S_nO_m + H_2O$ ⟶ $2H_2SO_4$
    (c) H_nIO_m ⟶ $H^+ + IO_4^- + 2H_2O$

44. How many grams of NCl_3 and HCl are produced if 1.00 gram of ammonia reacts *stoichiometrically* (completely, molecule for molecule, mole for mole) with 3.00 grams of chlorine gas in the reaction shown below?

$$NH_3 + 3Cl_2 \longrightarrow NCl_3 + 3HCl$$

How much of each reactant remains present after the reaction is over? Which reactant is the limiting species?

45. Suppose that 1.05 grams of HCl are obtained in the reaction described in the exercise above. Is this amount the maximum possible? What is the percent yield?

46. Balance the equation

$$\_\_ CH_4 + \_\_ H_2O \longrightarrow \_\_ CO + \_\_ H_2$$

and answer: (a) What mass of water is needed to react completely with 82.55 grams of methane? (b) How many grams of carbon monoxide and hydrogen are produced thereby?

47. Methanol can be manufactured by reaction of carbon monoxide and hydrogen. (a) Balance the equation

$$\_\_ CO + \_\_ H_2 \longrightarrow \_\_ CH_3OH$$

and then, from the stoichiometry, determine the limiting reactant when 10.0 grams of CO combine with 10.0 grams of H_2. (b) If the process occurs stoichiometrically, how many grams and moles of CH_3OH should be obtained? (c) How many grams of CO and H_2 should remain?

48. Suppose that 10.0 grams of CH_3OH are recovered from the reaction of CO and H_2 described just above. What is the percent yield?

3

Prototypical Reactions

3-1. The Nature of Chemical Reaction
3-2. Acids, Bases, and Salts
3-3. Reduction and Oxidation
3-4. Dissolution and Precipitation
3-5. Radical Reactions
3-6. Catalysis
    REVIEW AND GUIDE TO PROBLEMS
    EXERCISES

3-1. THE NATURE OF CHEMICAL REACTION

Chemistry has about it the air of the marketplace, a sense of buyers and sellers looking for mutual gain. One recognizes a peculiar economics to chemical reactions, whimsical in a way but telling nonetheless. There is a trading system complete with competition, profit and loss, conflict between short-term and long-term strategy, and even the equivalent of brokers to bring the participants together. And, most revealing of all, nothing is given away for free.

On offer are electrons—particularly the valence electrons of atoms and molecules, available for sharing or transfer from one entity to another. Species A, seeking a better use for some of its electrons, finds a species B with a complementary need. Often the quantity of electrons offered is only a small fraction of the total, but big things can happen. If a net benefit accrues to the parties involved (A, B, and their surroundings), then a reaction takes place. Old bonds are broken and new ones are formed; atoms are switched; electrons move. A molecule may ionize, or a solid material may dissolve in water. Perhaps an acid and base are neutralized to form a salt. Or a substance might burn or corrode, freeze or melt. Complex biological processes such as photosynthesis and metabolism may occur, proceeding according to an intricate sequence of elementary steps. Something happens. The **reactants** A and B become the **products** C and D.

The currency of these transactions is the thermodynamic *free energy*, a measure of a system's drive toward both lower internal energy and higher disorder, or entropy. We shall explore the great question of thermodynamic drive later on, in Chapter 14, but even now we should peek behind the curtain; for here, in a word, is the "why" of all chemical reactions, the reason-to-be, the motive force: It is a drive toward lower free energy, as if to fall from a higher valley down to a lower valley in the metaphorical landscape of Figure 3-1. Here, in a picture more literal than it first appears, are the energetic hills and valleys in which all reactions unfold.

To shed free energy is, above all, to relax. To shed free energy is to spread out, to fall down, to settle into a more stable state. To shed free energy is to do things the easy way; and, for matter and energy, the easy way is the natural way. Nature takes the path of least resistance.

Humpty-Dumpty's great fall tells it all. He falls from the wall and breaks into pieces, never to be mended. He goes from high gravitational energy to low, and he changes from a single, orderly arrangement into scattered, disordered bits and pieces. Lower energy. Higher entropy. Lower free energy.

FIGURE 3-1. The energetic ups and downs of an imaginary reaction. Reactants A and B eventually become products C and D, settling into a valley of lower free energy—but not without first crossing over the mountaintops. Sometimes they find themselves temporarily trapped, as intermediates, in a higher valley such as E or E′.

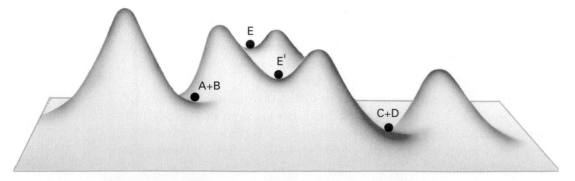

Molecules, like Humpty-Dumpty, also take the easy road, the road that leads to less energy and more disorder. Only if the reactants can decrease their free energy will a reaction occur on its own, for only then is the transformation profitable. The larger the drop in free energy, the more thoroughly do the reactants combine to form products. A lower free energy is the thermodynamic profit that nature demands to convert A and B into C and D.

Is such profit ever realized? Maybe, maybe not; there is still chemistry to be done. The expected change in free energy offers only the *possibility* of reaction, the expectation of gain or loss if indeed something happens. Species A and B, allowed the opportunity to become C and D, still need to carry out the transaction.

Here, during actual execution, arise the specific questions of reaction mechanism and kinetics, in particular *how* the changes are brought about and how fast. We ask how—not whether—the reactants will turn into the products; we ask what path the reaction will follow, what compounds will be made and unmade en route, how much time it will take. Often there is more than one way to go from here to there, just as there are different ways to go from valley to valley through the mountains staked out in Figure 3-1.

On the road, say, to becoming C and D, reactants A and B might first have to climb a mountain and then pass through an *intermediate* structure E, so that really we have

$$A + B \longrightarrow E \longrightarrow C + D$$

E lies over a peak higher than either the valley of the reactants (A and B) or the valley of the products (C and D), and this extra energy must be mustered before A and B can cross over into C and D. If E is hard to reach, a bottleneck may develop and the reaction will proceed slowly or not at all. But if an easier route over a lower mountaintop can be found, then faster conversion of reactants into products becomes possible. Such facilitation we call *catalysis*, using the term **catalyst** to mean some agent— a third party—that speeds a process without suffering any net change itself.

Catalysts are the brokers in the marketplace, often essential for making reactions happen. They are the matchmakers, guides, and escorts that bring molecules together and help them negotiate the pathways of chemical change. Compare, for example, the catalyzed and uncatalyzed reactions tracked symbolically in Figure 3-2. Which is easier? Which is faster? Look at the two roads and see.

Look still deeper, and realize that reactions (catalyzed or not) are intrinsically microscopic events. They occur electron by electron, bond by

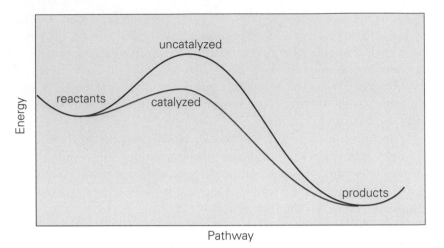

FIGURE 3-2. Catalysts work in many ways, but they share a common purpose: to open up a new, faster, easier route from reactants to products over a lower mountain of energy. Doing so, a catalyst changes neither the beginning nor end of the journey, but rather the road in between. It finds a less arduous way to take the molecules from the valley of the reactants to the valley of the products.

bond, molecule by molecule, so that, at these fine-grained levels of intimacy, the energetics of reaction can be dictated only by quantum mechanics: electron to electron. One molecule, with electrons to share, collides with another. Can the second species use those electrons profitably? Where might they go? Will the resulting products be more stable than the reactants? The answers lie rooted in the electron configurations of the particles, both in the electrons' energies and in their spatial distributions. Although we have, for now, just a rough but serviceable picture of octets and electron pairs, eventually we shall explain the intimate encounters between molecule and molecule in the language of quantum mechanics.

And then there is this to consider: Reactions occur not only molecule by molecule, but mole by mole as well. From molecule to mole, building up to almost unimaginably large numbers, a *macroscopic* change develops as the smoothed-out result of trillions upon trillions of individual events. About any single molecular incident we can never be sure. We know only the odds, not the outcome; and so we can say only, Maybe yes and maybe no. Whether or not an individual reaction occurs is always a matter of probability, a game of chance.

So, to react or not to react? Matter often finds itself at various probabilistic forks in the road where nature must make a decision: coin flips given names like Boltzmann distributions (see Chapters 10 and 14) and

Arrhenius factors (see Chapter 18). Will molecules A and B go to C and D? The odds might be 60% at one temperature, 93% at another temperature, 12% at yet another. Will C and D turn around and go back to A and B? Not likely at room temperature, but perhaps more probable 100 degrees lower.

Expect therefore to discover that quantum mechanics, thermodynamics, and kinetics are all microscopic games of chance—but games decided by incredibly many players, by uncountably large swarms of atoms and molecules. Thus constrained, chemical phenomena will succumb inevitably to the herdlike behavior of any large group. From chance and uncertainty at the molecular level, then, must come regularity and certainty at the macroscopic level.

We should at least be aware of these interrelated principles because in every reaction, however simple, rests all of chemistry. There is, first, the question of structure, specifically the arrangement of nuclei and electrons in reactants and products. There is the quantum mechanical problem of available electrons and a place to put them, and there is the further issue of thermodynamic potentiality. Kinetics and mechanism, too, determine whether one set of products (not necessarily the lowest in free energy) is made *fast* or whether some stronger, more stable set will emerge later from the competition. And everything is governed finally by probability and statistics; that much we must concede to the almost incomprehensibly large magnitude of the mole itself.

With such background we begin to explore various categories of chemical reactions, looking first to adopt some broadly useful classifications. Our aim at present is merely to see *what* happens in the most general sense—what goes in, what comes out, what differs from start to finish. We bear in mind the larger questions of equilibrium and free energy, rate, mechanism, and molecular structure, while deferring their detailed consideration to later chapters where the picture can be developed sufficiently.

3-2. ACIDS, BASES, AND SALTS

G. N. Lewis, early in the 1920s, defined an **acid** as any species capable of *accepting* a pair of electrons; in matching fashion he defined a **base** as a species able to *donate* an electron pair. With these generalizations, sweeping and incisive, we have a simple picture to describe what takes place in most chemical reactions: electrons given and taken, shared or transferred. So understood, much of chemistry falls under the heading of acid–base interactions.

We have already seen examples of Lewis acids and bases in Chapter 2, notable among them the reaction between boron trifluoride and ammonia:

$$\underset{\underset{F}{|}}{\overset{\overset{F}{|}}{F-B}} + \underset{\underset{H}{|}}{\overset{\overset{H}{|}}{:N-H}} \longrightarrow \underset{\underset{F}{|}}{\overset{\overset{F}{|}}{F-B}} - \underset{\underset{H}{|}}{\overset{\overset{H}{|}}{N-H}}$$

The base, $:NH_3$, has a lone pair suitable for donation to the octet-deficient (and consequently acidic) BF_3. Acid and base have complementary needs; each has what the other lacks and so they combine for mutual benefit. The union is called a **neutralization**, in which the product of the reaction—termed a "salt" complex—has the characteristics of neither the original acid nor base. The neutralized product emerges as a compound that enjoys a measure of chemical satisfaction, at least until some better opportunity comes along.

Figure 3-3 takes a fanciful view of the whole process, but a view not far from the truth either. Chemistry is the coming together of opposites,

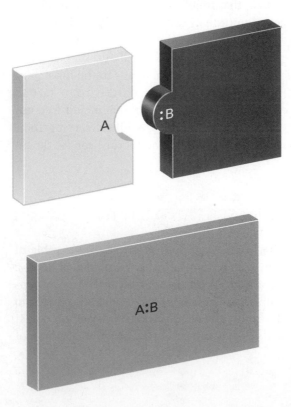

FIGURE 3-3. Haves and have-nots: Lewis acids and bases. The base B, heavily shaded, is rich in electrons. The acid A is not. One gives and the other takes; and together, forming a union of opposites as suggested below, they settle into a new structure more stable than before. The description applies in some way to most chemical reactions.

and neutralization is just the beginning. Lewis acids and bases give us our first brush with chemical yin and yang, certainly not the last.

Here is another. Consider the reaction

$$Ni^{2+} + 6H_2O \longrightarrow Ni(H_2O)_6^{2+}$$

in which six water molecules form a *coordination complex* with a nickel ion:

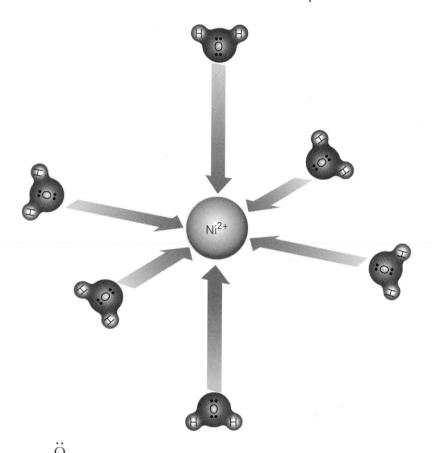

Each H $\ddot{O}$ H, acting as a Lewis base, offers a lone pair (on the oxygen) to the electron-receptive transition metal. The resulting octahedral arrangement of six *ligands* about the central atom is, by Lewis's definition, the salt of an acid–base reaction. Giver unites with taker.

There is a rich chemistry associated with transition-metal complexes, often recognized as brilliantly colored cationic, neutral, or anionic structures (Chapter 19). Lewis acid–base interactions bind the ligands to the central species through so-called ***coordinate covalent*** (or ***dative***) bonds, in which the electron pair is contributed entirely by the base. Commonly occurring ligands include water (H_2O), ammonia (NH_3), carbon monoxide (CO), the chloride ion (Cl^-), the cyanide ion

(CN⁻), and the hydroxide ion (OH⁻). They are all Lewis bases, each with a pair of electrons to give.

As perhaps the most common Lewis base of all, hydroxide is neutralized by the hydrogen ion to form water according to the reaction

$$H^+ + OH^- \longrightarrow H_2O$$

The cation H^+ suggests a naked hydrogen nucleus—an H atom stripped to its proton—and this positive ion clearly is a species crying out for electrons. It has none. With its electronic shell empty and its positive charge laid bare, a hydrogen ion welcomes negatively charged contributions from any source. The hydroxide ion serves admirably in this role as a Lewis base.

A proton on the loose is important. Indeed, H^+ is passed around so frequently in chemical reactions that it becomes worthy of careful attention and tracking. The hydrogen ion can serve, moreover, as the basis for an alternative definition of acids and bases, one advanced independently by Brønsted and Lowry at about the same time as Lewis. Their proposed interpretation of acidity and basicity: a giving and taking of H^+, from acid to base. A ***Brønsted-Lowry acid*** is a proton donor; a ***Brønsted-Lowry base*** is a proton acceptor.

This conception (Figure 3-4) is narrower than Lewis's model, for which it forms a limiting case. Since any species with a proton to *give* is

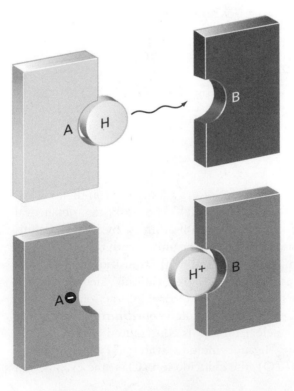

FIGURE 3-4. Haves and have-nots, another view: Brønsted-Lowry acids and bases. Here the acid HA has a movable proton (the H^+ ion), and the base B has a place to put it. One gives and the other takes; and, as molecules do, they settle into a new, more stable combination, going over to A⁻ and BH⁺. But take note: The real medium of exchange is still the electron, because Brønsted and Lowry's acid (a bare proton) is clearly also an acid in the Lewis sense. It accepts electrons.

necessarily one able to accept electrons as well, a Brønsted-Lowry acid acts simultaneously as a Lewis acid. In the same way, a species able to accept a proton can do so precisely because it has electrons to offer the positively charged particle H^+. The relationship is not reciprocal, though, for Lewis acids and bases do not require protons for their definition. Lewis put the onus entirely on the electrons, consistent with the idea that these are the subatomic particles truly at the heart of chemistry.

Yet despite its narrower scope, the Brønsted-Lowry picture of acids and bases does provide sharp insight into many reactions, since at times we can better follow the travels of a proton than the feathery movement of an electron. The hydrogen ion is, in some ways, a more substantial particle, one of the heavy bricks of a molecule. We might satisfyingly describe, for example, the neutralization reaction

$$HCl + NH_3 \longrightarrow NH_4^+ + Cl^-$$

as a transfer of H^+ from the proton-donating HCl (a Brønsted-Lowry acid) to the proton-accepting $:NH_3$ (a Brønsted-Lowry base) to form the ammonium ion NH_4^+ and the chloride ion Cl^-:

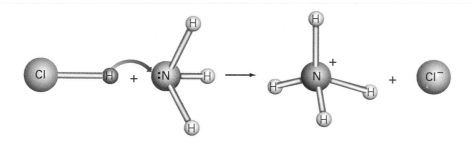

Ammonium chloride is thus the salt of reaction: a union of acid and base, brought together to produce the ionic compound NH_4Cl.

The Brønsted-Lowry description, by pointing to such a substantial change in the structure, captures something especially physical about an atomic rearrangement. Particularly for organic reactions, which we take up in Chapter 8, the model of proton transfer helps unify a variety of seemingly diverse processes.

In this same practical spirit, we come to appreciate the earlier and more restrictive model put forth by Arrhenius to describe acids and bases in *aqueous* (water-based) solutions. Here the chemistry is dominated by the formation of hydrogen and hydroxide ions, mediated by the presence of H_2O molecules. An *Arrhenius acid* is defined, appropriately, as a substance that yields H^+ ions when dissolved in water, whereas an *Arrhenius base* is something that produces OH^- ions in aqueous solution. Hydrochloric acid therefore is an Arrhenius acid by virtue of the H^+ ions it releases in the presence of water:

$$HCl \longrightarrow H^+ + Cl^-$$

Sodium hydroxide, a source of OH^-, is an Arrhenius base:

$$NaOH \longrightarrow Na^+ + OH^-$$

This **dissociation** (breakup) into ions is made possible by the water molecules, which subvert the covalent bonds between hydrogen and chlorine and the electrostatic bonds between the sodium and hydroxide ions, leaving a solution of separated anions and cations. Because water is highly polar, it has the electrical wherewithal to attract either anions or cations or other polar molecules. The H_2O molecule has, in addition, the lone pairs to compete effectively for protons.

Recognizing that capacity, we write the acid dissociation reaction explicitly as

$$\overset{\frown}{HCl} + H_2O \longrightarrow H_3O^+ + Cl^-$$

to represent the exchange of a proton. H^+ is ripped from the HCl molecule and attached to the H_2O molecule, producing the **hydronium ion** H_3O^+ in the process. The proton transfer is an acid–base reaction in the Brønsted-Lowry sense, just as (same thing) the acceptance by H^+ of a lone pair from H_2O is an acid–base interaction according to the broader Lewis definition. Lewis encompasses Brønsted-Lowry, and Brønsted and Lowry's model similarly subsumes that of Arrhenius, which remains nevertheless an apt description for the many acid–base reactions occurring in aqueous solution. The Arrhenius picture involves the water directly, using hydronium and hydroxide ions to mark the presence of acids and bases.

H_3O^+ is itself the primary acidic species in watery solution, and OH^- is the principal base. The two are neutralized to form water,

$$H_3O^+ + OH^- \longrightarrow 2H_2O$$

in the presence of whatever ions are left behind by the original acids and bases. A complete equation for the neutralization of, say, hydrochloric acid and sodium hydroxide would then be

$$H_3O^+ + Cl^-(aq) + Na^+(aq) + OH^- \longrightarrow 2H_2O + NaCl(aq)$$

where we use the symbol "aq" selectively to indicate that NaCl exists as the ions Na^+ and Cl^- dispersed in aqueous solution—no different on either side of the equation. Sodium chloride, dissociated in water, is the salt of this reaction between hydrochloric acid and sodium hydroxide; and, for once, the chemical term "salt" resonates with everyday experience: NaCl, in its solid form as an ionic crystal, is the main component of ordinary table salt. From such correspondences does some of the arcane nomenclature of chemistry originate.

There are further classifications. **Strong acids** produce relatively large amounts of H_3O^+ when introduced into water, yielding nearly one mole of hydronium ions per mole of acid (Figure 3-5a). Well-known examples include hydrochloric acid (HCl), sulfuric acid (H_2SO_4), and nitric acid (HNO_3). **Weak acids**, by contrast, are incompletely dissociated in aqueous solution (Figure 3-5b), and we shall return to these species in Chapter 16 better equipped to understand their equilibrium properties.

(a)   (b)

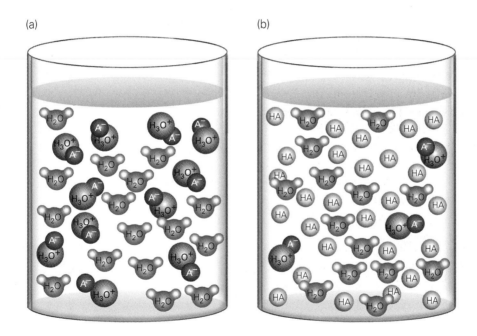

FIGURE 3-5. Arrhenius acids (HA), strong and weak. (a) A strong acid dissociates nearly completely in water, giving up one hydronium ion and one A^- for every HA molecule attacked by H_2O. (b) A weak acid resists dissociation more effectively, releasing very few hydronium ions into solution. Most of the original acid molecules keep their detachable proton.

Also of interest are acids with more than one dissociable hydrogen, termed **polyprotic**, for which the strength of acidity may vary from proton to proton. Sulfuric acid with its two ionizable hydrogens is a *diprotic* acid, while phosphoric acid (H_3PO_4) is *triprotic*. For a varied assortment of Arrhenius acids, see the listing in Appendix C (Table C-20). They come in many forms.

Common to all, however, is their production of H_3O^+ in water, testimony that a bare hydrogen nucleus (the proton) cannot survive as H^+ in the presence of H_2O molecules. Nature prefers neutrality, so charged particles do not remain unchaperoned for long. The stripped proton is best stabilized by the nearby presence of negative charges, supplied here by the excess electrons on water's oxygen. The interaction is called **hydration** or, more generally, **solvation** to include species other than water.

The precise nature of acidic hydration remains incompletely understood. A free proton starts as a bare speck of matter, tiny and easy to coordinate with the negative portion of the polar water molecule. Far smaller than a neutral H atom, the H^+ cation also can fit into more elaborate arrangements involving additional waters of hydration—such as $H_9O_4^+$, for example, in which four H_2O molecules solvate the ion. Some experimental findings do suggest that H_3O^+ is not the only solvated form of H^+, but our subsequent grasp of acid–base equilibria will be unimpaired by any revisions to the structure of hydronium. For simplicity, then, we shall pretend that one H_2O molecule attaches to one H^+ ion to produce one hydronium ion, H_3O^+:

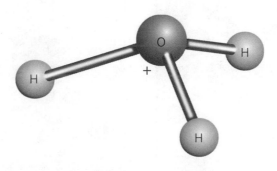

3-3. REDUCTION AND OXIDATION

Chemistry occurs at the margins. Already we see how the shift of a few valence electrons from one atom to another can make a molecule and, beyond that, how two molecules can coordinate their outer electrons still further in acid–base and other reactions. If the electrons abstain from interaction, there is no chemistry. No molecules, no reactions, no change.

Even the simplest models of chemical bonding and reactivity assign to the valence electrons this central role, whether by the attainment of octets or through the formation of molecular orbitals (Chapter 7). The Lewis conception of acids and bases, with all its power and generality, is nothing more than a qualitative picture of electrons on the move.

Electrons are always on the move. We acknowledge as much by saying, for instance, that an atom of sodium can lose its single valence electron to realize a noble-gas configuration:

$$Na \longrightarrow Na^+ + e^-$$

This ***ionization*** is a chemical reaction as well, and surely one of the most fundamental:

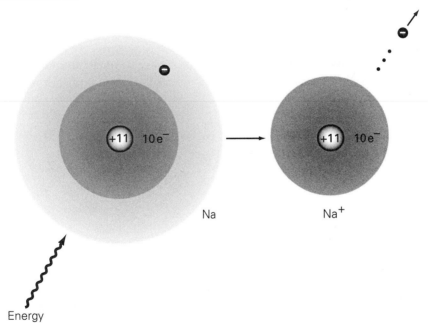

A neutral atom of sodium (Na) reacts to produce a monopositive cation (Na^+) and a free electron (e^-). Everything balances, as always it must: on the left, a sodium atom consisting of a nucleus and 11 electrons; on the right, a sodium ion consisting of a nucleus and 10 electrons, *plus* the liberated electron. Mass is conserved in a chemical reaction, and now charge is conserved. Total charge (here, zero) is neither created nor destroyed. It is a quantity conserved as strictly as momentum and energy.

Any loss of electrons is broadly called an ***oxidation***, a designation originally reserved for the burning of a substance in oxygen but now understood more generally. We say, according to this terminology, that the sodium atom is somehow "oxidized" to the sodium cation, yielding a species isoelectronic with neon and also an unattended electron.

Unattended initially . . . but not for long. The free electron, like the aforementioned H^+ ion, is poorly suited to exist on its own, and so the particle e^- will be either recaptured promptly by the Na^+ or else snared by some other electron-receptive species. Chlorine is a possibility, should a molecule of it happen to be present. Normally existing as a diatomic element, Cl_2 can acquire two electrons and subsequently split up into two chloride anions:

$$Cl_2 + 2e^- \longrightarrow 2Cl^-$$

Any such *gain* of electrons is termed a ***reduction***. What one party (Na) loses, the other party (Cl_2) gains.

Oxidation and reduction (Figure 3-6) are thus complementary processes; one feeds the other. Understand that the sodium atom does not simply give away its electron to all comers, despite our hunch that unconditional surrender might offer a quick way to make an octet. No. The "detachable" electron does not, by itself, fall off the atom like a dead

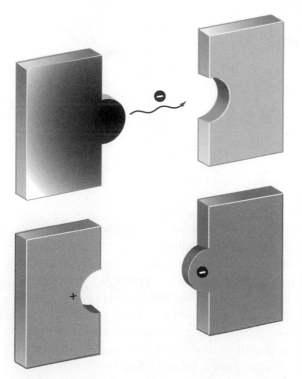

FIGURE 3-6. Haves and have-nots, yet again: oxidation and reduction. One species, shown with a heavily shaded tab, has electrons to give; the other has the capacity to take them. The exchange is similar in spirit to Lewis's acid–base chemistry, but restricted to processes involving a change in oxidation number.

leaf from a tree; it must be torn. Left alone, an atom of sodium will exist in the gas phase complete with valence electron and bereft of octet. Octet or no octet, sodium's outermost electron is a negative particle attracted to a positive nucleus, and there it will stay unless some stronger agent pulls it away. For ionization to occur, the metal exacts a price: energy.

The energy for ionization must come from somewhere. It might come from the outside in the form of heat or light, stripping off the electron as if by a hammer blow. Or it might come from the inside, supplied by another chemical entity and accompanied by a simultaneous oxidation and reduction—an *electron transfer*, in this case from sodium to chlorine. Indeed, the sodium atom suffers oxidation only because there is a nearby species able to accept the loose charge. More accurate would be to say that chlorine tears off sodium's valence electron in a strong collision, expending energy in the process but eventually gaining some of it back. For although it costs energy to ionize the sodium, this initial investment (known as the **ionization potential**) is repaid both by the energy gained when the chlorine atoms receive their electrons (measured as **electron affinity**) and when cation and anion come together as NaCl.* What transpires here is a reciprocal interaction—an oxidation–reduction, or reduction–oxidation, or just **redox reaction**—in which an electron lost is an electron gained. The Cl_2, by securing for itself a reduction, *causes* the oxidation of Na; it is said, accordingly, to be the **oxidizing agent**. So, too, is Na designated the **reducing agent**. By undergoing oxidation, it supplies the electron that initiates the reduction.

Putting together the oxidation and reduction reactions, properly balanced for atoms and electrons, we obtain an overall equation for the production of NaCl (the familiar "salt" again):

$$
\begin{array}{rcl}
2\,Na & \longrightarrow & 2\,Na^+ + 2\,e^- \\
Cl_2 + 2\,e^- & \longrightarrow & 2\,Cl^- \\
\hline
2\,Na + Cl_2 & \longrightarrow & 2\,NaCl
\end{array}
$$

It is close to an acid–base reaction of the Lewis type, since the key event remains the donation and acceptance of electrons. A reducing agent (Na) gives; an oxidizing agent (Cl_2) takes. The electrons, strictly speaking, do not travel as a *pair*, but such distinctions are minor in light of the essential similarity.

Bear in mind that "one electron lost and one electron gained" does not refer literally to a single, integral subatomic particle. Individual electrons are too elusive to be tracked so exactly, and the entire distribution of

---

*Energy must also be supplied to break the chlorine–chlorine bonds. All that matters in the end, though, is the net energy gain or loss associated with the entire sequence of events.

negative charge is smeared out over space. Oxidation means, more realistically, that some portion of the electron cloud shifts significantly from one atom to another. The charge transferred might correspond to one whole electron, or perhaps a half or a quarter or maybe 0.346 units. What counts is that the loss incurred on oxidation be compensated exactly by the gain enjoyed on reduction. The electron cannot simply disappear.

Understanding, then, the redox transfer of an electron as a relative loss and gain of charge (with the total strictly conserved), we may use a convenient accounting device known as the *oxidation state*. An atom's oxidation state is a positive or negative number, usually but not always an integer, designed to mirror local electronic changes relative to some normal condition. For an atom in elemental form, the oxidation state is 0. A singly charged atomic cation has the value +1; a singly charged anion, −1. For an ionized atom in general, the oxidation state is equal to the net charge (expressed as a multiple of the fundamental electronic value, -1.60×10^{-19} C). Thus in the formation of NaCl from Na and Cl_2, the oxidation state of chlorine is *reduced* in an algebraic sense (made smaller, more negative) from 0 in Cl_2 to −1 in the Cl^- anion. Sodium, by losing some negative charge, goes from 0 in the elemental form Na to +1 in the Na^+ cation. Its oxidation number increases; it becomes less negative.

Carried over from monatomic ions to covalently bonded species, oxidation numbers (Figure 3-7) give some idea of the flow and distribution of electrons between atoms of different electronegativities. Atoms in a molecule are assigned values that reflect both their usual bonding patterns and their relative abilities to attract electrons. Typically the oxidation state corresponds to the charge an atom would assume if the bonding were ionic rather than covalent—if, in other words, the atom had lost or gained the full electronic charge. The algebraic sum of all such oxidation numbers must equal the net charge of the species (zero for a neutral molecule), and the total for all species must remain constant during any reaction. That much we demand to guarantee conservation of charge.

Two electrons shy of an octet, oxygen is most often assigned an oxidation state of −2 (the exception arising in molecules such as hydrogen peroxide, H_2O_2, where it is −1). Hence in the neutral molecule CO, the carbon atom must be +2 to balance the oxygen's −2. In CO_2, by contrast, the two oxygens have a combined value of −4, and so the carbon is imagined to exist as C^{4+}. Carbon is a notable example of a species with multiple oxidation states, a feature that lends considerable chemical versatility to the atom.

Hydrogen's oxidation number is fixed at +1 in covalent compounds with nonmetals, such as HCl (H^+, Cl^-) and NH_3 (N^{3-}, H^+). When

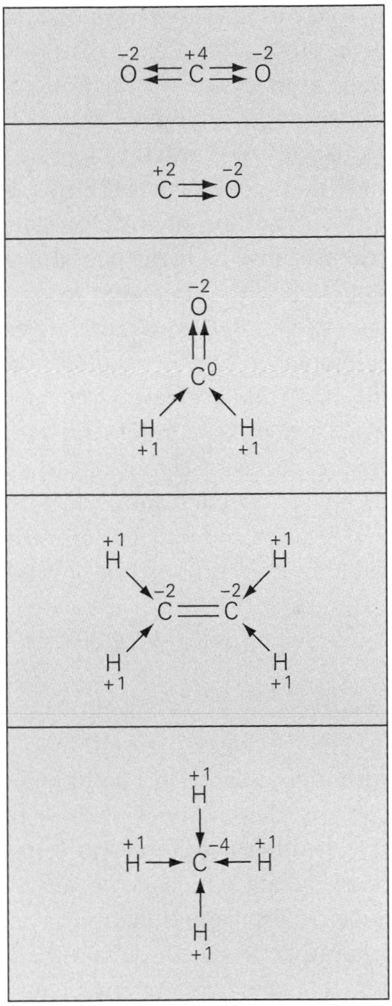

FIGURE 3-7. Oxidation states, a rough way to describe the drift of electrons in a molecule or polyatomic ion. The scheme: (1) Assume that the more electronegative atom takes the whole electron, as if a polar covalent bond were really an ionic bond. (2) Use an arrow to mark the flow of negative charge—one arrow for a single bond, two for a double, three for a triple. (3) For each polar bond, *increase* the oxidation state by 1 for the atom at the arrow's tail (where the electron departs) and *decrease* it by 1 at the tip (where the electron arrives). (4) Bonds between like atoms, shown without arrows, are unskewed and contribute nothing to the total.

bonding to highly electropositive metals, however, hydrogen attracts the electron more strongly than does the metal and consequently exists as the *hydride* form, H^-. Lithium hydride (Li^+H^-) is a good illustration.

In deciding between two covalently bonded atoms, we generally award the more electronegative species a negative oxidation state equal to its charge in a typical ionic compound. Halogen atoms such as chlorine and fluorine therefore are counted as -1.

These few rules bring together a wealth of chemical experience. Given the oxidation states, we have a useful shorthand to predict what happens when atoms combine—not a deep understanding of *why* they combine, but still a reliable guide to behavior. Oxidation states also allow us easily to identify redox reactions wherever they occur, sometimes with little outward evidence of the workings within. There need not be anything as macroscopically dramatic as a fire or even as microscopically conspicuous as a proton transfer to signal the change. Reduction and oxidation are marked rather by the ever-so-slight alteration in oxidation state that accompanies the transfer of an electron.

The rusting of iron, for one, proceeds as a seemingly quiet redox reaction in which the metal is oxidized from elemental Fe^0 to its Fe^{3+} state in Fe_2O_3. Molecular oxygen, acting as an oxidizing agent in the presence of water, brings about a very slow corrosion, albeit one where considerable energy is gradually liberated. Purely a kinetic consequence, this controlled release of energy over a long time makes the reaction appear deceptively quiet and cool. Viewed thermodynamically, though, rusting secures a significant lowering of the free energy. It is a chemically profitable transaction, even if it is a transformation slow to realize.

A faster and noisier demonstration of reduction and oxidation is provided by the combustion of methane,

$$CH_4 + 2O_2 \longrightarrow CO_2 + 2H_2O$$

where carbon is oxidized from C^{4-} in methane (CH_4) to C^{4+} in carbon dioxide. Oxygen is simultaneously reduced from 0 in O_2 to -2 in H_2O. Although usually unaccompanied by obvious electrical signs like ions or current, there is a net transfer of electrons nevertheless.

Another manifestation of redox appears in a *disproportionation* reaction, during which the same chemical entity is both oxidized and reduced. Disproportionation is an internal process, a form of self-redox, whereby one species acts as reducing agent while another (of like kind) serves as oxidizing agent. Thus two molecules of nitrogen dioxide can disproportionate in water to yield nitric and nitrous acids:

$$2NO_2 + 2H_2O \longrightarrow H_3O^+ + NO_3^- + HNO_2$$

One NO_2 gives up an electron to the other, with nitrogen going from N^{4+} in NO_2 to N^{5+} in NO_3^- (the nitrate anion) and N^{3+} in HNO_2 (nitrous acid).

Evidently all that distinguishes redox from Lewis acid–base reactions is the modification of oxidation state, a bookkeeper's distinction. Atoms undergoing redox must change their oxidation numbers; Lewis acids and bases need not. Interpreted broadly, the transfer of electrons in a redox reaction is conceptually the same as the donation of an electron pair from Lewis base to acid. There is a fundamental unity in all these processes, never to be overlooked.

3-4. DISSOLUTION AND PRECIPITATION

One substance *dissolves* in another when molecules or ions of some "guest" are dispersed individually throughout some "host," as in the acid, base, and salt solutions mentioned above. The ionic bonds in NaCl, for example, can be broken by the solvating action of water molecules, which surround and stabilize the ions to form an aqueous solution. Clusters of H_2O molecules use their electron-rich oxygens to attack the sodium cations, thereby enveloping each Na^+ within a protective shell of negative charge. By presenting, instead, their electron-poor (partially positive) hydrogen poles, the same waters are able to carry away the oppositely charged chloride anions. The story, in brief, is told in Figure 3-8: positive seeks out negative; negative seeks out positive.

Such a mixture of ions, electrically neutral overall, will conduct electricity and is called an **electrolytic solution**. Floating within the liquid is a population of hydrated anions and cations—charged particles—able to undergo directed motion in an electric field. Solutes that produce ions are classified as strong or weak **electrolytes** according to their degree of dissociation and additionally by their ability to carry a current. Aqueous solutions of acids and bases are themselves electrolytic, formed by mobile H_3O^+ and OH^- ions.

How shall we interpret dissolution? Again it is a variant form of Lewis acid–base interaction, liberally construed, through which a **solvent** species (the host) coordinates a portion of its electron cloud with some region on another species, a **solute** particle (the guest). The solute may be ionic or covalent, polar or nonpolar, solid, liquid, or gas. As for the solvent, it too may be solid, liquid, or gas; but of all the possibilities, certainly ordinary water stands out as the most common and generally useful host. For biochemical reactions in particular, water is the natural medium.

Coordination between solvent and solute produces a solvated particle, a so-called **solute–solvent complex** analogous to the salt in a neutralization reaction. The plot remains the same: Electron-giver finds electron-taker. They meet. They match. Each fills the need of the other. Solute–solvent interactions are weaker and more transient than covalent bonds, but the fundamental idea of electron donation and acceptance is preserved—even if somewhat tenuously and even if without any intimate intermingling of charge. Atoms, molecules, and ions can influence each other only through their electrons (directly or indirectly), and so dissolution takes its place along with acid–base and redox reactions as a specific realization of a thoroughly general phenomenon: the giving and taking of electrons.

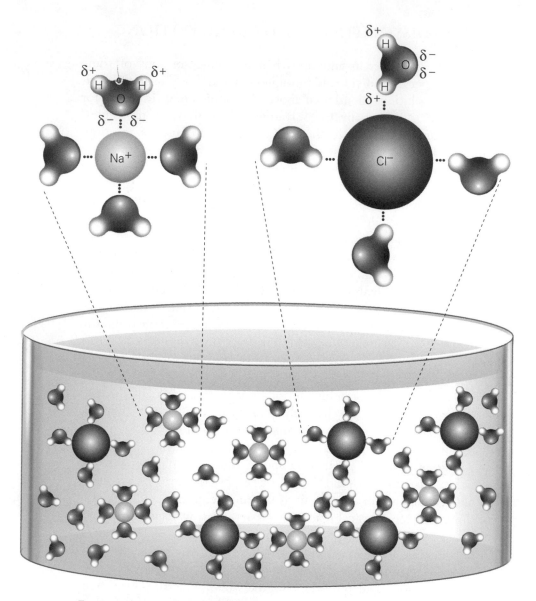

FIGURE 3-8. Solute and solvent in aqueous solution: positive attracts nega-tive. Molecules of the host, water, swarm around their ionic guests and isolate them inside solvated clusters (rendered schematically in two dimensions only). Regions rich in electrons are shaded heaviest.

Opposite to dissolution is **precipitation**, a process by which solids spontaneously "fall out" of solution. Consider first the aqueous solutions of two strong electrolytes such as sodium chloride and silver nitrate ($AgNO_3$). Introduced into separate vessels, each solid material is ionized and hydrated by water molecules to give the corresponding electrolytic solutions:

$$NaCl \longrightarrow Na^+(aq) + Cl^-(aq)$$

$$AgNO_3 \longrightarrow Ag^+(aq) + NO_3^-(aq)$$

Nothing further happens unless the two solutions are mixed together, whereupon a white solid instantly appears and separates from the solution. The *insoluble* material AgCl is produced as the Ag^+ cations form ionic bonds with the Cl^- anions, a process (Figure 3-9) described in full by the balanced equation

$$Ag^+(aq) + NO_3^-(aq) + Na^+(aq) + Cl^-(aq) \longrightarrow$$
$$AgCl(s) + Na^+(aq) + NO_3^-(aq)$$

What remains is a still-dissociated solution of sodium nitrate ($NaNO_3$) and a solid pile of silver chloride (labeled "s" in the equation).

On the one hand, we might regard this transaction as an **exchange**, or **metathesis**, reaction between NaCl and $AgNO_3$ in which the two compounds switch partners:

$$AgNO_3 + NaCl \longrightarrow AgCl + NaNO_3$$

On the other hand, recognizing that the sodium and nitrate species never leave the solution, we may disregard these so-called **spectator ions** and write a **net ionic equation** for the precipitation as

$$Ag^+(aq) + Cl^-(aq) \longrightarrow AgCl(s)$$

The spectator ions (Na^+ and NO_3^-) float by passively while the solid (AgCl) drops to the bottom.

Why some ions remain aloof while others unite is a matter for thermodynamics to decide. Ask if, in the end, the solvated ions are thermodynamically more stable than the precipitated solid form. Among the factors to consider (in Chapter 15) are the size of the ion, the distribution of charge, and the geometry of the solute–solvent complex. But always the means of attraction is the same: the affinity of positive for negative. Positive charge seeks negative charge, whether offered by the electrons on H_2O or by the electrons on the ion.

For silver and chlorine, we discover by experiment that the electrostatic attractions of an ionic crystal are evidently stronger than the forces of hydration. Prevailing over H_2O in this electrical competition, the Cl^-

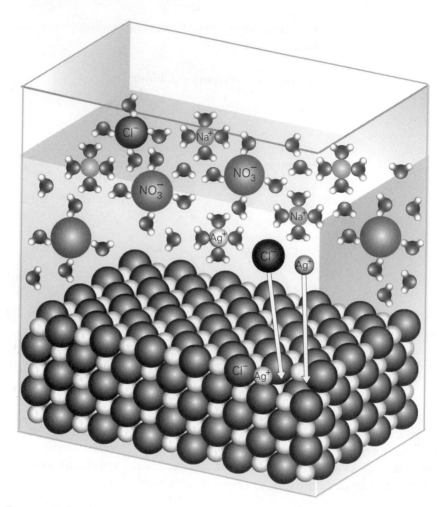

FIGURE 3-9. Precipitation, the reverse of dissociation. Solvated particles slip away from the water molecules and reassemble into a solid. The solid literally falls out of solution (hence the term "precipitation" from the Latin *praecipitare*, "to cast down"). Note that here, as in Figure 3-8, solvated clusters are pictured in a simplified two-dimensional arrangement not necessarily indicative of their true structure.

anions (Lewis bases) present their electrons to the Ag^+ cations (Lewis acids) to force the ionic solid AgCl out of solution. Left behind in solution are the hydrated Na^+ and NO_3^- ions, which remain better off energetically in the embrace of the water molecules.

3-5. RADICAL REACTIONS

Another important avenue of reaction involves the recombination of radicals. Containing an *unpaired* electron, a **radical** species offers half a covalent bond (one electron) to any potential partner. Typical of many radicals is a free atom of chlorine, $:\overset{..}{\underset{..}{Cl}}\cdot$, which we abbreviate as Cl· to emphasize the unpaired electron. From that single dot, moreover, we see that radical Cl· lacks the stabilizing octets of both the chloride ion and diatomic Cl_2 molecule. Cl· is an atom, of course—chemically legitimate and electrically neutral—yet it is poorly adapted for solitary existence.

With an electron configuration so obviously incomplete, the free radical Cl· cannot remain free for long. Expect it to form a covalent pair bond at the first opportunity, either by removing an atom from a nearby molecule or else by pairing with another radical. In a novel variation on the theme of Lewis acids and bases, each radical both donates and accepts an electron, this time to make a pair and form a covalent bond:

$$Cl\cdot + \cdot Cl \longrightarrow Cl_2$$

One radical begets another when it strips an atom from a paired-up molecule. Consider the hydroxyl radical, ·OH, which unlike the hydroxide ion :OH⁻ is a neutral species. Whereas hydroxide will attach a hydrogen *ion* (H^+) to produce water, the hydroxyl radical takes a whole hydrogen *atom* instead—electron as well as nucleus, H·. The ·OH species finds an appropriate molecule, rips out a hydrogen, and leaves behind a damaged structure with one unpaired electron. If we denote the hydrogen-containing victim by the general formula R—H (where R labels the rest of the molecule), then the reaction goes as

$$R\!-\!H + \cdot OH \longrightarrow H_2O + R\cdot$$

and leaves behind the radical fragment R·. Should the broken molecule be important to a living organism, such a reaction might damage cells and tissues.

Radicals are generated in diverse ways, with irradiation by light (**photolysis**: *photo*, "light"; *lysis*, "breaking apart") proving especially important in atmospheric reactions. Smog formation is thought to involve the production of atomic oxygen following the absorption of ultraviolet sunlight by nitrogen dioxide. The oxygen radical, highly reactive, then can attack a number of other airborne molecules to create still more chemical radicals. A variety of additional pollutants form in the ensuing recombinations.

The ease with which radicals propagate frequently leads to all manner of **chain reactions**, sometimes useful and sometimes destructive. Radical chain reactions (Figure 3-10) are used by chemists to link small molecules into extended structures called *polymers*. Such materials, ubiquitous as various plastics and fibers, are of obvious commercial importance. More malign, however, is the role played by radical chain

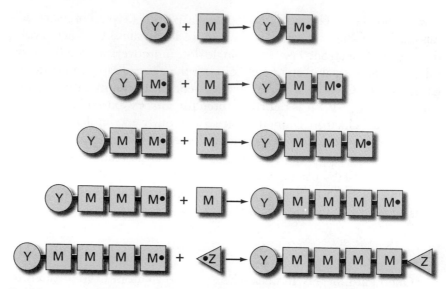

FIGURE 3-10. Free-radical polymerization. From a single small molecule M grows a long-chained molecule $Y-M_n-Z$, a *polymer* often containing thousands of repeating units. Attacked by a free radical $Y\cdot$, the first M molecule becomes the radical $Y-M\cdot$, which then attacks another M molecule to make a larger radical $Y-M-M\cdot$, followed by another, and another, and another, n times in all; and thus the chain grows until finally the free radical $Z\cdot$ terminates it. Example: When M is $CH_2=CHCl$ (vinyl chloride), the polymer produced is $Y(CH_2CHCl)_nZ$ (polyvinyl chloride). Y and Z take various forms.

reactions in the depletion of the ozone layer. Generated by the photolysis of chlorofluorocarbons (CFCs) in the upper atmosphere, a chlorine radical first attacks and destroys an ozone molecule (O_3):

$$Cl\cdot + O_3 \longrightarrow ClO\cdot + O_2$$

Recombination of two $ClO\cdot$ radicals,

$$2ClO\cdot \longrightarrow 2Cl\cdot + O_2$$

then regenerates the original $Cl\cdot$ and allows the process to begin again.

3-6. CATALYSIS

Bonds break and bonds form; reactants become products. Such are the ins and outs of a chemical reaction, present in all the examples cited above. Energy goes in, at first, when the reactants break apart, and energy comes out when the products finally pull together. There is, as suggested earlier, a refundable cost payable at the start: The reactants face an ***activation barrier***, a hill of energy to be climbed while bonds are first stretched or broken.

Climbing uphill, *spending* energy, the molecules arrange themselves at the summit into an unstable intermediate structure—a structure in crisis, a hybrid creation no longer reactants but not yet products. See it symbolized in Figure 3-11 as a ***transition state*** perched atop the mountain, teetering on the brink, convertible into either the old reactants or the new products. Once over the top, the reacting molecules then can roll down and recover the activation energy on their way to products. Reaction accomplished; thermodynamic profit earned. But since energy is needed first to build the critical transition state, the reactants must have a push up the hill to get going.

Sometimes the push comes from heat alone. Molecules in motion have an intrinsic kinetic energy (the higher the temperature, the faster they move), and so a reaction might be initiated just by brute force. Simply raise the temperature and let the reactants slam into each other; let them collide with enough force to break whatever bonds need to be

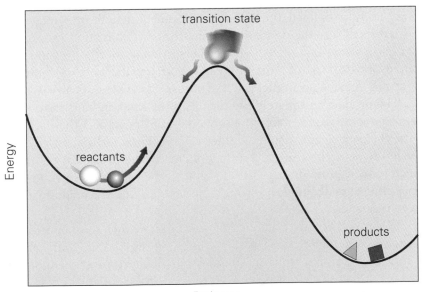

FIGURE 3-11. Separating reactants and products is an *activation barrier*, a mountain of energy—and poised at the summit is the *transition state*, ready to go either way. It does not endure for long. The transition state either falls forward into the valley of products or backward into the valley of reactants. A catalyst, by lowering the activation barrier, makes a transition state easier to form.

FIGURE 3-12. The production of ethane (CH_3—CH_3) from hydrogen (H_2) and ethylene (CH_2=CH_2), usually very slow, speeds dramatically when carried out over a metallic surface. Opening up a new pathway, the platinum catalyst allows hydrogen to combine more easily with ethylene. The reactants have a smaller activation barrier to overcome.

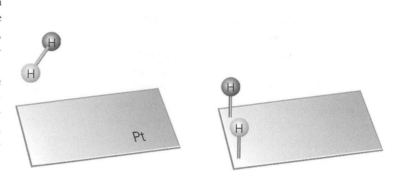

broken and to form whatever transition states need to be formed. For many reactions this thermal option proves sufficient; for others, including those in living cells, the requisite temperature might be impractically high or destructive to the system itself.

Where brute force fails, a catalyst can provide a subtle and more effective alternative. Rather than force the reactants over an impossibly high activation barrier, a catalyst instead opens up a new pathway with a smaller hill to climb. The assistance rendered is usually in the form of a more easily assembled transition state. Some catalysts integrate themselves directly into the structure of an intermediate (creating a short-lived catalyst–reactant complex), while others provide more of a supporting framework—a kind of atomic or molecular workbench available to the reactants. A single example, just one of many, is given in Figure 3-12.

The products ultimately produced are the same in kind and amount as in an uncatalyzed reaction, but now the transformation goes faster because the participants come together more economically. Less energy is needed to start the reaction and consequently the products appear that much sooner, often dramatically so. Unless the catalyst is somehow *poisoned*, it eventually emerges from the reaction unchanged and ready for more reactants.

When both catalyst and reactants coexist within the same state of matter (solid, liquid, or gas), the effect is described as **homogeneous catalysis**. Many reactions in solution, particularly those in biological systems, are catalyzed by acids and bases. Acid-catalyzed processes also are frequently used to produce various polymers, and free radicals play significant roles in homogeneous catalysis as well.

Heterogeneous catalysis, by contrast, involves materials in different phases, often with liquid or gaseous reactants interacting on the surface of a solid catalyst (see again, Figure 3-12). The reactants adhere to the surface, where they become better positioned to undergo a final

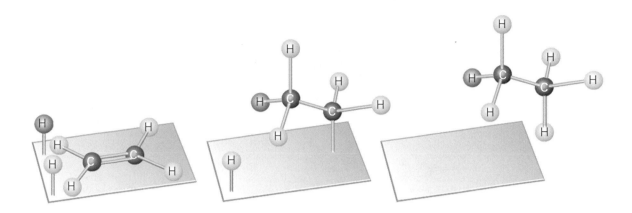

transformation. Once formed, the products must be released to permit further catalytic action.

Metals, especially, are used as heterogeneous catalysts to accelerate a broad range of reactions, and perhaps the most familiar example is the platinum–rhodium catalytic converter used to treat automotive emissions. The metallic platinum facilitates a rapid oxidation of carbon monoxide and assorted hydrocarbons to carbon dioxide and water, whereas the rhodium helps reduce gases such as NO and NO_2 to molecular nitrogen.

Of special catalytic interest are the *zeolites* (Figure 3-13), an extensive class of minerals manufactured both by nature and by chemists in the laboratory. These crystalline materials are constructed from aluminum, silicon, and oxygen atoms arranged into three-dimensional structures containing networks of channels and cavities. A charge imbalance develops whenever aluminum in the +3 oxidation state replaces Si^{4+}, and this difference of −1 must be compensated by a +1 somewhere else. Positively charged ions, usually cations of alkali or alkaline earth metals, enter the structure to make the zeolite electrically neutral.

Zeolites are porous materials with diverse properties, catalytic and otherwise. Ions can work their way through the channels, for instance, and this access permits the original cations to be exchanged with species introduced from the outside. Zeolites thus might act as *molecular sieves*, in which one ion is substituted for another as, say, when the Ca^{2+} cations in hard water are replaced by Na^+. Zeolites serve as drying agents as well, able to sequester water molecules in their cavities and thereby dehydrate a material passing through its pores. More generally, the ability of zeolites to accommodate various other molecules endows them with catalytic properties. Highly specific reactions, sometimes determined by molecular geometry and shape, are often favored within the close quarters of an internal cavity. Noteworthy here is the zeolitic catalysis of the conversion of methyl alcohol (CH_3OH) into gasoline.

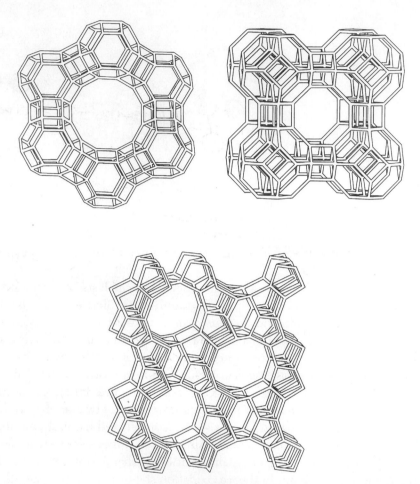

FIGURE 3-13.   Assorted zeolites, showing some of the pores and channels in the aluminosilicate structures.

Catalysis is an enormous field, critical both for understanding the details of chemical reactions and also for making them happen. Without the invention of suitable catalysts, many industrial processes would be either impossible or uneconomical. The refinement of petroleum and the production of petrochemicals, in particular, depend on catalysis. Nor could biochemical reactions take place without the aid of naturally occurring catalysts called *enzymes*. We shall see examples of enzymes in later chapters, and we shall better understand the mechanics of catalysis and transition states in the more thorough treatment of kinetics to come.

In the meantime, having sketched this qualitative picture of chemical structure and reactivity, let us return for a more detailed look into the microscopic world. Our attention shifts to light and matter, to quantum mechanics, and to how atoms and molecules come to exist.

REVIEW AND GUIDE TO PROBLEMS

Means. Motive. Opportunity. Without them, there would be no chemical reactions; there would be no change.

But change there is, everywhere—chemical reactions in every living cell, thousands of reactions, unrelenting; chemical reactions in every dead cell too, in every rock, in every molecule, in every bit of matter.

Why? Because, not least, matter *can* change. Matter has the simple *means* to interact, to push and to pull. Electrons are attracted to nuclei, and nuclei are attracted to electrons. Electrons are repelled by electrons; nuclei, by nuclei. No piece of matter stands alone. Every particle is a part of something else, subject to forces, alert to the presence of other particles. If not, we would walk through walls oblivious to the matter around us, for there would be no wall-to-body communication. Even more: There would be no bodies and no walls to begin with, for there would be no glue to hold together the atoms and molecules. There would be no atoms. There would be no molecules.

That atoms actually do exist, that molecules exist, that walls and bodies exist, that atoms and molecules and walls and bodies undergo change—for that, we have electromagnetic interactions to thank. Nuclei, electrons, atoms, molecules, and ions all communicate among themselves. They exert forces and respond to forces. They come together; they move apart. They make bonds; they break bonds.

But if matter's electrical structure gives it the means to change, then a drive toward self-improvement supplies the *motive* to change. Not a conscious desire, of course; not a motive in any volitional sense; rather a motive in the original sense of a drive, a tendency for something to happen.

There are, remember, lots of particles out there; and they have ample *opportunity* to smash together and thus expose themselves to the means of change, already acknowledged. Molecules meet. Molecules break. Maybe something new comes out of it; maybe not. Maybe a rearranged structure will show itself to be stronger and more stable than the original; if so, expect the new and improved molecule to persist.

And, with trillions upon trillions of such encounters, expect a kind of mob action as well. Expect the strong to drive out the weak, sooner or later, but inevitably. Nature relies on little more than dumb luck, blind chance, undirected stumbling; yet the results are so predictable as to suggest a driving force, a motive.

Means, motive, and opportunity. The means of change: the fundamental electromagnetic force, controlled by the laws of quantum mechanics. The motive for change: thermodynamic strength and stability, driven by the random but predictable motion of the mob. The opportunity for change: when molecules meet, when molecules break, when molecules rebuild; when molecules find a mechanism, a kinetic pathway.

Quantum mechanics, thermodynamics, and kinetics—the blueprint for change, the three fates ruling a chemical reaction. We shall deal with each in turn, systematically, step by step over the chapters following. First, however, let us review the broad classification of reactions sketched in the pages above. It will be the end of the beginning, the end of our introductory look into matter, energy, and change.

IN BRIEF: REACTIONS AND REACTIVITY

Chemical reactions are negotiated at the fringes of a molecule, in the valence region, where electrons dwell farthest from the nuclei. Away from the positive charge, pushed out by the electrons in the core, the valence electrons are most susceptible to outside influences. For metals, the exposed electrons are easiest to detach; for nonmetals, the vacancies are easiest to fill. Reactions occur at the margin. Electrons are given, taken, borrowed, shared. To follow the reaction, follow the electrons.

1. OXIDATION NUMBER. Where do they go? Where do the electrons go when atoms and molecules are rearranged? They go from the less electronegative to the more electronegative species, taking negative charge *away* from one region and bringing it to another.

Thus flow the electrons when the nonpolar molecules H_2 and Cl_2 combine as

$$H_2 + Cl_2 \longrightarrow 2HCl$$

to become the polar molecule $H^{\delta+}-Cl^{\delta-}$. Before the reaction, the electrons in both H_2 and Cl_2 are distributed symmetrically about the two nuclei. It is the normal state for these two elements, which exist by themselves as diatomic molecules. Neither end is more negative than the other. After the reaction, however, the distribution is different: The hydrogen end of HCl is slightly positive, and the chlorine end is slightly negative. The electrons spend more time near the chlorine, less near the hydrogen.

Now whenever a hydrogen nucleus finds itself surrounded by *fewer* electrons than in the elemental form H_2, we assert that H has become less negative (more positive) by comparison. How much more positive, we do not venture to say—but, be assured, H in HCl is positive compared with H in H_2. And if, in some other environment, a hydrogen nucleus were to gather proportionally *more* electrons, we would say that H has become more negative.

The same goes for chlorine; the same goes for any other element. An atom assumes a *negative oxidation state* if (compared with its elemental form) the nucleus attracts a greater following of electrons. When fewer electrons are present, the atom is in a correspondingly *positive oxidation state*. The H in HCl has an oxidation state of +1; the Cl in HCl has an oxidation state of −1.

+1 what? −1 what? Not the whole electron is transferred, not a full −1.6 × 10^{-19} coulombs, but rather some portion of the whole; the electron, drawn more to the chlorine, spends more time near that end. We say +1 and −1 *as if* the electron went all the way from one side to the other, *as if* the two species existed as the ions H$^+$ and Cl$^-$, but we don't mean it literally. We say +1 and −1 only to have a standard for comparison, a way to keep track of the flow. That was the thinking behind oxidation numbers, and here again are the rules for their assignment:

1. Establish a baseline for comparison. Any element in its natural, elemental state takes an oxidation number of 0. Whether that form be an atom (Ne), molecule (N$_2$), metallic solid (Na), or anything else, the electrons in an element display no bias toward one nucleus or the other.

2. Winner take all. In compounds containing two different elements, we arbitrarily grant the more electronegative species the maximum number of electrons available. The oxidation number thus becomes equal to the element's charge in simple ionic compounds.

3. Certain species exhibit only one oxidation state. Group I atoms take the value +1 in combination; Group II atoms take the value +2; Group III atoms, +3 (usually). Hydrogen gets +1, except when bonded to a less electronegative atom such as Li or Na (LiH, NaH) where it becomes −1. Oxygen usually bonds as −2. Fluorine is always −1.

4. There are exceptions. Fluorine, the most electronegative atom of all, prevails over runner-up oxygen in the O—F bond. O assumes a positive oxidation state when attached to F. Oxygen also loses its −2 state in compounds containing —O—O— linkages (peroxides).

5. The accounts must balance. All the oxidation numbers add to zero in a neutral molecule. For an ion, positive or negative, the sum is equal to the signed charge of the entire species.

2. SIMPLIFY. SIMPLIFY. For all the ways that chemicals can react (and there are many, many possibilities), one fundamental transaction stands out: the giving and taking of electrons. One party to the reaction offers a portion of its negative charge to another.

Originally it was Lewis who, with his idea of electron donors and electron acceptors, saw chemical reactivity as this union of opposites, this pairing of haves and have-nots. From Lewis we have since learned to look for the acids and bases, the reducers and oxidizers, the cations and anions, the givers and takers.

Where needed, we might distinguish further among combustion reactions, hydrolysis reactions, electrolysis reactions, metathesis reactions, disproportionation reactions, decomposition reactions, and all the rest, but first we ought to simplify. Better, at the start, to bring out the big similarities than to become entangled in the small differences.

Recall, then, our minimalist view of chemical give-and-take: acid–base, reduction–oxidation, dissolution and precipitation, radical reactions. See how the differences blur even among the members of this small set.

3. REDUCTION AND OXIDATION (REDOX). Electrons are transferred in such a way as to change the oxidation states of both giver and receiver.

The giver undergoes *oxidation*. It loses electrons. It gives up negative charge. Its oxidation number becomes more positive.

The taker undergoes *reduction*. It gains electrons. It takes up negative charge. Its oxidation number becomes more negative (algebraically smaller, reduced).

The taker, by taking, *causes* the giver to give. It (the taker) is the *oxidizing agent*, the party that brings about the oxidation.

The giver, by giving, *causes* the taker to take. It (the giver) is the *reducing agent*, the party that brings about the reduction.

Photosynthesis, metabolism, combustion, corrosion—all involve redox reactions, and there are still many more. Much of chemistry falls squarely in the category of oxidation and reduction.

4. ACIDS AND BASES. Much of chemistry also falls under the heading of acid–base reactions, variously defined. One structure (the acid) meets some other structure (the base), and the two unite in a *neutralization* reaction. They produce something new, a *salt*, a substance with the properties of neither the original acid nor base. Biochemical pathways, catalysis, the formation of polymers—again, just a small sample—all bring acids and bases together in some way.

Regard as a *Lewis acid* anything that can accept a pair of electrons: a *pair*, two electrons, precisely the makings of a new covalent bond. The bare, electron-deprived hydrogen ion, H^+ (a free proton), serves as a ready example, but the general conception is far broader. A *Lewis base*, complementary to the acid, then emerges as anything that can donate a pair to a Lewis acid. The base :B shares its two electrons with the acid A, and from this sharing arises a covalent bond:

$$A + :B \longrightarrow A:B$$

The bonded complex, A—B, is the salt of reaction.

Similar to redox, in a way? Yes, there is a shifting of electrons; there is a giver and receiver. Any differences? At least three:

1. The Lewis acid–base reaction, in its original sense, involves a *pair* of electrons, whereas the giver in a redox reaction may transfer any number of electrons to the receiver.

2. The acid–base reaction produces a united complex A—B, cemented by a new covalent bond. The partners in a redox process may, by contrast, be separated in space, as they are at the positive and negative electrodes of a battery. They send the electrons over a wire, never meeting each other valence to valence, never bonding directly and locally as A—B.

3. Although a redox reaction always causes a change in oxidation number, there is no similar requirement for a neutralization. Even so, the electrons still move when acid meets base; there is still a give-and-take.

Other forms of acidic give-and-take arise as well, each related to the Lewis picture. We recognize a **Brønsted-Lowry acid**, for instance, as any species that can donate H^+, a positive hydrogen ion. A **Brønsted-Lowry base** is any species that can accept H^+.

Dissolved in water, a hydrogen ion surrounded by H_2O molecules exists as a *hydronium ion* (represented, incompletely, by the formula H_3O^+). An **Arrhenius acid** produces hydronium ions when placed in water, and an **Arrhenius base** produces **hydroxide ions** (OH^-). Neutralization of an Arrhenius acid and base brings together H_3O^+ and OH^- to produce two molecules of H_2O.

5. DISSOLUTION. Guest meets host. Something (a *solute*) is dispersed into something else (a *solvent*) to create a *solution*: sugar in water, salt in water, toluene in benzene. Particles of solvent surround the solute; electron-rich species seek electron-poor species; positive meets negative—all told, a picture consistent with our scheme of give-and-take.

PRECIPITATION: Dissolution in reverse. Guest leaves host. Dissolved particles fall out of solution and reorganize themselves into a solid. How so? Electron-rich species seek electron-poor species. Positive meets negative. Haves give to have-nots.

6. RADICAL RECOMBINATION. Another giving and taking: this time, a union of two species with unpaired electrons, the *radicals* A· and ·B (also called *free radicals*). Half Lewis acid and half Lewis base at the same time, radical A· has just the material for half a covalent bond: one electron. Radical ·B has the other half. Apart, they exist with capacity unfulfilled. Together, as A:B, they enjoy the stability of a pair bond.

SAMPLE PROBLEMS

We begin with some practice in electronic bookkeeping, first using oxidation numbers as described above. Then, in Example 3-3, we introduce an alternative method: formal charge.

EXAMPLE 3-1. Oxidation Numbers: Winner Take All

PROBLEM: What is carbon's oxidation state in each of the following molecules: (a) Methane, CH_4. (b) Ethane, C_2H_6. (c) Ethylene, C_2H_4. (d) Acetylene, C_2H_2. (e) Dichloromethane, CH_2Cl_2. (f) Carbon monoxide, CO. (g) Carbon dioxide, CO_2.

SOLUTION: The oxidation numbers sum to zero in these neutral molecules. Assigning $+1$ to H, -1 to Cl, and -2 to O, we can deduce the value for C.

(a) *Methane, CH_4.* Since each hydrogen in CH_4 contributes $+1$ for a total of $+4$, carbon must be -4.

(b) *Ethane, C_2H_6.* The two carbons must balance a combined oxidation level of $+6$ from the six hydrogens. With the molecule arranged as H_3C-CH_3, the C atoms sit in identical environments and thus both carry the same oxidation number: -3.

(c) *Ethylene, C_2H_4.* Again the two carbons are indistinguishable, this time doubly bonded as $H_2C=CH_2$. Each takes an oxidation number of -2 to counteract its two attached hydrogens.

(d) *Acetylene, C_2H_2.* Now there are two fewer hydrogens, and the carbons are connected by a triple bond in the molecule $HC\equiv CH$. Carbon's oxidation number is -1.

(e) *Dichloromethane, CH_2Cl_2.* Here is methane with two hydrogens replaced by two chlorines. We count -1 for each Cl atom (a Group VII species, a halogen), $+1$ for each H atom, and therefore 0 for C. Carbon takes electrons from hydrogen and gives them to chlorine.

(f) *Carbon monoxide, CO.* O is -2. C is $+2$. Note the positive sign: carbon releases electrons to the more electronegative oxygen.

(g) *Carbon dioxide, CO_2.* Connected to one more oxygen than in CO, the carbon has been oxidized from $+2$ up to $+4$. It gives up more electrons.

What do we learn? We learn that carbon is a versatile element, able to bond to many different atoms (including itself). We learn that the various oxidation states reflect both the number of bonds (single, double, triple) and the hunger of the neighboring atom for electrons. Carbon is hungrier than hydrogen, and oxygen is hungrier than carbon.

And we learn, most important, that carbon-containing molecules have considerable potential for change. They can give and take electrons in different ways; they can be oxidized and reduced to various levels. They can go from one structure to another, up and down the ladder of oxidation states.

So versatile is carbon that it plays a dominant role in the chemistry of living systems. For more, see Chapter 8.

EXAMPLE 3-2. **Unusual Oxidation Numbers**

PROBLEM: Determine the oxidation number for each of the following atoms: (a) Oxygen in KO_2. (b) Oxygen in H_2O_2. (c) Oxygen in OF_2. (d) Fluorine in F_2. (e) Manganese in MnO_4^-.

SOLUTION: Some of the values are high; some are low; some are fractional; some are exceptional. Observe, in particular, how oxygen can depart from its usual -2 state when a structure contains an $-O-O-$ bond.

(a) *KO_2*, an ionic compound formed by the reaction of potassium metal with excess oxygen gas, contains the *superoxide* ion, O_2^-. Each oxygen in O_2^- takes an oxidation number of $-\frac{1}{2}$, with the K^+ ion retaining its usual value of $+1$.

(b) *HOOH*. Consider the environment around each oxygen. On one side is an H atom, which tends to give electrons; on the other side is an O atom, an equal partner which neither gives nor takes. Each hydrogen therefore remains at $+1$ while the bonded oxygens exist as -1 in the *peroxide* $-O-O-$ bond.

(c) *OF_2*. Fluorine, the most electronegative of the elements, always gets the electron; and so F always displays an oxidation number of -1. Oxygen, with an oxidation number of $+2$ in OF_2, gives electrons to each of the two attached fluorines. Overmatched, oxygen turns positive in the company of fluorine.

(d) *F_2*. The only atom that can stand up to fluorine is another fluorine. F and F, covalently bonded as F_2, fight their battle to a draw, with each atom receiving an equal share of the valence electrons. The oxidation state is 0, consistent with our prior definition: All elements in their natural state are at zero oxidation level (and the diatomic molecule, recall, is the elemental form of fluorine).

(e) *The permanganate ion, MnO_4^-*. Each of the four oxygens contributes -2, making a total of -8. Manganese must be assigned an oxidation number of $+7$ to conform with the overall ionic charge of -1.

The high oxidation state signals that Mn is ripe for reduction in MnO_4^-. Offered electrons, the permanganate ion usually takes them. It reacts. It gets reduced. It becomes something else. The oxidation number of manganese goes down—moves nearer to zero—as the atom adapts to an environment better approaching electrical neutrality. Nature, favoring neutrality, tends to smooth out the differences.

Reaction, indeed, is nature's way of smoothing all kinds of differences: differences in charge, differences in density, differences in pressure, differences in temperature, differences in energy. High values go down and low values come up; and there, in brief, is fulfilled the thermodynamic motive for change, as we shall discover soon enough (starting in Chapter 12).

For now, though, let us simply use the oxidation number in its stated role: as a benchmark, a way to assess a change, a way to recognize that a shift of electrons has occurred. But it is only a gimmick, a gimmick based on the arbitrary notion of winner take all. And, except in a monatomic ion, the oxidation number does not measure a real charge. True, the Mn in MnO_4^- is partially positive because electrons do go to the oxygens; but not seven full electrons, certainly not that many.

Clearly, we sometimes overstate the charge separation by insisting upon winner take all. So why, upon reflection, should we always credit the more electronegative atom with a whole electron? Suppose the two atoms differ only slightly in electronegativity; suppose the bond is nearly nonpolar. What then?

Example 3-3. Formal Charge: Share and Share Alike

What then? Try a new bookkeeping scheme, something we have not considered until now: *formal charge*. Assume, regardless of electronegativity, that the electrons in a bond A—B are shared equally. Assign half the bonding electrons to atom A and half to atom B. After that, credit all lone pairs fully to their respective atoms.

Doing so, we pretend that (1) A and B have equal rights to any electrons in the bond, but (2) they keep all other electrons exclusively to themselves. It is like a division of community property after marriage— a 50-50 split even if one partner may have contributed the greater share.

Whereas with oxidation numbers we make every bond an ionic bond, with formal charge we make every bond a *covalent* bond and a *nonpolar* one at that. We knowingly ignore all differences in electronegativity, seeking to devise a model complementary to oxidation numbers.

It will be good to have both. Although neither formal charge nor oxidation number corresponds to a real charge, together they help to bracket the range. Just as oxidation numbers overstate the polarity, formal charges understate it. Each prescription is a device constructed for our convenience; each reflects a mixture of chemical knowledge, quantum mechanical understanding, and an intuition born of long experience.

PROBLEM: Compute oxidation numbers and formal charges for the atoms in: (a) Nitrogen oxide, NO. (b) Sodium chloride, NaCl.

SOLUTION: (a) *NO*. We know, from Example 2-5(c), that NO cannot be represented by a single Lewis structure. With 11 electrons in the valence, an odd number, the molecule is unable to confer octets upon N and O simultaneously. We have both

Structure I: :Ṅ=Ö

and

Structure II: N̈=Ȯ:

to consider. Is one choice better than the other?

Yes. Start with structure I, and count electrons. For purposes of formal charge, N claims two electrons from the double bond (half of four) plus the three nonbonding electrons: five in all. But five, note, is exactly the number of valence electrons in neutral, *uncombined* nitrogen; and, if so, then N in structure I has neither more nor less negative charge than it would have as a free atom belonging to Group V. The formal charge is 0.

As is the formal charge for oxygen in structure I: also zero. Here the O atom has six valence electrons in this representation (two from the double bond, four from the two lone pairs), compared with six in the free atom. To obtain the formal charge, we subtract the number of valence electrons imputed to the bound atom (in the Lewis structure) from the number in the isolated atom:

Formal charge (O in NO, structure I) = 6 (e^- in free O) − 6 (e^- in NO) = 0

In structure II, the nitrogen has six valence electrons: two from the double bond, four from the two lone pairs. The formal charge of −1 (5 − 6 = −1) corresponds to one more electron compared with structure I. That electron comes, of course, from the oxygen, which now retains only five (two from the double bond, together with three nonbonding electrons). Oxygen's formal charge is therefore +1 (6 − 5 = 1):

Formal charge: 0   0   −1 +1

:Ṅ=Ö   N̈=Ȯ:

Structure: I   II

Now face reality: NO is a molecule, not a cluster of ions; and it is a molecule formed by two nonmetals not far apart in electronegativity. We expect from that union a polar bond, but not an exaggerated polarity; we expect some separation of charge, but not a huge disparity. Doesn't structure I, with its formally *neutral* atoms, appear more likely than the ionic structure II?

It does, although some contribution other than structure I must be at work as well. Surely the bond has a partially ionic quality, even if small, since NO does have a dipole moment. The molecule does have a positive and negative end, a separation of charge of the kind made possible by a structure like II. Structure I, however, is *better* than II, probably a more important contributor, a structure more in harmony with nature's tendency to suppress differences in charge.

The oxidation numbers, incidentally, are the same for both I and II (and equally distorted): $+2$ for N; -2 for O. They are far too high because we made a bad assumption. We assumed, wrongly, that NO is an ionic compound.

(b) *NaCl.* By contrast, NaCl manifestly *is* an ionic compound, and here the oxidation numbers correspond exactly to Na^+ and Cl^-. This time, it is the formal charge that misses the mark. By imagining that Na and Cl share their electrons equally, we incorrectly turn a strong ionic attraction into a nonpolar covalent bond.

Both Na and Cl have formal charges of 0 in $Na\!:\!\ddot{\underset{..}{Cl}}\!:$, as shown in the calculation below:

	Na	Cl
number of bonding electrons	1	1
number of nonbonding electrons	$+\,0$	$+\,6$
total valence electrons in NaCl	1	7
number of valence electrons in free atom	$-\,1$	$-\,7$
formal charge	0	0

EXAMPLE 3-4. Balancing a Redox Equation

We shall have much more to say about reduction and oxidation in Chapter 17, where a chemical transfer of electrons will be put fruitfully to work in an electric circuit. At present, though, be content with just the barest truths:

1. For every electron lost, there is an electron gained. There is both an oxidizing agent *and* a reducing agent; and unless both work together, nothing happens. One gives. The other takes.

2. Charge is conserved. The total number of electrons remains the same.

Good enough. Knowing what we do, we already have enough to develop a systematic, direct, almost surefire way of balancing simple redox equations.

PROBLEM: Balance the equation

$$I^-(aq) + Br_2(\ell) \longrightarrow I_2(s) + Br^-(aq)$$

explicitly as a redox process. How many moles of electrons are transferred from giver to taker?

SOLUTION: Yes—the equation is simple enough to balance just by looking, but look more closely to see that I^- is giving electrons to Br_2. Iodine, existing first as the iodide ion (I^-), *loses* electrons to become the iodine molecule, I_2. Bromine *gains* electrons to go from Br^0 in elemental bromine (a molecular liquid) to Br^- in the bromide ion.

Start with the oxidation reaction, pretending for a moment that it has some separate existence:

$$I^- \longrightarrow I_2 \qquad \text{(oxidation, unbalanced)}$$

To balance the number of iodine atoms on each side, we match the I_2 molecule with two iodide ions; thus:

$$2I^- \longrightarrow I_2 \qquad \text{(oxidation, balanced for mass)}$$

To balance the *charge*, we introduce two electrons on the right:

$$2I^- \longrightarrow I_2 + 2e^- \qquad \text{(oxidation, balanced for mass and charge)}$$

The electrons, represented by the symbol e^-, appear as if they are a product of reaction.

Conclusion so far: Two iodide ions ($2I^-$) combine to form one iodine molecule (I_2), losing two electrons in the process. But really nothing is lost, of course, since the two electrons are picked up by molecular bromine (Br_2) in its reduction to two bromide ions:

$$Br_2 \longrightarrow Br^- \qquad \text{(reduction, unbalanced)}$$

$$Br_2 \longrightarrow 2Br^- \qquad \text{(reduction, balanced for mass)}$$

$$2e^- + Br_2 \longrightarrow 2Br^- \qquad \text{(reduction, balanced for mass and charge)}$$

We put together, finally, the giver with the taker, so that the combined oxidation and reduction reactions add up to a single, concerted, balanced process:

$$
\begin{array}{ll}
2I^- \longrightarrow \quad I_2 \;+ 2e^- & \text{(oxidation)} \\
\underline{2e^- + Br_2 \longrightarrow 2Br^-} & \text{(reduction)} \\
2I^- + Br_2 \longrightarrow \quad I_2 \;+ 2Br^- & \text{(redox)}
\end{array}
$$

The two electrons, absent from the net equation, connect the redox partners and drive the reaction forward.

For more details and more elaborate examples, see Chapter 17.

EXAMPLE 3-5. Out of Many, Few: Recognizing Reactions

Give-and-take: an electron offered, a hydrogen ion displaced, a reaction underway. And small change: a bond broken, a bond formed, an oxidation number reduced, an electron shared, an electron transferred. Many reactions, few designs. We know what to look for.

PROBLEM: Classify each of the following reactions as either acid–base, reduction–oxidation, dissolution–precipitation, or radical production–recombination. Identify the giver and receiver of electrons.

(a) $2SO_2(g) + O_2(g) \longrightarrow 2SO_3(g)$

SOLUTION: Gaseous sulfur dioxide, an air pollutant, reacts with atmospheric oxygen to produce sulfur trioxide. Sulfur trioxide, SO_3, is a precursor to sulfuric acid, H_2SO_4, the major component of acid rain.

The reaction indicated is a redox process. Oxygen gains electrons, undergoing reduction from an oxidation state of 0 in O_2 to -2 in SO_3. Sulfur, which is oxidized from S^{4+} in SO_2 to S^{6+} in SO_3, supplies the electrons. Reducing agent: SO_2. Oxidizing agent: O_2.

(b) $O(g) + N_2O(g) \longrightarrow 2NO(g)$

SOLUTION: O is a free radical, an uncombined *atom* rather than the diatomic molecule O_2. It has two unpaired electrons.

From Examples 2-5(c) and 3-3(a), we know that nitric oxide, NO, is a radical as well. One of the 11 valence electrons is unpaired in the neutral molecule.

In simple terms, then, an oxygen radical offers two electrons to N_2O (nitrous oxide), whereupon two NO radicals are generated. The division of property is equal: one unpaired electron goes to each.

Oxygen atoms are produced in the atmosphere when ultraviolet light from the sun splits an O_2 molecule into O and O. These oxygens, reacting with N_2O (a pollutant), subsequently produce NO according to the reaction above. NO, in turn, consumes ozone in the further radical reactions

$$NO + O_3 \longrightarrow NO_2 + O_2$$

$$NO_2 + O \longrightarrow NO + O_2$$

following which the cycle renews itself with fresh NO.

(c) $Zn(OH)_2(s) + 2OH^-(aq) \longrightarrow [Zn(OH)_4]^{2-}(aq)$

SOLUTION: As written, the equation represents a net ionic reaction free of spectator cations. It shows that zinc hydroxide, dissolved in basic solution, forms the complex anion $[Zn(OH)_4]^{2-}$.

The Zn^{2+} ion in $Zn(OH)_2$ is a Lewis acid, and the OH^- ion is a Lewis base. OH^- gives Zn^{2+} a pair of electrons to produce the coordination complex $[Zn(OH)_4]^{2-}$. We have a Lewis acid–base reaction.

(d) $Zn(OH)_2(s) + 2H_3O^+(aq) \longrightarrow Zn^{2+}(aq) + 4H_2O(\ell)$

SOLUTION: The same substance, $Zn(OH)_2$, now turns around and acts as a *base*, releasing OH^- to be neutralized by the acid H_3O^+. Pulled off the zinc, each of the OH^- ions (a base) takes a proton from H_3O^+ (an acid) to form water:

$$H_3O^+(aq) + OH^-(aq) \longrightarrow 2H_2O(\ell)$$

Whereas in reaction (c) the Zn^{2+} component of zinc hydroxide acts as an acid, in reaction (d) the OH^- component acts as a base. $Zn(OH)_2$ is said to be *amphoteric*; it has the capacity to go both ways, to serve as either acid or base. Depending on what is offered, the same compound will either take electrons or give electrons. It has the facilities for both.

Say what you like. (1) $Zn(OH)_2$ is a Lewis base because each of its OH^- ligands has a pair of electrons to give. (2) $Zn(OH)_2$ is a Brønsted-Lowry base because each of the OH^- groups can take a proton. (3) $Zn(OH)_2$ is an Arrhenius base because it releases the hydroxide ion. Say what you like, but recognize that something is always lost and something is always gained.

(e) $CH_3COOH(aq) + NH_3(aq) \longrightarrow NH_4^+(aq) + CH_3COO^-(aq)$

SOLUTION: Another acid–base reaction, easily recognized as the shift of H^+ from CH_3COOH (acetic acid) to $:NH_3$ (ammonia, a base). The salt consists of NH_4^+ (ammonium) ions and CH_3COO^- (acetate) ions.

(f) $CH_3COO^-(aq) + NH_4^+(aq) \longrightarrow NH_3(aq) + CH_3COOH(aq)$

SOLUTION: The tables are turned. Acting as an *acid*, NH_4^+ gives a proton to the electron-rich, proton-poor CH_3COO^- anion—a *base*. For the on-again-off-again pair CH_3COOH/CH_3COO^-, the proton lost in (e) is regained in (f).

(g) $2\,Au(CN)_2^-(aq) + Zn(s) \longrightarrow Zn(CN)_4^{2-}(aq) + 2\,Au(s)$

SOLUTION: An elemental substance present on only one side of an equation (where its oxidation number is zero) warns us of a possible redox reaction. Zinc, notice, goes from Zn^0 to Zn^{2+}, while gold goes from Au^+ to Au^0. Zinc gives the electrons; gold takes them. Zn displaces Au from the cyanide complex, liberating pure gold.

(h) $Ag^+(aq) + NO_3^-(aq) + K^+(aq) + Br^-(aq) \longrightarrow$
$$AgBr(s) + K^+(aq) + NO_3^-(aq)$$

SOLUTION: Here is a simple precipitation reaction, cluttered needlessly by the redundant presence of the potassium and nitrate spectator ions. Since $K^+(aq)$ and $NO_3^-(aq)$ appear unchanged on both sides, we eliminate them to obtain the net ionic equation

$$Ag^+(aq) + Br^-(aq) \longrightarrow AgBr(s)$$

Our interpretation is now straightforward: With no change in oxidation number, a cation in solution reacts with an anion in solution to yield an insoluble solid. It is the very definition of a precipitation.

It is also the very definition of a Lewis acid–base reaction. The bromide ion gives a pair of electrons to the silver ion, thereby bringing together two opposites. They meet; they give and take; they merge into a salt, neutralized and no longer driven to react. Precipitation, we realize again, is little more than the encounter of a Lewis acid with a Lewis base . . . just as redox is a kind of acid–base reaction, just as radical recombination is a kind of acid–base reaction, just as any chemical change takes an electron from one partner and gives it to the other.

Let the description serve a purpose. If the appearance of a precipitate is the key event, call the process a precipitation reaction. If a point-to-point flow of electrons is critical (in a battery, say), use the term "redox." If redox seems too general for some gritty application (like the purification of metallic gold), call it something more specific. But always look for the common ground. Look for the electrons. Look for small changes.

EXAMPLE 3-6. Solutes and Solutions

Look, especially, at reactions in aqueous *solution*, because in seas and cells is where much of chemistry takes place. The first step is to recognize which compounds dissolve as molecules and which dissociate into ions.

PROBLEM: Classify each of the following substances as an electrolyte or nonelectrolyte: (a) Glucose, a sugar with formula $C_6H_{12}O_6$. (b) Ammonium hydroxide, NH_4OH. (c) Ammonium nitrate, NH_4NO_3. (d) Perchloric acid, $HClO_4$.

SOLUTION: Ionic compounds, Arrhenius acids, Arrhenius bases, and salts will produce electrolytic solutions.

(a) *Glucose*, $C_6H_{12}O_6$, is a covalently bonded molecule with no easily ionizable protons. It dissolves as a molecule, a nonelectrolyte.

(b) *Ammonium hydroxide*, NH_4OH, is an Arrhenius base. In water, it dissociates into NH_4^+ and OH^- ions. Electrolyte.

(c) *Ammonium nitrate*, used extensively in the manufacture of fertilizer, dissociates into NH_4^+ and NO_3^- ions. It is an electrolyte, the salt of nitric acid (HNO_3) and ammonium hydroxide (NH_4OH).

(d) *Perchloric acid*, a strong acid with one ionizable proton, dissociates into H^+ and ClO_4^-. Electrolyte.

Note also that $HClO_4$ is one of four related *oxyacids* spanning the range of oxidation states from Cl^+ to Cl^{7+}:

		Cl	O	H
hypochlorous acid	HClO	+1	−2	+1
chlorous acid	$HClO_2$	+3	−2	+1
chloric acid	$HClO_3$	+5	−2	+1
perchloric acid	$HClO_4$	+7	−2	+1

Existing in a relatively high oxidation state, the chlorine in each acid thus offers a target for reduction. It becomes a site at which fresh electrons may be imported.

EXAMPLE 3-7. Solution Stoichiometry

A last, practical note . . . To count molecules and moles in solution, we define a *concentration*: the quantity of solute (moles, grams, or molecules) contained in a stated volume of solution. In particular, the *molarity* (*M*) specifies the number of *moles* of solute dissolved per liter. A 1 *M* solution (read "one molar") contains 1 mole of solute dispersed throughout a total volume of 1 liter.

PROBLEM: Beaker A contains 50.0 mL of a 0.500 *M* solution of NaCl. Beaker B contains 50.0 mL of a 5.00 *M* solution of $AgNO_3$. (a) How many Na^+ ions are dissolved in the first beaker? (b) How many grams of AgCl precipitate from solution when 1.00 mL of the $AgNO_3$ solution is added to the 50.0 mL NaCl? Assume that the reaction proceeds stoichiometrically, with total consumption of the limiting reactant.

SOLUTION: Given the molarity, we know the number of moles in any volume of the liquid. 1 L of a 1 M solution contains 1 mole of solute. 0.5 L of a 1 M solution contains 0.5 mole of solute. 0.1 L of a 1 M solution contains 0.1 mole of solute. The relationship is straightforward:

$$\text{Moles} = \frac{\text{moles}}{\text{volume}} \times \text{volume}$$

(a) *How many Na^+ ions are present in 50.0 mL of a 0.500 M solution of NaCl?* Knowing both the molarity and volume, we compute first the number of moles and then (using Avogadro's number) the individual particles:

$$\frac{0.500 \text{ mol NaCl}}{L} \times \frac{1 \text{ mol Na}^+}{1 \text{ mol NaCl}} \times 50.0 \text{ mL} \times \frac{1 \text{ L}}{1000 \text{ mL}} \times$$

$$\frac{6.02 \times 10^{23} \text{ ions}}{\text{mol}}$$

$$= 1.51 \times 10^{22} \text{ Na}^+ \text{ ions}$$

There are 1.51×10^{22} Na^+ cations and 1.51×10^{22} Cl^- anions: 0.0250 mole of NaCl dissolved in a total volume of 0.050 L.

(b) *How many grams of AgCl will precipitate from solution upon addition of 1.00 mL of 5.00 M $AgNO_3$?* To the 0.0250 mole of NaCl, we evidently add 0.00500 mole of $AgNO_3$:

$$\frac{5.00 \text{ mol Ag}^+}{L} \times 1.00 \text{ mL} \times \frac{1 \text{ L}}{1000 \text{ mL}} = 5.00 \times 10^{-3} \text{ mol Ag}^+$$

The stoichiometry is 1:1, and the scarce Ag^+ ion serves as the limiting reactant:

$$\text{Ag}^+(aq) \quad + \quad \text{Cl}^-(aq) \quad \longrightarrow \quad \text{AgCl}(s)$$

$$5.00 \times 10^{-3} \quad 5.00 \times 10^{-3} \quad \quad 5.00 \times 10^{-3} \text{ mol}$$

$$5.00 \times 10^{-3} \text{ mol AgCl} \times \frac{143.32 \text{ g AgCl}}{\text{mol}} = 0.717 \text{ g AgCl}$$

Unaffected by the precipitation, the Na^+ and NO_3^- spectator ions stay dissolved in the resulting mixture.

EXERCISES

1. Determine chlorine's oxidation number in each of the following species:

    (a) ClO^-, the hypochlorite ion
    (b) ClO_2^-, the chlorite ion
    (c) ClO_3^-, the chlorate ion
    (d) ClO_4^-, the perchlorate ion

2. Determine sulfur's oxidation number in both SO_3^{2-} (the sulfite ion) and in SO_4^{2-} (the sulfate ion).

3. Determine nitrogen's oxidation state in each of the following species:

    (a) N_2O, dinitrogen oxide (nitrous oxide)
    (b) NO, nitrogen oxide (nitric oxide)
    (c) NO_2, nitrogen dioxide
    (d) NO_3, nitrogen trioxide
    (e) N_2O_3, dinitrogen trioxide
    (f) N_2O_4, dinitrogen tetroxide
    (g) N_2O_5, dinitrogen pentoxide

4. Determine the oxidation state of each atom in:

    (a) P_4, white phosphorus
    (b) P_4O_6, tetraphosphorus hexoxide
    (c) Cl_2O_7, dichlorine heptoxide
    (d) HNO_2, nitrous acid
    (e) HNO_3, nitric acid

5. In which ion does the phosphorus atom have the higher oxidation state . . . PO_4^{3-} (the phosphate ion) or PO_3^{3-} (the phosphite ion)?

6. In which ion or molecule does carbon attract the greatest proportional share of the electrons . . . CO_3^{2-} (the carbonate ion), CO (carbon monoxide), or C_3H_8 (propane)?

7. Compute oxidation numbers and formal charges for the atoms in:

   (a) NO_3^-, the nitrate ion
   (b) CO_3^{2-}, the carbonate ion
   (c) SO_2, sulfur dioxide

   In each ion or molecule, the oxygens are equidistant from a central atom. Can any of these systems be represented by a single Lewis structure?

8. Calculate the formal charge of each atom in these three structures proposed for the thiocyanate ion, NCS^-:

$$N{=}C{=}S^- \qquad C{=}S{=}N^- \qquad S{=}N{=}C^-$$

   Which structure is most likely to be correct?

9. Complete and balance each of the neutralization reactions given below, identifying acid, base, and salt as they appear in aqueous solution:

   (a) $HNO_3 + NaOH \longrightarrow \underline{} + \underline{}$
   (b) $CH_3COOH + KOH \longrightarrow \underline{} + \underline{}$
   (c) $H_3O^+ + OH^- \longrightarrow \underline{} + \underline{}$
   (d) $H_2SO_4 + \underline{} \longrightarrow NaHSO_4 + H_2O$

10. Complete and balance each of the neutralization reactions given below, identifying acid, base, and salt as they appear in aqueous solution:

    (a) $\underline{} + Ca(OH)_2 \longrightarrow Ca(HCO_2)_2 + H_2O$
    (b) $HCl + NH_4OH \longrightarrow \underline{} + \underline{}$
    (c) $H_2C_2O_4 + \underline{} \longrightarrow C_2O_4^{2-} + H_2O$
    (d) $H_3O^+ + \underline{} \longrightarrow CH_3NH_3^+ + H_2O$

11. Identify the acid and base in each pair:

    (a) Fe^{3+}, CN^-
    (b) NH_3, BF_3
    (c) H_3O^+, OH^-
    (d) Al^{3+}, OH^-
    (e) CH_3COO^-, CH_3COOH

    Which of them, if any, can be described only as a Lewis acid or Lewis base—and not, more narrowly, as either a Brønsted-Lowry or Arrhenius species as well?

12. Complete and balance:

(a) $CH_3COO^-(aq) + H_2O(\ell) \longrightarrow$ ___ + ___

(b) $NH_3(g) + H_2O(\ell) \longrightarrow$ ___ + ___

(c) $NH_3(g) +$ ___ $\longrightarrow NH_4Cl(s)$

(d) $CaCO_3(s) + HCl(aq) \longrightarrow CaCl_2(aq) + H_2O(\ell) +$ ___

13. Identify the oxidizing agent and reducing agent in each reaction:

(a) $Zn(s) + 2MnO_2(s) + H_2O(\ell) \longrightarrow Zn(OH)_2(s) + Mn_2O_3(s)$

(b) $N_2(g) + 3H_2(g) \longrightarrow 2NH_3(g)$

(c) $Zn(s) + CuSO_4(aq) \longrightarrow ZnSO_4(aq) + Cu(s)$

(d) $14H^+(aq) + Cr_2O_7^{2-}(aq) + 6I^-(aq) \longrightarrow 2Cr^{3+}(aq) + 7H_2O(\ell) + 3I_2(s)$

14. Identify the oxidizing agent and reducing agent in each reaction:

(a) $2IO_3^-(aq) + 5Cu(s) + 12H^+(aq) \longrightarrow I_2(s) + 5Cu^{2+}(aq) + 6H_2O(\ell)$

(b) $Cd(s) + NiO_2(s) + 2H_2O(\ell) \longrightarrow Cd(OH)_2(s) + Ni(OH)_2(s)$

(c) $Pb(s) + PbO_2(s) + 4H^+(aq) + 2SO_4^{2-}(aq) \longrightarrow 2PbSO_4(s) + 2H_2O(\ell)$

(d) $6I^-(aq) + 2Al^{3+}(aq) \longrightarrow 3I_2(s) + 2Al(s)$

15. Add the correct number of electrons to the appropriate side of each equation:

(a) $F_2(g) \longrightarrow 2F^-(aq)$

(b) $NO_3^-(aq) + 4H^+(aq) \longrightarrow NO(g) + 2H_2O(\ell)$

(c) $H_2(g) + 2OH^-(aq) \longrightarrow 2H_2O(\ell)$

(d) $Al(s) \longrightarrow Al^{3+}(aq)$

Which of the processes are oxidations? Which are reductions?

16. Balance these redox equations:

(a) ___ $Br^-(aq) +$ ___ $F_2(g) \longrightarrow$ ___ $Br_2(\ell) +$ ___ $F^-(aq)$

(b) ___ $Cr(s) +$ ___ $Cu^{2+}(aq) \longrightarrow$ ___ $Cr^{3+}(aq) +$ ___ $Cu(s)$

(c) ___ $Fe^{3+}(aq) +$ ___ $I^-(aq) \longrightarrow$ ___ $Fe^{2+}(aq) +$ ___ $I_2(s)$

How many moles of electrons are transferred in each process?

17. Write each equation explicitly as the sum of an oxidation reaction and a reduction reaction:

(a) $Zn(s) + 2H^+(aq) \longrightarrow Zn^{2+}(aq) + H_2(g)$

(b) $Cr_2O_7^{2-}(aq) + 14H^+(aq) + 6I^-(aq) \longrightarrow 2Cr^{3+}(aq) + 7H_2O(\ell)$
$+ 3I_2(s)$

(c) $H_2SO_3(aq) + 2Mn(s) + 4H^+(aq) \longrightarrow S(s) + 2Mn^{2+}(aq) +$
$3H_2O(\ell)$

(d) $2Ni^{2+}(aq) + 2H_2O(\ell) \longrightarrow 2Ni(s) + 4H^+(aq) + O_2(g)$

How many moles of electrons are transferred between reducing agent and oxidizing agent?

18. Identify the free radicals in each reaction:

(a) $OH + CO \longrightarrow H + CO_2$

(b) $CH_3 + OH \longrightarrow CH_3OH$

(c) $Br + CH_4 \longrightarrow HBr + CH_3$

(d) $OH + HBr \longrightarrow H_2O + Br$

19. How many valence electrons are contained in each of the free radicals found above? How many of the electrons are unpaired?

20. Complete each of the following radical reactions by supplying a likely species at the position indicated:

(a) $CH_3 + CH_3 \longrightarrow$ __

(b) $CH_3O + HBr \longrightarrow$ __ $+ Br$

Of the structures represented, which are free radicals? How many unpaired electrons does each contain?

21. Identify the ions produced when each of the following species is dissolved in water:

(a) $MgSO_4$, magnesium sulfate

(b) HCl, hydrochloric acid

(c) $Ca(OH)_2$, calcium hydroxide

(d) $NaCH_3COO$, sodium acetate

(e) NH_4OH, ammonium hydroxide

(f) $CaCO_3$, calcium carbonate

22. Simplify each equation to its net ionic equivalent:

(a) $NaCH_3COO(aq) + H_2O(\ell) \longrightarrow CH_3COOH(aq) + NaOH(aq)$

(b) $Na_2SO_4(aq) + Pb(NO_3)_2(aq) \longrightarrow PbSO_4(s) + 2NaNO_3(aq)$

(c) $Al_2(SO_4)_3(aq) + 6NaOH(aq) \longrightarrow 2Al(OH)_3(s) + 3Na_2SO_4(aq)$

23. Classify each of the reactions given in the previous exercise as acid–base, reduction–oxidation, dissolution–precipitation, or radical recombination:

 (a) $NaCH_3COO(aq) + H_2O(\ell) \longrightarrow CH_3COOH(aq) + NaOH(aq)$
 (b) $Na_2SO_4(aq) + Pb(NO_3)_2(aq) \longrightarrow PbSO_4(s) + 2NaNO_3(aq)$
 (c) $Al_2(SO_4)_3(aq) + 6NaOH(aq) \longrightarrow 2Al(OH)_3(s) + 3Na_2SO_4(aq)$

 Use more than one category if appropriate.

24. Although $Zn(OH)_2$ is insoluble in water, many other compounds containing zinc and hydroxide ions in different environments indeed *can* be dissolved. From the list of water-soluble ionic solids given below, choose any two compounds that will react to produce zinc hydroxide:

$$NaCl, \ Na_2SO_4, \ NaOH, \ KOH, \ ZnCl_2, \ ZnSO_4$$

25. Write, first, a complete and balanced equation for the precipitation reaction chosen in the preceding exercise; and then, eliminating the spectator ions, write the net ionic equation.

26. Compute the concentration (in *molarity*, moles per liter) of an aqueous solution that contains 5.00 g $ZnCl_2$ dissolved in a total volume of 200.0 mL.

27. (a) How many grams of $ZnCl_2$ must be dissolved in a total volume of 500.0 mL to produce an aqueous solution with concentration 0.3750 M? (b) How many Zn^{2+} ions are present in the 500.0 mL? (c) How many Cl^- ions?

28. Suppose we have 100 mL of a 1.00 M solution of $ZnCl_2$. What change in total volume must we achieve, by adding water, to dilute the concentration to 0.500 M? to 0.250 M? to 0.100 M?

29. (a) What mass of insoluble $Zn(OH)_2$ will precipitate from solution when 300.0 mL of 0.600 M $ZnCl_2$ are mixed with 700.0 mL of 0.250 M NaOH? (b) What ions stay dissolved despite the reaction? (c) Which of the reactants is limiting?

30. (a) What volume of 0.250 M NaOH must be added to 300.0 mL of 0.600 M $ZnCl_2$ in order to precipitate 5.00 g $Zn(OH)_2$? (b) How many moles of $ZnCl_2$ remain dissolved after the reaction?

4

Light and Matter—
Waves and/or Particles

4-1. A New World
4-2. Light: Electromagnetic Waves
      What, When, and Where
      The Electromagnetic Spectrum
      Diffraction and Interference
4-3. Matter: Particles of the Atom
4-4. Light as a Particle: Photons and the Photoelectric Effect
4-5. Matter as a Wave
      Atomic Spectra
      Standing Waves
      De Broglie Wavelength
4-6. The Ideas of Quantum Mechanics
      Probability
      Indeterminacy
      REVIEW AND GUIDE TO PROBLEMS
      EXERCISES

4-1. A NEW WORLD

Light and matter are strangely intertwined, having more in common than ordinary appearances might suggest. They share, in the most fundamental way, a connection so intimate that neither matter nor light can be understood properly without the other.

It is, at first, a connection that threatens to defy common sense. Light is something insubstantial, something that slips through the fingers. Matter is substance itself. Matter is stuff that can be held and weighed and molded into form. Matter is a stone. Matter is an electron, a proton, an atom. Our imagination, comfortable only with things it can visualize, readily conjures up the classic image of matter as a particle: a little billiard ball, solid and predictable. That is what we know from the world at large; that is what we expect.

What we expect, however, proves wrong. Our entire picture of the microscopic world must be revised accordingly, along with the whole notion of common sense. Common sense is merely the accumulated wisdom of *macroscopic* experience, and such judgment offers no guide to the small-scale behavior of electrons and light. Light is more substantial than you think. Electrons are less substantial than you think.

A new theory of waves and particles—**quantum mechanics**—came upon the world in the 1920s, when 30 years of puzzling observations finally demanded an explanation. The difficulties with light and matter could not be solved simply by patching up the old equations here and there, but rather an entire outlook had to be rethought. Existing ideas about matter and electric forces were tossed out, despite their success at describing every phenomenon known until the late 19th century. The rock-solid model of Newtonian particles in motion was gone, at least for particles as small as electrons.

Yet quantum mechanics, although revolutionary in every sense, could not sweep away everything in its path. Electrons aside, Newton's classical mechanics describes the motion of large objects straightforwardly and with unquestioned accuracy. Must we say that the theory is wrong for such things? Doesn't force move matter? Don't particles seem to move through space along well-defined paths? Our macroscopically trained intuition demands that such commonplaces be true for the big things. Quantum mechanics, as we shall see presently, is a slippery sort of model that leaves us guessing where the electron really is. Must there be two theories—one for electrons and one for billiard balls?

One theory may prevail over the other, but both ought to be reconciled if there is to be only one world. It was all the more convincing, then, to discover that the new quantum theory could *contain* the classical model as a subset. Quantum mechanics, which is designed on a small scale, blurs indiscernibly into classical mechanics when applied to matter of a certain size. Everything happens smoothly and consistently. There is no sudden crossover, no abrupt transition from a mysterious microworld to a familiar macroworld. For a billiard ball, quantum mechanics ultimately makes the same predictions as Newton's laws. The classical picture of big particles thus remains intact. For an electron, though, one is forced to use quantum mechanics exclusively and to deal with the microworld on its own terms.

Our point of departure is light and matter, for it was there that quantum mechanics originally came together. First we shall view light as a wave and matter as a particle, seeing the world as a scientist of the late 19th century. Then, confronted with some startling experimental evidence, we shall turn the tables and discover what quantum mechanics is about.

4-2. LIGHT: ELECTROMAGNETIC WAVES

Light, understood subjectively, is the agency by which matter is made visible to human consciousness. Construed more broadly, light is a manifestation of the basic electromagnetic force—a means to push and pull charged pieces of matter.

Light arises from charge in motion; light sets charge in motion. Where there is light, there is an electromagnetic field. Where there is an electromagnetic field, there is opportunity for a charged particle to move. Where there is a charged particle, there is opportunity to create light. Light and matter, even at this level of understanding, are already entangled.

The sunshine registered by our eyes is a form of light. The perceived warmth of the sun is another kind of light, as are the ultraviolet rays that burn the skin. Radio and television waves are further examples of the same influence. So are microwaves. So are X rays. All are *electromagnetic waves*.

A traveling electromagnetic wave is a mobile field of force. A wave of light sweeps through space, carrying forward an electric and magnetic field. Intangible and immaterial, these fields wash over a charged particle just as a wave of water might wash over a swimmer. The particle feels the force and moves in response; the swimmer, too, bobs up and down in waves of a different sort.

A wave—any kind of wave—is a *disturbance*. Something varies. It might be a wall of water, measurable in units of length. It might be a variation in air pressure, interpreted by our ears and brains as sound. It might be an electric and magnetic field, able to exert force on a charge. Whatever it is, the "something" that varies gives the wave its particular *amplitude*, or maximum extent of oscillation.

Light, in particular, is an electromagnetic disturbance of space itself, a disturbance by which an ever-changing electric and magnetic force is impressed onto the surroundings. The value of the field rises and falls at each point, repeating some basic pattern as the influence propagates through space. If a charged particle gets in the way, it absorbs energy from the field and rises and falls correspondingly. That, according to classical theory, is what happens when the sun shines.

What, When, and Where

The answers to seven questions will characterize an electromagnetic wave sufficiently for our purposes:

1. What varies? Light is a wave of electromagnetic force. Its amplitude comes from the electric field (E) and magnetic field (B).

2. What is the repeating motif? Most generally, the wave pattern is a sine curve of the kind illustrated in Figure 4-1. The field begins at zero, rises to a crest, comes down to zero again, falls into a trough, and finally climbs back to zero. The **cycle** then begins anew.

3. Over what distance does the pattern repeat? One **wavelength** marks the spatial separation between similar points in the cycle: crest to crest, trough to trough, or any such equivalent pair. Denoted by the Greek letter λ (lambda), wavelength has dimensions of length. Commonly used units include the kilometer (km),

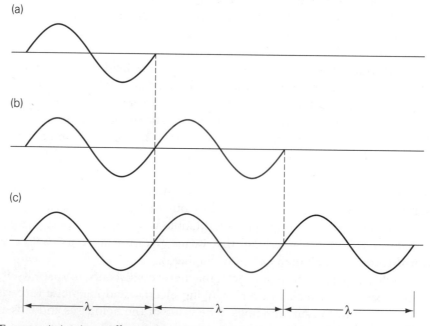

FIGURE 4-1. An oscillating disturbance (a wave), represented in its purest form: as a sine function. The wavelength λ marks the distance between corresponding points in successive cycles. (a) One cycle. (b) Two cycles. (c) Three cycles.

meter (m), centimeter (cm), millimeter (mm), micrometer (or micron, 1 μm = 10^{-6} m), nanometer (1 nm = 10^{-9} m), angstrom (1 Å = 10^{-10} m), picometer (1 pm = 10^{-12} m), and femtometer (1 fm = 10^{-15} m).

4. How often does the pattern repeat? At a given position in space, how many times per second does the wave reach its crest? *Frequency*, measured in ***hertz (Hz)*** and symbolized by ν (nu), indicates the number of completed cycles every second. A frequency of 1 Hz means that a crest occurs once per second, and so 1 Hz is expressed equivalently as 1 s^{-1}. Familiar multiples of the hertz include the kilohertz (1 kHz = 10^3 Hz), megahertz (1 MHz = 10^6 Hz), and gigahertz (1 GHz = 10^9 Hz).

5. How fast does the wave move forward? Recalling the rippling water, we realize that the swimmer not only bobs up and down but is also swept along by the traveling wave. A wave of electromagnetic field moves in much the same way, advancing at some speed c while completing a cycle (one wavelength) every $1/\nu$ seconds (its ***period***). Wavelength and frequency therefore are linked to speed by the relationship

$$\frac{\text{distance}}{\text{cycle}} \times \frac{\text{cycle}}{\text{time}} = \frac{\text{distance}}{\text{time}}$$

$$\text{Wavelength} \times \text{frequency} = \text{speed}$$

and by the resulting equation

$$\lambda\nu = c$$

The speed c, measured in vacuum, is equal to 2.99792458×10^8 m s^{-1} for all wavelengths and all frequencies.* Given either λ or ν, we can immediately determine the other.

6. Where is the wave *right now*? What fraction of a wavelength has been completed at this instant? Suppose we freeze the wave and inspect its leading edge. Is the cycle only beginning, so that the field is equal to zero? Has it just reached its first crest at one-quarter of a wavelength? Is the variation now halfway through the sine pattern? Or is the cycle three-quarters complete, with

---

*Note that light traveling through *matter* always has a speed less than c, and note further that the speed in a material also can vary depending on the wavelength. Since, however, the slowdown in air is only very slight, we shall assume the speed to be c everywhere.

the field in a trough? The fractional oscillation is called the ***phase*** of the wave. Phase tells an observer whether the disturbance is at a crest, trough, or anywhere in between. See Figure 4-2.

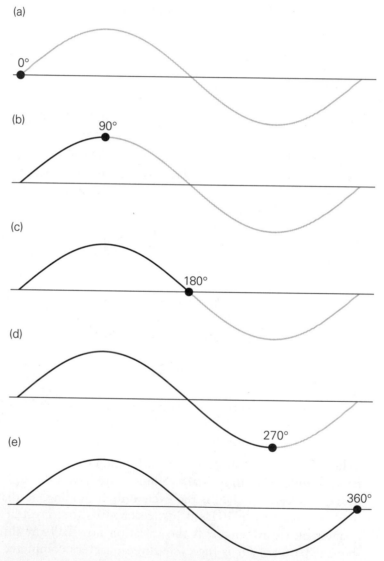

FIGURE 4-2. Spanning 360 degrees, the progress of a cycle is specified by its *phase*: (a) Oscillation begins; phase = 0°. (b) One-quarter of a cycle is completed; phase = 90°. (c) Half a cycle; phase = 180°. (d) Three-quarters of a cycle; phase = 270°. (e) One cycle ends and another begins; phase = 360°.

7. In what direction is the disturbance oriented? Light propagates as a *transverse* wave, meaning that the electric and magnetic fields are each perpendicular to the direction of travel. If we were to look straight into the oncoming wave shown in Figure 4-3, we would see that *E* and *B* oscillate in a plane perpendicular to our gaze. Up goes the electric field. Down goes the electric field. Left goes the magnetic field. Right goes the magnetic field. *E* and *B* are at right angles both to each other and to the forward direction of the wave. The electromagnetic force (and hence the electrical push or pull) thus points in a certain direction while it rises and falls so many times per second. That particular orientation is called the wave's **polarization**.

And so we have our electromagnetic wave, complete with amplitude, frequency, wavelength, speed, phase, and polarization. Thinking classically and macroscopically, we say that an oscillating electromagnetic force arises from the oscillating electric and magnetic field. A charged particle, sensing this force, is driven in sympathy. It is impelled back and forth in a certain direction at a certain rate, its movements related in some way to those of the field. The particle thereby acquires a new velocity and a new

FIGURE 4-3. Propagation and polarization of an electromagnetic wave. Mutually perpendicular, the electric (*E*) and magnetic (*B*) fields vary at right angles to the line of travel. The directions of *E* and *B* define the wave's *polarization*.

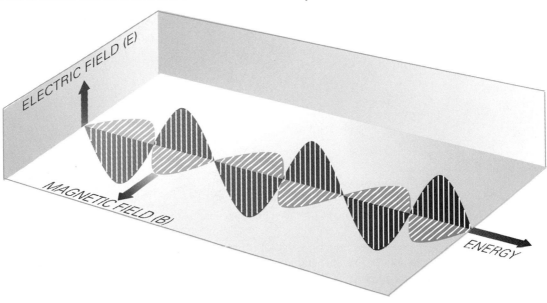

kinetic energy. Kinetic energy, as we know from Chapter 1, is always positive and is proportional to the square of the velocity.

Consequently, the energy of a classical wave must be related to the *square* of the disturbance—not E and B alone (which can be positive or negative), but rather $E^2 + B^2$. The field itself only makes the particle move, and the speed imparted will be equally fast for positive and negative forces. The energy comes, instead, from the square of the electric and magnetic fields. More precisely, we define the **intensity** (I) of a wave as the energy transported across a unit area per unit time, taking I as proportional to the square of the wavelike disturbance.

The Electromagnetic Spectrum

Light is light. All that distinguishes one form of light from another is the wavelength (or, equivalently, the frequency), from which comes the idea

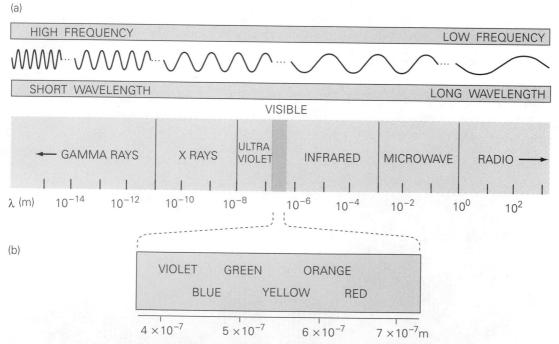

FIGURE 4-4. The electromagnetic spectrum. In vacuum, all electromagnetic disturbances propagate at the same speed but with different frequency, wavelength, and energy. (a) Overview of the entire electromagnetic spectrum. (b) The visible region alone. Frequencies increase in the order *red, orange, yellow, green, blue, indigo, violet* (ROYGBIV).

of color—more generally, the *electromagnetic spectrum*, laid out in Figure 4-4.

Radiation with wavelengths between 400 nm and 700 nm, for example, is able to deliver energy to certain molecules lining the retina. The electrons in these molecules take up the energy and produce an electrical response of their own. Various charged particles in the eye and brain are set in motion as a result, and ultimately we experience the subjective sensation of color. A wavelength of 400 nm is arbitrarily interpreted by the brain as violet, whereas a wavelength of 700 nm is seen as red. Between those endpoints we have nearly the entire *visible spectrum*, the colors of the rainbow.

At wavelengths lying just below the visible spectrum are *ultraviolet* waves (beyond the violet). Ultraviolet light, although undetected by the human eye, is similarly able to deliver energy to electrons in particular molecules. Small particles, for reasons soon to become apparent, absorb light selectively, and the valence electrons in chemical bonds are especially receptive to visible and ultraviolet radiation. Core electrons, by contrast, usually respond to *X rays*, for which λ ranges from 10^{-8} m to 10^{-11} m.

Above 800 nm lies the *infrared* portion of the spectrum, invisible to the eye but perceptible as heat. Infrared radiation induces vibrations in molecules, making bonds expand and contract, stretch and bend. Microwaves, even longer than infrared light, cause molecules to rotate.

The electromagnetic spectrum is completed by radio and television waves at the long end (λ greater than 1 m) and gamma rays at the short end (λ less than 10^{-11} m). Any wavelength is allowed, provided that $\lambda v = c$.

Diffraction and Interference

Consider what happens when one wave runs into another. Forgetting, temporarily, the abstraction of the electromagnetic field, let us disturb the waters of an imaginary pond with an imaginary stone. Circular waves emanate from the point of impact, with each successive wavefront rising and falling in rhythm. Every curved line in the left-hand side of Figure 4-5 represents such a wave at its crest.

Imagine next that the ripples are impeded, partly, by a slightly leaky breakwater—not an entirely solid obstacle, but a wall containing only two holes. Daughter waves are generated afresh at each opening, and two separate sets of ripples emerge on the right. The new waves spread out until finally crashing into a solid wall on the far right, along which we shall measure the energy delivered.

Each ripple arrives at the wall according to its own schedule, because

FIGURE 4-5. Waves impinging on a barrier with two openings. Regenerated at each hole, the ripples preserve any differences in phase while propagating forward to the far wall. An interference pattern develops wherever the waves recombine.

the waves generated at the two holes travel different distances to reach the same final position. Eventually they meet and deliver their combined energy to the barrier, point by point. How much energy is that?

The question goes straight to the heart of all wavelike behavior. What happens at the wall (and, more generally, what happens whenever any two waves cross paths) is a singular phenomenon that distinguishes a classical wave from a classical particle. When wave meets wave, the amplitudes combine to produce an effect called **interference**, an effect we need to examine with some care. Interference is what makes a wave a wave, and interference will prove essential to understanding quantum mechanics.

Start by following the paths of two arbitrary wavefronts, one from each opening:

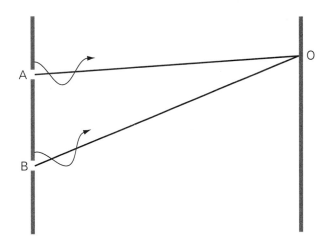

The geometry of **diffraction** (this spreading out of the waves) now becomes plain: A ray from the upper aperture (*AO*) meets a ray from the lower opening (*BO*) at some point *O* on the far wall. The wave on top (wave A) clearly has the shorter path in this triangular arrangement. We stipulate further that waves A and B start from their respective holes at the same time and at the same point in their cycles. They are *in phase*.

For simplicity, let both waves be at their crests when they begin to diffract away from the holes. Assume additionally that the shorter distance, *AO*, is some whole number of wavelengths. Any integer will do, since the wave motion repeats over and over again. If the displacement begins at a maximum, then it returns to a maximum at the beginning of every new cycle. Starting out at some value (say 1 unit), the wave always comes back to this same height after traveling for one wavelength. The value is 1 after one wavelength, 1 after two wavelengths, 1 after three wavelengths, 1 after one thousand wavelengths. If we had only this upper wave, then A would hit the wall at point *O* with a height of 1. The total intensity or energy, coming only from A, would be the square of the vertical displacement: 1^2, or simply 1 unit.

And wave B? Wave B evidently has a longer way to go, and this extra interval is all that distinguishes it from wave A. Ask, specifically, what is B's *phase* at point *O*, meaning at what fraction of a cycle does it complete the journey. The excess distance (and consequently the final phase) depends on both the angle θ and the separation *d* between the holes, as worked out in the diagram that follows:

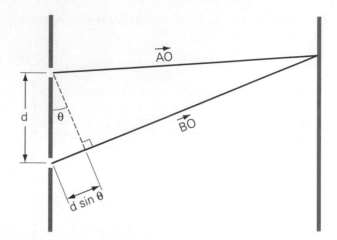

That difference in path is approximately $d \sin \theta$ for very long lines *AO* and *BO* (our stated condition), and we can control the outcome just by choosing a particular value of θ. Suppose, for instance, that d and θ are arranged so that wave A travels 1000.0 wavelengths and wave B travels 1000.5 wavelengths ($d \sin \theta = 0.5 \, \lambda$). Wave B would hit the wall exactly half a cycle after wave A, finding itself down while A was up. If, however, B were to travel 1001.0 wavelengths ($d \sin \theta = 1.0 \, \lambda$), it would arrive in the same condition as A. Both would be at a maximum, in phase, with their full heights restored.

Knowing how A and B behave separately, we now contemplate three experiments:

1. Close the lower hole but leave the upper hole open. Wave A arrives alone and transfers 1 unit of intensity, finishing at its crest after 1000 wavelengths:

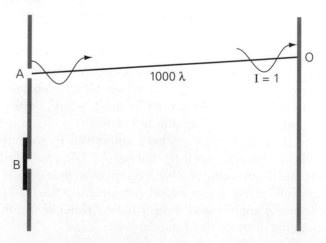

2. Close the upper hole and leave the lower hole open. B arrives alone, cresting after 1001 wavelengths and also bringing 1 unit of intensity:

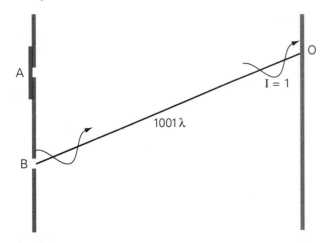

3. Open both holes and let A and B converge simultaneously on point *O*, differing by exactly 1 wavelength. Wave A makes a journey of 1000 wavelengths; B makes a journey of 1001 wavelengths:

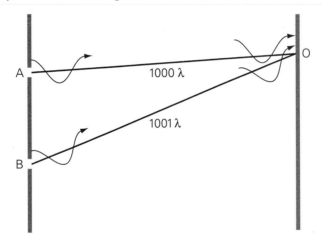

This last time, just before they meet, A is carrying 1 unit of energy and B too is carrying 1 unit of energy. Does 1 + 1 = 2?

For numbers, yes; for oscillating waves, no. The waves do not impinge on the walls separately. They combine with each other. They interfere. They meet—matched in phase, each at a crest, each with a height of 1 unit—and their individual heights *add* to create a combined height of 2 units. The joint energy therefore is $2^2 = 4$ units, not the 1 + 1 = 2 units that would follow from the separate arrival of two isolated waves. Here, with both A and B at a crest (Figure 4-6a), the intensity is enhanced. One

(a)

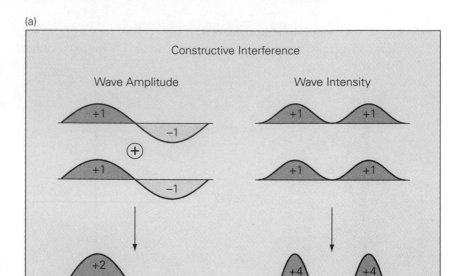

(b)

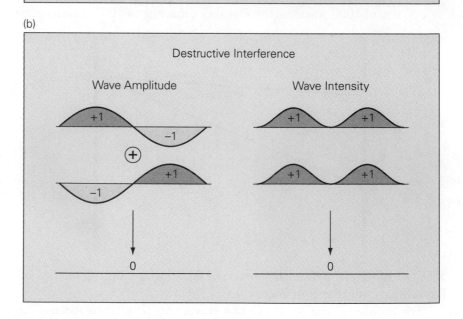

FIGURE 4-6. Interference. (a) Constructive interference: Two waves add *in phase*, crest to crest and trough to trough. The resulting intensity of the combined waves is greater than the value expected if the waves were to cross without interfering. (b) Destructive interference: *out-of-phase* combination of two waves, crest to trough. The resulting intensity of the combined waves is zero, even though each wave by itself has a positive intensity.

says that there has been ***constructive interference***, and the proper arithmetic is $1 + 1 = 4$. The amplitudes add ($1 + 1 = 2$), but the intensities interfere ($1 + 1 = 4$).

There can be ***destructive interference*** as well (Figure 4-6b), whereby the combined amplitude and intensity are diminished. If B travels an extra *half*-wavelength, for example, then it falls to a minimum (height $= -1$) just as A reaches a maximum (height $= 1$). The out-of-phase waves annihilate each other upon mixing (height $= 1 - 1 = 0$) and thus have no energy to impart at that point. Acting alone, each wave would have carried 1 unit of energy at either crest or trough. But now, coming together in precisely the worst way, they end up with nothing. Light plus light equals darkness, and the new arithmetic is $1 + 1 = 0$.

In fact, $1 + 1$ can add up to any number between 0 and 4 when two waves come together. It all depends on the difference in phase. Looking at the intensity along the entire wall, as in Figure 4-7, we discern the full interference pattern (also called the diffraction pattern) as a series of alternating maxima and minima. The maxima develop from fully constructive interference, at positions where the waves add crest to crest. The minima are the result of fully destructive, crest-to-trough interference. Phase differences other than 0 or $\lambda/2$ contribute to all the intermediate intensities between 0 and 4.

Despite the curious arithmetic, energy is neither created nor destroyed by the interfering waves. More energy does go to some spots

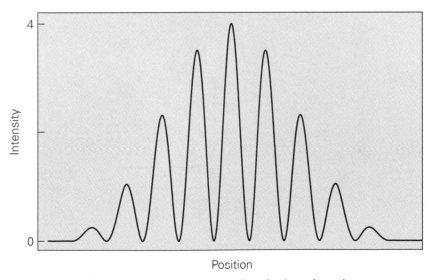

FIGURE 4-7. The interference pattern produced when the split waves recombine, marked by alternating bands of minimum and maximum intensity. Also called a *diffraction* pattern.

and less to others, but the net result is only a reshuffling. For every constructive meeting where $1 + 1 = 4$, there is a destructive encounter where $1 + 1 = 0$. The average is 2, reflecting the original energy carried by the waves separately. For every instance where the combined intensity is 3 units, there is a corresponding point where the total is 1 unit. Again the average is 2, as it is for every possible combination of constructive and destructive interference. Energy is conserved. It always is.

There, in brief, is the mark of a wave: interference. One disturbance interferes with another, and the resulting intensity can be more or less than the sum of its parts. No longer is $1 + 1$ guaranteed to be 2. Both the physical picture and mathematics of diffraction and interference are the same for water waves, electromagnetic waves, and similar oscillations of all kinds. For light, the pond of water becomes a region of space filled with an electromagnetic field of force. The ripples are the same, and the interference is the same.

A striking example of electromagnetic interference is provided by the *X-ray diffraction pattern* shown in Figure 4-8. Beams of X rays are

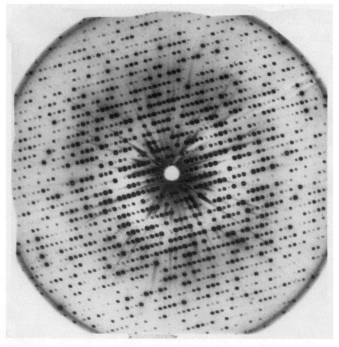

FIGURE 4-8. X rays, propagating as electromagnetic waves, suffer interference while passing through the latticework of a crystal. Scattered by the electrons in each atom, the waves recombine to produce a diffraction pattern consistent with the interatomic spacings. The pattern shown is for a crystal of the protein myoglobin.

directed onto a regularly spaced array of atoms in a crystal. Passing through, the electromagnetic radiation deflects (or *scatters*) off the electrons in the atoms, moving past these obstacles like the two sets of water waves. The diffracted X-ray intensity is determined by the spacing between atoms, just as the diffraction of water waves is controlled by the separation between the holes. From the X-ray diffraction pattern, one can thus determine the spacing between atoms in the crystal. It is a tool that reveals the microscopic structure of a crystal with impressive accuracy.

Interference, although remarkable, seems reasonable once we appreciate that the measurable quantity—the intensity, the energy—is proportional not to the wave's amplitude but rather to the amplitude's square. Still, waves appear special and set apart from particles. Two billiard balls, after all, never seem to mix themselves into a diffraction pattern as if they were two electromagnetic waves. If *billiard balls* A and B each have 1 unit of energy, we expect that the total energy will be 2 units however they might rattle about. The arithmetic for macroscopic particles is $1 + 1 = 2$.

Experience tells us that two billiard balls cannot produce a diffraction pattern. Picturing an electron as a billiard ball, one would draw on that experience to say . . . no, an electron should not produce an interference pattern. That kind of behavior is characteristic of waves, not particles. Isn't it?

But we have already been warned to abandon common sense when dealing with electrons. In consideration of that warning, let us reserve judgment on the whole issue of waves and particles and light and matter. We have not yet exhausted the characteristics of light, and we have still to observe how electrons really behave. It is appropriate, having arrived at this point, to look more closely at the atom and its components.

4-3. MATTER: PARTICLES OF THE ATOM

Although little was known about atomic structure before the 20th century, one attribute surely was taken for granted: substance. Atoms were expected to be particles of some sort, hard and indivisible. It was an idea that originated with the so-called atomistic philosophers of ancient Greece, who argued that a continuously changing world *should* be fashioned from a minimal set of unchanging parts. By saying "should" they were indulging in philosophy rather than experimental science, of course, but these early philosophers planted the seeds of atomic theory   the idea that matter is grainy and discontinous.

Those seeds lay dormant for more than two millennia, germinating only with the work of John Dalton at the beginning of the 19th century.

A scientist of that time would have reasoned that atoms (if they existed at all) might be invisible but certainly not insubstantial. Presumed to be the building blocks of matter, these unseen atoms eventually must amount to something tangible. One would expect them to share the characteristics of larger particles. They would have mass and possibly charge. They would move in response to forces. They would acquire momentum and energy. They would obey Newton's laws.

These expectations were borne out initially by the experiments of the 19th century. Stoichiometric observations, for instance, revealed that chemical combinations always involved whole entities—atoms of some sort, one corresponding to each element. Atoms seemed to join together into molecules, and such molecules transformed themselves into still new structures. Atomic and molecular weights were measured. An early form of the periodic table, developed by Dmitri Mendeleev, was known by the end of the century.

Seeing is believing, though, and the first visual traces of atoms and molecules in motion came with an understanding of *Brownian motion*. Viewed under a microscope, a small particle (a dust mote, say, or a grain of pollen) bounces around continually and randomly on the surface of a liquid. Albert Einstein showed, in 1905, that these random motions arise as the molecules of the liquid collide with the visible particle from all directions. Were we to imagine a volleyball batted around by a crowd of invisible players, the effect would be similar. The motion of the ball implies the presence of the players; the motion of the dust mote implies the presence of atoms and molecules. Both images are consistent with a particulate, billiard-ball kind of world.

Other experiments quickly revealed the atom to be not just one billiard ball, but many all at once: a composite structure built from smaller electrical particles. The electron was discovered in 1897 by J. J. Thomson, who used the properties of electromagnetic fields to measure the ratio of charge to mass (Figure 4-9). The current value, determined to high accuracy, is $-1.75881962 \times 10^{11}$ C kg^{-1}. Soon after, Robert Millikan was able to attach individual electrons to oil droplets and subsequently measure a droplet's rise and fall in an electric field. From this landmark experiment (Figure 4-10) he determined the mass and charge of the electron, now known to be $9.1093897 \times 10^{-31}$ kg and $-1.60217733 \times 10^{-19}$ C, respectively. It was all very particle-like. The electron had charge and mass, and the wispy particle responded to electric forces in the expected way.

The atom's other major constituent—the nucleus—proved to be a dense, positively charged object set amidst a diffuse sea of electrons. In a classic experiment performed in 1911 (Figure 4-11), Ernest Rutherford

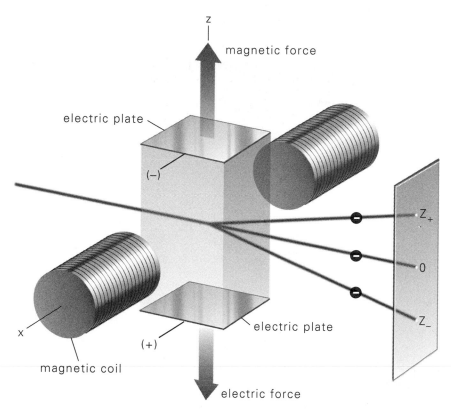

FIGURE 4-9. The electron, a charged particle, responds to both electric and magnetic forces, as discovered by J. J. Thomson in the late 19th century: A beam of electrons is deflected *down* (toward the positive plate) by the electric field and *up* by the magnetic field. Adjustment of the electric field enables the experimenter to aim the beam anywhere between points Z_+ and Z_-—and, from that, to determine the ratio of charge to mass: e/m.

and his associates bombarded a thin sheet of gold foil with a stream of microscopic "bullets" from a radioactive atom: *alpha* radiation, later identified as helium-4 nuclei stripped of electrons. Small, hard, and positively charged, most of these alpha particles passed right through the foil while suffering only minimal deflection. Unexpectedly, though, some were scattered through larger angles, and a few even ricocheted back directly at the observer. Rutherford, stunned by these latter results, likened the boomerangs to a 15-inch shell bouncing off a sheet of tissue paper.

Rutherford's explanation serves as a simple model even to this day: An atom exists largely as sparsely filled space, with its mass concentrated

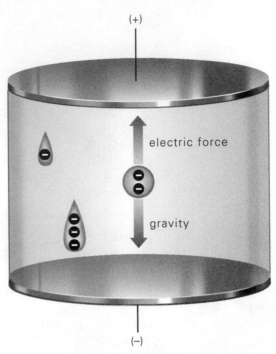

FIGURE 4-10. Robert Millikan's oil-drop experiment (1906), making use of Thomson's ratio e/m, reveals the values of e and m separately: e, the charge of an electron; m, the mass. The method: Fine droplets of oil, each given a negative charge, are pushed *upward* by an electric force while they simultaneously fall *downward* under the influence of gravity. Controlled variation of the electric field shows that the negative charge on each drop is always a multiple of a certain value e—the fundamental charge of the electron.

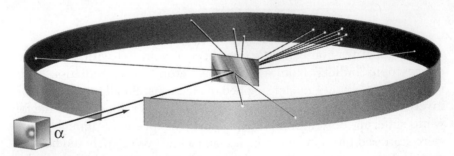

FIGURE 4-11. Ernest Rutherford (1911), firing his "15-inch shells" at a paper-thin sheet of gold foil, discovers the pit in the peach: the nucleus. The projectiles are really *alpha particles*, bare helium nuclei with a full charge of +2. Fired at the foil, these positively charged particles are repelled by the similarly positive nuclei set deep within the gold atoms. Some of the alpha particles bounce straight back after a head-on collision.

into a small but heavy nuclear kernel. The electrons swarm about the positively charged nucleus and give the atom its volume. So vast is the space and so tiny is an alpha particle, however, that relatively few of them ever approach a gold nucleus closely enough to alter their trajectories appreciably. On those occasions when an alpha particle does pass close by a nucleus, the electrostatic repulsion of positive against positive redirects the path. If the incoming particle hits the nucleus dead on, the collision sends it back whence it came.

Here, then, are the two basic parts of the atom as revealed by the experiments of Thomson, Millikan, and Rutherford: electrons and nucleus. Both appear to be particles. They have charge. They have mass. They are pushed and pulled by electromagnetic forces. They have momentum and energy. They collide with other particles and scatter. We might envision a billiard table or maybe a miniature solar system, with planetlike electrons orbiting about a nuclear sun in predictable paths.

Reasonable as it seems, this easy-to-grasp image has no basis in reality. The little solar system cannot endure. Classical mechanics and the laws of electromagnetism declare, unambiguously, that such an arrangement is impossible. A charged particle executing a curved trajectory loses energy while it moves, and hence an orbiting electron must eventually slow down and crash into the nucleus. If, by contrast, the electron tries to remain stationary, it is instantly pulled toward the positively charged nucleus by the attractive electric force. Either way, the planetary atom is doomed to collapse if the particles obey Newton's laws. Figure 4-12, icon of the nuclear age, gives entirely the wrong picture.

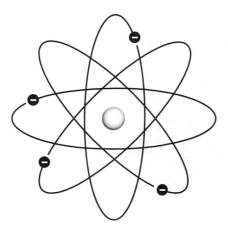

FIGURE 4-12. Popular image of the nuclear atom, grossly incorrect in its depiction of a miniature solar system. Electrons, unlike planets, do not orbit around an attractive center in well-defined paths. Such motion would drain the charged particles of energy and send them spiraling into the nucleus.

There must be another explanation, because atoms clearly do manage to exist—not as indivisible balls, admittedly, but as compound structures containing electrons and a nucleus. Something other than classical mechanics must enable the atom to assemble its parts into a stable structure.

4-4. LIGHT AS A PARTICLE: PHOTONS AND THE PHOTOELECTRIC EFFECT

Incomplete, too, is the picture of light as a wave. For despite exhibiting wavelike properties such as diffraction and interference, electromagnetic radiation also bears the marks of a particle.

A classical particle, stated crudely, is a kind of lump. It is a *whole* something, held together as a self-contained bundle of momentum and energy. It is a lump that can collide with other lumps and make them move. One lump hits another; momentum and energy are transferred; the target is set in motion. One particle knocks away another particle.

Consider, by comparison, the ***photoelectric effect*** (Figure 4-13): A beam of light (an electromagnetic *wave*?) shines on the surface of a metal. Momentum and energy are transferred, and some of the electrons bound in the metal are released. But are these electrons loosed from their bonds by the persistent up-and-down undulation of a light wave? Apparently not. They are knocked out abruptly by *particles* of light, one collision at a time. The effect is closer to the short, sharp rap between two billiard balls than to the crash of an ocean wave upon the shore.

FIGURE 4-13. Light delivers energy and momentum to matter, as evidenced by the *photoelectric effect*: Shining on the surface of a metal, a beam of light causes a current of electrons to flow.

Earlier we said that light behaves like an electromagnetic wave, and in most instances it does. Associated with light is, undeniably, an electromagnetic field that varies in space and time. The wave's intensity, proportional to the square of the amplitude, usually does control the energy transferred to any charged particles in its wake. Most telling, light routinely exhibits diffraction and interference just like any other wave.

Always, however, there is a tacit assumption in this treatment of light as a wave: There is presumed to be, in each manifestation, light of such high intensity that the individual particles of the electromagnetic field are blended together. It is the difference between looking at an entire beach and looking at a few grains of sand. With the photoelectric effect we begin to see the individual grains.

One grain of light (a *particle*, not a wave) knocks out one grain of matter, an electron. The light's momentum and energy are delivered to the electrons in discrete lumps. Expelled from the metal by these packets of energy, the liberated electrons (*photoelectrons*) subsequently flow as an electric current that can be exploited in such devices as burglar alarms and solar cells. More significant, though, this unexpected encounter between light and matter contains the proof that light acts as a particle.

Realize that if the ejected photoelectrons acquired their energy from *waves* of light, then the current could be switched on and off by the electromagnetic intensity alone. Some minimum amount of energy, called the **work function**, would be needed first to free an electron from its bonds in the metal and thereby initiate the current. Electrons would then flow unhindered once the light's intensity exceeded this threshold, regardless of the wave's frequency. Yet these predictions, devised for a wave, are contradicted by the experimental results.

Rather it is the frequency of the light that determines when the current starts to flow. Look at the graph shown in Figure 4-14. No photocurrent is induced until the wave's *frequency*, ν, surpasses a critical value ν_0, different for each metal. However high the intensity, the electrons remain bound if ν is less than ν_0. But let ν become greater than ν_0 (even with an intensity so faint as to be barely perceptible), and suddenly an electron will leave the metal with a certain kinetic energy. If, next, we increase the intensity while keeping the frequency constant, there will be *more* photoelectrons but they will not be more energetic.

Einstein, in another celebrated paper of 1905, explained these observations by proposing that light acts as a particle. Termed a **photon**, a "lump" of light carries an energy

$$E = h\nu$$

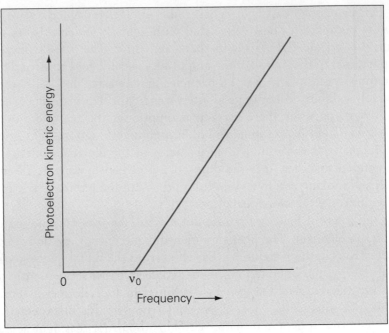

FIGURE 4-14. Not the intensity, but the *frequency* of the light (v) controls the photocurrent. Above a certain threshold frequency v_0, the kinetic energy of an ejected electron varies in direct proportion to v. Below the threshold, no current flows under any circumstances.

proportional to the frequency v, with v bearing its usual units of s^{-1}. The proportionality constant h, equal to $6.6260755 \times 10^{-34}$ J s, is **Planck's constant**, discovered a few years earlier by Max Planck in his analysis of the radiant energy emitted by a hot body.

Planck and Einstein, together, introduced the idea that energy is *quantized* (restricted into discrete little parcels). Planck, looking at matter, needed to explain the distribution of energy radiated by objects at high temperature. He could do so only by proposing that atoms absorb and emit energy in indivisible packets equal to hv—one **quantum** at a time—no more, no less. Einstein, looking at light, could explain the photoelectric effect only by saying that the electromagnetic field is similarly packaged into quanta of energy hv. Each view revealed a complementary aspect of the same phenomenon.

Light, which behaves macroscopically as a continuous electromagnetic field, is understood microscopically as an assembly of photons. Associated with each photon is a frequency v, a wavelength λ, and an energy hv. High frequencies (and thus short wavelengths) carry high energies; low frequencies and long wavelengths carry low energies. A photon with a wavelength of 4000 Å, for example, has twice the energy of a photon with $\lambda = 8000$ Å. Photons of blue light are more energetic than photons of red light, and X rays deliver more energy than microwaves.

Unlike a material particle, a photon has zero rest mass and moves at the speed of light. No massive object can travel as fast, and with this motion also comes the ability to convey momentum—although not in the sense we ordinarily think of mass × velocity. The massless photon's momentum depends, instead, on the wavelength, varying in inverse proportion to λ:

$$p = \frac{h}{\lambda}$$

Just as the energy of a photon is quantized, so is its momentum. Momentum is doled out in discrete parcels of h/λ, in the same way that energy is allocated in steps of $h\nu$. Notice how Planck's constant appears in both expressions, suggesting that h might be a characteristic amount associated with a quantum. Concerning the dimensions themselves, note that h (expressed in units of J s, or N m s) divided by λ (m) gives N s. A newton-second reduces properly to the basic units of momentum: kg m s^{-1}, mass × velocity.

To have momentum is to be a particle; to have a wavelength and to diffract is to be a wave. A photon therefore is both particle and wave together, its two natures linked by the reciprocal expressions $p = h/\lambda$ and $\lambda = h/p$. It is a particle with a wavelength. It can collide like a billiard ball and interfere like a wave.

With photons we now have a consistent explanation of the photoelectric effect. If light is a particle, then each photon delivers *one* lump of momentum (h/λ) and *one* lump of energy ($h\nu = hc/\lambda$) to the *one* electron it hits. We know, from Planck's discovery of quantized atomic energies, that part of the photon's energy (equal to the work function, $h\nu_0$) is needed just to free the electron from the forces holding it within the metal. After this initial investment, the difference between the photon energy and the work function becomes the kinetic energy (E_k) of the freed electron:

$$E_k = h\nu - h\nu_0$$

Energy is conserved. The electron absorbs a quantized dose of energy, part of which is used to escape from the metal. The ejected particle keeps the rest in the form of kinetic energy. The higher the photon's energy, the faster the photoelectron. The lower the photon's energy, the slower the photoelectron. Let there be too long a wavelength—a photon with too little energy—and no photoelectron appears at all.

Even when there is but one photon, a photoelectron still can be produced if the energy delivered is sufficiently high. This one photon knocks out one electron. A second photon, where present, knocks out a second electron. A third photon knocks out a third electron. If all the photons have the same frequency, then all the photoelectrons acquire the same energy.

A higher intensity, interpreted according to the particle model of light, means only that more photons contribute to the electromagnetic field. More photons produce more photoelectrons, and thus the photocurrent increases in proportion to the incident light's intensity. These are the experimental facts.

All of them, together, point to a picture of *particulate* light, at least under certain circumstances. Sometimes light behaves as a particle; other times, as a wave. If the light is of very low intensity, its photons become apparent one by one. When, by contrast, light seems to act as a continuous wave, we see a torrent of photons all at once. Enormous numbers of photons appear to merge into a continuous electromagnetic wave—an entire beach, so to speak.

What if, by putting together an electromagnetic field photon by photon, we try to build such a beach one grain of sand at a time? Suppose only a single photon of violet light ($\lambda = 400.0$ nm) is present initially, with an energy of

$$E = h\nu = \frac{hc}{\lambda} = \frac{(6.626 \times 10^{-34} \text{ J s})(2.998 \times 10^{8} \text{ m s}^{-1})}{400.0 \times 10^{-9} \text{ m}} =$$

$$4.966 \times 10^{-19} \text{ J}$$

After a second photon is added, the total energy is now exactly two quanta: 9.932×10^{-19} joule. It is not 6.000×10^{-19} joule or 8.123×10^{-19} joule or any number between one and two photon's worth of energy. Photons come one lump at a time. The lumps may be small, but they are lumps all the same.

When there are only a few photons, each stands out as an individual bundle of momentum and energy. But if there is too large a crowd (10^{19} violet photons, for instance, to give a total energy of 4.966 J), then we fail to notice any gaps between the allowed energies. There *are* gaps, but who can distinguish effectively between 10^{19} photons and 10^{19} photons plus one? The difference amounts to only 0.0000000000000000004966 joule out of 4.966, and the field blurs into a continuous wave.

4-5. MATTER AS A WAVE

A photon is a particle that can behave like a wave. Might an electron (or any other particle) exhibit wavelike characteristics as well?

If the question seems odd, remember that we have yet to understand how atoms can exist at all. Even the perseverance of simple hydrogen, with its lone electron and proton, is inexplicable under the old model of electromagnetic forces. A new way of thinking is required, and maybe the photon can provide some guidance.

Atomic Spectra

To start, note that an atom's mere existence is not its only unexplained property. There is also the *quantization of energy*, for which the element hydrogen serves as a prototypical example. Here the electron's energy is restricted to certain well-defined values: some particular number E_1, another number E_2, a third number E_3, and so on. These **energy levels** become apparent when the atom is irradiated with light, resulting in a transfer of electromagnetic energy to the bound electron.

Imagine that a photon, carrying a quantum $h\nu$, strikes an electron confined within an atom. If the electron truly were a billiard ball, it would absorb the energy and momentum of the photon—whatever those values might be. It would accept a red photon as readily as a blue photon, an infrared photon as well as an X-ray photon. Billiard balls take whatever they are given.

Electrons do not. An atom can accept a photon only if the quantum of electromagnetic energy, $h\nu$, is just enough to excite the electron from one energy level to another. Expressed in symbols, the condition for a *transition* between levels then becomes

$$\Delta E = E_i - E_j = h\nu$$

where E_i and E_j are two electronic energies. The atom absorbs and emits energy (a photon) only in discrete steps, a single quantum at a time. Think of rungs on a ladder, irregularly spaced (Figure 4-15).

By applying light of different frequencies, an experimenter obtains a **spectrum** of response: A distinct signal appears each time the photon energy successfully bridges two energy levels of the atom. The spectrum for hydrogen, first observed in the late 19th century, is a set of separated

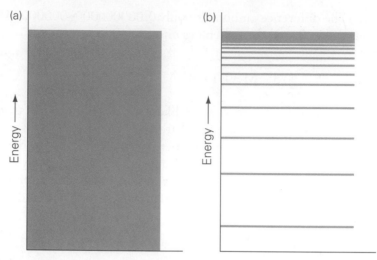

FIGURE 4-15. (a) In the macroworld, energies vary continuously. The spacing between one value and the next shrinks to nothing. (b) Not so in the microworld, where gaps punctuate the pattern all throughout—like open spaces on a ladder, offering no foothold to a particle. Microworld energies are quantized.

lines. Each frequency corresponds to a transition from one atomic state to another. An example is provided in Figure 4-16.

Other atoms and molecules yield more complicated spectra, but always there are discrete transitions. The energies of confined electrons are quantized.

Standing Waves

Quantization, by itself, is neither a new nor mysterious phenomenon in nature. It is an attribute of all **standing waves**, a phenomenon instinctively

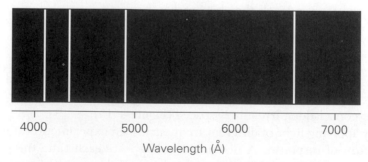

FIGURE 4-16. A portion of the spectrum of atomic hydrogen, showing four discrete transitions in the visible region. The atom emits a matching photon as it falls from a level of high energy to a level of low energy.

appreciated by any musician. A violinist knows that a vibrating string, held down at each end, is not free to oscillate at arbitrary frequencies. The sound has a certain pitch and timbre instead, shaped by fixed ***modes of vibration*** characteristic of the string's length and the instrument's construction. Unheard-of behavior for a particle, such quantization is a natural characteristic of any confined wave.

Standing waves, illustrated in Figure 4-17, are stationary disturbances in space. They *stand*. Crests and troughs remain fixed in place, growing and shrinking periodically as time passes. Always the net vibration is up

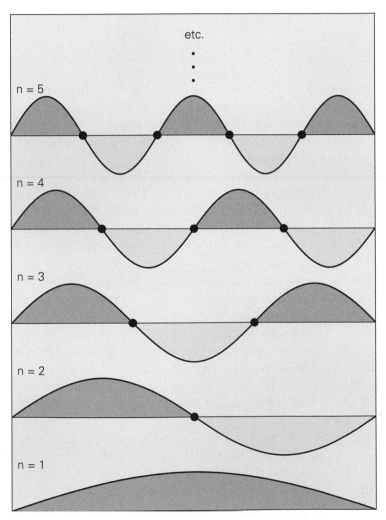

FIGURE 4-17. Standing waves: vibrations fixed in place between two points. Each mode contains *n* half-cycles, where the quantum number *n* takes the values 1, 2, 3, . . . , ∞. Nodes in the patterns are marked by large dots.

and down at each point in this view, and the pattern moves neither left nor right:

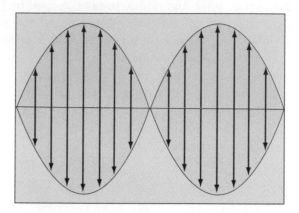

A standing wave develops when one or more half-cycles of oscillation ($\lambda/2$) are confined between the two endpoints. There must be an integral number of these half-wavelengths in order for the disturbance to vanish at each end. If the terminal displacements are nonzero, the stationary pattern disappears.

Wherever the displacement is zero, the wave is said to be at a **node**. Ignoring the clamping at each end, we see that a vibrating string is describable by its number of nodes. The more nodes, the more crests and troughs in the pattern. The frequency of vibration therefore increases with the number of nodes, growing in proportion as more half-waves are added to the oscillation.

See what we have. Lowest in the series, with zero nodes, is a standing wave containing only one half-cycle ($\lambda/2$). Next is a mode of vibration with one node, arising from one complete cycle ($2 \cdot \lambda/2 = \lambda$). Then comes a wave with two nodes ($3 \cdot \lambda/2$) and three nodes ($4 \cdot \lambda/2$), and so forth. There are no admissible patterns in between; the vibrations are quantized. The allowed frequencies (call them ν_1, ν_2, ν_3, . . .) are separated by gaps and labeled by a **principal quantum number** ($n = 1$, 2, 3, . . .), with $n - 1$ indicating the number of nodes: 0 for the first state, 1 for the second, 2 for the third.

These gaps in frequency always exist, but the *percent* change from one wave to another shrinks at high quantum numbers. Differences begin to blur as more and more half-wavelengths are confined within the same space. When, as already seen in Figure 4-17, there are just a few nodes, the peaks are well separated and easy to distinguish. Not so when we move from, say, 1,000,000 nodes to 1,000,001, for which the change is so subtle that the jump appears to be infinitesimally small. One mode of vibration blends continuously into the next at sufficiently high quantum numbers, and a similar kind of smoothing also occurs when the waves are allowed to spread out over larger distances. The quantized frequencies of a large system are uniformly smaller than those of a small system.

Thus the higher the quantum number and the longer the string, the less obvious is the quantization. Small size and low quantum numbers tend, in general, to emphasize the gaps in frequency, whereas large size and high quantum numbers tend to obscure them. It is the same for photons, which crowd together to create the illusion of an unbroken electromagnetic field. Such transitions between large and small suggest, for the first time, how quantized atomic energies might ultimately blend into the continuous fabric of bulk matter.

Now a hydrogen atom is no violin string, but its spectroscopic frequencies—and hence its energies—are indeed quantized. And, like a violin string, a hydrogen atom is a confined structure. Its electron is restricted to a limited region of space; its electron has discrete energies. So, too, are the effects of confinement and quantization apparent within a metal, where a photoelectron comes loose only when struck by a suitably energetic photon (strong enough to overcome the work function $h\nu_0$). Note also that once the electron is removed from its atomic moorings, any gaps between the energy levels become far smaller. Quantization is indeed *less* obvious under such conditions, and a free photoelectron does seem to run through a continuous range of kinetic energy. The analogy strongly suggests that a confined particle—an electron in an atom— might be quantized just like a confined wave.

What kind of wave, if any, is a question we have still to answer.

De Broglie Wavelength

If a particle is to behave like a wave, then it ought to have a wavelength. So said Louis de Broglie in 1924, adding that any kind of "matter wave" (whatever that may be) should also undergo diffraction and interference. The ability to interfere is, above all, what makes a wave a wave.

By analogy with the photon, that sometimes-particle-and-sometimes-wave, de Broglie suggested that an electron might have a characteristic wavelength given by

$$\lambda = \frac{h}{p} = \frac{h}{mv}$$

This expression, with the wavelength λ inversely proportional to the momentum p, is identical to that for a photon. The only accommodation for a material particle is the explicit substitution $p = mv$, where m

denotes the mass and v the velocity. The units for λ reduce properly to meters when mass is expressed in kilograms, velocity is expressed in meters per second, and Planck's constant is expressed in joule-seconds (recall that a joule, or newton-meter, has dimensions of kg m^2 s^{-2}).

Applying de Broglie's formula to an electron moving at 1.00×10^6 m s^{-1}, for example, we compute a wavelength of 7.27 Å (7.27×10^{-10} m):

$$\lambda = \frac{h}{mv} = \frac{6.626 \times 10^{-34}\ \text{kg m}^2\ \text{s}^{-1}}{(9.109 \times 10^{-31}\ \text{kg})(1.00 \times 10^6\ \text{m s}^{-1})} = 7.27 \times 10^{-10}\ \text{m}$$

A faster electron has more momentum and thus a shorter wavelength; a slower electron has a correspondingly longer wavelength.

The whole notion was an inspired guess. If true, then any material particle would have a wavelength and could, in principle, suffer interference. Electrons, protons, neutrons, atoms and molecules, billiard balls and boulders, all might produce diffraction patterns just like light waves.

The small particles do. Accelerated to speeds ranging from 10^6 m s^{-1} to 10^8 m s^{-1}, an electron has a not-so-small wavelength of approximately 0.1 Å to 10 Å. These distances are comparable to the spacing between atoms in crystals and, against all billiard-ball intuition, a beam of such electrons does diffract off the latticework of a suitable crystal. The first experimental proof came just a few years after de Broglie's proposal.

A typical electron diffraction pattern is reproduced in Figure 4-18. Realize that it is an *electron* beam, not an X ray, that generates the rings of intensity shown here. These are *particles*, particles with mass and momentum, but particles that behave strangely like waves with wavelength, frequency, and phase. They interfere. Neutrons do the same thing, on an even finer scale, owing to their larger mass and shorter de Broglie wavelengths ($\lambda = h/mv$). Small atoms and molecules also can be made to diffract.

Strange, but true: The diffraction of particles—a mystery of nature—has been exploited routinely, for decades, in the *electron microscope*. An important research tool for biology and materials science, this instrument uses an electron's short wavelength in place of visible light. Since a microscope can resolve objects no closer together than its illuminating wavelength, the angstrom-scale electronic wavelengths yield thousand-fold increases in magnification.

The wavelike diffraction of particles is a phenomenon unlike anything seen in the macroscopic world, where particles are particles and waves are waves. Thinking big, we accept without question that particles bounce and waves interfere. A ball simply drops into a pocket on the

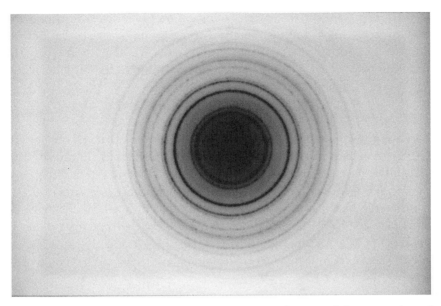

FIGURE 4-18. Electrons, like X rays, undergo interference while passing through a grid of atoms, as made clear in this electron diffraction pattern obtained from cubic zirconia. The alternating bands of light and dark beg the question: Are the electrons waves or particles? The fuzzy reality of quantum mechanics blurs the difference.

pool table, with no waviness and with no hint of interference. Now, abruptly forced to think small, we are unsure how even to distinguish a particle from a wave—if such a distinction matters at all.

Still, from the de Broglie formula

$$\lambda = \frac{h}{mv}$$

one can surmise why macroscopic objects move like ordinary particles: their wavelengths are nearly zero. Just consider that a kilogram-sized ball moving at 10 meters per second displays a wavelength of 6.63×10^{-35} m, a number splendid in its minuteness:

$$\lambda = 0.00000000000000000000000000000000663 \text{ m}$$

A distance so tiny, 25 powers of 10 smaller than a hydrogen atom, is beyond perception. No instrument can react to such a wavelength, and consequently the interference is never made visible. There are no 10^{-35}-m apertures through which a fist-sized ball can diffract.

We are guided, by this result, along another blended transition from small to large. Look for high quantum numbers and large sizes as before, but now look for a vanishingly small de Broglie wavelength as well. The transition back to everyday life proceeds according to the limit

$$\lambda \longrightarrow 0 \qquad \text{when} \qquad h \ll p \,,$$

thereby following a rule called the **correspondence principle**, a rule with far-reaching implications for quantum mechanics. Enunciated by Niels Bohr, the correspondence principle states that quantum effects and interference fade away whenever Planck's constant appears small relative to other physical quantities. Quantum mechanics will necessarily include factors of h in its equations, but all these expressions reduce to classical forms as the number 6.626×10^{-34} J s gradually loses importance. The h's disappear, and classical mechanics reappears. One world, after all.

4-6. THE IDEAS OF QUANTUM MECHANICS

Inexplicably stable atoms. Quantized energies and standing waves. Photons and electrons. The photoelectric effect and electron diffraction. Waves that deliver little packets of momentum and energy. Particles that diffract through crystals, apparently interfering with *something*—but who knows what. What is one to make of such things?

Quantum mechanics was born amidst this confusion, emerging first in the mid-1920s and developing rapidly over the next decade. The new mechanics was eventually to provide a mathematically consistent view of microscopic motion and energy, and it was to redefine our idea of a particle. The theory works for photons, electrons, atoms, molecules, semiconductors, superconductors, nuclei, and subnuclear particles. One of the enduring achievements of the 20th century, it is the basis of modern chemistry and physics.

Yet, deep down, nobody truly understands quantum mechanics. We use its equations confident of getting the correct answers, feeling certain about the "whats" but ever puzzled about the "hows." The quantum mechanical world gets stranger and stranger the harder one looks, until finally there is scarcely any handhold in ordinary reality. Readers encountering these ideas for the first time should appreciate and accept how discomforting that world will now seem.

It's also the only world we have.

Probability

Most of the weirdness of quantum mechanics stems from the ***wave–particle duality*** just described, by which light and matter exhibit properties of both wave and particle. Stripped bare, it is really the problem of diffraction and interference by *particles*. With what, we ask, can a particle interfere? How does a wavelike interference pattern develop when particles diffract through a pair of openings as in Figure 4-19?

More to the point, ask what happens when only *one* particle arrives at such a fork in the road. A water wave, remember, would simply spread itself over both openings and continue forward. Can a particle similarly split in two? Or must a lumpy photon or electron go through either one hole or the other? We wonder, naturally, what might emerge on the other side.

What emerges on the other side is assuredly a particle, a localized bundle of momentum and energy. Its arrival can be recorded on a piece of film or some other detector, and always the journey ends with a particle-like thud. The impact, whether made by photon or electron, is not spread over all space like a wave. Looking at the exposed film to the right of the barrier, an observer sees *one* spot at *one* location. One particle in and one particle out. There is not the faintest trace of a diffraction pattern. Electron or photon, the particle hits the film like a bullet.

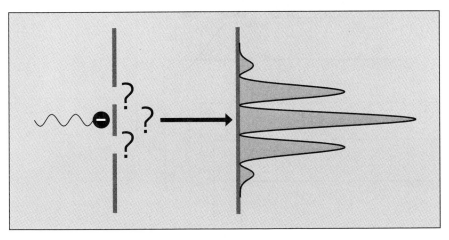

FIGURE 4-19. Quantum mechanical weirdness: How does an electron or photon (each a particle, in some sense) interfere with itself to produce a diffraction pattern? To do so, a lumpy particle would have to pass through both openings at the same time, as if it were a continuous water wave. But how can a particle be in two places at once? What happens when it reaches the two slits? Where does it go from there? Nobody knows.

Aim a second particle at the two openings, and again the film registers a distinct spot. And a third and a fourth and a fifth, and each time a new mark appears. Sometimes the particle lands where another has gone before; sometimes it falls in new territory. Bands of light and dark begin finally to develop on the film, and after repeated passages through the holes we eventually recognize the familiar diffraction pattern in Figure 4-20. It is a display of interference built up one grain at a time, formed from particles arriving seemingly at random.

Shooting an electron or photon through the openings is akin to throwing dice, for we cannot say with certainty where the particle will land. We have just the odds based on the final interference pattern. Some bands are more intense than others, and those are the locations

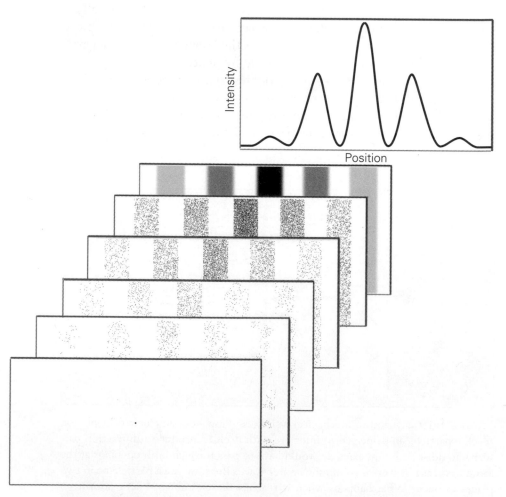

FIGURE 4-20. Strange but true: Shot through the holes one by one, as if in a time-lapse photograph, the electrons eventually do produce an interference pattern.

(a)

(b)

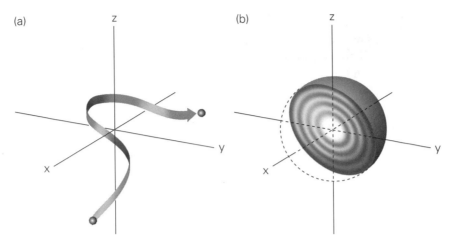

FIGURE 4-21. Two views of the world. (a) Classical mechanics: A particle, follow-ing a *path*, moves from point to point in predictable sequence. Its motion is deter-mined by the prevailing forces, and there is no doubt where it will go. (b) Quantum mechanics: no more paths, no more point-to-point trajectories, but a system of highly controlled guesswork instead. The classical path gives way to a statistical prob-ability—a map showing where the particle is likely (but not certain) to be found.

where the particle is more likely to be. We know where it *can* go, but not where it *will* go. Quantum mechanics, unlike classical mechanics, provides only a set of probabilities in lieu of a definite path (Figure 4-21).

Classical mechanics says: Now the body is here and one second later it will be there. The future is certain. Quantum mechanics says: There is a 10% chance it will be in this place, a 50% chance it will be in that place, and a 40% chance it will be in the other place. Make a measure-ment. Roll the dice.

And, quantum mechanics adds, if there are two options (like hole A and hole B), then expect to see evidence of interference. The film will record a diffraction pattern, generated one particle at a time.

When we see interference, we think of waves. We presume that some kind of amplitude (denote it here by ψ, the Greek letter psi) is os-cillating up and down. We remember the ripples in the water and the ripples in the electromagnetic field. We remember how waves from two sources, A and B, can combine amplitudes to become

$$\psi = \psi_A + \psi_B$$

We remember that the intensity, proportional to

$$(\psi_A + \psi_B)^2 = \psi_A^2 + \psi_B^2 + 2\psi_A\psi_B,$$

varies as the *square* of the amplitude (as ψ^2, not ψ), and that the combined quantity ψ^2 is not equal simply to $\psi_A^2 + \psi_B^2$. The whole differs from the sum of its parts. There is now an extra term, $2\psi_A\psi_B$, which creates the interference that renders $1 + 1$ unequal to 2. See it once more in Figure 4-22.

(a)

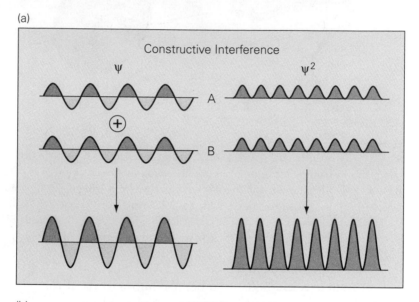

(b)

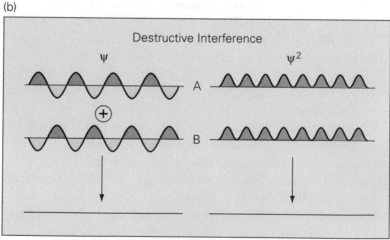

FIGURE 4-22. Constructive and destructive interference: a reminder. When wave amplitudes ψ_A and ψ_B combine, the joint intensity $(\psi_A + \psi_B)^2$ differs from the simple sum $\psi_A^2 + \psi_B^2$. It can be more (a), and it can be less (b); and from this interference of a wave function comes a diffraction pattern—true for water waves, light waves, photons, electrons. In quantum mechanics, the "intensity" ψ^2 gives the probability for finding the particle at a certain point.

Add two waves and square the sum . . . and suddenly there is interference. That, again, is what makes a wave a wave, and that is precisely the message of an electron diffraction pattern. The bands recorded on film (representing the probability of an electron's arrival point by point) suggest that some "wavelike" influence is detected as a squared quantity. Apparently one amplitude originates from hole A and another from hole B; ψ_A combines with ψ_B, and the detector responds to the quantity $(\psi_A + \psi_B)^2$.

We shall associate, accordingly, a *probability amplitude* with every particle, some suitable function ψ having a value at each position in space (as in Figure 4-23). This *wave function* bears all the characteristics of a wave amplitude and, most important, $\psi^2(x, y, z)$ communicates the likelihood of finding the particle at point (x, y, z). It is no more than a probability—a suggestion where to look, not a guarantee—but it is the best we can do.

So there is de Broglie's wave, identified finally as a probability amplitude. It is an odd kind of wave. By itself, ψ has no physical meaning and is not accessible experimentally. Carrying no energy, the wave function is measurable only as the *probability density* ψ^2. But wavelike it is, for one amplitude can combine with another to create interference. The diffracted electron, with an amplitude to be *simultaneously* at both hole A and hole B, has a wave function proportional to the combined amplitude

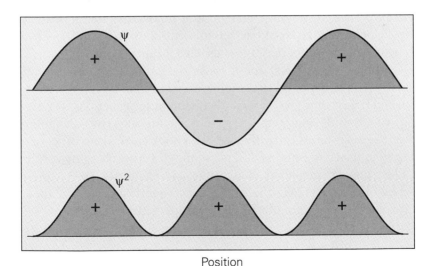

Position

FIGURE 4-23. An arbitrary example of a wave function (ψ) and its square (ψ^2), appropriate for a particle confined in one dimension—like a standing wave. The square of ψ, interpreted as a probability, is always positive, whereas the wave function itself can be either positive or negative.

$\psi_A + \psi_B$. Its probability density is proportional to $(\psi_A + \psi_B)^2$, and thence comes the interference—and there, with the baffling recognition that a particle can interfere with itself, is where quantum mechanics begins.

Indeterminacy

Honestly considered, the diffraction of a particle admits no ready explanation. If the electron definitely went through hole A, the wave function would be ψ_A and the probability distribution would be ψ_A^2. If the electron definitely went through hole B, the wave function would be ψ_B and the probability distribution would be ψ_B^2. The actual distribution, however, is not simply the sum of the two squares. There is interference.

Words and images fail to describe what happens. We cannot say that the particle "chose" either A or B, for then there would be no interference. Nor can we imagine that it split into two parts and wriggled through, because particles just do not do such things. An electron or photon is indivisible, yet somehow (in a manner impossible for us to picture) it seems to interact with both holes together. There is no easy way out of the confusion.

Suppose, though, we resolve to peek at the holes just as the particle begins to diffract. Surely we can see what happens at that critical moment and quickly clear up the mystery. Fine: Let us attempt to track an electron, perhaps by sending out a photon and waiting for an echo. Such a measuring technique would be reminiscent of ordinary radar, whereby a reflected electromagnetic wave is used to locate a target.

Think, for a moment, about radar detection. A flood of photons bounces off an aircraft and later returns, whereupon (given the photons' speed and transit time) we determine the point of reflection. The target, meanwhile, travels along unperturbed by the impact. The weak momentum of the radar beam is hardly likely to knock an airplane off course— no more so than a few grains of sand would disturb a charging elephant.

Electrons, far smaller than airplanes and elephants, are more easily knocked off course. Make the encounter as gentle as possible, then, by using only a single photon for the microscopic radar. This photon will be effective over a distance Δx approximately equal to its wavelength λ,

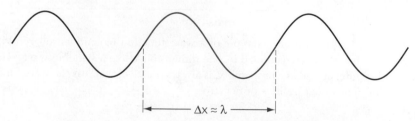

and the particle of light will hit the electron with momentum equal to h/λ:

$$\Delta p \approx \frac{h}{\lambda}$$

Jolted by the photon, the electron then changes momentum by some amount Δp, up to a maximum of h/λ.

A soft touch by one photon is the best compromise available. But even so, any measurement able to pin the electron down to one range of positions ($\Delta x = \lambda$) will irrevocably alter the momentum through an associated range ($\Delta p = h/\lambda$). Inevitably there are errors in both position and momentum such that

$$\Delta p \, \Delta x \gtrsim h$$

in rough accord* with the **indeterminacy principle** (or **uncertainty principle**) enunciated by Werner Heisenberg: The position and momentum of a particle cannot be measured simultaneously to unlimited accuracy. One quantity is traded against the other. If we adopt a short wavelength to decrease the uncertainty in x, then the uncertainty in p increases correspondingly. If, on the other hand, the measurement leaves the electron's momentum relatively undisturbed, then we must accept a fuzzy determination of position. This minimum product of the uncertainties, $\Delta x \, \Delta p$, is small but not zero; and no amount of technical improvement can ever make it less. It is a law of nature.

Constrained by the indeterminacy principle, we look at the electron emerging near the openings in Figure 4-24 and see (what else?) an electron. Our single-photon radar, employing a short wavelength, measures the position to high accuracy, and shows that each particle does indeed come through just *one* hole: upper or lower, not both at the same time. We pay for this careful determination of position, however, by forfeiting simultaneous knowledge of the electron's momentum. The energetic photon knocks the electron hopelessly off course, rendering any subsequent measurement of the momentum meaningless. Wherever the electron was going before, it is going there no longer.

The result: the two-hole interference pattern disappears. Having determined, for certain, that each electron went through one aperture or the other, we find that the detected intensity is frustratingly reduced to ψ_A^2 and ψ_B^2 separately. The interference is gone, destroyed by the act of measuring the position. A full diffraction pattern reappears only when we refrain from touching the electron at that most critical moment.

---

*A more thorough analysis, in which Δp and Δx are defined quantitatively, yields the inequality $\Delta p \, \Delta x \geq h/4\pi$. The underlying concepts remain the same.

(a)

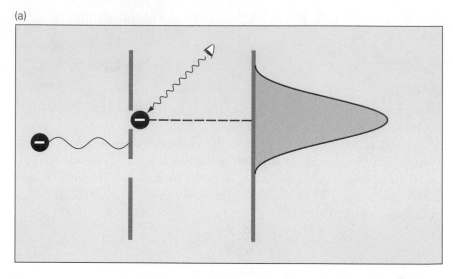

(b)

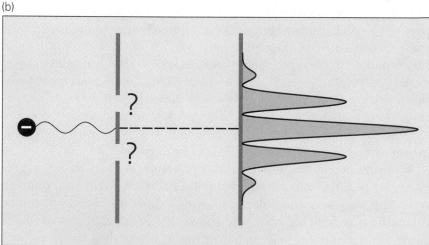

FIGURE 4-24. Curiosity kills the interference pattern. (a) Too precise a measurement of x (which hole?) renders measurement of p (what direction? what speed?) indeterminate. When the upper opening is illuminated, the diffracted intensity seems to arise from that source alone. (b) The double-slit diffraction pattern appears only when the electrons interact with both apertures undisturbed.

The indeterminacy principle is a fact of life in quantum mechanics, where it imposes limits on a guessing game of wave functions and probability. The microworld is a lightweight and jittery place in which things are easily disturbed, but at least now we know some of the rules. Forewarned, let us proceed in the next chapter to understand how quantum mechanics governs the formation of atoms and molecules.

REVIEW AND GUIDE TO PROBLEMS

Look to the sky and see a mechanical universe: a clockwork of stars and planets in orderly motion, wheels and gears turning, past and future all laid bare. We know where the earth was yesterday. We know where it is today. We know where it will be tomorrow.

Look into a single atom and see a universe too, but a universe in miniature: a quantum mechanical universe, a small world not nearly so certain. No longer can we say exactly where the electron goes; not yesterday, not today, not tomorrow. We know only the neighborhood it haunts. We know where the electron *should* be, where it is *likely* to be, where it *probably is*, but never with 100% certainty.

Kick a ball. It will move; it will accelerate in the direction kicked; it will obey the equation $F = ma$. Those are the laws of a classical mechanical universe.

Swing the earth around the sun, like a ball on a tether, and the force of gravity will hold the planet in a stable orbit. The earth can orbit forever in the vacuum of space, unencumbered by friction, retaining its original energy. Those are the laws of a classical mechanical universe.

Swing an electron around a proton, and let the electrical attraction do its work. Let a hydrogen atom take shape, as if the electron were a miniature earth orbiting a miniature sun.

Do it, and see how your hydrogen atom must collapse in a classical mechanical universe. There, proton and electron cannot stay apart forever. Like any other charged particle on a curved path, the classical electron will radiate energy and tire of the motion. The negative charge must spiral closer and closer to the positive nucleus, attracted more and more, drawn inward until finally the atom collapses. Those are the laws of a classical mechanical universe.

The atom has different laws to obey.

IN BRIEF: PARTICLES, WAVES, QUANTA

To justify the atom's existence, to make sense of the Lewis model, to explain the periodic table, to understand molecules, reactions, gases, liquids, solids—in short, to understand chemistry—we need to uncover the laws of the quantum universe.

Here, in the first of four related chapters, we began with the strange blurring of particles and waves in the atomic world.

1. A *particle* is a lump of energy and momentum, a lump found in one place at one time: here *or* there, one point or the other, not spread abroad. Be it bean or bullet, a particle has mass, charge, position, velocity, momentum, energy. It delivers its energy all at once, all in one place, the whole bundle with a localized ping or thud.

2. Not so a wave: A *wave* spreads through space, bringing force and energy to many points at the same time. The force comes not as a lump, not all at once, but as a recurring front that washes periodically over a shoreline. Delocalized over space and time, the disturbance keeps coming. The leading edge bobs up and down as the wave moves ahead.

3. *Light* is a wave, a wave of electromagnetic force. Traveling through space, light carries the force between charged particles. The electromagnetic field is, in the world of the atom, matter's messenger—the bringer of the force. It puts one charged particle in touch with another.

Stand still and let the field of an oncoming electromagnetic wave wash over an electron. Stand there, fixed in space, and measure the force imparted at that one point. The push-and-pull varies periodically with time, rising and falling in a repeating *cycle*: high one moment, low the next.

How high? Call the wave's largest displacement its *amplitude*, the strongest electromagnetic field available at any place and time. The field supplies force to a charged particle and thus delivers an electromagnetic energy: an energy proportional to (important) the *square* of the oscillating amplitude, not the amplitude itself. Up or down, positive or negative, the same height of wave delivers the same energy.

Next, count the number of electromagnetic waves cresting each second in this one place. The value obtained is the *frequency* of oscillation, ν, measured in units of *hertz* (1 Hz = one event per second, 1 s^{-1}). If 10^6 waves arrive every second, the frequency is 10^6 Hz. Each cycle is completed in $1/10^6$ s.

Now stop the motion. Freeze the wave and measure the spacing between ripples. From crest to crest is one *wavelength*, λ, the distance (in meters) over which the disturbance goes from high to low and back to high again: one full cycle.

Restart the wave. How fast are the wavefronts approaching? How many meters does a leading edge advance in one second? If some particular crest is here—now—where will it be one second later?

Observe: The crest moves a distance of one wavelength (λ) every time a cycle is completed. Since ν cycles are executed each second, the leading edge evidently moves forward at the rate of $\lambda\nu$, wavelength times frequency. Deduce from that the wave's *speed*, c, which follows as

$$\lambda\nu = c$$

The value $c = 3.00 \times 10^8$ m s^{-1} is the speed of light traveling in vacuum.

Light propagates at different frequencies and with different wavelengths, but always with the same speed in vacuum. Force is force; light is light. The same electromagnetic force is carried at the same speed, whether the frequency is low or high, the wavelength large or small. All wavelengths are possible, from radio waves (meters and kilometers and more) to X rays and beyond (angstroms and less). Radio waves, microwaves, infrared, visible, ultraviolet radiation, X rays, gamma rays— together they make the full rainbow of light, the *electromagnetic spectrum*.

4. Light, a wave, undergoes *interference*. Waves A and B meet, add their amplitudes, and build a new wave with combined amplitude $A + B$. But now the energy is proportional to

$$(A + B)^2 = A^2 + B^2 + 2AB$$

rather than simply $A^2 + B^2$. The difference is in the interference term $2AB$, which will be positive, negative, or zero depending on how the waves meet. *In phase* (crest to crest), the interference is *constructive*; combined, the two waves deliver more energy than they can separately. *Out of phase* (crest to trough), the interference is *destructive*; the joint energy is less. Alternations in intensity create an interference pattern (equivalently, a *diffraction pattern*) as the disturbances move in and out of step.

It is the signature of a wave. Not merely the up-and-down motion, but the ability to interfere is what distinguishes a wave from a classical particle. Particles keep to themselves. Waves blend. Waves spread. Waves interfere.

5. Thus was the world divided, particle and wave, until discoveries such as the *photoelectric effect* began to fuzz the boundary between the two. Light is a wave, we know. It has a wavelength, a frequency, a speed, a phase, a polarization, an amplitude, all the properties of a wave. It extends over space and varies with time. It turns corners; it casts shadows; it spreads out over obstacles. It diffracts. It interferes. For all these reasons, light is a wave; but sometimes, despite the waviness, light acts as a particle too. Sometimes, electromagnetic radiation seems to deliver a localized punch as pinglike as any particle.

Because it *is* a particle. Ultimately, the electromagnetic field *is* particulate, just like matter. Just like matter, light is fine-grained but not continuous; just like matter, light has its building blocks. Individual lumps of energy and momentum (*photons*) build up the field grain by grain; and, although limited by the size of *Planck's constant* ($h = 6.63 \times 10^{-34}$ J s), the small bundles are distinct as beans:

Energy: $E = h\nu$

Momentum: $p = h/\lambda$

One by one, photon by photon, the electromagnetic field fills its space like so many particles, even though large numbers create the illusion of a seamless web of light. Not until we address individual packets does the illusion disappear and the grain in the pattern reappear—just like matter, just like atoms, just like electrons.

6. But now we have matter itself to reconsider, for there is still another discovery: *particles* are not as grainy as we thought. Electrons, despite their lumpiness, suffer interference under certain conditions just as a wave would. Shot like bullets through, say, the fine latticework of a crystal, a beam of electrons will produce a diffraction pattern.

So what do we have, a wave-and-or-particle? A wavicle? Whatever it is, we have this one fact to face: What once was a particle, a hard kernel of energy and momentum, now seems to interfere like a wave; and, like a wave, the electron also has a characteristic dimension, the ***de Broglie wavelength***:

$$\lambda = \frac{h}{p} = \frac{h}{mv}$$

Like an unbound wave, too, a *free* electron—an electron left alone, an electron unconfined—can extend its presence over all space. It is a particle, with mass and momentum; yet somehow the free electron is a wave as well, with wavelength and infinite extension.

7. And, like a wave, it can be trapped. Like a wave clamped on a string, an electron bound into an atom can lose its freedom. Confined, pinned down, kept near a nucleus, the electron's de Broglie wave develops **nodes**: fixed points of zero amplitude. The once free wave becomes a ***standing wave***, retracing the same pattern back and forth. So it is for violin strings, and so it is for confined electrons.

Only certain standing-wave combinations are allowed, because only those arrangements with integral numbers of half-wavelengths can persist. A confined wave is therefore a *quantized* wave, trapped into a stationary pattern with discrete frequency and a fixed number of nodes. As if ascending and descending a ladder, a confined de Broglie wave can go only from one step to another, needing just the right energy (not too much and not too little) to bridge the gap. Similar to light, matter's energy is dished out in discrete bits called **quanta**.

We find, for atoms and molecules, a ladder of quantized states (E_1, E_2, . . . , E_∞); and to negotiate these steps we must give or take a photon with matching energy:

$$\Delta E = E_i - E_j = h\nu$$

To *absorb* a photon is to go up the ladder; to *emit* a photon is to go down. The systematic probing of energy levels is called **spectroscopy**.

8. Interference, quantization, the wave–particle duality—from observations of the small world one discovers the new **quantum mechanics**: the laws for particles that interfere like waves; for particles that allow themselves only certain energies; for particles that refuse to reveal their position and path.

Indeterminacy hovers over the quantum world. If we measure the electron's position within a range Δx, then any measurement of its momentum is burdened by a minimum uncertainty Δp (equivalent to $m\Delta v$) such that

$$\Delta p \, \Delta x \gtrsim h$$

The smaller the error in position, the larger the error in momentum. If we know exactly where the particle is, we have no idea how fast and in what direction it is going. To know x is to be ignorant of p, and to know p is to be ignorant of x. So says Heisenberg's **uncertainty principle (indeterminacy principle)**.

9. No longer do we give coordinates and velocity to a particle in the quantum world; we assign, instead, a **wave function** ψ that extends through space. It is a mathematical function that can travel, stand, oscillate, and interfere just like a wave. Yet with all these characteristics, ψ somehow lacks the physical presence of a water wave or an electromagnetic wave. It is an object like none we have seen so far, something abstract and intangible; ψ is a quantum mechanical **probability amplitude**.

A probability amplitude, to be treated as follows: Evaluate the *square* of the wave function, ψ^2, at any point in space, and behold: a number, a number that reflects the statistical probability of finding the electron at that point.

In the next chapters we shall apply these new ideas first to the hydrogen atom, then to all the atoms of the periodic table, and finally to molecules. But before moving on, let us work through an assortment of problems concerning waves, particles, and quanta.

SAMPLE PROBLEMS

EXAMPLE 4-1. Light: Wavelength and Frequency

PROBLEM: Imagine an atom in the grip of an electromagnetic wave, positioned so that the field reaches a maximum at the center of the nucleus at some instant (see drawing). Describe, very roughly, the variation of the field from the atom's center to edge under the following four frequencies: (a) 300 MHz. (b) 1 GHz. (c) 5×10^{14} Hz. (d) 3×10^{18} Hz.

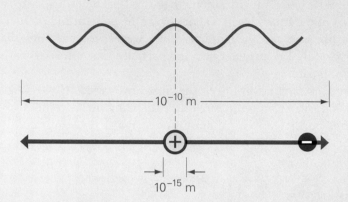

SOLUTION: Let our typical atom have a diameter of 1 Å (10^{-10} m), approximately the size of hydrogen. Taking this number as a standard, we then compare it to the electromagnetic wavelength. To do so, we use the relationship between frequency, wavelength, and speed: $\lambda \nu = c$.

(a) *300 MHz.* Light in this range falls into the *radiofrequency* portion of the electromagnetic spectrum, where wavelengths are meters and more. For 300 MHz (1 MHz = 10^6 s^{-1}), the corresponding wavelength is 1 m:

$$\lambda = \frac{c}{\nu} = \frac{3 \times 10^8 \text{ m s}^{-1}}{3 \times 10^2 \text{ MHz}} \times \frac{1 \text{ MHz}}{10^6 \text{ s}^{-1}} = 1 \text{ m}$$

Picture, then, a dot of electric charge 10^{-10} m in diameter (the atom) set within a field that stretches out to 1 m. The atom occupies a mere one part in ten billion, a speck amidst infinity, a speck so minute that the electromagnetic field stays effectively constant over the entire particle. We would need 10,000,000,000 such atoms all in a row to sample the disturbance over just one wavelength.

(b) *1 GHz* (=1000 MHz = 10^9 s^{-1}). At 0.3 m, this *microwave* wavelength still far exceeds the atomic diameter. The instantaneous field is constant over the atom, as before.

(c) 5×10^{14} *s*$^{-1}$. The frequency corresponds to a wavelength of 6×10^{-7} m (6000 Å), placing λ in the orange region of the *visible* spectrum. With dimension 6000 times greater than the atom, the light wave again remains constant over the small distance.

(d) 3×10^{18} *s*$^{-1}$. Finally, an atom-sized wavelength: $\lambda = 10^{-10}$ m (1 Å). The light falls within the *X-ray* portion of the spectrum, energetic enough to execute a complete cycle between one end of the atom and the other.

EXAMPLE 4-2. Light: Speed, Distance, Time

PROBLEM: How much time elapses when light travels the following distances? (a) Across a hydrogen atom. (b) Down the length of a football field. (c) From the earth to a communications satellite and back (50,000 miles). (d) From the sun to the earth (93,000,000 miles).

SOLUTION: Make repeated use of the relationship

$$\text{Time} = \frac{\text{distance (m)}}{\text{speed (m/s)}},$$

being careful to maintain a consistent set of units. To go from miles to meters in (c) and (d), for example, we use the conversion factor

$$1 \text{ mi} \times \frac{5280 \text{ ft}}{\text{mi}} \times \frac{12 \text{ in}}{\text{ft}} \times \frac{2.54 \text{ cm}}{\text{in}} \times \frac{1 \text{ m}}{100 \text{ cm}} = 1.609 \times 10^3 \text{ m}$$

to obtain the metric distances:

$$5.0 \times 10^4 \text{ mi} \times \frac{1.609 \times 10^3 \text{ m}}{\text{mi}} = 8.0 \times 10^7 \text{ m} \qquad \text{(satellite roundtrip)}$$

$$9.3 \times 10^7 \text{ mi} \times \frac{1.609 \times 10^3 \text{ m}}{\text{mi}} = 1.5 \times 10^{11} \text{ m} \qquad \text{(sun)}$$

For the atom, we shall assume 10^{-10} m; for the football field, 100 m.

(a) *Hydrogen atom*. With $c = 3.00 \times 10^8$ m s^{-1} (fast) and the distance only 10^{-10} m (small), the time is fantastically short: 3.3×10^{-19} s.

$$1.0 \times 10^{-10} \text{ m} \times \frac{1 \text{ s}}{3.00 \times 10^8 \text{ m}} = 3.3 \times 10^{-19} \text{ s}$$

Faster than even the swiftest chemical processes, light's trip across the atom takes almost "no time" at all.

(b) *Football field*. Here is a familiar distance, typical of the dimensions we find in the macroscopic world. Traversing the equivalent of 1,000,000,000,000 hydrogen atoms, the light covers the football field in only 3.3×10^{-7} s—less than one-millionth of a second, fast enough to seem instantaneous to quarterback and receiver.

(c) *Satellite bounce*: 0.27 s, clearly a measurable interval. The lag can produce an audible echo during a telephone call transmitted by satellite.

(d) *From the sun*: 500 s. Sunlight travels more than eight minutes before reaching the earth.

EXAMPLE 4-3. Interference: Peaks and Valleys

Remember the construction we used to understand interference, reproduced now in the following diagram:

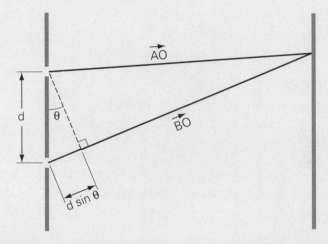

Arriving at the two openings, an electromagnetic wave splits into two outgoing signals that come together some distance to the right. Where they recombine crest to crest, the intensity is maximum; where they meet crest to trough, the intensity is minimum. The resulting interference creates a *diffraction pattern*, a series of alternating peaks and valleys in the intensity.

PROBLEM: Suppose the first maximum in a two-slit diffraction pattern corresponds to $\theta = 10°$. What is the spacing between the openings, *d*, if the wavelength is 1.00 Å?

SOLUTION: One wave travels a distance *AO*, and the other wave travels a distance *BO*. For constructive interference to develop, the two paths must differ by an integral number of wavelengths (*n*):

$$AO - BO = d \sin \theta = n\lambda \qquad (n = 1, 2, \ldots)$$

When they do, the diffracting waves recombine *in phase*.

We therefore set $n = 1$ (first maximum) and solve for d:

$$d = \frac{\lambda}{\sin \theta} = \frac{1.00 \text{ Å}}{\sin(10°)} = 5.76 \text{ Å}$$

Think of the implications. If somehow we could find a set of openings spaced between 1 and 10 Å, . . . and if we could produce electromagnetic radiation with a wavelength of approximately 1 Å, . . . and if we could measure a diffraction pattern, . . . and if we could determine θ and n for each peak, . . . and if we could analyze the many instances of interference, . . . then, knowing all that, we might be able to deduce the layout of the openings themselves.

We can do it, too, because: (1) Nature provides a ready-made supply of slits and openings: crystals, regular arrays of atoms spaced at distances of a few angstroms. Around each atom is a cloud of electrons able to receive and retransmit electromagnetic radiation. (2) X rays fall conveniently in the range needed to generate a diffraction pattern from a crystal. (3) With a suitably constructed instrument (an X-ray diffractometer), we can indeed produce and analyze such a pattern. The analysis is not quite so simple as suggested above, but still the idea is the same: Constructive interference arises only when two waves combine crest to crest. Using the diffraction pattern we can locate the atoms one by one, thereby mapping out the microscopic structure of the crystal—a remarkable accomplishment.

EXAMPLE 4-4. Luminous Billiard Balls

PROBLEM: Hit by an X-ray photon, a free electron starts to move. What happens to the photon? Does its wavelength increase, decrease, or remain the same?

SOLUTION: Diffracting through a crystal, a beam of light interferes like a wave. Pinging against an electron, a single photon recoils like a particle. Which is it? Wave? Particle? Wave and particle together?

Better to say: Light is put together with particle-like photons that blend together in a crowd to simulate a wave. One experiment may reveal one aspect; another experiment, a different aspect.

Here, apparently, the events suggest more the interactions of particles than waves, something like a collision of two billiard balls. A photon with momentum h/λ and energy $h\nu$ strikes an electron, delivering a blow in a given direction. The recoiling electron absorbs part of this energy and momentum while declining the rest, and (with the absorbed E and p) it accelerates to velocity v and momentum mv. The photon, meanwhile, bounces away with *less* energy and *less* momentum than before:

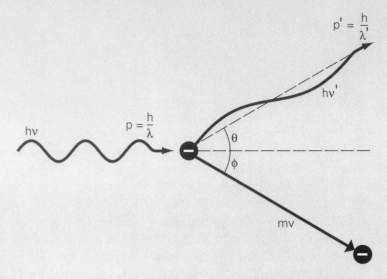

The result? Nothing is lost and nothing is gained. Energy is conserved. Momentum is conserved. The electron takes some of each but returns the rest. What the electron gains, the photon loses.

The scattered photon now has a longer wavelength than before, with its final energy and momentum both inversely proportional to this new value λ'. The longer the wavelength, the lower the energy and momentum.

Called the *Compton effect*, such a change in wavelength proves that a photon carries a quantized momentum h/λ in addition to its quantized energy $h\nu$. The discovery, coming in 1923, was timely. Not only did the Compton effect further establish the particulate properties of light, it also provided important evidence for the new quantum mechanics.

EXAMPLE 4-5. Photons and Photoelectrons

Once more, light in its particle-like incarnation . . .

PROBLEM: The work function of metallic cesium is 3.43×10^{-19} J. Calculate the velocity of the photoelectron produced by a photon of wavelength 525 nm.

SOLUTION: First we compute the photon's energy using the relationship

$$E = h\nu = \frac{hc}{\lambda} = \frac{(6.626 \times 10^{-34} \text{ J s})(2.998 \times 10^8 \text{ m s}^{-1})}{525 \times 10^{-9} \text{ m}}$$

$$= 3.78 \times 10^{-19} \text{ J}$$

If this energy were less than the work function ($E_0 = 3.43 \times 10^{-19}$ J), there would be no photoelectron. With E greater than E_0, however, we expect the ejected electron to have an excess kinetic energy

$$E_k = \frac{1}{2}mv^2 = E - E_0 = (3.78 - 3.43) \times 10^{-19} \text{ J} = 3.5 \times 10^{-20} \text{ J}$$

and a corresponding velocity

$$v = \sqrt{\frac{2E_k}{m}} = \sqrt{\frac{2 \times 3.5 \times 10^{-20} \text{ kg m}^2 \text{ s}^{-2}}{9.109 \times 10^{-31} \text{ kg}}}$$

$$= 2.8 \times 10^5 \text{ m s}^{-1}$$

The calculation amounts to a straightforward enforcement of energy conservation.

Note, incidentally, that only two of the three key quantities (photon energy, work function, kinetic energy) are independent, for we can always deduce the third from the other two. Measurement of the energies of both photon and photoelectron, for example, leads us directly to the work function. And, with that, we have a way to measure the binding energy of the electron in the metal—or, more generally, in any other structure where a photoelectron can be produced. From this idea comes an important experimental technique called *photoelectron spectroscopy* (Chapter 20).

EXAMPLE 4-6. De Broglie Wavelength

PROBLEM: Compute the de Broglie wavelength for the photoelectron described in Example 4-5.

SOLUTION: Unbound, the liberated electron has a de Broglie wavelength

$$\lambda = \frac{h}{mv} = \frac{6.626 \times 10^{-34} \text{ kg m}^2 \text{ s}^{-1}}{(9.109 \times 10^{-31} \text{ kg})(2.8 \times 10^5 \text{ m s}^{-1})}$$

$$= 2.6 \times 10^{-9} \text{ m}$$

Think about it. Is the wavelength comparable to the diameter of a hydrogen atom? Would electrons at this velocity produce a crystalline diffraction pattern?

EXAMPLE 4-7. It Went That Way (or Did It?)

PROBLEM: An electron passes through a vertical slit of length d, arranged as shown in the accompanying illustration. In what direction and with what momentum will it emerge if: (a) $d = 1.00 \times 10^{-11}$ m? (b) $d = 1.00$ m?

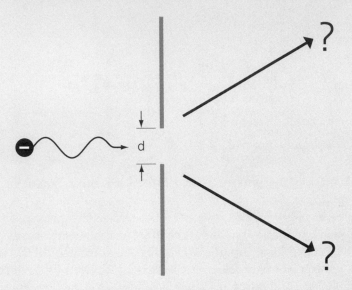

SOLUTION: Who knows? We can only estimate a range of possible momenta based on the uncertainty principle.

Exactly where we cannot say, but *somewhere* in that slit the electron must be found. Passing through the aperture, the particle will be pinned down at least that much; it will be confined at least to the region delineated by d. Consequently the uncertainty in position is d, the length of the opening in the vertical direction.

If so, then Heisenberg's indeterminacy principle tells us that the particle's vertical momentum (up and down the slit) is uncertain by roughly

$$\Delta p \sim \frac{h}{d}$$

When d is small, the electron's whereabouts are well defined; we know precisely where it is. At the same time, though, the range of allowable momenta in the vertical direction ($\sim h/d$) is correspondingly high. The particle might bounce up with a momentum h/d . . . or bounce down with a momentum $-h/d$ . . . or pass straight through with zero vertical momentum . . . or else take any route in between. Which way does it go on *this* trip? We simply do not know.

We know only that something unknowable happens. Something changes when electron or photon arrives at a narrow opening. The particle interacts with the slit abruptly and unpredictably, suffering a different change in direction each time. It goes this way and that, landing in some places more than others, scattering like buckshot over some angle θ. Eventually the random experiences of many particles condense into a highly predictable diffraction pattern.

(a) A *narrow slit* produces a wide diffraction pattern. The uncertainty in position is small, but the uncertainty in direction and speed is large. We know where the electron is, but not where it is going. Such are the circumstances for the 0.100 Å aperture, where the uncertainty in momentum is approximately

$$\Delta p = \frac{h}{d} = \frac{6.63 \times 10^{-34} \text{ kg m}^2 \text{ s}^{-1}}{1.00 \times 10^{-11} \text{ m}} = 6.63 \times 10^{-23} \text{ kg m s}^{-1}$$

The spread in p is large relative to the momentum a lightweight electron ($m = 9.11 \times 10^{-31}$ kg) typically can acquire.

(b) A *wide slit* (1.00 m) produces a narrow diffraction pattern. Here the uncertainty in position is enormous, but the uncertainty in momentum has been reduced from its value in (a) by 11 orders of magnitude: $\Delta p = 6.63 \times 10^{-34}$ kg m s^{-1}. The spread in velocity follows as $\Delta v = \Delta p / m = 7.28 \times 10^{-4}$ m s^{-1}.

Take your choice: Fix the position and forgo the momentum, or forfeit knowledge of position in exchange for knowledge of momentum. Such is the law of a quantum mechanical universe.

EXERCISES

1. Consider the relationship between a wave's *period* (the time taken to complete a cycle) and its *frequency* (the number of cycles completed per unit time). If, for example, a wave goes through 1 full cycle in 1 s (period = 1 second per cycle), the frequency is 1 s^{-1} (1 cycle per second). Now answer: What frequency corresponds to a period of 0.1 s? 0.01 s? 0.001 s? 10 s? 100 s? 1000 s?

2. What period corresponds to a wave frequency of 100 Hz? 1 kHz? 1 MHz? 1 GHz?

3. Monitoring a single point in space, an observer counts 7200 wave crests during the course of an hour. (a) What is the average frequency of oscillation, expressed in hertz? (b) What is the period, expressed in seconds?

4. An observer, measuring a wave at a single instant in time, notes that 101 successive crests are spread over a distance of 20 cm:

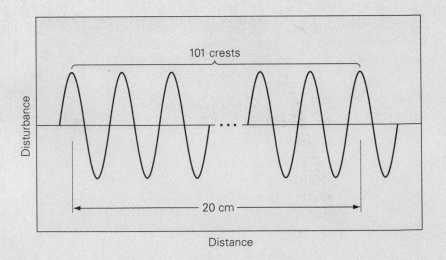

What is the average wavelength, expressed in meters?

5. A certain wave advances at a speed of 100 m s^{-1}, each trough separated by 2 m. Calculate the frequency and period of oscillation.

6. Suppose that 1000 wave crests arrive at a given point during 1 hour, each crest separated in space by 10 meters. What is the speed of the waves?

7. Below is a representation of a wave: an oscillating *disturbance*, a quantity of some sort moving up and down, up and down, varying regularly in space and time:

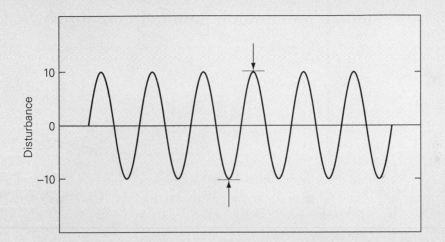

   Noting that the disturbance rises and falls by 20 units between trough and crest, draw a similar curve that shows the *energy* carried by the wave.

8. Using curves like the one in the exercise above, draw two waves that are *in phase*. Let each wave vary by 20 units from crest to trough, and let each wave oscillate with the same frequency and wavelength. Then: (a) Add the two disturbances together, drawing the combined wave and also the pattern of energy it carries. (b) What is the maximum energy delivered? (c) At what points of the wave is this maximum reached? (d) Where is the energy at a minimum?

9. Repeat the exercise, this time for two otherwise identical waves that are *out of phase* by exactly half a cycle: 20 units crest to trough, same frequency, same wavelength, same speed, but a phase difference of 180°. Again: (a) Add the two disturbances together, drawing the combined wave and also the pattern of energy it carries. (b) What is the maximum energy delivered? (c) At what points of the wave is this maximum reached? (d) Where is the energy at a minimum?

10. Refer to the diagram below, a representation of waves emerging from two narrow openings in a screen:

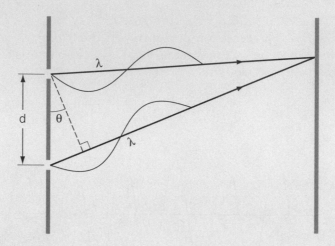

    Assume that the combined intensity is maximum when the angle θ is 30° and when d, the spacing between the slits, is 1.0×10^{-6} m. If so, what is the wavelength λ of the oscillation?

11. Estimate the slit-to-slit spacing, d, that will produce a diffraction maximum at $\theta = 20°$ for each of the following wavelengths:

    (a) 1 mm   (b) 1 μm   (c) 1 nm

12. Focus now on electromagnetic waves, describing briefly the *kind* of disturbance, the *speed* of the wave, and the range of possible frequencies and wavelengths.

13. For each of the frequencies stated, compute the distance traveled in vacuum by an electromagnetic wave propagating for 1.000 s:

    (a) 328 s^{-1} (b) 8.02×10^4 Hz   (c) 75 MHz
    (d) 9.7 GHz   (e) 12.8 kHz   (f) 60 Hz

14. For each of the electromagnetic frequencies stated, compute the corresponding wavelength and assign it to the appropriate portion of the spectrum:

    (a) 7.67×10^{15} Hz   (b) 770 kHz
    (c) 317 MHz   (d) 1.03 GHz
    (e) 5.00×10^{14} Hz   (f) 8.02×10^{13} Hz

15. For each of the electromagnetic wavelengths stated, compute the corresponding frequency and assign it to the appropriate portion of the spectrum:

    (a) 0.761 Å   (b) 4.01 mm   (c) 115 m
    (d) 5617 Å   (e) 218 nm   (f) 1.06 μm

16. For each of the electromagnetic frequencies stated, compute the corresponding photon energy and assign it to the appropriate portion of the spectrum:

    (a) 1 MHz   (b) 100 MHz   (c) 1 GHz
    (d) 10^{12} Hz   (e) 10^{15} Hz   (f) 10^{18} Hz

17. For each of the electromagnetic wavelengths stated, compute the corresponding photon energy and assign it to the appropriate portion of the spectrum:

    (a) 1 m   (b) 1 mm   (c) 1 μm
    (d) 1 nm   (e) 1 pm   (f) 1 fm

    Express the values both as joules per photon and as kilojoules per mole.

18. For each of the photon energies stated, compute the corresponding frequency, wavelength, and position in the electromagnetic spectrum:

    (a) 10^{-26} J   (b) 10^{-25} J   (c) 10^{-21} J
    (d) 10^{-20} J   (e) 10^{-19} J   (f) 10^{-16} J

19. A 60-watt source radiates 60 joules of electromagnetic energy per second. (a) How much energy is produced in 1 hour by a 60-watt bulb emitting light with a wavelength of 5000 Å? (b) How many photons is that?

20. How many photons are needed to supply 1 joule at each of the following wavelengths?

    (a) 1 km   (b) 1 m   (c) 1 mm   (d) 1 μm
    (e) 1 nm   (f) 1 pm   (g) 1 fm

21. The work function of metallic cesium is 3.43×10^{-19} J. What is the longest wavelength sufficient to eject a photoelectron?

22. Potassium has a work function equal to 3.69×10^{-19} J. How many photoelectrons, in principle, will be produced under each of the following illuminations? (a) Light with a wavelength of 538 nm delivers a total energy of 3.69×10^{-16} J. (b) Light with a wavelength of 500 nm delivers a total energy of 3.97×10^{-16} J. (c) Light with a wavelength of 600 nm delivers a total energy of 3.31×10^{-14} J.

23. Suppose that sodium, which has a work function equal to 4.41×10^{-19} J, is illuminated by light with a wavelength of 4390 Å. (a) With how much kinetic energy, momentum, and velocity will each photoelectron emerge? The mass of an electron is 9.11×10^{-31} kg. (b) How many such photoelectrons can be produced by a burst of light with a total energy of 5.00 J?

24. Again, try to produce a photocurrent from a piece of sodium (work function $= 4.41 \times 10^{-19}$ J)—this time, though, with light having a frequency of 6.63×10^{13} Hz. How many photoelectrons will be produced by an irradiation of 5000 J? Explain.

25. The mass of an electron is 9.11×10^{-31} kg. Compute the de Broglie wavelength of a photoelectron ejected from manganese (work function $= 6.6 \times 10^{-19}$ J) by one photon at each of the following wavelengths:

    (a) 3.0×10^{-7} m   (b) 2.8×10^{-7} m   (c) 2.5×10^{-7} m

    Explain the trend.

26. With what momentum must an electron move to have a de Broglie wavelength equal to 1 Å? What is the corresponding velocity?

27. The mass of a proton is 1.67×10^{-27} kg. With what momentum must a proton move to have a de Broglie wavelength equal to 1 Å? What is the corresponding velocity?

28. Starting from rest (zero velocity), an electron is accelerated through a voltage of 10^4 volts (1 V = 1 J C^{-1}). What velocity does the electron finally attain, and what is the resulting de Broglie wavelength?

29. Same experiment, but now with a proton: A proton is accelerated from rest through a potential difference of 10^4 volts. What is its final velocity and de Broglie wavelength?

30. Compute the characteristic wavelength for each of the following *macroscopic* systems:

(a) a 70-kg sprinter moving at 10 m s^{-1}
(b) a 10-g bullet fired at 250 m s^{-1}
(c) a 200-g ball thrown at 100 miles per hour
(d) a 4000-lb car traveling at 60 miles per hour.

31. Show that this statement is wrong:

    "According to the Heisenberg uncertainty principle, one can never know *either* the position or momentum of a particle to arbitrary accuracy."

    Indeed we can. Assume first, for the sake of argument, that we use technically perfect instruments, devices able to measure position and momentum to an infinite number of significant digits. Then, after restating the principle correctly, demonstrate a way to determine an electron's momentum with virtually no uncertainty. For help, go on to the next problem.

32. Suppose that, relative to some axis x, the position of an electron is known to fall somewhere between points x_1 and x_2, as below:

    Estimate the minimum uncertainty in the particle's momentum and velocity over each of the following intervals, assuming a mass of 9.11×10^{-31} kg:

    (a) $x_1 = 1.000$ m, $x_2 = 1.001$ m
    (b) $x_1 = -1.001$ m, $x_2 = -1.000$ m
    (c) $x_1 = 0$, $x_2 = 1$ mm
    (d) $x_1 = 0$, $x_2 = 1$ μm
    (e) $x_1 = 0$, $x_2 = 1$ nm

    Comment on the trend, and answer: When the uncertainty in momentum is, for practical purposes, so small as to be *zero*, is the indeterminacy principle violated in any way? Why must Δp be the same for (a), (b), and (c)?

33. Let a free electron (mass = 9.11×10^{-31} kg) pass through an opening of the kind described just above. If the particle's kinetic energy is 1.6×10^{-16} J before the encounter, what is its allowable range of momentum and velocity *after* becoming localized within each of the following intervals?

    (a) $\Delta x = 1$ mm   (b) $\Delta x = 1$ μm   (c) $\Delta x = 1$ nm

    Use the values of Δp already computed in the previous exercise.

34. Do the same now for a particle with greater mass: Take a proton (mass = 1.67×10^{-27} kg) with kinetic energy equal to 1.6×10^{-16} J, as before, and compute its allowable range of momentum and velocity once the position is pinned down to the same intervals:

    (a) $\Delta x = 1$ mm   (b) $\Delta x = 1$ µm   (c) $\Delta x = 1$ nm

    Compare the results for proton and electron. What stays the same? What changes?

35. Repeat the foregoing calculations for systems with even greater mass:

    (a) a helium atom, $m = 6.65 \times 10^{-27}$ kg
    (b) a nitrogen molecule, $m = 4.65 \times 10^{-26}$ kg
    (c) a sucrose molecule ($C_{12}H_{22}O_{11}$), $m = 5.68 \times 10^{-25}$ kg

36. Does the uncertainty principle impair measurements of macroscopic systems? Suppose that an observer uses infrared light (wavelength near 1 µm, say) to locate the position of a thrown ball. (a) What is the resulting uncertainty in velocity if the ball has a mass of 100 g? (b) If the measured velocity is 30 m s^{-1}, by what percentage is the value in doubt owing to disturbance of the ball's motion by the photon?

37. How is our image of the "size" of a particle affected by the entanglement of Δp and Δx? Picture a free electron, unbound, moving unimpeded in a universe all by itself—and moving, moreover, in a straight line with a constant, invariable, absolutely certain velocity and momentum. *Where* is it? We know the speed; we know the direction; but do we know anything about the position?

38. Same idea, but consider instead the other extreme: an electron confined to an atom, kept within a few angstroms of a nucleus. What can we say about the electron's momentum in such an arrangement?

5

Quantum Theory of the Hydrogen Atom

5-1. The Atom
5-2. The Wave Equation
5-3. Orbitals of a One-Electron Atom
Three Dimensions, Three Quantum Numbers
Sorting Out the Orbitals
Energy
The s Orbitals: Size and Shape
The p Orbitals: Shape
The p Orbitals: Orientation
The d Orbitals
5-4. Atomic Spectroscopy
5-5. A Fourth Quantum Number
REVIEW AND GUIDE TO PROBLEMS
EXERCISES

5-1. THE ATOM

We have now to consider the mechanics of the atom: the conditions of an atom's existence, its energies, what holds it together. The framework will be provided by quantum mechanics.

The quantum theory blends together our previously separate notions of particle and wave. An electron, a lump of matter that can diffract like a wave, must be treated as if it *were* a wave. It will be described by a ***wave function*** ψ and corresponding ***probability density*** ψ^2, from which we shall guess the particle's whereabouts from moment to moment.

Were the electron as big as a billiard ball, there would be no guess-work. A macroscopic object would unerringly follow the one (and only one) path laid down by Newton's laws, and we could always measure where it is, where it is going, and how fast it is moving at each instant. Now here, now there. We would see a classical particle, a lump with a path.

A quantum mechanical particle, by contrast, has no predetermined path, no single and unchangeable way to move when prodded by a force. And even if it did, the indeterminacy principle would prevent us from ever measuring the route. Remember that any measurement of position exacts a price, for to locate an electron is to change its momentum and thus to knock it off course. We shall make do instead with the wave function, seeking the electron within some region of space without ever specifying an exact address. It will be an electron that is maybe here, maybe there . . . ever changing, never certain.

Somewhat fuzzy around the edges, the quantum mechanical atom emerges nevertheless as a substantial and stable structure. It exists. It may flicker a bit, but still the atom persists. Like the blur of a propeller at high speed, an indeterminate cloud of electrons gives the atom a surprisingly solid aspect. Atom and molecule alike thereby acquire size, shape, and permanence within the constraints of probability and uncertainty. Probability and uncertainty, yes; total anarchy, no.

From the regularity of atomic structure comes the regularity of the periodic table and the regularity of chemical combination. It all begins with the quantum mechanical wave equation and its solutions.

5-2. THE WAVE EQUATION

An archer draws a bowstring and struggles to maintain a position. The taut line, disturbed from rest, pulls back in a direction opposite to the archer's draw. It resists with a strength derived from the elasticity, thickness, and tension of the string and bow.

Fighting against the archer's finger are forces seeking to restore the status quo. Drawn back, the string is held at a point of high potential energy. It can move. Given the opportunity, it *will* move. Once released and abandoned to the restoring force, the bowstring hurtles forward and converts potential energy into kinetic. Overshooting first one way and then the other, the string is jerked back and forth by the mechanical force. There is a whizzing sound, an audible buzz as the surrounding air mass moves in response. The string vibrates; it oscillates; it creates a wave.

Whether acting on bowstring or ocean wave, some force (equivalently, some change in potential energy) drives a mechanical oscillation. Combined with the appropriate law of motion, that agency is reflected in a *wave equation* that shows how the disturbance varies in space and time.

Solutions to any particular wave equation describe the shapes and energies of the allowed vibrations.

All waves and all wave equations are remarkably alike, differing only in details such as the kind of disturbance (an electromagnetic field? the Red Sea? a de Broglie amplitude?), the way it travels, and the specific forces involved. Choose, therefore, a suitable wave equation, plug in the pertinent force constants and conditions of confinement, and out come the shapes, energies, and quantization rules. De Broglie's waves should be no exception.

If so, then how are the electrons in atoms and molecules to be governed? Too lightweight for gravity to play a role, the charged particles respond almost entirely to electrical influences: the attraction of an electron to a nucleus, the repulsion of two electrons, and (in a molecule) the repulsion of one nucleus for another. The electrons have little more than their electrical potential energy and their kinetic energy. Atoms, molecules, solids, liquids, gases, chemical reactions, all of chemistry—everything—must be fashioned from these bits of charge and the electromagnetic tendencies that drive them.

Given the forces, we need a rule to understand how the electron's wave function, ψ, changes in space and time. Newton's laws no longer apply, since acceleration cannot be proportional to force in a world without path. It fell to Erwin Schrödinger, consciously guided by the classical laws, to reconfigure the old mechanics and fashion a specifically quantum mechanical equation of motion. Using de Broglie's relation between the electron's wavelength and momentum ($\lambda = h/p$), he incorporated the electrical energy into an ordinary looking wave equation and then solved for the standing waves of a confined electron.

It worked. The quantized energies demanded by the **Schrödinger equation** (1926) are precisely those measured for the hydrogen atom. Still more compelling, Schrödinger's wave function displayed all the peculiar properties demanded by the wave–particle duality. Here was a wave whose squared amplitude could be interpreted as the probability for finding the electron, an **electron density**. Here was a wave consistent with the uncertainty principle. Here was a wave that could undergo diffraction and interference. Here, finally, was a quantum mechanical wave that became a classical particle for large masses and large quantum numbers. Bohr's correspondence principle was built in from the start.

The mathematical details of the Schrödinger equation would take us far afield, and so we shall concentrate mostly on its solutions. Our goal is to understand the energy, shape, and orientation of the standing waves— to know, within limits, where the electron is likely to be and what bonds it is likely to form.

We start, as did Schrödinger, with the hydrogen atom (H, not the molecule H_2) and related one-electron ions such as He^+ and Li^{2+}. The solitary electron localized within each atom or ion is attracted to a nucleus containing Z protons, beginning with hydrogen ($Z = 1$).

5-3. ORBITALS OF A ONE-ELECTRON ATOM

Pin down a string, pluck it, and let the forces do their work. Clamped along a single line, the vibrations up and down must fit neatly between the endpoints: one lobe ($\lambda/2$), two lobes ($2 \cdot \lambda/2$), three lobes ($3 \cdot \lambda/2$), four lobes ($4 \cdot \lambda/2$), and on up, an additional half-wavelength ($\lambda/2$) squeezed in progressively for each mode. These standing waves, as we recall from Chapter 4 and see again in Figure 5-1, are fully described by just one quantum number, n, and an associated set of $n - 1$ nodes.

Look at them. They *fit*. Oscillating in place, they fill the space allotted. See especially how each stationary wave satisfies the condition

$$ L = n \frac{\lambda}{2} $$

since the n half-wavelengths are necessarily pinned down over a length L. The vibrational frequencies are limited to discrete values, and the intensity of the wave spreads out more and more evenly as n grows larger.

A confined wave is inevitably a quantized wave. Restricted to one dimension, the standing waves of a vibrating string need only the one quantum number n. Restricted to two dimensions, the standing waves of a vibrating drumhead need two quantum numbers. Bound in three dimensions by the forces within an atom, the standing waves of an electron need three quantum numbers.

Although an electron neither vibrates like a string nor quivers like a mass of gelatin, it does obey a wave equation and it does have a wave function. It may be a particle without a path, but still the electron must be found somewhere. Its wave function needs to fit, in some sense, around the nucleus, since only certain solutions of the Schrödinger equation can support themselves amidst the electromagnetic goings-on of the atom. Quantization emerges once again as a natural consequence of close confinement.

Three Dimensions, Three Quantum Numbers

Imagine sitting at the nucleus and looking out some distance r. Let that distance be the first dimension. Looking this way and that but always

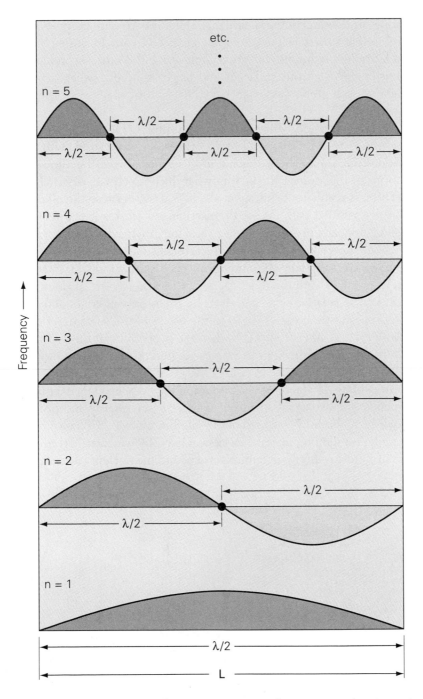

FIGURE 5-1. Confinement and quantization in one dimension: a vibrating string. Marked by a single quantum number n, each standing wave supports n half-cycles over the length of the string.

marking off the same interval, we map out a sphere with radius equal to r. Once on that sphere (Figure 5-2), we proceed to any location by specifying a latitude and longitude—two angles, θ and ϕ, the second and third dimensions. We go out to larger and larger spheres, sweep around the surface of each globe, and eventually cover all of space with these three dimensions.

Three dimensions, three quantum numbers. One quantum number corresponds to a radius, while two others are associated with latitude and longitude on a sphere. Each trio describes one of the allowed wave functions of the atom, called (somewhat deceptively) an **orbital**.

There is really no "orbiting" at all, of course, because the electron has no predetermined path. What the particle has, instead, is a territory in which it appears probabilistically, showing up maybe here and maybe there. One never knows for sure. An orbital ψ merely encloses some volume in space, assigning one number to each spot within. Is the electron here, at point 1? Roll the dice. Or is it somewhere else, at point 2? Roll the dice. The relative odds are given by the numbers $\psi^2(1)$ and $\psi^2(2)$ at the sites in question, each related to the *square* of the wave function ψ. If, say, $\psi^2(1) = 0.1$ and $\psi^2(2) = 0.5$, then the probability at the second site is five times greater. For every five times we find the electron at point 2, we expect to find it only once at point 1.

An orbital—a standing-wave solution of the Schrödinger equation— thus lends stability to the electron, conferring an energy and providing a noncollapsing home as mapped out by the function ψ^2. Collapse would be inevitable, remember, under the classical laws, by which an "orbiting" electron must radiate energy and ultimately fall into the nucleus (where potential

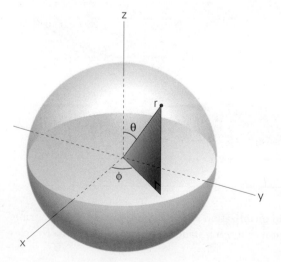

FIGURE 5-2. Distance and direction. To locate a point on a sphere, three polar coordinates (r, θ, ϕ) are better suited than three rectangular coordinates (x, y, z). The angles θ and ϕ give the latitude and longitude.

energy is lowest). Under quantum mechanics, however, the Schrödinger equation predicts a compromise: a balance between potential energy and kinetic energy, shaped so as to keep electron and nucleus apart, on the average, at some optimum distance.

Too close, and the momentum and kinetic energy of the electron grow too large. Too far, and it is the potential energy that becomes too high. Were we to picture the orbital as a simple de Broglie wave (from Chapter 4), then we would expect Heisenberg's uncertainty principle to regulate the corresponding spreads in position and momentum. Any narrowing of the position must be compensated by a large spread in velocity. The orbital offers a set of positions in which the total energy is optimized for stability.

The orbital offers no planned itinerary, though, no preset sequence of visits to be made within its territory. One says only that the electron "occupies" a particular orbital or, alternatively, that it exists in a particular **stationary state**. Using the shorthand language of quantum numbers, we then sketch the outlines of these various states as follows:

There is, first, the **principal quantum number**, which determines the energy and rough extent of each orbital. Denoted by the letter n, the principal quantum number takes the values 1, 2, 3, . . . up to $n = \infty$. Related closely to the radial dimension, n is analogous to the single quantum number needed to characterize a vibrating string.

Next there is the **azimuthal quantum number** (or **angular momentum quantum number**), ℓ, which establishes the orbital's shape. Is it a sphere, a dumbbell, a cloverleaf? We shall soon see. For the moment, note simply that values of ℓ are restricted to the integers 0, 1, 2, . . . up to $n - 1$. Only $\ell = 0$ is allowed if $n = 1$, for instance, whereas the range expands to include $\ell = 0, 1, 2$ if $n = 3$. Usually the number is replaced by a code letter, in keeping with a traditional (although arcane) nomenclature once associated with atomic spectroscopy:

$$\ell = 0 \longrightarrow s$$

$$\ell = 1 \longrightarrow p$$

$$\ell = 2 \longrightarrow d$$

$$\ell = 3 \longrightarrow f$$

$$\ell = 4 \longrightarrow g$$

The sequence then proceeds alphabetically.

Finally there is the **magnetic quantum number**, m_ℓ, which establishes the orientation of the orbital. It imposes a quantization of *direction*, a restriction on which way the orbital can point. With m_ℓ allowed the integral values $0, \pm 1, \pm 2, \ldots, \pm \ell$ ($2\ell + 1$ possibilities in all), this third quantum number creates $2\ell + 1$ additional choices for each value of ℓ. Hence an *s* orbital, with $\ell = 0$, can support just the single value $m_\ell = 0$, but a *p* orbital ($\ell = 1$) can exist in three varieties: $m_\ell = -1, 0, 1$. Similarly, there are five possible *d* orbitals ($m_\ell = -2, -1, 0, 1, 2$) and seven possible *f* orbitals ($m_\ell = -3, -2, -1, 0, 1, 2, 3$). We shall, in time, consider these various orientations along with the shapes.

Sorting Out the Orbitals

Working orbital by orbital, we build up the hierarchy of allowed states for a one-electron atom. Each wave function is specified by its own set of quantum numbers n, ℓ, and m_ℓ. Each is a possible home for the electron. Each has a size and shape, and each allows just *one* energy for an electron within its boundaries.

Start with $n = 1$, for which both ℓ and m_ℓ can only be 0 (ℓ since its maximum is limited to $n - 1$; m_ℓ since its maximum is limited to ℓ). The result is termed a 1*s* orbital, with "1" denoting the principal quantum number (or **shell**) and *s* denoting the azimuthal quantum number (or **subshell**).

Having exhausted the $n = 1$ shell, we proceed to $n = 2$. There is now a lone 2*s* orbital followed by three 2*p* orbitals. Sharing the same principal and azimuthal quantum numbers, the members of the *p* subshell span the values $m_\ell = -1, 0$, and 1. With that, the highest value of ℓ is reached and the $n = 2$ shell is closed. There can be no 2*d* orbitals, no 2*f* orbitals, no 2*g* orbitals, nothing in the second shell beyond 2*p*. When *n* is equal to 2, ℓ reaches no higher than 1.

The pattern continues for $n = 3$, where the shell expands to include one 3*s* orbital, three 3*p* orbitals, and five 3*d* orbitals. A set of seven *f* orbitals ($\ell = 3$) is then accommodated in the $n = 4$ shell, which houses also the 4*s*, 4*p*, and 4*d* subshells. The whole tableau is summarized in Figure 5-3.

All told, there are *n* subshells and n^2 orbitals in each shell. The first shell contains just an *s* orbital: one subshell. The second shell contains both an *s* orbital and a trio of *p* orbitals: two subshells, *s* and *p*. The third shell has an *s*, three *p*'s, and five *d*'s: three subshells. The fourth shell adds a seven-member *f* subshell beyond *s*, *p*, and *d*: four subshells. On and on we go, always finding *n* subshells in the *n*th shell.

One nucleus, one electron, and all those orbitals. Into which does the electron go, and what energy does it possess?

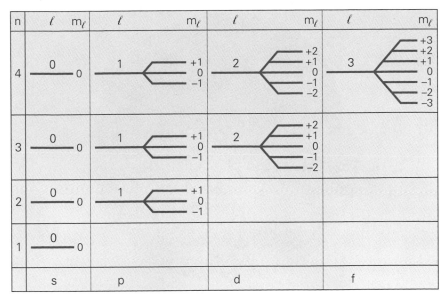

n	ℓ	m_ℓ	ℓ	m_ℓ	ℓ	m_ℓ	ℓ	m_ℓ
4	0	0	1	+1, 0, −1	2	+2, +1, 0, −1, −2	3	+3, +2, +1, 0, −1, −2, −3
3	0	0	1	+1, 0, −1	2	+2, +1, 0, −1, −2		
2	0	0	1	+1, 0, −1				
1	0	0						
	s		p		d		f	

FIGURE 5-3. Shells and subshells in the hydrogen atom, organized according to the spatial quantum numbers n, ℓ, and m_ℓ. Three dimensions; three quantum numbers.

Energy

The electrical attraction of negative for positive binds electrons into atoms, atoms into molecules, and atoms and molecules together into solids and liquids. Such attractions are nowhere more apparent than in a one-electron atom, where the positive charge of Z protons pulls unmolested on the electron. Nothing gets in the way. One bare charge faces another, from which develops a tug of forces governed by the Schrödinger equation.

The nuclear pull can depend only on the electron's distance from the center (along a radius), since no direction is more special than any other. Look out one way for some distance r; there is a pull. Turn around and look the other way; there is the same pull at the same distance r. Anywhere on that entire sphere of radius r (Figure 5-4), the attraction toward the center is always the same. An interaction of *one* electron for *one* nucleus therefore is said to have **spherical symmetry**.

This fundamental attraction of negative for positive, balanced by the kinetic energy, confines the electron to a particular volume around the nucleus. There arises a stationary state analogous to a standing wave, an *orbital* in which the electron is trapped. The higher the nuclear charge Z, the stronger the pull. The closer the electron, the stronger the pull.

And: the stronger the pull, the lower is the electrical potential energy. A body tugged by an attractive force tends to lower its potential

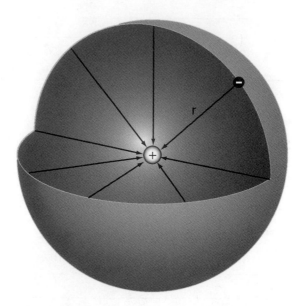

FIGURE 5-4.   Spherical symmetry. The electric force between two particles is the same in all directions—as here, in an atom of hydrogen. The attraction between the electron and nucleus depends only on the distance r between them.

energy. It moves. It comes in closer, like a bowstring after release, trading potential energy for kinetic. The stronger the force, the larger is the change in potential energy from point to point.

Putting these basic ideas together, we can now make sense of the electron's energy. First, a higher Z will always exert a stronger pull and thereby lower the total energy. Regardless of where the electron happens to be, it will be attracted more strongly to a larger positive charge than to a smaller one.

Second, we know from Chapter 1 that the Coulomb force between a single proton and a single electron varies as $1/r^2$. The electrical potential, with dimensions of force $\times$ distance, varies in turn as $1/r$ and creates a situation unique to a one-electron atom. The energies come to depend solely on the distance of the electron from the nucleus, not on the shape or orientation of the orbital. Only the principal quantum number (the shell) proves relevant, for it is only n that derives from the radial dimension r. The quantum numbers ℓ and m_ℓ, which relate instead to angles and directions, have no effect. Expect, accordingly, a $2s$ electron to have the same energy as a $2p$ electron. Likewise expect the energies of the $3s$, $3p$, and $3d$ orbitals to be all the same in a one-electron atom. But do not expect this kind of equality in an atom with many electrons; that very different problem we shall come to soon enough in the next chapter.

Third, realize that orbitals with lower values of n confine the electron closer to the nucleus. Lowest of all is $n = 1$, the most compact of the allowed territories. Subsequently, with each increment in n, the electron's average distance from the nucleus increases and the electrical pull becomes correspondingly weaker. The energy goes *up*.

The energy goes up until finally, for $n = \infty$, the electron finds itself at an infinite distance, $r = \infty$. Let us take that point, where the electrical potential (going as $1/r$) falls to zero, as a convenient origin for the electronic energy scale. We shall say that such an electron has a potential energy of zero. Attracted inward from infinity, the particle's energy then decreases with decreasing distance from the nucleus. The energy becomes more and more *negative*.

Here, then, is the exact relationship obtained by solving Schrödinger's equation for a one-electron atom: Confined to some orbital n, the electron acquires a quantized energy

$$E_n = -R_\infty \frac{Z^2}{n^2} = -(2.18 \times 10^{-18} \text{ J})\frac{Z^2}{n^2}$$

where the **Rydberg constant** ($R_\infty = 2.17987 \times 10^{-18}$ J) establishes a characteristic energy per electron. We confirm that this energy decreases (becomes more negative) as the nuclear charge increases, varying in direct proportion to Z^2. Any orbital of He$^+$ ($Z = 2$), for example, will have an energy four times more negative than an orbital of hydrogen ($Z = 1$) with the same value of n. We see also that E_n depends only on n (not ℓ or m_ℓ) and that the shell energy increases (becomes less negative) as n increases. The dependence goes as $1/n^2$, so that the allowed energies for hydrogen are $E_1 = -R_\infty$, $E_2 = -R_\infty/4$, $E_3 = -R_\infty/9$, $E_4 = -R_\infty/16$, . . . , $E_\infty = 0$.

Figure 5-5 reminds us that these energy levels are discrete. There are gaps between neighboring states, although the spacing shrinks as the principal quantum number becomes large. The energies blend into a quasi-continuous band toward the top of the diagram, tending ultimately toward the limit $E_\infty = 0$.

After that, the electron is on its own. It leaves the atom and becomes a particle possessing only kinetic energy, unconfined and unquantized. No longer influenced by the now-too-distant nucleus, the ionized electron moves unencumbered until, possibly, it strays too close to a positive charge. Then, with energy lowered by the electrical attraction, the formerly free particle is recaptured and confined to an orbital. A new atom or ion is born.

Such are the energetic options available to hydrogen and to any other one-electron atom as well. An electron in a stationary state persists in an

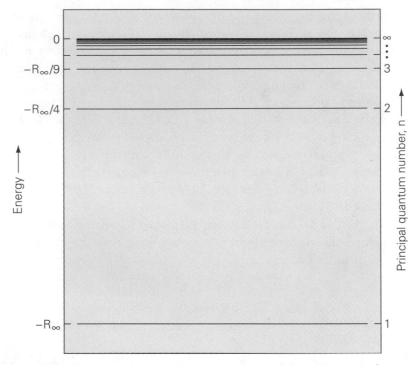

FIGURE 5-5. Energy levels of the hydrogen atom. Proportional to $1/n^2$, the discrete energies collapse into a continuum as the quantum number increases to infinity.

orbital with a fixed energy, roaming throughout some particular volume. Left to itself, the electron tends naturally toward the point of lowest energy and usually resides in the **ground state**: $n = 1$, the $1s$ orbital.

The s Orbitals: Size and Shape

Schrödinger's equation thus gives us a series of possible wave functions ψ, each characterized by quantum numbers n, ℓ, and m_ℓ and each corresponding to a particular orbital: first ψ_{1s}, then ψ_{2s}, then three varieties of ψ_{2p}, then ψ_{3s}, and so on, continuing up to $n = \infty$. Understand, once more, that any orbital is just a set of numbers, a set in which one particular value is assigned to every point in space. We go out some radius r, specify a latitude θ and longitude ϕ on the appropriate sphere, and then compute the value of ψ at that point. The result is a number, a plain number with a sign—maybe -0.50 or 0.34 or even 0—but still a number and not a physical record of the electron. The symbol ψ denotes only the wave function's amplitude (and a highly abstract, hard-to-interpret amplitude at that).

To find the particle, we need the equivalent of an intensity. We need the square of the amplitude, ψ^2, which functions as an electron density and which will now become our map of the atom. The electron turns up most often in regions where this always-positive number is large.

Start to explore the space around the nucleus, searching first for hydrogen's electron in either the $1s$ or $2s$ orbital. In what direction shall we look? It turns out not to matter, for an s orbital is always spherically symmetric. The wave function appears the same in all directions, just as the attraction of the nucleus for its one electron is independent of angle.

The left-hand column of Figure 5-6 shows how $\psi_{1s}(r)$ and $\psi_{1s}^2(r)$ vary with distance from the center. Here in the ground state, with n equal to 1, there are no nodes. The $1s$ orbital of hydrogen, like the lowest mode

FIGURE 5-6. The $1s$ and $2s$ orbitals of a hydrogen atom, plotted—irrespective of angle—against the electron's distance from the nucleus. Left: The $1s$ orbital contains no radial nodes. Right: The $2s$ orbital develops a node at the point where $\psi_{2s}(r)$ changes sign. The probability for finding the electron anywhere at this distance falls to zero, just as the amplitude of a standing wave vanishes at a node. Radial wave functions, $\psi(r)$, are shown at the top; the corresponding probability distributions, $\psi^2(r)$, below. Values are the same in all directions.

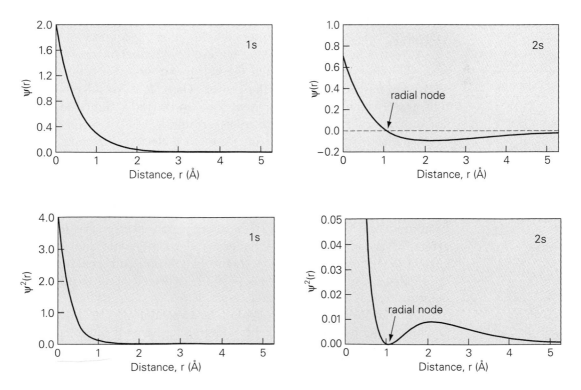

of a vibrating string, is everywhere positive. Not so for the 2s orbital, though, which develops the single radial node depicted in the right-hand column of Figure 5-6. The $n = 2$ wave function changes sign at that point (approximately 1 Å from the nucleus), crossing zero while going from positive to negative. Each successive s orbital then develops an additional node, thereby fixing the total number of nodes at $n - 1$ (again reminiscent of a string).

Look next to the heart of the atom, the nucleus, and discover from the curves a feature unique to s orbitals: always a finite—nonzero— value at the point $r = 0$, where the electron falls directly atop the nucleus. The probability $\psi^2(0)$ is greater for one orbital than the other (as witnessed by the different vertical scales), but neither value is zero. We learn, correspondingly, that an electron in the 1s or 2s orbital, indeed in any s orbital, is never excluded entirely from the nucleus. Only s electrons display this special characteristic, which allows them to *penetrate* close to the atom's core. No p, d, f, or higher electron is ever found at exactly $r = 0$.

The reason has to do with **angular momentum**. We know that a particle moving in a straight line has a *linear* momentum, and we know further that application of a force will change such momentum. A push or pull in a straight line makes the particle accelerate along that same straight line. Similarly, a body moving along a curve has a corresponding *angular* momentum that can be changed by a corresponding force as well—called a *torque*, a "twisting" force that alters the direction of motion.

Anyone who has ever swung a tethered ball understands what happens. The line stretches taut and the ball moves in a controlled circle, seemingly pushed away from the center. There is an angular momentum, from which comes an apparent force in the radial dimension. The fast-moving ball is repelled outward as a result.

A p electron has angular momentum. A d electron has angular momentum. An f or g or h electron has angular momentum, and hence electrons in all these orbitals are excluded from the center. Their wave functions are forced to zero at the nucleus.

This electronic angular momentum clearly cannot be the same angular momentum of a ball on a string, because an electron has no fixed orbit. Nevertheless, the correspondence exists: A quantized form of angular momentum prevents the p, d, and f electrons from penetrating to the nucleus. Only the s electrons ($\ell = 0$), deprived of all angular momentum, can approach the center.

Our hunt for the s electron must be expanded now to include all possible locations. The searchlight, which so far has only illuminated *one* radius in *one* direction, will be swept over the surface of every possible sphere

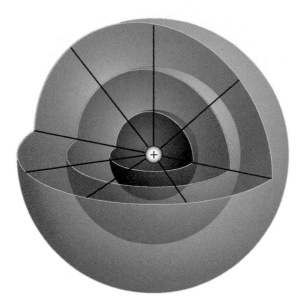

so that $\psi^2(r)$ can be accumulated from point to point. But because the probability is the same regardless of direction, we need only multiply $\psi^2(r)$ at each radius by the area of the associated sphere, $4\pi r^2$. And so is determined the ***radial probability distribution***

$$P = 4\pi r^2 \psi^2(r)$$

which provides the likelihood of finding the electron along a radius r in any direction:

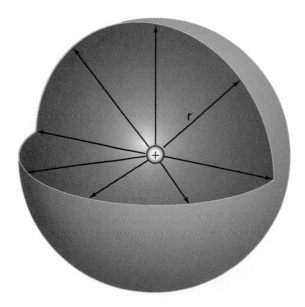

Look this way. Look that way. Always the same $\psi^2(r)$. The total radial probability, equal for all points on the surface, is boosted by the weighting factor $4\pi r^2$, more and more as the sphere expands.

That changes our perspective. Studying the plot of 1s radial density in Figure 5-7(a), we find the electron most frequently not at the nucleus

(a)

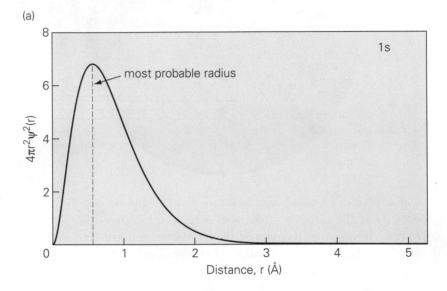

(b)

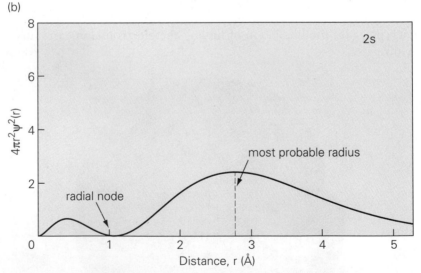

FIGURE 5-7. Radial density functions. Each curve $P = 4\pi r^2\psi^2(r)$ lays out the probability of finding the electron anywhere on a sphere of radius r. The peak gives the most probable radius, a rough measure of the size of an orbital. Orbitals: (a) 1s. (b) 2s. For a view from another perspective, see Figure 5-8.

itself but rather a finite distance away—anywhere on the surface of a sphere generated at some ***most probable radius***. Here is where the distribution peaks, and here is where the electron spends most of its time. Although the value of $\psi^2(r)$ is highest directly *at* the nucleus ($r = 0$), the area of the surrounding sphere ($4\pi r^2$) is vanishingly small. The nucleus may be a good place for an *s* electron to reside, but there is scarcely any room for a comfortable stay.

The optimum location offers, instead, a high value of $\psi^2(r)$ spread over a large area. This most favorable distance, which for a hydrogen 1*s* orbital is termed the ***Bohr radius*** ($a_0 = 0.52918$ Å), roughly establishes the atom's size. The electron subsequently moves to larger and larger distances as *n* increases, eventually finding itself completely detached (ionized) when *n* reaches infinity.

Even the 2*s* orbital of hydrogen, just one state above ground, is significantly expanded already. Its major hump occurs at 2.75 Å (Figure 5-7b), placing the more energetic 2*s* electron five times farther away than the 1*s*. Yet in neither the 1*s* nor the 2*s* nor any other orbital is there an abrupt cutoff past which the electron never ventures. There is just a continual weakening of the electron density, a smearing-out that leaves the atom a bit fuzzy around the edges.

Recognizing that fuzziness, let us give shape and substance to the orbitals first by visualizing the three-dimensional 1*s*, 2*s*, and 3*s* electron densities in cross section. We slice the sphere through its center, and in Figure 5-8 (top of next page) we track the probability by plotting ψ^2 at each point above the resulting plane. The tallest portions of the surface then correspond to those sites in the plane sampled most frequently by the electron.

Observe, right away, that the density decreases continuously out to infinity inside the nodeless 1*s* orbital. There is never any distance where the probability, however slight, is exactly zero. The much larger 2*s* orbital, by contrast, develops a nodal ring at the radius where ψ is zero, and the even larger 3*s* orbital displays *two* nodal rings (corresponding to the two radial nodes present when $n = 3$).

There, with such images, are rendered simple yet useful representations of the electron cloud. That done, let us simplify the picture still further by drawing a boundary surface around, say, 90% of the electron density (Figure 5-9). We choose an outermost surface over which the probability is uniform—always a sphere for *s* orbitals—and, by so doing, intentionally obscure the details within. What remains is an uncluttered representation that conveys roughly the size and shape of the electron density.

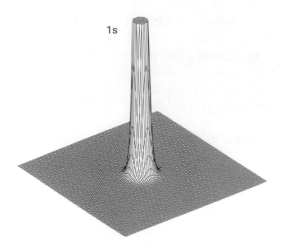

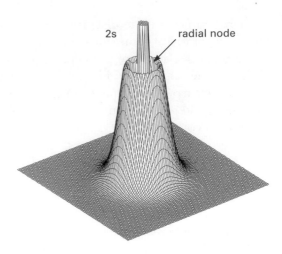

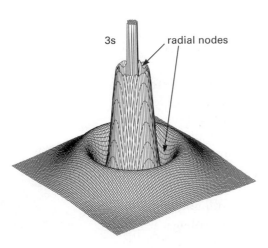

FIGURE 5-8. Finding a home: probability distributions for the 1*s*, 2*s*, and 3*s* orbitals, each evaluated in cross section over a 20 Å × 20 Å grid. The surfaces, shown truncated at the center, measure ψ^2 point by point over any plane taken through the nucleus. All such planes yield the same result, and together the probabilities combine to produce a spherically symmetric three-dimensional distribution. Except for the 1*s*, the *s* orbitals all have at least one nodal surface: an inner sphere (seen here as a ring) on which the probability falls to zero.

FIGURE 5-9. Size and shape of the first three *s* orbitals, shown in gross outline: The spherical surfaces enclose those regions where the electron spends 90% of its time. Point-to-point variations of the electron density (evident in Figure 5-8) are suppressed here for the sake of simplicity.

The p Orbitals: Shape

Only *s* electrons lack angular momentum, and therefore only *s* orbitals are spherically symmetric. For them, everything looks the same in all directions. An electron in any other orbital, however, acquires an angular momentum, and suddenly angles enter into the wave functions. Part of the electron density, losing its spherical symmetry in *one* of the three *p* orbitals, or *one* of the five *d* orbitals, or *one* of the seven *f* orbitals, now may appear different in different directions. Characteristic shapes develop according to the value of the angular momentum quantum number, ℓ.

Consider what it might mean for an orbital to depart from the monotony of a sphere. It means, perhaps, that atoms will form molecules in far greater variety than otherwise. Bear in mind that molecules are combinations of atoms—combinations of orbitals, we expect—and so atoms with varied orbitals should be able to produce more diverse molecular shapes and angles.

The orbitals on different atoms are wave functions, moreover, and (like other waves) they mix together and interfere. Being collections of algebraic numbers with signs, wave functions can combine either constructively or destructively when atoms come together. Added constructively as positive-to-positive, for example, two orbitals create more electron density in a region. An enhancement is obtained. A bond. Added destructively as positive-to-negative, they destroy electron density instead. The two alternatives are poles apart, and what matters most is this simple difference in the algebraic sign of ψ.

Anticipating the potential importance of signs and shapes, we first inspect the boundary surface of a typical *p* orbital. Shown in Figure 5-10(a), the dumbbell-shaped electron density looks like a teardrop reflected in a mirror. There is a lobe above, a lobe below, and nothing in between. The entire plane between the two teardrops is a nodal surface, a zone of zero electron density. Looking up from the nucleus, we see electrons; looking down, we see electrons; looking side to side in a horizontal plane, we see nothing. The empty plane is an **angular node**, and a given orbital always has ℓ directionally dependent nodal surfaces of various shapes: zero for *s* ($\ell = 0$), one for *p* ($\ell = 1$), two for *d* ($\ell = 2$), and three for *f* ($\ell = 3$).

We know nothing yet about the sign of ψ from this picture, since Figure 5-10(a) shows only the electron density ψ^2. Coming from the square of a wave function, these probabilities are necessarily all positive numbers. Let us move then to a representation of ψ itself, which will restore the missing information. In Figure 5-10(b) we see that the *p* orbital, before squaring, is a dumbbell formed from two spheres—one tangent to the other, one positive and one negative. Top and bottom differ in sign,

(a)   (b)

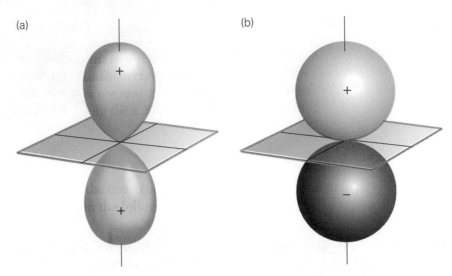

FIGURE 5-10. Shape of a *p* orbital ($\ell = 1$). (a) Square of the wave function, giving a simplified view of the electron density. The probability ψ^2 is exactly zero everywhere in a perpendicular nodal plane containing the nucleus at the origin. (b) Angular variation of the wave function before squaring, showing how the sign of ψ changes from positive to negative. Note that the algebraic sign of any wave function is purely a mathematical attribute, not to be mistaken for a positive or negative electric charge.

and the orbital must pass through zero to make that transition. The null plane between opposite-signed spheres is the angular node.

There is more to the orbital, certainly, for hidden below the surface is all manner of numerical variation. Recall that the density changes as the electron moves away from the nucleus, and with that dependence comes the possibility of *radial* nodal surfaces (each a sphere, independent of direction). The picture becomes sufficiently complicated, though, for us to forgo any detailed study, so we must be satisfied with two general observations:

1. The total collection of nodes, radial and angular together, is always equal to $n - 1$. Thus for a 2*p* orbital (where $n - 1 = 1$), there is just one node: an angular node demanded by the quantum number $\ell = 1$. In a 3*p* orbital, for which $n - 1 = 2$, there are two such surfaces: the $\ell = 1$ angular node as before, supplemented by one radial node.

2. Although the orbital expands as the principal quantum number increases, the basic shape is maintained from shell to shell. Ignoring the radial variation within, one sees the same lobes and the same algebraic signs. The electron just spends more time farther away from the nucleus in orbitals of higher *n*.

The p Orbitals: Orientation

The principal quantum number establishes the orbital's size, and the angular momentum quantum number establishes its shape. The third quantum number, m_ℓ, completes the picture by fixing the orientation in space.

An *s* orbital has but one orientation. It is a sphere, and nothing more needs to be said. The wave function points everywhere without discrimination, not knowing its left from its right. An individual *p* orbital, by contrast, lacks the symmetry of a sphere. It has angular momentum. It can be aimed at something.

Return to the image of a tethered ball being swung in a circle. In the unquantized world of heavy objects, nothing prevents us from swinging this ball wherever and however we desire. The rotation can be fast or slow, clockwise or counterclockwise, above the waist, below the waist, out front, in back, or in a plane tilted anywhere in between. A ball's angular momentum can be varied continuously.

A nucleus, exerting its electromagnetic pull, also swings an electron (in some indeterminate fashion), but here there is a difference. The angular momentum of an electron is *quantized*. If that ball were an electron in a *p* orbital, the nucleus would be able to swing it in ways corresponding to only three quantum numbers: up ($m_\ell = 1$), down ($m_\ell = -1$), and in the middle ($m_\ell = 0$).

Hence *p* orbitals come in threes, with one of many equivalent combinations depicted in Figure 5-11. Each wave function points in a certain direction, and each is perpendicular to the other two. We shall call them the p_x, p_y, and p_z orbitals according to their respective orientations.

Is there really a difference? For an atom all alone—isolated from other atoms, isolated from all electric and magnetic fields—the p_x, p_y, and p_z orbitals are indistinguishable and have the same energy. An orientation in space is meaningful only if there is something different to see. If not, one has no way to mark any particular axis. Who is to determine what we call *x*, *y*, and *z*? We can change the labels just by turning our heads.

The atom, remember, is held together by forces that look the same in all directions, and so its corporate electron density must preserve this symmetry as well. No direction is special. There can be no permanent lumps in the distribution; there can be no assertion that one portion of space is favored over another. Averaged over time, the total electron density of an isolated atom must be spherical.

The total electron density. Viewed *individually*, an orbital need not be spherically symmetric—and, with the exception of the *s* family, none is. The p_x points somewhere. The p_y points somewhere else. The p_z

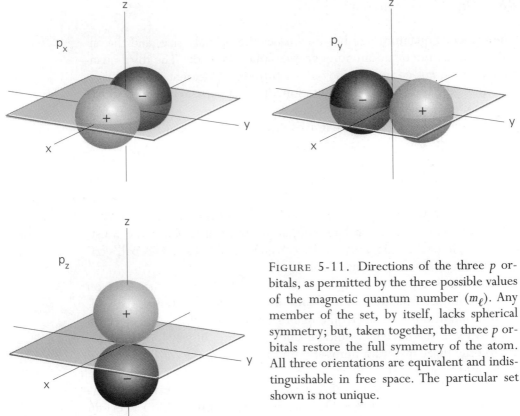

FIGURE 5-11. Directions of the three *p* orbitals, as permitted by the three possible values of the magnetic quantum number (m_ℓ). Any member of the set, by itself, lacks spherical symmetry; but, taken together, the three *p* orbitals restore the full symmetry of the atom. All three orientations are equivalent and indistinguishable in free space. The particular set shown is not unique.

points in yet another direction. Taken together, however, the full set of three orbitals effectively does cover a sphere. Taken together, the *arbitrarily* oriented trio of p_x, p_y, and p_z does preserve spherical symmetry.

But let there arise some distinguishing feature in space (a magnetic field, for instance, or the approach of another atom to form a molecule), and immediately the three *p* orbitals assert their individuality. Directions are no longer arbitrary, because now there is an objective measure of up, down, left, and right. The energies of p_x, p_y, and p_z change accordingly, and the directions matter.

The d Orbitals

The *d* wave functions ($\ell = 2$) come in groups of five, starting with the third shell. Labeled d_{z^2}, $d_{x^2-y^2}$, d_{xy}, d_{xz}, and d_{yz}, they display mostly cloverleaf shapes with orientations as shown in Figure 5-12. All five orbitals are equivalent in an isolated atom, despite the "dumbbell in a doughnut" appearance

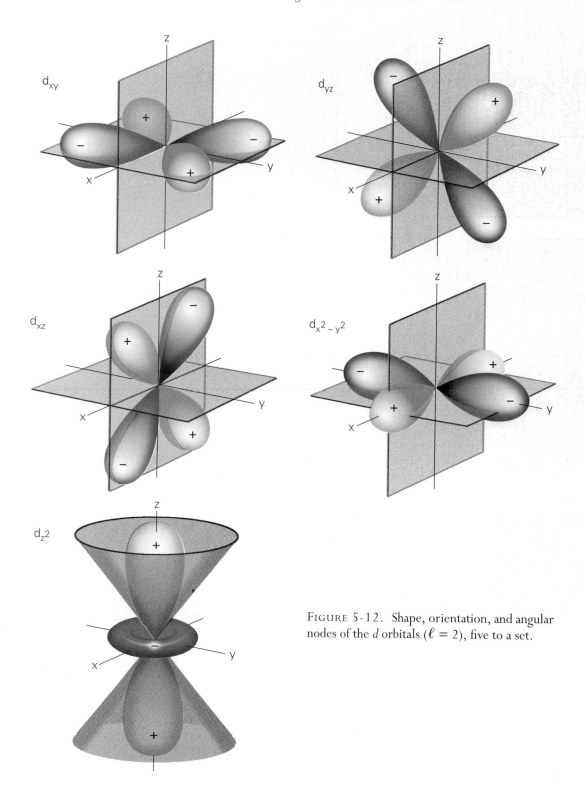

FIGURE 5-12. Shape, orientation, and angular nodes of the *d* orbitals ($\ell = 2$), five to a set.

of the d_{z^2}. Possessing two angular nodes, each d has more angular momentum than a p orbital.

Just like the p orbitals (just like *all* orbitals), the complete set of d orbitals preserves the symmetry of a sphere. Such symmetry may not be obvious from the picture, yet it is there nonetheless. It must be there, for no isolated atom can prefer one direction to another. A molecule can and does distribute its electrons nonspherically; an atom cannot.

The d orbitals play crucial roles in the chemistry of transition metals, as we shall discover in later chapters. Even more complicated and having still higher angular momentum are the f functions ($\ell = 3$), which assume seven orientations and which appear first in the $n = 4$ shell. These orbitals determine the properties of the lanthanide and actinide elements, shown as two special series at the bottom of the periodic table.

5-4. ATOMIC SPECTROSCOPY

Having drawn a picture of quantized energies and orbitals, we ask next how an electron goes from one state to another. It is a question answered best by *spectroscopy*, the systematic use of light waves to measure the energies of atoms and molecules.

To obtain the spectrum of an atomic gas, an experimenter first supplies energy to the system by means of a high-voltage electrical discharge. The excited atoms subsequently release the additional energy while returning to their original conditions. This energy is emitted as electromagnetic radiation of various wavelengths, and the different components are separated by a device called a *spectrograph*—similar to the way white light spreads into a rainbow when passed through a prism. The resulting *emission spectrum* (Figure 5-13) contains lines at distinct frequencies, each separate signal arising from a transition between two allowed states.

Introduced in the preceding chapter, hydrogen's discrete *line spectrum* compelled scientists of the early 20th century to develop the new mechanics of the atom. Here was evidence, completely unexpected, that an atom existed only in certain allowed states and could tolerate only certain energies. Atomic spectra provided early warning that nature was not continuous; that energy was negotiated only in certain fixed amounts; that an atom was, demonstrably, a ladder of energy levels to be climbed rung by rung, photon by photon.

An atom typically exists in its ground state, where energy is lowest. Left undisturbed, hydrogen's one electron occupies the $1s$ orbital and the atom persists as a stable, spherically symmetric ball of electron density.

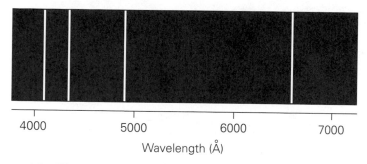

FIGURE 5-13. The emission spectrum of excited hydrogen atoms, recorded in the visible region and resolved by a spectrograph.

Suddenly, though, it is bathed in energy; suddenly it is disturbed by an incoming photon and becomes unstable. Something is about to happen.

Occupying the $n = 1$ shell, the electron has one allowed energy ($E_1 = -2.18 \times 10^{-18}$ J) coming from the quantization formula

$$E_n = -R_\infty \frac{Z^2}{n^2} = -(2.18 \times 10^{-18} \text{ J}) \frac{Z^2}{n^2}$$

If this energy is to change, then the electron must leave the first shell and go somewhere else. It cannot remain a stable $1s$ electron while having an energy other than the one prescribed. But to leave the $1s$ orbital, the excited electron must go to another shell with another discrete value of n. It has to jump from one stationary state to another, and to do so requires a photon with just the right energy. To go from $n = 1$ to $n = 2$, for example, demands an energy of $3R_\infty/4$ when $Z = 1$,

$$\Delta E = E_2 - E_1 = -R_\infty \left(\frac{1}{4} - \frac{1}{1} \right) = +\frac{3}{4} R_\infty$$

whereas a jump from $n = 1$ to $n = 3$ involves an energy of $8R_\infty/9$. If the photon's energy ($E = h\nu$) is too large or too small, then the electron has nowhere to land and the transition cannot occur. Energy must be conserved.

A transition (Figure 5-14) can proceed either up or down, as **absorption** or **emission**. An atom *absorbs* the energy of a photon when the interaction promotes an electron from a state of lower n to a state of higher n. The electron acquires more energy and settles into a new orbital, farther away from the nucleus. An atom *emits* a photon when the electron drops to an orbital lower in energy.

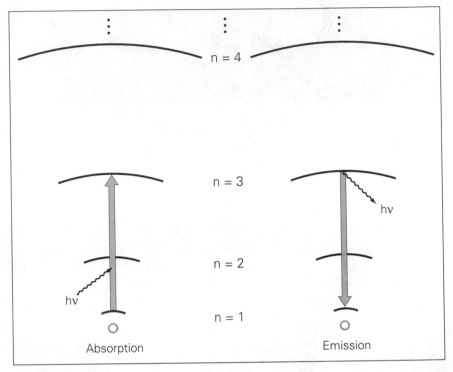

FIGURE 5-14. Transitions between states ("quantum leaps"): absorption and emission. An absorptive transition takes the electron to higher energy, farther from the nucleus. An emissive transition takes it to lower energy, closer to the nucleus.

Putting everything together for hydrogen ($Z = 1$), we obtain the expression

$$\Delta E = h\nu = -R_\infty\left(\frac{1}{n_f^2} - \frac{1}{n_i^2}\right)$$

where n_i and n_f denote, respectively, the initial and final quantum numbers. The change in energy is positive for absorption ($n_f > n_i$) and negative for emission ($n_f < n_i$).

Outlined in Figure 5-15, hydrogen's entire emission spectrum is contained within this one simple equation. The atoms, initially excited to all possible levels ($n_i = 2, 3, \ldots, \infty$), quickly tumble down to the states below and emit photons of energy ΔE. Patterns of spectral lines develop according to the quantum number of the final state, with the various series of emissions named after their discoverers. Most energetic are transitions terminating at the lowest level, $n_f = 1$, the bottom rung of the ladder. These signals fall in the ultraviolet spectrum and are called the **Lyman series**. The first four emissions to the $n_f = 2$ shell, by contrast,

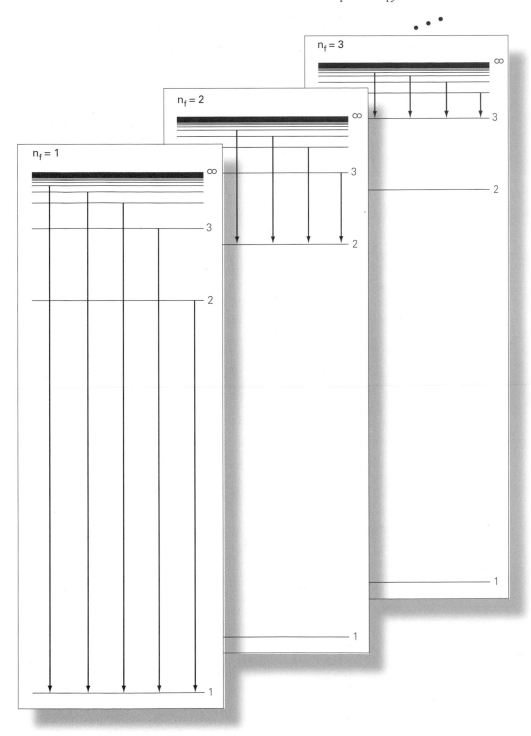

FIGURE 5-15. Pattern of emissions for an excited hydrogen atom. Each transition ends with the electron in a particular shell, giving rise to the Lyman series ($n_f = 1$), the Balmer series ($n_f = 2$), the Paschen series ($n_f = 3$), and so forth.

appear as visible light and are registered as the **Balmer series**:

$$\Delta E_{3 \longrightarrow 2} \quad 656\,\text{nm} \longrightarrow \text{red}$$

$$\Delta E_{4 \longrightarrow 2} \quad 486\,\text{nm} \longrightarrow \text{green}$$

$$\Delta E_{5 \longrightarrow 2} \quad 434\,\text{nm} \longrightarrow \text{blue}$$

$$\Delta E_{6 \longrightarrow 2} \quad 410\,\text{nm} \longrightarrow \text{violet}$$

From each pair of energy levels comes one transition. From each transition comes a separate color of light.

Atomic emission is by now a familiar part of the nighttime landscape, recognizable in the sodium streetlamp's yellow glow and the red-orange of the neon sign. Fireworks and flames, too, acquire color from atomic transitions, and even stars have their own emission spectra. All are quantized. All obey the Schrödinger equation.

5-5. A FOURTH QUANTUM NUMBER

Electricity and magnetism are two sides of a coin, for what we call "magnetism" is precisely the force produced by a charge in motion. Woven together into a unified electromagnetic force, this relationship (Figure 5-16) is the principle that drives the electric motor and the electromagnet. It is a fact of everyday life. An electric current circulating in a loop produces a magnetic field capable of acting on a piece of iron in its midst. Electrons move; magnetic forces develop; rods are pushed up and down; cranks turn.

Magnetism is thus a natural property of electrons and, by extension, of atoms as well. There are electrons in wires; there are electrons in atoms. There are currents in wires; there are currents in atoms. If the

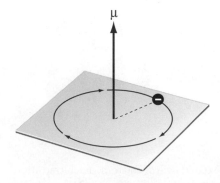

FIGURE 5-16. Electricity and magnetism. An electron circulating in a loop generates a *magnetic moment* μ, bringing with it the properties of a compass needle: a susceptibility to the force imposed by any other magnetic field. All charged particles with angular momentum behave similarly.

electron within an orbital is pictured as executing a tiny loop of quantum mechanical current, then one should expect magnetism to arise from this motion. It does. The magnetic field takes its strength and direction from the electron's orbital angular momentum, quantized according to ℓ and m_ℓ in the way shown in Figure 5-17. Such an effect is fully expected within the electrical context of the atom.

Completely unforeseen, however, was the discovery of a magnetism unrelated to electronic motion from point to point. Working in the 1920s, the physicists Otto Stern, Walther Gerlach, George Uhlenbeck, and Samuel Goudsmit identified a new form of angular momentum built into the particle itself. This angular momentum, dubbed **electron spin**, differs from the angular momentum of an electron moving about a nucleus, for an electron would have a spin (so called) even if it were standing

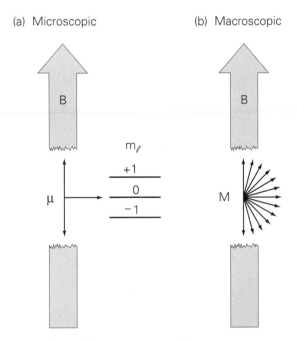

FIGURE 5-17. Orbital angular momentum and atomic magnetism. (a) Any electron (except *s*) acquires a magnetic moment while moving about the nucleus, but the direction of this "orbital" magnetism is restricted by the quantum number m_ℓ. For an electron with $\ell = 1$ (a *p* electron), the magnetic moment μ has only three sets of orientations in an external magnetic field *B*. Averaged over time, some of the microscopic magnets point one way ($m_\ell = 1$); some point another way ($m_\ell = 0$); and some, yet another ($m_\ell = -1$). There are no possibilities in between. (b) By contrast, a macroscopic magnet *M* suffers no restrictions on its energy or orientation. A compass needle can be made to point in any direction.

in one place. The particle undergoes, apparently, some kind of internal motion suggestive of a spinning top or rotating globe. The "spinning" charge produces a magnetic field just by being there.

The use of quotation marks emphasizes that this term *spin*, although long accepted, threatens to mislead. To say that the electron actually spins like a top is to go too far, since we know nothing about the particle's internal structure—or even if it has any. Nor does the electron's spin properly develop into an ordinary macroscopic rotation when the quantum numbers and masses become large. It remains purely a quantum mechanical phenomenon, made real by the experiment sketched in Figure 5-18.

That said, we shall accept the electron for what it is: a charged particle with an unspecified internal motion that generates spin angular momentum. All alone, it produces a magnetic field. So does a proton, and so does a neutron. Each has a spin. Each has some kind of internal distribution of charge. Even a neutron, despite its overall neutrality, is composed of still smaller electrical particles, and consequently even a neutron can be magnetic. Many heavier nuclei possess net spin and magnetism as well, having been built from spin-endowed protons and neutrons.

Like other forms of angular momentum, spin is quantized. Its magnitude is determined first by a ***spin angular momentum quantum number*** s, analogous to the orbital angular momentum quantum number ℓ. This quantized spin can point in only the $2s + 1$ directions prescribed by a related ***spin magnetic quantum number*** m_s, just as there are $2\ell + 1$

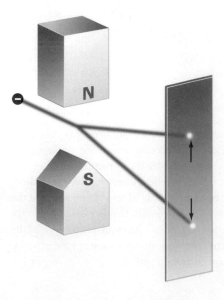

FIGURE 5-18. Spin angular momentum and atomic magnetism. Even without overt orbital angular momentum, an electron possesses a native "spin" angular momentum—built-in and unquenchable. The quantization becomes apparent in the *Stern-Gerlach experiment*, where a beam of atoms containing unpaired electrons $(s = \frac{1}{2})$ splits into two components $(m_s = -\frac{1}{2}, \frac{1}{2})$ when passing through a nonuniform magnetic field. The spin quantum numbers s and m_s are analogous to the orbital quantum numbers ℓ and m_ℓ.

values of m_ℓ. All that differs is the form of the numbers: They are half-integers, not integers, and they never become infinite. We discover, specifically, that an electron's spin is characterized simply by

$$s = \tfrac{1}{2}$$

rather than the series $\ell = 0, 1, 2, \ldots, \infty$ that could conceivably apply to orbital angular momentum.

Corresponding to $s = \tfrac{1}{2}$ are two orientations, loosely referred to as spin-up and spin-down:

$$m_s = \tfrac{1}{2} \quad \text{(up)}$$

$$m_s = -\tfrac{1}{2} \quad \text{(down)}$$

These quantum numbers obey the same rules as m_ℓ, ranging through the $2s + 1$ (two) values from s to $-s$. Were we to imagine, incorrectly, that the electron was a sphere spinning about its axis, then the two choices for m_s would designate clockwise and counterclockwise rotations.

Since s is always fixed at $\tfrac{1}{2}$ for an electron, we shall henceforth ignore it as redundant. Only the spin magnetic quantum number, m_s, with its two possible values of $\tfrac{1}{2}$ and $-\tfrac{1}{2}$, has anything further to tell us about the electron. This parameter joins n, ℓ, and m_ℓ as a fourth quantum number, and immediately in the next chapter we shall see that m_s is the key to explaining the periodic table.

Appearing in this ad hoc way, the spin seems unfortunately to have been tacked on as an afterthought, almost as if it were pulled from thin air. Actually it was. By necessity, the extra quantum number was appended to the wave function in just this arbitrary fashion when quantum mechanics was first developed. An unanticipated complication, spin was part of the picture nevertheless. Certain fine details of atomic spectra, for instance, could be explained only by the presence of spin. Its existence was an experimental fact, and the pertinent quantum numbers (although half-integral) were analogous to ℓ and m_ℓ. Since there were already three spatial dimensions and three quantum numbers, a fourth parameter was added to accommodate the newcomer.

Four dimensions, four quantum numbers. Before long, P. A. M. Dirac showed that all four quantum numbers flow naturally out of a *four-dimensional* system of coordinates: space and time together, all treated on the same footing. Einstein, as we shall see in Chapter 21, had already accorded equal status to space and time in his theory of relativity, wherein the four coordinates x, y, z, and t become part of one shared, interchangeable set. Applying Einstein's ideas to Schrödinger's equation,

Dirac produced a modified wave equation from which the electron auto-matically emerges with its proper spin. Spin therefore ceases to be an appendix, for the fourth quantum number is a normal part of a more in-clusive, four-dimensional world.

Still, the original three-plus-one picture of space and spin is adequate for almost all chemical phenomena. With this final quantum number, then, our description of the one-electron atom is complete: quantized en-ergies dependent only on n; spatial orbitals characterized by n, ℓ, and m_ℓ; and an additional spin angular momentum prescribed by m_s.

The next step leads directly into the heart of chemistry: the ele-ments. We must take the tools already developed—the Schrödinger equation, the one-electron orbitals, the spin—and try to understand how the atoms in the periodic table are held together.

REVIEW AND GUIDE TO PROBLEMS

One proton, one electron, and one electrical attraction to hold it all together—not much, perhaps, but enough to seed the universe with ordinary matter. Simplicity itself: the hydrogen atom, first of the elements.

It is pattern and prototype, cornerstone of the material world. Everything else derives from hydrogen, directly or indirectly. Everything else begins with this simple design.

From proton and electron comes a stable structure, a neutral atom, a structure able to exist and combine. The hydrogen atom finds size, shape, and energy in a quantum mechanical universe. It does not collapse. The electron finds a home.

IN BRIEF: THE HYDROGEN ATOM

1. Home for the electron is an *orbital*, a solution of the *Schrödinger equation*; home is a *wave function*, a three-dimensional pattern of ups, downs, and nodes analogous to the standing waves on a string. For the electron, an orbital is an allowed territory. It is a place to go, a step on the quantum ladder. Potential energy and kinetic energy are optimized. The atom finds a stable state, impossible under the laws of classical mechanics.

There is a mathematical sense and a physical sense. The orbital is, first, a mathematical function, a number assigned to each point in space. Pick a point; get a number. The number is the value of the wave function $\psi(x, y, z)$, and the squared value $\psi^2(x, y, z)$ lets us compute the probability of finding the electron close to the given site. We solve an equation, and we get a set of numbers.

A dry set of numbers, hopelessly abstract? . . . not when we see the numbers physically as a region in space, as a volume in which the electron may appear, as a three-dimensional *probability distribution*. There is size, shape, and orientation to that distribution; there is substance to the orbital; there is something to see. The orbital's territory may be near to the nucleus or far. The volume may be shaped like a sphere, a dumbbell, a cloverleaf, maybe something else. It may point this way or that or in no particular direction at all, but there is finally something to grasp. We draw the orbital, and we have a snapshot of the electron's whereabouts.

A snapshot, not a path. No path is sought or implied, despite the unfortunate term "orbital," which suggests the motion of a planet about the sun. The name is a mistake, an accident of history, the remnant of an old misunderstanding. Forget about the name and its misleading connotations, and remember this: An orbital is a point-by-point tabulation of probability, a tableau of possibilities rather than a trail for the electron to follow.

2. Three *quantum numbers* summarize the orbital's size, shape, and orientation: n, ℓ, and m_ℓ. The first, n, controls how the wave function changes at successive distances *away* from the nucleus, points taken along any radial line without regard to direction. Two other quantum numbers, ℓ and m_ℓ, then describe the angular variation of ψ *around* the nucleus independent of distance. There is a radius r, a latitude θ, and a longitude ϕ. Three dimensions, three quantum numbers.

Recall the meaning of shells and subshells. Orbitals with the same value of n belong to the same *shell*, from which we have the first shell ($n = 1$), the second shell ($n = 2$), the third shell ($n = 3$), and so forth. A shell, in turn, is organized into *subshells* of orbitals with the same value of ℓ (*spdf*); and within each subshell there is a further division according to m_ℓ.

3. The *principal quantum number* n, which assumes integral values from 1, 2, 3 through ∞, correlates with the orbital's size, energy, and total number of nodes ($n - 1$). The higher the n, the more likely is the electron to be found farther from the nucleus. Farther away, attracted less by the positive charge, an electron with high n is comparatively easy to remove. Its energy is higher than an electron with low n. It is held less tightly.

4. The *azimuthal quantum number* (or *angular momentum quantum number*) ℓ, restricted to the integers 0, 1, 2, . . . , $n - 1$, fixes the orbital's shape. An *s orbital* ($\ell = 0$) is shaped like a sphere, a *p orbital* ($\ell = 1$) like a dumbbell, and a *d orbital* ($\ell = 2$) either like a cloverleaf or like a dumbbell in a doughnut.

The different shapes correspond to different values of the electron's quantized angular momentum around the nucleus. A higher number for ℓ means higher angular momentum, analogous to the way a diver's body spins faster when tucked in tighter. And a higher angular momentum means more *angular nodes* in the wave function, more combinations of angles where $\psi(\theta, \phi)$ changes sign. An orbital with azimuthal quantum number ℓ has ℓ such angular nodes, each a surface with a certain shape and orientation: zero angular nodes for an *s* orbital (which is a sphere and thus insensitive to direction); one angular node for a *p* orbital (the plane separating the dumbbell's positive lobe from the negative lobe); two angular nodes for a *d* orbital (two planes dividing four lobes).

5. The *magnetic quantum number* m_ℓ, taking the values 0, ± 1, ± 2, . . . , $\pm \ell$, determines the orientation of the orbital. An *s* orbital, a perfect sphere, points everywhere the same and has but one orientation. Its angular quantum numbers are $\ell = 0$ and $m_\ell = 0$. A *p* orbital ($\ell = 1$, $m_\ell = -1, 0, 1$) has, by contrast, three possible directions while a *d*

orbital ($\ell = 2$, $m_\ell = -2, -1, 0, 1, 2$) has five. In general, there are $2\ell + 1$ admissible orientations within a subshell of fixed ℓ.

Can we tell which way is up? No, not yet. To the electron in hydrogen, all directions look the same. All orbitals with given values of n and ℓ have identical energy and are equivalent in every way, notwithstanding their different values of m_ℓ. The one-electron atom enjoys the symmetry of a sphere, having the same view in all directions.

So does the earth, at least if we ignore its natural magnetic field. Although the residents of New York may assert, very strongly, that all their buildings point from the ground *up*, the citizens of Sydney justifiably make the same claim for their own structures. Neither viewpoint is right, and neither is wrong. Without a compass, there is no special direction on which everyone can agree.

Give them a compass, though, and then they do have an external marker to point the way—the direction in which the magnetized needle aligns. Then there would be a difference; then New York and Sydney could agree on what was up and what was down.

So would an atom. For an atom, space loses its sameness in the presence of, say, an electric field, a magnetic field, another electron, another nucleus, another atom, another molecule. Let there be some marker, and then suddenly the orientations of the orbitals become meaningful.

6. One electron, but an infinity of orbitals. Each orbital represents a possibility for the electron, a potential home, a place where it *can* stay but not where it *must* stay. There is always some other place to go.

Imagine a guest arriving at an empty hotel, finding himself with the run of the house. Low in energy, as travelers often are, a tired visitor will usually choose a room on the ground floor. Some few guests, perhaps with extra energy, might occasionally climb one, two, three flights of stairs or more, but surely the lowest levels will be favored by most. If we imagine further that there is one room on the ground (first) floor, four rooms on the second floor, nine rooms on the third floor, and so on, then we have a parable for how an electron comes to stay in a hydrogen atom.

Lowest in energy is the ***ground state*** orbital, the $1s$. Next are the four orbitals of the second shell ($2s$, $2p_x$, $2p_y$, $2p_z$), all with the same energy absent any marker in space to say otherwise. Then come the nine orbitals of the third shell ($3s$, $3p_x$, $3p_y$, $3p_z$, $3d_{z^2}$, $3d_{x^2-y^2}$, $3d_{xy}$, $3d_{xz}$, $3d_{yz}$), and the 16 orbitals of the fourth shell, and, in general, the n^2 orbitals of the nth shell.

7. ENERGY LEVELS. Each orbital in the same n shell has the same energy, no matter what the value of ℓ and m_ℓ:

$$E_n = - R_\infty \frac{Z^2}{n^2}$$

The *Rydberg constant* ($R_\infty = 2.18 \times 10^{-18}$ J) pertains to a single electron. For a mole, we multiply by Avogadro's number to convert R_∞ into 1.31×10^3 kJ mol^{-1}.

E_n is expressed as a negative quantity, consistent with the energy lost when a free electron comes under the attractive sway of the nucleus. The stronger the binding, the more negative the energy. The more negative the energy, the harder it is to tear off the electron.

So: An electron close to the nucleus feels a correspondingly strong Coulomb attraction, balanced by a kinetic energy that keeps the atom from collapsing. Proportional to $1/n^2$, the strength of binding increases as n decreases. E_n grows more negative.

And also: A large positive charge attracts an electron more strongly. Proportional to Z^2, the binding becomes tighter as the atomic number increases. E_n grows more negative.

There is little else. Just these two parameters, Z and n, embrace almost all the energetic possibilities for the one-electron atoms.

8. SPECTROSCOPY. Like weary travelers, most electrons do fall into the ground state, where energy is lowest; but, like our fictitious guests, they also have the run of the house. Gaining energy they can go up, and losing energy they can go down (although not below the ground state, no, because there is no basement; the atom does not collapse). This freedom up and down they always have, provided they pay their way energetically. Any photon absorbed or emitted must match the spacing between levels.

The requisite energy might come from the surroundings, perhaps from lightning bolts or ultraviolet sunlight or just the normal bumps, jolts, and collisions suffered by the particles. Or it can be delivered systematically in the laboratory under an experimental procedure called *spectroscopy*. Either way, the atom remains a set of possibilities to be realized, a ladder of energies for the electron to ascend and descend.

9. Finally, a quantum surprise: the *spin* of the electron. We know that the electron has an *orbital* angular momentum by virtue of its indeterminate motion about the nucleus—like a quantum mechanical earth about a quantum mechanical sun. But there is more than that. The electron has an additional angular momentum, a so-called *spin* angular momentum that reminds us of a ball rotating about its own axis—like the earth turning daily from day to night.

No, not really. The electron has no internal structure that we know of, so it has nothing about which to spin; and, even if it did, the angular momentum thus generated would not be what we expect. Electron spin is rather an angular momentum intrinsic to the particle, structure or no structure. It represents a fourth quantum number, part of a four-dimensional world that marries space and time (about which, see Chapter 21).

What the electron clearly does have, irrespective of origin, is this **spin angular momentum**, marked by quantum numbers s and m_s analogous to ℓ and m_ℓ. Since, however, s for an electron always has the magnitude $\frac{1}{2}$ and never changes, we have only to deal with the two possible *orientations* of m_s: spin "up" ($m_s = \frac{1}{2}$) and spin "down" ($m_s = -\frac{1}{2}$). Our final set of four quantum numbers then includes n, ℓ, m_ℓ, and m_s, three quantum numbers for the spatial coordinates and one for the spin.

Spin—an interesting surprise, a quirk of nature? Maybe, but here is no mere curiosity. The electron spin, we shall find, takes on transcendent importance, because it is the spin, this twofold possibility of up or down, that creates the chemical world. The spin is what shapes all atoms heavier than hydrogen; the spin is what forces the valence electrons away from the core; the spin is what allows atoms to combine into molecules. We shall see why in the next chapter.

The whole periodic table is yet to come. Meanwhile, here are some problems dealing just with space quantization in the one-electron atom.

SAMPLE PROBLEMS

EXAMPLE 5-1. Quantum Numbers and Orbitals

PROBLEM: Which of the following are impossible combinations of n and ℓ: $1p$, $2s$, $3f$, $3g$, $4f$, $5g$?

SOLUTION: The orbitals are designated as $n\ell$, with the *spdf* letter code used for ℓ:

ORBITAL	n	ℓ	LEGAL?
$1p$	1	1	no
$2s$	2	0	yes
$3f$	3	3	no
$3g$	3	4	no
$4f$	4	3	yes
$5g$	5	4	yes

Since ℓ is restricted to the values 0, 1, 2, . . . , $n - 1$, the $1p$ ($\ell = n$), $3f$ ($\ell = n$), and $3g$ ($\ell > n$) orbitals do not exist.

EXAMPLE 5-2. Shells and Subshells

PROBLEM: How many orbitals are contained within the fifth shell? Name them and specify quantum numbers for each.

SOLUTION: Follow the rules. With $n = 5$, the azimuthal quantum number ($\ell = 0, 1, \ldots, n - 1$) ranges from 0 to 4. In each subshell there are then $2\ell + 1$ orbitals with magnetic quantum numbers $m_\ell = 0, \pm 1, \pm 2, \ldots, \pm \ell$.

The s subshell, as always, supports only one magnetic orientation. The p subshell, as always, supports three. The d subshell, five. The f subshell, seven. The g subshell, nine. Quantum numbers are summarized in the table below:

ORBITAL	n	ℓ					m_ℓ				
5s	5	0					0				
5p	5	1				−1	0	1			
5d	5	2			−2	−1	0	1	2		
5f	5	3		−3	−2	−1	0	1	2	3	
5g	5	4	−4	−3	−2	−1	0	1	2	3	4

Count them up: $1 + 3 + 5 + 7 + 9 = 25 = 5^2$. There are always n subshells and n^2 orbitals in a shell.

EXAMPLE 5-3. Orbitals, Quantum Numbers, Nodes

PROBLEM: What is the total number of nodes contained within each orbital in the fifth shell? How many are angular nodes, and how many are radial nodes?

SOLUTION: The total number of nodes is $n - 1$, and the number of angular nodes is ℓ. The number of radial nodes therefore must be $n - \ell - 1$:

$$\text{ANGULAR} \quad + \quad \text{RADIAL} \quad = \quad \text{TOTAL}$$
$$\ell \qquad\qquad n - \ell - 1 \qquad\qquad n - 1$$

Tabulated by subshell, the results are:

ORBITAL	n	ℓ	ANGULAR NODES	RADIAL NODES	TOTAL
5s	5	0	0	4	4
5p	5	1	1	3	4
5d	5	2	2	2	4
5f	5	3	3	1	4
5g	5	4	4	0	4

Each orbital in the fifth shell contains four nodes, split between angular and radial. The higher the angular momentum, the more angular nodes in the wave function.

Nodal distribution is independent of m_ℓ. Each of the three $5p$ orbitals, for example, contains one angular node and three radial nodes. They differ only in orientation, not shape.

EXAMPLE 5-4. More Nodes

PROBLEM: How many radial nodes are contained in the $1s$, $2s$, $3s$, $4s$, and $5s$ orbitals?

SOLUTION: We have just a simple repetition of the exercise above, but there is a point worth making: s orbitals are spherically symmetric. They have no angular nodes.

Since each s orbital is free of angular nodes ($\ell = 0$), the full complement of $n - 1$ nodes is exclusively radial:

ORBITAL	n	ℓ	ANGULAR NODES	RADIAL NODES	TOTAL
$1s$	1	0	0	0	0
$2s$	2	0	0	1	1
$3s$	3	0	0	2	2
$4s$	4	0	0	3	3
$5s$	5	0	0	4	4

Notice how the number of nodes increases with orbital energy.

EXAMPLE 5-5. Two Protons Are Stronger Than One

PROBLEM: Consider the one-electron species H, He$^+$, and Li^{2+} in their ground states. (a) Which atom has the smallest radius? (b) Compute the ionization energy for each.

SOLUTION: Start by noting that: (1) H has a nuclear charge of $+1$ ($Z = 1$), He$^+$ a nuclear charge of $+2$, and Li^{2+} a nuclear charge of $+3$. (2) The ground state for each species is the $1s$ orbital. (3) The orbital energies

$$E_n = -R_\infty \frac{Z^2}{n^2}$$

depend strictly on Z and n. With that, we have all the information we need.

(a) *Radius*. In H, one proton attracts one electron. In He$^+$, two protons attract one electron. In Li^{2+}, three protons attract one electron. The progressively stronger attraction pulls the $1s$ orbital in more tightly, leaving Li^{2+} as the smallest atom of the three. The radii increase in the order Li^{2+}, He$^+$, H.

(b) *Ionization energy*. To ionize the atom, we must supply sufficient energy to promote the electron from $n = 1$ to $n = \infty$. The ionization energy is therefore

$$E_\infty - E_1 = -R_\infty Z^2 \left(\frac{1}{\infty^2} - \frac{1}{1^2} \right) = +R_\infty Z^2$$

where $R_\infty = 2.18 \times 10^{-18}$ J/atom (or 1.31×10^3 kJ/mol). The sign is positive. We put energy *into* the atom to tear off its electron.

For H, the requisite energy is R_∞. For He$^+$, where $Z = 2$, the amount is four times greater (2^2); for Li^{2+}, nine times greater (3^2). The numbers scale directly with Z^2:

	ΔE	PER ATOM	PER MOLE
H	R_∞	2.18×10^{-18} J	1.31×10^3 kJ
He$^+$	$4R_\infty$	8.72×10^{-18} J	5.24×10^3 kJ
Li^{2+}	$9R_\infty$	1.96×10^{-17} J	1.18×10^4 kJ

EXAMPLE 5-6. Capture

PROBLEM: A free electron moves with velocity $v = 1.00 \times 10^6$ m s^{-1} before being captured by a hydrogen nucleus. Eventually it lands in the 1*s* orbital. What is the change in energy?

SOLUTION: The free electron's energy, entirely kinetic, is initially $\frac{1}{2}mv^2$. The bound electron's energy is $E_n = -R_\infty(Z^2/n^2)$. Trapped by the nucleus, then, the electron suffers a *loss* equal to

$$\Delta E = E_{bound} - E_{unbound} = -R_\infty \frac{Z^2}{n^2} - \frac{1}{2}mv^2$$

Substituting the quantities $Z = 1$, $n = 1$, $m = 9.11 \times 10^{-31}$ kg, $v = 1.00 \times 10^6$ m s^{-1}, and $R_\infty = 2.18 \times 10^{-18}$ J, we have

$$\Delta E = -R_\infty - \frac{1}{2}mv^2$$

$$= -2.18 \times 10^{-18} \text{ J} - \frac{1}{2}(9.11 \times 10^{-31} \text{ kg}) \times (1.00 \times 10^6 \text{ m s}^{-1})^2$$

$$= -2.18 \times 10^{-18} \text{ J} - 4.56 \times 10^{-19} \text{ J} = -2.64 \times 10^{-18} \text{ J}$$

Where does the energy go?

EXAMPLE 5-7. Line Spectra

PROBLEM: Compute the longest wavelength in the Lyman series for both H and He$^+$.

SOLUTION: The Lyman series includes all emissions where the electron falls from an initial shell n_i to the $1s$ ground state (designated the final state, $n_f = 1$). Lowest in energy and longest in wavelength is the transition from $n_i = 2$ to $n_f = 1$, for which

$$\Delta E = -R_\infty Z^2 \left(\frac{1}{n_f^2} - \frac{1}{n_i^2} \right) = -R_\infty Z^2 \left(1 - \frac{1}{4} \right) = -\frac{3}{4} R_\infty Z^2$$

and simultaneously

$$\Delta E = h\nu = \frac{hc}{\lambda}$$

Equating the two magnitudes, we then have

$$\frac{hc}{\lambda} = \frac{3}{4} R_\infty Z^2$$

with a corresponding wavelength

$$\lambda = \frac{4hc}{3R_\infty Z^2} = \frac{4 \times (6.63 \times 10^{-34} \text{ J s}) \times (3.00 \times 10^8 \text{ m s}^{-1})}{3 \times (2.18 \times 10^{-18} \text{ J})Z^2}$$

$$= 1.217 \times 10^{-7} \text{ m}/Z^2 = 121.7 \text{ nm}/Z^2$$

For H ($Z = 1$), the wavelength to three significant figures is 122 nm. For He$^+$, where $Z = 2$, the wavelength is 30.4 nm—shorter and more energetic.

EXERCISES

1. How many quantum numbers suffice to describe the standing waves of a vibrating string? What values do they take, and what physical information do they convey?

2. Sketch the first five standing waves that can develop on a string of length 0.72 m, its two ends pinned in place. Arrange the patterns in order of increasing frequency, indicating the wavelength and number of nodes for each.

3. The standing wave of lowest frequency is called the *fundamental* mode, or *first harmonic*, of a vibrating system. The second harmonic corresponds to the state just above the fundamental vibration; the third harmonic, to the state after that; the fourth, the fifth, and so on. (a) In principle, how many harmonics might be sustained on a vibrating string? (b) If the fundamental mode vibrates with the frequency ν, what are the frequencies of the second, third, and fourth harmonics? Illustrate the relationships with a sketch.

4. Suppose that the fundamental frequency of a certain 100-cm string is 440 Hz, as, say, the A above middle C on a piano. (a) Will a longer string of the same material vibrate at a lower or higher fundamental frequency? (b) What length will produce a fundamental frequency of 880 Hz (one octave higher)? (c) What length will produce a fundamental frequency of 220 Hz (one octave lower)?

5. Explore, by example, the spacing between successive standing waves, and observe: If the fundamental frequency of a string has the value ν, its second harmonic has the value 2ν—making for a difference of 100%:

$$\frac{2\nu - \nu}{\nu} \times 100\% = \frac{\nu}{\nu} \times 100\% = 100\%$$

The percentage change from the second to third harmonic, however, falls to 50%, and the change from third to fourth drops further to 33%. (a) By what percentages do the fourth and fifth harmonics differ? (b) The fifth and sixth? (c) The seventh and eighth? (d) The hundredth and hundredth plus one? (e) The millionth and millionth plus one? Use a rough sketch to make the point.

6. Taking into account the results of the last exercise, consider: How might we expect the proportional spacing between levels of a quantized *atom* to change as the energy increases?

7. (a) How many spatial quantum numbers describe a one-electron atom? (b) What information do they provide about the size, shape, and orientation of the electron distribution? (c) Which of the quantum numbers is most reminiscent of the description of a vibrating string?

8. Suppose that we know exactly *where* an electron is at some instant. Can we determine, using Schrödinger's equation, where the electron will be immediately thereafter? How does the Heisenberg indeterminacy principle limit our knowledge?

9. Explain the physical significance of a one-electron orbital, distinguishing between the amplitude (ψ) and the squared amplitude (ψ^2) of the wave function. What information, in particular, does the quantity ψ^2 carry?

10. For each of the following one-electron orbitals, specify the quantum numbers n and ℓ:

    (a) $3p$ (b) $2s$ (c) $4p$ (d) $1s$ (e) $2p$

11. For each of the following one-electron orbitals, specify the quantum numbers n and ℓ:

    (a) $3d$ (b) $4d$ (c) $3s$ (d) $4s$ (e) $4f$

12. For each combination of quantum numbers n and ℓ, give the name of the corresponding subshell:

    (a) $n = 3, \ell = 2$
    (b) $n = 2, \ell = 1$
    (c) $n = 4, \ell = 3$
    (d) $n = 1, \ell = 0$

13. For each combination of quantum numbers n and ℓ, give the name of the corresponding subshell:

    (a) $n = 2, \ell = 0$
    (b) $n = 6, \ell = 5$
    (c) $n = 3, \ell = 1$
    (d) $n = 3, \ell = 0$

14. List the possible values of ℓ belonging to each of the first five shells.

15. List the possible values of m_ℓ belonging to each of these subshells:

    (a) $1s$ (b) $2s$ (c) $2p$ (d) $3s$ (e) $3p$

16. List the possible values of m_ℓ belonging to each of these subshells:

    (a) $3d$ (b) $4s$ (c) $4p$ (d) $4d$ (e) $4f$

17. Sketch, for each orbital in the first three shells, the shape of ψ.

18. Consider a hydrogen atom with its electron in the $3p$ subshell. (a) In the absence of a magnetic field, how many distinct values of energy are available to this single $3p$ electron? Is that energy less than, greater than, or equal to the energy of a $3s$ electron? (b) How many distinguishable energies may the $3p$ orbitals assume when a magnetic field is indeed present? (c) Including the effects of electron spin, into how many different levels will the p orbitals divide themselves in the presence of a magnetic field?

19. The "spin" of an electron: (a) Does the electron literally spin about some axis? (b) What information does the quantum number m_s convey? (c) From where does it arise? (d) What values may it take for a single electron?

20. Which of the following combinations of quantum numbers n and ℓ are forbidden?

    (a) $1d$ (b) $5d$ (c) $5f$ (d) $5h$ (e) $6h$ (f) $8d$

    For each violation, state the reason why.

21. Name the subshell corresponding to each set of quantum numbers below:

    (a) $n = 3$, $\ell = 2$, $m_\ell = 1$
    (b) $n = 2$, $\ell = 1$, $m_\ell = 0$
    (c) $n = 4$, $\ell = 3$, $m_\ell = -2$
    (d) $n = 1$, $\ell = 0$, $m_\ell = 0$

22. Name the subshell corresponding to each set of quantum numbers below:

    (a) $n = 2$, $\ell = 0$, $m_\ell = 0$
    (b) $n = 6$, $\ell = 5$, $m_\ell = 5$
    (c) $n = 3$, $\ell = 1$, $m_\ell = -1$
    (d) $n = 3$, $\ell = 0$, $m_\ell = 0$

23. Which of these combinations of n, ℓ, m_ℓ, and m_s are impossible?

    (a) $n = 4$, $\ell = 2$, $m_\ell = 1$, $m_s = 0$
    (b) $n = 2$, $\ell = 1$, $m_\ell = 2$, $m_s = \frac{1}{2}$
    (c) $n = 5$, $\ell = 5$, $m_\ell = -2$, $m_s = -\frac{1}{2}$
    (d) $n = 1$, $\ell = 0$, $m_\ell = -1$, $m_s = \frac{1}{2}$

    For each violation, state the reasons why.

24. Which of these combinations of n, ℓ, m_ℓ, and m_s are impossible?

    (a) $n = 3$, $\ell = 2$, $m_\ell = 2$, $m_s = 1$
    (b) $n = 6$, $\ell = -1$, $m_\ell = 5$, $m_s = \frac{3}{2}$
    (c) $n = 3$, $\ell = 0$, $m_\ell = 1$, $m_s = \frac{1}{2}$
    (d) $n = 3$, $\ell = \frac{1}{2}$, $m_\ell = 0$, $m_s = -\frac{1}{2}$

    For each violation, state the reasons why.

25. How many nodes are contained within each orbital in the fourth shell? How many are angular nodes, and how many are radial nodes?

26. Does a hydrogen atom expand or contract when its electron moves from the ground state to an excited state? Explain the terms "expand" and "contract" in their proper quantum mechanical context.

27. Spectroscopic transitions: (a) What amount of energy must be supplied to excite a hydrogen electron from its ground state to the next higher level? (b) Is visible light sufficiently energetic? (c) Can the transition be induced by collision with a helium atom moving at 1000 m s^{-1}? Assume that all the kinetic energy of the helium atom is transferred to the hydrogen.

28. Compute the difference in energy between the fourth and fifth shells of a hydrogen atom. Is the comparable spacing more or less in a He$^+$ ion?

29. Between what two states of a hydrogen atom may a photon of frequency 6.169×10^{14} s^{-1} be absorbed or emitted?

30. Calculate energy and wavelength for each of the following transitions in a hydrogen atom:

    (a) $1s \longrightarrow 2p$
    (b) $5f \longrightarrow 4d$
    (c) $3d \longrightarrow n = \infty$
    (d) $n = \infty \longrightarrow 1s$

    Indicate whether the process is absorptive or emissive.

31. Calculate energy and wavelength for each of the following transitions in a He^+ ion:

    (a) $1s \longrightarrow 2p$
    (b) $5f \longrightarrow 4d$
    (c) $3d \longrightarrow n = \infty$
    (d) $n = \infty \longrightarrow 1s$

    How do the results compare with those obtained for H in the preceding example? Explain.

32. Excited by electromagnetic radiation with a wavelength of 827 Å, an electron is stripped from a ground-state hydrogen atom. (a) How much kinetic energy does the liberated electron possess after ionization? (b) What is its velocity?

33. Suppose that a free electron (velocity 1.00×10^6 m s^{-1}) comes under the influence of a hydrogen nucleus. How much energy is released when the electron settles into each of the following orbitals?

    (a) $3d$ (b) $3p$ (c) $3s$ (d) $2p$ (e) $2s$ (f) $1s$

34. This time, let the same free electron ($v = 1.00 \times 10^6$ m s^{-1}) come under the influence of a lithium nucleus. How much energy does the electron emit now as it falls into each of the following orbitals?

    (a) $3d$ (b) $3p$ (c) $3s$ (d) $2p$ (e) $2s$ (f) $1s$

35. The *Paschen series* arises when an emission terminates in the third shell of a one-electron atom. Compute the energy and wavelength of the photon emitted for the first four such transitions in a hydrogen atom:

$$n_i = 4 \longrightarrow n_f = 3$$
$$n_i = 5 \longrightarrow n_f = 3$$
$$n_i = 6 \longrightarrow n_f = 3$$
$$n_i = 7 \longrightarrow n_f = 3$$

    In what portion of the electromagnetic spectrum do the transitions fall?

36. Compute the energy and wavelength of the most energetic transition in the Paschen series for He^+ (see above). Is the change in energy larger or smaller than the comparable transition in a hydrogen atom?

37. Compute the energy and wavelength of the least energetic transition in the Paschen series for He^+. Is the change in energy larger or smaller than the comparable transition in a hydrogen atom?

38. The most energetic emission in the *Pfund series* for hydrogen appears at a wavelength of 2.279 μm. In what shell does the transition terminate?

39. Using results from the preceding exercise, calculate the energies and wavelengths of the first four lines in the Pfund series for Be^{3+}. Where in the electromagnetic spectrum do the signals lie?

40. The *Brackett series* of one-electron emissions terminates in the level $n = 4$. Compute the energy and wavelength of the first four Brackett transitions for H, He^+, and Li^{2+}.

41. Calculate, in turn, the longest wavelength sufficient to ionize H, He^+, and Li^{2+}.

42. A xenon atom with kinetic energy 2.28×10^{-18} J collides with a hydrogen atom in its ground state, leaving the hydrogen electron in the $n = 3$ level after the collision. With what speed does the xenon atom recoil away?

43. The ionization energy of a certain one-electron atom is 3.28×10^4 kJ mol^{-1}. How many protons are contained in its nucleus?

44. A series of four spectral lines is obtained from the one-electron atom described in the preceding exercise. The first three emissions have frequencies of 3.93×10^{14} Hz, 6.63×10^{14} Hz, and 8.56×10^{14} Hz. (a) In what shell do the emissions terminate? (b) Compute the energy, frequency, and wavelength of the fourth line in the series.

45. The ionization energy of an unknown atom is 3.28×10^3 kJ mol^{-1}. Does the atom contain one electron or more than one?

6

Periodic Properties
of the Elements

6-1. Orbitals in a Many-Electron Atom
   The Pauli Exclusion Principle
   Penetration and Shielding
   Electron—Electron Repulsion
6-2. Electron Configurations: Building Up the Atoms
   The First Shell: Hydrogen and Helium
   The Second Shell: From Lithium to Neon
   The Third Shell (s and p): From Sodium to Argon
   d Orbitals: The Transition Elements
   f Orbitals: Lanthanides and Actinides
   Organization of the Periodic Table
6-3. Valence Properties and the Periodic Law
   Across
   Down
   Atomic Size
   Ionization Energy
   Electron Affinity
   REVIEW AND GUIDE TO PROBLEMS
   EXERCISES

Quantum mechanics provides more than just a tidy accounting of the hydrogen atom. Beyond that revealing description—with its discrete orbitals and energies, its unraveled emission spectra—the quantum theory offers a new way to look at the world in miniature. A detailed understanding of the simplest atom, hydrogen, is only the first fruit of that worldview.

   Our task now is to come to a similar understanding of atoms with more than one electron: many-electron atoms; in short, the rest of the atomic universe. The one-electron atom will serve as a guide throughout.

6-1. ORBITALS IN A MANY-ELECTRON ATOM

Solving Schrödinger's equation for a one-electron atom is a tricky piece of work, but still the procedure can be carried out and carried out exactly. No approximations need be made to obtain the orbitals. No terms are thrown away; no simplifications are introduced. The hydrogenic wave functions are mathematically impeccable.

Not true for any other atom, though, for hydrogen stands apart as the only example where exact, analytical solution is possible. Add just one more electron and already there develops a fatal complication. The Schrödinger equation, although valid over the entire atomic and molecular world, remains insoluble unless some compromises are made.

Why is hydrogen different from all other atoms? H stands out, clearly, for the simplicity of its electrical interactions, notably for the absence of any competing electrons. One electron is attracted to one nucleus, with nothing else to intervene. There is no second electron to interpose a charge between the first electron and the nucleus, nor is there any repulsion between like-charged particles. Unhindered by additional interactions, hydrogen's lone electron has the nucleus all to itself. The pull depends on one distance only, equal in all directions.

That simplicity of forces is abruptly destroyed upon entrance of a second electron, as occurs first in helium (Figure 6-1). Here each electron is subjected to two interactions: an attractive pull toward the nucleus, plus a repulsive push away from the other negative charge. Electron 1's potential energy depends, consequently, not just on its own radial distance from the nucleus, but on the position of a third body as well. Suddenly there is an electron–electron repulsion to worry about, and this additional Coulomb interaction is controlled by the whereabouts of electron 2. Close together, the electrons repel more strongly. Far apart, less so. The electrical energy scales inversely with the distance r, varying as $1/r$.

Electron 1 therefore might undergo different repulsions even while its own attraction to the nucleus is fixed at a given radius. The potential energy of electron 1 depends on the positions of two other particles, nucleus *plus* electron 2, and the added complexity makes exact solution hopeless. Although computers can be enlisted to provide approximate (and sometimes highly accurate) answers, one can never write down a simple form using pencil and paper. Still worse, the mutual entanglement of electrons 1 and 2 takes away the familiar orbitals and quantum numbers of the hydrogen atom.

But consider this proposal as a way out. Suppose we pretend initially that the two electrons in helium move independently of each other. Here is number 1, and here is number 2. Number 1 buzzes around the nucleus, attracted by the positive charge; number 2 buzzes around the same

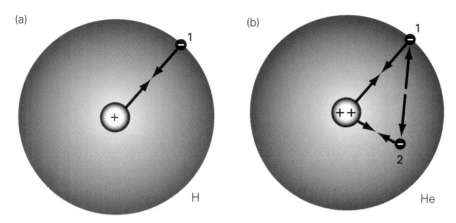

FIGURE 6-1. Electrical interactions in atoms. (a) In hydrogen, the attraction between the electron and proton is between two bodies only. The potential energy depends on just one distance. (b) Beginning with helium, repulsions as well as attractions combine to shape the many-electron atom. Every electron repels every other.

nucleus, also attracted by the positive charge. The whole picture dissolves into a whirl of rapid motion, a blurred cloud of electrons in which the separation between 1 and 2 is constantly changing. Imagine, then, that each electron reacts to the other simply as a diffuse cloud of charge, rather than as a point.

If so, then a remarkable simplification has been effected. Those point-to-point variations of electron–electron repulsion are smoothed away, replaced by an average push coming from all positions. Consequent to that smoothing—that everywhere-at-once blurriness—any previous sensitivity to a third body is apparently lost for each electron. The independence of a one-electron atom is restored, thereby turning the big atom into a loose confederation of individual electrons with a common nucleus at the center.

The image (Figure 6-2) is of a somewhat unsociable apartment building, to be contrasted with the enforced community of a military barracks. Although the apartment dwellers are aware of (and possibly seek to avoid) one another, each resident still moves in a largely personal realm. Adjustments are made, but such minimal coordination is insufficient to merge together all their separate lives.

So shall the structure of atoms fit this image. Each electron, presented with a centralized $1/r$ interaction over the course of time, acts as if it were in its own compartmentalized atom. An electron thus conceived can then be described using orbitals and quantum numbers similar to those already established for the hydrogen atom. It is a problem we can solve, a tempting possibility.

It is, moreover, a physically reasonable picture of a complex system, a model certainly worth putting to the test. If the assumptions are invalid,

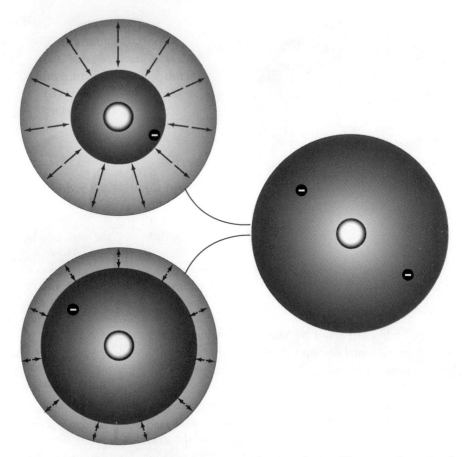

FIGURE 6-2. Many atoms in one? A mathematical simplification, the *orbital model* treats a many-electron atom as an association of one-electron atoms. The argument comes down to this: Each electron moves through its own space, attracted to a central nucleus while repelled—equally in all directions—by a cloud formed from all the other electrons.

then contradictory experimental results will discredit the hypothesis soon enough. But should the plan actually have merit, we shall acquire a compelling way to understand the atoms of the periodic table: as populated by quasi-independent electrons confined to hydrogen-like orbitals. There will be shells, subshells, and quantum numbers as before, subject to whatever modifications are needed to represent the averaged electron–electron repulsions.

One wonders, appropriately, if such a view conforms to reality. Often it does, if reality is taken to be those facts learned through observation and measurement. An imposing body of data suggests that a model of quasi-independent electrons is qualitatively suitable for many elements, at least in their ground states. This orbital picture accounts for the observed periodicities and also helps to explain how atoms react and form bonds. The scheme is approximate and admittedly imperfect, but the inevitable exceptions detract little from its larger chemical significance.

Let us assume, accordingly, that a many-electron atom has a full set of hydrogen-like orbitals at its disposal (1s, 2s, . . .), governed by the usual quantization rules for n, ℓ, and m_ℓ. Into these orbitals will go the atom's electrons, each deposited into a particular region of space with a certain fixed energy. Confined to an orbital, an individual electron then suffers some average repulsion owing to the presence of all the other negative charges. That will be our working model of the atom.

The Pauli Exclusion Principle

Immediately we ask how many of an atom's electrons can be placed into a single orbital. The answer is dictated not by Schrödinger's equation but by an additional law called the **exclusion principle**, first revealed by Wolfgang Pauli in the mid-1920s.

The Pauli exclusion principle provides quantum mechanical underpinning to a familiar fact of life: Two objects cannot be in the same place at the same time. It explains why fingers do not easily go through walls, why buildings do not fall to the center of the earth, and why atoms and molecules exist in their present forms.

Framed at the atomic level, the exclusion principle states that *no two electrons in an atom can have the same four quantum numbers*. They may share one quantum number or two or three, but not all four. Two electrons never reside in exactly the same state, occupying both the same region in space and having the same spin. They exclude each other.

The exclusion principle limits an orbital to a maximum of two electrons, and then only if the two spins are opposite. Placed into a common orbital, the particles start out with identical values of n, ℓ, and m_ℓ. To avoid duplicating the fourth quantum number, one spin must be up ($m_s = \frac{1}{2}$) and the other must be down ($m_s = -\frac{1}{2}$). Except for that, there is no way to force another electron into a single orbital without violating the Pauli principle.

Try to put three electrons into, say, a $2p_z$ orbital. Let the first particle enter with its spin up, so as to acquire the quantum numbers $n = 2$, $\ell = 1$, $m_\ell = 0$, and $m_s = \frac{1}{2}$. The second goes in with a different spin magnetic quantum number, $m_s = -\frac{1}{2}$, in order to comply with the law. These two electrons, represented by up and down arrows ($\uparrow\downarrow$), are now said to form a *spin pair*. Each member of the pair moves throughout a common volume of space with a different spin, and together they exhaust all possible arrangements of n, ℓ, m_ℓ, and m_s. Henceforth the $2p_z$ orbital is closed to new entrants, since any additional electron has only the same two combinations of quantum numbers to offer: $(1, 1, 0, \frac{1}{2})$ or $(1, 1, 0, -\frac{1}{2})$. Both sets are already in use, and so the filled orbital resists further incursion.

Thus established, the exclusion principle prevents electrons from piling into one region of space without limit. Nature has arranged things that way for reasons unknown, but the rules are enforced wherever there are electrons: maximum occupancy of two per orbital, paired spins only. It is a kind of zoning ordinance for ordinary matter.

Exclusion is responsible ultimately for matter's incompressibility, for the determined resistance encountered by a head banging against a wall. Too many electrons in the skull try to share intimate space with the electrons in the wall. Eventually something has to give.

Penetration and Shielding

Geography is key to an electron as it roams through the space of an orbital. Always at issue is the matter of location, framed in terms of distances between charges, of near and far, of where one particle is relative to another. It is a question best posed in two parts. Ask first how close an electron comes to the nucleus, and then ask how far that negative charge stays away from all others. Two by two, the charged particles make their contributions to the electrical energy. The negative-to-positive encounter between electron and nucleus is attractive and therefore energy lowering; the negative-to-negative encounter between electron and electron is repulsive, or energy raising. Uphill or downhill; a push or a pull. One road is easy, and the other is hard. Together, they make an atom.

For one-electron atoms, everything is easy. There is no repulsion between electrons, and the attractive pull of the nucleus is equally strong in all directions. The orbital energies, which scale roughly with the average electron–nucleus distance, depend only on the principal quantum numbers. Shape is irrelevant. All orbitals in a particular n shell have the same energy, with no distinction made between s, p, d, and f. The $1s$ has a certain energy E_1; the four $n = 2$ orbitals ($2s$, $2p_x$, $2p_y$, $2p_z$) have a different value E_2; and the nine $n = 3$ orbitals ($3s$, $3p_x$, $3p_y$, $3p_z$, $3d_{xy}$, $3d_{xz}$, $3d_{yz}$, $3d_{z^2}$, $3d_{x^2-y^2}$) have yet another value E_3. Wave functions such as these, equal in energy, are termed **degenerate**.

Degeneracy is usually a sign of symmetry, a sign that something stays the same despite changes elsewhere—as does the uniquely simple Coulomb energy in a one-electron atom, which varies equally as $1/r$ in all directions around a central point (the nucleus). That special $1/r$ invariance is lost with a second electron, however, and the accompanying degeneracy is lost along with it. No longer do the $2s$ and $2p$ orbitals have the same energy in a many-electron atom; no longer are the five 4*d* orbitals degenerate with the seven 4*f*'s.

Electrical circumstances clearly are different in the many-electron atom, where now an electron in one orbital is forced to adjust for the presence of all the rest. Each negative particle competes for a share of the nuclear attraction, and each faces some measure of electron–electron repulsions while so doing. The orbital energy is determined by the interplay between these two tendencies.

Particularly fruitful to this understanding are the related notions of **penetration** and **shielding** (Figure 6-3), which offer qualitative guides to "inner" and "outer" charge density. Penetration, on the one hand, is a measure of how close an electron gets to the nucleus. It is a view from the inside. Shielding indicates how thoroughly some other electron, farther away, is blocked from the nucleus and repelled by its fellows. It represents the outsider's perspective.

Picture, as an example, the relationship between two electrons in a single atom, one close to the nucleus and the other distant. A suitable choice might be an excited helium atom in which the 1*s* and 5*p* orbitals contain one electron apiece. Right away we know that these two electrons will meet only rarely, for the 5*p* resides far from the nucleus while the 1*s* remains tightly bound. Consider how each negative charge then feels the nuclear pull.

For the 1*s* electron, securely on the inside, that pull is the pull of *two* protons—the full nuclear charge of helium. Exposed to this uncovered

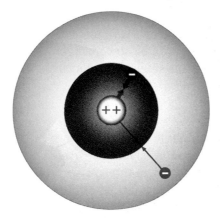

FIGURE 6-3. Penetration and shielding. The attractive positive charge looms larger for the electron nearer to the nucleus, the insider. This inner electron *penetrates* deeper into the core, gaining a stronger attraction for itself. At the same time, it *shields* the electron outside from the full nuclear charge.

nucleus with $Z = 2$, the inner electron is attracted more strongly than it would be in hydrogen. The orbital energy decreases relative to a hydrogen $1s$ state, and the included volume draws in closer. The $1s$ electron is said to *penetrate* deep into the atom's electronic core; it gets ahead of the other electron and enjoys a disproportionately large nuclear attraction.

For the $5p$ electron, the pull of the positive charge is diluted by the intervening $1s$ electron wrapped around the nucleus. Looking from afar, the outer electron sees two charges one atop the other: the first, $+2$, from the nucleus; the second, -1, from the inner electron. The effect from a distance is to present a net charge of $+1$ to the $5p$ electron ($2 - 1 = 1$), as if the nucleus were really hydrogen rather than helium. Thus not only does the $5p$ fail to penetrate, but it also incurs the additional penalty of being *shielded* (or *screened*) from the full positive charge by the $1s$ electron whirring close about the nucleus. Offered this attenuated attraction, the shielded electron moves higher in energy and expands its orbital volume accordingly.

To penetrate, then, is to get in close and decrease your own energy. To be shielded is to be locked out by the presence of other, more tightly bound electrons. Penetration and shielding—the insider's view and the outsider's view of the nucleus—jointly affect the energetics of a many-electron atom, combining both attractions and repulsions. They come together in a single parameter called the ***effective nuclear charge***, Z_{eff}, which seeks to estimate how much residual positive charge is available to a particular electron.

Under the present scenario, for instance, Z_{eff} would be approximately 2 for the inner electron and close to 1 for the outer electron. For neutral atoms in general, Z_{eff} falls between the atomic number Z (envision the innermost electron at a bare nucleus) and 1 (picture the outermost electron, assuming it to be screened completely by $Z - 1$ negative charges). The definition, rendered graphically in Figure 6-4, is

$$Z_{\text{eff}} = Z - S$$

where S denotes the average number of inner electrons responsible for the shielding of some outer electron. Realize that S, representing an electron density, need not be an integer.

What more can we say about penetration and shielding, given our knowledge of orbital shapes? Most strongly penetrating are the s functions, which alone display nonzero probability at the nucleus. All other electrons are thrust farther away because they possess angular momentum: f more so than d; d more than p; p more than s. Electrons in the p, d, and f subshells, finding themselves increasingly distant from the nucleus, then respond to a smaller and smaller effective nuclear charge as

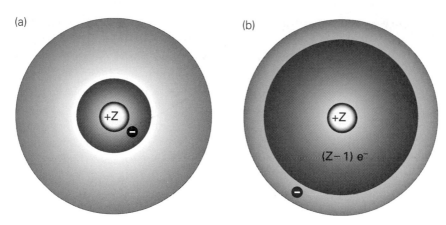

FIGURE 6-4. The effective nuclear charge: $Z_{eff} = Z - S$. Screened by a repulsive cloud in between, a shielded electron interacts with a nuclear charge apparently reduced from its full value. Z is the atomic number. S is the shielding factor, symbolized by the colored circle around the nucleus. (a) Z_{eff} approaches Z for an electron deep in the core, close to the nucleus. (b) Z_{eff} approaches 1 for the outermost electron in the valence.

well. The attraction is necessarily weakened, and the orbital energy increases from *s* to *p* to *d* to *f*.

Figure 6-5 makes the point clear by showing radial distribution functions for *s* and *p* orbitals within a single shell. Recall (from Chapter 5) that

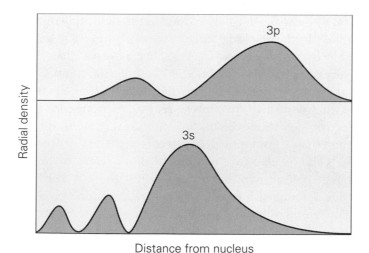

FIGURE 6-5. The lower the angular momentum, the more closely an electron approaches the nucleus. Its radial density peaks at a smaller distance, as seen here for the 3*s* and 3*p* orbitals of a sodium atom. The *s* electron, penetrating more, is less shielded. The *p* electron, penetrating less, is more shielded.

these curves predict the total electron density at any distance, regardless of direction. The *s* electron, which penetrates deeper than the others, accumulates extra humps of probability close to the positive charge. It gets in closer. The *s* penetrates better than the *p* or, put the other way, the *p* is better *shielded* than the *s*. Similarly a *p* electron penetrates more strongly than a *d* electron, thereby leaving the five *d* orbitals more highly shielded than the three *p* functions. Expect, therefore, that orbitals in a given *n* shell will be ordered energetically as $E_s < E_p < E_d < E_f$. The nearer to the nucleus, the lower the energy.

Those are the arguments concerning subshells belonging to a certain principal quantum number (why, for example, E_{2s} is less than E_{2p}). Jumping next from shell to shell, we see additional changes brought on by the overall expansion of the electron distribution. Orbitals of lower *n*, built closer to the nucleus, usually penetrate more effectively than those of higher *n*. E_{2p} should be less than E_{3s} according to this reasoning, since orbitals in the second shell tend to be smaller than ones in the third. Such is typically the trend, and usually the energies do increase with *n*—but still there are noteworthy exceptions, which we shall soon uncover during our analysis of atomic structure and the periodic table.

Electron–Electron Repulsion

Before filling the orbitals, let us delve further into the issue of electron versus electron. Being in an *imagined* one-electron orbital does not allow an electron to escape repulsion, and indeed some of this repulsion is already implicit in the shielding effect. We say, somewhat loosely, that a shielded electron is attracted to a smaller positive charge. What we mean, expressed more sharply, is that the full attraction is reduced by negative-to-negative repulsions involving various other electrons.

Since energy-raising encounters are most pronounced in regions where electron densities overlap, the repulsions come to depend on orbital size and shape. Sometimes the effect is slight and easily overlooked, as for the 1*s* and 5*p* electrons mentioned above. Here the *n* = 1 shell is concentrated far nearer to the nucleus than is the *n* = 5 shell, and so a considerable amount of interelectron separation is enforced by this difference alone. Adding to the separation are the different territories covered by these orbitals, with the *p* electron permanently excluded from the immediate nuclear vicinity (where its wave function vanishes). Kept apart by such restrictions, the two particles incur minimal repulsions.

More repulsive than a 1*s*–5*p* interaction would be an encounter between 1*s* and 5*s*, for then the similarly shaped *s* densities (both spherical)

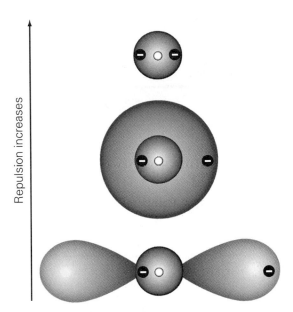

FIGURE 6-6. Orbitals similar in size and shape occupy space in common. Electrons in partially overlapping orbitals suffer correspondingly greater repulsion, all the more as the common region grows larger.

would be better able to overlap. Still more intrusive would be a $1s$–$4s$ interaction, exceeded in turn by a $1s$–$3s$ and then a $1s$–$2s$ repulsion. With each drop in the principal quantum number, the second electron's cloud becomes progressively smaller and hence better matched to the $1s$. The effect, illustrated in Figure 6-6, is strongest when both electrons jointly occupy the same orbital, forming a spin pair in the $1s$.

The reason is plain. Two electrons in a single orbital travel through exactly the same space, and thus two paired electrons suffer the most repulsion. With one spin up and the other spin down, their three spatial quantum numbers (n, ℓ, m_ℓ) are identical. They meet *each other* more frequently than anyone else, and this increased repulsion is the price paid for double occupancy of the orbital. It costs energy to form spin pairs.

That cost is worth paying if the alternatives are even more expensive. Helium's two electrons, for instance, readily enter the $1s$ orbital as a spin pair (as do the first two electrons of all atoms in their ground states). It is an energetic bargain: The increased $1s$–$1s$ electron–electron repulsion is small compared with the much larger gap in energy between the first and second shells (Figure 6-7). Were helium to exist as $1s2s$, the atom would spend far more energy in promoting an electron from $n = 1$ to $n = 2$ than it could save by avoiding double occupancy of the $1s$.

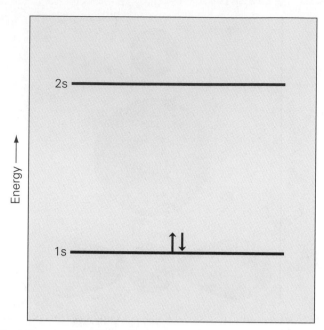

FIGURE 6-7. Atoms in the same orbital pay a price to be together, but the energy exacted is usually less than the gap between shells. Helium's electrons both occupy the 1*s* orbital in the atom's ground state—one electron with spin up, one with spin down. All other configurations lie at higher energies.

Yet double occupancy is not merely the lesser of two evils, for there are genuine advantages as well. The second electron is a net winner. By entering a portion of space already occupied by one electron, it positions itself closer to the nucleus than otherwise possible. Two electrons in the 1*s* orbital shield each other least of all, precisely because they do claim common ground. Since neither particle, on the average, penetrates closer to the nucleus, the second electron enjoys the increased nuclear attraction coming from a higher Z_{eff}. For a 1*s* electron, the benefit of a filled $n = 1$ shell is worth the price paid to share space.

Consider what might happen, though, when electrons are distributed over a number of orbitals with *identical* energies, as can occur whenever ℓ is unequal to zero. For although *s*, *p*, *d*, and *f* orbitals all have different energies in a complex atom, there is still no differentiation within the various subshells. Each of the three *p* functions has the same energy. Each of the five *d* functions has the same energy. Each of the seven *f* functions has the same energy. So they must, because these orbitals differ only in direction rather than shape. Lacking some special marker (such as a magnetic field) to distinguish *z* from *x* from *y*, we can go from p_z to p_x to p_y just by rotating the axes. That sameness is evidence of an underlying symmetry, from which arises a corresponding degeneracy once again—here a degeneracy in the subshell energies. No energy can depend on a set of arbitrary labels on a diagram, and so one value must be the same as another.

Look at the choices available to two electrons entering a p subshell, laid out systematically in Figure 6-8: $p_x p_x$, $p_x p_z$, $p_x p_y$, $p_z p_z$, $p_y p_z$, $p_y p_y$, one spin up and one spin down ($\uparrow\downarrow$), one spin down and one spin up ($\downarrow\uparrow$), and (Pauli permitting) both down ($\downarrow\downarrow$) or both up ($\uparrow\uparrow$). Some of these

FIGURE 6-8. Two electrons, each with a spin, may distribute themselves among three degenerate p orbitals in various ways. Some configurations have lower energies than others.

possibilities, obviously higher in energy than the rest, stand out instantly as unlikely candidates for the ground state. Note, in particular, how an arrangement such as __ __ ↑↓ will force the electrons into gratuitously close quarters:

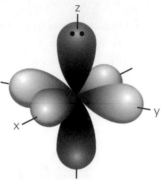

The spin pair creates needless repulsions, since each of the unoccupied p orbitals lies at the same energy and is freely accessible at no extra cost. To put two electrons into p_z, say, rather than one apiece into p_x and p_y, is to do things the hard way. It is energetically easier for an atom to send the two electrons off in different directions:

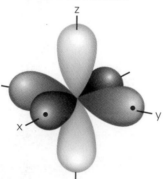

Electrons thus fill a set of degenerate orbitals in the same way that unacquainted passengers would fill the seats of a bus: one at a time, pairing up only after all slots have been occupied. They stay out of each other's way and effectively minimize the repulsive interactions. Electrons, too, can further increase their avoidance by entering the different orbitals with parallel spins, either all up or all down. Arrangement I, for example, lies at a lower energy than arrangement II:

Keeping their spins parallel lessens the chance that the two electrons will meet at the same point in space.

Because: Even to share this *one* quantum number (the spin) runs counter to the spirit of the exclusion principle. Nature says: If you have the same spin, stay apart wherever possible—in different orbitals, of course, but you can still do better than that. Take care that your positions never actually coincide. Keep away. Avoid close calls. The less repulsion, the better.

These observations find formal expression in **Hund's rule**, which states that an atom in its ground state tends to maximize both the number of unpaired electrons and the number of spins in the same direction. Distributed over a group of orbitals equal in energy, electrons enter one at a time with parallel spins. Pairing is avoided until each degenerate orbital contains one electron.

6-2. ELECTRON CONFIGURATIONS: BUILDING UP THE ATOMS

What remains is to fill the orbitals with electrons, one by one from the bottom up. Our conceptual tools, developed just above, include:

1. The Pauli exclusion principle, which restricts orbital occupancy to no more than two electrons, one with spin up ($m_s = \frac{1}{2}$) and the other with spin down ($m_s = -\frac{1}{2}$).
2. Qualitative ideas of penetration, shielding, and electron–electron repulsion, which suggest how the levels might be ordered.
3. Hund's rule, which accounts for the filling of degenerate orbitals.

Understand, however, that these tools did not spring fully grown early in the 20th century, but were developed empirically to help organize a mass of experimental data. First came spectroscopic observations, through which the principal features of each atom became known. Quantities such as electronic energy, angular momentum, and spin were measured by atomic spectroscopy. Careful observation (not unbounded imagination, not mathematical speculation) revealed the number and spacing of the quantized energy levels, together with the number of unpaired electrons in the various atoms. *Then* came the model. *Then* came the orbitals; then came the exclusion principle; then came the rules.

With that admonition, we go on to explore the ground states of many-electron atoms. It will be a process of building up (called the **Aufbau principle**, from the German), and it will be an exercise that mimics nature's own progressive formation of the atoms: Add a proton to the nucleus; add an electron outside. Maximize the attractions and minimize the repulsions.

The First Shell: Hydrogen and Helium

Lowest in energy and hence first to be filled is the $1s$ orbital, into which goes one electron for hydrogen (↑) and two for helium (↑↓). Using superscripts to indicate the orbital population, we can express these two **electron configurations** compactly as

$$
\begin{array}{lll}
\text{H} & 1s^1 & \uparrow \\
\text{He} & 1s^2 & \uparrow\downarrow
\end{array}
$$

while recognizing that helium's up-down pairing is implicit in the notation $1s^2$.

With such symbols, we reinterpret Lewis's dot diagrams (H· and He:) in a quantum mechanical context. Hydrogen's one dot means something very specific now. It means an electron in the $1s$ orbital, to be found most often on a sphere of radius 0.529 Å. It means an electron with zero angular momentum—an electron able, in turn, to penetrate deeply toward the nucleus. It means an electron bound to that nucleus with an energy of -2.18×10^{-18} J, trapped in the lowest of all possible orbitals and therefore hardest to remove. Clearly this atom or any other will be toughest in its ground state, most resistant there to disruption and consequently most likely to survive.

Hydrogen's one dot means, also, an electron with an unpaired spin. The electron becomes a tiny magnet able to line up in a magnetic field. Acting like a quantized compass needle, it can point either up ($m_s = \frac{1}{2}$) or down ($m_s = -\frac{1}{2}$) in the field. This property, called **paramagnetism**, is demonstrated by any atom having at least one unpaired electron spin.

Along comes a second electron (together with a second proton in the nucleus), and a neutral atom of helium is formed. Now one spin is up and the other is down, leaving the first electron to cancel the magnetism of the second. The helium atom, with two electrons in the $1s$ orbital, is termed **diamagnetic** rather than paramagnetic. Its electrons do not line up in a magnetic field, for the *net* spin angular momentum is zero. Individually magnetic, the electrons are nonmagnetic as a pair.

We already know why helium prefers its $1s^2$ configuration to alternatives such as $1s^1 2s^1$ or $1s^1 2p^1$. The atom's total energy, a balance of attraction and repulsion, is lowest with two electrons in the $1s$ orbital. So into the $1s$ they both go, filling up the first shell and excluding any other occupants. This closed-shell configuration, He:, then stands as a completed chapter in the buildup. To accept an electron from another atom entails the difficulty of opening a new shell, and to surrender an electron means tearing away a tightly bound particle from a hard-to-disturb, filled shell. Unable to find a better deal energetically, helium remains chemically aloof. It is the first of the noble gases.

The Second Shell: From Lithium to Neon

With $1s$ completely filled, the $2s$ orbital becomes the lowermost level available. Lithium's third electron enters this orbital to establish a $1s^2 2s^1$ configuration, followed by beryllium ($Z = 4$) with its $1s^2 2s^2$ configuration. Lithium and beryllium, starting anew in the second shell, thus repeat the pattern of hydrogen and helium by populating an s orbital. Like hydrogen, a lithium atom contains one unpaired electron and is paramagnetic. Like helium, beryllium is diamagnetic owing to its fully paired-up spins.

But, unlike helium, the chapter is not closed with the filling of beryllium's $2s$ orbital. Still accessible (at a relatively modest cost in energy) are the three p orbitals of the $n = 2$ shell, which together can accommodate six electrons. Six electrons: one additional electron for boron ($Z = 5$), two for carbon ($Z = 6$), three for nitrogen ($Z = 7$), four for oxygen ($Z = 8$), five for fluorine ($Z = 9$), and six for neon ($Z = 10$). Filing into the degenerate orbitals one by one, they spread out according to Hund's rule to produce the sequence

	$n = 1$	$n = 2$	$2p_x$	$2p_y$	$2p_z$
B	$1s^2$	$2s^2 2p^1$	↑	—	—
C	$1s^2$	$2s^2 2p^2$	↑	↑	—
N	$1s^2$	$2s^2 2p^3$	↑	↑	↑
O	$1s^2$	$2s^2 2p^4$	↑↓	↑	↑
F	$1s^2$	$2s^2 2p^5$	↑↓	↑↓	↑
Ne	$1s^2$	$2s^2 2p^6$	↑↓	↑↓	↑↓

where x, y, and z denote *any* three perpendicular axes. Recognize, again, that what we call x, y, and z is irrelevant, since the p-orbital energies are all the same in an isolated atom. Without imposing any unwarranted special direction, we simply label the three sublevels arbitrarily and adopt the abbreviated notation p^1, p^2, . . . , p^6.

The rest is taken care of by Hund's rule. There are three equivalent p orbitals to be filled, and the favored configurations are those best able to minimize electron–electron repulsions. Each atom so manufactured—except neon—has at least one unpaired spin and is paramagnetic as a result.

Upon formation of neon, two electrons reside in the $2s$ orbital and six in the $2p$. There are, altogether, eight electrons in the $n = 2$ shell, from which comes the stable configuration that G. N. Lewis called an octet. Neon, with its fully occupied second shell, is the second of the noble gases.

The Third Shell (s and p): From Sodium to Argon

The series lithium through neon, over which the second shell is opened and closed, constitutes the second row of the periodic table (after hydrogen and helium):

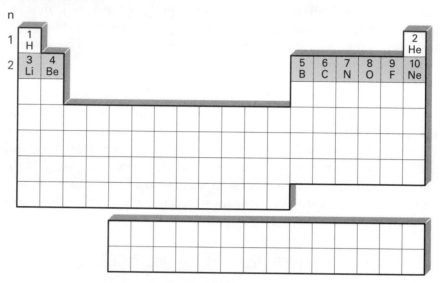

The Aufbau process then continues with the filling of the 3s and 3p orbitals, beginning with sodium ($Z = 11$) and terminating at argon ($Z = 18$), the next noble gas. These atoms are all built upon a 10-electron, neonlike core containing five doubly occupied inner orbitals. Using the symbol [Ne] to denote this $1s^2 2s^2 2p^6$ core density, we then run through eight new configurations,

$$
\begin{array}{ll}
\text{Na} & [\text{Ne}]3s^1 \\
\text{Mg} & [\text{Ne}]3s^2 \\
\text{Al} & [\text{Ne}]3s^2 3p^1 \\
\text{Si} & [\text{Ne}]3s^2 3p^2 \\
\text{P} & [\text{Ne}]3s^2 3p^3 \\
\text{S} & [\text{Ne}]3s^2 3p^4 \\
\text{Cl} & [\text{Ne}]3s^2 3p^5 \\
\text{Ar} & [\text{Ne}]3s^2 3p^6
\end{array}
$$

in a reprise of the sequence just described for $n = 2$.

Any differences between the $n = 3$ and $n = 2$ atoms are outweighed by the similarities. The heavier species do contain [Ne] rather than [He] cores, but, far more important, the pattern of occupancy beyond the inner shells is identical across the two series. Lithium has one valence electron, an *s* in the second shell; sodium has one valence electron, an *s* in the third shell. Fluorine has seven valence electrons in a $2s^2 2p^5$ configuration; chlorine has seven valence electrons in a $3s^2 3p^5$ configuration. Neon has an octet; argon has an octet.

There, in recurring patterns such as these, is the origin of chemical periodicity. In the rhythmic return of the valence configurations we find the reason why sodium has properties similar to lithium, why chlorine has properties similar to fluorine, why argon has properties similar to neon. It is outside the core—outside, where the orbitals contain vacancies—that chemistry takes place. It is these outermost electrons, the chemically active *valence* electrons, that are shared and passed from atom to atom. It is here, outside the core, where atoms in the same family match up, electron for electron and orbital for orbital.

As now: The eight third-shell elements, sodium through argon, line up directly beneath the eight second-shell elements. They form a new row:

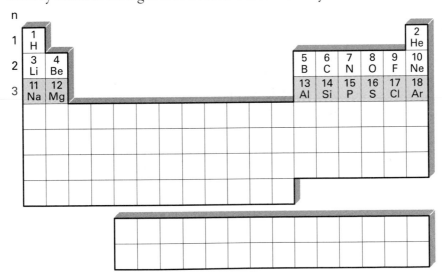

d Orbitals: The Transition Elements

From hydrogen through argon, where orbital energies go as

$$E_{1s} < E_{2s} < E_{2p} < E_{3s} < E_{3p}$$

our simple expectations concerning penetration and shielding are apparently fulfilled. Energy, we said, should increase with *n*, since an electron's average distance from the nucleus goes up from one shell to the next. The farther away, the weaker the attraction. The weaker the attraction, the higher the energy. We argued also that orbital energies within a single shell should increase in the order $s < p < d < f$, consistent with a progressive decrease in penetration and a corresponding increase in shielding. A $3p$ electron, for example, is kept farther from the nucleus and therefore at higher energy than a $3s$ electron.

Next in line (by this reasoning) would come the $3d$ orbital, set to be filled just after argon. The pattern is abruptly broken, however, when the $4s$ makes an unscheduled appearance in potassium ($Z = 19$) and calcium ($Z = 20$) to begin the periodic table's fourth row. These configurations, shown experimentally to be

$$K \qquad [Ar]4s^1$$
$$Ca \qquad [Ar]4s^2,$$

suggest that E_{4s} lies below E_{3d} despite having the higher principal quantum number n. Indeed n tells only part of the story, because shielding (see above) is determined also by the angular momentum quantum number ℓ—or, pictured physically, by the *shape* of the orbital. And, in this complicated environment of some 19 and 20 electrons, the $3d$ orbital's higher angular momentum proves decisive. The $4s$ electron, better able to penetrate than the strongly shielded $3d$, garners the higher Z_{eff} and gets the lower energy. The difference is small but sufficient, enabling the $4s$ orbital to fill before the $3d$ in potassium and calcium.

This inversion is short-lived, though, and E_{3d} again falls beneath E_{4s} beginning with the very next element (scandium, $Z = 21$), where the levels stand as

$$4s \quad \underline{}$$
$$3d \quad \underline{} \;\; \underline{} \;\; \underline{} \;\; \underline{} \;\; \underline{}$$

Nevertheless there is still an anomaly, for the observed configuration

$$Sc \qquad [Ar]4s^2 3d^1$$

contains a filled $4s$ subshell and only one d electron, not the $[Ar]3d^3$ configuration

$$4s \quad \underline{}$$
$$3d \quad \uparrow \;\; \uparrow \;\; \uparrow \;\; \underline{} \;\; \underline{}$$

expected if all three valence electrons entered the lower-lying $3d$ subshell. Yet despite the slightly lower energy of the $3d$ orbitals, the electrons fill the levels seemingly out of turn: one in the $3d$ (downstairs) and two in the $4s$ (upstairs) to yield the upside-down configuration

$$4s \quad \uparrow\downarrow$$
$$3d \quad \uparrow \;\; \underline{} \;\; \underline{} \;\; \underline{} \;\; \underline{}$$

This ordering is firmly established by ionization experiments, which show that a $4s$ electron—not a $3d$—is first to be stripped from a scandium atom.

We learn something. We learn that an atom's *total* energy is not equal simply to the sum of its orbital energies. That naive expectation is compromised by the electron–electron repulsions, which force each electron to recognize the existence of every other. Since no electron can truly be independent in a many-electron atom, no orbital is truly isolated and set apart. The orbitals are only approximations, reflecting a simplified picture of electron behavior and hence always subject to revision. An atom's complete energy, inclusive of repulsions, is determined instead by all the electrical interactions.

Scandium therefore does what it does for the sake of the lowest total energy, not to respect any idealized layout of orbitals. The two electrons in the $4s$ subshell stay where they are, content to decline the offer of a lower orbital energy in the $3d$. The $4s$ electrons, as individuals, become weaker and easier to detach; but the atom, as a whole, grows stronger. The energy of all the electrons together is lower.

Such occasional surprises occur throughout the periodic table, particularly when small differences in energy are involved (as here between $4s$ and $3d$). They are in no way distressing, given the streamlined assumptions of the orbital model. The picture of a loosely connected community of electrons continues to hold.

Comforted by that thought, we discover the same kind of anomalies while proceeding to organize the d electrons into the first 10 **transition elements**. The atoms thus formed, appearing as a bulge in the row just begun, are:

		$4s$	$3d$				
Sc	$[Ar]4s^2 3d^1$	↑↓	↑	—	—	—	—
Ti	$[Ar]4s^2 3d^2$	↑↓	↑	↑	—	—	—
V	$[Ar]4s^2 3d^3$	↑↓	↑	↑	↑	—	—
Cr	$[Ar]4s^1 3d^5$	↑	↑	↑	↑	↑	↑
Mn	$[Ar]4s^2 3d^5$	↑↓	↑	↑	↑	↑	↑
Fe	$[Ar]4s^2 3d^6$	↑↓	↑↓	↑	↑	↑	↑
Co	$[Ar]4s^2 3d^7$	↑↓	↑↓	↑↓	↑	↑	↑
Ni	$[Ar]4s^2 3d^8$	↑↓	↑↓	↑↓	↑↓	↑	↑
Cu	$[Ar]4s^1 3d^{10}$	↑	↑↓	↑↓	↑↓	↑↓	↑↓
Zn	$[Ar]4s^2 3d^{10}$	↑↓	↑↓	↑↓	↑↓	↑↓	↑↓

The five d sublevels are populated according to Hund's rule, but notice how both chromium and copper transfer one electron out of the $4s$

orbital. This rearrangement, which creates a half-filled subshell ($3d^5$) for Cr and a filled $3d$ subshell for Cu ($3d^{10}$), decreases the total energy by reducing electron–electron repulsions.

Zinc, with its $3d^{10}$ configuration, closes the third shell. At this point, too, the filled $3d$ orbitals have fallen so far below the $4s$ as to become part of the *core*, not the valence. Next to be filled in the valence shell ($n = 4$) are the $4p$ orbitals, starting with gallium ($Z = 31$, $[Ar]4s^2 3d^{10} 4p^1$) and ending with the noble gas krypton ($Z = 36$, $[Ar]4s^2 3d^{10} 4p^6$). After that, there begins a similar 18-atom row (rubidium through xenon) in which the s, d, and p orbitals are again filled in succession. The sequence opens with the $5s$, continues with 10 new ($4d$) transition metals from yttrium through cadmium, and concludes with the buildup of the $5p$ subshell between indium and xenon:

Following xenon, another noble gas, a new row commences with the filling of the $6s$ orbital in cesium and barium.

f Orbitals: Lanthanides and Actinides

By now we have been through the first 56 elements, hydrogen through barium, observing the orbital energies to increase in the order

$$E_{1s} < E_{2s} < E_{2p} < E_{3s} < E_{3p} < E_{4s} < E_{3d} < E_{4p} < E_{5s} < E_{4d} < E_{5p} < E_{6s}$$

Conspicuously absent, however, have been the seven $4f$ functions, which stay well shielded and far from the nucleus throughout. Every other electron, it seems, gets in closer and enjoys a lower energy. The $4f$ subshell, handicapped energetically by both high angular momentum and a far-flung radial electron density, must cede its place to orbitals in the fifth and sixth shells. Unable to compete with more penetrating *s* and *p* electrons, the 4*f* orbitals remain higher in energy than the 5*s*, the 5*p*, and even the 6*s*. They stay unfilled until the 5*d* electrons, strongly shielded themselves, enter the picture.

Here is the sequence: The 57th element, lanthanum (with configuration [Xe]$6s^2 5d^1$), accepts one 5*d* electron to initiate a third series of transition elements, but immediately this buildup is interrupted. The seven 4*f* orbitals, having finally fallen sufficiently low in energy, intervene and receive their full complement of electrons. Fourteen **rare-earth**, or **lanthanide elements** are formed in turn, from cerium ($Z = 58$, [Xe]$6s^2 4f^1 5d^1$) through lutetium ($Z = 71$, [Xe]$6s^2 4f^{14} 5d^1$). The lanthanides, their chemistry determined by the *f* orbitals, are segregated at the bottom of the table into a supplementary row.

With the 4*f* subshell filled at element number 71, the remaining 5*d* transition elements are completed. The interrupted series resumes in the row starting with cesium ($n = 6$), where it picks up at hafnium ($Z = 72$, [Xe]$6s^2 4f^{14} 5d^2$) and continues to mercury ($Z = 80$, [Xe]$6s^2 4f^{14} 5d^{10}$). A last noble gas, radon ($Z = 86$), is formed upon full population of the 6*p* orbitals:

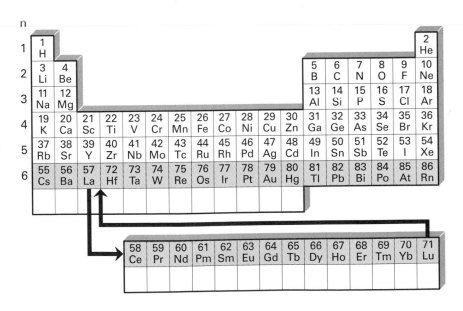

A new row starts with the 7*s* electrons of francium and radium, after which the 6*d* and 5*f* orbitals become active from actinium ($Z = 89$, [Rn]$7s^2 6d^1$) through the recently discovered elements number 110, 111, and 112. Included in this run are the radioactive **actinide elements** (separated, like the lanthanides, from the body of the periodic table). The actinide series extends from thorium ($Z = 90$) to lawrencium ($Z = 103$), over which the seven 5*f* orbitals are filled:

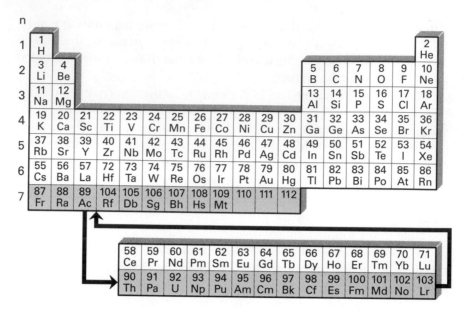

Most of these last atoms do not occur naturally, but instead must be generated by artificial transmutation of the nucleus. The nuclear reactions so induced, described in Chapter 21, bring about a change in the number of protons.

Organization of the Periodic Table

This, to recapitulate, is the genealogy of the periodic table: From hydrogen to helium, from lithium to neon, from sodium to argon . . . and, correspondingly, from the $1s$ orbital to the $2s$ and $2p$, to the $3s$ and $3p$, $4s$, $3d$, and beyond. This is our model, fully realized, of the chemical elements, a model of atoms, electrons, and orbitals. It is a model of electrons shielded and deshielded, attracted and repelled, bumped up and down in energy; a model complicated in its details but manageable overall. The order of filling, a summary of the entire building-up process, is sketched in the mnemonic diagram of Figure 6-9.

Still more useful is the block diagram of the periodic table shown in Figure 6-10, less a mnemonic than a real guide to organization and understanding. Within these boxes, the broad features of the electron configurations fall into place. Laid out plainly are the recurring patterns common to the buildup, grouped according to the pertinent valence subshells.

There is the *s block*, consisting of the first two columns, for which the chemistry is determined by one or two electrons in a valence s orbital. The s block brings together lithium and cesium, with analogous $2s^1$ and $6s^1$ configurations; and it brings together calcium and strontium as well, with $4s^2$ and $5s^2$ configurations. These ns^1 and ns^2 species are the alkali

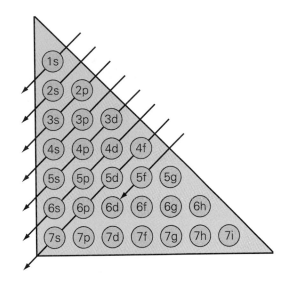

FIGURE 6-9. The building-up sequence: an aid to memory. Starting at the upper left-hand vertex, follow the diagonal arrows to recall the order (1s, 2s, 2p, 3s, . . .) in which the orbitals usually fill.

metals and alkaline earth metals of Groups I and II, having one and two electrons to lose. Surrendering them, the s-block atoms acquire oxidation states of +1 and +2 in their respective compounds and thereby achieve noble-gas configurations.

Then there is the **p block**, standing off to the right. A rectangle six columns across and five rows deep, the p block cordons off the valence p orbitals. At the top, the second-shell atoms add electrons to create the configurations $2p^1$ through $2p^6$; at the bottom, with lower-lying d and f subshells already completed, the 6p orbitals are similarly filled from $6p^1$ to $6p^6$. Pooling their p electrons with the two s electrons already present

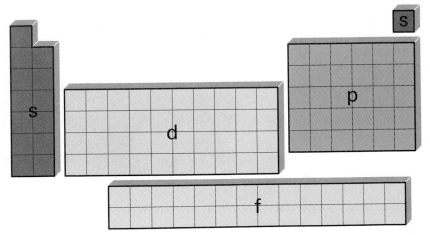

FIGURE 6-10. The periodic table, partitioned into s, p, d, and f blocks. Each block takes its name from the last valence subshell to be filled during the Aufbau process.

(same n), the atoms in each row of the p block exhibit varied oxidation states and combining capacities.

Terminating the p block are the noble gases, unreactive except for some xenon-fluorine and krypton-fluorine compounds. The noble gases are satisfied energetically with what they have, whether it be 2 s electrons in the valence (He), or 8 s and p electrons (Ne, Ar), or 18 s, p, and d electrons (Kr, Xe), or 32 s, p, d, and f electrons (Rn). All the shells are filled.

Next is the **d block**, stretched out between the s group at the left and the p at the right. Here the d electrons present themselves in three rows of transition metals—three rows with a rich chemistry involving many oxidation states and punctuated by bold colors. Deriving their special properties from the d orbitals, the transition elements are set apart from the aforementioned **representative** or **main-group elements** of the s and p blocks.

Finally there are the two 14-element rows of the **f block**, in which the lanthanides and actinides are displayed as auxiliary periods. For these atoms it is the exotic chemistry of the f orbitals that proves determinative, and the lanthanide elements, in particular, really do behave as a block. Existing close together in nature, they share similar properties and are difficult to separate. The actinides, which decay radioactively into atoms with lighter nuclei, have a complex and more varied chemistry.

6-3. VALENCE PROPERTIES AND THE PERIODIC LAW

The periodic table existed in rudimentary form long before chemists understood basic atomic structure, to say nothing of electron configurations. What they observed, instead, was a regularity in the *properties* of the elements, a recurrence that we know today to arise from the electrons within.

But first there were only properties: properties such as boiling point and melting point, atomic weight, density, color, combining capacity, compounds formed. Next, thanks to the separate insights of Mendeleev and Meyer around 1870, there was *organization*: a table of the elements, arranged by increasing atomic weight. And even though the ordering should have been by atomic number (an unimaginable parameter at the time), still it was an eye-opening display. The first periodic table brought into focus, on a single sheet of paper, the commonality of lithium and sodium and potassium; of magnesium and calcium; of fluorine and chlorine; of all the various groups. It codified the **periodic law**, which asserts that elemental chemical and physical properties reappear at regular intervals according to atomic number.

Soon after, the periodic law was demonstrated convincingly with the prediction and discovery of several new elements. Mendeleev, faced with a hole in his table, would call boldly for some previously unheard-of substance to be brought to light. One such element X, for instance, he expected to be an easily melting metal with a relative atomic mass of 68 and a density of 6.0 g cm^{-3}. Among its compounds would be an oxide with empirical formula X_2O_3 and density 5.5 g cm^{-3}, as well as a chloride having formula XCl_3. This "eka-aluminum" had to exist, argued Mendeleev, in order to fill the gap in a group already composed of boron, aluminum, indium, and thallium.

Eka-aluminum, our present-day gallium, was found just a few years after Mendeleev's statement of the periodic law, along with other new elements similarly predicted. Observed and expected properties were in striking agreement, and the cycle of observation–hypothesis–experiment was crowned each time with a most dramatic discovery: a new atom, a new particle of matter.

The periodic law, impressively confirmed, is experimental proof that there is regularity in matter and order among the elements. Nevertheless, for all the drama of prediction and discovery, an *unexplained* periodic table leaves something to be desired. The rows and columns are neat, orderly, and practical, yes, but why? An uninterpreted periodic table is more of a puzzle than a solution. After organization must come explanation.

The explanation, we know, lies in the electrical and quantum mechanical properties of the valence electrons. Chemistry is the giving and taking of electrons. Chemistry is the merging of atoms into molecules through the sharing of electrons; chemistry is the transformation of one molecule into another, mediated also by the shuffle of electrons. Chemistry is the electrical association of atoms and molecules into solid, liquid, and gas.

A chemical transformation may hinge on whether some electron is easy or hard to move from here to there, or whether this or that ion is large or small relative to some other. These are underlying microscopic questions that are answered macroscopically in the flash of a chemical reaction—hard questions that now can be addressed using the ideas of penetration, shielding, and effective charge.

To do so, we follow the trail of the valence electron configurations, moving across the rows and down the columns of the periodic table.

Across

Nuclear charge increases steadily and uniformly across a row, one proton at a time. Lithium, beryllium, boron, . . . ; $Z = 3, 4, 5, ...$ Sodium, magnesium, aluminum, . . . ; $Z = 11, 12, 13, ...$ Each new proton is accompanied by a matching electron, and entire subshells are opened and closed along the way.

Herded into a common shell or subshell, the valence electrons find themselves poorly shielded while the nuclear charge grows from left to right over each period. Start with lithium, for example, where the $2s$ electron interacts with a three-proton nucleus through a screen of two $1s$ electrons:

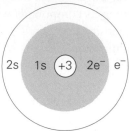

Such shielding, relatively strong, leaves the valence electron drawn toward a value of Z_{eff} close to $+1$ (1.3, after more detailed calculation). What, then, should we expect for beryllium, which must place its fourth and final electron in the very same $2s$ orbital?

Both of beryllium's $2s$ electrons are held at the same average distance from the nucleus, and so neither can penetrate to the disadvantage of the other:

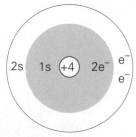

Aside from their mutual repulsion, these two electrons jointly see a *four*-proton nucleus screened through the same doubly-negative $1s^2$ core. The positive charge looks even stronger as a result, for the valence $2s$ electrons are unable to shield each other efficiently. Z_{eff} increases. The rise is not all the way to $+2$ (since $2s$–$2s$ repulsions do push up the energy), but still to a value greater than lithium's. The number is approximately 1.9.

Whereas lithium's one valence electron is governed by a Z_{eff} close to $+1$, beryllium's two outer electrons respond to a charge nearly equal to $+2$. The difference comes from the incomplete shielding existing within the same orbital, and its effect is to lower the Be $2s$ orbital energy compared with Li. A larger effective charge enhances the attraction of electron for nucleus. The orbital is drawn in more tightly. It shrinks.

The pattern continues across the row, with Z_{eff} increasing from atom to atom as the nucleus steadily acquires protons. First there is boron,

(a)

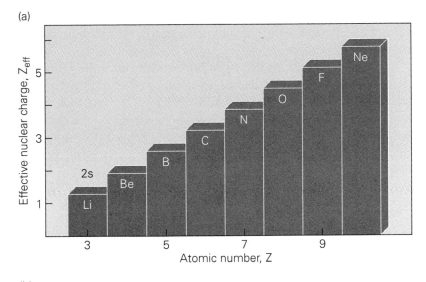

(b)

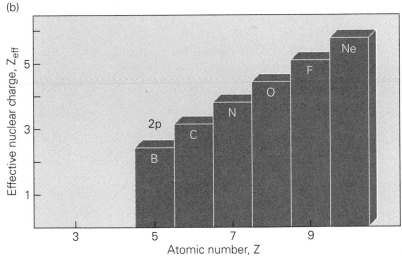

FIGURE 6-11. Effective nuclear charge, computed for atoms in the second row. Z_{eff} increases from left to right across the period. (a) $2s$. (b) $2p$.

where three second-shell electrons (poorly shielded in a $1s^2 2s^2 2p^1$ configuration) face a five-proton nucleus. Then comes carbon, with four valence electrons facing a six-proton nucleus, followed by nitrogen, oxygen, fluorine, and neon. The trend, typical of the representative elements, becomes plain from the data plotted in Figure 6-11.

Down

Having gone left to right, we now move top to bottom: down a column, as from lithium to sodium to potassium, where the descent is from

[He]$2s^1$ to [Ne]$3s^1$ to [Ar]$4s^1$. From s^1 to s^1 to s^1, it is an electronic realization of the periodic law. Atoms periodically renew their configurations and start over again.

Valence configurations therefore remain the same going down a column, each supported by an additional completed shell in the core. The inner electrons, lower in energy and nearer to the nucleus, mount a united electrical front to shield the valence electrons on the outside.

Observe: A two-electron $1s^2$ core shields a three-proton nucleus in lithium, yielding (after mathematical analysis) $Z_{\text{eff}} = 1.3$ for the valence $2s$ electron:

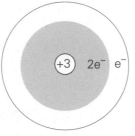

A 10-electron $1s^2 2s^2 2p^6$ core shields an 11-proton nucleus in sodium, yielding $Z_{\text{eff}} = 2.5$ for the valence $3s$ electron:

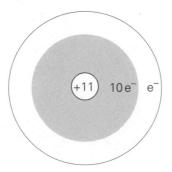

An 18-electron $1s^2 2s^2 2p^6 3s^2 3p^6$ core shields a 19-proton nucleus in potassium, yielding $Z_{\text{eff}} = 3.5$ for the valence $4s$ electron:

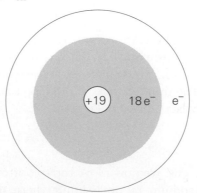

Eight protons are added in each instance, but still the outermost *s* electron is left with a relatively small Z_{eff}—far less than the real nuclear charge, *Z*. The underlying core shells add bulk to the atom, and they screen the valence electrons in rough proportion.

Thus the nuclear charge transmitted to the valence shell grows only slowly down a column. It is, moreover, an increasingly distant positive charge, because the valence density is displaced farther and farther from the nucleus with each jump in *n*. The orbital moves to higher energy as a result.

Atomic Size

Two broad tendencies emerge from the analysis above, each carrying its own set of expectations (Figure 6-12).

First, *effective nuclear charge increases across a row.* Orbital energies decrease accordingly, with the valence electrons attracted by a growing positive charge. Bound tighter and closer to the core, an outer electron becomes progressively more difficult to remove. The density shifts closer to the nucleus, and the atom shrinks. We expect smaller atoms, harder to ionize.

Second, *the principal quantum number increases down a column.* Filled out by an expanding core, the atom gets bigger and can shed its far-off valence electrons more readily. We expect larger and looser atoms, easier to ionize.

Compare now these expectations with experimental reality, beginning with measurements of atomic radii. An atom has no fixed size, of course, given the diffuse quality of its electron density, but there are reasonable

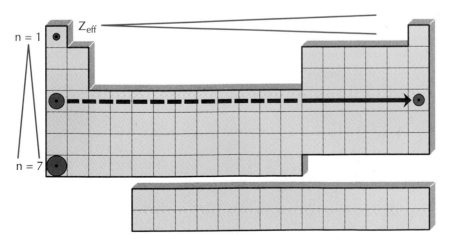

FIGURE 6-12. Periodicity at a glance: Effective nuclear charge increases across a row; principal quantum number increases down a column. Atoms shrink from left to right and expand from top to bottom. Properties recur from period to period.

ways to estimate a volume nevertheless. Picture the atom as a hard sphere (like our oft-considered billiard ball), and imagine that two of these balls are brought together, just touching. It is a crude model of two atoms in a homonuclear crystal or diatomic molecule, with one impenetrable ball simply tangent to another. Then, measuring the true nucleus–nucleus distance as D, we obtain a radius for the sphere by taking $R = D/2$, an **atomic radius**:

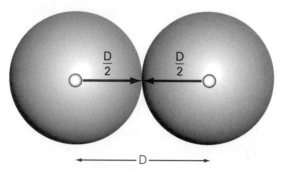

Atomic radii, so determined, remain sensibly constant in different compounds, as if the atoms really did combine as hard spheres. The radius marks out a certain inviolable space around the nucleus, a space ordinarily not breached by a second species. This distance-of-closest-approach establishes an effective minimum volume of electrons for an atom or ion.

Radii for most of the elements are displayed in Figure 6-13, where we see our expectations borne out. Across any one period, the atoms get smaller. From one period to the next, each set of bars is displaced

FIGURE 6-13. Atomic radii, arranged according to row: from the first period (H, He) at the far left to the sixth (Cs through Rn) at the far right. Size generally decreases across a row and increases down a column. For a closer look at the three series of transition metals, see Figure 6-15.

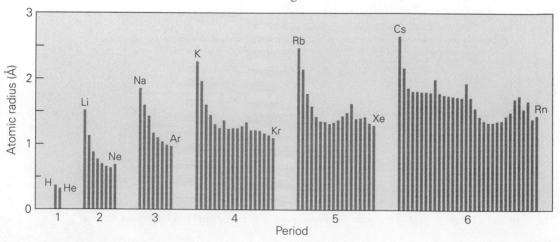

upward to higher radius. There are minor irregularities in spots, usually reflecting the special circumstance of a half-filled or filled subshell, but overall the pattern is clear.

One particularly striking deviation—interesting for its association with the f orbitals—takes the name **lanthanide contraction**. To illustrate, consider the electron configurations of two pairs of transition metals: one pair in the second series; the other just below, in the third series. Yttrium/lanthanum and zirconium/hafnium, which line up as

Y	$[Kr]5s^2 4d^1$	Zr	$[Kr]5s^2 4d^2$
La	$[Xe]6s^2 5d^1$	Hf	$[Xe]6s^2 4f^{14} 5d^2$

are good examples. Since yttrium and lanthanum have analogous $ns^2(n-1)d^1$ valence configurations, differing only in n, we expect the heavier atom to be bigger. It is. Lanthanum, bolstered by its filled $4d$ and $5p$ orbitals, has a larger radius than yttrium. Yet hafnium, with a $6s^2 5d^2$ configuration, has almost the same radius as zirconium with its $5s^2 4d^2$ configuration. The reason lies in the anomalously poor shielding provided by hafnium's $4f^{14}$ subshell, which annuls the usual down-the-column radial expansion.

Consider: Even though fully occupied, the oddly shaped $4f$ orbitals (Figure 6-14) are unable to offer effective screening to the $5d$ electrons.

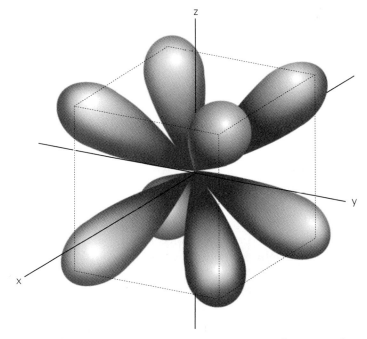

FIGURE 6-14. One of the f orbitals, f_{xyz}, taken from a set of seven. High in angular momentum, an f electron ($\ell = 3$) is thrust far from the nucleus and provides scant shielding to other electrons.

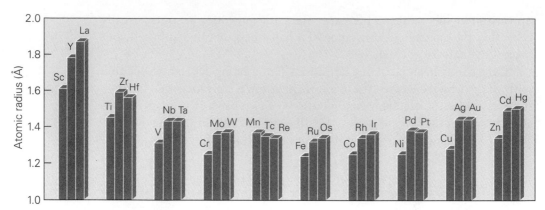

FIGURE 6-15. The lanthanide contraction. Poor shielding from a filled $4f^{14}$ core leaves the sixth-period transition metals with anomalously small atomic radii. Each group of three bars shows radii for a triad of elements, from Sc/Y/La at the far left to Zn/Cd/Hg at the far right. Note how the third bar in each set, representing a sixth-period transition metal, is scarcely taller and sometimes even shorter than the second (a fifth-period metal)—despite the higher principal quantum number. Only lanthanum, which has no intervening $4f^{14}$ subshell, runs counter to the trend.

The valence shell in hafnium, inadequately buffered from a higher nuclear charge, is pulled in more tightly than otherwise. And now that hafnium and zirconium have nearly identical radii as well as analogous valence configurations, the two atoms become even more similar in their chemical behavior. The lanthanide contraction, summarized in Figure 6-15, partially blurs the distinction between the second and third series of transition elements.

Ionization Energy

An electron bound to a nucleus is just that: bound, bound by electrical forces. Trapped in a well, such an electron has given up energy to combine with the atom. To detach it thereafter requires a compensatory repayment of energy. Some electrons go easy and others go hard, but none goes for free. The price demanded is equal approximately to the energy of the orbital into which the electron has fallen.

One can measure by experiment a binding energy for each electron, defined (for some atom X in a gas) as the minimum energy needed to strip the electron from its orbital and move it beyond the reach of the nuclear attraction—to infinity, we say, or to wherever the binding potential falls effectively to zero. The quantity so determined, known as the orbital *ionization energy*, is analogous to the work function needed to eject a photoelectron from a metal.

Equivalently expressed as either joules per atom or kilojoules per mole, the ionization energy is different for each orbital. Inner-shell electrons, sunk deep into the core and subject to a high Z_{eff}, lie low in energy and are difficult to remove. Highly energetic X rays are needed to overcome the fierce attraction these electrons feel for a nearly exposed nucleus. Valence electrons, farthest from the nucleus and highest in energy, come off with greater ease and are loosened by appropriately longer wavelengths (typically visible or ultraviolet radiation, with photon energies 1000 to 10,000 times smaller than those of X rays).

The *first ionization energy*, I_1, is the minimum energy required to detach the most weakly bound electron, usually from the highest valence orbital. Shorn of this outermost negative charge, a correspondingly *smaller* cloud of $Z - 1$ electrons is then left to face a nucleus containing Z protons. The effective nuclear charge is enhanced for the electrons that remain, and the monopositive ion, X^+, shrinks in response:

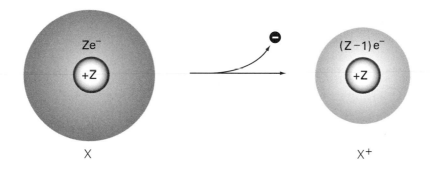

Each orbital within the ion drops to lower energy, reflecting both the increased nuclear attraction and also the reduced repulsion among the fewer electrons.

Removal of a second electron demands considerably more energy. I_2, the *second ionization energy*, exceeds I_1 even if both first and second electrons come from the same orbital, as would happen, for example, when beryllium's $2s^2$ electrons are taken off in succession. Realize that when neutral Be ([He]$2s^2$) is ionized to produce Be$^+$ ([He]$2s^1$), the $2s$ orbital left behind is *not* the same $2s$ orbital found in the original atom. An electron is gone, and everything is different. Attractions are different; repulsions are different; Z_{eff} is different; size is different; energy is different. With the cation now smaller and more tightly constructed than the neutral atom, the second ionization is harder to execute than the first. So goes the trend for consecutive ionizations in all atoms, growing progressively tougher from I_1 to I_2 to I_3 to I_4 and more, all the way down to the last electron clinging tenaciously to a stripped nucleus.

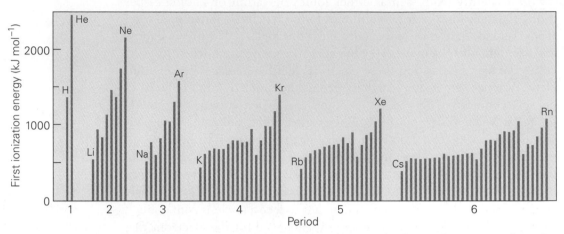

FIGURE 6-16. Atomic ionization energies, arranged according to row: from the first period (H, He) at the far left to the sixth (Cs through Rn) at the far right. The strength of binding increases across a row and decreases down a column.

We can use experimental values of I_1, the smallest of the ionization energies, to test the model of electrons in orbitals. Inspecting the data in Figure 6-16, we find quick confirmation of our two major predictions: I_1 increases across the rows and decreases down the columns. Progressively deshielded and therefore subjected to a rising Z_{eff}, the valence electrons are bound tighter and tighter across a period. It becomes all the harder to remove the first electron, and so I_1 increases from left to right: from the metals (which tend to lose electrons) to the nonmetals (which tend to gain them). Top to bottom, though, the ionizable electron moves farther away and rises to higher energy. The additional expense needed for ionization goes down, since the electron is already that much closer to leaving the atom.

Superimposed on these general trends are a few bumps in the patterns, mostly easy to explain. Boron's $2p$ electron is ionized more readily than beryllium's $2s$, for instance, owing to the increase in shielding incurred upon opening up the p subshell. The filled $2s$ orbital in boron casts a sphere of negative charge around the nucleus, a repulsive filter through which the $2p$ electron experiences a diluted nuclear attraction. Pushed to higher energy, the shielded p electron is more loosely bound and easier to remove. I_1 dips in response, decreasing 11% between beryllium and boron (but rising again for carbon and nitrogen).

From nitrogen to oxygen, too, there is a small dip. Electrostatic repulsions increase when oxygen's last electron is forced to share a p orbital in a $[He]2s^2 2p^4$ configuration. The orbital energy is raised by these additional repulsions, and I_1 falls as expected.

Electron Affinity

Most neutral atoms (and all positive ions) are able to accommodate one additional electron within their existing orbital structure. Successful attachment of an electron to a particular atom X yields an anion X⁻, accompanied by the release of energy. It is the reverse of ionization.

The energy evolved, measured for gaseous species, is termed the **electron affinity** (*EA*). The more negative the affinity, the greater the energy produced and the stronger the attachment. A positive value, by contrast, means that energy must be *applied* to force the electron onto an unwilling host. The intended ion is unstable and does not endure.

Rather if an electron is to be captured, then the transaction must not cost energy. To turn Ne into Ne⁻, say, would require a ninth electron to be inserted beyond the filled $1s^2 2s^2 2p^6$ core into an unreceptive $3s$ orbital, significantly higher in energy. Neon resists such addition, as do the other noble gases, and the electron affinity is appropriately positive.

Nor do the alkaline earth metals accept additional electrons, for just the same reason. The atoms of Group II, configured as ns^2, have vacancies only in the higher-lying np orbitals. Lacking sufficient nuclear charge to hold the extra electron, ions such as Be⁻ and Mg⁻ do not freely exist.

Among those atoms with favorable affinities (the majority of the elements), the addition is easiest toward the right-hand side of the periodic table. The halogens, especially, standing just one electron short of a noble-gas configuration, benefit from an extra negative charge. This last electron transforms the valence from $ns^2 np^5$ to $ns^2 np^6$, leaving it isoelectronic with the noble gas just following. Thus F readily becomes F⁻, enjoying the same electronic arrangement as Ne and lowering its energy by over 300 kJ per mole.

The anion, although lower in energy than its parent atom, proves to be the larger species. For to add an electron is, unavoidably, to pump more negative charge into orbitals far from the nucleus. More electrons take up more space, and more electrons generate more repulsions as well:

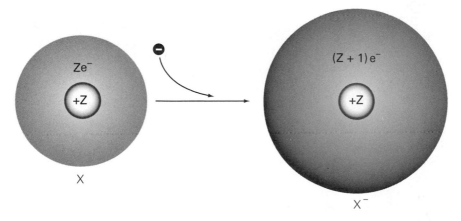

Plotted in Figure 6-17, the experimental data roughly confirm that electron affinities become stronger—more negative—going from column to column across the periodic table. Notable exceptions include the elements of Groups II and VIII (already mentioned), as well as the elements of Group V. Nitrogen, for one, has no affinity for an extra electron within its half-filled $2p^3$ subshell ($\uparrow\!\_\ \uparrow\!\_\ \uparrow\!\_$). Forced to pair up in any of these three orbitals, a new electron would increase the repulsions disproportionately and make the cost of occupancy far too high.

Paralleling a similar trend in ionization energies, but less pronounced, there is a tendency for electron affinity to weaken as the atoms grow larger down a group. Imagine the electron approaching from infinity, to be trapped ultimately by the valence orbital highest in energy. The energy released during that fall is the electron affinity, and hence *EA* depends on how soon the descent is terminated. The larger the atomic radius, the higher the orbital energy and the shorter the fall. The shorter the fall, the less negative is the electron affinity. So it is for iodine (−295 kJ/mol) compared with fluorine (−328 kJ/mol), since iodine's $5p$ orbitals extend to greater distances than fluorine's more compact $2p$ subshell.

FIGURE 6-17. Electron affinities of the main-group elements, arranged according to column: from Group I (H through Cs) at the far left to Group VIII (He through Rn) at the far right. Atoms with negative affinities are able to add a valence electron. The more negative the affinity, the more energy is released upon attachment. Bars pointing upward represent atoms with no affinity for an extra electron in the gas phase—typically species with filled s and p subshells.

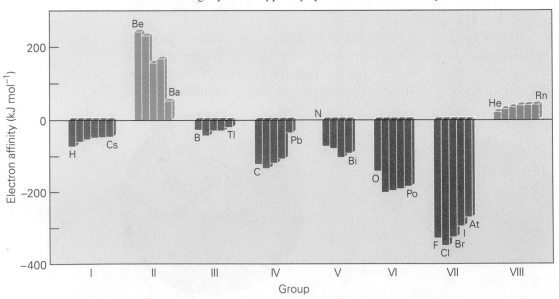

But note that this modest down-the-column trend in Figure 6-17 is not without exception, for chlorine's electron affinity (−349 kJ/mol) exceeds fluorine's despite the larger atomic radius in chlorine. Indeed fluorine is so compact that an added electron exacerbates the repulsions inside the crowded $2p$ subshell. Chlorine, with its more diffuse $3p$ orbitals, is better able to spread out the extra charge and thus mitigate some of these repulsions.

We began our pursuit of the quantum theory with light and matter, turning waves into particles and particles into waves. It seemed, for a time, that matter threatened to disappear into a nebulous description of waves and probabilities, hovering uncertainly, neither here nor there.

But now we have rebuilt our atoms into substantial and predictable quantum mechanical objects. Fashioned from electrons and nuclei into several dozen elements, the atoms of the periodic table step forth as the interchangeable parts of everyday matter. They are the building blocks, the basic ingredients of the chemical mix. They are building blocks equipped with quantum mechanical hooks and hinges; with electrons that can come and go; with orbitals that can add, subtract, and interfere.

From this machinery will come molecules and the chemical bond.

REVIEW AND GUIDE TO PROBLEMS

It should be easy, one would think, to go from the hydrogen atom to a similar model for helium: (1) Add one proton to the nucleus and one electron outside. (2) Sort through the electrical interactions. (3) Write the appropriate Schrödinger equation. (4) *Solve* the Schrödinger equation. (5) Analyze the results.

Easy to say is not easy to do, however, because we can never solve the equation for helium cleanly and exactly. The repulsion between two electrons, absent in hydrogen, shatters all previous simplicity. Just this one complication, the byplay between electron 1 and electron 2, makes exact solution impossible. Two electrons—not fifty, not thirty, not ten, but merely two—and already we are stuck. Step 4 is the problem.

It gets worse. Lithium's three electrons pose an even greater challenge, not to mention beryllium and boron and all the heavier atoms. Hydrogen, with its rigorously correct orbitals and quantum numbers, proves to be a special case, never to be duplicated.

Do we give up? No; we need a working model, and so we make allowances. Accepting the reality of electron repulsion, we assess its importance. We devise alternatives to the full Schrödinger equation. We use the lessons of the hydrogen atom wherever possible. We work around the problem. We do not ask for perfection. We compromise.

The results, remember, can be pretty good.

IN BRIEF: THE MANY-ELECTRON ATOM

1. A single electron, like the one in hydrogen, tends naturally toward the nucleus, where attraction is high and potential energy is low. Any electron, isolated from all others, would do the same. Opposites attract.

Each electron in a many-electron atom would (if it could) cluster tightly about the nucleus, enjoying there the lowest energy, taking care of itself alone, indifferent to the rest. All the electrons would, if they could, huddle near the positive charge, but they cannot. They are not independent actors. They are in competition. Space is limited, and they have each other to fight.

Electrons *repel*. They push apart. The closer they come, the harder they push. To be close is to endure a high electrostatic energy, to fight uphill, to do things the hard way. Electrons stay apart wherever possible. Nature takes the easy way.

Nature also enforces the ***Pauli exclusion principle***, which states: No two electrons in the same atom may occupy the same quantum state, lest they be identical in both space and spin. If the spins are the same (both up or both down), then the electrons cannot claim precisely the same region in space. If they share the same space, their spins must be opposite.

All electrons, then, may compete for the positive charge, but not all can win. They repel and they exclude. Rather than cluster about the nucleus, they are forced to spread out. An atom takes shape from the inside out as the electrons make room for each other.

2. If so, then how are the repulsions balanced against the attractions? How do the electrons spread out, and where do they go?

There, unfortunately, intrudes step 4 into our efforts to understand: the hopelessly complicated Schrödinger equation, made insoluble by the point-to-point repulsion of the electrons.

But step 4 is *our* problem, not nature's problem, because electrons do what they do regardless of the mathematics. Looking on, we must use the tools at hand; and so, if one equation is too hard, we cautiously devise another—a simpler model, perhaps, yet one still grounded in reality. If the less ambitious plan works, then later we can make adjustments and refinements as needed.

3. Try the *orbital approximation*: Replace the herky-jerky electron repulsion with a smooth, symmetric force field averaged over space and time. Imagine each electron to move, on the average, through a blurred cloud of negative charge presented by all the other electrons. Make it simple. Let the particle be repelled not by each electron separately, but by this one smooth and centralized force instead.

Each electron in our model now moves independently of the rest. There is no bias in any one direction, only a composite force (different for every electron) that pushes the negative charge away from the nucleus. To the individual electron, these imagined circumstances recreate the two-body simplicity of a hydrogen atom.

Good for us, because that is a problem we know how to solve: the one-electron atom, with its exact orbital solutions and quantum numbers. Viewed this way, the many-electron atom becomes an apartment complex for quasi-independent particles, a structure where every electron has its own orbital with its own energy, territory, spin, and quantum numbers n, ℓ, m_ℓ, and m_s. A given electron enjoys just one attraction (to the nucleus) and suffers just one smeared-out repulsion. That is our approximation.

4. PENETRATION AND SHIELDING. Thus arise the orbitals of a many-electron atom, but not all orbitals are created equal. No longer does each member of a shell have the same energy; no longer does the energy depend only on the principal quantum number n. In hydrogen, the $2s$ and $2p$ orbitals have the same energy. They are **degenerate**. In helium, no more.

In helium, an electron in the 2s orbital comes closer to the nucleus than does an electron in a 2p orbital. The s electron has less angular momentum than the p electron. It **penetrates** more effectively. It spends more time nearer to the positive charge. Its energy is lower.

The 2p electron is **screened**, or **shielded**, from the nucleus by the more intrusive 2s electron. The 2p (with higher angular momentum) is pushed away and made to face a weakened, more distant positive charge. The shielded electron's energy moves higher, as if the nucleus offered not its full charge Z but a smaller **effective nuclear charge**, Z_{eff}, reduced by some amount S:

$$Z_{eff} = Z - S \qquad (S > 0)$$

For each shell (same n, different ℓ), we then discover that orbital energies increase in the order $E_{ns} < E_{np} < E_{nd} < E_{nf}$. Electrons with higher angular momentum are cast farther from the nucleus. They penetrate less. They are screened more. Their energies go up.

5. BUILDING UP. Many-electron atoms differ in how their orbitals are filled, where the one-electron energies stand, and where the territories lie—all matters to be determined by electron–electron interactions and the Pauli principle. The effects are subtle, interwoven, sometimes inconsistent, yet still we manage to identify several broad influences.

First, orbitals owe their different energies to the interplay of penetration and shielding. Within a shell, energy increases with ℓ. From shell to shell, energy usually increases with n (although there are exceptions, notably where E_{ns} lies close to $E_{(n-1)d}$).

Second, the Pauli exclusion principle restricts an orbital to a maximum of two electrons, and only if the two particles have opposite spins. An orbital may be vacant (__), singly occupied ($\uparrow$ or $\downarrow$), or doubly occupied ($\uparrow\downarrow$), but nothing else. There are no other options.

Third, double occupancy costs energy. To place two electrons in one orbital is to force them into close quarters, where they repel more and suffer accordingly. Recognizing the costs of spin pairing, **Hund's rule** states: An atom in its ground state tends to maximize the number of unpaired electrons in degenerate orbitals, keeping them parallel if possible. Within a d subshell, for example, the arrangement $\uparrow$ $\uparrow$ __ __ __ is favored over both the arrangement $\uparrow\downarrow$ __ __ __ __ and the arrangement $\uparrow$ $\downarrow$ __ __ __.

Fourth, the **electron configuration** (overall orbital occupancy) aims to minimize the energy of the entire atom. The order of filling is usually

$$1s < 2s < 2p < 3s < 3p < 4s < 3d < 4p < 5s < 4d < 5p < 6s$$

for the ground state.

6. THE PERIODIC LAW. There is a rhythm to the elements, a rhythm of structure and properties. The atoms come and the atoms go, but always the electrons return to their courses. The same valence configurations recur as old shells are closed and new ones opened.

We see the pattern in the *periodic table*, where the atoms fall into the *s block, p block, d block*, or *f block* depending on which valence subshell remains active. The alkali metals (ns^1) lie in the s block, the halogens (ns^2np^5) in the p block, the transition metals in the d block, and the lanthanides in the f block.

Down the columns of the table are *groups* of elements having similar valence configurations. The same subshells are filled in the same order, with only the principal quantum number differing from row to row. Simplest are the alkali metals of Group I,

$$
\begin{array}{ll}
\text{Li} & \text{[He]}2s^1 \\
\text{Na} & \text{[Ne]}3s^1 \\
\text{K} & \text{[Ar]}4s^1 \\
\text{Rb} & \text{[Kr]}5s^1 \\
\text{Cs} & \text{[Xe]}6s^1 \\
\text{Fr} & \text{[Rn]}7s^1
\end{array}
$$

constructed to give each member a single s electron. Top to bottom, atoms in the group grow larger as more and more completed shells are tucked into the core. Pushed away from the nucleus, the valence electrons become easier to ionize and correspondingly harder to attach.

Across the rows are *periods* of elements in which the various subshells are filled, sequences such as

$$
\begin{array}{cccccccc}
\text{Li} & \text{Be} & \text{B} & \text{C} & \text{N} & \text{O} & \text{F} & \text{Ne} \\
2s^1 & 2s^2 & 2s^22p^1 & 2s^22p^2 & 2s^22p^3 & 2s^22p^4 & 2s^22p^5 & 2s^22p^6
\end{array}
$$

built around a common core (here, $[\text{He}] = 1s^2$). Be aware also of the transition metals, where the d orbitals receive their electrons; and the lanthanides, where the f orbitals are first opened.

Across a row, the atoms generally grow smaller as one electron follows another into the same shell. Since *intra*shell shielding ($2s$–$2p$, say) is less effective than *inter*shell shielding ($2s$–$3s$), the incoming electrons are exposed to a proportionately higher Z_{eff} going across the row. There is a tightening of the structure, a contraction; and (mostly) the atoms become harder to ionize.

So now, at last, we have Lewis's dot diagrams, shells, and octets recast into quantum mechanical language. Here are the elements, the atoms, the literal stuff of matter broken down into electronic wave functions—abstract, maybe, but also a mirror of very real material properties.

And more: Different across the rows and similar down the columns, the atomic structures testify to thrift and regularity in nature. In the rhythm of the configurations we see a discipline, a simplicity, and yet still a limitless capacity for variation. Much is done with little.

SAMPLE PROBLEMS

EXAMPLE 6-1. Shielding In Helium

PROBLEM: Without attempting a detailed calculation, estimate the first ionization energy for a ground-state helium atom. Determine the lowest and highest values possible.

SOLUTION: Suppose it were a hydrogen atom. Then, using the formula from Chapter 5, we know that the orbitals have energies

$$E_n = -R_\infty \frac{Z^2}{n^2}$$

where $R_\infty = -1312$ kJ mol^{-1}. The energy needed to ionize the one electron from hydrogen's $1s$ orbital is therefore R_∞:

$$I_1 = E_\infty - E_1 = 0 - (-R_\infty) = R_\infty$$

Now consider helium and its *two* $1s^2$ electrons, attracted to a nucleus containing two protons ($Z = 2$). Competing for the same nuclear charge, the electrons screen one another. They repel. Exactly how much we cannot say, but we certainly can establish upper and lower limits for the ionization energy.

Lower limit: Pretend that each $1s$ electron screens the other completely, interposing a full charge of -1 between its orbital mate and the nucleus. If so, the screened electron sees roughly an effective nuclear charge of $+1$ ($Z_{eff} = 2 - 1$); and, if it does, we have a pseudo-hydrogen atom. The lowest possible ionization energy is R_∞, the value expected for a single electron bound to a nucleus with $Z_{eff} = 1$.

Upper limit: At the other extreme, pretend that there is no screening at all. Assume that each electron interacts with the *full* nuclear charge so as to make $Z_{eff} = 2$. The ionization energy, proportional to Z_{eff}^2, then quadruples to $4R_\infty$.

Helium's first ionization energy thus falls between R_∞ (1312 kJ mol^{-1}) and $4R_\infty$ (5248 kJ mol^{-1}). Where in between? . . . probably in the middle, since the two $1s$ electrons screen each other in roughly equal measure. They are, after all, in the same orbital; they are caught in the same region of space, with the same energy, with the same angular momentum. If electron 1 screens electron 2 by some amount S, then electron 2 should screen electron 1 to the same extent. Venturing a guess, we predict an ionization energy of approximately $2R_\infty$ (2624 kJ mol^{-1}).

The experimental value is 2372 kJ mol^{-1}.

EXAMPLE 6-2. Spin and Electron Repulsion

PROBLEM: Which of the following p^2 configurations should realize the lowest energy? Which should realize the highest? Which are forbidden?

(a) ↑↑ __ __

(b) ↑↓ __ __

(c) ↑ ↓ __

(d) ↑ ↑ __

SOLUTION: (a) is impossible, a clear violation of the Pauli exclusion principle. Two electrons can occupy the same orbital only if their spins are oppositely directed.

The doubly occupied orbital in (b) does satisfy the exclusion principle, although at the highest cost in energy. Forced to share the same space, the two electrons suffer the most repulsion.

Of the two arrangements remaining, one of them, (c), places two electrons with opposite spins into two different orbitals; whereas the other, (d), also places the electrons into two orbitals but with parallel spins. The question then reduces to: Is the balance between attractions and repulsions greater in ↑ ↓ __ (c) or ↑ ↑ __ (d)?

Hund's rule tells us that the *parallel* configuration (↑ ↑ __) will exist at the lower energy, in part because of demands made once again by the Pauli principle: Two electrons with the same spin cannot be in the same place at the same time. True, parallel electrons already comply with the exclusion principle because they have different values of the spatial quantum number m_ℓ; but, also true, the duplication of spin further encourages the electrons to stay apart wherever possible. Their motion is said to be "correlated" so as to avoid repulsive encounters and to seek the attraction of the nucleus instead. Very subtle (and still a subject for research), but the balance proves more favorable for parallel electrons than for those in an antiparallel configuration such as ↑ ↓ __.

Summing up: Energy is lowest in (d), intermediate in (c), and highest in (b). Choice (a) is forbidden to electrons.

EXAMPLE 6-3. Spin and Magnetism: A Closer Look

Endowed with spin, an electron produces a magnetic field. The magnetism is always there, a fixed property of the charged particle. The electron may stay in one place, not circulating, yet still the magnetic field persists. Electrons are natural magnets.

A spin has direction, and so does the accompanying magnetism. Up (↑) or down (↓), the electron's magnetic field is quantized according to the two possible values of m_s. Put together two electrons with parallel spins; the fields add. Put together two electrons with opposite spins; they subtract. No net magnetism remains.

When one or more electrons exist *unpaired* (for example, ↑ ↑ __ rather than ↑↓ __ __), a system is termed **paramagnetic**. The atom or molecule or ion is drawn into a magnetic field just as a compass needle lines up in the field of the earth.

Where all the electrons are *paired* (↑↓ __ __ rather than ↑ ↑ __), a system becomes **diamagnetic**. Opposed one to the other, the electrons have no net magnetism to give. An external magnetic field holds no attraction for them; they are repelled, pushed out by any field.

PROBLEM: Which of the following species are paramagnetic in their ground state? Which are diamagnetic? (a) He. (b) Be. (c) C. (d) Na. (e) Na^+.

SOLUTION: For each, we write the electron configuration according to the procedure detailed above. *Aufbau principle*: Orbitals are filled in the order 1s, 2s, 2p, 3s, 3p, 4s, 3d, 4p, ... *Pauli principle*: Maximum orbital occupancy is two electrons, but only if the spins are opposite. *Hund's rule*: Inside each subshell, electrons preferentially occupy the orbitals one at a time with parallel spins before pairing.

(a) *He, 2 electrons.* With two electrons paired in the 1s orbital (↑↓), helium's configuration is $1s^2$. Diamagnetic.

(b) *Be, 4 electrons.* Diamagnetic. The configuration is $1s^2 2s^2$, again with no unpaired electrons:

2s   ↑↓

1s   ↑↓

(c) *C, 6 electrons.* The additional two electrons enter the 2p subshell one at a time, unpaired, consistent with Hund's rule:

2p   ↑ ↑ __

2s   ↑↓

1s   ↑↓

The $1s^2 2s^2 2p^2$ configuration renders carbon paramagnetic.

(d) *Na, 11 electrons*. With first and second shells both filled, sodium's lone valence electron makes it paramagnetic. The full configuration is $1s^2 2s^2 2p^6 3s^1$:

$3s$	↑
$2p$	↑↓ ↑↓ ↑↓
$2s$	↑↓
$1s$	↑↓

(e) *Na⁺, 10 electrons*. The sodium ion is isoelectronic with neon and consequently has the closed-shell configuration $1s^2 2s^2 2p^6$. Diamagnetic.

QUESTION: How, experimentally, can we tell whether a substance is diamagnetic or paramagnetic?

ANSWER: As follows . . . (1) Paramagnetic systems are pulled *into* a magnetic field; diamagnetic systems are pushed *out*. (2) To be pushed or pulled is to endure a force, no different from the way that gravity pulls an object earthward and gives it a weight. (3) Using a special balance (a *Gouy balance*), we weigh the material in a magnetic field. A paramagnetic system, drawn into the field, betrays itself by an apparently greater weight than usual. Careful measurement of the magnetic force reveals the number of unpaired electrons.

EXAMPLE 6-4. More Configurations

PROBLEM: Write electron configurations for the following atoms: (a) Zr. (b) Xe. (c) Pb.

SOLUTION: We can always count electrons and then follow the usual filling sequence ($1s$, $2s$, $2p$, $3s$, . . .), but here the structures are becoming inconveniently large. Better to use the periodic table, which (with a few exceptions) telegraphs the configuration just by the atom's position.

In the row beginning with Li, the $2s$ and $2p$ subshells are filled in turn. The p block starts at Group III, with the element boron.

In the row beginning with Na, the $3s$ and $3p$ subshells are similarly filled. Aluminum is the first p-block atom.

In the row beginning with K, the $4s$, $3d$, and $4p$ subshells take on two, ten, and six electrons respectively. Scandium initiates the d-electron transition metals, and Ga reopens the p block.

So it goes, each atom falling neatly into place. Elements with identical valence configurations line up one below the other, grouped into the recurring patterns that give meaning to the table.

(a) *Zirconium, Zr (Z = 40)*. Find it in the fifth row, just under titanium (Ti) in the *d* block. The valence contains two $5s$ and two $4d$ electrons in the arrangement $5s^2 4d^2$, paralleling titanium's $4s^2 3d^2$ configuration. To specify the zirconium atom completely, use the shorthand form $[Kr]5s^2 4d^2$ (where

$$[Kr] = 1s^2 2s^2 2p^6 3s^2 3p^6 4s^2 3d^{10} 4p^6$$

represents the closed-shell configuration of krypton, the closest noble gas in the table).

(b) *Xenon, Xe (Z = 54)*. A noble gas, xenon concludes the fifth period. Its 54 electrons are distributed as

$$1s^2 2s^2 2p^6 3s^2 3p^6 4s^2 3d^{10} 4p^6 5s^2 4d^{10} 5p^6$$

or, in abbreviated notation, as $[Kr]5s^2 4d^{10} 5p^6$. From Kr to Xe we see the completion of the $5s$, $4d$, and $5p$ subshells, a total of 18 electrons in all.

(c) *Lead, Pb (Z = 82)*. The 82nd element, Pb sits in the *p* block at the intersection of the sixth row with Group IV. Having 28 electrons more than xenon, lead's $[Xe]6s^2 4f^{14} 5d^{10} 6p^2$ configuration includes completed $4f^{14}$ and $5d^{10}$ shells. Note the interposition of the lanthanide series in the sixth row, over which the seven $4f$ orbitals receive their 14 electrons.

EXAMPLE 6–5. Atomic Radii

PROBLEM: Which atom in each pair has the smaller radius? (a) Be, N. (b) B, Br. (c) C, Si. (d) Al, Ca.

SOLUTION: For elements in the *s* and *p* blocks, we rely on two broad guidelines. (1) Atomic size generally decreases across a row, consistent with a progressive deshielding of the valence. Left to right, Z_{eff} grows larger while the *ns* and *np* subshells are filled. The atom contracts. (2) Top to bottom, the principal quantum number increases. Radii expand as the new valence shell is layered upon a larger and larger core.

(a) *Be and N*. Locate both atoms in the second row ($n = 2$), with Be to the left of N. Beryllium's valence configuration is $2s^2$; nitrogen's is $2s^2 2p^3$. Poorly shielded by the *p* electrons, the nitrogen atom contracts under a higher Z_{eff}. N has a smaller radius than Be.

(b) *B and Br*. Boron, a second-row element, has the valence configuration $2s^2 2p^1$. Bromine, in the fourth row, has the configuration

$4s^2 4p^5$. Here the difference in principal quantum numbers proves decisive, making B smaller than Br.

(c) *C and Si.* Carbon $(2s^2 2p^2)$ sits just above silicon $(3s^2 3p^2)$ in Group IV. C has a smaller radius than Si.

It matters. Carbon is the key element in living organisms; silicon is the key element in sand.

(d) *Al and Ca.* Al $(3s^2 3p^1)$ is both above and to the right of Ca $(4s^2)$. Aluminum is smaller than calcium.

EXAMPLE 6-6. Ionic Radii

PROBLEM: Which atom or ion in each pair has the smaller radius? (a) Li, Li^+. (b) F, F^-. (c) O^{2-}, F^-. (d) Na^+, Mg^{2+}.

SOLUTION: Short by one electron, a monopositive cation is smaller than its parent atom. The remaining $Z - 1$ negative charges are attracted more strongly by the Z protons. By contrast, an anion is larger than the neutral atom. The negative ion needs more space to accommodate the extra electron, a repulsive presence.

(a) *Li and Li^+.* Simply an atom and its cation. Li^+ is smaller than Li.

(b) *F and F^-.* Atom and anion. F is smaller than F^-.

(c) *O^{2-} and F^-.* Trickier. The anions are isoelectronic, each with eight valence electrons in the closed-shell neon configuration $2s^2 2p^6$. The oxygen nucleus, however, has only eight protons to attract the electrons, whereas the fluorine nucleus has nine. The fluoride ion has the smaller radius.

(d) *Na^+ and Mg^{2+}.* Here we have two isoelectronic cations, each stripped down to a neon configuration. The magnesium nucleus, with one more proton than sodium, exerts the stronger pull. Mg^{2+} is smaller than Na^+.

EXAMPLE 6-7. Ionization Energy

PROBLEM: Which atom or ion in each pair has the larger ionization energy? (a) C, N. (b) N, O. (c) O, S. (d) Na, Na^+.

SOLUTION: The same trends apply. Across: Shielding decreases. Z_{eff} increases. Atoms become smaller and more difficult to ionize. Down: Atoms grow larger. The valence electrons, more distant now, are easier to remove.

(a) *C and N.* Nitrogen, the smaller atom, has the higher effective nuclear charge and hence the higher ionization energy.

(b) *N and O.* An exception to the rule. Despite standing to the left of oxygen, nitrogen has the larger ionization energy thanks to its half-filled $2p$ subshell:

$2p$ ↑ ↑ ↑   ↑↓ ↑ ↑

$2s$ ↑↓   ↑↓

       N   O

Oxygen's *paired p* electrons bring with them additional repulsions, leaving the oxygen atom more loosely bound and therefore easier to ionize than nitrogen. The effect is slight.

(c) *O and S.* Oxygen, situated just above sulfur in Group VI, is smaller and more difficult to ionize.

(d) *Na and* Na^+. To ionize a cation is difficult at best, since prior removal of an electron from a neutral atom always produces a more tightly bound species (in which Z protons attract only $Z - 1$ electrons). Here, however, the effect is even more pronounced because Na^+ enjoys the special stability of a neon configuration. The sodium ion is far harder to ionize than the sodium atom.

EXAMPLE 6-8. Electron Affinity

PROBLEM: Which atom in each pair has the stronger electron affinity? (a) S, Cl. (b) Be, B.

SOLUTION: Trends for electron affinity are not as sharply drawn as those for radius and ionization energy, but still we ask the same kinds of questions. Which species has the higher effective nuclear charge? Which system can better accommodate the added electron? Must the electron be added to a singly occupied orbital, so as to worsen the electron–electron repulsion? Will an extra electron *open* a subshell (say from ns^2 to ns^2np^1) or *close* a subshell (as from ns^2np^5 to ns^2np^6)? How does the attached electron alter the balance between attractions and repulsions in the system?

(a) *S and Cl.* The elements are side by side in the third row, chlorine to the right of sulfur. Chlorine's higher Z_{eff} is better able to bind the electron in Cl^-, and, moreover, the added charge goes to complete the octet: $3s^2 3p^6$. Cl has the stronger (more negative) electron affinity.

(b) *Be and B.* To accept an electron, beryllium $(1s^2 2s^2)$ would have to open up its $2p$ subshell and then hold a total of five electrons with only four protons. Boron, by contrast, already has a p electron $(1s^2 2s^2 2p^1)$ and thus can accommodate the extra charge more easily. Whereas beryllium's electron affinity is strongly positive (binding *costs* energy), boron's affinity is weakly negative. The B atom has a capacity, albeit small, to accept one more electron.

1. Beyond hydrogen: (a) How does an orbital in a many-electron atom differ from an orbital in a one-electron atom? (b) In what sense does an electron "occupy" an orbital? (c) What circumstances limit the accuracy of the orbital model?

2. A lithium atom has the electronic configuration $1s^2 2s^1$. What does this notation signify?

3. Does the $2s$ electron in a lithium atom interact with the lithium nucleus in precisely the same way as do the $1s$ electrons? Do all three electrons respond to the same effective nuclear charge?

4. Describe, briefly, the interplay of electron–electron repulsion, penetration, and shielding in a many-electron atom such as lithium. In what way does the effective nuclear charge, Z_{eff}, reflect these influences?

5. Consulting only the periodic table, make a rough estimate of the effective nuclear charge acting on the three electrons in a lithium atom. Give a range of possible values for each orbital, good enough to restrict Z_{eff} to "no less than this number but no more than that number."

6. Do the same for the beryllium atom, Be. (a) Write the configuration. (b) Estimate the magnitude of Z_{eff} sensed by each electron. (c) Compare beryllium's values to lithium's values, taking into account any differences in penetration and shielding.

7. According to the analysis of the preceding exercise, which atom—lithium or beryllium—should have the higher ionization energy? Why?

8. State the Pauli exclusion principle.

9. Each of the following electron configurations violates the Pauli exclusion principle:

   (a) ↑↓↑↓
       $1s$

   (b) ↑↓   ↑↑   __ __ __
       $1s$ $2s$ $2p$

Why? Are these states absolutely forbidden or simply unlikely? May they exist under *any* circumstances?

10. State Hund's rule. What does it say about the electrostatic repulsion suffered by electrons in the same subshell?

11. Which one of the following configurations satisfies Hund's rule?

(a) ↑↓   ↑↓   ↑↓ ↓   —
    $1s$ $2s$ $2p$

(b) ↑↓   ↑↓   ↑ ↑ ↑
    $1s$ $2s$ $2p$

(c) ↑↓   ↑↓   ↑ ↓ ↑
    $1s$ $2s$ $2p$

Why? Are the other two configurations absolutely forbidden or merely unfavorable in some way? May they exist under certain circumstances?

12. According to Hund's rule, which of the following d^3 configurations has the lowest energy?

(a) ↑ ↓ ↑ — —

(b) ↑↓ ↑ — — —

(c) ↑ ↑ ↑ — —

(d) ↓ ↑ ↓ — —

13. Use Hund's rule to predict the f^4 configuration having the lowest energy:

? ? ? ?

— — — — — — —
f

Is an f^4 system likely to be paramagnetic?

14. Distribute nine electrons over seven f orbitals, again in the arrangement predicted by Hund's rule:

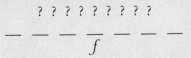

? ? ? ? ? ? ? ? ?

— — — — — — —
f

How many electrons remain unpaired?

15. Again, but this time for four electrons over five *d* orbitals:

$$? \quad ? \quad ? \quad ?$$

$$\underline{\quad} \; \underline{\quad} \; \underline{\quad} \; \underline{\quad} \; \underline{\quad}$$
$$d$$

How many electrons remain unpaired?

16. Suppose, however, there is a vacant *s* orbital close in energy to the *d* orbitals. How will the same four electrons occupy the manifold of orbitals below?

$$? \quad\quad ? \quad\quad ? \quad\quad ?$$

$$\underline{\quad} \quad\quad \underline{\quad} \; \underline{\quad} \; \underline{\quad} \; \underline{\quad} \; \underline{\quad}$$
$$ns \quad\quad\quad (n-1)d$$

How many electrons remain unpaired? Explain.

17. Which of these two configurations—$4s^2 3d^4$ or $4s^1 3d^5$—has the lower energy?

$4s^2 3d^4$:   ↑↓   ↑ ↑ ↑ ↑ __
          $4s$ $3d$

$4s^1 3d^5$:   ↑   ↑ ↑ ↑ ↑ ↑
          $4s$ $3d$

Why?

18. Similar: Which of these two configurations—$4s^2 3d^9$ or $4s^1 3d^{10}$—has the lower energy?

$4s^2 3d^9$:   ↑↓   ↑↓ ↑↓ ↑↓ ↑↓ ↑
          $4s$ $3d$

$4s^1 3d^{10}$:   ↑   ↑↓ ↑↓ ↑↓ ↑↓ ↑↓
          $4s$ $3d$

19. Bearing in mind the results of the preceding two exercises, write full electron configurations for Cr and Cu in their ground states.

20. (a) Write ground-state electron configurations for Mn and Zn. (b) How many *d* electrons remain unpaired in each atom?

21. Write ground-state electron configurations for the following atoms:

    (a) Be, B, C, O, Ne

    (b) Mg, Al, Si, S, Ar

    (c) Ca, Ga, Ge, Se, Kr

    To what block of the periodic table (s, p, d, or f) does each of these elements belong?

22. Write ground-state electron configurations for the following atoms:

    (a) Sc, Ti, V, Fe, Co, Ni

    (b) Y, Zr, Nb, Ru, Rh, Pd

    (c) La, Hf, Ta, Os, Ir, Pt

    To what block of the periodic table (s, p, d, or f) does each of these elements belong?

23. Write the ground-state electron configuration of seaborgium (Sg), element number 106. To what block of the periodic table does it belong?

24. What element, first reported in 1999, should fall just below radon in the periodic table, thus becoming the next noble gas? State its atomic number and write its configuration in the ground state.

25. Write ground-state electron configurations for each ion:

    (a) O^{2-}, F^-, Na^+, Mg^{2+}

    (b) S^{2-}, Cl^-, K^+, Ca^{2+}

26. Write ground-state electron configurations for each ion:

    (a) Sc^{3+}, Ti^{4+}, V^{5+}

    (b) Y^{3+}, Zr^{4+}, Nb^{5+}

    (c) La^{3+}, Hf^{4+}, Ta^{5+}

27. Identify the atom from the configuration stated:

    (a) $[Ne]3s^2 3p^3$

    (b) $1s^2 2s^2 2p^6 3s^2 3p^6 4s^2 3d^{10} 4p^6$

    (c) $[Kr]5s^2 4d^{10} 5p^1$

28. Identify the atom from the configuration stated:

    (a) $[Xe]6s^2$

    (b) $[Xe]6s^25d^1$

    (c) $[Xe]6s^24f^15d^1$

29. Identify the ion (X) from the configuration stated:

    (a) $X^+ = 1s^22s^22p^63s^2$

    (b) $X^{2+} = 1s^22s^22p^63s^23p^6$

    (c) $X^- = 1s^22s^22p^63s^23p^64s^23d^{10}4p^6$

30. Identify the ion (X) from the configuration stated:

    (a) $X^{3+} = [Ne]$

    (b) $X^{2+} = [Ar]3d^2$

    (c) $X^+ = [Rn]$

31. None of these full configurations corresponds to a ground-state atom:

    (a) $1s^22s^23p^1$

    (b) $2s^23s^23p^5$

    (c) $1s^22s^33p^3$

    Explain why.

32. The interatomic distances in H_2 and F_2 are 0.7414 Å and 1.4119 Å, respectively. What are the atomic radii of H and F, and what (approximately) is the expected distance between H and F in a molecule of hydrogen fluoride, HF? Compare this estimate with the measured bond length of 0.9169 Å.

33. The distance between H and I in the molecule HI is 1.6090 Å. Use both this value and information from the preceding exercise to estimate the atomic radius of iodine.

34. Using only the periodic table, arrange each set of atoms in order of increasing radius:

    (a) Rb, Cs, Li

    (b) B, Li, F

    (c) Cl, F, Ba

    (d) Rb, Be, K

    (e) Cl, Al, Ba

35. Arrange each set of atoms and ions in order of increasing radius:

    (a) O, O^-, O^{2-}
    (b) Li^+, Li, Be^{2+}
    (c) Na^+, Mg^{2+}, Al^{3+}
    (d) F^-, Br^-, O^{2-}
    (e) Cs^+, Ba^{2+}, Al^{3+}

36. Arrange each set of atoms and ions in order of increasing electron affinity:

    (a) Li, Li^-, F
    (b) F, O, N
    (c) S, Cl, Ca
    (d) I, Ba, Tl
    (e) Br, At, I

    Which of the species above are unlikely to have any affinity at all for an extra electron? Why?

37. Pick the largest ionization energy from each set:

    (a) $I_1(Li)$, $I_2(Li)$, $I_1(Be)$
    (b) $I_2(Cs)$, $I_1(Rb)$, $I_7(Na)$
    (c) $I_1(Y)$, $I_1(Zr)$, $I_3(Zr)$

38. Consider, in summary, how the Pauli exclusion principle shapes every aspect of chemistry, building up the periodic table with this one demand: Electrons may not occupy the same quantum state, but rather each electron must find a state for itself—even if that other state lies at a higher energy. What kind of chemistry would we have if the Pauli exclusion principle were repealed? Could we exist?

7

Covalent Bonding
and Molecular Orbitals

7-1. From Atoms to Molecules
7-2. Diatomic Molecules
       The p Orbitals: σ and π
       Building Up
       Heteronuclear Diatomic Systems
7-3. Bonds in Molecules
7-4. Molecular Geometry
       Hybridization
7-5. Delocalization
       REVIEW AND GUIDE TO PROBLEMS
       EXERCISES

Graced by perfect symmetry, a solitary atom is a small masterwork of matter. It can stand alone, stable and self-contained, held together in exquisite balance.

Yet rarely is an atom a finished work, destined to remain complete and entire. An atom *can* stand alone, but usually it does not—at least not when other atoms are present, at least not under ordinary earthly conditions, at least not for long. Atoms combine. They intermingle their electrons and merge into larger structures. Except for a few noble gases, free atoms are hard to find. Look first for a molecule.

7-1. FROM ATOMS TO MOLECULES

Nature draws no fundamental distinction between atom and molecule. Each is an assembly of charged particles, brought together by electrical interactions. The particles—whatever they are, whatever their number—submit to the same electromagnetic forces and to the same laws of quantum mechanics. Atom or molecule, they obey the Schrödinger equation. Atom or molecule, they are describable by a wave function. Atom or molecule, they are governed by the laws of thermodynamics.

Mostly it is a difference in degree: an atom has only one nucleus; a molecule has two or more. But with extra nuclei come certain special features, and these new aspects turn into subtly important details. To understand them properly is to understand what makes a molecule a molecule.

Foremost is the loss of spherical symmetry. A molecule, having gained nuclei, loses an atom's invariance to direction and acquires new-found possibilities in the process. Things look different in different directions. Energies, wave functions, and electrons follow suit.

Think only of the simplest example, H—H, to appreciate why. Here space is indelibly marked by the line joining the two nuclei, which defines, once and for all, what was previously an arbitrary z-axis. There is one special direction plain to see, and (owing to this pointer in space) the molecule is denied the *s-p-d-f* orbitals of the hydrogen atom. Those familiar wave functions derive solely from the $1/r$ Coulomb potential and spherical symmetry of the atom. Expect something different for the hydrogen molecule, something that matches its own symmetry—the symmetry of a cylinder, not a sphere.

There are other changes as well. The H_2 species, typical of molecules, can move in ways impossible for an atom. It can vibrate. The two nuclei, attracted to the electrons and repelled by each other, move rapidly back and forth with quantized energies. It can rotate. The molecule tumbles and spins like a top, again with quantized energies. These motions (Figure 7-1), said to be additional *degrees of freedom*, are unavailable to a solitary nucleus.

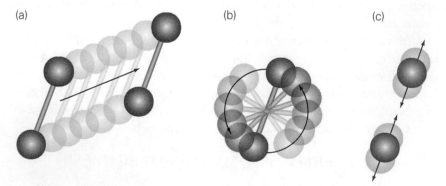

(a)   (b)   (c)

FIGURE 7-1. Three ways for a molecule to move: (a) *Translation*. The structure goes from one point in space to another, changing neither its orientation nor its internuclear distances. (b) *Rotation*. The molecule spins about an axis. (c) *Vibration*. The internuclear distances expand and contract.

Now, knowing what we do, let us try to imagine how a hydrogen molecule, H:H, might be formed from two separated atoms, H· and ·H. Let the atoms start far apart and alone, with each hydrogen accommodating its one electron in a very atomlike, very separate $1s$ orbital. The first of these $1s$ wave functions we shall call ψ_A; the second, equivalent in every way, we shall call ψ_B. Negative charge is distributed spherically and identically around each nucleus, and there is no common electron density in between:

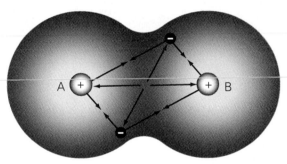

Imagine next that the atoms come slowly together, bringing their particles close enough for electrical cross-interactions to take hold. The electron originally near nucleus A is simultaneously attracted to nucleus B, while the electron on B is similarly attracted to A. Electron 1 starts to repel electron 2, and nucleus A starts to repel nucleus B. A composite system of two electrons and two nuclei is born:

New attractions and new repulsions develop in this way, leaving the energy to be reckoned anew for each distance. Look at the change in potential energy, point by point, as events unfold along the curve of Figure 7-2. Cross-attractions between electrons and nuclei draw the nuclei together, but repulsions work to push them apart. As long as attractions continue to exceed repulsions at any distance, the net change is favorable and the atoms are encouraged to move still closer. Nearer and nearer they come, sharing electrons ever more intimately, until eventually the repulsions dominate. The nuclei are too close, and the electrons are too close. Further compression can only raise the energy and force the atoms apart. A molecule of H_2 thus takes shape at some optimal separation, held together—in quantum mechanical fashion—by the attraction of two nuclei for two electrons. Tugged constantly in one direction and then the other, it is a jittery connection in which the nuclei vibrate back and forth as if joined by a spring. We speak therefore of an *average* bond distance to acknowledge such oscillations (typically small) about the point of lowest energy.

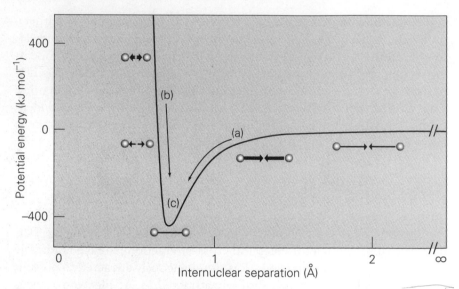

FIGURE 7-2. The potential energy of two hydrogen atoms, plotted for internuclear separations ranging from zero to infinity. (a) Where the curve slopes downward from right to left, the interaction is attractive. By coming together, the particles decrease their potential energy. (b) Where the curve slopes downward from left to right, the interaction is repulsive. The particles decrease their energy only by moving apart. (c) Where the curve is flat, at the bottom of the well, there is no tendency either to attract or repel. The structure is stable.

Cast in quantum mechanical language, this merger of hydrogen atoms translates into a mixing of atomic orbitals. One wave of probability amplitude, ψ_A, meets another, ψ_B. They come together. They interfere. A covalent bond, viewed quantum mechanically, then appears as yet another example of the strange arithmetic that follows when waves combine: 1 plus 1 proves to be more (or less) than 2. The bond is cemented when two wave functions, previously separate, come together constructively. The amplitudes ψ_A and ψ_B mix, as waves, and create something bigger than either of them could make alone.

It happens roughly in the following way. Far apart, there are two isolated atomic wave functions, ψ_A *and* ψ_B, sketched below in radial form:

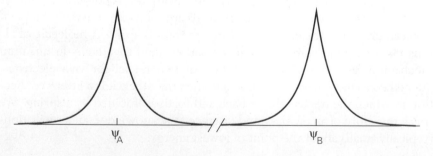

Close together, ψ_A and ψ_B can mix either destructively to yield a combined molecular wave function proportional to the difference,

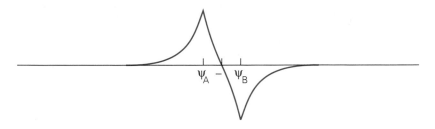

or—more promising—they can mix constructively to yield something proportional to the sum:

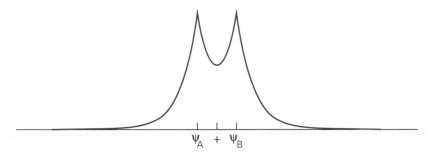

For when the two orbitals do reinforce each other, the joint density

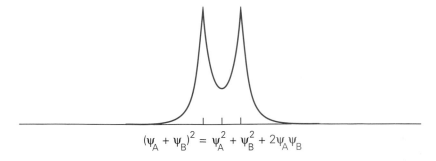

becomes bigger than ever before. The combined probability, we recall from Chapter 4, is not simply the sum of ψ_A^2 and ψ_B^2, which would follow from straightforward addition of two independent electron clouds. There is more. The electron probability becomes $\psi_A^2 + \psi_B^2$ (atom plus atom) *plus* $2\psi_A\psi_B$, where this new contribution describes the additional density concentrated between the nuclei. Simultaneously attracted to an increased negative charge, the nuclei draw together more closely. With stronger attractions, too, the energy of the molecule is lower than the energy of the uncombined atoms. A bond develops.

A ***molecular orbital*** for H_2 is thus formed, arising from a combination of the original atomic orbitals. This extended wave function becomes an orbital like any other, except now the electrons roam freely around two nuclei rather than one. The molecular orbital will play the same role for a molecule that an atomic orbital plays for an atom. It will provide an option for an electron to acquire a particular energy, a particular angular momentum, and a particular territory within a molecule.

That said, remember that any orbital—atomic or molecular—is no more than an approximation, flawed by an idealized picture of electron–electron repulsions. An orbital model offers us a good beginning, but not the final word on chemical bonding. But remember also that we have done well with atomic orbitals, and perhaps we shall have similar success with molecules. The idea, after all, is simple enough: Since molecules come from atoms, molecular wave functions should somehow come from atomic wave functions. A straightforward mixing of ψ_A and ψ_B into a molecular orbital is a plausible way to start.

With its roots in the atoms, then, this first molecular orbital is a wave function appropriate for the two electrons and two nuclei together. Its combined density, enhanced in the center (see Figure 7-3a), further

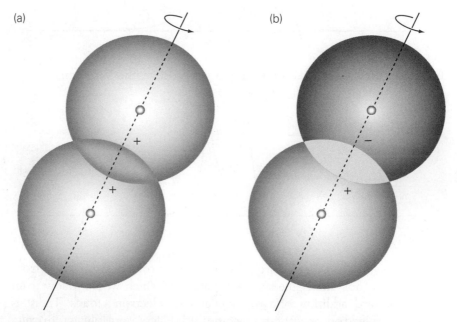

FIGURE 7-3. Mixing of the $1s$ wave functions from two hydrogen atoms: an angular view. The overlap, made directly along the line of approach, is termed *sigma* (σ). Electrons appear with equal probability at any angle around the bond axis. (a) A σ bonding combination, formed by constructive addition of the two orbitals. Electron density between the nuclei is enhanced. (b) A σ^* antibonding combination, formed by destructive subtraction of the orbitals. Density in the shared region decreases.

conforms to the symmetry of the molecule: the same left and right; the same all about the line joining the nuclei. We demand no less. The electrons must be distributed uniformly from side to side, because there is no discernible left and right in the entire structure. Nucleus A is indistinguishable from nucleus B, and so the electrons are divided without preference. Nor can there be any differentiation about the axis of the H—H bond, just as one looks down a pipe and expects to see only sameness all around.

This electron-enhancing combination, $\psi_A + \psi_B$, symmetric about the internuclear axis, is termed a ***sigma bonding orbital*** (σ, a Greek *s*). Resulting from the direct ***overlap*** of two atomic *s* orbitals along the line joining A and B, the σ bond blends the atoms together with the most favorable constructive interference. A σ electron, moreover, has zero angular momentum around the bond axis (and hence no angular node), in the same way that an *s* electron has zero angular momentum about an atom's center. So should it be, for the molecule's σ_{1s} orbital derives explicitly from the nodeless *s* orbitals of the atoms.

Having reached such understanding, we have still to deal with the possibility of destructive interference—as would arise when ψ_A and ψ_B are subtracted, not added. Here the whole is less than the sum of its parts, and here the combined density goes as $\psi_A^2 + \psi_B^2$ *minus* $2\psi_A\psi_B$. What was gained earlier from the extra density, $2\psi_A\psi_B$, is lost measure for measure. *Less* negative charge accumulates between the nuclei (Figure 7-3b), and the attraction is weakened correspondingly. Energy increases. Destructive interference pushes the atoms apart, creating an effect exactly opposite to a bond.

The result is a ***sigma antibonding orbital*** (σ^*), the destructive counterpart to the bonding combination. Although the overlap is still cylindrically symmetric about the unique axis (like the σ), the orbitals meet positive-to-negative and are cut perpendicularly by a nodal plane. The antibonding orbital expels electrons from its midst, leaving the nuclei exposed to even stronger positive–positive repulsions.

Constructive interference (bonding) or destructive interference (antibonding) . . . which will it be? We have, in the end, transformed two separate atomic orbitals into two molecular orbitals, producing the pattern shown in Figure 7-4. Averaged together, the energies of the new orbitals are approximately equal to those of the 1s atomic states. The σ_{1s} bonding level drops to some extent, whereas the σ^*_{1s} antibonding level rises by that same amount and usually slightly more (owing, in part, to the increased internuclear repulsion). Into these orbitals must go the molecule's two electrons, to be filled according to the same rules observed for the atomic *Aufbau* process in the previous chapter.

Molecular hydrogen (H_2), with two electrons to distribute, deploys them both into the bonding level as a spin pair. By so doing, the molecule

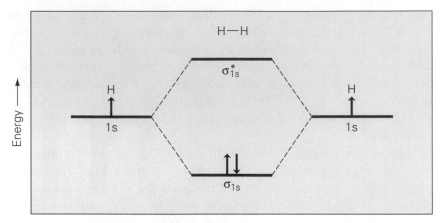

FIGURE 7-4. When two wave functions mix, one combination goes down in energy while the other goes up. If more electrons go into bonding than into antibonding orbitals, then the molecule will be more stable than the atoms apart. Hydrogen passes the test. H_2 exists in preference to H + H.

falls to an energy below that of its separated atoms. This new corporate structure, H_2, holds the two electrons more tightly than H and H alone, leaving the molecule stronger and better adapted to the bruising encounters of the microscopic world. Survival of the fittest is the rule there as well.

Unfit for survival is the nonexistent He_2 molecule (Figure 7-5), which could form only if helium atoms were able to share electrons profitably.

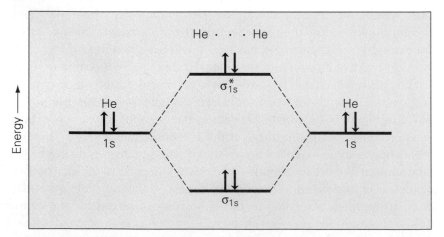

FIGURE 7-5. A molecule of He_2 has no energetic reason to exist: Two of its electrons would be in the bonding orbital, but two electrons would also be in the antibonding orbital. The result is a wash—no molecule.

They cannot. Coming together with four electrons, two helium atoms can (like hydrogen) mix their $1s$ orbitals up and down to produce one bonding and one antibonding level. In contrast to hydrogen, though, the new combinations lend no added stability. Two electrons go into the bonding orbital, lower in energy; but another two electrons also go into the antibonding orbital, higher in energy. The Pauli exclusion principle enforces a separation. There is no net gain and consequently no energetic advantage to be won from a merger.

No energetic advantage; no molecule. Electrons and nuclei band together only if there is new strength to be had. The molecule must be tougher than its component atoms on their own—tougher not by choice but by chance, simply because weaker species cannot long endure in a violent world of collisions and sudden change. Only the more robust structures, well settled in energy, remain intact and retain all their electrons. It is a fact of microscopic life.

We go forward now to molecules of greater complexity, but the added complexity is mostly in the details. Hydrogen and helium have already taught us two of the most fundamental lessons, valid for even the largest species: (1) Covalent bonds arise from the constructive overlap of wave functions, accompanied by an enhancement of negative charge between the nuclei. (2) Atoms stay united only if the larger structure brings more stability than they could otherwise realize.

H_2, for example, assigned a configuration $(\sigma_{1s})^2$, is a quantum-mechanically perfect realization of G. N. Lewis's *pair bond*: two electrons shared equally between two nuclei. Supplementing that original description, we add only that the pair occupies a *molecular* orbital lower in energy than the original two atomic orbitals. Bear that unifying picture in mind as we proceed to systems with more electrons, with more orbitals, with one-electron bonds, with two-electron bonds, with three-electron bonds, with hybridized bonds, with localized bonds, with delocalized bonds, or with any of the myriad variations to be expected from millions of molecules.

The next step, suitably modest, involves only a few examples from all those millions. We undertake a building up of the homonuclear diatomic molecules formed from second-row elements, beginning with Li_2. What follows is the molecular equivalent of the atomic Aufbau principle—but with two nuclei instead of one, with molecular orbitals rather than atomic, and with cylindrical symmetry rather than spherical.

7-2. DIATOMIC MOLECULES

Understand first that orbitals mix best when they are similar in shape and energy. Atoms bringing together well-matched levels ($1s$ and $1s$, for instance) will produce the most widely separated σ and σ* combinations. The bonding orbital drops to its lowest possible level, and the antibonding orbital rises to its highest. Knowing that relationship, we shall suppose initially (Figure 7-6) that the atomic orbitals come together one pair at a time: first-shell orbitals combine with first-shell orbitals; second-shell

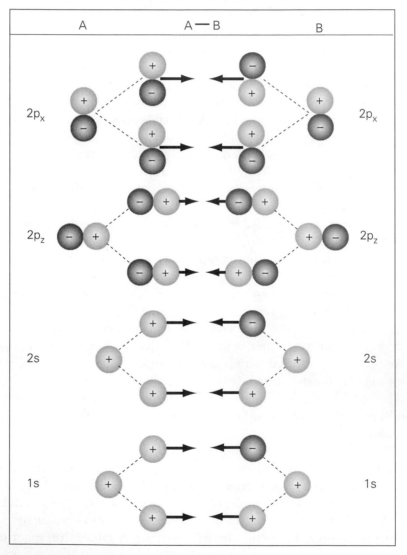

FIGURE 7-6. A simplified scheme for turning atomic orbitals into diatomic molecular orbitals: The $1s$ function on one atom mixes with the $1s$ function on the other. The $2s$ mixes with the $2s$; the $2p_z$ mixes with the $2p_z$; the $2p_x$ mixes with the $2p_x$; the $2p_y$ mixes with the $2p_y$. Orbitals in the y direction, not shown, stand perpendicular to the plane of the page.

with second-shell; s with s; p with p. It is a simple, not-too-unrealistic way to approach a complicated problem.

Start with two atoms of lithium, A and B, each bearing three electrons configured as $1s^2 2s^1$. Let them come together and interfere, and let them weave that interference into a molecule. The two $1s$ atomic orbitals are reshaped into a sigma bonding ($\sigma_{1s} \propto 1s_A + 1s_B$) and sigma antibonding ($\sigma_{1s}^* \propto 1s_A - 1s_B$) combination. The $2s$ orbitals similarly transform into ($\sigma_{2s} \propto 2s_A + 2s_B$) and ($\sigma_{2s}^* \propto 2s_A - 2s_B$), as depicted in Figure 7-7.

From four atomic orbitals thus come four molecular orbitals, divided into two sets of bonding antibonding pairs. Clearly we see the continuation of a pattern already begun by hydrogen and helium. Formally called a ***linear combination of atomic orbitals*** into molecular orbitals, such up-and-down mixing provides a consistent framework for our developing image of covalent bonding. Molecules, we insist, come from atoms. If atoms are precursors to molecules, then molecular orbitals should reasonably be related to atomic orbitals. What better, more logical way is there to express that dependence than to add and subtract atomic wave functions in varying amounts? It is a scheme we can extend to molecules of any size, and always we can count on one universal rule: n atomic orbitals, when mixed, will generate n molecular orbitals. If n orbitals go in, then n orbitals come out.

As for Li—Li, four molecular orbitals stand ready to receive a combined pool of six electrons. Of the first four electrons, one pair goes into

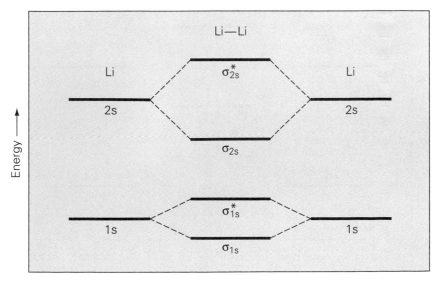

FIGURE 7-7. Molecular orbitals appropriate for Li_2. The σ level in each pair goes down in energy about the same amount as the σ^* goes up.

σ_{1s} and the other goes into σ_{1s}^*. The result so far: nothing, no bonding. These $1s$ electrons dwell deep in the atoms' cores, buried close to their nuclei underneath the valence shell. Unable to come sufficiently close, the $1s$ atomic orbitals fail to overlap and therefore contribute little to the eventual bonding. That job will be done, instead, by the two second-shell valence electrons, which interact at the outermost contours of the atoms and make the bond.

Those last two electrons enter the σ_{2s} bonding orbital, out of which develops the actual covalent linkage. A pair of electrons is shared, via head-on sigma overlap, in an orbital formed by constructive combination of the atomic $2s$ functions. The valence configuration is $(\sigma_{2s})^2$. It is a single bond, legitimately represented by the dot diagram Li:Li, which we now interpret as denoting two paired electrons in a bonding molecular orbital.

From that specific observation (Figure 7-8) let us define a more general quantity, the **bond order**, as $\frac{1}{2}$ the net number of bonding electrons in any arrangement of molecular orbitals:

$$\text{Bond order} = \tfrac{1}{2} \, (\text{no. of bonding electrons} - \text{no. of antibonding electrons})$$

With one Lewis-like bonding pair, Li_2 has a bond order of 1. The molecule is, unsurprisingly, analogous to H_2 in electronic structure, but its

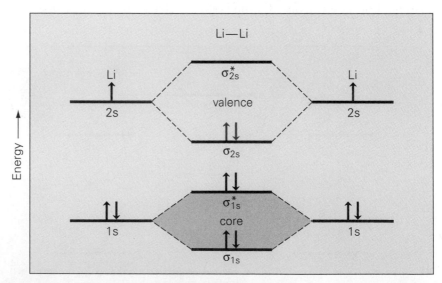

FIGURE 7-8. Electronic configuration of the Li_2 molecule, showing a net distribution of two electrons in bonding orbitals: a single bond, with bond order equal to 1. The valence levels are rendered distinct from the core.

bond is longer and weaker. Picture rather how two hydrogen atoms, un-encumbered by filled inner shells, should be able to approach more closely and enjoy a correspondingly stronger overlap. The effect is evident in the **bond energy** or **bond dissociation energy**, which we take to be the amount needed to tear apart one mole of the species in gaseous form. For hydrogen (436 kJ mol^{-1}) and dilithium (110 kJ mol^{-1}), the difference is nearly fourfold.*

Just as lithium mimics hydrogen, so does beryllium follow helium. With eight electrons to be deposited into the four molecular orbitals, the bond order is zero. The filled antibonding levels annul the overlap created by the filled bonding levels.

Two atoms of beryllium do not bond together covalently under normal conditions. The bond energy (less than 10 kJ mol^{-1}) is exceedingly weak, and the Be_2 molecule has a tenuous existence.

The p Orbitals: σ and π

Past beryllium, the $2p$ orbitals dominate the bonding from boron through neon. Each atom bears a set of three degenerate p functions, utterly indistinguishable under spherical symmetry. Differing only in orientation, they share the same shape and angular properties. Each orbital displays two lobes opposite in sign, divided by a nodal plane; together, the full set (p_x, p_y, p_z) is oriented arbitrarily along any three perpendicular axes. For the atom, it is a world without direction.

Not for the molecule. For the molecule, the unique direction of approach defines a unique direction for the p orbitals as well. Only in *one* orientation (call it z) can the p functions meet head-on in a sigmalike overlap, symmetrically disposed about the bond axis:

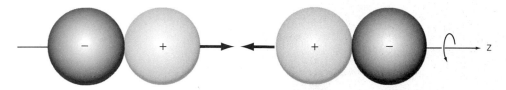

These $2p_z$ atomic orbitals interfere to produce a low-energy sigma bonding combination (*in* phase, σ_{2p_z}) and a corresponding sigma anti-bonding combination (*out* of phase, $\sigma^*_{2p_z}$), both shown in Figure 7-9. Notice also that the bonding orbital is nodeless between the nuclei,

---

*Dilithium, incidentally, is more stable than two separated lithium atoms, but not more stable than lithium *metal*, the element's normal state. We shall deal with metals and other states of matter beginning in Chapter 9.

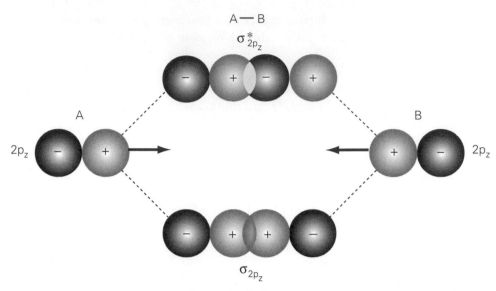

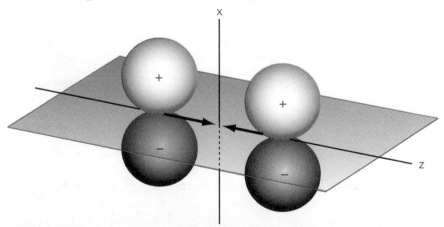

FIGURE 7-9. Mixing of $2p_z$ orbitals from two atoms to create a σ_{2p_z} and a $\sigma^*_{2p_z}$ molecular orbital—formed, like all sigma functions, by head-on overlap cylindrically symmetric about the internuclear axis.

because the wave functions add together with matching signs. Positive lobe meets positive lobe, or (same thing) negative lobe meets negative lobe.

The other *p* orbitals, meanwhile, are constrained to *side-to-side* overlap once the p_z orientation is locked into place:

Neither the *x* orbital (arbitrarily chosen) nor the equivalent *y* orbital can meet its counterpart straightaway along the internuclear axis. Standing perpendicular to the internuclear axis, for instance, $2p_{xA}$ and $2p_{xB}$ interact above and below their nodal planes to form the bonding and antibonding functions depicted in Figure 7-10. This mixing, called *pi overlap*, leaves the bonding orbital split by a nodal plane, thereby preserving the single angular node of the atomic *p* orbitals. The arrangement produces a less direct, less concentrated electron sharing than the highly efficient σ geometry. Typically the π interaction creates an orbital of higher energy and hence a weaker bond.

The resulting combinations are labeled **pi bonding orbitals** ($\pi_{2p_x} \propto 2p_{xA} + 2p_{xB}$) and **pi antibonding orbitals** ($\pi^*_{2p_x} \propto 2p_{xA} - 2p_{xB}$). To this pair is added an equivalent set in the *y* direction (π_{2p_y} and $\pi^*_{2p_y}$), thus giving the diatomic molecule a total of eight valence levels with expected energies

$$\sigma_{2s} < \sigma^*_{2s} < \sigma_{2p_z} < \pi_{2p_x} = \pi_{2p_y} < \pi^*_{2p_x} = \pi^*_{2p_y} < \sigma^*_{2p_z}$$

The π and π* functions are twofold degenerate, as they must be to honor the symmetry of the molecule. Since no special features exist to

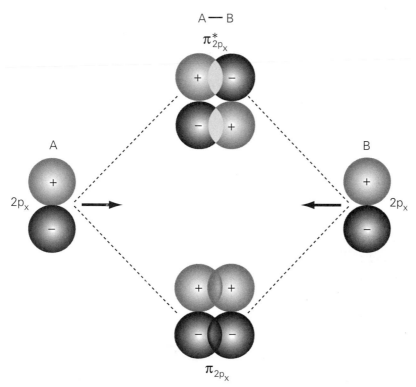

FIGURE 7-10. Pi overlap: $2p_x$ orbitals combine side to side, overlapping above and below a nodal plane. The bonding orbital is called π_{2p_x}; the antibonding orbital is called $\pi^*_{2p_x}$.

differentiate x from y, everything looks the same perpendicular to the z-axis (which connects the nuclei). Consequently all energies must be identical in the x-y plane.

This presumed ordering of the molecular levels (sketched in Figure 7-11a) follows the pattern observed for the uncombined atoms, in

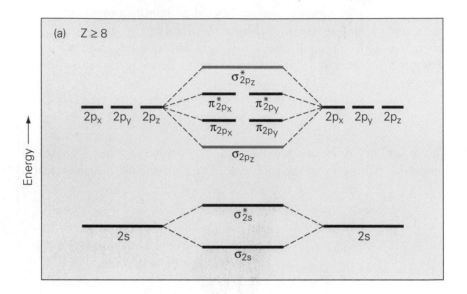

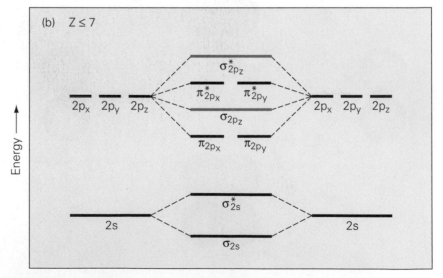

FIGURE 7-11. Valence energy levels for diatomic molecules in the second row. There are two possible orderings: (a) The σ_{2p_z} orbital lies below the π_{2p}, as it does for O_2 and F_2. (b) The σ_{2p_z} orbital lies above the π_{2p}, as observed for Li_2 through N_2.

which $E_{2s} < E_{2p}$. The scheme places, accordingly, the *s*-like molecular orbitals (σ_{2s} and σ_{2s}^*) underneath the six combinations derived from *p* functions. Bonding orbitals lie below their antibonding counterparts within each of these two sets, and (among the *p*-derived group) the σ falls beneath the π. The arrangement reflects the simple assumptions made above, beginning with our decision to combine *s* only with *s*, and *p* only with *p*.

Additional refinement is needed, however, to bring this model fully into line with experiment. For we see, looking more closely at the bonding patterns, that both the σ_{2s} and σ_{2p_z} orbitals lay claim to some of the same internuclear territory. Made from 2*s* and $2p_z$ atomic orbitals, respectively, the σ_{2s} and σ_{2p_z} combinations are sufficiently close in shape and energy to overlap and mix together themselves—despite our original assumptions. Each takes a bit from the other, and they combine to form two new bonding functions. The σ_{2s} (with a tinge of $2p_z$) goes down in energy and the σ_{2p_z} (with a tinge of 2*s*) goes up, producing a different splitting for each diatomic molecule. We need not consider the details further except to note that, at the extreme, the σ_{2p_z} is pushed *above* the π energies to yield the pattern sketched in Figure 7-11(b):

$$\sigma_{2s} < \sigma_{2s}^* < \pi_{2p_x} = \pi_{2p_y} < \sigma_{2p_z} < \pi_{2p_x}^* = \pi_{2p_y}^* < \sigma_{2p_z}^*$$

Some molecules are built according to the one pattern; some according to the other. We shall soon see examples of each.

Subtle? Earlier experience with atoms (Chapter 6) teaches us to expect just this kind of inversion in orbital models. It is the price paid for a simplified picture.

Building Up

Electrons follow both the "normal" and "inverted" filling sequences during formation of the second-period diatomic molecules. Figure 7-12 provides the results at a glance.

The series continues with B_2, which (see Figure 7-12) arranges its six valence electrons into the inverted, pi-before-sigma configuration $(\sigma_{2s})^2(\sigma_{2s}^*)^2(\pi_{2p})^2$. Occupation of the degenerate π bonding orbitals conforms to Hund's rule, with one electron in each slot and with the two spins parallel. From this electron structure, written explicitly as $(\pi_{2p_x})^1(\pi_{2p_y})^1$, there develops a paramagnetic molecule having a bond order equal to 1. It is a single bond, attributable to a net sharing of two electrons in the π system.

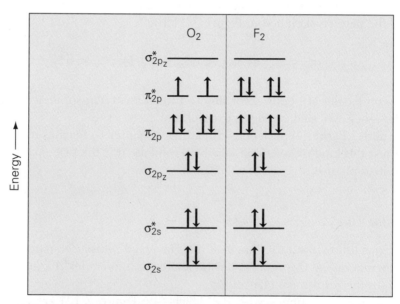

FIGURE 7-12. Valence configurations for the second-row diatomics: Li_2 through F_2.

C_2, with filled π orbitals, is no longer paramagnetic, but is bound instead by two spin pairs. Its configuration is $(\sigma_{2s})^2(\sigma_{2s}^*)^2(\pi_{2p})^4$, and its bond order is 2.

Nitrogen, N_2, still following the pattern of Figure 7-11(b), adopts the configuration $(\sigma_{2s})^2(\sigma_{2s}^*)^2(\pi_{2p})^4(\sigma_{2p_z})^2$. Three bonding pairs (see the drawing below) are distributed among the σ_{2p_z}, π_{2p_x}, and π_{2p_y} orbitals, giving the molecule a bond order of 3:

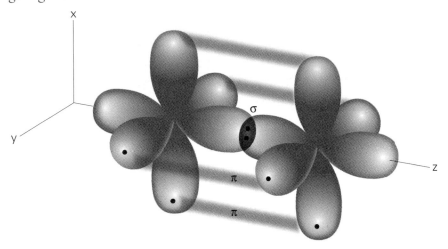

The connection is a classic *triple bond*, represented originally in Chapter 2 by a dot structure in which each atom shares three pairs of electrons and keeps one pair to itself:

$$:N:::N:$$

These 10 electrons, we now realize, are better understood as filling five orbitals of three distinct kinds:

1. A σ_{2p_z} orbital, which connects the nuclei through head-on σ overlap.

2. Two π orbitals, which yield the second and third bonds. The π overlap ties the nuclei together above and below.

3. A *nonbonding* 2s atomic orbital on each nitrogen, corresponding to the lone pairs in the Lewis structure $:N:::N:$. With both σ_{2s} and σ_{2s}^* orbitals fully occupied, the 2s electrons make scant contribution to the bonding.

Through this correspondence comes a satisfying link between molecular orbitals and Lewis's earlier, much simpler picture of electron pairs and octets.

There is more, too, because molecular orbitals can explain phenomena that dots cannot, a telling example being the paramagnetism of molecular oxygen. In O_2, where the levels switch over to the sigma-before-pi pattern of Figure 7-11(a), the configuration becomes $(\sigma_{2s})^2(\sigma_{2s}^*)^2(\sigma_{2p_z})^2(\pi_{2p})^4(\pi_{2p}^*)^2$. Consistent with the molecule's known paramagnetism, there are two *unpaired* electrons in the degenerate π^*_{2p} orbitals:

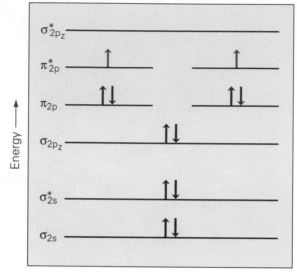

Coexisting with these two antibonding electrons are six in *p*-derived bonding orbitals (two in the σ_{2p_z}, four in the π_{2p}) to make a net bond order of 2.

Oxygen therefore does get a double bond, on balance, but not a double bond in which all of the electrons are paired spin-up and spin-down. Recalling the Lewis diagram

$$:\ddot{O}::\ddot{O}:$$

drawn originally in Chapter 2, we admit that the simple dot picture fails to account for the molecule's magnetic properties. True enough. Our model of bonding continues to evolve.

Heteronuclear Diatomic Systems

Indistinguishable end to end and around the bond, a *homonuclear* diatomic molecule inherits considerable symmetry from its atoms. Each nucleus lays equal claim to the electrons, and so the wave function is biased neither left nor right. Rather the electrons are shared impartially along a purely covalent bond, without buildup of excess charge near one nucleus or the other. We say, using the terminology introduced in Chapter 2, that there is no difference in electronegativity and also no dipole moment. There is no left-hand side. There is no right-hand side.

But left and right do exist in a *heteronuclear* diatomic molecule (Figure 7-13), where two different nuclei make each end instantly

(a)

(b)

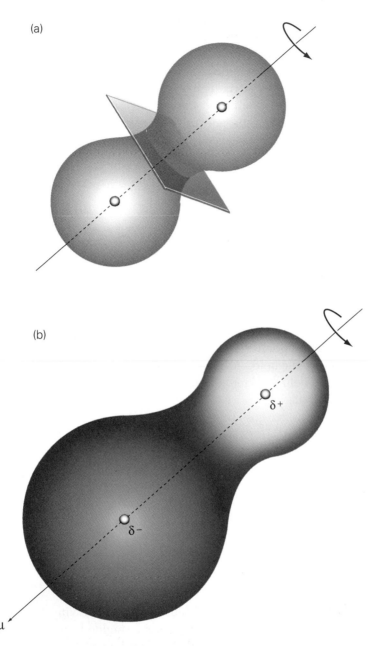

FIGURE 7-13. The difference between heteronuclear and homonuclear bonding:
(a) In a homonuclear molecule, always nonpolar, neither side accumulates more
electrons than the other. The structure appears as if reflected in a mirror, indistinguishable from end to end. (b) In a heteronuclear molecule, one of the atoms takes
a larger share of the electrons. An excess of negative charge builds up at that end to
produce a dipole moment (μ)—although the molecule retains, still, the cylindrical symmetry common to all diatomics.

recognizable. And, with these different nuclei and different atomic orbitals, the approaching atoms compete unequally for a share of the total pool of electrons. The electron density is skewed inevitably to one side. One atom attracts proportionally more negative charge and thus makes its own end partially negative. The other site, denied part of its share, is left partially positive. From that separation of *two* charges, Q and $-Q$, there develops a **dipole moment** (*di*-pole: two charges, equal and opposite) and hence a *polar* molecule.

Defined as

$$\mu = Qd$$

where d denotes the distance between the charges, the dipole moment is a directed quantity that points from the positive to the negative charge:

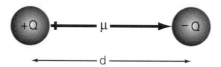

Indeed it really is a pointer, because a polar molecule will align itself end to end in an electric field. The positive portion points toward the negative pole of the field, making μ a property susceptible to both measurement and manipulation.

We shall consider just one specific molecule—hydrogen fluoride, HF—to suggest how electrons come to be shared asymmetrically in heteronuclear bonds. Let the species H· and ·F̈: approach with configurations $1s^1$ and $1s^2 2s^2 2p^5$, and assume for simplicity that the primary mixing occurs constructively between hydrogen's $1s$ orbital and fluorine's $2p_z$:

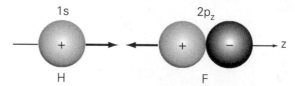

Admittedly it is an unequal partnership. Fluorine, carrying a high effective nuclear charge, initially holds its poorly shielded $2p_z$ electron in a level lower in energy than hydrogen's $1s$. The fluorine and hydrogen orbitals then mix according to the scheme illustrated in Figure 7-14, from which develops a lopsided distribution of electrons biased toward the fluorine. Lower to begin with, the fluorine $2p_z$ orbital sinks still more and becomes the dominant contributor to the σ bonding function. With both electrons trapped in this fluorine-weighted molecular orbital, the

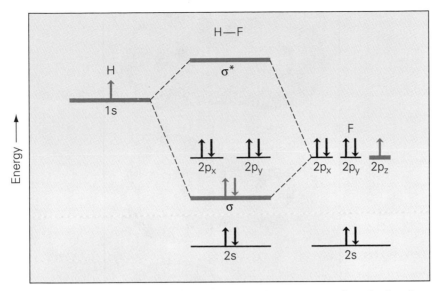

FIGURE 7-14. Formation of molecular orbitals between H and F. The low-lying fluorine 2s atomic orbital, unable to mix effectively with hydrogen's 1s, remains nonbonding, as do two of the three p orbitals. The bonding σ orbital favors the fluorine site, occupied by the more electronegative atom.

molecule HF enjoys a more negative energy than the uncombined atoms. Together, the bonding electrons spend more time in the attractive environment of the deshielded fluorine nucleus.

The structure is manifestly different side to side, but still indistinguishable around the bond. Although drawn toward the more electronegative atom, the electrons continue to occupy space *between* the nuclei. Wrapped uniformly around the axis, this reinforced internuclear charge density enforces the cylindrical symmetry while binding together the molecule. It is an unequal bond, but covalent all the same; it is a *polar* covalent σ bond fashioned from *shared* electrons.

Now imagine again, as we did in Chapter 2, what happens when one atom's electron-attracting power starts to overwhelm the other. More and more negative charge shifts to one end, and the electron cloud between the nuclei is stretched increasingly thin. Steadily the molecule becomes more polar (Figure 7-15), until the sharing effectively disappears and ions are formed. With internuclear overlap so drastically reduced, the covalent bond then gives way to an unvarnished electrostatic attraction of positive charge for negative charge.

No longer connected by a line of shared electrons, the ions reacquire the spherical symmetry of the original atoms. They are on their own, free to interact electrostatically with whatever charged particles come

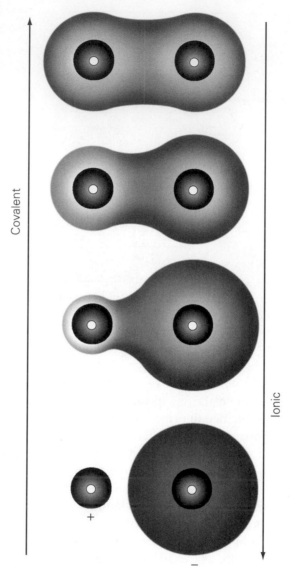

FIGURE 7-15. From one extreme to another. In the panel at the top, two atoms share the electrons equally in a nonpolar covalent bond. At the bottom of the diagram, two ions interact noncovalently by an electrostatic Coulomb interaction. No electrons are shared; no electrons appear in the space between the nuclei. In the middle, there is a continuous range of polar covalent bonds—part covalent, part ionic.

along—in any direction, at any distance, in any number. Usually the individual molecule ceases to exist under such conditions, and the ions condense en masse into an extended solid like NaCl, a collective *state of matter* (to reappear, in more detail, in Chapter 9). See Figure 7-16.

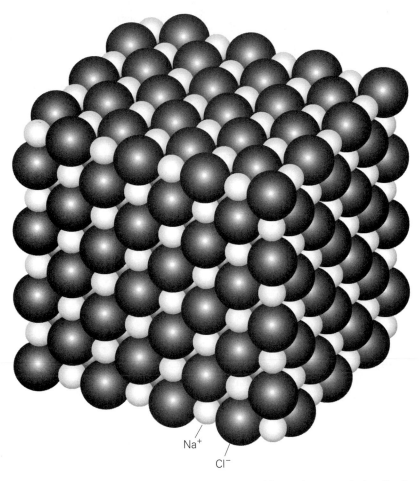

Na⁺

Cl⁻

FIGURE 7-16. NaCl, an ionic crystal, holds itself together mostly by Coulomb interactions—but not completely. Even here, a certain amount of shared electron density persists between the ions.

7-3. BONDS IN MOLECULES

Somewhere between the one-to-one partnership of H_2 and, say, the communal interactions of 10^{23} ions in NaCl lie our millions of molecules. Consider them: covalently bonded molecules, built from atoms and stitched together with shared electrons; structures with size, shape, and electrical characteristics; structures that can vibrate and rotate; structures that can yield color and stimulate taste; structures that can carry out the business of life. All different, yet all built from the same plan. Common to all is the covalent chemical bond.

We can measure bond energies. We can measure bond lengths. We can measure bond angles. We can measure bond vibrational frequencies. We can measure dipole moments. Then, having done so, we would find remarkably little variation in the numbers from molecule to molecule. C—H bond lengths, for example, fall close to an average value of 1.09 Å in all kinds of environments. A distance may be 1.08 Å in this molecule and 1.11 Å in that, but such differences remain small.

One is tempted to view the bond as an interchangeable part for a polyatomic molecule, able to connect any two nuclei in some standard way. A gross oversimplification, certainly, yet the trends above suggest that bonding might be largely a local, nucleus-to-nucleus affair, similar to that found in diatomic systems. Each of these quasi-autonomous bonds would then acquire substance and integrity of its own within the molecule's overall organization. It would be a spoke in a wheel.

Making that argument (Figure 7-17), we concede that a molecule is just so many charged particles, brought together and abandoned to the fundamental electromagnetic force. Nuclei and electrons will fall where they may, subject only to the laws of quantum mechanics and thermodynamics. What the data imply, however, is that often the electrons do fall precisely between the nuclei, site by site, in generic arrangements corresponding to our idea of covalent bonds. If so, then we already have our prototype: the diatomic molecule, with its constructively overlapping orbitals and (in many instances) its two Lewis-like electrons per bond.

FIGURE 7-17. Bonds in molecules: a chemist's view of matter. On the one hand, molecules are integrated structures in which electrons have free access to every site. On the other hand, molecules often behave as if they were modularly constructed from snap-together parts: from *pair bonds*, local accumulations of electron density between two nuclei. Even in very different molecules, the highlighted A—B bonds share nearly the same length, energy, polarity, and other properties.

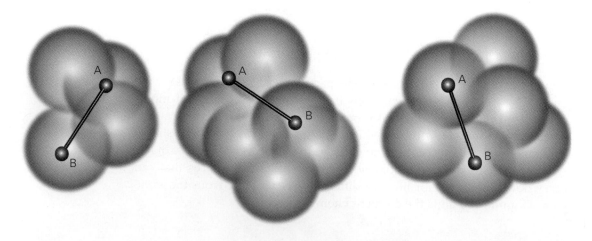

It is an incisive, simplifying view of molecular architecture, appearing all the more powerful when applied to large and complicated structures. Sometimes it may be too simple a view (and we shall take notice where necessary), but the notion of bonds-in-molecules rightfully dominates the modern conception of chemistry. The proof is in the properties; the proof is in the data.

Take some of the key properties, starting with those parameters that bear on a bond's strength. First there is the *bond energy*, defined earlier as the quantity of energy needed to remove a covalently bonded atom to infinity. Typical values, collected in Appendix C (Table C-13), amount to several hundred kilojoules per mole for bonds in their most common molecular settings.

Since strong linkages have large bond energies, such bonds tend to be less reactive. At approximately 950 kJ mol^{-1}, for instance, a nitrogen–nitrogen triple bond ranks as a particularly strong connection, hard to break under normal conditions. Just think of molecular nitrogen, :N≡N:, which makes up 80% of ordinary air and which we inhale without chemical consequence. Sometimes the best chemistry is no chemistry at all.

Ask again: What makes a bond strong? Shared electrons. The more negative charge between the nuclei, the greater the attraction. Hence doubly bonded nitrogen, N=N, connected by fewer electrons than N≡N, has a bond energy of only some 400 kJ mol^{-1}. Weaker still, at an average value near 160 kJ mol^{-1}, is the singly bonded linkage, N—N.

Although bond energy generally does increase with bond order, the scaling is by no means uniform. N—N is not energetically equivalent to one-half of N=N, nor is the double bond equivalent to two-thirds of a triple bond. Total strength is determined by too many other factors, among which are the aforementioned differences between σ and π overlap. Nor do these correlations hold between different pairs of atoms. There is no reason, therefore, why a C—F single bond (484 kJ mol^{-1}) should not be stronger than a N=N double bond (409 kJ mol^{-1}). Each connection arises from its own particular combination of attraction, repulsion, effective charges, orbital energies, and overlap.

Strong bonds are *short* bonds too, as borne out by the average distances also available for inspection in Appendix C (Table C-13). From single to double to triple bond, nitrogen nuclei draw progressively closer together: from 1.47 Å in N—N to 1.24 Å in N=N to 1.10 Å in N≡N linkages. The same pattern holds for C—C, C=C, and C≡C; the same pattern holds for C—N, C=N, and C≡N; the same pattern holds for multiple bonds in general.

So it goes for the strength of a bond. Next we turn to the charge distribution and polarity, looking to test this same notion of the bond as an interchangeable part. To address the issue is to inquire, once again, into the differing abilities of atoms to compete for electrons.

Electronegativity, introduced in Chapter 2 as an atom's ability to attract electrons, provides a useful—although rough—way to predict polarity. It is an empirical system by which electron-withdrawing power is rated on a dimensionless scale, with 4.0 assigned to the most competitive element: fluorine.

By now we can understand why. Poorly shielded fluorine, needing one electron to complete its second shell, attracts incoming negative charge better than any other element, including the other halogens. As a small species with a large Z_{eff}, the fluorine atom resists ionization but will readily accept an additional electron. Contrast that receptivity, though, with the ease of *detachment* exhibited by an atom like cesium, a large structure bearing a single valence electron, strongly shielded. Size, shielding, and valence combine to make cesium one of the least electronegative of the elements, with a rating of 0.7. Hydrogen, which can both give and take an electron, lies between these extremes with a rating of 2.1.

The electronegativity scale, devised expressly to reproduce known properties, uses a combination of ionization energy (I) and electron affinity (EA) to rank the atoms under average conditions. A species with low I and low EA (making it easy to lose an electron, hard to gain one) will usually transfer negative charge to a species with high EA and high I (where the charge is easy to attach, but hard to lose). Atoms in the first category, such as the metals, have low electronegativities. Atoms in the second class, notably the halogens, have high electronegativities. Values typically increase up the columns and across the rows, as is apparent in Figure 7-18.

FIGURE 7-18. Relative electronegativity of the elements. Least electronegative are the alkali metals, particularly the larger atoms. Best able to attract an electron are the halogens, led by fluorine (which tops the scale at 4.0).

H 2.1																	
Li 1.0	Be 1.5											B 2.0	C 2.5	N 3.0	O 3.5	F 4.0
Na 0.9	Mg 1.2											Al 1.5	Si 1.8	P 2.1	S 2.5	Cl 3.0
K 0.8	Ca 1.0	Sc 1.3	Ti 1.5	V 1.6	Cr 1.6	Mn 1.5	Fe 1.8	Co 1.9	Ni 1.9	Cu 1.9	Zn 1.6	Ga 1.6	Ge 1.8	As 2.0	Se 2.4	Br 2.8
Rb 0.8	Sr 1.0	Y 1.2	Zr 1.4	Nb 1.6	Mo 1.8	Tc 1.9	Ru 2.2	Rh 2.2	Pd 2.2	Ag 1.9	Cd 1.7	In 1.7	Sn 1.8	Sb 1.9	Te 2.1	I 2.5
Cs 0.7	Ba 0.9	La 1.0	Hf 1.3	Ta 1.5	W 1.7	Re 1.9	Os 2.2	Ir 2.2	Pt 2.2	Au 2.4	Hg 1.9	Tl 1.8	Pb 1.9	Bi 1.9	Po 2.0	At 2.2

I   II {◄————————— Transition metals ————————►} III   IV   V   VI   VII

Electronegativity, a *relative* quantity, is best taken as a difference between two atoms. A larger difference means a more polar bond. Possibilities range from purely covalent (as in H_2, where $2.1 - 2.1 = 0$) to highly ionic bonds (such as in NaF; $4.0 - 0.9 = 3.1$). Polar covalent bonds of various compositions are found in between.

7-4. MOLECULAR GEOMETRY

Whereas atoms are spherical, molecules are angular. Supporting any molecule is a craggy skeleton of nuclei, laid out with bends and points and planes to which the electrons must conform. There is, as the imagined image in Figure 7-19 reminds us, always a framework to the structure—be it a line, a triangle, a tetrahedron, a hexagon, an octahedron, or some combination of geometric figures. Holding everything together are the electrons between the nuclei.

To account for molecular shape is the next challenge for our model of bonding, beginning briefly with the question of *why*. We wonder why, for example, CO_2 is linear yet H_2O is bent; and why CH_4 is tetrahedral rather than square; and why C_6H_6 is arranged as a hexagon rather than a prism.

Why? Because that shape, whatever it is, best allows those nuclei and electrons to lower their combined energies. If such an answer seems merely to repeat the question, then refer back to the formation of H_2 as portrayed in Figure 7-2. There, with only two nuclei, the potential energy depends solely on one variable: the internuclear distance. Adjusting to the changing force, the budding molecule samples different structures until settling finally around the point of minimum energy. That is where H_2 finds itself least disturbed; that is "why" H_2 has an average bond length of 0.74 Å.

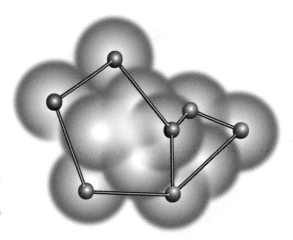

FIGURE 7-19. Nooks and crannies. An atom is round, but a molecule is angular. It has a framework, a skeleton, corners, joints, bends. It has a shape.

For polyatomic molecules there are more variables, and with more variables comes a complicated dependence of energy on distances and angles (Figure 7-20). No longer can all questions be answered by a single curve on a flat sheet of paper. One needs, instead, to understand exactly how a molecule's total potential energy depends on each of the geometric variables. It is a hard and compelling problem, different for every structure.

Still the shortest, simplest, most obvious answer remains correct even for the most intricate molecular combinations: nuclei and electrons fall blindly into an arrangement where their potential energy is lowest. Such was already our thinking in Chapter 2, where we sought to explain molecular geometry by the VSEPR model. The idea then, plausible but incomplete, was that pairs of electrons would spread out to minimize repulsions about a central atom. Now, having developed a workable picture of orbitals, we can go considerably beyond that first treatment.

It will be, for the moment, a localized picture involving identifiable bonds between pairs of nuclei—an approach called ***valence bond theory***.

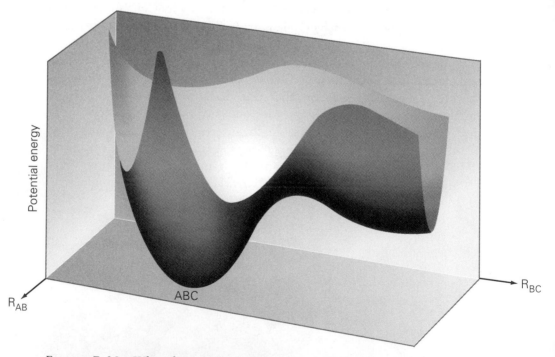

FIGURE 7-20. When three or more atoms come together, their potential energy changes instantaneously with the geometric arrangement. The energy depends on every internuclear distance and angle, as seen here in this *potential energy surface* for the head-on approach of molecule AB and atom C. The linear molecule ABC lies at a minimum on the surface.

Building on the notion of bonds-in-molecules, we shall attribute each "valence bond" to constructive overlap between orbitals on adjacent atoms. Every connection will be assigned a two-site, nucleus-to-nucleus bond held together by a pair of electrons with opposite spin. The plan is to reconstruct the molecule piece by piece, one covalent linkage at a time.

We start by looking at some facts.

Hybridization

Fact: Methane (CH_4) is a tetrahedral molecule:

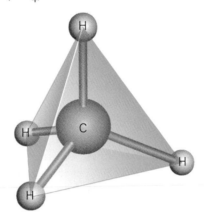

Four bonds, equal in length and strength, radiate from the central carbon. Each H—C—H angle is 109.5°, as in a regular tetrahedron. Each carbon–hydrogen distance is 1.09 Å, consistent with a bond order of 1.

Fact: Ethylene (C_2H_4) is a planar molecule:

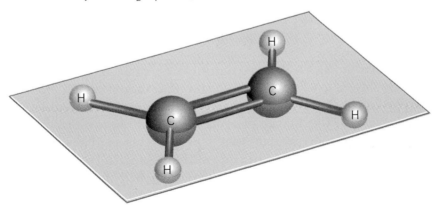

Its four carbon–hydrogen bonds, also equivalent, are of length 1.09 Å, again suggesting a bond order of 1. The carbon–carbon connection, spanning a length of 1.34 Å, has a bond order of 2, as in a C=C double bond. Each of the four C—C—H bond angles is equal and close to 120°.

Fact: Acetylene (C_2H_2) is a linear molecule:

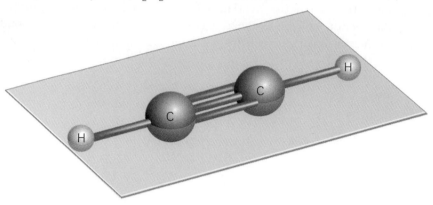

The two carbon–hydrogen bonds, indistinguishable left and right, have lengths of 1.06 Å and hence appear to be C—H single bonds. The 1.20 Å carbon–carbon link has the character of a C≡C triple bond.

Three molecules, three shapes. One kind of atom (carbon) forms bonds to another kind of atom (hydrogen) in three distinct ways to produce three distinct angular structures. Yet methane and ethylene and acetylene, if broken apart, would each revert to spherical carbon and hydrogen atoms with configurations of $1s^2 2s^2 2p^2$ and $1s^1$. From identical atomic ingredients must come, apparently, three different molecules with new shape and symmetry. From atomic carbon's three p orbitals— mutually perpendicular, with 90° angles all around—must come tetrahedra, triangles, and lines.

Why not? Atomic carbon is not molecular carbon, and the electrons in a molecule need not respect the orbitals of an atom. Those old orbitals were the orbitals of a space having no north, south, east, and west. The new orbitals, set amidst the angles and bends of the nuclei, are the orbitals of tetrahedra, triangles, and lines.

Ask how methane's electrons can possibly hold the molecule together in a tetrahedral configuration. Ask how one carbon atom, originally equipped with a $2s$ valence orbital and three $2p$ functions, can make four indistinguishable C—H bonds. The question is not *why* (we already know why, broadly considered), but *how*—how, in a nuts-and-bolts sense, the electrons can arrange themselves into the most satisfactory structure.

One route might be to reshape the valence shell of carbon into four equivalent orbitals, each directed toward one corner of a tetrahedron, each able to make a single bond. Such adjustment, if possible, would produce the observed structure of methane, and it would teach us the difference between five spherical atoms (C, H, H, H, H) and one tetrahedral molecule (CH_4). Can our valence bond model account for that?

It can. Just let the old $2s$, $2p_x$, $2p_y$, and $2p_z$ orbitals of the free carbon be recombined to produce four **hybrid atomic orbitals** appropriate for the carbon bound *inside the molecule*. We mix and match, free to do so, proceeding according to this quantum mechanical recipe: (1) Take four carbon orbitals ($2s$, $2p_x$, $2p_y$, $2p_z$). (2) Mix thoroughly. (3) Combine them in the way prescribed in Figure 7-21, and draw out four identically shaped hybrid functions differing only in orientation: four ***sp³ hybrids*** (read "ess-pee-three") derived from one s and three p functions.

Each of these degenerate sp^3 combinations is now a 25%–75% composite of s and p characteristics, blended together as intimately and inseparably as a mixture of hot and cold water. No longer does carbon have three p orbitals and one s; it has four identical hybrid orbitals instead, merged together seamlessly to point in four new directions. Each sp^3 hybrid is part s and part p, with energy intermediate between the original $2s$ and $2p$ levels:

$$2p \quad \uparrow \quad \uparrow \quad — \qquad\qquad \uparrow \quad \uparrow \quad \uparrow \quad \uparrow \quad sp^3$$

$$2s \quad \uparrow\downarrow$$

free C atom carbon in CH_4

Into every one of these four sp^3 orbitals goes one electron apiece, as if the carbon atom were configured as $\cdot\dot{C}\cdot$ rather than $:\dot{C}\cdot$. Then, in the

FIGURE 7-21. Construction of four sp^3 hybrid orbitals from four combinations of the s, p_x, p_y, and p_z orbitals on a single atom. An additional small lobe is omitted from each function for the sake of clarity: (1) $s + p_x + p_y + p_z$. (2) $s - p_x - p_y + p_z$. (3) $s - p_x + p_y - p_z$. (4) $s + p_x - p_y - p_z$. Degenerate in energy, the hybrid orbitals point to the corners of a regular tetrahedron.

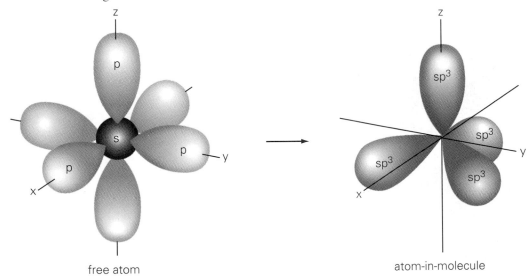

free atom atom-in-molecule

presence of four hydrogen atoms, the sp^3 carbon forms four ordinary
sigma bonds to produce tetrahedral CH_4:

$$H$$
$$\overset{\cdot}{\underset{\cdot}{H \cdot \; \cdot C \cdot \; \cdot H}}$$
$$\overset{\cdot}{H}$$

It is a scenario consistent with a VSEPR model. The sp^3 hybridization
leaves CH_4 with four shared pairs of electrons around the carbon, accommodated in localized orbitals and separated to the largest extent
possible.

Realize, too, that hybridization is just another variation on the persistent theme of orbital combination. A hybrid orbital is a mixture of
two or more atomic orbitals centered on one nucleus. It is, like all orbitals, a mathematical device used to explain the reality of molecular
geometry *after the fact*. Surely a molecule does not "know" it is hybridized, any more than an atom knows it has perpendicular p orbitals.
The molecule only knows, so to speak, how best to arrange electrons
and nuclei for lowest energy. Looking at the final arrangement, *we* devise

FIGURE 7-22. Formation of three sp^2 hybrid orbitals from a mixture of s, p_x, and p_y
orbitals. An additional small lobe is omitted from each function for the sake of
clarity: (1) $s + \sqrt{2}p_y$. (2) $s + \sqrt{3/2}p_x - \sqrt{1/2}p_y$. (3) $s - \sqrt{3/2}p_x - \sqrt{1/2}p_y$. The degenerate combinations lie together in a plane, spaced at intervals of $120°$. The
original p_z orbital, standing perpendicular to the sp^2 plane, remains unhybridized.

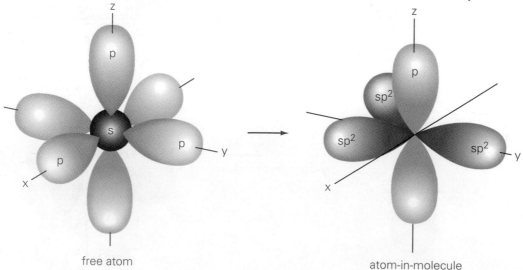

free atom

atom-in-molecule

our "as if" picture, saying: as *if* the electrons were housed individually in orbitals; as *if* these bonding orbitals were hybrid blends of *s* and *p*.

On to ethylene, then, where the bond angles are 120°. Here the after-the-fact recipe calls for one *s* orbital and only two *p* orbitals, from which are formed the three ***sp²*** *hybrid orbitals* pictured in Figure 7-22. Spaced 120° apart like three spokes of a wheel, the degenerate *sp²* functions all lie in one plane. Perpendicular to this plane is the leftover *p* function, which always remains part of the set whether occupied or not.

Two *sp²*-hybridized carbon atoms can engineer one *sp²*–*sp²* sigma bond in the head-on arrangement shown below:

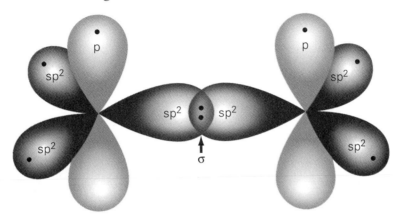

Held together by that first bond, the two atoms then make a π bond between the unhybridized *p* orbitals sticking out of the plane. The π overlap, running above and below like two ribbons, locks the four remaining *sp²* orbitals into a common plane:

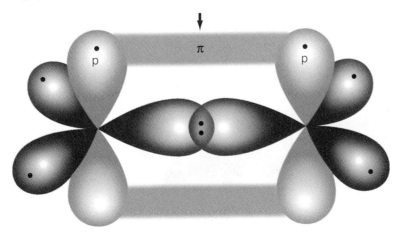

Each half-filled hybrid, having one electron to share, concludes a σ bond with a hydrogen 1*s* orbital:

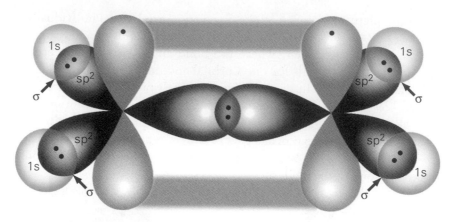

A planar ethylene molecule is thus assembled, joined together by a C=C double bond. The first carbon–carbon linkage, stronger of the two, arises from sp^2–sp^2 σ overlap. The second connection, more easily broken, is attributable to p–p π overlap. It is this π bond that freezes the molecule into a planar geometry, and (as we shall explore further in the next chapter) presents a tempting target for electron-hungry Lewis acids.

Finally there is acetylene, H—C≡C—H, in which the four atoms are laid out all in a row. The linear geometry traces to ***sp hybrid orbitals***, formed in what has become a predictable scheme. One *s* orbital plus one *p* orbital first makes two *sp* hybrids for each carbon atom, leaving two *p* orbitals still unmixed:

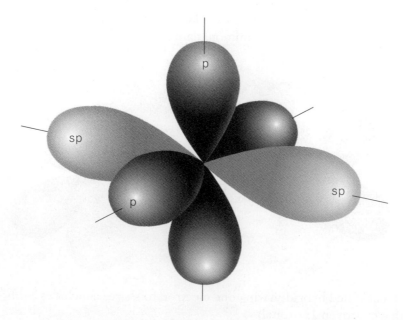

Pointing in opposite directions, these twin *sp* hybrids have the same shape and energy. Each can make a sigma bond. And so, with *two* unhybridized *p* orbitals per carbon atom, the molecule is connected by one *sp–sp* σ bond and *two p–p* π bonds encircling the carbon–carbon axis:

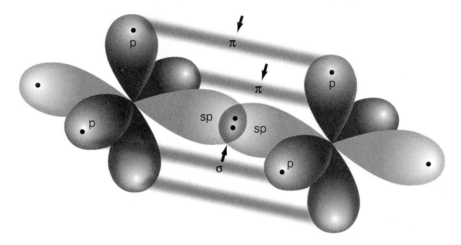

One hydrogen atom is attached, by means of sigma overlap, to the receptive *sp* orbital jutting out from each end. The result is a linear molecule, held together in the center by a triple bond:

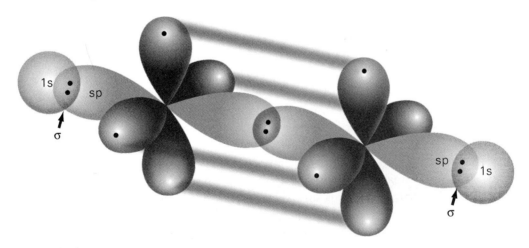

7-5. DELOCALIZATION

This chapter will end in the spirit it began, with a reaffirmation of our original conception of a molecule. Putting aside any notions about valence bonds or electron pairs, we shall again permit the charged particles to fall where they may. We shall treat a molecule as a unified assembly of electrons and nuclei, trusting the Schrödinger equation to reveal the orbitals within. The result will be a true *molecular* orbital theory, as opposed to a site-by-site valence bond approach involving pairs of electrons. From the Schrödinger equation will come molecular orbitals for the entire structure, and we shall fill them according to the Pauli exclusion principle and Hund's rule.

Anything can happen. There might be, depending on the molecule, filled bonding and antibonding orbitals; or lone pairs in nonoverlapping, *nonbonding* orbitals; or half-occupied orbitals of all kinds, plus unoccupied orbitals as well. Sometimes there will be evidence for two-electron bonds between pairs of nuclei; sometimes not.

The proposal should sound familiar, for it is the very same procedure we adopted with the diatomic systems. Those molecules taught us the basics of bonding, notably: (1) Bonds arise from constructive interference between electronic wave functions. Electron density between the nuclei is enhanced by this overlap, and the electrical energy is lowered correspondingly. (2) Various combinations of atomic orbitals from different atoms can provide useful approximations to molecular orbitals. (3) Orbital overlap is facilitated by such factors as proximity in space, similarity in shape, and similarity in energy. (4) Bonding molecular orbitals tend to concentrate electrons between the nuclei, whereas antibonding orbitals tend to dilute the internuclear charge density.

We took these observations and, consistent with measured properties, developed the idea of a bond as an entity in itself. The bond became a part within a whole—a generic connection within a molecule—and that bond-by-bond scheme worked surprisingly well. We have just seen how a valence bond model, with its locally shared pairs of electrons, brings consistency to much of the molecular world. The examples cited above are testimony to its success.

Nevertheless, the success is incomplete. There are molecules that resist a one-bond-at-a-time approach. There are molecules (like oxygen) with unpaired electrons. There are highly symmetric molecules in which the charge density is spread evenly throughout the entire structure. For these many molecules, a sensible model must allow the electrons to range freely over all the nuclei. For them, the molecular orbitals must truly be "molecular" and not preordained to favor specific sites.

Benzene (C_6H_6) will serve as our prototype. A planar molecule, benzene's most striking geometric feature is the hexagonal arrangement of its six carbon atoms. All carbon–carbon bonds within the six-membered ring are equal (1.40 Å), and the angles are 120° all around. The structure bears, unmistakably, the marks of sp^2 hybridization.

A network of sp^2–sp^2 carbon–carbon bonds holds benzene together, with one unhybridized p orbital protruding perpendicularly at each vertex of the hexagon:

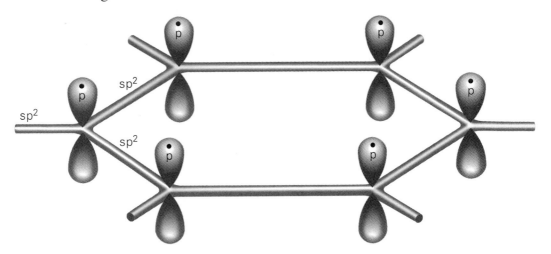

Constrained by this framework, a given carbon atom can contribute three electrons toward the formation of three ordinary σ bonds: the first, to a carbon on one side; the second, to a carbon on the other side; the third, to hydrogen. One p electron remains.

Together the six sp^2 carbons accumulate six residual p electrons, which clearly are destined for some sort of π bonding. The p electrons can, for example, group themselves into three double bonds and three single bonds as in the following structure I (seen from above):

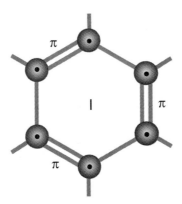

But for that species we should expect an alternation of a long C—C bond (1.47 Å) with a short C=C bond (1.34 Å). Instead there are six *equal* carbon–carbon bonds, each held at a middling distance of 1.40 Å. The proposed bonding arrangement is inconsistent with benzene's known structure.

Something is obviously wrong. By arbitrarily drawing three double bonds, we have shown no regard for the molecule's symmetry; we have introduced special directions and bonds into an environment where such distinctions are meaningless. Equally valid (or, better, equally *invalid*) would be an alternative arrangement,

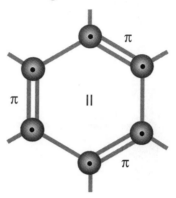

in which the C—C and C=C bonds are displaced by one position. And look: structures I and II are really the same, differing only by a rotation of 60°. Tilt the page by 60°, and the first drawing turns into the second. From that, we discover what it means to be a regular hexagon.

If benzene is to be a regular hexagon, then its molecular orbitals must reflect the symmetry of the entire ring. There can be no preferential overlap between different pairs of carbons; there can be no favoritism for this side or that side. That kind of preference leads inevitably to the joint appearance of benzene I and benzene II, neither of which is any good by itself. Were we to adopt such asymmetrical thinking, then we would be forced into the following argument: Benzene's "true" electronic structure lies midway between two extremes (call them *resonance structures* I and II), blended together in some hybrid fashion. Neither I nor II exists by itself, we would say, but rather each contributes equally to make a unitary molecule, symmetric and undivided. The actual molecule would be, again, like a lukewarm mixture impossible to disentangle—a so-called *resonance hybrid* of two equivalent forms, both present simultaneously.

Thus runs the "resonance" description of bonding in benzene. The terminology is unfortunate (suggesting, perhaps, a nonexistent alternation between forms I and II), but the resonance picture does preserve the often useful idea of localized bonds. It can be invoked whenever a molecule requires more than one Lewis structure for a complete description.

Resonance structures serve most usefully, though, as a symptom that something is wrong. The existence of alternative bonding patterns for benzene is a warning that a model of localized π bonds is entirely out of place. Arising from the symmetry of a hexagon, structures I and II confirm that all six electrons demand unbiased access to the whole ring. The two possibilities remind us that a p orbital cannot overlap with one neighbor while ignoring the other. Better we should represent the molecule as

where the circle symbolizes the electrons' freedom to roam from p to p to p. We should say, more accurately, that electrons in benzene's π system are **delocalized**. They are *not* allocated, two at a time, to specific pair bonds; rather they move freely throughout the ring, lost in the symmetry of the hexagon.

The procedure henceforth is the same as for diatomic molecules, differing only in mathematical detail. In brief: Take benzene's six unhybridized p orbitals and combine them—all of them—into six π molecular orbitals. Then take the six available electrons and fill the levels in the usual way, as depicted in Figure 7-23. Lowest in energy is the nodeless bonding orbital, formed by constructive overlap between every pair of p functions. Electrons confined to this state are equally likely to be found near equivalent points on the hexagon, and consequently the charge density forms a uniform ring above and below the plane. Compare that pattern now with the highest antibonding orbital, in which there is destructive interference (and a corresponding node) at each point of overlap. In between are four molecular orbitals laid out as two degenerate pairs. Each of these intermediate combinations has its own arrangement of bonding, antibonding, and nonbonding connections, as is apparent in the diagrams.

Benzene's ground state takes shape when its three bonding molecular orbitals are completely filled, leaving the delocalized structure with a lower energy than either of the two hypothetical resonance forms. The result is a net bonding interaction, which acts symmetrically around the ring to create a *delocalized π system* independent of the σ framework. Between any two carbons the π bond order is effectively $\frac{1}{2}$ (equivalent to three full π bonds for the entire structure), whence symmetric benzene

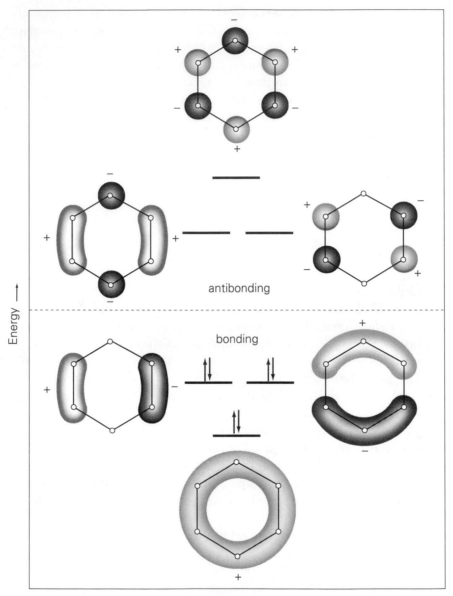

FIGURE 7-23. The delocalized π orbitals of benzene, viewed from above the molecular plane. Contrasting regions mark points where *p* orbitals combine out of phase, positive against negative. Six electrons fill the three bonding levels.

gets its intermediate bond distances. One full σ connection plus approximately half the normal π overlap holds together each pair of carbons. All six electrons are held as common property within the delocalized system of π molecular orbitals.

There we shall end our brief survey of covalent bonding, having exposed the broad principles governing molecular architecture. Nowhere, however, are those principles better illustrated than in the richly abundant compounds of carbon, which begin with methane and ethylene and acetylene and benzene and culminate with DNA and the proteins. That will be our subject for the next chapter, where we begin to see what nature can do with this most powerful and versatile tool: the covalent bond.

REVIEW AND GUIDE TO PROBLEMS

From particle to wave to wave function . . . from the free electron to the bound electron . . . from the hydrogen atom to the helium atom . . . from the helium atom to the elements . . . and now from the elements to the compounds.

Through it all, some things never change. Nuclei repel nuclei. Nuclei attract electrons. Electrons repel electrons. Nuclei and electrons obey the Schrödinger equation and Pauli principle when they interact. They conserve energy. They conserve momentum. They mix and mingle into atoms and molecules.

An atom: electrons trapped about a single nucleus, the same all around. From hydrogen to helium and beyond, there persists this single architecture. A large sphere of negative charge wraps itself around a small, heavy kernel of positive charge.

A molecule: electrons bound by two nuclei or more, no longer a sphere. A molecule is a structure with shape and direction, lines and planes, corners and angles. It can rotate in one direction or another. Its nuclei can vibrate. It has new ways to move.

IN BRIEF: MOLECULES AND CHEMICAL BONDING

1. When assemblies of electrons and nuclei—atoms—come together, they attract and repel. Far apart, they attract. Close together, they repel. Somewhere in between, they find the right balance and make a molecule.

Somewhere in between, as if settled in a valley, the attractions just match the repulsions; and here, where the potential energy is lowest, the arrangement is strongest and most stable. A stable molecule breathes in and out continually about this point of lowest energy, held in check by the attractions and repulsions. The structure neither implodes nor explodes.

2. What holds the electrons and nuclei together, in controlled oscillation about a set of average bond distances and angles? With what cord is a molecule bound?

With the same cord that binds all matter: the electromagnetic interaction, the attraction of positive for negative. An electron shared between two nuclei becomes the rope in a quantum mechanical tug-of-war. Pulling on the same electron, the nuclei are brought together. They have something in common; they share a connection, a *covalent bond*.

3. Far apart, two atoms. A and B. Close together, one molecule. AB. The atoms approach; their electrons mingle; their wave functions combine. Probability amplitudes ψ_A and ψ_B interfere one with the other, and the two atoms blend into a united molecule.

When ψ_A *adds* to ψ_B, the interference is constructive. There is an enhancement of the charge density, a gain beyond just the simple mixing of A's electrons with B's electrons. The density between the nuclei now becomes proportional not to $\psi_A^2 + \psi_B^2$, but to

$$\psi_A^2 + \psi_B^2 + 2\psi_A\psi_B$$

There is extra negative charge in the middle, a new source of attraction, a **bonding** relationship.

When ψ_A *subtracts* from ψ_B, the interference is destructive. The probability falls to

$$\psi_A^2 + \psi_B^2 - 2\psi_A\psi_B$$

as electrons are expelled from the space between the nuclei. The nuclei are pushed apart, victims of an **antibonding** interaction.

When ψ_A *fails* to interfere with ψ_B (a **nonbonding** relationship), there is neither gain nor loss. Atom A is here, atom B is there; and, like strangers in an elevator, they fail to mix. A's electrons stay close to A, and B's electrons stay close to B. The combined electron density

$$\psi_A^2 + \psi_B^2$$

brings nothing extra to the space between the nuclei. No added attraction draws them together, and no new repulsion pushes them apart.

4. The finished work is a molecule, and there it stands for us to describe: nuclei, electrons, attractions, repulsions, and a Schrödinger equation finally to pull the structure together. The problem, of course, is that we cannot solve the equation in its pristine form. Once again, we need to make the **orbital approximation**.

So imagine, as with the many-electron atom, that each electron moves nearly independently of the rest; that it sees the others only as an averaged-out repulsive force; that it has its own energy, territory, and spin; that it occupies, therefore, its own **molecular orbital**. Let the electrons fill a set of these molecular orbitals, just as they would fill a set of atomic orbitals—from low energy to high, building up in agreement with the Pauli principle and Hund's rule. It was a good model for the atom, and it is a good model for the molecule.

5. ATOMS, MOLECULES, ORBITALS. How shall we construct the orbitals? Yes, *we*. *We* must devise approximate wave functions to suit the

symmetry and properties of the molecule. Nature does not make models; we do.

Good molecular orbitals should, first, reflect something of the molecule's parentage: atoms. Molecular orbitals should mix together the electrons found originally on the free atoms, just as we envisioned the generic wave functions ψ_A and ψ_B to interfere constructively and destructively. The usual approach is to add and subtract the atomic orbitals in various proportions, taking *linear combinations of atomic orbitals* to form the molecular wave functions.

Many models, many options. We might, for example, specify *delocalized molecular orbitals* to cover the entire structure, especially when there is considerable symmetry (as in benzene, with its hexagonal ring). Or we might, symmetry permitting, use the *valence bond approach* to construct localized orbitals between pairs of nuclei. Doing so, we argue that such pairings have actual meaning for a molecule, notably because bond properties (length, dissociation energy, orientation, polarity, and others) vary so little from system to system. To use the valence bond approach, then, is to assert outright that electrons build up preferentially in the region between bonded nuclei, just as Lewis long ago maintained.

6. And this added flexibility too: We can use *hybrid atomic orbitals* to match a given molecular shape, mixing together a set of atomic orbitals on the same atom. Four tetrahedrally oriented *sp^3 hybrids* (109.5°) arise from the mixing of one *s* orbital with three *p* orbitals; three trigonally oriented *sp^2 hybrids* (120°) come from a 1:2 mix; two antiparallel *sp hybrids* (180°) come from a 1:1 combination. Reshaped and redirected, these new combinations then reach out and mix with orbitals on other atoms.

7. OVERLAP. Atomic orbitals that meet head to head will produce a *sigma (σ) molecular orbital*, a barrel-like combination of the two contributing wave functions. When the atomic orbitals add, they make a σ bonding orbital. When they subtract, we have a σ* antibonding combination. Either way, the sigma orbital retains the symmetry of a cylinder. The cylinder's axis is the line joining atom A with atom B:

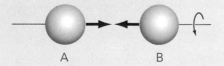

A   B

Weaker and less direct is the interaction in a *pi (π) molecular orbital*, occasioned when orbitals overlap side to side. It is a circumstance unavailable to *s* functions but common among *p* orbitals, suggested even by the appearance of the letter π: two upright legs, parallel, side by side.

Atomic orbital A stands next to atomic orbital B, both perpendicular to the line between the nuclei, the two surfaces overlapping above and below:

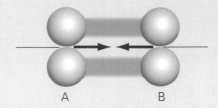

8. ELECTRONS, ORBITALS, BONDS. The molecular orbitals, each a possible home for up to two electrons, stand ready to be filled. Lowest in energy are the bonding combinations; highest, the antibonding. The electrons occupy the levels according to the usual rules of the Aufbau principle.

Some enter bonding orbitals, whereas others (depending on the pattern of energy levels) might occupy antibonding or nonbonding orbitals. Electrons in bonding combinations bring nuclei together. Electrons in antibonding combinations push them apart. Electrons in nonbonding combinations make no difference. The total energy thus realized will decide whether the molecule is stronger or weaker than the separated atoms. The more electrons in bonding orbitals, the lower the energy.

For that, we define the **bond order** as half the excess of bonding electrons over antibonding electrons. Two electrons (net) in bonding orbitals produce a bond order of 1, equivalent to a single bond. Four such electrons yield a bond order of 2: a double bond, shorter and stronger than a single bond. Six bonding electrons give a bond order of 3, shorter and stronger still.

Bond order thereby reconnects us to the Lewis picture of single, double, and triple bonds, but in a way more general than before. No longer do we restrict a covalent bond to just two electrons between two nuclei, because now we have new possibilities to acknowledge. In some bonds there may be one electron between two nuclei; in others, five. In some bonds the order may be fractional; in others, integral. In some bonds there may be two nuclei; in others, more.

In the end, we envision the electrons as either (1) localized between pairs of nuclei (valence bonds), or (2) dispersed throughout the molecule (delocalized molecular orbitals). That is our model, and the electron distribution so obtained paints a picture of the molecule for us. It shows us where the electrons go, where the molecule is positive and negative, and whether there is a dipole moment overall. It gives us the energy of the molecule and the strength of the bonds. It lets us predict the properties of the structure, all within the context of the orbital picture.

SAMPLE PROBLEMS

Start small: two identical atoms, a tissue of electrons between them, split symmetrically as if by a mirror. Recognize in that arrangement a homonuclear diatomic molecule, nonpolar, its bond perfectly covalent. Any structure like He_2^+ (Example 7-1) or F_2 (Example 7-2) is really just a pure covalent bond in a self-contained package, and with these simple systems we begin anew.

EXAMPLE 7-1. When Less Is More

PROBLEM: Predict a ground-state configuration and bond order for He_2^+. Is the ion paramagnetic?

SOLUTION: Does He_2^+ even exist? We have already seen that the neutral molecule, He_2, does not. He_2 would have two bonding electrons, two antibonding electrons, and a bond order of zero—no molecule. Should the ion be different?

Yes. Accommodating one fewer electron, He_2^+ acquires the configuration $(\sigma_{1s})^2(\sigma_{1s}^*)^1$ and a bond order of $\frac{1}{2}$:

$$\sigma_{1s}^* \quad \uparrow \qquad \qquad \text{Bond order} = \tfrac{1}{2}(2-1) = \tfrac{1}{2}$$
$$\sigma_{1s} \quad \uparrow\downarrow$$

The two σ_{1s} electrons enjoy lower energy than they would inside a neutral helium atom, enough to offset the rise in energy suffered by the lone antibonding electron in σ_{1s}^*.

For two helium nuclei, less is more: three electrons are more effective than four. Unable to share four electrons profitably, the nuclei (a total charge of +4) manage instead with only three. A fourth electron—an *antibonding* electron—would be a repulsive presence, and hence its absence helps bring the nuclei together.

The He_2^+ ion, made paramagnetic by the unpaired σ_{1s}^* electron, does exist as a stable structure. Compared with H_2, its dissociation energy (251 kJ/mole) is 60% smaller and its bond length 45% larger. Roughly speaking, we have half a bond: one bonding electron delocalized between two nuclei.

EXAMPLE 7-2. Homonuclear Bonding in the Second Row

PROBLEM: Predict a ground-state configuration and bond order for F_2^+, F_2, and F_2^-.

SOLUTION: We take the same approach as in Example 7-1, this time using the filling sequence

$$\sigma_{2s} < \sigma_{2s}^* < \sigma_{2p_z} < \pi_{2p_x} = \pi_{2p_y} < \pi_{2p_x}^* = \pi_{2p_y}^* < \sigma_{2p_z}^*$$

for the valence molecular orbitals. Into these orbitals must go 13 electrons for F_2^+, 14 for F_2, and 15 for F_2^- (seven from each neutral F atom, give or take one for the ionic charge). The configurations, populated in accord with the exclusion principle and Hund's rule, then become:

$$F_2^+ \qquad (\sigma_{2s})^2(\sigma_{2s}^*)^2(\sigma_{2p_z})^2(\pi_{2p})^4(\pi_{2p}^*)^3$$

$$F_2 \qquad (\sigma_{2s})^2(\sigma_{2s}^*)^2(\sigma_{2p_z})^2(\pi_{2p})^4(\pi_{2p}^*)^4$$

$$F_2^- \qquad (\sigma_{2s})^2(\sigma_{2s}^*)^2(\sigma_{2p_z})^2(\pi_{2p})^4(\pi_{2p}^*)^4(\sigma_{2p_z}^*)^1$$

	F_2^+	F_2	F_2^-
$\sigma_{2p_z}^*$	—	—	↑
π_{2p}^*	↑↓ ↑	↑↓ ↑↓	↑↓ ↑↓
π_{2p}	↑↓ ↑↓	↑↓ ↑↓	↑↓ ↑↓
σ_{2p_z}	↑↓	↑↓	↑↓
σ_{2s}^*	↑↓	↑↓	↑↓
σ_{2s}	↑↓	↑↓	↑↓

F_2^+ and F_2^- are paramagnetic, one unpaired electron in each. F_2 is diamagnetic. For the bond orders, we have:

$$F_2^+ \qquad \tfrac{1}{2}(8 - 5) = \tfrac{3}{2}$$

$$F_2 \qquad \tfrac{1}{2}(8 - 6) = 1$$

$$F_2^- \qquad \tfrac{1}{2}(8 - 7) = \tfrac{1}{2}$$

The model thus predicts that F_2^+, with the fewest antibonding electrons, will make the strongest and shortest bond. Expect dissociation energies to increase and bond lengths to decrease in the order F_2^- before F_2 before F_2^+.

Example 7-3. Dipole Moments

Next comes a *heteronuclear diatomic* molecule, symmetric all around but skewed end to end. One of the atoms, better equipped to receive electrons, draws a disproportionate share of the negative charge. The winner of this tug-of-war is the more ***electronegative*** species, made that way by a favorable combination of radius, effective nuclear charge, and valence configuration.

The result: a polar bond, negative at one end, positive at the other. There is a separation of charge. A ***dipole moment*** arises.

Problem: Which of the following molecules has the largest dipole moment . . . HF, ClF, HCl, or NO?

Solution: For each structure AB, we compute the difference in electronegativity between the two atoms $(EN_B - EN_A)$. The larger the difference, the more skewed is the electron density:

AB	EN_A	EN_B	$EN_B - EN_A$
→ HF	2.1	4.0	1.9 ←
ClF	3.0	4.0	1.0
HCl	2.1	3.0	0.9
NO	3.0	3.5	0.5

Hydrogen fluoride is substantially more polar than the other three.

Recall that the dipole moment μ is defined as

$$\mu = Qd$$

where d is the distance (in meters, m) between charges Q and $-Q$. If we express Q as a fractional electronic charge δ such that $Q = \delta e$ (with $e = 1.602 \times 10^{-19}$ C), then μ has units of *debye* (1 D = 3.336×10^{-30} C m) in the table of experimental data below:

Molecule	Dipole Moment (D)
HF	1.826
ClF	0.888
HCl	1.109
NO	0.159

Note how the dipole moments loosely follow the trend in electronegativity differences.

EXAMPLE 7-4. Ionic Character

Knowing a molecule's dipole moment, we can now estimate the *ionic character* of the bond—the extent to which electrons are separated rather than shared.

PROBLEM: Assume, for simplicity, that a structure exists literally in the form $A^{\delta+}B^{\delta-}$, with point charges of δ and $-\delta$ at each end. A charge $\delta = 1$ then would indicate a shift of one full electron ($Q = \delta e = e$), the mark of a 100% ionic bond (atom A gives all it has to B). The value $\delta = 0$, however, should signal no net movement of charge at all, suggesting a purely covalent bond instead (A and B give and take in equal measure). In between, the bond is part covalent and part ionic.

Using the data below, estimate the ionic character for each bond:

AB	DIPOLE MOMENT (D)	BOND LENGTH (nm)
HCl	1.109	0.127
NO	0.159	0.115
NaCl	9.001	0.236

SOLUTION: Working backward from the definition of dipole moment given in Example 7-3, we determine the point charge δ as

$$\mu = Qd = \delta e d$$

$$\delta = \frac{\mu}{ed}$$

For HCl, the calculation goes as

$$\delta = \frac{1.109 \text{ D}}{(1.602 \times 10^{-19} \text{ C})(0.127 \times 10^{-9} \text{ m})} \times \frac{3.336 \times 10^{-30} \text{ C m}}{1 \text{ D}}$$

$$= 0.182$$

and likewise we compute δ for all three bonds:

AB	DIPOLE MOMENT (D)	BOND LENGTH (nm)	δ
HCl	1.109	0.127	0.182
NO	0.159	0.115	0.029
NaCl	9.001	0.236	0.794

Hydrochloric acid's dipole moment thus is consistent with point charges of approximately ± 0.2 *e*, as if an electron spent an extra 20% of its time close by the chlorine.

Nitric oxide, by contrast, displays a far smaller separation of charge, amounting to only 3% ionic character. The molecule is less polar than HCl, and the electrons spend correspondingly more time between the nuclei.

At the other extreme is sodium chloride, a system with 80% ionic character. Nearly a full electronic charge moves from Na to Cl, a shift large enough to justify the label "ionic" bond. Even so, there is still considerable electron density (20%) *between* the nuclei. Some covalent character always persists in most ionic bonds, if only slightly.

EXAMPLE 7-5. Hybrid Atomic Orbitals

Moving on to polyatomic systems, we revisit the water molecule, H_2O, the greatest solvent on earth.

The structure is already familiar from Chapter 2: a bent molecule, polar, with single bonds between O and H and with four pairs of valence electrons arranged in a Lewis diagram as

$$\ddot{\text{O}}$$
$$\text{H} \quad \text{H}$$

The H—O—H bond angle is 104.5°.

Two of the electron pairs are bonding, and two are nonbonding. The VSEPR model puts these four pairs at the vertices of a tetrahedron, spreading them out to minimize the electrostatic repulsions. Strong lone-pair repulsions, however, squeeze the two bonding pairs closer, shrinking the bond angle from 109.5° (tetrahedral) to 104.5°.

How might we go beyond this VSEPR picture, which ascribes the structure entirely to electrostatic repulsion?

PROBLEM: Use the valence bond method to describe the structure of H_2O.

SOLUTION: The valence bond model, derived from the Schrödinger equation, is properly quantum mechanical. It includes both attractions and repulsions, and it uses wave functions to describe the electrons.

We begin with the facts of water's existence—most important, the molecule's near-tetrahedral bond angle of 104.5°. A tetrahedral geometry instantly suggests sp^3 hybridization, a role for which oxygen is well suited. Its $2s$ and $2p$ valence orbitals will mix to produce four tetrahedral sp^3 hybrids, going from a free configuration of $2s^2 2p^4$ to an in-the-molecule configuration of $(sp^3)^6$ in which all four sp^3 levels are equivalent:

$2p$ ⇅ ↑ ↑   ⇅ ⇅ ↑ ↑ sp^3

$2s$ ⇅

free O atom   oxygen in H_2O

Two of the four sp^3 hybrids, singly occupied, make σ bonds with one hydrogen apiece: one sp^3 electron from the oxygen, one $1s$ electron from a hydrogen. The other two sp^3 orbitals, doubly occupied, house the lone pairs:

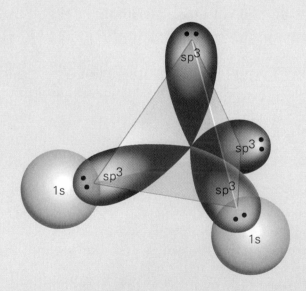

EXAMPLE 7-6. sp^2 Hybridization: The Orbital Left Behind

PROBLEM: Why is BF_3 a Lewis acid?

SOLUTION: Because it has an empty valence orbital. The central boron has a reserve capacity to attract and hold electrons.

When we first met BF_3 in Chapter 2, we found a trigonal boron two electrons shy of an octet:

$$:\ddot{F}:$$
$$|$$
$$B$$
$$:\ddot{F}:\quad:\ddot{F}:$$

Faced with the octet-deficient structure, we argued that boron trifluoride would benefit from the addition of two electrons; that the central boron would, in some way, profit from a full complement of eight. Absent in our reasoning, though, was any idea *where* these new electrons might go.

The valence bond model now supplies the hows and wheres: sp^2 hybridization around the boron, consistent with the 120° bond angles. Suppose, accordingly, that boron's $2s$ orbital and *two* of its $2p$ orbitals mix to produce three equivalent sp^2 combinations:

$2p$ ↿   —   —   — $2p$

                                ↿   ↿   ↿ sp^2

$2s$ ↿⇂

      free B atom   boron in BF_3

Each of the sp^2 atomic orbitals, holding one of boron's three valence electrons, then overlaps with a singly occupied sp^3 orbital on a fluorine atom (which, having realized four electron pairs, is assumed to be tetrahedral). Three B—F sigma bonds are formed, and there are no electrons left over.

But an *orbital* is left over. The unhybridized p function on boron, never occupied, juts up perpendicular to the plane of the molecule:

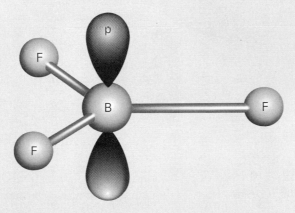

It stands there, empty, like a house without people—just as accommodating, its presence just as marked. Filled or empty, the orbital remains mathematically in place. It stakes out a territory, a possible home for two electrons.

QUESTION: True, but all atoms have empty orbitals above the ground state. Should we then expect one of the fluorines to accept a pair of electrons in, say, its vacant $3s$ orbital? Why is boron the only acid site?

ANSWER: All atoms and molecules *do* have unfilled orbitals, but usually these orbitals stand too high in energy (especially when they lie outside a filled shell). Only orbitals close to the valence energies can engage the electrons effectively.

EXAMPLE 7-7. Hybridization of *d* Orbitals

Here is something new: hybrid orbitals built from d functions. Later, in Chapter 19, we shall use such combinations to describe the structure of transition-metal complexes.

Start with a d^2sp^3 mixture. Take two d orbitals, one s, and three p. Add and subtract them in various ways. Draw out six equivalent *d^2sp^3 hybrid orbitals*, each pointing toward one of the corners of an octahedron:

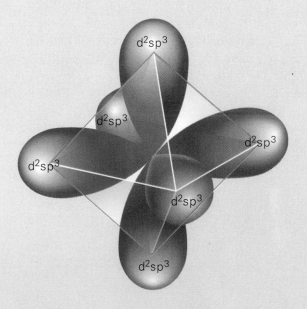

And another, the dsp^3: one d, one s, and three p orbitals, mixed into five *dsp^3 hybrid orbitals*. Together, they support a trigonal bipyramid:

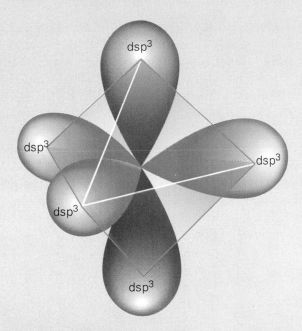

Note that the three *equatorial* combinations, spaced 120° apart in the trigonal plane, are not strictly equivalent to the two *axial* combinations directed up and down.

PROBLEM: Suggest a hybridization scheme for each of the following molecules: (a) SF_6. (b) PCl_5.

SOLUTION: Since we are given no structural information, we shall use the VSEPR model to predict the geometry from the Lewis diagram. After that, we can choose a set of hybrids to fit the shape of the molecule.

(a) *SF_6*. Sulfur falls in the third row, in Group VI just below oxygen. Its six valence electrons are configured as $3s^2 3p^4$ in the free atom.

The six fluorine atoms, meanwhile, contribute seven valence electrons apiece, giving SF_6 a grand total of 48 and a Lewis structure of

Each fluorine atom, with a conventional octet, is surrounded by four pairs of electrons in tetrahedral (sp^3) geometry. Around the central sulfur atom, however, an *expanded* octet (compare with SF_4, Example 2-5a) holds *six* pairs of electrons.

VSEPR theory predicts an octahedral geometry for the six-coordinate structure as a whole: six equivalent S—F bonds pointing to the six vertices of a regular octahedron, with 90° bond angles about the sulfur.

Clearly, sulfur cannot support this octahedron using s and p orbitals alone, but it doesn't have to. It has nearby d orbitals, unoccupied in the free atom but available to the molecule. Sufficiently low in energy, the d orbitals give the bonded sulfur atom more ways to make bonds.

To form six octahedral bonds, then, sulfur needs only to put together the six octahedral orbitals present in a d^2sp^3 configuration. Singly occupied, each d^2sp^3 hybrid contains one of sulfur's six valence electrons; and then, one at a time as in Figure R7-1, the d^2sp^3 orbitals make σ bonds with a singly occupied sp^3 orbital on each fluorine.

Thus configured, six valence bond orbitals (12 electrons) surround the sulfur atom in SF_6. Lewis's octet, normally arising from a filled

FIGURE R7-1. Bonding in SF_6. Sulfur, hybridized as d^2sp^3, forms σ bonds with six sp^3 fluorine atoms. The molecular geometry is octahedral.

ns^2np^6 configuration, expands to 12 once the two d orbitals participate.

(b) *PCl$_5$*. Forty valence electrons (five from P, seven from each Cl) are distributed over the Lewis structure as

$$
\begin{array}{c}
:\ddot{\text{C}}\text{l}: \quad :\ddot{\text{C}}\text{l}: \\
| \quad \diagup \\
:\ddot{\text{C}}\text{l} - \text{P} \\
| \quad \diagdown \\
:\ddot{\text{C}}\text{l}: \quad :\ddot{\text{C}}\text{l}:
\end{array}
$$

Presented with five pairs of electrons around the central atom, VSEPR theory predicts that the structure will be trigonal bipyramidal.

Phosphorus, a third-row element, has the nearby d orbitals needed to form five dsp^3 hybrids in the correct orientation. Each of the five hybrids makes a σ bond with a singly occupied sp^3 orbital on a chlorine (Figure R7-2).

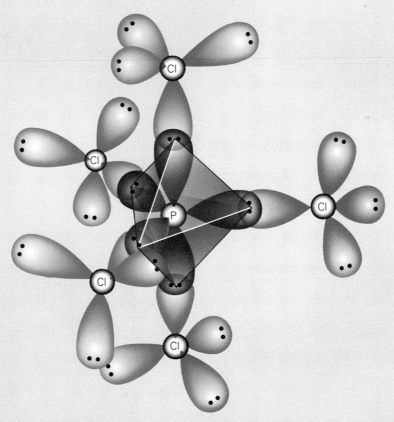

FIGURE R7-2. Bonding in PCl$_5$. The trigonal bipyramidal geometry is attributed to a phosphorus atom hybridized as dsp^3.

EXAMPLE 7-8. Delocalization

PROBLEM: Consider the *allyl cation*, $CH_2CHCH_2^+$, shown below as a skeleton of carbon and hydrogen nuclei:

$$
\begin{array}{c}
\text{H} \\
| \\
\text{H} \quad\quad \text{C} \quad\quad \text{H} \\
\text{C} \overset{+}{} \text{C} \\
\text{H} \quad\quad\quad\quad \text{H}
\end{array}
$$

Assumptions: (1) The eight atoms lie in a single plane. (2) The H—C—H bond angles are all nearly 120°, as is the central C—C—C angle. (3) Both carbon–carbon bond lengths are the same, falling between the usual values for C—C single and C=C double bonds.

Propose a model for the bonding.

SOLUTION: 120° angles, atoms in a plane, and carbon–carbon bonds intermediate between single and double—it sounds like benzene.

Benzene, remember, brings together our valence-bond and delocalized-orbital pictures in one molecule. Its structure includes both a planar skeleton of localized, sp^2-hybridized σ bonds *and* a delocalized zone of π electrons above and below. Allylic systems do the same.

Sixteen valence electrons contribute to the allyl cation, and for Lewis structures we have either (double bond to the left),

$$ \text{I:} \quad\quad CH_2\!\!=\!\!CH\!-\!\overset{+}{C}H_2 $$

or (double bond to the right):

$$ \text{II:} \quad\quad \overset{+}{C}H_2\!-\!CH\!\!=\!\!CH_2 $$

But I and II are really the same, because nowhere in this molecule is there any meaningful left and right. There is only CH_2 on one end, CH_2 on the other, CH in between, and two identical carbon–carbon bonds throughout. Flip the page and confirm: left becomes right and right becomes left. No single Lewis diagram can ever respect that symmetry.

Better, then, that we represent the allyl cation as a *delocalized* structure, with some of its electrons shared by all three carbons:

$$ [CH_2 \cdots CH \cdots CH_2]^+ $$

There are σ orbitals, and there are π orbitals. From sp^2-hybridized carbon atoms will come, first, the planar 120° network of C—C and C—H sigma bonds that we need. Ten of the 16 electrons go into the five

C—H bonds, while 4 go into the two C—C sigma bonds. Two electrons and three unhybridized *p* orbitals remain:

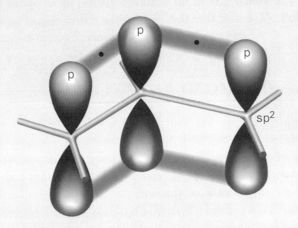

Perpendicular to the plane, these three *p* orbitals then combine to make a delocalized home for the last two electrons. Looking right and left, the central *p* sees its partners as identical twins; and thus, preserving the symmetry, the three atomic *p* orbitals mix themselves into three π molecular orbitals.

They mix constructively, in phase, to produce a π *bonding* combination with no nodes perpendicular to the molecular plane:

They mix destructively, all three out of phase, to produce a π *antibonding* combination with two nodes:

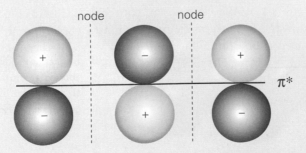

And they mix indifferently, excluding the middle p, to produce a π *nonbonding* combination with one node:

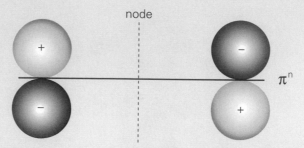

From three atomic p orbitals we now have three molecular π orbitals, ordered energetically according to the number of nodes. Lowest is the constructively formed bonding orbital, which falls below the energy of the unbound atoms. Here the electron density is concentrated in the middle of the molecule, peaking near the second carbon and pulling the three atoms together as a result. Next is the nonbonding combination, equal in energy to the free atoms. It neither enhances nor diminishes the electron density in the center of the molecule. Highest in energy, finally, is the antibonding orbital, which pushes the three atomic centers apart.

Housed solely in the bonding orbital, the two π electrons therefore manage to lower the energy of the cation:

$$\pi^* \; \underline{}$$

$$\pi^n \; \underline{}$$

$$\pi \;\; \underline{\uparrow\downarrow}$$

The structure is stable, diamagnetic, and symmetric. Left and right, above and below, the π electrons are reflected in each plane as if through a mirror. The two carbon–carbon distances are the same.

With bond order equal to 1 over all three carbons, each p–p pi linkage acts as half a conventional pair bond; and so, together with the sp^2–sp^2 sigma connection, the total bond order ($\sigma + \pi$) between adjacent carbons is 1.5. We saw the same outcome for benzene, and we shall see it manifested soon (next chapter) among a very general, very important class of carbon-containing molecules—so-called *conjugated* systems such as $CH_2{=}CH{-}CH{=}CH_2$, in which $C{-}C$ single bonds and $C{=}C$ double bonds alternate in the Lewis structure.

EXERCISES

1. *Modus operandi*: If electrons repel electrons and if nuclei repel nuclei, then by what means does a molecule hold together?

2. *Raison d'être*: Why should molecules exist at all? What advantage might a molecule have over an atom?

3. Describe, in broad terms, the coming together of atoms as a mixing of wave functions. Show how bonding and antibonding combinations may arise from the interaction of two hydrogen atoms.

4. Same thought, but now in more detail: (a) What physical interpretation is given to a wave function and its square—the *probability amplitude* (ψ) and the *electron density* (ψ^2)? (b) Can the mixing of two wave functions (ψ_A and ψ_B) both diminish and enhance the electron density between the nuclei?

5. (a) In what ways is a molecular orbital similar to an atomic orbital? In what ways is it different? (b) What thinking lies behind the method of *linear combination of atomic orbitals (LCAO)*?

6. (a) What is a "hybrid" orbital? What theoretical purpose does it serve? (b) Is the hybrid wave function an *atomic* orbital (assigned to one nuclear site) or a *molecular* orbital (assigned to two or more)? Is it special in any way?

7. (a) What characteristics are common to the electron distributions of all homonuclear diatomic molecules? (b) May a homonuclear diatomic molecule ever possess a permanent dipole moment? Why or why not?

8. Show, on a diagram of atomic and molecular energy levels, the combinations that arise when *s* orbitals from two identical atoms are mixed.

9. Similar: Sketch the bonding and antibonding orbitals that arise from the combination of *p* orbitals on two atoms of the same type.

10. (a) What characteristics are common to the electron distributions of all heteronuclear diatomic molecules? (b) May a heteronuclear

diatomic molecule possess a permanent dipole moment? Why or why not?

11. (a) Sketch, roughly, the energy levels that arise when s orbitals from two different kinds of atoms are mixed. (b) In what ways do the σ molecular orbitals in a heteronuclear system differ from those in a homonuclear system?

12. Similar: (a) Sketch the energy levels that arise from the combination of p orbitals on two atoms of different type. (b) In what ways do the π molecular orbitals in a heteronuclear system differ from those in a homonuclear system?

13. Removal of one electron from a hydrogen molecule produces the H_2^+ ion: (a) Why is H_2^+ different from every other molecule? (b) In what way is it similar to a hydrogen *atom*? In what way is it different?

14. Use molecular orbital arguments to assess the stability of H_2^+.

15. (a) How many electrons occupy bonding orbitals in the H_2^+ ion? How many occupy antibonding orbitals? (b) What is the bond order? (c) Is the ion paramagnetic?

16. (a) In which molecule is the H—H bond longer . . . H_2^+ or H_2? (b) In which is it stronger?

17. (a) Write the ground-state configuration of the molecule B_2, using the appropriate building-up sequence. Show the orbital populations on a diagram. (b) Is the system paramagnetic? (c) What is its bond order?

18. Suppose, contrary to observation, that the σ_{2p_z} molecular orbital in B_2 were lower than the π_{2p} orbitals. (a) If so, would B_2 be paramagnetic? (b) What bond order would it have?

19. Predict the ground-state configuration, bond order, and magnetic properties of B_2^+, B_2, and B_2^-. Show the orbital populations on a diagram, and rank the three structures in order of increasing dissociation energy and decreasing bond length.

20. Predict the ground-state configuration, bond order, and magnetic properties of C_2^+, C_2, and C_2^-. Show the orbital populations on a diagram, and rank the three structures in order of increasing dissociation energy and decreasing bond length.

21. Predict the ground-state configuration, bond order, and magnetic properties of O_2^+, O_2, and O_2^-. Show the orbital populations on a diagram, and rank the three structures in order of increasing dissociation energy and decreasing bond length.

22. Identify the homonuclear diatomic molecule, A—A, from the valence configuration stated:

    (a) $(\sigma_{2s})^2(\sigma_{2s}^*)^2$
    (b) $(\sigma_{2s})^2$
    (c) $(\sigma_{2s})^2(\sigma_{2s}^*)^2(\pi_{2p})^4(\sigma_{2p_z})^2$

    Which molecules are paramagnetic? Show the orbital populations on a diagram, and rank the three structures in order of increasing dissociation energy and decreasing bond length.

23. (a) Using the electronegativity scale as a guide, rank the following bonds in probable order of increasing polarity: NO, BeN, PN. (b) Which end is partially negative in each linkage?

24. Dipole moments and bond lengths for the molecules HF, HCl, HBr, and HI are given below:

AB	Dipole Moment (D)	Bond Length (Å)
HF	1.826	0.917
HCl	1.109	1.275
HBr	0.827	1.415
HI	0.448	1.609

    (a) Estimate the fractional ionic character of each polar covalent bond. (b) Which end of the molecule is partially negative?

25. Dipole moments and bond lengths for several bromine-containing molecules are given below:

AB	Dipole Moment (D)	Bond Length (Å)
IBr	0.726	2.469
BrO	1.76	1.717
TlBr	4.49	2.618
LiBr	7.268	2.170
NaBr	9.118	2.502
KBr	10.628	2.821

    (a) Estimate the fractional ionic character of the bonds. (b) In which molecules is the bromine atom partially negative? (Note how elements that normally form ionic compounds—such as KBr, NaBr, and LiBr—may, under certain conditions, stick together as molecules in the gas phase).

26. Dipole moments and bond lengths for several hydrogen-containing molecules are given below:

AB	DIPOLE MOMENT (D)	BOND LENGTH (Å)
HI	0.448	1.609
NH	1.39	1.036
LiH	5.884	1.595

(a) Estimate the fractional ionic character of the bonds. (b) In which molecule is the hydrogen atom partially negative?

27. BeF_2 exists in the gas phase as a linear molecule. The two Be—F bonds have the same length. (a) Is the shape consistent with the VSEPR model? Draw the Lewis structure. (b) Is each Be—F bond polar? If so, which end is partially negative? (c) Does the molecule possess a dipole moment overall?

28. Use a valence bond approach to describe the hybridization and bonding in BeF_2.

29. The ammonia molecule, NH_3, is a pyramidal structure in which the three N—H bonds all have the same length. The nitrogen atom sits atop the pyramid, and the hydrogen atoms form the base. (a) Is ammonia's shape consistent with the VSEPR model? Draw the Lewis structure. (b) Is each N—H bond polar? If so, which end is partially negative? (c) Does the molecule possess a dipole moment overall?

30. Fluorine is the most electronegative element of all, yet the molecule NF_3 has just a small dipole moment. Compare the electronic structure with NH_3, and suggest a reason why the polarity might be so low in NF_3.

31. Use a valence bond approach to describe the hybridization and bonding in NH_3.

32. (a) Use the VSEPR description to predict the shape of an ammonium ion, NH_4^+. (b) Propose a hybridization scheme consistent with the VSEPR geometry.

33. Sulfur trioxide, SO_3, is a planar molecule. The three oxygen atoms, equidistant from the sulfur atom, occupy the vertices of an equilateral

triangle. (a) Is the shape consistent with the VSEPR model? Draw the Lewis structure, taking into account the possibility of resonance forms. (b) Is each S—O bond polar? If so, which end is partially negative? (c) Does the molecule possess a dipole moment overall?

34. Use a valence bond approach to describe the hybridization and bonding in SO_3.

35. (a) Use the VSEPR description to predict the shape of a sulfite ion, SO_3^{2-}. (b) Propose a hybridization scheme consistent with the VSEPR geometry.

36. Describe the structure, hybridization, and bonding of the product formed by reaction of BF_3 (a Lewis acid) and NH_3 (a Lewis base).

37. (a) Use a valence bond approach to describe the geometry, hybridization, and bonding in a molecule of ethane, C_2H_6. How are the atoms connected? (b) Which bonds are polar? (c) Is the molecule polar?

38. (a) How are the atoms connected in CH_2ClCH_2OH, a form of ethane in which two hydrogens are replaced by Cl and OH. (b) Use a valence bond approach to describe the geometry, hybridization, and bonding. (c) Is the molecule polar?

39. Use a valence bond approach to describe the geometry, hybridization, and bonding in the molecule CCl_3CCH. Start with the Lewis structure, and then identify the single bonds, multiple bonds, σ bonds, and π bonds present in the molecule.

40. The five fluorine atoms in PF_5 occupy the vertices of a trigonal bipyramid. Phosphorus sits at the center. (a) Is the shape consistent with the VSEPR model? Draw the Lewis structure. (b) Is the structure consistent with either sp, sp^2, or sp^3 hybridization at the phosphorus? (c) Use an appropriate valence bond approach to describe the hybridization and bonding in PF_5.

41. Nitrogen, although directly above phosphorus in Group V, does *not* bond with five fluorines to form NF_5, the analog of PF_5. According to valence bond theory, what feature distinguishes phosphorus from nitrogen? Why is nitrogen unable to expand its octet?

42. Arsenic and antimony lie below phosphorus in Group V. Will these atoms, like phosphorus, be able to bond with five fluorines?

43. (a) Use both the VSEPR description and a valence bond approach to predict the shape of SbF_5, antimony pentafluoride. (b) Which of antimony's atomic orbitals contribute to the hybridization? Specify the shells and subshells.

44. One of the following structures—C_2H_4, CO_3^{2-}, CH_4—is a candidate for a *delocalized* π molecular orbital treatment, a picture in which the electrons no longer are restricted to pair bonds between designated nuclei. (a) Which is it? (b) Draw a set of Lewis structures (resonance forms) to make the point.

45. Challenge: The molecule 1,3-butadiene, represented in part by the Lewis structure below, contains a sequence of C=C, C—C, and C=C bonds:

Measurements show that the distance between the central carbons is 1.47 Å, whereas the spacing between the outer carbons is only 1.35 Å at each end; thus:

$$\underbrace{\text{C=C}}_{\text{short}}\underbrace{\text{—C}}_{\text{long}}\underbrace{\text{=C}}_{\text{short}}$$

Try to account for this structure using a picture of delocalized molecular orbitals, and assume—for simplicity—that the nodal patterns are similar to the standing waves on a vibrating string (Chapters 4 and 5). The following set of questions may help: (a) How many unhybridized p orbitals are available for combination into molecular orbitals? (b) How many nodes are present in the various π molecular orbitals, and how do the different energies compare? (c) For each π combination, show (on a rough sketch) where the electrons are expected to concentrate. (d) How many electrons are distributed throughout the π system? Which molecular orbitals are occupied? (e) Examine carefully the distribution of electrons in the occupied orbitals, and show how the model predicts (qualitatively) the short-long-short alternation of bond lengths.

8

Some Organic
and Biochemical Species
and Reactions

8-1. Why Carbon?
8-2. Hydrocarbon Structure and Bonding
Isomerism Among the Alkanes
Double and Triple Bonds: A Review
Geometric Isomerism
8-3. Functional Groups
8-4. Biopolymers
8-5. Summary of Organic Reactions
REVIEW AND GUIDE TO PROBLEMS
EXERCISES

8-1. WHY CARBON?

Although only a small component of the earth's crust, the element carbon boasts a versatility unmatched by any other atom. It enters into limitless combinations, from one-carbon methane to giant structures of enormous molecular weight. Millions of compounds contain carbon; far fewer, a shrinking minority, do not.

The sixth element is a fact of life, all pervasive: Carbon dominates the chemistry of life; carbon gives the cell its molecular machinery, its infrastructure; carbon provides framework and support for life's molecules. And carbon, driving a huge chemical industry, brings chemistry not just to biological life but to economic life as well: fossil fuels, petrochemicals, plastics; in short, money and material, all from carbon. A small atom, it shows up in interesting places.

The chemistry of carbon is the domain of **organic chemistry**, a name

261

chosen originally to describe the many carbon-containing molecules found in living organisms (broadened now to include all sources, alive or not). Closely related is **biochemistry**, which focuses on those particular molecules and reactions involved directly in the workings of life. Together, organic chemistry and biochemistry are so significant—so rich is their display of molecular formation and transformation—that we do well to survey both fields early on, even if only in broad outline. Let these molecules of carbon be our first reward for the study of chemical bonding just undertaken.

Several factors combine to make carbon what it is. Sitting atop Group IV, the atom has four valence electrons and thus can bond covalently with up to four additional partners. No other second-row element possesses a bonding capacity so extensive, nor can the Group IV semimetals (silicon and germanium) form double and triple bonds with the same ease. Owing partly to a small radius and middle-of-the-road electronegativity, however, the carbon atom makes multiple bonds with a variety of species—especially other carbons. It uses this self-linking capacity to build chains and rings of diverse size and shape, sometimes to create intricate and extended three-dimensional structures. The four-way bonds are comparatively strong as well, and so carbon molecules can endure under conditions where tetravalent silicon compounds (otherwise similar) would be unstable. Yet the carbon bonds are still *weak* enough to provide ample capacity for reaction and change, and here is where organic molecules excel: in their marriage of sturdy carbon–hydrogen skeletons with special reactive sites called *functional groups*, from which new structures blossom in almost infinite variety. We shall consider, accordingly, first the molecular frameworks and then the functional groups to gain some appreciation for carbon's possibilities.

8-2. HYDROCARBON STRUCTURE AND BONDING

Built entirely from carbon and hydrogen, the **hydrocarbons** provide a foundation for most organic molecules. Simplest are the **alkanes**, C_nH_{2n+2}, which contain n carbon atoms connected by single bonds only. Methane (CH_4), with one carbon, is the most elementary example, followed by ethane (C_2H_6), propane (C_3H_8), butane (C_4H_{10}), pentane (C_5H_{12}), hexane (C_6H_{14}), heptane (C_7H_{16}), octane (C_8H_{18}), and so forth. These molecules are termed **saturated hydrocarbons** because the carbon valences are exhausted; each has a full complement of four covalent bonds with four other atoms and therefore can accommodate no more.

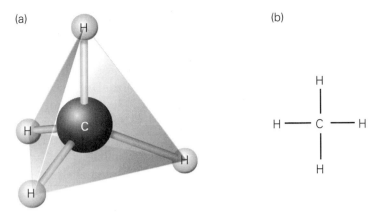

FIGURE 8-1. Tetrahedral carbon, reduced to its simplest form: a molecule of methane, CH_4. (a) The four hydrogen nuclei occupy the vertices of a regular tetrahedron, forming a four-sided figure with an equilateral triangle for each face. Carbon sits at the center, equidistant from the hydrogens. The H—C—H angles are all 109.5°. (b) A simplified projection in two dimensions.

Unsaturated hydrocarbons, by contrast, contain one or more multiple carbon–carbon bonds, either C=C or C≡C, which can be opened up and used to attach new atoms. The **alkenes** incorporate at least one *double* bond in their structures, whereas the **alkynes** contain one or more *triple* bonds. Ethylene, or ethene ($H_2C=CH_2$), is the parent alkene; acetylene, or ethyne (HC≡CH), is the parent alkyne.

Isomerism Among the Alkanes

The geometry of four-coordinate carbon is dominated by the tetrahedron. In methane there are four covalent C—H bonds directed toward the corners of a regular tetrahedron, a structure consistent with both the VSEPR model and also the picture of sp^3 hybrid orbitals developed in Chapter 7. Each bond has equal length since one hydrogen is the same as another, and each H—C—H angle is similarly fixed at 109.5°. This highly symmetric arrangement, pictured in Figure 8-1, defines the basic configuration of organic carbon. It must be appreciated thoroughly if more complex molecular structures are to be understood.

Imagine, then, grasping a tetrahedral methane molecule by any of the C—H bonds, so that the carbon and remaining three hydrogens form a pyramid. At the pyramid's apex is the carbon, near our imaginary handle, with the hydrogens arranged below in an equilateral triangle:

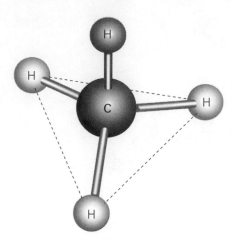

Looking at the H atoms spaced 120° apart at the base of the pyramid, we see a pattern akin to a three-spoke wheel.

Now ask: Does a tetrahedrally bonded carbon, by itself, offer any opportunity for *isomerism*—alternative arrangements of the same atoms? Can the atoms be reconfigured into a new molecule with the same chemical formula but different structure?

Methane, no. No isomers are possible for methane with its four indistinguishable hydrogens, so to explore further we must introduce different substituents. Suppose a chlorine atom replaces one of the hydrogens to yield CH_3Cl, as pictured below; what then?

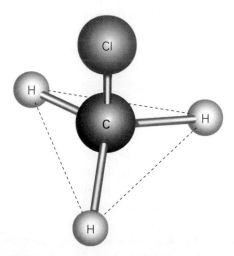

Nothing new. We grab the C—Cl bond, which is now unique among the four bonds, and we continue to look down at three identical (and hence interchangeable) hydrogens. No distinguishable rearrangements are possible on the equilateral triangle, as before, and again there are no isomers. Nor do any isomers exist for CH_2Cl_2 or $CHCl_3$, since the symmetry is still too high.

But if all *four* attached atoms are different (try it with H, Cl, Br, and I), then suddenly there is the possibility of two independent structures—one the mirror image of the other—which under these conditions cannot be superimposed:

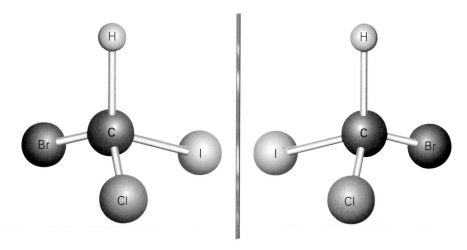

Were we to pick up each molecule by, say, the C—H bond, then the Cl-Br-I sequence would appear clockwise in one arrangement and counterclockwise in the other. The two structures are undeniably different, and no amount of manipulation will ever bring them into agreement.

These two nonsuperimposable mirror images, suggestive of a pair of hands, are called **enantiomers**. Just look: A left hand, placed palm down on a mirror, produces an image of the right hand appearing palm up; yet right and left never can be matched one over the other. Enantiomeric molecules, possessing a handedness of their own, are appropriately called **chiral** (from the Greek word for "hand"), and a carbon to which four different groups are attached is termed a *chiral center* or, more precisely, a *stereogenic atom*.

Listen to what the word says. First, *stereo*: borrowed from the Greek to suggest solidity, volume, three-dimensionality. Second, *genic*: Greek again, meaning "producing" or "causing." From that, a stereogenic atom: an atom able to support two distinct spatial arrangements of its attached groups. Where a molecule contains one stereogenic carbon, and only one, we find that the structure has a nonsuperimposable mirror image: an enantiomer. Some enantiomeric molecules, though, have no stereogenic atom, and some molecules with more than one "chiral" center are not chiral at all; their reflections are identical to the originals. Only the mirror can tell us for sure.

Now *enantiomers*, like hands, do distinguish left from right; and they make the distinction clear when they interact with plane-polarized light—light where the electric field (described in Chapter 4) vibrates in just one plane. If polarized waves pass through a solution of chiral molecules, the plane of polarization rotates either left or right as judged by an observer looking into the oncoming beam. One enantiomer, the *dextrorotatory* form (+), rotates it to the right (clockwise); the other enantiomer, the *levorotatory* form (−), rotates it to the left (counterclockwise). In recognition of those preferences, the two enantiomers—nonsuperimposable mirror images—are given the alternative name **optical isomers**.

The properties of optical isomers are identical in other respects, except in relation to molecules also endowed with the property of handedness. Subtle differences become important. Enzymes, for example, have chiral structures that admit only one of two possible optical isomers. The catalysis so critical to life thus can be impaired if an enantiomeric molecule is replaced by its ill-fitting mirror image.

Remarkable, too: one carbon, four attachments, and already a subtlety in structure . . . but the story is just beginning. There are still more kinds of isomers to find, as we soon discover on proceeding up the chain from methane to ethane:

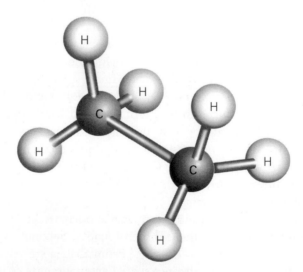

Here the molecule has two tetrahedral carbons, assembled so that two pyramidal methyl groups (CH_3) are connected by a C—C single bond. Looking down the C—C axis (rendering the molecule using a sticklike *Newman projection*), we see two H-H-H wheels, one at each end, able to rotate about the interconnecting single bond:

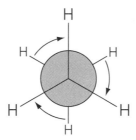

This freedom creates an infinity of possible **conformations**, distinguished by the way one wheel of hydrogens is oriented relative to the other. Out of all those angles, however, emerge two principal conformational isomers, or **conformers** for short: **eclipsed** (0°), where the hydrogens coincide; and **staggered** (60°), where one set is interleaved between the other:

eclipsed   staggered

Less favored of the two is the eclipsed arrangement, because now the opposing groups are closer together and consequently more exposed to electrostatic repulsion. Energy is lower for the staggered conformation, in which the two ends of the molecule suffer minimal interference. Between staggered and eclipsed lies a continuous range of potential energy, sketched in Figure 8-2.

The methyl groups do not rotate unhindered about the C—C single bond. It takes a push to spin the CH_3 groups around the C—C axis, to go from staggered to eclipsed to staggered again—although not much of a push, and the small hurdle is overcome easily enough to permit essentially free rotation at room temperature. Only at low temperatures, where a molecule may possess insufficient energy to climb that hill, are the staggered conformations routinely frozen in place.

Beyond two carbons there are potentially many additional conformers, since in larger hydrocarbons there are more C—C single bonds about which rotations can take place. Remember that each carbon in an alkane is approximately tetrahedral, and so do not be misled by a straight-looking molecular formula like $CH_3CH_2CH_2CH_2CH_2CH_2CH_2CH_3$ into

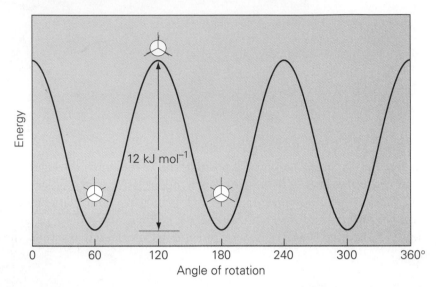

FIGURE 8-2. Barriers to rotation: With every twist of 60°, ethane alternates between a staggered (low-energy) and an eclipsed (high-energy) conformation. The graph shows how the potential energy rises and falls as the CH_3 groups rotate about the molecule's C—C single bond.

thinking that the eight carbons in "linear" octane literally form a straight chain. Rather it is a complicated geometric structure,

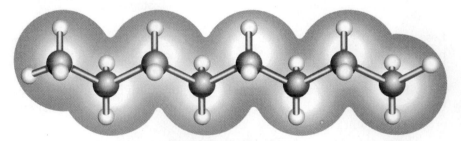

with the carbons zigzagging up and down at the tetrahedral angle, giving rise to a serpentine molecule that twists and turns to the extent permitted by the various rotational barriers. Conformational isomers of all kinds abound among the alkanes.

Structural isomers abound as well: molecules with the same set of atoms, but connected in different ways. The carbons are no longer linked one to the other to form a simple chain, but instead they acquire branches off the main line. Bonds break and form again at different points while the atoms regroup. It is a change more substantive and permanent than for conformational isomers, where the atoms merely rotate about intact covalent bonds.

Structural isomers become possible first among four carbons, where we observe how the molecule butane (connected as C—C—C—C) can develop a branch and isomerize into isobutane:

$$CH_3—CH_2—CH_2—CH_3$$

butane

$$\overset{\overset{\displaystyle CH_3}{|}}{CH_3—CH—CH_3}$$

isobutane

Although the two molecules share the formula C_4H_{10}, their structures are clearly different. The branched alkane, appearing as if one CH_3 (methyl) group and one H atom have moved from their original sites in butane, contains a central carbon bonded to three rather than only two other carbons. It is rounder than the unbranched form.

With more carbons come more structural isomers. Whereas among four carbons we have only one choice for the displaced CH_3, in pentane (five carbons) there are two branched isomers in addition to the straight chain:

$$CH_3—CH_2—CH_2—CH_2—CH_3$$

pentane

$$\overset{\overset{\displaystyle CH_3}{|}}{CH_3—CH—CH_2—CH_3} \qquad \overset{\overset{\displaystyle CH_3}{|}}{\underset{\underset{\displaystyle CH_3}{|}}{CH_3—C—CH_3}}$$

isopentane   neopentane

The three molecules, all having the formula C_5H_{12}, are structurally distinct, each with different shape and properties. Neopentane, in particular, completes the rounding trend begun by isobutane.

See the difference in Figure 8-3: With four methyl groups attached to one center, the arrangement of five carbons in neopentane closely approximates a sphere (Figure 8-3b). Hence two neopentanes—coming together just like two large balls—can touch only at one point, in contrast to molecules of pentane, which can nestle along the entire length of the chain (Figure 8-3a). The linear molecules are better able to attract each other electrically, and so the "stickier" pentane boils at a higher temperature than neopentane. More energy is needed to break up the weak attractions between the straight chains in the liquid state. We shall return to this point in the next chapter, when the question of intermolecular interactions is taken up in more detail.

(a)

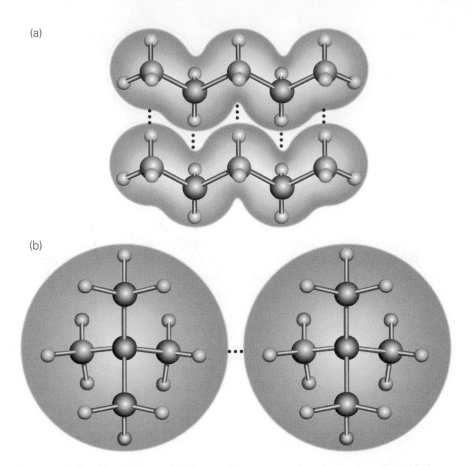

(b)

FIGURE 8-3. Two structural isomers of pentane, each with the same formula but a different shape: (a) Normal pentane, an unbranched hydrocarbon. The molecules make contact at many points along their five-carbon chains. (b) Neopentane, a tetrahedral structure like methane, looms bulky and round. Molecules of neopentane approach each other only over a limited area.

Structural isomers proliferate rapidly as the molecules become larger. Hexane (C_6H_{14}) shows five variations; heptane (C_7H_{16}) has nine; octane (C_8H_{18}) has 18, and the number quickly escalates into the thousands and millions. In addition, there are cyclic alkanes *(cycloalkanes)* in which the carbon atoms are connected at the ends to create rings, cages, and other closed structures. The menu of hydrocarbons grows bigger and bigger, and almost all conceivable arrangements consistent with carbon's valence—straight chains, branched chains, cyclic forms— are found in *petroleum*, the principal source of many raw alkanes. Crude oil is a complex soup of hydrocarbons from which molecules of various shapes and weights can be separated, or *refined*, on the basis of their different boiling points. One at a time, the different components boil away and are collected separately as the temperature is raised.

Double and Triple Bonds: A Review

Molecular geometry is inevitably altered whenever two carbon atoms share more than one pair of electrons. Allowed only four bonds altogether, a carbon forfeits one additional covalent opportunity for every extra bond it makes to another carbon. Any C atom participating in a C=C double bond therefore has a free valence of only two, which shrinks to one in a C≡C triple bond.

   The tetrahedron is gone. All the atoms affected by the multiple bond lie instead in a single plane, held in place without free rotation—as, for example, in ethylene:

$$\begin{array}{ccc} \text{H} & & \text{H} \\ \diagdown & & \diagup \\ & \text{C}=\text{C} & \\ \diagup & & \diagdown \\ \text{H} & & \text{H} \end{array}$$

Here the 120° H—C—H and H—C—C angles are attributed to sp^2-hybridized carbon (Chapter 7), from which comes a planar σ framework for the molecule:

$$\begin{array}{ccc} \text{H} & & \text{H} \\ \diagdown & & \diagup \\ & \text{C}-\text{C} & \\ \diagup & & \diagdown \\ \text{H} & & \text{H} \end{array}$$

Holding together any of these C—H or C—C sigma bonds is a covalent, head-on sharing of electrons, distributed symmetrically around the internuclear axis.

   Such single bonds, by themselves, continue to permit free rotation. But still to be accommodated is the fourth valence electron from each carbon atom, left untouched after formation of the three sp^2 orbitals. We recall that a second C—C bond can no longer be head to head, since this slot is already filled by the σ electrons. Rather, the remaining two electrons (one per carbon) are distributed above and below the plane of the molecule, to form the π bond we first met in Chapter 7 and see again now in Figure 8-4. Although weaker than a C—C σ bond, the π bond is nonetheless strong enough to eliminate free rotation and thereby keep the molecule rigid. Together, the two bonds hold the carbons closer and tighter than would a single bond alone. And, as we shall note presently, the π bond—which, with its abundance of exposed, accessible

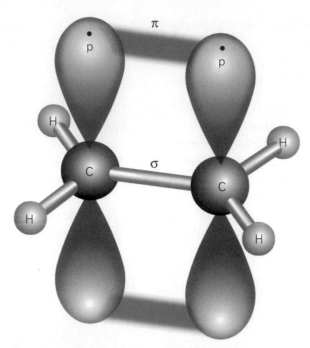

FIGURE 8-4. Valence bond description of ethylene: an sp^2–sp^2 σ bond locked in place above and below by a p–p π bond. The exposed π electrons allow the double bond to act as a Lewis base.

electrons, is surely a Lewis base—provides both a source of electrons for other species and also a spare bonding capacity for the original molecule. This "extra" C—C bond can later be tapped and its electrons used to attach new atoms.

The model of σ and π bonding recurs throughout organic chemistry, notably in the class of cyclic and planar molecules called ***aromatic***. Benzene (C_6H_6), a hydrocarbon consisting of a hexagonal ring of carbons, is the prototype and most well known of the aromatic structures. Introduced in Chapter 7, benzene's six-membered ring takes shape from a planar framework of sp^2 C—C and C—H sigma bonds, with an unhybridized p orbital at each vertex (Figure 8-5). Every carbon in the ring supplies one π electron, and so there is no gap at any site. The six π electrons wander freely in *delocalized* molecular orbitals above and below, thus able to fill (completely) a full set of three bonding levels.

It is a special characteristic, this delocalization, one that both defines the aromatic molecules and lends special stability to them. Not every planar ring can enjoy such stability, because full aromatic delocalization also demands an uninterrupted system of p orbitals that can house $4n + 2$ pi electrons to best effect ($n = 0, 1, 2, \ldots$). At a minimum, though,

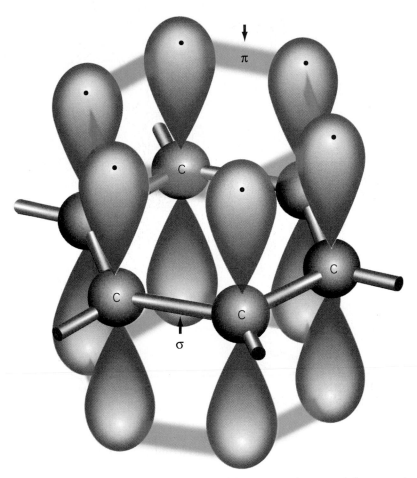

FIGURE 8-5. Valence bond description of benzene: a hexagonal framework of sp^2–sp^2 σ bonds set between two rings of delocalized π electrons. The six hydrogen atoms are omitted for clarity.

the presence of delocalization is often signaled by "resonance" Lewis structures in which single bonds alternate with double bonds—as in benzene I and II, for example, shown below:

I   II

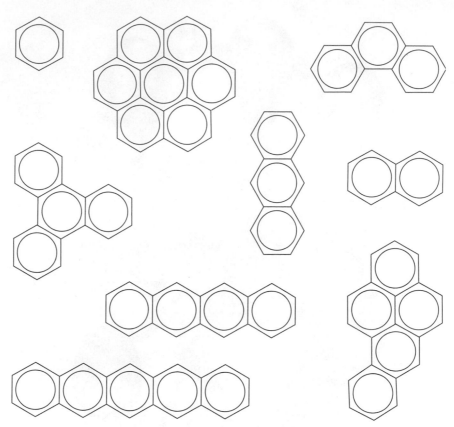

FIGURE 8-6. Aromatic rings. These and similar molecules offer their π electrons an optimally configured network of delocalized molecular orbitals, symbolized by the circles. Each vertex depicts a carbon atom, with any attached hydrogen omitted.

The sequence of C=C followed by C—C followed by C=C creates a continuous π system, since each carbon then has one electron to contribute to an uninterrupted sequence of p orbitals. Examples of some other aromatic rings are provided in Figure 8-6.

Nor is delocalization confined to rings. Similarly extended orbitals appear also in **conjugated** alkenes such as 1,3-butadiene (CH_2=CH—CH=CH_2), where alternation in the Lewis structure again creates an unbroken string of p orbitals. The effect is not quite the same as in benzene, which is aromatic, but still there is a shared network of electrons. Figure 8-7 offers a sketch of the π system in this unbranched alkene.

Turning finally to the alkynes, we find many of these same features carried over to the C≡C triple bond. The principal modification is the reduction of carbon's free valence to one, mandated now that three pairs of electrons are tied up in a triple bond. Acetylene (H:C⋮⋮C:H), with its *sp* hybridization, is typical. As before, there is a C—C σ bond in the middle and a C—H σ bond at each end, oriented this time at 180° to

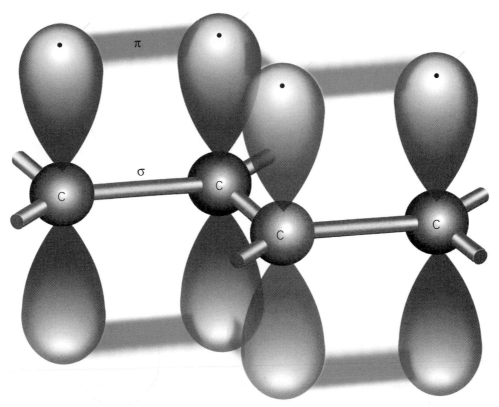

FIGURE 8-7. Delocalization of π electrons in the conjugated alkene $CH_2=CH-CH=CH_2$, typical of a Lewis structure where $C-C$ single bonds alternate with $C=C$ double bonds. Hydrogen atoms are omitted for clarity.

produce a straight molecule (really straight, note, not like the slithery alkane chains). The second and third carbon–carbon bonds, shown in Figure 8-8, are of π symmetry, nominally perpendicular to each other but actually blending into a cylindrical cloud of charge wrapped around

FIGURE 8-8. Valence bond description of acetylene: an *sp–sp* σ bond locked in place by two *p–p* π bonds. The π electrons encircle the axis.

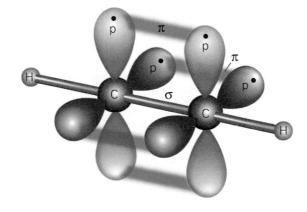

the C—C axis. Triple bonds, too, if alternated with single bonds, develop regions of delocalized electrons, just as do aromatics and conjugated alkenes.

Geometric Isomerism

No carbon involved in a multiple bond can be stereogenic, because the primary requirement to form bonds with four different groups is violated right from the start. Around C=C double bonds, however, there arises a new opportunity for isomerism in the placement of the associated four single bonds. Consider the two molecules

$$
\begin{array}{ccc}
\mathrm{H}\!\diagdown \quad \diagup\mathrm{H} & & \mathrm{H}\!\diagdown \quad \diagup\mathrm{Cl}\\
\mathrm{C}{=}\mathrm{C} & \text{and} & \mathrm{C}{=}\mathrm{C}\\
\mathrm{Cl}\!\diagup \quad \diagdown\mathrm{Cl} & & \mathrm{Cl}\!\diagup \quad \diagdown\mathrm{H}\\
\textit{cis} & & \textit{trans}
\end{array}
$$

which differ only in whether the chlorines are attached on the same side of the double bond *(cis)* or opposite sides *(trans)*. These rearranged structures, called **geometric isomers**, are not freely converted one into the other, but rather the two hydrogens and two chlorines are held in place by the rigid double bond. Bonds must be broken to go from cis to trans.

   Similar variations appear when different groups are attached to rings as well, opening up still more geometric possibilities. Into the mix go tetrahedra, triangles, and lines, not to mention hexagons, octagons, icosahedra and more; out come isomers of all types—optical, structural, and geometric, among others. Add to that the influence of free rotation, hindered rotation, and bonds both σ and π, and soon the range of combination looks inexhaustible. That is what makes carbon so special, and therein lies the richness of organic chemistry.

8-3. FUNCTIONAL GROUPS

If this richness seems, somehow, still to be merely a static richness of shape and structure, it is because we have yet to consider *functional groups* and the capacity for change they bring. A saturated hydrocarbon is, in the end, just that: saturated. With its valence monopolized by hydrogen, a pure alkane has little opportunity for transformation. There is no room to add more hydrogen, and the existing bonds are hard to break. The alkanes, for all their diversity, really cannot do much on their own, and the few options available are easy to enumerate: (1) They burn in oxygen to produce carbon dioxide and water. This oxidative process (combustion) is one we have seen once before (Chapter 3) and will see again. (2) They can be made to isomerize into various branched structures, an important consideration in the production of gasoline. (3) They will react with free radicals such as Cl·, also described in Chapter 3. (4) They can be broken down, or "cracked," into smaller hydrocarbons during the refinement of petroleum. There are some other reactions, too, but alkanes still exist mostly to provide support and structure for the real agents of change, the functional groups—or, put simply, those parts of the molecule that are *not* hydrogen atoms.

For as soon as a C—H bond gives way to anything else, the local character of the structure is abruptly altered. Whatever that new group may be, it will be different from hydrogen. Perhaps it will donate more electrons to the carbon, making the site more negative; or perhaps it will withdraw electrons, leaving the carbon more positive. It might be a Lewis acid; it might be a Lewis base. It might be big; it might be small. It might be anything at all, but it will be different. The functional group (Figure 8-9) will present a new aspect to both the host and neighboring molecules.

We have already encountered functional groups in the form of double and triple carbon–carbon bonds, where the "anything but hydrogen" role is filled by another carbon. The π bonds in alkenes and alkynes, which bring additional electrons to particular locations on the carbon skeleton, create new functional capability, new possibilities for reaction. These sites act as electron-rich Lewis bases able to coordinate their

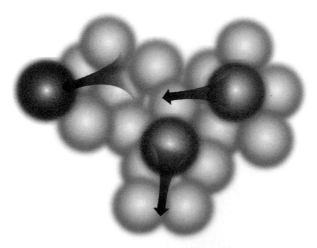

FIGURE 8-9. Functional groups and modular reactivity: a conceptual view. Interchangeable parts such as C=C, C≡C, OH, Cl, C=O, COOH, and other functional groups bring new, specially tailored chemical properties wherever they go. Some attract electrons toward a site; others expel them.

accessible π electrons with suitable Lewis acids. What might have been a functionally dead alkane linkage instead becomes a target for reaction, a hot spot.

A selection of key functional groups is displayed in Table 8-1, among which are such entities as halides, alcohols, ethers, ketones, aldehydes, carboxylic acids, anhydrides, esters, amines, and amides. These small species are both reactive and portable from molecule to molecule; they

TABLE 8-1. Selected Functional Groups

Group	Structure[a]
alcohol (hydroxyl)	$-$OH
aldehyde	$\underset{\displaystyle -\overset{\textstyle\|}{\text{C}}-\text{H}}{\overset{\textstyle\text{O}}{}}$
alkene	$\text{C}=\text{C}$
alkyne	$-\text{C}\equiv\text{C}-$
amide	$\underset{\displaystyle -\overset{\textstyle\|}{\text{C}}-\text{N}-}{\overset{\textstyle\text{O}}{}}$
amine	$-\text{N}-$
anhydride	$-\overset{\text{O}}{\overset{\|}{\text{C}}}-\text{O}-\overset{\text{O}}{\overset{\|}{\text{C}}}-$
carboxylic acid	$-\overset{\text{O}}{\overset{\|}{\text{C}}}-\text{OH}$
ester	$-\overset{\text{O}}{\overset{\|}{\text{C}}}-\text{O}-$
ether	$-\text{O}-$
halide, acyl	$-\overset{\text{O}}{\overset{\|}{\text{C}}}-\text{X}$ (X = F, Cl, Br, I)
halide, alkyl	$-\text{X}$ (X = F, Cl, Br, I)
ketone	$-\overset{\text{O}}{\overset{\|}{\text{C}}}-$
nitrile	$-\text{C}\equiv\text{N}$
nitro	$-\text{NO}_2$
sulfide	$-\text{S}-$
thiol	$-\text{SH}$

[a]Groups attach at the open positions indicated. True angles are not necessarily shown.

attach to various parts of a hydrocarbon and impart some special character to that region. Replace an H with an OH, say, and instantly there appears an *alcohol* at a given site. It is only the OH (hydroxyl) group—just those two atoms—that gives an alcohol its alcoholic properties, and the rest of the molecule is, in some sense, hardly more than a hydrocarbon appendage. We shall denote that rest-of-molecule for the time being as R, deferring consideration of its structural details. We do so because R—OH has properties determined first by the functional group alone (OH), whether the entire molecule be CH_3OH (methyl alcohol, or methanol), CH_3CH_2OH (ethyl alcohol, or ethanol), or a structure far more elaborate.

The group R does play a role, however, in how those alcoholic properties are manifested. Is R so big and bulky that other molecules cannot come close enough to interact with the OH site? Does R give or take electrons from the hydroxyl group? If the OH were to leave, what kind of existence would R face on its own? Answers to such questions, posed symbolically in Figure 8-10, are what organic chemistry is all about: knowing how the overall molecular environment affects the way in which a functional group expresses its unique properties.

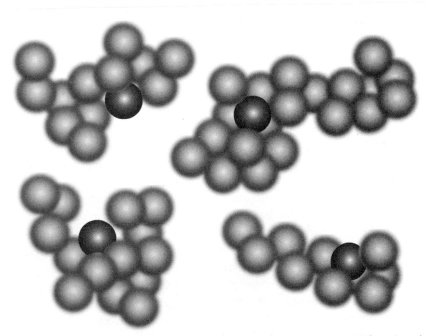

FIGURE 8-10. The way in which a functional group performs its function depends, in large measure, on the local environment: the size and shape of the host molecule, the presence of neighboring groups, the electron density at the site. The same functional group (colored ball) may operate differently in different settings, as suggested in these stylized renderings.

We consider now a few examples of other such groups, along with some of the reactions they undergo. First in line are the *alkyl halides*, R—X, where X denotes any halogen: CH_3F (methyl fluoride), CH_3Cl (methyl chloride), CH_3Br (methyl bromide), CH_3I (methyl iodide), CH_3CH_2Cl (ethyl chloride), and so on.

HALIDES: $X^- = F^-$, Cl^-, Br^-, I^-. Bonded as R—X, the halogen atom or ion brings to the site a charge density more substantial than that of hydrogen—up to eight valence electrons—of which it displays three lone pairs to the world outside. Already rich in electrons, it takes up residence as a Lewis base. The strongly electronegative halogen then withdraws additional charge from the carbon, skewing the distribution to create a polar bond partially negative at the X end: $C^{\delta+}-X^{\delta-}$. The X atom makes a difference just by being there, and that difference is enough to turn the site into a target for other molecules. Something can happen.

Substitution will occur, for instance, when one halogen replaces another or when some other electron-rich species comes along—a Lewis base, anionic or neutral, maybe OH^- or CN^- or NH_3. Under appropriate conditions the halogen X can be removed from R, giving way to a more aggressive group Y able to take its place:

$$R-X + Y \longrightarrow R-Y + X^-$$

Such transformations are among the most important organic reactions.

Both X: and Y: act here as **nucleophiles**, electron-rich species seeking a less negative, more positive (more nucleus-like) environment for themselves. These "nucleus loving" nucleophiles are in competition, too, since they both fight for accommodation at the same partially positive carbon site. They are, understand, ordinary Lewis bases with electrons to give, but we introduce the special term *nucleophilicity* to emphasize the competitive aspect of the substitution. A more powerful nucleophile is a better kinetic competitor, able to substitute *faster* than another, even if it is not necessarily a stronger base. In the short run, the stronger nucleophile finishes the race sooner and locks out the weaker nucleophile. It finds an easier pathway to go from reactants to products, a route that demands less of a "start-up" cost to stretch and break the first bonds (in the sense of an *activation energy*, touched on in Chapter 3). But in the long run—if given the chance—the stronger Lewis base would win the final competition by forming molecules with the lowest energy. That we shall see as we wend our way through thermodynamics and kinetics beginning in Chapter 12.

For now, though, we use the alkyl halides simply to glimpse the main features of this process of **nucleophilic substitution**, looking for principles that apply broadly to many other organic reactions. We start with the general transformation

$$R—X + Y \longrightarrow R—Y + X^-$$

and note that each participant has a decisive role to play: the group *coming* (Y), the group *leaving* (X), and the group *staying* (R). First, Y must have something to offer R. The offering should be something that X does not already supply, something that will favor the formation of structure R—Y over structure R—X. Second, X must be able to go off and exist on its own in a state of reasonably low energy. Third, Y must make intimate contact with R—X if the transaction is to be completed. A large, bulky R might prevent a sufficiently close approach.

Then there is this additional consideration: Departing as an anion (X^-), the halogen might leave behind a carbon cation, or **carbocation** (R^+), to balance the X^-. And since nature favors neutral species over ions, a successful breakup will then depend largely on the ability of the two charged structures, R^+ and X^-, to survive. Questions to ask now about the carbocation include: For how long must it endure? Are there opportunities to disperse the charge through delocalization into nearby aromatic rings or conjugated bonds? Are there other portions of the molecule, less electronegative than the exposed C^+, able to pump electrons toward that site and thus mitigate the positive charge? Alkane carbons, for example, donate electrons more readily than hydrogens, so a heavily alkyl-substituted carbocation such as

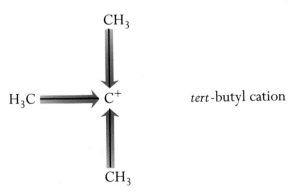

tert-butyl cation

is better able to exist alone than a simple CH_3^+ cation (which has only hydrogen atoms bonded to the C^+).

Where a carbocation is hard to sustain, as it would be during the conversion of CH_3Br into CH_3OH, the substitution reaction must proceed through some other route. Under such conditions, the OH^- can attack the tetrahedral $-\overset{|}{\underset{|}{C}}-Br$ site from behind to eject a bromide ion. It

(a)

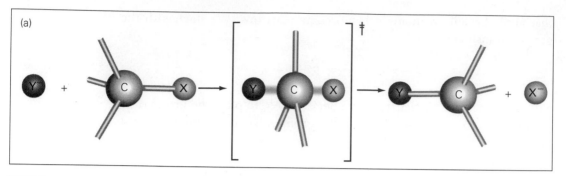

(b)

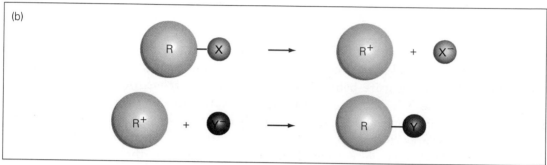

FIGURE 8-11. Nucleophilic substitution. (a) Substitution brought about forcibly by a collision of two molecules: the S_N2 mechanism. An attacking anion (Y^-) strikes the carbon atom from behind, displacing the leaving group (X^-) from the site. The system passes through a short-lived *transition state* (‡) in which five groups are simultaneously attached to one carbon. (b) Substitution facilitated by a unilateral departure: the S_N1 mechanism. The leaving group leaves, exposing a carbocation (R^+) open to attack by another group with electrons to give.

all occurs in one concerted motion, never allowing the dubious CH_3^+ cation to form. This sequence, illustrated symbolically in Figure 8-11(a), is called the **S_N2** mechanism (Substitution, Nucleophilic, bimolecular— *two* species come together at once).

Under circumstances more favorable for carbocations, say when $(CH_3)_3C-Br$ (the reactant R—X) is transformed into $(CH_3)_3C-OH$ (the product R—Y),

$$CH_3-\underset{\underset{CH_3}{|}}{\overset{\overset{CH_3}{|}}{C}}-Br \ + \ OH^- \ \longrightarrow \ CH_3-\underset{\underset{CH_3}{|}}{\overset{\overset{CH_3}{|}}{C}}-OH \ + \ Br^-$$

the substitution reaction proceeds instead by the two steps shown in Figure 8-11(b). First, the leaving group X^- departs on its own to leave behind the carbocation R^+ (here, the comparatively robust *tert*-butyl cation). The incoming group Y⁻, finding a sudden vacancy for its excess electrons, then attacks the carbocation to complete the substitution. Designated S_N1, this mechanism is a unimolecular (one molecule) process; there is no collision between the incoming nucleophile and the original reactant to force the substitution.

GROUPS CONTAINING C=O. Moving beyond the halides, we come next to the versatile *carbonyl* group, C=O, noted especially for its incorporation into *ketones* and *aldehydes*:

$$
\begin{array}{cc}
\overset{\displaystyle O}{\overset{\|}{R-C-R'}} & \overset{\displaystyle O}{\overset{\|}{R-C-H}} \\[6pt]
\text{ketone} & \text{aldehyde}
\end{array}
$$

Common to these compounds is the C=O double bond (again σ and π, similar to C=C), which makes the site unsaturated. Hybridized as sp^2, the carbonyl's π system contains exposed, accessible, *bondable* electrons that can be diverted to other partners:

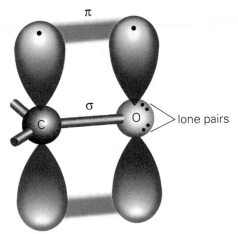

This reserve bonding capacity, in addition to the two lone pairs on oxygen, enables new species to break into the double bond and attach themselves to carbon and oxygen. We say that two incoming groups *add across the double bond*, such as when the carbonyl site is attacked simultaneously by a Lewis acid (often H⁺) and some Lewis base denoted by Y⁻:

$$
Y^- \;+\; {\textstyle \diagdown \atop \diagup}C{=}O \;+\; H^+ \quad\longrightarrow\quad Y-\overset{|}{\underset{|}{C}}-OH
$$

The π bond is ruptured, enabling Y⁻ to react with the carbon while H⁺ attaches to the oxygen.

It happens that way because the carbonyl linkage is significantly polar, made so by the difference in electronegativity between C and O. The strongly electronegative oxygen drains electrons away from carbon, growing partially negative while leaving carbon partially positive. The $O^{\delta-}$ then acts as a Lewis base (a nucleophile), offering its excess electrons to an incoming Lewis acid such as H^+. At the other end is electron-deficient $C^{\delta+}$, a Lewis acid (an electrophile, an "electron lover") that mates with the electrons carried by Y^-.

Positive to negative; negative to positive. Electron-rich meets electron-poor; Lewis base meets Lewis acid; nucleophile meets electrophile. See it all throughout chemistry; always a merging of opposites, different in name only. Crudely put: Look for where the electrons are and where they aren't, and then match up the haves with the have-nots.

CARBOXYLIC ACIDS: R—COOH. These compounds are acids in the Brønsted-Lowry sense, functioning as donors of H^+. The carboxylic acid

$$\overset{\overset{\textstyle O}{\|}}{R-C-OH}$$

carboxylic acid

releases a proton to leave behind a *carboxylate* anion (R—COO⁻), a newly created base, which then must find a way to sustain the unbalanced negative charge:

$$\overset{\overset{\textstyle O}{\|}}{R-C-OH} \longrightarrow R-\overset{\overset{\textstyle O}{\cdot}}{C}\!\!\cdot\!\!\overset{-}{O} \;+\; H^+$$

Our first question, once again, is: Can R—COO⁻ support itself and its extra electron?

Indeed it can; in part, because a carboxylate anion is able to spread the charge throughout the delocalized π orbitals that link O to C to O. With the true bonding pattern neither $O—C=O^-$ nor $O=C—O^-$ but rather something in between (represented by the broken line), the carboxylate structure allows the electron access to the whole π system, just as in the aromatics and conjugated alkenes noted earlier. Alcohols, by contrast, usually cannot spread out the charge created when R—OH loses a proton to become R—O⁻. Carboxylic acids, better able to leave behind a sustainable anion, consequently are the more effective proton donors. They enjoy other advantages as well.

CONDENSED FUNCTIONAL GROUPS. What else can functional groups do? They can serve as intermolecular connectors, joining one molecule or fragment to another. Often the link is made by removal of H_2O in a so-called **condensation reaction** (not to be confused with gas-to-liquid condensation). Typical of organic condensation is the formation of an *ester* from a carboxylic acid and an alcohol,

$$\begin{matrix} & O & & & & & O \\ & \parallel & & & & & \parallel \\ R-C-OH & + & HO-R' & \longrightarrow & R-C-O-R' & + & H_2O \\ \text{acid} & & \text{alcohol} & & \text{ester} & & \text{water} \end{matrix}$$

and the formation of an *anhydride* from two carboxylic acids:

$$\begin{matrix} & O & & & & & O & & O \\ & \parallel & & & & & \parallel & & \parallel \\ R-C-OH & + & HO-C-R' & \longrightarrow & R-C-O-C-R' & + & H_2O \\ \text{acid} & & \text{acid} & & \text{anhydride} & & \text{water} \end{matrix}$$

There are far more possibilities than these, but the general capability is clear and the implications are far-reaching. Condensed functional groups, found everywhere, become the joints in composite structures—in chains built from groups of molecules—and nowhere are these links more crucial than in the biological polymers: life's macromolecules, the proteins and nucleic acids.

8-4. BIOPOLYMERS

From protons and electrons to atoms to molecules, we build up now to some structures of real personal interest: ourselves, and other living organisms. It happens carbon by carbon, one connection at a time.

Start with a single carbon, and attach to it (1) a hydrogen atom, H; (2) an *amino* group, NH_2; (3) a carboxylic acid group, COOH; and (4) something else (a *side chain*, denoted R), from which is formed an **amino acid**:

$$\begin{matrix} & R & O \\ & | & \parallel \\ H_2N-&C-&C-OH \\ & | \\ & H \end{matrix}$$

Such molecules, properly fitted out with the appropriate side chains, are the universal building blocks for proteins. They are the makings of dog hair, eagle feathers, skin and bone, life itself.

Realize, first, that an amino acid is both acid (COOH) and base (NH_2) simultaneously, despite the *acid* in its name. A proton can be passed back and forth between the carboxylic acid and the basic nitrogen of the amino group, as symbolized by a bidirectional arrow in the transformation

$$H_2N-\overset{\overset{\displaystyle R}{|}}{\underset{\underset{\displaystyle H}{|}}{C}}-\overset{\overset{\displaystyle O}{\|}}{C}-OH \quad \rightleftharpoons \quad H_3\overset{+}{N}-\overset{\overset{\displaystyle R}{|}}{\underset{\underset{\displaystyle H}{|}}{C}}-\overset{\overset{\displaystyle O}{|}}{C}\overset{-}{=}O$$

The structure at the right, termed a ***zwitterion***, is a doubly ionized but electrically neutral species (a *dipolar* ion) often found under typical biological conditions. At a minimum, then, an amino acid might play host to this internal acid–base reaction, but (read on) there is far more to its chemistry than that.

For variety, we look to the side chains. Since the components H, NH_2, and COOH are common to all amino acids, only the R group can distinguish one from another. Each side chain comes with its own functional capability, from which nature (ever thrifty) selects some 20 examples to fashion proteins for every living organism on earth. The assortment of R ranges from a single hydrogen atom (glycine) or methyl group (alanine) to more complicated entities involving aromatic rings, sulfur atoms, additional carboxylic acid and amino groups, hydroxyls, and various other organic functional groups.

Many of these side chains appear in Table 8-2, where we discover that all the biological amino acids except glycine are chiral. Observe, for each: Four different groups are attached to the central carbon, allowing for either of two mirror-image configurations. The enantiomers are designated L (Latin *levo*, "left") and D (Latin *dextro*, "right"):

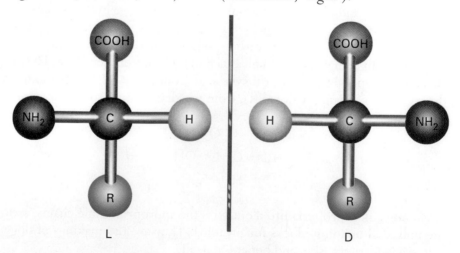

For proteins, the choice is L. Biological catalysts, as we asserted earlier, recognize left and right, and so a D enantiomer interacts with an L site as poorly as a right hand fits into a left-hand glove. One optical isomer is not the same as another, at least not in the chemistry of life.

TABLE 8-2. A Selection of Amino Acids

Amino Acid	Symbol	Side Chain	Functional Group
glycine	Gly	$-H$	hydrogen
alanine	Ala	$-CH_3$	alkyl
valine	Val	$-CH$ with CH_3 above and CH_3 below	alkyl
leucine	Leu	$-CH_2-CH$ with CH_3 above and CH_3 below	alkyl
isoleucine	Ile	$-CH$ with CH_3 above and CH_2CH_3 below	alkyl
serine	Ser	$-CH_2OH$	alcohol
threonine	Thr	$-CH$ with OH above and CH_3 below	alcohol
lysine	Lys	$-CH_2CH_2CH_2CH_2NH_2$	amine (basic)
aspartic acid	Asp	$-CH_2COOH$	carboxylic acid
glutamic acid	Glu	$-CH_2CH_2COOH$	carboxylic acid
glutamine	Gln	$-CH_2CH_2\overset{\displaystyle O}{\overset{\|}{C}}-NH_2$	amide
cysteine	Cys	$-CH_2SH$	thiol
methionine	Met	$-CH_2CH_2SCH_3$	sulfide
phenylalanine	Phe	$-CH_2-\bigcirc$	aromatic
tyrosine	Tyr	$-CH_2-\bigcirc-OH$	aromatic

Few as they are, these 20-odd amino acids join together to form the innumerable proteins needed for life—the structural proteins that make up skin and muscle and tissue, the hormones that control growth and metabolism, the enzymes that catalyze biochemical reactions. The characteristic linkage, a **peptide bond**, is made between the NH_2 group on one amino acid and the COOH group on another, accompanied by the loss of H_2O. It is a standard condensation reaction, which occurs over and over again in organic and biological processes.

The road to a protein begins therefore with a single step, the condensation of two amino acids into a dipeptide:

$$H_2N-\underset{\underset{H}{|}}{\overset{\overset{R}{|}}{C}}-\overset{\overset{O}{\|}}{C}-OH \;\; + \;\; H-\underset{\underset{H}{|}}{\overset{\overset{R'}{|}}{N}}-\underset{H}{\overset{\overset{O}{\|}}{C}}-OH \;\; \longrightarrow$$

$$H_2N-\underset{\underset{H}{|}}{\overset{\overset{R}{|}}{C}}-\overset{\overset{O}{\|}}{C}-\underset{H}{N}-\underset{\underset{H}{|}}{\overset{\overset{R'}{|}}{C}}-\overset{\overset{O}{\|}}{C}-OH \;\; + \;\; H_2O$$

Held together by a $-\overset{\overset{O}{\|}}{C}-\underset{H}{N}-$ linkage, the dipeptide retains an amino group at one end and a carboxylic acid at the other. So endowed, it can condense further with new amino acids and eventually grow (outward from each end) into a long chain.

From two amino acids comes a dipeptide; from three comes a tripeptide; from many comes a polypeptide; and from a polypeptide consisting of hundreds or thousands of amino acids comes a protein. Yet however complicated the protein may be, the structure (Figure 8-12) unravels to reveal one or more chains of amino acids joined sequentially by peptide bonds. This **primary structure**—the backbone of amino acids—is just a sequence of elementary peptide linkages between two simple functional groups: NH_2 and COOH. With the skeleton firmly in place, regions of a protein first assume ordered three-dimensional shapes to establish a **secondary structure** (usually some sort of helix), and then the protein folds into a still more complicated **tertiary structure**. The amino acid side chains come into play throughout, helping to determine the overall shape of the protein and hence its biological function.

We shall return briefly to the elaborate shapes of proteins in the next chapter. For the moment, though, let us remain at the primary level of amino acid sequence, viewing the protein as a chain of amino acids joined by simple peptide linkages. It is a view consistent with our understanding of the role played by functional groups in organic reactions.

The primary sequence for every protein is specific, different, and sometimes unforgiving. Although certain mistakes can be tolerated, there are structures where the misplacement of even one amino acid in a thousand will render a protein nonfunctional, perhaps fatally so. With proteins of all sorts constantly manufactured in living cells, one wonders how the process works at all. So complex are the final products that it seems they can only be derived from a detailed pattern.

They are. The molecular blueprint is, as mentioned previously,

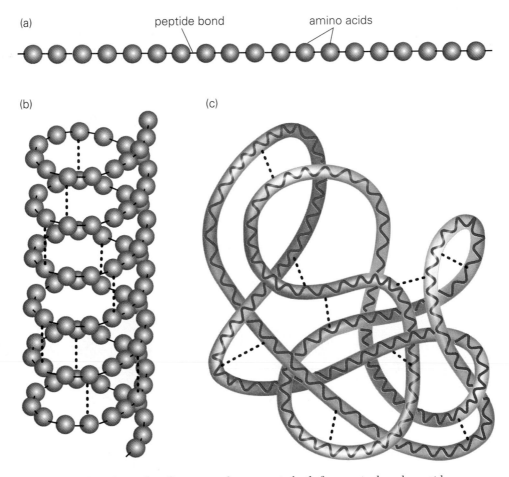

FIGURE 8-12. Hierarchy of structure for a protein built from a single polypeptide chain. (a) Primary: the linear sequence of amino acids, each molecule linked covalently to the next. (b) Secondary: the coordination of nearby amino acids by noncovalent bonds (dotted lines; see Chapter 9). The interactions often produce either a helical or sheetlike structure. (c) Tertiary: the folding together of amino acids from distant points on the polypeptide chain, again by noncovalent interactions.

deoxyribonucleic acid, the biopolymer familiarly known as DNA. It too has a three-dimensional structure (the celebrated double helix), although once again we limit our immediate focus to the linear sequence of the building blocks. The monomeric units are no longer amino acids but rather **nucleotides** assembled from the three components pictured in Figure 8-13: a molecule of phosphoric acid (H_3PO_4), a deoxyribose molecule (a variety of sugar), and one of the four nitrogen-containing bases designated adenine, guanine, cytosine, and thymine. The related *ribonucleic acid*, abbreviated RNA, is built from similar ingredients except for the substitutions of ribose for deoxyribose and uracil for thymine.

The nucleotide shown in Figure 8-13 may look complicated, but

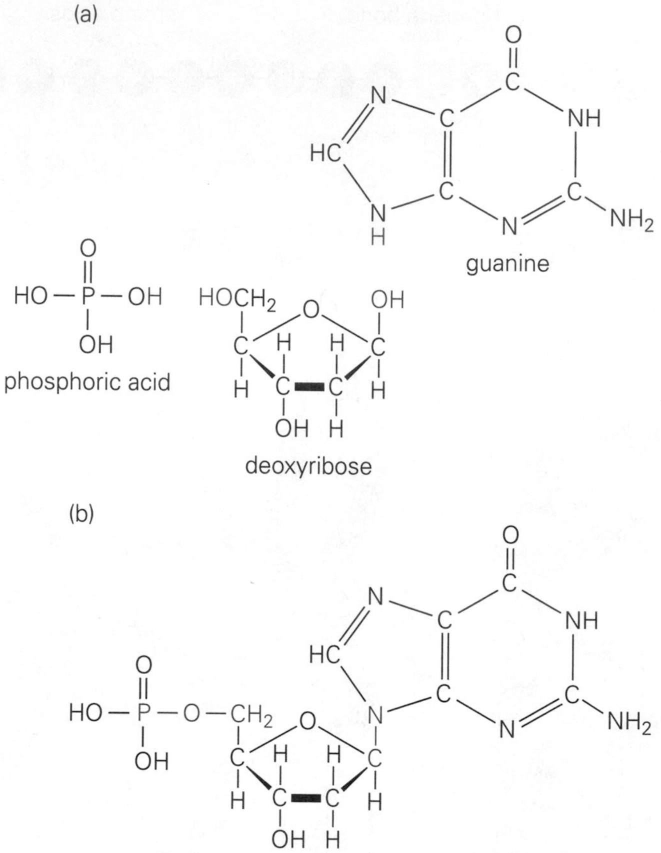

(a)

guanine

phosphoric acid

deoxyribose

(b)

FIGURE 8-13. Structure of a nucleotide. The example shown here contains de-oxyribose (a sugar) and the base guanine, one of four such nitrogen-containing molecules found in DNA. (a) The nucleotide's three components: phosphoric acid, sugar, base. (b) The assembled structure, held together by condensation linkages.

really it is just three molecules joined at two positions by simple conden-sation reactions. A P—O—CH$_2$ linkage develops when H$_2$O is removed from the CH$_2$OH group of the sugar and a P—OH group of the phos-phoric acid. Similarly, a C—N bond forms upon loss of water from the N—H of the nitrogen base and a C—OH fragment of the sugar. Following that, one nucleotide is joined to another through an additional condensation reaction—between a P—OH and a CH$_2$OH, as before—to yield first a dinucleotide and then a trinucleotide and eventually a full-fledged polymer: the polynucleotide, or nucleic acid. The buildup is, in its essence, scarcely different from protein formation, nor is anything fundamentally changed in most other biochemical polymerizations. Large biological molecules are all assembled one block at a time, ce-mented together by the removal of one molecule of water at each link.

The alternating pattern of sugar–phosphate–sugar–phosphate establishes the backbone for a strand of DNA, with the nitrogen bases jutting out as depicted in Figure 8-14. And here, in the various arrangements of the bases adenine (A), guanine (G), cytosine (C), and thymine (T), is

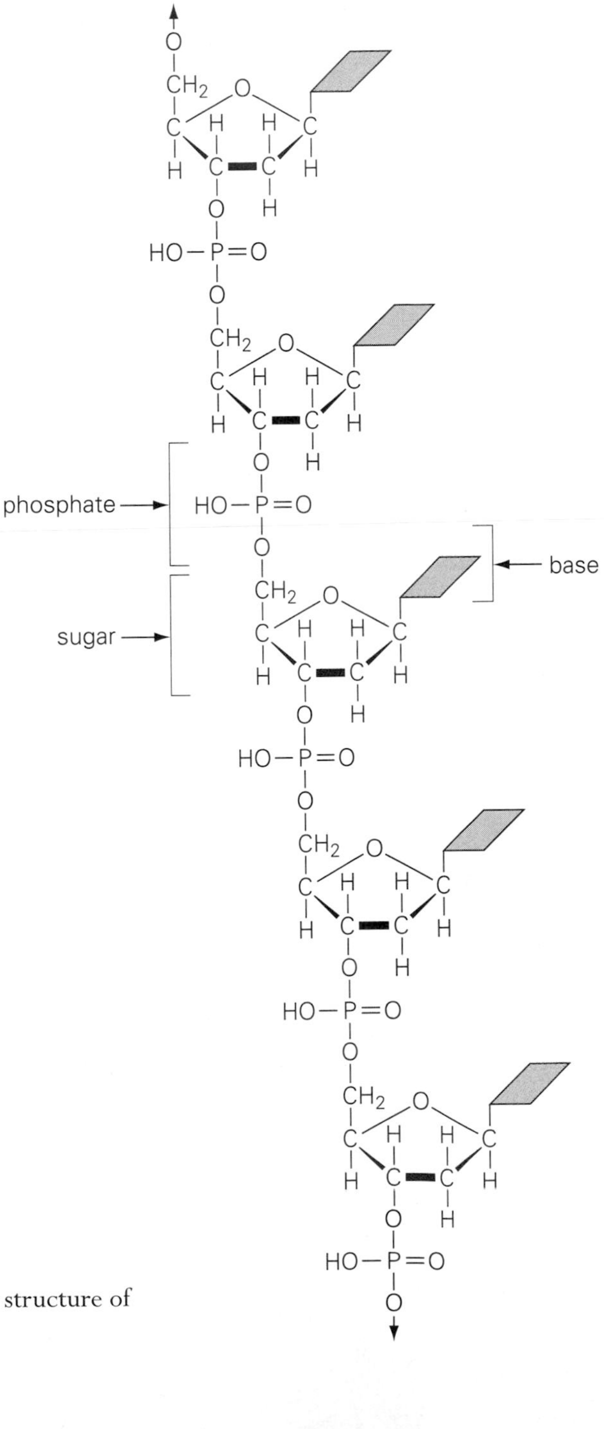

FIGURE 8-14. Primary structure of a polynucleotide chain.

written the genetic code: markers for the amino acids needed to assemble all conceivable proteins for all living organisms. Apart from a few minor deviations, it is a universal code for life as we know it. The recipes for the specific proteins may differ (and different proteins are, finally, what distinguishes a beagle from an eagle), but the instructions are written in the same language.

For with scant exception, we do regard DNA as the common language of life. The letters of its alphabet are formed from the nitrogen bases, taken three at a time to represent each of the amino acids. The letters GCC, for example, correspond to the amino acid alanine, whereas GGA translates to glycine. Every amino acid matches up with at least one three-letter word, called a *codon*, and thus the detailed instructions for making a protein are written on a strand of nucleic acid using a genetic alphabet of only four letters: A, G, C, and T, the four nitrogen bases.

If, as in Figure 8-15, some hypothetical segment were to read ACC–GCC–AGC–GGA, then out would come (after considerable cellular activity, involving RNA as well as DNA) the amino acid sequence threonine–alanine–serine–glycine. It is a small example, offered here with little detailed explanation, but it is of no small consequence. By following such patterns, organisms are able to develop from a single cell and to grow, sustain, and renew themselves. Millions of condensation reactions between organic functional groups make that possible.

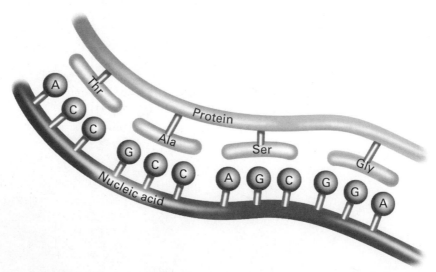

FIGURE 8-15. Living systems build proteins using a molecular template: the nitrogen bases in DNA and RNA, taken three at a time. Each amino acid recognizes its own particular triplet of bases (a codon) in the sequence.

8-5. SUMMARY OF ORGANIC REACTIONS

We have moved swiftly from methane to DNA—from swamp gas to the architecture of life—mindful always of the commonality and chemical simplicity underlying these many compounds. Analyzed sensibly, even the most complicated molecules possess a spare elegance of form and structure. Nature's tricks are relatively few, but they are used over and over again to spectacular effect.

Along the way, here and in Chapter 3, we have noted some of the reactions undergone by organic molecules: acid–base, oxidation, free radical, condensation, and a number of others in passing. Now, having attended to both skeletal structure and the properties of functional groups, we pause to organize these ideas into a concise (if incomplete) model of organic reactivity.

All the conclusions drawn in Chapter 3 still hold. Chemical reactions, organic as well as inorganic, hinge mostly on the sharing or transfer of electrons. The broad categories of acid–base, reduction–oxidation, dissolution, and radical reactions—particularly the supercategory of Lewis acid–base—apply with the same force to methane and DNA as to sodium chloride and iron. Nevertheless, organic molecules often give us compelling reason to look at the larger movements: the protons that are transferred, the water molecules that leave, the hydrogen atoms that add. Organic reactions are understood most vividly by a microscopic observer who keeps track of entire atoms or groups while, at the same time, understanding the electronic motives underlying the various shifts.

Already we have seen several examples of reactions, portrayed abstractly and conceptually in Figure 8-16. Consider the following types of change, all very tangible and structural:

1. *Addition* reactions. Atoms are added across a double bond, as when ethylene is converted into ethane by the addition of H_2, or when HY adds to a carbonyl group. The reserve bonding capacity of the electron-rich unsaturated bond is used to attach additional atoms.

2. *Elimination* reactions. The reverse of addition: when something leaves and is not replaced. An atom or other species is lost, and the structure closes up to form new bonds.

3. *Condensation* reactions. Two molecules join together, as when a carboxylic acid and an alcohol react to produce an ester. A molecule of H_2O or some other by-product is liberated during the process.

4. *Rearrangement* reactions. Part of a molecule, maybe a hydride

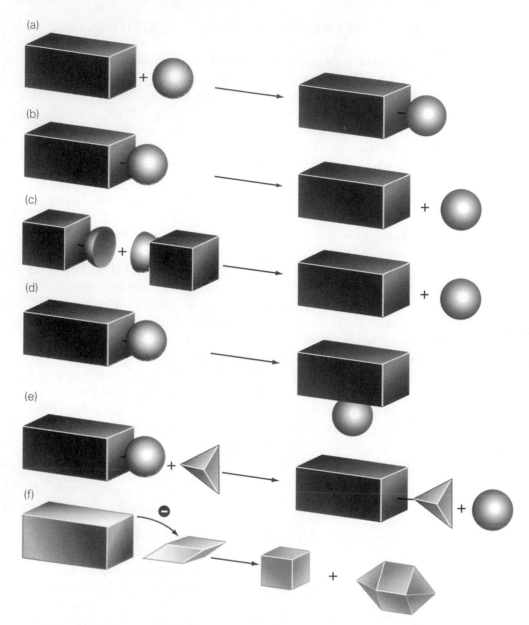

FIGURE 8-16. A summary of the principal organic reactions. (a) Addition.
(b) Elimination. (c) Condensation. (d) Rearrangement. (e) Substitution. (f) Reduction–
oxidation.

ion (H:⁻) or methyl group or larger fragment, leaves its moorings and travels to a new site on the same molecule. Bonds are
broken. Bonds are formed. A functional group may undergo a
chemical change as well.

If, during a rearrangement, there are no changes in the number and kind of atoms, then the process is an ***isomerization*** reaction. Isomers differ only in the spatial arrangements of the atoms.

5. ***Substitution*** reactions. Some atom or group of atoms takes the place of another, as when hydroxide replaces bromide in the conversion of CH_3CH_2Br into CH_3CH_2OH.

6. ***Reduction–oxidation*** reactions. Electrons are transferred, certainly, but for organic molecules there is often more solid, structural evidence of change: a movement of atoms. Oxidation usually entails the removal of hydrogen from a functional group, whereas reduction is marked by its addition. An alcohol, for instance, is oxidized to a carbonyl group when atoms of hydrogen are taken from the hydroxyl oxygen and adjacent carbon:

$$-\overset{|}{\underset{|}{C}}-OH \quad + \quad \text{oxidant} \quad \longrightarrow$$
$$\overset{H}{}$$

$$\overset{\diagdown}{\underset{\diagup}{C}}=O \quad + \quad \text{reduced oxidant} \quad + \quad \text{by-products}$$

The hydroxyl, stripped of hydrogen, finds itself with extra bondable electrons that go toward a second C=O bond. A legitimate oxidation takes place (electrons are given away, oxidation numbers increase), but to some eyes the *structural* change may appear more obvious than in, say, the oxidation of iron from Fe^{2+} to Fe^{3+}.

Although a few illustrations will never capture the full wealth of organic reactions, we use these images, more modestly, to savor the gritty texture of microscopic encounter—to expose the nuts and bolts, to witness the tearing down and building up, to take the stonemason's view of chemistry. At times, the pieces even seem to move less like molecules and more like everyday large objects bumping this way and that. In a sense, they are: Considerations of sheer bulk often do determine the course of reaction, by restricting a molecule's reactive sites to those incoming species small enough to fit. Size and shape are important.

In the end, however, it is all an illusion; all these large-scale movements of atoms are controlled by the electrons, and so we must return to our persistent theme of electronic give and take. Whatever happens, happens for the benefit of the electrons. Always it is *electrophilic* addition or *nucleophilic* addition, or electrophilic elimination or nucleophilic elimination, or electrophilic substitution or nucleophilic substitution, or some other movement to accommodate the electrons. Look for where the electrons are and where they aren't, we recall, and that will tell us what will react with what.

Review and Guide to Problems

From modest beginnings, the most complex and varied structures take shape. With a small selection of stones, a mason builds a wall, a bridge, a house, a cathedral; with a small selection of atoms, nature builds a hydrocarbon, a nucleic acid, a protein, a membrane. House and cathedral, hydrocarbon and protein—each structure has its own form and function, manifestly so different, yet different only from a distance. Up close, they all look the same. Up close, one sees only stone against stone, atom against atom, electron against electron.

And there, up close, we marvel at both the extravagance and frugality of carbon-based chemistry: an extravagant variety of structure and reaction, a frugal use of atoms and bonds. Frugal, too, is the approach we have taken in our first look at this extravagantly rich field.

IN BRIEF: ORGANIC STRUCTURE AND REACTIVITY

1. Carbon forms four bonds: single bonds, double bonds, triple bonds, localized bonds, delocalized bonds, bonds to fellow carbons, bonds to hydrogen, oxygen, nitrogen, sulfur, and others . . . all with a valence of 4 at each carbon. See a carbon; count to 4.

Four single bonds, as in methane:

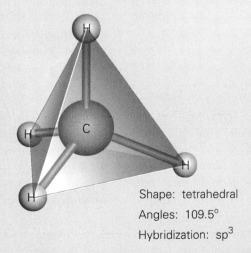

Shape: tetrahedral
Angles: 109.5°
Hybridization: sp^3

Or one double bond and two single bonds, as in ethylene (ethene):

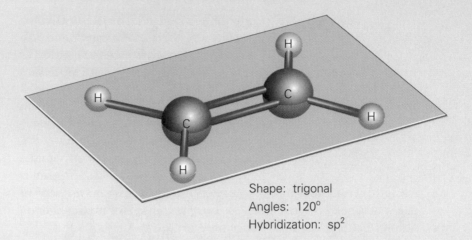

Shape: trigonal
Angles: 120°
Hybridization: sp^2

Or one triple bond and one single bond, as in acetylene (ethyne):

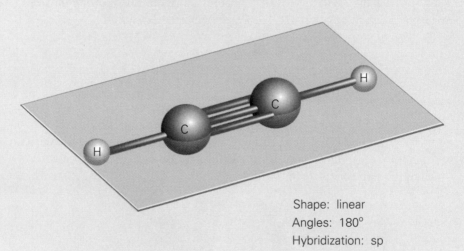

Shape: linear
Angles: 180°
Hybridization: sp

Carbon, with a valence of 4, has four electrons to give and take.

From these beginnings, the **hydrocarbons** (C and H only) grow and grow, from one carbon to two, three, four, more. There are the **alkanes** (C_nH_{2n+2}), with C—C single bonds throughout; the **alkenes**, with at least one C=C double bond; and the **alkynes**, with one C≡C triple bond or more. There are ring-shaped **cycloalkanes** containing all single bonds, and there are ring-shaped **aromatics** with delocalized π systems above and below.

2. Organic molecules frequently exist as **isomers**, different arrangements of the same set of atoms. **Structural isomers**, for example, differ in the sequence of carbon atoms, allowing us to distinguish between normal (unbranched) and branched —C—C—C—C— chains:

NORMAL

$$H-\underset{\underset{H}{|}}{\overset{\overset{H}{|}}{C}}-\underset{\underset{H}{|}}{\overset{\overset{H}{|}}{C}}-\underset{\underset{H}{|}}{\overset{\overset{H}{|}}{C}}-\underset{\underset{H}{|}}{\overset{\overset{H}{|}}{C}}-H$$

BRANCHED

$$H-\underset{\underset{H}{|}}{\overset{\overset{H}{|}}{C}}-\underset{\underset{H}{|}}{\overset{\overset{CH_3}{|}}{C}}-\underset{\underset{H}{|}}{\overset{\overset{H}{|}}{C}}-H$$

$CH_3CH_2CH_2CH_3$

C_4H_{10} (butane)

$$CH_3-\underset{\overset{|}{CH}}{\overset{CH_3}{\overset{|}{}}}-CH_3$$

C_4H_{10} (isobutane)

Each pair of formulas, left and right, represents the same two structural isomers.

For *geometric isomers*, the difference arises in the placement of atoms about a C=C double bond, either *cis* (same side) or *trans* (across):

$$\underset{Cl}{\overset{H}{\diagdown}}C=C\underset{Cl}{\overset{H}{\diagup}}$$

cis

$$\underset{Cl}{\overset{H}{\diagdown}}C=C\underset{H}{\overset{Cl}{\diagup}}$$

trans

For *optical isomers*, the variation shows up when a structure of sufficiently low symmetry is reflected through a mirror:

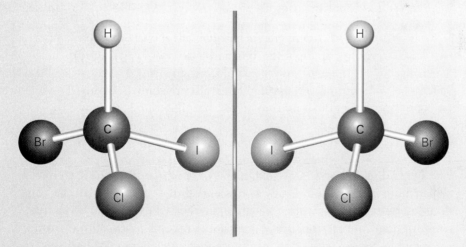

The two optical isomers are nonsuperimposable mirror images *(enantiomers)*, like right and left hands, and any molecule having an enantiomeric twin is aptly called *chiral* (from the Greek word for "hand"). Optical isomers exert opposite effects on polarized light, rotating the electromagnetic field either to the left or right.

And, finally, still more isomers: ***conformational isomers***, structures in which bonds need not be broken to go from one form to another. Different conformations develop when a group of atoms rotates about a C—C single bond, as in the passage from staggered to eclipsed to staggered conformations shown below for ethane:

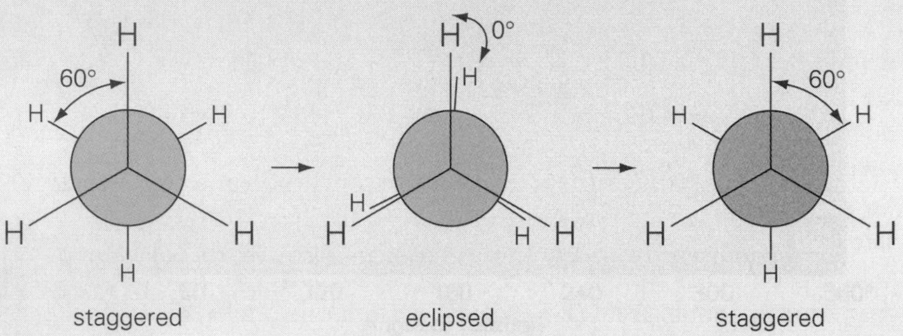

| staggered | eclipsed | staggered |

3. FUNCTIONAL GROUPS. Pure alkanes are ***saturated hydrocarbons***, able to react in only limited ways. Each carbon has its valence tied up in some combination of four C—C and C—H single bonds—four *strong* bonds, four bonds that are hard to break, four bonds that are difficult to change once they are in place.

A double bond, by contrast, is more amenable to modification. Hybridized as sp^2, the C=C bond has weakly bound π electrons that can be diverted to other purposes, as when ethylene's double bond is broken to accept two additional hydrogen atoms:

$$CH_2=CH_2 \ + \ H_2 \longrightarrow CH_3-CH_3$$

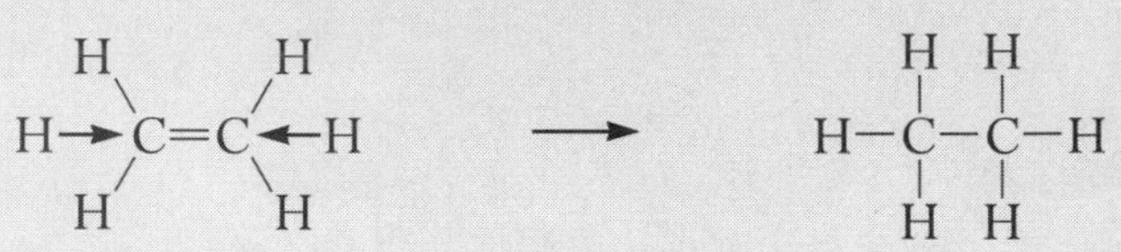

The C=C double bond is, we say, ***unsaturated***, because the system can redistribute electrons into new bonds and thus accept new atoms. The double bond is a reservoir of negative charge, a Lewis base, a potential site for chemical reaction. It is our first example of a ***functional group***.

Double bonds, triple bonds, and various other functional groups replace hydrogen within the hydrocarbon framework and thereby break the chemical sameness of the molecule. They open the way for subtle modifications of the structure. They give electrons and they take electrons. They create reactive sites where electron-rich species come together with electron-poor species. The chemistry of organic molecules is the chemistry of the functional groups.

Among the most important of these groups are C=C double bonds, C≡C triple bonds, halides, alcohols, ketones, aldehydes, carboxylic acids, esters, anhydrides, and amines. For structural formulas, see Table 8-1 in the first part of the chapter.

4. REACTIONS. Question: What does an organic molecule do with its functional groups? Answer: everything. It oxidizes them. It reduces them. It moves them. It swaps them. It gives to them. It takes from them.

EXAMPLE: *Addition*—where new atoms invade a double bond, rip open the π system, and use the liberated electrons to form single bonds. We have seen the process unfold in Section 3 just above, in the reaction of H_2 with $CH_2{=}CH_2$ to produce $CH_3{-}CH_3$:

$$
\begin{array}{ccc}
\overset{\displaystyle H}{\underset{\displaystyle H}{\diagup}} \text{C}{=}\text{C} \overset{\displaystyle H}{\underset{\displaystyle H}{\diagdown}} & \longrightarrow & \underset{\displaystyle H\ \ H}{\overset{\displaystyle H\ \ H}{H{-}\text{C}{-}\text{C}{-}H}}
\end{array}
$$

EXAMPLE: *Elimination*—the reverse of addition. A molecule releases one or more groups, as when ethyl alcohol loses H_2O to become ethylene:

$$CH_3{-}CH_2OH \longrightarrow CH_2{=}CH_2 \ + \ H_2O$$

$$
\underset{\displaystyle H\ \ OH}{\overset{\displaystyle H\ \ H}{H{-}\text{C}{-}\text{C}{-}H}} \longrightarrow \overset{\displaystyle H}{\diagup}\text{C}{=}\text{C}\overset{\displaystyle H}{\diagdown} \ + \ H_2O
$$

EXAMPLE: *Condensation*—the fusing of two molecules, brought about when each partner loses a small piece of itself. Most common is the ejection of H_2O, a prominent architectural theme in the structure of proteins, nucleic acids, and other polymers (large molecules built from repeating units). Proteins, for instance, are born when individual amino acids join together, two at a time, to produce a series of *peptide* linkages. The COOH group of one amino acid reacts with the NH_2 group of another, expelling H_2O in the process:

$$
\underset{\displaystyle H}{H_2N{-}\overset{\displaystyle R}{\text{C}}{-}\overset{\displaystyle O}{\text{C}}{-}OH} \ + \ \underset{\displaystyle H}{H_2N{-}\overset{\displaystyle R'}{\text{C}}{-}\overset{\displaystyle O}{\text{C}}{-}OH} \longrightarrow
$$

$$
\underset{\displaystyle H}{H_2N{-}\overset{\displaystyle R}{\text{C}}{-}\overset{\displaystyle O}{\text{C}}{-}}\underset{\displaystyle H}{N{-}}\underset{\displaystyle H}{\overset{\displaystyle R'}{\text{C}}{-}\overset{\displaystyle O}{\text{C}}{-}OH} \ + \ H_2O
$$

EXAMPLE: **Substitution**—where one group enters and another group leaves, such as when OH replaces Br in the reaction

$$CH_3Br + OH^- \longrightarrow CH_3OH + Br^-$$

See the same process repeated in many ways in many molecules, always with the same questions to be resolved. Is the site electron-rich or electron-poor? Which species (the group attacking or the group defending) is better able to attach itself to the molecule? Which species, detached from the molecule, is better able to exist on its own?

EXAMPLE: **Rearrangement**—where groups migrate from one part of a structure to another during the reaction, as does a hydride ion $(H\!:^-)$ in the two carbocations below:

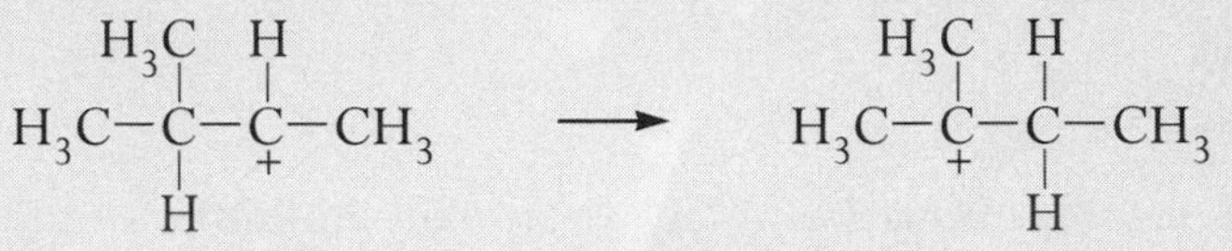

If there are no additions, eliminations, substitutions, or other changes to the functional groups, then reactant and product are isomers: the same set of atoms arranged in two different ways.

EXAMPLE: **Redox**—where one of the carbons undergoes a change in oxidation number (Chapter 3). In, say, the chromium-assisted oxidation of an alcohol to an aldehyde,

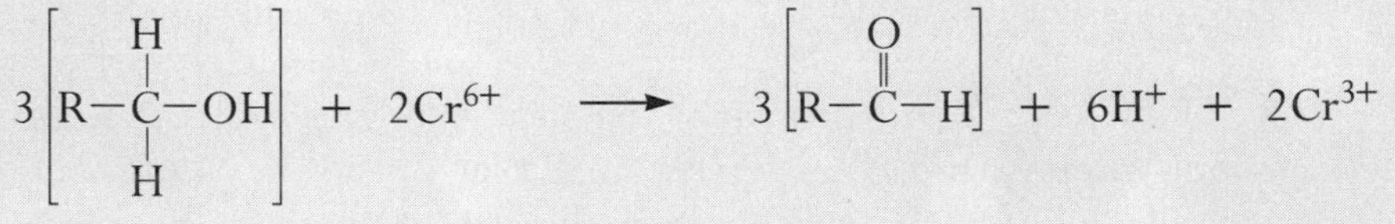

$$3\begin{bmatrix} & H & \\ & | & \\ R-&C&-OH \\ & | & \\ & H & \end{bmatrix} + 2Cr^{6+} \longrightarrow 3\begin{bmatrix} & O & \\ & \| & \\ R-&C&-H \end{bmatrix} + 6H^+ + 2Cr^{3+}$$

    alcohol   aldehyde

the electrons flow from RCH_2OH to Cr^{6+}, accompanied by the loss of $2H^+$. The alcohol RCH_2OH, which loses the electrons, is oxidized. The oxidizing agent Cr^{6+}, which gains them, is reduced. Its oxidation number goes down.

SAMPLE PROBLEMS

EXAMPLE 8-1. Structural Isomers

PROBLEM: How many structural isomers are possible for the molecule C_6H_{14}? Draw them.

SOLUTION: Recognize here an *alkane*, C_6H_{14}, a hydrocarbon that conforms to the generic formula C_nH_{2n+2}. The parent molecule is therefore hexane, a chain of six single-bonded carbons:

$$CH_3-CH_2-CH_2-CH_2-CH_2-CH_3$$

$$1 \quad 2 \quad 3 \quad 4 \quad 5 \quad 6$$

We start by breaking the molecule into a five-carbon chain (pentane) and a CH_2 fragment,

$$CH_3-CH_2-CH_2-CH_2-CH_3 \qquad CH_2$$

$$1 \quad 2 \quad 3 \quad 4 \quad 5$$

and then reassembling the pieces however we can.

There are, at first, only two choices, because the severed carbon can be reattached at just two distinct points on the pentane chain: carbon 2 or carbon 3, no more. Not carbon 1, because to do so would simply regenerate normal hexane with its unbranched chain of six carbons. Not carbon 5, because carbon 5 is exactly the same as carbon 1 (1 becomes 5 and 5 becomes 1 when we turn our heads). Likewise not carbon 4, because 4 is equivalent to 2. Hence the only two possibilities for the five-carbon chain are

2-methylpentane	3-methylpentane

$$\begin{array}{c} CH_3 \\ | \\ CH_3-C-CH_2-CH_2-CH_3 \\ | \\ H \end{array} \qquad \begin{array}{c} CH_3 \\ | \\ CH_3-CH_2-C-CH_2-CH_3 \\ | \\ H \end{array}$$

$$1 \quad 2 \quad 3 \quad 4 \quad 5 \qquad\qquad 1 \quad 2 \quad 3 \quad 4 \quad 5$$

and, using a systematic notation (see Appendix A), we shall call these isomers *2-methylpentane* and *3-methylpentane*. In each variation, *one* methyl group (CH_3) is substituted for *one* hydrogen at *one* numbered position on a pentane chain.

Repeating the procedure now for a *four*-carbon (butane) chain with *two* methyl groups, we find two new isomers:

2,3-dimethylbutane	2,2-dimethylbutane

$$\begin{array}{c} CH_3 \quad H \\ | \qquad | \\ CH_3-C-\!\!-C-CH_3 \\ | \qquad | \\ H \quad\; CH_3 \end{array} \qquad \begin{array}{c} CH_3 \\ | \\ CH_3-C-CH_2-CH_3 \\ | \\ CH_3 \end{array}$$

$$1 \quad 2 \quad 3 \quad 4 \qquad\qquad 1 \quad 2 \quad 3 \quad 4$$

And no more. Unique variations are no longer possible either for the four-carbon chain or anything smaller.

There are five structural isomers in all: normal hexane (the unbranched chain), two methyl-substituted forms of pentane, two dimethyl-substituted forms of butane.

EXAMPLE 8-2. Substituted Benzenes

PROBLEM: In the molecule dichlorobenzene, chlorine replaces hydrogen at two positions on the six-membered ring. How many isomers are possible?

SOLUTION: Three. After we attach one chlorine to carbon 1, the second chlorine finds only three unique positions on the ring: at carbon 2, carbon 3, or carbon 4.

ortho   meta   para

Position 6 is equivalent to position 2; position 5 is equivalent to position 3.

The three isomers are called *ortho* (1,2-dichlorobenzene), *meta* (1,3-dichlorobenzene), and *para* (1,4-dichlorobenzene).

EXAMPLE 8-3. Geometric Isomers: cis and trans

PROBLEM: Draw two geometric isomers for the molecule 2-butene. The empirical formula is C_4H_8.

SOLUTION: The name *2-butene* signals that we have a four-carbon structure with a double bond between carbons 2 and 3. Translation: (1) The stem *but* means *butane* (four carbons). (2) The suffix *ene* (alkene) means a double bond. (3) The numerical prefix *2* tells us that the double bond starts at carbon 2.

Adding the requisite hydrogens, we discover four groups (CH_3, H, H, CH_3) attached to the four available positions of a C=C bond:

$$CH_3-CH=CH-CH_3$$

1   2   3   4

CH_3 and H are connected to carbon 2; CH_3 and H are connected to carbon 3.

There are two geometric possibilities. The two methyls will be found either on the same side of the double bond, as in *cis*-2-butene:

$$\begin{array}{ccc} H & & H \\ \diagdown & & \diagup \\ & C{=}C & \\ \diagup & & \diagdown \\ CH_3 & & CH_3 \end{array} \qquad \text{cis}$$

Or on opposite sides, as in *trans*-2-butene:

$$\begin{array}{ccc} H & & CH_3 \\ \diagdown & & \diagup \\ & C{=}C & \\ \diagup & & \diagdown \\ CH_3 & & H \end{array} \qquad \text{trans}$$

EXAMPLE 8-4. Stereogenic Centers

PROBLEM: The amino acid *threonine*

$$\begin{array}{c} \quad\; H \quad O \\ \quad\; | \quad\;\; \| \\ H_2N{-}C{-}C{-}OH \\ \quad\; | \\ H_3C{-}C{-}OH \\ \quad\quad\; | \\ \quad\quad\; H \end{array}$$

contains two stereogenic carbons. Identify them.

SOLUTION: To be stereogenic, an atom must be able to assume two spatially distinct configurations about its center. A carbon singly bonded to four *different* substituents will meet the test.

Concerning the amino acids, recall that the following four groups are always attached to a tetrahedral carbon at the molecule's center: (1) an amino group, NH_2; (2) a carboxylic acid group, COOH; (3) a hydrogen atom, H; and (4) a side chain, R. With the sole exception of glycine (where R = H), all four groups are indeed different, and hence the central carbon is stereogenic. This site, marked below with an asterisk, is called the α carbon (to identify it as the position adjacent to the acid group):

$$\begin{array}{c} \quad\; H \quad O \\ \quad\; | \quad\;\; \| \\ H_2N{-}C^*{-}C{-}OH \\ \quad\; | \\ \quad\; R \end{array} \qquad\qquad R \;=\; \begin{array}{c} | \\ H_3C{-}C{-}OH \\ | \\ H \end{array}$$

Of the three carbons remaining, we eliminate the carboxyl $-\overset{\overset{\displaystyle O}{\|}}{C}-OH$ (because of the C=O double bond) and the methyl $-CH_3$ on the side chain (because of the three hydrogens). The carbon at the center of the side chain, however, does make bonds to four different groups: (1) a methyl group, CH_3; (2) a hydroxyl group, OH; (3) a hydrogen atom, H; and (4) the rest of the molecule. Marked again with an asterisk, this carbon furnishes threonine with a second stereogenic site:

$$H_2N-\overset{\overset{\displaystyle H}{|}}{\underset{\underset{\displaystyle H_3C-\overset{\overset{\displaystyle }{|}}{\underset{\underset{\displaystyle H}{|}}{C^*}}-OH}{|}}{C^*}}-\overset{\overset{\displaystyle O}{\|}}{C}-OH$$

EXAMPLE 8-5. The Longest Molecule Begins with the First Linkage

PROBLEM: Write a structure for the tetrapeptide Thr–Ala–Ser–Gly.

SOLUTION: Nature uses the same connection over and over again to build polypeptides and proteins: a condensation reaction between the amino group (NH_2) of one amino acid and the carboxylic acid group (COOH) of another:

$$H_2N-\overset{\overset{\displaystyle R}{|}}{\underset{\underset{\displaystyle H}{|}}{C}}-\overset{\overset{\displaystyle O}{\|}}{C}-OH \;+\; H-\overset{\overset{\displaystyle R'}{|}}{\underset{\underset{\displaystyle H}{|}}{\underset{H}{N}}}-\overset{\overset{\displaystyle O}{\|}}{\underset{\underset{\displaystyle H}{|}}{C}}-OH \longrightarrow$$

$$H_2N-\overset{\overset{\displaystyle R}{|}}{\underset{\underset{\displaystyle H}{|}}{C}}-\overset{\overset{\displaystyle O}{\|}}{C}-\overset{}{\underset{\underset{\displaystyle H}{|}}{N}}-\overset{\overset{\displaystyle R'}{|}}{\underset{\underset{\displaystyle H}{|}}{C}}-\overset{\overset{\displaystyle O}{\|}}{C}-OH \;+\; H_2O$$

Formation of the peptide CO(NH) linkage, an example of an **amide** group, proceeds with the loss of H_2O from COOH and H_2N.

The tetrapeptide thus becomes

$$H_2N-\overset{\overset{\displaystyle R_{Thr}}{|}}{\underset{\underset{\displaystyle H}{|}}{C}}-\overset{\overset{\displaystyle O}{\|}}{C}-\overset{}{\underset{\underset{\displaystyle H}{|}}{N}}-\overset{\overset{\displaystyle R_{Ala}}{|}}{\underset{\underset{\displaystyle H}{|}}{C}}-\overset{\overset{\displaystyle O}{\|}}{C}-\overset{}{\underset{\underset{\displaystyle H}{|}}{N}}-\overset{\overset{\displaystyle R_{Ser}}{|}}{\underset{\underset{\displaystyle H}{|}}{C}}-\overset{\overset{\displaystyle O}{\|}}{C}-\overset{}{\underset{\underset{\displaystyle H}{|}}{N}}-\overset{\overset{\displaystyle R_{Gly}}{|}}{\underset{\underset{\displaystyle H}{|}}{C}}-\overset{\overset{\displaystyle O}{\|}}{C}-OH$$

with the following side chains:

$$\text{Threonine:} \quad R_{Thr} = \quad H_3C-\overset{\overset{\displaystyle H}{|}}{\underset{|}{C}}-OH$$

$$\text{Alanine:} \quad R_{Ala} = \quad -CH_3$$

$$\text{Serine:} \quad R_{Ser} = \quad H-\overset{\overset{\displaystyle OH}{|}}{\underset{|}{C}}-H$$

$$\text{Glycine:} \quad R_{Gly} = \quad -H$$

Structures for the side chains are taken from Table 8-2, available in the first part of the chapter.

EXAMPLE 8-6. More Condensation

PROBLEM: The molecule represented below is an ester, formed by the reaction of an acid and alcohol:

$$CH_3-\overset{\overset{\displaystyle O}{\|}}{C}-O-CH_2CH_3$$

ethyl acetate

What acid? What alcohol?

SOLUTION: An ester linkage develops when H—OH is lost between a carboxylic acid (R—COOH) and an alcohol (R′—OH). From the general reaction

$$R-\overset{\overset{\displaystyle O}{\|}}{C}-OH \ + \ HO-R' \ \longrightarrow \ R-\overset{\overset{\displaystyle O}{\|}}{C}-O-R' \ + \ H_2O$$

we deduce that R = CH_3 (for the acid) and R′ = CH_2CH_3 (for the alcohol). The primary reactants therefore are CH_3COOH and CH_3CH_2OH, acetic acid and ethyl alcohol (ethanol):

$$CH_3-\overset{\overset{\displaystyle O}{\|}}{C}-OH \ + \ HO-CH_2CH_3 \ \longrightarrow$$

acetic acid   ethyl alcohol

$$CH_3-\overset{\overset{\displaystyle O}{\|}}{C}-O-CH_2CH_3 \ + \ H_2O$$

ethyl acetate   water

Acetic acid is the principal ingredient in vinegar; ethyl alcohol is found in many beverages.

EXAMPLE 8-7. Organic Acids

PROBLEM: Which carboxylic acid is more likely to release a proton—CH_3COOH (acetic acid) or CF_3COOH (trifluoroacetic acid)?

SOLUTION: Trifluoroacetic acid is the stronger acid. Far more electronegative than hydrogen, the fluorine atoms effectively withdraw electrons from the carboxylate anion after H^+ leaves:

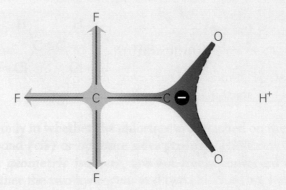

Better dispersal of the negative charge gives the trifluoroacetate a lower energy than the hydrogen-only acetate ion (CH_3COO^-), allowing CF_3COOH more freedom to shed its proton.

We shall take up the question of acids and bases again in Chapter 16.

EXAMPLE 8-8. Recognizing Reactions

PROBLEM: Classify each of the following reactions as either addition, elimination, condensation, substitution, rearrangement, or reduction–oxidation.

(a) *Hydrolysis of an acyl chloride: reaction with water*

$$CH_3\overset{\overset{\displaystyle O}{\|}}{C}-Cl \ + \ H_2O \ \longrightarrow \ CH_3\overset{\overset{\displaystyle O}{\|}}{C}-OH \ + \ HCl$$

SOLUTION: Nucleophilic substitution. OH^- replaces Cl^- at the partially positive end of the carbonyl bond. Note that the reactant CH_3COCl (acetyl chloride) is an example of an *acyl halide* (or *acid halide*), a compound with the structure

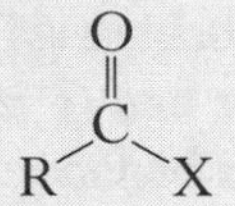

$$
\underset{R}{\overset{\displaystyle O}{\underset{|}{\overset{||}{C}}}}\!\!-\!X
\qquad X = F,\ Cl,\ Br,\ I
$$

The product of substitution, CH_3COOH, is acetic acid.

(b) *Formation of an anhydride from two acids*

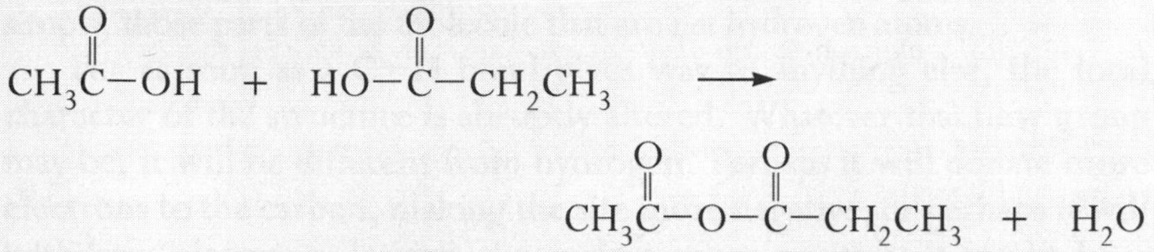

$$
CH_3\overset{\displaystyle O}{\overset{||}{C}}-OH \; + \; HO-\overset{\displaystyle O}{\overset{||}{C}}-CH_2CH_3 \;\longrightarrow
$$

$$
CH_3\overset{\displaystyle O}{\overset{||}{C}}-O-\overset{\displaystyle O}{\overset{||}{C}}-CH_2CH_3 \; + \; H_2O
$$

SOLUTION: A joining of two molecules, sealed by the loss of water: a condensation reaction similar to ester formation.

(c) *Hydrohalogenation*

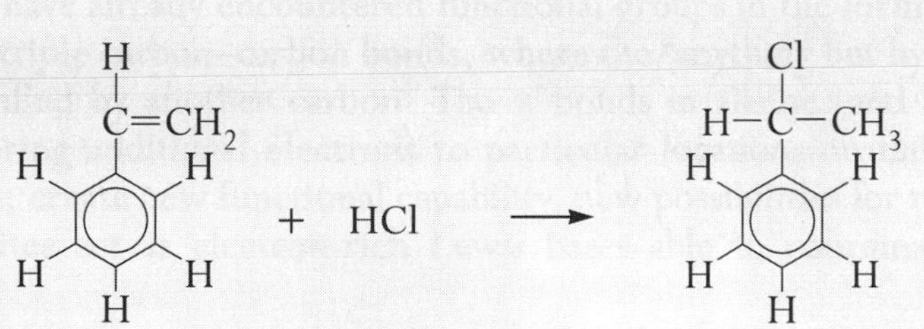

$$
\text{(structure)} \; + \; HCl \;\longrightarrow\; \text{(structure)}
$$

SOLUTION: An addition of H^+ and Cl^- across the C=C double bond. Each of the two highlighted carbons, formerly unsaturated, is now *saturated* with four single bonds apiece.

(d) *Dehydrohalogenation*

$$
CH_3CHBrCH_3 \; + \; KOH \;\longrightarrow\; CH_3CH{=}CH_2 \; + \; KBr \; + \; H_2O
$$

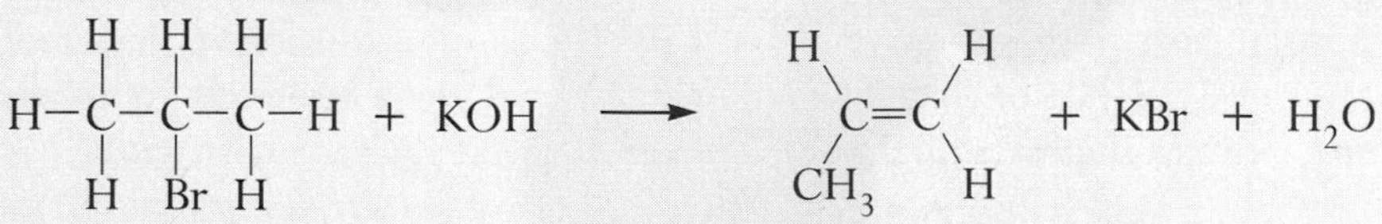

$$
\underset{\underset{H}{|}\;\underset{Br}{|}\;\underset{H}{|}}{H-C-C-C-H} \; + \; KOH \;\longrightarrow\; \overset{H}{\underset{CH_3}{C}}{=}\overset{H}{\underset{H}{C}} \; + \; KBr \; + \; H_2O
$$

SOLUTION: Elimination of HBr from an alkyl bromide, resulting in the formation of an alkene. See the process as opposite to the hydrohalogenation reaction in (c), where an alkyl halide is produced by addition across a double bond.

(e) *Combustion of a hydrocarbon*

$$CH_4 + 2O_2 \longrightarrow CO_2 + 2H_2O + heat$$

SOLUTION: This reaction we have already seen in Chapter 3: oxidation of carbon from C^{4-} in methane to C^{4+} in carbon dioxide; reduction of elemental oxygen to O^{2-} in water.

Methane is the principal component of natural gas.

(f) *Biological conversion of an alcohol into an aldehyde*

Called a *coenzyme*, NAD (nicotinamide adenine dinucleotide) participates in many biological reactions as an assistant to an enyzme. It helps the enzyme function properly as a catalyst.

SOLUTION: Even without specifying the structure of NAD^+, we recognize that reduction and oxidation are underway. Losing two hydrogens along with two electrons, ethanol (CH_3CH_2OH) is oxidized during the conversion into acetaldehyde (CH_3CHO):

ethanol   acetaldehyde

NAD^+ is reduced.

(g) *Conversion of 1-butene into 2-butene*

1-butene   *cis*-2-butene

SOLUTION: A rearrangement, specifically a transformation between two structural isomers. In 1-butene, the double bond falls between the first and second carbons; in 2-butene, between the second and third. Each molecule otherwise contains four carbon atoms and eight hydrogen atoms.

QUESTION: From Example 8-3 above, we know that 2-butene exists in both cis and trans forms. Does 1-butene similarly exist as two geometric isomers?

ANSWER: No. Since the same group (H) already occupies three positions around the C=C bond, no new structures can be created by rearrangement of the four substituents in 1-butene.

Suppose, for example, we were to move the ethyl group (CH_2CH_3) from the lower right corner to the upper left:

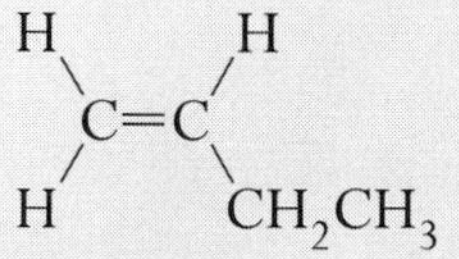

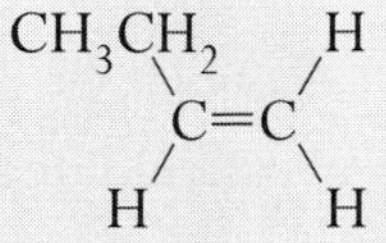

$$
\begin{array}{ccc}
H & & H \\
\ \ \diagdown & & \diagup \\
& C=C & \\
\diagup & & \diagdown \\
H & & CH_2CH_3
\end{array}
\qquad\qquad
\begin{array}{ccc}
CH_3CH_2 & & H \\
\ \ \ \ \ \diagdown & & \diagup \\
& C=C & \\
\diagup & & \diagdown \\
H & & H
\end{array}
$$

Except for a meaningless rotation in the plane, the two representations are the same.

EXERCISES

1. Start by adding the missing hydrogen atoms to a hydrocarbon skeleton, as in the following example:

   CARBON FRAMEWORK   FULL MOLECULE

   C—C=C

   $$\underset{\overset{\displaystyle |}{H}}{\overset{\displaystyle H}{H-C}}-\underset{}{\overset{\displaystyle H}{C}}=\underset{H}{\overset{\displaystyle H}{C}}$$

   Do so for each of the partial structures given below, and classify each molecule as saturated or unsaturated:

   (a) C—C≡C—C

   (b) C—C—C—C

   (c) C—C=C—C

   What are the molecular formulas?

2. Predict the bond angles and hybridization around each carbon in the hydrocarbons (a), (b), and (c) treated above.

3. Add the missing hydrogens, and predict the bond angles and hybridization around each carbon:

   $$\underset{\overset{\displaystyle |}{\overset{\displaystyle C}{\underset{|}{C}}}}{C}-C-C-\overset{\overset{\displaystyle C}{|}}{C}-C=\overset{\overset{\displaystyle C}{|}}{C}-C-C≡C$$

   What is the molecular formula?

4. Although both benzene (C_6H_6) and cyclohexane (C_6H_{12}) are built from closed rings of six carbons, the two molecules have different structures and properties. To see why, add the hydrogens and describe the hybridization around each carbon in C_6H_6 and C_6H_{12}:

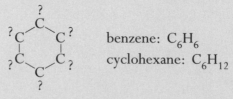

   benzene: C_6H_6

   cyclohexane: C_6H_{12}

Compare: Benzene is a planar molecule with a set of delocalized π orbitals perpendicular to the ring. Is cyclohexane planar? Does cyclohexane support a delocalized π system?

5. Which molecules in the list below contain π bonds? In which of them are the electrons delocalized over three or more carbon sites?

(a) CH_2CHCH_3

(b) CH_2CH_2

(c) $CH_3CH_2CH_2CH_3$

(d) $CH_2CHCH_2CH_3$

(e) $CH_2CHCHCHCHCH_2$

6. Again: Which of these molecules contain π bonds? In which of them are the electrons delocalized over three or more carbon sites?

(a) CHCH

(b) $CHCCH_2CH_2CH_3$

(c) CHCCCH

(d) $CHCCH_2CCH$

(e) C_3H_8

7. Heptane (C_7H_{16}) has nine structural isomers. Draw them.

8. Calculate the weight percent of carbon and hydrogen in each of heptane's nine structural isomers.

9. Octane (C_8H_{18}) has 18 structural isomers. Draw as many as you can.

10. Calculate the weight percent of carbon and hydrogen in each of octane's 18 structural isomers.

11. Rearrange this branched alkane into a straight chain:

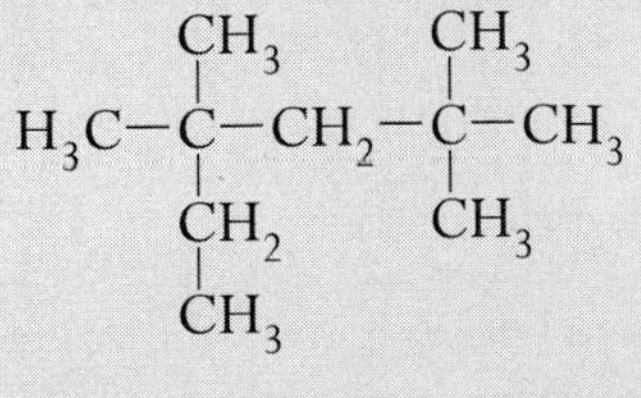

$$H_3C-\overset{\overset{\displaystyle CH_3}{|}}{\underset{\underset{\displaystyle CH_3}{|}}{C}}-CH_2-\overset{\overset{\displaystyle CH_3}{|}}{\underset{\underset{\displaystyle CH_3}{|}}{C}}-CH_3$$

12. Identify the stereogenic carbons in the molecule drawn below:

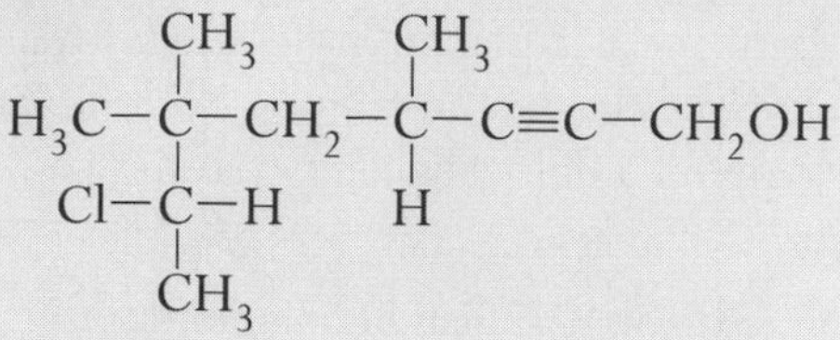

$$\begin{array}{ccc}
& \underset{|}{CH_3} & \underset{|}{CH_3} \\
H_3C-\underset{\underset{|}{\overset{|}{Cl-C-H}}}{C}-CH_2-\underset{\underset{|}{H}}{C}-C\equiv C-CH_2OH \\
& \underset{|}{CH_3} &
\end{array}$$

13. Which molecules contain a stereogenic center?

(a)

$CH_3-CH_2-CH_2-CH_2-CH_3$

(b)
$$\begin{array}{c}
\underset{|}{CH_3} \\
CH_3-CH-CH_2-CH_3
\end{array}$$

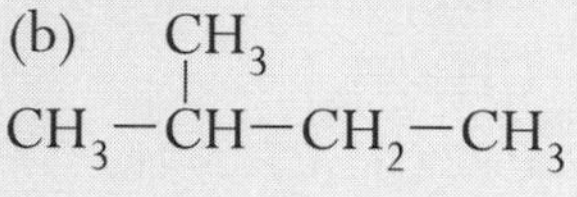

(c)
$$\begin{array}{c}
\underset{|}{CH_3} \\
CH_3-\underset{\underset{|}{CH_3}}{C}-CH_3
\end{array}$$

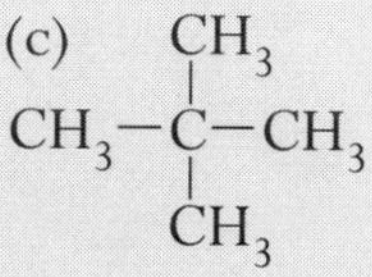

(d)

$CH_2=CH-CHCl-CH_2-CH_3$

(e)
$$\begin{array}{c}
\underset{|}{CH_2OH} \\
CH_3-CH-CH_2-CH_3
\end{array}$$

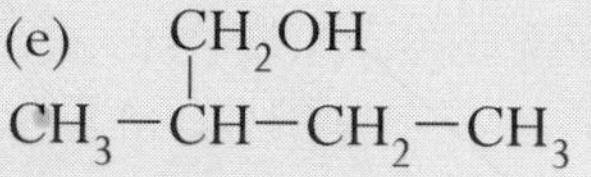

(f)
$$\begin{array}{c}
\underset{|}{Cl} \\
CH_3-\underset{\underset{|}{Br}}{C}-CH_3
\end{array}$$

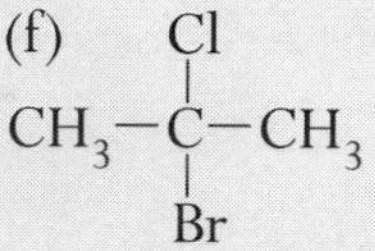

14. Amino acids conform to the general pattern

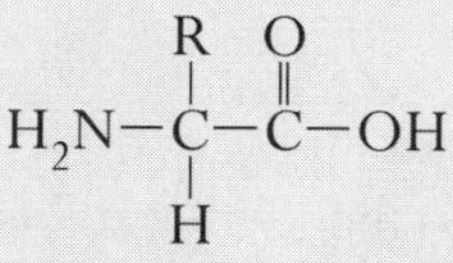

$$\begin{array}{c}
R \quad O \\
\underset{\underset{|}{H}}{\underset{|}{H_2N-C-C-OH}}
\end{array}$$

where R is a side chain specific to each molecule. Indicate, for each side chain listed, whether the molecule is optically active; and, if appropriate, draw the two enantiomers:

(a) glycine, $R = H$

(b) alanine, $R = CH_3$

(c) serine, $R = CH_2OH$

(d) cysteine, $R = CH_2SH$

15. Consider the two geometric isomers of 1,2-dibromoethene:

$$\begin{array}{cc}
\underset{Br}{\overset{H}{\diagdown}} C=C \underset{Br}{\overset{H}{\diagup}} & \underset{H}{\overset{Br}{\diagdown}} C=C \underset{Br}{\overset{H}{\diagup}}
\end{array}$$

Which is the cis form? Which is the trans form? Which of the two has a dipole moment?

16. Do these two structural formulas represent two distinct isomers of dibromoethene?

$$\underset{Br}{\overset{H}{\diagdown}}C=C\underset{Br}{\overset{H}{\diagup}} \qquad \underset{H}{\overset{Br}{\diagdown}}C=C\underset{H}{\overset{Br}{\diagup}}$$

How about these?

$$\underset{Br}{\overset{Br}{\diagdown}}C=C\underset{H}{\overset{H}{\diagup}} \qquad \underset{H}{\overset{Br}{\diagdown}}C=C\underset{H}{\overset{Br}{\diagup}}$$

If so, what *kind* of isomers are they . . . optical, structural, conformational, geometric? Explain the differences.

17. Draw cis and trans geometric isomers for 2-pentene:

$$CH_3-CH=CH-CH_2-CH_3$$

18. Shown below is a representation of 6-methyl-4-propyl-*trans*-2-nonene:

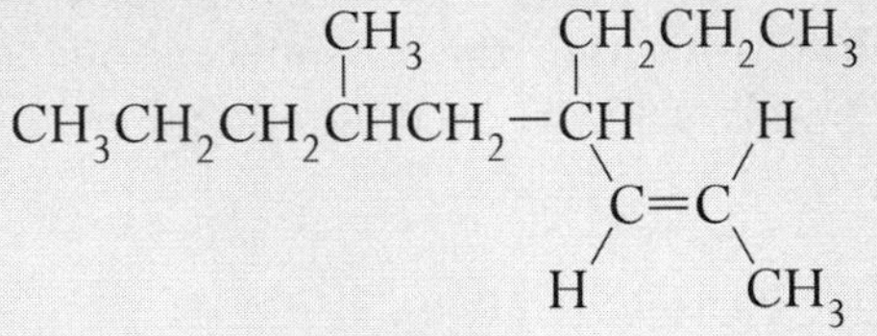

$$CH_3CH_2CH_2\overset{\overset{\displaystyle CH_3}{|}}{CH}CH_2-\overset{\overset{\displaystyle CH_2CH_2CH_3}{|}}{CH}\diagdown\underset{\underset{H}{\diagup}C=C\underset{CH_3}{\overset{H}{\diagdown}}}{}$$

Draw the cis isomer.

19. Which of the following molecules may exist as *conformational isomers*?

    (a) $H_2C=CH_2$
    (b) $H_2C=CH-CH_3$
    (c) $H_2C=CH-CH_2-CH_3$
    (d) $HC\equiv CH$
    (e) $HC\equiv C-CH_2-CH_3$

20. Redraw the structural formula shown below, making every connection and every bond explicit:

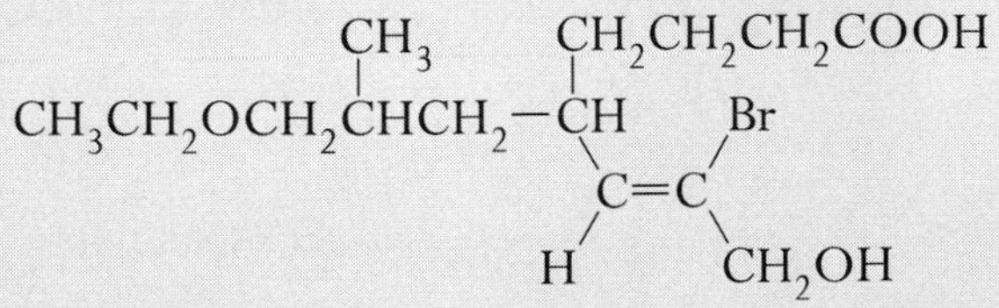

$$CH_3CH_2OCH_2\overset{\overset{\displaystyle CH_3}{|}}{CH}CH_2-\overset{\overset{\displaystyle CH_2CH_2CH_2COOH}{|}}{CH}\diagdown\underset{\underset{H}{\diagup}C=C\underset{CH_2OH}{\overset{Br}{\diagdown}}}{}$$

Recognize, for example, that atoms in the group CH_2OH are connected in sequence as

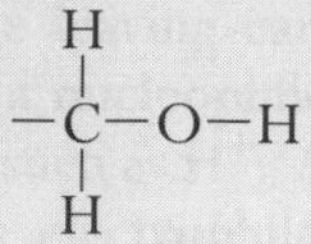

$$-\overset{\displaystyle H}{\underset{\displaystyle H}{\overset{|}{\underset{|}{C}}}}-O-H$$

After that, pick out the functional groups on the molecule and identify each group by name.

21. Same idea. Pick out the functional groups on the molecule displayed below:

$$CH_3COCH_2\overset{\displaystyle \overset{CH_3}{|}}{CHCH_2}-\overset{\displaystyle \overset{CH_2CH_2CH_2COOCH_3}{|}}{\underset{\displaystyle \underset{|}{C}=\underset{|}{C}}{CH}}$$

$$\underset{H \qquad CHO}{C=C}$$

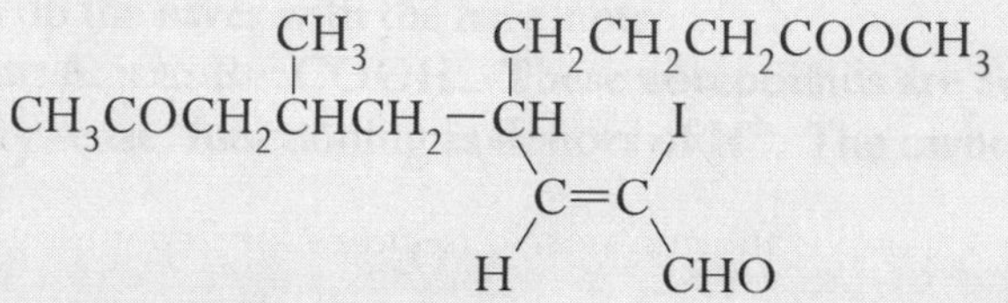

Redraw the formula to show the connections explicitly, and identify each group by name.

22. Once more. Pick out the functional groups on the molecule displayed below, rewriting the formula as needed to show the sequence of bonds. Identify each functional group by name:

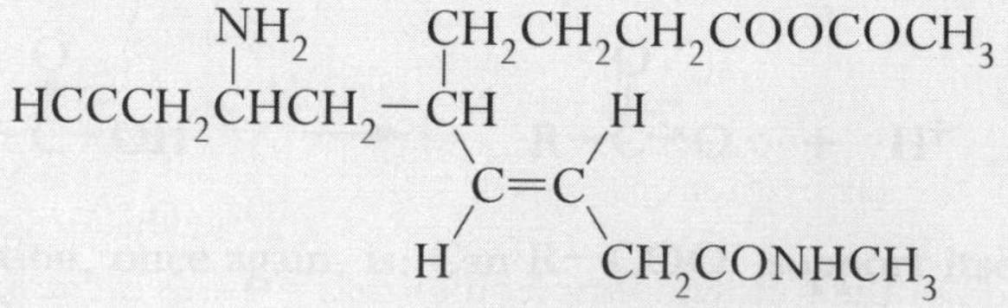

$$HCCCH_2\overset{\displaystyle \overset{NH_2}{|}}{CHCH_2}-\overset{\displaystyle \overset{CH_2CH_2CH_2COOCOCH_3}{|}}{CH}$$

$$\underset{H \qquad CH_2CONHCH_3}{C=C}$$

23. Look carefully at these two structural formulas:

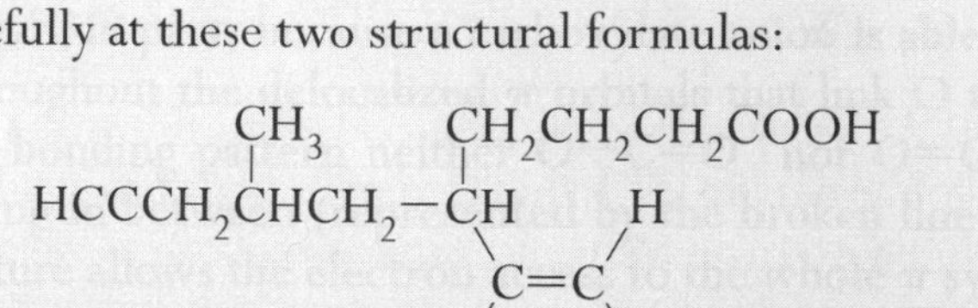

$$HCCCH_2\overset{\displaystyle \overset{CH_3}{|}}{CHCH_2}-\overset{\displaystyle \overset{CH_2CH_2CH_2COOH}{|}}{CH}$$

$$\underset{H \qquad CH_2CONHCH_3}{C=C}$$

Molecule I

$$HCCCH_2\overset{\displaystyle \overset{CH_3}{|}}{CHCH_2}-\overset{\displaystyle \overset{CH_2CH_2CH_2COOCOCH_3}{|}}{CH}$$

$$\underset{H \qquad CH_2CONHCH_3}{C=C}$$

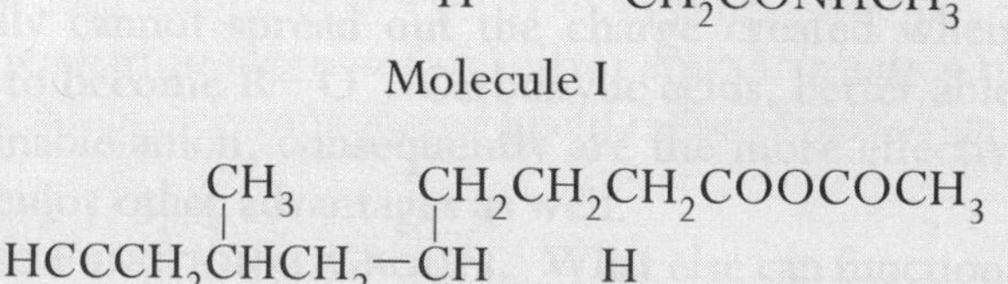

Molecule II

Complicated? Yes—but for all their complexity, Molecules I and II differ at just a single site. Find this one point of difference, and identify the groups involved.

24. Having spotted the difference between Molecules I and II in the example above, identify now the Mystery Molecule X that will react with Molecule I to produce Molecule II:

    Molecule I + Mystery Molecule X $\longrightarrow$ Molecule II + Molecule Y

    What *kind* of reaction is it? What by-product, Molecule Y, accompanies the formation of Molecule II?

25. Similar. Look closely at the two structures, and then decide how to convert Molecule I into Molecule II. First the molecules:

$$\begin{array}{cc} \underset{\underset{\displaystyle H}{|}}{\overset{\overset{\displaystyle H}{|}}{H_2N-C}}\overset{\overset{\displaystyle O}{\|}}{-C}-OH & \underset{\underset{\displaystyle H}{|}}{\overset{\overset{\displaystyle H}{|}}{H_2N-C}}\overset{\overset{\displaystyle O}{\|}}{-C}-\underset{\underset{\displaystyle H}{|}}{N}-\underset{\underset{\displaystyle CH_2OH}{|}}{\overset{\overset{\displaystyle H}{|}}{C}}\overset{\overset{\displaystyle O}{\|}}{-C}-OH \\ \text{Molecule I} & \text{Molecule II} \end{array}$$

Then the reaction:

    Molecule I + Molecule X $\longrightarrow$ Molecule II + Molecule Y

What kind of reaction takes place? What by-product is generated?

26. Same thing. The molecules:

$$\underset{\underset{\displaystyle H}{|}}{\overset{\overset{\displaystyle H}{|}}{H_2N-C}}\overset{\overset{\displaystyle O}{\|}}{-C}-\underset{\underset{\displaystyle H}{|}}{N}-\underset{\underset{\displaystyle CH_2OH}{|}}{\overset{\overset{\displaystyle H}{|}}{C}}\overset{\overset{\displaystyle O}{\|}}{-C}-OH$$

Molecule I

$$\underset{\underset{\displaystyle H}{|}}{\overset{\overset{\displaystyle H}{|}}{H_2N-C}}\overset{\overset{\displaystyle O}{\|}}{-C}-\underset{\underset{\displaystyle H}{|}}{N}-\underset{\underset{\displaystyle CH_2OH}{|}}{\overset{\overset{\displaystyle H}{|}}{C}}\overset{\overset{\displaystyle O}{\|}}{-C}-\underset{\underset{\displaystyle H}{|}}{N}-\underset{\underset{\displaystyle H}{|}}{\overset{\overset{\displaystyle H}{|}}{C}}\overset{\overset{\displaystyle O}{\|}}{-C}-OH$$

Molecule II

The reaction:

    Molecule I + Molecule X $\longrightarrow$ Molecule II + Molecule Y

Identify Molecule X, Molecule Y, and the kind of reaction. What role do these processes play in biological chemistry?

27. Again. For each pair of Molecules I and II, determine the structures of Molecules X and Y in the reaction

    Molecule I + Molecule X ⟶ Molecule II + Molecule Y

    Here are the structures:

MOLECULE I	MOLECULE II
(a) $CH_3CH_2CHOHCOOH$	$CH_3CH_2CHOHCOOCH_2CH_2CCl_3$
(b) CH_3COOH	$CH_3COOCOCH_2CH_3$
(c) $CH_3CH_2CH_2CH_2OH$	$CH_3CH_2CH_2CH_2OOCCH_3$

    What are the reactions?

28. Consider now a *substitution* reaction of the kind

    Molecule I + X ⟶ Molecule II + Y

    Given the various combinations of I, X, II, and Y, fill in the blanks in the table below:

I	X	II	Y
(a) CH_3Cl	H_2O	__	__
(b) $CH_3COCH_2CH_2CH_2Br$	F^-	__	__
(c) CH_3Br	CH_3CH_2OH	__	HBr
(d) CH_3COCl	__	CH_3COOCH_3	HCl

29. Add the maximum amount of molecular hydrogen to each unsaturated molecule below and write the structural formula of the product:

    (a) CH_2CH_2

    (b) $CHCH$

    (c) $CH_3CH_2CH_2CHCHCH_3$

    (d) $CH_2CHCHCH_2$

    (e) CH_3CCH

    How many moles of H_2 will react completely with one mole of each hydrocarbon?

30. For each of the first five alkanes (methane, ethane, propane, butane, pentane), write a balanced equation for its complete combustion to carbon dioxide and water:

    $$\underline{}\ C_nH_{2n+2} + \underline{}\ O_2 \longrightarrow \underline{}\ CO_2 + \underline{}\ H_2O$$

31. Classify each transformation as rearrangement, redox, substitution, addition, elimination, or condensation, using more than one category where appropriate:

    (a) $C_6H_5CH_2OH$ becomes C_6H_5CHO

    (b) $CH_3CH_2CH_2OH$ becomes $CH_3CHOHCH_3$

    (c) $CH_3CH_2CH_2OH$ becomes $CH_3CH_2CH_2F$

    (d) $CH_3CH_2CH_2OH$ becomes $CH_3CH_2CH_2OCOCH_3$

    For clarity, redraw the formulas to make the key connections explicit.

9

States of Matter

9-1. The World Within
9-2. Traditional Classifications
9-3. Noncovalent Interactions
9-4. Order and Disorder in Gases, Liquids, and Solids
9-5. Symmetry: Crystals and Quasicrystals
9-6. The Solid State
  Network Crystals
  Ionic Crystals
  Molecular Crystals
  Metallic Crystals
9-7. Clusters
9-8. Liquid Crystals
9-9. Polymers
  Biopolymers
  Synthetic Polymers
  REVIEW AND GUIDE TO PROBLEMS
  EXERCISES

9-1. THE WORLD WITHIN

We live in what seems to be a very real world, a macroscopic world of large, touchable objects. Through our senses comes an unschooled perception of matter as smooth and continuous: a rush of wind, a raindrop, a frozen lake, a dog, a meadow. Only with great difficulty—only by looking very closely, by using instruments, by analyzing experiments—have we become aware of the agitated microscopic realm underlying our own. It is understandably a strange place, for it is a world of protons and electrons, a world of atoms, molecules, ions, and all the other particles unimaginably small. Yet these are the particles that ultimately make up the texture of everyday experience, and these are the particles that make up the immediate reality of physical science.

Without an appreciation of atoms and molecules, chemistry would be little more than magic. It would have remained where it was before the modern atomic hypothesis began to emerge in the 19th century, appearing mostly as a series of intriguing but incomprehensible phenomena. To each observation there would be attached a specific question: Why do these two clear liquids turn blue when mixed? Why should this solid fall out of that solution? Why do these gases explode when subjected to a spark, yet not those? With a microscopic explanation, however, these questions become not only fewer but also less specific. The general rules governing matter and its transformations start to appear, and previously unrelated events become part of a general pattern.

We have begun cautiously to negotiate the unfamiliar terrain of the microscopic world, seeking first to understand chemistry's fundamental particles as individuals. From previous chapters we know something about the structure of atoms and how atoms combine into molecules; we have an idea of chemical reactivity and a dawning awareness of organic and biochemical processes. But now this inquiry must go beyond the level of the mere individual—beyond one atom, or one or two molecules, or one-on-one collisions in chemical reactions. For although the atoms and molecules do act as individuals, they do so within a larger "society" and are subject always to constraints. The collective context within which the microscopic particles interact is provided by the various *states* of matter—solid, liquid, gas, and a number of less precisely defined aggregates—and it is here that the microscopic becomes macroscopic. It is here that one mole of unseen H_2O molecules freezes into a cube of ice; it is here that countless N_2 and O_2 molecules turn unmistakably into a gust of wind.

To bring the macroscopic into harmony with the microscopic will be our single-minded pursuit, and now we shall take some tentative steps forward by meeting the states of matter. It can be no more than an introduction, aimed mostly at showing how atoms and molecules cooperate to form the **condensed phases** (solids and liquids), but this overview will offer at least a glimpse beneath the surface. After a brief pause to absorb some essential concepts and language (Section 9-2), let us then see how the invisible particles of the microscopic world come together to create the macroscopic world of ordinary matter.

9-2. TRADITIONAL CLASSIFICATIONS

Matter, large and small, is physically apprehensible, material, concrete *stuff*. It takes up space. It has mass. It is tangible.

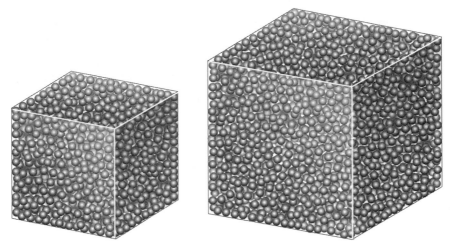

FIGURE 9-1. Density: mass per unit volume, suggested here by the packing of small spheres into a cube. Both cubes have the same density, even though one has a larger volume. Each box contains the same number of spheres in a given space.

The most elementary characterization of bulk matter is provided by its **density**, defined as mass per unit volume. Liquid water, for example, has a density of approximately 1 g/mL, meaning simply that the molecules in 1 gram of H_2O occupy 1 cubic centimeter as a liquid. Density (Figure 9-1) is thus understood to be an *intensive* property, independent of the amount of material. Whereas 100 grams of water fill a volume of 100 mL, 10 grams of water take up only 10 mL. The density is the same.

A sample of matter with uniform properties throughout is said to be **homogeneous**, consisting of a single **phase**. There is no variation in such parameters as density, color, and melting or boiling temperature from point to point within a phase. Pure oxygen gas exists as a single phase, as does a well-stirred solution of sucrose (table sugar) or sodium chloride (table salt) in water. But if ice cubes are added to either solution, as in one of the panels of Figure 9-2, then there is introduced a second phase: the *solid* form of water, with physical properties different from the liquid. The presence of more than one physically distinct phase makes the material **heterogeneous**, and the various properties are uniform only within the individual phases.

If a homogeneous, single-phase material cannot be separated into different components, then it is called a **substance**. Elements in their unadulterated forms exist as pure substances, having only atoms with the same atomic number. Macroscopic amounts of *compounds*, too, can be pure substances if not mixed with molecules of other types. When, however, a separation can be accomplished without breaking any chemical bonds (for instance, by selectively boiling away liquids with different boiling points), then the material is a **mixture**.

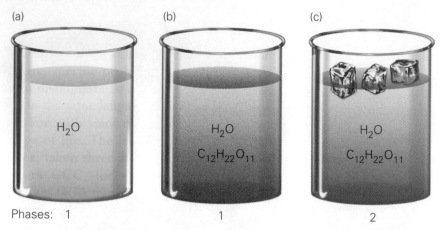

FIGURE 9-2. Phases. (a) Pure water: one phase, homogeneous. (b) A solution of sucrose in water, uniform throughout: one phase, homogeneous. (c) A solution of sucrose in water plus three blocks of ice: two phases (solution and solid), heterogeneous. The three ice cubes count for just one phase.

Crude oil, to take an especially complex example, is a mixture of very many organic hydrocarbons. Air is a mixture of nitrogen, oxygen, carbon dioxide, argon, and other gases, each of which would exist as a pure substance by itself. Sugar in water is a mixture as well, formed by the combination of one pure substance (sucrose) with another (water) to create a homogeneous phase. Here the water can be removed either by rapid boiling or slow evaporation to leave behind a solid phase of sucrose. The residual sugar is a substance, a chemical compound with molecular formula $C_{12}H_{22}O_{11}$, which can be broken down further only by disruption of the covalent bonds within the molecules. Sucrose recovered from the solution therefore is not a *physically* separable mixture of carbon, hydrogen, and oxygen atoms; it is a macroscopic quantity of *chemically* bonded $C_{12}H_{22}O_{11}$ molecules sitting in a heap at the bottom of a cup.

Note, incidentally, that sugar ground up with a solid material (say sodium chloride) also may constitute a mixture, but not an intimate one since the mixing does not occur microscopically. Instead there will be relatively large granules of each solid compound, interspersed to some extent and blended into a coarse-grained mix without any direct particle–particle interactions between sugar and salt. Such a material is best described as a heterogeneous mixture.

For intimate, microscopically *homogeneous* mixtures like sugar water or salt water or aqueous sodium hydroxide, we adopt the term **solution** from Chapter 3. Realize, though, that solutions are not restricted to solid **solutes** dispersed in liquid **solvents**, for there are also solutions consisting entirely of gases, or gases in liquids, or even intimately mixed solids in solids. All that matters is how thoroughly the particles are intermingled at the microscopic level.

Solutions are characterized by the **concentration** of the solute in the solvent, usually expressed as a **molarity** (mol/L, or mol L^{-1}, or just M) and symbolized by square brackets. Hence a 1.00 M aqueous solution of sodium chloride contains 58.4 g NaCl mixed in sufficient water to yield a volume of 1.00 L; we say that [NaCl] = 1.00 M. Solute concentrations are also expressed as **mole fractions** (X), defined for each component as the number of moles of substance divided by the total number of moles in the mixture. In a gaseous solution containing, for example, 1.00 mole of oxygen and 2.00 moles of hydrogen, the mole fractions would be

$$X_{O_2} = \frac{1.00 \text{ mol}}{3.00 \text{ mol}} = 0.333$$

$$X_{H_2} = \frac{2.00 \text{ mol}}{3.00 \text{ mol}} = 0.667$$

This definition ensures that a complete set of mole fractions for any mixture always adds up to 1.

9-3. NONCOVALENT INTERACTIONS

Viewed from a human perspective, the ordinary matter of the earth is astonishingly varied in form and organization. Everything is different. Every rock or chemical substance or living thing appears to be unique.

Such diversity, marvelous in itself, seems even more remarkable when one realizes that all this variety comes from just two reusable parts: electrons and nuclei. For if we could microscopically disassemble a sample of matter (as if it were an automobile engine), then at some point we would have only the electrons and nuclei. Broken down piece by piece, all earthly objects prove to be rearrangements of these two electrical particles. Much of our science—physics, chemistry, biology, geology, medicine, and the rest—is an attempt to understand how electrons and nuclei organize themselves into larger structures, mostly along the lines laid down in Figure 9-3.

First there is the atom. An atom arises when some number of electrons, Z, congregate about a nucleus with Z protons. Outside the nucleus, it is entirely an electromagnetic affair. The negatively charged electrons are repelled by their own kind but attracted to the positively charged nucleus. Governed by the laws of quantum mechanics (the Schrödinger equation, the Heisenberg uncertainty principle, the Pauli

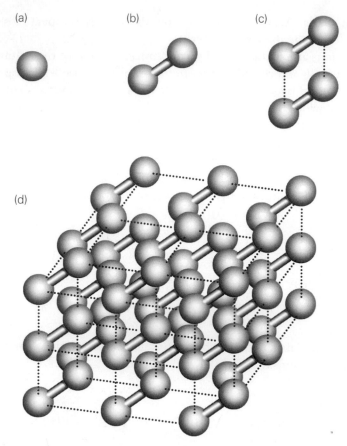

FIGURE 9-3. Hierarchy of matter. (a) An atom, built from electrons and a nucleus. (b) A covalent molecule, built from atoms. The bonded nuclei share their electrons intimately. (c) A small cluster, built from molecules. Linked weakly by a noncovalent interaction, each molecule retains its own electrons. (d) A phase: molecules aggregated in bulk, woven together by noncovalent interactions.

exclusion principle), the atom holds together by electromagnetic interactions. There is no other fundamental force at play.

A molecule develops in much the same way, except now there are two or more nuclei competing for the electrons. The nuclei attract the electrons; the electrons repel the electrons; the nuclei repel the nuclei. A molecule may differ markedly from an atom in details such as shape, size, and energy levels, but there remains this profound similarity. Atoms and molecules (ions too) are fundamentally electrical objects, deriving shape, volume, and chemical substance from the electrons around their nuclei. We say that atoms fashion themselves into molecules by means of "chemical" bonds, but really the attraction is just one peculiar example of the electromagnetic force.

And beyond atoms and molecules? Beyond atoms and molecules,

the rules remain the same: electromagnetic forces and quantum mechanics. Opposite charges attract; like charges repel. When atoms and molecules move up to a higher level of organization, they can do so only by using electromagnetic forces. No other mechanism is available.

Atom meets atom. Molecule meets molecule. Molecule meets atom. Atom meets ion. Ion meets ion. Ion meets molecule. So run the assorted possibilities for noncovalent interactions, but always it is the meeting of one electrical entity with another. Always it is the same kind of force, and always it is the same law of motion. What differs is just the specific way the force is applied, because here the species cannot commingle their electrons as intimately as in the formation of an atom or molecule. These objects already *are* atoms and molecules, and (provided there is no chemical reaction) they do not allow other electrons to mix into their structures. Instead they approach and interact as recognizably separate species, all the while maintaining their chemical integrity. Although the electron clouds suffer some distortion, the interacting species remain intact as atoms and molecules throughout. It is a friendship, not a marriage.

These meetings, in diverse manifestations, are the **noncovalent interactions** (or **nonbonded interactions**) that enable matter to organize on a larger scale. Often they are dubbed *physical* interactions to indicate that electrons are not exchanged (as in a so-called chemical bond), but such terminology should never obscure the fundamental similarity between covalent and noncovalent bonds. They are both electrical phenomena.

Already familiar to us are **ion–ion** interactions (Figure 9-4), which

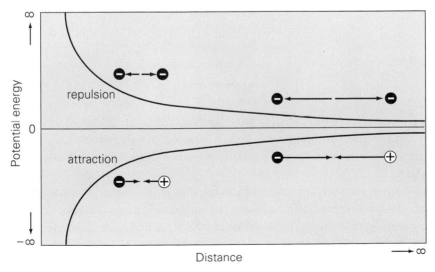

FIGURE 9-4. Coulomb interaction between two ions, strongest and longest-reaching of the noncovalent bonds. Opposite charges attract; like charges repel. The potential energy falls off as $1/r$.

give rise to the ionic bonds that sew together materials such as NaCl. Each ion behaves like a simple charged particle, out of which comes the usual combination of forces: anion attracts cation; anion repels anion; cation repels cation. See it as an implementation of Coulomb's law between particles with charges q_1 and q_2 (Chapter 1), so that the electric force varies as

$$F \propto \frac{q_1 q_2}{r^2}$$

for two ions separated by a distance r. The potential energy, with units of force times distance, then goes as $1/r$—inversely with the distance alone, not the square. As a result, energies of interaction (attractive or repulsive) are cut in half when r is doubled and reduced to a fourth when r is quadrupled. This falling away with distance is comparatively gradual, making ion–ion interactions the most far-reaching of the noncovalent couplings. They are the strongest as well, on a par with covalent bonds.

Next we have noncovalent interactions involving molecules with permanent dipole moments, sketched in Figure 9-5. Some molecules, recall, develop an imbalance of charge when electrons spend more time in certain regions than others. These *polar* molecules have partially negative and partially positive portions, which provide opportunity for further interactions. The two electrical poles present a *dipole moment* (μ), which we defined in Chapter 7 as

$$\mu = Qd$$

for equal and opposite charges, Q and $-Q$, separated by a small distance d:

$$Q \xmapsto{\;\;d\;\;} -Q$$

In an **ion–dipole** interaction, for example, the positive end of a molecular dipole moment might be pulled toward the negative charge of an anion. Or, equivalently, the partially negative end might be attracted to a cation.

Either way, the attraction of ion for dipole is reduced compared with an ion–ion interaction, because the dipole's charge (Figure 9-5a) is smeared over a finite region rather than packed into a single point. Ion–dipole energies are about 10 to 20 times less than ion–ion energies, and the ion–dipole interaction has a shorter range as well. Picture what happens: The closely spaced charges of the dipole—equal and opposite—start to seem like no charge at all the farther away one looks. Go

(a)

(b)

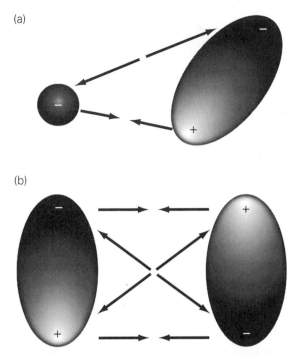

FIGURE 9-5. Ions and dipoles. (a) Ion–dipole interaction. One end of the dipole attracts the lone charge; the other end repels it. Weaker than an ion–ion coupling (Figure 9-4) but stronger than a dipole–dipole coupling, the potential energy of the ion–dipole interaction decreases as $1/r^2$. (b) Dipole–dipole interaction. Positive and negative ends of the dipoles align for lowest energy, proportional to $1/r^3$.

far enough away, and the opposite charges appear to merge into a single charge of zero: Q and $-Q$ together, one atop the other. The dipole shrinks to a neutral *point* like railroad tracks converging in the distance, and the electrostatic energy it imparts to an ion falls off as $1/r^2$. The drop in energy is fourfold for every doubling of the interval r between ion and dipole.

Dipoles also interact with other dipoles. An attractive **dipole–dipole** force (Figure 9-5b) develops whenever the negative portion of one dipole aligns with the positive portion of its partner. Even weaker than ion–dipole effects, dipole–dipole energies are typically 100 times less than ion–ion bonds. The range is shorter too, since now there are two dipoles to shrink in the distance. A dipole–dipole energy falls away as $1/r^3$, diminishing eightfold as the separation is doubled. By comparison, the dependence is $1/r^2$ for ion–dipole energies and $1/r$ for ion–ion energies.

The interaction between polar species, although weak under most circumstances, proves exceptionally strong in a **hydrogen bond**, which

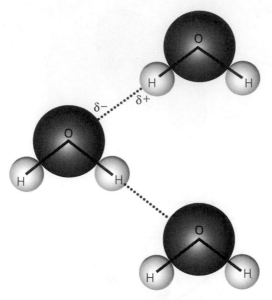

FIGURE 9-6. Formation of noncovalent "hydrogen" bonds, effective for hydrogen atoms already bonded to oxygen, nitrogen, or fluorine. The $H^{\delta+}$ end of one dipole, strongly polarized, attracts the $\delta-$ lone pair of another, pulling the three affected atoms tightly into a straight line.

comes about when a hydrogen atom is covalently attached to a strongly electronegative atom: O, N, or F (Figure 9-6). In water, for instance, the highly electronegative oxygen withdraws charge from the hydrogen, which is then free to associate with a lone pair from an oxygen on another molecule. The interaction is particularly effective because the strongly positive hydrogen atom, having no core electrons to shield its nucleus, provides a good link between two partially negative atoms. The exposed $H^{\delta+}$ is small enough here to fit into a tight space and thus connect the two molecules through a nearly linear $O^{\delta-}—H^{\delta+}\cdots O^{\delta-}$ bridge. Hydrogen bonds are typically the sturdiest of the intermolecular interactions (about 10 times stronger than direct dipole–dipole mechanisms), and they are a familiar sight in the molecular world. Among other roles, hydrogen bonds help cement the three-dimensional structure of biological macromolecules—as we shall see later in this chapter.

Completing the catalog are interactions involving *induced* dipoles (Figure 9-7), which take hold momentarily even in species lacking any fixed separation of charge. It is a quantum mechanical effect, rooted in the vagaries of an electron cloud. Remember that electrons have no well-defined paths; they follow no predetermined route in their motion about the nuclei. We rely, instead, on positions *averaged* over time to conjure up a picture of atom or molecule, but always there are slight fluctuations, slight flickerings in the distribution.

Go so far as to imagine the electrons as a swarm of bees buzzing unpredictably about the nucleus, never the same one moment to the next. Anything can happen. For an instant, the swarm might distort so that

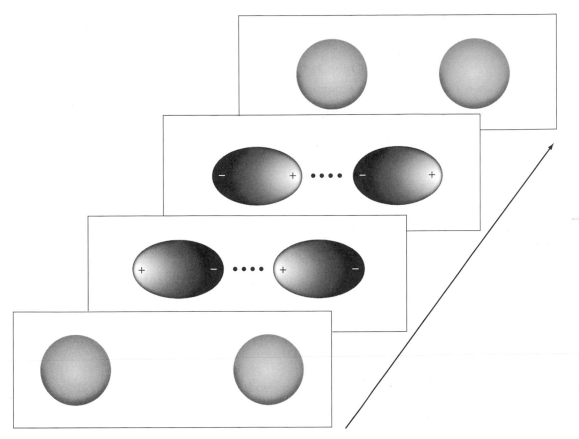

FIGURE 9-7. Interaction between induced dipoles: a fleeting alignment of transient dipoles, induced by momentary fluctuation of the electron density. Weak and limited in range, the attraction falls away as $1/r^6$. Its influence is strongest for large, heavy systems with easily distorted electron clouds. (The interaction is called the London dispersion force—after Fritz London, who explained the effect theoretically.)

one end grows slightly negative and the other slightly positive: a temporary dipole, a sphere turned into an egg. This newly created dipole moment, putting its negative end forward, then will approach another nonpolar molecule and push aside some of the electrons it encounters there. The electrons of the second retreat from the electrons of the first; and, as before, a sphere becomes an egg. The repulsion *induces* a temporary dipole in the second molecule.

There are two dipole moments now, and they can attract one another ever so briefly before the charge distributions fluctuate once more. While it lasts, this **London dispersion force** (named for physicist Fritz

London) engineers a weak dipole–dipole attraction between the mole-cules. The effect diminishes rapidly with distance, dropping as $1/r^6$. Energy sinks by a factor of 64 for every doubling of the separation.

Dispersion forces are most effective when an electron distribution is highly *polarizable*, meaning that the electrons are easily pushed aside. The bigger the cloud, the larger are its fluctuations and the more suscep-tible are its distant, loosely bound electrons to induced dipoles. Small atoms and molecules, with their comparatively few electrons all bound tightly around the nuclei, are difficult to polarize.

9-4. ORDER AND DISORDER
IN GASES, LIQUIDS, AND SOLIDS

Matter, held together by these various interactions, arranges itself into the three broadly familiar (if sometimes ambiguously defined) categories of solid, liquid, and gas. Each aggregated state is distinguished by certain microscopic and macroscopic characteristics, some of which are incor-porated into well-known and commonsensical definitions.

The simplest phase of matter is a **gas**, which we shall take up in de-tail beginning in the next chapter. Particles in a gaseous state are in con-stant motion, never in any one place for long. Here one moment and gone the next, they move rapidly and randomly inside their container. Whatever interactions exist between the microscopic constituents are fleeting at best, and so each particle travels alone with no regard for its neighbor. The result is a low-density, loosely knit phase composed mainly of empty space, summed up best by the name itself: *gas*, derived from the Greek word for "chaos." If we consider a state of matter to be a kind of atomic and molecular fabric, then a gas is a threadbare and thin piece of cloth indeed.

This loose confederation of gaseous particles produces a highly squeezable macroscopic phase (Figure 9-8), one that is easily expanded or compressed under external pressure. A gas has neither fixed shape nor volume. With no significant interparticle interactions to hold the material together, it simply takes the shape of its container. A gas occu-pies whatever space is provided, expanding and contracting as needed. Compression is easy because there is ample empty space to fill, and ex-pansion into a larger volume is unrestricted. The particles just fly off in all directions until they encounter the walls of a container.

Chemical reactions do take place in the gas phase; gaseous molecules do come together and transform themselves into new species. Never-theless, many of the properties that define the gaseous state itself (pres-sure, density, and temperature, among others) are usually indifferent to

FIGURE 9-8. A gas, most tenuous of the three normal states of matter: Barely interacting, the particles fill whatever space they have available. The gas expands and compresses easily, adjusting its density to fit the container.

the chemical identity of the particles. Viewed mechanically, one gas looks much like another: a disorganized collection of generic, undifferentiated, featureless *particles*. The particles travel mostly by themselves and have scant opportunity to affect one another as chemically distinct individuals. An argon atom, a CO_2 molecule, an H_2O molecule, whatever you like—in a gas, each seems merely to be an undistinguished pinprick of mass flying off in the distance.

Liquids, by contrast, are both more idiosyncratic and more organized. Here the particles no longer behave generically, but instead they are attracted and repelled as single atoms, molecules, or ions with individual characteristics: shape, volume, charge distribution, dipole moment, all the rest. Intermolecular relationships in a liquid are more specific, more selective, more discriminating than in a gas.

For proof, take (from Chapter 3) our picture of the solvated clusters in a liquid solution, where molecules of solvent surround and carry away the solute. What is it, after all, that allows the microscopic particles to associate intimately in a condensed phase? What brings them together? Why are no two liquids exactly the same? In a word, call it this: *influence*, the reciprocated action of one uniquely crafted electrical entity on another.

In a gas, the particles have the potential to interact weakly but not the opportunity; they move too fast. In a liquid, the same particles have less energy and correspondingly more chance to engage. Motion, although present, is slow enough to permit the molecules or ions to form weakly bound clusters. The various types of electrical attractions, discussed above, hold these clusters together and impart a certain cohesiveness to the liquid state. The particles communicate.

They do so in whatever ways they can, driven always by the electrical affinity of positive for negative. Polar molecules, for instance, stick together by aligning their ready-made δ+ and δ− regions. The dipole–dipole interaction is a relatively weak link, shattered by heat when the liquid boils, but still the association endures far longer than any conceivable connection in the faster-moving gas phase. Ions dissolved in polar solvents are similarly solvated by ion–dipole interactions, different in detail but again with the same result: small clusters of particles, clumped so that positive meets negative and negative meets positive. And even without a permanent dipole moment, a molecule can enjoy (temporarily) the weak dispersion attractions that come from induced dipole–dipole couplings.

Any means to an end, we see, for from all these interactions emerges something that transcends the individual differences: a liquid; a state of matter; a new kind of association; a step away from the chaos of a gas. A liquid is a liquid—not a gas, not a solid—because of a microscopic organizational scheme special to the liquid *state*. A liquid exhibits **short-range order**, falling between the helter-skelter of a gas and the sleepy regularity of a crystalline solid. Groups of associated particles drift through the liquid phase, but there persists no long-lasting connection between one group and another. On a scale larger than an individual cluster, no particular regularity is apparent. That very ambiguity makes a liquid what it is.

Even the local order in a liquid is impermanent. Clusters break up and form again; clusters drift randomly throughout the phase. Were we to remove one or more molecules from the liquid and then look away, changes in the pattern would inevitably ensue. At first there might appear a hole where once the missing particles resided, but that memory is destined to be short-lived. The scene portrayed in Figure 9-9 will evolve continuously as the remaining particles jostle about in unceasing motion, and eventually little will remain of the original picture.

Weak molecule–molecule attractions and short-range order give to the macroscopic liquid a fixed volume, although not a fixed shape. Like a gas, a liquid flows; its particles stick together to some extent, but not so strongly as to make the material rigid. Like a gas, a liquid needs the walls of a container to give it shape—but *unlike* a gas, it does not rush to fill an expanse completely. A liquid has a finite volume, and there it stays at the bottom of a vessel. Held together by intermolecular forces, the fluid is incapable of infinite expansion. Nor can a liquid be compressed as easily as a gas, because interparticle distances are already far smaller and there is scarcely any space for the molecules to squeeze closer together.

Solids, we know, have both fixed volume and fixed shape, reflecting a highly ordered microscopic state woven into a dense macroscopic fabric. The particles remain mostly in place, vibrating only slightly around their average positions. There is no flying about and hardly any jostling or drifting, for the intermolecular forces hold the particles together with

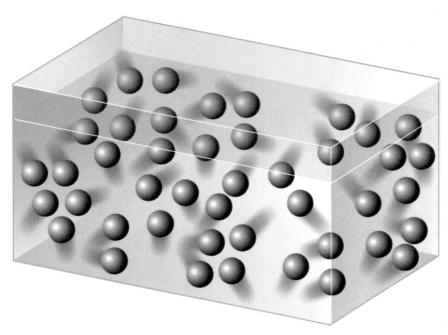

FIGURE 9-9. Short-range order in a liquid: Clusters come together, break up, and form again. Doing so, the molecules organize themselves locally but maintain no regularity over long distances.

more stability and permanence than in a liquid. If now an atom or molecule were to be removed, the vacancy created would be obvious and persistent.

Closest in character to a liquid is an ***amorphous solid*** (or ***glass***), in which only *local* order and very slow motions prevail. Such a material, exemplified by ordinary window glass, resembles a frozen liquid in its lack of long-range order. Although individual clusters of particles do exist within a glass (and these are precisely the associations that establish *short-range* order), there is otherwise no connection between one ordered region and another. The atoms and molecules stay in their places, but no real regularity is apparent over long distances. Given the disposition of atoms in one location, we can say nothing about the configuration at another position. Everything is decided at random, and there are no rules. An amorphous solid (Figure 9-10a) thus offers what amounts to a stop-action view of a liquid: evidence of organization here and there, yet without a master plan to impose order on the system as a whole.

A ***crystalline solid***, on the other hand, does follow a plan and does obey some rules. It exhibits ***long-range order*** in its microscopic structure, with different regions of the crystal showing the same arrangements of the particles. There is no random variation over the material,

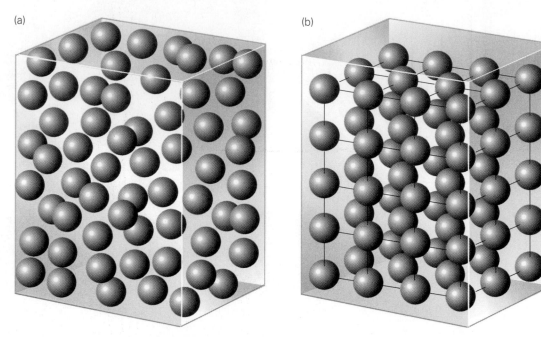

FIGURE 9-10. Solid phases. (a) An amorphous solid, similar to a frozen liquid with short-range order only. (b) A crystalline solid, the most regular form of matter. It offers short-range order, long-range order, and a fixed address and orientation for every particle.

nor does the structure change appreciably with time. An ideal crystal (Figure 9-10b) is the most ordered and regular state of matter, arguably the most beautiful, and to understand it further we shall need to become better aware of symmetry and its far-reaching implications. This we do by first considering a familiar macroscopic activity with unsuspected microscopic import: tiling a floor.

9-5. SYMMETRY: CRYSTALS AND QUASICRYSTALS

Suppose one wishes to cover a floor completely with tiles of a single shape, leaving no gaps at any point. What are the options?

Surprisingly few. A rectangular tile will do the job, as will a square, rhombus, or any other parallelogram. Each of these shapes can be fitted together flush on all sides, resulting in the uniform coverage shown in Figure 9-11(a). Triangles are similarly effective and so are hexagons (which produce a honeycomb pattern), but otherwise there are no more options among the regular (equilateral) figures. Neither the pentagons

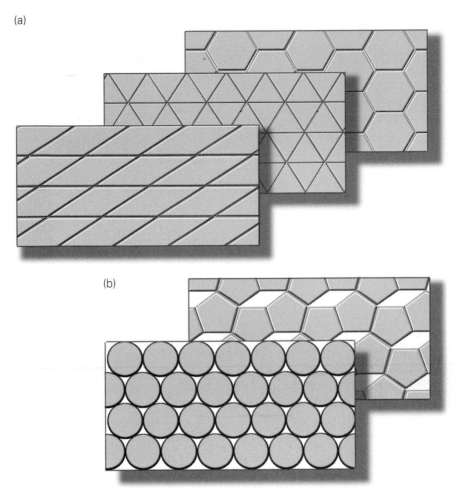

FIGURE 9-11. Tiling a floor. (a) Parallelograms, equilateral triangles, and regular hexagons will cover a space leaving no gaps. (b) Circles, pentagons, and all other regular polygons beyond hexagons will not.

drawn in Figure 9-11(b) nor heptagons nor any other regular polygon can cover the floor without leaving spaces; circles clearly are useless, able to touch only at one point and unable to come closer.

These tiles on the floor bring out a fundamental fact of nature and life: symmetry. The uniform coverage creates what is called ***translational symmetry***, through which a pattern is repeated in space at regular intervals. As a result of this symmetry, each mosaic in Figure 9-11(a) derives from one simple rule which links together all its tiles. With that rule (Figure 9-12) we can reach any particular tile from any other tile simply by walking off a set number of steps in two independent directions. There is a fixed spacing between adjacent tiles, and the entire layout becomes

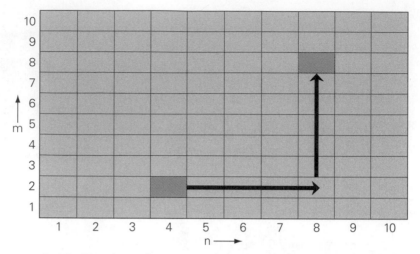

FIGURE 9-12. Translational symmetry. Like houses on a grid, the address of every tile in a mosaic derives from a rule: *n* steps in one direction, *m* steps in the other.

periodic as a result. Periodicity and the accompanying translational symmetry are possible only when the coverage is uniform—which, we agree, will certainly occur with parallelograms, equilateral triangles, and regular hexagons (but not pentagons and not heptagons and not any other regular polygon).

Translational symmetry means, in effect, that a traveler starts at a point *A* and then goes a set distance to reach some new point *B* only to find that it looks identical to *A*. Additional displacements of the same magnitude take the traveler from *B* to *C* and then from *C* to *D* and on forever, yet with never any apparent change in the environment. Interstate highways, according to this definition, display a species of translational symmetry. A driver can discern no differences between highway rest stops spaced 50 miles apart, perceiving instead only an indistinguishable succession of identical fast-food restaurants separated by a constant interval.

Symmetry, in general, describes an *invariance* to some particular operation. Walking from tile to tile is a symmetry operation, as is driving 50 miles from rest stop *A* to rest stop *B* and then to rest stop *C*. At the end of each maneuver, the scene appears unchanged and thus we say that the environment is "invariant" or "symmetric" to that operation. We close our eyes, do something, and then look around to find that everything still looks the same. If anything happened, how would we know?

Rotations (Figure 9-13) are another form of symmetry operation. If, say, we rotate a square by 90°, it still appears to be the same square in the same orientation. The four corners are all indistinguishable, and with eyes closed we have no way of knowing which has been interchanged with which. A square therefore has **rotational symmetry**—*fourfold* rotational

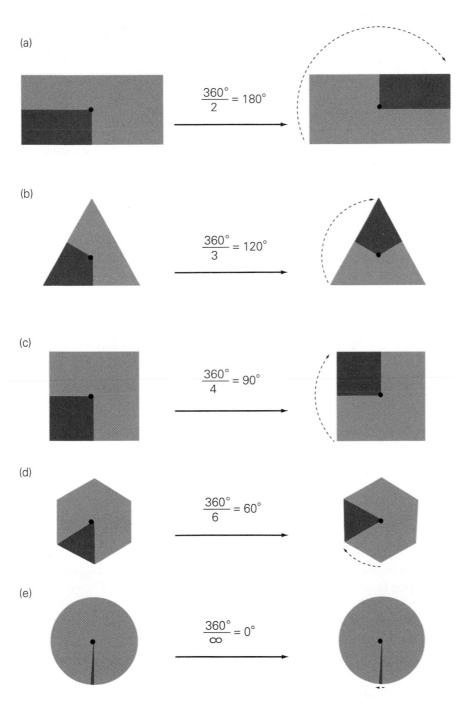

FIGURE 9-13. Rotational symmetry. (a) A rectangle has twofold symmetry. Rotation by 180° leaves it looking the same as before, whereas two such rotations (180° + 180° = 360°) return the figure to its original position. (b) An equilateral triangle has a threefold axis of symmetry. (c) A square: fourfold symmetry. (d) A hexagon: sixfold. (e) A circle: infinite.

symmetry, to be exact, since four successive turns by 90° (a 360° rotation) bring it back to its original position:

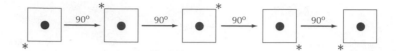

A rectangle or general parallelogram possesses just twofold rotational symmetry, for these figures can only be benignly rotated by 180°. The equilateral triangle has a threefold axis allowing for 120° twists, and the regular hexagon has a sixfold axis which accommodates 60° rotations. A diatomic molecule, endowed with cylindrical symmetry, acquires an infinite axis of rotation; the structure is invariant to even the smallest rotation about the bond.

Our lesson from laying tiles is that only twofold, threefold, fourfold, or sixfold rotational invariance is consistent with translational symmetry. To comply, we need either one of our three prototypes—a four-sided figure with two pairs of parallel sides (a parallelogram), a regular three-sided figure, a regular six-sided figure—or some variant that possesses no "forbidden" symmetry axis such as fivefold or sevenfold. Anything else destroys the periodicity and leaves gaps in the pattern. This lesson then carries over directly to the formation of crystals in three dimensions, which can be visualized also as a tiling operation. Now, however, the tiles grow up into boxes with opposite faces parallel, able to fit together seamlessly in all directions. Where rotational symmetry is present (and usually it is), the axis naturally must be either twofold, threefold, fourfold, or sixfold if the box is to serve as a flush three-dimensional tile.

Simplest to understand is a cube, which is invariant to 90° rotations about an axis perpendicular to each face. Picture a large number of cubic boxes, constructed maybe of some transparent material that allows us to see through the framework. The cubes first line up neatly in one and two dimensions (just as square tiles fit together in a plane) and then stack up layer upon layer to create the three-dimensional network of boxes shown in Figure 9-14.

Points corresponding to the corners of the cubes make up a *lattice*, a mental construct useful for picturing the translational symmetry. Each cubic box, called a *unit cell*, is the basic repeating block of the periodic structure. We know in general, from laying tiles, that infinite repetition is possible if the faces of the unit cells are parallelograms, equilateral triangles, or regular hexagons. Restriction of the lattice to twofold, threefold, fourfold, or sixfold rotational symmetry follows as a result.

Six parameters specify a unit cell: the lengths of the three edges (a, b, c) and the angles between these edges (α, the angle between b and c; β, between a and c; γ, between a and b). Only seven types of unit cell,

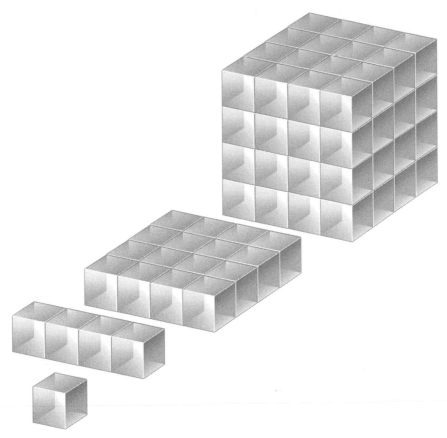

FIGURE 9-14. Assembly of a cubic lattice in one, two, and three dimensions.

listed in Table 9-1 and sketched in Figure 9-15, are compatible with translational symmetry in a crystal. We have just addressed the cubic cell, for which $a = b = c$ and $\alpha = \beta = \gamma = 90°$; the six others range from *triclinic* (a general figure with unequal edges and angles) to the highly symmetric *hexagonal* cell.

Now a lattice is a mental device—a regular arrangement of math-

TABLE 9-1. Unit-Cell Constants for the Seven Crystal Systems

System	Edges	Angles
cubic	$a = b = c$	$\alpha = \beta = \gamma = 90°$
tetragonal	$a = b \neq c$	$\alpha = \beta = \gamma = 90°$
orthorhombic	$a \neq b \neq c$	$\alpha = \beta = \gamma = 90°$
hexagonal	$a = b \neq c$	$\alpha = \beta = 90°; \gamma = 120°$
monoclinic	$a \neq b \neq c$	$\alpha = \gamma = 90°; \beta \neq 90°$
trigonal	$a = b = c$	$\alpha = \beta = \gamma \neq 90°$
triclinic	$a \neq b \neq c$	$\alpha \neq \beta \neq \gamma \neq 90°$

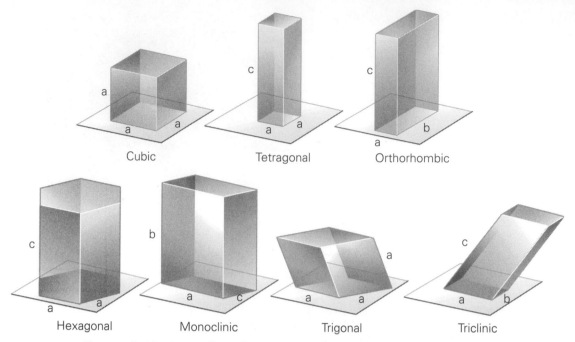

Cubic	Tetragonal	Orthorhombic

Hexagonal	Monoclinic	Trigonal	Triclinic

FIGURE 9-15. Unit cells in the seven crystal systems.

ematical points—but a crystal is something real. Suppose, therefore, that atoms exist either at the lattice points or in their vicinity, so that immediately the translational symmetry of the lattice is stamped onto the particles themselves, as below:

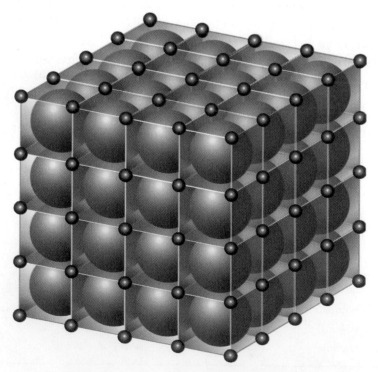

Every unit cell has within it the same configuration of atoms, and hence a fully periodic crystal structure comes to be built upon the imaginary scaffolding of the lattice. We could even assemble the crystal by placing an identical arrangement of atoms in each unit cell and then stacking the boxes. The framework of unit cells defines the translational symmetry, and then the actual atoms fill the space at regular intervals to make a periodic crystal.

Microscopic turns spectacularly into macroscopic when crystals are formed. The symmetry of the lattice is preserved as unit cell is added to unit cell, and the crystal grows larger and larger while adhering to the same elementary pattern. Eventually a crystal big enough to see and touch is produced, with sharply etched faces that mirror the fundamental symmetry within. The rules of symmetry, moreover, apply with equal force to systems large and small, and so the constraints imposed on ordinary flooring tiles similarly govern the microscopic architecture of the crystalline state.

Imagine the surprise, then, when researchers in the 1980s discovered apparently crystalline materials with *fivefold* rotational symmetry, as in Figure 9-16. Dubbed **quasicrystals**, these structures display long-range order yet possess no translational symmetry. Lengthy and ostensibly regular rows of atoms attest to the presence of long-range order; but although quasicrystals (like ordinary crystals) do obey a set of rules from

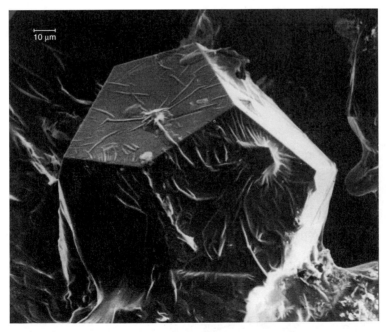

FIGURE 9-16. A grain of quasicrystalline aluminum-copper-iron, made visible by the scanning electron microscope. A fivefold axis of rotation is plain to see.

which a pattern can be generated, the pattern that ensues (Figure 9-17) is not periodic. It never repeats itself, and there is no translational symmetry. The fivefold quasicrystal cannot be built by repetition of only one unit cell, just as regular pentagons cannot cover a surface uniformly.

We see why by looking carefully at displacements in both pentagonal and hexagonal environments. Under *sixfold* symmetry, the entire array first can be shifted along the line *AB* (see below) and then along *BC* to leave the lattice seemingly unchanged:

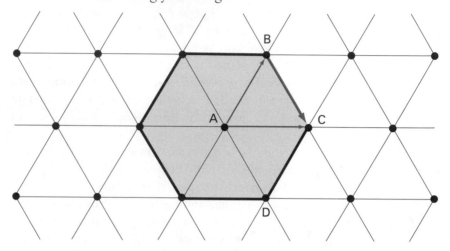

Precisely the same operation is also achieved by a single displacement in the direction *AC*, and indeed the sixfold rotational symmetry guarantees that *any* series of displacements along the lattice directions will coincide with the existing grid. Each atom slips into an equivalent position.

In the fivefold environment, by contrast, successive displacements along *AB* and *BC* fail to bring the pentagon back onto itself:

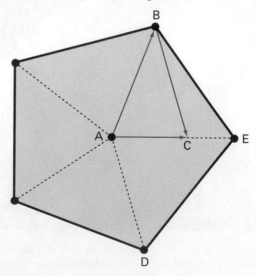

(a)

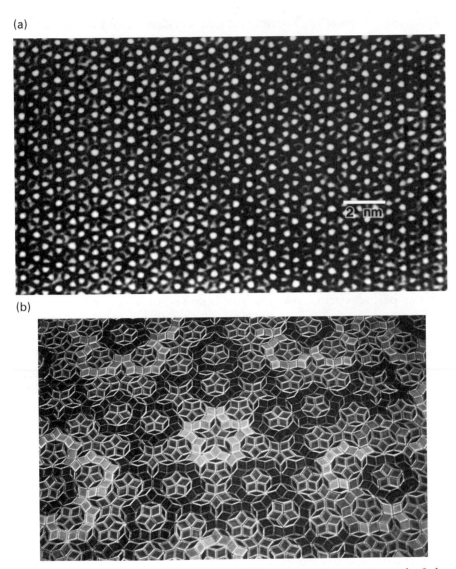

(b)

FIGURE 9-17. Science and art. (a) Fivefold symmetry in a quasicrystal of aluminum, manganese, and silicon, visualized by electron microscopy. (b) A Penrose tiling, named for physicist Roger Penrose. The patterns display fivefold rotational symmetry but are not translationally periodic—despite the tiling's adherence to a fixed architectural rule in two dimensions.

Even though the second shift, from *B* to *C*, lies parallel to an allowed translation (*AD*), still the fivefold symmetry is broken. The excursion creates an isosceles—not equilateral—triangle, which leaves the third leg (*AC*) short of a legitimate lattice position (*E*). Translational symmetry thus is not commensurate with fivefold rotational symmetry, and the arrangement is not translationally periodic.

Lacking translational symmetry, quasicrystals are not crystals in the conventional sense. Possessing long-range order, they also are not amorphous materials. Rather quasicrystals are a new and controversial solid phase about which important questions still remain. How do they form? Why do they form? What symmetries are possible beyond those already observed, which include fivefold, sevenfold, eightfold, tenfold, twelvefold, and twentyfold. There are questions concerning composition and properties. For example, many of the newly discovered quasicrystals contain aluminum—why? There is considerable interest, too, in exploring the elastic and electric properties of various quasicrystalline alloys. It is a field just beginning.

9-6. THE SOLID STATE

Nature, through the constraints of symmetry, lays down the law concerning periodic structures. The law, however, only prescribes the general *appearance* of a crystal; it does not dictate by what method that structure is to be erected. Atoms, molecules, and ions are free to use whatever means necessary to make a crystal, so long as the result conforms to the rules of symmetry. The association may involve covalent bonds, ionic bonds, nonbonded interactions, or some other modality, and it is through these different options that the solid state achieves its diversity. Here, following, are a few of the major variations.

Network Crystals

Also called **covalent crystals**, these materials arise when individual atoms crystallize into a structure resembling an infinitely extended molecule. Regular covalent bonds bind the atoms together to create a tightly interconnected crystalline lattice. Network solids typically are brittle, hard, and difficult to melt owing to the strong bonds between the atoms. *Diamond*, the hardest substance known, is the most familiar example of a covalent crystal.

Every carbon atom in diamond forms four single C—C bonds, one to each of four neighbors so as to make a tetrahedron. From this basic arrangement, which we have come to expect for four-coordinate carbon, grows an ever-expanding network. Each sp^3 carbon at a corner of the original tetrahedron fills its valence with three more carbons to complete another tetrahedron, and then the newly attached atoms make still more bonds, and thus develops the continuous network of C—C bonds pictured in Figure 9-18(a). What results is a macroscopic aggregate of covalently bonded carbon atoms, or something rather like a gigantic molecule.

(a)

(b)

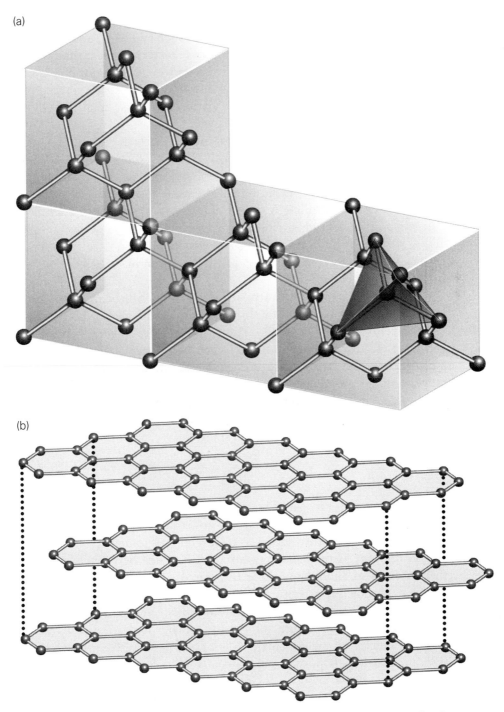

FIGURE 9-18. Two of carbon's three allotropes. (a) Diamond: a covalently bonded network crystal formed from tetrahedral carbon. (b) Graphite: a layered structure containing sheets of hexagonally bonded carbon stacked one atop the other.

Macroscopic properties follow directly from the bonding pattern. Diamond, melting at 3550°C or higher, is tough and hard, and so it finds frequent industrial application as a cutting material. Yet although diamond derives considerable strength from its web of covalent bonds, the interdependence of the network makes for a brittle structure as well. The crystal loses integrity and can be shattered if even a small part of the whole is fractured.

In addition to diamond, elemental carbon exists in two other forms, or **allotropes**: *graphite* and the recently discovered *buckminsterfullerene*, which we shall address presently. Looking for the moment just at graphite (Figure 9-18b), here we find two-dimensional layers of atoms rather than the space-filling tetrahedral network of diamond. Each layer consists now of hexagonal rings of sp^2 carbons, fused together at 120° angles to form a continuous sheet suggestive of chicken wire. The sheets are stacked one upon the other and held loosely in place by noncovalent interactions.

The weaker nonbonded attractions between adjacent layers allow the sheets to slide back and forth, making graphite a soft, slippery substance useful as a lubricant. Graphite's six-membered rings, with delocalized π electrons similar to those of benzene (Chapter 7), also make it a **conductor** of electricity. The π electrons are mobile and therefore able to respond to an electric field and carry a current along the layers. Diamond, with a far different electronic structure, is an **insulator**, a poor conductor of electricity—but still a good conductor of *heat* over its interconnected three-dimensional framework.

Ionic Crystals

Here the microscopic particles are ions, and the electrostatic forces holding together the crystal are largely noncovalent. Solid NaCl, already encountered many times, serves as the archetypal ionic crystal.

So extreme is the electronegativity difference between Na and Cl that there is an almost complete transfer of charge. Sodium's valence electron is effectively captured by the chlorine atom to create Na^+ cations and Cl^- anions. With hardly any electronic charge shared between the nuclei, the sodium and chloride ions behave to some extent as simple electrical particles—little balls of charge. Positive attracts negative, positive repels positive, and negative repels negative. The closer the particles, the stronger is the interaction.

Sodium and chloride ions come together in a delicate balance to form a crystal. The positively and negatively charged particles arrange themselves just right; the attractions bring them together, but the repulsions keep them far enough apart to avoid a collapse. It works every

time. The crystal (Figure 9-19) is strong and highly symmetric, adopting
a cubic unit cell which is well suited for this electrical balancing act.

Cubic symmetry is typical for the ionic crystals that form between
the metals of Groups I and II and the nonmetals of Groups VI and VII.
These materials are generally hard, brittle, and nonconducting, and they
possess high melting temperatures for reasons similar to those elaborated for network solids. Ionic crystals clearly are network solids themselves, differing only to the extent that the valence electrons are
monopolized by the anions rather than shared.

Ionic crystals do have something that covalent crystals lack: ions.
When sodium chloride melts at 801°C, the molten liquid contains mobile Na^+ cations and Cl^- anions able to flow as an electric current. And
just as liberating is a spoonful of water, which can quickly break up a
strongly bound ionic crystal of NaCl into a salty solution, seemingly with

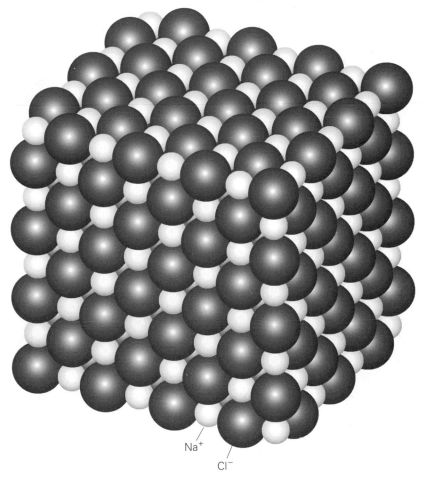

Na^+
Cl^-

FIGURE 9-19. Sodium chloride, a cubic crystal.

little effort. The forces of solvation are surely not to be underestimated, for they can quietly disrupt the same ionic bonds that otherwise resist melting at temperatures up to hundreds and thousands of degrees. Not to be dismissed as "mere" electrical attractions between the ions and polar water molecules, these noncovalent interactions are able to dismantle a crystal of salt within a matter of seconds. They demand a measure of respect.

Molecular Crystals

Nonbonded interactions provide the glue that holds together molecular crystals. Contained within these solids are intact covalent molecules, which are set free upon melting or dissolution. An ordered arrangement shapes up when the weak couplings (weaker than the covalent bonds) bring the molecules into contact, again in just the right balance between attraction and repulsion. Water forms a molecular crystal; so does benzene and so does sucrose. The noble gases, although monatomic, qualify as well.

Nonpolar structures such as oxygen and argon have at their disposal only the London dispersion force, which induces short-lived dipole moments in the otherwise balanced electron clouds. Polar molecules, by comparison, come together more directly through an alignment of their permanent dipole moments, but always we find just one simple rule for all: the rule of electrical attraction; positive to negative, and negative to positive. One way or another, the molecules fall into place.

Water is a good example. The partially positive hydrogen in one H_2O molecule forms noncovalent hydrogen bonds with the negatively biased oxygen in another H_2O molecule. From this preferential association there grows an intricately symmetric crystal of ice, as depicted in Figure 9-20. Other molecules crystallize differently, each with their own interactions and their own constraints.

The properties of molecular crystals depend, accordingly, on the strength of the particular nonbonded interactions responsible for the ordering. Hydrogen-bonded solids such as ice melt at far higher temperatures than do nonpolar solids such as nitrogen and helium, which are held together only by the dispersion force. But compared with the ferociously strong ionic and network solids, of course, even the strongest molecular crystals are soft and easy to melt. Strength is relative.

Metallic Crystals

Until now we have seen metals only as individual actors, most often in solution, where each atom surrenders its valence electrons in a one-to-

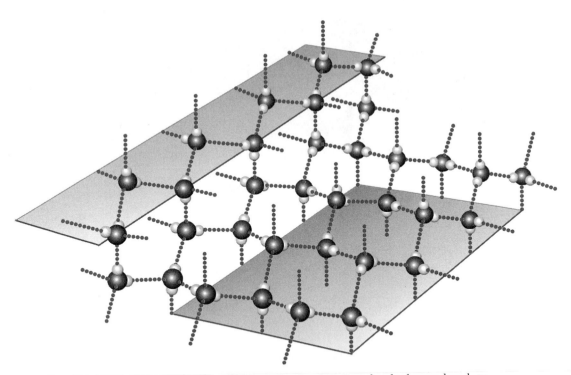

FIGURE 9-20. A crystal of ice. Each molecule of H_2O makes hydrogen bonds to its four nearest neighbors.

one chemical exchange. We know well how sodium, a Group I alkali metal, gives up its lone valence electron to nonmetallic chlorine to produce sodium chloride, an ionic compound. A transition metal also has electrons to offer (*d* as well as *s*), and the loss of negative charge leaves the atom partially positive in many covalent compounds. That, in short, is our microscopic picture of a metal: an atom that loses electrons; an atom that would rather be a cation.

The word *metal* suggests nothing of this microscopic transfer of electrons, but the old familiar name does evoke a set of macroscopic properties common to the metallic elements. The physical attributes are what one routinely expects from shiny pieces of metal—good conduction of electricity and heat, malleability, luster—and all these characteristics originate in the peculiar organization of metal atoms in the solid phase. The easy-to-remove valence electrons are, scant surprise, what makes a metal unusual as a collective state of matter.

Unusual, because a metallic crystal blends together aspects of both solid and gas: an ordered lattice, like a solid; a chaotically disordered mob,

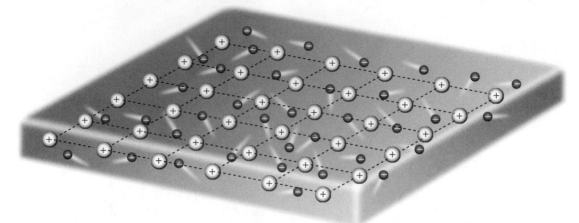

FIGURE 9-21. A metallic crystal, envisioned as a lattice of cations held together by a paste of shared valence electrons. Common to all the atoms, the delocalized electrons roam freely over the whole structure and make the metal a conductor of heat and electricity.

like a gas. Fixed into lattice positions are the ionized cores of the metal atoms (Na^+, for example), while swirling around in the open spaces are *all* the free electrons everywhere at once. Each atom loses electrons from its valence, and these electrons subsequently join a communal pool of negative charge. Shared equally by all the positive ions, the swarm of electrons is reminiscent of a gas in its freedom of movement.

These roaming electrons give us a delocalized system on a grand scale, a community wherein we cannot assign any member to a particular atomic core. Nevertheless we do know that negative will attract positive, as always, and so it does: The negatively charged sea of electrons becomes an adhesive that binds the cations into a lattice like the one drawn in Figure 9-21. A strong and collective binding develops throughout the solid material, similar in some ways to a covalent network. The mobile electrons can be marshalled into a current (hence the high electrical conductivity), and they can transport heat efficiently throughout the metal (hence the high thermal conductivity).

The metallic crystal solidifies as a latticework of cations, cemented together by a grout of electrons. Symmetry typically is high, often cubic or hexagonal, and the ions are frequently arranged in ***close-packed*** configurations with 12 immediate neighbors and minimal free space. Here is the same arrangement we might find if tennis balls were crammed into a carton, row upon row, with the balls in one layer falling into the gaps left in the layer below. Close packing, illustrated in Figure 9-22, affords the most efficient use of space. Note, however, that not all metals are close-

(a)   (b)

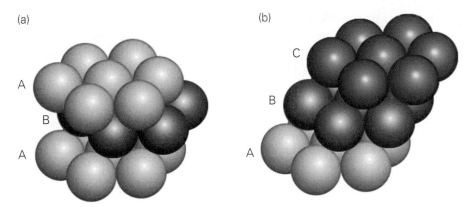

FIGURE 9-22. Close packing, layer by layer: (a) Hexagonal. Each sphere in the first layer (A) touches six others. Spheres in the second layer (B) fall into the hollows left in the first. Spheres in the third layer line up with those in the first. The pattern is ABABAB . . . . (b) Cubic. Spheres in the third layer are offset from those in the first, giving a pattern ABCABC . . . .

packed, and among the important exceptions are our familiar sodium (with eight neighbors) and tin, with only six.

Before dividing the world exclusively into conducting metals and insulating nonmetals, let us take into account two intermediate classifications: semimetals and semiconductors. *Semimetals* (or ***metalloids***) such as antimony, arsenic, and boron straddle the border between metals and nonmetals; they sit on a zigzag line that divides the periodic table into the categories of electron losers and electron gainers:

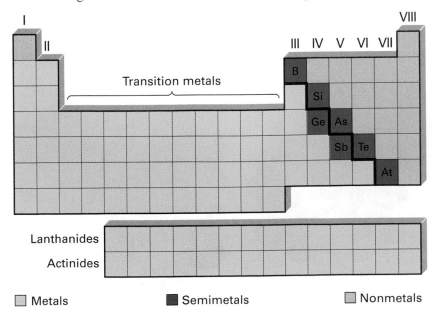

In some ways metallic and in other ways nonmetallic, semimetals are testimony to the ambiguity inherent in chemical taxonomy. In similar fashion, a ***semiconductor*** like silicon or germanium or gallium arsenide falls into a gray area between a true metallic conductor and a nonmetallic insulator.

A *semiconductor*, we say: sometimes yes, sometimes no. Silicon, analogous to carbon in electronic structure and bonding, also forms network solids and is (usually) nonconducting for the same reasons that apply to diamond. Still, there is a critical distinction. Silicon crystals sometimes *can* conduct electricity, owing to subtle differences in their electronic energies compared with carbon. Under the right circumstances—when the temperature is increased, for instance, or when certain impurities are present—*some* of silicon's networked electrons can acquire the enhanced mobility needed to carry a current. Discovery and manipulation of just these circumstances has created the entire modern electronics industry.

9-7. CLUSTERS

Hovering between large and small is the ***cluster*** (Figure 9-23), a state of matter that falls outside the traditional divisions of solid, liquid, and gas. More than a molecule yet less than a bulk material, a cluster endures as a finite-sized aggregate of atoms with properties intermediate between microscopic and macroscopic. It is a gray state of matter—something in

FIGURE 9-23. More than a molecule but less than a bulk liquid or solid, a *cluster* (nanocrystal) has properties all its own. Shown here are stable clusters of 13, 55, and 147 noble-gas atoms, held in place by London dispersion interactions. Nanocrystals vary widely in shape, size, composition, and mode of interaction.

between, not sharply delineated like a millimeter-long crystal or a beaker of liquid. Four atoms or forty might make a cluster, or possibly even several hundred. If sufficiently small, the structure usually behaves more or less like a molecule; if sufficiently large, a cluster starts to exhibit properties of the bulk state. Most interesting is the zone where it is neither here nor there.

At this point we already have some acquaintance with clusters, recognizing them as the stuck-together species that create short-range order and cohesiveness in liquid phases. Solvated molecules or ions are clusters, as are the groups of weakly interacting particles found in pure liquids. Yet although these structures are familiar to an extent, until now they have always existed for us within the larger context of an entire macroscopic phase. Never have we imagined a cluster to stand in isolation, as *just* a cluster and not as one family unit among many in a liquid or solid. But clusters can indeed survive by themselves (if, sometimes, only in the laboratory), and their special properties are currently a matter of intense investigation.

The questions go directly to our understanding of matter and to the very meaning of macroscopic. How big is big? When does a group of associated particles start to become a macroscopic phase? By what mechanism is such a transition accomplished? In what ways do the properties of clusters differ from bulk materials? What structures do small clusters assume, and why?

There are commercial and practical considerations as well. Consider the possibilities if, say, we could mimic the properties of a bulk metal with a cluster of several hundred or several thousand atoms. Expensive materials like the platinum and rhodium used in automotive catalytic converters could be replaced at a fraction of the cost. Catalytic chemistry in general, semiconductors and electronics, lasers, superconductors, and other materials and processes might eventually be enhanced by such capabilities.

Clusters are not limited to just the weakly bound collective structures found in liquids. In various forms they are held together in all the usual ways—through covalent or metallic bonds, perhaps, or through dipole–dipole and assorted other noncovalent interactions. There are clusters of carbon, niobium, sodium, and argon, to name a few, and these varied species exhibit all manner of size and shape and properties. Some are similar to either isolated molecules or bulk states; others are extraordinarily different. Understand, though, that both the atoms within the structure and the means by which they attract one another are entirely ordinary. What makes the cluster so different is not its atoms but rather its unique not-quite-macroscopic aspect.

If, as we have suggested earlier, a state of matter is a kind of molecular society, then bulk material might be a large city while an isolated cluster would be a small town. In the city, most residents are packed deep inside and know only their nearest neighbors; in a small town, everybody knows everybody else and there is far less layering of the population from the inside out. It is a more interactive, more *informed* arrangement lacking the compartmentalization of the city.

Picture, then, the urban environment of a crystal, with its lattice of internal unit cells. There is a sharp distinction between the inner and outer reaches of the structure. Only well *inside* the crystal is the apparently infinite translational symmetry made manifest, for the periodicity begins to break down near the surface. On the surface, clearly, there is no more periodicity (the crystal just stops), and so properties on the surface necessarily differ from those in the interior.

A cluster, by contrast, might be *all* or at least mostly surface. With fewer atoms to accommodate, it is not forced to crowd as many residents into the interior. In structures of 13 argon atoms, for example, 12 surface atoms surround only one in the center, and even for larger argon clusters there are appreciable numbers existing directly on the surface: 42 (surface)/55 (whole cluster) and 92/147, say, just two possible architectures among many. Compare that with a full mole of frozen argon, where atoms on the surface constitute a minuscule fraction of the bulk solid, and we realize that the properties of the crystal and *nanocrystal* (the very small cluster) are bound to differ.

Melting points may be different. Clusters containing a few tens of atoms typically melt more readily than the bulk material, a phenomenon attributable to the dominant role played by the surface. Atoms on the surface interact only with themselves and those one layer below, thereby exposing themselves to binding forces different from those that control the interior. Clusters, with their disproportionately large surfaces, show considerable variation in the strength of binding.

Optical, electrical, and magnetic properties also may differ, as demonstrated by aggregates of mercury. Normally a liquid metal—a metal with free valence electrons but no fixed lattice of cations—mercury retains its metallic bonding and persists as a conductor down to clusters of a few hundred atoms. Reduced to only a few atoms, however, the mode of association changes to weak dispersion interactions, and the smaller structure becomes an insulator.

Properties of clusters often vary sharply when additional atoms are added. Whereas the attachment of one additional atom to a crystal of 10^{23} particles can scarcely be noticed, the effect stands out far more in a group of only a few dozen. There are, in fact, certain geometric structures possible with some numbers and not with others—particularly the *icosahedron* (Figure 9-24), which is of especial importance for noncovalent clusters such as argon. Going from 12 to 13 to 14 or from 54 to 55 to 56 atoms makes a big difference in these instances. It can mean the

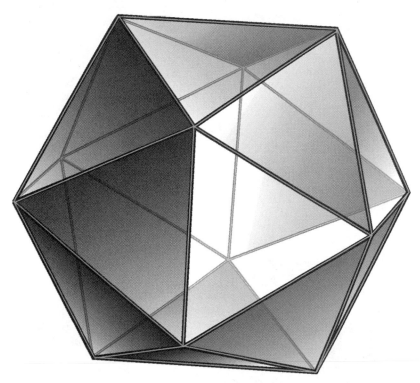

FIGURE 9-24. A regular icosahedron. Displaying 20 triangular faces, the structure contains 30 edges and 12 vertices.

difference between a wobbly arrangement on the one hand, and a stable icosahedral structure on the other.

The icosahedron is an extraordinary piece of geometry, a solid figure with 20 triangular faces, 30 edges, and 12 vertices. Its high symmetry dominates the intermediate world of clusters, not only for noncovalent aggregates like Ar_n but also for carbon's third state: C_{60}—*buckminsterfullerene*—which, undiscovered until 1985, is now known to occur naturally in the soot of burning hydrocarbons. C_{60} (Figure 9-25) has become the parent structure for a growing class of man-made compounds called *fullerenes*, which often incorporate various metal atoms to yield novel electrical and magnetic properties.

In contrast to argon, carbon atoms usually do not form clusters of arbitrary sizes. Carbon condenses, instead, mostly in well-defined aggregates of 60 atoms, although other closed structures do exist (notably C_{70}, but also many more). The predominant C_{60} configuration is a truncated icosahedron that forms when the figure's original 12 vertices are sliced away. Its architecture is none other than R. Buckminster Fuller's geodesic dome (whence the name buckminsterfullerene) as well as the stitching pattern of an ordinary soccer ball. The soccer-ball arrangement

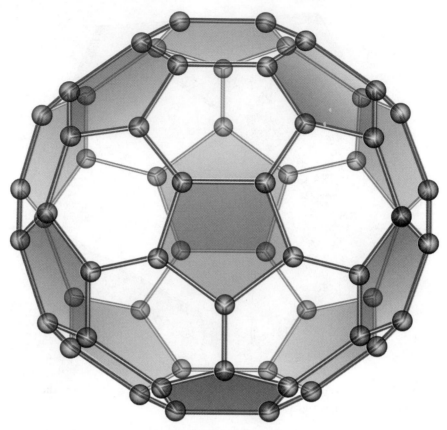

FIGURE 9-25. Carbon's third allotrope: carbon-60, known as buckminster-fullerene. The soccer-ball network of 12 pentagons (shaded panels) and 20 hexagons (open spaces) arises when the 12 vertices of an icosahedron are cut off. One carbon atom sits at each of the 60 exposed corners.

has 20 hexagonal and 12 pentagonal faces, with one carbon atom at each of its 60 vertices to create a closed, surface-only cluster. Every carbon is bonded to three others by ordinary covalent interactions, and the fused hexagons and pentagons appear as alternating single and double bonds throughout the Lewis structure.

9-8. LIQUID CRYSTALS

Another "in-between" state of matter is the *liquid crystal*, which combines some of the ordering of a crystal with some of the freedom of a liquid. The result is a homogeneous phase with properties distinct from either traditional state. Today these properties are exploited in electronic devices such as wristwatch and calculator displays, screens in laptop computers, and instrument panels.

Consider, first, the manifest differences between liquids and crystals. Atoms in a crystal reside at fixed locations, maybe jiggling a bit but never straying far. There they sit, held together by some balance of attractions and repulsions, and thus the solid maintains its shape and cannot flow. We even know the whereabouts of every atom, because the entire lattice is built according to definite rules of symmetry.

Particles in a liquid have no assigned places, but are free instead to roam throughout the container. They can move in all directions, tumble and spin, come together to form clusters and move apart. Although less organized than a solid, the liquid holds together well enough to permit the macroscopic fluid to flow.

Such are the differences between the ideal crystal and the ideal liquid. Now a liquid crystal, being neither ideal liquid nor ideal crystal, sits midway between the two. Its molecules are no longer confined to fixed locations, and consequently the material can flow like a liquid. Still, the flow is not unrestricted—no unlimited spinning and tumbling—so some remnant of crystal-like ordering is preserved.

Stiff rodlike molecules, for example, are prone to organize into liquid crystalline phases. We might start out with a regular crystal built from long and thin molecules of the sort shown in Figure 9-26, in which initially the rods line up rigidly in set positions. Then heat is applied to melt the crystal, and the molecules start to come loose. They leave their original locations, *flowing* to some extent, but their movements are constrained by the crowd. A given molecule is hemmed in by its neighbors, forced to line up with the others and unable to rotate freely along all axes. Stuck in one orientation, a molecule in the middle (and there is always a molecule in the middle) has no room to spin about or tumble in every direction. What develops is a homogeneous herd of rigid rods, flowing with nearly the same rotational orientations but now without any *translational* order. For although the molecules are carried along

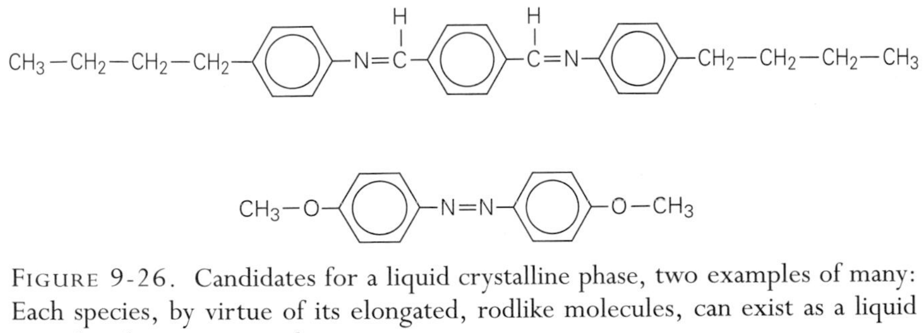

FIGURE 9-26. Candidates for a liquid crystalline phase, two examples of many: Each species, by virtue of its elongated, rodlike molecules, can exist as a liquid crystal under certain conditions.

with the flow, they do not travel in orderly rows; their centers do not line up or fall into any regular pattern. The liquid crystal is almost like a disorganized military unit marching in broken ranks, yet not marching so haphazardly as to allow each soldier to meander off at random or turn cartwheels.

Translationally disordered like a liquid but rotationally ordered like a crystal, this hybrid phase is pictured in Figure 9-27. It is called a *nematic* liquid crystal, one of many possibilities. There are, in addition, *smectic* liquid crystals (which include some forms of soap structures in water), *discotics*, *cholesterics*, *hexatics*, *columnars*, and others. Each has its own special brand of order and disorder. Some of the phases are layered; some arise from molecules with shapes different from long rods (notably the discotic phase, with disklike molecules). There is considerable diversity, but already with our consideration of a nematic we have the essential character of a liquid crystal.

Any liquid crystal exists in a fragile state. It arises when a crystal does not fully melt (or, equivalently, when a liquid does not fully freeze). There is enough heat to disrupt the crystal lattice and release the molecules from their positions, yet not enough energy to permit tumbling in all directions. Only at higher temperatures can the molecules break free of the crowd. Stirred up by sufficient heat, the long rods then brush each other aside and tumble without further hindrance. At that point, with the molecular marchers having broken ranks completely, the liquid crystal becomes an ordinary liquid.

9-9. POLYMERS

We close with a sketch of matter in its most complicated form—the **macromolecules**, or **polymers**—and then begin a new chapter with a detailed examination of gases, the simplest phase of all.

Polymers are large (very large) molecules built by linking together small units called **monomers**, an act of simple repetition that produces structures of great subtlety and complexity. Properties vary widely among the macromolecules, and small changes in a monomer usually produce big changes in the polymer that results. Replace H by F in a molecule of $CH_2=CH_2$, for example, and polyethylene becomes Teflon, a different material with different physical attributes. Or, more telling, alter the nucleotides in a person's DNA even slightly and the person may die.

They are important molecules. There are, to begin, biological macromolecules such as proteins (built from amino acids), nucleic acids (built from nucleotides), and polysaccharides (built from sugars). Take them away and we cannot live at all. Added to these are a long list of synthetic polymers without which we cannot live the way we currently do: polyethylene,

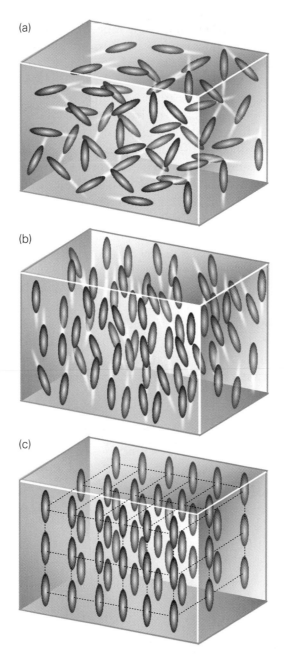

FIGURE 9-27. Liquids and liquid crystals. (a) Molecules in a liquid fall victim to both rotational and translational disorder over long distances. Given time, each molecule flows through all points of the liquid and tumbles randomly in every direction. (b) Not so in a liquid crystal, where molecules are restricted rotationally (around a point) but not translationally (point to point). Their orientations—although not their positions—stand in some fixed relationship. (c) In a crystal, translational order is enforced as well. The molecules take up preassigned positions and orientations.

polypropylene, polystyrene, and many more, most developed since the middle of the 20th century. From plastic films to fibers to home furnishings to automotive parts to toys to storage containers to electronic products to artificial limbs to replacement cardiac valves—wherever we turn, we find macromolecules at the economic heart of modern life.

Are macromolecules different from other molecules? Yes and no. Polymers can be huge, with thousands of atoms and with molecular weights into the millions, but still they arise from the same covalent and noncovalent interactions that govern every other molecule. Like all molecules, too, they can organize into phases, and indeed polymers do exist as amorphous and partially crystalline solids, solutions, melts, and other forms. Yet polymers are no ordinary molecules, even so, and they fall into a special class for perhaps the most obvious of reasons. They are big.

Imagine, on the one hand, a few dozen oranges piling up neatly in a basket, each adjusting to the other as an organized assembly takes shape. Now contrast that scene with an image of tree trunks being tossed in a heap, or a tangle of enormously long snakes, or a fleet of battleships trying simultaneously to negotiate a small harbor. Size makes a difference. There is a difference in the extent of cooperation and coordination demanded when the pieces coming together are so large.

Giant molecules, like tree trunks and tangled snakes,

face special constraints when forced to act in concert. The way they pack depends intricately on size and shape, different for every polymer. Are the macromolecules long and thin? Are the chains flexible? Are they able to fold? Do they fold randomly or do they fall into predictable patterns? Are the molecules spherical? Are groups of atoms able to rotate freely? Are there bulky side chains that get in the way of orderly packing? Do the molecules consist of one chain or many? Are all chains the same length? Are all chains the same kind? Are individual molecules *cross-linked*, their chains connected one to another through a network of covalent bonds?

Such distinctions, exploited in the manufacture of synthetic polymers, produce materials with uniquely crafted physical properties—ranging from the soft plastic of contact lenses to the ultra-strong polycarbonates used in bulletproof windows.

Even acting alone, however, a polymer is no ordinary molecule, for a polymer may be able to organize *itself* in a way impossible for a small species. Granted sufficient size and flexibility, a single macromolecule sometimes can fashion itself—unaided—into an ordered structure molded by noncovalent interactions between different parts of the same molecule. A polymer then becomes its own state of matter, organized internally without the assistance of any other parties.

For the supreme examples of macromolecular assembly, we look first to nature and not art. We start with the architecture of proteins and nucleic acids, paying especial attention to the self-organization of single polymers.

Biopolymers

We have seen before, in Chapter 8, illustrations of two important classes of biological macromolecules: the proteins and nucleic acids, formed by condensation reactions involving, respectively, amino acids and nucleotides. Recall from that first description how the monomers join together when a group of atoms is removed at each link, as when two amino acids make a peptide bond:

$$\underset{\underset{H}{|}}{\overset{\overset{R}{|}}{H_2N-C}}-\overset{\overset{O}{\|}}{C}-OH \;\; + \;\; \underset{\underset{H}{|}}{\overset{\overset{R'}{|}}{H-N-C}}-\overset{\overset{O}{\|}}{C}-OH \;\; \longrightarrow$$

$$\underset{\underset{H}{|}}{\overset{\overset{R}{|}}{H_2N-C}}-\overset{\overset{O}{\|}}{C}-\underset{\underset{H}{|}}{N}-\underset{\underset{H}{|}}{\overset{\overset{R'}{|}}{C}}-\overset{\overset{O}{\|}}{C}-OH \;\; + \;\; H_2O$$

The connecting $-\overset{\overset{O}{\|}}{C}-\underset{\underset{H}{|}}{N}-$ bridge, also called an *amide linkage,* develops when OH from the COOH group on one amino acid "condenses" with H from the NH_2 group on an adjacent residue to produce H_2O. Repeated condensation reactions eventually produce a long molecular chain of connected monomers, drawn from an assortment of 20 different amino acids (for the proteins) and five different nucleotides (for DNA and RNA). And that, so far, is what we have: a molecule, a big molecule bonded together covalently like any other.

But we are prepared already to see this kind of big, flexible molecule develop into something else—into a structure organized from within, a patterned structure governed by *non*covalent interactions between atoms not directly bonded. These weaker forces, acting randomly but with unerring effect, mold strings of monomers into the three-dimensional machinery needed precisely to support life.

The story is one best told with pictures, and the accompanying illustrations should give some idea of the manifold possibilities. Common to many biopolymer structures is the ubiquitous hydrogen bond, which brings different portions of the molecule together by means of a specialized noncovalent interaction. A polypeptide protein chain, for instance, organizes into a helix by forming hydrogen bonds between amide N—H groups and carbonyl oxygen atoms (Figure 9-28). The electronegative nitrogen withdraws electrons from the hydrogen, leaving the H atom on one amino acid partially positive and therefore able to form a bridge to the oxygen on another:

$$\overset{\displaystyle \delta- \quad \delta+ \qquad \delta- \quad \delta+}{N-H\cdots O=C}$$

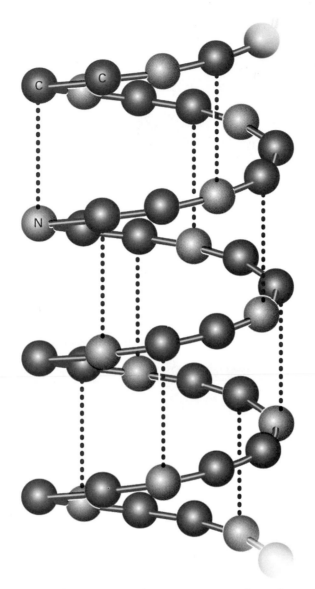

FIGURE 9-28. Secondary structure of a protein; an example: Hydrogen bonds bring together C=O and N—H groups on nearby amino acids and thus twist a polypeptide chain into a helix. For simplicity, only the helical backbone (-C-C-N-C-C-N-) is shown. C-C and N are rendered in contrasting colors, with R groups and all other atoms omitted.

This level of association gives rise to the *secondary structure* of the protein, to be distinguished from the simple sequence of amino acids that establishes the *primary structure*. Additional interactions allow the protein to fold into an even more complex configuration, thereby establishing its *tertiary structure*.

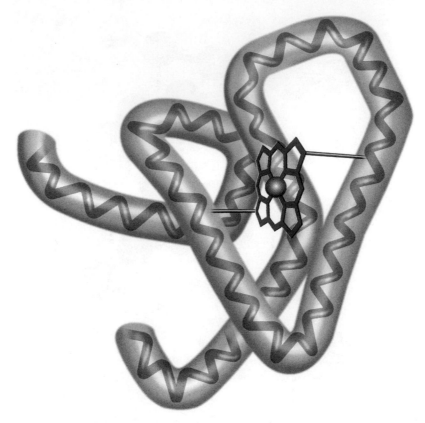

FIGURE 9-29. The tertiary structure of a protein; an example: myoglobin. Interactions between distant amino acids cause the protein to fold into an elaborate pattern. Myoglobin's internal cavity enables it to store and transport O_2 molecules.

Tertiary structures are exquisitely complicated, but never for adornment alone. Some function is always served by the geometry. Myoglobin, for one, is a *globular* protein, so named for its approximately spherical shape. Myoglobin's role in biological cells is to store oxygen, which it carries in a protected internal pocket perfectly suited for this role (Figure 9-29). Form and function are inseparable.

Globular proteins typically have sundry kinks, bends, and cavities to accommodate smaller atoms and molecules of predetermined shape. Enzymes, in particular, often do their catalytic work by discriminating in favor of certain shapes. Only the target molecule will fit into the enzyme's active site, where it can be held in place while a suitable intermediate is assembled.

Certainly the most celebrated macromolecular structure of all is the double helix of DNA, shown in Figure 9-30. The three-dimensional shape is maintained once again by hydrogen bonds, here between the nitrogen bases as shown in the illustration. Adenine is coordinated to thymine (A–T) whereas cytosine is coordinated to guanine (C–G), and these are the links that bring together two separate strands of DNA into an interwoven helix. For every adenine on one strand there is a complementary thymine on the other; for every cytosine there is similarly a guanine.

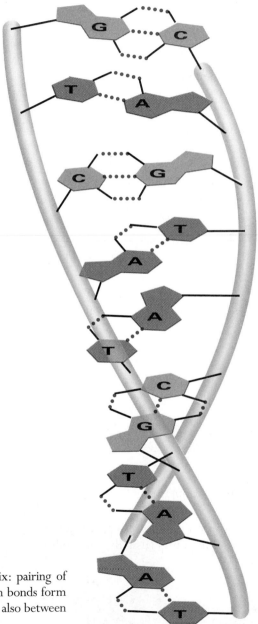

FIGURE 9-30. The double helix: pairing of nitrogen bases in DNA. Hydrogen bonds form between adenine and thymine and also between cytosine and guanine.

The hydrogen bonds in DNA, like so much else in nature, are just right: strong enough to keep the helix intact amidst the normal bumping and jolting of the molecular world, yet weak enough to let the coils unwrap during the manufacture of a protein. In this way is the genetic blueprint stored in compact and protected form within a living cell, one of the happy chemical circumstances that make life on earth possible.

Synthetic Polymers

The challenge for the polymer chemist is now to improve life, both intellectually and economically, by taking a lesson from nature—by combining small molecules into large molecules with dramatically new properties. First from monomer to polymer, step by step, and then from single macromolecule to aggregated state, the macroscopic properties follow inevitably from the microscopic structure. The relationships are not simple, but maybe nowhere else is the link between microscopic and macroscopic more alluringly intricate.

Take, for example, the small molecule ethylene as a monomer,

$$\begin{array}{c} H \\ \diagdown \\ \diagup \\ H \end{array} C = C \begin{array}{c} H \\ \diagup \\ \diagdown \\ H \end{array}$$

and polymerize it to form *poly*ethylene by repeated *addition* reactions (described in Sections 3-5 and 8-5):

$$\cdots \; + \; \begin{array}{c} H \\ \diagdown \\ \diagup \\ H \end{array} C = C \begin{array}{c} H \\ \diagup \\ \diagdown \\ H \end{array} \; + \; \begin{array}{c} H \\ \diagdown \\ \diagup \\ H \end{array} C = C \begin{array}{c} H \\ \diagup \\ \diagdown \\ H \end{array} \; + \; \cdots \; \longrightarrow$$

$$\cdots \; \begin{array}{c} H\;H\;H\;H \\ |\;\;|\;\;|\;\;| \\ -C-C-C-C- \\ |\;\;|\;\;|\;\;| \\ H\;H\;H\;H \end{array} \; \cdots$$

With each opening of a C=C double bond, two electrons are freed to make new single bonds with two other ethylene molecules; and so, monomer after monomer, the process continues until the chain contains n repeating units (usually hundreds or thousands):

$$n \; \begin{array}{c} H \\ \diagdown \\ \diagup \\ H \end{array} C = C \begin{array}{c} H \\ \diagup \\ \diagdown \\ H \end{array} \; \longrightarrow \; \left[\begin{array}{c} H\;\;H \\ |\;\;\;| \\ -C-C- \\ |\;\;\;| \\ H\;\;H \end{array} \right]_n$$

The reaction terminates at some point, and we have a single molecule of polyethylene. *One* macromolecule.

Suppose it were a "perfect" macromolecule of polyethylene, just as written: a single continuous chain of repeating $-CH_2-CH_2-$ monomers, zigzagging tetrahedrally from one end to the other like a fantastically large version of the alkanes we saw in Chapter 8. Suppose, further, that all the macromolecules in the sample were built from the same number of monomers, n, and suppose that we wished to crystallize the collective system from a fluid phase. What might happen?

If the conditions are as we describe, and if we cool the liquid very slowly and carefully, then—ideally—the material would form a covalent crystal, reasonably well ordered although packed with very large molecules. It would form a crystal with uniform density and other physical properties.

Does it? Again, yes and no. Crystalline portions of linear polyethylene do form, and the molecules pack in such a way as to produce *high-density polyethylene (HDPE)*, generally a strong and hard plastic with density greater than 0.96 g cm^{-3}. HDPE finds application in pipes, bottles, molded containers, drums, toys, and other products where strength is needed. But the crystal structure is not nearly so regular as that arising from a small molecule, because an orderly occupation of lattice points is problematic for molecules of this size. Nor does the material usually exist as a single crystal patterned according to a single lattice. The long, twisting chains are difficult to separate as the crystal grows, and the final configuration varies greatly with the speed of crystallization. Some chains find themselves disrupted at the surface of a rapidly advancing crystal (see below), where ordinarily they would fold back and reenter the lattice:

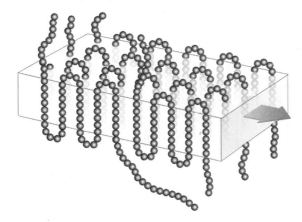

Here, however, another chain slips into position instead, and the ordering is impaired. Left out, portions of the excluded chains either clump together in amorphous regions or else start adjacent crystalline domains of their own (Figure 9-31).

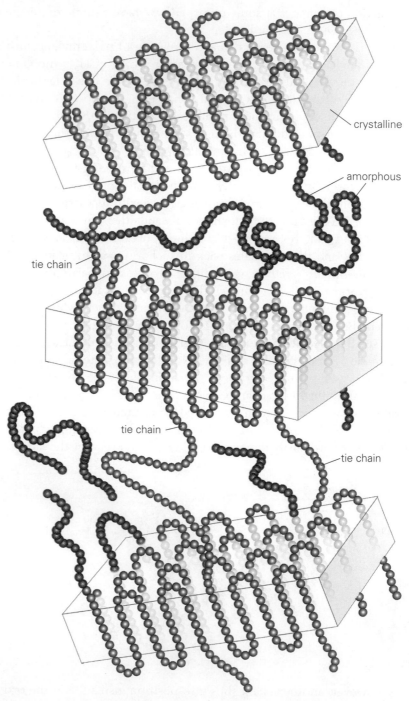

FIGURE 9-31. Alternation of crystalline and amorphous regions in a rapidly cooled melt of semicrystalline polyethylene.

This sandwich structure, where present, gives the material mechanical properties different from those of an ordered single crystal. Not just *one* crystal develops here, but rather we find a stack of imperfect crystals in which molecules are ordered differently from one layer to the next. Chains in one layer (the "tie chains" in Figure 9-31) thus help tie together domains on either side, lending structural integrity to this *semicrystalline* material as a whole. The effect is somewhat like the strengthening achieved in plywood, where sheets of wood are cemented together with grains at opposing angles.

Amorphous regions interspersed with the crystalline portions provide further support. With their molecules distributed over a range of directions, the noncrystalline layers prove strong, resilient, and particularly well suited to absorb stress. They melt over a range of temperatures different from those in the semicrystalline regions. Local order, made possible by noncovalent interactions within individual chains, is present here and there as well, despite the lack of overall crystallinity.

Other outcomes are also possible, especially when we drop our assumptions that all the macromolecules are linear polyethylenes of exactly the same length. Mixtures of different chains—some long, some short—will exhibit different densities and different degrees of crystallinity. Typically, the relative molecular mass for HDPE varies around an average value of 1,000,000.

Moreover, not all polymerization reactions of ethylene result in pure unbranched polyethylene of the kind

$$-\overset{\displaystyle H}{\underset{\displaystyle H}{C}}-\overset{\displaystyle H}{\underset{\displaystyle H}{C}}-\overset{\displaystyle H}{\underset{\displaystyle H}{C}}-\overset{\displaystyle H}{\underset{\displaystyle H}{C}}-\overset{\displaystyle H}{\underset{\displaystyle H}{C}}-\overset{\displaystyle H}{\underset{\displaystyle H}{C}}-\overset{\displaystyle H}{\underset{\displaystyle H}{C}}-\overset{\displaystyle H}{\underset{\displaystyle H}{C}}-\overset{\displaystyle H}{\underset{\displaystyle H}{C}}-\overset{\displaystyle H}{\underset{\displaystyle H}{C}}-\overset{\displaystyle H}{\underset{\displaystyle H}{C}}-$$

that we have specified. Some produce molecules in which hydrocarbon side chains substitute for hydrogen atoms at various points along the main chain; as below, say:

$$-\overset{H}{\underset{H}{C}}-\overset{H}{\underset{H}{C}}-\overset{H}{\underset{H}{C}}-\overset{H}{\underset{H}{C}}-\overset{H}{\underset{H}{C}}-\overset{H}{\underset{H}{C}}-\overset{H}{\underset{H}{C}}-\overset{H}{\underset{CH_2}{C}}-\overset{H}{\underset{H}{C}}-\overset{H}{\underset{H}{C}}-\overset{H}{\underset{CH_2}{C}}-$$

with side chains $-CH_2-CH_2-CH_3$ and $-CH_3$

The effect is to inhibit crystallization, a difficult feat made still more difficult by the added hindrance of the side chains. The branched chains no longer can disentangle sufficiently to occupy a lattice. They freeze in

TABLE 9-2. Selected Polymers

Polymer	Structure
polyethylene	$\left[\text{CH}_2-\text{CH}_2\right]_n$
polyvinyl chloride (PVC)	$\left[\text{CH}_2-\text{CH}(\text{Cl})\right]_n$
polytetrafluoroethylene (Teflon)	$\left[\text{CF}_2-\text{CF}_2\right]_n$
polystyrene	$\left[\text{CH}_2-\text{CH}(\text{C}_6\text{H}_5)\right]_n$
polypropylene	$\left[\text{CH}_2-\text{CH}(\text{CH}_3)\right]_n$
polybutadiene	$\left[\text{CH}_2-\text{CH}=\text{CH}-\text{CH}_2\right]_n$
polyisoprene	$\left[\text{CH}_2-\text{C}(\text{CH}_3)=\text{CH}-\text{CH}_2\right]_n$
polyethylene terephthalate (PET)	$\left[\overset{\text{O}}{\text{C}}-\text{C}_6\text{H}_4-\overset{\text{O}}{\text{C}}-\text{O}-\text{CH}_2-\text{CH}_2-\text{O}\right]_n$
polycarbonate	$\left[\text{O}-\text{C}_6\text{H}_4-\text{C}(\text{CH}_3)_2-\text{C}_6\text{H}_4-\text{O}-\overset{\text{O}}{\text{C}}\right]_n$
polyurethane	$\left[\overset{\text{O}}{\text{C}}-\overset{\text{H}}{\text{N}}-\text{R}-\overset{\text{H}}{\text{N}}-\overset{\text{O}}{\text{C}}-\text{O}-\text{R}'-\text{O}\right]_n$
nylon 66	$\left[\overset{\text{O}}{\text{C}}-(\text{CH}_2)_4-\overset{\text{O}}{\text{C}}-\overset{\text{H}}{\text{N}}-(\text{CH}_2)_6-\overset{\text{H}}{\text{N}}\right]_n$
poly(p-phenylenediamine terephthalamide) (Kevlar)	$\left[\overset{\text{H}}{\text{N}}-\text{C}_6\text{H}_4-\overset{\text{H}}{\text{N}}-\overset{\text{O}}{\text{C}}-\text{C}_6\text{H}_4-\overset{\text{O}}{\text{C}}\right]_n$

place, trapped in an amorphous solid phase

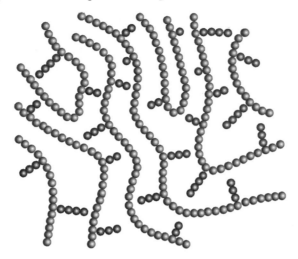

where the packing is less dense than for unbranched molecules. Branches on one macromolecule prevent another chain from approaching as closely as before. The material that forms, generically called *low-density polyethylene*, has a density less than 0.94 g cm^{-3} and is used for soft packaging, plastic bags, squeeze bottles, films, plastic sheets, and so on. Its average molecular mass is typically around 10,000.

The list continues. Low-density polyethylene, high-density polyethylene, polyvinyl chloride, polystyrene, polypropylene, polyethylene terephthalate, polyurethane, nylon, *and* nature's own DNA, RNA, myoglobin, glycogen, starches, cellulose, and many others not named in Table 9-2—polymers all. Each polymer is the complex result of a simple process using simple parts. Each polymer arises from different monomers. Each polymer has its own encumbrances, its own constraints, its own microscopic properties made macroscopic. Each polymer has its own internal ordering and noncovalent associations, highly specific, which can turn even a single macromolecule into a corporate state of matter. Each macromolecule also has its own external relationships with other macromolecules as part of a larger bulk state. And each, of course, is shaped by the same forces and laws that shape all matter.

REVIEW AND GUIDE TO PROBLEMS

Scenes from a formal recital, just before and just after the performance comes to a close.

Scene 1: The audience is quiet, attentive, orderly. Row by row, one behind the other, each listener occupies a seat in a symmetric array. See the arrangement as a kind of lattice, a fixed network of points labeled Row A/Seat 12, Row G/Seat 1, Row X/Seat 5, Row This/Seat That. Individuals fidget here and there, but otherwise they stay in their assigned positions. We know where to find them.

They are not free. The concertgoers are mindful of each other, and they behave as befits a formal gathering: no talking, no kicking, no standing, no walking. They stay tightly connected throughout the performance. They interact.

Scene 2: The performance ends and the spell is broken. The audience applauds, stands, and leaves the hall. No longer are they constrained by a latticework of seats. Although still confined to the building, they are more mobile.

They continue to interact, yet less so. Knots of people form and break up as acquaintances meet, chat for awhile, and then part. The crowd mills about; shoulders brush against shoulders; bottlenecks form. It is a looser interaction, less enduring than before. Neighbor is still aware of neighbor, but the contacts are briefer and weaker.

Scene 3: The crowd leaves the theater and disperses into the night. The people are on their own now, free to move without constraint. They make little contact one with the other. They see each other only from a distance, unable even to distinguish young from old, tall from short, man from woman. They do not interact.

Solid. Liquid. Gas.

IN BRIEF: NONCOVALENT INTERACTIONS AND CONDENSED PHASES

1. Molecules together are different from molecules apart, just as any true group is more than simply the sum of its members. Acting in concert, no longer free, molecules must adjust their positions, orientations, energies, and electrons to suit their neighbors. They assemble themselves into collective *states of matter*, into *phases*—into such broadly defined categories as solid, liquid, and gas, and also into such

in-between states as liquid crystal and cluster. They are constrained. They interact.

Electrical interactions thus grant to electrons, nuclei, atoms, molecules, and ions the power to *organize*. Electrical interactions enable the particles of matter to stick together and form corporate structures. It is the fundamental electromagnetic force, always, that binds together the atom, the molecule, the cluster, the crystal.

For atoms, we find structures in which the electrical attractions and repulsions are maintained in quantum mechanical balance about the nucleus. We use the Schrödinger equation to understand the energy of the atom, the distribution of its electrons, the tendency to react or not to react.

For molecules, we employ the ready term "covalent bond" to capture an even more complex balance of attractions and repulsions, but still we deal with the same fundamental electromagnetic force. The same laws and equations of quantum mechanics still apply.

And for atoms *not* covalently bonded, for groups of molecules, for all those structures where atoms, molecules, and ions stick together without intimately pooling their electrons—for all the various *states of matter*, the fundamental electromagnetic force remains the same too, and so do the laws of quantum mechanics. The differences are in the details.

2. Details such as: What kind of charge distribution is interacting with what? A point charge with a point charge? A point charge with a dipole? A dipole with a dipole? A nonpolar cloud of electrons with a dipole? One nonpolar cloud with another?

Details such as: Is the interaction strong or weak? How does the binding energy compare with the strength of a covalent bond? Is the attraction transient or enduring, flickering or persistent? For how long is the contact maintained?

Details such as: By how much does the coupling decrease as the particles move apart? Does the potential energy scale with distance as $1/r$, $1/r^2$, $1/r^3$, $1/r^4$, $1/r^5$, $1/r^6$? Does the direction of approach affect the strength of interaction?

In details such as these, we see, are some of the differences between oil and water, sand and salt, blood and bile explained.

3. INTERACTIONS. Here, in summary, are the principal noncovalent forces that control the states of matter. Recognize in each of them a meeting of electrical opposites.

Ion–ion: the strongest and most far-reaching of all the noncovalent couplings, a Coulomb interaction between two charged particles. One charge is here and another is there, each sufficiently compact to be considered a mere point in space. Unlike charges attract, like charges repel, and, significantly, every ion has an effect on every other. Even those from far away, from one end of a crystal to another, are drawn into the network of attractions and repulsions. The potential energy falls off only slowly with distance (as $1/r$), decreasing by just half when the separation is doubled.

Example of an ion–ion interaction: Na^+ and Cl^-. Potential energy is approximately 550 kJ mol^{-1} at 2.5 Å, and 275 kJ mol^{-1} at 5.0 Å.

Ion–dipole: an ion (a point charge) interacts with a dipole (a *two-point* charge), such as when Na^+ encounters H_2O. With the charge now separated into positive and negative ends, one point of the dipole attracts the ion while the other point repels it:

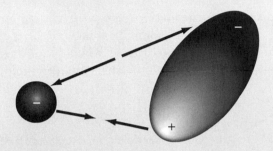

Pushed and pulled, the ion salvages only a weaker, less direct attraction than otherwise possible, acquiring a potential energy that falls away as $1/r^2$. A value near 20 kJ mol^{-1} is typical for a separation of 4 Å.

Dipole–dipole: arising from the mutual attraction of two dipoles. Weaker still than the ion–dipole coupling, the dipole–dipole potential energy decreases as $1/r^3$. Not just distance, moreover, but orientation is critical here as well. Lowest energies are realized when the two dipoles line up positive to negative, as pictured below:

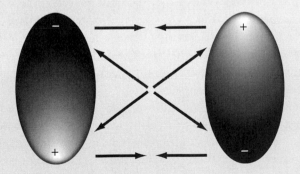

Dipole–dipole couplings amount to only about 2 kJ mol^{-1} at distances of a few angstroms.

Hydrogen bond: an electrostatic interaction in which a hydrogen atom acts as a bridge between two strongly electronegative atoms: either N,

O, or F. Covalently bonded as, say, $O^{\delta-}$—$H^{\delta+}$, the partially positive hydrogen will attract the accumulated negative charge of a lone pair on another electronegative atom, say N:

$$\begin{array}{ccc} \delta- & \delta+ & \delta- \\ O-\!\!\!\!\! & H\cdots\!\!\!\! & N \end{array}$$

The small hydrogen atom brings the oxygen and nitrogen close together, thereby lowering the potential energy by approximately 20 kJ mol^{-1}. Linear arrangements of the three atoms produce the strongest connections.

London dispersion interaction: a transient attraction between *induced* dipoles, arising from a momentary skewing of the charge in an electron cloud. The London interaction, which requires no permanent dipole moment, permits even atoms and nonpolar molecules to associate.

Example: Ar and Ar. Potential energy: 2 kJ mol^{-1} or less at 5 Å. Distance dependence: $1/r^6$.

4. SOLID, LIQUID, GAS. Particles *can* interact. Particles *do* interact. Electrical forces are the means, and the diverse states of matter are the ends. From solid to liquid to gas, atoms and molecules go from states of order to states of disorder, from slow motion to frenzied motion, from strong interactions to weak, from connected groups to disconnected individuals.

Bound the tightest are the solids, existing in both crystalline and amorphous forms. A *crystal* develops when the atoms occupy positions in a symmetric *lattice*, whereby a single *unit cell*, reproduced over and over, generates a regular, predictable structure—like bricks in a wall, seats in a theater, or tiles on a floor. And, just so, we know where to find every atom according to some rule, or *symmetry operation*. We travel *m* steps of length *a* in one direction, *n* steps of length *b* in another direction, *p* steps of length *c* in a third direction, and, however great the journey, we find the local environment seemingly unchanged. There is *long-range order*.

Suppose, though, that our rules only apply over limited distances, perhaps just a few steps in any direction. Starting out near atom A, for example, we first take one step in direction *a*, two steps in direction *b*, one step in direction *c*, and then look around. If we find ourselves standing near another atom A, then we know that the structure is symmetric in this small neighborhood. But five steps away, or ten steps away, or fifty steps away, the view may not be the same; farther away, there may be an atom B where we would otherwise expect an atom A. If so, then the structure lacks long-range order. It is either an *amorphous solid* (a *glass*) or a liquid, a system ordered locally but not globally. The particles clump together into finite clusters, but the organization does not extend to the whole phase.

The difference between an amorphous solid and a liquid? *Mobility*. The particles of a liquid have more freedom and energy to move, to flow, to adapt to the shape of their container. A liquid plays host to a shifting, fluid phase in which clusters continually form, break up, and form again somewhere else. An amorphous solid is, instead, more like a frozen liquid, retaining a fixed shape, flowing slowly if at all, its clusters persisting over longer times.

The difference, then, between a liquid and a gas? Again: more mobility; more energy; less order (or, better said, *no* order). Well separated, the particles of a gas move rapidly and find little time or opportunity to interact.

5. EXAMPLES OF CRYSTALLINE SOLIDS: covalent crystals, metallic crystals, ionic crystals, molecular crystals.

A *covalent crystal (network crystal)* is a giant molecule, an unbroken network of atoms linked by covalent bonds. Example: diamond, one of the elemental forms of carbon.

Another kind of extended solid is the *metallic crystal*, a state in which the nuclei are held in a lattice but the valence electrons are free to wander. Electronic mobility makes a metal a good conductor of electricity and heat.

In an *ionic crystal*, by contrast, the electrons remain localized near individual nuclei, and the ions are held together by Coulomb forces. In a *molecular crystal*, the electrons stay bound to individual *molecules*, and these intact molecules then associate by hydrogen bonds, dipole–dipole interactions, or London dispersion interactions, as appropriate.

6. EXAMPLES OF IN-BETWEEN STATES OF MATTER: isolated clusters, liquid crystals, polymers.

Two particles, three particles, ten, two hundred—a single *cluster* is a structure caught between microscopic and macroscopic, with properties different from both a molecule and a bulk material. Held together by any means available (covalent bonds, ionic bonds, hydrogen bonds, dipole–dipole interactions, dispersion forces), the cluster, or *nanocrystal*, is distinguished by a large surface area relative to its interior.

A *liquid crystal* combines the rotational order of a crystal with the translational disorder of a liquid. Although the material flows like a liquid, the molecules are unable to tumble randomly in all directions.

A *polymer* is an overgrown molecule, a *macromolecule*, built from a limited set of smaller *monomers* such as amino acids or nucleotides. The macromolecule is often flexible enough to interact with itself, twisting and folding to make noncovalent bonds between atoms on different parts of the chain. Interactions *between* macromolecules give bulk polymers special properties.

In each form of association, we recognize a means to an end. The means: interparticle interactions. The end: a state of matter. Solid? Liquid? Liquid crystal? Gas? Which state and why?

Save those questions for the next five chapters, where we shall see how macroscopic conditions and the laws of thermodynamics jointly determine the microscopic behavior of the particles. Put pressure on a gas, for instance, and eventually the particles will be squeezed into a more ordered state, a liquid. Or apply heat to a crystal, and observe how the particles break away from the ordered arrangement of the lattice as the solid melts . . . but all that is still to come.

At the moment, there are other problems to solve.

SAMPLE PROBLEMS

To go from the macroscopic to the microscopic, from the continuous to the discrete, from grams to atoms—we start with the *density* of a substance, defined as the ratio of mass to volume:

$$\text{Density} = \frac{\text{mass}}{\text{volume}}$$

A macroscopic quantity, the density tells us how much mass is packed into a certain volume. A *microscopic* quantity as well, density also tells us how many particles contribute to that mass.

EXAMPLE 9-1. Spreading Out: Solid to Liquid to Gas

PROBLEM: Densities for nine materials in their normal states are given below. For each, compute: (a) The number of atoms or molecules in a volume of 1 L. (b) The volume, in milliliters, occupied by 1 mole of substance.

SUBSTANCE	SYMBOL	STATE	DENSITY (g/mL)
diamond	C	solid	3.51
graphite	C	solid	2.25
gold	Au	solid	19.31
mercury	Hg	liquid	13.55
water	H_2O	liquid	1.00
bromine	Br_2	liquid	3.12
oxygen	O_2	gas	1.43×10^{-3}
fluorine	F_2	gas	1.70×10^{-3}
hydrogen	H_2	gas	8.99×10^{-5}

SOLUTION: Remember that 1 milliliter (mL) corresponds to a volume of 1 cubic centimeter (cm^3), equivalent to a cube with dimensions 1 cm $\times$ 1 cm $\times$ 1 cm. A volume of 1 liter (L) similarly defines a cube 10 cm on a side ($1000 \ cm^3 = 1000 \ mL = 1 \ L$).

(a) *Particle density: atoms per liter.* We go from grams to moles to atoms and, simultaneously, from milliliters to liters. For diamond, the calculation takes the following form:

$$\frac{3.51 \text{ g}}{\text{mL}} \times \frac{1 \text{ mol}}{12.011 \text{ g}} \times \frac{6.02 \times 10^{23} \text{ atoms}}{\text{mol}} \times \frac{1000 \text{ mL}}{\text{L}} =$$

$$\frac{1.76 \times 10^{26} \text{ atoms}}{\text{L}}$$

(b) *Molar volume.* Use here the inverse density (volume/mass), again employing the same relationship between grams and moles:

$$\frac{1 \text{ mL}}{3.51 \text{ g}} \times \frac{12.011 \text{ g}}{\text{mol}} = \frac{3.42 \text{ mL}}{\text{mol}} \qquad \text{(diamond)}$$

The completed table now becomes:

SUBSTANCE	SYMBOL	STATE	g/mL	particles/L	mL/mol
diamond	C	solid	3.51	1.76×10^{26}	3.42
graphite	C	solid	2.25	1.13×10^{26}	5.34
gold	Au	solid	19.31	5.90×10^{25}	10.20
mercury	Hg	liquid	13.55	4.07×10^{25}	14.80
water	H_2O	liquid	1.00	3.34×10^{25}	18.0
bromine	Br_2	liquid	3.12	1.18×10^{25}	51.2
oxygen	O_2	gas	0.00143	2.69×10^{22}	22.4×10^3
fluorine	F_2	gas	0.00170	2.69×10^{22}	22.4×10^3
hydrogen	H_2	gas	0.0000899	2.68×10^{22}	22.4×10^3

Note, especially, how the three gases stand out from the liquids and solids.

First point: The condensed phases clearly do pack more particles into the same space, just as we expect. Whereas a few tens of cubic centimeters suffice to hold a mole of solid or liquid, the same amount of gas fills a volume of fully 22.4 *liters* (22,400 cm^3). The expansion is approximately thousandfold.

Second point: The condensed phases have their own distinct properties, one different from the next. Different materials pack together in different ways. The particle densities are different. The molar volumes are different. The atoms and molecules have diverse sizes, shapes, electron distributions, and dipole moments—and they interact accordingly. They interact *significantly* in the condensed phases, and they interact *differently*. Some structures mesh closely as solids and liquids; others do not.

But a gas is a gas is a gas. The particle densities (2.7×10^{22} molecules per liter) and molar volumes (22.4 L) seem scarcely to depend on the particular molecule, suggesting to us that all gases fill their space in the same way. Perhaps they should, too: In a volume so large, with distances between the molecules so great, with opportunity for intermolecular interactions so scant—what chance is there for either oxygen or fluorine or hydrogen to assert its individual chemical identity? Should not each molecule appear just as a speck of mass in the distance, a molecular billiard ball with no other distinguishing characteristics? Bear that image in mind when we take up the model of the *ideal gas* in Chapter 10.

EXAMPLE 9-2. Counting Particles in Solution

Since so much of chemistry takes place in the solution phase, we need to consider carefully the relationships between mass, volume, particles, and concentration.

PROBLEM: 3.00 g of sucrose ($C_{12}H_{22}O_{11}$) are dissolved in sufficient water to yield 15.0 mL of solution. (a) What is the molarity of the solution? (b) How many molecules are contained in each mL? (c) How does the concentration change when the original volume (15.0 mL) is diluted to 20.0 mL?

SOLUTION: *Molarity* (*M*) is defined as the number of moles of solute dissolved per liter of solution (the total volume to include both solvent and solute after mixing).

(a) *3.00 g sucrose/15.0 mL solution.* Go from grams to moles and from milliliters to liters, using 342.3 g for the molar mass of $C_{12}H_{22}O_{11}$:

$$\frac{3.00 \text{ g}}{15.0 \text{ mL}} \times \frac{1 \text{ mol}}{342.3 \text{ g}} \times \frac{1000 \text{ mL}}{\text{L}} = \frac{0.584 \text{ mol}}{\text{L}} = 0.584 \text{ } M$$

(b) *How many molecules in 1 mL?* Go from moles to molecules and from 1 L to 1 mL:

$$\frac{0.584 \text{ mol}}{\text{L}} \times \frac{6.02 \times 10^{23} \text{ molecules}}{\text{mol}} \times \frac{1 \text{ L}}{1000 \text{ mL}} =$$

$$\frac{3.52 \times 10^{20} \text{ molecules}}{\text{mL}}$$

(c) *Volume = 20.0 mL.* The mass of solute stays the same, but the concentration falls as the volume of solution decreases:

$$\frac{3.00 \text{ g}}{20.0 \text{ mL}} \times \frac{1 \text{ mol}}{342.3 \text{ g}} \times \frac{1000 \text{ mL}}{\text{L}} = 0.438 \ M$$

EXAMPLE 9-3. Symmetry Elements

In this example and the two following, we review the relationship between crystalline symmetry and structure.

PROBLEM: Think of the digits

800008800008800008800008

not as a number, but as a graphical object—an array of shapes and forms, something like a crystal in one dimension, and ponder: What is the unit cell from which this object is built? How many cells are strung together? What elements of symmetry exist *within* each unit cell? What elements of symmetry exist within each separate digit?

SOLUTION: As the smallest repeating unit, the unit cell stands out more clearly when presented like this:

800008 800008 800008 800008

The *translational symmetry* arises, we see, from fourfold repetition of the six-digit unit cell 800008. No other group will do. Not 800, because 800 would generate the array 800800800 . . . . Not 800008800008, because 800008800008 can be broken down further into the identical units 800008 and 800008. Not larger, not smaller, but just right: 800008 is the minimal building block of the array.

But notice, too, that there are additional elements of symmetry in the unit cell itself, namely a mirrorlike plane through the center:

800 | 008

See the six digits as divided by a mirror that juts out perpendicular to the page, and compare left and right: One position to the left of center is a 0; one position to the right of center is a 0. *Two* positions to the left of center is a 0; two positions to the right of center is a 0. Three positions to the left is an 8; three positions to the right is an 8. The larger array 800008 thus appears when a smaller object, 800, is reflected in a mirror.

And keep looking, because mirror planes cut through the individual digits as well: two through the 8 and two through the oval-shaped 0, horizontally and vertically.

The 8 we can generate from a small half-oval ( ), reflecting the object

left–right (to produce ◯) and then up–down (to produce 8).

The 0 we can generate likewise from a quarter-oval ( ),

again reflecting left–right and up–down in the sequence shown below:

Now, put everything together: (1) Construct the digit 8 by two suitable reflections of a half-oval. (2) Make a 0 by two reflections of a quarter-oval, and then repeat to create the block 800. (3) Reflect 800 in a mirror to produce the unit cell 800008. (4) Reproduce the unit cell four times to yield the full structure (and be glad to discover the symmetry and simplicity inherent in the formidable array 800008800008800008800008).

QUESTION: Yet certainly the symmetry is imperfect? After all, there is a single 8 at each end but paired 88s everywhere else. Aren't the endpoints different from the middle?

ANSWER: Yes, the symmetry is imperfect. Yes, the endpoints are different. Yes, the structure must end somewhere. At some point the symmetry must be broken.

The endpoints of a linear array become the *surface* of a three-dimensional array, and here we recall a crucial difference between macrocrystals (big arrays) and nanocrystals (small clusters): In the nanocrystal, there is an overweighting of the surface relative to the interior. For the present example, just think how the importance of the endpoints (the surface) diminishes as the one-dimensional numerical crystal grows larger:

800008

800008800008800008800008

800008800008800008800008800008800008800008800008800008

EXAMPLE 9-4. Crystals, Symmetry, and Density

Consider now the structure of sodium chloride, which crystallizes into a *face-centered cubic* lattice of the kind shown in Figure R9-1(a). The unit cell, a cube of dimension $a = 5.64$ Å, contains 14 lattice points: one at each of the eight corners, plus one each at the center of all six faces. Cl^- ions occupy the corner and face positions; Na^+ ions fall *between* the lattice points, along the edges and at the center of the cube as shown in Figure R9-1(b).

PROBLEM: Compute the density of crystalline NaCl.

SOLUTION: We can do so, remarkably, because the crystal is built according to a plan. Give us the unit cell, and we know the entire crystal; and give us, in particular, the volume of the unit cell and the mass contained within, and we know the density everywhere. The symmetry of the structure guarantees it.

First, the mass. Studying Figure R9-1(b), we count *one-eighth* of a chloride ion at each corner (a point where eight unit cells come together), and *one-half* of a chloride ion on each face (where two unit cells

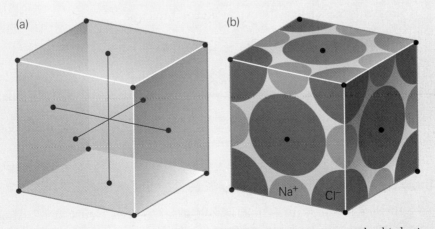

FIGURE R9-1. Cubic structure of crystalline NaCl. (a) Face-centered cubic lattice. Each unit cell has eight points at the corners and six on the faces. (b) Placement of sodium and chloride ions in the lattice.

come together). There is also, at the center of the cube, one whole and unshared sodium ion; and, finally, one-quarter of a sodium ion along each edge (where four cells come together). The occupancies for the unit cell therefore are

$$\frac{\frac{1}{4} \text{ Na}^+}{\text{edge}} \times 12 \text{ edges} + \frac{1 \text{ Na}^+}{\text{center}} \times 1 \text{ center} = 4 \text{ Na}^+$$

$$\frac{\frac{1}{8} \text{ Cl}^-}{\text{corner}} \times 8 \text{ corners} + \frac{\frac{1}{2} \text{ Cl}^-}{\text{face}} \times 6 \text{ faces} = 4 \text{ Cl}^-$$

and the total mass, expressed in atomic mass units (u), is

$$4 \text{ Na} + 4 \text{ Cl} \equiv 4 \times 22.9898 \text{ u} + 4 \times 35.453 \text{ u} = 233.77 \text{ u}$$

Recalling (from Chapter 2) that the atomic mass unit is defined so that

$$1 \text{ atom } {}^{12}\text{C} \equiv 12.0000 \text{ u}$$

$$1 \text{ mole } {}^{12}\text{C} \equiv 12.0000 \text{ g} \equiv 6.02 \times 10^{23} \text{ atoms}$$

we then use Avogadro's number to obtain the amount in grams:

$$233.77 \text{ u} \times \frac{1 \text{ g}}{6.02 \times 10^{23} \text{ u}} = 3.88 \times 10^{-22} \text{ g}$$

Since this mass falls within a cubic volume equal to

$$V = a^3 = (5.64 \text{ Å})^3 = (5.64 \times 10^{-8} \text{ cm})^3 = 1.79 \times 10^{-22} \text{ cm}^3$$

the density must be

$$\text{Density} = \frac{\text{mass}}{\text{volume}} = \frac{3.88 \times 10^{-22} \text{ g}}{(5.64 \times 10^{-8} \text{ cm})^3} = 2.16 \text{ g cm}^{-3}$$

EXAMPLE 9-5. More Cubic Lattices

Two further forms of the cubic lattice are shown in Figure R9-2: the *primitive* cell and the **body-centered** cell. The primitive form has lattice points only at the corners; the body-centered cell contains an additional point at the center of the cube.

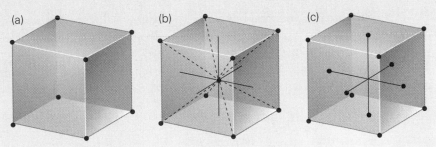

FIGURE R9-2. Cubic lattices. (a) Primitive. (b) Body-centered. (c) Face-centered.

PROBLEM: How many atoms are included in each cubic cell? (a) Primitive. (b) Body-centered. (c) Face-centered.

SOLUTION: Eight cubic cells meet at a single, shared corner; four meet along an edge; two meet at a face. Hence each cell claims one-eighth of a corner atom, one-fourth of an edge atom, one-half of a face atom, and one whole atom at the body center.

(a) *Primitive.* Eight corners are occupied, each contributing one-eighth of an atom:

$$\text{Total:} \quad 8 \times \tfrac{1}{8} = 1$$

(b) *Body-centered.* In addition to the eight corners (one atom), there is one whole atom at the center:

$$\text{Total:} \quad (8 \times \tfrac{1}{8}) + 1 = 2$$

(c) *Face-centered.* This structure we already know from Example 9-4: eight corners (one atom) and six faces (three atoms).

$$\text{Total:} \quad (8 \times \tfrac{1}{8}) + (6 \times \tfrac{1}{2}) = 4$$

EXAMPLE 9-6. Noncovalent Interactions

PROBLEM: What are the principal interactions holding together each of the following phases? (a) Ice, $H_2O(s)$. (b) Water, $H_2O(\ell)$. (c) Steam, $H_2O(g)$. (d) A hydrated potassium ion, $K^+(aq)$, in an aqueous solution of KCl. (e) Liquid nitrogen, $N_2(\ell)$. (f) Solid carbon monoxide, $CO(s)$. (g) Solid potassium bromide, $KBr(s)$.

SOLUTION: Ask: Is the species neutral or charged? Polar or nonpolar? Are there opportunities for hydrogen bonds?

(a) *Solid water.* A neutral but polar structure, water interacts most effectively by means of relatively strong hydrogen bonds. A partially positive hydrogen on one water molecule coordinates with the partially negative oxygen on another, arranged so that each H_2O makes four such connections in ice:

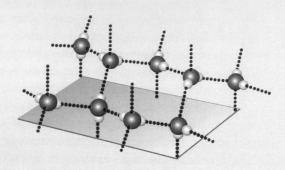

(b) *Liquid water.* Same molecule, same interactions: hydrogen bonds. In the liquid phase, however, the molecules move faster, more energetically, more independently. Hydrogen-bonded clusters of water molecules arise and fall and arise again, leaving the liquid with no definite shape.

(c) *Gaseous water.* Same molecule, same inherent capability for intermolecular associations—but, in the gas phase, no time to make them. Since the molecules travel too rapidly for the interactions to be well expressed, the phase is held together only by the walls of its container.

(d) *Hydrated K^+.* Here we have a cation dissolved in a polar liquid, solvated by **ion–dipole interactions**. Water molecules cluster about the potassium ion, turning their negative oxygen sites inward to face the positive charge on K^+.

(e) *Liquid nitrogen.* Neutral and nonpolar, molecular nitrogen has only the weak **London dispersion interactions** at its disposal.

(f) *Solid carbon monoxide.* The slightly polar CO molecules interact via weak **dipole–dipole interactions**.

(g) *Solid potassium bromide.* An ionic crystal, KBr is held together by electrostatic **ion–ion interactions** between K^+ and Br^-.

EXAMPLE 9-7. Strength Amidst Weakness

PROBLEM: Which of the following substances should we expect to boil at the higher temperature? (a) He or Ar. (b) H_2O or CH_4. (c) C_2H_6 or C_2H_5OH. (d) $CH_3CH_2CH_3$ or $CH_3CH_2CH_2CH_2CH_2CH_2CH_2CH_3$. (e) HF or HCl. (f) NaCl or HCl.

SOLUTION: A liquid boils only when its molecules have enough energy to break free of their attractive forces. The stronger the interaction, the more energy is needed. The more energy is needed, the higher will be the boiling temperature.

Now, think of the factors that would make a weak interaction strong. For polar species, a large dipole moment. For nonpolar species, a large cloud of electrons susceptible to induced polarization. For all species, a large surface area offering many points of contact.

(a) *He or Ar.* The larger atom, argon, with the more polarizable electron cloud, is more readily distorted into a temporary dipole. Compared with helium, argon's electrons are bound less tightly around the nucleus; and thus, more susceptible to fluctuation, they respond with larger induced dipole moments. Argon normally boils at $-185.7°C$. Helium boils at $-268.9°C$. Both elements are gases at room temperature.

(b) H_2O *or* CH_4. Water is polar; methane is nonpolar. Liquid water, knitted together by strong hydrogen bonds, boils at $100.0°C$. Liquefied methane, with only induced dipole–dipole interactions to hold the phase together, boils at $-161.5°C$.

(c) C_2H_6 *or* C_2H_5OH. Ethanol, made polar by the hydroxyl OH group, boils at $78.2°C$. Nonpolar ethane, boiling at $-88.6°C$, normally exists as a gas.

(d) $CH_3CH_2CH_3$ *or* $CH_3CH_2CH_2CH_2CH_2CH_2CH_2CH_3$. Both propane (C_3H_8) and octane (C_8H_{18}) are nonpolar, but octane's longer chain and larger electron cloud offer more opportunity for effective London interactions. Octane starts to boil at $125.6°C$; propane, at $-42.1°C$.

(e) *HF or HCl.* Fluorine, more electronegative than chlorine, gives HF the larger dipole moment and allows it to form hydrogen bonds. Hydrogen fluoride boils at $19.5°C$. Hydrogen chloride boils at $-84.9°C$.

(f) *NaCl or HCl.* Sodium chloride, our prototype ionic compound, does not even *melt* until $801°C$, so strong are the electrostatic forces. Under the influence, still, of these powerful forces, the molten liquid subsequently holds together until finally boiling at a temperature of $1413°C$. Covalent HCl, with its low boiling temperature of $-84.9°C$, exists as a gas at room temperature—a dangerous, very corrosive, very toxic gas.

EXAMPLE 9-8. Lattice Energy

The strength of an ionic crystal is reflected in its **lattice energy**, the energy needed to transform one mole of solid material into gaseous ions. Strong ionic bonds produce a correspondingly high lattice energy.

Consider, then, the tangle of electrostatic interactions in sodium chloride: Sodium ion number 1 is attracted to chloride ion number 1, and also to number 2, and to number 3, and to number 4, and so on; and sodium ion number 1 is *repelled* by sodium ion number 2, and number 3, and number 4, and so on; and, likewise, chloride ion number 1 is re-pelled by chloride ion number 2, and number 3, and number 4, and so on. Count them all, and see how an ionic crystal becomes the sum of its electrostatic interactions, ion by ion, everywhere throughout the solid. So far-reaching is the Coulomb interaction that every pair of ions con-tributes to the binding. The lattice energy is the sum of all these pairwise contributions.

For simplicity, though, let us use just one well-chosen ionic bond to gauge the strength of the entire crystal: the attraction between adjacent cation and anion. It is the shortest and strongest connection, a telltale measure for the system as a whole.

PROBLEM: Which of the following ionic crystals should we expect to have the higher lattice energy? (a) NaCl or MgO. (b) NaCl or KBr.

SOLUTION: Recall (from Chapter 1) that the Coulomb force be-tween charges q_1 and q_2 is proportional to $q_1 q_2 / r^2$, where r is the dis-tance between the particles. The potential energy, related to force times distance, goes as $q_1 q_2 / r$.

Which means: (1) Ions with larger charges enjoy a proportionately stronger interaction. (2) Ions separated by a smaller distance interact more strongly as well. If we assume that oppositely charged ions are packed so close as to be just touching, then species with small radii will be attracted more than those with large radii.

(a) *NaCl or MgO.* Na^+ and Cl^- are monopositive ions; Mg^{2+} and O^{2-} are dipositive. O^{2-} is, moreover, a smaller ion than Cl^-, since chlorine (a third-row element) supports an additional shell of electrons. The magne-sium ion, for its part, is also smaller than the sodium ion. On grounds of both charge and size, magnesium oxide should have the higher lattice en-ergy, and it does: approximately 3800 kJ mol^{-1} for MgO; 788 kJ mol^{-1} for NaCl.

(b) *NaCl or KBr.* Since all the ions are monopositive, any difference in binding should be attributable to the spacing between the charges. Na^+ and Cl^- are the smaller ions, and sodium chloride does have the higher lattice energy: 788 kJ mol^{-1} compared with 682 kJ mol^{-1}.

EXERCISES

1. The three principal states of matter are often defined according to their macroscopic properties:

   Solid: a form of matter with definite shape and volume.

   Liquid: a form of matter with definite volume but indefinite shape.

   Gas: a form of matter with neither definite shape nor volume.

   Expand these descriptions to include the *microscopic* character of solids, liquids, and gases.

2. Explain the difference between long-range order and short-range (local) order, using solid, liquid, and gas as examples.

3. Criticize the following statement from the text:

   > "We know the whereabouts of every atom in a crystal, because the entire lattice is built according to definite rules of symmetry."

   Do we indeed know the location of every atom exactly? Reconcile the statement with the indeterminacy principle.

4. How does a quasicrystal differ from a usual crystal?

5. Briefly describe the properties and ordering of a *cluster* (nanocrystal).

6. Describe, similarly, the properties and ordering of a liquid crystal. Between what two principal states of matter does it fall?

7. What noncovalent forces hold together the various states of matter? Why do solids, liquids, and gases differ in their degrees of order?

8. Review the source and operation of the key noncovalent interactions:

   (a) ion–ion
   (b) ion–dipole
   (c) dipole–dipole
   (d) hydrogen bond
   (e) London dispersion

   How does each effect vary with distance? In what ways are the different interactions similar?

9. Molecule X is nonpolar. Molecule Y is polar. Both have approximately the same mass. One of them, however, exists as a liquid at room temperature; the other, as a gas. Which is which?

10. Molecule X has a molar mass of 150 g. Molecule Y has a molar mass of 1200 g. Both are nonpolar; both exist as liquids at room temperature. Which probably has the higher boiling point?

11. Molecule X is a long, unbranched hydrocarbon. Molecule Y is a highly branched isomer of molecule X, nearly spherical. Both are liquids. Which probably has the higher boiling point?

12. Temperature is a measure of how fast the molecules are moving. Pressure is a measure of the external force pressing on a material. (a) What combination of temperature (high? low?) and pressure (high? low?) will favor the formation of a solid? (b) What combination will favor the formation of a gas?

13. States of matter differ in their *compressibility*—the ease with which they change their volume in response to pressure. A substance with a high compressibility deforms readily under pressure. A substance with a low compressibility, hard to squeeze, resists a change in volume. How do solid, liquid, and gas compare in relative compressibility? Explain why.

14. Sodium fluoride (NaF) crystallizes in a face-centered cubic unit cell, just like NaCl. The unit cell is 4.6342 Å on an edge. Sketch the arrangement of the atoms, and compute the density of NaF in g/cm^3.

15. AgCl, which has a density of 5.5710 g/cm^3, also crystallizes in a face-centered cubic cell. Compute the cell dimensions.

16. A certain ionic compound, AgX, forms a face-centered cubic crystal with a unit cell dimension of 5.7745 Å and a density of 6.4772 g/cm^3. Identify the unknown element X.

17. Metallic sodium (Na) crystallizes in a body-centered lattice with a unit cell dimension of 4.2856 Å. (a) How many atoms are contained within each unit cell? (b) Sketch the arrangement and compute the density in g/cm^3.

18. A certain element crystallizes in a body-centered cubic unit cell, 2.8664 Å on a side. The density is 7.86 g/cm^3. What is the element?

19. Given the densities, compute the molar volumes of the elements listed below:

	SUBSTANCE	DENSITY (g/mL)
(a)	Ag	10.50
(b)	B	2.34
(c)	Bi	9.80
(d)	K	0.86
(e)	Ne	0.00090

For each element, how many atoms are contained in a cube 1 mm on a side?

20. Although chlorine, bromine, and iodine—all halogens—have similar chemical properties, each element exists as a different state of matter at room temperature. One is a solid; one is a liquid; one is a gas. Which is which, and why?

21. The molar volumes of chlorine, bromine, and iodine are given below:

	SUBSTANCE	MOLAR VOLUME (L)
(a)	Cl_2	22.06
(b)	Br_2	0.0512
(c)	I_2	0.0515

Compute the corresponding densities, and answer: Why are bromine and iodine so similar? Why is chlorine so different?

22. (a) Using only the information derived from the preceding two exercises, predict the normal state of fluorine—solid, liquid, or gas. (b) Estimate the density and molar volume of F_2 at room temperature.

23. Calculate the concentration of sodium ions in a 463-mL aqueous solution containing 36.4 g Na_2SO_4. Express the result as a *molarity*, M: moles of solute per liter of solution.

24. How many sulfate ions are contained in 1.00 mL of the solution described in the preceding exercise?

25. Compute the mole fraction of both dextrose (solute) and water (solvent) in a 1.00 M solution. Notice that neither the molar mass of dextrose nor the volume of solution is needed to solve the problem. Why?

26. The densities of methanol, ethanol, and water at 20°C are given below:

SUBSTANCE	DENSITY $(g\ mL^{-1})$
CH_3OH	0.7914
CH_3CH_2OH	0.7893
H_2O	0.9982

    Calculate the mole fraction of each component in the following mixtures:

    (a) 10.0 mL CH_3OH + 10.0 mL CH_3CH_2OH
    (b) 10.0 mL CH_3OH + 10.0 mL H_2O
    (c) 10.0 mL CH_3CH_2OH + 10.0 mL H_2O
    (d) 10.0 mL CH_3OH + 10.0 mL CH_3CH_2OH + 10.0 mL H_2O
    (e) 72.0 mL CH_3OH + 1.05 L CH_3CH_2OH + 413.0 mL H_2O

27. 2.00 g Li_2SO_4 are dissolved in sufficient water to produce a solution volume of 50.0 mL. (a) Compute the molarity. (b) Compute the mole fractions of Li^+, SO_4^{2-}, and H_2O.

28. Imagine now that the solution in the preceding exercise is diluted to a total volume of 1.00 L. The mass of dissolved solute stays the same. (a) Compute the molarity. (b) Compute the mole fractions of Li^+, SO_4^{2-}, and H_2O.

29. Beaker A, containing a 0.500 *M* solution of Li_2SO_4, holds 50.0 mL of the liquid. Beaker B, containing a 0.250 *M* solution of the same compound, holds 100.0 mL. What final concentration is attained when the two solutions are mixed together?

30. Arrange each set of ionic solids in order of increasing lattice energy:

    (a) NaCl, KCl, LiCl, RbCl
    (b) AgBr, AgI, AgCl, AgF
    (c) CaS, MgS, NaCl

31. Describe the noncovalent forces that hold together each of the following solids:

    (a) calcium carbonate, $CaCO_3$
    (b) elemental platinum, Pt
    (c) frozen nitrogen, N_2
    (d) ice

32. Kevlar, a polymer of exceptional strength, has the chemical structure shown below:

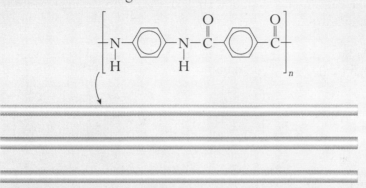

The solid material, a *fiber*, is obtained from liquid crystalline solutions in which the rodlike molecules are kept nearly parallel. This parallel ordering carries over into threadlike fibers with extended chains that run the length of the structure:

(a) By what noncovalent forces are the long molecules held parallel in the fiber? Which functional groups play a key role? (b) Kevlar can stop a bullet. Low-density polyethylene cannot. Why not?

33. In 1839, Charles Goodyear discovered a way to make *vulcanized rubber* by heating the sticky sap of the *Hevea brasiliensis* tree with sulfur:

natural rubber

vulcanized rubber

The reaction produces a certain number of sulfur cross-links between individual macromolecules. (a) Only a dimeric portion (formed from two monomers) is shown for each macromolecule in the equation above. From what monomer does natural rubber derive? Draw the chemical structure. (b) What noncovalent forces bring together different macromolecules in natural rubber? (c) Are the cross-links in vulcanized rubber covalent or noncovalent? (d) Vulcanized rubber is considerably more elastic than natural rubber. When stretched, the vulcanized compound tends to restore its original position. Why might that happen?

34. Which polymer—nylon 66 or polystyrene—is likely to have a higher degree of crystallinity?

nylon 66   polystyrene

35. Arrange the following substances in order of increasing melting point:

$$Ag, \ CO_2, \ Ne, \ He, \ H_2O$$

Specify the dominant noncovalent forces in each system.

36. Specify the noncovalent forces that hold together each of the following liquids:

(a) methanol, CH_3OH
(b) iodine, I_2
(c) tetrachloromethane (carbon tetrachloride), CCl_4
(d) trichloroiodomethane, CCl_3I
(e) water

37. Arrange the following substances in order of increasing boiling point:

$$CH_3OH, \ CF_4, \ CH_4$$

Specify the dominant intermolecular forces in each system.

38. Pick the compound with the higher boiling point:

(a) CH_3CH_3, $CH_3CH_2CH_2CH_2CH_2CH_2CH_2CH_2CH_2CH_3$
(b) $CH_3CH_2CH_3$, $CH_3CH_2CH_2OH$
(c) $C(CH_3)_4$, $CH_3CH_2CH_2CH_2CH_3$

39. Pick the substance more likely to be a gas at room temperature and atmospheric pressure:

    (a) C_2H_4, C_2H_2BrCl
    (b) C_2H_6, $C_{12}H_{26}$
    (c) NH_3, NH_4Cl

40. Is a polymer ever likely to be a gas at room temperature and atmospheric pressure?

41. In what way is a metal like a gas?

42. In what way are all gases alike?

10

Macroscopic to Microscopic— Gases and Kinetic Theory

10-1. The Ideal Gas
10-2. Equation of State
   Measurement of Pressure: Barometer and Manometer
   Pressure: Boyle's Law and Compressibility
   Boyle's Law in Action
   Temperature: Charles's Law
   Number of Moles: Avogadro's Law
   Summing Up: The Ideal Gas Law
10-3. Kinetic Theory
   The Statistics of Pressure
   Energy, Temperature, and Motion
   Collisions
   Thermal Energy
   Speed and Temperature
   The Boltzmann Distribution
   REVIEW AND GUIDE TO PROBLEMS
   EXERCISES

10-1. THE IDEAL GAS

The gaseous state is the simplest, loosest, least constrained form of matter we know. With its territory vast and its inhabitants few, a gas consists mostly of empty space. The gas particles, whether individual atoms or molecules, are on their own. Usually far apart, they travel independently at high speeds like cars on a lonely highway.

Such is the model of an *ideal* gas, ideal not in some aesthetic sense but rather in its idealized conception of noninteracting particles. To simplify the physical picture of a gas, we imagine each microscopic constituent as a point-sized object with a definite mass yet no volume. The point masses fly about the container, bounce off its walls, and travel in straight lines before one collides with another—an event that does little to compromise the presumption of ideality. Energy is simply reshuffled during such a collision, and the two particles go off in new directions to resume their motion unhindered. Except for these brief "elastic" encounters, no particle influences the path of any other.

The ideal gas is an approximation, a theoretical description of a general state and not of any specific substance. We make this approximation knowing well that molecules have shapes, volumes, and electrical characteristics, and knowing also that intermolecular forces certainly exist. In conceiving of ideal gas behavior, however, we do not stubbornly ignore reality. The assumptions of the model are reasonable given the manifest nature of a gas, namely its sparse population and the generally large distances between neighbors. All gases actually do behave ideally at sufficiently low density, typically where the temperature is high and the pressure is low—conditions that tend to increase the distances between the particles.

That makes sense. Returning to our rural road, for instance, we neither know nor care about the make and model of the car in the distance. Insofar as that vehicle affects our own driving (if at all), its shape or size or color is neither relevant nor even recognizable. From that distance all cars look the same, and so do molecules in a rarefied gas.

When prevailing conditions do not support ideal behavior, we shall modify the original assumptions and explore the relationships between intermolecular interactions and changes in state (from gas to liquid and from liquid to solid). In the meantime, the ideal gas model suggests an appealingly simple way to understand the behavior of an important state of matter.

And gases are indeed important to chemistry and to life. Air is a gas, consisting of 78.1% nitrogen, 20.95% oxygen, 0.93% argon, and trace quantities of several additional components. Carbon dioxide, another atmospheric gas present in small amounts, is a key ingredient in photosynthesis. Gases find many technological uses as well, beginning with the internal combustion engine of the automobile. We would be moved to study gases for practical reasons alone, such is their importance, but there is an additional, less tangible reward: This tenuous state of matter, characterized by a handful of simple variables, offers a direct connection between the macroscopic and the microscopic. The gas will demonstrate how the smooth, predictable relationships of macroscopic experience emerge from a molecular world of fantastic complexity. The microscopic origin of everyday phenomena is our recurrent and ever-compelling theme.

10-2. EQUATION OF STATE

Let us begin with the macroscopic properties, recognizing a gas as a fluid form of matter with no fixed shape and generally with large distances between the constituent particles. A gas fills the boundaries of a container and thus acquires a shape only by virtue of its confinement.

Surprisingly few variables—just the pressure (P), volume (V), temperature (T), and number of moles (n)—suffice to describe the gaseous state. Masked by this apparent simplicity, though, is a collection of maybe 10^{23} particles in random motion, each of which obeys Newton's laws. Somewhere, a connection must exist between macroscopic and microscopic.

Already we have an intuitive sense of the macroscopic quantities. **Pressure** tells us how much force a gas exerts on the walls of its container, familiar enough to anyone who has ever handled a balloon or inflated a tire. The concise mathematical definition of pressure (Figure 10-1) is force per unit area ($P = F/A$), and so the appropriate SI unit, the **pascal (Pa)**, has dimensions of N m^{-2}. Commonly used units of pressure include the **atmosphere** (1 **atm** = 1.01325×10^5 Pa) and the **torr** (1 atm = 760 torr). Volume, the spatial extent of the container, needs no further elaboration, of course, and neither does the *amount* of substance (number of moles). We are left, finally, with temperature as a familiar measure of hot and cold, an indicator that heat can flow between

FIGURE 10-1. Pressure: the ratio of force to area. (a) A force of 1 newton imposed over 1 square meter exerts a pressure of 1 pascal. (b) The same weight (1 N) distributed over twice the area (2 m^2) produces half the pressure (0.5 Pa). Standard atmospheric pressure is slightly greater than 100,000 Pa.

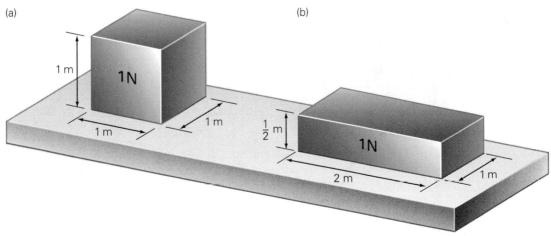

(a)

1 m

1N

1 m

1 m

$\frac{1}{2}$ m

(b)

1N

2 m

1 m

two bodies. A more mechanistic and quantitative understanding of this most important parameter will come shortly as well.

Measurement of Pressure: Barometer and Manometer

Macroscopic pressure aptly describes a pressing, an exertion of force such as the kind applied by the earth's atmosphere to every square meter of the surface. Force due to gravity is what we recognize as weight, and air has a substantial weight arising from all its molecules pressing down on the surface. This force pressing down can, in turn, be used to push something else up, say a column of mercury in a glass tube. From that simple idea comes Torricelli's 17th-century **barometer**, sketched in Figure 10-2(a).

A barometer is made from an evacuated glass tube, sealed at the top but with its open end sitting in a dish of mercury. Air presses down on the mercury and drives the liquid up into the empty tube. The height attained by the column provides a measure of the force per unit area—the

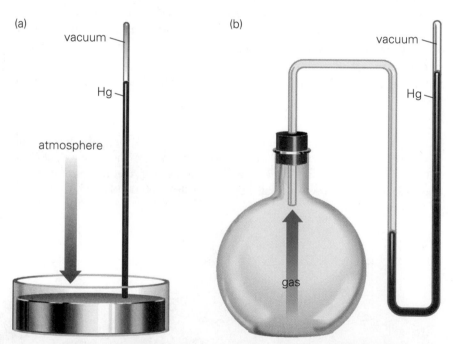

FIGURE 10-2. (a) A simple barometer, designed to measure atmospheric pressure. The weight of the air pushes mercury up the evacuated glass tube, forcing the column to a height proportional to the prevailing pressure. Standard atmospheric pressure (1 atm) corresponds to 760 millimeters of mercury at 0°C (760 torr). (b) A manometer, useful for measuring pressure in a closed system. For a confined gas, the pressure is proportional to the difference in height between the two mercury columns.

pressure. Under normal conditions the barometer reads near 760 mm of Hg (760 torr at 0°C), from which is taken the definition of the standard atmosphere mentioned above.

Gas pressure inside a closed vessel is measured with a **manometer** (Figure 10-2b), a device that uses the same principle as a barometer but in a slightly different application. A U-shaped tube filled with mercury is exposed to the pressure of a confined gas, typically through an arrangement of tubes and valves. Gas entering one arm of the manometer presses down and thereby forces the mercury in the other arm to rise. The pressure is taken as the difference in height between the two columns of mercury.

Pressure: Boyle's Law and Compressibility

Armed with manometer, thermometer, balance, and volumetrically calibrated glassware, the experimenter's first task is to observe how gases respond to different conditions. That done, the next step is to search for an equation connecting the macroscopic variables P, V, T, and n.

In a careful experiment, an investigator changes just a single variable and notes the effect on another variable. A good place to begin is with the relationship between pressure and volume, as measured for a fixed quantity of gas (constant n) at a fixed temperature. Provided that the density of gas is low enough to permit ideal behavior, a curve of P versus V takes the shape shown in Figure 10-3. The relationship (known as

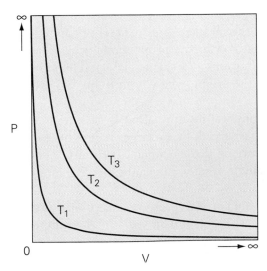

FIGURE 10-3. Boyle's law: For a fixed amount of gas at constant temperature, the product of pressure and volume is constant. Each hyperbolic curve, called an *isotherm*, conforms to the equation $PV = C$. The constant C depends on both the temperature and number of moles, and the isotherms are drawn so that $T_3 > T_2 > T_1$.

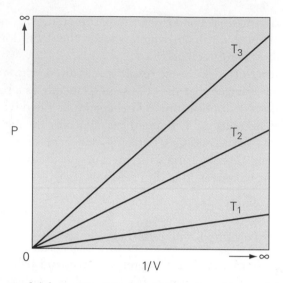

FIGURE 10-4. Boyle's law again, this time expressed as a linear relationship: A plot of pressure (P) versus the *reciprocal* of volume ($1/V$) yields the straight line $P = C/V$. The pressure goes down as the gas expands, falling away to zero as the volume approaches infinity ($1/V = 0$). Temperatures match those used in Figure 10-3, with $T_3 > T_2 > T_1$.

Boyle's law, after the physicist who first reported it in 1662) is described by the equation for a hyperbola,

$$PV = \text{constant}$$

which indicates that P and V are inversely proportional. Alternatively, we recognize that P is *directly* proportional to $1/V$ and obtain a straight line by plotting P against $1/V$ (Figure 10-4). The intercept of this line for an ideal gas is at the origin, where both P and $1/V$ are zero and therefore V is infinite.

Again, the physical interpretation is easy to accept. As the volume gets bigger, the gas spreads over a larger region of space and consequently exerts less pressure on the container. To say that P goes to zero as V goes to infinity is really to restate, in quantitative fashion, our picture of a gas itself.

The connection between pressure and volume is straightforward. See it again in Figure 10-5: Squeeze the gas . . . the volume decreases . . . the pressure increases. Expand the gas . . . the volume increases . . . the pressure decreases. If the temperature is kept constant, then the product PV will always be equal to the same number, and the P-V curve, known as an **isotherm** (iso, "equal", therm, "heat"), will be a hyperbola. The curve stays hyperbolic even at a higher temperature, although here the constant

FIGURE 10-5. Pressure and volume. (a) When a gas contracts, the particles crowd into a smaller volume and press against a reduced area: the shrunken walls of the container. Pressure goes up. (b) When the system expands into a larger container, the gas exerts its force over a wider area. The pressure drops.

in Boyle's law is found experimentally to be a larger number. This new isotherm measured at higher temperature lies above the original one, as evident from the curves already plotted in Figure 10-3.

All gases tend toward ideal behavior at very low pressure and density, where intermolecular interactions become less and less important. To gauge the effect, we define the molar density (moles per liter) as

$$\rho = \frac{n}{V}$$

and consider what a plot of PV versus ρ would look like for an ideal gas with fixed T and n: a horizontal line, since $P \times V$ will always be the same number regardless of the volume. Any deviation of the actual value of PV from this horizontal line thus provides immediate proof of nonideality, as apparent in Figure 10-6.

The curves in Figure 10-6 tell us that real gases surely do depart from the ideal limit at higher pressures. Nevertheless, for all gases (hydrogen, oxygen, sulfur dioxide, or any other) the product PV converges on the same numerical value as the density goes to zero. The model of the ideal gas clearly takes on universal significance under these conditions, and that attribute alone is worthy of appreciation. It means that

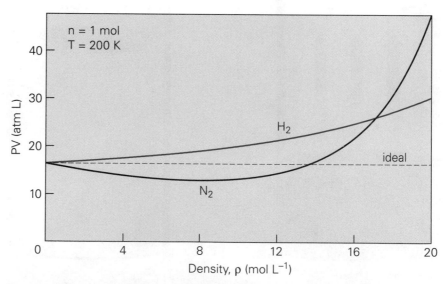

FIGURE 10-6. For an ideal gas obeying Boyle's law, a plot of PV versus density ($\rho = n/V$) is flat across the entire range—as it will be also for *any* gas, regardless of structure, whenever ρ goes to zero. Reason: Where both density and pressure are low, the volume is high and the particles remain far apart. They behave ideally, the same for each gas. Deviations from ideality set in when the particles crowd together under high density and begin to interact. Each nonideal gas is different.

we do not need one set of rules for nitrogen and another for oxygen, but rather a single pattern is common to all.

An important property here is **compressibility**, which measures how the volume of gas changes in response to pressure. Filled mostly with empty space, a gas at low pressure is easy to squeeze. One needs only to increase the external pressure on the container to force the gas into less volume; such a system is said, appropriately, to be highly compressible. Given so much empty space the particles have ample room to adjust to the smaller container. A gas at higher density, however, is already "presqueezed" and so tends to resist further compression. Too much pressure forces the particles uncomfortably close together.

Instances of high compressibility (where the pressure and density are low to begin with) and low compressibility (high pressure, high density) are pictured conceptually in Figure 10-7 and mathematically in the isotherm below:

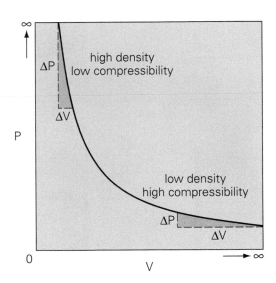

Observe how the compressibility, expressing essentially the ratio of volume change (ΔV) to pressure change (ΔP), is related to the reciprocal of the isotherm's slope at a given pressure. We interpret the inverse slope, $\Delta V/\Delta P$, as "change in volume per change in pressure" and define a compressibility (γ) at constant temperature as

$$\gamma = -\frac{1}{V}\left(\frac{\Delta V}{\Delta P}\right)$$

for very small values of ΔP. The negative sign makes γ a positive number since $\Delta V/\Delta P$ by itself is negative; the volume goes *down* when the pressure goes up.

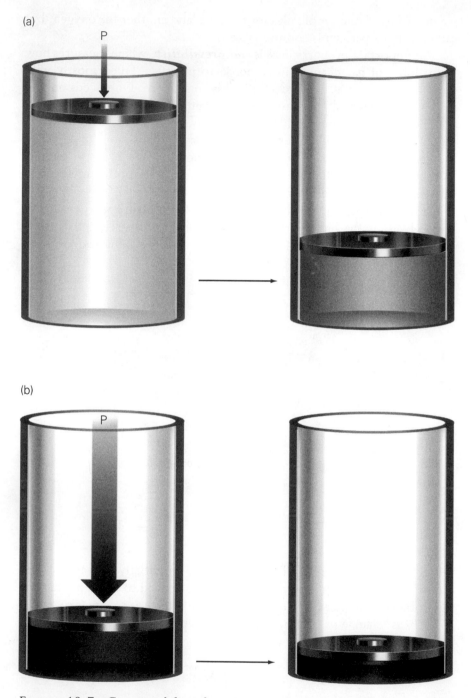

FIGURE 10-7. Compressibility: the sensitivity of volume to pressure. (a) With room to spare, a dilute gas at low pressure and high volume compresses easily. A small change in pressure produces a large change in volume. (b) A dense gas, already squeezed into a small volume at high pressure, resists further compression. A large change in pressure produces only a small change in volume.

Boyle's Law in Action

Breathing, an activity of undeniable importance, offers us a ready demonstration of Boyle's law. Our lungs (Figure 10-8) function as a flexible container, sucking in and expelling air in response to changes in pressure and volume. The trick is to maintain a proper balance between the pressure inside and outside the lungs. If the internal pressure is much less than atmospheric pressure, then the lungs are crushed from without. If the internal pressure greatly exceeds the external, then they explode from within. Neither eventuality is desirable, but fortunately a simple mechanism has evolved to make breathing automatic and safe under normal conditions.

The size of the chest cavity is controlled in humans by a large muscle, the diaphragm, which descends during inhalation and effectively increases the volume available to the lungs. Internal pressure then decreases correspondingly, and air rushes in to reestablish the necessary balance. The procedure is reversed during exhalation.

When the lungs are used under abnormal conditions, as during deep-sea diving, an awareness of Boyle's law helps to ensure against an unhappy outcome. External pressure increases steadily with depth, adding approximately

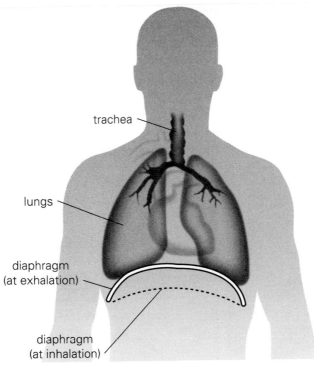

trachea

lungs

diaphragm
(at exhalation)

diaphragm
(at inhalation)

FIGURE 10-8. Breathing and Boyle's law. Regulated by the diaphragm, the available volume demands a suitable pressure to keep the lungs in balance with the atmosphere outside. Air rushes in and out as the lungs expand and contract.

1 atmosphere every 10 meters, and the conditions needed to maintain the balance of pressure change accordingly. A diver, having attained pressure equilibrium at a certain depth, must be careful to breathe normally when rising to the surface. External pressure diminishes during the ascent, and therefore the lungs must enlarge to maintain a constant value of PV. Holding the breath can be disastrous now, for the gas trapped in the lungs will expand inexorably and perhaps explosively. Boyle's law demands that the volume increase according to the ratio of the two pressures, P_{high}/P_{low}.

The danger is greater closer to the surface, where relative changes in pressure are higher. Going from 2 atm at 10 m to 1 atm on the surface, for example, causes a halving of the pressure, compared with only a 10% change between 90 m and 80 m (where 10 atm goes to 9 atm). Since compressibility decreases at higher pressures, the expansion of the lungs is less pronounced at the greater depths.

Temperature: Charles's Law

Temperature is a subtle notion demanding interpretation at many levels, and its multifaceted character makes this variable an important bridge between macroscopic and microscopic. Already we have a practical understanding of temperature as a measure of relative hot and cold, of the tendency of a hot body to warm up a cold body. It is a notion grounded in daily experience, commonsensical if a bit vague, and the effects implied by Figure 10-9 are plain to see. A cold hand, clasped around a cup of hot liquid, warms up while the cup cools down. The temperature of

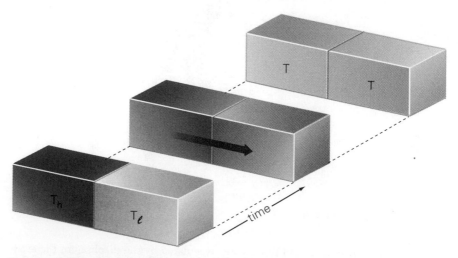

FIGURE 10-9. Heat flows from a region of high temperature (T_h) to a region of low temperature (T_ℓ) until T is intermediate and uniform throughout.

the hand increases; the temperature of the cup decreases; eventually both are equal. All very macroscopic, it seems, as are the *mechanical* consequences of heat in connection with temperature as well: A column of mercury rises and falls in a thermometer. Strips of metal inside a thermostat expand and contract as the air grows hot and cold. The volume of a tire changes with the seasons. Aware of such effects, we routinely accept the idea that changes in volume can form the basis of a temperature scale.

Most materials do undergo a change in volume when heated, usually by expanding as the temperature increases. Gases, with their free-form construction and noninteracting particles, respond with especial facility. And, as with measurements of pressure and volume, gases at low pressure display universal behavior when the temperature is varied.

The result, determined by experiment, is straightforward: At constant P and n, gas volume is linearly dependent on temperature. If volume is plotted against the temperature in °C (Figure 10-10), the data fall on a straight line that extrapolates to zero volume at −273.15°C. *All gases exhibit this pattern, provided they are investigated in a region where ideal behavior can be expected. An ideal gas—any ideal gas—shrinks uniformly as the temperature is lowered.*

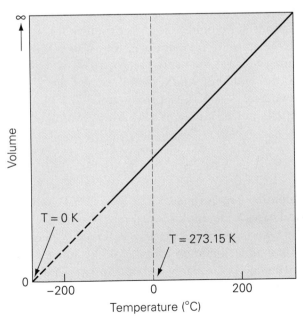

FIGURE 10-10. Charles's law: For a fixed amount of gas at constant pressure, the volume varies directly with temperature. Heated to a higher temperature, a gas expands. Cooled to a lower temperature, it shrinks. The volume extrapolates to zero at $T = -273.15$°C, thus defining the zero point for an absolute scale of temperature: 0 K.

The universal *V-T* line immediately suggests the possibility of an **absolute temperature** scale, one that can provide a measure of heat without being tied to a specific substance. So conceived, the unit of absolute temperature is the **kelvin** (abbreviated K, not °K), a dimension with the same size as one Celsius degree. "Zero" on the absolute scale, equivalent to −273.15°C, is the temperature at which an ideal gas would be thermally shrunk to zero volume. That is the end of the line, beyond which the volume would become (impossibly) negative. Absolute zero is the coldest it gets.

The ideal gas model supposes that the individual particles are volumeless, noninteracting point masses. Since the particles neither take up space nor influence each other, there is nothing to prevent a thermal compression all the way down to zero volume. We assert, as a result, that at fixed values of *n* and *P* the volume of an ideal gas varies in direct proportion to its kelvin temperature (**Charles's law**):

$$\frac{V}{T} = \text{constant}$$

If the absolute temperature is doubled, the volume doubles. If the absolute temperature is halved, the volume is halved.

Observe this one caution, though: *T* must be expressed in kelvins for the direct relationship to hold. Say a gas occupies 1 L at 273 K. Then at 546 K, twice the absolute temperature, the volume will increase to 2 L under constant pressure; but were we to work the problem using the Celsius scale, the numbers would come out wrong: 273 K is approximately 0°C whereas 546 K is approximately 273°C. The ratio of the two temperatures in Celsius is not 2:1. Always we need to use the conversion

$$T_K = T_{°C} + 273.15$$

when implementing Charles's law.

Now, ideal gases aside, look toward reality and ask: What happens in the real world where molecules do have volume and where they interact as well? To a first approximation (notably under low pressure), ideal gas behavior is maintained reasonably well as the temperature decreases. At some point during cooling, however, intermolecular interactions will become important and the gas will cease to be a gas. It will undergo a phase transition, changing into a more ordered state of matter such as a liquid or solid. Still, if the *V* versus *T* data *before* the phase transition are extrapolated to zero volume (as shown earlier in Figure 10-10)—or, put another way, if the gas behaves ideally and miraculously down to the very bitter, very cold end—then the line must always intercept the temperature axis at 0 K. Such is the macroscopic meaning of absolute temperature in the context of gases. All gases exhibit this property; the behavior is universal.

Number of Moles: Avogadro's Law

Experiments in the 19th century demonstrated yet another universal characteristic of the gaseous state: *Equal volumes of gases contain equal numbers of particles* at a given temperature and pressure. Hence the volume scales directly with the number of moles, so that

$$\frac{V}{n} = \text{constant}$$

at fixed values of T and P (**Avogadro's law**). One mole of any gas occupies the same volume as one mole of any other gas; two moles of gas particles require exactly twice as much volume; three moles require three times the volume.

Historically, establishment of the direct proportionality between V and n was an important step in understanding the connection between stoichiometry and the existence of discrete atoms and molecules. In hindsight we recognize that this relationship, too, is another statement of ideal gas behavior under the appropriate conditions. Its message (Figure 10-11) is that one particle in an ideal gas is <u>mechanically</u> no different from any other.

FIGURE 10-11. Avogadro's law: For a gas at constant pressure and temperature, the volume is directly proportional to the number of moles—or, put another way, equal volumes contain equal numbers of particles.

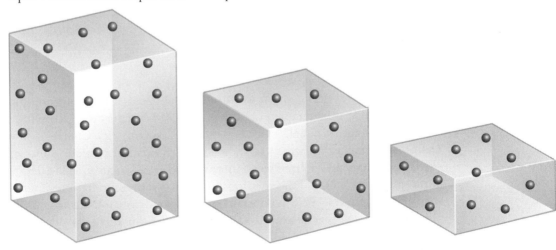

Chemically different, yes. Molecules of hydrogen, helium, nitrogen, oxygen, and carbon dioxide undergo manifestly different reactions in the gas phase. But mechanically different? No. The same number of hydrogen, helium, nitrogen, oxygen, and carbon dioxide molecules will (ideally) exert the same pressure, occupy the same volume, and exist at the same temperature under the same conditions. An ideal gas behaves as if its particles have no volume, no structure, and therefore no distinguishing features. Mechanically, they are all alike. To have *no* structure is, after all, to have the same structure.

Summing Up: The Ideal Gas Law

We now combine all the results above into one *equation of state* for the ideal gas:

$$\frac{PV}{nT} = R$$

R is the **universal gas constant**, the same for all ideal gases. Its dimensions are (pressure × volume)/(mol K), and its specific units must be consistent with those used for P and V.

Observe that the numerator, PV, has units of energy (namely, force × length). It takes work to expand or contract a gas against an applied pressure. With pressure defined as force per unit area, PV then becomes

$$PV = \text{pressure} \times \text{volume} = \frac{\text{force}}{(\text{length})^2} \times (\text{length})^3 = \text{force} \times \text{length}$$

The experimentally determined value of R is 0.0820578 atm L mol^{-1} K^{-1} or, expressed in joules, 8.31451 J mol^{-1} K^{-1} (8.31451×10^{-3} kJ mol^{-1} K^{-1}).

The resulting equation of state,

$$PV = nRT$$

ties together everything we know about ideal gases at the macroscopic level. Within it are contained (1) Boyle's law ($PV = $ constant), for which Boyle's constant is equal to nRT at fixed T and n; (2) Charles's law ($V/T = $ constant), with the constant equal to nR/P for fixed P and n; (3) Avogadro's law ($V/n = $ constant), with the constant equal to RT/P at fixed T and P; and (4) the relationship between pressure and temperature at fixed n and V ($P/T = $ constant), mentioned now for the first time.

Equipped with a measured value for R, we can solve the ideal gas equation for any unknown variable, given the values of the other three:

$$P = \frac{nRT}{V} \qquad V = \frac{nRT}{P} \qquad n = \frac{PV}{RT} \qquad T = \frac{PV}{nR}$$

Furthermore, recognizing that the number of moles (n) is just the total mass of gas (m_t) divided by the molar mass ($\mathcal{M}$), we may also write

$$PV = \frac{m_t}{\mathcal{M}} RT$$

and solve for either m_t or $\mathcal{M}$:

$$m_t = \frac{\mathcal{M}PV}{RT} \qquad \mathcal{M} = \frac{m_t RT}{PV}$$

The molar density ($\rho = n/V$, in units of mol L^{-1}) follows directly as

$$\rho = \frac{P}{RT}$$

Here, in the equation $PV = nRT$, is thus a remarkably compact description of the entire state, requiring only a consistent set of units for P, V, n, R, and T. From it we shall compute, as an important first example, the molar volume at **standard temperature and pressure (STP)**, defining these benchmark conditions as $T = 273.15$ K (0°C) and $P = 1.000$ atm (atmospheric pressure):

$$V = \frac{nRT}{P} = \frac{(1.000 \text{ mol})(0.08206 \text{ atm L mol}^{-1} \text{ K}^{-1})(273.15 \text{ K})}{1.000 \text{ atm}} =$$

$$22.41 \text{ L}$$

This result, again, carries universal significance. It asserts that one mole of *any* ideal gas occupies 22.41 L at STP, filling a cube approximately 28 cm on a side. The universal molar volume derives from the absence of inter-molecular interactions, and its relatively large size demonstrates the vast emptiness of a gas. Consider that 1 mole of liquid water, with a density of approximately 1 g/mL, occupies a volume of only 18 mL. As a gas, however, the same molecules spread out over a volume of 22,410 mL, expanding by a factor of nearly 1250. The difference is all empty space, confirming for us that intermolecular interactions are negligible.

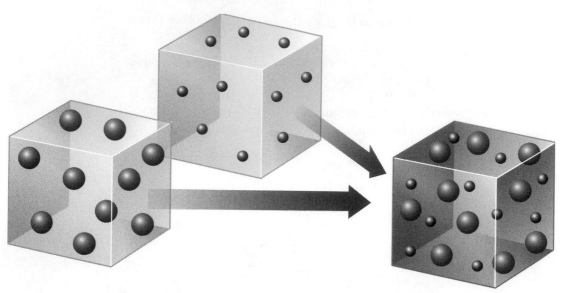

FIGURE 10-12. Dalton's law of partial pressures: Each ideal gas in a mixture exerts a partial pressure independent of any other component, as if it were the only one present. Since all such particles behave alike, the pressure at a given volume and temperature is determined solely by the number of molecules, not the kind. The various partial pressures add to produce the total.

The generic nature of the ideal gas equation also justifies **Dalton's law of partial pressures**, illustrated in Figure 10-12: If several gases are mixed together, the total pressure is the sum of the partial pressures from the different components. Realize that $PV = nRT$ stipulates that pressure depends only on volume, temperature, and the *number* of moles—not their chemical identities. Whether arising from oxygen or hydrogen or neon, a "partial" pressure is just the pressure that would be exerted by some component if it alone were present. Since ideal behavior presupposes a collection of independent particles, we then expect that each gas fills the container uniformly and generates its own pressure without interference. Consequently for a mixture of K components (with molar quantities $n_1, n_2, \ldots, n_K$), Dalton's law becomes

$$P_{\text{total}} = P_1 + P_2 + \cdots + P_K = (n_1 + n_2 + \cdots + n_K)\frac{RT}{V} = \frac{nRT}{V}$$

where

$$n = n_1 + n_2 + \cdots + n_K$$

gives the total number (in moles) of generic particles. The partial pressure of a given component, P_i, is related to this total pressure simply as

$$P_i = X_i P_{total}$$

for which the **mole fraction**

$$X_i = \frac{n_i}{n}$$

has the same form defined in Chapter 9.

10-3. KINETIC THEORY

We turn now to things microscopic, to understanding how the properties of an ideal gas derive from its atoms and molecules.

A gas, we know, is a collection of many particles—*many* particles; not two or three, not several hundred, but rather something like 100,000,000,000,000,000,000,000—moving about independently under the dictates of Newton's laws. From this chaos must somehow come the strikingly simple equation of state ($PV = nRT$) that describes the highly regular properties of the bulk fluid.

A given particle is characterized by certain microscopic variables such as mass (m) and velocity (denoted by a lowercase v, not to be confused with the uppercase V used for volume). To solve Newton's equations, one needs to begin with six pieces of information for each particle: the x-, y-, and z-coordinates of its position and velocity (speed and direction) at any one instant in time, as set out in Figure 10-13. For just one or two particles, the problem is simple enough to solve by hand. Even for a few hundred or a few thousand, a computer can still keep track of all the independent paths. But before long, no computer will be big and fast enough to deal with so many variables. All the world's most powerful computers working together for millions of years would not succeed in solving this problem by brute force alone. Are we out of luck?

Not at all. In fact, we should be grateful precisely because the microscopic system is so enormously complicated. For this very complexity permits us to deal with *average* quantities—with *statistical* instead of individual behavior—and therefore frees us from the distracting details. Statistics may be unreliable if there are only a few dozen particles, say, or a few hundred, but they are razor sharp when applied to Avogadro's number. The statistical behavior of huge groups becomes extremely regular, since deviations from the average occur with vanishingly low probabilities. Insurance companies, for example, might not know who, specifically, out

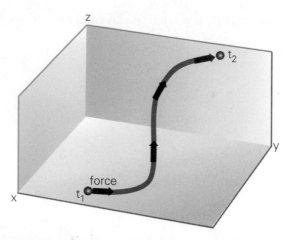

FIGURE 10-13. For any particle in classical motion, its next move follows inevitably and exactly from (1) its current condition (position and velocity) and (2) the force acting on it at just that instant (t_1). Problem solved? No; to solve Newton's equations for a macroscopic system is a practical impossibility, requiring an immense amount of advance information: some 10^{23} individual sets of coordinates and velocities.

of a group of 20 people will be absent next year, but they can forecast with dead certainty the mortality rate for a country of 250 million people. There is predictive power in large numbers.

The Statistics of Pressure

Take a mole of ideal gas particles, each with mass m and velocity v, flying about within a cubical box of volume V and continually bouncing off the walls. Again and again a particle hits a wall and imparts a force; each time, the wall responds with a force of its own and the particle ricochets away.

Observing these collisions at short intervals (imagine taking a snapshot every 10^{-12} seconds), we detect a sequence of discrete blips of the kind pictured in the first panel of Figure 10-14. One collision . . . pause . . . another collision . . . pause . . . three collisions in rapid succession . . . a constant pinging against the wall. Some impacts are hard and some soft, but most cluster about some average value.

If, instead, we were to make a prolonged exposure (in effect, averaging over time and over particles), discrete blips would no longer be resolved. What appears in their place is just an average of many hits, a smoothing out of the highs and lows to produce a continuous blur. The pinging becomes a humming, and that persistent hum is perceived as a macroscopic pressure. A steady force presses against the wall.

This pressure depends, arguably, both on the force delivered with each impact and on the number of collisions per unit area. Hence an expression of the form

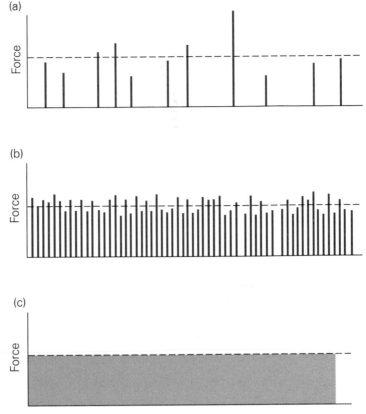

FIGURE 10-14. Pings and pressure. (a) Detected in rapid succession, the collisions register as distinct blips. Each impact delivers a small dose of momentum to the wall, different from one moment to the next. (b, c) Recorded over longer periods, as if in a time exposure, the individual collisions blur into a smooth force.

$$P = \frac{\text{force}}{\text{collison}} \times \frac{\text{number of collisions}}{\text{area}},$$

which has the correct units of force per area, is a reasonable starting point.

The first factor, the force associated with each ping, should depend on the particle's mass and velocity. The more massive an object, the harder it will hit; the faster it travels, the harder it will hit as well. One needs only to imagine a boxer's gloved hand "pinging" one's face to appreciate the relationship between force, mass, and velocity.

Remember (from Chapter 1) that **momentum** is precisely the quantity mass × velocity, *mv*, and remember too that Newton's second law states that *force is equal to the time rate of change of momentum*. And, in the box, a particle's momentum changes every time it hits the wall. A particle that collides head-on with momentum mv, for instance, will bounce straight back with momentum $-mv$ and thus undergo a reversal equal to $-2mv$:

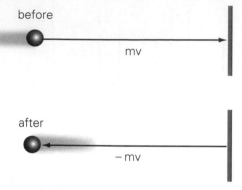

before

mv

after

$-mv$

The wall, meanwhile, suffers a change equal and opposite ($+2mv$) so as to conserve the total momentum:

$$\text{particle} \quad \text{wall} \quad \text{total}$$
$$-2mv \; + \; 2mv \; = \quad 0$$

If such an encounter occurs over a time interval Δt, the force delivered to the wall is $2mv/\Delta t$.

Recognizing that this force comes from momentum transferred per unit time, we can revise the aforementioned expression for pressure to read

$$P = \frac{\text{momentum transfer}}{\text{collision}} \times \frac{\text{number of collisions}}{\text{time} \cdot \text{area}}$$

The macroscopic pressure (force/area) develops as individual particles, uncountably many, impinge on the wall.

The second factor above, the collision rate per unit area, evidently must be proportional to particle velocity, since objects moving more rapidly will arrive at the wall with greater frequency. It should similarly scale with the number of particles (N), which we shall usually take as Avogadro's number (N_0) for convenience. The time between collisions, too, should be proportional to the distance covered, allowing us to substitute length·area (the volume of the box) for the quantity time·area in the denominator. The hit rate then becomes inversely proportional to the volume, going as $1/V$: A particle takes less time to travel between the walls of a smaller box, and a wall with a smaller area endures more impacts when N is fixed. Working together, as summarized in Figure 10-15, the three contributions combine as vN/V.

Length · Area = time · area
length time

COLLISION RATE

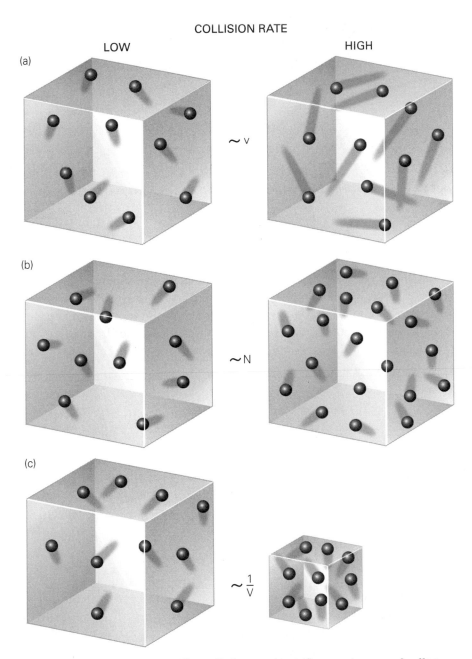

FIGURE 10-15. Bouncing off a wall: factors that influence the rate of collisions per unit area. (a) *Velocity*. The greater the speed, the more hits per second. Covering more ground in less time, a fast particle arrives at the wall sooner than a slow particle. (b) *Number*. The more particles available, the more of them can hit the wall. The collision rate increases in direct proportion. (c) *Volume*. The smaller the volume, the more impacts occur over a given area. The cross-sectional rate increases as the size of the container decreases.

For simplicity, then, take $2mv$ as roughly the momentum transferred per collision and vN_0/V as the collision rate per mole. Reasoning in this heuristic fashion—trying just to conjure up some plausible form with the correct dimensions—we arrive at

$$P = 2mv \cdot \frac{vN_0}{V} = \frac{2mv^2N_0}{V}$$

A quick (but crucial) inspection reveals that the units of mv^2/V do work out to force per area as needed, giving us a legitimate pressure:

$$\frac{mv^2}{V} \sim \frac{\text{kg m}^2 \text{ s}^{-2}}{\text{m}^3} = \frac{\text{kg m s}^{-2}}{\text{m}^2} = \text{N m}^{-2}$$

Although just a guess, this estimate comes surprisingly close to the exact relationship obtained from a more careful derivation. Here the next steps would be to apply statistical averaging and to make amends for the various shortcuts (such as assuming that all particles hit straight on, which fails to consider the collisions properly in three dimensions). That done, the final and correct expression,

$$P = \frac{1}{3} \frac{m \langle v^2 \rangle N_0}{V}$$

differs in only two details from the approximate solution. First, the constant $\frac{1}{3}$ appears when the original factor of 2 is multiplied by $\frac{1}{6}$—a necessary adjustment, since a particle is equally likely to hit any one of the six walls in a cubical box. Second, the squared speed for a single particle, v^2, is replaced by the **mean-square speed** $\langle v^2 \rangle$ of the entire assembly (the angle brackets mean "take the average").

This one number, $\langle v^2 \rangle$, the average of v^2 over all the particles, arises from statistical reasoning. Unable to track the individual speeds, we finesse the problem by substituting one key value to represent the entire distribution. With that, from the microscopic complexity of a mole of gas particles, comes macroscopic simplicity: one number, a pressure.

Energy, Temperature, and Motion

Upon straightforward rearrangement of the expression for pressure, we next obtain a microscopic representation of Boyle's law,

$$PV = \frac{2}{3}\left(\frac{m\langle v^2 \rangle}{2}\right) N_0 = \frac{2}{3} N_0 \langle \epsilon_k \rangle$$

$N_0 \langle \epsilon \rangle = E_k$

$\frac{2}{3} N_0 \langle \epsilon_k \rangle = PV$

$\frac{2}{3} E_k = PV$

$E_k = \frac{3}{2} PV$

in which the constant quantity $m\langle v^2 \rangle / 2$ stands out instantly as the *average kinetic energy per particle*, symbolized by a Greek epsilon in angle brackets: $\langle \epsilon_k \rangle$. To see why, recall that the combination

$$\tfrac{1}{2} \times \text{mass} \times (\text{velocity})^2$$

corresponds to kinetic energy, the mechanical energy associated with a moving body. Understanding that $N_0 \langle \epsilon_k \rangle$ must be equal to E_k, the *total* kinetic energy of one mole, we then have a macroscopic expression

$$E_k = \tfrac{3}{2}PV$$

to go with the microscopic equation involving ϵ_k.

Everything is consistent so far, since PV has been shown earlier to have units of energy. The final step is to use the ideal gas law to replace PV with nRT, obtaining now the truly remarkable result

$$E_k = \tfrac{3}{2}nRT$$

and with it our first quantitative connection between macroscopic and microscopic.

Put into words: *Temperature, a macroscopic variable, proves to be a direct measure of the internal kinetic energy of the moving gas particles.* Hot and cold take on a new, sharper meaning. A hot gas (Figure 10-16) is one where the particles, on the average, move faster than in a cold gas. The temperature is higher.

To say, for instance, that a mole of nitrogen gas has a temperature of 273.15 K is to note that the combined kinetic energy of the $6.0221 \times$

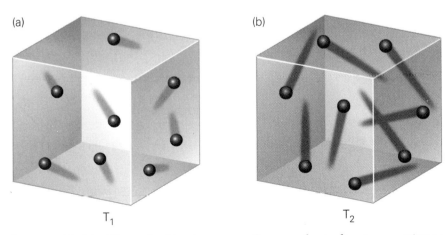

(a)   (b)

T_1 T_2

FIGURE 10-16. The molar kinetic energy of a gas scales in direct proportion to its temperature: $E_k = \tfrac{3}{2}RT$. Particles at high temperature (panel b) move at average speeds greater than those at low temperature (panel a). $T_2 > T_1$.

10^{23} N_2 molecules is exactly 3.4067 kJ ($\frac{3}{2}RT$):

$$E_k = \frac{3}{2} nRT = \frac{3}{2} \times 1.0000 \text{ mol} \times \frac{8.3145 \times 10^{-3} \text{ kJ}}{\text{mol K}} \times 273.15 \text{ K}$$

$$= 3.4067 \text{ kJ}$$

And, going further, we can manipulate the definition of E_k to obtain a representative *speed* for each particle in the gas:

1. Take one mole of particles (N_0 in all), each with average kinetic energy

$$\langle \epsilon_k \rangle = \tfrac{1}{2}m\langle v^2 \rangle$$

   to produce a total kinetic energy of

$$E_k = N_0 \langle \epsilon_k \rangle = \tfrac{1}{2}N_0 m \langle v^2 \rangle = \tfrac{3}{2}RT$$

2. Recognize that N_0 (one mole) times m (the mass of one particle) is equal to $\mathcal{M}$ (the mass of one mole). Substitute $\mathcal{M} = N_0 m$ into the expression above to obtain

$$\tfrac{1}{2}\mathcal{M}\langle v^2 \rangle = \tfrac{3}{2}RT$$

   and isolate the mean-square speed, $\langle v^2 \rangle$, on the left-hand side:

$$\langle v^2 \rangle = \frac{3RT}{\mathcal{M}}$$

3. Take the square root of both sides, thereby defining the **root-mean-square speed**, v_{rms}, as

$$v_{rms} = \sqrt{\langle v^2 \rangle} = \sqrt{\frac{3RT}{\mathcal{M}}}$$

   for a given temperature (T) and molar mass ($\mathcal{M}$). This one number, v_{rms}, the square root of the quantity $\langle v^2 \rangle$, is itself firmly rooted in the average kinetic energy. See Figure 10-17.

Understand, first, what v_{rms} means mathematically. By example: If we have five numbers (say $v = 1, 2, 3, 4, 5$) then

$$v_{rms} = \sqrt{\tfrac{1}{5}(1^2 + 2^2 + 3^2 + 4^2 + 5^2)} = 3.32$$

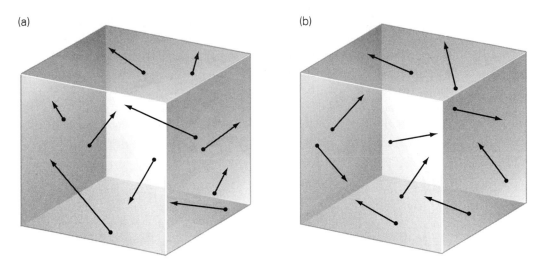

FIGURE 10-17.   (a) Particles in a gas move through a range of velocities distributed about an average. The arrows suggest instantaneous speed and direction. (b) Use of the root-mean-square velocity simplifies the picture. A single number, v_{rms}, represents the speed of each particle.

whereas the simple average, $\langle v \rangle$, is about 10% smaller:

$$\langle v \rangle = \tfrac{1}{5}(1 + 2 + 3 + 4 + 5) = 3$$

But understand, even more, what v_{rms} signifies physically: a single, deliberately chosen speed to associate with every particle in the gas, as if they were all moving at this one rate. We take the value v_{rms}, square it, multiply by $\mathcal{M}/2$, and we recover, finally, the molar kinetic energy $\tfrac{3}{2}RT$. The root-mean-square speed thus tells us, again by example, that our typical room-temperature N_2 molecule travels at approximately 500 m s^{-1} (1100 miles per hour):

$$R = 8.3145 \text{ J mol}^{-1} \text{ K}^{-1} = 8.3145 \text{ kg m}^2 \text{ s}^{-2} \text{ mol}^{-1} \text{ K}^{-1}$$

$$T = 273.15 \text{ K}$$

$$\mathcal{M} = 28.0134 \text{ g mol}^{-1} = 28.0134 \times 10^{-3} \text{ kg mol}^{-1}$$

$$v_{rms} = \sqrt{\frac{3 \, (8.3145 \text{ kg m}^2 \text{ s}^{-2} \text{ mol}^{-1} \text{ K}^{-1})(273.15 \text{ K})}{0.0280134 \text{ kg mol}^{-1}}} = 493.17 \text{ m s}^{-1}$$

Were the temperature to be doubled, the speed would be multiplied by $\sqrt{2}$ since v_{rms} goes as $\sqrt{T}$. The "hotter" molecules move faster.

We should also expect *lighter* molecules to move faster, and indeed they do—v_{rms} goes inversely with the square root of the molar mass, $1/\sqrt{\mathcal{M}}$. Given the same kinetic energy, a less massive particle acquires a greater velocity than a more massive particle. The rms speed of H_2, for example, with a molar mass of 2.0159 g mol^{-1}, is nearly four times that of N_2 (the exact factor being 3.73, from the ratio $\sqrt{28.0134/2.0159}$). A mole of each gas has the same total kinetic energy at the same temperature, but the lighter hydrogen molecules must move faster to compensate for their lower mass.

Collisions

We have remained silent concerning interparticle collisions even as the kinetic picture has become more detailed, focusing instead on activity at the walls. Accidents in mid-container do happen, however, and these direct particle-to-particle encounters have macroscopic consequences. Realize that under normal temperature and pressure, a small molecule such as oxygen or nitrogen might travel 50 to 100 nanometers from collision to collision. Distances on this scale are large relative to the dimensions of the particle (roughly a few hundred molecular diameters), yet tiny compared with the size of the container.

Expressed quantitatively, the average distance between collisions is called the **mean free path**; it is obtained as the product of the average speed and the average time between collisions. Mean free path decreases under any conditions that crowd the particles closer together.

In making its way from one end of the container to another, therefore, a molecule cannot avoid suffering repeated collisions. Its direction of motion changes with each hit (Figure 10-18), and the path degenerates

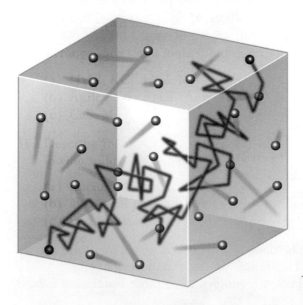

FIGURE 10-18. Repeated collisions hinder the progress of a particle through a gas, resulting in a slow, zigzag *diffusion* rather than rapid transit in a straight line. The *mean free path* specifies the average distance between collisions.

into a zigzag pattern. What would have been a rapid trip "as the crow flies" is corrupted into a random meandering—like the purposeless wanderings of a drunkard. The particle's travel time through the container increases considerably as it works its way forward, and the motion that arises from such continual jostling is known as **diffusion**.

The faster a particle moves between collisions, the faster it will traverse some stated net distance. Qualitatively, then, we expect the rate of diffusion to show the same dependence as the root-mean-square speed, going inversely as the square root of the molar mass. Lighter particles with smaller diameters move faster, have longer mean free paths, and reach their destinations sooner than heavier particles. A similar $1/\sqrt{\mathcal{M}}$ dependence also applies to the phenomenon of **effusion**, by which gas particles escape through a small hole in the container (**Graham's law**, Figure 10-19). Molecules low in mass arrive at the opening with greater frequency.

Thermal Energy

The total kinetic energy of one mole is $\frac{3}{2}RT$, but what is the average energy per particle? As always, Avogadro's number provides the link between an individual particle and a mole. Defining, first, **Boltzmann's constant**

FIGURE 10-19. Effusion. Particles impinge upon a small hole and escape from the container. Lighter particles, traveling faster, are more likely to pass through. The rate of effusion varies as $1/\sqrt{\mathcal{M}}$.

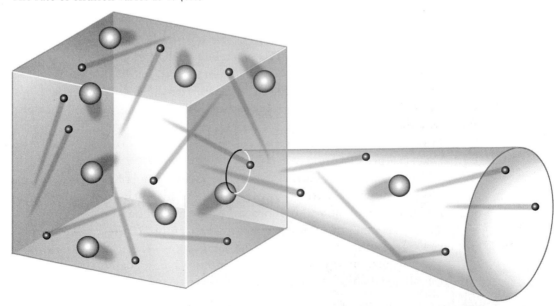

($k_B = 1.38066 \times 10^{-23}$ J K^{-1}) as the universal gas constant divided by Avogadro's number,

$$\langle \epsilon_k \rangle N_0 = E_k \qquad\qquad k_B = \frac{R}{N_0}$$

and recalling that $\langle \epsilon_k \rangle$ is equal to E_k / N_0, we obtain

$$\langle \epsilon_k \rangle = \tfrac{3}{2} k_B T$$

for the average kinetic energy per particle (having done nothing more than divide $\tfrac{3}{2}RT$ by N_0). That said, henceforth we may speak interchangeably of particle energy and molar energy, sometimes saying $\tfrac{3}{2}RT$ per mole, sometimes saying $\tfrac{3}{2}k_B T$ per particle.

More important, though, is to understand $\tfrac{3}{2}RT$ ($\tfrac{3}{2}k_B T$) as the system's **thermal energy**, the energy associated with a definite temperature. This microscopic/macroscopic expression for E_k and ϵ_k, which relates average energy to temperature, is but one specific realization of a more general, wide-ranging physical principle. Called the **equipartition theorem**, the larger law regulates the energy of all mechanical systems obeying Newton's laws of classical physics.

Equipartition says that a molecule gains thermal energy in the amount $\tfrac{1}{2}k_B T$ for each *degree of freedom*, a degree of freedom being merely any way something can move. The molecules of an ideal gas are taken as point masses with no internal structure, so all they can do is undergo translational motion in three dimensions. Simply stated, they *move*. Allowed $\tfrac{1}{2}k_B T$ for movement along each axis in space (Figure 10-20), they acquire a total thermal energy of $\tfrac{3}{2}k_B T$. The average value per particle is the same in all directions.

3° of Freedom

Of course it is. Should thermal energy be larger along the x-axis and smaller along the y-axis, different north and south? No—not unless there is some objective marker (like an electric field) to define north and south, to give meaning to our arbitrary labels x, y, z. To be otherwise would be to violate all that we understand about the symmetry of forces in nature. For thermal energy, the rule is simply this: share and share alike. Nature plays no favorites.

The concept of thermal energy pervades chemistry, physics, and biology. It supplies a rough answer to questions such as why hydrogen molecules exist at room temperature, why helium does not liquefy above 4.2 K, and why atoms are ripped apart into fragments in the interior of the sun.

Thermal energy is nature's minimum wage, a guaranteed allotment for all mechanical systems. It's just there. Atoms and molecules at a

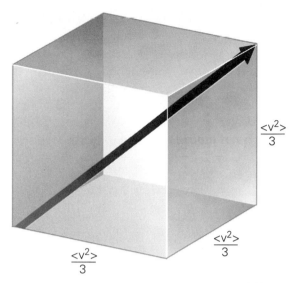

FIGURE 10-20. Equipartition of energy. Proportional to $\langle v^2 \rangle / 3$, the average kinetic energy in each of three dimensions is the same: $\frac{1}{2}k_B T$. Total thermal energy is $\frac{3}{2}k_B T$, the sum of contributions in three perpendicular directions.

stated temperature naturally jiggle about with (approximately) $k_B T$'s worth of energy, and so all other processes must compete with this motion. Thermal energy can tear down what might otherwise be built up. If a molecule is to be formed, it must be able to withstand the disruptions of $k_B T$. If a bond is to be broken by heat alone, then the thermal energy (Figure 10-21) must be large enough to overcome the force of attraction.

Strength and weakness are always interpreted relative to the prevailing temperature. Two hydrogen atoms, as we saw in Chapter 7, form a covalently bonded hydrogen molecule with a bond energy exceeding 400 kJ mol^{-1}. Such a molecule is sturdy enough to persist at room temperature, where $\frac{3}{2}RT$ is less than 4 kJ mol^{-1}. The atoms remain strongly attached, stirred up only slightly by thermal energy and needing far more than 4 kJ to break apart one mole.

But no structure can survive a determined thermal assault. H_2 is surely doomed to be ripped in two at some higher temperature, presumably near the point where its thermal energy becomes comparable to the bond energy. Pairs of helium atoms, by contrast, are far less persistent. Helium atoms do not form covalent bonds, but instead rely on much weaker intermolecular forces to hold together. Since He–He attractions amount to only about 0.1 kJ mol^{-1}, the atoms cannot resist the thermal energy unless the temperature is very low. Helium, kept apart by thermal motion, remains a monatomic gas all the way down to 4.2 K.

Thermal energy serves as a benchmark, determining whether any proposed event is feasible at a given temperature. A quick survey of temperatures might begin in the sun's interior, where at 10^7 K the thermal energy is so enormous that hydrogen nuclei undergo thermonuclear fusion reactions to form helium. Considerably cooler is the 6000 K surface of the sun, where a thermal energy of 75 kJ mol^{-1} permits the existence of some molecular hydrogen.

FIGURE 10-21. Atoms, molecules, and other assemblies of matter grow increasingly fragile at high temperatures, more likely to blow apart in a high-speed collision. To survive, a structure must withstand the buffeting that comes from its *thermal energy*—the natural kinetic energy of the particles, proportional to $k_B T$. Only at comparatively low temperatures are atoms and molecules able to stay intact.

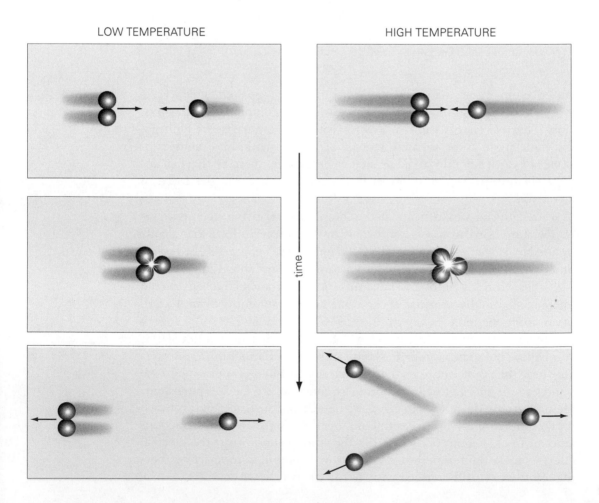

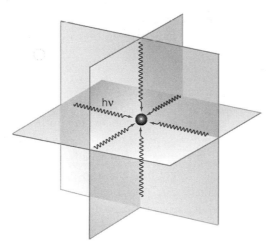

FIGURE 10-22. Laser cooling. Pinned in place by the opposing momenta of several photons, an atom loses much of its thermal energy. It slows to a crawl, and the temperature of the system drops nearly to absolute zero.

Tungsten, used as a filament for incandescent lightbulbs, melts near 3400 K. Here we know from experience that the heat generated by a bulb is insufficient to melt the wire. Continuing down to more familiar temperatures, we note that water boils at 373 K (thermal energy = 4.7 kJ mol^{-1}) and freezes at 273 K (3.4 kJ mol^{-1}). This is our own world, one ideally suited for chemistry and for life. It is cool enough for molecules to exist, but still warm enough to permit chemical change.

Going lower, we find that carbon dioxide goes from solid to gas at 195 K (2.4 kJ mol^{-1}), nitrogen liquefies at 77 K (1 kJ mol^{-1}), and helium liquefies at 4.2 K (0.05 kJ mol^{-1}). To drop further requires some effort, but modern research is pushing the limits. Helium-3, the lighter isotope of helium, liquefies at 0.3 K, and can be cooled to the millikelvin range (10^{-3} K) by diluting it with helium-4. Even more remarkable are the microkelvin temperatures (10^{-6} K) recently achieved through *laser cooling* (Figure 10-22), whereby photons bombard atoms from different directions and so bring them (almost) to a standstill. It is a dramatic manifestation of light acting as a particle, able to deliver momentum to another particle.

Currently the lowest temperatures registered are in the nanokelvin range (10^{-9} K), where thermal energies are on the order of 10^{-11} kJ mol^{-1}. These systems, the coldest observed, contain highly ordered arrangements of nuclear spins—small magnets associated with certain nuclei, just as spin and magnetism are associated with electrons.

The unreachable limit is absolute zero, 0 K, at which the thermal energy is zero and all translational motion ceases. No mechanical system can be colder. A gas, for one, may have neither negative volume nor negative thermal energy.

Speed and Temperature

Let us begin to tie up some loose ends concerning both thermal energy and the kinetic theory. Although we are pleased to know the root-mean-square speed in an ideal gas, this quantity v_{rms} is still only a single number—a very meaningful number, but clearly not a comprehensive microscopic description. Even more valuable, if possible, would be a tabulation showing how the individual speeds are actually distributed, as if to say:

Out of 100 particles, 3 have a speed between 0 and 200 m s^{-1};

out of 100 particles, 10 have a speed between 200 and 400 m s^{-1};

out of 100 particles, 57 have a speed between 400 and 600 m s^{-1};

out of 100 particles, 24 have a speed between 600 and 800 m s^{-1};

out of 100 particles, 6 have a speed between 800 and 1000 m s^{-1};

and, out of all that, to determine that v_{rms} has the value such-and-such.

Now, individual particles in a gas do have available to them a broad range of speeds and energies. Any *one* atom or molecule can race ahead, lag behind, or cruise at any velocity whatsoever. A large collection of particles, though, has no such freedom, for there the system is constrained by the laws of large numbers. A certain average speed must always be maintained among the whole group. But how does that average value (that *temperature*) come about? How fast is an individual particle likely to be moving?

Right away we venture a tentative answer to the second question: A given particle is likely to be moving at or near the average speed of the whole ensemble. Remember that no particle is truly alone in the gas, despite having to travel some distance before encountering another. Accidents happen. The container is big, but inevitably there are collisions—the same collisions that are responsible for gaseous diffusion, mentioned just above. These encounters continually change the speeds and directions of the particles, thereby smoothing out the range of possibilities.

Thus before a collision, particle A might be moving much faster than particle B; after the collision, however, B might be the faster of the two, having acquired some of A's kinetic energy from the impact. The scene is no different from the events on a pool table, except now the table is really a three-dimensional box containing fantastically many balls. With so many collisions and with so many ricochets and so many changes in speed, the particles all end up traveling at about the same rate. The more an individual speed deviates from that *most probable* value, the less likely it is to occur. Relatively few particles will be moving at the extremes; most will cluster instead about a narrow range.

Nor, again, can there be a preferred direction of motion—not north or south, east or west, *x*, *y*, or *z*. No axis is special. Do we ever see a gas balloon spontaneously develop a bulge at one point, as if a crowd of particles were to move suddenly in that direction? Never. The leveling effect of countless random collisions prevents such an odd occurrence.

Search, then, for the simplest distribution of speeds that meets our basic requirements, namely:

1. Most particles travel at a certain smoothed-out average value.
2. There is no dependence on direction.

Condition 2 we can satisfy by choosing a suitable function of v^2 (the *square* of the velocity), since here is an appropriate quantity possessing magnitude but no specified direction. A speed v to the right (positive) has the same value of v^2 as a speed v to the left (negative). So, too, for a speed v toward the northeast, northwest, southeast, southwest, or in any direction at all; v^2 is identical everywhere along a sphere of radius v:

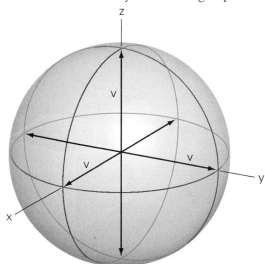

What we need, next, to satisfy condition 1 is a function of v^2 that is zero at both very low and very high speeds, and rises to a single maximum in between—near the popular average value.

What will it be? Forgoing a detailed derivation, we shall simply state the result; but, even so, one should think of the intellectual effort involved here: to understand how macroscopic order comes out of microscopic chaos; to analyze, somehow, the random, disconnected motions of countless tiny particles and to summarize them in a single chart. Yet that was precisely the problem solved by 19th-century physicists James Clerk Maxwell and Ludwig Boltzmann, whose combined efforts produced the ***Maxwell-Boltzmann distribution***:

$$F(v) = Kv^2 e^{-mv^2/2k_B T} \qquad K = 4\pi \left(\frac{m}{2\pi k_B T} \right)^{3/2}$$

This curve, plotted in Figure 10-23, has just the form we need: zero at the endpoints, peaked in the middle, dependent only on v^2. The value of K, a constant, is determined by the mass and temperature. The symbol e denotes the base of the natural logarithm (2.718281828), which we represent alternatively by the symbol *exp* (where exp a is equivalent to e^a; see Appendix B).

The Maxwell-Boltzmann distribution provides a complete summary of all possible speeds, to be read now in the following way. (1) Pick a point, a speed v on the horizontal axis, and locate the corresponding point $F(v)$ on the vertical axis. (2) Ask: For some very small increment

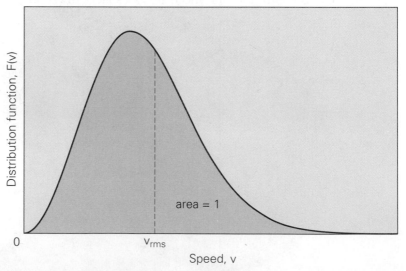

FIGURE 10-23. Distribution of speed for particles in an ideal gas (the Maxwell-Boltzmann distribution). The proportion of molecules moving at a given speed v in any direction is plotted against v. Some particles lag and some particles lead, but most travel at about the same speed. Since all fractions $F(v)\Delta v$ must sum to 1 over the full range of v, the area beneath the curve always has the same value: 1.

Δv, what fraction of the particles will be traveling at speeds between v and $v + \Delta v$? (3) The answer is $F(v)\Delta v$. To find it, we multiply the small interval Δv by the value $F(v)$, a single number. If $F(v)$ at a certain speed v_2 is then, say, twice the value at another speed v_1, we learn that twice as many particles are moving at v_2:

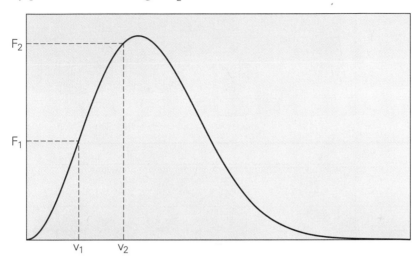

All such ratios stand fixed for a given mass and temperature; and we know, moreover, that the sum of probabilities across the entire curve will always add to 1, since all the particles must eventually be included in the census.

Look closely at the particular distribution in Figure 10-23. The curve begins at zero, rises to a maximum, tapers to zero at large v, and is sensitive only to v^2. A few particles travel very slowly; a few travel very fast. Most move at approximately the same speed, and directions are irrelevant. Observing further, we also see that v_{rms}, although not quite at the peak of the distribution, is nonetheless close to the value observed for most of the particles. When the temperature is increased (Figure 10-24), the whole distribution shifts to the right—as does v_{rms}, consistent with its already known dependence on $\sqrt{T}$:

$$ v_{rms} = \sqrt{\frac{3RT}{M}} = \sqrt{\frac{3k_B T}{m}} $$

A higher fraction of the total is moving faster at the higher temperature, with the exponential factor $\exp(-mv^2/2k_B T)$ going up correspondingly as the exponent $(-mv^2/2k_B T)$ becomes less negative.

Yet the area beneath the curve is forever constant. The unchanging area represents the sum of all the products $F(v)\Delta v$, and since every particle

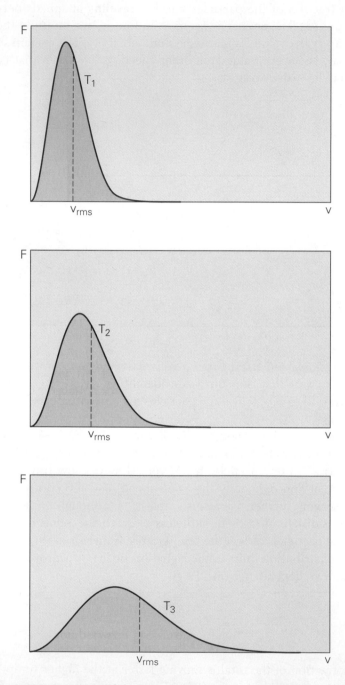

Figure 10-24. Variation of the Maxwell-Boltzmann distribution with temperature, presented so that $T_3 > T_2 > T_1$. As the temperature increases, more particles move faster. The entire curve—peak and tail—shifts to the right while preserving its unit area.

must be included in the counting, the individual populations always add up to the total number of particles.

The Boltzmann Distribution

Look carefully, once more, at the exponential factor in the Maxwell-Boltzmann distribution: $\exp(-mv^2/2k_BT)$. Contained within it are two familiar quantities, the kinetic energy of an individual particle ($\epsilon = mv^2/2$) and the approximate thermal energy (k_BT). Written in the form

$$\exp\left(-\frac{\epsilon}{k_BT}\right) = \frac{1}{\exp\left(\dfrac{\epsilon}{k_BT}\right)},$$

this function (called the **Boltzmann factor**) has a significance that transcends its present role in the kinetic theory of gases. It happens that ϵ can be *any* kind of atomic or molecular energy—not just the translational energy of gas particles, but anything at all. The variable ϵ might pertain to the energy needed to make a molecule rotate or vibrate; it might designate a difference between electronic energy levels; it might be the energy needed to flip an electron spin. Anything.

And, it happens as well, the Boltzmann factor is proportional to the probability, $P(\epsilon)$, that a particle will adopt the specific energy ϵ given the prevailing thermal energy:

$$P(\epsilon) \propto \exp\left(-\frac{\epsilon}{k_BT}\right)$$

The distribution of speeds in an ideal gas, presented just above, thus proves to be merely one example of a much more general statistical result. Maxwell, in fact, first used reasoning similar to ours to guess the general form that $F(v)$ must take for an ideal gas, and then he applied the requisite mathematics to obtain the exact expression. Not until some years later did Boltzmann recognize the far-reaching significance of the quantity $\exp(-\epsilon/k_BT)$. He showed how it was no accident that Maxwell's exponential factor depends on the ratio of particle energy to thermal energy. The Maxwell-Boltzmann distribution of speeds, $F(v)$, is really just the probabilistic Boltzmann function

$$\exp\left(-\frac{\epsilon}{k_BT}\right) = \exp\left(-\frac{mv^2}{2k_BT}\right)$$

for kinetic energy in some arbitrary direction, multiplied by a speed factor v^2 and a normalization constant K. Together, K and v^2 scale the probability to include contributions from every angle and also to ensure that the various fractions add to 1.

We need not follow Boltzmann's complex mathematical arguments to appreciate the broad meaning of the Boltzmann distribution, to which we shall return in other contexts. Observe, for example, how $\exp(-\epsilon/k_BT)$ goes from a maximum of unity (when ϵ is very much less than k_BT and hence ϵ/k_BT approaches zero) to a minimum of zero (when ϵ/k_BT goes to infinity):

$$\epsilon \ll k_BT \longrightarrow \exp\left(-\frac{\epsilon}{k_BT}\right) \approx \exp(-0) = 1$$

$$\epsilon \gg k_BT \longrightarrow \exp\left(-\frac{\epsilon}{k_BT}\right) \approx \exp(-\infty) = 0$$

If ϵ is small compared with k_BT, as suggested in the first part of Figure 10-25, then the thermal energy is large enough to kick the particle

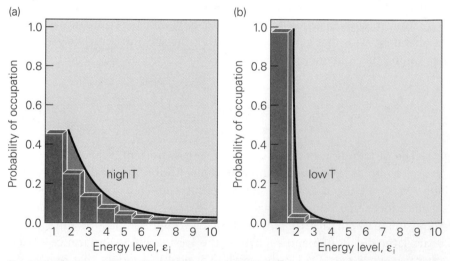

FIGURE 10-25. The Boltzmann factor, $\exp(-\epsilon_i/k_BT)$, determines whether a particle is likely to have the energy ϵ_i. A simple exponential distribution develops when each of these energies can be realized in only one way. (a) High temperature: When excitation energies are small compared with k_BT, the system has sufficient thermal energy to promote its particles from the ground state (ϵ_1) to excited states ϵ_2, ϵ_3, and higher. The distribution of energy is spread broadly over the various levels. (b) Low temperature: Excitation energies are relatively large compared with k_BT, and now most of the particles stay trapped in the ground state.

up to that particular level. If, by contrast, ϵ is high and T is low, then it becomes difficult to achieve the required thermal boost. That is what we mean, mathematically, when we say that all energies must be compared relative to $k_B T$. That, too, is why atoms and molecules tend to exist in their ground states at room temperature, where $k_B T$ is insufficient to propel the system up its quantized ladder of energies. The distribution of energy under such circumstances comes to resemble the distribution of money: the poor are many, the rich few.

This very basic relationship is a statistical result of great power and generality, applicable far beyond the ideal gas, for its quantitative aspect gives exact meaning to the whole notion of thermal energy. The Boltzmann factor shows how the likelihood of an event changes with temperature, and it appears all throughout chemistry. It shows up in statistical thermodynamics (kinetic theory being but one example), kinetics, quantum mechanics, and indeed wherever there are questions of energy and temperature. Always the Boltzmann factor is the answer to a most important question: How many microscopic particles (atoms, molecules, electrons, nuclei, whatever you like) will have such-and-such an energy?

Our attention shifts now to nonideal gases, where the particles are not point masses but rather are molecules that must accommodate each other in some sense. The interactions that develop will eventually take us from the gaseous to the liquid state.

REVIEW AND GUIDE TO PROBLEMS

Intangible, always . . . invisible, often . . . formless, elusive, a challenge to the senses—a gas seems to lack substance, as if it were scarcely matter at all. Thin air, we say. Vapor. Wind. In everyday language the gaseous state becomes a metaphor for nothingness, for unreality.

But no: Air is thin, but not empty; wind is invisible, but not immaterial. Air is the sum of its molecules, mostly nitrogen and oxygen, always in motion, bearers of momentum and energy. From the momentum of N_2 and O_2, we feel the force of the wind; and from the chemical energy of oxygen, we draw breath. There is substance amidst all the empty space.

IN BRIEF: THE IDEAL GAS AND KINETIC THEORY

Like all states of matter, a gas has its external face and its internal face: the macroscopic properties, very few; the particles that produce them, very many. Inside the gas vessel are particles in constant motion, spread far apart, unable to interact, restrained only by the walls.

1. MACROSCOPIC QUESTIONS. How big a container? How much mass does it hold? How many moles and how many particles? How much force do they exert? What is the temperature?

Ask. Measure. Define. Define the macroscopic properties, the external face of the gaseous state: volume, amount, pressure, temperature. Volume, V: the capacity of the container, filled completely and uniformly by the gas. Amount, n (moles): the number of particles pinging about the box, each impact delivering a dose of momentum and force to one of the walls. Pressure, P: the average force per unit area generated by all those pings against the walls. Temperature, T: a measure of the kinetic energy of the particles, related microscopically to the average speed.

P, V, n, and T. Four macroscopic properties, four numbers to be measured. With them, we know all there is to know about the macroscopic gas: how much space it occupies (V), how much stuff it contains (n), how much force it exerts (P), and how much kinetic energy it carries (proportional to T).

2. GAS LAWS. Make some changes. Double the volume, but leave n and T the same—what happens to the pressure? Halve the temperature, but maintain constant P and n —what happens to the volume? Triple the number of moles at constant P and V—what happens to the temperature?

The answers, discovered over two centuries of experimentation, are condensed into one law, the *ideal gas equation of state*:

$$PV = nRT$$

in which R, the **universal gas constant**, has the value 8.31451 J mol^{-1} K^{-1} (0.0820578 atm L mol^{-1} K^{-1}).

Valid for all gases at sufficiently low pressures, the ideal gas law subsumes

Boyle's law: $PV = $ constant (at fixed n, T)

Charles's law: $V/T = $ constant (at fixed n, P)

Avogadro's law: $V/n = $ constant (at fixed P, T)

and is supplemented by

Dalton's law of partial pressures: the pressure exerted by a gaseous mixture is the sum of the individual pressures of the separate components.

3. IDEALITY. All gases behave alike at low densities. Oxygen, nitrogen, carbon monoxide, carbon dioxide, methane, and helium all obey the same laws. All respond in the same way to changes in pressure, volume, temperature, and number of moles. Never mind that carbon dioxide and helium are so different chemically, because such differences are revealed only during intimate encounters and only over sufficient time. Only then, by making sustained contact, can two molecules interact truly as *molecules* rather than featureless billiard balls. Only then, by sticking together, can they express their individual sizes, shapes, and polarities. If not, one molecule has nothing to distinguish itself from another.

Thus under 1 atm of pressure, at 273 K, in a volume of 22.4 L, there are 6.02×10^{23} molecules of gaseous CO_2. $PV = nRT$. Under the same conditions, there are also 6.02×10^{23} atoms of helium. $PV = nRT$. And 6.02×10^{23} molecules of methane, and oxygen, and nitrogen, and any other "ideal" gas as well.

$PV = nRT$. Not $PV = nRT$ for one molecule and $PV = $ Something Else for another, but $PV = nRT$ for all molecules. Why? Because the molecules are far apart. They move fast and resist association. They look alike. They act alike. There is, after all, only one way *not* to interact, and so all noninteracting particles are alike in their very lack of interaction.

Different from everything we have seen to this point, the ideal gas is simply the sum of its noninteracting parts; no more, no less. Different from a wave: Waves interfere; gas particles do not. Gas particles add their energies; waves do not. Different from an atom too: Electrons attract and repel from a distance; gas particles do not. Gas particles add their energies; electrons do not. Different from a molecule, different from a cluster, different from a liquid, different from a solid—different, unique, because in an ideal gas there are no interactions to which the individual particles must submit. Everywhere else, in all states where interactions do exist, the collective structure is either more or less than its individual members. Suppress the interactions, and the result is an ideal gas.

Every particle in the ideal gas is an independent agent, acting alone. Oxygen, nitrogen, carbon monoxide, carbon dioxide, methane, and helium mix together without interaction and without distinction except for mass and speed. Each component, by itself, obeys the same ideal gas law. Each component has its own pressure, as if it were the only substance present. Each component fills the same volume, again as if it were all alone.

4. MICROSCOPIC ANSWERS. No more wave functions, then, and no more atoms and molecules, but only these: particles, huge numbers of generic gas particles, pinpricks of mass stripped of all chemical identity. Heavy enough and fast enough to behave like billiard balls, the particles move in predictable paths as they fly about in all directions. Their de Broglie wavelengths are small, and so the particles follow the beaten tracks of classical mechanics. We know, in principle, where to find them.

Principle, however, gets us only so far, for here is a problem that we shall never solve completely. Not now and not ever can we successfully track the individual paths of a mole-sized collection of billiard balls. There are just too many to consider.

But we can do something better, something far more useful: we can study their behavior as a group. Rather than worry about each particle's whereabouts and energy, we look instead for statistically meaningful average values—the *average* kinetic energy, the *average* speed, the *average* time and distance between collisions. With numbers so large, the averages are deadly accurate and the deviations small. Nothing is gained, in the end, by trying to follow the motion of each particle separately.

Ask not: How fast is particle 328 traveling? When will particle 802 hit the left wall? How much momentum will particle 6,054,094,720 deliver to particle 115,660,652,002,138? Those problems are formidable to solve, and, more important, offer scant benefit for the effort required.

Ask, rather: What is the average kinetic energy of all the particles combined? How many particles have energies within 10% of the average?

How many have speeds within 20% of the average? What is the average pressure exerted on the walls? Those are the sensible questions, and those are the questions we can answer statistically.

To do so is to apply ***statistical mechanics***: a statistical, probabilistic approach to understanding the motion of large groups of anything, be they electrons, nuclei, atoms, molecules, or generic gas particles. Statistical reasoning offers a powerful way to extract simple, regular macroscopic properties from the chaotic complexity of the microworld.

The kinetic theory of the ideal gas, a late 19th-century conception, was the first application of statistical mechanics. A hundred years later, it still stands.

5. KINETIC THEORY. Recall the principal results obtained from a statistical treatment of the ideal gas.

First, we learn the *why* of Boyle's law; we discover, microscopically, why PV is constant. Macroscopic pressure accumulates as particle after particle collides with the walls, with each transfer of momentum described by the Newtonian laws of motion. Averaged over many collisions, the relationship between pressure and volume then becomes

$$PV = \tfrac{2}{3}E_k$$

for one mole of gas. E_k, the combined kinetic energy of all the particles, is constant if n and T are held steady.

Second, temperature is directly proportional to kinetic energy:

$$E_k = \tfrac{3}{2}nRT$$

We call this quantity the ***thermal energy*** ($\tfrac{3}{2}RT$ per mole), the kinetic energy generated by random motion. A given amount of thermal energy is associated with a given temperature.

Third, particles move faster in a gas at higher temperature. Recognizing that the average kinetic energy is proportional to the mean-square speed, $\langle v^2 \rangle$, we express the ***root-mean-square speed*** as

$$v_{\text{rms}} = \sqrt{\frac{3RT}{\mathcal{M}}}$$

where $\mathcal{M}$ is the molar mass. Lighter, more energetic particles move faster on the average, since the average speed is proportional to both $\sqrt{T}$ and $1/\sqrt{\mathcal{M}}$.

Fourth, the ***equipartition principle***: nature plays no favorites. Thermal energy is apportioned equitably along the three axes of motion: $\tfrac{3}{2}RT$ per mole, allocated as $\tfrac{1}{2}RT$ in each direction. For a single particle, the average thermal energy is similarly distributed as $\frac{1}{2}k_{B}T + \frac{1}{2}k_{B}T + \frac{1}{2}k_{B}T$ (where k_{B}, the **Boltzmann constant**, is equal to R divided by Avogadro's number). On the average, no particle gets more than its fair share of the total energy. Nor does energy build up unduly in any one direction.

Fifth, speed and thermal energy conform to the ***Maxwell-Boltzmann distribution***,

$$F(v) = Kv^2 \exp\left(-\frac{mv^2}{2k_{B}T}\right) \qquad K = 4\pi\left(\frac{m}{2\pi k_{B}T}\right)^{3/2}$$

which we interpret piece by piece: (1) The Boltzmann factor, $\exp(-mv^2/2k_{B}T)$, is proportional to the number of particles whose kinetic energy ($\frac{1}{2}mv^2$) arises from motion in a specific direction. Note, however, that all particles with the same speed have the same kinetic energy and the same Boltzmann factor regardless of their line of motion. The atom moving at 1000 m s^{-1} along the x-axis has the same values of $\frac{1}{2}mv^2$ and $\exp(-mv^2/2k_{B}T)$ as the atom moving at 1000 m s^{-1} along the y-axis or z-axis. (2) The velocity factor v^2 accounts, in part, for all these different directions of motion that will produce a given speed and energy. (3) The constant K ensures that all particles are properly tallied in the census.

The quantity $F(v)\Delta v$ then gives us the fraction of particles traveling randomly at speeds confined to a narrow range between v and $v + \Delta v$. Some particles, very few, move very slowly. Some particles, very few, move very fast. Most of them, caught up in the crowd, move at speeds near the peak of the curve (a number not far different from the root-mean-square value, v_{rms}, our preferred benchmark).

The distribution depends on temperature. The higher the temperature, the more energetic are the particles. Taken over large populations, the average becomes $\frac{3}{2}k_{B}T$ per particle.

6. QUESTIONS YET TO COME. Why do molecules react under some conditions but not others? When energy is reshuffled during a reaction, where does it go? How is a fixed quantity of energy allocated among all the different degrees of freedom in atoms and molecules? How much goes to the electrons? How much is invested in the vibration of bonds and the rotation of molecules? How much goes into translational kinetic energy? Why does heat flow from a body at high temperature to a body at low temperature? Why does the Boltzmann distribution have its peculiar form?

Such questions we shall soon explore, mostly in Chapters 12 through 14, but already the kinetic theory gives us a clue: Where numbers are large, macroscopic outcomes are few. Nature is blind. She plays no favorites. Electrons, nuclei, atoms, and molecules stumble along the easiest, most likely route from A to B.

SAMPLE PROBLEMS

Begin with the four questions:

1. What is the pressure? P
2. What is the volume? V
3. What is the amount? n
4. What is the temperature? T

These four numbers—P, V, n, T—provide a complete macroscopic description of an ideal gas. We need to specify a value for each of them. First, consider a typical problem for which Boyle's law applies.

EXAMPLE 10-1. Boyle's Law: Pressure Up, Volume Down

PROBLEM: 300. mL of gaseous CO_2 are confined in a vessel initially at atmospheric pressure and a temperature of 273 K (STP, standard temperature and pressure). The pressure in the vessel is controlled by raising and lowering a plunger, and the gas is assumed to behave ideally under the prevailing conditions. Compute the volume occupied at a pressure of 3.00 atm and a temperature of 273 K.

SOLUTION: We are told that the number of moles is the same for each of the two states specified—or so we infer, since the gas is said to be "confined" in a vessel. If nothing comes in and nothing goes out, then $n_1 = n_2$. Similarly, the stated temperature is fixed at $T_1 = T_2 = 273$ K.

The problem thus describes a straightforward *isothermal compression*, a squeezing of the gas at constant temperature. Pressure increases and volume decreases. Since two of the variables (n and T) are fixed, Boyle's law requires that PV must have the same value for states 1 and 2:

$$P_1 V_1 = P_2 V_2$$

Given the information

$$P_1 = 1.00 \text{ atm} \qquad P_2 = 3.00 \text{ atm}$$

$$V_1 = 300. \text{ mL} \qquad V_2 = ?$$

we solve

$$V_2 = V_1 \frac{P_1}{P_2} = 300. \text{ mL} \times \frac{1.00 \text{ atm}}{3.00 \text{ atm}} = 100. \text{ mL}$$

Be aware, foremost, of the correct direction of change. Here the pressure is increased, so the volume must decrease. If, sadly, we were to compute a *larger* volume at higher pressure, then we would know immediately that the ratio is upside-down.

Message: There are only two choices. Pick the correct one.

EXAMPLE 10-2. 1 atm = 760 torr

The following demonstration emphasizes the importance of keeping the units straight in any application of Boyle's law.

PROBLEM: After an isothermal increase in pressure from 380 torr to 1.0 atm, an ideal gas is found to have a volume of 2.0 L. What was its original volume?

SOLUTION: We are given

$$P_1 = 380 \text{ torr} \qquad P_2 = 1.0 \text{ atm}$$

$$V_1 = ? \qquad V_2 = 2.0 \text{ L}$$

The point is simply to use one set of units for the pressure, so as not to mix torr with atm:

$$P_1 = 380 \text{ torr} \times \frac{1 \text{ atm}}{760 \text{ torr}} = 0.50 \text{ atm}$$

Since the pressure has doubled, the volume must be half the initial value calculated below:

$$V_1 = 2.0 \text{ L} \times \frac{1.0 \text{ atm}}{0.50 \text{ atm}} = 4.0 \text{ L}$$

EXAMPLE 10-3. Charles's Law: Temperature Up, Volume Up

Problems involving a direct application of Charles's law,

$$\frac{V_1}{T_1} = \frac{V_2}{T_2}$$

are solved using the same straightforward ratio approach as for Boyle's law. Remember, though, that T must be expressed in kelvins for this relationship to hold.

PROBLEM: A sample of He gas occupies 1.00 L at 10.0°C. Compute the volume at 405 K.

SOLUTION: The units are inconsistent: T_1 is specified in °C; T_2 is specified in K.

We need, then, to make the proper conversion,

$$T_K = T_{°C} + 273.15$$

$$T_1 = 10.0 + 273.15 = 283.2 \text{ K}$$

after which the rest is easy:

$$V_1 = 1.00 \text{ L} \qquad\qquad V_2 = ?$$

$$T_1 = 283.2 \text{ K} \qquad\qquad T_2 = 405 \text{ K}$$

Temperature up, volume up:

$$V_2 = V_1 \frac{T_2}{T_1} = 1.00 \text{ L} \times \frac{405 \text{ K}}{283.2 \text{ K}} = 1.43 \text{ L}$$

EXAMPLE 10-4. The Ideal Gas Law

Bear in mind three crucial requirements: (1) Use a consistent set of units. (2) Make sure that the units of R match those of P, V, T, and n. (3) Keep the units straight. Little can go wrong if these simple rules are followed.

Our task is to determine one of the state variables given values of the other three.

PROBLEM: The pressure in a 1.000-L bulb containing nitrogen gas is found to be 556 torr at 298 K. How many molecules are contained in the vessel?

SOLUTION: We know P, V, and T. By using $PV = nRT$ to calculate n, the number of moles, we can then determine the number of molecules:

$$P = 556 \text{ torr} \times \frac{1 \text{ atm}}{760 \text{ torr}} = 0.732 \text{ atm}$$

$$V = 1.000 \text{ L}$$

$$T = 298 \text{ K}$$

$$n = \frac{PV}{RT} = \frac{(0.732 \text{ atm})(1.000 \text{ L})}{(0.08206 \text{ atm L mol}^{-1} \text{ K}^{-1})(298 \text{ K})} = 2.99 \times 10^{-2} \text{ mol}$$

The final step is to compute N, the number of molecules, making use of Avogadro's number ($N_0 = 6.02 \times 10^{23}$):

$$N = n \times N_0 = (2.99 \times 10^{-2} \text{ mol}) \times \frac{6.02 \times 10^{23} \text{ molecules}}{\text{mol}}$$

$$= 1.80 \times 10^{22} \text{ molecules}$$

There are 18,000,000,000,000,000,000,000 molecules of N_2.

ALTERNATIVE SOLUTION: Knowing that one mole of ideal gas particles at STP occupies 22.41 liters, we can solve this problem simply as a change between two states. This approach permits use of the ratio method and thus eliminates the gas constant.

REFERENCE STATE	UNKNOWN STATE
$P_1 = 1.000$ atm	$P_2 = 0.732$ atm
$V_1 = 22.41$ L	$V_2 = 1.000$ L
$T_1 = 273.15$ K	$T_2 = 298$ K
$n_1 = 1.000$ mol	$n_2 = ?$

Consider each variable in turn:

1. *Change the pressure while holding the volume and temperature constant.* n is directly proportional to P. Since state 2 is at lower pressure compared with state 1 (STP), we expect likewise that state 2 will contain fewer moles. Accordingly, we ensure that the ratio of pressures is less than 1—in this case, P_2/P_1.

2. *Change the volume while holding the pressure and temperature constant.* n is directly proportional to V. Since the volume of state 2 is smaller than 22.41 liters, we know that n_2 must be less than 1 mole. The appropriate ratio here is V_2/V_1.

3. *Change the temperature while holding the pressure and volume constant.* n is inversely proportional to T. With the temperature in the bulb higher than standard, the ratio once again must be less than 1. Its value is T_1/T_2.

Putting everything together, we have

$$n_2 = n_1 \times \frac{P_2}{P_1} \times \frac{V_2}{V_1} \times \frac{T_1}{T_2}$$

$$= 1.00 \text{ mol} \times \frac{0.732 \text{ atm}}{1.000 \text{ atm}} \times \frac{1.000 \text{ L}}{22.41 \text{ L}} \times \frac{273.15 \text{ K}}{298 \text{ K}}$$

$$= 2.99 \times 10^{-2} \text{ mol}$$

EXAMPLE 10-5. Density

PROBLEM: A sample of Ar gas is maintained at a pressure of 4.053×10^5 Pa and a temperature of $-20.50°C$. What is the molar density?

SOLUTION: Recall that the molar density ($\rho = n/V = P/RT$) is determined solely by pressure and temperature:

$$P = (4.053 \times 10^5 \text{ Pa}) \times \frac{1 \text{ atm}}{1.01325 \times 10^5 \text{ Pa}} = 4.000 \text{ atm}$$

$$T = -20.50 + 273.15 = 252.65 \text{ K}$$

$$\rho = \frac{n}{V} = \frac{P}{RT} = \frac{4.000 \text{ atm}}{(0.08206 \text{ atm L mol}^{-1} \text{ K}^{-1})(252.65 \text{ K})}$$

$$= 0.1929 \text{ mol L}^{-1}$$

EXAMPLE 10-6. Molar Mass

PROBLEM: 5.000 g He gas occupy a volume of 28.00 L at 2.000 atm and 546.30 K. Compute the molar mass, $\mathcal{M}$, of the helium.

SOLUTION: Denoting the total mass by m_t, we use the ideal gas equation with the substitution $n = m_t/\mathcal{M}$:

$$m_t = 5.000 \text{ g}$$

$$P = 2.000 \text{ atm}$$

$$V = 28.00 \text{ L}$$

$$T = 546.30 \text{ K}$$

$$\mathcal{M} = \frac{m_t RT}{PV} = \frac{(5.000 \text{ g})(0.08206 \text{ atm L mol}^{-1} \text{ K}^{-1})(546.30 \text{ K})}{(2.000 \text{ atm})(28.00 \text{ L})}$$

$$= 4.003 \text{ g mol}^{-1}$$

ALTERNATIVE SOLUTION: Compare the conditions relative to STP and then use the known molar volume of 22.41 L mol^{-1}.

1. Determine the number of moles that would be present under standard conditions:

$$28.00 \text{ L} \times \frac{1 \text{ mol}}{22.41 \text{ L}} = 1.249 \text{ mol at STP}$$

2. Correct the number of moles for nonstandard pressure. Since P is exactly twice standard pressure, n must be doubled.

3. Correct the number of moles for nonstandard temperature. Since T is exactly twice standard temperature, n must be halved.

4. Compute: $n = 1.249 \text{ mol} \times 2 \times \frac{1}{2} = 1.249 \text{ mol}$

5. The molar mass is therefore

$$\frac{5.000 \text{ g}}{1.249 \text{ mol}} = 4.003 \text{ g mol}^{-1}$$

EXAMPLE 10-7. Mixtures: Dalton's Law of Partial Pressures

PROBLEM: Sally fills a rigid 1.00-L vessel with O_2 gas at a pressure of 1.25×10^5 Pa and a temperature of 500 K. Maintaining constant temperature and volume, she then adds just enough He gas to raise the total pressure to 3.75×10^5 Pa. What are the mole fractions of He and O_2 in the mixture?

SOLUTION: Provided that the gases behave ideally, it makes no difference whether we have O_2 molecules or He atoms or any other species in the vessel. We are concerned only with the *number* of particles at a given pressure, volume, and temperature.

Here, with volume and temperature held constant, the pressure is directly proportional to the number of moles:

$$\frac{P}{n} = \frac{RT}{V} = \text{constant} \qquad (T, \ V \text{ fixed})$$

If n doubles, so does P. If n triples, so does P. If n quadruples, so does P. More particles exert more pressure.

Since the pressure triples upon addition of the helium, we conclude that there are two He atoms (two moles) present for each O_2 molecule (one mole) in the mixture:

$$n_{\text{He}} = 2n_{O_2}$$

The mole fractions, good to within three significant figures, are $\frac{2}{3}$ for He and $\frac{1}{3}$ for O_2:

$$X_{He} = \frac{n_{He}}{n_{He} + n_{O_2}} = \frac{2n_{O_2}}{2n_{O_2} + n_{O_2}} = 0.667$$

$$X_{O_2} = \frac{1}{1 + 2} = 0.333$$

EXAMPLE 10-8. Pressure, Temperature, Energy

PROBLEM: Dick heats a confined system of Ar gas sufficiently to double its pressure. If the volume remains the same, how much kinetic energy does the gas gain or lose? Express the result symbolically.

SOLUTION: Understanding that the energy is directly proportional to temperature, we look first to the relationship between P and T:

$$\frac{P}{T} = \frac{nR}{V} = \text{constant} \qquad (n, \ V \text{ fixed})$$

Any doubling of the pressure (at constant n and V) is accompanied by a doubling of the absolute temperature ($T_2 = 2T_1$), and hence the translational kinetic energy doubles in response:

$$E_k = \tfrac{3}{2}nRT_2 = \tfrac{3}{2}nR(2T_1)$$

Moving faster, the molecules hit the walls with greater momentum and exert greater pressure. The increase in total energy is $\tfrac{3}{2}nRT_1$.

EXAMPLE 10-9. An Isothermal Compression

PROBLEM: 11.2 L of propane gas (C_3H_8) at STP are compressed isothermally to 5.60 L. (a) Compute the number of moles in the confined system. (b) Compute the pressure after the reduction in volume. (c) Compute the kinetic energy before and after compression.

SOLUTION: The key word is *isothermal*—constant temperature. Energy, sensitive only to a change in temperature, is unchanged by the isothermal compression.

(a) *Number of moles.* Since 1.00 mol occupies a volume of 22.4 L at STP ($T = 273$ K, $P = 1$ atm), there is evidently 0.500 mol in the container before compression:

$$\frac{1 \text{ mol (STP)}}{22.4 \text{ L}} \times 11.2 \text{ L} = 0.500 \text{ mol}$$

(b) *Pressure.* Confined within a smaller volume at the same temperature, the same number of molecules will generate a higher pressure:

$$1.00 \text{ atm (STP)} \times \frac{11.2 \text{ L}}{5.60 \text{ L}} = 2.00 \text{ atm (after)}$$

(c) *Kinetic energy.* The molecules move with the same average energy before and after the isothermal compression:

$$E_k = \tfrac{3}{2}nRT = \tfrac{3}{2} \times 0.500 \text{ mol} \times \frac{8.3145 \times 10^{-3} \text{ kJ}}{\text{mol K}} \times 273 \text{ K} = 1.70 \text{ kJ}$$

EXAMPLE 10-10. Temperature, Mass, Speed

Answer the following question using a minimum of numerical computation.

PROBLEM: Nitrogen molecules travel at a root-mean-square speed of approximately 500 m s^{-1} at STP. Suppose that v_{rms} for unknown particle X is equal to 476 m s^{-1} at 400 K. What species is X likely to be . . . CO_2, He, or H_2O?

SOLUTION: The root-mean-square speed

$$v_{rms} = \sqrt{\frac{3RT}{\mathcal{M}}}$$

is proportional to $\sqrt{T/\mathcal{M}}$, where $\mathcal{M}$ is the molar mass and T is the absolute temperature. Hence if T were to go from 273 K (at STP) to 400 K, then *nitrogen's* v_{rms} would increase by the factor $\sqrt{400/273}$:

$$500 \text{ m s}^{-1} \times \frac{\sqrt{400}}{\sqrt{273}} \approx 605 \text{ m s}^{-1} \qquad \text{(N}_2 \text{ at the higher temperature)}$$

An approximate value of 600 m s^{-1} will suffice for our rough calculation.

For X at this temperature, we now have two possibilities to consider. Either: X is less massive than N$_2$ ($\mathcal{M} = 28.0$ g mol^{-1}), so that its v_{rms} will be greater than 600 m s^{-1}. Or: X is more massive than N$_2$, so that its v_{rms} will be less than 600 m s^{-1}.

Answer: X is *more* massive than N$_2$. Neither He ($\mathcal{M} = 4$ g mol^{-1}) nor H$_2$O ($\mathcal{M} = 18$ g mol^{-1}) has a molar mass consistent with the data specified. CO$_2$, however, has just the right molar mass (44 g mol^{-1}) to yield a value close to the stated v_{rms}:

$$600 \text{ m s}^{-1} \times \frac{\sqrt{28}}{\sqrt{44}} \approx 479 \text{ m s}^{-1} \qquad (CO_2 \text{ at } 400 \text{ K})$$

EXAMPLE 10-11. Maxwell-Boltzmann Distribution

For the following calculation, note that the mass of an individual molecule of N_2 (m) is 4.65×10^{-26} kg:

$$\frac{28.0134 \text{ g } N_2}{\text{mol}} \times \frac{1 \text{ mol}}{6.022 \times 10^{23} \text{ molecules}} \times \frac{1 \text{ kg}}{1000 \text{ g}}$$

$$= 4.65 \times 10^{-26} \text{ kg}$$

PROBLEM: Consider a sample of nitrogen gas at 273 K. (a) For every 1000 molecules traveling at a speed of 500 m s^{-1}, how many travel at a speed of 1000 m s^{-1}? (b) What is the equivalent ratio at 373 K?

SOLUTION: The Maxwell-Boltzmann distribution

$$F(v) = Kv^2 \exp\left(-\frac{mv^2}{2k_BT}\right) \qquad K = 4\pi \left(\frac{m}{2\pi k_BT}\right)^{3/2}$$

tells us what fraction of the population is traveling with a given speed, v, at a particular temperature. The count includes particles moving in all possible directions.

To compare the distribution at two speeds, v_1 and v_2, we need to compute the ratio

$$\frac{F(v_2)}{F(v_1)} = \frac{v_2^2}{v_1^2} \exp\left[-\frac{m(v_2^2 - v_1^2)}{2k_BT}\right]$$

into which we insert the following values:

$$v_1 = 500 \text{ m s}^{-1}$$

$$v_2 = 1000 \text{ m s}^{-1}$$

$$m = 4.65 \times 10^{-26} \text{ kg}$$

$$k_B = 1.38 \times 10^{-23} \text{ J K}^{-1}$$

$$T = 273 \text{ K}, 373 \text{ K}$$

The exponent, with units of energy in both numerator and denominator, is properly dimensionless:

$$\frac{mv^2}{k_\mathrm{B}T} \sim \frac{\mathrm{kg\ m^2\ s^{-2}}}{\mathrm{J\ K^{-1}\ K}} \sim \frac{\mathrm{J}}{\mathrm{J}} \sim 1$$

(a) *273 K.* The ratio is

$$\frac{F(v_2)}{F(v_1)} = \frac{1000^2}{500^2} \exp\left[-4.65 \times 10^{-26} \times \frac{1000^2 - 500^2}{2 \times 1.38 \times 10^{-23} \times 273}\right]$$

$$= 0.0391 = \frac{39.1}{1000}$$

For every 1000 molecules traveling at 500 m s^{-1}, there are approximately 39 traveling at 1000 m s^{-1}.

(b) *373 K.* $F(v_2)/F(v_1)$ increases to 135/1000 at the higher temperature, as more and more of the molecules are pushed to higher speeds.

EXERCISES

1. Suppose that two objects have the same mass (1 kg) and same volume (1 m³) but different shapes: object A, with dimensions 100 cm × 100 cm × 100 cm; and object B, with dimensions 1000 cm × 100 cm × 10 cm. (a) In the orientations shown below, which object exerts the greater pressure on the ground?

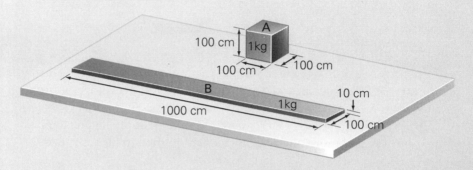

First, answer "A" or "B" without doing an explicit numerical calculation. (b) Now solve the problem to three significant figures: Compute the two pressures, remembering that a body is pulled toward the earth by a gravitational force (weight) equal to

$$F = mg,$$

where m is the mass and

$$g = 9.81 \text{ m s}^{-2}$$

is the acceleration due to gravity. Note, from that, how the SI unit of pressure, the *pascal*, follows naturally as

$$1 \text{ Pa} = 1 \text{ N m}^{-2}$$

Note also that the same letter of the alphabet denotes both the unit "meters" and the general quantity "mass" in the three expressions above. Context determines the correct meaning.

2. Take object B from the preceding exercise (mass = 1 kg, length = 1000 cm, width = 100 cm, height = 10 cm) and compute the pressure it exerts on the ground in three different orientations: (a) Lying flat, so that the 1000 × 100 side presses down. (b) Standing on its

long edge, so that the 1000 × 10 side presses down. (c) Standing on its short edge, so that the 100 × 10 side presses down. See the sketch below:

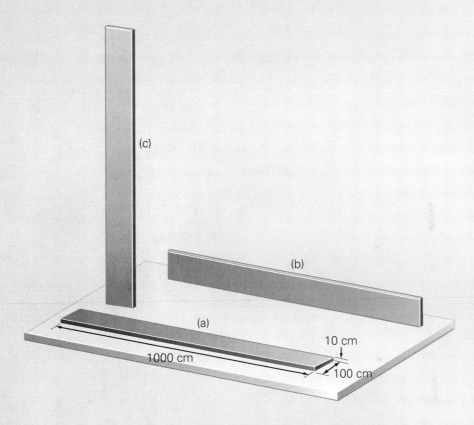

3. Explain, using words and not numbers, the term *atmospheric pressure*. What does it mean for the atmospheric pressure to have such-and-such a value?

4. Normal atmospheric pressure is approximately 100,000 Pa (1 *bar*), close enough so that 1 *atmosphere* (atm) is defined as 101.325 *kilopascals* (kPa):

$$1 \text{ atm} = 1.01325 \times 10^5 \text{ Pa} = 101.325 \text{ kPa}$$

What force does a circular column of air, 1.00 cm in diameter, exert on the ground below to produce a pressure of 1 atm? Assume that the column extends from ground level to outer space, and remember that

$$1 \text{ Pa} = 1 \text{ N m}^{-2}$$

5. Convert the normal value of atmospheric pressure

$$1 \text{ atm} = 1.01325 \times 10^5 \text{ Pa} = 101.325 \text{ kPa}$$

into dimensions of *pounds per square inch* (lb/in^2), noting that a mass of 1.00 kg exerts a force of 2.21 lb near the earth's surface.

6. The manufacturer's instructions call for a certain tire to be inflated to a pressure of 30 lb/in^2. What pressure must exist *inside* the tire to give a reading of 30 lb/in^2 on a pressure gauge? Use the results of the previous exercise.

7. Describe the operation of a barometer, paying particular attention to the statement "1 atmosphere of pressure corresponds to 760 millimeters of mercury at 0°C (760 torr)."

8. Now use the relationship

$$1 \text{ atm} = 101.325 \text{ kPa} = 760 \text{ mm Hg} \qquad \text{(at } 0°\text{C)}$$

to determine the density of liquid mercury in g/cm^3. Do it this way:

STEP 1. In a barometer, the full force of the atmosphere presses down upon a dish of mercury and forces the liquid up a tube. Let h denote the height attained by the mercury, and let A denote the area of the tube in cross section:

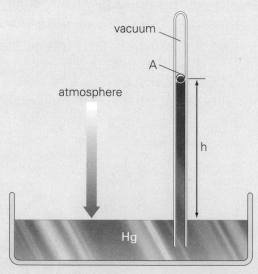

STEP 2. The volume of mercury in the tube is thus Ah, area multiplied by height. The *mass* contained within that volume depends solely on the density ρ, as yet unknown:

$$\text{Mass} = \text{density} \times \text{volume} = \rho Ah$$

STEP 3. Using the standard acceleration due to gravity ($g = 9.81$ m s^{-2}), express the force exerted upon the given mass:

$$F = mg = \rho Ahg$$

STEP 4. Divide force by area to obtain the pressure:

$$P = \frac{F}{A} = \frac{\rho Ahg}{A} = \rho gh$$

Take note: Given two of the three variables (P, ρ, h), the third is automatically determined. Also, the area of the tube does not affect the height of the column.

9. Water has a density of approximately 1.00 g mL^{-1}. To what height will a column of water rise when subjected to a pressure of 1.00 atm? State the result in *feet*.

10. Supply the missing values:

(a) 0.816 atm = ___ torr

(b) 712.3 torr = ___ atm

(c) 712.3 torr = ___ kPa

(d) 24.32 atm = ___ kPa

(e) 7659.1 Pa = ___ atm

(f) 7659.1 Pa = ___ torr

11. Suppose that a fixed quantity of gas has a pressure of 1 atm and a volume of 1 L. What will the volume be when the pressure is changed isothermally to (a) 10 atm, (b) 100 atm, (c) 0.1 atm, (d) 0.01 atm? Assume that pressure and volume obey Boyle's law.

12. One mole of helium gas occupies a volume of 22.4 L at STP. When the external pressure is reduced to 616 torr, what will be the new volume if the temperature is held constant?

13. A sudden change in pressure causes a fixed amount of nitrogen gas to change its volume from 435.1 mL to 172.9 mL. By what factor has the pressure increased or decreased? Assume that the temperature remains constant.

14. A fixed quantity of neon gas, initially occupying 1.0 L, is pressurized from 1.0 atm to 54.6 atm at constant temperature. What is the new volume?

15. Supply the missing values:

    (a) $0.0°C =$ ___ K
    (b) 0 K $=$ ___ $°C$
    (c) 315.75 K $=$ ___ $°C$
    (d) $-123.6°C =$ ___ K

16. To convert between the Celsius and Fahrenheit temperature scales, we use the points

$$0°C = 32°F$$

$$100°C = 212°F$$

    to establish a linear relationship between the two:

$$°C = \tfrac{5}{9}(°F - 32) \qquad °F = \tfrac{9}{5}(°C) + 32$$

    Supply the missing values:

    (a) $342.45°C =$ ___ $°F$
    (b) $98.6°F =$ ___ $°C$
    (c) $-10.6°C =$ ___ $°F$
    (d) $-40.0°F =$ ___ $°C$

17. 1.00 L of air is cooled from $0°C$ to $-50°C$ under a constant pressure of 1.00 atm. What is the new volume? Assume that the amount of gas is fixed and that Charles's law applies.

18. One mole of helium gas, held originally at STP, is allowed to double its volume at constant pressure. What is the new temperature?

19. A fixed amount of argon gas is heated from $400°C$ to $500°C$, attaining a final volume of 2.38 L under constant pressure. What was the original volume?

20. A fixed amount of krypton gas is cooled from 400 K to 300 K under constant pressure. If the final volume is 344 cm^3, what was the initial volume?

21. How many gas particles are contained in each of the following samples at STP?

    (a) 22.4 L CO_2
    (b) 22.4 L N_2

(c) 22.4 L O_2

(d) 22.4 L of air

Assume that the gases behave ideally.

22. How many atoms of helium are contained in a 22.4-liter vessel under each set of conditions?

    (a) STP

    (b) $P = 2.00$ atm, $T = 273$ K

    (c) $P = 0.500$ atm, $T = 273$ K

    (d) $P = 1.00$ atm, $T = 546$ K

    (e) $P = 1.00$ atm, $T = 136.5$ K

23. How many molecules of CO gas are present under each set of conditions?

    (a) $P = 1.00$ atm, $V = 22.4$ L, $T = 273$ K

    (b) $P = 1.00$ atm, $V = 44.8$ L, $T = 273$ K

    (c) $P = 1.00$ atm, $V = 11.2$ L, $T = 273$ K

24. As a way to remember the value of the universal gas constant, R, simply recall this one number: 22.4 liters per mole. Meaning: One mole of an ideal gas occupies a volume of 22.4 L at STP, obeying the ideal gas equation of state

    $$PV = nRT$$

    Compute, from that, the value of R in units of atm L mol^{-1} K^{-1}.

25. Take the value of R determined in the preceding exercise and reexpress it in the following units:

    (a) torr L mol^{-1} K^{-1}

    (b) Pa m^3 mol^{-1} K^{-1}

    (c) J mol^{-1} K^{-1}

26. (a) How much volume is occupied by 3.28 moles of methane gas (CH_4) at a temperature of 312.4°C and a pressure of 589.1 torr? (b) If both the volume and temperature are halved, what happens to the pressure?

27. Gaseous ammonia (NH_3) is confined to a volume of 2.00 L at a temperature of 400°F and a pressure of 1.00 atm. If the temperature is increased to 800°F at constant volume, what happens to the pressure?

28. (a) How many moles of ethane (C_2H_6) are contained in a volume of 12.56 L at a pressure of 2.944 atm and a temperature of 298.2 K? (b) If the volume is halved while the pressure remains constant, what happens to the temperature?

29. One mole of nitrogen is added to a 22.4-L container already containing oxygen at STP. What pressure would then be needed for the mixture to sustain a temperature of 405 K?

30. Suppose that 160 grams of methane, confined to a volume of 1.00 L, are made to react with 480 grams of oxygen:

$$CH_4(g) + 2O_2(g) \longrightarrow CO_2(g) + 2H_2O(g)$$

(a) How many moles of methane and oxygen remain after the combustion? (b) How many moles of carbon dioxide and water are produced? (c) What total pressure do the gases exert, after reaction, when the system is cooled to a temperature of 125°C?

31. A gaseous mixture containing 2.00 g CO_2, 2.00 g H_2O, and 2.00 g CH_4 is kept in a 1.00-L bulb at 10.0 atm. What is the temperature?

32. A mixture of helium and nitrogen exerts a total pressure of 1.17 atm in a volume of 1.00 L at 285 K. (a) If the partial pressure of N_2 is 1.00 atm, what is the mole fraction and partial pressure of He? (b) How many moles of each component are present?

33. Calculate the density (mol L^{-1}) of a sample of N_2 gas maintained at −10.0°C and 2.50 atm.

34. How many grams of Cl_2 gas will fill a volume of 3.00 L at a temperature of 48.6°C and a pressure of 976 torr?

35. 2.14 grams of an undetermined homonuclear diatomic gas, X_2, occupy a volume of 1.50 L at STP. What is the gas?

36. An unknown compound contains 81.7% carbon and 18.3% hydrogen by weight. If 34.84 grams of this material occupy a volume of 17.7 L at STP, what is the molecular formula?

37. Compute the total translational kinetic energy resident in one mole of each ideal gas at 150 K, 300 K, and 450 K:

    (a) H_2 (b) He   (c) Ne   (d) Ar

38. Compute the total translational kinetic energy resident in each quantity of H_2 gas at 150 K, 300 K, and 450 K:

    (a) 1.008 g (b) 2.016 g (c) 4.032 g (d) 6.048 g (e) 8.064 g

39. Compute the average translational kinetic energy per H_2 *molecule* in each quantity of hydrogen gas at 150 K, 300 K, and 450 K:

    (a) 1.008 g (b) 2.016 g (c) 4.032 g (d) 6.048 g (e) 8.064 g

    Do the values depend on n, the number of moles?

40. (a) Compute the root-mean-square speed of a hydrogen molecule at 150 K, 300 K, and 450 K. (b) Is v_{rms} directly proportional to temperature? Why or why not?

41. Repeat the last series of calculations now for a different system, argon, 20 times more massive than molecular hydrogen. Start by computing the total translational kinetic energy resident in each quantity of Ar gas at 150 K, 300 K, and 450 K:

    (a) 19.97 g   (b) 39.95 g   (c) 79.90 g   (d) 119.84 g
    (e) 159.79 g

    Compare the values with those for hydrogen at the same temperatures.

42. Compute the average translational kinetic energy per Ar atom in each quantity of Ar gas at 150 K, 300 K, and 450 K:

    (a) 19.97 g   (b) 39.95 g   (c) 79.90 g   (d) 119.84 g
    (e) 159.79 g

    Do the argon atoms possess more or less kinetic energy than hydrogen molecules at the same temperature?

43. (a) Compute the root-mean-square speed of an argon atom at 150 K, 300 K, 450 K. (b) Do the argon atoms move faster or slower than hydrogen molecules at the same temperatures? By what factor do the root-mean-square speeds differ?

44. Suppose that argon atoms could only move in two dimensions, as if

they were confined to a flat surface rather than a three-dimensional box. (a) What total translational kinetic energy would then be resident in one mole of gas at 150 K, 300 K, and 450 K? (b) What average kinetic energy would each argon atom have at 150 K, 300 K, and 450 K? (c) What would be the root-mean-square speed at 150 K, 300 K, and 450 K? (d) In what ways do these values differ from those computed in the three exercises above?

45. Pick the system with the higher root-mean-square speed (if indeed there is any difference):

(a) 3 L H_2 at 100 atm and 700 K . . . or 3 L H_2 at 1 atm and 700 K

(b) 1 L He at 350 K . . . or 100 L He at 350 K

(c) Ar at 400 K . . . or Xe at 400 K

(d) CO_2 at 500 K . . . or Kr at 1000 K

46. Which particles escape (effuse) faster from a small opening under the same conditions of temperature and pressure?

(a) H_2 or He

(b) CO_2 or C_3H_8

(c) N_2 or C_2H_6

(d) N_2 or CO

(e) N_2 or HCN

Use Graham's law to calculate the ratio of the effusion rates in each case:

$$\frac{\text{Rate A}}{\text{Rate B}} = \frac{\sqrt{\text{mass B}}}{\sqrt{\text{mass A}}}$$

47. When molecules collide, do they collide with enough violence to shatter bonds? Sometimes yes; sometimes no. Take H_2 as an example: The dissociation energy of a hydrogen molecule is 436 kJ mol^{-1}. (a) At what temperature, approximately, does the thermal energy begin to approximate the bond energy? (b) Is a hydrogen molecule likely to dissociate at STP?

48. Some particles move at this speed; others, at that speed. Compute the root-mean-square speed for helium at 50 K, 300 K, and 600 K, and then sketch the Maxwell-Boltzmann distribution at each temperature. Indicate the position of v_{rms} on the curves.

49. Pick the more likely particle speed in each of the following sets:

    (a) He at 50 K: 50 m s^{-1} . . . or 500 m s^{-1}
    (b) He at 50 K: 500 m s^{-1} . . . or 1000 m s^{-1}
    (c) He at 300 K: 50 m s^{-1} . . . or 1500 m s^{-1}
    (d) He at 300 K: 1500 m s^{-1} . . . or 2000 m s^{-1}
    (e) He at 600 K: 1000 m s^{-1} . . . or 2000 m s^{-1}
    (f) He at 600 K: 2000 m s^{-1} . . . or 4000 m s^{-1}

50. Consider a sample of neon gas at STP. For every million molecules traveling at the root-mean-square speed, approximately how many travel at the following speeds?

    (a) $0.1v_{rms}$ (b) $0.9v_{rms}$ (c) $2v_{rms}$ (d) $3v_{rms}$

11

Disorder–Order and Phase Transitions

11-1. Intermolecular Potential

11-2. Real Equations of State

11-3. Real Isotherms: Critical Temperature and Phase Transitions

11-4. Phase Equilibria
  Vapor Pressure of a Pure Liquid
  Vapor Pressure of a Solution

11-5. Phase Diagrams

11-6. A Survey of Disorder–Order Transitions
  REVIEW AND GUIDE TO PROBLEMS
  EXERCISES

A gas deviates from ideal behavior when its constituents begin to interact. No longer idealized as volumeless particles, the atoms and molecules in a "real" gas must be credited not just with mass but also with size and shape and, most of all, electrical characteristics. Having such properties—taking up space, responding to nearby electric charges—forces a molecule to recognize and adjust to its neighbors whenever they are near. The model of the ideal gas breaks down once these mutual interactions become significant, typically under conditions of high density.

Real gases display collective effects impossible among noninteracting particles. Interactions lead first to new equations of state and eventually to *changes* of state, or **phase transitions**. Intermolecular forces are the glue that holds together liquids and solids, the condensed forms of matter.

11-1. INTERMOLECULAR POTENTIAL

Atoms and molecules exert forces. They attract and repel in diverse ways, ever moving about to balance the pushes and pulls. And typically they stumble downward: down toward lower potential energy. With a lowering of the energy usually comes added strength and stability, and already we know such directional flow to be characteristic of many natural processes. Water runs downhill; the dust settles; a battery discharges and dies.

Two objects will attract each other if they can decrease their combined energy by moving closer. So too will they be repelled—pushed apart—if moving away is energetically more profitable. The outcome is governed by the distance between them, and this whole notion of attraction and repulsion is made quantitative through a function called the **intermolecular potential** or **interatomic potential**. The potential function $U(r)$, sketched in Figure 11-1, lays out the mutual energy of two atoms or molecules separated by distances r ranging from zero to infinity.

At infinite separation, the two bodies have no means of communication. Unable to interact effectively, they maintain their kinetic energies and nothing more. The combined energy at infinity is simply the sum of the two separate energies, and we use this value to establish an arbitrary zero. $U(r)$ then becomes a measure of how much the energy *changes* while the objects are brought in from a large distance.

As the distance decreases, attractive forces begin to take hold and the potential energy goes down. Generally it keeps going down, with the attraction strengthening as the gap is closed, but eventually some minimum energy U_0 is reached at a separation r_0. Come any closer and the repulsive forces begin to dominate. The most stable position is at this **equilibrium distance** r_0—at the bottom of the so-called potential well. Here is where the attractive and repulsive forces are balanced to yield the lowest energy, like a spring at rest.

Chemical bonding arises from a particularly strong expression of the electrical interaction, a coming-together in which electrons are intimately mingled to make a molecule. Recall (from Chapter 7) how two approaching hydrogen atoms follow the potential energy curve shown in Figure 11-1(a). Coming in from infinity, the atoms begin to share electrons and form a covalent bond. At any separation along the way, each atom contributes its one electron to an incipient molecule containing two hydrogen nuclei. The negatively charged particles are attracted by both nuclei simultaneously, and eventually neither electron retains any preference for its original site.

This sharing proves optimal at an H–H distance of 0.74 Å, where the energy of the H_2 molecule compared with the separated H atoms is lower (more negative) by over 400 kilojoules per mole. When compressed further, however, the molecule moves to higher energy as the net interaction starts to turn repulsive. What, at moderate distances, was a mutually beneficial *sharing* of electrons suddenly becomes an intrusion. The repulsive interactions between like charges (nucleus against

(a)

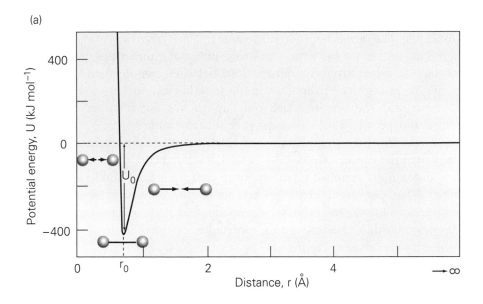

(b)

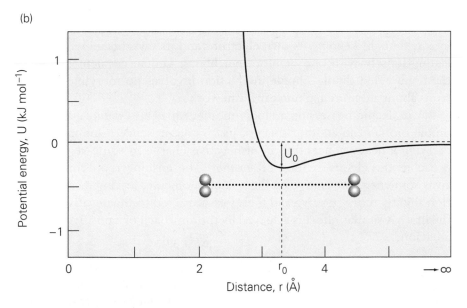

FIGURE 11-1. Atoms and molecules attract and repel, searching blindly for the lowest potential energy. Far apart, they draw together. Up close, they move away. Somewhere in the middle, caught in a *potential well*, they find stability. (a) Interatomic potential for two H atoms, showing a hydrogen molecule stabilized by more than 400 kJ mol^{-1}. A covalent H—H bond forms at an equilibrium distance of less than 1 Å. (b) The intermolecular potential for two hydrogen *molecules* gives rise to a weak dimer at a separation of just over 3 Å. Dispersion interactions between H_2 and H_2 carve out a shallow well of only 0.3 kJ mol^{-1}, shown here on an appropriately enlarged scale.

nucleus, electrons against electrons) control the energy at the shorter distances. The atoms begin to move apart.

Such retreat evokes almost an image of social interactions gone awry, as they do when attractive interactions between people turn repulsive owing to proximity. Although coming together might be good, any attraction will remain attractive only up to a point. Neither electron clouds nor people want to occupy precisely the same space.

Returning now to molecules, we know from Chapter 9 that a variety of noncovalent interactions exist as well, each deriving from Coulomb's law stated in its simplest form: opposite charges attract and like charges repel. That persistent electrical thread is common to intermolecular interactions in general, whether the encounters involve dipoles, induced dipoles, pointlike charges, atoms, molecules, or ions. Some of the possibilities are recalled in Figure 11-2.

Differences in electronegativity, for example, leave electrons distributed unevenly over parts of a molecule, making some regions slightly positive and others slightly negative. The resulting separation of charge produces a dipole moment, and these *polar* molecules then can lower their energies by lining up positive to negative. Each molecule remains intact, though, keeping its own electrons and its own nuclei while associating loosely with one or more neighbors. Unlike covalent bonding, the nonbonded dipole—dipole interaction involves no merging of electrons about a communal nuclear framework.

A molecule possessing a dipole moment may also *induce* a transient dipole moment in another species, just by its presence. One molecule, carrying a heavy concentration of negative charge in some region, approaches the electron cloud of another. The ensuing repulsion pushes away some negative charge on the second molecule, leaving its near portion slightly more positive and therefore attracted to the negative end of the first. A similar effect is achieved by the approach of a positive or negative ion.

FIGURE 11-2. Selected means of intermolecular interaction. Each panel shows (top to bottom) a sequence of three snapshots as the interacting particles, endowed with thermal energy, come together and move apart. (a) Between polar molecules: alignment of permanent dipole moments, positive to negative. Other low-energy arrangements are possible as well. (b) Between polar and nonpolar molecules (or atoms): induction and alignment of a transient dipole moment. The permanent dipole moment of one molecule distorts, briefly and weakly, the electron cloud of the other. (c) Between atoms or nonpolar molecules: the London dispersion interaction. A fragile attraction develops from the alignment of induced dipole moments in each structure.

(a)

(b)

(c)

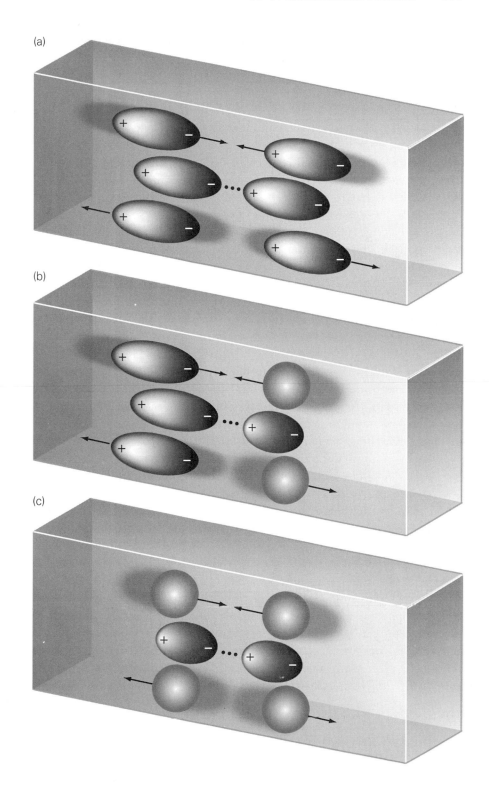

Remember, too, that certain noncovalent interactions develop even without a permanent dipole or ionic charge. Our same hydrogen molecule, symmetric and nonpolar, provides a ready example. H_2 possesses neither a dipole moment nor a net charge, nor can it form additional chemical bonds since its capacity to share electrons is exhausted. Yet a molecule of hydrogen remains an electrical object and thus retains the ability to interact with other charge distributions. Picture H_2 just as a symmetric cloud of negatively charged electrons surrounding two positively charged nuclei. Averaged over time the dipole moment is, assuredly, zero—but the electron cloud is not frozen in space and time. Rather it undergoes continuous and rapid fluctuation, so that for brief intervals the charge may appear skewed and a dipole moment may arise instantaneously. The momentary fluctuation creates a weak dipole that is able to induce other dipole moments of similar kind, if only for an instant.

This weakly attractive interaction, called the **London dispersion force**, is highly fragile and falls away as the sixth power of the distance (decreasing 64-fold in energy when the separation is only doubled). With the potential well, U_0, shallower than 1 kJ mol^{-1} for H_2–H_2 and He–He interactions, the effect is small but ultimately perceptible even by the least gregarious atoms and molecules. The London force permits the liquefaction of H_2, He, Ar, and other weakly interacting atomic or molecular species at low temperature, where the attractions eventually prevail over the disruptive thermal energy $k_B T$.*

The bigger and "floppier" the electron cloud, the more easily it is *polarized* to yield an instantaneous dipole. Noncovalent interactions consequently are more effective between larger atoms—stronger, say, for argon than for helium. Argon atoms, with many more electrons and a correspondingly greater spatial extent, move in a potential well of approximately −1 kJ mol^{-1}, compared to −0.1 kJ mol^{-1} for atoms of helium.

Shape is another consideration. Since the dispersion force operates only at very close range, molecules need to maximize their points of contact. Spherical molecules such as neopentane (Figure 11-3a), which can touch only at a single point, therefore interact less strongly than a pair of elongated molecules able to attract each other over a large area (Figure 11-3b). Atoms, always spherical, have the most difficulty generating substantial dispersion interactions.

---

*We use $k_B T$ and RT interchangeably for the thermal energy. The microscopic value $k_B T$ gives the energy per atom, molecule, bond, and so forth; the macroscopic quantity RT expresses the same energy per mole.

(a)

(b)

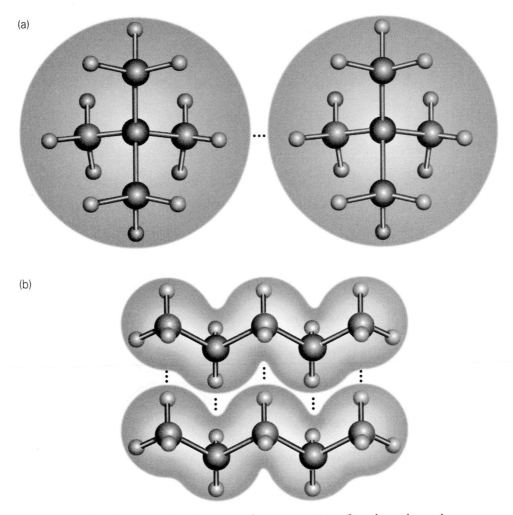

FIGURE 11-3. The strength of a noncovalent interaction often depends on the shapes of the molecules involved. (a) Bulky, round species find scant opportunity to approach closely and cohere, as noted previously for neopentane (Chapters 8 and 9). (b) Molecules of normal pentane, interacting at multiple points along their chains, enjoy a stronger attraction.

11-2. REAL EQUATIONS OF STATE

Looking at Figure 11-4, we now understand exactly what is implied by the model of the ideal gas: The intermolecular potential function, $U(r)$, must be zero everywhere. Ideal gas particles have the same energy regardless of the distance between them. They do not interact.

Isotherms for ideal gases (plots of pressure versus volume at constant

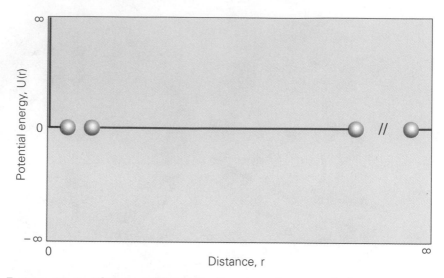

FIGURE 11-4. The potential energy of two particles in an ideal gas: zero. At no distance is there the slightest tendency for the particles to attract, repel, or otherwise alter their motion—unless they collide and simply bounce away. Since the potential energy is everywhere the same, its influence vanishes. One particle has no effect on the other.

temperature) are hyperbolas, since *PV* is always constant. That much we know both from macroscopic observations and from the kinetic theory discussed in the previous chapter. Remember: Once $U(r)$ is assumed to be zero, the pressure can be derived solely by considering collisions of the particles with the enclosing walls. The ideal gas equation of state inevitably follows.

Ideal isotherms, which trace an unbroken progression from low density (low pressure, high volume) to high density (high pressure, low volume), owe their appearance to the absence of an intermolecular potential. Lacking any way to interact, the particles of an ideal gas can only squeeze together or move apart when the pressure is varied. All that changes is the density. An ideal gas (Figure 11-5) stays a gas no matter how tightly its particles are compressed.

Molecules in real gases (Figure 11-6) respond differently under pressure. They have a real intermolecular potential, through which the energy is changed when the molecules squeeze closer together. Thermal energy permitting, they will stick together at distances near r_0 so as to lower their energy by approximately U_0. A gas thus starts to express its *non*ideality at temperatures where the depth of the potential well is comparable to the prevailing thermal energy, *RT*. The illusion of ideality vanishes as the gas cools.

FIGURE 11-5. Squeeze together the particles of an ideal gas, and the result is simply this: the same ideal gas, compressed more densely into a smaller volume. Whether near to another or far, each particle moves as if it were the only one present. It exerts no influence. It accepts none. It makes no accommodations and no alliances. Its potential energy is everywhere zero.

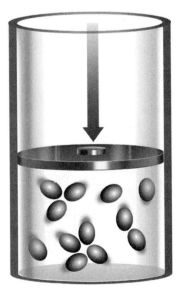

FIGURE 11-6. Squeeze together the molecules of a real gas, and eventually the pressure becomes too great to bear. Attractions take hold. Clusters form. The position of one molecule affects the potential energy of another.

What we need is a *real* equation of state to account for these various effects. How, for instance, would pressure be affected by the tendency of molecules to attract at certain distances? One possibility, not unreasonable, is to suppose that such stickiness will rob molecules of some momentum and also interfere with their arrivals at the wall. For if a particular molecule suddenly finds itself occupied with other molecules, then surely it has less time and energy to expend banging into the wall. The ideal pressure (nRT/V) will be reduced as a result, and so we may tentatively write

$$P = \frac{nRT}{V} - \text{some correction for intermolecular interactions}$$

to represent the net pressure exerted by a real gas.

The volume, on the other hand, should be corrected to account for the repulsive part of the potential, which derives from the nonzero volume of the molecules themselves. Repulsion means, after all, that two electron clouds of finite size cannot be forced into the same region of space. So if one mole of real molecules—packed together with no gaps—were to occupy some irreducible volume b, then the empty space within the container would necessarily be decreased by just that amount. For n moles, then, the free volume available to the molecules must be no greater than

$$V_{\text{free}} = V - nb,$$

an amount corresponding to the empty container's volume (V) diminished by nb.

Van der Waals, in the 19th century, adopted this model of an excluded volume and argued further that the pressure correction (attributed to *two* interacting particles) should depend on the *square* of the density, going as an^2/V^2. He said: Since one molecule (one density factor, n/V) must first collide with another molecule (a second density factor, n/V), the probability of interaction will be proportional to n^2/V^2. Taking a as an empirical proportionality constant to characterize a particular molecule, he then proposed that

$$P = \frac{nRT}{V - nb} - \frac{an^2}{V^2}$$

in a real gas. For a view in pictures, see Figure 11-7.

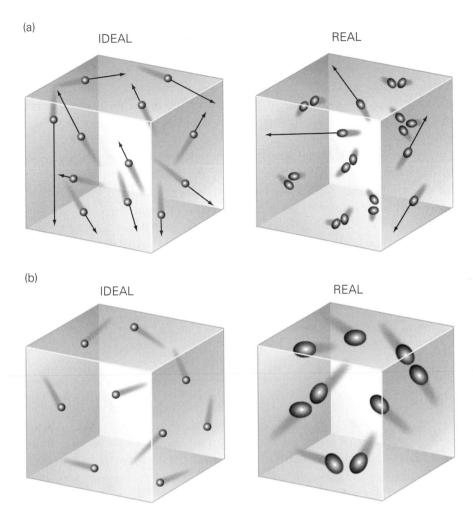

FIGURE 11-7. All gases behave ideally at sufficiently low density, and all ideal gases look alike: structureless, volumeless particles uninfluenced by the presence of others. But at sufficiently high density, every gas becomes a *real* gas; and every real gas is different: a unique assembly of interacting molecules. (a) Able to attract and repel, real molecules clump together to form clusters. Collisions with the walls grow softer and less frequent, and the pressure drops correspondingly. (b) A real molecule takes up space—space from which all other molecules are now excluded. The volume available to a real gas is smaller than the volume of its container.

Notice how the two terms work in opposite directions. The first, which corrects for the excluded volume nb, produces a pressure greater than the ideal value:

$$\frac{nRT}{V - nb} > \frac{nRT}{V}$$

Confined within a smaller effective volume $(V - nb)$, the molecules hit the walls more frequently and generate a consequently higher pressure.

The second term, $-an^2/V^2$, reduces the pressure. The stickier the attraction (as assessed by the coefficient a, a positive number), the lower the pressure. The denser the gas, too, the lower the pressure. To have more particles crowded into a smaller volume is to have more opportunity for interaction.

These two parameters a and b, obtained by fitting experimental data to the expression above, follow roughly the pattern one would expect for different gases. Both, for example, are bigger for argon than for helium—both the attractive coefficient a, which reflects the enhanced dispersion interactions of the larger atom, and similarly the size correction b. Given numerical values of a and b determined for different molecules (see Table C-14 in Appendix C), we start to appreciate the effects of molecular size and interactions in real gases.

Rearranged to resemble the more familiar $PV = nRT$, the **van der Waals equation**

$$\left(P + \frac{an^2}{V^2}\right)(V - nb) = nRT$$

is but one example of an equation of state for a real gas. As one of the earliest, it is of historical interest and is easily understood qualitatively. Many other possible equations of state have been investigated as well, each making different assumptions about the intermolecular potential. These various equations can be derived using statistical methods similar in spirit to the kinetic theory we discussed, but the mathematical formalism becomes considerably more complicated. Our purposes are served amply, however, just by an awareness of what goes into a real equation of state and what its general form must be. To that end, the van der Waals equation stands as an instructive prototype even after 100 years.

11-3. REAL ISOTHERMS: CRITICAL TEMPERATURE AND PHASE TRANSITIONS

The van der Waals equation reduces to the ideal gas equation when the density is low, under conditions where relatively few moles are spread over a large volume. With $V \gg n$, the correction terms an^2/V^2 and nb are both small compared with P and V, respectively, and we recover $PV = nRT$. So we must, since gases do behave ideally at low densities and high temperatures.

High-temperature isotherms computed from a real equation of state therefore remain the same smooth hyperbolas already seen for an ideal gas. The molecules stay far apart, and their high thermal energy wipes out any attractive interactions. The intermolecular forces are not yet expressed.

But the isotherms change visibly as the temperature is lowered to the point where interactions take hold. The variables start to obey the real equation of state (under which *PV* is *not* equal to a constant), and so the isotherms lose their hyperbolic shape and begin to display the additional structure shown in Figure 11-8. The former smoothness, so evident for an ideal gas, proves to be characteristic only of a vanishing intermolecular potential.

Eventually at some temperature T_c, called the **critical temperature**, the isotherm acquires a stark new feature: it develops a horizontal jog when the curvature suddenly changes. At lower temperatures still, the slope flattens out to form a plateau as seen in the diagram. Corners and edges start to appear. Such drastic alterations in the isotherms are symptoms of an intermolecular potential able fully to exert its influence, a potential made real by the drop in thermal energy suffered with falling temperature. The onset of critical behavior, a clear sign that the gas is no

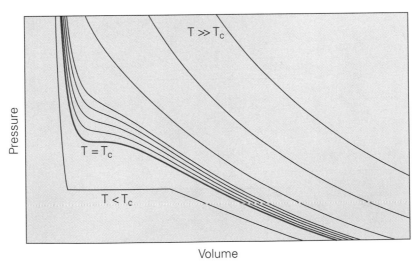

FIGURE 11-8. No longer governed by the equation $PV = nRT$, isotherms for a real gas cease to be hyperbolic once reality sets in. Note these two features: (1) Only at the highest temperatures (where interactions remain unexpressed) will a *P-V* plot yield a hyperbola. (2) As the gas cools, its isotherms become progressively distorted until eventually they change curvature at a certain *critical temperature* T_c. Application of suitable pressure below T_c will force a condensation of the gas.

longer ideal, cannot occur until RT becomes comparable to U_0. Until then, the molecules are too energetic (Figure 11-9).

The stronger the potential, the sturdier is the association. The stronger the interaction, the higher is the critical temperature at which the molecules can clump together. They may associate two at a time to form *dimers*, three at a time to form *trimers*, or arrange themselves into larger clusters, but in any event *something* starts to happen at the critical temperature—some kind of collective, or *cooperative* phenomenon. It is a crisis, a turning point.

The critical temperature of an ideal gas? Zero. The particles never associate preferentially; there are no interactions. The isotherms stay the same.

The critical temperature of helium? 5 K. Held together by the barest of threads, the helium atoms cannot cooperate unless the thermal energy is minimal as well. So weak is the dispersion force that clusters can survive only among the coldest, most sluggish collections of these atoms.

Argon? Approximately 150 K. Its London interactions are stronger than helium's; its potential well is deeper; its T_c is higher.

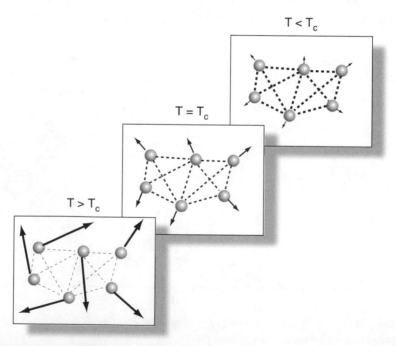

FIGURE 11-9.   Above T_c, molecules have the *means* to interact but not the opportunity. Their thermal energy, still too high, quickly tears apart any budding clusters, and so the gas remains a gas. Not until the system cools to its critical temperature do the interactions (broken lines) begin sufficiently to take hold. Each gas behaves differently.

Now, given this understanding, let us move along a real isotherm below the critical temperature, beginning at the lower right where we clearly have a gas at low pressure and high volume:

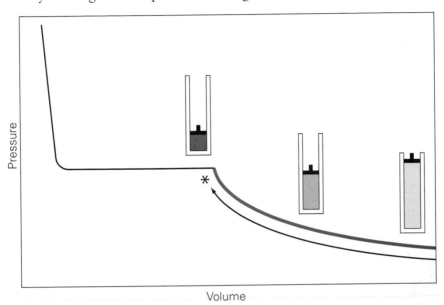

Compression along the isotherm appears normal at first, as the gas squeezes down to a lower volume at higher pressure. Abruptly, though, the flat portion of the isotherm is encountered, whereupon the pressure stays constant:

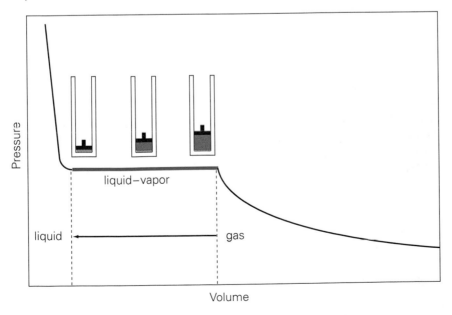

The volume of gas continues to decrease, but the pressure remains the same and thus the compressibility becomes infinite.

A *change in state* (Figure 11-10) is occurring along the flat section of the isotherm. Molecules are being squeezed out of the gas into a liquid. The pressure is constant during this **condensation**, because both the number of moles of gas *and* the gas volume are decreasing. Water molecules, pressed into close association, are falling out of the vapor while the gas shrinks with each droplet. In the meantime, the much denser liquid occupies only a very small volume compared with the original capacity of the gas.

As soon as that first droplet appears, the liquid provides a reservoir to accept molecules from the gas phase. Further squeezing of the gas just produces more liquid while, above the surface, a dwindling supply of gas molecules continues to exert the same pressure in a smaller volume. At last all the molecules are forced into the liquid phase, and the isotherm begins to climb upward again.

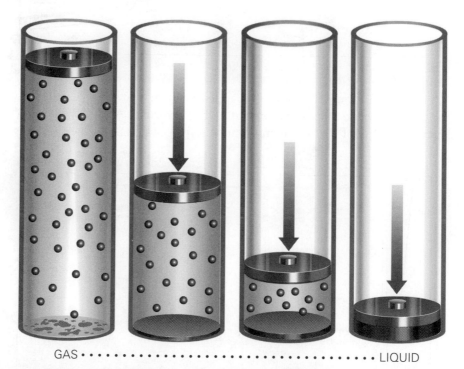

GAS · LIQUID

FIGURE 11-10. Cooled below the critical temperature, the molecules of a gas have both the means and opportunity to cooperate. Squeezed hard enough, they stick together and stay together. They associate to the point where the gas is no longer a gas. They condense into a liquid. The transition occurs along the flat part of the isotherm in Figure 11-8, under constant pressure.

Having been completely liquefied, the system is much less compressible than before. Average distances between molecules are far smaller now, so the material requires considerably more squeezing to yield even a small change in volume. The slope of the isotherm grows steeper.

We have gone from a *disordered* gaseous state to a more ordered liquid state, passing through a region where both phases exist simultaneously. The many molecules in a gas are in random motion, with many possible arrangements and with only the most fleeting of associations. Molecules in a liquid, by contrast, tend to stick together locally, binding themselves briefly into weakly interacting clusters before succumbing to the thermal energy. In this sense the gas is disordered relative to the liquid—but only comparatively so, for the liquid itself possesses no long-range order. The interactions come and go, and the molecules never occupy fixed positions as they would in a solid. Although the liquid is unquestionably more ordered than the gas, the two states continue to have something in common. We shall see evidence of this commonality in our upcoming discussion of supercritical fluids.

With the terms *order* and *disorder*, we also anticipate a thermodynamic quantity called *entropy*, which will be examined at length over the next three chapters. Entropy is, in one interpretation, a measure of disorder, or the number of ways a system of microscopic particles can be arranged to yield the same macroscopic properties. Notably it is the drive toward increased global entropy—toward greater disorder everywhere—that controls the direction of chemical change. We shall explore in detail, later, the relationship between energy and entropy, and see how such change comes about.

The gas–liquid condensation is, moreover, merely the first example of a broad class of **disorder–order transitions**. The freezing transition between liquid and solid is another, and we shall mention several more toward the end of this chapter. Most disorder–order transitions bear little resemblance, superficially, to changes in state such as condensation–evaporation and freezing–melting, yet they are profoundly similar nevertheless. All share many of the same characteristics at a deeper level.

11-4. PHASE EQUILIBRIA

The coexistence of gas and liquid during the phase transition is an example of **equilibrium**, recognizable macroscopically as a condition that seems never to change. An observer looks coarsely and sees only constancy: a pool of liquid sitting under a column of gas, with the pressure of

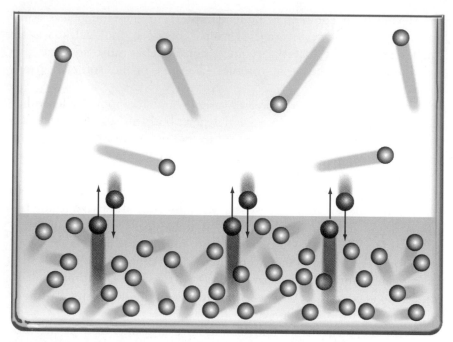

FIGURE 11-11. Dynamic equilibrium. Some molecules go this way, and some go that way; but in due course they reach a stalemate. The rate of condensation becomes equal to the rate of evaporation, and the macroscopic picture ceases to change. A steady vapor pressure is maintained above the liquid.

the gas always the same. Viewed microscopically, however, the gas–liquid equilibrium depicted in Figure 11-11 is not a static landscape, unchanging in time. It is rather an intense competition fought to a draw. The most energetic molecules are continually escaping from the liquid and evaporating into the gas; they, in turn, are replaced by a matching flow of molecules condensing from the gas into the liquid to preserve the balance. There is a *dynamic* opposition of forces, equally weighted, and thus is the pressure maintained constant.

Vapor Pressure of a Pure Liquid

A gas coexisting with its liquid is customarily called a ***vapor***, and the constant pressure existing above the liquid is termed the ***vapor pressure***. We may read the vapor pressure directly from the isotherm for a phase transition, taking it simply as the value along the flat portion of the *P-V* curve.

Vapor pressure is a fixed property of the equilibrium system, dependent only on the temperature, and in Chapter 12 we shall recognize and define this quantity properly as an *equilibrium constant*. It takes the same value for a given temperature no matter how much gas, no matter how much liquid, no matter what the volume. The isotherm, and hence the vapor pressure, moves lower as the temperature decreases.

The vapor pressure of water, displayed in Figure 11-12, goes from 23.8 torr at 25°C to 760 torr (1 atm) at 100°C. At 100°C, with the vapor having reached atmospheric pressure, water begins to boil. Any liquid will similarly boil when its vapor pressure exceeds the ambient pressure. Molecules under these conditions are violently expelled from the liquid, able now to overcome the force of the atmosphere pressing down on the surface. At lower pressure, say on a mountaintop, the boiling point is lower than normal. At higher pressure, as in a pressure cooker, it is greater.

To boil a liquid under any conditions, one must supply sufficient thermal energy to overcome the attractions of its intermolecular potential. Strongly interacting polar species such as water, for example, demand more energy and consequently boil at a higher temperature than liquid argon or helium. Indeed we can often predict qualitative trends in boiling points just by considering the expected strength of the intermolecular forces. Ion–ion interactions are strongest of all, followed by ion–dipole interactions, dipole–dipole interactions, and finally by the dispersion interactions.

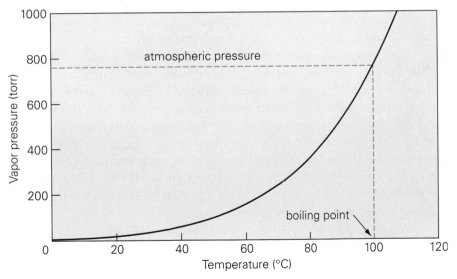

FIGURE 11-12. The vapor pressure of water, measured at temperatures ranging from 0°C to just over 100°C. Wherever liquid and vapor are in equilibrium, the pressure is constant and fixed by the temperature alone—not by the volume of either phase.

Liquid water, with its polar molecules networked together by hydrogen bonds, boils at an appropriately high temperature.

Related to boiling point (but construed more broadly) is the aforementioned critical temperature, T_c, which acquires a crisp meaning independent of pressure. Realize that above the critical temperature, regardless of the pressure and density, no liquid can exist in equilibrium with its vapor. Squeeze as one might, the gas will never condense. The thermal energy is so high compared with U_0 that the gas molecules cannot cooperate.

Is, then, water's critical temperature above or below its boiling point of 100°C? Evidently above—if not, we would never see water and steam together in the same kettle. *Any* amount of liquid, if found to coexist with its vapor, proves irrefutably that the ambient temperature is below T_c.

T_c for water, incidentally, is 647 K (374°C). For others, see Appendix C (Table C-14).

Vapor Pressure of a Solution

Our picture so far has been of pure states: one species distributed over pure liquid and pure vapor, with molecules ascending and descending from one phase to the other. Just as some energetic particles in the liquid will break free of the surface and rise to the vapor, so will certain gaseous particles crash down to the liquid. At equilibrium, the opposing rates are equal and the vapor pressure holds steady.

Now expand that picture to allow for two or more species in the liquid phase, as in systems like water/ethanol or water/sucrose or water/sodium chloride. Such mixtures, familiar already from Chapters 3 and 9, are *solutions*, formed when a particular solute is dispersed uniformly throughout a solvent. Solutions, found everywhere from the seas to the blood to the laboratory vessel, have special phase properties that compel our interest.

Confronted by any new and hard problem, a scientist will try to simplify the circumstances—to replace the real by the ideal, if only temporarily. As we did for the gas, let us do the same now for the liquid and devise a model of the ***ideal solution***.

What would make a solution ideal? Whereas ideality for a gas arises from a complete lack of intermolecular interaction, ideality for a solution implies total *uniformity* of interaction. In an ideal gas, the intermolecular potential is zero for all particles at all distances. No molecule affects any other by its presence or absence. In an ideal solution, subtly different, the intermolecular potential is imagined to be *identical* for every species. The molecules do influence each other, but always in the same way.

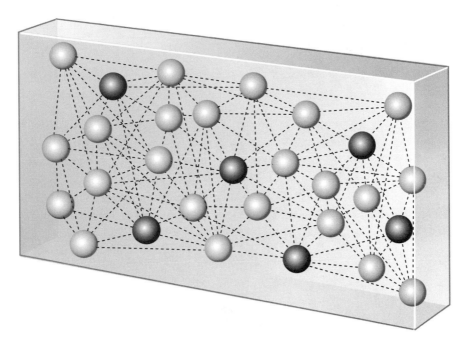

FIGURE 11-13. The ideal solution: a hypothetical system in which solvent–solvent, solute–solvent, and solute–solute potentials all vary in the same way. Solute particles are rendered with darker shading, both in this illustration and subsequently. Dashed lines of equal weight (as here) suggest interactions having the same mechanism, strength, and sensitivity to distance.

Solvent–solvent interactions, solute–solvent interactions, and solute–solute interactions are indistinguishable in an ideal solution (Figure 11-13). Adding a solute neither enhances nor diminishes the attractions and repulsions normally present in the pure solvent; the solute is assimilated into the solvent without disturbance. That, at least, is the simplifying assumption, but that is also the behavior actually exhibited by dilute solutions. With fewer particles of solute available to cause distortions, our picture of ideality becomes more reasonable.

If so, then we know what to expect. Suppose, first, that the solute is itself a *volatile* liquid, able to expel molecules into a vapor phase without difficulty. Introduced into a volatile—but ideal—solvent, such a solute will do what it normally does: vaporize. Interactions with the solvent leave the solute particles no more or less tightly clustered than before. Nor is the vaporization of the solvent aided or impaired by interaction with the ideal solute. Each species of molecule is free to come and go independently, as if no other were present (Figure 11-14).

The vapor properties then depend only on the number of particles, not the kind. The volatile *solute* (call it component 2) evaporates and

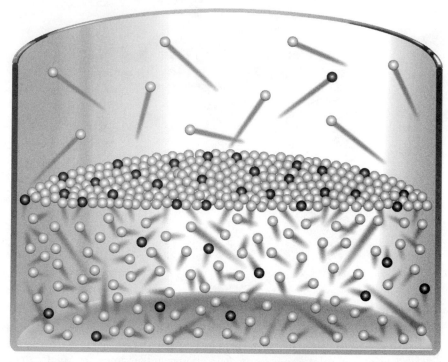

FIGURE 11-14. Evaporation of a volatile component from an ideal solution: by the numbers. To escape, a molecule needs the requisite energy and also a vacant launching point along the phase boundary. Will it find such an opening? Will it vaporize? Chance will decide: If all molecules really do interact in the same way, then slots at the boundary should fall to each component in proportion to population.

condenses according to its numerical representation in the solution, given by the mole fraction

$$X_2 = \frac{\text{number of moles of solute}}{\text{total number of moles}} = \frac{n_2}{n_1 + n_2}$$

If its vapor pressure is P_2° in a pure liquid ($X_2 = 1$), then the value in solution is reduced proportionally to

$$P_2 = X_2 P_2^\circ$$

Similarly unhindered, the volatile *solvent* (component 1) evaporates and condenses according to its own mole fraction

$$X_1 = \frac{\text{number of moles of solvent}}{\text{total number of moles}} = \frac{n_1}{n_1 + n_2}$$

and equilibrates at the corresponding vapor pressure

$$P_1 = X_1 P_1^\circ$$

Each component establishes a reduced vapor pressure consistent with its fractional population in the ideal solution. It is a straightforward apportionment according to the numbers.

Proceed next to a nonvolatile solute, a species with little or no vapor pressure. Sucrose in water is a good example, one where we know not to expect a cloud of sugar vapor above a cup of sweet liquid. The dissolved sugar molecules stay dissolved in the liquid while the water molecules move back and forth between liquid and vapor.

But a nonvolatile solute is not wholly a benign presence, for it does interfere with the emigration of the H_2O species. Sugar molecules, distributed with a mole fraction X_2, take up space like so much debris. A water molecule seeking to escape the liquid might find itself blocked by one of these hulking particles and consequently unable to vaporize. The solvent's vapor pressure is lowered as a result. Fewer water molecules get to the surface, and fewer water molecules get out.

The more abundant the solute, the more likely a blockage of the kind pictured in Figure 11-15 is to occur. Take, for the moment, the most naive view of the statistics: If one out of ten molecules is sucrose ($X_2 = 0.1$), then water will fail to vaporize one out of every ten attempts. It will succeed the other nine times out of ten, consistent with its own mole fraction ($X_1 = 0.9$). If, instead, four out of ten molecules are sucrose ($X_2 = 0.4$), then 40% of the encounters at the surface are unfavorable while 60% ($X_1 = 0.6$) are favorable . . . and so on for all the other mole fractions, according to this simplest of interpretations. We find, therefore, that the solvent's vapor pressure is lowered again to

$$P_1 = X_1 P_1^\circ$$

in what amounts to a definition of **Raoult's law** for any volatile component in an ideal solution. The result tells us specifically here that (1) pure solvent ($X_1 = 1$) has its usual vapor pressure P_1°; (2) pure nonvolatile solute has a vapor pressure of zero; and (3) all mixtures vaporize at pressures in between, always lower than P_1°. Although reality is never so simple as we have imagined, Raoult's law is valid nevertheless for sufficiently low concentrations of solute—where the solution behaves ideally.

Conceived as flotsam and jetsam, the nonvolatile solute particles lose their chemical identity in the hypothetically ideal solution. Whether hindered by sucrose molecules or dextrose molecules or chloride ions, the water molecules respond to featureless, undifferentiated *particles* that function only to impede vaporization. The reduction in vapor pressure, determined just by the concentration of such particles, is then said (aptly or not) to be a ***colligative property***, a property of the "collection" rather than the individual.

Such collectivity means that an ionic solute differs from a molecular solute principally in the number of particles produced upon dissolution. Whereas one mole of sucrose produces one mole of dissolved particles ($C_{12}H_{22}O_{11}$ molecules), one mole of NaCl produces *two* moles of dissolved particles (one mole each of Na^+ and Cl^- ions). Mole for mole,

FIGURE 11-15. Unable to escape the liquid, a nonvolatile solute lowers the vapor pressure of its solvent. Particles of solute, distinguished by dark shading, occupy space on the surface and block the exit of solvent molecules. The higher the concentration, the greater the obstruction. Fewer solvent molecules escape to establish a vapor pressure in the space above.

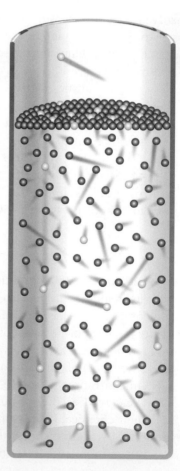

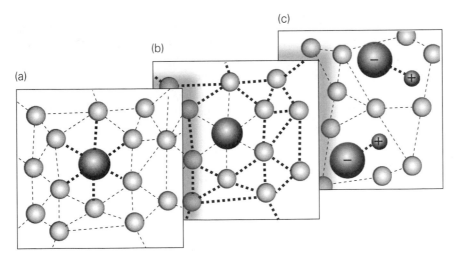

FIGURE 11-16. Real solutions. (a) Negative deviation: Solute–solvent interactions, represented by the thicker dashed lines, are stronger than solvent–solvent interactions. Bound more tightly to the solute than to each other, solvent molecules vaporize less than predicted by Raoult's law. (b) Positive deviation: Solute–solvent interactions are weaker than solvent–solvent interactions. Particles of solute, where present, disrupt the clusters of solvent molecules and increase the vapor pressure accordingly. (c) Ion pairing: Anions and cations associate preferentially, effectively reducing the number of free particles in solution.

the sodium chloride exerts twice the colligative effect—if the solution is ideal.

It is a big *if*. Dropping the pretense of ideality, we concede that solutes are not just particles; that solutes do assert themselves as chemically unique individuals; that solute and solvent do interact in distinguishable ways. When they do, we should consider a number of more realistic models (Figure 11-16) involving real solutes and real solvents:

1. The solute has a strong affinity for its solvent, so strong that the solute–solvent interaction exceeds the usual solvent–solvent binding. Some of the solvent molecules, held more tightly by the solute, are unduly handicapped when they try to vaporize. The vapor pressure then falls below Raoult's prediction for an ideal solution, less than $X_1 P_1^\circ$. Call that a *negative deviation*.

2. Particles of a solute, having little affinity for the solvent, come between the solvent molecules and disrupt the relationships normally present. Disturbed by the interference, clusters of solvent molecules are weakened and hence more likely to vaporize. The vapor pressure rises above the ideal benchmark of $X_1 P_1^\circ$, making for a *positive deviation*.

3. The anions and cations of an ionic solute, able to sustain strong attractions, sometimes pair up as oppositely charged particles are wont to do. Each *ion pairing* of, say, Na^+ with Cl^- effectively replaces two species with one. The result is fewer independent particles in solution, something less than the stoichiometric maximum. Any colligative properties of the solution are reduced correspondingly.

Deviations of this sort occur all the time, but still we have much to gain from the simple image of an ideal solution. It is a model worth keeping, even if real solutions are nuanced and idiosyncratic. The ideal solution, like the ideal gas, allows us to be "almost" right with little effort—a position not to be scorned.

Accepting those limitations, we come now to understand two more colligative properties: *boiling-point elevation* and *freezing-point depression*. Undifferentiated *particles*, we have seen, clog the solvent and interfere with its normal comings and goings. Crowds of particles alter the equilibrium between liquid and vapor. A certain number of particles (be they atoms or molecules or ions), will lower the vapor pressure by so many torr. With its vapor pressure held down, the solvent then must expel more and more molecules to keep pace with those species returning—at a rate undiminished—from the gas above. To produce a vapor pressure of 1 atmosphere and thence to boil, the solution is forced to work harder than the formerly pure solvent. It will boil at a higher temperature.

The increase amounts to

$$\Delta T_b = K_b m$$

where m is not the molarity but rather the solution *molality* (moles of dissolved solute per *kilogram* of solvent). K_b is the *molal boiling-point-elevation constant*, characteristic only of the solvent and equal to $0.51°C\ m^{-1}$ for water. One mole of ideal solute particles added to one kilogram of water will increase the boiling temperature by approximately half a degree.

The same dissolved particles also will *lower the freezing point* to the extent

$$\Delta T_f = K_f m$$

where K_f, the *molal freezing-point-depression constant*, has the value $1.86°C\ m^{-1}$ for water. The phenomenon is long familiar: Seawater freezes at a lower temperature than fresh water; calcium chloride ($CaCl_2$) melts the ice on a frozen highway; ethylene glycol ($C_2H_6O_2$) added to water produces automotive antifreeze.

The depressive effect comes again from the clutter created by the solute particles, which now are prevented from entering the solid phase of the freezing solvent. Most solutions exclude solute when they freeze, creating solid regions consisting of pure solvent while confining the solute to the liquid phase. Unable to cross the liquid–solid boundary, the dissolved particles effectively lower the rate at which solvent molecules enter the solid (Figure 11-17).

FIGURE 11-17. Depression of the freezing point in solution, also by the numbers: Barred entry to the solid phase of the *solvent*, the darkly shaded solute particles hinder the traffic of solvent molecules across the liquid–solid boundary. Similar to vapor-pressure lowering and boiling-point elevation, the effect is *colligative*. It depends on the number of dissolved solute particles, not the kind.

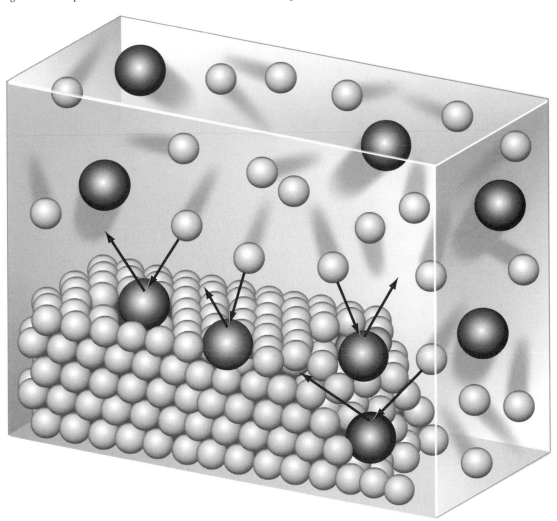

Meanwhile, the reverse flow from solid to liquid continues unaffected by the dissolved solute outside, since the melting solvent can sweep away any solute particles temporarily resident on its surface. To equalize the rates and establish phase equilibrium, the solid must cut back its supply of molecules to the liquid. The necessary balance between freezing and melting is established at an appropriately lower temperature.

11-5. PHASE DIAGRAMS

Complicated by solutes and solvents, by ions and molecules, and by ideal and real interactions, phase equilibria become increasingly individualistic in their details. We need inquire no further, but instead we shall develop a practical scheme to tabulate all the phases possible under any conditions.

Although P-V isotherms persuasively chart the progress of a given phase transition, even more data can be summarized in a compact ***phase diagram*** of the sort shown in Figure 11-18. Here the axes are pressure

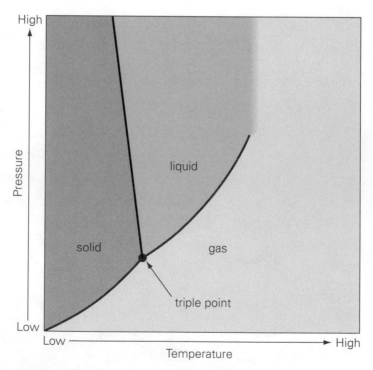

FIGURE 11-18. Schematic phase diagram for water, showing combinations of temperature and pressure under which solid, liquid, and gaseous phases can exist—either alone, as in the open regions; or in equilibrium, as along the solid–liquid, liquid–gas, and solid–gas lines. All three phases simultaneously maintain a dynamic equilibrium at the *triple point*, where the three coexistence lines intersect.

and temperature, the two variables most accessible to manipulation in the laboratory.

The distinct parts of a typical three-phase diagram fall into three classes: (1) open areas, (2) lines, and (3) the point where the three lines intersect. For each combination of pressure and temperature, represented on the diagram as an ordered pair (P, T), the diagram registers the number and kind of phases present. Volume is irrelevant in such a presentation, since the designated phases will exist in a system of any size.

OPEN AREAS CORRESPOND TO SINGLE PHASES, indicated for pure water as solid, liquid, and gas:

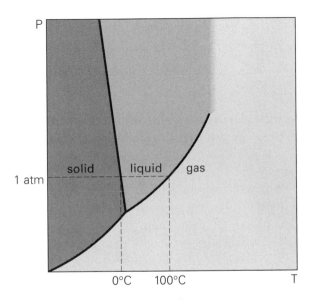

Although water actually can form more than one solid phase, the simplified diagram shown here pertains only to the most common variety of ice (known as ice-1). Looking first for some familiar points, we quickly confirm that water at 1 atm is a solid below 0°C, a liquid between 0°C and 100°C, and a gas above 100°C.

Note that the solid phase appears in the diagram precisely where it should, under conditions tending toward high pressure and low temperature. The gaseous state, just as naturally, favors conditions of low pressure and high temperature.

LINES REPRESENT COEXISTENCE REGIONS, along which two states of matter are present together. The line itself becomes emblematic of the *interface*, or phase boundary, through which molecules pass back and forth to maintain the equilibrium. Any point on the liquid–gas line, for instance, marks some combination of temperature and pressure that supports a liquid–vapor equilibrium. Most obvious is the example of water–steam at 1 atm and 100°C:

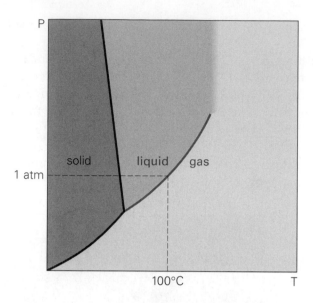

Vapor pressure, the parameter characterizing the liquid–gas equilibrium, corresponds directly to the various pressure values along the liquid–gas line. We simply read the pressure associated with a given temperature at which both phases are present.

A POINT OF INTERSECTION SHOWS THREE PHASES EXISTING SIMULTANEOUSLY. Termed a **triple point**, this special set of conditions produces a three-way equilibrium between solid, liquid, and gas. The triple point for water is $0.01°C$ and 4.6 torr, while for carbon dioxide it is $-56.6°C$ and 5.1 atm:

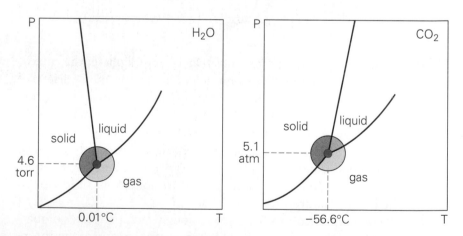

Having undertaken this quick orientation, we can now move more confidently around the phase diagram. Particularly interesting is behavior near the critical temperature T_c, above which there can be no liquid–vapor equilibrium. That temperature, paired with a critical pressure P_c, appears at a certain **critical point** which abruptly terminates the liquid–gas line at the upper right:

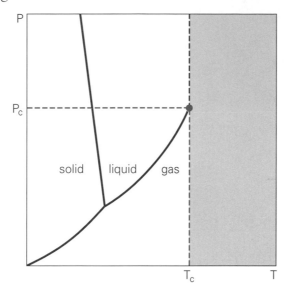

A vertical line (the isotherm $T = T_c$) connects the critical point to the temperature axis and thereby delineates the start of the shaded **supercritical temperature** region.

To the right of the critical isotherm there is no coexistence of liquid and gas. The single phase that does exist is best called a **fluid**, a state that sometimes looks like a liquid and sometimes looks like a gas, depending on the density. As a start, we can agree on at least one manifestly obvious characteristic: it *flows*.

Fluids beyond the critical point have a number of unusual properties, and in their liquidlike incarnations are especially useful as solvents. Supercritical carbon dioxide is used to extract caffeine from coffee beans, for example, and certain supercritical organic hydrocarbons are able to dissolve coal.

Many fluids display exotic properties in the immediate vicinity of the critical point as well. *Critical opalescence*, for one, can be a striking effect. With the isothermal compressibility infinite at (T_c, P_c), a critical fluid may exhibit large fluctuations in density from point to point. Light shining on the material is scattered differently at each position, and so a normally clear substance may either become turbid or else display a variety of flickering colors.

But is it a liquid or a gas? To compound the ambiguity, the phase diagram evidently shows a way to go, imperceptibly, from gas to liquid by detouring around the critical point. Starting with a normal subcritical gas, we follow the roundabout path shown in Figure 11-19 instead of

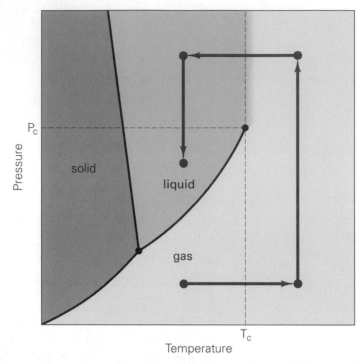

FIGURE 11-19. A change from gas to liquid, routed around the critical point to avoid a discontinuous phase transition. There is gas at the beginning and liquid at the end, but no sign of condensation or any phase boundary in between. All throughout, gas and liquid never exist together in equilibrium.

crossing the gas–liquid line directly. Here the temperature is first raised above T_c, after which the fluid is compressed to higher and higher density until the critical pressure is exceeded. Finally, cooled below T_c and decompressed somewhat, the material clearly becomes a liquid occupying a legitimate point in the liquid-only region. Along the way, too, the fluid looks increasingly like a liquid as the pressure is elevated, yet (oddly) no phase transition is ever apparent. At no point does a boundary between liquid and gas appear, and never is there any hint of discontinuity. No droplets form.

To locate a genuine phase transition brought about by isothermal compression, we need only to draw a vertical temperature line that crosses the gas–liquid coexistence line. All vertical lines on the *P-T* diagram correspond to isotherms, of course, and we have already noted the position of one such marker: the critical isotherm. Squeezing at some other (lower) temperature then takes us *up* the appropriate isotherm, going from low to high pressure as the gas gives way to the liquid:

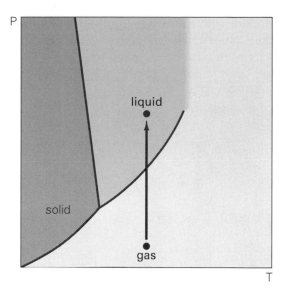

In addition to the liquid–vapor equilibrium, we have also to consider liquid–solid (*freezing* and *melting*) and solid–gas (*sublimation* and *deposition*) transitions. The analysis remains the same: If any of these changes is to be induced at constant temperature, then at least one isotherm must cross the designated coexistence line. Sometimes, though, that "vertical" requirement proves to be impossible—just note how liquid water cannot freeze isothermally at, say, 400 K however high the pressure, since the 400 K isotherm connects only the liquid and gaseous regions:

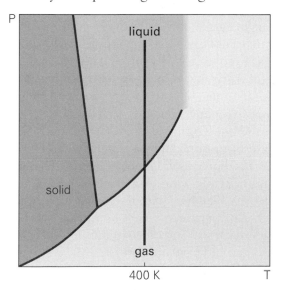

If, instead, the transition occurs *isobarically* (at constant pressure), then the phase change proceeds along a horizontal *isobar* between two regions. Sublimation from solid to gas, in particular, may take place on any horizontal line beneath the triple point (where the liquid phase, absent from this part of the diagram, cannot exist):

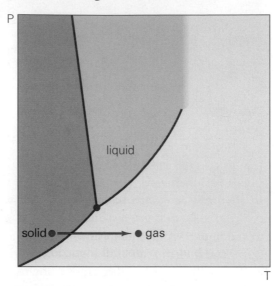

Observe, however, that no critical point exists for either the liquid–solid or solid–gas transition, because there is no pathway to go continuously from one phase to the other. The difference between both liquid–solid and solid–gas is more pronounced in some respects than between liquid and gas. A transformation between gas and liquid, we have remarked, is a transformation merely between extreme disorder and some partial, short-range, not-very-permanent order. To go from either liquid to solid or gas to solid is, by strong contrast, to end up in a highly ordered, rigid state. A gas and a liquid are close enough structurally to be indistinguishable at times, able to merge into an ambiguous fluid state when their densities become equal.

Apropos of the issue of liquid–solid transitions, we discover finally that water's phase diagram exhibits a significant anomaly highlighted in Figure 11-20(a): The liquid–solid line has a negative slope, extending up right to left from the triple point. Water, unlike most materials, forms a solid of lower density than its liquid; it expands upon freezing and hence ice floats. The crystal structure of ice is dominated by hydrogen bonding (Chapter 9), secured by strong electrostatic interactions between partially negative oxygen and partially positive hydrogen atoms in molecules of H_2O. The packing produces a comparatively open structure (Figure 11-21) in which each water molecule interacts with only its four nearest neighbors. Upon melting, the hydrogen bonds suffer a partial collapse and the density increases by about 9%.

If ice at $0°C$ is squeezed isothermally, it tends naturally toward a

(a)

(b)

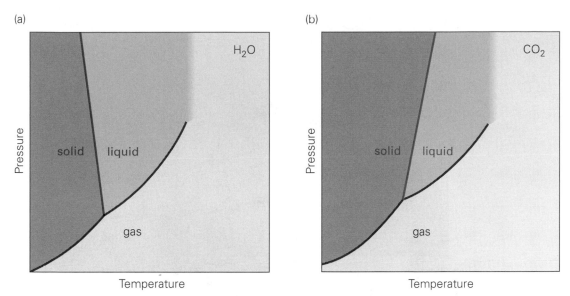

FIGURE 11-20. Comparison of the phase diagrams for H_2O and CO_2 (not to scale). (a) The line between water and ice slopes upward, anomalously, from right to left since the solid phase has lower density. Ice liquefies under pressure. (b) For carbon dioxide, the liquid–solid line slopes upward from left to right—normal behavior for any solid more dense than its corresponding liquid. Solid carbon dioxide remains a solid when subjected to higher pressure.

state of higher density: *liquid* water. Ice therefore melts under pressure, and the anomalous transition appears on the phase diagram as a vertical line beginning in the solid region and crossing into the liquid:

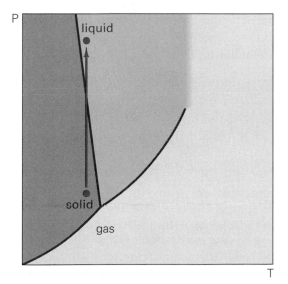

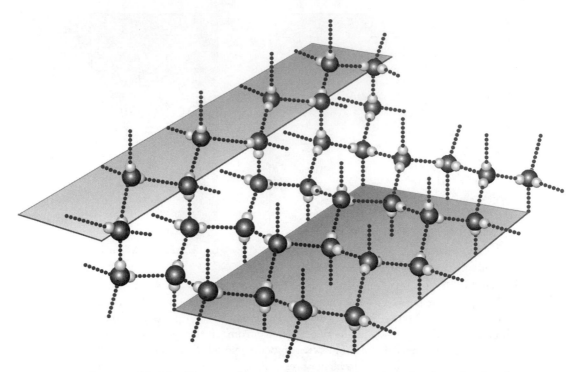

FIGURE 11-21. The crystal structure of ice. Each H_2O molecule makes four hydrogen bonds to its four nearest neighbors, creating an open structure less dense than liquid water at the same temperature.

Carbon dioxide, on the other hand, has a normal liquid—solid line sloping upward from left to right, as shown in Figure 11-20(b). Since the solid form of CO_2 is already the denser phase, additional squeezing produces a denser solid but no liquid.

11-6. A SURVEY OF DISORDER—ORDER TRANSITIONS

Few phenomena better exemplify the link between macroscopic and microscopic than a change of state. Allowed to communicate through an intermolecular potential, molecules respond to each other and act cooperatively. Interactions bring about phase changes.

But microscopic communication need not be limited to noncovalent interactions, and microscopic particles need not be restricted to atoms and molecules. Phase transitions encompass far more than ice melting or iodine subliming or helium condensing; these changes of state, already so varied, are themselves subordinated to a larger class of changes known as disorder–order transitions.

Magnetism offers a good example. Here the role of microscopic particle is played by neither molecule nor atom, but instead by a *magnetic moment* arising from one or more unpaired electrons. Electrons, as we recall from Chapter 5, have an attribute called *spin* that endows certain atoms with magnetic properties. Just as a compass needle aligns itself with the earth's magnetic field, so too does an atomic magnetic moment respond to an external magnetic field. The energy of the magnet depends on its orientation.

If the magnetization of a slab of iron seems distant from the melting of an ice cube, remember that interparticle interactions provide the key. And remember that these interactions automatically arise whenever energy depends on angle or distance, for such a dependence forces a microscopic particle to adjust its own position relative to its neighbor.

Within our block of iron there are more than 10^{23} atomic moments, each producing a magnetic field and each acquiring energy from the fields produced everywhere else. The particles interact, and the magnet-to-magnet relationship (Figure 11-22a) provides the equivalent of an intermolecular potential. Energy depends on position. Magnetic dipoles separated by different distances or tilted in different ways incur different potential energies.

Now if the internal magnetic fields (from the unpaired electrons) are weak compared with some larger external field H (imposed from elsewhere,

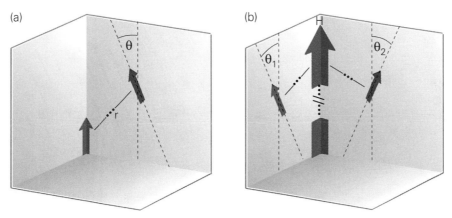

(a)

(b)

FIGURE 11-22. (a) Responding to a magnetic field from a neighboring particle, the energy of a second magnetic dipole depends both on the separation (r) and on the orientation (θ) relative to its partner. (b) Overwhelmed by a much stronger external field (H), the two magnetic dipoles respond separately—as disconnected individuals, not a coupled pair. Each microscopic magnet has its own potential energy, varying with its own orientation (θ_1 and θ_2) and with the strength of H.

as in Figure 11-22b), then the internal interactions cannot be expressed. The interactions similarly disappear when the interparticle magnetic energies are small compared with the thermal energy. Either way, the iron atoms will behave like the aloof particles of an ideal gas, and their orientation and energy will be determined individually. We describe that state macroscopically as *paramagnetic* (Figure 11-23a).

If, by contrast, both H and T are low enough to allow the coupling *between* magnetic moments to emerge, then the microscopic constituents will act cooperatively. Aligning preferentially to enjoy the lowest energy, the moments will form an ordered, *ferromagnetic* state (Figure 11-23b) with a macroscopic magnetic moment M. This transformation between paramagnetic and ferromagnetic states, just like the transformation between gas and liquid, is a *disorder—order transition* made possible by interactions and ultimately driven by larger considerations of entropy. There is a satisfying correspondence between P, V, T, and $U(r)$ on the one hand, and M, H, T, and the internal magnetic energy on the other. The same general theory, with its isotherms and critical points and all the rest, applies to both.

The more phenomena we can describe consistently with one model, the better—and there are many more. A small sample might include:

LIQUID—LIQUID CRYSTAL. The liquid crystal, discussed in Chapter 9, is a rotationally ordered liquid phase, typically formed by elongated molecules. It flows like a liquid, but under the right conditions the

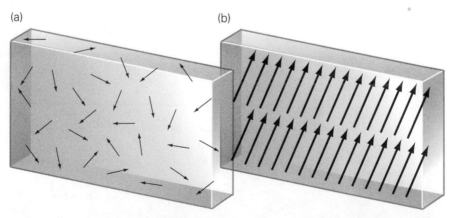

(a)   (b)

FIGURE 11-23. Magnetic order and disorder. (a) Lacking a sufficiently strong field to establish direction, the microscopic magnetic moments of a paramagnetic system fail to organize. Each particle is too weakly magnetic to command the orientation of its neighbor. (b) Armed with a strong magnetic field of its own, each particle in a ferromagnetic system wields influence over its neighbors. The microscopic magnets organize internally, aligning themselves one against the other to realize the lowest energy. No external field is needed.

molecules can move together while maintaining approximately the same angular orientation. The positions of the molecules overall remain random, so the liquid crystal still lacks the long-range order of a solid:

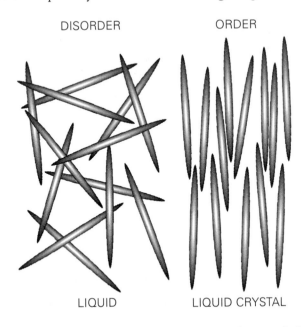

DISORDER ORDER

LIQUID LIQUID CRYSTAL

MONOMER–POLYMER. Small units, such as ethylene molecules, combine to form a macromolecule such as polyethylene. The polymer, allowed fewer possible arrangements than the separated monomers, is the ordered phase:

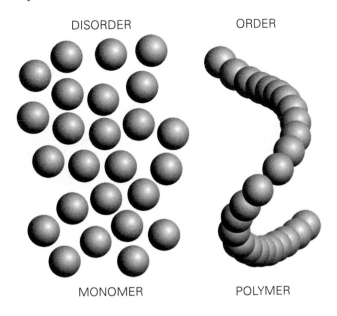

DISORDER ORDER

MONOMER POLYMER

ALLOY–BINARY SYSTEM. The alloy is a random mixture of atoms; the binary system separates the atoms into two distinct zones:

DISORDER   ORDER

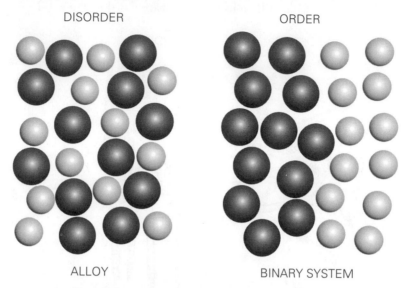

ALLOY   BINARY SYSTEM

RANDOM COIL–HELIX. Noncovalent interactions within a protein create an ordered internal conformation, without which biological activity ceases. A protein, when *denatured*, loses this conformation and exists as a disordered, fluctuating chain. Transitions between unfolded and various folded forms (random coil–helix) are disorder–order phenomena:

DISORDER   ORDER

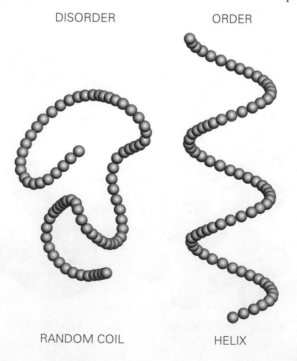

RANDOM COIL   HELIX

INCOHERENT LIGHT–LASER LIGHT. Light from an ordinary source is disordered. The waves emerge at different times with different phases of oscillation, so that some reach their crests while others reach their troughs. Laser light, by comparison, is coherent; all the waves are synchronized:

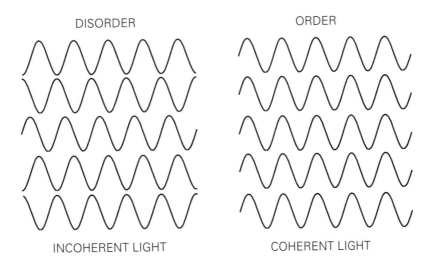

DISORDER   ORDER

INCOHERENT LIGHT   COHERENT LIGHT

CONDUCTOR–SUPERCONDUCTOR. Superconductivity describes a condition of zero electrical resistance in a material (not just a weak resistance, but virtually nothing at all—zero). Metallic superconductors have critical temperatures near 4.2 K, below which electrons are able to form specially correlated pairs that move unimpeded through the material. Normal resistance develops in the disordered state at higher temperature:

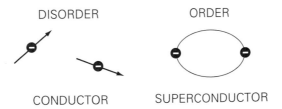

DISORDER   ORDER

CONDUCTOR   SUPERCONDUCTOR

See how far we have come: What began with the ideal gas is now thoroughly entwined with intermolecular interactions, states of matter, phase transitions, and disorder–order transitions of diverse kinds—all the way from Boyle's law to superconductivity. Next comes the even larger question of *change* in general, of how energy and entropy determine the transformation of matter. Looking at the dividing lines on a phase diagram, we already have a hint of what is yet to come: equilibrium and thermodynamics. As before, the implications will extend far beyond such early encounters as condensation, vapor pressure, and liquid–gas coexistence.

REVIEW AND GUIDE TO PROBLEMS

Electron by electron, atom by atom, molecule by molecule, we have drawn a picture of matter as in a still photograph—an image of small parts formed into finished structures, like bricks and mortar made into buildings. By experiment and model we have studied the architecture of matter on a small scale, looking first at atoms and molecules and phases one by one, seeing each as a completed work. How, we asked, is a molecule of CO_2 put together? Is elemental titanium paramagnetic or diamagnetic? How are the ions arranged in a crystal of $NaCl$? What is the difference between the liquid and gaseous states of water?

But until this chapter, we have not asked such questions as: How and why does water *change* from liquid to gas and from gas to liquid? Under what conditions will a reaction occur? How and why do substances react at all?

Our molecular world stands on the threshold of change. We have come to a crossroads.

We have always known that matter has the power to change. Henceforth, it will.

IN BRIEF: THE DISORDER–ORDER TRANSITION, A MODEL FOR CHANGE

Freezing, melting, evaporation, condensation: *phase transitions*, changes in the *state* of matter, natural changes familiar to everyone. We began with the transition from gas to liquid, a good example upon which to build a later appreciation of equilibrium, thermodynamics, and kinetics in general.

1. Before the change: a gas, presumably ideal. The molecules behave as if they are without shape and volume. H_2O or CO_2 or He or CH_4? No difference. One ideal gas is just like another. The particles fill the same container, exert the same pressure, exist at the same temperature. They move fast, randomly, and resist association. Disorder is total.

2. After the change: a liquid. Now the molecules have time for each other. They travel more slowly. They stick together. They form locally ordered structures, and each liquid is different. In one, the molecules are big; in another, small. One kind of molecule is polar; another is nonpolar. One molecule forms strong associations; another does not. This liquid boils at a low temperature; that one, at a high temperature. This one is viscous and dense; that one is thin and light. Close up, water is not the same as helium; close up, carbon dioxide is not the same as methane.

3. What happened? The independent particles of the gas came together, interacted, and became the clustered molecules of the liquid. They slowed down and became more organized. They lost energy. They gained order.

4. How did the molecules come together? They availed themselves of the same kinds of nonbonded interactions that we examined in Chapter 9: dipole–dipole, London forces (involving induced dipoles), hydrogen bonds, each interaction appropriate for a particular structure.

As an ideal gas, unable to interact, the particles behaved as if there were no potential energy between them. Near or far, they suffered neither attraction nor repulsion. Their own kinetic energy was so high that it overwhelmed what little particle–particle interactions there may have been.

But then, moving closer and closer, the gas particles abruptly began to attract one another. The *intermolecular potential* (the variation of the potential energy with particle–particle distance) started to assert itself. There was a weak attraction at large distances, a strong repulsion at short distances, and an accommodating well of minimum energy at some point in between. The well eventually loomed large enough to trap the particles into groups of two, three, and more. The gas became a liquid.

5. Why now? Under what sudden compulsion, present now and not before, was the gas transformed into a liquid?

It was a matter of opportunity. The gas particles always had the power to interact, but they lacked the opportunity to use those powers. Although the attractive potential well was always there, the rapidly moving particles had too much energy ever to stick together.

Then something changed; something from without, something external to the system. Maybe the surroundings became cooler, drawing energy away from the particles and thus slowing them down. Or perhaps the pressure on the gas was increased (as by, say, the weight of a piston) and the particles were squeezed into a smaller volume. But whatever it was, some new demand was imposed on the gas by its surroundings; and, evidently, these new conditions brought the molecules close enough for the dormant interactions to become active.

6. Is the gas no longer ideal at this point? Correct; reality has set in. The squeezed gas or the cold gas is no longer ideal. There are interactions in play between the particles, enforced at each distance by the intermolecular potential. PV is not equal to nRT.

Instead, some new equation of state describes the relationship between P, V, n, and T: a *real equation of state*, which now includes the effect of the intermolecular potential.

7. What form does the equation take? *The* equation? Ask, rather, *which* equation. There are many forms possible, just as there are many individual examples of real gases. Unlike the one and only ideal gas,

each real gas is real in its own way. They are like Tolstoy's unhappy families.

As a particularly instructive example, though, recall the **van der Waals equation of state**, one of the first attempts to describe real gases. It uses two parameters, *a* and *b*, to account for (*a*) the effect of intermolecular attractions on the pressure, and (*b*) the reduction of free volume attributable to the nonzero volume of the molecules. We shall see how, briefly, in Examples 11-1 and 11-2 below.

8. Is there one special combination of pressure and temperature at which a certain gas will condense? No; instead there is a range of suitable conditions, and we can summarize them on a single **phase diagram**—a chart that shows whether a substance exists as solid, liquid, or gas over all possible values for *P* and *T*.

Some general conditions: At low temperatures, where the particles move slowly, the pressure required for condensation usually is relatively low. At high temperatures, where the particles move fast and are spread far, the corresponding condensation pressure tends to be high.

There are limits. Higher than the **critical temperature** T_c, no amount of pressure will ever suffice to condense the gas. The particles are too energetic to be forced together. Liquefaction by compression becomes possible only when the hot fluid cools down to (at least) T_c, whereupon application of a certain **critical pressure** P_c will press the gas into a liquid.

Pushed beyond this **critical point** (P_c, T_c), a gas becomes a **supercritical fluid** with special properties. Liquid and vapor never appear together once the temperature exceeds T_c. There is no discernible boundary, or **interface**, between the two phases; there is only one seamless, homogeneous fluid state instead.

Below the critical point, however, the vapor may coexist with its associated liquid. At still lower temperatures, a liquid–solid transition usually develops as well. There are also particular combinations of *T* and *P* called **triple points**, where solid, liquid, and gas coexist simultaneously in dynamic equilibrium.

9. Dynamic equilibrium? See it as a battle back and forth with no winner. Molecules condense from gas to liquid at the same rate as molecules evaporate from liquid to gas. If 1,500,288,311,292,856 molecules fall down from the gas each second, then another 1,500,288,311,292,856 simultaneously rise up from the liquid to take their place. The status quo is preserved.

Thus is established **equilibrium**, a term taken from the Latin meaning an "equal balance" between opposing forces. We look at the macroscopic face of the system and see no apparent change—just a steady **vapor pressure** above the liquid, a single, constant pressure determined only by the temperature.

Yet this apparent constancy masks an underlying *dynamic* equilibrium, a *vigorously active* opposition of forces (from the Greek). We look into the microscopic guts of the system and see an unending, stalemated struggle between two opposing reactions: condensation and vaporization.

10. What if the liquid contains more than one component? How does a solution (solvent and solute together) come to equilibrium?

In the same way. Molecules evaporate, molecules condense, and eventually they reach a stalemate. If two or more volatile liquids are mixed together, then each component establishes its own vapor pressure as if it were the only one present.

We have, on the one hand, an ideal gas (low density), and an ***ideal solution*** (low concentration) on the other. Just as ideal gases conform to Dalton's law of partial pressures, so do ideal solutions conform to ***Raoult's law***,

$$P_i = X_i P_i^\circ$$

which states: The vapor pressure P_i of any volatile component i is proportional to its mole fraction in the mixture. The resulting pressure is reduced from the value P_i° that would exist over a pure liquid containing i alone.

In an ideal gas, remember, all the particles behave alike because they fail to interact. In an ideal solution, where the particles clearly do interact, solute and solvent continue to look alike as long as they interact in the same way. Let them interact in the same way, after all, and there will be nothing to distinguish this molecule from that molecule—until finally they begin to crowd one another at high concentrations, whereupon deviations from ideality should set in.

Until then, however, only the total number of dissolved species is important, not the kind. Raoult's law, which depends on the mole fraction alone, apportions a system's vapor pressure strictly by the numbers, treating it as a property of the overall collection (a ***colligative property***).

Where there is present a *nonvolatile* solute, unable to evaporate, the dissolved particles interfere with vaporization and freezing of the solvent. The higher the concentration of solute, the lower the freezing point and the higher the boiling point. Colligative properties again, both ***freezing-point depression*** (ΔT_f) and ***boiling-point elevation*** (ΔT_b) are proportional to the molality of solution (m):

$$\Delta T_f = K_f m \qquad \text{(understood as a decrease)}$$

$$\Delta T_b = K_b m \qquad \text{(understood as an increase)}$$

K_f and K_b are, respectively, the molal freezing-point-depression and boiling-point-elevation constants, two experimentally determined numbers different for each solvent. The symbol m, in a dual usage, represents both the numerical value of the molality (as in the equation $\Delta T_b = K_b m$) and the *units* of molality as well (moles of solute per kilogram of solution).

11. LOOKING AHEAD. With every equilibrium between liquid and vapor, solid and liquid, reactants and products, we have the plot line for all chemical reactions. Reactants become products, and products turn back into reactants. The forward reaction competes with the reverse reaction; time passes; eventually a dynamic equilibrium is reached. At that point, the relative amounts of products and reactants cease to change.

Question: Are there more products than reactants at equilibrium? Answer: to come from an understanding of thermodynamics (Chapters 12 through 14). The tendency is to go from high energy to low energy, and from high order to low order.

Question: How long will the reaction take to reach equilibrium? Answer: to come from an understanding of kinetics (Chapter 18). Different molecules interact in different ways. As in any trip, the time elapsed depends on the road traveled.

SAMPLE PROBLEMS

EXAMPLE 11-1. Toward Reality

PROBLEM: Take the values

$$a = 3.00 \text{ atm L}^2 \text{ mol}^{-2}$$

$$b = 3.00 \times 10^{-2} \text{ L mol}^{-1}$$

as typical parameters for the van der Waals equation:

$$P = \frac{nRT}{V - nb} - \frac{an^2}{V^2}$$

Suppose now that 1.00 mole of a real gas having these parameters is confined within a volume of 22.40 L. Compared with an ideal gas, what pressure would this real gas generate at temperatures of 77 K, 273 K, and 373 K?

SOLUTION: Our model of the ideal gas was always just that: a model, an approximation justified only under certain conditions. We now begin to assess the accuracy of those assumptions, first under relatively gentle pressures.

To compute the van der Waals pressure we substitute the effective volume,

$$V_{eff} = V - nb = 22.40 \text{ L} - (1.00 \text{ mol})(0.0300 \text{ L mol}^{-1}) = 22.37 \text{ L}$$

and the "*a*" correction,

$$a\frac{n^2}{V^2} = (3.00 \text{ atm L}^2 \text{ mol}^{-2})\frac{(1.00 \text{ mol})^2}{(22.40 \text{ L})^2} = 0.00598 \text{ atm}$$

into the equation

$$P = \frac{nRT}{V_{eff}} - a\frac{n^2}{V^2}$$

Since neither V_{eff} nor an^2/V^2 changes with temperature, only the quantity nRT needs to be recomputed for each value of T.

In what follows, both below and in Example 11-2, note that we shall sometimes carry an extra (nonsignificant) digit in order to capture the small variations.

(a) *77 K.* The ideal pressure is

$$P = \frac{nRT}{V} = \frac{(1.00 \text{ mol})(0.08206 \text{ atm L mol}^{-1} \text{ K}^{-1})(77 \text{ K})}{22.40 \text{ L}} = 0.282 \text{ atm}$$

and the van der Waals pressure (P_{vdW}) is 0.276 atm, deviating from ideality by approximately 2%:

$$P_{vdW} = \frac{(1.00 \text{ mol})(0.08206 \text{ atm L mol}^{-1} \text{ K}^{-1})(77 \text{ K})}{22.37 \text{ L}} - 0.00598 \text{ atm}$$

$$= 0.276 \text{ atm}$$

Nitrogen liquefies at this temperature.

(b) *273 K.* Water freezes. Ideal pressure = 1.00 atm (STP). Van der Waals pressure = 0.995 atm, a deviation of only 0.5%. The gas is warmer; the molecules move faster; the interactions are suppressed considerably.

(c) *373 K.* Water boils at this temperature. Ideal pressure = 1.366 atm. Van der Waals pressure = 1.362 atm. The trend toward ideality continues as the particles move even faster. The deviation falls to 0.3%.

EXAMPLE 11-2. Toward a Harsher Reality

PROBLEM: Using the same parameters (a, b) as in Example 11-1, compute ideal and van der Waals pressures under the following conditions:

$$n = 1.00 \text{ mol}$$

$$V = 1.00 \text{ L}$$

$$T = 77 \text{ K, } 273 \text{ K, } 373 \text{ K}$$

SOLUTION: It is a harsher reality because the same amount of gas is now squeezed into less than 5% of the original volume. Forced into closer quarters, the particles interact appreciably more.

Here, in a much smaller total volume, the same excluded volume of particles ($nb = 0.0300$ L) makes a proportionally greater impact. The nominal volume, $V = 1.00$ L, is reduced by fully 3% to an effective volume of 0.97 L:

$$V_{\text{eff}} = V - nb = 1.00 \text{ L} - (1.00 \text{ mol})(0.0300 \text{ L mol}^{-1}) = 0.97 \text{ L}$$

Meanwhile, the correction for attractions (an^2/V^2) increases 500-fold in the smaller container:

$$a\frac{n^2}{V^2} = (3.00 \text{ atm L}^2 \text{ mol}^{-2})\frac{(1.00 \text{ mol})^2}{(1.00 \text{ L})^2} = 3.00 \text{ atm}$$

The higher densities (and stronger interactions) thus are responsible for pronounced deviations from ideality at each temperature:

T (K)	P_{ideal} (atm)	P_{vdW} (atm)	DEVIATION (%)
77	6.3	3.5	44
273	22.4	20.1	10
373	30.6	28.6	7

Nowhere over this range do we have a truly ideal gas, although the interactions do begin to fail at increasing temperature.

QUESTION: If so, are we wrong to believe that a gas can ever be ideal?

ANSWER: Not necessarily, but certainly we must consider both the strength of the interactions and the macroscopic conditions. The question is not whether PV equals nRT in general, but whether PV equals nRT for some particular set of P, V, n, and T.

The answer varies case by case. The lower the density, the more likely it is that the gas behaves ideally. Low pressure, high temperature, small amount, large volume, *and* an intermolecular potential energy small compared with $k_B T$—all five of these factors tend to encourage ideality.

EXAMPLE 11-3. Critical Temperature

The particles of an ideal gas, unable to interact, will never condense into a liquid. No matter how high the pressure, no matter how low the temperature, the fictional ideal gas remains a gas. A vanishing intermolecular potential ensures that the system's critical temperature is exactly zero. Without interactions, there is nothing to hold the particles together.

A real gas, however, will always condense over some range of temperature and pressure. Slowed down sufficiently (at the critical temperature T_c or below), the particles begin to interact and stick together. Adequate pressure will then force a condensation.

PROBLEM: Arrange the following substances in order of decreasing critical temperature: Ar, H_2, Kr, CCl_3F.

SOLUTION: There is no certain way to predict T_c, but we can make at least this one assertion: The stronger the interactions, the higher the critical temperature. Strongly interacting particles should be able to come together even at relatively high temperatures, despite their high kinetic energies and consequent rapid motions. By contrast, weakly interacting particles need to shed thermal energy and slow down more before they can cooperate. The microscopic connections are more fragile.

Of the four substances given, only CCl_3F (trichlorofluoromethane) is a polar molecule, and so it alone can interact via a set of comparatively strong dipole–dipole forces. We expect (or rather, we *guess*) that CCl_3F will have the highest T_c of the set.

The remaining three substances, which include two atoms (Ar, Kr) and one homonuclear diatomic molecule (H_2), can interact only by way of *induced* dipole–dipole interactions (the London dispersion force). Far weaker than the attraction between permanent dipole moments, the induced interaction increases roughly with the size of the particle's electron cloud. Heavier atoms or molecules—those having *more* electrons to polarize, more *distant* electrons to polarize, hence more *weakly bound* electrons to polarize—are more likely to enjoy strong dispersion interactions. Using atomic or molecular mass as a rough gauge of polarizability, we would then predict T_c to decrease in the order Kr (36 electrons), Ar (18 electrons), and H_2 (2 electrons).

Experimental values are given in the table below, together with the boiling points (T_b) and corresponding van der Waals parameters (a, b):

SUBSTANCE	T_c (K)	T_b (K)	a (atm L^2 mol^{-2})	b (L mol^{-1})
CCl_3F	471.2	296.9	14.68	0.1111
Kr	209.4	119.9	2.33	0.0396
Ar	150.9	87.3	1.36	0.0320
H_2	33.0	20.3	0.25	0.0265

Overall, the trends go in the directions we predict. (1) Intermolecular potential energy is strongest for polar molecules and weakest for atoms and nonpolar molecules. (2) Other factors being equal, the effects are stronger for the larger species and weaker for the smaller species. (3) Critical point, boiling point, and the sticking coefficient (van der Waals a) all follow suit, decreasing with decreasing strength of interaction. (4) The imputed particle volume (van der Waals b) similarly decreases with decreasing mass.

Still, a warning: Ours is a very rough guess, not to be applied heedlessly. Real gases are more peculiar and less predictable than we yet suspect.

EXAMPLE 11-4. Vapor Pressure

PROBLEM: One mole of liquid water (1.00 mol) is allowed to equilibrate with its vapor inside a one-liter vessel (1.00 L). The vapor pressure at 20.00°C is then found to be 17.5 torr. (a) How many moles of water exist in the vapor state? Assume ideal behavior. (b) Suppose that the volume of the vessel is doubled to 2.00 L. What is the new vapor pressure? (c) How many moles of water vapor are present in the larger container?

SOLUTION: Visualize the initial circumstances described above: a small amount of liquid water (mass 18.0 g, density 1.00 g mL^{-1}, volume 18.0 mL) inside a much larger container (1000 mL), which we shall assume to be empty of water vapor at the start. Initially, all the H_2O molecules are in the liquid; not a single one is in the gas:

But circumstances change. One by one, the most energetic of the liquid molecules break free of their intermolecular interactions and escape to the space above. A gaseous phase forms above the surface of the liquid, fed by a steady stream of highly energetic escapees from the more ordered state below.

Then, with the passage of time, there develops a return traffic. Some of the *least* energetic gaseous molecules, straying too close to the surface, are sucked down into the liquid, easy prey for the attractions within. The rate of condensation is slow at first, but it increases steadily as the vapor phase becomes more and more crowded. Up and down the molecules continue to go until eventually there is an equal balance, an *equilibrium* between liquid and vapor. The pressure that prevails above the liquid is the system's equilibrium vapor pressure.

That vapor pressure, remember, depends only on the temperature, itself a measure of the average energy of the moving particles. Higher temperature means higher energy; higher energy means that more particles escape to the gas; more particles in the gas means higher pressure. Volume has nothing to do with it.

(a) *How many moles of water exist in the vapor state?* Once we understand the circumstances, we can solve the problem by a straightforward application of the ideal gas equation:

$$P = 17.5 \text{ torr} \times \frac{1 \text{ atm}}{760 \text{ torr}} = 0.0230(3) \text{ atm}$$

$$V = 1.00 \text{ L}$$

$$T = 20.00°C = 293.15 \text{ K}$$

$$R = 0.08206 \text{ atm L mol}^{-1} \text{ K}^{-1}$$

$$n = \frac{PV}{RT} = 9.57 \times 10^{-4} \text{ mol}$$

There are 9.57×10^{-4} moles of H_2O in the vapor phase, existing in equilibrium with $(1.00 - 9.57 \times 10^{-4})$ moles in the liquid phase—not the same molecules from moment to moment, understand, but always divided between vapor and liquid in the same proportion.

(b) *What vapor pressure will be established in a two-liter container at 20°C?* The same as in a one-liter container: 17.5 torr. The vapor pressure depends only on temperature.

(c) *How many moles of water vapor are present in the larger vessel?* More molecules must evaporate into the larger volume if the same pressure is to be exerted. Since the volume doubles, so must the number of moles. Whereas before there were 9.57×10^{-4} moles of vapor, now there are twice as many: 1.91×10^{-3} mol. Any less, and there cannot be equilibrium between liquid and vapor inside the container.

QUESTION: But suppose that there actually are fewer than 1.91×10^{-3} moles of liquid water in the 2.00-L container. What happens then?

ANSWER: The liquid dries up. There is no equilibrium. All the liquid evaporates, and the gas exists alone at whatever pressure is appropriate for the number of moles it contains. There can be no back and forth between a vapor and a residual liquid, because the reservoir of liquid is insufficient to support the equilibrium. The gas phase is thirsty. It will accept more water, if offered.

QUESTION: What if indeed we were to add another drop of liquid, and then another, and another, and another?

ANSWER: Evaporation will continue until the equilibrium vapor pressure is finally established and a reservoir of liquid finally appears. After that, the population of the vapor phase cannot be increased at the given temperature. Adding more liquid will only create a bigger puddle.

EXAMPLE 11-5. Dew

PROBLEM: Water's vapor pressure falls from 17.5 torr at 20.00°C to 9.2 torr at 10.00°C. If equilibrium is first established at 20.00°C in a volume of 1.00 L (as in Example 11-4a), then approximately how many milliliters of water will condense when the temperature is lowered to 10.00°C? Assume that liquid water has a density of 1.00 g mL^{-1}.

SOLUTION: From Example 11-4, we already know that 9.57×10^{-4} moles of water are present under the initial conditions stated ($P = 17.5$ torr, $V = 1.00$ L, $T = 20.00°C$). At the lower temperature, though, only 5.2×10^{-4} moles can be supported in equilibrium with a liquid reservoir:

$$P = 9.2 \text{ torr} \times \frac{1 \text{ atm}}{760 \text{ torr}} = 0.012 \text{ atm}$$

$$V = 1.00 \text{ L}$$

$$T = 10.00°C = 283.15 \text{ K}$$

$$R = 0.08206 \text{ atm L mol}^{-1} \text{ K}^{-1}$$

$$n = \frac{PV}{RT} = 5.2 \times 10^{-4} \text{ mol}$$

The difference, 4.4×10^{-4} mol, fills a liquid volume of just 0.0079 mL:

$$
\begin{array}{ll}
9.57 \times 10^{-4} \text{ mol} & (T = 20°C) \\
-5.2 \times 10^{-4} \text{ mol} & (T = 10°C) \\
\hline
4.4 \times 10^{-4} \text{ mol} & (\text{condensate})
\end{array}
$$

$$(4.4 \times 10^{-4} \text{ mol}) \times \frac{18.0 \text{ g}}{\text{mol}} \times \frac{1 \text{ mL}}{1.00 \text{ g}} = 7.9 \times 10^{-3} \text{ mL}$$

EXAMPLE 11-6. Vapor Pressure of an Ideal Solution

PROBLEM: The vapor pressure of a water/sucrose solution ($C_{12}H_{22}O_{11}$, table sugar) is found to be 16.1 torr at 20°C, lower by 1.4 torr compared with pure water at the same temperature. The solution, presumed to be ideal, contains 500. g of water. Compute the molality.

SOLUTION: Here is a mixture in which n_2 moles of sucrose (unknown) are dissolved in a known amount of solvent, 500. g, whence we derive n_1:

$$n_1 = 500. \text{ g } H_2O \times \frac{1 \text{ mol}}{18.0 \text{ g}} = 27.8 \text{ mol } H_2O$$

To learn the value of n_2, we write the mole fraction of solvent as

$$X_1 = \frac{n_1}{n_1 + n_2}$$

and then rearrange the equation to isolate n_2 on the left-hand side:

$$X_1(n_1 + n_2) = n_1$$

$$n_2 = \frac{n_1(1 - X_1)}{X_1}$$

We need to determine X_1.

Raoult's law, appropriate for an ideal solution, provides a way. Recall: Vapor pressure of a solvent (P_1) is proportional to its mole fraction, our desired X_1. The exact relationship is

$$P_1 = X_1 P_1^\circ$$

where P_1° denotes the vapor pressure of component 1 in pure form (water).

Solving for the mole fraction of water (component 1), we obtain

$$X_1 = \frac{P_1}{P_1^\circ} = \frac{16.1 \text{ torr}}{17.5 \text{ torr}} = 0.920$$

and thus

$$n_2 = \frac{n_1(1 - X_1)}{X_1} = \frac{(27.8 \text{ mol}) \, (1 - 0.920)}{0.920} = 2.41(7) \text{ mol}$$

for the number of moles of dissolved sucrose. The molality, defined as the moles of solute per kilogram of solvent, quickly follows:

$$\text{Molality} = \frac{2.417 \text{ mol sucrose}}{0.500 \text{ kg H}_2\text{O}} = 4.83 \ m$$

QUESTION: Not once did the molecular weight of sucrose appear in the calculation. According to this method, won't we derive the same molality for any other molecular solute under the stated conditions?

ANSWER: Yes. Vapor-pressure lowering is a colligative property, dependent on the number but not the kind of dissolved particles—provided, of course, that the solution is ideal. Since all interacting particles wield the same influence in an ideal solution, how can we then expect a different result for a different solute? They all behave alike.

EXAMPLE 11-7. Shattered Ideals

PROBLEM: Water's normal freezing point at 1 atm is 273 K. Comment on the data presented below for aqueous solutions of sucrose, sodium chloride, and potassium sulfate:

SOLUTE	MOLALITY (m)	FREEZING POINT (°C, at 1 atm)
$C_{12}H_{22}O_{11}$	1.00×10^{-3}	-1.86×10^{-3}
$C_{12}H_{22}O_{11}$	1.00×10^{-2}	-1.86×10^{-2}
$C_{12}H_{22}O_{11}$	1.00×10^{-1}	-1.86×10^{-1}
NaCl	1.00×10^{-3}	-3.66×10^{-3}
NaCl	1.00×10^{-2}	-3.61×10^{-2}
NaCl	1.00×10^{-1}	-3.48×10^{-1}
K_2SO_4	1.00×10^{-3}	-5.28×10^{-3}
K_2SO_4	1.00×10^{-2}	-5.02×10^{-2}
K_2SO_4	1.00×10^{-1}	-4.32×10^{-1}

For reference, note the value of the molal freezing-point-depression constant for water: $K_f = 1.86°C \ m^{-1}$.

SOLUTION: We know that a dissolved solute will depress the normal freezing point of water, but by how much?

For an ideal solution, the decrease is given by

$$\Delta T_f = K_f m$$

where the molality, m, includes *all* dissolved solute particles regardless of chemical type.

Taking, for example, a 0.100 m solution of K_2SO_4, we would expect to find an ion concentration of 0.300 m (0.200 m from K^+, 0.100 m from SO_4^{2-}) and hence a freezing point lower by

$$\Delta T_f = K_f m = (1.86°C\ m^{-1})(0.300\ m) = 0.558°C$$

But clearly we are mistaken, because the last line in the table shows T_f to be depressed by only 0.432°C. The solution is not perfectly ideal.

There is *ion pairing*. Positive K^+ ions and negative SO_4^{2-} ions associate preferentially in the solution, thereby contradicting our idealistic assumption that interactions are the same everywhere. The outcome is a decrease in the effective number of ions available to depress the freezing point, as if (see below) every mole of K_2SO_4 produced 2.32 moles of ions rather than 3.

For if we rewrite our equation as

$$\Delta T_f = jK_f m$$

to include a correction factor j, then we find a real-to-ideal ratio of 2.32/3.00:

$$j = \frac{2.32}{3.00} = 0.774$$

The extent to which j deviates from 1 is thus a measure of the solution's deviation from ideality:

Solute	Molality (m)	Correction Factor
$C_{12}H_{22}O_{11}$	1.00×10^{-3}	1.000
$C_{12}H_{22}O_{11}$	1.00×10^{-2}	1.000
$C_{12}H_{22}O_{11}$	1.00×10^{-1}	1.000
NaCl	1.00×10^{-3}	0.984
NaCl	1.00×10^{-2}	0.970
NaCl	1.00×10^{-1}	0.936
K_2SO_4	1.00×10^{-3}	0.946
K_2SO_4	1.00×10^{-2}	0.900
K_2SO_4	1.00×10^{-1}	0.774

Molecular sucrose, a nonelectrolyte, apparently behaves ideally over the entire range given. It produces no ions upon dissolution.

Sodium chloride, nearly ideal in the most dilute solution (0.001 *m*), deviates from ideality more and more as its concentration increases and ion pairing therefore becomes more likely.

For potassium sulfate, the deviations are consistently stronger at all concentrations. The larger charge on the SO_4^{2-} ion produces a stronger electric field and consequently a bigger drop in energy for the paired ions.

Be careful about what you call ideal.

EXAMPLE 11-8. Colligative Properties and Molar Mass

Reality aside, we proceed to use boiling-point data for a computation of molar mass under ideal conditions. Similar methods can be applied to other colligative properties as well, since all such effects derive from the same cause. The idea is simply to determine the number of moles associated with a given mass.

PROBLEM: A 1.00-g sample of an unknown molecular compound $C_nH_{2n}O_n$, dissolved in 100. g of water, produces a boiling-point elevation of 0.029°C ($K_b = 0.51$°C m^{-1}). Compute the molar mass and molecular formula, assuming the solution to be ideal.

SOLUTION: First, determine the apparent molality of the solution:

$$m = \frac{\Delta T_b}{K_b} = \frac{0.029°C}{0.51°C \ m^{-1}} = 0.057 \ m = 0.057 \ \text{mol kg}^{-1}$$

Second, determine the number of moles of solute represented by the 1.00 g:

$$\frac{0.057 \ \text{mol solute}}{\text{kg solvent}} \times \frac{1 \ \text{kg}}{1000 \ \text{g}} \times 100. \ \text{g solvent} = 0.0057 \ \text{mol solute}$$

Third, determine the molar mass:

$$\mathcal{M} = \frac{1.00 \ \text{g}}{0.0057 \ \text{mol}} = 1.8 \times 10^2 \ \text{g mol}^{-1}$$

The result is consistent with a molecular formula $C_6H_{12}O_6$, suggesting the sugar glucose.

EXAMPLE 11-9. Phase Diagrams

A quick review of phase transitions in water.

PROBLEM: Describe what happens during the following processes, shown below (not to scale) on a phase diagram:

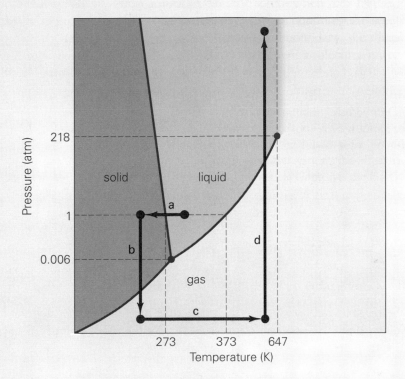

(a) Water at atmospheric pressure is cooled isobarically from 300 K to 250 K. (b) Continuing from the point just reached, the system's pressure is decreased isothermally from 1 atm to 0.0002 atm. (c) Next, temperature is increased isobarically from 250 K to 600 K. (d) Last, the system is compressed isothermally to a pressure of 1000 atm.

SOLUTION: Follow the four stages marked on the accompanying diagram. Horizontal lines correspond to the isobaric processes; vertical lines correspond to the isothermal processes.

(a) *(1 atm, 300 K)* ⟶ *(1 atm, 250 K)*. We make ice cubes. Starting from the liquid-only region, the system becomes steadily cooler and begins to freeze at a temperature of 273 K. Equilibrium between solid and liquid is established at this point, marked by the coexistence of water and ice. Upon further cooling (leftward on the diagram), the liquid disappears and only the solid remains.

(b) *(1 atm, 250 K)* ⟶ *(0.0002 atm, 250 K)*. The pressure above the ice is diminished below the triple point (0.0060 atm = 4.6 torr), and the system undergoes *sublimation*; it passes directly from solid to gas with no appearance of liquid. First, there develops an equilibrium between solid and gas when P and T conform exactly to a point on the solid–gas coexistence line. After that, the solid disappears completely as the pressure continues to decrease.

(c) *(0.0002 atm, 250 K)* ⟶ *(0.0002 atm, 600 K).* The gas becomes warmer, but no boundary lines are crossed and hence no changes of state occur.

(d) *(0.0002 atm, 600 K)* ⟶ *(1000 atm, 600 K).* Squeezed isothermally, the gas condenses at the point where the system crosses the gas–liquid line. During the condensation, liquid and vapor coexist in equilibrium; afterward, with all the vapor gone, the liquid is further pressurized but undergoes no change in state.

Note, incidentally, that the system remains subcritical throughout the entire compression, which takes place some 50 K below T_c. Had the process been initiated above 647 K, then no amount of squeezing would have ever resulted in equilibrium between liquid and vapor. Above T_c, there is no longer a liquid–vapor coexistence line to cross.

EXAMPLE 11–10. Frost

PROBLEM: Why does frost, not dew, appear on the ground during a cold winter's morning?

SOLUTION: Look again at the phase diagram for water, which shows that a liquid phase exists only at pressures above the triple point (4.6 torr, 0.01°C):

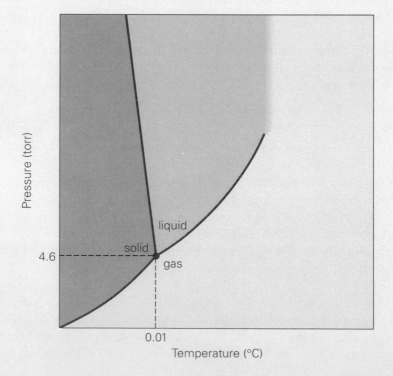

To maintain such equilibrium, water vapor in the air must exert a pressure no less than 4.6 torr on the liquid below. Anything lower is insufficient.

Pressed too gently, no solid can hold together. Below the triple point, especially, the pressure is so gentle that the solid bursts directly into a gas without passing through any liquid phase in between. The equilibrium that ensues is between solid (frost) and gas, never between liquid and gas. There is no liquid below the triple point.

Not the total pressure of the atmosphere, but rather the partial pressure of the water vapor is decisive here—and thus on cold, dry mornings when the air contains little moisture, frost is deposited on the ground.

EXERCISES

1. Under what circumstances will a gas undergo a transition to a liquid?

2. Sketch the interparticle potential for an ideal gas.

3. Is an ideal gas able to liquefy under any conditions at all? What causes a gas to deviate from ideality?

4. Sketch (a) the interatomic potential for two nitrogen atoms and (b) the intermolecular potential for two nitrogen molecules. What kinds of interactions give rise to each curve? Why is the dip in one curve so much deeper than in the other?

5. Review the principal noncovalent interactions between molecules. Which interactions are strongest? Which are most able to rob a gas of its ideality?

6. Under what conditions of temperature, pressure, and density is a gas *always* ideal? Under what conditions is a gas *never* ideal?

7. Values for van der Waals parameters (a, b) are given below for a set of three molecules:

SUBSTANCE	a (atm L^2 mol^{-2})	b (L mol^{-1})
Kr	2.33	0.0396
Ar	1.36	0.0320
H_2	0.25	0.0265

Insert a and b into the van der Waals equation,

$$P = \frac{nRT}{V - nb} - \frac{an^2}{V^2}$$

and then, using a density of 1.00 mol L^{-1}, compute the pressure of each gas at the following temperatures: (a) 300 K. (b) 1000 K. Compare the results with the corresponding ideal pressures.

8. Repeat the exercise above for H_2 gas ($a = 0.25$ atm L^2 mol^{-2}, $b = 0.0265$ L mol^{-1}), this time using a density of 10.0 mol L^{-1} at each temperature: (a) 300 K. (b) 1000 K. How do the van der Waals pressures compare with the ideal pressures now?

9. Explain the terms *critical temperature* and *critical pressure*.

10. Differentiate between a substance's critical temperature and its normal boiling temperature. In what ways are the two parameters related?

11. Pick the molecule with the higher critical temperature in each pair:

    (a) CH_3F, CH_4
    (b) CH_4, He
    (c) C_3H_8, $C_{18}H_{38}$
    (d) Ne, Xe
    (e) CO, H_2O

12. NH_3 boils at a higher temperature than PH_3, despite the similarity in structure and despite the larger mass of PH_3. Why?

13. Water has a critical temperature of 374°C and a critical pressure of 218 atm. In what state (solid, liquid, gas, or something else) will H_2O exist at 700°C under each of the following pressures?

    (a) 1 atm   (b) 10 atm   (c) 100 atm   (d) 1000 atm

14. The critical point for carbon dioxide is reached at 304 K and 73 atm, and the triple point stands at −56.6°C and 5.1 atm. The solid form is more dense than the liquid. (a) Without consulting the text, sketch the phase diagram for CO_2. (b) What phase or phases can exist at STP? (c) Ice cubes floating in water are a familiar sight at normal atmospheric pressure. Will Dry Ice cubes (frozen CO_2) ever be found floating in liquid carbon dioxide under the same conditions?

15. Describe how a state of *dynamic equilibrium* is reached and maintained between two phases.

16. Explain the following statement: "The vapor pressure of water at 15°C is 12.8 torr."

17. Again: The vapor pressure of water at 15°C is 12.8 torr. (a) Under what conditions will water boil at 15°C? (b) Under what conditions will gaseous water exert a pressure *less* than 12.8 torr at 15°C?

18. The vapor pressure of a substance is independent of volume. Why?

19. The vapor pressure of water at 15°C is 12.8 torr, and the density of the liquid is approximately 1.00 g mL^{-1}. (a) 10.0 mL of $H_2O(\ell)$

are left to evaporate inside a 5.00-L container at 288 K. What partial pressure of water is eventually established? (b) How much liquid remains? Assume ideal gas behavior.

20. Let the same 10.0 mL of water now be placed inside a 100-L container, a capacity 20 times greater than the volume of the vessel in the previous exercise. Compute the partial pressure of water and the amount of liquid that remains at 288 K.

21. Once more, although this time for 10.0 mL of $H_2O(\ell)$ in a volume of 1000 L: (a) Are there enough molecules to sustain a dynamic equilibrium between a liquid and vapor phase? (b) What partial pressure of water is established? (c) How much liquid, if any, remains?

22. As before, choose a volume of 1000 L and a temperature of 288 K— but suppose now that the partial gas pressure of water is only 6.1 torr, less than half the equilibrium vapor pressure of 12.8 torr. (a) Is any liquid present? (b) If not, then below what temperature will the gas begin to condense? (c) What volume of liquid will condense at a temperature near 273 K? Data relevant to the calculation will be found in Appendix C.

23. In summary: 10.0 mL of liquid water are left to evaporate at 288 K in a closed container. (a) What is the largest volume able to support a phase equilibrium between liquid and vapor? (b) Does the vapor pressure depend upon the presence of air or any other gas in the container? Why or why not?

24. Are the results of these last five exercises in any way inconsistent with the assertion that vapor pressure is independent of volume?

25. A gas is said to be "ideal" when its interparticle interactions are unexpressed. How, then, can a solution ever be called ideal if interactions of some sort are always present?

26. Compute the vapor pressure of water above each of these solutions at 15°C, using information either from previous exercises or from Appendix C:

    (a) 10.0 g $C_6H_{12}O_6$ (glucose) in 100.0 g H_2O
    (b) 10.0 g NaCl in 100.0 g H_2O
    (c) 10.0 g $NaNO_3$ in 100.0 g H_2O
    (d) 10.0 g Na_2SO_4 in 100.0 g H_2O

Assume that Raoult's law applies. Is there a pattern to the results? Why or why not?

27. By what percentage is the vapor pressure of water reduced above each of these solutions?

    (a) 18.0 g $C_6H_{12}O_6$ in 100.0 g H_2O
    (b) 2.92 g NaCl in 100.0 g H_2O
    (c) 4.25 g $NaNO_3$ in 100.0 g H_2O
    (d) 4.73 g Na_2SO_4 in 100.0 g H_2O

    Assume, again, that Raoult's law applies. Is there a pattern now to the results? Why?

28. How many grams of each solute must be added to 500. g of water to lower the vapor pressure from 50.0 torr to 49.0 torr?

    (a) $NaC_2H_3O_2$
    (b) $NaClO_3$
    (c) $NaClO_4$
    (d) $C_{12}H_{22}O_{11}$
    (e) $C_6H_{12}O_6$
    Assume ideal behavior.

29. Analysis of an unknown compound (Y) reveals the following composition by mass:

    $$\begin{array}{ll} C: & 38.7\% \\ H: & 9.7\% \\ O: & 51.6\% \end{array}$$

    When 5.00 grams of this substance are dissolved in 125.0 mL of ethanol (C_2H_5OH), the vapor pressure of ethanol above the solution is 57.15 torr at 25°C. (a) Given that pure ethanol has a density of 0.7873 g mL^{-1} and a vapor pressure of 59.3 torr at 25°C, what is the molar mass of Y? Assume ideal behavior. (b) What is its molecular formula?

30. Calculate the partial pressure (at 25°C) of both ethanol and water above a solution, presumably ideal, containing equal volumes of each component. Use the data below:

COMPONENT	DENSITY (g mL^{-1})	VAPOR PRESSURE (torr)
H_2O	0.9971	23.8
C_2H_5OH	0.7873	59.3

31. Compute the freezing point and boiling point of each solution:

    (a) 3.52 g KCl in 100.0 g H_2O

    (b) 5.72 g $Ce_2(SO_4)_3$ in 100.0 g H_2O

    (c) 11.2 g $NaC_2H_3O_2$ in 100.0 g H_2O

    (d) 30.0 g $C_{12}H_{22}O_{11}$ in 100.0 g H_2O

    Assume that the systems behave ideally.

32. Pick the aqueous solution with the lower freezing point, assuming each system to behave ideally:

    (a) 0.010 *m* $Pb(NO_3)_2$, 0.012 *m* KCl

    (b) 0.050 *m* $C_6H_{12}O_6$, 0.020 *m* $NaNO_3$

    (c) 0.030 *m* $C_3H_8O_3$,   0.040 *m* $C_2H_6O_2$

    (d) 0.010 *m* Li_2SO_4,   0.010 *m* $MgSO_4$

    Which solution in each pair has the higher boiling point?

33. (a) If a certain aqueous solution boils at 101.70°C, at what temperature does it freeze? Assume ideal behavior. (b) How many grams of ethylene glycol ($C_2H_6O_2$) must be added to 100 grams of water to produce such a solution?

34. Pure benzene (C_6H_6) has a freezing point of 5.5°C and a boiling point of 80.1°C. When 10.0 g of anthracene ($C_{14}H_{10}$) are dissolved in 500. g of benzene, however, the freezing point drops to 4.95°C and the boiling point rises to 80.38°C. Determine, to two significant figures, the values of the molal freezing-point-depression and boiling-point-elevation constants, K_f and K_b. Assume ideal behavior.

35. (a) In a theoretically *ideal* solution containing 1.00 kg of water, how many moles of solute particles would reduce the freezing point by 5.00°C? (b) Will the requisite amount of, say, $NaNO_3$ in a *real* solution be substantially larger or smaller? (c) How about ethylene glycol, $C_2H_6O_2$?

36. Which solute should produce the greater deviation from ideality at a given concentration?

    (a) $NaHCO_3$ or $C_6H_{12}O_6$
    (b) LiF or Li_2SO_4
    (c) KCl or $Ce_2(SO_4)_3$

37. Which of the following *real* aqueous solutions will have the higher boiling point? Which will have the lower freezing point?

    (a) 1.5 *m* KCl or 1.0 *m* K_2SO_4
    (b) 0.5 *m* KCl or 1.0 *m* $C_{12}H_{22}O_{11}$

12

Equilibrium—
The Stable State

12-1. The Nature of Equilibrium

12-2. Energy, Entropy, and the Drive
to Equilibrium: A Preview

12-3. Approaches to Equilibrium: A Preview

12-4. The Equilibrium Constant and the Law of Mass Action

12-5. Vapor Pressure and the Equilibrium Constant

12-6. Stressed Equilibria

12-7. Solving Problems Using the Equilibrium Constant
REVIEW AND GUIDE TO PROBLEMS
EXERCISES

To be in equilibrium is to be in a state of delicate balance, poised between opposing forces so as to appear at rest. Two arms wrestling on a tabletop, neither able to budge the other, are in a state of equilibrium. So too is a picture hanging on a wall, where the downward pull of gravity is balanced exactly by the upward pull of the wire. Coffee and cream, well mixed, is another example, as is ice in water. In all the natural world there are really only two options: to be in equilibrium, or else to be approaching it. If you are in equilibrium, there you will stay unless disturbed by some outside influence. If you are out of equilibrium, as indeed are all living organisms, you will inevitably get there. Everything does; it's only a matter of time.

12-1. THE NATURE OF EQUILIBRIUM

A chemical system in equilibrium presents an unchanging macroscopic face. Gross properties such as pressure, temperature, mass, concentration, color, and all the rest remain constant in time and uniform in space, revealing nothing new to the outside observer. Time stands still. Looking at a system in equilibrium, we have no idea how or when it came to be. All memory of previous conditions is lost, and without change there is no way to mark time. We know only those few macroscopic properties characteristic of the final state, yet these variables alone provide an unambiguous recipe for its construction. The equilibrium state can be duplicated—no questions asked—merely by preparing a system with the appropriate macroscopic values. Past history is irrelevant.

So much for the macroscopic picture. At the *microscopic* level, we have already seen (in Chapter 11) that the seemingly static face of equilibrium belies an ongoing but stalemated competition between opposing processes. With a liquid–gas equilibrium, for example, the number of molecules evaporating from the liquid is balanced precisely by those condensing from the gas. A constant vapor pressure is maintained as a result. Such is the fate, ultimately, of any change that can go both ways, either forward (——▶) or in reverse (◀——). A **dynamic equilibrium**, defined symbolically in Figure 12-1, is attained when the forward and reverse rates become equal.

For generality, we take an arbitrary chemical process

$$a\mathrm{A} + b\mathrm{B} \rightleftharpoons c\mathrm{C} + d\mathrm{D}$$

and write its stoichiometric equation with a bidirectional double arrow, emblematic of a two-way street: Reactants (*a* moles of A, *b* moles of B)

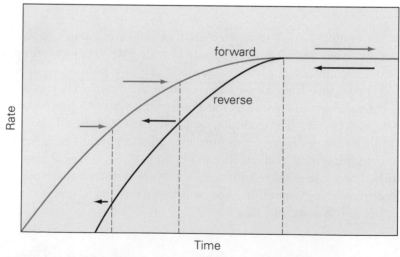

FIGURE 12-1. Approach to a *dynamic equilibrium* between opposing processes, realized when rates in the forward (——▶) and reverse (◀——) directions finally become equal. After that, nothing changes. The flow of time effectively stops, and all macroscopic properties remain constant.

go to products (*c* moles of C, *d* moles of D), and products go back to reactants. Back and forth, this way and that—two reactions in competition—and, in the end, all the participants come to equilibrium.

There is nothing special about two reactants and two products, and our observations are in no way limited to a particular stoichiometry or class of reaction. Whatever we say, we say for all chemical reactions in all their variety. The attainment of equilibrium is a universal imperative, a compulsion of nature to smooth out the highs and lows wherever and whatever they may be. Superficial differences between one process and another melt away.

Here, by way of anticipation, are a few specific transformations that already embrace a wealth of chemical possibility:

1. $2NO_2(g) \rightleftharpoons N_2O_4(g)$
2. $H_2O(\ell) \rightleftharpoons H_2O(g)$
3. $CaCO_3(s) \rightleftharpoons CaO(s) + CO_2(g)$
4. $CH_3COOH(aq) + H_2O(\ell) \rightleftharpoons H_3O^+(aq) + CH_3COO^-(aq)$
5. $Ag^+(aq) + Cl^-(aq) \rightleftharpoons AgCl(s)$

A short list, but in it are such varied examples as (1) *dimerization–dissociation* (the formation and subsequent breakup of one molecule from two), in this case a **homogeneous equilibrium** within a single (gas) phase; (2) a **heterogeneous equilibrium** between two phases (*evaporation–condensation*), where properties are uniform in each phase but change abruptly at the interface; (3) *decomposition* of one substance into two others, another heterogeneous system (solid and gas); (4) a homogeneous *acid–base* equilibrium in solution; and (5) *precipitation–dissolution*, a heterogeneous equilibrium between a solution and solid phase. Different as they seem, all these reactions are governed by the same laws, and there will be just one universal equation to fix the ratios of concentrations in each instance.

12-2. ENERGY, ENTROPY, AND THE DRIVE TO EQUILIBRIUM: A PREVIEW

Equilibrium is about change or, more appropriately, about the *end* of change and the beginning of a new, more stable state. Consider: A system that reaches equilibrium must be "better off" in some way than it was before. If not, then why would it be there? Why would it stay there? Why would it always come to the same point and stop?

There is, we shall see, a tendency in nature that acts as a driving force behind all transformations, and the outcome is always some unchanging, final, *predictable* state of equilibrium—the end of the line, the point where the driving force is satisfied and no further improvement can be effected. Expect no surprises. The nature of the equilibrium state is predestined by the never-violated laws of thermodynamics; here, at equilibrium, is where everything comes to rest.

We use these metaphors, for now, only as stand-ins for the more rigorous language of thermodynamics, the study of energy in its many forms. Equilibrium is bound up deeply with thermodynamic quantities like internal energy, work and heat, enthalpy, entropy and internal disorder, free energy, and others. What are such things, we ask, and how do they determine the inevitable sameness of equilibrium?

The answers will come over the next several chapters, during which many interrelated ideas must be woven together. We cannot grasp everything at once, though, so our initial efforts will concern mostly the destination—the equilibrium condition itself—as we look to portray that final state simply for what it is, not necessarily how and why it came to be. Even so, we should still like some small, early notion of what drives the process overall, while leaving the details to await later elaboration. Let us start slowly and patiently, then, using broad strokes to sketch the tendencies that impel a system to equilibrium.

Figure 12-2 has much to say about equilibrium and thermodynamics. It provides a symbolic record of the progress of a general transformation,

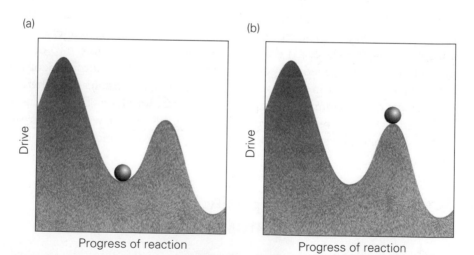

FIGURE 12-2. Like a ball rolling down a hillside, a chemical reaction moves ahead under the urging of a certain "driving force": a tendency toward a more uniform dispersal of energy; or, in broad terms, a more disordered universe. (a) Trapped in a valley, its drive at a minimum, the system resists small displacements to either side. It persists in a state of local equilibrium, able to fall further only if pushed sufficiently hard. (b) Poised at the summit, an unstable system cannot withstand even the smallest disturbance.

and the hills and valleys of the curve represent the ups and downs of the natural tendency toward change. Down is easy; up is hard. A ball might roll downhill into the first valley at the left, where (provided it is not moving too fast) it may eventually come to rest. The bottom of that valley, a *local minimum* in the curve, is a point of equilibrium. The ball sits there. In response to a small push, it rocks back and forth a bit and settles down again into equilibrium. We say, accordingly, that a system in equilibrium is locally **stable** to small changes, to small *perturbations*.

Suppose that the ball were poised motionless at the *top* of the central hill, for that too is possible. Here again there is no change, but this condition is no equilibrium since it cannot withstand a small perturbation. The slightest push to the left or right will send the ball down into one of the valleys. What we have instead is an **unstable state**, one that remains unchanged only if left absolutely untouched and undisturbed.

One good push over the top of the hill and the ball comes to rest in the second valley, thereby attaining a new equilibrium at a deeper level. Any particular equilibrium, we see, typically is just a local phenomenon and not really an irrevocable, final stage. True finality is elusive, for there is likely to be an even lower equilibrium under another set of conditions. The upper valley, for example, might represent a mixture of mostly unreacted H_2 and O_2, and the lower valley primarily H_2O. Water, situated at a deeper thermodynamic level, is undeniably favored over hydrogen and oxygen alone, but the uncatalyzed formation of H_2O proceeds with excruciating slowness. Absent some impetus to propel them on their way, the hydrogen and oxygen molecules will remain unchanged, perhaps indefinitely, without producing much water. They stand in equilibrium with respect to their nuclei and electrons, each structure a stable molecule. Application of a spark to the mixture, however, sends the reactants over the barrier and allows the water to form. When it does, we speak of a new equilibrium between H_2, O_2, and H_2O.

Someone else, seeking to challenge us, might argue that the hydrogen molecules are themselves unstable compared with the helium nuclei that can be formed by nuclear fusion. Such processes indeed do occur in the sun at extraordinarily high temperature and pressure, and so the remark emphasizes again that all equilibria are local. Conceding that point, we then take a stubbornly practical, earthbound attitude and insist: Under the conditions presently specified, for any reasonable interpretation, the H_2–O_2–H_2O system is in equilibrium. If we choose not to perturb the state drastically, there it will remain.

But what about this driving force, this apparent compulsion of nature always to seek equilibrium? We shall lay the drive bare in the next two chapters, but even now we can ask: Is it *heat*? Must a system give off heat to effect a change? Clearly not, for otherwise there would be no

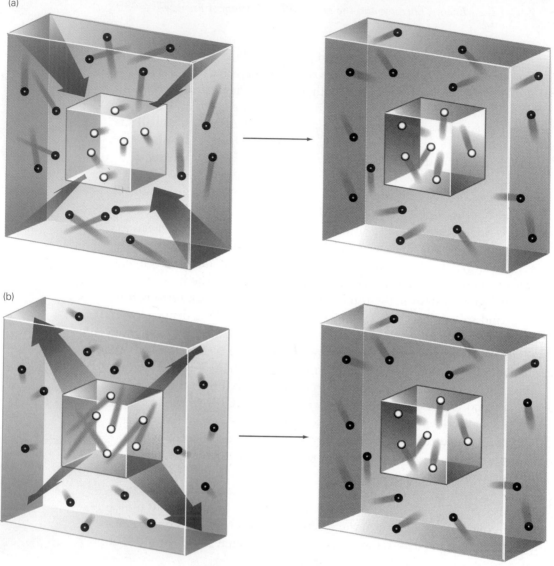

FIGURE 12-3. Movement of heat between a chemical system (inner box) and its surroundings (outer box). (a) Heat flows *into* the system during an endothermic process. The reacting molecules, gaining thermal energy at the expense of those outside, grow more agitated and possibly more disordered. (b) Flowing *out* of the system, heat from an exothermic reaction stirs up molecules in the surroundings.

endothermic reactions—processes (as in Figure 12-3a) that go forward on their own by absorbing heat from the environment, unaided and unforced. And, we know, there *are* endothermic processes. Some systems absorb heat naturally. It happens. The melting of ice, certainly the most natural of changes, is convincing proof of that.

So, no . . . not heat. Heat may flow into a system, as it does when a block of ice absorbs heat endothermically to disrupt the intermolecular forces and free the water molecules. Or (just as naturally) heat may flow out, as it does in an **exothermic** process such as freezing. Here the attractive interactions bring the molecules together with an attendant release of heat to the outside (Figure 12-3b). Exothermicity and endothermicity alone cannot tell the whole story.

Yet if endothermicity is, in some way, "bad"—a difficult climb uphill—perhaps there is some other "good" to be found in the messy breakup of a melting solid. Remember: Ice going to liquid water tends to disorder the system (Chapter 11), and an inclination toward disorder might be just as natural as a tendency toward lower energy. True? Things get messier as time marches on, and everything appears to run down. Maybe it is just easier to be disorganized, since surely there are many more ways to arrange things haphazardly than carefully. To be neat means to accept only one or a very few ordered arrangements, whereas to be sloppy opens up an almost unlimited range of choice. Figure 12-4, sketchy as it is, suggests as much: that a liquid is a more forgiving, more disordered, more probable state to stumble upon *by chance* than a solid, with its rigidly fixed pattern and far fewer alternative configurations.

Disorder—a multiplicity of possible arrangements—is the signature of entropy, and it will be global entropy that emerges as the impetus to go from one equilibrium to another. Understand that an equilibrium state is determined by only a few macroscopic variables, but that there are innumerable *microscopic* configurations all able to reproduce the required macroscopic values. Recalling the ideal gas, for instance, we discover that there is not just one possible set of molecular speeds and positions associated with a given pressure. Instead there are many (many!) different yet perfectly equivalent configurations consistent with that one macroscopic value. Expect, accordingly, that the drive toward a final state will depend on the multiplicity of such microscopic arrangements.

Getting there is a matter of blind chance. Any transformation proceeds inexorably toward the outcome most likely in a statistical sense, namely the one that offers the most choice and hence the greatest disorder for the universe. Getting to equilibrium becomes, therefore, a probabilistic imperative rather than the result of purposeful guidance. The driving force proves to be a quantity called the "free" energy, which combines changes in both local energy (Chapter 13) *and* local entropy (Chapter 14) into a single measure of entropy everywhere: global disorder, always increasing.

We shall learn that the hills and valleys of Figure 12-2 are really the hills and valleys of the free energy, and it will be in one such valley that

(a)

(b)

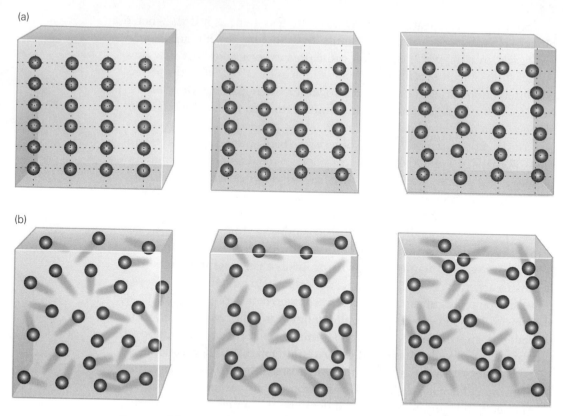

FIGURE 12-4. Order and disorder. (a) The atoms in a crystal are allowed only a limited set of positions, beyond which the crystal ceases to be a crystal. (b) A liquid, more tolerant of microscopic rearrangement, is a looser, less ordered state of matter. Many different internal configurations are consistent with a single set of external properties.

reactants and products (our rolling ball) come to equilibrium. Thermodynamics will enable us not only to locate that valley, but also to understand what it means for free energy to be at a minimum and global entropy to be at a maximum.

12-3. APPROACHES TO EQUILIBRIUM: A PREVIEW

Merely looking at the landscape of Figure 12-2 tells us nothing about how a rolling ball might actually negotiate the hills and valleys. Without knowing something about the ball itself—its initial velocity, its size,

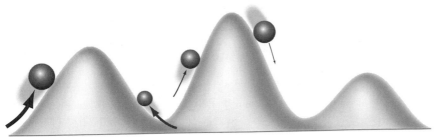

FIGURE 12-5. Questions of kinetics: How much of a push is needed to start the ball rolling? How likely is the ball to clear the first peak? How fast will it travel downhill? What forces might slow the motion? When will the ball come to rest?

shape, mass, density, texture—we cannot predict *how* it will come to equilibrium. We can assert only that there are two potential equilibrium states, one lower than the other, and eventually the ball will be at rest. But when? And how will it get there? Will it bounce around violently at first or else settle down quietly? Will it overcome the barrier between the upper and lower valleys?

These questions (Figure 12-5) are the province of *kinetics*, the study of the rates and pathways of chemical transformations. The bare mountain range, with no ball rolling around, represents only a thermodynamic potentiality or, put simply, what *can* happen. The specific route by which the ball really gets to equilibrium is controlled by kinetics, and that is a separate story which we shall address more fully later on, in Chapter 18.

Nevertheless let us anticipate, ever so briefly, how systems approach their equilibrium states, where all macroscopic variables are constant in time and uniform in space. Just as we need information about the ball on the hill, here we need information about the participants in a chemical reaction and also about the general conditions: initial concentrations of the species, temperature, pressure, and so on. All these factors contribute to the course of chemical change, as does the specific mechanism by which the reactant molecules are converted into product molecules.

There are three general approaches to equilibrium, which we mention in turn:

The concentration of each species may reach its equilibrium value *monotonically*, as in Figure 12-6, growing or shrinking in just one direction. If a concentration is too high initially, it decreases continuously until the final value is attained. If too low initially, then it increases uniformly on the way to equilibrium.

Monotonic processes, although extremely common, provide an incomplete picture of the kinetic possibilities. Chemical reactions may be *oscillatory* as well, with the concentrations cycling back and forth rhythmically

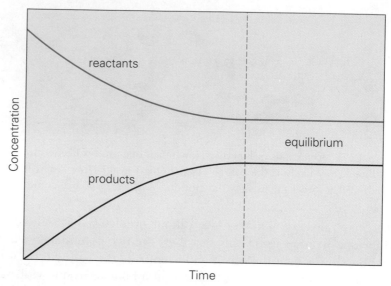

FIGURE 12-6. Monotonic approach to equilibrium (schematic): Concentrations of reactants decrease. Concentrations of products increase. The change is in one direction only; there is no turning back.

about the equilibrium values (Figure 12-7). Rather than come to equilibrium in some steady fashion, the system swings between extremes. A concentration may start out too high, then decrease so much as to become too low, and then shoot back up again. A cyclic sequence ensues, although the swings eventually become smaller and smaller as equilibrium is achieved over time. Oscillatory reactions occur frequently in biological

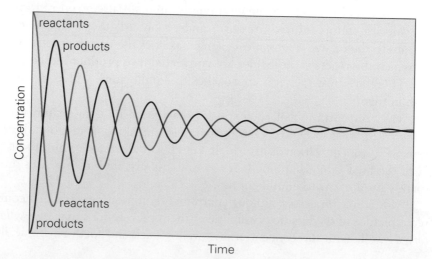

FIGURE 12-7. Oscillatory approach to equilibrium (schematic).

processes, many of which (sleep patterns and heartbeat, among others) are intrinsically rhythmic. Living organisms are, of course, the most conspicuous examples of systems temporarily out of equilibrium.

Beyond regular oscillations, reactions may proceed *chaotically*. Understood only recently, chaos in chemistry has a distinctive meaning that transcends mere randomness. A system may embark on a simple and apparently deterministic course yet still obtain a chaotic outcome, depending on the reaction mechanism and initial conditions. Patterns of concentrations become complex and unpredictable while displaying inordinate sensitivity to the slightest change in starting point. Still there is method in the seeming madness portrayed in Figure 12-8, and a rich variety of chaotic behavior has been described using mathematics and computer simulations.

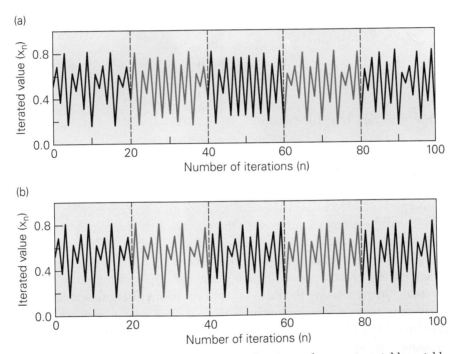

FIGURE 12-8. Chaos, illustrated by example: A simple equation yields a richly complex outcome, sensitive to the slightest change in conditions. Here, a series of numbers is computed according to the following recipe: (1) Pick a starting value x_0 and a certain parameter A. (2) Calculate the next number, x_1, as $x_1 = Ax_0(1 - x_0)^2$. (3) Use x_1 to calculate x_2 as $Ax_1(1 - x_1)^2$. (4) Use x_2 to calculate x_3; use x_3 to calculate x_4; use x_4 to calculate x_5, and so forth. Now look at the results obtained for two slightly different values of x_0, each with $A = 5.50$: (a) 0.5000 and (b) 0.5001. The small change in starting point (from 0.5000 to 0.5001) produces a noticeably different sequence that grows progressively larger after about 20 iterations. Similar effects appear everywhere in nature, from chemical reactions to biological rhythms to weather patterns.

Turbulence, the kind that arises when cream is poured into a cup of coffee, is an example of chaos. Swirling, complicated structures form and dissipate in space and time, never the same in any two cups. Chemical reactions, too, may develop similar structures in a chaotic regime.

12-4. THE EQUILIBRIUM CONSTANT AND THE LAW OF MASS ACTION

Putting aside the larger questions of why and how, we now take up the more pragmatic issue of *what*—what we can say quantitatively about a chemical system in equilibrium.

Such understanding begins with the *law of mass action*, elucidated by C. M. Guldberg and P. Waage in the 1860s: At equilibrium, the *reaction quotient*

$$Q = \frac{[C]^c [D]^d}{[A]^a [B]^b}$$

for an arbitrary transformation

$$a A + b B \rightleftharpoons c C + d D$$

is equal to a constant, $Q = K(T)$, dependent only on the temperature and nothing else. The concentration of each species X, represented as [X], appears in the expression raised to the power of its stoichiometric coefficient.

$K(T)$ is termed the *equilibrium constant*. It is, for a given reaction at a given temperature, one single *number* that can be determined experimentally by measuring the ratio of concentrations at equilibrium. It is a number of tremendous importance, since it conforms to a universal requirement imposed on all chemical systems at equilibrium—on homogeneous, single-phase equilibria; on heterogeneous systems; on phase equilibria; on dimerization; on dissociation; on decomposition; on acid–base neutralization; on precipitation—on all the ways that molecules can associate and disassociate.

The equilibrium constant is an expression akin to an equation of state, for it sets forth a fixed relationship among the macroscopic variables. Just as $PV = nRT$ connects P, V, n, and T for all ideal gases, so does K connect the concentrations of reactants and products for all types of reactions. Again we acquire a universal description covering many phenomena at once, an idea to be respected and nurtured.

Given a balanced chemical equation, we now have an immutable constraint on the proportions of reactants and products at equilibrium. Take, for example, the simple reaction A $\rightleftharpoons$ B, for which the equilibrium constant is $K = [B]/[A]$. Suppose that measurement reveals the value $K = 1$. It follows immediately that reactants and products will be present in equal concentrations at equilibrium, whether that be 1 mole per liter or 10 moles per liter or 100 moles per liter. And not only is the ratio fixed but, equipped with knowledge of the initial conditions, we can pinpoint the absolute amounts as well. Thus if the reaction begins with $[A] = 1$ mol L^{-1} and $[B] = 0$, the equilibrium concentrations must be $[A] = [B] = 0.5$ mol L^{-1}.

Observe: not all B and no A, but *some* B and *some* A. Reactants need not be consumed totally once a process attains equilibrium. The amounts left over (Figure 12-9) are determined by the magnitude of K. If $K \gg 1$, the system will consist mostly of products; the reaction is said to proceed "to the right." If $K \ll 1$, then it will be the reactants that dominate at equilibrium; the reaction goes to the left. Whether the equilibrium favors products or reactants is determined by the thermodynamic driving force alluded to above, notably the difference in free energy between products and reactants.

The first hint of a driving force appears in the reaction quotient itself. Away from equilibrium, certainly, the value of Q is unrestricted since we can always prepare an arbitrary mixture of products and reactants. We just dump everything together. But such a mixture is unlikely to be already in balance, and so—if allowed the opportunity—it must do what all out-of-equilibrium systems must do: get there. Either the

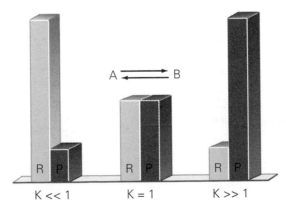

FIGURE 12-9. The value of K determines the "position" of an equilibrium, tilting the system either toward reactants (R) or products (P). Left: When the equilibrium constant is much less than 1, the final mixture contains mostly reactants. The equilibrium lies to the left. Center: When K is equal to 1, products and reactants coexist in equal concentrations for the process A $\rightleftharpoons$ B. Right: When K is much larger than 1, products prevail over reactants in the equilibrium mixture. The reaction goes to the right.

process will consume more reactants and make more products, or it will reconvert products into reactants. Given enough energy and time, the forward and reverse reactions will come to equilibrium. Doing nothing is not a permanent option.

Q can assume arbitrary values, but K cannot. K is *one* number, the target for the equilibrium state at a given temperature. If Q exceeds K at any instant, then there are too many products and not enough reactants. The reaction quotient has too big a numerator and too small a denominator. In response to that imbalance, the reverse reaction (products ⟶ reactants) gains the upper hand and overtakes the formation of products. If, by contrast, Q is less than K, there is initially a disproportionate excess of reactants over products. The forward reaction is then favored in order to increase Q. Only when Q becomes equal to K is equilibrium reached, whereupon the concentrations cease to change.

Rendered graphically in Figure 12-10, the three possibilities

$$Q < K \qquad \text{forward (make more products)}$$

$$Q = K \qquad \text{equilibrium (stay as you are)}$$

$$Q > K \qquad \text{reverse (make more reactants)}$$

are directional markers on the way to equilibrium for any chemical system.

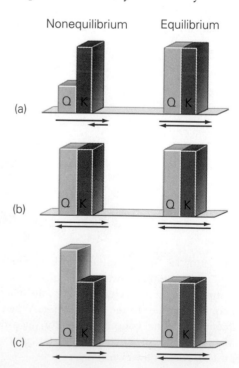

FIGURE 12-10. The reaction quotient, Q, may assume any value for an out-of-equilibrium mixture, but the equilibrium constant may not: K is fixed by the stoichiometry, temperature, and structure of the molecules involved. (a) The reaction moves to the right whenever Q is less than K. Products appear and reactants disappear until the reaction quotient rises to its equilibrium value. (b) When Q is equal to K, the process stands at equilibrium with no further drive toward either products or reactants. (c) When Q is greater than K, the reaction goes to the left. Products become reactants.

Both the reaction quotient and equilibrium constant should, it seems, be related to the free energy, the force that drives the transformation. Indeed they are, as we shall soon see in Chapter 14.

12-5. VAPOR PRESSURE AND THE EQUILIBRIUM CONSTANT

The vaporization of a liquid affords a simple example of heterogeneous equilibrium. Consider the phase transformation

$$H_2O(\ell) \rightleftharpoons H_2O(g)$$

and its equilibrium constant,

$$K(T) = \frac{P_{H_2O}}{[H_2O(\ell)]}$$

where P_{H_2O} denotes the partial pressure of the water vapor. Note that pressures are used preferentially in lieu of concentrations for gas-phase components, and note further that the concentrations of pure liquids and solids are essentially constant and so contribute no information to K.* We can, alternatively, multiply left and right by $[H_2O(\ell)]$ (which is just a constant roughly equal to 55.6 mol L^{-1}), and use instead the modified expression

$$K_p(T) = K(T)[H_2O(\ell)] = P_{H_2O} \qquad \text{(at equilibrium)}$$

The subscript p reminds us that this redefined equilibrium constant, K_p, is expressed in terms of pressure, easily convertible to gaseous concentration (n/V) through the relationship

$$[\text{gas}] = \frac{n}{V} = \frac{P}{RT}$$

According to the definition of $K_p(T)$, the equilibrium pressure above any liquid is constant at fixed temperature. Such constancy, too, is fully

---

*Strictly speaking, equilibrium constants are made dimensionless by using *activities*—the dimensionless ratios $[X]/c°$ and $P_X/P°$—in place of concentrations $[X]$ and pressures P_X in ideal systems. With the reference states $c°$ and $P°$ taken as 1 mol L^{-1} for solutions and 1 atm for gases, the net effect is to cancel the units when concentrations and pressures are measured in mol L^{-1} and atm, respectively. The reference concentration $c°$ for a pure liquid or solid is just the material's density, and hence the corresponding activity is unity.

(a)   (b)

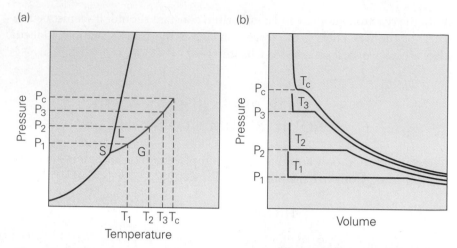

FIGURE 12-11. Vapor pressure. The pressure P exerted by a vapor in equilibrium with its liquid is, by definition, the equilibrium constant K_p for the transformation *liquid* $\rightleftharpoons$ *gas*. (a) Values of $K_p(T)$ at different temperatures correlate with points on the liquid–vapor line of a phase diagram. The coexistence line terminates at the critical point (T_c, P_c), where T_c denotes the critical temperature and P_c the corresponding critical pressure (see Chapter 11). (b) The same temperature dependence is obtained from a series of subcritical isotherms. The vapor pressure at each temperature corresponds to the flat portion of the curve.

consistent with our earlier identification (in Chapter 11) of vapor pressure as the flat, unchanging section of a subcritical *P-V* isotherm. With this realization, we appreciate now that vapor pressure itself is an equilibrium constant, a single *number* characteristic of the liquid–vapor system at equilibrium. It will be the same number regardless of where, when, and how the equilibrium arises. It will be the same number regardless of how much or how little liquid is present. It will be the same number—a *constant* of the equilibrium—as long as the temperature is fixed. And, both from the liquid–gas coexistence line of the phase diagram and from the isotherms shown in Figure 12-11, we can see exactly how K_p *increases* with temperature for the vaporization reaction.

Look more closely at this relationship between the equilibrium constant and temperature, interpreting it according to the picture of vaporization offered by Figure 12-12. Vaporization, an endothermic reaction, demands an influx of heat as if the process

$$H_2O(\ell) + \text{heat} \rightleftharpoons H_2O(g)$$

is fueled from left to right. Thermal energy (heat) must be supplied to break apart the intermolecular attractions of the liquid, and hence the gaseous product is favored at higher temperatures.

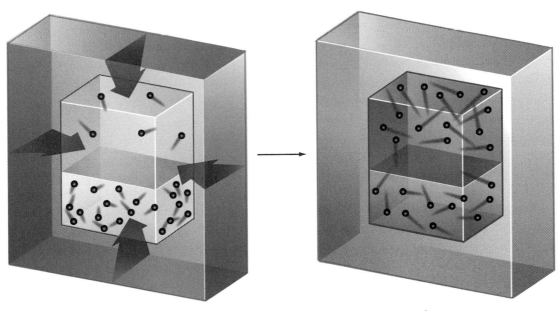

FIGURE 12-12. Vaporization, an endothermic process, requires energy to disrupt the structure of a liquid. As the temperature increases, more molecules acquire the thermal energy needed to escape from liquid to vapor. The equilibrium constant (vapor pressure) goes up.

Already, from the treatment in Chapter 11, we know what happens: Agitated by the elevated thermal energy, more molecules leave the liquid and the vapor pressure rises along with the temperature. Higher temperature means more thermal energy. More thermal energy means more gas. More gas means less liquid. This equilibrium constant for vaporization, $K_p(T)$, increases correspondingly with temperature—as will the equilibrium constant, evidently, for any endothermic reaction.

Condensation, on the other hand, is an exothermic process, accompanied by a loss of thermal energy while order is created in the liquid. Raising the temperature (pumping in thermal energy) promotes not condensation but rather its opposite: vaporization. For the molecules to condense, they must slow down and spend time together. To add heat, though, is to stir up the particles and thereby to inhibit the production of liquid. Raising the temperature encourages less liquid (less product) to form, and so raising the temperature fixes a smaller equilibrium constant for the exothermic reaction—a smaller *number*, a smaller ratio of products to reactants.

The mathematical representation follows directly. Write the conversion of gas into liquid as

$$H_2O(g) \rightleftharpoons H_2O(\ell) + \text{heat}$$

so that the exothermic condensation is viewed as the forward reaction. The equilibrium constant for the condensation (call it K_p' to distinguish it from K_p of vaporization) is then

$$K_p' = \frac{1}{P_{H_2O}} = \frac{1}{K_p}$$

One constant is the reciprocal of the other. K_p' for condensation goes down with temperature while K_p for evaporation goes up.

In general, the equilibrium constant for a "reverse" transformation

$$cC + dD \rightleftharpoons aA + bB$$

stands in a reciprocal relationship to that for a "forward" transformation

$$aA + bB \rightleftharpoons cC + dD$$

Products and reactants switch places in numerator and denominator, and so K flips over. If one direction is exothermic, the other direction is endothermic. K for the exothermic reaction decreases with temperature; K for the endothermic reaction increases with temperature.

12-6. STRESSED EQUILIBRIA

Say that equilibrium has been attained for the following process, a gas-phase synthesis of ammonia from nitrogen and hydrogen:

$$N_2(g) + 3H_2(g) \rightleftharpoons 2NH_3(g)$$

With the system at equilibrium, the reaction quotient

$$Q = \frac{P_{NH_3}^2}{P_{N_2} P_{H_2}^3} = K_p(T)$$

is equal, by definition, to $K_p(T)$ at the prevailing temperature. Constrained by the law of mass action, the three partial pressures hold constant and no further change is evident.

Despite outward appearances, however, the system remains *capable* of change and will respond appropriately when confronted with an external stress. Picture what happens when the pressure in the reaction vessel is doubled, as if by a sudden halving of the volume (Figure 12-13). Each partial pressure doubles at the same time, immediately rendering the reaction quotient

$$Q = \frac{2^2}{2 \cdot 2^3} K_p(T) = \frac{1}{4} K_p(T)$$

less than $K_p(T)$ and thus throwing the process out of equilibrium. With each pressure twice what it once was, the condition imposed by $K_p(T)$ is no longer fulfilled and the system must find its way back to equilibrium. The reaction is forced to the right: NH_3 forms at the expense of N_2 and H_2 until Q is restored to its equilibrium value of $K_p(T)$.

$K_p(T)$, dependent only on temperature, does not change its value under the higher pressure. There is only one equilibrium constant. What does change are the partial pressures of reactants and products, by which a new equilibrium mixture is established with more ammonia and proportionately less nitrogen and hydrogen. Although the new pressures are

FIGURE 12-13. Stress and response: the effect of pressure on a gas-phase equilibrium modeled after the N_2–H_2–NH_3 reaction. Assume, by analogy, that four molecules of reactants (dark shading) combine to form two molecules of product (light shading). (a) The system is initially at equilibrium. (b) An abrupt reduction of the volume increases the pressure, leaving the reaction quotient less than its equilibrium value. (c) The system restores equilibrium by dissipating the excess pressure. It reduces the number of gaseous molecules and hence the pressure, converting the more numerous reactants into the less numerous products. Four darkly shaded particles (reactants) are replaced by two lightly shaded particles each time the reaction goes forward.

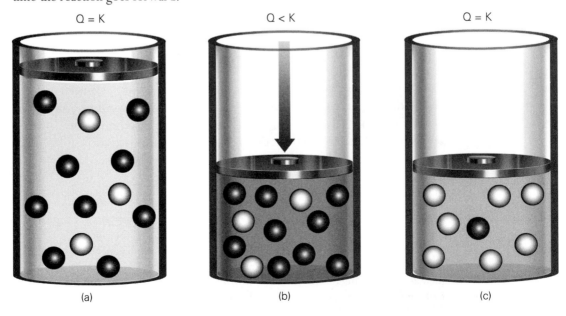

(a)   (b)   (c)

different individually, they remain connected through the equilibrium constant and they satisfy the same numerical requirement demanded by $K_p(T)$. Out of an infinite number of combinations consistent with the constant value of $K_p(T)$, the system under stress is forced from one specific set to another. We say that the "position" of the equilibrium shifts toward the products.

The response of the N_2–H_2–NH_3 reaction to an increase in pressure conforms to a general principle first enunciated by Le Châtelier: *A system perturbed from equilibrium will shift its equilibrium position so as to relieve the applied stress.* Here the stress is additional pressure (the molecules are squeezed closer together), and the gaseous system dissipates that pressure by reducing the number of molecules present. From fewer molecules will come fewer crashes into the walls and, in turn, lower pressure.

Forward progress (reactants ⟶ products) is demanded in this example, because four molecules on the left react to form only two molecules on the right:

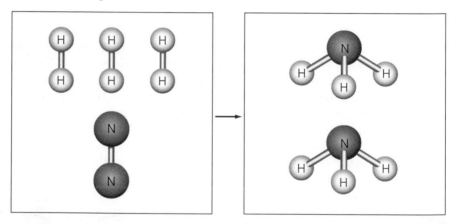

Were there *fewer* molecules of reactants than products, then an *elevated* reaction quotient would force the reverse direction to dominate at higher pressures. Were there equal numbers of gaseous molecules on both sides of the equation, the equilibrium position would be independent of pressure. Responding to the stress of increased pressure, the system simply shifts in the direction that will best remove molecules from the gas phase—to the right, to the left, or not at all. The three options are summarized in Figure 12-14.

FIGURE 12-14. Gas-phase equilibria and the stress of increased pressure: which way to go? (a) Forced out of equilibrium, a reaction shifts to the right if gaseous reactants (dark shading) outnumber gaseous products in the balanced equation. (b) Where reactants and products exist in equal stoichiometric proportions, the transformation is unaffected by a change in pressure. (c) If products outnumber reactants, an increase in pressure drives the reaction to the left.

EQUILIBRIUM   STRESS   EQUILIBRIUM

(a)

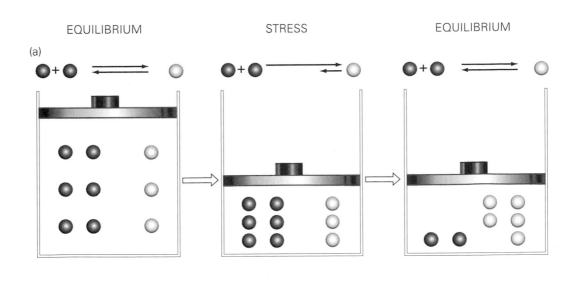

(b)

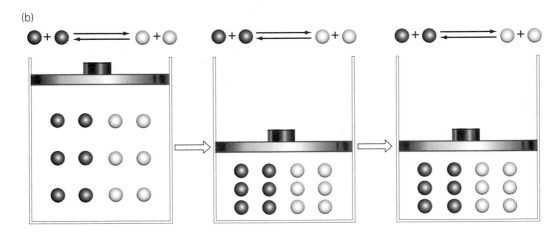

(c)

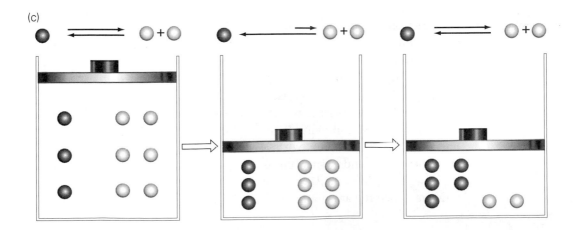

Changes in temperature are another source of stress, but now the value of the equilibrium constant itself is affected (as we have just demonstrated for vapor pressure). Consider, at this point, the exothermic dimerization reaction

$$2NO_2(g) \rightleftharpoons N_2O_4(g) + \text{heat}$$

which releases 57.2 kJ of heat with the formation of each mole of N_2O_4. Since $K_p(T)$ for an exothermic process decreases with temperature, a sudden increase in temperature will leave the instantaneous reaction quotient too *large* compared with the new equilibrium constant. Hence the reverse reaction, which proceeds endothermically by absorbing 57.2 kJ per mole of N_2O_4 dissociated, gains the advantage at a higher temperature:

$$N_2O_4(g) + \text{heat} \longrightarrow 2NO_2(g)$$

A new equilibrium at the smaller $K_p(T)$ is subsequently established in favor of NO_2.

The effect, again, is explained qualitatively by Le Châtelier's principle, with the stress of a higher temperature understood as the imposition of additional heat. Finding itself suddenly too hot, the out-of-equilibrium system acts to cool down. The extra heat (Figure 12-15) feeds the endothermic dissociation process and, doing so, enforces compliance with the new equilibrium constant.

Pressure, temperature, and Le Châtelier come together in the Haber-Bosch synthesis of ammonia (1911), which combines fundamental principles of equilibrium, thermodynamics, and kinetics in a process of considerable economic importance. The goal—to produce ammonia from readily available nitrogen and hydrogen gas—is thermodynamically favorable since the free energy of the product is substantially lower than that of the reactants. Thermodynamic feasibility and a correspondingly large equilibrium constant are of little use, however, if the kinetic pathway for the reaction is too slow. Such is the problem when one mole of N_2 and three moles of H_2 are brought together. Nothing happens, although thermodynamics unequivocally guarantees a big payoff of NH_3 . . . eventually. The challenge is to turn "eventually" into *today*.

A partial solution is to raise the temperature to between 400°C and 500°C and use a catalyst to speed the reaction further. At the higher temperature, faster-moving molecules smash into each other with higher energies and break bonds with greater facility. Many reactions proceed at higher rates under such circumstances, as we shall see in our later treatment of kinetics. Regrettably, though, a thermodynamic difficulty emerges since the ammonia synthesis is exothermic and consequently its

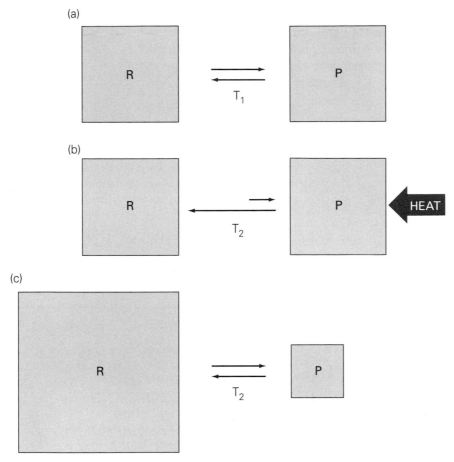

FIGURE 12-15. Stress and response: temperature. The example is for a reaction exothermic in the forward direction and endothermic in the reverse. $K(T)$, the equilibrium constant, decreases with temperature. (a) The system stands initially at equilibrium, its concentrations conforming to some ratio $K(T_1)$. (b) The stress imposed by a higher temperature T_2 throws the process out of equilibrium, making the reaction quotient suddenly greater than the new, reduced equilibrium constant $K(T_2)$. There are too many products (P) and too few reactants (R), and the system dissipates its excess thermal energy by running in reverse until (c) a new equilibrium is reached at the higher temperature. Products become reactants. Heat is consumed. Q shrinks.

equilibrium constant decreases with temperature. Less ammonia is actually produced, albeit faster.

The full solution is to exploit the pressure dependence of the reaction, as set forth above, to shift the equilibrium to the right at pressures between 200 atm and 600 atm. If ammonia is then removed as it forms, the reaction now suffers an additional stress: loss of product, which

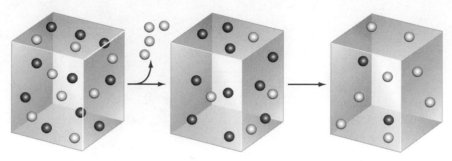

FIGURE 12-16. Stress and response: loss of product. For a process already at equilibrium, any tampering with the concentrations will destroy the balance. Removal of a product (light shading) forces the system to produce more in response.

makes Q less than K. Le Châtelier's principle requires that the stress be relieved by producing even more NH_3.

So it happens. Pulled continually away from equilibrium, never allowed to rest, the system under stress chases the departing material. The reaction goes forward, urged on by a perpetual imbalance between products and reactants. See Figure 12-16 for a reminder in pictures.

12-7. SOLVING PROBLEMS USING THE EQUILIBRIUM CONSTANT

Equilibrium computations tend to be straightforward, if sometimes tedious algebraically. Beginning with a balanced equation and the initial concentrations, one typically wants to know how much of everything is present at equilibrium. The procedure generally entails three steps: (1) Set up the reaction quotient and note the direction of change. (2) Adjust the concentrations (or pressures) in the reaction quotient to reflect the extent of the change. (3) Solve the algebraic equation that results. The complexity of this equation is governed largely by the stoichiometry of the reaction, which determines the exponents in the equilibrium expression. When mathematical difficulties develop, simplifications and approximations often can be used to facilitate quick solutions—many times with little loss in accuracy.

As a general (not overly simple) example, suppose that at 100°C we have an equilibrium mixture of NO_2 and N_2O_4 in which the partial pressures of the two components are 3.12 atm and 1.50 atm, respectively. Let the volume of the vessel undergo a 10-fold isothermal expansion. What will the partial pressures be when equilibrium is reestablished?

We begin by using the data to compute K_p for the dimerization reaction $2NO_2(g) \rightleftharpoons N_2O_4(g)$:

$$K_p = \frac{P_{N_2O_4}}{P_{NO_2}^2} = \frac{1.50}{(3.12)^2} = 0.154 \qquad \text{(at equilibrium)}$$

Increasing the volume by a factor of 10 reduces $P_{N_2O_4}$ from 1.50 atm to 0.150 atm and P_{NO_2} from 3.12 atm to 0.312 atm. The reaction quotient

$$Q = \frac{0.150}{(0.312)^2} = 1.54 \qquad \text{(out of equilibrium)}$$

is now too high, and equilibrium can be restored only if the reverse reaction accelerates to form NO_2 at the expense of N_2O_4. K_p—the target value—holds constant at 0.154 because the temperature never changes (remember: we are examining an *isothermal* expansion).

If x denotes the decrease in $P_{N_2O_4}$, then P_{NO_2} must increase by $2x$ since 2 moles of NO_2 are produced whenever 1 mole of N_2O_4 dissociates:

	$2NO_2$	$\rightleftharpoons$	N_2O_4
initial pressure	0.312		0.150
change in pressure	$2x$		$-x$
equilibrium pressure	$0.312 + 2x$		$0.150 - x$

K_p at the new equilibrium point is therefore given by

$$K_p = \frac{0.150 - x}{(0.312 + 2x)^2} = 0.154$$

which reduces to a quadratic equation in x:

$$0.616x^2 + 1.192x - 0.135 = 0$$

Using the quadratic formula (see Appendix B), we obtain $x = 0.1073$. The second root of the equation, $x = -2.0427$, we reject because it shows the change going in the wrong direction. The negative sign means that the pressure of N_2O_4 would increase, in violation of Le Châtelier's principle.

Upon reestablishment of equilibrium, N_2O_4 exerts a partial pressure of 0.043 atm $(0.150 - x)$ while NO_2 contributes 0.527 atm $(0.312 + 2x)$. The reaction quotient

$$Q = \frac{0.0427}{(0.527)^2} = 0.154$$

once again equals K_p, and so the change in equilibrium position leads to a 70% increase of the partial pressure exerted by NO_2.

Note, finally, that any problem involving gases can be cast in terms of molar concentrations rather than pressures. For gases behaving ideally, we just use the equation

$$[\text{gas}] = \frac{n}{V} = \frac{P}{RT} = P(RT)^{-1}$$

to obtain the relationship between K_c (K expressed in concentrations) and K_p (K expressed in pressures). The conversion goes, for example, as

$$K_c = \frac{[C]^c[D]^d}{[A]^a[B]^b} = \frac{P_C^c \, P_D^d}{P_A^a \, P_B^b} \, (RT)^{a+b-c-d}$$

$$= K_p(RT)^{a+b-c-d}$$

for the reaction

$$a\text{A} + b\text{B} \rightleftharpoons c\text{C} + d\text{D}$$

The general rule is thus

$$K_c = K_p(RT)^{-\Delta n}$$

where Δn is the difference between the number of moles of gaseous products and the number of moles of gaseous reactants in the balanced equation:

$$\Delta n = n_{\text{gaseous products}} - n_{\text{gaseous reactants}}$$

If Δn is zero, then K_p and K_c have the same value.

To sum up: All reactions come to equilibrium. Having addressed the question of *what* it is, we now turn to *why* it is. The "why" is wrapped up in thermodynamics, in the interrelationship of energy and entropy. Our agenda is to examine first one and then the other: energy in the chapter following, and entropy (or disorder) immediately thereafter.

REVIEW AND GUIDE TO PROBLEMS

Equilibrium: everywhere different, everywhere the same. Look for differences, and find similarities instead.

Somewhere in the world, liquid water is turning into vapor. The vapor phase grows rapidly at first, enriched by a flow of energetic molecules from the liquid, but competition soon develops. A reverse reaction sets in. More and more molecules, lower in energy, start to return from vapor to liquid until finally a balance is struck between the forward process (evaporation) and the reverse (condensation). In the end, as many molecules come as go. The rates of evaporation and condensation become equal; a steady vapor pressure is exerted; a state of dynamic equilibrium takes hold. The change is over, at least for now.

Somewhere else, a blast of compressed air fills a tire and starts another system on the road to equilibrium. Outside the tire, molecules in the air pepper the rubber continually with energy and momentum. Inside, the turbulent gas swirls around for a time, its particles colliding both with each other and with the inner walls. They bounce around randomly and chaotically, exchanging energy like so many billiard balls, until finally the turbulence is gone. The molecules continue to collide, but to an external observer the scene is quiet: a constant pressure, a constant volume, and a constant temperature (equal to the temperature outside). In the end, molecular energies are distributed according to the Boltzmann distribution, and a state of dynamic equilibrium takes hold. The change is over, at least for now.

Somewhere else again, in a commercial reactor, vast quantities of nitrogen gas and hydrogen gas are reacting to form ammonia, also in a process that goes both ways. Reactants smash together, break into pieces, and reassemble themselves into products, all the while fighting against the reverse transformation. At times, the products accumulate rapidly; at times, slowly. In the end, products and reactants come to a dynamic equilibrium whereby the concentrations of N_2, H_2, and NH_3 cease to change. The change is over, at least for now.

We see, then, that all transformations of matter, like everything else, have a beginning, middle, and end. They begin because something in the world is out of balance—because there is too much water in the lake and not enough in the air, or because there are too many speedy molecules in the tire and not enough sluggish ones, or because there is too much hydrogen in the reactor and not enough ammonia, or even (tragically) because Humpty-Dumpty is sitting too high on the wall and thus is ripe for a fall. But behind every change, however different it may appear, there always lurks an imbalance in need of correction.

For every change too, whatever course it takes, the general effect is always the same in its microscopic aspect: *energy is redistributed*. How? Particles collide. Molecules attract and repel. Atoms move. Bonds vibrate. Molecules rotate. Heat goes in and heat goes out. Clusters come and go. Covalent bonds are broken and remade in new ways. Take the broadest possible view, and see in it the essence of every possible change: a redistribution of energy into new forms of motion, like capital flowing from one investment to another.

And when such redistribution is complete, the process is at equilibrium and can progress no further. The imbalance is gone. Nature is satisfied, at least for now.

IN BRIEF: CHEMICAL EQUILIBRIUM
AND THE LAW OF MASS ACTION

Realize that we began this chapter not at the beginning but at the end, at equilibrium, with the reaction already over, the driving force spent, all outward change vanished. Still to come, piece by piece, is an understanding of the beginning and middle of a process—the thermodynamic forces that drive it, the redistribution of energy that ensues, the rates and mechanisms of the various reactions. For now, however, we choose to see the equilibrium state simply for what it is: an accomplished fact, a condition of the present without past or future.

1. A system in equilibrium loses all sense of time, for without change there is no time. Once brought to equilibrium, all the macroscopic properties remain constant. Pressure, temperature, volume, number of molecules, concentration, color—all have the same values yesterday, today, and tomorrow.

Shall we arrange for an ideal gas to be in equilibrium at standard temperature and pressure? Here is a plan in three steps: (1) Place one mole of particles in a container of capacity 22.4 L. (2) Put the gas in thermal contact with some large object at a temperature of 273 K (the atmosphere will do), and let heat flow in or out. (3) Wait.

Wait for a time, and then measure the properties: $P = 1$ atm, $V = 22.4$ L, $n = 1$ mol, $T = 273$ K; steady as a rock, never changing. PV has become equal to nRT, and thus the gas has reached an equilibrium where the pressure and temperature are appropriate for the volume and amount. To get to equilibrium, the particles passed energy back and forth until finally their average speed was consistent with a temperature of 273 K. And there they stay, in equilibrium, dead to the world outside.

Inside the system, there is animation and change everywhere.

Energy moves continually from particle to particle. Heat flows into and out of the surroundings, forcing some particles to speed up or slow down. There are collisions, ricochets, and bottlenecks all the time. But whatever happens microscopically, we fail to notice macroscopically. Whatever the particles do among themselves, they do so without affecting the macroscopic outcome. Same temperature. Same pressure. Same volume. Same amount.

Anywhere and anytime, the result is the same. Equilibrium is a unique macroscopic condition, recognizable instantly by the constant, predictable values of the variables. How the system got there is of no consequence, and how the system stays there is of no consequence. It is enough simply to *be* there. Does *PV* equal *nRT*? Yes? Then our ideal gas is, by definition, in equilibrium, to remain that way until provoked by a change from the outside. It is in a state of *thermal equilibrium*, a state that we shall be able to examine more sharply in Chapter 14.

2. CHEMICAL EQUILIBRIUM. Now devise a system in *chemical* equilibrium, where reactants become products and products become reactants. No longer do we have mere billiard balls bouncing around a box, juggling speed and energy while remaining otherwise intact. Instead, there is action of more consequence among the atoms and molecules: movements of electrons, smashing of bonds, making and breaking of clusters, all the creative destruction that accompanies the building up of new molecules and phases.

Throw together the reactants, and wait. Eventually the system comes to chemical equilibrium, all concentrations steady. Some of the reactants may remain unconverted, true, but still the reaction is finished. Never mind that molecules are being shattered and rebuilt every second, because in the end there is not one speck of products more or less. Forget the microscopic frenzy, because macroscopically the game is over. The process has reached equilibrium, and that is as far as it can go.

Try it again. Start with another mixture of reactants, perhaps one that is already seeded with some of the expected products. Mix them together, and wait. Some time later, there again develops a state of equilibrium whereupon the concentrations cease to change. Inside, the reactants furiously become products and (just as furiously) the products become reactants. Outside, we declare the reaction to be over.

3. THE LAW OF MASS ACTION. Just as an equilibrated gas obeys an equation of state, so does an equilibrated reaction: Its concentrations *at equilibrium* conform to the **law of mass action**, constrained by the value of the **equilibrium constant**

$$K(T) = \frac{[C]^c [D]^d}{[A]^a [B]^b}$$

for the model reaction

$$a\mathrm{A} + b\mathrm{B} \rightleftharpoons c\mathrm{C} + d\mathrm{D}$$

However it happens, the concentrations of products and reactants at equilibrium will satisfy the ratios demanded by the equilibrium constant $K(T)$, a number dependent on temperature alone. For a chemical system in equilibrium, the definition of K is the equivalent of $PV = nRT$ for a gas—an equation of state, a connection between the macroscopic variables. It tells us all there is to know.

Here is what to do. First, determine (by experiment) the stoichiometry and hence the balanced equation for the reaction. Second, measure the equilibrium concentrations of all reactants and products at a given temperature. Third, compute the value of $K(T)$ by inserting those numbers into the mass-action expression. Then: Choose a new set of initial concentrations; solve an equation; make a prediction. Predict the amount of products and reactants present at the new equilibrium, using $K(T)$ and the initial concentrations to see into the future.

We can do it. Knowing both the value of $K(T)$ and the concentrations present at the *start* of reaction, we automatically know the concentrations at the end. The end is predetermined thermodynamically from the start, as we shall see more profoundly in the two chapters following. So let us be patient, as always, because the reward will be a richer, deeper understanding of things, but we should understand at least this much right now: Sooner or later, all reactions come to an end; and when they do, we know what to expect. That information is ours for the price of a mathematical exercise, the solution of an algebraic equation involving $K(T)$ and the initial concentrations.

4. REACTION QUOTIENT. *At* equilibrium, the concentrations must satisfy the one and only value of the equilibrium constant, $K(T)$. *Away* from equilibrium, they can be whatever they like. To address that infinite range of possibilities, we define (by analogy to K) the generalized ***reaction quotient***

$$Q = \frac{[\mathrm{C}]^c[\mathrm{D}]^d}{[\mathrm{A}]^a[\mathrm{B}]^b}$$

for a system as it progresses toward equilibrium.

When Q is greater than K, there are too many products and too few reactants. Nature demands that some of the products be reconverted into reactants.

When Q is less than K, there are too many reactants and too few products. The system will not attain equilibrium until the reactants produce more products.

When Q is equal to K, the process is over. Forward and reverse reactions are in balance. Equilibrium prevails.

5. LE CHÂTELIER'S PRINCIPLE. Equilibrium prevails, yes, but only as long as the prevailing conditions prevail. Change anything, and the balance is gone. A new equilibrium must be negotiated.

Suppose, for example, that we add more products to the equilibrium mixture. Confronted with too much product to support, the system's reaction quotient then goes abruptly from $Q = K$ (equilibrium) to $Q > K$ (out of equilibrium). The response? Get rid of the extra products. Convert the products back into reactants, and do so until the reaction quotient has been cut down to the value $Q = K$.

Suppose, instead, that we add more *reactants* to the mixture, so as to depress the reaction quotient *below* its equilibrium value. The response now? Get rid of the excess reactants. Make more products.

Or suppose we change the temperature to a point where $K(T)$ itself becomes greater, thereby allowing the system to support proportionally more product at equilibrium. Suddenly there is excess reactant to dispose of and more product to make.

Whatever stress is imposed, whether from a change in concentration, temperature, or pressure, the system responds in a way to alleviate that stress: to rid itself of reactants, to rid itself of products, to absorb heat, to evolve heat, to make more, to make less, to return to equilibrium by whatever means available.

Thus states the **principle of Le Châtelier**.

6. We conclude with a set of examples designed to illustrate the practical aspects of equilibrium computations: setting up the equilibrium constant, using the initial conditions to construct an equation for the concentrations, making approximations, solving for the concentrations, using pressures instead of concentrations, and so forth.

But before we do, a warning concerning the symbol K: it will vary. Sometimes we shall write it as $K(T)$; sometimes, just as K or K_{eq}. At other times, it will be K_c (for concentration) or K_p (for pressure) or, in Chapter 15, K_{sp} (for solubility product) or, in Chapter 16, K_a (for acid), K_b (for base), and K_w (for water); or sometimes as simply a vapor pressure, as in Chapter 11 . . . but always a change in name only, with the same meaning in the end. Equilibrium: everywhere different, everywhere the same.

SAMPLE PROBLEMS

EXAMPLE 12-1. The Equilibrium Constant

Three rules to remember: (1) The concentrations of pure condensed phases (solids and liquids) are constant. Do not include them in the expression for K. (2) For gases, pressures are preferred to concentrations.

(3) The exponent assigned to each concentration or pressure is equal to the coefficient of that substance in the balanced equation.

PROBLEM: For each of the reactions indicated, write an expression for the equilibrium constant.

(a) *The ionization of acetic acid*

$$CH_3COOH(aq) + H_2O(\ell) \rightleftharpoons CH_3COO^-(aq) + H_3O^+(aq)$$

SOLUTION: Rule 1 reminds us that the concentration of a pure liquid—not a solution, but simply the unmixed liquid itself—is constant, forever fixed by the density of the substance. For water in particular, where the density is roughly 1 g mL^{-1} (hence 1000 g L^{-1}), the molarity stays unchanged at 55.6 M throughout the entire reaction:

$$\frac{1000 \text{ g}}{L} \times \frac{1 \text{ mol}}{18.0 \text{ g}} = 55.6 \text{ mol L}^{-1} = 55.6 \text{ } M$$

It tells us nothing new. Were we to include this 55.6 in the equilibrium constant, the expression would still depend only on the concentrations of acetic acid (CH_3COOH), the acetate anion (CH_3COO^-), and the hydronium ion (H_3O^+):

$$K = \frac{[CH_3COO^-][H_3O^+]}{[H_2O][CH_3COOH]} = \frac{[CH_3COO^-][H_3O^+]}{(55.6)[CH_3COOH]}$$

Instead, we opt to multiply left and right by 55.6 and obtain the *acid ionization constant*, K_a, in its standard form:

$$K_a = 55.6 \text{ } K = \frac{[CH_3COO^-][H_3O^+]}{[CH_3COOH]}$$

(b) *The decomposition of SiCl$_4$ to produce elemental silicon*, carried out at high temperature in the presence of H$_2$:

$$SiCl_4(g) + 2H_2(g) \rightleftharpoons Si(s) + 4HCl(g)$$

This process is used to produce pure material for electronic applications.

SOLUTION: Omitting, again, the constant concentration of the pure condensed phase (solid silicon), we express K for the gaseous components in terms of either pressure or concentration:

$$K_p(T) = \frac{P_{HCl}^4}{P_{SiCl_4}\, P_{H_2}^2}$$

$$K_c(T) = \frac{[HCl]^4}{[SiCl_4][H_2]^2}$$

(c) *Dissolution and precipitation of $Mg(OH)_2$ in water*

$$Mg(OH)_2(s) \rightleftharpoons Mg^{2+}(aq) + 2\,OH^-(aq)$$

SOLUTION: As before, the solid phase does not appear in the equilibrium expression:

$$K_{sp} = [Mg^{2+}][OH^-]^2$$

K_{sp} is called the *solubility-product constant*.

EXAMPLE 12-2. Computing an Equilibrium Constant

PROBLEM: Ammonia, a weak base, stimulates the production of ammonium ion and hydroxide ion in aqueous solution:

$$NH_3(aq) + H_2O(\ell) \rightleftharpoons NH_4^+(aq) + OH^-(aq)$$

Compute the equilibrium constant at 25°C, given that a 1.0 *M* solution of ammonia yields $[OH^-] = 0.0042$ *M* in equilibrium at this temperature. Assume that OH^- and NH_4^+ arise from no other source but the reaction of ammonia and water.

SOLUTION: Here we have the *base ionization constant*

$$K_b = \frac{[NH_4^+][OH^-]}{[NH_3]}$$

to consider, together with three other pieces of information:

First, we know that the *initial* concentration of NH_3—the concentration at the start of reaction, before anything else happens—is 1.0 *M*.

Second, from the wording of the problem, we assume that the initial concentrations of both OH^- and NH_4^+ are zero.

Third, the concentration of OH^- at equilibrium is 0.0042 *M*.

Now, from the balanced equation we know that reaction of *one* NH_3 molecule produces *one* NH_4^+ ion and *one* OH^- ion. And since ammonium and hydroxide are said to arise exclusively from the process above, then we also know that

$$[NH_4^+] = [OH^-] = 0.0042 \ M \qquad \text{(at equilibrium)}$$

Because: For every OH^- ion produced by reaction of NH_3, there must be one NH_4^+ ion produced as well—and, similarly, there must be one molecule of NH_3 consumed at the same time. The final concentration of NH_3 is therefore

$$[NH_3] = 1.0 \ M - 0.0042 \ M = 0.9958 \ M \qquad \text{(at equilibrium)}$$

and the equilibrium constant is

$$K_b = \frac{[NH_4^+][OH^-]}{[NH_3]} = \frac{(0.0042)(0.0042)}{0.9958} = 1.8 \times 10^{-5}$$

QUESTION: Scarcely any NH_3 seems to have reacted, a mere 0.4% of the original concentration. Since 0.9958 (namely, $1.0 - 0.0042$) is so close to 1.0, would it be acceptable to ignore that small difference when solving the problem?

ANSWER: Yes. Such a shortcut would make the calculation simpler without changing the final result (a good thing), and we shall use similar tricks wherever possible. Note that the simplified procedure yields the same value within our stated limits of two significant figures:

$$K_b = \frac{(0.0042)(0.0042)}{1.0 - 0.0042} \approx \frac{(0.0042)(0.0042)}{1.0} = 1.8 \times 10^{-5}$$

EXAMPLE 12-3. **Manipulating K: Reverse Reactions**

Algebraic fact: If K is the equilibrium constant for a process written in the forward direction,

$$\text{reactants} \rightleftharpoons \text{products} \qquad K$$

then $1/K$ is the corresponding value for the same process run in reverse:

$$\text{products} \rightleftharpoons \text{reactants} \qquad K' = 1/K$$

PROBLEM: Carbon dioxide and hydrogen react to form carbon monoxide and water according to the equation

$$CO_2(g) + H_2(g) \rightleftharpoons H_2O(\ell) + CO(g) \qquad K_p(25°C) = 3.22 \times 10^{-4}$$

Determine the value of the equilibrium constant for the reverse transformation,

$$H_2O(\ell) + CO(g) \rightleftharpoons CO_2(g) + H_2(g) \qquad K_p'(25°C) = ?$$

SOLUTION: We need only write down the two expressions and compare:

$$K_p = \frac{P_{CO}}{P_{CO_2} P_{H_2}} \qquad \frac{P_{CO_2} P_{H_2}}{P_{CO}} = K_p'$$

The numbers are reciprocals, one the inverse of the other, giving us

$$K_p K_p' = 1$$

and

$$K_p' = \frac{1}{K_p} = \frac{1}{3.22 \times 10^{-4}} = 3.11 \times 10^3$$

Meaning: The equilibrium lies to the left, driven in that direction by a large equilibrium constant. The tendency to produce $CO_2(g)$ and $H_2(g)$ far exceeds the tendency to produce $CO(g)$ and $H_2O(\ell)$.

EXAMPLE 12-4. **Manipulating K: Concentration and Pressure**

Algebraic fact: The relationship between an equilibrium constant expressed in concentrations (K_c) and an equilibrium constant expressed in pressures (K_p) is

$$K_c = K_p(RT)^{-\Delta n}$$

where Δn denotes the change in the number of gaseous moles during the reaction:

$$\Delta n = n_{\text{gaseous products}} - n_{\text{gaseous reactants}}$$

PROBLEM: Given that K_p for the reaction

$$CO_2(g) + H_2(g) \rightleftharpoons H_2O(\ell) + CO(g) \qquad K_p(25°C) = 3.22 \times 10^{-4}$$

has the value 3.22×10^{-4} (Example 12-3), compute the corresponding value for the equilibrium constant expressed in concentrations, K_c. Assume that the gases behave ideally.

SOLUTION: For an ideal gas X, the relationship between pressure (P_X) and concentration ($n_X/V = [X]$) is

$$P_X = \frac{n_X}{V} RT = [X]RT$$

Inserting $[X]RT$ for the pressure of each component X, we first transform K_p into K_c:

$$K_p = \frac{P_{CO}}{P_{CO_2} \, P_{H_2}} = \frac{[CO]RT}{[CO_2]RT \, [H_2]RT} = K_c\left(\frac{1}{RT}\right)$$

We then supply values for R (0.0821 atm L mol^{-1} K^{-1}) and T ($25°C = 298$ K):

$$K_c = K_p RT = (3.22 \times 10^{-4})(0.0821 \times 298) = 7.88 \times 10^{-3}$$

QUESTION: Doesn't RT have units of atm L mol^{-1}?

ANSWER: We write the numbers without units, a seemingly sloppy practice, but we justify the omission as follows: Each pressure in K_p is understood implicitly to be the dimensionless ratio $P_X/P°$, where P_X denotes the pressure of gas X (in atm) and $P°$ is a reference pressure of exactly 1 atm. Similarly, each concentration in K_c is taken as the dimensionless ratio $[X]/c°$, where $c°$ represents a reference concentration of 1 mol L^{-1}.

Making the appropriate substitutions, we find that both K_p and K_c are dimensionless as well—and also that RT is really $c°RT/P°$, a dimensionless number with magnitude RT. For the example at hand, the relationship works out to be

$$K_c = K_p \frac{c°RT}{P°} = K_p\left(\frac{1 \text{ mol L}^{-1}}{1 \text{ atm}}\right)RT = K_p RT \quad \text{(dimensionless)}$$

The conversion factor so obtained, RT, thus conforms to the general rule for gas-phase reactions stated above:

$$K_c = K_p(RT)^{-\Delta n}$$

Here, specifically, we have one mole of gas on the right (CO), two moles on the left ($CO_2 + H_2$), and hence a change given by

$$\Delta n = 1 - 2 = -1$$

The result is

$$K_c = K_p RT$$

EXAMPLE 12-5. Manipulating K: Multiplication of Chemical Equations

Algebraic fact: When all the coefficients in a balanced equation are multiplied by a constant factor j, the equilibrium constant (originally K) becomes K^j.

PROBLEM: Using the information given in Example 12-3, compute K_p for the reaction shown below:

$$2CO_2(g) + 2H_2(g) \rightleftharpoons 2H_2O(\ell) + 2CO(g)$$

SOLUTION: The equation as written differs from the form in Example 12-3,

$$CO_2(g) + H_2(g) \rightleftharpoons H_2O(\ell) + CO(g)$$

only by a uniform factor $j = 2$. There are now *two* moles of CO_2, not one; *two* moles of H_2, not one; *two* moles of H_2O, not one; *two* moles of CO, not one.

Before, the equilibrium constant was

$$K_p = \frac{P_{CO}}{P_{CO_2} P_{H_2}} = 3.22 \times 10^{-4}$$

Now, with all coefficients doubled, the expression becomes

$$K'_p = \frac{P^2_{CO}}{P^2_{CO_2} P^2_{H_2}} = K^2_p = (3.22 \times 10^{-4})^2 = 1.04 \times 10^{-7}$$

The scaling factor ($j = 2$) multiplies the exponent on each of the pressures, and the effect is to square the entire expression.

EXAMPLE 12-6. Manipulating *K*: Addition of Chemical Equations

Algebraic fact: When balanced equations are added together, the equilibrium constant for the combined process is equal to the product of the equilibrium constants for each step.

PROBLEM: The equilibrium constant K_1 for the ionization of acetic acid (reaction 1; see below) is equal to 1.76×10^{-5} at 25°C. The equilibrium constant K_2 for the autoionization of water (reaction 2) has the value 1.0×10^{-14} at the same temperature:

$$\text{(1)} \quad CH_3COOH(aq) + H_2O(\ell) \rightleftharpoons CH_3COO^-(aq) + H_3O^+(aq)$$
$$K_1 = 1.76 \times 10^{-5}$$

$$\text{(2)} \quad H_2O(\ell) + H_2O(\ell) \rightleftharpoons H_3O^+(aq) + OH^-(aq)$$
$$K_2 = 1.0 \times 10^{-14}$$

Compute the value of the equilibrium constant for reaction 3, the neutralization of acetic acid by hydroxide ion:

$$\text{(3)} \quad CH_3COOH(aq) + OH^-(aq) \rightleftharpoons CH_3COO^-(aq) + H_2O(\ell)$$
$$K_3 = ?$$

SOLUTION: First, notice that the neutralization reaction (3) can be written as the sum of reaction 1 and the reverse of reaction 2 (call it 2'):

$$\text{(1)} \quad CH_3COOH(aq) + H_2O(\ell) \rightleftharpoons CH_3COO^-(aq) + H_3O^+(aq)$$
$$+ \quad \text{(2')} \ H_3O^+(aq) + OH^-(aq) \rightleftharpoons 2H_2O(\ell)$$
$$\overline{\text{(3)} \quad CH_3COOH(aq) + OH^-(aq) \rightleftharpoons CH_3COO^-(aq) + H_2O(\ell)}$$

Next, observe that the equilibrium expression for the neutralization reaction, K_3, is exactly equal to the product of the equilibrium expressions for reactions 1 and 2':

$$\frac{[CH_3COO^-]}{[CH_3COOH][OH^-]} = \frac{[CH_3COO^-][H_3O^+]}{[CH_3COOH]} \times \frac{1}{[H_3O^+][OH^-]}$$
$$K_3 \qquad = \qquad K_1 \qquad \times \qquad K_{2'}$$

Finally, put everything together. We know, from the information provided, that K_1 is equal to 1.76×10^{-5}; and we also know, from Example 12-3, that

$K_{2'}$ is equal to $1/K_2$. The result is

$$K_3 = K_1 K_{2'} = \frac{K_1}{K_2} = \frac{1.76 \times 10^{-5}}{1.0 \times 10^{-14}} = 1.8 \times 10^9$$

EXAMPLE 12-7. Appraising K

PROBLEM: Look again at the reactions described in Example 12-6, and comment on the size of K in each case:

(1) $CH_3COOH(aq) + H_2O(\ell) \rightleftharpoons CH_3COO^-(aq) + H_3O^+(aq)$

$$K_1 = 1.76 \times 10^{-5}$$

(2) $H_2O(\ell) + H_2O(\ell) \rightleftharpoons H_3O^+(aq) + OH^-(aq)$

$$K_2 = 1.0 \times 10^{-14}$$

(3) $CH_3COOH(aq) + OH^-(aq) \rightleftharpoons CH_3COO^-(aq) + H_2O(\ell)$

$$K_3 = 1.8 \times 10^9$$

SOLUTION: A large value ($K \gg 1$) means that far more products than reactants are present at equilibrium, so that the reaction proceeds to the right. A small value ($K \ll 1$) means that the reaction goes to the left, with relatively few products evident in the final mixture. An intermediate value of K (near 1) means that the reaction comes to equilibrium with roughly equal amounts of products and reactants at the end.

The first reaction, the ionization of acetic acid, has a small equilibrium constant: 0.0000176. Relatively few CH_3COOH molecules lose a proton, and consequently the equilibrium mixture consists mostly of unionized CH_3COOH. There is just a smattering of CH_3COO^- and H_3O^+.

The second reaction, the self-ionization of water, proceeds with an even smaller equilibrium constant: 0.000000000000010. Only one water molecule in many millions will give up a proton to produce hydronium and hydroxide ions.

The third reaction, the neutralization of acid and base, comes to a different end. Here the equilibrium constant is huge (1,800,000,000), large enough to guarantee that the limiting reactant is entirely consumed. The reaction therefore goes to completion, with the push forward coming from the enormous value for $K_{2'}$ that governs the reuniting of H_3O^+ and OH^- (reaction $2'$, the reverse of 2):

(2') $H_3O^+(aq) + OH^-(aq) \rightleftharpoons 2H_2O(\ell)$ $K_{2'} = 1/K_2 = 1.0 \times 10^{14}$

QUESTION: Haven't we previously assumed that *all* reactions go to completion? We always claimed to predict the limiting stoichiometry of reaction from the coefficients of the balanced equation alone.

ANSWER: Yes. We have, until this chapter, naively assumed that all reactions proceed completely to the right, as if the equilibrium constant were infinitely large. Many times, too, K is indeed big enough to warrant that view, but now we know better than to make such an assumption without specific knowledge. We understand, instead, that: When a process comes to an end, the ratio of products to reactants is fixed by the value of the equilibrium constant. Only when K is exceedingly large is the reaction's stoichiometric promise completely fulfilled.

EXAMPLE 12-8. Appraising Q

PROBLEM: Expressed in concentrations, the equilibrium constant for the reaction

$$H_2(g) + I_2(g) \rightleftharpoons 2HI(g)$$

is 50.5 at 448°C. Suppose that a vessel at this temperature contains 5.00 M H_2, 3.00 M I_2, and 1.00 M HI. Is the mixture in equilibrium?

SOLUTION: Evaluate the reaction quotient, Q, and compare with $K = 50.5$:

$$Q = \frac{[HI]^2}{[H_2][I_2]} = \frac{(1.00)^2}{(5.00)(3.00)} = 0.0667 < K$$

With Q smaller than K, there are too many reactants and too few products. To attain equilibrium, the system must consume H_2 and I_2 to produce additional HI.

EXAMPLE 12-9. Using K

PROBLEM: Determine the concentrations of H_2, I_2, and HI when the mixture described in Example 12-8 comes to equilibrium.

SOLUTION: The initial concentrations are

$$H_2(g) \ + \ I_2(g) \ \rightleftharpoons \ 2HI(g)$$

$$5.00 \ M \quad 3.00 \ M \qquad 1.00 \ M$$

Now, assume that some unknown concentration of I_2 (call it x) is consumed during the approach to equilibrium. If so, then the stoichiometry of the reaction assures us that exactly twice as much HI ($2x$) will be produced. Likewise, the concentration of H_2 will decrease by x since one mole of H_2 reacts with one mole of I_2. At equilibrium, then, we have

$$H_2(g) \ + \ I_2(g) \ \rightleftharpoons \ 2HI(g)$$
$$5.00 - x \quad 3.00 - x \qquad 1.00 + 2x$$

and a corresponding expression for K:

$$K = \frac{[HI]^2}{[H_2][I_2]} = \frac{(1.00 + 2x)^2}{(5.00 - x)(3.00 - x)} = 50.5$$

We have only to solve for x.

To do so, first multiply both sides of the equation by the expression in the denominator:

$$(1.00 + 2x)^2 = 50.5(5.00 - x)(3.00 - x)$$

Second, expand the expressions in parentheses:

$$4x^2 + 4.00x + 1.00 = 50.5(x^2 - 8.00x + 15.0)$$

$$4x^2 + 4.00x + 1.00 = 50.5x^2 - 404x + 757.5$$

Third, collect the terms:

$$46.5x^2 - 408x + 756.5 = 0$$

Fourth, recognize that we have a quadratic equation of the form

$$ax^2 + bx + c = 0$$

with coefficients

$$a = 46.5$$

$$b = -408$$

$$c = 756.5$$

Fifth, solve the equation using the quadratic formula (see Appendix B):

$$x = \frac{-b \pm \sqrt{b^2 - 4ac}}{2a} = \frac{408 \pm 160.48}{93.0}$$

There are two solutions, as required for any quadratic equation:

$$x_+ = 6.11 \; M$$

$$x_- = 2.66 \; M$$

The first of these, x_+, we reject because it is *greater* than the limiting initial concentration of I_2, which was only 3.00 M. To consume more than the original amount is to do the impossible.

The second number, x_-, is the physically meaningful solution, and from it we obtain the concentrations at equilibrium:

$$[H_2] = 5.00 - x_- \;\; = 2.34 \; M$$

$$[I_2] = 3.00 - x_- \;\; = 0.34 \; M$$

$$[HI] = 1.00 + 2x_- = 6.32 \; M$$

To check the arithmetic, finally, we should substitute these values back into the expression for K (retaining additional digits to avoid a round-off error):

$$K = \frac{(6.3229)^2}{(2.3385)(0.3385)} = 50.5$$

EXAMPLE 12-10. Using K: A Shortcut

Sometimes there is an easier way.

PROBLEM: The equilibrium constant for the ionization of nitrous acid is 0.00045 at 25°C:

$$HNO_2(aq) \rightleftharpoons NO_2^-(aq) + H^+(aq) \qquad K_a = 0.00045$$

What is the equilibrium concentration of HNO_2 in a solution initially 1.0 M?

SOLUTION: The initial concentrations are

$$HNO_2 \rightleftharpoons NO_2^- + H^+$$
$$1.0 \; M \qquad 0 \qquad 0$$

If HNO_2 ionizes to the extent x, then the equilibrium concentrations are

$$HNO_2 \rightleftharpoons NO_2^- + H^+$$
$$1.0 - x \qquad x \qquad x$$

and the expression for K_a becomes

$$K_a = \frac{[H^+][NO_2^-]}{[HNO_2]} = \frac{x^2}{1.0 - x} = 0.00045$$

If we solve the resulting quadratic equation, as in Example 12-9, the correct solution proves to be $x = 0.021$ M:

$$x^2 = (0.00045)(1.0 - x)$$

$$x^2 + 0.00045x - 0.00045 = 0$$

$$x = 0.021, \quad -0.021$$

This time, however, there is a simpler way to get the same answer. We see from the equation

$$\frac{x^2}{1.0 - x} = 0.00045$$

that x is likely to be small compared with 1, since the equilibrium constant is itself so small. If true, then we can assume that

$$1.0 - x \approx 1.0$$

and hence

$$x^2 = 0.00045$$

A simple square-root operation gives us back again the value $x = 0.021$ M.

QUESTION: Would this trick work if the initial concentration were smaller, say 0.010 M rather than 1.0 M?

ANSWER: Less likely, for then the equilibrium equation would be

$$\frac{x^2}{0.010 - x} = 0.00045$$

and we would be forced to assume that x is small compared with 0.01. Is it? Is $0.010 - x \approx 0$? The results this time are $x = 0.0021$ for the quick solution and $x = 0.0019$ for the solution obtained from the full quadratic equation, an error greater than 10%. Since the initial concentration, 0.010 *M*, is only five times the size of *x*, the shortcut becomes less accurate.

EXAMPLE 12-11. Avoiding *K*: Evolution of a Gas

PROBLEM: Consider the reaction of sodium carbonate with hydrochloric acid:

$$Na_2CO_3(aq) + 2HCl(aq) \longrightarrow 2NaCl(aq) + H_2O(\ell) + CO_2(g)$$

Allowed to occur in the open air, this process goes to completion. Why?

SOLUTION: The reactants and products never come to a stable equilibrium. One of the products, carbon dioxide, is a gas that escapes into the open air as soon as it appears. This perpetual loss of product encourages the reactants to produce still more material in response, a consequence of Le Châtelier's principle. The system remains always just shy of equilibrium.

If initiated in a *closed* vessel, with all products and reactants in a forced coexistence, a reaction that produces a gas can progress only as far as its equilibrium constant allows. Exposed to the air, however, it goes to completion instead. Under the permanent stress of having too many reactants and too few products ($Q < K$), the process continues until there is nothing left to react. Spared the need to compute the equilibrium values, we determine the amount of products by stoichiometry alone (as first described in Chapter 2). The magnitude of *K*, large or small, is irrelevant.

EXAMPLE 12-12. Stress: Changing the Mix

Another demonstration of Le Châtelier's principle.

PROBLEM: Under the conditions stated in Example 12-9, we found that a mixture of $H_2(g)$, $I_2(g)$, and $HI(g)$ can exist at equilibrium (448°C) in the following concentrations:

$$H_2(g) + I_2(g) \rightleftharpoons 2HI(g) \qquad K = 50.5$$

$$2.34\ M \quad 0.34\ M \qquad 6.32\ M$$

What would happen if the concentration of HI were suddenly increased to 10.00 *M*?

SOLUTION: There would be a move toward a new equilibrium, because the reaction quotient (Q) now exceeds the equilibrium constant ($K = 50.5$):

$$Q = \frac{[HI]^2}{[H_2][I_2]} = \frac{(10.00)^2}{(2.34)(0.34)} = 126 > K$$

The extra HI disturbs the equilibrium and puts the reaction under stress. To relieve the stress, the system must destroy enough HI to bring the reaction quotient down to its equilibrium value of 50.5.

The procedure is the same as in Example 12-9. First, we note the initial conditions:

$$H_2(g) \ + \ I_2(g) \ \rightleftharpoons \ 2HI(g)$$

$$2.34 \ M \quad 0.34 \ M \qquad 10.00 \ M$$

Second, the conditions at equilibrium:

$$H_2(g) \ + \ I_2(g) \ \rightleftharpoons \ 2HI(g)$$

$$2.34 + x \quad 0.34 + x \qquad 10.00 - 2x$$

Third, we solve for x in the equation for K:

$$K = \frac{[HI]^2}{[H_2][I_2]} = \frac{(10.00 - 2x)^2}{(2.34 + x)(0.34 + x)} = 50.5$$

$$46.5x^2 + 175.34x - 59.8222 = 0$$

$$x_+ = 0.315$$

$$x_- = -4.09$$

Of the two possible roots, we reject x_- because it indicates negative concentrations for both H_2 and I_2. The acceptable solution comes rather from x_+, which produces equilibrium concentrations of

$$[H_2] = 2.34 + x_+ \quad = 2.66 \ M$$

$$[I_2] = 0.34 + x_+ \quad = 0.66 \ M$$

$$[HI] = 10.00 - 2x_+ = 9.37 \ M$$

The stress of having too much HI is thus relieved, and the reaction is stalemated once again at a new equilibrium.

EXAMPLE 12-13. Stress: Pressure

More Le Châtelier.

PROBLEM: For each of the reactions written below, indicate whether a reduction in volume will shift the equilibrium toward products, toward reactants, or not at all. Assume that each process occurs in a closed vessel.

(a) $CO(g) + 3H_2(g) \rightleftharpoons CH_4(g) + H_2O(g)$

(b) $CO_2(g) + C(s) \rightleftharpoons 2CO(g)$

(c) $I_2(g) + Br_2(g) \rightleftharpoons 2IBr(g)$

(d) $2H_2O(g) \rightleftharpoons 2H_2(g) + O_2(g)$

SOLUTION: The stress comes from an increase in pressure, occasioned by a decrease in volume. Confined to a smaller volume, the gas exerts more pressure; and for a system already at equilibrium, any additional pressure creates a stress that must be relieved. Since gas pressure rises and falls with the number of moles, the reaction shifts in the direction toward fewer molecules and hence lower pressure.

(a) $CO(g) + 3H_2(g) \rightleftharpoons CH_4(g) + H_2O(g)$. More products. There are four moles of gas on the left, two on the right. Subjected to higher pressure, the reaction moves toward the right. A higher proportion of products exists at the new equilibrium.

(b) $CO_2(g) + C(s) \rightleftharpoons 2CO(g)$. Of the two reactants, only one—carbon dioxide—is a gas. With one mole of gaseous reactants but two moles of gaseous products, the perturbed system moves to produce more reactants. It goes to the left.

(c) $I_2(g) + Br_2(g) \rightleftharpoons 2IBr(g)$. No effect. There are two moles of gaseous reactants and two moles of gaseous products.

(d) $2H_2O(g) \rightleftharpoons 2H_2(g) + O_2(g)$. More reactants. Pressure is dissipated whenever three molecules of product are converted into only two molecules of reactant.

QUESTION: Does the value of K change in response to higher pressure?

ANSWER: No. K depends only on the temperature, not pressure. The individual amount of each component changes, but the collectively determined ratios at equilibrium all remain the same.

EXAMPLE 12-14. When Pressure Is Not Stressful

PROBLEM: Suppose that a gaseous mixture of ammonia, hydrogen, and nitrogen comes to equilibrium at a total pressure of 1 atm. In what direction would the equilibrium shift if helium gas were forced into the system at a pressure of 10 atm. Assume that temperature and volume remain the same.

SOLUTION: The equilibrium would not shift at all, because helium has no effect on the partial pressures of N_2, H_2, and NH_3 in the reaction

$$N_2(g) + 3H_2(g) \rightleftharpoons 2NH_3(g)$$

Dalton's law of partial pressures ensures that each component in the mixture is responsible for its own partial pressure, which is determined individually by three variables: (1) Volume (unchanged). (2) Temperature (unchanged). (3) The number of moles of that component's *own kind* (unchanged, because helium reacts with neither nitrogen nor hydrogen nor ammonia). The three partial pressures (N_2, H_2, NH_3) that actually determine the reaction quotient stay the same.

It is a system under higher pressure but not greater stress. The equilibrium remains undisturbed.

EXAMPLE 12-15. Stress: Temperature and the Equilibrium Constant

A final look at the ammonia reaction.

PROBLEM: The equilibrium constant for the transformation

$$N_2(g) + 3H_2(g) \rightleftharpoons 2NH_3(g)$$

goes from $K_c(T) = 0.50$ at 400°C to $K_c(T) = 0.16$ at 450°C. Do the surroundings grow warmer or cooler when ammonia is produced by this process?

SOLUTION: If the reaction releases heat (is exothermic), then the surroundings will grow warmer. If the reaction consumes heat from its environment (is endothermic), the surroundings will grow colder. Which is it?

See how the equilibrium constant behaves: $K_c(T)$ is smaller at the higher temperature. If a mixture in equilibrium at 400°C were suddenly heated to 450°C, then the system would find itself with too many products and too few reactants. Whereas the original equilibrium mixture had only to satisfy the constraint $K_c(400°C) = 0.50$, the new conditions call for new proportions: $K_c(450°C) = 0.16$. Products must be reconverted into reactants if the revised, smaller target is to be met.

At the same time, the system at higher temperature is under stress from additional thermal energy—more heat—and so it must dissipate the heat and thereby relieve the stress. The smaller equilibrium constant tells us, then, that the reverse reaction (which is favored at the higher temperature) evidently *consumes* heat. To go backward, from NH_3 back to N_2 and H_2, is to take the endothermic route. The reverse reaction sucks heat out of the surroundings, thus relieving the stress of a higher temperature.

Hence the forward reaction, the *production* of ammonia, is exothermic:

$$N_2(g) + 3H_2(g) \rightleftharpoons 2NH_3(g) + heat$$

When N_2 and H_2 react, heat is produced along with NH_3. The surroundings grow hotter as they absorb the heat of reaction; and the hotter the environment grows, the more difficult it becomes for the chemical reaction to continue dumping out heat.

Specific result? Less ammonia, and consequently less heat, is produced at the higher temperature. The equilibrium constant is smaller.

General conclusion? For an exothermic reaction, K decreases with T.

EXERCISES

1. What does it mean for a reaction to "go to completion"? Under what special circumstances will the limiting reactant be completely consumed?

2. How and when does a reaction come to an end? Does the process simply stop? What action occurs on the molecular level? What activity is visible macroscopically?

3. Describe chemical equilibrium as a dynamic phenomenon. How *final* is the final state of equilibrium, so called? Once a system reaches equilibrium, is its state still susceptible to external manipulation?

4. Suppose we know the balanced equation and equilibrium constant for a certain reaction at a certain temperature: a number, $K(T)$. From that number, nothing more, can we compute the concentration of each component present at equilibrium? If not, what further information is required? What can we say for sure, even without additional information?

5. Consider the generalized reaction

$$a\text{A} + b\text{B} \rightleftharpoons c\text{C} + d\text{D}$$

and its equilibrium expression

$$K(T) = \frac{[\text{C}]^c[\text{D}]^d}{[\text{A}]^a[\text{B}]^b}$$

Is there a unique set of equilibrium concentrations, some *one* particular combination of $[\text{A}]_{eq}$, $[\text{B}]_{eq}$, $[\text{C}]_{eq}$, $[\text{D}]_{eq}$ and no other?

6. Look again at the mass-action expression in the exercise above. The notation $K(T)$ reminds us explicitly that the equilibrium constant, K, is a function of temperature, yet nowhere in the equation itself does the temperature appear. Concentrations, yes; temperature, no. In what way, then, is K a single-valued function of T, and why is it not a function of concentration?

7. Which of the following actions will alter the numerical value of K?

(a) The concentrations of reactants and products are changed.
(b) The volume of the system is changed.
(c) The pressure of the system is changed.
(d) The temperature of the system is changed.
(e) A catalyst is added to the system.

8. Imagine some arbitrary exothermic reaction

$$2A(g) + 4B(g) \rightleftharpoons C(g) + D(g) + \text{heat} \qquad K = 1$$

at equilibrium in the gas phase; and, just as arbitrarily, let K be 1. Now examine, in turn, each of the following stresses to the system:

(a) The temperature is increased.
(b) The temperature is decreased.
(c) The pressure is increased.
(d) The pressure is decreased.

In which direction, if any, does the equilibrium shift? When equilibrium is restored, are there more products or less than before? Is K more than 1, less than 1, or equal to 1 at the new equilibrium?

9. Repeat the preceding exercise, this time for an endothermic reaction of the type

$$2A(g) + 4B(g) + \text{heat} \rightleftharpoons C(g) + D(g) \qquad K = 1$$

Same stresses:

(a) The temperature is increased.
(b) The temperature is decreased.
(c) The pressure is increased.
(d) The pressure is decreased.

Same questions: In which direction, if any, does the equilibrium shift under each set of conditions? When equilibrium is restored, are there more products or less than before? Is K more than 1, less than 1, or equal to 1 at the new equilibrium?

10. In the Born-Haber synthesis of ammonia at high temperature and high pressure, a catalyst is used to make the reaction

$$N_2(g) + 3H_2(g) \rightleftharpoons 2NH_3(g)$$

go faster. Now consider: Does *faster* also mean *more product*? Does the catalyst change either the equilibrium constant itself or the composition of the final mixture?

11. Assume that we know both the form and numerical value of the equilibrium constant for some reaction, as well as the stoichiometry

of the process. If we measure the final concentrations at equilibrium, can we then determine what those concentrations were at the *start* of reaction?

12. Distinguish between the equilibrium constant and the reaction quotient.

13. Classify each equilibrium as homogeneous or heterogeneous:

    (a) $NaCl(s) \rightleftharpoons Na^+(aq) + Cl^-(aq)$
    (b) $CF_3COOH(aq) + H_2O(\ell) \rightleftharpoons CF_3COO^-(aq) + H_3O^+(aq)$
    (c) $SiO_2(\ell) + 2C(s) \rightleftharpoons Si(\ell) + 2CO(g)$
    (d) $2NO(g) + O_2(g) \rightleftharpoons 2NO_2(g)$
    (e) $Fe(s) + 2Ag^+(aq) \rightleftharpoons Fe^{2+}(aq) + 2Ag(s)$
    (f) $C_2H_5OH(\ell) \rightleftharpoons C_2H_5OH(g)$

14. Classify each equilibrium as homogeneous or heterogeneous:

    (a) $Ag^+(aq) + 2NH_3(aq) \rightleftharpoons Ag(NH_3)_2^+(aq)$
    (b) $Cl(g) + O_3(g) \rightleftharpoons ClO(g) + O_2(g)$
    (c) $4Fe^{2+}(aq) + O_2(g) + 4H_2O(\ell) + 2nH_2O(\ell) \rightleftharpoons$
    $2Fe_2O_3 \cdot nH_2O(s) + 8H^+(aq)$
    (d) $NH_4Cl(aq) + NaOH(aq) \rightleftharpoons NH_3(g) + H_2O(\ell) + NaCl(aq)$
    (e) $H_2O(\ell) \rightleftharpoons H_2O(s)$
    (f) $2NH_3(aq) + OCl^-(aq) \rightleftharpoons N_2H_4(aq) + Cl^-(aq) + H_2O(\ell)$

15. Write equilibrium expressions for the following reactions in terms of both concentration (K_c) and pressure (K_p):

    (a) $CO(g) + 2H_2(g) \rightleftharpoons CH_3OH(g)$
    (b) $CH_3OH(g) + \frac{3}{2}O_2(g) \rightleftharpoons CO_2(g) + 2H_2O(g)$
    (c) $B_2H_6(g) + 3O_2(g) \rightleftharpoons B_2O_3(s) + 3H_2O(g)$
    (d) $4H_3BO_3(s) \rightleftharpoons H_2B_4O_7(s) + 5H_2O(g)$

    Show, explicitly, how K_c and K_p are related by some power of RT, and then calculate the ratio K_c/K_p for each process at 600°C.

16. More of the same. Write the mass-action expressions (K_c, K_p) and compute K_c/K_p at 600°C:

    (a) $C_2H_5OH(g) + 3O_2(g) \rightleftharpoons 2CO_2(g) + 3H_2O(g)$
    (b) $CH_4(g) + 2O_2(g) \rightleftharpoons CO_2(g) + 2H_2O(g)$
    (c) $Cl(g) + O_3(g) \rightleftharpoons ClO(g) + O_2(g)$
    (d) $NH_4NO_3(s) \rightleftharpoons N_2O(g) + 2H_2O(g)$

17. Three different reactions,

$$(1)\ Na_2CO_3(s) \rightleftharpoons Na_2O(s) + CO_2(g)$$

$$(2)\ CaCO_3(s) \rightleftharpoons CaO(s) + CO_2(g)$$

$$(3)\ CO_2(s) \rightleftharpoons CO_2(g)$$

share the same algebraic expression for K:

$$K_p = P_{CO_2}$$

How can that be? Interpret the statement, "the equilibrium constant, K_p, is equal to the pressure of carbon dioxide."

18. Here are two reactions, each generating water as a by-product:

(a) $4NH_3(g) + 5O_2(g) \rightleftharpoons 4NO(g) + 6H_2O(g)$
(b) $Cu^{2+}(aq) + 2NH_2OH(aq) \rightleftharpoons$
$$Cu(s) + N_2(g) + 2H_2O(\ell) + 2H^+(aq)$$

Write the corresponding expressions for K, and note that H_2O appears in the numerator of just one of them, not both. Why?

19. Write K_c for each reaction:

(a) $2NH_3(aq) + OCl^-(aq) \rightleftharpoons N_2H_4(aq) + Cl^-(aq) + H_2O(\ell)$
(b) $N_2H_4(aq) + H_2O(\ell) \rightleftharpoons N_2H_5^+(aq) + OH^-(aq)$
(c) $SO_2(g) + H_2O(\ell) \rightleftharpoons H_2SO_3(aq)$
(d) $H_2SO_3(aq) + H_2O(\ell) \rightleftharpoons HSO_3^-(aq) + H_3O^+(aq)$
(e) $HSO_3^-(aq) + H_2O(\ell) \rightleftharpoons SO_3^{2-}(aq) + H_3O^+(aq)$
(f) $HSO_3^-(aq) + H_2O(\ell) \rightleftharpoons H_2SO_3(aq) + OH^-(aq)$

20. Write K_c for each reaction:

(a) $Fe^{3+}(aq) + 6CN^-(aq) \rightleftharpoons Fe(CN)_6^{3-}(aq)$
(b) $BaSO_4(s) \rightleftharpoons Ba^{2+}(aq) + SO_4^{2-}(aq)$
(c) $Ca_3(PO_4)_2(s) \rightleftharpoons 3Ca^{2+}(aq) + 2PO_4^{3-}(aq)$
(d) $Ca^{2+}(aq) + 2F^-(aq) \rightleftharpoons CaF_2(s)$
(e) $H_2O(g) \rightleftharpoons H_2O(\ell)$

21. The equilibrium constant for the reaction

$$(1)\ 4HCl(g) + O_2(g) \rightleftharpoons 2Cl_2(g) + 2H_2O(g)$$

is $K_p = 13.3$ at 753 K. Compute the numerical value of K_p when the same reaction is written in the following ways:

(2) $HCl(g) + \frac{1}{4}O_2(g) \rightleftharpoons \frac{1}{2}Cl_2(g) + \frac{1}{2}H_2O(g)$

(3) $2HCl(g) + \frac{1}{2}O_2(g) \rightleftharpoons Cl_2(g) + H_2O(g)$

Write the mass-action law for each of the three balanced equations, and explain why there is no inconsistency. Do the proportions of products and reactants at equilibrium depend in any way on our choice of equations (1), (2), or (3)?

22. The equilibrium constant K_p for the process

$$2NO(g) + Br_2(g) \rightleftharpoons 2NOBr(g)$$

is 2.40 at 100°C. Write the mass-action expression for the reverse process,

$$2NOBr(g) \rightleftharpoons 2NO(g) + Br_2(g)$$

and determine the corresponding value of K_p at the same temperature.

23. The equilibrium constant for the self-ionization of water,

$$H_2O(\ell) + H_2O(\ell) \rightleftharpoons H_3O^+(aq) + OH^-(aq)$$

is 1.0×10^{-14} at 25°C. Write the mass-action expression for the neutralization,

$$H_3O^+(aq) + OH^-(aq) \rightleftharpoons 2H_2O(\ell)$$

and calculate the corresponding value of $K(25°C)$.

24. Given equilibrium constants for reactions 1 and 2, compute K for reaction 3:

(1) $H_2O(\ell) + H_2O(\ell) \rightleftharpoons H_3O^+(aq) + OH^-(aq)$

$$K_1 = 1.0 \times 10^{-14}$$

(2) $NH_4^+(aq) + H_2O(\ell) \rightleftharpoons NH_3(aq) + H_3O^+(aq)$

$$K_2 = 5.7 \times 10^{-10}$$

(3) $NH_3(aq) + H_2O(\ell) \rightleftharpoons NH_4^+(aq) + OH^-(aq)$

$$K_3 = ?$$

Calculate, as well, K for reaction 4:

(4) $NH_4^+(aq) + OH^-(aq) \rightleftharpoons NH_3(aq) + H_2O(\ell)$

$$K_4 = ?$$

25. The equilibrium constant for the neutralization of acetic acid is 1.8×10^9 at 25°C:

    (1) $CH_3COOH(aq) + OH^-(aq) \rightleftharpoons CH_3COO^-(aq) + H_2O(\ell)$

    $$K_1 = 1.8 \times 10^9$$

    Combine this process with the aforementioned self-ionization of water,

    (2) $H_2O(\ell) + H_2O(\ell) \rightleftharpoons H_3O^+(aq) + OH^-(aq)$

    $$K_2 = 1.0 \times 10^{-14}$$

    to compute K for the ionization of acetic acid:

    (3) $CH_3COOH(aq) + H_2O(\ell) \rightleftharpoons CH_3COO^-(aq) + H_3O^+(aq)$

    $$K_3 = ?$$

26. Chloroacetic acid dissociates in aqueous solution to produce the chloroacetate and hydronium ions:

    $$CH_2ClCOOH(aq) + H_2O(\ell) \rightleftharpoons CH_2ClCOO^-(aq) + H_3O^+(aq)$$

    Compute K at 25°C, given the following set of equilibrium concentrations:

    $$[CH_2ClCOOH] = 0.0888 \ M$$

    $$[CH_2ClCOO^-] = 0.0112 \ M$$

    $$[H_3O^+] = 0.0112 \ M$$

27. Begin with an aqueous solution of acetic acid (CH_3COOH) and an aqueous solution of chloroacetic acid ($CH_2ClCOOH$), each initially at the same concentration. At equilibrium, which of the two mixtures will contain the greater concentration of hydronium ions? Use the equilibrium constants derived in the two exercises preceding.

28. The following equilibrium concentrations were measured for aqueous solutions of benzoic acid ($HC_7H_5O_2$) and formic acid ($HCHO_2$) at 25°C:

Benzoic Acid	Formic Acid
$[HC_7H_5O_2] = 2.0 \times 10^{-1} \ M$	$[HCHO_2] = 2.78 \times 10^{-2} \ M$
$[C_7H_5O_2^-] = 3.6 \times 10^{-3} \ M$	$[CHO_2^-] = 2.24 \times 10^{-3} \ M$
$[H_3O^+] = 3.6 \times 10^{-3} \ M$	$[H_3O^+] = 2.24 \times 10^{-3} \ M$

    Calculate K for each system. Which acid is stronger?

29. Into a 1.00-L vessel are introduced 1.00 mol $PCl_3(g)$, 1.00 mol $PCl_5(g)$, and 1.00 mol $Cl_2(g)$. Maintained at 523 K, the system reacts according to the equation

$$PCl_5(g) \rightleftharpoons PCl_3(g) + Cl_2(g)$$

and eventually reaches equilibrium. The concentration of $Cl_2(g)$ then becomes fixed at 1.20 mol L^{-1} from that point on. (a) Compute the equilibrium concentrations of the other two components. (b) Compute K_c at 523 K. (c) Compute K_p at 523 K.

30. The equilibrium constant for the dissociation of hydrochloric acid,

$$HCl(aq) + H_2O(\ell) \rightleftharpoons H_3O^+(aq) + Cl^-(aq)$$

is approximately 10^7 at room temperature. Calculate the equilibrium concentrations of HCl, H_3O^+, and Cl^- under each set of conditions: (a) 1.00 mol HCl is dissolved in a total volume of 0.10 L H_2O at 25°C. (b) 1.00 mol HCl is dissolved in a total volume of 1.00 L H_2O at 25°C. (c) 1.00 mol HCl is dissolved in a total volume of 10.00 L H_2O at 25°C.

31. Repeat the exercise above, but with this one difference: Compute the number of *moles* of H_3O^+ and Cl^- present at equilibrium when (a) 1.00 mol HCl is dissolved in a total volume of 0.10 L H_2O at 25°C; (b) 1.00 mol HCl is dissolved in a total volume of 1.00 L H_2O at 25°C; (c) 1.00 mol HCl is dissolved in a total volume of 10.00 L H_2O at 25°C.

32. The equilibrium constant for the dissociation of hydrazoic acid,

$$HN_3(aq) + H_2O(\ell) \rightleftharpoons H_3O^+(aq) + N_3^-(aq)$$

is 1.9×10^{-5} at 25°C. Calculate the equilibrium concentrations of HN_3, H_3O^+, and N_3^- under each set of conditions: (a) 1.00 mol HN_3 is dissolved in a total volume of 0.10 L H_2O at 25°C. (b) 1.00 mol HN_3 is dissolved in a total volume of 1.00 L H_2O at 25°C. (c) 1.00 mol HN_3 is dissolved in a total volume of 10.00 L H_2O at 25°C.

33. Repeat the exercise above, but with this one difference: Compute the number of *moles* of H_3O^+ and N_3^- present at equilibrium when (a) 1.00 mol HN_3 is dissolved in a total volume of 0.10 L H_2O at 25°C; (b) 1.00 mol HN_3 is dissolved in a total volume of 1.00 L H_2O at 25°C; (c) 1.00 mol HN_3 is dissolved in a total volume of 10.00 L H_2O at 25°C.

34. Summarize the results of the previous four exercises to show, qualitatively, how the equilibrium attained by a strong acid (HCl) differs from the equilibrium attained by a weak acid (HN_3).

35. A question perhaps not to be answered right away, but begging to be asked just the same: Hydrochloric acid, which dissociates into H^+ and Cl^-, is a strong acid. Its equilibrium constant for the ionization is large. Hydrazoic acid, which dissociates into H^+ and N_3^-, is a weak acid. The equilibrium constant is small. What's the difference? One acid produces a mononegative anion, and so does the other. Chloride, Cl^-; nitride, N_3^-. Why is HCl strong and HN_3^- weak? . . . Perplexed? Chapter 16 addresses the question of relative strength and weakness among acids and bases.

36. The equilibrium constant for the reaction

$$Hg_2Cl_2(s) \rightleftharpoons Hg_2^{2+}(aq) + 2Cl^-(aq)$$

is $K(25°C) = 1.5 \times 10^{-18}$. Calculate the equilibrium concentrations of each ion in an aqueous solution at 25°C.

37. The equilibrium constant for the reaction

$$AgCl(s) \rightleftharpoons Ag^+(aq) + Cl^-(aq)$$

is $K(25°C) = 1.8 \times 10^{-10}$. Calculate, for each ion, the number of moles able to equilibrate with solid AgCl in solutions having the following volumes:

(a) 1.00 L   (b) 1.00×10^{-3} L   (c) 1.00×10^3 L

38. At 25°C, the equilibrium constant for the reaction

$$CO_2(g) + H_2(g) \rightleftharpoons CO(g) + H_2O(\ell)$$

is $K_p = 3.2 \times 10^{-4}$. Assume that carbon dioxide, hydrogen, and carbon monoxide are confined in a closed volume at 25°C, each gas exerting a partial pressure of 1.00 atm at the start of reaction. (a) Compute the reaction quotient. Will the evolution toward equilibrium generate more products or less products? (b) Write algebraic expressions to represent the *initial* pressure, the *change* in pressure, and the *final* pressure of each substance. (c) Solve the resulting equation for the equilibrium pressure of each gaseous component.

39. Allow the reaction

$$CO_2(g) + H_2(g) \rightleftharpoons CO(g) + H_2O(\ell) \qquad K_p(25°C) = 3.2 \times 10^{-4}$$

to reach equilibrium so that the final pressures are

$$P_{CO} = 3.2 \times 10^{-4} \text{ atm}$$
$$P_{CO_2} = 1.0 \text{ atm}$$
$$P_{H_2} = 1.0 \text{ atm}$$

at 25°C. Then halve the volume. (a) Calculate the reaction quotient. In which direction must the process move to restore the proper mass-action ratio? (b) Calculate the partial pressure of each component after the system regains its equilibrium.

40. Suppose that gaseous Br_2, I_2, and IBr all come to equilibrium at 350 K in a closed vessel ($V = 100.0$ L), whereupon the amounts of each component in the reaction

$$I_2(g) + Br_2(g) \rightleftharpoons 2\,IBr(g)$$

become fixed at the following values:

$$Br_2 \quad 0.250 \text{ mol}$$
$$I_2 \quad 1.000 \text{ mol}$$
$$IBr \quad 8.972 \text{ mol}$$

(a) Calculate K_c and K_p. (b) Are both numbers the same? Does K for this reaction depend upon volume in any way?

41. Same reaction and same conditions as above, but this time the equilibrium quantities are

$$Br_2 \quad 1.85 \text{ mol}$$
$$I_2 \quad 6.76 \text{ mol}$$
$$IBr \quad 63.44 \text{ mol}$$

(a) Recalculate K_c and K_p. (b) Is there any change? Why or why not?

42. Now let the same three gases be present in equal amounts:

$$Br_2 \quad 20.00 \text{ mol}$$
$$I_2 \quad 20.00 \text{ mol}$$
$$IBr \quad 20.00 \text{ mol}$$

(a) Is the mixture in equilibrium at 350 K? (b) If not, will the reaction proceed to the right (formation of IBr) or to the left (formation of I_2 and Br_2)? Use the equilibrium constant previously computed to make the assessment.

43. Stay with the process

$$I_2(g) + Br_2(g) \rightleftharpoons 2IBr(g)$$

at 350 K, and consider three different mixtures of reactants and products:

	MIXTURE 1	MIXTURE 2	MIXTURE 3
Br_2	10.00 mol	1.00 mol	1000.00 mol
I_2	10.00 mol	100.00 mol	0.10 mol
IBr	10.00 mol	10.00 mol	10.00 mol

Using the value of K derived in the exercises above, note first that each mixture is out of equilibrium; and note further that each mixture has the same reaction quotient (what is it?). Now observe: The three systems all begin with the same amount of product (10.00 mol IBr), but at equilibrium the final amounts of IBr will differ widely—even though each mixture starts out with the same reaction quotient. Explain why.

44. Taking again the three out-of-equilibrium systems described in the previous exercise, describe how each of the following external actions will affect the final equilibrium attained:

    (a) Decrease the volume of the container.
    (b) Increase the volume of the container.
    (c) Decrease the temperature.
    (d) Increase the temperature.
    (e) Remove IBr.
    (f) Add IBr.
    (g) Remove Br_2.
    (h) Add Br_2.
    (i) Remove I_2.
    (j) Add I_2.

    Will there be more, less, or the same amounts of Br_2, I_2, and IBr? If a shift in position occurs, will the reaction move toward the right or toward the left? Will the equilibrium constant increase, decrease, or remain the same? Note that the process is endothermic in the forward direction.

45. Upon reaching equilibrium, each of these gas-phase reactions is suddenly disturbed by a reduction in the system's volume:

    (1) $2NO(g) + 2H_2(g) \rightleftharpoons N_2(g) + 2H_2O(g)$

    (2) $2NO(g) \rightleftharpoons N_2(g) + O_2(g)$

    (3) $2NOBr(g) \rightleftharpoons 2NO(g) + Br_2(g)$

(a) Which processes (1, 2, or 3) are driven out of equilibrium?
(b) In what direction will the equilibrium shift in each of the perturbed systems? Will there be more or less product when equilibrium is reestablished?

46. Something new to consider: the difference between (1) a true state of equilibrium and (2) a nonequilibrium *steady state*.

    On the one hand, a system in equilibrium is self-sustaining. It supports itself, all alone, presumably forever. The counterbalanced processes persist with no help from the outside, and all macroscopic properties cease to change.

    Not quite the same for a steady state: A system in a nonequilibrium steady state is *not* self-sufficient, even though its macroscopic properties do seem to remain constant. The steadiness comes at a price, because the system maintains itself only by actively exchanging matter and energy with the outside. Take away its fuel, hold back its waste, close its entrances, block its exits—leave the system to itself, unsupported and alone—and the steady state dies.

    Which of these states, then, are truly in equilibrium, and which are actively maintained by forces outside the system? Which will persist indefinitely, and which will eventually give out?

    (a) Two arms wrestling on a tabletop, neither able to budge the other.
    (b) A chemical reaction in a closed vessel, with all concentrations conforming to their mass-action ratios.
    (c) A process in which fresh reactants are added just as the newly formed products escape. The concentrations of reactants and products never deviate from their equilibrium ratios.
    (d) A living cell operating at a fixed temperature with fixed concentrations of certain ions.

13

Energy, Heat, and Chemical Change

13-1. Work and Heat
13-2. The First Law of Thermodynamics
13-3. Energy and Enthalpy
13-4. Thermochemical Computations
13-5. Exothermic and Endothermic Processes
REVIEW AND GUIDE TO PROBLEMS
EXERCISES

Not without irony is industrial society preoccupied with energy and its conservation—ironic because energy stands as one of the great constants of the universe, never to be augmented or diminished. Its totality is conserved, strictly and without exception, throughout all changes however extreme. There is no more or less energy in the universe today than yesterday, nor will there be tomorrow. No new energy is ever created and no old energy is ever destroyed.

But energy constantly *switches* from one form to another, by processes both natural and man-made. It appears and reappears in a variety of guises: kinetic energy one moment, potential the next; now electrical, now thermal; even, most profoundly, as mass itself. Although the total quantity stays fixed, its distribution is ever-changing and always a matter of compelling importance. Understanding energy and its transformations, harnessing its capacity for useful work, and predicting the outcome of reactions are the goals of chemical thermodynamics.

13-1. WORK AND HEAT

Energy is an ability to do work, to move something. We do work on a body (Figure 13-1) by pushing it over some distance; the harder and

461

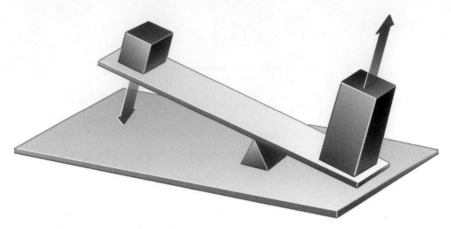

FIGURE 13-1. Work: the application of force over distance, illustrated by a simple lever. Movement of a mass downward, instigated by a force, provides the work to raise a heavier mass (at the right) against gravity. For a fuller discussion of work and energy, see the parallel treatment in Chapter 1.

longer the push, the more work is done. Hence the work expended in a given direction has dimensions of force (a push) times distance (a displacement),

$$w = Fd$$

as we saw first in Chapter 1. Here the natural units are kg m^2 s^{-2}, or joules (J), understandable directly from Newton's second law of motion. Recall what it says: A mechanical force is equal to mass (kg) $\times$ acceleration (m s^{-2}). Force multiplied by distance gives us kg m^2 s^{-2}.

The work is positive (done *on* the particle) if the motion is in the same direction as the push, and negative (done *by* the particle) if the motion is in the opposite direction. If *we* move the particle, then *we* do the work and call it positive. If the particle moves us, then *it* does the work and we call the result negative.

Just a few basic forces in nature give rise to all kinds of work and energy. To raise an object against the force of gravity requires gravitational work: the mass of the object (m) times the acceleration due to gravity (g) times the change in height (h). To move a charged particle in an electric field requires electrical work: the charge (q) times the difference in electrical potential (or voltage, V). To sustain a body in motion requires kinetic energy: $\frac{1}{2}mv^2$, where v is the magnitude of the velocity—the speed.

Electrons in a molecule have both kinetic energy and potential energy, the latter arising from electrical forces and governed by quantum mechanics. Potential energy, by its definition (Chapter 1), is the energy inherent in a body's position, and we saw already in Chapter 11 how whole molecules are attracted or repelled according to their *intermolecular* potential energy at various distances.

The list goes on and on. There is elastic energy in a spring, radiant energy in light, chemical energy in a reaction, energy in an atomic nucleus, and energy in the mass of a proton. All different, yet fundamentally all the same. They are all interconvertible and indestructible.

An example: Consider, as in Figure 13-2, the exchange of potential and kinetic energy that enables a pendulum to swing back and forth. The bob accelerates on the way down, slows on the way up, pauses momentarily at the top of its arc, and then repeats the whole sequence in the opposite direction. Motionless at the two extremes, the pendulum at these turning points has no kinetic energy; all its energy is stored as gravitational potential (mgh), a currency fully convertible to the kinetic form $\frac{1}{2}mv^2$. At every point on the trajectory thereafter, part of the stored potential energy is traded for the kinetic energy of motion. The conversion is complete at the low point of the arc, where the bob moves with maximum speed and with zero potential energy. At that location and all others, however, the sum of the kinetic and potential energies is always the

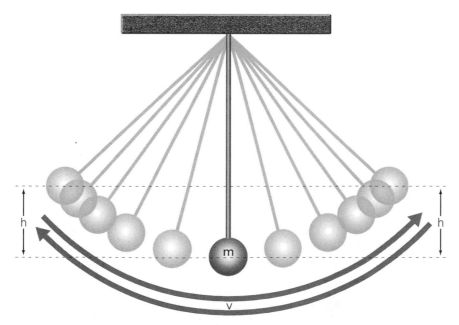

FIGURE 13-2. The motion of a pendulum. Kinetic and potential energy are interconverted during each swing of the bob, from top (all potential) to bottom (all kinetic). Dissipative forces eventually bring the orderly macroscopic motion to a halt, redistributing the original energy into the disorderly motion of the microscopic particles.

same. The total energy is conserved, and so we expect the pendulum to oscillate forever with undiminished amplitude.

Yet it never happens that way. Pendulums in the real world continually rub against the air, and inevitably they slow down and stop. All the original energy, all of that supposedly constant sum of kinetic and potential energy, is eventually given up to the atoms within. Slowly but inevitably, the moving object disperses the entire amount. Not for long do the atoms all work together, like oarsmen pulling in the same direction; instead, little by little, the pendulum dribbles its energy into the chaotically disordered motion of countless unseen particles. The resulting transfer of thermal energy—heat—produces a rise in temperature, reflecting a change in the internal kinetic energy.

With that, too, we come to an intellectual crossroads between macroscopic and microscopic. Viewed mechanically, in the sense of Figure 13-3, a flow of heat reduces to just ordinary work invested in the random motion of atoms and molecules. And certainly it does, but now ask yourself this one question just the same: Must we be so literal, so insistent on a fantastically complicated microscopic picture of heat, temperature, and molecular motion? Perhaps not. As an alternative, a macroscopic and *particle-free* sense of heat predates even the atomic theory and requires, moreover, no microscopic justification for its basis. Since thermodynamics makes no

FIGURE 13-3. A flow of heat, pictured mechanically as inciting the random movement of electrons, atoms, and molecules: microscopic work, disorganized and unseen. Unlike macroscopic work, where all parts move the same way, no external motion develops from this uncoordinated internal agitation.

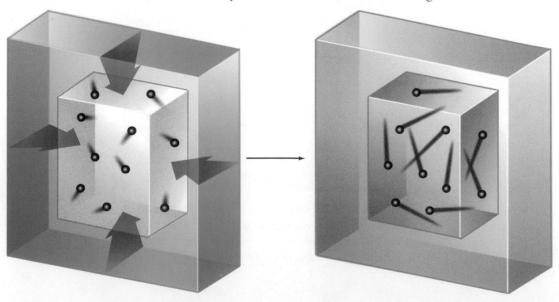

mention of atoms and molecules, the macroscopic theory would stand unaffected even by the complete demise of quantum mechanics—and therein lies its power and generality. We can gain some of that power ourselves by relaxing, if only temporarily, our belief in a particulate view of matter. So let that be our plan for the moment: to approach heat as a macroscopic phenomenon.

Realize that work and heat began as two separate ideas. Work, on the one hand, was perceived as a mechanical process involving the application of force over a distance. Work evoked images of pulleys pulling and gears turning, water falling, projectiles flying. Heat, on the other hand, was a nonmechanical, almost intangible effect uniquely associated with temperature. Heat meant warm and cold. Heat was a "flowing" within matter, something invisible.

Joule's great accomplishment was to measure the *mechanical equivalence* of work and heat, and to prove (Figure 13-4) that one is convertible into the other. His experiment, in the 1840s, showed quantitatively how

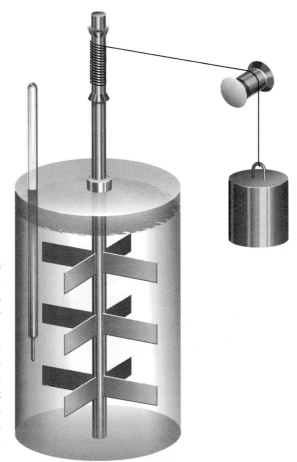

FIGURE 13-4. Work is convertible into heat, as demonstrated by Joule's experiment: beating leads to heating. Even though the water is thermally isolated, its temperature will rise when a paddlewheel powered by gravity churns the liquid. The increase results from mechanical work done on the molecules, not from any contact with an object at higher temperature. An organized external motion (work) turns into disorganized internal motion (heat).

the application of ordinary mechanical work to an object acts to raise its temperature, despite the absence of any obvious source of heat. The equivalence is expressed through the exact relationship

$$1 \text{ cal} \equiv 4.184 \text{ J}$$

$$\text{Heat} \equiv \text{work}$$

where the calorie (cal) is the amount of heat needed to increase the temperature of 1.00 g of water from 14.5°C to 15.5°C.

Heat should be understood not as a property of an object, but rather as a mechanism by which energy is exchanged. Heat *flows* from one body to another, from a region of higher temperature to one of lower temperature. The transfer of energy continues until the temperature is uniform throughout, at which point equilibrium (specifically, *thermal* equilibrium) is achieved.

Temperature, by controlling the direction and duration of the heat exchange, thus becomes a kind of macroscopic "potential" for energy transfer, playing a role suggested by the gradation of color in Figure 13-5. Wherever a difference in temperature exists, there is a tendency for heat to flow and ultimately to establish equilibrium. We know, consequently, that one mole of gas at fixed temperature, pressure, and volume must already be in a state of thermal equilibrium. No portion of the gas is hotter or colder than any other. The temperature is the same everywhere, and there is no flow of heat. The thermal potential (the ability to redistribute thermal energy) is everywhere equal. Nothing changes.

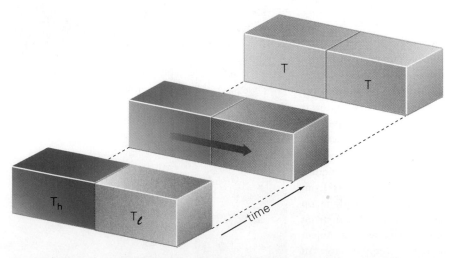

FIGURE 13-5. A difference in temperature points the way for a transfer of energy. The hotter body, at a higher temperature T_h, cedes energy to the colder body (T_ℓ) until the two temperatures become equal. The result: thermal equilibrium.

Heat and temperature are but one manifestation of a far more comprehensive understanding, for all natural processes are similarly governed by potentials. Change, in general, occurs only when there is an imbalance in some potential, and the process stops once that difference is erased. Current in a wire, for example, flows from a point of high voltage to a point of low voltage. As the battery runs down, the potential difference erodes and so the current stops. If the battery happens to be an electrochemical cell (a device we shall explore in a later chapter), then it uses the free energy of a redox reaction to establish the required electrical potential. Free energy is itself a generalized potential for chemical change. When the free energies of reactants and products become equal, a reaction stands at equilibrium and progresses no more.

It seems suddenly to be a macroscopic world of potentials and flows, a universe wholly comprehensible without molecules. The Industrial Revolution, no less, was launched in such a world, a place where heat was seen strictly as a macroscopic phenomenon. Thermodynamics alone (energy, work, and heat; not molecules) powered the steam engine and built the railroads.

Still, there *are* molecules. Molecular agitation does give rise to thermal energy, and heat is just a transfer of energy from molecule to molecule. Be glad, therefore, that the molecular interpretation agrees completely with the macroscopic description afforded by thermodynamics. A reconciliation is brought about by a discipline called statistical mechanics, of which the kinetic theory of gases (Chapter 10) served as our first example.

Thinking statistically, we can calculate thermodynamic quantities by appealing to the laws of large numbers. The outcome is preordained. We know that the molecules will submit—eventually—to the constraints that govern all large assemblies at equilibrium. There is never a conflict, never a discrepancy. Harmony prevails between the microscopic (statistical) and the macroscopic (thermodynamic) world, offering us an enduringly beautiful way to draw simplicity out of complexity.

13-2. THE FIRST LAW OF THERMODYNAMICS

Thermodynamics distinguishes between system and surroundings (Figure 13-6), with the **system** taken as some specific part of the world under observation—a liter of water, 15 grams of copper, a mole of hydrogen gas, a pendulum, or any other object of interest that *we* choose. The **surroundings** (everything else) provide an environment available to the system for exchanges of heat and work, say a glass beaker and the air around. Together, system and surroundings take the imposing name **universe**.

To the system is ascribed an **internal energy** E, a well-defined *property of the equilibrium state*. The internal energy of one mole of ideal gas, for

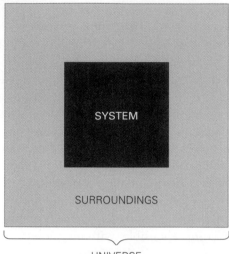

FIGURE 13-6. System, surroundings, and universe in thermodynamics. The boundaries of the system are defined by the observer.

example, we know to be $\frac{3}{2}RT$ (Chapter 10). Given values for the pressure and volume, we immediately deduce the temperature through the equation of state ($PV = nRT$) and thence the internal energy. Temperature, a single macroscopic variable, fixes the internal energy of the gas once and for all. Whatever may be done to the gas—whether it suffer changes in pressure and volume, whether it be expanded or compressed, warmed or cooled—it undergoes no net change of internal energy *if*, at the end, the temperature is the same as at the beginning. What happens in the interim is irrelevant once the initial state is restored; we say, accordingly, that E is a **state function**, specified here entirely by a macroscopic **state variable** (the temperature).

Internal energy is classified as an **extensive property**, for it scales directly with the extent or quantity of matter. The more stuff, the more energy within: $\frac{3}{2}RT$ for one mole of ideal gas, $3RT$ for two moles, $\frac{9}{2}RT$ for three moles, $\frac{3}{2}nRT$ for n moles. Volume and mass, too, are extensive properties, whereas size-independent variables such as temperature, pressure, and density are considered **intensive properties**. The distinction is made pictorially in Figure 13-7.

Countless observations have shown that any change in a system's internal energy, ΔE, occurs only through the exchange of heat and work with the surroundings. The **first law of thermodynamics** summarizes this accumulated experience in one compact equation,

$$\Delta E = q + w$$

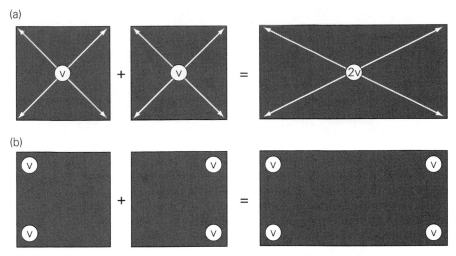

FIGURE 13-7. The difference between extensive and intensive properties: (a) The value (*v*) of an *extensive* property requires the entire system for its definition, increasing in direct proportion to mass and volume. Examples include internal energy, enthalpy, and entropy. (b) An *intensive* property holds its value at each point within a system, independent of the total amount of material. Examples include density and pressure.

where q denotes the heat and w denotes the work. Heat flowing *into* the system and work done *on* the system by the surroundings are considered to be positive. Heat flowing *out* of the system and work done *by* the system are algebraically negative. One state ($E = E_1$) can be converted into another state ($E = E_2$) only by a combination of heat and work exactly equal to the energy difference, $\Delta E = E_2 - E_1$.

The first law thus is a symbolic statement of the conservation of energy, an energetic balance sheet of the kind rendered in Figure 13-8. A given change in energy is deemed traceable to clearly identifiable sources, and this simple equation demands that all the ins and outs of heat and work add up to the net change in internal energy. Energy is neither created nor destroyed, only shuffled about.

We say "simple" but the conservation of energy is no small thing. It is a profound fact of nature, rooted in a basic symmetry of the universe: time. Physical laws are the same yesterday, today, and tomorrow, and therefore natural processes depend not on absolute times (t_1 = July 4, 1776; 6:00:00 P.M. and t_2 = July 4, 1776; 6:00:45 P.M.) but only on time *differences* ($\Delta t = t_2 - t_1 = 45$ seconds). From this symmetry in time follows inevitably (but not obviously) the conservation of energy; the one demands the other. Even more, experience and analysis show that *some* quantity is always conserved wherever a symmetry exists.

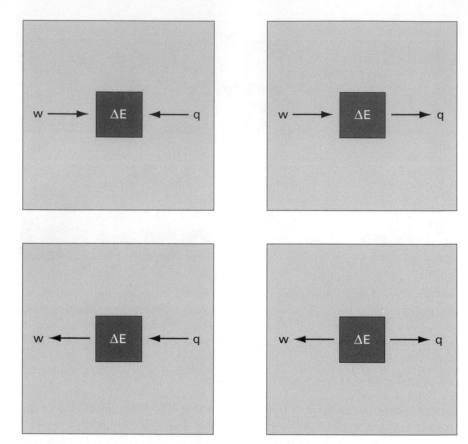

FIGURE 13-8. The first law of thermodynamics: energy is conserved. A change in internal energy ($\Delta E = q + w$) results entirely from work done on or by the system (inner box), together with any heat that flows in or out. The energy of the universe remains constant throughout all exchanges of heat and work between system and surroundings.

That, apparently, is how the world works. Momentum and angular momentum, for instance, are conserved precisely because physics is invariant to absolute positions and directions. Newton's laws are the same in Jerusalem and Berlin, left and right, up and down. And they are also the same on Sunday and Monday, from which comes the conservation of energy.

Duly impressed by the conservation laws, we must nevertheless understand that internal *energy* is a conserved quantity (a state function) but heat and work individually are not. Flows of heat and work are events specific to particular processes. The distinction is important.

We begin to appreciate that distinction by considering the isothermal expansion of an ideal gas, expecting the system eventually to satisfy Boyle's law at constant temperature ($PV =$ constant). Clearly the change in internal energy ($\Delta E = \frac{3}{2}R \, \Delta T$) is zero for such a transformation, since the temperature never changes. From the resulting first-law equation,

$$\Delta E = q + w = 0$$

we then conclude that

$$q = -w$$

and offer the following interpretation: The work done *by* the gas ($-w$) in expanding isothermally against an external pressure is supplied by the heat (q) coming *in* from the surroundings. This relationship holds no matter how the expansion is carried out, because the net change in state (ΔE) must always be zero. Constant temperature can be maintained only if the heat coming in matches the work going out.

But the actual work done depends very much on the path taken, even if the internal energy does not. Imagine a quantity of gas confined in a cylinder and pressurized externally by a weighted piston (Figure 13-9). Initially the gas is in equilibrium, with its internal pressure P_1 equal to the external pressure P_{ext}. The equation of state, $P_1 V_1 = nRT$, is in force and all is quiet. Then suddenly we halve the external pressure, instantly rendering the volume too small for the applied pressure (which is now $P_1/2$, half the original equilibrium pressure). Thrown out of equilibrium, the gas expands until its final *internal* pressure P_2 is equal to the reduced external pressure *and* its new volume V_2 is consistent with the equation of state. Boyle's law, $P_2 V_2 = P_1 V_1$, is satisfied when

$$P_2 = \frac{P_1}{2}$$

$$V_2 = 2V_1$$

and in this way a new equilibrium is reached after some possibly turbulent expansion.

Getting there took work. While recovering its equilibrium, the gas had to push up the weighted piston and expand against the external pressure. Recall from Chapter 10 that pressure × volume has units of work, and so the work done by the gas is just the constant external pressure times the change in volume—in this instance equal to

$$w = -P_{ext}(V_2 - V_1) = -\tfrac{1}{2}P_1(2V_1 - V_1) = -\tfrac{1}{2}P_1V_1$$

The negative sign is used to indicate work done by the system.

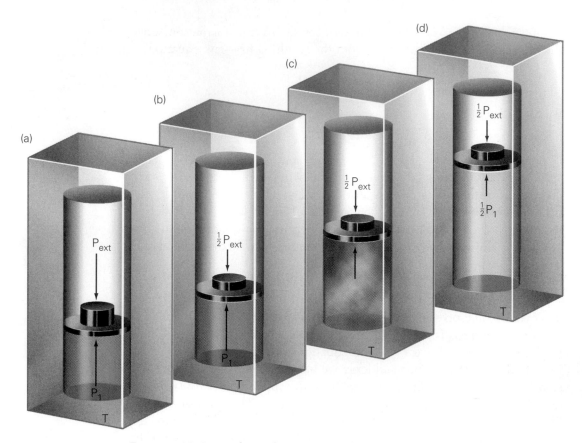

FIGURE 13-9. Isothermal expansion of a gas, performed suddenly and irreversibly. Pressure is proportional to the length of the arrow in the direction shown. (a) The gas, held at constant temperature in a heat bath, stands initially in equilibrium. Its internal pressure is equal to the external pressure imposed by a weighted piston. (b) Weights are removed from the piston to cut P_{ext} in half, jarring the gas temporarily out of equilibrium. The original internal pressure is too high, and an adjustment must now take place. (c) The gas begins an uncontrolled, turbulent expansion against the piston, drawing heat from the surroundings to do the work. (d) A new equilibrium is finally established at the lower pressure, with the final volume stabilizing at twice its original value.

Now an abrupt halving of the external pressure is a particularly violent way to expand a gas. The sudden change destroys the equilibrium, *invalidates* the law $PV = nRT$, and sends the system scrambling for a new resting position. Suppose, instead, we proceed in the gentlest way possible by making only infinitesimal changes in P_{ext}. Very s-l-o-w-l-y we reduce the external pressure and allow the confined gas to readjust its internal pressure at every small step. There is no turbulence and no

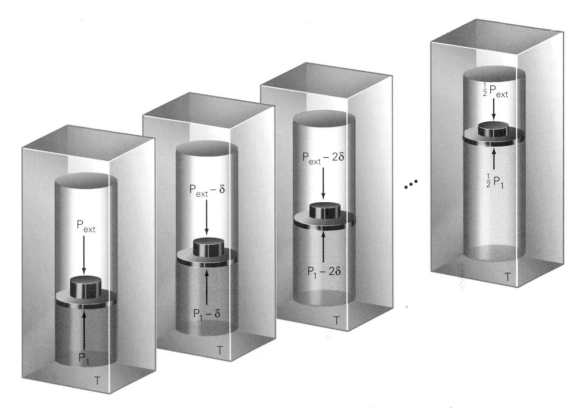

FIGURE 13-10. Isothermal expansion of a gas, performed reversibly in a series of small steps. At each stage, the reduction in external pressure (δ) is gentle enough to keep the gas always in a state of equilibrium. No turbulence develops, and eventually the gas reaches the same final state as it does for the uncontrolled expansion shown in Figure 13-9. ΔE is equal to zero for both these isothermal processes ($\Delta T = 0$), but q and w are different for each.

sudden loss of equilibrium, since the gas in Figure 13-10 is never more than infinitesimally displaced from a point of true equilibrium. The ideal gas law is never violated, and therefore the instantaneous internal pressure decreases inversely with volume as

$$P = \frac{nRT}{V}$$

At any moment, too, the pressure can be increased ever so slightly to return the system to its last equilibrium state.

Such a sequence of gradual, never-really-out-of-equilibrium changes is termed **reversible**, and we can easily devise a reversible pathway along which the volume is doubled isothermally. Computing the $P \Delta V$ work at each tiny step, however, and then adding everything up, we discover that the total work differs from that accomplished by the sudden expansion. The result* is

$$w_{rev} = -nRT \ln\left(\frac{V_2}{V_1}\right) = -P_1 V_1 \ln\left(\frac{2V_1}{V_1}\right) = -P_1 V_1 \ln 2$$

where the subscript *rev* is used to denote reversibility. Here the gas does *more* work than during the sudden expansion, approximately $0.7 P_1 V_1$ compared with $0.5 P_1 V_1$, even though its net change in internal energy is still zero. The reversible heat inflow, q_{rev}, which powers the isothermal expansion and is linked to w_{rev} by the first law, follows immediately as

$$q_{rev} = -w_{rev} = +nRT \ln\left(\frac{V_2}{V_1}\right) = P_1 V_1 \ln 2$$

It differs from w_{rev} only in sign.

For any number of conceivable pathways, we similarly find variable amounts of work and heat exchanged. Unlike internal energy, work is not determined by destination alone; it is determined by the road taken. *How* the system goes from state 1 to state 2 (Figure 13-11) makes all the difference.

Thus neither work nor heat are functions of state; they are functions of path. The first law asserts, though, that the sum of q and w is always a path-independent state function—a conserved quantity, ΔE, the change in internal energy.

Heat, while clearly not path-independent, nevertheless does figure in another important thermodynamic quantity which is indeed a function of state. Designated S, this new variable is associated with a reversible heat flow, q_{rev}, at constant temperature. Changes in S are given by the expression

$$\Delta S = \frac{q_{rev}}{T}$$

in which ΔS ($S_2 - S_1$) denotes the difference between two states. But, in contrast to q_{rev} alone, division by the absolute (Kelvin) temperature

*The solution requires a straightforward application of integral calculus, namely integration of the internal pressure over the change in volume. The integral to be evaluated is $\int (nRT/V)dV$ between the limits V_1 and $V_2 = 2V_1$.

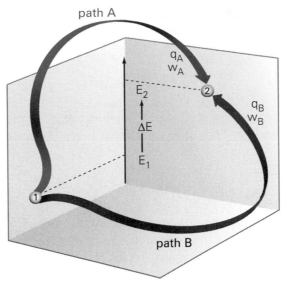

FIGURE 13-11. Internal energy is independent of path, but heat and work are not. Although the difference in energy between states 1 and 2 never changes, the specific combination of heat and work varies with the route taken. E is a function of state; q and w are functions of path.

makes the extensive function S a property of state. It scales with the quantity of matter and is utterly indifferent to the specific route taken between states. It is characteristic of the state itself, on a par with the internal energy.

And what is S? Surprisingly, perhaps, S is the entropy—the same entropy that we are coming to associate with microscopic disorder. Yet S in the present context is a macroscopic thermodynamic quantity lacking any microscopic interpretation. It makes a statement about temperature, heat flows, and state functions cast in the particle-free language of thermodynamics. An early triumph of statistical mechanics was to show that this state function is precisely the macroscopic aspect of microscopic entropy. In Chapter 14 we shall understand this relationship ourselves, using the isothermal expansion of an ideal gas as a key example. The expression derived above for q_{rev} will help unify these macroscopic and microscopic interpretations of entropy.

13-3. ENERGY AND ENTHALPY

Experimental conditions dictate the specific ways in which heat and work can be exchanged. The work done by a chemical system, for example, is often limited to $P \, \Delta V$ effects—typically the expansion and contraction of gases under fixed external pressure—for which the first law can be recast as

$$\Delta E = q - P \, \Delta V$$

Immediately we see that any change in energy at constant volume is attributable entirely to heat. The work term is zero (since $\Delta V = 0$), leaving just

$$\Delta E = q_V$$

where q_V denotes the heat flow at constant volume. Knowing this requirement, we can then measure ΔE for a reaction by monitoring a temperature change within a rigidly enclosed vessel, the so-called *bomb calorimeter* illustrated in Figure 13-12. The change in temperature is related to the heat flow by the **molar heat capacity at constant volume** (a substance-specific property designated c_V and possessing units of

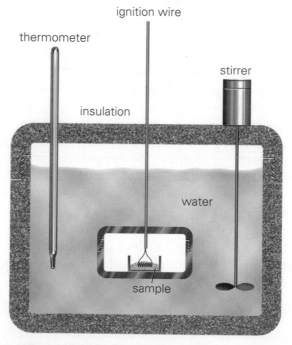

FIGURE 13-12. A bomb calorimeter, designed to measure a flow of heat at constant volume. The reaction takes place in a rigid chamber (center), and a corresponding change in temperature is registered in the water outside.

J mol^{-1} K^{-1}). A material's heat capacity indicates how much heat is needed to raise the temperature of one mole by one degree; for n moles, the corresponding relationship is

$$q_V = nc_V \, \Delta T$$

Alternatively, we may use the **specific heat capacity**, c_s, which measures the heat per gram rather than per mole. Consider that 10 g of liquid water, with a specific heat capacity of 4.184 J g^{-1} K^{-1}, demands 418.4 J for a 10-K rise in temperature. A quantity ten times as large, 100 g, therefore requires ten times as much (4184 J = 4.184 kJ), a trend entirely consistent with the extensive nature of heat flow.

    Constant-volume conditions sometimes exist naturally, such as when a reaction takes place entirely in solution. Since liquids generally have low compressibilities, $P \, \Delta V$ work usually is negligible and heat then becomes the primary mechanism for changing the system's internal energy. Explicit control of volume, however, tends to be difficult experimentally, and chemists usually prefer to manipulate the temperature and pressure instead. Many reactions are run in open vessels under constant pressure (normally atmospheric), and almost all the biochemical processes occurring in living organisms are both isobaric and isothermal. So rather than use the internal energy, which is best understood at constant volume, we should like a new state function better suited to constant pressure. That wish is easily fulfilled in the form of a slightly modified energy function called the **enthalpy** (H), developed especially for the conditions shown in Figure 13-13.

    Enthalpy is easy to define because the first law allows considerable flexibility in the construction of state functions. The guidelines are simply to conserve energy and to respect the extensive or intensive nature of any property. Within these constraints we are free to propose the function

$$H = E + PV$$

and to express the change in enthalpy at constant pressure as

$$\Delta H = \Delta E + P \, \Delta V = q - P \, \Delta V + P \, \Delta V \equiv q_P$$

Again the first law ($\Delta E = q - P \, \Delta V$) has been used to eliminate $P \, \Delta V$, and the symbol q_P has been introduced by analogy to q_V. The result of the rearrangement is now clear: ΔH is the heat transferred at constant pressure.

    Enthalpy is a legitimate state function, formed deliberately as a

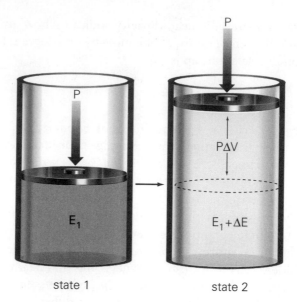

FIGURE 13-13. A change in *enthalpy*, ΔH, registers the heat transferred under a constant pressure P. There are two contributions to ΔH: (1) ΔE, the change in internal energy. (2) $P \, \Delta V$, the work done when the system expands or contracts. So defined, enthalpy is a state function just like energy. ΔH becomes equal to ΔE for processes undertaken at constant volume and pressure.

combination of E and the state variables P and V. Like the internal energy, enthalpy has units of joules. Like the internal energy, its changes can be measured by heat transfer under certain conditions. And, where the changes in volume are slight, ΔH and ΔE are nearly identical. Enthalpy is only a convenient way to express the energy, an alternative definition particularly well adapted for standard laboratory conditions.

We measure ΔH by observing the temperature change in a calorimeter at constant pressure, sometimes with just the simple apparatus sketched in Figure 13-14. The relevant molar heat capacity, dubbed c_{P}, then gives the heat processed by n moles at constant pressure (usually atmospheric) as

$$q_{\mathrm{P}} = nc_{\mathrm{P}} \, \Delta T$$

Note, in addition, that neither c_{P} nor c_{V} is necessarily a single number. Since heat capacities may vary with both the temperature and phase (solid, liquid, gas) of a material, we use just those values appropriate for a given range of temperature.

Another consideration: Only *changes* in enthalpy or energy are

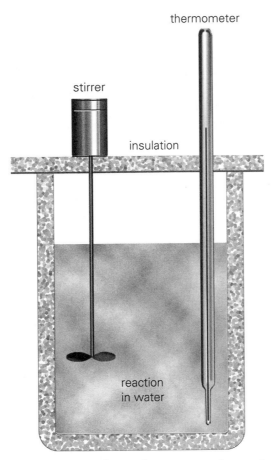

FIGURE 13-14. A calorimeter designed to measure changes in temperature at constant pressure. The loose-fitting top ensures that the reaction occurs at 1 atm.

meaningful, for there is no true zero point to provide an absolute scale. We cannot say that substance A has an enthalpy of 10 joules and substance B has an enthalpy of 12 joules. What we can say—all we can say—is that the enthalpy of A is 2 joules lower than the enthalpy of B. Whether $H_A = 146$ J and $H_B = 148$ J, or $H_A = 1031$ J and $H_B = 1033$ J, or $H_A = 10$ J and $H_B = 12$ J is unknowable and irrelevant. Behaving as a state function sensitive only to differences, enthalpy can be referred to an arbitrary zero chosen solely for our convenience.

The convention is to assign enthalpy values of *zero* to elements in their standard states, defined for each element as the form in which it naturally exists under atmospheric pressure at some specified temperature (often 25°C, but not necessarily). The **standard enthalpy of formation** (ΔH_f°) is then taken to be the change in enthalpy accompanying the formation, *from the appropriate elements*, of one mole of a compound in its standard state under standard conditions. By standard conditions we shall understand implicitly a pressure of 1 atm for a gas, a concentration of 1 mol L^{-1} for a solution, and the normally existing phase for a pure solid or liquid.

Standard enthalpies of formation have been measured and tabulated for a wide range of compounds. With the existing data we can predict the heat evolved or absorbed for countless reactions, both actual and hypothetical. Such calculations form the basis of **thermochemistry**, to which we now turn our attention.

13-4. THERMOCHEMICAL COMPUTATIONS

Take an arbitrary chemical reaction

$$R \longrightarrow P$$

in which some set of reactants (R) goes to some set of products (P). Is the process exothermic? Endothermic? How much heat is involved? How can we tell without measuring it?

We need to know the difference in enthalpy between products and reactants, or the generic **enthalpy of reaction** (ΔH_r, often called simply ΔH). The difference, moreover, is *all* we need, because enthalpy is a function of state. The details of how R makes its way to P are of no concern. Suppose, therefore, that we know nothing about the reaction as written, but that we do know the enthalpy change for each of the following two processes involving certain other substances "E":

$$R \longrightarrow E \qquad \Delta H = \Delta H_{RE}$$

$$E \longrightarrow P \qquad \Delta H = \Delta H_{EP}$$

The rules for state functions permit us to imagine an indirect path from R to P, which begins with R $\longrightarrow$ E and ends with E $\longrightarrow$ P. The enthalpy is changed first by ΔH_{RE} (when the intermediates E are produced) and then by ΔH_{EP} (when they are consumed). The net result is R $\longrightarrow$ P, from which the enthalpy of reaction automatically follows as

$$\Delta H_{RP} = \Delta H_{RE} + \Delta H_{EP}$$

The problem is solved.

This addition of ΔH_{RE} and ΔH_{EP} to yield ΔH_{RP}, made possible by the path-independence of the enthalpy, is an example of **Hess's law**, which states: The overall enthalpy of a process is the sum of the enthalpies from *any* series of other reactions that, taken together, transform the desired reactants into the desired products. Whatever the path chosen, the end result always remains the same. There may be a thousand different ways to convert R into P, but the net change in enthalpy will still be the same for each one: the sum of the enthalpies at each step. "All's well that ends well" thus becomes nature's motto for any function of state.

Hess's law means, right away, that the enthalpy change for some arbitrary reaction can be obtained merely by subtracting the enthalpies of formation of the reactants from the enthalpies of formation of the products. To see why, we need only identify "E" with the component elements of R, so as to equate R $\longrightarrow$ E with the decomposition of the reactants to their elemental units (usually atoms). The rest follows straightforwardly from the definition of a heat of formation.

Track the process now as it unfolds in Figure 13-15. Since decomposition is the reverse of formation, the enthalpy change for

$$\text{reactants} \longrightarrow \text{elements}$$

is just (note the sign reversal) $-\Delta H_f(R)$. But if the overall equation

$$\text{reactants} \longrightarrow \text{products}$$

is stoichiometrically balanced, then these are the same elements needed to form the products in exactly the right quantity. All the ingredients are there. Consequently, the enthalpy change for

$$\text{elements} \longrightarrow \text{products}$$

must be $\Delta H_f(P)$—positive sign this time—and the overall transformation proceeds with

$$\Delta H_r = \Delta H_f(\text{products}) - \Delta H_f(\text{reactants})$$

The path from reactants to products is complete, and the net enthalpy change is fixed by the individual heats of formation.

For compounds in their standard states, the **standard enthalpy of reaction** then becomes

$$\Delta H_r^\circ = \Sigma\, \Delta H_f^\circ(\text{products}) - \Sigma\, \Delta H_f^\circ(\text{reactants})$$

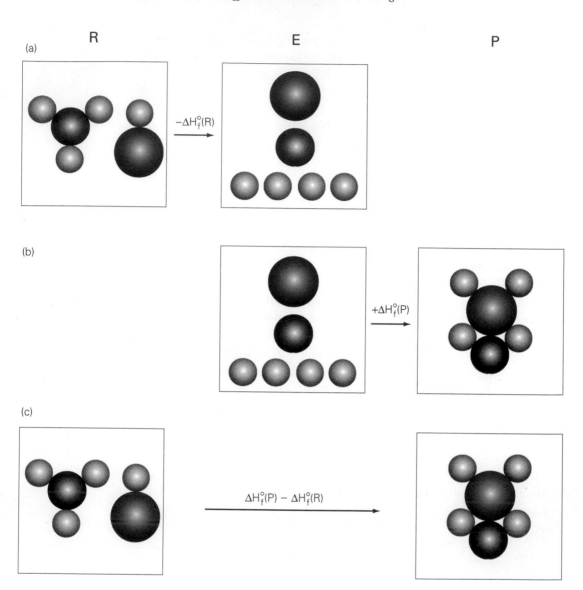

FIGURE 13-15. The problem is to determine the difference in standard enthalpy between an arbitrary set of reactants (R) and products (P). Hess's law allows any such transformation to be imagined in two steps: (a) The reactants break apart into their component elements (E), undergoing a change in enthalpy equal to $-\Delta H_f^\circ(R)$. (b) The products take shape from the very same elements just released, this time with an enthalpy change equal to $\Delta H_f^\circ(P)$. Since mass is conserved, all the atoms contained in the reactants find their way into the products. Nothing is left over. (c) The net change in enthalpy, a function of state, is thus fixed: $\Delta H_f^\circ(P) - \Delta H_f^\circ(R)$, the sum of the changes for the two steps.

where the large Greek sigma (Σ) means "form the sum." Extensive tables of standard enthalpies of formation facilitate ready calculation of ΔH_r° for most conceivable transformations. Just be aware that the tabulated values report the heat per *mole* of compound formed, leaving us to remember that enthalpy is an extensive property. Hence for a reaction

$$a\text{A} + b\text{B} \longrightarrow c\text{C} + d\text{D}$$

the quantities in the table must be multiplied by the proper stoichiometric coefficients to obtain the correct total heat:

$$\Delta H_r^{\circ} = c \; \Delta H_f^{\circ}(\text{C}) + d \; \Delta H_f^{\circ}(\text{D}) - a \; \Delta H_f^{\circ}(\text{A}) - b \; \Delta H_f^{\circ}(\text{B})$$

Then, if needed, ΔH_r for a reaction at a nonstandard temperature can be obtained by use of the molar heat capacity c_p.

13-5. EXOTHERMIC AND ENDOTHERMIC PROCESSES

We conclude with a brief look at the thermochemistry of a few representative reactions. Remember that ΔH is negative for an exothermic process, where heat is given off to the surroundings as high-enthalpy reactants are converted into low-enthalpy products (Figure 13-16a). An endothermic transformation, by contrast, sucks heat *in* from the surroundings in order to raise the reactants to a higher level of enthalpy. With $\Delta H > 0$ for the endothermic reaction, the surroundings become colder and the temperature decreases correspondingly (Figure 13-16b).

Examine first the dissociation of gas-phase molecular hydrogen into atoms:

$$\tfrac{1}{2}\text{H}_2(g) \longrightarrow \text{H}(g) \qquad \Delta H_r^{\circ} = +218 \text{ kJ}$$

The process is endothermic, requiring 218 kJ to form one mole of atomic hydrogen in the gas phase.

Neither the sign nor the magnitude of ΔH_r° is new to us. We already know, first, that H_2 is the normal form of hydrogen and might well have the lower enthalpy. Since energy must be supplied to break the H—H bonds, the reaction is expected to be endothermic. We know further that the intermolecular potential of H_2, discussed in Chapters 7 and 11, shows a minimum of -436 kJ mol^{-1} when measured as an enthalpy. This quantity—the difference in enthalpy between one mole of $\text{H}_2(g)$ and

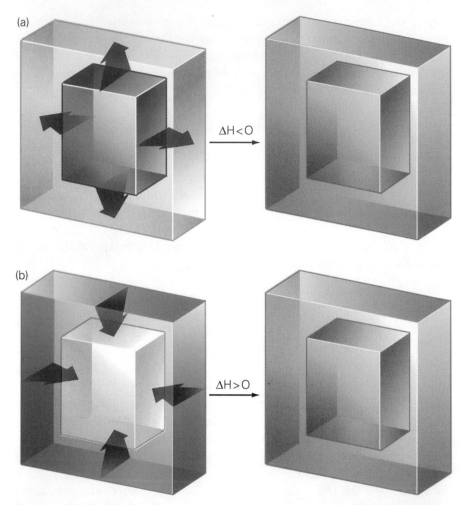

FIGURE 13-16. (a) Heat flows *out* of the system during an exothermic reaction. The system grows cooler; the surroundings grow warmer. (b) For an endothermic process, the reverse: heat flows *into* the system. Partially deprived of thermal energy, the surroundings drop to a lower temperature.

two moles of hydrogen atoms—gives molecular hydrogen the advantage over atomic hydrogen. It means that one mole of H_2 molecules (the normal state) is assembled exothermically from its component atoms according to the equation

$$2 H(g) \longrightarrow H_2(g) \qquad \Delta H_r^\circ = -436 \text{ kJ}$$

Conclusion: Since dissociation is the reverse of formation, the enthalpy change for the desired reaction will have the same magnitude but the opposite sign. The equations also differ by a stoichiometric factor of 2, consistent with the overall heats of reaction (−436 kJ versus +218).

Thermochemical measurement thus can indicate how *strong* a bond is, showing in this case that it takes 436 kJ to break the H—H bonds in one mole of H_2. Other bonds may be stronger or weaker, depending on the degree to which the electrons are shared. N_2 molecules, for instance, have a dissociation energy of 946 kJ mol^{-1}, more than twice that of H_2. Nitrogen shares three pairs of electrons in the structure

$$:N:::N: \qquad\qquad N\equiv N$$

whereas hydrogen shares only one pair,

$$H:H \qquad\qquad H—H$$

and the strength of the triple bond is reflected in the higher dissociation energy.

Attractive forces *between* molecules are considerably weaker than covalent bonds, with the difference readily apparent in the **enthalpy of vaporization** for water:

$$H_2O(\ell) \longrightarrow H_2O(g) \qquad \Delta H^\circ_{vap} = 44.0 \text{ kJ}$$

Compared with an H—H bond, it takes only one-tenth the energy to disrupt the nonbonded interactions in water. Liquid water ($\Delta H^\circ_f = -285.8$ kJ mol^{-1}) is transformed to vapor ($\Delta H^\circ_f = -241.8$ kJ mol^{-1}) at 25°C, absorbing 44.0 kilojoules per mole in the process. Here is a quantitative expression of thermal energy and its relation to the intermolecular potential, and here too is a simple demonstration of Hess's law (−241.8 + 285.8 = 44.0).

Understand, as in Figure 13-17, that the enthalpy of vaporization goes entirely toward tearing away the molecules in the liquid, *not* toward raising the temperature. Temperature remains constant during the phase transition while the excess heat is absorbed to separate the molecules. The difference shows up in the higher enthalpy of the vapor.

No energy can be lost—ever—so 44 kilojoules of enthalpy are likewise returned to the surroundings during condensation, the reverse process. The energy generated by the exothermic condensation is the amount by which the intermolecular potential is lowered in the liquid. Temperature again remains constant throughout the phase transition, this time as the molecules come together into a more ordered phase.

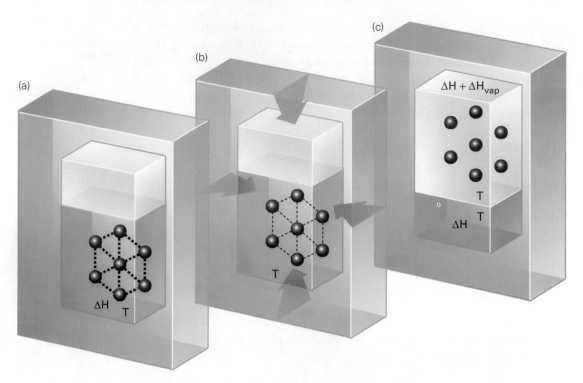

FIGURE 13-17. A liquid and vapor in equilibrium have the same temperature, but the vapor is hotter: its enthalpy is higher. The difference, called the *enthalpy of vaporization*, goes to disrupt the intermolecular interactions of the liquid. (a) Before vaporization: a liquid at its boiling temperature, with noncovalent clusters still intact. (b) During vaporization: Heat flows in from the surroundings, providing just enough energy to pull apart the clusters. Temperature remains constant as the phase transition proceeds. (c) After vaporization: Freed from the liquid, particles in the vapor now carry the extra enthalpy (ΔH_{vap}) applied previously to separate them.

Turn next to the heat available from some hydrocarbon fuels, starting with these standard enthalpies of formation:

$$CH_4 \quad \text{(methane)} \qquad -74.8 \text{ kJ mol}^{-1}$$
$$C_2H_6 \quad \text{(ethane)} \qquad -84.7 \text{ kJ mol}^{-1}$$
$$C_2H_4 \quad \text{(ethylene)} \qquad +52.3 \text{ kJ mol}^{-1}$$
$$C_2H_2 \quad \text{(acetylene)} \quad +226.7 \text{ kJ mol}^{-1}$$

Both methane and ethane, formed exothermically relative to their elements, have limited ability to participate in heat-producing reactions. For if these compounds are to serve as *reactants* in an exothermic process, the products must be still lower in enthalpy. Heat cannot be emitted otherwise, yet the options are limited because CH_4 and C_2H_6

are already so low on the enthalpy scale. Nevertheless they do react with oxygen, evolving substantial heat through **combustion** while producing carbon dioxide and water. Methane, for one, yields nearly 900 kilojoules per mole:

$$CH_4(g) + 2O_2(g) \longrightarrow CO_2(g) + 2H_2O(\ell) \quad \Delta H_r^\circ = -890.3 \text{ kJ mol}^{-1}$$

Ethylene and acetylene, by comparison, are higher in enthalpy relative to elemental carbon and hydrogen, and consequently these unsaturated hydrocarbons react with many other compounds to form products lower in energy. Combustion with oxygen is especially productive, releasing over 1400 kilojoules per mole of ethylene burned:

$$C_2H_4(g) + 3O_2(g) \longrightarrow 2CO_2(g) + 2H_2O(\ell) \quad \Delta H_r^\circ = -1410.9 \text{ kJ mol}^{-1}$$

Observe that both ethylene ($H_2C=CH_2$) and acetylene ($HC\equiv CH$) contain multiple bonds, in contrast to methane and ethane. Such differences in bonding show up in the different heats of combustion.

A fuel need not even make a big bang to be effective, as the sugar glucose ($C_6H_{12}O_6$) so clearly demonstrates. With its standard heat of formation equal to -1274 kJ mol^{-1}, glucose is enthalpically lower than just about everything else—but *not* lower than the total enthalpy contributed by the molecules of water and carbon dioxide produced in a combustion reaction. A small amount of $C_6H_{12}O_6$, oxidized exothermically in a series of steps, is thereby able to supply energy for biochemical processes, among which are included temperature regulation, maintenance of ion gradients, generation of nerve impulses, and general metabolism. Perspiration and radiative cooling allow the body to dissipate the excess heat produced by the exothermic reaction.

Thermochemistry, governed by the first law, can tell us only so much. Although knowledge of the heat absorbed or evolved in a reaction is undeniably important, its real significance is only local. *We* care, but the universe does not. The combined energy of system and surroundings is conserved no matter what happens.

A process can be exothermic and release enthalpy; a process can be endothermic and consume enthalpy. Both occur in nature, so enthalpy and the first law represent just part of the story. Heats of reaction, by themselves, cannot tell us whether some transformation will or will not take place.

Entropy—something the universe does care about—stands behind *that* decision. It is the basis for the second law of thermodynamics, the "arrow of time."

REVIEW AND GUIDE TO PROBLEMS

Ironic, too, is the human preoccupation with material things, as if matter has a critical role to play in the universe. It doesn't. Energy does.

Matter is dust. Electrons, protons, neutrons, atoms, molecules, parakeets, mountains, galaxies—all of them, nothing but dust: specks of mass and charge all in a heap, specks that exist solely to be pushed around. Matter's lot is only to suffer forces and to *move*, not to pose like a statue.

For whatever matter may be, it is no statue. Nothing made of matter stands perfectly still; everything is in motion. Inside the mountain, the atoms jiggle. Inside the atoms, the electrons jiggle and the nuclei jiggle. Inside the nuclei, the protons and neutrons jiggle. Everything jiggles, because everything is driven about by one form of energy or another. The particles of matter are pawns in the game. The fundamental forces, with energy as their agent, move the pieces.

Why, then, should we worry about the matter when we can track the energy instead?

IN BRIEF: THE FIRST LAW OF THERMODYNAMICS

A scientific understanding of the world comes entirely from observation and experiment, not philosophy, not metaphysics. We look, we learn, and we take what we get; and from the study of **thermodynamics** (energy and its transformations), we learn nature's great secret: that all the energy there ever was and ever will be is here right now; that energy is neither created nor destroyed; that energy can only be passed around from one mode of motion to another; that one particle's loss is another particle's gain.

Thus the first law of thermodynamics, as we now have it: In the redistribution of energy, nothing is lost or gained. Energy is conserved.

And the second law, to come in the next chapter: Energy is distributed fairly, spread out over as many modes of motion as possible.

For the moment, though, first things first—a review of the first law, its mechanics, its implications, its practical consequences.

1. WORK AND ENERGY. **Work** is the service bought when energy is spent. Energy, stored throughout the universe in various forms, confers upon matter the ability to do work. Energy is the currency tendered in exchange for motion.

Like dollars and pesos, marks and yen, energy is a freely convertible currency. Whether posing as electrical energy or gravitational energy or chemical energy or mechanical energy or nuclear energy or mass energy,

energy is energy. One form is transformed into the other, and all kinds of energy carry within them the same ability: to move matter, to do work.

To move a mass under the agency of a force is to do work. For the specific case where a constant force F (measured in newtons, N) acts over a distance d (in meters, m), the work done is given by

$$w = Fd$$

when the displacement is parallel to the force. Analogous to energy, work has units of joules (N m, or kg m^2 s^{-2}).

Take note: (1) The stronger the force, the greater the work. (2) The greater the displacement, the greater the work. (3) No displacement ($d = 0$), no work—no matter how strong the force. (4) For motion in the same direction as the force, a mass works *with* the force and the work is positive (done *on* the mass). (5) Moving against the force, the mass fights back with a force of its own to go in the opposite direction. The work under those circumstances, done *by* the mass, is negative.

There are as many kinds of work as there are ways to move matter. There is the work done by falling water, whereby gravitational potential energy is converted into the motion of an electrical turbine. There is the work done by a chemical reaction, whether aimless and destructive (as in an explosion) or controlled and directed (as in an engine). There is the work done by a battery, through which the energy of a redox reaction is converted into a flow of electricity.

There is, as well, the mechanical work done by a gas expanding or contracting under an external pressure, equal to

$$w = -P_{\text{ext}} \, \Delta V$$

where ΔV is the change in volume accomplished under a constant pressure P_{ext}. When $\Delta V > 0$, the gas pushes out and thus does the work itself (w is negative). When $\Delta V < 0$, the gas is pushed inward by the pressure and suffers the work done (w is positive). Either way, there is a force exerted over a distance, as we confirm by looking at the units of pressure and volume:

$$\text{Pressure} = \frac{\text{force}}{\text{area}} \sim \text{N m}^{-2}$$

$$\text{Volume} = \text{area} \times \text{length} \sim \text{m}^3$$

$$\text{Pressure} \times \text{volume} = \text{force} \times \text{length} \sim \text{N m} = \text{J}$$

Pressure times volume is the same as force times distance.

Why the interest in *PV* work? Because *PV* work is often the only visible mechanical work that accompanies a chemical reaction, especially those processes that occur in solution under constant atmospheric pressure.

2. HEAT. When all the pistons have pumped and all the cranks have turned . . . when all the electrons have flowed . . . when all the gases have pushed and pulled and when all the work is done, there is usually something more: a change in temperature. Somewhere it is warmer and somewhere it is colder. There has been a flow of **heat**.

Or so we say. Finding no further evidence of obvious work (things moving), we say that "heat" flows—as if heat were some fluidlike essence permeating matter, some invisible property that can emerge during a process. But we say so only because our vision is too weak to see the microscopic *work* done inside the matter, where particles speed up and slow down continually as energy comes and goes.

Heat, interpreted this way, becomes simply a flow of energy invested in the motion of the particles. There is work done just as surely as a moving piston does work, but it is work too small to see and too disorderly always to understand in every microscopic detail. So we coin the term *heat*, thereby subsuming all the microscopic pistons and cranks under one parameter, a temperature.

Yet heat and work forever remain two sides of the same coin. The invisible work that we call heat can be converted into the visible work that we see every day, and it was indeed this discovery (the equivalence of heat and work) that launched both the Industrial Revolution and the science of thermodynamics in the 19th century.

3. THE FIRST LAW. Somewhere, we have our **system** of interest (a sample of water, say) set amidst its **surroundings** (a glass beaker plus the atmosphere plus everything else in the universe). Then something happens. State R (reactants) turns into state P (products). There is a change, and we set out to investigate.

First, we hunt for all signs of visible work done on or by the system, looking for such clues as flying timber or changes in volume or other telltale movements of matter. We compute the number of joules expended on each activity, add the values together, and record the result as *w*: the total work done.

Second, we track down any changes in temperature that may have occurred in system and surroundings during the transformation from R to P. Knowing that different materials have different **heat capacities** (Examples 13-5, 13-6, 13-8), we can compute the number of joules transferred during each change in temperature:

Heat = amount of material × heat capacity × change in temperature

That done, we add the results together and record the sum as q: the total heat exchanged.

Finally, we add w and q and make a startling discovery: The sum (call it ΔE, the change in the system's internal energy) is always the same. When state R becomes state P, the total work done and heat absorbed by the system is inevitably

$$\Delta E = q + w$$

and that number never varies. Energy is conserved. Any new energy that goes into the system can only come from the surroundings, and any energy that leaves the system can only return to those very same surroundings. Nothing is lost; nothing is gained. A constant amount of energy is shared between system and surroundings, passed back and forth through various combinations of work and heat.

4. A FUNCTION OF STATE. When R goes directly to P, the change in system energy is a certain value ΔE. Whether the transformation takes one second or one year to accomplish, the value remains the same. And even if R goes first to X and only then to P, the internal energy still changes by the identical amount ΔE. Or from R to Y to P, or from R to Z to Y to X to P, or from R to P to R to P, or any way at all—as long as we start at R and end at P, ΔE is the same. The change in internal energy depends only on the starting and ending points, R and P, not on the route taken.

This single focus on beginning and end, to the exclusion of everything in the middle, defines the internal energy as a ***function of state***. Fixed by the identity of the state itself, a state's energy depends only on what that state *is* and not how it was prepared. In an ideal gas, for example, the internal kinetic energy is proportional solely to temperature and amount (Chapter 10), going as

$$E = \tfrac{3}{2}nRT$$

If we know the temperature and number of moles, we know the energy of the gas. Nothing else is relevant.

The consequences are general and deep. State R has an internal energy E_R, regardless of how it came to be; and state P likewise has its own internal energy, E_P. When R changes into P, the net change in energy is simply

$$\Delta E = E_P - E_R = q + w$$

and the number of joules is always the same no matter how the transformation is carried out. Although the amounts of heat and work may differ between one route and another, the total change in energy never varies. Heat and work are not state functions, but energy is; and energy is always conserved. Thus declares the first law of thermodynamics, a fact of nature.

5. STATE FUNCTIONS AND HESS'S LAW. There are other functions of state as well, notably the *enthalpy*

$$H = E + PV$$

which is constructed entirely from E, P, and V, all quantities of state. The enthalpy, recall, corresponds to the heat exchanged at constant pressure (q_P), which we derive as

$$\Delta H = \Delta E + P\,\Delta V = q + w + P\,\Delta V = q - P\,\Delta V + P\,\Delta V \equiv q_P$$

when the work done, w, is equal to $-P\,\Delta V$. We shall use ΔH as a convenient measure of the heat absorbed by the system (in an **endothermic** reaction) or produced by the system (in an **exothermic** reaction), because so many chemical processes are run under constant atmospheric pressure.

Four functions of state—energy (E), enthalpy (H), entropy (S), and free energy (G, Chapter 14)—account for most chemical changes, and all these thermodynamic quantities are pleasingly simple to manipulate. Once their values are measured under an agreed-on set of **standard conditions** ($P = 1$ atm; all substances in their normal phases, typically at $T = 25°C$), the numbers can be added and subtracted to give us the *standard differences* that exist between states R and P:

$$\Delta E° = E_P° - E_R° \qquad \text{internal energy}$$

$$\Delta H° = H_P° - H_R° \qquad \text{enthalpy}$$

$$\Delta S° = S_P° - S_R° \qquad \text{entropy}$$

$$\Delta G° = G_P° - G_R° \qquad \text{free energy}$$

And to compute those standard differences, we have **Hess's law**: When a reaction proceeds in a sequence of steps 1, 2, 3, . . . , the change in any state function ($\Delta H°$, for instance) is equal to the sum of the changes at each step,

$$\Delta H° = \Delta H_1° + \Delta H_2° + \Delta H_3° + \cdots$$

so that the net change is the same regardless of the path taken.

Put another way: The change in standard enthalpy (or any other state function) caused by the reaction

$$aA + bB \longrightarrow cC + dD$$

is equal to

$$\Delta H° = [c\ \Delta H_f°(C) + d\ \Delta H_f°(D)] - [a\ \Delta H_f°(A) + b\ \Delta H_f°(B)]$$

sum of products   −   sum of reactants

where $\Delta H_f°(A)$, $\Delta H_f°(B)$, $\Delta H_f°(C)$, $\Delta H_f°(D)$ are the standard heats of formation of substances A, B, C, D. The **standard heat (enthalpy) of formation**, in particular, measures the change in enthalpy undergone when a compound is assembled from its elements under standard conditions. By defining the standard heat of formation as zero for all elements, we have a convenient (if arbitrary) reference point that leaves intact the differences measured between any two states.

SAMPLE PROBLEMS

The first six examples will concentrate on the relationship between internal energy, enthalpy, heat, and work in an especially simple system: a monatomic gas. For each problem, imagine that we start with 1.00 mol of helium in a state of thermal equilibrium (state 1), confined by a piston to a volume of 22.4 L at STP:

$$P_1 = 1.00 \text{ atm}$$

$$V_1 = 22.4 \text{ L}$$

$$n_1 = 1.00 \text{ mol}$$

$$T_1 = 273 \text{ K}$$

Then a change will take place. There might be a flow of heat into the system (the gas) or out to the surroundings; there might be an expansion of the gas; there might be a contraction; maybe a change in temperature, maybe not; but something will happen to disturb the equilibrium.

Whatever happens, happens. In the end, the gas will find itself in a new state of equilibrium (state 2) with a new set of variables (P_2, V_2, n_2, T_2) for us to determine:

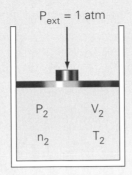

$P_{ext} = 1$ atm

P_2 V_2

n_2 T_2

All throughout the process, too, unless stated otherwise, the weighted piston will press down on the gas so as to apply a constant external pressure, P_{ext}, equal to 1 atm.

EXAMPLE 13-1. A Zero-Sum Game

PROBLEM: Sufficient heat is supplied to double the volume and temperature of the gas while maintaining an internal pressure of 1.00 atm. What is the change in internal energy? How much work is done if the external pressure remains constant at 1.00 atm? How much heat is added? What is the change in enthalpy?

SOLUTION: Helium, a monatomic gas, has only three possible modes of motion: The atoms can move straight ahead in each of three perpendicular directions, accumulating an internal energy equal to $\frac{3}{2}nRT$. If helium could do anything else—like vibrate or rotate—then the system would acquire additional thermal energy for each new *degree of freedom* (Chapter 10), but isolated atoms have no such opportunities apart from electronic excitation (which we shall ignore for now). Any change in internal energy is invested, therefore, entirely in the kinetic energy of the atoms, so that

$$\Delta E = E_2 - E_1 = \tfrac{3}{2}n_2RT_2 - \tfrac{3}{2}n_1RT_1$$

Since the only variation comes from the temperature ($T_2 = 546$ K, $T_1 = 273$ K), the change in internal energy follows directly as

$$\Delta E = \tfrac{3}{2}nR(T_2 - T_1) = \tfrac{3}{2}(1.00 \text{ mol})(8.3145 \text{ J mol}^{-1} \text{ K}^{-1})(546 \text{ K} - 273 \text{ K})$$

$$= 3.40 \times 10^3 \text{ J}$$

The system's energy is higher at the higher temperature, corresponding to a positive change in our system-centered view of the process. Energy goes *into* the gas.

Next, we compute the work done *by* the gas as it pushes against a constant external pressure (P_{ext} = 1.00 atm) during the expansion from V_1 = 22.4 L to V_2 = 44.8 L:

$$w = -P_{ext}(V_2 - V_1) = (-1.00 \text{ atm})(44.8 \text{ L} - 22.4 \text{ L}) = -22.4 \text{ atm L}$$

Recalling that R (the gas constant) is equal to both 0.08206 atm L mol^{-1} K^{-1} and 8.3145 J mol^{-1} K^{-1}, we then quickly convert from atmosphere-liters into joules:

$$w = -22.4 \text{ atm L} \times \frac{8.3145 \text{ J}}{0.08206 \text{ atm L}} = -2.27 \times 10^3 \text{ J}$$

The negative sign makes explicit that the gas spends some of its own energy to raise the piston.

Enter now the first law of thermodynamics: energy is conserved. If the system *paid* 2270 joules to finance the expansion, then it must have *received* 5670 joules as heat from the surroundings to earn a net energetic profit of 3400 joules:

$$\Delta E = q + w \quad = 3.40 \times 10^3 \text{ J}$$

$$q_P = \Delta E - w = (3.40 \times 10^3 \text{ J}) - (-2.27 \times 10^3 \text{ J}) = 5.67 \times 10^3 \text{ J}$$

This quantity of heat is precisely the change in enthalpy, ΔH, which we have already shown to be the heat transferred under constant pressure.

Summary: The gas is 3400 joules richer, having paid out 2270 while taking in 5670. The rest of the world is 3400 joules poorer, having paid out 5670 while taking in 2270. The universe is neither energetically richer nor poorer; it plays a zero-sum game.

EXAMPLE 13-2. Any Means to an End

PROBLEM: Compute ΔE for a doubling of the temperature at constant volume, carried out with the piston held rigidly in place. How much heat must be supplied?

SOLUTION: We know, from Example 13-1, that $\Delta E = 3.40 \times 10^3$ J for any process in which one mole of an ideal gas goes from 273 K to 546 K. The internal energy depends solely on the temperature and number of moles. It is a function of state.

But since there is now no change in volume, the *work* done to realize that same doubling of energy is zero—not the 2.27×10^3 J calculated before. At no time does the piston move. No expansion. No contraction. No mechanical work:

$$\Delta V = 0$$

$$w = 0$$

The first law then tells us that the change in energy (at constant volume) comes entirely from the heat absorbed:

$$\Delta E = q + w = q + 0 = q$$

$$q_V = 3.40 \times 10^3 \text{ J}$$

QUESTION: Does this quantity of heat correspond to the change in enthalpy as well?

ANSWER: No. The enthalpy measures the heat transferred at constant pressure, not constant volume.

QUESTION: If so, what is ΔH here?

ANSWER: Look what happens to the pressure of the gas inside the fixed volume. *P* doubles along with *T*, going from

$$P_1 = 1.00 \text{ atm}$$

$$V_1 = 22.4 \text{ L}$$

$$T_1 = 273 \text{ K}$$

to

$$P_2 = 2.00 \text{ atm}$$

$$V_2 = 22.4 \text{ L}$$

$$T_2 = 546 \text{ K}$$

These *internal* pressures, P_1 and P_2, are the values we use to compute the change in enthalpy, a state function.

For states 1 and 2, the enthalpies are

$$H_1 = E_1 + P_1 V_1$$

$$H_2 = E_2 + P_2 V_2$$

and the difference is thus

$$\Delta H = H_2 - H_1 = (E_2 - E_1) + (P_2 V_2 - P_1 V_1)$$

$$= \Delta E + (P_2 V_2 - P_1 V_1)$$

from start to finish. Equivalently, we use the ideal gas equation ($PV = nRT$) to write

$$\Delta H = \Delta E + nR(T_2 - T_1) = \Delta E + nR\,\Delta T$$

and obtain the net change in enthalpy as 5.67×10^3 J, the same as in Example 13-1:

$$\Delta H = (3.40 \times 10^3\ \text{J}) + (1.00\ \text{mol})(8.3145\ \text{J mol K}^{-1})(273\ \text{K})$$

$$= 5.67 \times 10^3\ \text{J}$$

QUESTION: Why are ΔE and ΔH the same for both processes? In Example 13-1, the system expanded against an external pressure and did work. In Example 13-2, the gas was heated at constant volume and did no mechanical work at all. Aren't the final states different?

ANSWER: Both systems underwent the same change in temperature, from $T_1 = 273$ K to $T_2 = 546$ K. Moreover, both systems underwent the same change in PV, from $P_1 V_1 = 22.4$ atm L to $P_2 V_2 = 44.8$ atm L. *How* they did so has no effect on either ΔE or ΔH. Energy and enthalpy are functions of state.

EXAMPLE 13-3. Properties of State: All's Well That Ends Well

PROBLEM: The system now undergoes a transformation in three steps. In step 1, the temperature is halved and the volume is halved. In step 2, the volume is quadrupled isothermally. In step 3, the temperature is doubled and the volume is halved. If the amount of gas remains constant throughout, what will ΔE and ΔH be at the end of the process?

SOLUTION: Zero; nothing; nil. $\Delta E = 0$. $\Delta H = 0$. The system completes a round trip from state 1 back to state 1, starting and finishing with the state variables P, V, n, and T unchanged.

How do we know? From the wording of the question, we can deduce the values of T, V, and n at each stage; and from the gas law ($PV/T =$ constant), we can easily compute the accompanying pressure as shown below:

$$
\begin{array}{cccc}
\text{STEP 1} & \text{STEP 2} & \text{STEP 3} & \\
T_1 \longrightarrow & \dfrac{T_1}{2} \longrightarrow & \dfrac{T_1}{2} \longrightarrow & T_1 \\[2ex]
V_1 \longrightarrow & \dfrac{V_1}{2} \longrightarrow & 2V_1 \longrightarrow & V_1 \\[2ex]
n_1 \longrightarrow & n_1 \longrightarrow & n_1 \longrightarrow & n_1 \\[2ex]
P_1 \longrightarrow & P_1 \longrightarrow & \dfrac{P_1}{4} \longrightarrow & P_1
\end{array}
$$

Energy and enthalpy are both functions of state, their values fixed forever by the interconnected state variables P, V, n, and T. If, at the end, those variables are unchanged, then the energy and enthalpy will be unchanged as well. Along the way, heat will flow in and out; the gas will expand; the gas will contract; work will be done—yet after all that, the state functions E and H must end where they began, as if the system rode up and down an elevator and finally got off on the original floor.

QUESTION: Have we solved the round-trip problem for all time? What would happen, say, if the system returned to its original state by following a route completely different from the one just described?

ANSWER: Devise any route you like, but there will still be no net change in E and H once the gas returns home. Energy, enthalpy, entropy, free energy, and all the other functions of state have no ties to the past. They are properties of the present and the present alone.

QUESTION: What about work and heat?

ANSWER: Different. Work and heat, unlike the internal energy, are not state functions. They do retain memories from the past, and they do depend very much on the route traveled. Here, where the state undergoes a round trip ($\Delta E = 0$), the first law guarantees only that the heat flowing into the system will be exactly equal to the work done by the gas:

$$\Delta E = q + w = 0$$

The heat (q) is therefore equal to the negative of the work $(-w)$, but how much work is that? The answer demands an examination of the entire trip. Different pathways require different amounts of work and heat to return the system to its starting point.

There is a difference, after all, between climbing up and down a mountain and climbing up and down a flight of stairs. Both leave the climber in the same gravitational state at the end, but one requires manifestly more work than the other.

EXAMPLE 13-4. Extensive and Intensive Properties

PROBLEM: What happens when the amount of He gas is doubled to 2.00 mol at constant volume and temperature? Compute P_2 and ΔE.

SOLUTION: Start with what we know about states 1 and 2:

STATE 1	STATE 2
$P_1 = 1.00$ atm	$P_2 = ?$
$V_1 = 22.4$ L	$V_2 = 22.4$ L
$T_1 = 273$ K	$T_2 = 273$ K
$n_1 = 1.00$ mol	$n_2 = 2.00$ mol

The first detail is to determine the internal pressure, P_2, by routine application of the gas laws. With twice as many atoms now packed into the same volume at the same temperature, the pressure doubles from 1.00 atm to 2.00 atm:

$$P_2 = P_1 \times \frac{n_2}{n_1} = 1.00 \text{ atm} \times \frac{2.00 \text{ mol}}{1.00 \text{ mol}} = 2.00 \text{ atm}$$

The internal energy, which depends directly on the number of moles, then doubles as well, despite the constant temperature. The extra mole contributes an extra $\frac{3}{2}RT$ to the total energy:

$$\Delta E = E_2 - E_1 = \tfrac{3}{2}n_2 RT_2 - \tfrac{3}{2}n_1 RT_1 \qquad (T_1 = T_2)$$

$$= \tfrac{3}{2}RT_1(n_2 - n_1) = \tfrac{3}{2}RT_1(1.00 \text{ mol})$$

$$= \tfrac{3}{2}(8.3145 \text{ J mol}^{-1} \text{ K}^{-1})(273 \text{ K})(1.00 \text{ mol}) = 3.40 \times 10^3 \text{ J}$$

We prove thereby that the internal energy is an **extensive property**, going up and down with the amount of substance. If one mole has energy E, then two moles have energy $2E$. More particles, more energy.

Temperature, by contrast, is an **intensive property**. It reflects not the combined energy of all the particles but rather the *average* value—the total energy divided by the number in the sample. Since state 2 has both twice the energy and also twice the number of moles found in state 1, the average energy per particle (proportional to temperature) remains unchanged.

QUESTION: What effect does the higher internal pressure, P_2, have on the internal energy of state 2?

ANSWER: No direct effect. P_2 plays a role only insofar as it influences the values of T_2 and n_2 through the equation of state, $PV = nRT$.

EXAMPLE 13-5. Heat Capacity at Constant Volume

PROBLEM: Suppose that the helium gas absorbs 12.472 J of heat at constant volume. What is the change in temperature?

SOLUTION: The first law reduces from

$$\Delta E = q + w$$

simply to

$$\Delta E = q = 12.472 \text{ J}$$

in the absence of any work done. With the volume fixed so that $P_{ext} \Delta V$ (equal to $-w$) is zero, the influx of heat goes entirely into the kinetic energy of the atoms:

$$\Delta E = \frac{3}{2} nR \Delta T$$

$$\Delta T = \frac{2 \Delta E}{3nR} = \frac{2(12.472 \text{ J})}{3(1.00 \text{ mol})(8.3145 \text{ J mol}^{-1} \text{ K}^{-1})} = 1.00 \text{ K}$$

Recognizing that 12.472 J has the same magnitude as $\frac{3}{2}R$, let us see now what we have: an increase of 1 degree caused by the addition of 12.472 joules ($\frac{3}{2}R$) to 1 mole of helium at constant volume. But that description corresponds perfectly to the meaning of c_V, the **molar heat**

capacity at constant volume, in the expression

$$q_V = n c_V \Delta T$$

For n moles of a monatomic gas, then, we derive the universal result

$$\Delta E_V = \tfrac{3}{2} nR \, \Delta T = q_V$$

$$c_V = \frac{q_V}{n \, \Delta T} = \tfrac{3}{2} R$$

where the subscript V indicates constant volume throughout the process.

QUESTION: Is c_V larger or smaller for a polyatomic gas?

ANSWER: Larger, because a molecule has more ways to move than a single atom. A molecule has bonds that can vibrate and axes that can rotate; and the bigger the molecule, the richer the possibilities. More heat is needed to stimulate the additional degrees of freedom.

EXAMPLE 13-6. Heat Capacity at Constant Pressure

PROBLEM: By how much does the temperature of helium rise if 20.786 joules of heat are added to the system at constant pressure?

SOLUTION: Delivered at constant pressure, the influx of heat brings about a change in enthalpy

$$\Delta H = \Delta E + \Delta(PV) = q_P$$

which proves equivalent to

$$q_P = \Delta E + \Delta(nRT)$$

upon application of the ideal gas equation, $PV = nRT$. We know furthermore that (1) ΔE is equal to $\tfrac{3}{2} \Delta(nRT)$, and (2) $\Delta(nRT)$ becomes $nR \, \Delta T$ for constant n; and so we have

$$q_P = \Delta E + nR \, \Delta T = \tfrac{3}{2} nR \, \Delta T + nR \, \Delta T = \tfrac{5}{2} nR \, \Delta T$$

after making these substitutions. The resulting expression defines the **molar heat capacity at constant pressure**, c_P, for a monatomic gas:

$$q_P = nc_P \, \Delta T$$

$$c_P = \tfrac{5}{2}R$$

The last step is to compute the rise in temperature, using the data stated above ($q_P = 20.786$ J, $n = 1.00$ mol):

$$\Delta T = \frac{q_P}{nc_P} = \frac{q_P}{n(5R/2)} = \frac{20.786 \text{ J}}{(1.00 \text{ mol})(2.5 \times 8.3145 \text{ J mol}^{-1} \text{ K}^{-1})}$$

$$= 1.00 \text{ K}$$

The 20.786 joules added to the system is exactly $\tfrac{5}{2}R$, just enough to raise the temperature of one mole of helium by one degree.

EXAMPLE 13-7. Condensed Phases: Where $\Delta E \approx \Delta H$ (Usually)

PROBLEM: The molar heat capacity for liquid water is nearly constant over the range 0°C to 100°C. Using the approximate value $c_P = 75.3$ J mol^{-1} K^{-1} and the densities given below, compute the difference in internal energy (per mole) between water at 0°C and water at 90°C:

T (°C)	DENSITY AT 1 atm (g mL^{-1})
0	0.99984
30	0.99565
60	0.98320
90	0.96535

SOLUTION: The densities show explicitly that liquid water, typical of condensed phases, is far less squeezable than a gas. Since the volume responds relatively little to changes in pressure and temperature, the opportunity to do *PV* work is limited. Between 0°C and 90°C, for instance, the entire expansion in molar volume amounts to only 3.5%:

$$V(0°\text{C}) = \frac{1 \text{ mL}}{0.99984 \text{ g}} \times \frac{18.015 \text{ g}}{\text{mol}} \times \frac{1 \text{ L}}{1000 \text{ mL}} \times 1 \text{ mol} = 0.018018 \text{ L}$$

$$V(90°\text{C}) = \frac{1 \text{ mL}}{0.96535 \text{ g}} \times \frac{18.015 \text{ g}}{\text{mol}} \times \frac{1 \text{ L}}{1000 \text{ mL}} \times 1 \text{ mol} = 0.018662 \text{ L}$$

The increase in volume, $\Delta V = V(90°\text{C}) - V(0°\text{C})$, is equal to 0.000644 L.

Now we may choose any path we like to compute the internal energy, provided that the system starts with a temperature of 273 K (0°C) and ends at a temperature of 363 K (90°C). Our simplest option, given the information supplied, is then to arrange for an absorption of heat at a constant pressure of 1 atm, such that state 1 (at low temperature) is transformed into state 2 (at high temperature):

STATE 1   STATE 2

$P_1 = 1.00$ atm $P_2 = 1.00$ atm

$V_1 = 0.018018$ L $V_2 = 0.018662$ L

$n_1 = 1.00$ mol $n_2 = 1.00$ mol

$T_1 = 273$ K $T_2 = 363$ K

Knowing the molar heat capacity at constant pressure, c_p, we are able immediately to compute (to two significant figures) the heat absorbed under constant pressure:

$$q_P = nc_p \, \Delta T = (1.00 \text{ mol})(75.3 \text{ J mol}^{-1} \text{ K}^{-1})(363 \text{ K} - 273 \text{ K}) = 6800 \text{ J}$$

But this quantity of heat is also the change in enthalpy, ΔH, which is related to the change in internal energy by the definition

$$H = E + PV$$

We have, from that, the relationship

$$\Delta E = \Delta H - P_{\text{ext}} \, \Delta V$$

for systems doing only PV work under a constant external pressure.

With both internal pressure and external pressure in equilibrium at 1 atm, the work done by the expanding liquid is only 0.0653 joules, almost negligible in comparison with a ΔH more than 100,000 times larger:

$$P_{\text{ext}} \, \Delta V = (1.00 \text{ atm})(0.000644 \text{ L}) \times \frac{8.3145 \text{ J}}{0.08206 \text{ atm L}} = 0.0653 \text{ J}$$

$$\Delta E = \Delta H - P_{\text{ext}} \, \Delta V = 6800 \text{ J} - 0.0653 \text{ J} \approx 6800 \text{ J}$$

Conclusion: Unless the external pressure is very high, we generally find nearly equal values of ΔE and ΔH for systems in condensed phases.

Enthalpy and energy differ only by the $P \, \Delta V$ work done by the system, and a solid or liquid has little freedom to expand and contract. It does correspondingly little work, unless forced by a pressure far greater than atmospheric.

QUESTION: To simplify the calculation, we assumed that c_P stays constant over the range of temperatures considered. How might we refine our procedure to account for the variation of the heat capacity with temperature?

ANSWER: No matter how strong the variation, there should always be some interval in temperature, δT, small enough so that $c_P(T)$ and $c_P(T + \delta T)$ are essentially the same. If so, then we can safely compute the total heat transferred as a sum of N tiny steps

$$q_P = q_1 + q_2 + \cdots + q_N$$

in which c_P stays constant at each stage i:

$$q_i = n c_P(T_i) \, \delta T_i$$

The procedure is more streamlined than it sounds, because for infinitesimally small intervals δT_i, the step-by-step summation lends itself automatically to the techniques of integral calculus. We shall nowhere undertake such an integration, but merely the possibility of doing so is a powerful option to hold.

EXAMPLE 13-8. Calorimetry

Realize just how convenient it is to have the constant-pressure relationship

$$q_P = \Delta H = n c_P \, \Delta T$$

between the heat absorbed by a substance and the change in temperature thereby produced. With it, an experimenter can systematically measure the heat produced or consumed by any chemical reaction.

The first step is to pick a material that will act as a source or sink of heat, as needed. Water is often a good choice.

Second, we measure c_P experimentally for this heat reservoir over a range of temperatures.

Third, we select a chemical process for which the enthalpy of reaction is unknown. The transformation will either absorb heat from the surroundings (be endothermic) or give heat to the surroundings (be exothermic).

Fourth, we place the heat bath in thermal contact with the reacting system. One way is to run the process in solution, designating the solvent as the heat bath. A Styrofoam cup filled with water, held under atmospheric pressure, makes a homely but surprisingly effective constant-pressure calorimeter.

Fifth, we measure the change in the solution's temperature consequent to the reaction of a known amount of substance. If the temperature goes up, then the reaction evidently produces heat; it is exothermic. If the temperature goes down, then the dissolved system draws heat away from its surroundings (the solvent, water) to fuel an endothermic transformation.

PROBLEM: A precipitate of AgCl forms when solutions of $AgNO_3$ and NaCl are mixed together, in accord with the overall reaction

$$AgNO_3(aq) + NaCl(aq) \longrightarrow AgCl(s) + NaNO_3(aq)$$

and the net ionic reaction

$$Ag^+(aq) + Cl^-(aq) \longrightarrow AgCl(s)$$

When 25.0 mL of 0.100 M $AgNO_3$ and 25.0 mL of 0.100 M NaCl are stirred together in a constant-pressure calorimeter, the temperature of the solution rises from 25.000°C to 25.784°C. Compute the molar change in enthalpy, ΔH, making the following assumptions: (1) The combined solution has a volume of 50.0 mL and a mass of 50.0 g. (2) The heat of reaction is taken up entirely by the water, with no loss either to the walls of the calorimeter or to any other portion of the surroundings. (3) The molar heat capacity of water is 75.3 J mol^{-1} K^{-1} at 1 atm.

SOLUTION: We observe that a certain mass of water (50.0 g) absorbs a certain quantity of heat at constant pressure (q_P), accompanied by a rise in temperature of 0.784 degrees. The reaction, which increases the temperature of the surroundings, is clearly exothermic; it releases heat. But how much? How many joules are produced by how many moles?

First, we compute the heat absorbed by the water:

$$q_P = nc_P\, \Delta T = \left(50.0 \text{ g} \times \frac{1 \text{ mol}}{18.015 \text{ g}}\right) \times \frac{75.3 \text{ J}}{\text{mol K}} \times 0.784 \text{ K} = 163.85 \text{ J}$$

Next, finding that 0.00250 mol Ag^+ reacts with 0.00250 mol Cl^-, we assume that 0.00250 mol AgCl is formed stoichiometrically:

$$\frac{0.100 \text{ mol AgNO}_3}{L} \times 0.0250 \text{ L} \times \frac{1 \text{ mol Ag}^+}{1 \text{ mol AgNO}_3} = 0.00250 \text{ mol Ag}^+$$

$$\frac{0.100 \text{ mol NaCl}}{L} \times 0.0250 \text{ L} \times \frac{1 \text{ mol Cl}^-}{1 \text{ mol NaCl}} = 0.00250 \text{ mol Cl}^-$$

$$0.00250 \text{ Ag}^+(aq) + 0.00250 \text{ Cl}^-(aq) \longrightarrow 0.00250 \text{ AgCl}(s)$$

Putting together, finally, the quantities q_P (163.85 J) and n (0.00250 mol), we obtain the change in enthalpy per mole of substance formed. Since the reaction is exothermic, the sign of ΔH is negative:

$$\Delta H = \frac{-163.85 \text{ J}}{0.00250 \text{ mol}} \times \frac{1 \text{ kJ}}{1000 \text{ J}} = -65.5 \text{ kJ mol}^{-1}$$

The result is good to three significant figures.

QUESTION: What is water's **specific heat capacity**? How would we modify our calculation to use the specific heat capacity rather than the molar heat capacity?

ANSWER: The modification is only slight. The molar heat capacity expresses the change in enthalpy per mole; the specific heat capacity expresses it per gram. They differ by a factor equal to the molar mass:

$$\frac{75.3 \text{ J}}{\text{mol K}} \times \frac{1 \text{ mol}}{18.015 \text{ g}} = \frac{4.18 \text{ J}}{\text{g K}}$$

Molar heat capacity ÷ molar mass = specific heat

The equation for the heat then becomes

$$q_P = \text{mass} \times \text{specific heat} \times \text{change in temperature}$$
$$\text{J} \quad \text{g} \quad \text{J g}^{-1} \text{ K}^{-1} \quad\quad\quad \text{K}$$

QUESTION: Should we revise our assumption concerning a perfectly loss-free calorimeter? Surely no real piece of equipment would function this way.

ANSWER: Yes, we should. A more accurate determination would take into account the heat capacity of the calorimeter itself. An empirical correction factor, the *calorimeter constant*, is best used to adjust the observed change in temperature. Every calorimeter has its own calorimeter constant.

EXAMPLE 13-9. Hess's Law

Again we exploit the defining property of a function of state: insensitivity to path. The difference in enthalpy between reactants and products is the same regardless of what happens along the way.

PROBLEM: Using the results from Example 13-8 and the formation data provided below, compute the standard heat of formation for $AgCl(s)$:

SUBSTANCE	ΔH_f° (kJ mol^{-1})
$Ag(s)$	0
$Ag^+(aq)$	105.6
$Cl_2(g)$	0
$Cl^-(aq)$	−167.2

SOLUTION: Our goal is to determine ΔH_1° for the reaction

$$(1) \quad Ag(s) + \tfrac{1}{2}Cl_2(g) \longrightarrow AgCl(s) \qquad \Delta H_1^\circ = ?$$

at a temperature of 25°C and a pressure of 1 atm. This quantity, the heat absorbed or released per mole of $AgCl(s)$, conforms precisely to the definition of ΔH_f°—the change in enthalpy obtained upon formation of a compound from its elements under standard conditions.

From Example 13-8, we already know ΔH_r° for the precipitation of silver chloride at 25°C and 1 atm:

$$(2) \quad Ag^+(aq) + Cl^-(aq) \longrightarrow AgCl(s) \qquad \Delta H_2^\circ = -65.5 \text{ kJ mol}^{-1}$$

From the formation data supplied, we also know ΔH_r° for reactions (3) and (4) below:

$$(3) \quad Ag(s) \longrightarrow Ag^+(aq) + e^- \qquad \Delta H_3^\circ = 105.6 \text{ kJ mol}^{-1}$$

$$(4) \quad \tfrac{1}{2}Cl_2(g) + e^- \longrightarrow Cl^-(aq) \qquad \Delta H_4^\circ = -167.2 \text{ kJ mol}^{-1}$$

And since (1) is equivalent to the sum of reactions (2), (3), and (4), Hess's law ensures that

$$\Delta H_1^\circ = \Delta H_2^\circ + \Delta H_3^\circ + \Delta H_4^\circ = -65.5 + 105.6 - 167.2$$
$$= -127.1 \text{ kJ mol}^{-1}$$

By imagining this arbitrary (but mathematically convenient) route from reactants to products, we thus determine silver chloride's standard heat of formation without resorting to additional calorimetry. The number obtained, -127.1 kJ mol^{-1}, is identical to the value given in Appendix C (Table C-16).

EXAMPLE 13-10. Hess's Law Again

PROBLEM: By a method similar to the one used in Example 13-9, compute ΔH_f° for $CH_4(g)$ given the following data:

$$\text{(1) } C(s) + O_2(g) \longrightarrow CO_2(g) \qquad \Delta H_1^\circ = -393.5 \text{ kJ mol}^{-1}$$

$$\text{(2) } H_2(g) + \tfrac{1}{2}O_2(g) \longrightarrow H_2O(\ell) \qquad \Delta H_2^\circ = -285.8 \text{ kJ mol}^{-1}$$

$$\text{(3) } CH_4(g) + 2O_2(g) \longrightarrow CO_2(g) + 2H_2O(\ell) \quad \Delta H_3^\circ = -890.3 \text{ kJ mol}^{-1}$$

SOLUTION: The formation reaction for $CH_4(g)$,

$$C(s) + 2H_2(g) \longrightarrow CH_4(g)$$

is enthalpically equivalent to a sum of three known processes: reaction (1) plus twice reaction (2) plus the reverse of reaction (3).

$$
\begin{array}{ll}
C(s) + O_2(g) \longrightarrow CO_2(g) & \Delta H^\circ = \Delta H_1^\circ \\
2 \times [H_2(g) + \tfrac{1}{2}O_2(g) \longrightarrow H_2O(\ell)] & \Delta H^\circ = 2\,\Delta H_2^\circ \\
CO_2(g) + 2H_2O(\ell) \longrightarrow CH_4(g) + 2O_2(g) & \Delta H^\circ = -\Delta H_3^\circ \\
\hline
C(s) + 2H_2(g) \longrightarrow CH_4(g) & \Delta H^\circ \equiv \Delta H_f^\circ[CH_4(g)]
\end{array}
$$

The combined change in enthalpy then follows from Hess's law:

$$\Delta H_f^\circ[CH_4(g)] = \Delta H_1^\circ + 2\,\Delta H_2^\circ - \Delta H_3^\circ$$

$$= -393.5 + 2(-285.8) - (-890.3) = -74.8 \text{ kJ mol}^{-1}$$

Note the rules for manipulating the enthalpies. First, multiplication of all the stoichiometric coefficients by k results in a corresponding scaling of the enthalpy by k. Why? Because H is an extensive property; it depends on the amount of substance. ΔH becomes $k\,\Delta H$. Second, reversal of the sense of reaction similarly reverses the sign of ΔH. Why? Because if heat is absorbed (or released) during the assembly of the products, then exactly the same amount of heat must be released (or absorbed) during their disassembly. ΔH becomes $-\Delta H$.

EXAMPLE 13-11. Hess's Law: Manipulating the Heats of Formation

PROBLEM: Given that the standard heat of formation for $CH_4(g)$ is -74.8 kJ mol^{-1} (Example 13-10), compute $\Delta H°$ for the following reaction:

$$CH_4(g) + 2O_2(g) \longrightarrow CO_2(g) + 2H_2O(\ell)$$

SOLUTION: Here we can use the alternative form of Hess's law, which states that

$$\Delta H° = [c \, \Delta H_f°(C) + d \, \Delta H_f°(D)] - [a \, \Delta H_f°(A) + b \, \Delta H_f°(B)]$$

for any general reaction

$$aA + bB \longrightarrow cC + dD$$

The relevant heats of formation, taken either from Example 13-10 or Appendix C (Table C-16), are

SUBSTANCE	$\Delta H_f°$ (kJ mol^{-1})
$CH_4(g)$	-74.8
$O_2(g)$	0.0
$CO_2(g)$	-393.5
$H_2O(\ell)$	-285.8

and the overall change in enthalpy per mole of CH_4 is therefore

$$\Delta H° = \Delta H_f°[CO_2(g)] + 2 \, \Delta H_f°[H_2O(\ell)] - \Delta H_f°[CH_4(g)] - 2 \, \Delta H_f°[O_2(g)]$$

$$= -393.5 + 2(-285.8) - (-74.8) - 2(0) = -890.3 \text{ kJ mol}^{-1}$$

$\Delta H°$ is the same as that given for reaction (3) in Example 13-10, as we expect. The net change in enthalpy is independent of the path taken.

EXERCISES

1. Classify each of the following properties as either extensive or intensive:

    (a) pressure
    (b) temperature
    (c) number of moles
    (d) volume
    (e) concentration
    (f) density
    (g) molar volume

2. The molar heat capacity (the number of joules absorbed per degree *per mole*) is an intensive property, the same at all points in a material. But consider: Is it always so? Do intensive properties actually stay intensive no matter how small the scale? Would heat capacity or any other macroscopic quantity remain uniform if it pertained to only a handful of particles? Reformulate the definition of an intensive property to make clear exactly what we mean.

3. The lesson from above: Thermodynamics is the law of the large. To speak of a thermodynamic property is to admit that very many particles—so many that we need not even count them—contribute to the observed macroscopic behavior. Otherwise, it makes no sense. So ask: How many is enough? How small is too small? How many atoms of helium, say, are contained within each of the following cubic volumes at STP?

    (a) 1 cm × 1 cm × 1 cm
    (b) 1 mm × 1 mm × 1 mm
    (c) 1 μm × 1 μm × 1 μm
    (d) 10 nm × 10 nm × 10 nm

    Hint: Take care with the units. One liter denotes the volume contained within a cube 10 cm on a side, so that

    $$1 \text{ L} = 0.1 \text{ m} \times 0.1 \text{ m} \times 0.1 \text{ m} = 0.001 \text{ m}^3$$

4. What do you think now, after solving the last problem? When is it appropriate to speak thermodynamically, and when is it not?

5. Which of the following quantities are functions of state? Which, by contrast, depend upon the path taken between two states?

   (a) internal energy
   (b) enthalpy
   (c) heat
   (d) work
   (e) distance traveled between cities
   (f) minimum distance traveled between cities

6. Are there circumstances under which either work or heat can be a function of state? Give an example.

7. Suppose that one mole of an ideal gas is bottled under a pressure of 100 atm and a temperature of 1000 K; and suppose further that the container is a perfect insulator: No additional heat flows in; none flows out. No work is done *on* the system; no work is done *by* the system. If the gas is thus isolated from the world, left alone in its own private universe, what will be its temperature, pressure, and internal energy in the year 3000? Invoke the first law of thermodynamics.

8. Suppose, this time, that our ideal gas is sealed in a container with rigid, immovable walls impervious to the flow of matter but not to the flow of heat. Let thermal energy flow freely between the gas inside ($P = 100$ atm, $T = 1000$ K, $n = 1$ mol) and the surroundings outside ($P = 1$ atm, $T = 273$ K), and wait. . . . (a) Calculate the volume and internal energy of the gas at the start. (b) In which direction does heat flow—from the gas to the world or from the world to the gas? (c) Calculate the temperature, pressure, volume, and number of moles of gas at the end—after the system equilibrates with its surroundings. (d) By how much does the internal energy of the gas change? (e) On the average, are the gas particles moving faster or slower when the process is over?

9. Once more: An ideal gas is placed, as in the exercise above, in a rigid container that conducts heat but not matter. Outside, the world is at STP; inside, there is one mole of gas at $P = 1$ atm and $T = 173$ K. Heat starts to flow. (a) From where to where? (b) What are the values of temperature, pressure, volume, and number of moles of gas at the end—after the system equilibrates with its surroundings? (c) By how much does the internal energy of the gas change? (d) On the average, are the gas particles moving faster or slower after the process

is over? (e) Taking helium as a specific example, calculate the root-mean-square speeds before and after. (f) Do the same for neon. Why are the speeds different from those of helium at the same temperature, even though both systems have the same internal energy?

10. Which system has the greater thermal energy (enthalpy) overall? Which has the greater thermal energy per molecule?

    (a) a swimming pool at 80°F or the Atlantic Ocean at 33°F
    (b) 1 mL water at 100°C or 1 mL water at 0°C
    (c) 1 mL water at 10°C or 1 L water at 10°C
    (d) 1 L gaseous water at 1000 K or the entire universe at 5 K

11. Look again. Pick out those pairs in which there will be a flow of heat from one part of the system to another. From where to where?

    (a) a swimming pool at 80°F in contact with the ocean at 33°F
    (b) 1 mL water at 100°C in contact with 1 mL water at 0°C
    (c) 1 mL water at 10°C in contact with 1 L water at 10°C
    (d) 1 L gaseous water at 1000 K in contact with the entire universe at 5 K

    In which systems will heat *not* flow? Why?

12. (a) Which system has the greater thermal energy—one mole of liquid water at 100°C or one mole of water vapor at 100°C? (b) If one phase is indeed "hotter" than the other, then why is there no difference in temperature? Where does the energy go, if not into faster motion of the molecules?

13. Start to distinguish between internal energy (E) and enthalpy ($H = E + PV$): (a) The density of water at its normal boiling point is 0.95840 g mL^{-1}, and the difference in enthalpy between liquid and vapor is 40.7 kJ mol^{-1} at 100°C and 1 atm. Calculate the molar volume of water and water vapor at 100°C. (b) Calculate the change in internal energy, ΔE, when one mole of water boils. (c) Similarly, calculate the change in internal energy when one mole of steam condenses at 100°C. (d) Which number has the larger absolute value during condensation: ΔE or ΔH? Why?

14. Water and ice differ in enthalpy by 6.0 kJ mol^{-1} at 0°C and 1 atm. The density of the liquid at 0°C is 0.99984 g mL^{-1}, whereas the density of the solid is 0.917 g mL^{-1}. (a) Calculate the molar volume of water and ice at 0°C. Which is larger? (b) Calculate the change in internal energy, ΔE, when one mole of ice melts. (c) Calculate the

change in internal energy when one mole of water freezes. (d) Which number has the larger absolute value during freezing: ΔE or ΔH? Why?

15. Why is the difference between ΔH and ΔE so much greater for a liquid–vapor transition than for a liquid–solid transition?

16. Water, which expands when it freezes, is unusual; most other substances contract. Their molecules squeeze closer together in the solid phase. The volume decreases. Suppose, then, that ΔE for some normal material is -100 J during a freezing transition. (a) Which of the values below is most probable for the associated ΔH?

$$-101 \text{ J}, \quad -99 \text{ J}, \quad 0, \quad 99 \text{ J}, \quad \text{or} \quad 101 \text{ J}$$

(b) If so, then what is the correct value for ΔH during the melting transition?

17. Pursue further the related notions of expansion and contraction, work and heat, energy and enthalpy—beginning, just for fun, with an exercise involving the units of pressure, volume, and work-energy. Prove that the ratio

$$\frac{\text{force} \times \text{length}}{\text{pressure}}$$

has dimensions of volume, and compute the number of liters represented by each of the following quantities:

(a) 1 N m atm^{-1}
(b) 1 kg m^2 s^{-2} atm^{-1}
(c) 1 g cm^2 s^{-2} torr^{-1}
(d) 1 J Pa^{-1}

18. The temperature of a confined gas is 300 K. The temperature outside is 270 K. (a) Is the gas able to expand isothermally? (b) Is the gas able to contract isothermally?

19. A gaseous system expands against the pressure of the atmosphere while maintaining a constant temperature. Work is done. (a) Is the work done by the system on the surroundings, or by the surroundings on the system? (b) Suppose that w has a magnitude of 100 J. Give it a proper algebraic sign. (c) Consider just these 100 joules of work for the moment, nothing else: Does the mechanical expansion,

by itself, raise or lower the system's internal energy? (d) Now account for the thermal contribution: Is there a flow of heat? If so, which party supplies it—the system or the surroundings? (e) Does the heat flux, by itself, raise or lower the system's internal energy? By how many joules? (f) Compute ΔE and ΔH for both system and surroundings.

20. Do it again, but this time in the opposite direction: A gaseous system *contracts* under a constant pressure while maintaining a constant internal temperature. Accompanying the process is a heat flow of 100 J. (a) In which direction is the heat transferred? What is the algebraic sign of q? (b) How much work is done? How many joules? Is the sign positive or negative? (c) Compute ΔE and ΔH for both system and surroundings.

21. One mole of an ideal gas at STP absorbs 520 joules of heat from its surroundings, reaching a final temperature of 298 K. The process occurs under constant external pressure. (a) Compute the change in internal energy, ΔE, for the system. (b) How much work is done? (c) Does the volume of the gas increase, decrease, or remain the same? Is the change consistent with the ideal gas law? (d) Compute the change in enthalpy, ΔH. (e) Is the process exothermic or endothermic?

22. One mole of neon gas, initially at STP, is held in a rigid container at constant volume—where, interacting with the surroundings, the gas manages to increase its temperature from 273 K to 298 K. (a) Calculate q, w, ΔE, and ΔH. (b) In which direction does the heat flow?

23. Argon gas, initially at STP, suffers an abrupt increase of pressure to 100 atm followed by a sudden reduction of its volume to 50 cm^3. Then the gas is heated very slowly to 700 K, whereupon it undergoes 100 turbulent cycles of expansion and compression throughout which the pressure fluctuates wildly. Finally, the gas is cooled and stabilized at a pressure of 1.5 atm and a temperature of 273 K. (a) Calculate ΔE, the net change in internal energy from start to finish. (b) Has the system undergone a reversible transformation?

24. Heat capacity: The heat flows in and the heat flows out, but how much here and how much there? Say that a certain substance X has a constant-pressure heat capacity of 50 J mol^{-1} K^{-1}, the same at all temperatures. By contrast, the heat capacity of some other substance Y

varies over a range of temperature: 50 J mol^{-1} K^{-1} between 0 K and 200 K, but 100 J mol^{-1} K^{-1} everywhere else. (a) How many joules will substance X absorb while going from 100 K to 300 K? (b) How many joules does Y need to realize the same change in temperature?

25. For substance X, c_p is 100 J mol^{-1} K^{-1}. For substance Y, c_p is 200 J mol^{-1} K^{-1}. (a) Which substance—X or Y—suffers the greater change in temperature for the same amount of heat flowing in or out? (b) Why, now that we mention it, should different substances have different heat capacities? (c) Why do all ideal monatomic gases have the same heat capacity?

26. Which probably has the higher heat capacity?

    (a) helium gas or propane gas (C_3H_8)
    (b) methane gas (CH_4) or liquid octane (C_8H_{18})
    (c) neon gas or chlorine gas (Cl_2)
    (d) DNA or water
    (e) gaseous ammonia (NH_3) or gaseous hydrogen (H_2)

    How is the incoming thermal energy invested in these various systems?

27. (a) How much heat (how many joules) must 100 grams of neon absorb for the temperature to rise by 100 degrees at constant volume? (b) How much at constant pressure?

28. Imagine that a mass of 1.00 kg falls from a height of 10.0 m into a bucket containing 1.00 L of water. (a) By how much would the temperature rise if the water absorbs all the body's kinetic energy as heat? Assume a density of 1.00 g mL^{-1} and a heat capacity (c_p) of 75.3 J mol^{-1} K^{-1}. (b) What would be the change in temperature if the bucket held only 0.100 L of water?

29. A mass of 1.00 kg falls from a height of 10.0 m into a bucket containing 1.00 L of ethanol (C_2H_5OH). By how much would the temperature change if the liquid absorbs all the energy of the falling body as heat? The heat capacity (c_p) and density of ethanol are, respectively, 111.5 J mol^{-1} K^{-1} and 0.789 g mL^{-1} at 25°C.

30. The molar heat capacity of methanol (CH_3OH) is 81.6 J mol^{-1} K^{-1} at 25°C. (a) What is the *specific heat* of methanol, expressed in

$J g^{-1} K^{-1}$? (b) Suppose that the temperature of x grams of methanol increases by $1.17°C$ upon absorbing 313 joules of heat. Compute x.

31. The specific heat of silver is $0.235 J g^{-1} K^{-1}$. (a) Compute the heat capacity of 10.0 g of silver (the number of joules needed to raise the temperature of ten grams by one degree). (b) How many joules of heat must 10.0 g of silver absorb to increase its temperature from $20.0°C$ to $30.3°C$?

32. Use the data in Appendix C to calculate $\Delta H°$ for each of the reactions below. State whether the transformation is exothermic or endothermic.

    (a) $K(s) \longrightarrow K(g)$
    (b) $Mg^{2+}(aq) + 2Cl^-(aq) \longrightarrow MgCl_2(s)$
    (c) $NaOH(s) \longrightarrow Na^+(aq) + OH^-(aq)$
    (d) $CH_4(g) + 2O_2(g) \longrightarrow CO_2(g) + 2H_2O(\ell)$
    (e) $N_2(g) + 3H_2(g) \longrightarrow 2NH_3(g)$
    (f) $H_2O_2(\ell) \longrightarrow H_2O(\ell) + \frac{1}{2}O_2(g)$

33. Use the data in Appendix C to calculate $\Delta H°$ for each of the reactions below. State whether the transformation is exothermic or endothermic.

    (a) $S(s) + O_2(g) \longrightarrow SO_2(g)$
    (b) $H^+(aq) + OH^-(aq) \longrightarrow H_2O(\ell)$
    (c) $CH_4(g) \longrightarrow C(s) + 2H_2(g)$
    (d) $H_2O(g) \longrightarrow H_2(g) + \frac{1}{2}O_2(g)$
    (e) $2O_3(g) \longrightarrow 3O_2(g)$
    (f) $F_2(g) \longrightarrow 2F(g)$

34. Use the data in Appendix C to calculate $\Delta H°$ for each of the reactions below. State whether the transformation is exothermic or endothermic.

    (a) $C_4H_{10}(\ell) \longrightarrow C_4H_{10}(g)$
    (b) $C_4H_{10}(g) \longrightarrow C_4H_{10}(\ell)$
    (c) $C_6H_6(\ell) \longrightarrow C_6H_6(g)$
    (d) $C_6H_6(g) \longrightarrow C_6H_6(\ell)$
    (e) $I_2(s) \longrightarrow I_2(g)$
    (f) $I_2(g) \longrightarrow I_2(s)$

35. The standard enthalpy of reaction for the decomposition of 2 mol N_2O is -164.2 kJ:

$$2N_2O(g) \longrightarrow 2N_2(g) + O_2(g) \qquad \Delta H° = -164.2 \text{ kJ}$$

Compute the standard heat of formation $(\Delta H_f°)$ per mole of $N_2O(g)$.

36. Use the data in Appendix C to calculate $\Delta H°$ for each transformation:

(a) $H_2(g) + F_2(g) \longrightarrow 2HF(g)$
(b) $2C(s) + 2H_2(g) \longrightarrow C_2H_4(g)$
(c) $SO_3(g) \longrightarrow S(s) + \frac{3}{2}O_2(g)$
(d) $2NaCl(s) \longrightarrow 2Na(s) + Cl_2(g)$

37. Carrying over the results from the previous exercise, compute $\Delta H°$ for the formation of 1.00 g of product in each reaction:

(a) $H_2(g) + F_2(g) \longrightarrow 2HF(g)$
(b) $2C(s) + 2H_2(g) \longrightarrow C_2H_4(g)$
(c) $S(s) + \frac{3}{2}O_2(g) \longrightarrow SO_3(g)$
(d) $2Na(s) + Cl_2(g) \longrightarrow 2NaCl(s)$

38. (a) Write a balanced equation for the combustion of propane,

$$\_\_ C_3H_8(g) + \_\_ O_2(g) \longrightarrow \_\_ CO_2(g) + \_\_ H_2O(\ell)$$

and calculate, using the data in Appendix C, the standard enthalpy change for the reaction. (b) How many joules of heat are produced by the burning of 1.00 g of propane?

39. (a) Write a balanced equation for the combustion of butane,

$$\_\_ C_4H_{10}(g) + \_\_ O_2(g) \longrightarrow \_\_ CO_2(g) + \_\_ H_2O(\ell)$$

and calculate, using the data in Appendix C, the standard enthalpy change for the reaction. (b) How many joules of heat are produced by the burning of 1.00 g of butane? (c) Which fuel—propane or butane—yields more heat per gram?

40. Complete combustion of sucrose $(C_{12}H_{22}O_{11})$ to carbon dioxide and water releases 5644 kJ of enthalpy in the reaction shown below:

$$C_{12}H_{22}O_{11}(s) + 12O_2(g) \longrightarrow 12CO_2(g) + 11H_2O(\ell)$$
$$\Delta H° = -5644 \text{ kJ}$$

(a) Compute the standard enthalpy of formation for sucrose.

(b) Compute the enthalpy of formation liberated per gram of the sugar.

41. (a) First, use data from Appendix C to compute the enthalpy of reaction for the dissolution of sodium chloride:

$$NaCl(s) \longrightarrow Na^+(aq) + Cl^-(aq)$$

(b) Next, suppose that 10.00 grams are dissolved in 50.00 milliliters of water originally at a temperature of 25.0°C. What is the temperature of the water immediately after the reaction? Assume a density of 1.00 g mL^{-1} and a heat capacity (c_p) of 75.3 J mol^{-1} K^{-1}.

42. Study the formation data tabulated below:

SUBSTANCE	ΔH_f° (kJ mol^{-1})
$NaNO_3(aq)$	−445.1
$Na^+(aq)$	−240.1
$Ag^+(aq)$	+105.6
$AgNO_3(s)$	−124.4

(a) Compute ΔH_f° for the hydrated nitrate ion, NO_3^- (aq).

(b) Compute the reaction enthalpy for the dissolution of $AgNO_3$:

$$AgNO_3(s) \longrightarrow Ag^+(aq) + NO_3^-(aq) \qquad \Delta H^\circ = ?$$

(c) If 10.00 grams of silver nitrate are dissolved in 200.0 milliliters of water originally at 25.00°C, what is the temperature of the water after the reaction? Assume a density of 1.00 g mL^{-1} and a heat capacity (c_p) of 75.3 J mol^{-1} K^{-1}.

14

Free Energy and the Direction of Change

14-1. Statistical Inevitability
14-2. Distributions, Microstates, and Disorder
   Positional Disorder
   Energetic Disorder
14-3. Entropy
14-4. The Second Law of Thermodynamics
14-5. Free Energy
14-6. Standard Free Energy and Entropy
14-7. The Free Energy of Reaction
14-8. Free Energy and the Equilibrium Constant
14-9. Thermal Equilibrium and the Meaning of Temperature
   REVIEW AND GUIDE TO PROBLEMS
   EXERCISES

14-1. STATISTICAL INEVITABILITY

Beneath the quiet face of bulk matter is a hidden substructure of atoms and molecules, vast and unsettled, a miniature universe of small particles in motion. It is a regime governed by the laws of large numbers, by distributions and averages . . . by statistics. Statistical constraints, most of all, give matter its smooth and familiar aspect, leaving it sparsely characterized by pressure, volume, temperature, number of moles, and just a few other simple properties. But although these limited attributes aptly define some particular macroscopic condition (call it a **macrostate**), we know that such description provides only a coarse summary of an ever-changing microworld. Where are the molecules? How fast are they moving? What are their energies? What, in other words, are all the detailed

489

arrangements at the microscopic level—the **microstates**—that can produce this one particular macrostate?

Looking into the organization of ordinary matter, we come upon a small world so complex as to defy comprehension. Even the number of molecules in a thimbleful of water, exceeding 10^{23}, is already beyond any human experience; how much more so, then, is the multiplicity of microstates—all the different possibilities for particle positions, velocities, and energies. It is an unexpected picture of infinity, encountered not in the large but in the small. Yet infinite it is. No less than the stars in the sky, the microstates of just these few milliliters are virtually uncountable.

The consequences are clear: When a group becomes large enough, there is typically only one way it can behave. We can hardly imagine, let us say, flipping a coin 100 billion times and having it come up heads each throw. No. The only plausible result will be within a hairsbreadth of 50 billion heads and 50 billion tails, give or take maybe a million. No other outcome has anything but a vanishing probability, including a deviation not "too" drastic like 51 billion heads and 49 billion tails. And although such rare events might someday occur (if only one waits long enough), that day is scarcely to be expected. For when the numbers get really large, as they do with molecules, even the lifetime of the universe is not long enough to beat the odds.

Nor can chemistry beat the odds. Chemistry runs on statistics, by necessity, as we saw first in the kinetic theory of gases. Always there is too much of everything—too many particles, too many arrangements, too many degrees of freedom—to permit otherwise. Chemical changes are driven by statistical inevitability, formally embodied in the concept of entropy with its notion of disorder. The visible world, rendered smooth and simple, is the inescapable result.

We have come to appreciate, if only roughly, the power of large numbers and the tendency toward increasing disorder in the universe. To these rudimentary ideas we now lend some substance.

14-2. DISTRIBUTIONS, MICROSTATES, AND DISORDER

Flip a coin once and it will come up heads or tails; those are the only two choices. Flip it twice and there are four conceivable outcomes,

HH

HT TH

TT

which may be classified according to the number of heads: one way of getting zero heads (TT), two ways of getting one head (HT and TH), and one way of getting two heads (HH). This tabulation is expressed concisely as a **distribution**, by which we arrange the results according to total heads regardless of permutations such as HT and TH. Here the three broad categories (zero, one, and two heads) are distributed in the ratio 1:2:1, with the most likely outcome being one head and one tail.

For three tosses, the eight possibilities

HHH

HHT   HTH   THH

TTH   THT   HTT

TTT

are distributed as 1:3:3:1 for zero, one, two, and three heads. The pattern for four throws is 1:4:6:4:1, and once again the largest number of combinations is at the center of the distribution (two heads and two tails). Thus it continues, with the graphs in Figure 14-1 growing more and more peaked as the number of throws (N) increases. Long before N approaches 10^{11} or 10^{17} or 10^{23}, however, the distribution squeezes into a sharp spike as just one class of outcome (nearly equal numbers of heads and tails) becomes overwhelmingly probable. Nothing else comes close.

What we have is a simple model for microstates and macrostates. The macrostate: how many heads. The microstates: all the equally acceptable ways of getting that number. When $N = 4$, for example, the most likely macrostate is two heads and two tails, which can arise from any one of the six microstates HHTT, HTHT, HTTH, THHT, THTH, TTHH. The least likely macrostates are all heads and all tails, which can only come from one microstate apiece: HHHH or TTTT.

Positional Disorder

Picture now a container of gas divided into left and right halves by an imaginary partition, and let the N particles fly about in random motion. For a macrostate, we ask how many molecules are in the left half (L) and how many are in the right (R)—totals only. For the microstates, though, we want to assign each particle to a stated half of the container (particle 1 on the left, particle 2 on the right, particle 3 on the right . . .).

It amounts again to a coin flip. Each molecule has a 50% chance of being left or right at any time, and so the same statistical distribution applies

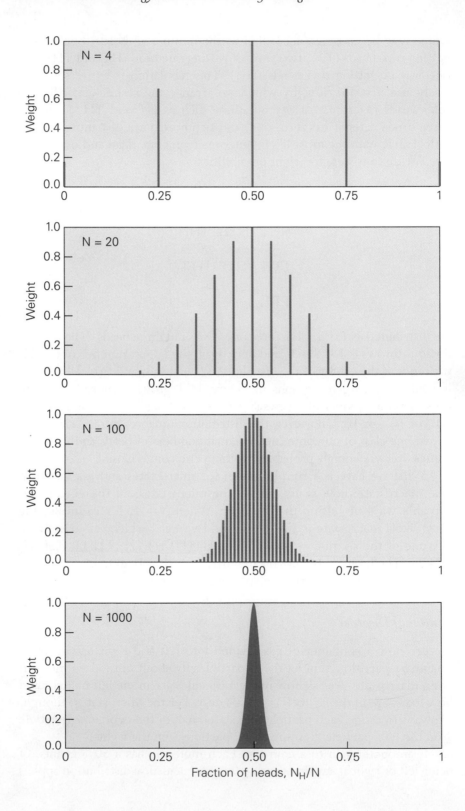

FIGURE 14-1 (opposite page). Distribution of heads and tails for a coin tossed N times, drawn to show the relative importance (the weight) attached to each group of outcomes: the different ways in which a certain number of heads can turn up. A given series produces N_H heads and $N - N_H$ tails, with the largest set containing equal numbers of each ($N_H = N/2$). All the weights are compared consistently on a scale where this most likely outcome—always in the center of the plot—has the value 1. The distribution sharpens to a spike as N grows large.

as before. The *least* likely macrostate (Figure 14-2) has all the particles on one side or another, an outcome for which there is just one acceptable partitioning. With, say, $N = 4$, the only microstate that puts all the molecules at the left is LLLL. The *most* likely macrostate (Figure 14-3) is an equal split between left and right, where we have the same six microstates as for the coins: LLRR, LRLR, LRRL, RLLR, RLRL, RRLL.

When N gets to be Avogadro's number, the most likely macrostate (half on the left, half on the right) is supported by so many microstates that no appreciably different arrangement is statistically probable. That condition is the equilibrium point for the gas, and that is the final macrostate which any nonequilibrium distribution will inevitably attain. Once in equilibrium, the N particles are free to place themselves in any of the myriad microstates consistent with $N/2$ particles on the left and $N/2$ on the right. Each arrangement yields the same pressure, the same volume, the same temperature, the same number of moles. Viewing the gas macroscopically, we have no way either to keep track of individual

FIGURE 14-2. In how many ways can four numbered particles occupy just half the volume available to them? Only one: Each particle must be somewhere in the designated zone (here, on the left) and nowhere else.

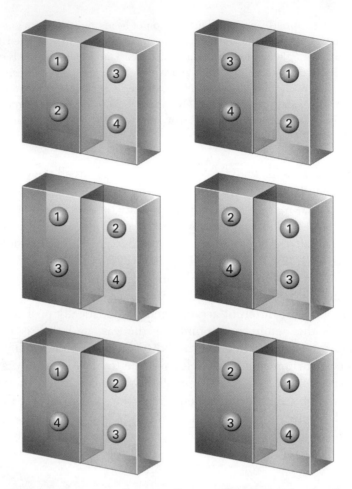

FIGURE 14-3. In how many ways can four numbered particles distribute themselves equally throughout the container? Six: There are six microstates, each containing two of the four particles on the left and two on the right. A 1:1 mixture, always the most likely distribution, becomes overwhelmingly probable as N increases.

particles or to make distinctions of any kind. To us, all the equilibrium microstates look the same.

We say, accordingly, that the gas is *positionally disordered*. At equilibrium, the particles have available to them the largest number of equally acceptable configurations; any set of those positions is as good as any other, and each set is equally accessible. Given that freedom, the gas naturally spreads out in the most random, most likely fashion as it comes to equilibrium. To be *out* of equilibrium, by contrast, is to be in a state with fewer compatible arrangements and hence less choice.

"Less choice," when pushed to extremes, might mean all the particles are forced into one small portion of the container. "Maximum choice" (equilibrium) means the particles are free to roam. Which outcome is inevitable for a gas, given sufficient time? Only one: the randomized state of equilibrium.

Conditions of positional order and disorder are easy to visualize. Two molecules are more disordered than one molecule. Gases are more disordered than liquids, and liquids are more disordered than solids. A perfect crystal, in which the atoms are allowed only one set of locations, is the most ordered state of all.

That said, a crystal strikes us as a statistically unlikely state—a surprise, an oddity, something to be excused. Granted little or no freedom of position, a crystalline lattice (even an imperfect, real-life example) is an arrangement that would be wildly improbable for a liquid or gas. Still, despite the discouraging statistics, crystals do exist. What keeps the structure locked into a state of such extreme order, able to beat the odds?

Lower energy. In particular, the lower energy afforded by intermolecular forces. Various attractions (recall from Chapter 9) hold the atoms and molecules in place, and with these associations comes a reduction of the energy. The crystal gives up a disordered structure in exchange for the lower energy it enjoys when the particles stick together.

Even then, the ordered arrangement achieved in Figure 14-4 persists only as long as conditions allow. Just think of the crystal's fate when the temperature is raised. Its molecules move faster and faster as they acquire more thermal energy; they can hold on to each other no longer; they are forced, irresistibly, into a state of increasing disorder as the temperature rises. Having absorbed sufficient heat, the solid will melt into a liquid and then vaporize into a gas.

Two powerful tendencies are at play within the system: the drive toward lower energy, and the drive toward decreased order. They work at cross purposes. Both are always present, although sometimes one dominates the other. Compromises must be made. A gas, for instance, balances its high energy against a wildly chaotic disorder, whereas a solid purchases its high order with a compensating low energy.

Energy is high in this system and low in that system. Order decreases here and increases there. If we can understand the global interplay of energy and disorder *everywhere*—both for a system and its surroundings together—then we shall be able to understand how and why matter is transformed. The work remaining in this chapter is largely to firm up the qualitative picture developed so far, which has already taken us from coin flips to gas particles and now to positional disorder in general. Energy is the next step.

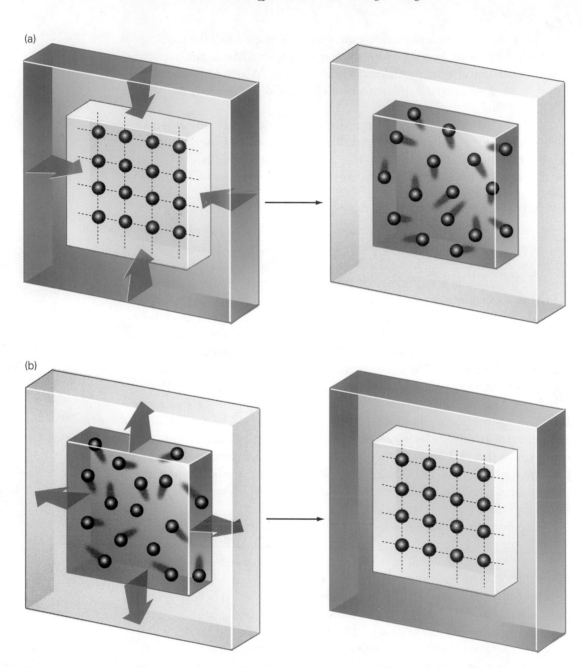

FIGURE 14-4. Overcoming a natural tendency toward disorder, a crystal maintains its symmetry only when the thermal energy is low. (a) An inflow of heat leaves the particles restless and disordered. (b) An outflow of heat—a cooling—allows the atoms to settle into their lattice positions.

Energetic Disorder

Distributions arise whenever an outcome is in doubt. Coin flips are distributed over various combinations of heads and tails. Examination scores are distributed over a bell curve. Molecules are distributed in space. Energy, too, is distributed over a range of values.

By extending the idea of a distribution to energy itself, we move still closer to appreciating the meaning of disorder and the related concept of entropy. We have, in fact, already taken a first step at the end of Chapter 10, where the distribution of particle speeds in a gas was considered. Let us return now to one mole of an ideal gas, but this time with better awareness of distributions and disorder.

There are 6.02×10^{23} particles (N_0), each of which may assume an energy from some set of m allowed values: $\epsilon_1, \epsilon_2, \ldots, \epsilon_m$. If such a gas can be characterized by a temperature, then immediately we know there is a fixed total kinetic energy

$$E_k = \tfrac{3}{2}RT$$

with a corresponding average particle energy, $\langle \epsilon \rangle$, equal to $\tfrac{3}{2}k_B T$. This relationship was one of the most productive results of the kinetic theory.

The aim is to determine how the individual energies are distributed about their average value. We ask how many particles are going this fast and how many are going that fast, looking for a spread in energy analogous to a spread in position. This same question was posed in our discussion of the Maxwell-Boltzmann and Boltzmann distributions in Chapter 10, and it comes back to us now as the central issue in statistical thermodynamics.

Here are the requirements: There will be n_1 molecules with ϵ_1; n_2 with ϵ_2; n_3 with ϵ_3, and so on throughout the whole range. We want to know the values of these numbers n_i; we want, in effect, to take a census of the energies in the way suggested in Figure 14-5. Not all conceivable distributions $n_1, n_2, \ldots, n_m$ are allowed, however, for there are two restrictions to be satisfied. First, since every particle must be accounted for, the individual populations must add up to the total, N_0:

$$n_1 + n_2 + \cdots + n_m = N_0$$

Second, the particle energies must add up to some well-defined total energy, E_k, so that we have

$$\epsilon_1 n_1 + \epsilon_2 n_2 + \cdots + \epsilon_m n_m = E_k$$

These limitations are easy to accept, but telling in their consequences. Suppose (to take an unrealistically small example) there are 1000 particles and the total energy is 10,000 units. Could 999 particles have zero energy

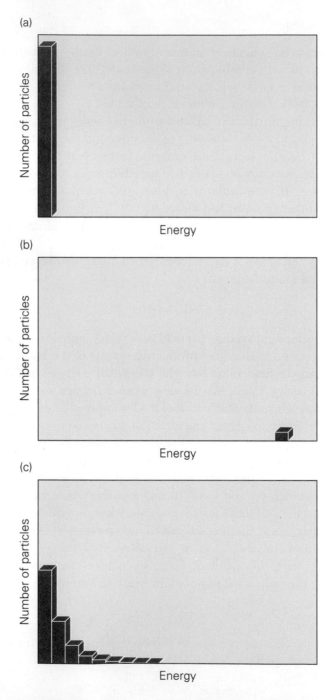

FIGURE 14-5. How will a certain amount of energy be divided among a group of particles? Which distribution will prevail? By how much? Here are three possibilities: (a) Egalitarian: each particle gets the same small allocation. (b) Feudal: one particle has all the energy; the others have none. (c) Boltzmann: more poor than rich, but something for everybody. Statistics will decide.

and one particle possess 10,000 units? Yes. Is that likely? No. Such a distribution is the energetic equivalent of crowding all the molecules into a corner. It can be done, but only in a limited number of ways—1000 possibilities, to be exact. Particle 1 might have $\epsilon = 10,000$ (and all the rest would have $\epsilon = 0$), *or* particle 2 might have $\epsilon = 10,000$, *or* particle 3 might have $\epsilon = 10,000$, . . . *or* particle 1000 might have $\epsilon = 10,000$. Each of these 1000 options is one *microstate*, and each microstate is an equivalent way to realize the macroscopic value $E_k = 10,000$.

How about assigning 100 units of energy to each of 100 particles, thereby also achieving the required value of E_k? Clearly this new distribution is preferable to one in which all the energy is crammed into one particle. Here, at least, there are more possible ways of shuffling the total quantity around—resulting in more microstates, more spread. We can measure that spread by counting microstates (as above) and come up with a number.

Nothing prevents us from making such a count. The procedure soon becomes tedious, though, so let us forgo the arithmetic and simply agree that the relevant number must exist. Call it W.

Does that particular value of W represent the most microstates *ever* possible? To know for sure, we envision going through every conceivable set of values for the n_i and asking two sets of questions. First, does this distribution (so many particles have energy ϵ_1, so many have ϵ_2, so many have ϵ_3, . . .) satisfy the constraints? Do all the populations add up to the total number of particles? Does the total energy add up to the right value? If not, we throw it out and try another distribution. Second (if the proposed distribution is otherwise acceptable), what is W? What is its actual numerical value?

W is the key quantity. W is disorder expressed in quantitative form. Nature's goal at equilibrium is to *maximize* W, the number of microstates, while adhering to the constraints on the populations. It is a contest of sorts, and (except for those two restrictions) the whole counting procedure is not unlike flipping coins. There is, in the end, a value of W recorded for each legal distribution, and the largest one—W_{max}—is the winner. And, with that, the system is granted its overwhelmingly most probable set of microstates, an almost inconceivably large number. Each of the W_{max} states is macroscopically indistinguishable, and together they create the unchanging uniformity of macroscopic equilibrium.

The winner (discovered, admittedly, by mathematics more elegant than the hopeless method of trial and error suggested above) is the Boltzmann distribution,

$$\frac{n_i}{n_j} = \exp\left(-\frac{\epsilon_i - \epsilon_j}{k_B T}\right)$$

first brought out in Chapter 10 and illustrated again in Figure 14-6. Its meaning, as before: Given any two energies ϵ_i and ϵ_j differing by $\Delta\epsilon$, the ratio of populations at equilibrium is $\exp(-\Delta\epsilon/k_BT)$, where k_B denotes the *Boltzmann constant*. Temperature is uniform throughout.

Thus originates not only the Boltzmann distribution for kinetic energy in a gas, but the Boltzmann distribution for electronic energy levels and the Boltzmann distribution for magnetic energy levels and the Boltzmann distribution for *any* set of energies at a fixed absolute temperature T. It is a

(a)

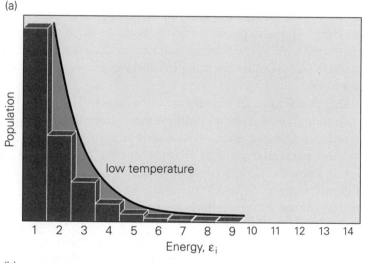

(b)

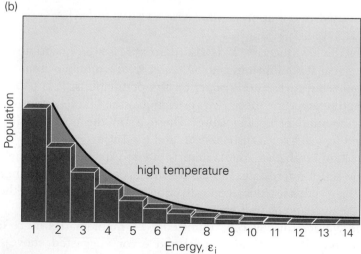

FIGURE 14-6. The winner: the Boltzmann distribution, whereby energy is parcelled out from the ground up. Occupancies of the energy levels are fixed by the temperature. (a) Between any two states differing by $\Delta\epsilon = \epsilon_i - \epsilon_j$, the ratio of populations is $\exp(-\Delta\epsilon/k_BT)$. (b) At a higher temperature, the distribution skews toward higher energy. The particles have more thermal energy to share.

statistical outcome demanded by the *one* set of microstates with the most votes. The number of accessible microstates is maximal at equilibrium.

14-3. ENTROPY

All along we knew that the total energy had to be constant, strictly conserved. Now, having grasped the Boltzmann distribution, we see where it goes. Some energy goes to particle 1 (or, more generally, energy level 1); maybe a little more goes to particle 2; perhaps a little less goes to particle 3. The ratios at equilibrium, determined by the temperature, are set by the exponential factors.

The more interesting question, therefore, is not how *much* energy, but rather *to what end* it is applied. Nature constantly divides and redivides a fixed energetic pie, distributing different amounts to all eligible recipients. Energy ebbs and flows, moving from atom to atom and place to place. Not before the entire universe comes to equilibrium (and forever ceases to change) will such transfers of energy disappear. Until then, it is the way in which energy is allocated—its distribution—that commands our interest. That distribution, too, is what comes to be identified with entropy and disorder.

W, through its count of microscopic configurations, gives quantitative meaning to the notion of order and disorder. An ordered system has relatively few microstates available to it and a correspondingly limited range of statistical options. A disordered system, by contrast, is able to express its macroscopic properties in many equivalent forms. Low *W* means high order; high *W* means low order.

Placing restrictions on molecular positions creates a manifestly more ordered arrangement, characterized in turn by a smaller value of *W*. We counted, for example, far fewer ways to stuff a gas into half its normal volume than to disperse it uniformly throughout. By similar thinking we concluded that a (disordered) liquid has more allowed microscopic configurations than its (ordered) solid form, in which the atoms might be held fixed in a crystalline lattice. And further: A gas obviously has more microscopic possibilities than a liquid, and two moles of gas more possibilities than one, and in this way the relationship between microstates and disorder becomes increasingly apparent.

Any quantitative definition of entropy, then, must depend explicitly on the number of microstates, going up and down with *W*. In addition, entropy is an extensive property which needs to increase as the sum

$$S = S_1 + S_2$$

when two individual systems are joined. The combined number of microstates, on the other hand, transforms into a *product* of independent factors,

$$W = W_1 \times W_2$$

a result we can verify again by flipping coins. If one coin has 2 choices (H and T; $W_1 = 2$), then two coins have 4 choices (HH, HT, TH, TT; $W = 4$):

$$W = W_1 \times W_2 = 2 \times 2 = 4$$

Three coins have 8 choices ($2 \times 2 \times 2$), and four coins have 16 choices ($2 \times 2 \times 2 \times 2$). All these requirements now must be reconciled in a consistent microscopic interpretation.

This way: Understood in statistical terms, entropy is defined as

$$S = k_B \ln W$$

so as to establish the desired link with the number of microstates. To begin, the natural logarithm ensures that S is properly additive since

$$\ln W_1 W_2 = \ln W_1 + \ln W_2$$

Furthermore, $\ln W$ does everything that W does—it goes up when W goes up and it goes down when W goes down—and in so doing it reflects the idea of spread, or disorder. Finally, Boltzmann's proportionality constant ($k_B = 1.38 \times 10^{-23}$ J K^{-1}) gives entropy dimensions of energy (equivalently, heat) divided by absolute temperature.

Carved in stone (literally) on Boltzmann's tomb in Vienna, the statistical definition of entropy is a conceptual breakthrough and one of the most profound equations ever devised. But plugging numbers into it is another story, and usually a hopeless task. For how do we actually calculate $k_B \ln W$ for large systems? The path is blocked by the same insuperable mathematical difficulties that made statistical averaging necessary in the first place.

Entropy becomes something practical, though, when the microscopic definition is connected to the macroscopic thermodynamic function of the same name. Recall the description (in Chapter 13) of an extensive function of state S, for which a small change is defined as

$$\Delta S = \frac{q_{rev}}{T}$$

This macroscopic entropy, so called, deals with heat flow in a reversible process at constant temperature, and superficially it seems to have little

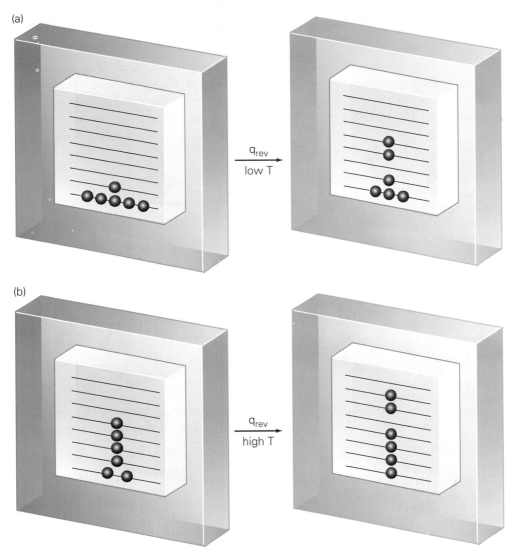

FIGURE 14-7. An *influx* of heat gives a system more options and more choice: more states to occupy, more ways to allot its energy microscopically. Turned around, a *loss* of heat leaves the particles with a more ordered, more tightly bunched distribution of energy. Either way, the change in entropy ($\Delta S = q_{rev}/T$) is determined by the heat transferred reversibly at a given temperature. (a) A system at low temperature, well organized at the start, reacts strongly to even a small flow of heat. Any redistribution of energy stands out clearly. (b) At high temperature, small disturbances are less noticeable. Proportionally more heat is needed to shake up a system already endowed with a high thermal energy.

in common with microstates and disorder. Yet if the two entropies *were* the same, then we could use straightforward measurements of heat and temperature to determine S.

They are indeed the same, because it is precisely this flow of heat into internal, "hidden" modes (Figure 14-7) that causes the redistribution of energy and the consequent change in entropy. We shall show the connection explicitly for one specific example (the isothermal expansion of an ideal gas) and subsequently accept the result as valid in general.

The task, set out below, is to compute ΔS as a gas expands from V_1 to $V_2 = 2V_1$ at constant temperature, during which the entropy goes from $S_1 = k_B \ln W_1$ to $S_2 = k_B \ln W_2$:

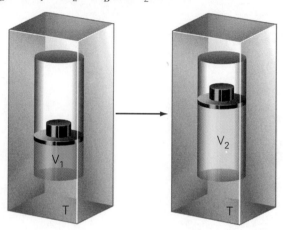

Doubling the volume makes available to each particle exactly twice the number of positions, so that $W_2 = 2W_1$. The change in entropy, *per particle*, follows immediately as

$$\Delta S = S_2 - S_1$$

$$= k_B \ln W_2 - k_B \ln W_1 = k_B \ln\left(\frac{W_2}{W_1}\right)$$

$$= k_B \ln\left(\frac{2W_1}{W_1}\right)$$

$$= k_B \ln 2$$

according to the rules for manipulating logarithms (Appendix B). This, then, the result for $N = 1$: $k_B \ln 2$.

When $N = 2$, the ratio W_2/W_1 is the product of two separate factors (2×2) because each particle moves independently of the other. For $N = 3$, the ratio is $2 \times 2 \times 2$ and thus we see how the pattern develops: 2^N for N particles.

The microstates for one mole ($N = N_0$) go, accordingly, as

$$\frac{W_2}{W_1} = 2^{N_0}$$

and the change in statistical entropy becomes

$$\Delta S = k_B \ln(2^{N_0}) = N_0 k_B \ln 2$$

Remembering (from Chapter 10) that the ideal gas constant is derivable from Boltzmann's constant as $R = N_0 k_B$, we then arrive at

$$\Delta S = R \ln 2 \qquad \text{(statistics)}$$

for one mole. But this is precisely the thermodynamic result obtained in Chapter 13 for the quantity q_{rev}/T, where we showed that

$$q_{rev} = RT \ln 2$$

and therefore

$$\Delta S = \frac{q_{rev}}{T} = R \ln 2 \qquad \text{(thermodynamics)}$$

for a reversible isothermal expansion. The job is done.

Entropy is entropy. It has a powerful microscopic interpretation related to microstates and disorder. It has an equally powerful macroscopic interpretation related to reversible heat flow. It has units of $J\ K^{-1}$, heat divided by temperature. It can be measured calorimetrically.

Note that entropy is a function of state, so any change is independent of process. ΔS remains the same whether the expansion is reversible or irreversible, gentle or violent. How, otherwise, could we have failed to specify step by step the route from V_1 to V_2 during our statistical enumeration of the microstates? Heat, by contrast, is not a state function, and only the very special quantity q_{rev}/T (involving *reversible* heat) is path-independent. Consequently we have not an equation for ΔS in general, but rather an inequality

$$\Delta S \geq \frac{q}{T}$$

that guarantees a minimal entropy change (equal to q/T) just for a reversible process. For all other changes, ΔS is greater than q/T and is not directly related to a heat flow. If an entropy change is to be measured calorimetrically, then heat must be made to flow reversibly for the equality to apply in the expression above.

That happens naturally during a phase transition, as we discovered in Chapter 13. When a solid melts, for instance, the temperature remains constant at T_m while heat is absorbed to disrupt the intermolecular forces. This heat is delivered reversibly, moreover, since such a transition is reversible by its very nature. Solid and liquid phases are always present simultaneously, *in equilibrium*, and hence the system is never more than infinitesimally displaced from a state of true equilibrium. If a little more heat is added, then a bit more solid melts. If some heat is withdrawn, then some more liquid freezes—with equilibrium between the phases maintained all the while, as in Figure 14-8. It is a clear example of reversibility.

The reversible heat exchanged during melting under constant pressure is a measurable, calorimetric quantity called the **enthalpy of melting** (or **enthalpy of fusion**, same thing). Given ΔH_{melt}, we have at once the **entropy of melting**

$$\Delta S_{melt} = \frac{\Delta H_{melt}}{T_m}$$

and, with that, a numerical value to measure the disordering of the solid. It is a positive quantity, consistent with an increase in entropy as the solid melts. Going the other way, we know that the entropy of freezing is equal to $-\Delta S_{melt}$ (because $\Delta H_{freeze} = -\Delta H_{melt}$). Entropy decreases when the liquid freezes to form an ordered solid, going down just as much as it

FIGURE 14-8. Entropy of melting. Maintaining thermal equilibrium with its liquid phase, a solid absorbs the requisite enthalpy of melting (ΔH_{melt}) at a constant melting temperature, T_m. The change in entropy during the phase transition is $\Delta H_{melt}/T_m$.

previously went up. So, too, must we find for any function of state, since *differences* are always fixed regardless of the path taken.

14-4. THE SECOND LAW OF THERMODYNAMICS

With the ***second law of thermodynamics***, nature puts unyielding limits on the way energy and matter can be rearranged. The terms: For a change to occur with no outside intervention (to be, we say, ***spontaneous***), the combined entropy of system and surroundings cannot decrease. At best, the entropy of the universe remains constant—but only for a certain class of idealized, reversible transformations that never happen in the real world. Otherwise, the total entropy goes *up* for all real processes. It goes up and up until a maximum is reached, and then the change is over. Equilibrium is attained.

Thus the second law of thermodynamics demands that all things must come to an end at some time; that the universe must eventually stumble into its statistically most likely set of microstates; that everything must come to equilibrium. Concentrations of reactants and products become constant. The ball settles down in the valley. The pendulum stops.

Spontaneous change (not necessarily *fast* change, but simply unaided, naturally occurring change) is permitted only if it leaves the universe more disordered than before. Every process must contribute its share to the ever-increasing total entropy. The universe might not care about its total *energy*, which is constant, but it does want that energy dispersed over as many microstates as possible. Like moisture seeping into a sponge, energy (Figure 14-9) spreads inexorably to atoms and molecules everywhere. The entropy of the universe never decreases.

Locally, yes; globally, no. Nature will tolerate localized reductions in entropy, provided they are paid for elsewhere. Liquids, after all, do freeze at certain temperatures and pressures despite the lower entropy of the solid. Allowance is made because freezing is an exothermic process, and the heat of reaction serves to disorder the surroundings and thereby compensate for the generation of local order. Although the system in Figure 14-10 (the ice cube) is more ordered, the surroundings (the air and everything else) are even more *disordered* after absorbing the additional thermal energy. The outside world is stirred up just enough to ensure that

$$\Delta S_{univ} = \Delta S_{sys} + \Delta S_{surr} > 0$$

in accordance with the second law.

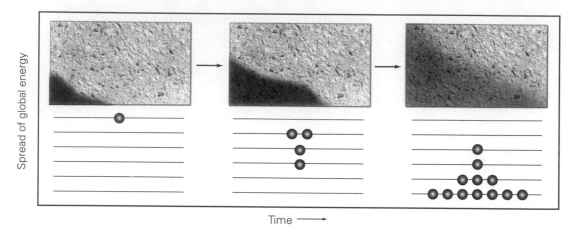

FIGURE 14-9. Energy makes its way microscopically into nature's uncountably many degrees of freedom: the quantized states of every electron, atom, and molecule in the universe. With time, the fixed pool of global energy dribbles into more and more levels—never to go back; never to revisit an earlier, more cramped distribution.

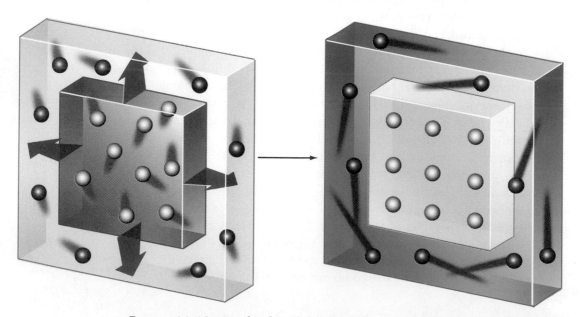

FIGURE 14-10. Local order; global disorder. As the liquid freezes, it releases energy to the surroundings. The universe as a whole realizes a larger spread in energy, even if the system (arbitrarily defined) does not.

A net increase in the entropy of the universe is the criterion for any spontaneous change. Without a positive ΔS_{univ}, A will not go to B. B will go to A instead.

14-5. FREE ENERGY

Dealing constantly with the entropy of the universe has its disadvantages. More convenient would be some other quantity—some alternative combination of state functions—that depends, explicitly, only on the system but which incorporates ΔS_{surr} implicitly and automatically. Such thermodynamic quantities exist. Generically called *free energy* functions, they are adapted to various experimental conditions: constant temperature and volume, constant pressure and temperature, and a number of other combinations.

Free energy, suitably defined, supplies the driving force for chemical change. It determines both the direction of the reaction and the composition of the system at equilibrium. Of particular interest is the **Gibbs free energy**,

$$G = H - TS$$

which is designed so that its change under constant pressure and temperature is given by

$$\Delta G = \Delta H - T\,\Delta S$$

Here ΔH and ΔS refer to the system alone (to be understood tacitly as ΔH_{sys} and ΔS_{sys}), but note that system and surroundings remain in equilibrium with each other throughout. Consider the relationship portrayed in Figure 14-11: Since both system and surroundings are at the same temperature and pressure, we associate ΔH (normal sign) with a reversible heat flow *into* the system and similarly interpret $-\Delta H$ (opposite sign) as a reversible *outflow* to the surroundings. System and surroundings are linked.

Something happens inside the system—a change in enthalpy—and the surroundings suffer a sympathetic change in entropy resulting from the flow of heat. Any shift of enthalpy occasioned by the system (ΔH_{sys}, or simply ΔH) brings order or disorder to the surroundings. Our macroscopic formula for the entropy ($\Delta S = q_{rev}/T$) tells us at once that this change is

$$\Delta S_{surr} = -\frac{\Delta H_{sys}}{T} \equiv -\frac{\Delta H}{T}$$

for the surroundings, and we see it appear when the system-based equation

$$\Delta G = \Delta H - T \, \Delta S$$

is divided by $-T$:

$$-\frac{\Delta G}{T} = -\frac{\Delta H}{T} + \Delta S_{sys} = \Delta S_{surr} + \Delta S_{sys} = \Delta S_{univ}$$

The resulting expression,

$$\Delta G = -T \, \Delta S_{univ}$$

connects ΔG (a quantity pertaining to the system alone) with the entropy of the universe.

A change in G just for the *system* therefore carries within it all the

FIGURE 14-11.   Changes in the Gibbs free energy, registered at constant temperature and pressure as the sum of two contributions: $\Delta G = \Delta H - T \, \Delta S$. (a) With system and surroundings in thermal equilibrium, the system's change in enthalpy (ΔH) affects the entropy of the surroundings by an amount $-\Delta H/T$. The enthalpy $-\Delta H$ (with sign reversed) represents here a reversible flow of heat from system to surroundings at constant temperature. (b) With the inclusion of ΔS for the system alone, the Gibbs function encompasses the full change in entropy inflicted on the universe: $\Delta G = -T \, \Delta S_{univ}$.

(a)

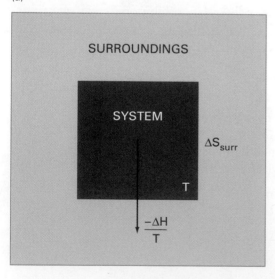

(b)

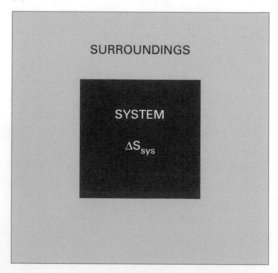

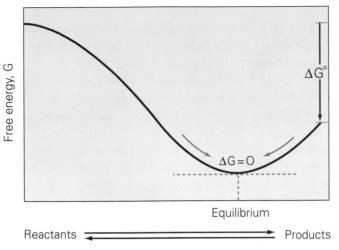

FIGURE 14-12. The hillsides and valley represent a system's drive to minimize its free energy and, by so doing, to maximize the entropy of the universe. Reactants and products follow the slope downward, moving at each point in the direction where the instantaneous change in free energy, ΔG, becomes less steep. The hypothetical curve shows how G might vary for mixtures ranging from all reactants (at the left) to all products (at the right). Somewhere in the middle is a point of equilibrium, with just enough products and reactants to yield a minimum in the free energy: a valley. To climb out of the valley on either side would be to raise the free energy.

requisite information about the entropy of the universe, on the assumption that T is constant, P is constant, and only $P\,\Delta V$ work is done. If a system under these conditions is driven toward minimum free energy, then the universe is automatically driven toward maximum entropy. A large negative value for ΔG means a large positive value for ΔS_{univ}.

Reacting systems seek the lowest free energy as if rolling down a hill, stopping only when their drive is spent. A process follows the slope of free energy downward, moving in whichever direction leads to a decrease in G: toward products if the instantaneous change ΔG is negative; toward reactants if not. And since the proportions of reactants and products vary continuously during the course of reaction, ΔG itself keeps changing until no further reduction is possible. Equilibrium is attained at the point where ΔG is zero, the bottom of the free energy valley pictured in Figure 14-12. There, at rest, the system is no longer able to lower its free energy by making small changes to the concentrations.

With that, the universe is satisfied; it can acquire no more entropy, at least not from the process at hand. An irreversible change has taken place which leaves the world forever different. The entropy gained can never be taken back.

Reactions do not proceed uphill. No spontaneous transformation can occur if the products are higher in free energy than the reactants. To climb the free energy hill is to bring order to the universe, and that has never happened before. We suppose it never will.

Whether or not G decreases—and, in return, brings disorder to the universe—depends on the interplay between ΔH, ΔS, and T. Going down in enthalpy is generally good for spontaneity, since a negative value for ΔH contributes directly toward a negative value for ΔG. Going up in entropy is similarly good, for the system acts on its own to disorder the universe. Neither ΔH nor ΔS can determine spontaneity by itself, however, owing to the weighting factor T in the expression $\Delta G = \Delta H - T\,\Delta S$. A reaction might be very exothermic yet still not proceed if $T\,\Delta S$ is large and negative (so that—note the minus sign—the contribution of $-T\,\Delta S$ to ΔG is positive). Similarly, a large increase in system entropy can be annulled by a large endothermicity.

The role of temperature is summarized compactly in a four-line table such as

ΔH	ΔS	ΔG	DIRECTION OF G
$-$	$+$	$-$	always decreases
$+$	$-$	$+$	always increases
$+$	$+$	T	decreases at high temperature
$-$	$-$	T	decreases at low temperature

There are just two certainties: (1) If enthalpy decreases and entropy increases, the change in free energy can only be negative. Both influences are favorable and the process occurs spontaneously. Reactants go to products; the battery discharges; the stone falls; the gas fills its container. (2) If enthalpy increases and entropy decreases, then everything is going in the wrong direction. ΔG is always positive and the transformation will not take place spontaneously.

The third option ($H\uparrow$, $S\uparrow$) depends on the temperature. A positive (bad) ΔH can be overcome by a positive (good) $T\,\Delta S$ if both T and ΔS are sufficiently large. This possibility of an entropic rescue is the reason why endothermic reactions are not strictly forbidden. Despite the uphill climb in enthalpy, there may be a compensating gain in entropy that allows the system's free energy to decrease. The enthalpy–entropy balance shifts in favor of entropy at higher temperatures, where a large $-T\,\Delta S$ term swamps ΔH and makes enthalpy irrelevant:

$$\Delta G = \Delta H - T\,\Delta S \approx -T\,\Delta S \qquad |T\,\Delta S| \gg |\Delta H|$$

Thermal energy is greater at elevated temperatures as well, and so endothermic reactions are, not surprisingly, similarly favored by Le Châtelier's principle under such conditions. The extra heat goes into the formation of more products, effectively relieving a stress on the system as the process

$$\text{reactants} + \text{heat} \rightleftharpoons \text{products}$$

moves from left to right (see Chapter 12).

The final possibility ($H\downarrow$, $S\downarrow$) is also dictated by temperature. An order-*inducing* reaction ($\Delta S < 0$) will occur in the system if there is sufficient exothermicity to disorder the surroundings through a large and negative ΔH. But now the temperature must be relatively low, so that the unfavorable $-T\,\Delta S$ term (which is positive because ΔS is negative) does not overpower the favorable ΔH. We ask, specifically, for a temperature at which

$$\Delta G = \Delta H - T\,\Delta S \approx \Delta H \qquad \left|T\,\Delta S\right| \ll \left|\Delta H\right|$$

when ΔH is negative. The tradeoff is between order in the system and disorder in the surroundings; and, for that, we find: Too high a temperature will reduce the entropy created in the surroundings, because

$$\Delta S_{\text{surr}} = -\frac{\Delta H}{T}$$

decreases as *T* increases. Again Le Châtelier's principle is consistent with the demands for a decrease in free energy, since exothermic reactions are naturally favored at low temperatures. The heat produced by the reaction

$$\text{reactants} \rightleftharpoons \text{products} + \text{heat}$$

is dissipated into the surroundings, and a large drop in enthalpy (a negative ΔH) becomes the primary criterion for spontaneity as $T\,\Delta S$ approaches zero.

14-6. STANDARD FREE ENERGY AND ENTROPY

Free energy is a thermodynamic function of state, just like enthalpy with its ups and downs. It is, first of all, a quantity for which there is no fixed zero point—no absolute measure of *G*, no way to know exactly how many joules of free energy are contained within a substance. Only differences between two states are relevant. A ***standard free energy of formation***, ΔG_f°, thus may be defined for any compound in its conventional standard state, following exactly the procedure established for enthalpy in Chapter 13.

Standard states, we recall, imply gas pressures of 1 atm, solution concentrations of 1 mol L^{-1}, and the normally existing phase at some stated reference temperature (often 25°C). The standard value ΔG_f° is simply the net change brought about when one mole of compound is formed from its elements under these conditions.

Changes in free energy obey a generalized Hess's law (Chapter 13), as do all state functions. Accordingly, ΔG° for any reaction is obtained by subtracting the combined ΔG_f° of the reactants from the combined ΔG_f° of the products (always with proper accounting for number of moles). A given free energy of formation, ΔG_f°, therefore offers a measure of a compound's thermodynamic stability under standard conditions, providing a rough gauge of possible reactivity. The lower the free energy of formation, the less likely a substance is to undergo change; there exist relatively few potential products with even *less* free energy. Conversely, reactants high on the free energy scale are relatively unstable and find proportionately more profitable transformations.

What is true for enthalpy and free energy—standard states, Hess's law, independence of path—is true for entropy as well, since S too is a state function. There is, though, one modification sufficiently notable to be elevated to the ***third law of thermodynamics***, which asserts: Entropy indeed *can* be measured on an absolute scale. There really is something identifiable as a state of zero entropy, namely a perfectly ordered crystal at 0 K (shown symbolically in Figure 14-13). With only this one arrangement permitted ($W = 1$), the entropy vanishes and a recognizable zero point is established:

$$S = k_B \ln W = k_B \ln 1 = 0$$

There exists, in contrast to energy, a knowable amount of entropy in a substance defined relative to a state of perfect order.

Elements are no exception. An element does not have *zero* entropy under standard conditions, for it is already disordered by comparison with a hypothetical state existing at absolute zero. We tabulate therefore not ΔS° but rather S°, reserving the symbol Δ for entropy changes during reactions.

Take, as an example, the entropy involved in the formation of one mole of gaseous NO_2 from its elements:

$$\tfrac{1}{2}N_2(g) + O_2(g) \longrightarrow NO_2(g)$$

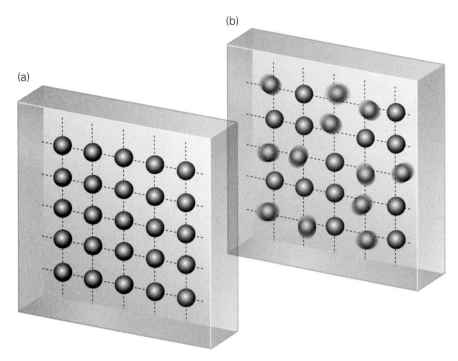

FIGURE 14-13. (a) The third law of thermodynamics establishes the zero point for a scale of absolute entropy: a perfectly ordered crystal at 0 K. The system has one microstate and zero entropy. (b) At all other temperatures, the substance is disordered relative to its ideal state at absolute zero. The entropy is positive.

Standard entropies, available from thermodynamic tables (such as Table C-16 in Appendix C), show that $S°$ is 191.5 J mol^{-1} K^{-1} for nitrogen gas, 205.0 J mol^{-1} K^{-1} for oxygen gas, and 240.0 J mol^{-1} K^{-1} for nitrogen dioxide. Bear in mind that these values reflect the change incurred as a substance is heated from a hypothetically perfect crystal (at 0 K) to standard conditions. $\Delta S°$ for the formation reaction pertains, instead, to the change in entropy relative to the elements as they exist at 25°C and 1 atm. That value, analogous to a quantity like $\Delta H_f°$, follows directly as

$$\Delta S° = S°[NO_2(g)] - \tfrac{1}{2}S°[N_2(g)] - S°[O_2(g)] = -60.8 \text{ J mol}^{-1} \text{ K}^{-1}$$

by extension of Hess's law. All we have done is to subtract reactant entropies from product entropies, a legitimate operation for any function of state.

A few representative standard entropies, culled from the data in Table C-16, will enable us now to attach numbers to this picture of entropy and disorder. We see, for example, that oxygen gas (205.0 J mol^{-1} K^{-1}) and chlorine gas (223.0 J mol^{-1} K^{-1}) have similar values, roughly 200 to 225 J mol^{-1} K^{-1}. Both substances are diatomic gases of comparable size and molar mass, and their entropies are more a reflection of the gaseous state itself than of individual molecular characteristics. Water vapor (188.7 J mol^{-1} K^{-1}) also falls into the same range, whereas liquid water displays a much lower entropy (70.0 J mol^{-1} K^{-1}) consistent with a more ordered phase. Solids are lowest of all, as evidenced by graphite with its $S°$ of only 5.7 J mol^{-1} K^{-1}.

14-7. THE FREE ENERGY OF REACTION

Given thermodynamic data for various compounds, we can compute changes in free energy and entropy for an almost limitless number of reactions, real and imagined. The rules are exactly the same as for thermochemical calculations of enthalpy changes:

$$\Delta G° = \Sigma\ \Delta G_f°(\text{products}) - \Sigma\ \Delta G_f°(\text{reactants})$$

$$\Delta S° = \Sigma\ S°(\text{products}) - \Sigma\ S°(\text{reactants})$$

$$\Delta H° = \Sigma\ \Delta H_f°(\text{products}) - \Sigma\ \Delta H_f°(\text{reactants})$$

A good starting point is the familiar equilibrium between liquid water and water vapor,

$$H_2O(\ell) \rightleftharpoons H_2O(g) \qquad \begin{array}{l} \Delta H° = 44.0 \text{ kJ mol}^{-1} \\ \Delta S° = 118.7 \text{ J mol}^{-1}\text{ K}^{-1} \end{array}$$

where the standard enthalpy of vaporization is $+44.0$ kJ mol^{-1}. Here at 25°C, under an external pressure of 1 atm, a mole of liquid water needs 44 kilojoules of additional enthalpy to escape into the vapor phase. The gas, having absorbed this heat, then will have climbed *up* the enthalpy hill relative to the liquid—a step in the wrong direction for a spontaneous change. But this same gas, more disordered now than the liquid, will have moved to a higher entropy as well, and so the transformation's direction must ultimately be arbitrated by the temperature. Temperature will determine whether the "good" represented by a positive $\Delta S°$ is good enough to outweigh the "bad" represented by a positive $\Delta H°$.

In this instance (at 298 K), the good is not good enough. The net change in free energy,

$$\Delta G^\circ = \Delta H^\circ - T \Delta S^\circ$$

$$= 44.0 \text{ kJ mol}^{-1} - (298.15 \text{ K})(0.1187 \text{ kJ mol}^{-1} \text{ K}^{-1})$$

$$= 8.6 \text{ kJ mol}^{-1}$$

is positive, meaning that such a transformation would (impossibly) bring order to the universe as a whole. Evidently the increased *local* disorder in the gas is insufficient to overcome its higher enthalpy, and hence the vaporization of liquid water is greatly suppressed at this temperature. The spontaneous direction of change is toward condensation, not evaporation.

Water's equilibrium vapor pressure, remember, is just 23.8 torr at 25°C. Only at higher temperatures does entropy become the dominant consideration, making ΔG negative and driving forward the vaporization to higher pressures. Not until 100°C does water come to equilibrium with its vapor at 1 atm.

Also dependent on temperature, albeit for another reason, is the dimerization reaction

$$2NO_2(g) \rightleftharpoons N_2O_4(g) \qquad \begin{array}{l} \Delta H^\circ = -57.2 \text{ kJ mol}^{-1} \\ \Delta S^\circ = -175.8 \text{ J mol}^{-1} \text{ K}^{-1} \end{array}$$

where both enthalpy and entropy decrease in the forward direction. The process is exothermic owing to the formation of new chemical bonds (good, make dimers), but unfavorable entropically since one mole of gaseous N_2O_4 is less disordered than two moles of gaseous NO_2 (bad, make monomers). To go forward, the system needs a temperature low enough to mitigate the $T \Delta S$ component of the free energy and, at the same time, permit the surroundings to absorb the heat produced during the reaction.

Again we have a reaffirmation of Le Châtelier's principle. Decreasing the temperature—removing heat—encourages an exothermic reaction to move forward. Even more, a lower temperature may reduce the unfavorable $T \Delta S$ term to the point where the free energy of $N_2O_4(g)$ is less than that of $NO_2(g) + NO_2(g)$. At the right temperature, where $\Delta H - T \Delta S$ becomes negative, the dimerization will proceed spontaneously.

When everything is truly right, of course, temperature is no longer a consideration for spontaneity, as in the combustion of carbon:

$$C(s) + O_2(g) \rightleftharpoons CO_2(g) \qquad \begin{array}{l} \Delta H^\circ = -393.5 \text{ kJ mol}^{-1} \\ \Delta S^\circ = 2.9 \text{ J mol}^{-1} \text{ K}^{-1} \end{array}$$

This reaction is always favorable thermodynamically, for enthalpy decreases while CO_2 gas is formed *and* entropy increases while the order of the solid is destroyed. Regardless of the temperature, then, one mole of carbon dioxide gas will be thermodynamically more stable than one mole of graphite and one mole of oxygen gas. The free energy of the gaseous CO_2 is always lower, and consequently the reactants will always tend to go spontaneously toward products—if they can get there.

Getting there is not always easy, nor necessarily fast. Although the products may sit in a valley lower in free energy than the reactants, often there is a hill to climb in between, as below:

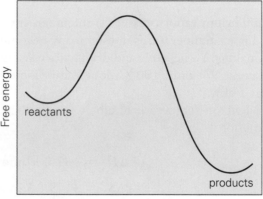

Course of reaction

Reactants must come together violently, perhaps forming transient structures on their way to making products, and such processes can cost free energy.

These issues belong properly to the study of reaction kinetics (to be taken up in Chapter 18), and here temperature will have yet another important role to play. Without some heat invested at the outset to push graphite and oxygen over the barrier, the combustion cannot take place no matter how much the free energy may be lowered eventually.

Finally there are those reactions for which nothing is right, where enthalpy goes up and entropy goes down. The endothermic formation of ozone,

$$\tfrac{3}{2}O_2(g) \rightleftharpoons O_3(g) \qquad \begin{aligned} \Delta H^\circ &= 142.7 \text{ kJ mol}^{-1} \\ \Delta S^\circ &= -68.7 \text{ J mol}^{-1} \text{ K}^{-1} \end{aligned}$$

with its accompanying decrease in entropy, concludes our survey of the four ways in which enthalpy and entropy can contribute to the free energy. Uphill all the way, this conversion of oxygen into ozone will never proceed of its own accord.

14-8. FREE ENERGY AND THE EQUILIBRIUM CONSTANT

With pressures and concentrations adjusted continually during an approach to equilibrium, there is always some up-to-the-moment difference in free energy, ΔG, between reactants and products:

$$\Delta G = G(\text{products}) - G(\text{reactants})$$

And this difference, as we have seen, is our sought-after driving force, the thermodynamic motive for reaction. A nonzero ΔG creates an imbalance that nature is compelled to erase.

The imbalance arises when a combined mass of products with one free energy, G_P, confronts a combined mass of reactants with a different free energy, G_R. It is an imbalance as intolerable as an uneven surface of water would be in a pond; and, just as all ponds eventually subside and become flat, so too must any disparity in G be smoothed away. It will be smoothed away by chemical reaction.

Out of equilibrium, a system cannot rest. Driven by either too many products or too many reactants, a reaction will proceed until there are just enough moles of everything to guarantee equal free energy for all. If G_P is higher than G_R, then products will be converted into reactants. If G_P is lower than G_R, then reactants will be converted into products. They will go back and forth as needed, doing whatever it takes until finally there is no difference between the two; until finally there is no tendency to prefer one over the other; until finally there is a dynamic equilibrium between reactants and products. Then, only then, will the imbalance be erased and will ΔG be equal to zero.

Thus we regard any instantaneous difference in G (per mole) as a measure of *chemical potential*, an unambiguous indicator of a reaction's direction. In the same way that heat flows from a hot object to a cold object until the temperatures are equal, matter flows (is converted) from a mixture at high G to a mixture at low G until the free energy is everywhere equal. When ΔG is negative, the reactants sit higher than the products and the reaction rolls to the right. When ΔG is positive, circumstances are reversed: The products occupy a higher level of free energy, and so the process goes to the left as products are consumed to reduce ΔG. Equilibrium is attained finally at the point where $\Delta G = 0$, when the free energies of reactants and products stand equal.

Now note, significantly, that this ΔG (without the superscript $^\circ$) is not the same as ΔG°, the difference measured between standard states. Rather, the standard difference in free energy

$$\Delta G^\circ = G^\circ(\text{products}) - G^\circ(\text{reactants})$$

applies only to materials having gas pressures of 1 atm and solution concentrations of 1 mol L^{-1}, whereas actual pressures and concentrations are by no means restricted to these values. Actual pressures and concentrations will be what they will be, regardless of the arbitrary conditions we choose for standard measurements. We need to relate, therefore, the standard conditions of the thermodynamic tables (which, conveniently, we have) to the prevailing, arbitrary conditions of a real reaction (which we must accept).

Indeed, should there not be a connection between ΔG, $\Delta G°$, and the equilibrium constant? For we know, first, that the final pressures and concentrations at equilibrium are constrained *from the start* by the mass-action expression (K), and we understand too that the reaction will be driven to equilibrium by a difference in free energy (ΔG). Furthermore, the reaction quotient

$$Q = \frac{[C]^c[D]^d}{[A]^a[B]^b}$$

for the general process

$$aA + bB \rightleftharpoons cC + dD$$

is itself a directional indicator (Chapter 12). Remember: If Q is less than K (too many reactants), the transformation goes forward to produce more products. If Q is greater than K (too many products), the transformation goes backward to regenerate more reactants. If Q is equal to K, a balance is struck and there is equilibrium between products and reactants.

We expect, as a result, some relationship between how *far* a system is from equilibrium and how *strong* is its tendency to get there. The further away, the harder the drive. If there is a large difference in free energy between reactants and products, then—at that moment—the instantaneous pressures and concentrations surely must be far from their foreordained equilibrium ratios. Such reactants and products have a long way to go before reaching equilibrium (and reach equilibrium they must), so the system works correspondingly harder to correct the imbalance. What remains, finally, is to fit the standard free energy ($\Delta G°$) into the picture.

The proper dependence (obtained by a modest use of calculus) is given by the expression

$$\Delta G = \Delta G° + RT \ln Q$$

in which the reaction quotient Q brings together the standard and nonstandard free energies. Only when $Q = 1$ (so that $\ln Q = 0$) are reactants and products all at unit pressure and concentration, thereby conforming to standard conditions and making ΔG equal to $\Delta G°$. The term $\ln Q$ tracks the difference between standard and nonstandard free energy.

It is, moreover, only when ΔG becomes zero that the driving force is satisfied and equilibrium is established, with $Q = K$ (Figure 14-14). What follows is

$$\Delta G° = -RT \ln K$$

or

$$K = \exp\left(-\frac{\Delta G°}{RT}\right),$$

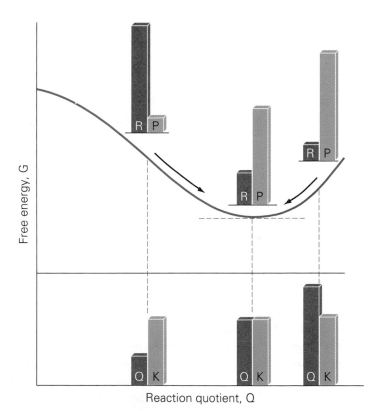

FIGURE 14-14. Variation of G with concentration, considered in and out of equilibrium: (1) Away from equilibrium, the reaction quotient determines the direction of spontaneous change: to the right if there are too many reactants ($Q < K$); to the left if there are too many products ($Q > K$). (2) At equilibrium ($Q = K$), where the curve is flat, no drive propels the system either to the right or left: $\Delta G = 0$. Products (P) and reactants (R) coexist with the same free energy, in proportions fixed by the difference in free energy under standard conditions, $\Delta G°$. The deeper the valley, the more the equilibrium ratio tilts in favor of the products.

from which comes a concise summary of the relationship between enthalpy, entropy, free energy, and equilibrium. We have, in one equation, a way to calculate the equilibrium constant using only standard free energies (conveniently tabulated, once and for all) and the temperature.

In an exponential dependence reminiscent of a Boltzmann distribution, the expression for K relates the eventual position of equilibrium to the thermodynamic properties of the original compounds. Just by knowing something about the *uncombined* reactants and products—their standard free energies of formation—we derive a thermodynamic quantity characteristic of the entire process: the standard free energy of reaction, $\Delta G°$. And from $\Delta G°$ now comes the equilibrium constant, the final word on the whole transformation. When $\Delta G°$ is positive, the reaction is nonspontaneous and K is less than 1. When $\Delta G°$ is negative, the reaction is spontaneous and K is greater than 1. From that we see how a large driving force in one direction skews the equilibrium accordingly, producing a suitably large K to match a large drop in free energy.

14-9. THERMAL EQUILIBRIUM AND THE MEANING OF TEMPERATURE

With heightened awareness, let us conclude by revisiting the notion of temperature as it appeared first for the ideal gas. This summary treatment will help bring together the key ideas behind equilibrium, thermodynamics, and statistical mechanics developed since Chapter 10.

The question is disarmingly simple: What does it mean to have a temperature? The answer, rooted in the statistics of large numbers, holds implications for all processes that occur in the natural world. Wherever there are atoms, wherever there are molecules, wherever particles come together in large numbers—their disposition is determined by temperature and the eventual attainment of thermal equilibrium.

A single particle has no temperature. Temperature is an intrinsically macroscopic property, a parameter that summarizes in one value the behavior of enormously large assemblies. For an ideal gas, we saw that the temperature provides a direct measure of the *average* kinetic energy of all the particles. Expressed more sharply, temperature came to acquire meaning really through the Boltzmann distribution

$$\frac{n_i}{n_j} = \exp\left(-\frac{\epsilon_i - \epsilon_j}{k_B T}\right)$$

which lays out exactly how the energy is distributed microscopically. It says: If n_i particles have an energy ϵ_i and another n_j particles have the energy ϵ_j, then the statistical ratio n_i/n_j is determined entirely by the energy difference $\Delta\epsilon$ and the temperature (through the exponent $-\Delta\epsilon/k_BT$). Shown once more in Figure 14-15, the Boltzmann distribution is maintained between all possible differences in energy, $\epsilon_i - \epsilon_j$. The temperature is fixed.

Reflect again on what it means for an ideal gas to have well-defined and persistent macroscopic properties: pressure, volume, temperature, number of moles, energy, and so on. How can we characterize a system with an unchanging macroscopic aspect? What shall we say about a system able to change, yet apparently unwilling?

It is, by definition, a system in equilibrium. To say that a confined sample of ideal gas possesses a definite temperature is to assert that it is in equilibrium. It is in ***thermal equilibrium***, able to share thermal energy among its own particles and (usually) with the surroundings outside.

The system looks quiet, but there is a dynamic character to its equilibrium. Gas particles, in constant motion, alter their velocities through

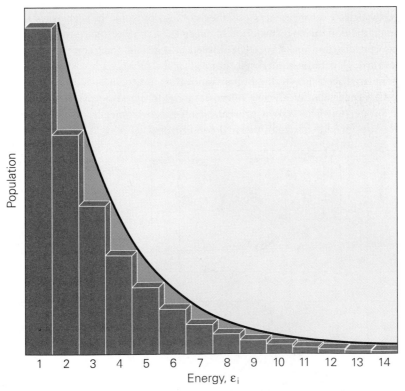

FIGURE 14-15. The Boltzmann distribution of energies at a fixed temperature, computed for a set of equally spaced levels.

collisions, and the microscopic distribution of energies undergoes continuous revision as a result. But—and here is where the laws of large numbers prevail—only some of the possible new distributions have nonzero probability. All the others are as unlikely as turning up 100 billion heads and zero tails.

The only acceptable choices prove to be those arrangements consistent with the peculiar ratios of the Boltzmann distribution. For reasons of probability alone, the system wanders through the restricted set of microstates describable by the exponential factors given above. Each of these microstates is manifested macroscopically by the same temperature, regardless of how the particles are flying about in Figure 14-16. An observer therefore sees only the unchanging macroscopic property and little suspects any variation in the underlying microscopic configurations. The temperature remains constant while the proper distribution of energies is maintained.

The attainment of thermal equilibrium is an inevitable consequence of

FIGURE 14-16. Thermal equilibrium, interpreted dynamically and microscopically: Particles exchange energy as they collide, but only in such ways that will maintain the Boltzmann distribution of energies. For a gas, the result is a Maxwell-Boltzmann distribution of speeds enforced at constant temperature. The system passes through countless microstates one after the other, each with its energy levels populated according to the Boltzmann ratios. Yet to an outside observer there is no difference, because every microstate yields the same macrostate with the same properties. The system presents itself macroscopically as unchanging and timeless, trapped in a state of thermal equilibrium.

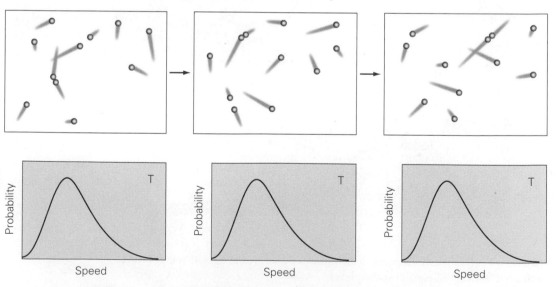

the statistics of large groups. Precisely such a foregone statistical conclusion applies both to the particles of an ideal gas and to *all* other equilibrium systems. There is nothing magical about the Boltzmann distribution; it simply allows the internal energies to be spread over the broadest range of possibilities. It ensures that the levels are filled from the ground up, with the thermal energy (proportional to $k_B T$) determining the assignment. At temperatures where $k_B T$ is small compared with $\Delta\epsilon$, most particles huddle in the very lowest energy levels. More of them are promoted to higher levels as the temperature increases, but always the exponentially decreasing distribution favors the lower energies.

The relationship between temperature and equilibrium now can be stated still more strongly: Absent an equilibrium Boltzmann distribution, the system does not even *have* a temperature. Temperature is a defining property of the equilibrium state.

A system with a temperature is in thermal equilibrium, and there it stays because escape is a statistical improbability. If, owing to some random fluctuation, the system ever gets out of equilibrium, statistical constraints pull it right back in.

Equilibrium for an ideal gas means that the particles are distributed uniformly throughout the volume, with their speeds falling on a Maxwell-Boltzmann curve appropriate for the prescribed temperature. Were we to disturb the balance by preparing an improbable microstate, the system would stumble back to equilibrium through a succession of increasingly probable configurations. Suppose, for example, that the volume was suddenly doubled so that all the particles found themselves in one half of the container. Such an arrangement is so odd, so hugely improbable compared with all other choices, that it cannot last. Before long, the gas spreads out to fill the expanded container and recover its equilibrium. Once again, PV becomes equal to nRT.

A change in temperature attendant upon the addition or withdrawal of thermal energy (heat) has a similar effect. Internal speeds sooner or later must be redistributed along whatever Maxwell-Boltzmann curve is ordained at the new temperature. No other outcome is reasonable.

Thermal equilibrium for any other system (not just a gas) is similarly characterized by a constant temperature, a Boltzmann distribution of energies, and a set of *most probable* microstates. Two bodies, one at a higher temperature than the other, come to thermal equilibrium when heat is allowed to flow from the hotter body to the colder.

Let Figure 14-17 point the way: Initially each object has its own equilibrium distribution of internal energy and corresponding temperature, chosen arbitrarily to make T_2 greater than T_1. Then they are brought together so that energy is transferred from the hot body to the cold. The colder atoms start to move faster; the hotter atoms start to move slower; and eventually

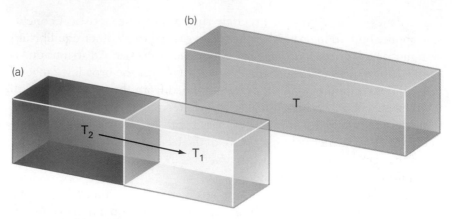

FIGURE 14-17. Thermal equilibrium, interpreted macroscopically: (a) Heat flows from a body at high temperature (T_2) to a body at low temperature (T_1), as if driven by a difference in thermal potential. (b) At equilibrium, the difference in potential is erased and the two temperatures stand equal. The final value T, lying somewhere between T_1 and T_2, establishes a new Boltzmann equilibrium.

the temperature becomes the same throughout. The new equilibrium temperature, T, stands intermediate between T_1 and T_2, and the internal energies of the composite body conform to a corresponding Boltzmann distribution. The inevitable equilibration is a direct consequence of molecular statistics and the unyielding demands of large numbers.

With these powerful ideas and just a few examples, we see how the maximization of global entropy dictates the drive to equilibrium for processes of all kinds. For *thermal equilibrium*: the flow of heat from one body to another until the temperature is uniform. For *chemical equilibrium*: the flow of matter between reactants and products until the free energies and concentrations, regulated by the law of mass action, hold steady. For dynamic equilibrium in general, where a flow of some sort or another is eventually stabilized and a system appears at rest.

———

Someone once summarized thermodynamics by paraphrasing the three laws in fatalistic but not necessarily inaccurate fashion. (1) You can't win; at best you might break even. (2) You can break even only at absolute zero. (3) You can never reach absolute zero. Total energy of the universe is conserved and total entropy grows. Those are the laws.

In the meantime, we can create *local* order and disorder through countless chemical reactions. To these possibilities we now turn, taking up such manifestations of thermodynamics and equilibrium as guests and hosts, solutes and solutions, acids and bases, and electrochemistry.

REVIEW AND GUIDE TO PROBLEMS

Think about the chemical world we have come to know: small, not large; grainy, not smooth; always in motion, never at rest; driven by energy, not matter. See it as a world of chance, accident, and unremitting change. See it as an ever-shifting flow of energy from one vessel to the next, a sea of infinite possibility. See it, figuratively, as a hidden world of strings to be plucked, of gears to be turned, of molecules to be built up, of molecules to be torn down. For that, really, is what we have: an inexhaustible store of energy to be shuffled around without cease and without loss.

Quantum by quantum, little packets of energy trickle into all the options the universe has to offer: into the electronic levels of every atom and molecule; into the vibrations of a bond; into the rotation of a molecule; into the intermolecular toughness of a solid; into the kinetic rush of a gas; into the strong binding of a nucleus; into the photons of an electromagnetic field; into this mode, into that mode; give a little here, take a little there; spread it around; share the energy among all.

Every particle has its effect on every other. They attract; they repel; they make new structures; they break old ones—but, above all, they redistribute their energy. They redistribute their energy among whatever states are available.

And which states are those? Which molecules get built? Which levels get filled? How does nature choose to distribute a fixed supply of energy?

Not by design, but by chance. Nature awards the energy by lottery, and the winners at equilibrium are simply those nuclei, atoms, molecules, and phases that hold the most tickets. Nature plays no favorites. The winners win because they can rearrange the same amount of energy in the most ways while still making the same product. They win because they offer the most microstates. They win because they outnumber their competition.

Such is the message of the second law of thermodynamics: not how *much* energy you have, but rather how you use it is what matters most.

IN BRIEF: THE SECOND LAW OF THERMODYNAMICS

1. MICROSTATES AND MACROSTATES. Let there be equilibrium, and let the atoms and molecules do what they will. To us, the effect is all the same; because to us, the macroscopic face of the system never changes: same pressure, same volume, same temperature, same energy,

same enthalpy, same everything. To be in equilibrium is to be always the same. To be in equilibrium is to be locked into a single **macrostate**, stuck with a fixed set of macroscopic properties.

But inside the system, hidden from view, the particles fight constantly to maintain their equilibrium. They bang into each other, and they bang into the walls around them. Heat goes in. Heat goes out. One molecule speeds up, and another slows down. This one gains energy; that one loses. The action never stops, and no two scenes are alike.

Standing outside, however, we never suspect what goes on within, because the energy-swapping particles conspire to fool us. They manage to stay in equilibrium, always yielding up the same macroscopic properties to unwitting observers unable to tell the difference. Throughout each of the system's changing **microstates**, flickering one after the other like the frames of a movie, the particles still produce the same overall effect. At equilibrium, every conceivable snapshot of the microworld blurs into the same smoothed-out macrostate.

2. WINNERS AND LOSERS. Take some arbitrary set of reactants (sodium chloride and water) and define a system. Mix the reactants together in a suitable container which, together with everything else in the world, will become the surroundings. Give the reactants ample time and opportunity to react; let them come to equilibrium; and, finally, ask: In a blind universe, ruled by chance, which is more likely—these particular reactants or these particular products? Solid NaCl and liquid H_2O, uncombined . . . or a solution of hydrated Na^+ and Cl^- ions? Who wins?

We have, on the one hand, the universe that *is*. Divided into three parts, it contains: (1) Some definite number of Na^+ and Cl^- ions locked into a lattice, held there by a balance of attractive and repulsive electric forces. (2) Some definite number of H_2O molecules moving throughout a liquid phase, grouped loosely into clusters that come and go. (3) Everything in the surroundings outside. Together, the world's total energy is distributed over a certain number of levels in system and surroundings. The energy goes into a certain number of microstates, W_{before}, each of which is indistinguishable macroscopically.

Then there is the universe that *might* be, a new world with new products and new surroundings, consisting of: (1) Some definite number of Na^+ and Cl^- ions caught within clusters of water molecules, floating in solution, no longer bound into a crystal. (2) Every particle in the surroundings outside, countless innocent bystanders, each changed forever by any heat that may flow when sodium chloride and water come together. In the new world, the same total energy would be redistributed over a new set of levels in system and surroundings. There would be a revised number of equilibrium microstates, W_{after}, each of them belonging, again, to a single macrostate.

Will this new world come to pass? Maybe, maybe not. If W_{after} is greater than W_{before} . . . if the energy of the universe is spread more widely as a result . . . if, in other words, there are simply more ways to produce a universe with salt water than without—then, yes, we shall have salt water. At equilibrium, we shall have more products than reactants; we shall have an equilibrium constant greater than 1. We shall have a more disordered universe, a universe closer to a final equilibrium in which all energetic imbalances would be erased.

And for this one reason, just this, we shall have that more disordered world: because it would be easier. The more ways there are to stumble into the same condition, the more likely it is for that condition to prevail. If trillions upon trillions of doors all lead to room A but only three doors lead to room B, to which room will a randomly knocking visitor be admitted?

Thus the **second law of thermodynamics**: The universe evolves toward a state of ever-increasing disorder. For a process to occur spontaneously, there must be a broader distribution of energy after the change. The entropy of the universe can never decrease. All isolated systems run down and eventually come to equilibrium.

3. THE MICROSCOPIC VIEW: STATISTICAL ENTROPY. Our general measure of disorder is the **entropy**, S, defined as

$$S = k_B \ln W$$

for a system with W microstates. S, a state function describing an extensive property, acquires units of $J \ K^{-1}$ from the Boltzmann constant, k_B (which we can replace by $R = N_0 k_B$ to express the entropy per mole as $R \ln W$).

Note, right away, that if there is only one microstate, then the entropy ($S = k_B \ln 1$) is exactly zero; and from this definition we draw out the **third law of thermodynamics**: The entropy of a perfect crystal at 0 K is zero, consistent with a single arrangement of the atoms. At any other temperature, the absolute entropy of that same system is always positive because there is always more than one way to distribute the energy. The more microstates, the greater the entropy.

4. THE MACROSCOPIC VIEW: THERMODYNAMIC ENTROPY. Any property that we interpret microscopically and statistically we can also describe macroscopically. Remember, then, how the entropy first appeared in Chapter 13 with its thermodynamic form and thermodynamic meaning: as a macroscopic function of state, S, for which any change is given by

$$\Delta S = \frac{q_{rev}}{T}$$

The symbol q_{rev} denotes the heat transferred during a hypothetically reversible process at a temperature T.

Macroscopic S claims only to be a function of state representing an extensive property. It makes no mention of particles, no mention of disorder, no mention of anything but heat and temperature; yet, just the same, thermodynamic entropy proves to be merely statistical entropy in another guise. The S in $S = k_B \ln W$ is the same S as in $\Delta S = q_{rev}/T$, just as the P in $PV = nRT$ is the same P as in the kinetic theory of pressure. There is only one entropy, whether we choose to frame it microscopically or macroscopically.

Thermodynamic S thus has both the same units and the same numerical value as statistical S ($k_B \ln W$), but it has this one great advantage as well: we can measure it. We can measure heat calorimetrically, whereas we cannot actually count microstates except in the simplest of systems (like perfect crystals at 0 K, or two or three gas molecules at a time, or other contrivances of the sort illustrated in Examples 14-1, 14-2, and 14-4).

5. THE GIBBS FREE ENERGY. Defined as $G = H - TS$ for *system* parameters H, T, and S, the **Gibbs free energy** is our shorthand way to monitor the entropy of the universe. It gives us, for any process occurring at constant temperature and pressure, a change in system free energy equal to

$$\Delta G = \Delta H - T \Delta S$$

or, put another way, a global change in entropy equal to

$$\Delta S_{univ} = -\frac{\Delta G}{T} = -\frac{\Delta H}{T} + \Delta S$$

To recall why, review the meaning of the two terms on the right: (1) $-\Delta H/T$, corresponding to the entropy change of the surroundings; and (2) ΔS, denoting the entropy change of the system.

First, recognize $-\Delta H$ (with the minus sign) as the heat flow registered in the *surroundings* while system and surroundings maintain thermal equilibrium at temperature T. Hence $-\Delta H/T$ is a thermodynamic entropy of the form q_{rev}/T (describing the entropy change of the surroundings, ΔS_{surr}) and also a statistical entropy, since this same flow of heat will bring either order or disorder to the surroundings. If the reaction is exothermic ($\Delta H < 0$), then heat goes *into* the surroundings and stirs up the particles. The flow has a disordering effect, reflected in a positive value for ΔS_{surr}. If, by contrast, the reaction is endothermic ($\Delta H > 0$), then heat goes out of the surroundings and into the system. The loss of heat leaves the surroundings more ordered, and the sign of ΔS_{surr} is correspondingly negative.

As for ΔS, the second term, remember that we have already defined it as the entropy of the system, leaving us with the result first mentioned above:

$$\Delta S_{univ} = -\frac{\Delta G}{T} = \Delta S_{surr} + \Delta S_{sys}$$

So: To increase the entropy of the universe and thereby comply with the second law, a process must produce products having lower free energy than the reactants. For a proposed happening to qualify as *spontaneous*, ΔG for the system must be negative. If not, the change cannot occur on its own.

We define and measure, accordingly, the **standard free energy of formation** for a compound (ΔG_f°) and also the **standard entropy** (where S° is now the difference in absolute entropy between a perfect crystal at 0 K and the substance under standard conditions). Then, recognizing that ΔG_f°, ΔH_f°, and S° are all proper functions of state, we use the same summation prescribed by Hess's law (Chapter 13) to compute net changes for an arbitrary reaction. We do so, finally, and examine the difference in free energy. If ΔG° is negative (the products have a lower free energy), then the reaction is guaranteed to leave the universe a more disordered place. The change *can* happen. Whether it actually *does* happen is a story still to unfold in Chapter 18.

6. FREE ENERGY AND THE EQUILIBRIUM CONSTANT. What shall we make of the enormously useful relationship $\Delta G^{\circ} = -RT \ln K$? It tells us, to begin, that the drive to minimize a system's free energy (to decrease the enthalpy, to increase the entropy) determines the eventual equilibrium state. It tells us that proportionally more products are formed when the drop in free energy is large. It tells us that there can be big winners and little winners, according to how much universal disorder is created.

Put together what we have, this time relying on the statistical meaning of ΔG° for the transformation

$$\text{reactants} \rightleftharpoons \text{products}$$

under standard conditions. Realize, first of all, that since $-\Delta G^{\circ}/T$ is equal to ΔS_{univ}°, the standard change in free energy should be related to the dispersal of energy into all the microstates of the universe. It is. If W_{before} and W_{after} denote, respectively, the number of global microstates before the change (with reactants present) and after the change (with products present), the relationship per mole becomes

$$-\frac{\Delta G°}{T} = \Delta S°_{univ} = R \ln W_{after} - R \ln W_{before}$$

Combining the logarithms and dividing by R, we then obtain

$$-\frac{\Delta G°}{RT} = \ln \left(\frac{W_{after}}{W_{before}} \right) \equiv \ln K$$

and there we have our connection: The equilibrium constant

$$K = \frac{W_{after}}{W_{before}} = \frac{\text{no. of global microstates with products}}{\text{no. of global microstates with reactants}}$$

is a statistical measure of winners and losers. If a world containing products can deploy more microstates than before, then there will be products. K will be greater than 1. The more microstates available, the more products will be formed.

SAMPLE PROBLEMS

EXAMPLE 14-1. Counting Microstates in a Simple System

PROBLEM: Suppose that a certain particle has only two allowed energies ($\epsilon = 0$ and $\epsilon = 1$), and suppose further that a system contains three such noninteracting particles. If the total energy sums to $E = 2$, how many microstates are possible?

SOLUTION: Two of the particles must occupy the level $\epsilon = 1$, thus leaving the third particle to assume the value $\epsilon = 0$. There is no other way for the three bodies to produce the total energy $E = 2$. If only one of them were to have $\epsilon = 1$, the total would be too low; if all three were to have $\epsilon = 1$, the total would be too high.

Let us say, then, that we have a *distribution* such that

$$n_0 = 1 \qquad \text{(the number of particles having } \epsilon = 0)$$

$$n_1 = 2 \qquad \text{(the number of particles having } \epsilon = 1)$$

and now ask: How many ways can we arrange the three particles consistent with this distribution?

Answer: three. Labeling the particles a, b, and c, we convince ourselves that the odd one out (the one with $\epsilon = 0$) can only be a *or* b *or* c as shown below:

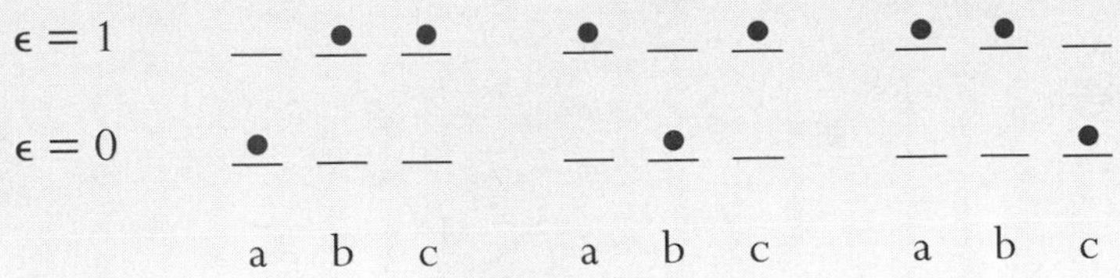

Each of these three arrangements is a distinct microstate. $W = 3$.

EXAMPLE 14-2. Picking a Winner

Still working by trial and error, here we explore a system with two possible distributions. One is more probable than the other.

PROBLEM: Three particles (a, b, c) combine to produce a total energy $E = 3$. If each particle has three allowable energies ($\epsilon = 0, 1, 2$), which distribution will contain the most microstates?

SOLUTION: Two overall distributions are consistent with a total energy of 3 units:

DISTRIBUTION	n_0	n_1	n_2	COMMENT
I	0	3	0	Three particles have $\epsilon = 1$.
II	1	1	1	One particle has $\epsilon = 1$.

Distribution I, in which all three particles are assigned 1 unit of energy, is simple enough. There is only one way to do it; hence only one microstate:

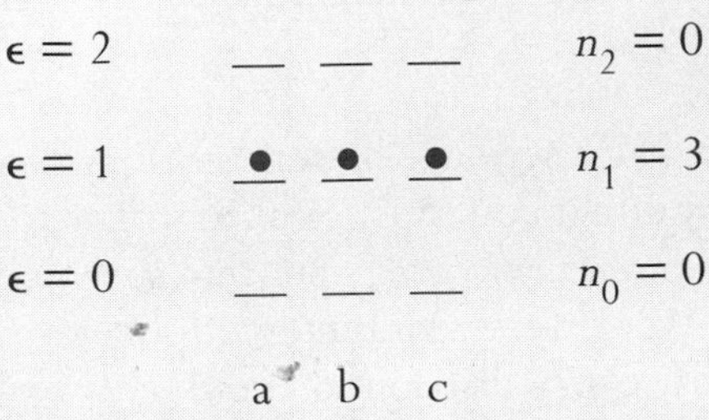

Distribution II, however, gives us a sixfold greater set of choices. First we pick a *single* particle to have $\epsilon = 0$, finding exactly three options:

either a or b or c, just as in Example 14-1. For each of these three possibilities, we then dispose of the other two particles in a way that will produce the total energy $E = 3$. One particle must go into the level $\epsilon = 1$; the other, into the level $\epsilon = 2$. There are six microstates in all ($W = 6$):

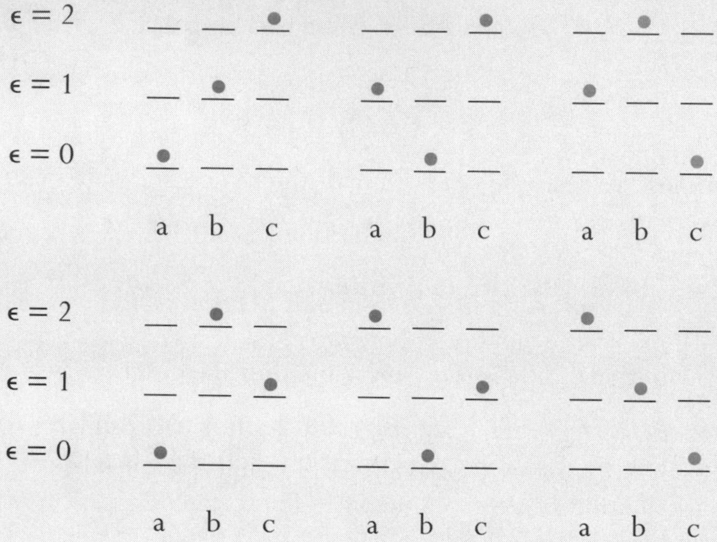

Distribution II, offering six ways to arrange the individual energies, is therefore six times more probable than distribution I with its lone microstate. Realize, though, that this narrow victory is just the beginning of a growing statistical dominance, because the advantage explodes to enormous proportions in larger, more complicated systems. With more particles, there are more and more possible distributions; and, before long, the winning set of microstates outnumbers all competitors by a fantastic margin.

QUESTION: Our trial-and-error approach is clearly inefficient. Will this method succeed in systems much larger than a few particles and a few levels?

ANSWER: No, but there are better, more systematic ways to proceed than simply by direct counting. Realize that we are solving a known problem in *combinatorial arithmetic*, choosing specified groups of objects from a larger set. Were we to continue with bigger systems, we would quickly discover that every distribution conforms to the same counting pattern. There is a general mathematical expression for W, valid for all distributions.

EXAMPLE 14-3. Atoms, Molecules, Phases

PROBLEM: Which system has the higher entropy? (a) A molecule of methane (CH_4) or a molecule of ethane (C_2H_6) at the same temperature. (b) Gaseous CO_2 or solid CO_2 (Dry Ice). (c) Water at 10°C or water at 90°C. (d) Diamond or graphite at the same temperature. (e) One mole of AgCl or two, again at the same temperature. (f) One mole of helium gas in state 1 (P, V, T) or state 2 ($P/2$, $2V$, T).

SOLUTION: Our rules of thumb are: (1) Entropy is an extensive property. The more material, the more entropy. (2) Disorder usually increases with both temperature and volume. (3) Gases are more disordered than liquids, and liquids are more disordered than solids. (4) Other factors being the same, a molecule with more degrees of freedom will have more ways to disperse its energy. The more complicated structure should have the higher entropy.

(a) *Methane or ethane.* Ethane (C_2H_6) has more atoms, more bonds, more electrons, more ways to rotate, more ways to vibrate, more of everything than methane (CH_4). Ethane has the higher entropy.

(b) *Gaseous or solid carbon dioxide.* The gas is more disordered than the solid.

(c) *Water at 10°C or water at 90°C.* The water at the higher temperature has more freedom of movement and more energy to disperse in a slightly larger volume. Its entropy is higher.

(d) *Diamond or graphite.* Graphite, with its sliding sheets of carbon atoms, has more freedom than diamond, a network crystal with every atom locked rigidly in place. Graphite has the higher entropy.

(e) *One mole of AgCl or two.* Two moles have twice the entropy of one.

(f) *Helium in state 1 (P, V, T) or helium in state 2 ($P/2$, $2V$, T).* The larger volume is responsible for greater positional disorder and thus higher entropy in state 2. Note that the total energy, which depends on n and T, is the same in both states.

Remember also: A gas will always expand spontaneously into a vacuum, moving irreversibly toward a state of greater disorder. Fortunately, too, for we need never fear a loss of oxygen caused by spontaneous contraction of the air around us.

EXAMPLE 14-4. Fear of Fluctuations

Notwithstanding the assurances of Example 14-3(f), how justified are we in feeling secure from sudden asphyxiation? Anything is possible, after all, so why not a wild fluctuation in the density of the air? The following calculation should help to allay such fears.

PROBLEM: Let there be N molecules of O_2 in the two-chambered container shown in the diagram:

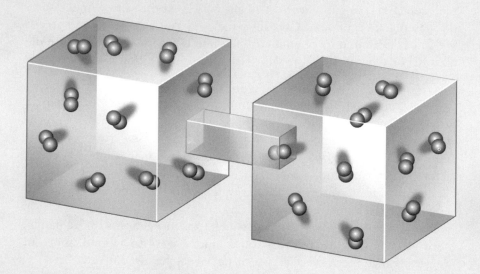

If each particle has unfettered access to both compartments through the connecting bridge, what is the probability that all N particles will be found on the left? Compute the numerical probability for $N = 1$, 10, and 100.

SOLUTION: Bouncing around randomly, each molecule has an equal probability of turning up in either compartment ($p = \frac{1}{2}$). Two molecules, neither interfering with the other, then have a combined probability equal to $\frac{1}{2} \times \frac{1}{2}$; three molecules, $\frac{1}{2} \times \frac{1}{2} \times \frac{1}{2}$; four molecules, $\frac{1}{2} \times \frac{1}{2} \times \frac{1}{2} \times \frac{1}{2}$; and so on, leading us to the general result

$$p = \frac{1}{2} \times \frac{1}{2} \overset{N \text{ times}}{\times \cdots \times} \frac{1}{2} = \frac{1}{2^N}$$

for N independent particles.

When there is only one molecule, it will be on the left one out of every two times that we look.

With 10 molecules, the probability falls to about one in a thousand ($p = 1/2^{10} = 1/1024$)—rarer than before, true, but still frequent enough to notice. Were we to check once per second, we would find the right-hand compartment empty every 17 minutes or so.

Already with 100 molecules, though, the probability of this extreme fluctuation all but vanishes: $p = 1/2^{100} = 1/(1.3 \times 10^{30})$, one occurrence in roughly every 1,000,000,000,000,000,000,000,000,000,000 trials. Making one observation per second, we would wait for over 10^{22} years, a span substantially greater than the estimated age of the universe.

Even then, a hundred molecules still would amount to only slightly more than 10^{-22} mole; a speck, practically nothing at all. Beyond that, try (and fail) to imagine the probability for finding a mere thousand or million or billion molecules clumped together unduly, let alone Avogadro's number.

Large numbers, huge numbers, enormous numbers, numbers that numb the mind and defy all understanding—those are the numbers that support the second law, and those are the numbers that impel the universe forward.

EXAMPLE 14-5. Where Predictions Can Fail

PROBLEM: In which of the following reactions should we expect the products to have a higher combined entropy than the reactants?

(a) $N_2(g) + 3H_2(g) \longrightarrow 2NH_3(g)$

SOLUTION: When four moles of gas become two moles, the total disorder of the system decreases along with the missing particles. The reactants should have the higher entropy.

(b) $C(s) + H_2O(g) \longrightarrow H_2(g) + CO(g)$

SOLUTION: One mole of solid and one mole of gas become two moles of gas. The entropy of graphite, a solid, is so low that we expect the products to have higher entropy.

(c) $CO(g) + H_2O(g) \longrightarrow CO_2(g) + H_2(g)$

SOLUTION: With two moles of gaseous reactants and two moles of gaseous products, similar in molecular structure, we have scant reason to predict one way or the other. In Example 14-6 we shall consult a table of standard entropies to resolve the question experimentally.

(d) $NaCl(s) \longrightarrow Na^+(aq) + Cl^-(aq)$

SOLUTION: A rigidly bonded ionic solid dissolves in water, surrendering its ions to a strongly polar solvent. The hydrated ions, surrounded by water molecules, then are dispersed throughout the solution. More order? Less order? The same?

Collapse of a crystal, dispersal into a liquid, more freedom of movement—it sounds surely like an increase in overall entropy, except for this one order-inducing wrinkle: the formation of hydrated clusters. Clustered around the ions, the water molecules give up some of their freedom. They become more ordered. They lose some of their original entropy.

As for how much is lost and how much is gained, we had better not guess. Every system has its idiosyncrasies, and the competing effects are too subtly interwoven to permit casual prediction. Better we should do an experiment and see what really happens, rather than cling to overly simple notions that no longer apply.

The finding, incidentally, is that entropy indeed does increase when NaCl is dissolved in water (Example 14-6d). Although relatively small, the gain is sufficient to prove that the disordering of the solute outweighs the ordering of the solvent in this one particular instance.

But not for all dissolutions. See the reaction below.

$$\text{(e)} \quad MgCl_2(s) \longrightarrow Mg^{2+}(aq) + 2Cl^-(aq)$$

SOLUTION: Another breakup of an ionic solid, and again we make the same inconclusive arguments as for NaCl. But we shall soon see, in the very next example, that for $MgCl_2$ the change in entropy is negative, not positive. Magnesium chloride is a *different* structure, with ions of different size and charge, and here the dissolved particles do have a lower entropy than the solvent and ionic solid taken separately.

Conclusion: Our rules of thumb become unreliable when competing factors produce a small net change in entropy. Subtle differences prove decisive.

EXAMPLE 14-6. Standard Entropies

PROBLEM: Compute $\Delta S°$ for each of the reactions given in Example 14-5, consulting Appendix C for any relevant thermodynamic data.

SOLUTION: First, look up in Table C-16 the standard entropy, $S°$, for each reactant and product. The value tabulated is the difference in absolute entropy between the substance at 298 K and a hypothetically perfect crystal at 0 K. The third law specifies that all entropy vanishes at absolute zero.

Next, combine the standard entropies according to the procedure we developed in Chapter 13 to implement a summation based on Hess's law:

$$aA + bB \longrightarrow cC + dD$$

$$\Delta S° = [cS°(C) + dS°(D)] - [aS°(A) + bS°(B)]$$

$$\text{sum of products} \quad - \quad \text{sum of reactants}$$

Entropy, a state function, obeys the same summation rule as does the enthalpy.

(a) $N_2(g) + 3H_2(g) \longrightarrow 2NH_3(g)$

$$\Delta S° = \quad 2S°[NH_3(g)] \quad - \quad S°[N_2(g)] \quad - \quad 3S°[H_2(g)]$$

$$= 2 \text{ mol} \times \frac{192.3 \text{ J}}{\text{mol K}} - 1 \text{ mol} \times \frac{191.5 \text{ J}}{\text{mol K}} - 3 \text{ mol} \times \frac{130.6 \text{ J}}{\text{mol K}}$$

$$= -198.7 \text{ J K}^{-1}$$

The entropy is lower owing to the loss of two moles of gas, consistent with our prediction in Example 14-5(a). Note also the similarity in $S°$ for $NH_3(g)$ and $N_2(g)$.

Will the lower entropy of the products prevent this reaction from occurring spontaneously? Not necessarily. The ordering suggested by a negative $\Delta S°$ pertains only to the system (the mixture of gases), whereas the criterion for spontaneity is an increase in *global* entropy (system plus surroundings). If the reaction is exothermic, then a strongly negative ΔH might disorder the surroundings sufficiently to ensure a net drop in the Gibbs free energy. For that, see Example 14-7.

(b) $C(s) + H_2O(g) \longrightarrow H_2(g) + CO(g)$

$$\Delta S° = S°[H_2(g)] + S°[CO(g)] - S°[C(s, \text{ graphite})] - S°[H_2O(g)]$$

$$= 130.6 + 197.7 - 5.7 - 188.7 = 133.9 \text{ J K}^{-1}$$

The low entropy of graphite ($5.7 \text{ J mol}^{-1} \text{ K}^{-1}$) allows the net change to be positive.

(c) $CO(g) + H_2O(g) \longrightarrow CO_2(g) + H_2(g)$

$$\Delta S° = S°[CO_2(g)] + S°[H_2(g)] - S°[CO(g)] - S°[H_2O(g)]$$

$$= 213.6 + 130.6 - 197.7 - 188.7 = -42.2 \text{ J K}^{-1}$$

Here, with equal numbers of gaseous reactants and products, we were expecting either a small increase or decrease in entropy. So it is: a small decrease, -42.2 J K^{-1}.

(d) $NaCl(s) \longrightarrow Na^+(aq) + Cl^-(aq)$

$$\Delta S° = S°[Na^+(aq)] + S°[Cl^-(aq)] - S°[NaCl(s)]$$

$$= 59.0 + 56.5 - 72.1 = 43.4 \text{ J K}^{-1}$$

A modest increase in entropy accompanies the dissolution of sodium chloride.

$$(e) \ MgCl_2(s) \longrightarrow Mg^{2+}(aq) + 2Cl^-(aq)$$

$$\Delta S° = S°[Mg^{2+}(aq)] + 2S°[Cl^-(aq)] - S°[MgCl_2(s)]$$

$$= -138.1 + 2(56.5) - 89.6 = -114.7 \ J \ K^{-1}$$

The Mg^{2+} ion, smaller than Na^+ and with a larger charge, attracts the partially negative oxygen of a water molecule with a stronger electric field. The effect is to increase the extent of hydration around the magnesium, thereby bringing more order to the solution.

QUESTION: How might we determine the standard change in free energy, $\Delta G°$, for each of these reactions?

ANSWER: If the process occurs at 25°C, we may use the free energies of formation tabulated in Appendix C (Table C-16) to compute $\Delta G°$, recognizing that any *element* is assigned the arbitrary value $\Delta G_f° = 0$. In the ammonia synthesis, for example, straightforward application of a generalized Hess's law yields $\Delta G° = -33.0$ kJ for the overall reaction:

$$N_2(g) + 3H_2(g) \longrightarrow 2NH_3(g)$$

$$\Delta G° = 2 \ \Delta G_f°[NH_3(g)] - \Delta G_f°[N_2(g)] - 3 \ \Delta G_f°[H_2(g)]$$

$$= 2 \ mol \times \frac{-16.5 \ kJ}{mol} - 0 - 3(0)$$

$$= -33.0 \ kJ$$

Alternatively, as in Example 14-7 immediately below, we can construct the function

$$\Delta G° = \Delta H° - T \ \Delta S°$$

at any desired temperature, often by using values of $\Delta H°$ and $\Delta S°$ determined at 1 atm and 25°C.

Our justification for so doing: Although $\Delta G°$ varies very strongly with temperature, the individual contributions $\Delta H°$ and $\Delta S°$ typically do not. Most of the temperature dependence comes from the explicit factor T in the definition of G, not from the slow variation of $\Delta H°$ and $\Delta S°$. Provided that the *pressure* remains standard at 1 atm, we shall then consider the tabulated values of ΔH° and ΔS° (which were determined at 25°C) to be valid over a wider range of temperature. It is an approximation that can be corrected later where appropriate.

EXAMPLE 14-7. Free Energy, Temperature, and Spontaneity

PROBLEM: Is the reaction

$$N_2(g) + 3H_2(g) \rightleftharpoons 2NH_3(g)$$

spontaneous in the forward direction at 25°C? Estimate the highest temperature at which spontaneity is preserved, assuming that ΔH° and ΔS° remain roughly constant throughout.

SOLUTION: From the standard enthalpies of formation listed in Table C-16, we first compute ΔH° for the reaction at 25°C:

$$\Delta H^\circ = 2\ \Delta H_f^\circ[NH_3(g)] - \Delta H_f^\circ[N_2(g)] - 3\ \Delta H_f^\circ[H_2(g)]$$

$$= (2\ \text{mol})(-46.1\ \text{kJ mol}^{-1}) - 0 - 3(0) = -92.2\ \text{kJ}$$

The forward transformation is evidently exothermic, releasing 92.2 kilojoules to the surroundings for every two moles of ammonia produced. For the process to occur spontaneously, this heat must disorder the surroundings enough to compensate for the system's *decrease* in order, which we know to be

$$\Delta S^\circ = -198.7\ \text{J K}^{-1} = -0.1987\ \text{kJ K}^{-1}$$

from Example 14-6(a).

Does it? Will 92.2 kilojoules dumped outside ($-\Delta H^\circ$) suffice to overcome the system's ordering at 298.15 K ($\Delta S^\circ = -0.1987$ kJ K^{-1})? We combine ΔH° and ΔS° to obtain the standard difference in free energy, finding here the same number as derived previously from tabulated values of ΔG_f°:

$$\Delta G^\circ = \Delta H^\circ - T\,\Delta S^\circ = -92.2\ \text{kJ} - (298.15\ \text{K})(-0.1987\ \text{kJ K}^{-1})$$

$$= -92.2\ \text{kJ} + 59.24\ \text{kJ} = -33.0\ \text{kJ}$$

The negative sign shows that the reaction is spontaneous at this relatively low temperature, with the favorable ΔH term (negative) not yet overwhelmed by the unfavorable $-T\,\Delta S$ term (positive).

The crossover ($\Delta G° = 0$) is reached when

$$\Delta H° = T \, \Delta S°$$

so that we have

$$T = \frac{\Delta H°}{\Delta S°} = \frac{-92.2 \text{ kJ}}{-0.1987 \text{ kJ K}^{-1}} = 464 \text{ K}$$

as the highest possible temperature for a spontaneous reaction.

EXAMPLE 14-8. Free Energy and the Equilibrium Constant

PROBLEM: Compute the equilibrium constant for the reaction

$$N_2(g) + 3H_2(g) \rightleftharpoons 2NH_3(g)$$

at 298 K, 464 K, and 1000 K.

SOLUTION: We use the relationship

$$K = \exp\left(-\frac{\Delta G°}{RT}\right)$$

at each temperature, expecting the equilibrium constant to decrease as the change in free energy becomes less and less negative.

At 298 K, where $\Delta G° = -33.0$ kJ (Example 14-7), the dimensionless exponent is

$$-\frac{\Delta G°}{RT} = \frac{-(-33.0 \text{ kJ})}{(8.3145 \text{ J/K})(298 \text{ K})} \times \frac{1000 \text{ J}}{\text{kJ}} = 13.32$$

and the equilibrium constant is approximately

$$K(298 \text{ K}) = \exp(13.32) = 6.1 \times 10^5$$

At 464 K, just at the point where the reaction ceases to be spontaneous, there is no thermodynamic preference for either products or reactants. The change in free energy is zero, and the equilibrium constant becomes equal to unity:

$$K(464 \text{ K}) = \exp(0) = 1$$

At 1000 K, we estimate that

$$\Delta G°(1000\ K) \approx \Delta H°(298\ K) - T\ \Delta S°(298\ K) = +106.5\ kJ$$

and hence

$$K(1000\ K) = \exp(-12.81) \approx 3 \times 10^{-6}$$

The process is now spontaneous in the reverse direction, with reactants favored over products.

EXAMPLE 14-9. Driving to Equilibrium

PROBLEM: Suppose that hydrogen, nitrogen, and ammonia gas are mixed together in a closed vessel, each component having a partial pressure of 4.0 atm. Is the mixture in equilibrium at 25°C? If not, how much free energy is still to be released?

SOLUTION: The reaction quotient

$$Q = \frac{P_{NH_3}^2}{P_{N_2} P_{H_2}^3} = \frac{(4.0)^2}{(4.0)(4.0)^3} = 0.0625$$

differs substantially from the equilibrium constant at 25°C ($K = 6.1 \times 10^5$, Example 14-8). There are still too many reactants and too few products. The mixture is not yet at equilibrium.

As a measure of the driving force, we have the instantaneous difference in free energy

$$\Delta G = \Delta G° + RT \ln Q$$

$$= -33.0\ kJ + (0.0083145\ kJ\ K^{-1})(298\ K)(\ln 0.0625) = -39.9\ kJ$$

It is, notice, a *nonstandard* G (no superscript °), because the pressures deviate from the standard values of 1 atm. Had all the partial pressures been 1 atm instead, the free energy would have been

$$\Delta G = \Delta G° + RT \ln 1 = \Delta G° = -33.0\ kJ$$

as we calculated before. Made lower by excess reactants, however, the actual reaction quotient ($Q = 0.0625$) drives the process forward to produce a mixture increasingly rich in products.

But not without limit: As products appear and reactants disappear (making Q larger), the driving force ΔG simultaneously shrinks more and more until finally vanishing at equilibrium. At that point, with $\Delta G = 0$ and $Q = K$, the free energies of reactants and products are equal. There is no drive to produce one over the other. The system is at rest, brought into conformance with the equilibrium constant

$$K_p = \exp\left(-\frac{\Delta G^\circ}{RT}\right)$$

Nothing more can be done.

EXAMPLE 14-10. At Equilibrium: Entropy of a Phase Transition

PROBLEM: Use *enthalpy* data to estimate the entropy of vaporization for water boiling at 1 atm.

SOLUTION: Consider what it means for the process

$$H_2O(\ell) \rightleftharpoons H_2O(g)$$

to be in dynamic equilibrium. Since liquid and vapor have the same free energy ($\Delta G = 0$), there is no tendency to favor one phase over the other. The system has done all it can do. Everything is in balance; the system is at rest.

From the equilibrium condition

$$\Delta G = \Delta H - T\,\Delta S = 0$$

we quickly derive an equation for the difference in entropy between the phases, expressing ΔS_{vap} in terms of the enthalpy of vaporization (ΔH_{vap}) and the boiling temperature ($T_b = 373$ K):

$$\Delta S_{vap} = \frac{\Delta H_{vap}}{T_b} = \frac{\Delta H_f[H_2O(g)] - \Delta H_f[H_2O(\ell)]}{T_b}$$

Insertion of formation data tabulated at 25°C (our usual simplification) then yields the approximate value

$$\Delta S_{vap} = \frac{[-241.8 - (-285.8)]\ \text{kJ mol}^{-1}}{373\ \text{K}} = \frac{44.0\ \text{kJ mol}^{-1}}{373\ \text{K}}$$

$$= 118\ \text{J mol}^{-1}\ \text{K}^{-1}$$

Had we used numbers measured *at* the boiling temperature (where $\Delta H_{vap} = 40.7$ kJ mol^{-1}), the result would have been 109 J mol^{-1} K^{-1}. The error is 8%.

QUESTION: The vapor is higher in both enthalpy and entropy than the liquid, yet at equilibrium there is no difference in temperature between the two phases. Why, despite having more enthalpy, does the vapor not have a higher temperature?

ANSWER: The enthalpy of vaporization goes toward tearing apart the liquid, not toward making the newly freed gas particles move faster. Yes, the vapor is hotter because it absorbs this additional heat, ΔH_{vap}; but no, its temperature is not higher since the heat is spent entirely on separating the particles.

Where, then, did the energy go and where is it now? It went into the intermolecular potential. ΔH_{vap} was used to overcome the intermolecular attractions, and there it stays: invested as potential energy, energy of position, the difference between interacting and noninteracting particles. Enthalpy buys entropy, and entropy makes a gas.

EXERCISES

1. Differentiate between a *spontaneous* process and a *fast* process. Does one have anything to do with the other?

2. Why are some transformations spontaneous and others not? What makes a reaction spontaneous? Is a nonspontaneous process physically impossible?

3. Which of the following processes are spontaneous? (a) An apple falling from a tree. (b) A ball rolling uphill. (c) The segregation of N_2 and O_2 molecules into separate halves of a room. (d) The intermingling of ethanol and water molecules in solution. (e) The discharge of a battery. (f) A flow of heat from a hand at 300 K to a hand at 301 K.

4. (a) Microstates: List the 36 outcomes possible when a gambler throws a pair of dice, completing the sequence begun below:

   | Die 1: ● | Die 1: ● | Die 1: ● | . . . |
   | Die 2: ● | Die 2: ●● | Die 2: ●●● | . . . |

   (b) Macrostates: Which total (2, 3, . . . , 12) is most likely to appear as the sum of the two dice? How many of the 36 microstates contribute to this most probable macrostate? Which macrostates (totals) are *least* likely? How many of the 36 microstates contribute to each of the least likely macrostates?

5. (a) Microstates: How many outcomes are possible when a gambler throws four dice? (b) Macrostates: Which of the totals (4, 5, . . . , 24) is most likely to appear? Which are least likely? How many microstates contribute to each of the least likely macrostates?

6. (a) Microstates: How many outcomes are possible when a gambler throws 10 trillion dice? (b) Macrostates: What are the odds, approximately, that a given throw will produce the most probable macrostate? Which macrostates are least likely? How many microstates contribute to each? (c) In what ways do reacting molecules resemble tossed dice?

7. Give a microscopic explanation of entropy. Why is entropy familiarly associated with disorder? What kind of disorder? How is disorder related to spontaneity?

8. This is work: a big, visible piston compressing a big, visible volume of gas. We can see the moving parts all working together. This is heat: tiny, invisible particles being knocked about in random directions as energy comes and goes . . . tiny, invisible particles doing tiny bits of disorganized, *microscopic* work. Is there perhaps a practical difference, then, between work and heat?

9. Continue the thought: In principle we can convert work (macroscopic work) into heat (microscopic work) with perfect efficiency, so that not even a picojoule goes astray. We cannot convert *heat*, however, into useful work with the same efficiency: 100 joules of heat might be transformed into 50 joules of work, or 75 joules, or 90 joules, or 99 joules, or even 99.999999999999 joules of work, but never into the full 100 joules. Some part of the energy is always squandered in the form of new heat. Why should that be?

10. The facts of life described above give us a macroscopic formulation of the second law:

    Heat cannot be converted into work with 100% efficiency.

    Think about the underlying *microscopic* cause of this asymmetry between work and heat, and argue that the picture is consistent with our molecular view of the second law:

    The universe evolves spontaneously toward a state of increasing global disorder.

11. So what's all the fuss about "saving" energy? Consider this modest proposal instead: Since the first law of thermodynamics guarantees us a fixed and interconvertible supply of global energy, suppose that we find a way to recapture and recycle *all* our spent energy (note the emphasis). The result would be a perpetual-motion machine, a device that could run forever without resupply of energy. Sounds good. Unfortunately, it will never work (and people have tried for centuries to make it work). What is the fatal flaw in the argument? What does the second law of thermodynamics have to say?

12. Who says that order cannot arise spontaneously in the universe? Suppose that Billy, acting entirely on his own, puts away his toys and tidies up his room. Nobody forces him; nobody helps him. Has Billy managed somehow to violate the second law of thermodynamics?

13. Humpty-Dumpty, sitting on a wall, suffers such a great fall that not even all the king's horses and all the king's men can put disordered Humpty together again. Really? If they want, the king's horses and

men can indeed put him together again, never once violating the second law of thermodynamics. Explain how.

14. Now, if Billy can clean up his room (and if—maybe—a properly equipped, hard-working group of horses and men can put Humpty-Dumpty back on the wall), shouldn't we be able to recharge a dead battery? For that bit of work, specifically, see Chapter 17; but in preparation, think about the issue in general and ask: What is the difference between local equilibrium and global equilibrium? Under what terms will nature permit the maintenance of *local order*?

15. Take a familiar example of local order tolerated amidst an ever in-creasing global disorder: a refrigerator. A refrigerator takes heat from a cold object (the milk inside) and dumps it into a warm object (the air outside), rather than allowing heat to flow spontaneously from high temperature to low. How can it do that? What price must be paid?

16. Biff has no air-conditioner in his house, but he does have a refrigerator. He also knows nothing about thermodynamics. What happens when he tries to cool the room by propping open the refrigerator door?

17. Calcium oxide and carbon dioxide combine spontaneously at tem-peratures below a certain threshold,

$$CaO(s) + CO_2(g) \longrightarrow CaCO_3(s)$$

even though the product (calcium carbonate) has a lower entropy than the reactants. (a) Is the reaction exothermic or endothermic? Answer without consulting Appendix C or any other set of data. (b) Is the entropy of the surroundings higher or lower after the re-action? (c) Is the entropy of the universe higher or lower?

18. Again, without consulting Appendix C: (a) Is the decomposition of calcium carbonate, the reverse of the process described just above, an endothermic or exothermic reaction?

$$CaCO_3(s) \longrightarrow CaO(s) + CO_2(g)$$

(b) Is the change in entropy positive or negative for the system? (c) Is the process spontaneous at high temperature or low temperature?

19. The decomposition of ammonium nitrate to form dinitrogen oxide is spontaneous at all temperatures:

$$NH_4NO_3(s) \longrightarrow N_2O(g) + 2H_2O(g)$$

(a) Is the reaction exothermic or endothermic? Answer without consulting Appendix C or any other set of data. (b) Are the products higher or lower in entropy than the reactants? (c) The reverse reaction,

$$N_2O(g) + 2H_2O(g) \longrightarrow NH_4NO_3(s)$$

is nonspontaneous at all temperatures. Does it never occur?

20. A rigid one-liter vessel is filled rapidly with one mole of an ideal gas. The container is then thermally isolated from the rest of the world (no heat flows in or out), but the system is kept at atmospheric pressure by a movable piston. (a) Does the gas come to equilibrium *immediately* after the container is sealed off? (b) By what means do the particles reach a final state of thermal equilibrium? (c) What temperature is established when they do?

21. One mole of gas, confined within a rigid, insulating, one-liter vessel at atmospheric pressure, comes to an internal thermal equilibrium (as described in the exercise above). After that, the insulation is broken and the container comes into thermal contact with the outside world at STP. (a) When contact is first established, is the gas in thermal equilibrium with its surroundings? (b) By what means do the particles attain thermal equilibrium with the external world? In what direction does the energy flow? (c) Does the gas have a proper temperature while it seeks equilibrium with the surroundings? (d) What temperature does the gas have when overall thermal equilibrium is finally secured? Assume that the surroundings are so vast as to have an infinite heat capacity: However much heat they emit or absorb, they remain at the same temperature.

22. Pick the system with the higher entropy at 25°C:

    (a) 100 mL water . . . or 200 mL water
    (b) 100 mL water . . . or 100 mL water vapor at 1 atm
    (c) 1 mol argon at 1 atm . . . or 1 mol argon at 2 atm
    (d) 1 L CH_4 at 1 atm . . . or 1 L CH_4 at 2 atm

23. Pick the system with the higher molar entropy:

    (a) liquid butane (C_4H_{10}) at 25°C . . . or liquid octane (C_8H_{18}) at 25°C
    (b) methane at 250 K . . . or methane at 300 K
    (c) the ice cube . . . or the drink
    (d) elemental oxygen at 0 K . . . or elemental titanium at 0 K

24. (a) State the third law of thermodynamics. (b) ΔH_f° is arbitrarily assigned a null value for any element in its standard state. Why are

standard entropies, $S°$, not treated the same way? Since entropy, like enthalpy, is a function of state, why should we bother to establish an *absolute* scale rather than a relative scale of values?

25. Use the data in Appendix C to compute the standard entropy change for each reaction at 25°C:

    (a) $H_2(g) + \frac{1}{2}O_2(g) \longrightarrow H_2O(\ell)$
    (b) $H_2(g) + \frac{1}{2}O_2(g) \longrightarrow H_2O(g)$
    (c) $CH_4(g) + 2O_2(g) \longrightarrow CO_2(g) + 2H_2O(\ell)$
    (d) $H_2(g) + I_2(g) \longrightarrow 2HI(g)$

26. Use the data in Appendix C to compute the standard entropy change for each reaction at 25°C:

    (a) $2NO(g) + O_2(g) \longrightarrow 2NO_2(g)$
    (b) $2Fe_2O_3(s) \longrightarrow 4Fe(s) + 3O_2(g)$
    (c) $I_2(s) \longrightarrow I_2(g)$
    (d) $2Na^+(aq) + SO_4^{2-}(aq) \longrightarrow Na_2SO_4(s)$

27. Use the data in Appendix C to compute the standard free energy change for each transformation at 25°C:

    (a) $H_2O(\ell) \longrightarrow H_2O(g)$
    (b) $H_2O(g) \longrightarrow H_2O(\ell)$
    (c) $CCl_4(\ell) \longrightarrow CCl_4(g)$
    (d) $CCl_4(g) \longrightarrow CCl_4(\ell)$
    (e) $CH_3OH(\ell) \longrightarrow CH_3OH(g)$
    (f) $CH_3OH(g) \longrightarrow CH_3OH(\ell)$

    What is the normal state of each substance (H_2O, CCl_4, CH_3OH) at 1 atm and 25°C?

28. Use the data in Appendix C to compute the standard free energy change for each transformation at 25°C:

    (a) $C(s, graphite) \longrightarrow C(s, diamond)$
    (b) $C(s, diamond) \longrightarrow C(s, graphite)$
    (c) $C_6H_6(\ell) \longrightarrow C_6H_6(g)$
    (d) $C_6H_6(g) \longrightarrow C_6H_6(\ell)$
    (e) $C_4H_{10}(\ell) \longrightarrow C_4H_{10}(g)$
    (f) $C_4H_{10}(g) \longrightarrow C_4H_{10}(\ell)$

What is the normal state of each substance (C, C_6H_6, C_4H_{10}) at 1 atm and 25°C?

29. Use the data in Appendix C to estimate the difference in Gibbs free energy between $C_2H_5OH(\ell)$ and $C_2H_5OH(g)$ under each set of conditions:

    (a) $T = 25°C$, $P = 1$ atm
    (b) $T = 78.2°C$ (the normal boiling point), $P = 1$ atm

    For simplicity, assume that $\Delta H°$ and $\Delta S°$ do not vary with temperature.

30. Use *enthalpy* data from Appendix C to estimate the entropy of vaporization for benzene (C_6H_6), which boils at 80.1°C. Ignore any dependence of the enthalpy on temperature.

31. Use enthalpy data from Appendix C to estimate the entropy of vaporization for elemental bromine (Br_2), which boils at 58.78°C. Ignore any dependence of the enthalpy on temperature.

32. Use enthalpy data from Appendix C to estimate the entropy of vaporization for mercury, which boils at 356.6°C. Ignore any dependence of the enthalpy on temperature.

33. Calculate the standard change in free energy, $\Delta G°$, for each reaction:

    (a) $NaCl(s) \longrightarrow Na^+(aq) + Cl^-(aq)$
    (b) $NaBr(s) \longrightarrow Na^+(aq) + Br^-(aq)$
    (c) $NaNO_3(s) \longrightarrow Na^+(aq) + NO_3^-(aq)$
    (d) $NaOH(s) \longrightarrow Na^+(aq) + OH^-(aq)$
    (e) $AgCl(s) \longrightarrow Ag^+(aq) + Cl^-(aq)$

    Which of the five compounds is least soluble in water at 25°C?

34. Calculate $\Delta H°$ and $\Delta S°$ for these same dissolutions:

    (a) $NaCl(s) \longrightarrow Na^+(aq) + Cl^-(aq)$
    (b) $NaBr(s) \longrightarrow Na^+(aq) + Br^-(aq)$
    (c) $NaNO_3(s) \longrightarrow Na^+(aq) + NO_3^-(aq)$
    (d) $NaOH(s) \longrightarrow Na^+(aq) + OH^-(aq)$
    (e) $AgCl(s) \longrightarrow Ag^+(aq) + Cl^-(aq)$

    Which of the reactions are exothermic and which are endothermic? In which reactions do the products have more entropy than the reactants?

35. Calculate $\Delta H°$, $\Delta S°$, and $\Delta G°$ at 25°C for each of the following reactions:

    (a) $C_2H_4(g) + H_2(g) \longrightarrow C_2H_6(g)$
    (b) $C_4H_{10}(g) + \frac{13}{2}O_2(g) \longrightarrow 4CO_2(g) + 5H_2O(\ell)$
    (c) $C_4H_{10}(\ell) + \frac{13}{2}O_2(g) \longrightarrow 4CO_2(g) + 5H_2O(\ell)$
    (d) $C_2H_5OH(\ell) + 3O_2(g) \longrightarrow 2CO_2(g) + 3H_2O(\ell)$
    (e) $CH_4(g) + 2O_2(g) \longrightarrow CO_2(g) + 2H_2O(\ell)$

36. Over what range of temperature is each of these reactions spontaneous?

    (a) $C_2H_2(g) + 2H_2(g) \longrightarrow C_2H_6(g)$
    (b) $H_2(g) + I_2(g) \longrightarrow 2HI(g)$
    (c) $2O_3(g) \longrightarrow 3O_2(g)$
    (d) $2CO_2(g) + 4H_2O(g) \longrightarrow 2CH_3OH(g) + 3O_2(g)$

37. Over what range of temperature is each of these reactions spontaneous?

    (a) $2NO(g) + O_2(g) \longrightarrow 2NO_2(g)$
    (b) $2NOCl(g) \longrightarrow 2NO(g) + Cl_2(g)$
    (c) $CaCO_3(s) \longrightarrow CaO(s) + CO_2(g)$
    (d) $H_2O_2(\ell) \longrightarrow H_2O(\ell) + \frac{1}{2}O_2(g)$

38. (a) Use the relationship between $\Delta G°$ and K to compute an equilibrium constant for the reaction

    $$2HBr(g) + F_2(g) \rightleftharpoons 2HF(g) + Br_2(g)$$

    at 300 K, 600 K, and 900 K. Neglect any temperature variation in $\Delta H°$ and $\Delta S°$. (b) Does the yield at equilibrium change noticeably at the higher temperatures? (c) Does the yield increase or decrease at higher pressure? (d) Does the yield increase or decrease in the presence of a catalyst?

39. Assume that neither $\Delta H°$ nor $\Delta S°$ changes with temperature for the reaction

    $$N_2(g) + 3F_2(g) \rightleftharpoons 2NF_3(g)$$

(a) At what temperature is the equilibrium constant equal to 1? (b) At what temperature is K equal to 10? (c) At what temperature is K equal to 0.1?

40. (a) Use thermodynamic data from Appendix C to determine whether the following reaction is spontaneous at 298 K:

$$NaHCO_3(s) \rightleftharpoons NaOH(s) + CO_2(g)$$

(b) Will the equilibrium pressure of CO_2 be higher or lower at 700 K?

41. Suppose that carbon dioxide exerts a pressure of 1.00 atm in a heterogeneous reaction with calcium carbonate and calcium oxide at 25°C:

$$CaCO_3(s) \rightleftharpoons CaO(s) + CO_2(g)$$

(a) Compute both the equilibrium constant and the reaction quotient at 25°C. Is the system in equilibrium at 25°C? (b) If not, then at what temperature would 1 atm of $CO_2(g)$ coexist in equilibrium with CaO(s) and $CaCO_3(s)$?

42. A one-liter vessel contains one mole each of the gases N_2O_4 and NO_2, which react as

$$N_2O_4(g) \rightleftharpoons 2NO_2(g)$$

The temperature is 298 K. (a) Calculate the reaction quotient and equilibrium constant. Is the mixture in equilibrium at 298 K? (b) If not, how much free energy is still to be released? In which direction will the reaction proceed?

43. The equilibrium constant for the ionization of acetic acid,

$$CH_3COOH(aq) \rightleftharpoons H^+(aq) + CH_3COO^-(aq)$$

is 1.8×10^{-5} at 25°C. Estimate $\Delta G°$.

44. Say that a mixture of oxygen, sulfur dioxide, and sulfur trioxide stands at equilibrium:

$$2SO_2(g) + O_2(g) \rightleftharpoons 2SO_3(g)$$

The temperature is 25°C. (a) Calculate ΔG and $\Delta G°$. (b) Will ΔG increase, decrease, or remain the same if the partial pressure of oxygen is increased above its equilibrium value? (c) Will $\Delta G°$, the standard free energy difference, be affected by similar changes in the partial pressures?

45. One more reminder, near the end, of microstates and macro-states—really the key to understanding equilibrium, entropy, and free energy. Consider: A system consists of three particles (a, b, c) with a total energy of three (3) units among them. Each particle has four allowed energies ($\epsilon_1 = 0$, $\epsilon_2 = 1$, $\epsilon_3 = 2$, $\epsilon_4 = 3$). (a) There are, in all, 10 microstates consistent with the constraints as stated, grouped into three distributions. Work them out. (b) Which distri-bution contains the most microstates? (c) In what way is this model drastically different from a real chemical system?

46. System, surroundings, statistics: Observing gas particles to be dis-tributed uniformly throughout a container, we attributed this homo-geneity of phase to a drive toward higher entropy. We asserted simply that a uniform mixture was by far the most likely macrostate, a statistical certainty impossible to beat. But what shall we say about substances that do *not* mix well—substances like benzene and water, which spontaneously segregate into two distinct phases when shaken together? Do these *immiscible* liquids somehow manage to defy the statistics? How can the separation of a mixture lead to greater en-tropy in the universe?

15

Making Accommodations—
Solubility and
Molecular Recognition

15-1. Thermodynamics: Prologue to Reaction
15-2. Solutes and Solutions
  Order, Disorder, and Equilibrium
  The Solubility Product
  Selective Precipitation
  Space, Time, and Equilibrium
  Enthalpy and Entropy
  Structure and Solubility
  Hydrophilic and Hydrophobic Effects
15-3. Guests and Hosts
  Molecular Recognition
  Thermodynamics of Binding
  REVIEW AND GUIDE TO PROBLEMS
  EXERCISES

15-1. THERMODYNAMICS:
PROLOGUE TO REACTION

Nothing stands still. Ours is a universe of ceaseless motion, driven along by a shifting flow of energy. It is energy that moves matter; it is energy that effects change.

And it is thermodynamics, with but a few austere laws, that regulates the transfer of energy and thereby governs all spontaneous change. Blindly and impartially, thermodynamics lays down the rules.

Rule 1: Energy is conserved. Total motive power is, has been, and continues to be fixed and unchanging. Only in the *redistribution* of energy is there any freedom to reshape the material world.

Rule 2: Time marches on. Matter moves. Energy is shared and exchanged among uncountably many recipients, always tending toward a broader and broader distribution. Inexorably and irrevocably, global entropy increases as nature's fixed endowment of energy is divided ever more finely. Whatever happens, happens. There is no turning back.

And since there is no turning back, we have a way to distinguish yesterday from today. We observe that, today, the universe is more disordered than it was yesterday. Its total energy, still unaltered, is spread out further. Any spontaneous change—the motion of a planet, the burning of gasoline, the rusting of iron—contributes to that growing disorder. There are no exceptions, not even among the most complex and ordered pieces of machinery found anywhere: living organisms.

How, then, do we know whether some chemical change occurs spontaneously, with the products coming *after* the reactants? Again, we observe that the universe *with the products* is more disordered than it was with the reactants alone. The system's local free energy, G, decreases as

$$\Delta G = G_P - G_R = \Delta H - T\,\Delta S < 0$$

upon going from reactants to products, matched invariably by an increase in entropy for system and surroundings combined.

To react is thus to correct an unbalanced distribution of free energy. Reactants A and B, standing higher in free energy than products C and D, roll down a hill of "chemical potential" just as a ball rolls down a hill of gravitational potential. Matter transforms itself from high G to low G, always disordering the universe irreversibly in the process. Differences in free energy are eventually leveled out, and equilibrium prevails thereafter. Where no differences exist, matter cannot move and matter cannot change. That is what thermodynamics has to tell us.

Our aim now is to bring thermodynamics to bear specifically on the molecular world, where a particle will interact according to its own size, shape, and electrical characteristics. Fashioned quantum mechanically from charged bits of matter, the atoms and molecules are creations of the electromagnetic interaction. Electric forces make them move; electric forces give them energy; electric forces enable them to change . . . and thermodynamics legislates what is possible and what is impossible.

Here, and in the two chapters following, we shall begin to connect molecular structure with thermodynamic potentiality, looking to understand how microscopic attributes combine to determine ΔG° and hence the equilibrium constant. The emphasis shifts back to elementary chemical reactions—to such fundamental transactions as the transfer of an electron, the movement of a hydrogen ion, and the coordination of electrons in noncovalent interactions. These processes, introduced first in

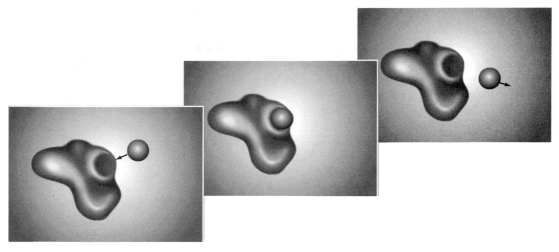

FIGURE 15-1. Guest and host: paradigm for a wide-ranging chemistry of weak, noncovalent, often highly specific interactions. The relationship is between independent associates, with each partner free to come and go.

Chapter 3, are the leitmotifs of all reactions. Recycled and reused, they enable molecules to carry out the most complex maneuvers one step at a time.

We start with the chemistry of noncovalent interactions, embracing everything from the formation of a solution to the formation of an enzyme–substrate complex. Think of it as a chemistry of accommodation, a chemistry in which one species (the guest) is taken up within some other structure (the host). Such association encourages a curiously transient kind of relationship, a chemistry of temporary accommodation whereby the guest is often later released to seek new partners. Linked by noncovalent interactions, guests and hosts share a relationship less intimate than in a covalent bond. They come and go.

They come and go with considerable variety too. The guest in Figure 15-1 might be a solute surrounded by solvent particles in a liquid. It might be an ion or amino acid bound to an enzyme; it might be a xylene molecule within a zeolite, a hormone bound to a receptor, or even an antigenic protein recognized by an antibody. But all such manifestations are answerable to the same basic laws, and so we need pose only one set of questions for each reaction. Questions such as: What aspects of microscopic structure enable these species to interact? Where are the electrons? Where *aren't* the electrons? Where and how is the enthalpy changed? Does the local entropy increase or decrease? Is the free energy dominated by either enthalpy or entropy? In what direction does the free energy flow? Where, in the end, will the equilibrium lie?

15-2. SOLUTES AND SOLUTIONS

Dropped into water, a sugar cube grows soft and soon disappears into solution. Molecules of water first erode the crystal, loosing the noncovalent bonds between one highly polar molecule of sucrose ($C_{12}H_{22}O_{11}$)

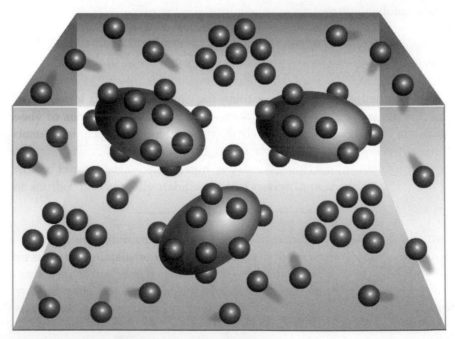

and another. Groups of H_2O, offering noncovalent bonds of their own, then surround each particle of solute and carry it away, victors in a thermodynamic competition. The sugar dissolves. A host—a shifting, fluid host of solvent clusters (Figure 15-2)—has been assembled just for the occasion, the *solvation* of a guest. And with water acting as the solvent,

FIGURE 15-2. Assault on a solid: solvation. The solvent, rendered as small spheres, plays host to a solute in this schematic view of a solution. Molecules of solvent take apart the erstwhile solid and trap its particles inside solvated clusters. The solvent host, readily adaptable, rearranges itself as needed to embrace (and later to release) its guest.

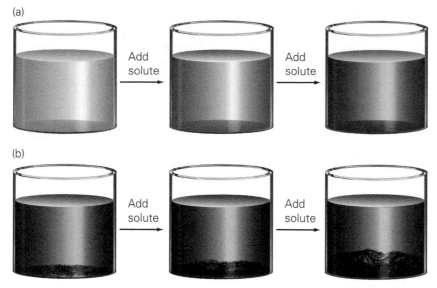

FIGURE 15-3. Equilibrium in solution, from one phase to two: (a) Below the saturation point, no solid appears even as the concentration increases. The particles, completely dissolved, are dispersed uniformly throughout the solvent, maintaining everywhere the same temperature. The system persists in a state of homogeneous equilibrium, instituted for this one phase only: the solution. (b) At the saturation point, a second phase—undissolved solute—enters into a heterogeneous equilibrium with the solution. Particles travel continually from solid to solution and back again to solid, but the concentration of dissolved solute never changes.

we speak specifically of a *hydration* of the solute; it is an especially versatile accommodation made possible by the wide-ranging affinity of the polar H_2O molecule for other polar species.

Stirred into solution, the hydrated sucrose molecules spread themselves uniformly over a homogeneous liquid phase. Eventually no part of the solution is more or less sweet than any other, and eventually the temperature is the same throughout. There the mixture stands in thermal equilibrium, homogeneous over the one phase.

Add more sugar, as in Figure 15-3, and the solution becomes more concentrated until finally no additional solute can be accommodated. With a newly formed pile of undissolved sucrose at the bottom of the cup, the water's sweetness then ceases to change. A heterogeneous equilibrium sets in between the hydrated sucrose and the sucrose in the solid state. The solution is **saturated**.

Order, Disorder, and Equilibrium

We behold, in any saturated solution, an equilibrium between a disordered and ordered phase, conceptually no different from the vapor–liquid transformation of Chapter 11. The disordered phase: the solvated particles floating free, liberated from their close association in the solid. The ordered phase: the undissolved solute, bound up in a crystal. Like vapor–liquid and liquid–solid systems, the heterogeneous equilibrium between solute and solution develops from a disorder–order transition going both ways.

Adding sugar to water should remind us, therefore, of a gas subjected to isothermal compression. Squeezed along an isotherm, the gas undergoes a rise in pressure until suddenly droplets of liquid begin to appear and equilibrium takes hold. Arising from a delicate balance between vapor and liquid, the equilibrium vapor pressure then remains fixed for as long as the two phases continue to coexist. So, too, will a solution accept more and more solute until the onset of its own disorder–order transition, signaled by the abrupt precipitation of a solid phase. Having increased steadily to the point of saturation, the concentration of dissolved solute now holds constant.

This relationship between solid and solution (Figure 15-4) is a dynamic equilibrium, just as before. Solute molecules continually leave the solid and enter the solution, counterbalanced by a return traffic from solution to solid. At equilibrium, where no difference in free energy separates solid and solution, the opposing rates are equal and the system

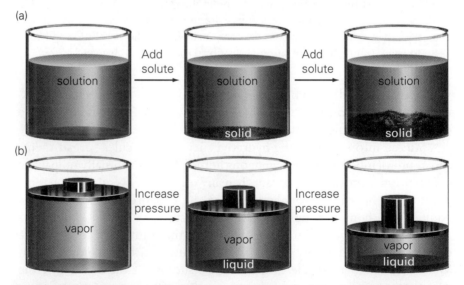

FIGURE 15-4. Order and disorder, a comparison of solution–solid equilibria and vapor–liquid equilibria: (a) A solid *precipitates* (literally, "falls down") from solution once the concentration reaches its saturation point. (b) A vapor similarly condenses at sufficiently high pressure. In each system, a dynamic equilibrium is maintained between an ordered phase and a disordered phase. Both the concentration of solute and the pressure of the vapor remain constant for as long as the heterogeneous equilibrium continues.

appears to be at rest. There is no net tendency for matter to flow from one point to another.

Analogous to the vapor pressure above a pure liquid, the concentration of a saturated solution holds steady in the presence of its undissolved solid. From one phase to another, precipitation is matched by dissolution. The balance is maintained molecule for molecule and mole for mole, unaffected by how much or how little undissolved material is on hand. Even one grain of solid is proof that the solution is saturated; even one grain is sufficient to sustain the equilibrium and preserve the fixed concentration.

For a single dissolved species, moreover, this unwavering concentration serves as an equilibrium constant, acting again like a vapor pressure. We cast the transformation in general as

$$A(s) \rightleftharpoons A(aq)$$

where $A(s)$ and $A(aq)$ denote, respectively, the solid and aqueous states of some nondissociating molecule, and we write

$$K' = \frac{[A(aq)]}{[A(s)]}$$

for the equilibrium constant. Disregarding the never-changing solid concentration, $[A(s)]$, we then obtain

$$K = K'[A(s)] = [A(aq)]$$

at saturation.

The concentration of a saturated solution, $[A(aq)]$, thus is equal to the equilibrium constant for the solid–solution transition. If, instead, the actual concentration of A is less than K, then the low value of $[A(aq)]$ serves to depress the reaction quotient, Q, below its equilibrium threshold. No longer is there any undissolved solid. More solute can yet be added to this out-of-equilibrium solution, and so the reaction tends to the right until finally $[A(aq)]$ grows equal to K. At that point, with equilibrium established, the first grain of solid precipitates from the solution.

If the solute concentration ever becomes *greater* than K (by partial evaporation of solvent, for example), then the reverse reaction

$$A(s) \longleftarrow A(aq)$$

will force the excess solid out of solution until $Q = K$. Equilibrium returns following that precipitation, but note that the number K itself is never subject to negotiation. Its value is invariant, always the same at a given temperature, and the concentration of a saturated solution is forever fixed.

Realize that no such restriction applies to a *gaseous* solute dissolved in a liquid (Figure 15-5), since here the position of equilibrium depends further on the pressure. The balance is between gas-phase molecules (the disordered state) and solution-phase molecules (now the ordered state). More gas can be forced into the liquid simply by pressing down on the liquid's surface, since increased pressure will drive the equilibrium leftward, from gas to liquid, in the transformation

$$A(aq) \rightleftharpoons A(g)$$

We then rearrange the equilibrium expression

$$K = \frac{P_A}{[A(aq)]}$$

as

$$P_A = K[A(aq)]$$

to draw out a statement of **Henry's law**: The dissolved concentration of a gas is proportional to the pressure of that gas over the solution. At

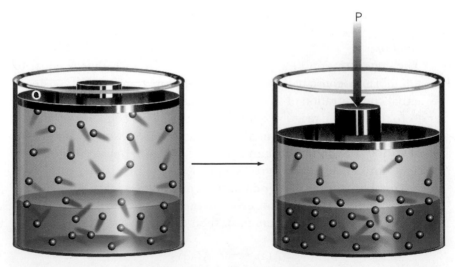

FIGURE 15-5. Pressed into solution, a gaseous solute attains a concentration given by *Henry's law*: in direct proportion to the gas pressure above the liquid. Any increase in the partial pressure shifts the equilibrium in favor of the solution.

higher pressure, more gas molecules are squeezed into the liquid—as when CO_2 is dissolved under high pressure to produce a carbonated beverage.

The Solubility Product

Henceforth we shall be concerned mainly with solids in liquids, where the concentration of a saturated solution is fixed by its equilibrium constant. Particularly important is the so-called **solubility-product constant**

$$K_{sp} = [A^{m+}]^n [B^{n-}]^m$$

associated with the equilibrium between an ionic solid ($A_n B_m$) and its hydrated ions (A^{m+} and B^{n-}):

$$A_n B_m(s) \rightleftharpoons nA^{m+}(aq) + mB^{n-}(aq)$$

Use of K_{sp} is most reliable for highly dilute, nearly ideal solutions (where solute–solvent, solvent–solvent, and solute–solute interactions are all assumed to be the same, as described in Chapter 11).

Examples of solubility products at 25°C include

$$AgCl(s) \rightleftharpoons Ag^+(aq) + Cl^-(aq)$$

$$K_{sp} = [Ag^+][Cl^-] = 1.8 \times 10^{-10}$$

and

$$PbCl_2(s) \rightleftharpoons Pb^{2+}(aq) + 2Cl^-(aq)$$

$$K_{sp} = [Pb^{2+}][Cl^-]^2 = 1.6 \times 10^{-5}$$

Varying only with temperature, these small numbers are ordinary equilibrium constants: products of ionic concentrations, each raised to the appropriate stoichiometric power.

They are not solubilities. A **solubility** is the amount of material that will dissolve to produce a saturated solution, often measured as grams or moles of solute per liter of solution. The solubility-product constant, by comparison, is an expression of the constraints imposed upon a solid and its dissociated ions at equilibrium. K_{sp} is not a statement about, say, the ability of Ag^+ *or* Cl^- to dissolve under any circumstances, but rather about the ability of these two hydrated ions to coexist specifically with undissolved $AgCl$, an ionic solid.

A low K_{sp} (barely over 10^{-10} for AgCl) shows that prospects are bleak for Ag^+ and Cl^- together in the same aqueous solution. How bleak? We denote both $[Ag^+]$ and $[Cl^-]$ by x so that

$$K_{sp} = x^2 = 1.8 \times 10^{-10}$$

and, taking the square root, we quickly determine that x is just 1.34×10^{-5} M—a very small dissolved concentration, coming from a correspondingly small equilibrium constant. The direction of spontaneous change is largely from solution to solid, driven by a substantial decrease in free energy upon precipitation, not dissolution.

Although here the thermodynamics clearly favors the solid, such judgment need not extend to solids in general. The contest for lower free energy is won sometimes by the solid and sometimes by the solution, subject to individual criteria that we shall soon consider. For the moment, let us assert only that one mole of crystalline AgCl has (owing to the size and disposition of its ions) a far lower free energy than one mole of the same Ag^+ and Cl^- ions in aqueous solution. The flow of matter is overwhelmingly toward the solid, reversed only when hydrated Ag^+ and Cl^- fall below their small equilibrium concentrations in the watery phase outside.

Nevertheless, a low K_{sp} for the AgCl equilibrium says nothing about the solubility of Ag^+ and Cl^- under altered conditions. Quite the contrary, for certain other ionic solids such as $AgNO_3$ and NaCl are readily soluble in water. Fully 1.2 *kilograms* of $AgNO_3$ will dissolve in one liter of water at $0°C$, consistent with a K_{sp} 11 orders of magnitude greater than for AgCl. Almost as soluble is NaCl, with a value of 0.357 kg L^{-1} at $0°C$.

Different materials make for different free energies. With the right partners and in the right environments, each of these dissolved ions (Ag^+ and Cl^-) evidently can enjoy an equilibrium shifted strongly toward the solution phase. Put Ag^+ and Cl^- together, though, and AgCl(s) will precipitate in enormous quantity; formation of that one particular solid, AgCl, disorders the universe more profoundly than hydration of the Ag^+ and Cl^- ions in a shared solution. To destroy a crystal of AgCl by hydration entails a significant climb upward in free energy, whereas to attack a sample of a more soluble material is to drop downward (or perhaps to climb a smaller hill).

We learn, accordingly, that the fate of any ion in solution is linked to all the other species simultaneously resident. Individual concentrations and solubilities are always contingent upon specific conditions. It is only the product of concentrations, K_{sp}, that remains inviolable.

Just because Cl^- *can* form a 1:1 compound with Ag^+, for instance, is no guarantee that a solution must always contain equal numbers of these two ions. Suppose we add 1.3×10^{-5} mol AgCl to 1.0 L of a solution already 1.0 M in AgNO$_3$. Will this small quantity of AgCl dissolve to form a saturated solution, as it would when introduced into pure water?

No. The dissolution of AgCl will be constrained by a large population of Ag$^+$ already present in solution, utterly indistinguishable from any other silver ions and granted equal rights at equilibrium. To this initial Ag$^+$ concentration of 1.0 M (from the AgNO$_3$), AgCl then contributes an additional x moles per liter of both Ag$^+$ and Cl$^-$ as it dissolves; and, when it does, the whole mixture is compelled finally to satisfy K_{sp} at equilibrium:

$$[\text{Ag}^+] = 1.0 + x$$

$$[\text{Cl}^-] = x$$

$$K_{sp} = [\text{Ag}^+][\text{Cl}^-] = (1.0 + x)x = 1.8 \times 10^{-10}$$

Since x must be a small number (no greater, certainly, than the 1.3×10^{-5} mol L^{-1} on hand), we can safely assume that $1.0 + x$ is very nearly 1.0 M. That done, the equation is solved: $x = 1.8 \times 10^{-10}$ M. Dissolved silver ions, most of them coming from AgNO$_3$, outnumber chloride ions in the final solution by a factor approaching 10^{10}.

The solubility of AgCl, small even in pure water, is therefore depressed by five orders of magnitude in a 1.0 M solution of AgNO$_3$ (Figure 15-6), where the massive presence of Ag$^+$ shifts the equilibrium drastically toward solid AgCl. Called the **common-ion effect**, such behavior offers a straightforward demonstration of Le Châtelier's principle: The system relieves itself of the stress suffered by having too much product (in this instance, dissolved Ag$^+$). It is, as we shall see in Chapter 16, a frequent consideration in solutions of acids and bases.

Selective Precipitation

Similar effects ensue whenever two or more equilibria compete for the same ions. Observe what happens, for example, when chloride and iodide anions are simultaneously exposed to silver cations, so that now there are two possible reactions with two very different equilibrium constants:

$$\text{AgCl}(s) \rightleftharpoons \text{Ag}^+(aq) + \text{Cl}^-(aq) \qquad K_{sp}(\text{AgCl}) = 1.8 \times 10^{-10}$$

$$\text{AgI}(s) \rightleftharpoons \text{Ag}^+(aq) + \text{I}^-(aq) \qquad K_{sp}(\text{AgI}) = 8.3 \times 10^{-17}$$

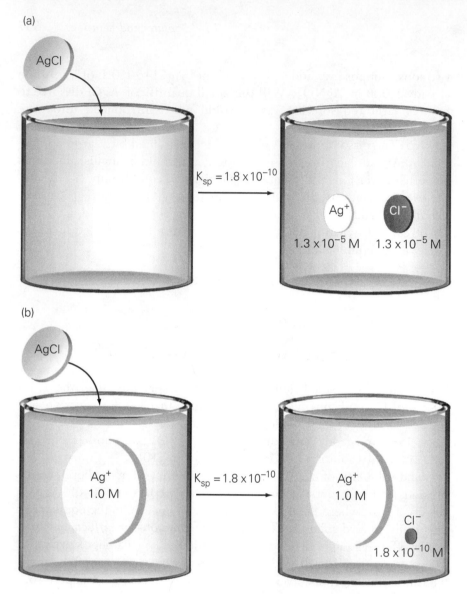

FIGURE 15-6. Common-ion effect. The presence of one solute alters the solubility of another, as shown here for AgCl and $AgNO_3$: (a) Alone, AgCl dissolves in water to a maximum of 1.3×10^{-5} mol L^{-1}. (b) Added to 1.0 M $AgNO_3$, chloride ions no longer dissolve as they do in pure water. The solution, already rich in Ag^+, can scarcely accept any additional contribution from AgCl, and the solubility of silver chloride plummets in response. Reason: K_{sp}, the equilibrium constant for $[Ag^+]$ and $[Cl^-]$, is forever fixed at 1.8×10^{-10}. More Ag^+ from $AgNO_3$ means less Cl^- from AgCl.

To a solution initially containing both NaCl and NaI (try it with $[Cl^-] = [I^-] = 0.010$ M), we add small doses of a suitably concentrated $AgNO_3$ solution, monitoring carefully the process pictured in Figure 15-7. Ag^+ ions enter the solution continually, drop by drop, but neither Cl^- nor I^- is induced to precipitate until its ion product first reaches the appropriate

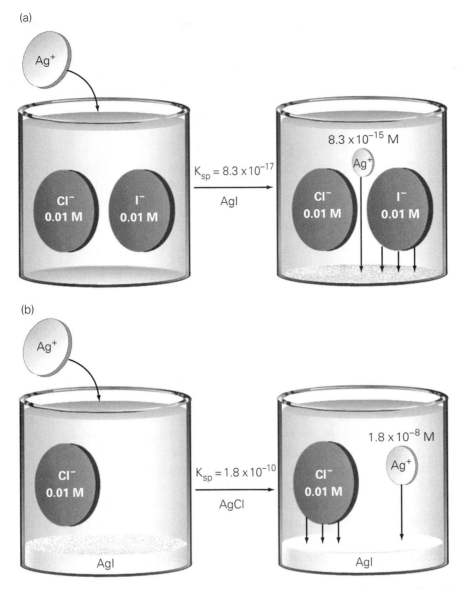

(a)

$K_{sp} = 8.3 \times 10^{-17}$

AgI

8.3×10^{-15} M

(b)

$K_{sp} = 1.8 \times 10^{-10}$

AgCl

1.8×10^{-8} M

FIGURE 15-7. Selective precipitation. Present in equal numbers, the anions Cl^- and I^- compete with unequal results for the same Ag^+ cation: (a) AgI precipitates first from solution, driven out by its vanishingly small equilibrium constant. (b) With a K_{sp} 2 million times greater, AgCl resists precipitation until all the I^- is gone. The equilibrium constant for AgCl, though small, is nevertheless large compared with the value for AgI.

K_{sp}. Anything less, and the solution remains *under* the threshold for heterogeneous equilibrium; and it stays there, unsaturated, for as long as the reaction quotient lies below its own K_{sp}. Upon attaining that target level, however, the system settles into equilibrium and a solid phase forms. K_{sp} acts as a limit to dissolution, a thermodynamic barrier to further coexistence.

For chloride, the threshold for precipitation is met at a silver concentration of

$$[Ag^+] = \frac{K_{sp}(AgCl)}{[Cl^-]} = \frac{1.8 \times 10^{-10}}{0.010} = 1.8 \times 10^{-8} \; M$$

whereas for iodide the limit is $8.3 \times 10^{-15} \; M$:

$$[Ag^+] = \frac{K_{sp}(AgI)}{[I^-]} = \frac{8.3 \times 10^{-17}}{0.010} = 8.3 \times 10^{-15} \; M$$

This large disparity, which conforms to the ratio of equilibrium constants, ensures that AgI(s) will appear after introduction of just enough Ag^+ to make $[Ag^+] = 8.3 \times 10^{-15} \; M$. Silver chloride, meanwhile, remains below its equilibrium threshold—unprecipitated—until 2 million times more silver is added, by which point the iodide has already vanished.

Here, evidently, are two anions structurally capable of binding to the same cation, but thermodynamically they compete on unequal terms. The salt with the smaller equilibrium constant (AgI) is destined to precipitate first, allowing us to manipulate the equilibria and separate one ion from the other. It is an example of *selective precipitation*, an important technique for the purification of mixtures.

Space, Time, and Equilibrium

Not to be overlooked are equilibria within the solution phase itself, made manifest by a uniformity over space and time. We see how the temperature of an equilibrated solution is everywhere constant, as are its concentrations and all the other macroscopic attributes. Maintained by particle-to-particle contacts within the solution, this balance will develop whether or not a solid phase is present. For despite the unremitting motion of the particles—indeed *because* of such motion—a single phase in thermal equilibrium betrays no net flow of matter, no unjustifiably preferred direction, no spontaneous change.

The reason: Solute and solvent species continually come together, exchange energy, and move apart as they sustain a Boltzmann distribution at the prevailing temperature. Essentially no different from particles colliding in a gas, solute and solvent maximize their positional disorder by spreading uniformly over the volume. Only then is the solution truly at rest; only then is it in chemical and thermal equilibrium, with free energy everywhere the same.

But, like any other equilibrium, the delicate balance can be disturbed by imposition of a stress. Let there arise some spatial nonuniformity in, say, the concentrations, and immediately the equilibrium portrayed in Figure 15-8 is destroyed. Variations of G appear from point to point, making the free energy suddenly high over here and low over there. It is a signal for matter to move—an indication that some kind of *potential* is out of balance, just as a difference in temperature is a signal for heat to flow.

There is an imbalance in free energy. The free energy per mole of substance, understood precisely as this *chemical potential*, no longer is distributed evenly throughout the available space. To restore equilibrium, matter must flow from high potential to low potential until any differences are erased.

It happens all the time, particularly within living cells. Enclosed by its membrane, a cell is a bag of ions and molecules dissolved in aqueous solution. Outside is another kind of liquid, usually a solution at some different concentration.

Now, between inside and outside matter must flow. The cell must take in raw materials selectively and relieve itself of waste, keeping what it needs and discarding the rest. Survival depends on such discrimination. What this chemical factory cannot accept, clearly, is a fatal equilibrium with the outside world, whereby it would merge into a homogenized solution having the same concentration inside and out.

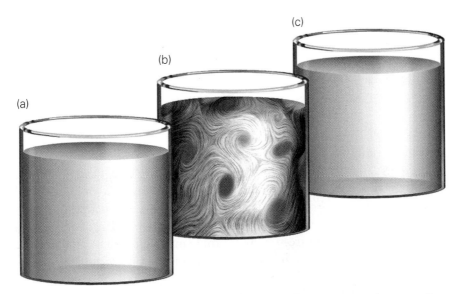

FIGURE 15-8. (a) At rest, matter in equilibrium has no net tendency to flow from point to point. Everywhere in the solution, the free energy is the same. (b) Disturbed from equilibrium, the system moves to restore the balance. Matter flows spontaneously from high concentration to low. (c) Equilibrium is reestablished and, once again, there are no differences in chemical potential. The concentration is uniform throughout.

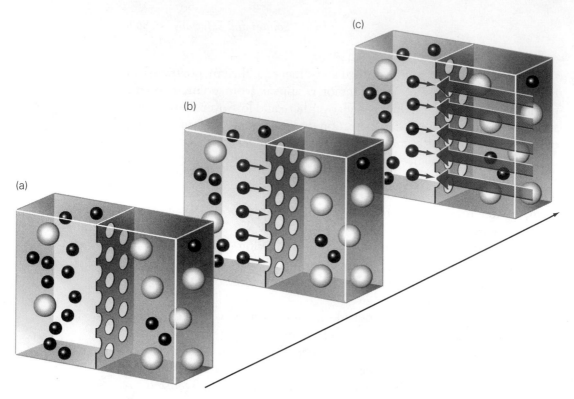

FIGURE 15-9. Osmotic flow across a membrane. The right-hand compartment in each panel represents a cell. (a) Open to water (small particles) but not sodium ions (large particles), the semipermeable membrane encloses a cellular solution where [Na$^+$] is higher than in the fluid outside. But to do so is to live dangerously in a world ruled by the second law of thermodynamics, and immediately the cell must fight for its separate osmotic equilibrium. (b) Triggered by the unequal concentrations, water rushes in as nature moves to establish a broader equilibrium between inside and outside. (c) Resisting the dilution, the cell spends energy to hold back the water with a counteracting *osmotic pressure*.

Protection comes in the form of a special *semipermeable membrane* (Figure 15-9), through which some particles pass and others do not. Often the discrimination is made according to size, with only small species able to traverse the membrane. So it happens that water, life's critical little solvent, typically flows in and out, while larger molecules such as proteins are confronted by an impermeable barrier. The blocked species, unable simply to diffuse across the membrane, must be actively carried across by other means.

Suppose, then, that the fluid outside a cell supports a certain concentration of sodium ions [Na$^+$], a value lower than found inside the cell. Richer in H$_2$O and poorer in Na$^+$, the extracellular solution then stands at a different level of free energy compared with the solution inside. It is at a different chemical potential, out of equilibrium with respect to the cellular contents.

Not for long. Water molecules soon flow through the membrane into the cell, moving to equalize the free energy between solvent outside and solution inside. The process, a form of diffusion, is called *osmosis*, and it is a phenomenon dependent only on the number of solute particles—a colligative effect like vapor-pressure lowering, boiling-point elevation, and freezing-point depression (all described earlier in Chapter 11). Solvent flows toward a locally high concentration of solute, responding more to the *number* of generic particles than to distinct chemical entities. It flows in order to equalize the concentrations and hence the free energy per mole; it flows in order to establish a more comprehensive equilibrium.

The cell, taking on this osmotic flow, swells as the incoming water dilutes the solution within. Its contents weigh heavier on the membrane, and the internal pressure increases correspondingly. This rise in pressure (call it Π) is again a colligative property dependent on how *many* moles of solute are to be diluted. Obeying an ideal gas–like equation of state,

$$\Pi = \frac{n}{V}\, RT$$

the effect scales with the original concentration, n/V. The greater the disparity between inside and out, the more solvent threatens to rush in.

Erasing a difference in chemical potential, the water will insist on equilibrium even to the point of bursting open the cell. In response, the cell might impose a counterbalancing *osmotic pressure* exactly equal to Π so as to stop the flow and go on living. Pushing against the incoming water, such pressure would supply just enough energy to render further inflow of solvent thermodynamically unnecessary (making, as we say, the solutions inside and outside *isotonic*).

A plant cell does just that: It resists the intrusion with a strong, cellulose-based cell wall that pushes back against the water until a livable standoff is attained. Animal cells, lacking the stiff wall, regulate the osmotic pressure by actively pumping Na^+ out across the membrane. One way or another, though, the chemical potentials are renegotiated and matter is set to rest.

Enthalpy and Entropy

From solid to solution, let us track the net change in free energy that governs the equilibrium between dissolution and precipitation.

We have, on one side, the uncombined reactants: (1) a solid *solute*, presented as an ordered, low-enthalpy array of ions or molecules; and (2) a liquid *solvent*, a loosely knit, variable arrangement of small clusters in constant flow, shifting about with no long-range order.

On the other side of the arrow is the product: a *solution*, a uniform dispersion of solute throughout the solvent. Each dissolved particle is accommodated within a cluster of solvent particles, kept there by some suitable form of electrical interaction.

Permitted (by Hess's law) to compute changes of state along an arbitrary path, we now imagine the dissolution to follow the three steps set forth in Figure 15-10. First, the solid is taken apart into independent, noninteracting particles. Second, the solvent clusters are forced open sufficiently to admit the solute; and, third, solute and solvent particles come together to form the solution. This fictitious but thermodynamically acceptable path will take us from reactants to product, and for each event there will be a recognizable change in the free energy.

Consider the changes in enthalpy and entropy, step by step. Ion by ion or molecule by molecule, the ordered structure of the solute is dismantled in step 1. Such disassembly is presumably a move toward higher entropy ($\Delta S_1 > 0$), but the disordering comes also at the expense of higher enthalpy. It takes work to overcome the forces of attraction. Energy must be supplied to destroy the crystal—the same **lattice energy** (or **lattice enthalpy**) that once was released when the particles came together into the solid. Enthalpy increases ($\Delta H_1 > 0$) while the solute–solute interactions are undone.

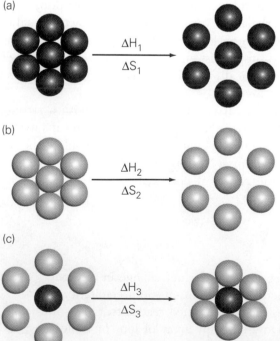

FIGURE 15-10. Dissolution of a solute, pictured in three stages: (a) Disassembly of the solid. (b) Preparation of the host solvent. (c) Reception of the solute by the solvent.

Step 2 likewise involves a pulling apart of interacting species, so both enthalpy and entropy increase again as solvent is separated from solvent ($\Delta H_2 > 0$ and $\Delta S_2 > 0$). Not until step 3, when picked-apart solute is housed within spread-out solvent, is the enthalpy first able to be lowered. Here particles are allowed at last to come together, to approach along the attractive portion of their intermolecular potential so that ΔH_3 usually turns negative. Finally the system receives a payback for the bonds broken in steps 1 and 2.

And entropy? The sign of the entropy change in step 3 arises, more subtly, from a compromise between the order-inducing formation of a cluster and the disorder induced by random dispersal of the solute. Many times (not always) it is this disorderly dispersal that wins out, thus making ΔS_3 positive.

Applying Hess's law, we combine these three imaginary steps into a net **enthalpy of solution**,

$$\Delta H_{soln} = \Delta H_1 + \Delta H_2 + \Delta H_3$$

an **entropy of solution**,

$$\Delta S_{soln} = \Delta S_1 + \Delta S_2 + \Delta S_3$$

and a **free energy of solution**:

$$\Delta G_{soln} = \Delta H_{soln} - T \, \Delta S_{soln}$$

Therein is contained a complete thermodynamic accounting for any dissolution, from which also comes a value for the equilibrium constant (provided that the *standard* free energy change, ΔG°_{soln}, is equated to $-RT \ln K$).*

The numbers vary from substance to substance. Since entropy often increases at each stage, ΔS_{soln} proves to be positive overall for many systems. So it should be, we guess, given that an ordered solid is broken into bits and redistributed over a much larger volume. Still undecided, however, is the crucial algebraic sign of ΔG_{soln}, despite any expectations we may have concerning the entropy. That determination—and hence the direction of spontaneous change—will be influenced by the heat of solution, ΔH_{soln}, with its two positive contributions (ΔH_1 and ΔH_2) and

*Caution: Standard conditions demand 1 M concentrations, difficult to achieve for sparingly soluble substances. A high concentration, moreover, usually renders the solution *nonideal*. Anions and cations will start to interact as pairs well before such conditions are attained (see Chapter 11).

one potentially negative contribution, ΔH_3. Which will it be?

Either one is possible. If the exothermic coming-together shows itself to be stronger than the endothermic tearing-apart, then the net change in enthalpy is negative and so is the free energy:

$$\Delta G_{soln} = \Delta H_{soln} - T\,\Delta S_{soln} < 0 \qquad (\Delta H_{soln} < 0, \; \Delta S_{soln} > 0)$$

Dissolution will proceed spontaneously at any temperature, accompanied by an evolution of heat. The surroundings grow warmer.

But the overall reaction turns endothermic when ΔH_3 is insufficiently negative, going into energetic deficit if the new associations are weaker than the old. Forced to absorb heat, the solution leaves the surroundings colder and more ordered. The requisite disordering of the universe now must occur entirely inside the solution.

Spontaneous dissolution then proceeds only if the $T\,\Delta S_{soln}$ term, which speaks to the disordering of the system, is large enough to produce a negative ΔG_{soln}. Often, too, it is just this increase in entropy that proves decisive, by investing the absorbed heat into a warmer and more disordered solution. An appropriately large increase in entropy can carry a system over a not-too-large hill of higher enthalpy (as, for instance, it does with strongly soluble NaCl, where the standard enthalpy of solution is endothermic by just under 4 kJ mol^{-1}).

Can a substance dissolve spontaneously if ΔS_{soln} is negative, thus bringing increased *order* to the solution? Sometimes it can, but not when the reaction is endothermic. Instead, the system must disorder the surroundings with an exothermic flow of heat large enough to make ΔG_{soln} negative. The magnitude of ΔH_{soln} (a negative number) must be greater than the magnitude of $-T\,\Delta S_{soln}$ (a positive number).

Structure and Solubility

With thermodynamic principles in hand, we ask simply: Will it dissolve? Will this solute dissolve in that solvent? It is a reasonable question, plainly posed; and it is an urgently practical question for anyone working in the laboratory.

Simple answers are not always forthcoming. Solubility emerges as the outcome of a complicated relationship between solute and solvent, not easy to predict, sometimes hard to explain. For specific numbers we must turn to experimental measurements (such as those abstracted in Tables C-17 through C-19 of Appendix C); but for the broadest of guidelines, never to be applied dogmatically, we look again to our picture of solute–solvent interactions.

Start with an image (Figure 15-11) of solvent molecules swarming

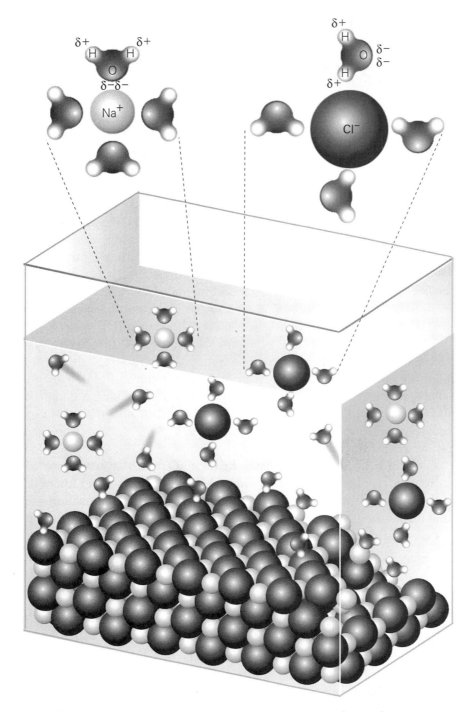

FIGURE 15-11. A polar solvent attacks an ionic solute, replacing the ion–ion interactions of the solid with the ion–dipole interactions of the solution. By offering this roughly similar kind of attraction, the solvating molecules can insert the ions into their own intermolecular network at little cost in enthalpy. The result, portrayed here schematically, is to release the particles from an ordered lattice into a free-form liquid, typically with an increase in entropy.

over the exposed surface of a solute, nibbling away at the edges. For water and sodium chloride, typical of ionic solutes, the partially negative oxygens attack the sodium cations, while simultaneously the electropositive hydrogens orient themselves toward the chloride anions. One set of electrostatic attractions is replaced by another, with the ion–ion interactions of the crystal supplanted by the ion–dipole interactions of the solution. The outlying Na^+ and Cl^- ions are carried away, hydrated amidst floating clusters of H_2O molecules, and one first bit of solid is thereby dissolved. Thereupon, with fresh crystalline surface laid bare, the dissolution continues to the extent permitted by the equilibrium constant.

Never mind that the overall dissolution of NaCl is slightly endothermic. Never mind that the new solute–solvent interactions cannot quite compensate for the original lattice energy. Any shortfall in enthalpy is small enough to be overcome by a local increase in entropy, and so this particular ionic crystal goes readily into aqueous solution. The process works—better for NaCl than AgCl, undeniably, but still it works.

It works because *this* solute and *this* solvent can interact effectively in solution. Similar in their electrical origins, the solute–solvent interactions are not conspicuously different from either the solvent–solvent or solute–solute interactions. In the face of that rough equality, it will be an increase in solution entropy that drives the dissolution forward. One species must spread out and mix with another; if not, there is no thermodynamic profit to be had.

Such mixing is only made easier by the similarity between solute and solvent. Strongly polar, water has the electrical capacity to extract Na^+ and Cl^- ions from the crystal and to hold them (almost as securely) in modified solvent clusters. Charge-bearing species themselves, the ions are integrated into water's polar environment without severely disrupting the existing solvent–solvent interactions.

Water molecules do not avoid the ions, nor does the polar solvent completely distort its own network of hydrogen-bonded connections to accommodate them. Rather, solute and solvent behave almost interchangeably; and so, to some extent, one particle can substitute unobtrusively for another. They are well matched.

No such electrical matching is available for sodium chloride in a strictly nonpolar environment, of the kind presented by carbon tetrachloride or benzene. Able to interact only by the weak London dispersion force (Chapter 9), molecules of these solvents cannot exert a comparably strong grip on the ions. Here the solvated clusters, standing high in enthalpy, are of flimsy construction compared with the strong environment of the undissolved crystal. No amount of entropic consideration suffices, then, to disperse NaCl appreciably into CCl_4 or C_6H_6.

The hoary old adage is *like dissolves like*. Polar solvents work best on polar or ionic solutes; nonpolar solvents work best on nonpolar molecules. Thus we find that sucrose, its polarity enhanced by the polar OH groups, dissolves easily in water. The intermolecular hydrogen bonds of the sugar crystal are replaced efficiently by hydrogen bonds to H_2O molecules in solution. The more of these polar functional groups that are present, too, the better an organic molecule dissolves in water. Pure hydrocarbons, largely nonpolar owing to their high symmetry and to the low difference in electronegativity between C and H, mix poorly with water.

Not all polar solutes dissolve readily in all polar solvents, of course, nor do all nonpolar solutes dissolve in all nonpolar solvents. Yet despite the abundant exceptions and subtleties overlooked by this simple model, "like dissolves like" remains a sensible rule of thumb. Similar structures are best able to mix together in solution, substituting one for the other without occasioning any gross distortion.

Even so, sensible is still not enough: for why should water-soluble NaCl be more "like" H_2O than the almost insoluble AgCl, an ionic solid of similar composition? Apparently all ionic solids are not quantitatively alike, and apparently all ions bearing the same charge are not interchangeable in solution. Should they be? Think, instead, of the differences: There are differences in lattice energies, causing one solid to be bound more tightly than another. There are differences in electronegativity. There are differences in solvation enthalpies; there are differences in solvation entropies. Even the solubility of potassium chloride, KCl, differs from that of sodium chloride, notwithstanding potassium's position directly below sodium in Group I. Other structural influences come into play.

Size is important. Smaller, more compact ions (with correspondingly greater charge densities) attract water molecules more tenaciously. Solvent molecules are drawn from greater distances by the stronger electric field of a small ion. More of them come, and they bind more tightly.

They bind in layers, from the inside out, so that we speak of *hydration shells* developing around the solute. There is, first, an inner layer of water molecules directly coordinated to the ion, held close by the strongest forces. Here, not surprisingly, solutes with the most surface area can support the most water molecules. All the details are not yet known, but this much is clear: the bigger the ion, the more room it has for primary coordination. A small ion like Li^+, for instance, can house fewer waters in its innermost shell than a comparatively large ion like Rb^+. Look past that first sphere of hydration, however, because beyond the primary layer is where a small cation collects its additional waters.

Packing the same charge into a reduced volume, the small ion attracts more solvent molecules to the secondary positions. Additional waters pile randomly onto the first layer, pulled in by a stronger field; the more compact the charge, the greater the pull. A smaller ion has the electrical resources to attract more H_2O molecules, in total, than a larger ion.

From solute to solution, we have now a reversed trend in ionic radius: The effective radius of a solvated cation *decreases* going down the group, consistent with a shrinking hydration sphere around an increasingly diffuse charge. No longer is Na^+ smaller than K^+, as it is in the solid state; in solution, the sodium ion becomes the larger of the two, dragging along more solvent in its wake. Expect, accordingly, that Na^+ will enjoy a substantially negative ΔH_{soln} owing to this added binding, but realize also that the extra waters bring increased order around the cation. More order means higher entropy, and so any decrease in solution enthalpy (good for spontaneity) may also be accompanied by an unfavorable reduction of ΔS_{soln}.

Charge plays a similarly ambiguous role, thoroughly entangled in the tug between enthalpy and entropy. The higher the charge, the stronger the attraction between ion and water. A charge's sign is significant as well, if only indirectly, for cations tend to be smaller than anions of similar electronic structure (recall Chapter 6). Around these smaller species there develops, again, a more ordered, more tightly bound shell of water molecules.

Hydrophilic and Hydrophobic Effects

Water, water everywhere—covering most of the earth's surface, providing some 60% to 90% of the mass of living tissue, and everywhere dominating our treatment of solubility.

Water, a giver and taker of hydrogen ions, is also acid and base simultaneously, and it sets the standard for acid–base interaction in general (about which, much more in Chapter 16). A natural field for chemical reaction, it is our primary solvent and life's natural medium.

Water stands alone, quantitatively different from all other liquids. Small and highly polar, it is an especially strong solvent able to interact with both positive and negative solutes. For a cation or an electron-poor atom, the H⟨Ö⟩H molecule has, first, an oxygenic lone pair to provide a ready concentration of exposed negative charge. Standing ahead of the V-shaped molecule's two small hydrogen atoms, the highly electronegative oxygen is then well positioned to approach the positive charge of its partner:

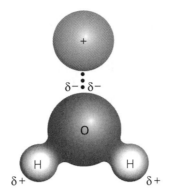

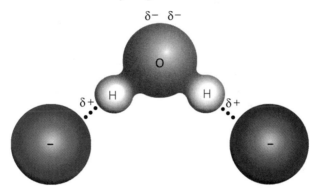

And for an *anion* or any other electron-rich structure, water has its partially positive hydrogens to offer. The protruding $H^{\delta+}$ poles, very small
and very accessible, are ideally disposed to form noncovalent bonds

with a diverse array of anions and polar molecules (including water itself).

Either way, whether presenting a partially positive hydrogen or partially negative oxygen, water delivers its charge efficiently. The solvating
molecules stay out of each other's way to an extent impossible for larger
species, thus allowing more of them to participate. The intermolecular
bonds are closer and tighter as a result.

So special, indeed, is water's role as a solvent that we classify everything else according to its relationship to the H_2O molecule: **hydrophilic**
(water-loving) structures, typically polar or ionic; and **hydrophobic**
(water-fearing) structures, mostly nonpolar. Ethanol and sucrose, both
carrying the polar OH group, are hydrophilic. Benzene and octane are
hydrophobic. Ethanol mixes with and dissolves in water. Benzene shies
from water.

To put benzene in water is to create more order, not less.
Confronted with a hydrophobic entity, molecules of water will build a
cagelike network around the interloper. The solvent's own intermolecular

attractions are disrupted by the intrusion, and soon *more* hydrogen bonds are formed between H_2O molecules in an effort to quarantine the nonpolar solute. Entropy decreases.

Then, too, there are molecules that approach water with apparent ambivalence, acting in part hydrophilic and in part hydrophobic. Described as **amphipathic**, such structures often contain a small, polar head group attached to a long hydrocarbon tail. The head, implementing its own version of "like dissolves like," seeks water. The tail, bulky and nonpolar, avoids water.

So now there is an option. Water-loving and water-fearing at the same time, a sufficiently large amphipathic molecule can satisfy both tendencies. One part interacts with water while the other turns away. An order-inducing disruption of the hydrogen bonds need never occur. "Unlike" might dissolve "like" after all.

Here is an example: Anions of long-chained hydrocarbon acids (*fatty acids*) such as

$$CH_3CH_2CH_2CH_2CH_2CH_2CH_2CH_2CH_2CH_2CH_2CH_2CH_2CH_2CH_2COO^-$$

are nonpolar except for the carboxylate group at one end. Introduced into water, dozens or hundreds of these amphipathic species will come together in the arrangement pictured in Figure 15-12—a *micelle*.

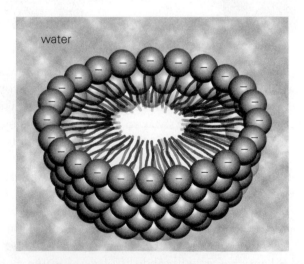

FIGURE 15-12. The hydrocarbon chains point inward; the carboxylate groups point outward; and together the fatty-acid molecules organize themselves into a *micelle*: a small sac, hydrophilic on the outside and hydrophobic on the inside. Soluble in water, some micelles will hold nonpolar molecules protected in their hydrophobic interiors. They offer a way to disperse otherwise insoluble materials in water.

Pointing inward, the individual hydrocarbon tails organize themselves to produce a sphere with (1) a water-excluding internal compartment, but also (2) a hydrophilic exterior studded with polar COO^- groups. The construction allows a micelle to trap nonpolar molecules within its hydrophobic interior, whereas the water-soluble surface makes the entire container compatible with an aqueous environment.

Ordinary soap works in just this fashion, by surrounding and carrying off largely hydrophobic grease in water-soluble micelles. It is a way to disperse, if not actually to dissolve, a nonpolar solute in water.

Consider in the same spirit a *phospholipid* molecule, a structure with two tails, formed when two fatty acids plus glycerol attach themselves to a substituted phosphate group. The example below is typical:

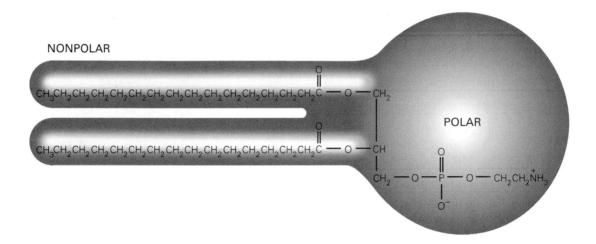

With such construction we see, once more, how an ambivalence toward water can create a unique assembly, this time a component critical to the maintenance of life: a cell membrane. For upon exposure to water, groups of phospholipid molecules act in concert to satisfy the electrical needs of both head and tail. The organization is similar to the formation of a micelle, except here the bulkier (dual-chained) fatty acids limit the maneuverability of a phospholipid.

The self-assembly is sketched in Figure 15-13. Note how the hydrophobic hydrocarbon chains again turn inwards, effectively shielding one another from the water. At the same time, the hydrophilic end containing the phosphate group is thrust outward—toward the water, its preferred environment.

The resulting two-tiered arrangement produces a lipid *bilayer*, a structure found in most semipermeable cell membranes. Mentioned a few pages earlier, the membrane encloses a cell while allowing small

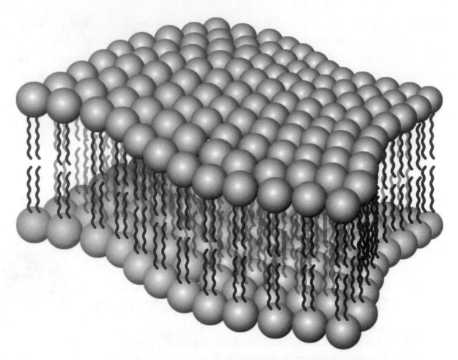

FIGURE 15-13. Like fatty acids, phospholipid molecules are amphipathic—partly hydrophobic, partly hydrophilic. The phosphate groups on the outside seek water, and the hydrocarbon chains turn away. A lipid bilayer, key component of most biological membranes, takes shape.

molecules to pass back and forth. Special *membrane proteins* stationed within the bilayer assist in the active transport of larger species.

The membrane, elaborately constructed and still incompletely understood, becomes the cell's protective fence in a watery environment: polar on either side, nonpolar in between. By combining both hydrophobic and hydrophilic tendencies, the bilayer separates the aqueous solution inside from the aqueous solution outside.

Even beyond the membrane, important as it is, life's workings depend critically on the tension between hydrophobic and hydrophilic interactions. Think of life as primarily a water-based chemistry of diverse species in delicate interplay. There are ions and polar molecules to be kept in their proper places, whether inside or outside a cell. There are also nonpolar molecules and water-insoluble fats, which (like it or not) must function amidst a predominantly aqueous environment. There are, most important, amphipathic molecules well represented by the various membranes and also by the proteins, which owe their structure in water primarily to the hydrophobic effect. The water-avoiding portions of a protein turn in; the water-seeking portions jut out; and together they fold into the intricate three-dimensional structures essential to life. The entire arrangement—thousands and thousands of atoms shaped into some unique macromolecular construction—is secured by noncovalent interactions, often intramolecular hydrogen bonds, as we discovered in Chapter 9.

15-3. GUESTS AND HOSTS

Hydrogen bonds, dipole–dipole interactions, ion–dipole interactions, dispersion forces; weak couplings, all of them . . . weak only as individuals, however, yet strong and purposeful as a collective. They are the nuts and bolts of organized matter, animate and inanimate alike; they are the agents of order. Each a variation on the theme "positive attracts negative," these noncovalent forces shepherd both large molecules and individual particles into ordered communities: into liquids, into crystals, into the tertiary structure of a protein, into the double helix of DNA. Without that binding, we would all be vapor.

But without *selective* binding, we would be dead just the same. Without *specificity* in biological structures, there could be no life. A protein, remember, does not fold haphazardly; rather it falls unerringly into a specific, ordered configuration. Shaped uniquely by the thermodynamics of weak bonds, an enzyme does not respond indiscriminately to any molecule that happens by. It will accommodate, at best, one or two or three properly shaped *substrates*, often only one. It is a tailored structure, a host ready to receive some special guest.

An ordinary fluid solvent (water, methanol, benzene) is, by contrast, a far more accommodating host, willing to accept a broader clientele. Water dissolves NaCl. Water dissolves sucrose. Water mixes with polar methanol. Water dissolves more compounds than we need to enumerate. Admittedly not all substances dissolve to the same extent, but still the general criteria for acceptance are generous.

For each solute, a traditional solvent does its thermodynamic best to minimize local enthalpy and maximize local entropy. It puts together a host to suit the particular solute at hand—a host adapted to the size, shape, and charge distribution of each newly arrived guest. Thus 16 water molecules might hydrate one kind of ion (a small cation, say), whereas only 10 might be mustered for another (a larger cation), and just 8 for another (an anion).

Paid for by changes in both enthalpy and entropy, this spontaneous reorganization of the solvent creates something entirely new: a host that was not there before, a host assembled on demand. Solvent molecules come together specially to meet this one particular solute, fashioning a different host for every guest.

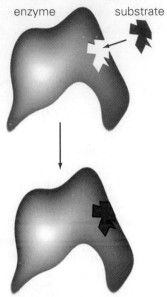

enzyme   substrate

enzyme–substrate complex

Figure 15-14. Lock and key: the specific, sometimes one-of-a-kind relationship between enzyme (host) and substrate (guest). Weak noncovalent bonds bind the substrate temporarily to the active site of the enzyme, shown here as a small opening in a large, complex structure. Simple but apt, the model is broadened considerably by two principal extensions: (1) Some enzymes undergo a limited structural modification to accept a substrate. The lock, so called, is *induced* to fit the key. (2) Often a bound substrate exists not in its ground state but rather as a *transition state*, an energized structure poised for immediate reaction. To understand the role played by the transition state in kinetics and catalysis, look back to Chapter 3 and ahead to Chapter 18.

Not so for the enzyme, a more discerning host. There it stands, already in place, with an individually sculpted *active site* of the kind shown schematically in Figure 15-14. Little or no assembly is required. The structure is largely **preorganized**.

The enzyme is, for every guest, nearly the same host. There are exceptions, but typically a biological catalyst accepts only those substrates that match either the existing architecture or an easily realized modification of the active site. One enzyme might bind strongly with a potassium ion, yet weakly with a sodium ion; another only with the amino acid tyrosine, not histidine; a third, with something else. An enzyme is a lock waiting for a key. It *recognizes* other species; it accepts and rejects guests according to prearranged features of design.

(a)

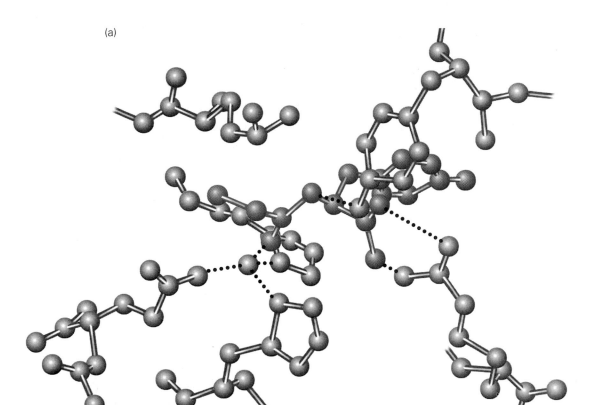

FIGURE 15-15. Molecular recognition: binding of a substrate to an enzyme. (a) A three-dimensional view of glycyltyrosine (the substrate, a dipeptide) joined to carboxypeptidase A at the active site. Although the structure of the enzyme changes somewhat during the course of binding, the final configuration appears made to order. The substrate fits snugly into the active site of the enzyme, held in place by hydrogen bonds and other electrostatic interactions. (b) A simplified rendering of the relationship between enzyme and substrate. One structure matches the other.

(b)

Molecular Recognition

How subtle—how subtle to find, in a universe of unrelenting disorder, matched structures like enzyme and substrate, structures that come together like hand and glove. In such pairings we discern the ultimate in selectivity and local ordering: the exercise of **molecular recognition**. One microscopic entity identifies and selects another, to the exclusion of all else. Enzymes do it; antibodies do it; zeolites and molecular sieves do it; and, ever since the 1960s, chemists do it with systems of their own devising.

1. MOLECULES RECOGNIZE MOLECULES BY POLARITY. That much we already know; we know that like dissolves like, and we know that hydrophilic excludes hydrophobic. But look further at the enzyme–substrate binding shown in Figure 15-15 (p. 557) to appreciate fully what can develop from an intricate network of weak interactions: a discriminating, selective chemistry critical to the maintenance of life. Pictured here, merely one example, are the very specific, very directional couplings that yoke together guest and host for some definite purpose.

The niche created in the enzyme appears shaped to fit. Few substrates, it seems, could nestle so snugly into this particular active site, so well matched to the structure of the host. Attractions and repulsions fall into place. The partially negative oxygens and partially positive hydrogens match up exactly where they should on guest and host. Not too long and not too short, the three-atom hydrogen bonds align to form the strongest links. Each enzyme, offering its own idiosyncratic array of polar and nonpolar bonds, binds only those select molecules with a complementary structure.

2. MOLECULES RECOGNIZE MOLECULES BY CHIRALITY. Nearly all amino acids (tyrosine, for instance) can exist as either of two optical isomers, the L and D enantiomers, which we encountered originally in Chapter 8:

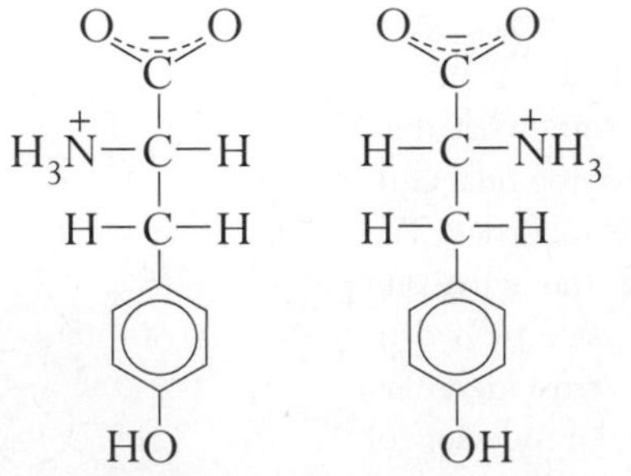

L-tyrosine   D-tyrosine

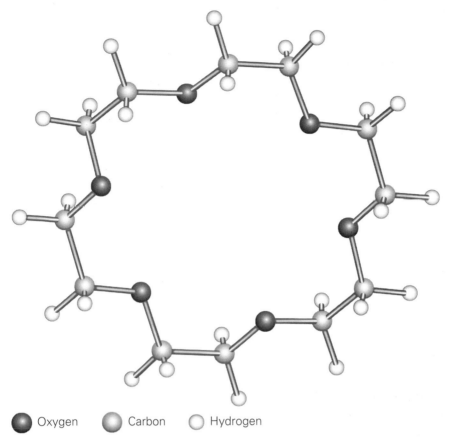

● Oxygen ● Carbon ○ Hydrogen

FIGURE 15-16. Molecular recognition: 18-crown-6. The coronet-shaped cyclic ether, $(OCH_2CH_2)_6$, has an open cavity at its center. Dotted with electron-rich oxygen atoms, the cavity has enough electrical wherewithal to hold an electron-poor guest—but optimally only a guest of just the right size. A potassium ion will fit well; a rubidium ion will not.

One form is the mirror image of the other, and the two cannot be super-imposed. D-tyrosine is to L-tyrosine just as right hand is to left hand. So, too, does the active site of any biological enzyme present itself as a left-hand glove, matched only to an L amino acid. Try (perversely) to insert a species of D chirality into the L environment of the enzyme—to no avail. Everything suddenly is wrong. The hydrogen bonds are misplaced; the fit is awkward; the binding is ineffective. A chiral structure recog-nizes a chiral molecule.

3. MOLECULES RECOGNIZE MOLECULES BY SIZE. Here is an example of a synthetic host, one of many developed in the laboratory to mimic the ion-binding capabilities of proteins, antibiotics, and various other natural systems. The molecule shown in Figure 15-16 is a cyclic ether, a *crown* ether, so named for its crownlike three-dimensional appearance.

A cleverly constructed doughnut, the structure

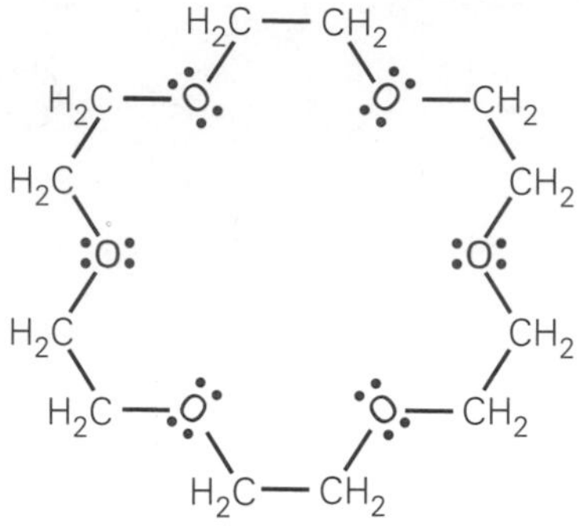

bristles with nonpolar CH_2 groups on the outside and electron-rich oxygens on the inside. In the middle is a hole, 2.7 Å in diameter, exposed directly to the hydrophilic (polar) inner surface.

Into this hole the oxygen in each $C\overset{\ddot{O}}{\frown}C$ linkage directs one of its lone pairs, offering the negative charge to a potential guest. But again, not just any guest; it must be a guest of a certain size, not too large and not too small, a guest able to fit snugly into the hole. Too small, and the electrostatic attraction is weakened by distance. Too large, and the guest cannot fit.

The potassium ion, K^+, like Goldilocks, finds the opening just right:

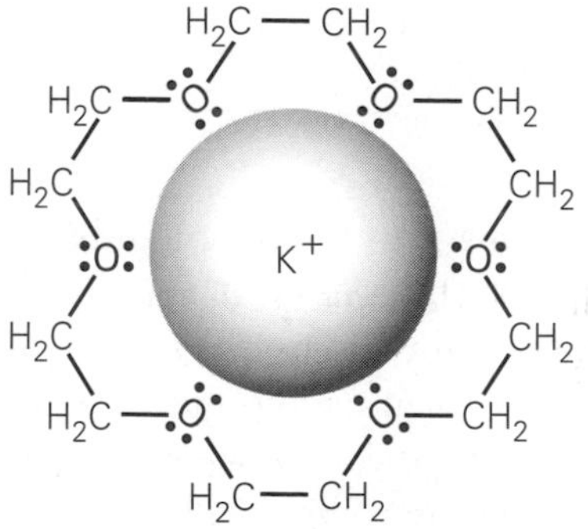

K^+ binds better than either the smaller Na^+ ion or the larger Rb^+ ion. More potassium will be present in the cavities at equilibrium.

Bound to 18-crown-6, the K^+ now enjoys a kind of "solvation" by a preorganized, amphipathic host. The cation is sheltered inside, coordinated to the polar oxygens, while the surface outside remains nonpolar and hence suited to a hydrophobic milieu. This entire package then can be sent through the hydrophobic interior of a cell membrane, into territory where an unescorted potassium ion would be retarded by the mismatch in polarity.

We acquire, through this molecular tinkering, an intriguing way to transport drugs and other materials into and out of cells, similar to the natural practices of certain bacteria. Such microorganisms produce host structures reminiscent of the crown ethers, and some of these cavernous molecules can traverse membranes and act as antibiotics. They accelerate (and consequently disrupt) the passage of K^+ and Na^+ across the membranes of other cells, forcing the afflicted systems to correct the imbalance at a high cost in metabolic energy.

4. MOLECULES RECOGNIZE MOLECULES BY SHAPE. Of course they do, and any sufficiently complex host implicitly demands as much. How, otherwise, could a substrate ever fit securely into the nooks and crannies of an enzymatic active site? Surely histidine will not dock effectively with an enzyme specially tailored for tyrosine—not given histidine's inopportunely arranged atoms and, ipso facto, its wrong *shape*.

But shape-selective reactions can recognize variations well above the level of gross morphology, going even so far as to distinguish among the three isomers of xylene:

ortho   meta   para

With each molecule displaying two methyl groups on a benzene ring, the differences lie only in the disposition of the CH_3 substituents: at twelve o'clock and two o'clock in *ortho*-xylene; at twelve o'clock and four o'clock in *meta*-xylene; at twelve o'clock and six o'clock in *para*-xylene. Everything else is the same, and the isomers all have roughly the same size.

Of the three molecules, though, *para*-xylene is particularly useful as a raw material for the production of polyester fibers and plastics. The placement of the methyl groups proves critical to subsequent reactivity, and economic considerations dictate that the para isomer be isolated in sufficiently attractive quantities. A shape-selective host, a zeolite called ZSM-5, provides the means.

Recollect, from Chapter 3, that zeolites (Figure 15-17) are porous, channel-containing structures built from aluminum, silicon, and oxygen

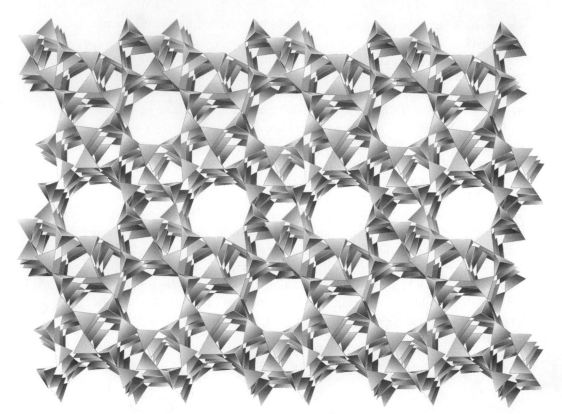

FIGURE 15-17. Architecture of the zeolite ZSM-5, showing a mostly silicate (SiO_2) framework shot through with pores and cavities. First synthesized in 1963, ZSM-5 finds commercial application as a molecular sieve and a catalyst. A form of ZSM-5 is used to convert methanol into gasoline.

atoms in richly varied combinations. Their cavities and passageways provide snug accommodation for molecules undergoing various reactions and, still more, the zeolite can incorporate detachable H^+ ions in its framework. These acidic protons often will trigger catalytic processes within the pores—one such process being the production of *ortho*-xylene, *meta*-xylene, and *para*-xylene from two molecules each of toluene ($C_6H_5-CH_3$, a benzene molecule substituted with only one, not two, methyl groups).

Now at first the acid-catalyzed reaction between two zeolite-bound toluene molecules (Figure 15-18) produces all three isomers of xylene, with only 25% having the hoped-for para configuration. Yet that distribution is soon to change. The para isomer, slimmer owing to the linear arrangement of its two methyl groups, slips out of the pores 1000 times faster than either ortho or meta. Differing not so much in total volume

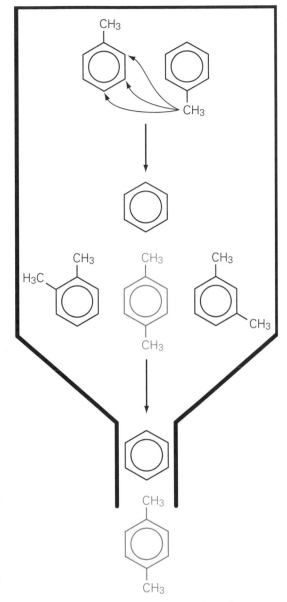

FIGURE 15-18. Molecular recognition: shape-selective catalysis. Passing easily through the channels of ZSM-5, the comparatively slim *para*-xylene molecule is produced at the expense of the bulkier ortho and meta isomers. The catalyzed reaction converts toluene largely into a mixture of benzene and *para*-xylene.

(size) as in *shape*, the thinner molecule fails to meet the binding requirements of the host. It departs.

The two other isomers, meanwhile, are eventually converted into *para*-xylene during their prolonged confinement. One at a time, these new products emerge from the pores until a substantial quantity of pure *para*-xylene is harvested. A shape-selective host makes it possible.

Thermodynamics of Binding

The relationship between guest and host is expressed concisely through the generalized equation

$$A + B \rightleftharpoons AB$$

and corresponding equilibrium constant

$$K = \frac{[AB]}{[A][B]}$$

A denotes the host; B, the guest; AB, the host–guest complex.

To be a good guest, we see, is merely to offer a large K; if so, the bound structures AB are guaranteed to outnumber the unbound species when the system attains equilibrium. To be, more ambitiously, a *preferred* guest is to offer a K significantly larger than all potential rivals. For whenever two guests, B_1 and B_2, vie for the same host, the eventual winner is determined by the respective equilibrium constants K_1 and K_2. Thus if we allow B_1 and B_2 to compete in equal numbers, so that $[B_1] = [B_2]$, the ratio of final concentrations follows straightforwardly as

$$\frac{[AB_1]}{[AB_2]} = \frac{K_1[A][B_1]}{K_2[A][B_2]} = \frac{K_1}{K_2}$$

The larger K always prevails at equilibrium, even to the point where one guest (producing a K much, much greater than any other) acquires a monopoly.

This ratio K_1/K_2 becomes the quantitative expression of what we have called selective binding, as understood both for ordinary solvation (selective precipitation) and for host–guest associations. The message is clear: In the end—however long it may take—the guest offering the largest K will occupy the most sites. It is a victory of statistics, resulting from the thermodynamic drive toward the lowest local free energy and the highest global entropy.

It is also nothing new, since all reactions obey the same thermodynamic law. This reaction, or any other, will come to equilibrium upon minimization of G, leaving the standard free energy change as

$$\Delta G^\circ = -RT \ln K = \Delta H^\circ - T \, \Delta S^\circ$$

A stable, intimately associated host–guest complex develops when ΔG° is large and negative, just as any similarly favored product would predominate at equilibrium. And, as always, we find $\Delta G°$ to be a compromise between $\Delta H°$ and $\Delta S°$: a temperature-mediated compromise between lower enthalpy (stronger bonds) and higher entropy (*broken* bonds, microscopic disorder).

In what directions, then, do we expect enthalpy and entropy to change when host and guest unite? The process, imagined in Figure 15-19, begins when a potential guest is removed from one environment and placed in another—going from, let us say, a K^+ ion in a crystal to an ion sequestered in 18-crown-6. Bonds first must be broken to free the guest from its original associations, a demand no different from the ΔH_1 step required for any solute. This initial investment of enthalpy is subsequently recovered, presumably at a profit, when the liberated guest forms stronger bonds with its new host (a step equivalent to our former ΔH_3).

Since the host, assembled long ago, already has a prearranged cavity waiting for the guest, there is little need for further enlargement of the receptive structure. The enthalpy-consuming "solvent expansion" step (ΔH_2) is thereby minimized or avoided entirely, and so we obtain the overall enthalpy of binding, $\Delta H°$, principally from ΔH_1 and ΔH_3: a *net*

Step 1: ΔH_1 +

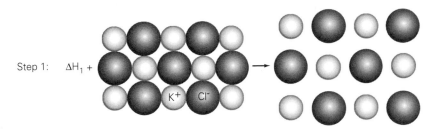

Step 2:

Step 3:

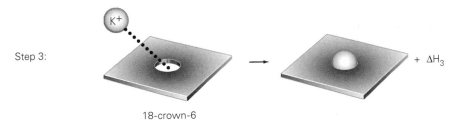

18-crown-6

FIGURE 15-19. The advantages of a preorganized host, pictured schematically for the sequestration of K^+ in 18-crown-6: no assembly required. Unlike a fluid solvent, the host molecule is ready to receive its guest. The usual endothermic "expansion" of the solvent (step 2) becomes unnecessary.

change, reflecting the breaking of old bonds and the formation of new ones. To replace a weak set of former connections by a strong new host–guest complex is the best way to ensure an exothermic binding.

Exothermicity is precisely what guest and host need to cement their union, for they are unlikely to be aided by an increase in entropy. The principal step, after all, is the creation of *more* order as two species become one, unmitigated by any further expansion of the host structure. The guest is captured by its preassembled host not so much for the sake of higher entropy, but instead for the lower enthalpy derived from a strong new set of attractions in the AB complex. There are exceptions, but usually entropy does not help.

Or does it? True, entropy generally decreases upon formation of the host–guest complex, but think how bad the loss *could* have been. Imagine the cost in entropy if the host had to be improvised on the spot—as in an ordinary fluid solvent. What would become of $\Delta G°$ if the $T \Delta S°$ term included the additional entropy required to organize the atoms of 18-crown-6 around the potassium?

Do an experiment. Try to dissolve potassium ions in a solvent like tetrahydrofuran,

$$\begin{array}{c} \text{O} \\ \text{H}_2\text{C} \quad \text{CH}_2 \\ \text{H}_2\text{C}-\text{CH}_2 \end{array}$$

where each molecule offers a single $\text{C}\overset{\overset{..}{\text{O}}..}{\diagup\diagdown}\text{C}$ ether linkage of the sort found all together in 18-crown-6. The effect is as if we had disassembled the crown ether into its electron-rich components, leaving them to reorganize later in the presence of an electron-deficient solute. Acting separately, the tetrahydrofuran molecules clearly have the electrical wherewithal to attract K^+; are they able, though, to do so cooperatively? Can they form themselves into the required new clusters?

They cannot. Too much entropy is lost if they do; too much additional order must be created. Any enthalpy released upon solvation is insufficient to remove K^+ from either an ionic crystal or an aqueous solution. The necessary reorganization cannot be financed by the available free energy.

Yet for a preorganized host—the crown ether, already assembled—the entropic cost never becomes an issue. The ordered structure already exists, and so the universe has already exacted its payment in increased disorder. Somewhere and sometime, something was stirred up just enough when our present 18-crown-6 molecule was formed. *Where* is irrelevant; *when* is irrelevant; even *how much* stirring is irrelevant. It is sufficient merely that the host molecule is here today.

So now, even if the host must adjust itself slightly to accommodate the guest (and such adjustments often do occur), the actual increase in local entropy is relatively modest. The potassium ion finds, therefore, that most of the entropic charges have been prepaid, leaving only a current bill for enthalpy. It is then up to K^+, or any aspiring guest, to find the host offering the strongest ties.

REVIEW AND GUIDE TO PROBLEMS

All the richness and breadth of chemistry begins with simple transactions, few enough to count on one hand. Solvents capture solutes; guests bind to hosts; bases neutralize acids; reductants reduce oxidants; this one gives; that one takes. In so many different ways, with so many different shadings, with such variety of result, atoms and molecules and ions all do the same thing however they react: They redistribute energy. They move matter. They do chemistry.

We began in this chapter with solutes, solvents, guests, hosts, and the chemistry of weak noncovalent bonds in general—the most tenuous of chemical embraces, yet the key also to most of life's processes.

IN BRIEF: SOLUTION EQUILIBRIA—
A CHEMISTRY OF ACCOMMODATION

1. NONCOVALENT ASSOCIATIONS. Weak bonds foster a notably gentle kind of chemistry, a chemistry dependent more on proximity and influence than on sudden, violent, or radical change. Rather than seeing molecules A and B emerge from a shattering collision as molecule C (completely new), instead we have A and B sticking together as AB while keeping much of their old character intact. Whether mediated by dipole–dipole forces or London dispersion forces or some other noncovalent interaction, the encounter produces a meeting but not a merging: a close friendship but not quite a marriage, an interpenetration of electric fields but not a mixing of electron density. Partners A and B maintain separate households, each holding onto its own electrons and its own identity. When a sodium ion is dissolved in water, it stays a sodium ion. When a hormone molecule recognizes a receptor, it stays a hormone molecule. When an enzyme binds a substrate, it stays an enzyme. The attachments are sometimes enduring and sometimes not, but there is always at least the possibility of release and separation if a thermodynamically better opportunity comes along.

2. BREAKING UP AND MAKING UP. We imagine that three steps accompany the integration of A into B, not in any strict sequence but logically as follows: (1) A–A interactions are overcome, allowing the bulk substance A to break up into its constituent particles. (2) B–B interactions are loosened sufficiently for the B particles to move apart and make room for A. (3) A and B come together.

Step 1, the breakup of the guest (solute), trades enthalpy for entropy.

Pulling particles apart costs energy, but giving them more space creates new possibilities and hence greater disorder. ΔH_1 is positive. ΔS_1 is positive.

Step 2, the expansion of the host (solvent), is thermodynamically similar to step 1: an expenditure of enthalpy, a gain in entropy. For a pre-organized host such as a crown ether, little or no expansion is required. For a more conventional solvent, however, there usually must be some rearrangement of the particles to accommodate the solute. ΔH_2 is positive. ΔS_2 is positive.

Step 3, the final mixing, brings the particles together, typically with a release of enthalpy as weak noncovalent bonds are made between solute and solvent. If so, ΔH_3 is negative. ΔS_3 is often positive, consistent with the freedom gained when a solute is dispersed throughout a solution.

The combined enthalpy, entropy, and free energy of solution all hinge, therefore, on the relative importance of the breakup, expansion, and mixing stages at a given temperature:

$$\Delta H_{soln} = \Delta H_1 + \Delta H_2 + \Delta H_3$$

$$\Delta S_{soln} = \Delta S_1 + \Delta S_2 + \Delta S_3$$

$$\Delta G_{soln} = \Delta H_{soln} - T\,\Delta S_{soln}$$

The outcome is determined by a tangled interplay among competing factors, many times with a small net difference as the result. Every solute, every ion, every molecule has its own structure and suffers its own thermodynamic fate. We learn that fate through the value of the equilibrium constant.

3. SOLUBILITY AND EQUILIBRIA. The equilibrium is between a substance in its undissolved and dissolved phases, represented in general by an equilibrium constant

$$K = \frac{[A(\text{dissolved})]}{[A(\text{undissolved})]}$$

for the symbolic transformation

$$A(\text{undissolved}) \rightleftharpoons A(\text{dissolved})$$

On the left, we have a solute in its native phase (crystalline sucrose, say, or crystalline sodium chloride, or liquid toluene, or gaseous carbon dioxide, or any other solid, liquid, or gas you like). On the right, we have a solute dispersed in a solvent (sucrose in water, sodium chloride in water, toluene in acetone, carbon dioxide in water, to name just a few).

The difference is simply this: The dissolved phase contains particles of A surrounded by particles of B; the undissolved phase contains particles of A alone. For certain concentrations under certain conditions, the dissolved and undissolved phases will coexist with the same free energy, balanced together in a dynamic equilibrium. When they do, the solution is *saturated*.

Concerning products, reactants, free energy, and equilibrium there is nothing fundamentally new to say, and so we recall the old lesson from Chapter 14. Namely: The equilibrium constant, K, exceeds unity only when the products sit at lower standard free energy than the reactants, thereby fulfilling the promise of a more disordered universe. The greater the drop in $\Delta G°$, the greater the proportion of products at equilibrium. For specific numbers, look to the specific species involved in the reaction.

Except for changes in name, then, the broad interpretation will remain the same whether the system contains either a gas dissolved in a liquid,

$$CO_2(g) \rightleftharpoons CO_2(aq) \qquad K = \frac{[CO_2(aq)]}{P_{CO_2}}$$

or a molecular solid dissolved in a liquid,

$$C_{12}H_{22}O_{11}(s) \rightleftharpoons C_{12}H_{22}O_{11}(aq) \qquad K = [C_{12}H_{22}O_{11}(aq)]$$

or an ionic solid dissolved in a liquid (with *solubility-product constant K_{sp}*),

$$PbCl_2(s) \rightleftharpoons Pb^{2+}(aq) + 2Cl^-(aq) \qquad K_{sp} = [Pb^{2+}(aq)][Cl^-(aq)]^2$$

or something else dissolved in something else. The names change. The symbols change. The numbers change. The meanings do not.

Just as in any other reaction, moreover, it takes *two* to make a dynamic equilibrium. The system needs both a "before" state and an "after" state; it needs a reactant and a product, a dissolved phase and an undissolved phase. If not, then exactly what shall we say is in equilibrium with what? If we can still add solute to the solution, then where is the undissolved phase that we allege to be in balance with the dissolved phase? Where is the back-and-forth between the two?

Rather, the solution only attains this heterogeneous kind of order–disorder equilibrium upon *saturation*—when, in a system perfectly balanced, not a speck more of solute can be dissolved; when there is a supply of undissolved material present to support traffic in both directions between two phases.

4. OSMOTIC EQUILIBRIUM. Saturated or not, a solution comes to equilibrium in whatever ways it can: in its distribution of energy, in its distribution of matter. A solution always tends toward both thermal equilibrium (uniformity of temperature) and osmotic equilibrium (uniformity of concentration).

Thermal equilibrium: where the temperature is the same from point to point. Otherwise, there would be an unsustainable variation of thermal energy over different parts of the sample. Heat would flow from the hotter regions to the colder regions until all was level.

Osmotic equilibrium: where the concentrations are the same from point to point. Otherwise, there would be an unsustainable variation of free energy (or *chemical potential*) over different parts of the sample. Matter would flow from points of high potential to points of low potential until all was level.

We know. Experience teaches that the first sip of wine is as cool and sweet as the last; that the cream spreads uniformly throughout the coffee; that the liquid settles down and eventually ceases to swirl. But remember, again, that without some kind of back-and-forth there can be no equilibrium; and so to attain osmotic equilibrium, there must be a free flow of matter. The system must permit matter to move from regions of high concentration to regions of low concentration.

A glass of wine does precisely that, but a cell enclosed by a *semipermeable membrane* does not. The membrane allows only certain species (such as water) to pass, while rejecting others (like sodium ions) as too large. And like much of life, perhaps, this action is necessary yet risky—necessary, because organisms can exist only when matter is segregated into highly organized structures; risky, too, because such discrimination runs counter to the natural tendency to establish equilibrium. What, after all, will prevent water from flowing back and forth to equalize the concentrations inside and out, heedlessly allowing cells to explode or shrivel for the sake of a larger equilibrium?

Answer: energy. A cell must push back if it wishes to defy nature's drive to stamp out imbalances wherever they arise. To do so, the cell exerts a counterpressure (an **osmotic pressure**, Π) to stanch the flow and maintain the concentrations necessary for survival. It costs energy to generate this osmotic pressure, of course, but such is the price paid to avoid premature equilibrium. Finding the energy to live is the true cost of living, for to be alive is to be defiantly out of equilibrium.

The price rises and falls with the number of particles. A colligative property, the osmotic pressure

$$\Pi = \frac{n}{V} RT$$

is determined by the concentration ($c = n/V$) independently of the nature and kind of particles in solution. We shall see, in some of the examples to follow, just how high that pressure can be under normal physiological conditions.

SAMPLE PROBLEMS

EXAMPLE 15-1. Equilibrium: Solubility and Solubility Product

PROBLEM: Attempting to dissolve 0.25 g $PbCl_2$ in 50.0 mL of water, a chemist finds that all but 0.03 g of the solid goes into solution at a temperature of 25°C. (a) Compute the solubility of $PbCl_2$ in water at this temperature, reporting the result in units of g L^{-1}. (b) Use the data to estimate the value of the solubility-product constant, $K_{sp}(25°C)$, for lead chloride.

SOLUTION: The presence of undissolved solid tells us, right away, that the solution is saturated, unable to accept additional solute. From that, we infer an equilibrium

$$PbCl_2(s) \rightleftharpoons Pb^{2+}(aq) + 2Cl^-(aq)$$

between $PbCl_2(s)$, $Pb^{2+}(aq)$, and $Cl^-(aq)$, governed by the solubility-product constant

$$K_{sp} = [Pb^{2+}][Cl^-]^2$$

Note the power of 2 on the chloride concentration.

(a) *Solubility.* We define a ***solubility*** as the amount of solute needed to produce a given volume of saturated solution. Now if 0.03 g $PbCl_2$ remains undissolved in a volume of 50.0 mL, then evidently 0.22 g of the original 0.25 must have gone into solution. The solubility is simply (0.22 g)/(50.0 mL), which we convert straightforwardly into units of g L^{-1}:

$$\frac{0.22 \text{ g}}{50.0 \text{ mL}} \times \frac{1000 \text{ mL}}{L} = 4.4 \text{ g } L^{-1}$$

(b) *Solubility-product constant.* Having determined that 4.4 g $PbCl_2$ will produce one liter of saturated solution, we also know the molarity of both $Pb^{2+}(aq)$ and $Cl^-(aq)$ from the reaction stoichiometry:

$$\frac{4.4 \text{ g PbCl}_2}{L} \times \frac{1 \text{ mol PbCl}_2}{278.1 \text{ g}} \times \frac{1 \text{ mol Pb}^{2+}}{1 \text{ mol PbCl}_2} = 0.0158 \text{ mol Pb}^{2+} \text{ L}^{-1}$$

$$\frac{4.4 \text{ g PbCl}_2}{L} \times \frac{1 \text{ mol PbCl}_2}{278.1 \text{ g}} \times \frac{2 \text{ mol Cl}^-}{1 \text{ mol PbCl}_2} = 0.0316 \text{ mol Cl}^- \text{ L}^{-1}$$

Since these are the concentrations existing *at equilibrium*, the solubility-product constant takes the value

$$K_{sp} = [Pb^{2+}][Cl^-]^2 = (0.0158)(0.0316)^2 = 1.6 \times 10^{-5}$$

to within two significant figures.

QUESTION: How would these results differ if our chemist had tried to dissolve 50 grams of lead chloride in the same volume of water?

ANSWER: Except for a larger pile of undissolved $PbCl_2$ (49.78 g), the results would not differ at all. There would be the same 0.22 g $PbCl_2$ in solution, producing the same equilibrium concentrations of 0.016 M (Pb^{2+}) and 0.032 M (Cl^-) and the same value of K_{sp}. The equilibrium is determined not by the *undissolved* material, but by the concentration of the dissolved material instead.

EXAMPLE 15-2. From Equilibrium to Equilibrium

PROBLEM: (a) Given the solubility-product constant reported in Example 15-1 [$K_{sp}(25°C) = 1.6 \times 10^{-5}$], compute the amount of $PbCl_2$ that will dissolve in 0.100 L of water. (b) If the volume of this saturated solution is suddenly doubled to 0.200 L, how much additional $PbCl_2$ must be dissolved to reestablish the disorder–order equilibrium? (c) Compute the reaction quotient for the out-of-equilibrium mixture that exists at the instant the volume is doubled.

SOLUTION: For every mole of $PbCl_2$ that dissolves, the solution takes up one mole of Pb^{2+} and two moles of Cl^-. Let x and $2x$ denote, then, the respective equilibrium concentrations of Pb^{2+} and Cl^-, so that we can write

$$PbCl_2(s) \rightleftharpoons Pb^{2+}(aq) + 2Cl^-(aq)$$

$$ x \phantom{Pb^{2+}(aq) + } 2x$$

and solve the resulting equation for x:

$$K_{sp} = [Pb^{2+}][Cl^-]^2 = x(2x)^2 = 1.6 \times 10^{-5}$$

$$4x^3 = 1.6 \times 10^{-5}$$

$$x = 0.016 \ M$$

The molarities are the same as in Example 15-1. Understand why: In any solution containing only Pb^{2+} and Cl^-—nothing else—the concentrations at the saturation point are dictated by K_{sp} and are independent of volume:

$$[Pb^{2+}] = \quad x \quad = 0.016 \ \text{mol L}^{-1}$$

$$[Cl^-] = 2x \ = 0.032 \ \text{mol L}^{-1}$$

$$K_{sp} = 4x^3 = 1.6 \times 10^{-5}$$

Grams can change, and moles can change, and liters can change; but moles per liter cannot. See below.

(a) *How many grams will dissolve in 0.100 L?* Since the solubility of lead chloride is 4.4 g L^{-1} (0.016 M), we quickly verify that 0.44 g will dissolve in 100 mL (exactly twice the amount found in Example 15-1 for 50 mL):

$$\frac{4.4 \ \text{g}}{\text{L}} \times 0.100 \ \text{L} = 0.44 \ \text{g}$$

(b) *How many additional grams will dissolve in 0.200 L?* If 0.44 g is needed to saturate a solution of 100 mL, then the solution will demand 0.88 g to maintain heterogeneous equilibrium in a volume twice as large. The new solution therefore can accommodate an additional 0.44 g of solute in its volume of 200 mL, all the while maintaining a constant molar concentration of dissolved $PbCl_2$(aq): $x = 0.016 \ M$.

(c) *What is the reaction quotient at the instant the volume is doubled?* Starting with molarities of

$$x = 0.016 \ \text{mol L}^{-1} = [Pb^{2+} \ \text{at equilibrium}]$$

$$2x = 0.032 \ \text{mol L}^{-1} = [Cl^- \ \text{at equilibrium}]$$

and the equilibrium constant

$$K_{sp} = x(2x)^2 = 4x^3 = 1.6 \times 10^{-5}$$

in a saturated solution, we abruptly halve the concentrations by doubling the volume:

$$y = \frac{x}{2} = 0.0080 \text{ mol L}^{-1} = [\text{Pb}^{2+} \text{ out of equilibrium}]$$

$$2y = x = 0.016 \text{ mol L}^{-1} = [\text{Cl}^- \text{ out of equilibrium}]$$

The reaction quotient, evaluated at these nonequilibrium concentrations y and $2y$, thus falls to one-eighth of its equilibrium value upon going from $4x^3$ to $4y^3$:

$$Q = [\text{Pb}^{2+}][\text{Cl}^-]^2 = y(2y)^2 = 4y^3$$

$$= 4y^3 = \frac{x^3}{2} = \frac{K_{sp}}{8}$$

$$= 2.0 \times 10^{-6}$$

Dissolution of the additional solute specified in part (b) will restore the concentrations needed to reestablish equilibrium.

EXAMPLE 15-3. Common-Ion Effect

PROBLEM: Calculate the amount of $PbCl_2$ that will dissolve in a 0.100 M solution of KCl.

SOLUTION: Here there are two sources of chloride ion at the start: (1) potassium chloride (KCl), already dissolved at a concentration of 0.100 M; and (2) lead chloride ($PbCl_2$), *waiting* to be dissolved at some concentration x.

All the chloride comes, at first, from the preexisting solution of KCl,

$$[\text{Cl}^-] = 0.100 \; M$$

and there is no initial concentration of lead:

$$[\text{Pb}^{2+}] = 0$$

Then the $PbCl_2$ dissociates into x moles of Pb^{2+} and $2x$ additional moles of Cl^- per liter of solution, eventually settling into an equilibrium

$$PbCl_2(s) \rightleftharpoons Pb^{2+}(aq) + 2Cl^-(aq)$$
$$ x \phantom{Pb^{2+}(aq) +} 0.100 + 2x$$

where the ion concentrations conform to the same unchanging, preset value of K_{sp}:

$$[Pb^{2+}][Cl^-]^2 = K_{sp}$$

$$x(0.100 + 2x)^2 = 1.6 \times 10^{-5}$$

Solving for x, we shall have our answer.

Rather than work with the complicated cubic equation, however, we make our usual approximation and assert that x (small to begin with) should presumably be small compared with 0.100. If so, then the term $(0.100 + 2x)^2$ can be replaced initially by $(0.100)^2$ to yield the approximate equation

$$x(0.100)^2 \approx 1.6 \times 10^{-5}$$

$$x = 0.0016 \ M$$

The shortcut proves good enough for a quick estimate, correct to within 6%.

Conclusion: We have an example of the **common-ion effect** (equivalently, Le Châtelier's principle), where the common ion Cl^- depresses the solubility of $PbCl_2$ compared with its value in pure water. The equilibrium constant never changes, but the dissolved concentration of Pb^{2+} falls 10-fold (from 0.016 M to only 0.0016 M) as the reaction shifts to the left. With most of the chloride concentration coming from KCl, the contribution of $PbCl_2$ to the equilibrium mixture drops correspondingly.

EXAMPLE 15-4. Selective Precipitation

PROBLEM: NaCl and NaI are mixed together to produce a solution 0.010 M in both Cl^- and I^-. Will a precipitate form if 0.0010 mole of $Pb(NO_3)_2$ is added to 1.0 L of the NaCl/NaI solution? Will either Cl^- or I^- remain dissolved?

SOLUTION: The relevant equilibria at 25°C (see Table C-19 in Appendix C) are

$$PbCl_2(s) \rightleftharpoons Pb^{2+}(aq) + 2Cl^-(aq) \qquad K_{sp}(PbCl_2) = 1.6 \times 10^{-5}$$

$$PbI_2(s) \rightleftharpoons Pb^{2+}(aq) + 2I^-(aq) \qquad K_{sp}(PbI_2) = 8.5 \times 10^{-9}$$

and the initial concentrations are

$$[Cl^-] = 0.010 \ M$$

$$[I^-] = 0.010 \ M$$

in a volume of 1.0 L. To this mixture is then added 0.0010 mole of Pb^{2+}, producing an initial Pb^{2+} concentration of 0.0010 M (where we assume, for simplicity, that the total volume is unchanged):

$$[Pb^{2+}] = \frac{0.0010 \ \text{mol}}{1.0 \ \text{L}} = 0.0010 \ M$$

From here we compute the reaction quotient first for the salt with the lesser solubility, PbI_2, finding the ion product already in excess of its value at equilibrium:

$$Q = [Pb^{2+}][I^-]^2 = (0.0010)(0.010)^2 = 1.0 \times 10^{-7}$$

To restore the proper balance, enough PbI_2 must precipitate from solution so that $[Pb^{2+}][I^-]^2$ becomes equal to the solubility-product constant

$$K_{sp}(PbI_2) = 8.5 \times 10^{-9}$$

For $PbCl_2$, more soluble than PbI_2, the reaction quotient

$$Q = [Pb^{2+}][Cl^-]^2 = (0.0010)(0.010)^2 = 1.0 \times 10^{-7}$$

remains well below the equilibrium value

$$K_{sp}(PbCl_2) = 1.6 \times 10^{-5}$$

The chloride ions stay in solution until the concentration of Pb^{2+} exceeds a certain threshold, whereupon precipitation begins. That point is reached when

$$[Pb^{2+}]_{precip} = \frac{K_{sp}(PbCl_2)}{[Cl^-]^2} = \frac{1.6 \times 10^{-5}}{(0.010)^2} = 0.16 \ M$$

EXAMPLE 15-5. Osmotic Pressure

PROBLEM: Compute the osmotic pressure of a 0.16 M solution of NaCl at 25°C.

SOLUTION: We use the relationship

$$\Pi = cRT$$

in which c denotes the concentration (moles per liter) of any and all dissolved particles regardless of structure. Osmotic pressure is a colligative property, dependent only on the total number of ions or molecules.

Since one mole of NaCl produces two moles of ions, the relevant concentration becomes 0.32 M and the osmotic pressure follows as

$$\Pi = cRT = (0.32 \text{ mol L}^{-1})(0.0821 \text{ atm L mol}^{-1} \text{ K}^{-1})(298 \text{ K}) = 7.8 \text{ atm}$$

If the value seems large, realize that 7.8 atm is close to the usual osmotic pressure of blood cells at 25°C. A 0.16 M solution of NaCl is therefore *isotonic* with the blood, neither drawing water from the cells nor forcing it into them. A solution with greater osmotic pressure is said to be *hypertonic*, more concentrated than the blood. A solution with lesser pressure is *hypotonic*.

Isotonic mixtures of NaCl find medical application as physiological saline solutions, useful for intravenous delivery.

EXAMPLE 15-6. Molar Mass from Osmotic Pressure

Osmotic pressure, like other colligative properties, allows us to determine the molar mass of a dissolved substance.

PROBLEM: A solution containing 5.00 g of sucrose in 50.0 mL of water generates an osmotic pressure of 7.15 atm at 25°C. Calculate the molar mass of sucrose.

SOLUTION: Write the concentration as

$$c = \frac{n}{V} = \frac{m/\mathcal{M}}{V}$$

where m is the mass in grams (5.00) and $\mathcal{M}$ is the molar mass (unknown). Then substitute into the equation

$$\Pi = \frac{n}{V}RT = \frac{mRT}{\mathcal{M}V}$$

and rearrange to solve for $\mathcal{M}$:

$$\mathcal{M} = \frac{mRT}{\Pi V} = \frac{(5.00 \text{ g})(0.0821 \text{ atm L mol}^{-1} \text{ K}^{-1})(298 \text{ K})}{(7.15 \text{ atm})(0.0500 \text{ L})} = 342 \text{ g mol}^{-1}$$

The molecular formula for sucrose is $C_{12}H_{22}O_{11}$.

EXAMPLE 15-7. Henry's Law: Gases in Liquids

PROBLEM: The partial pressure of CO_2 above a 0.031 M aqueous solution of carbon dioxide is 1.0 atm at 25°C. Under what pressure would the dissolved concentration of CO_2 rise to 1.0 M?

SOLUTION: For a gas in equilibrium with a solution, Henry's law tells us that the partial pressure above the liquid is proportional to the concentration of dissolved gas:

$$P_A = K[A(aq)]$$

K, a form of the *Henry's law constant*, is simply the equilibrium constant for the process

$$A(aq) \rightleftharpoons A(g)$$

Its value for CO_2 in water, from the data given, is evidently 32 atm M^{-1}:

$$K(25°C) = \frac{P_{CO_2}}{[CO_2(aq)]} = \frac{1.0 \text{ atm}}{0.031 \ M} = 32 \text{ atm } M^{-1}$$

As the gas presses harder on the liquid, more gas molecules go into solution. To reach a concentration of 1.0 M, for example, we need a gas pressure of 32 atm:

$$P_{CO_2} = K[CO_2(aq)] = (32 \text{ atm } M^{-1})(1.0 \ M) = 32 \text{ atm}$$

Bottled under high CO_2 pressure, a carbonated beverage will release much of its dissolved carbon dioxide when it comes to a new equilibrium at 1 atm.

EXAMPLE 15-8. Breaking Up: Lattice Energy

Before a solution can be put together, a solute first must be taken apart. The price paid is the **lattice energy**, accounted for in the enthalpy change we call ΔH_1.

PROBLEM: Which substance has the higher lattice energy? (a) NaCl or $MgCl_2$. (b) NaCl or KCl.

SOLUTION: We addressed the issue of lattice energy in Chapter 9 (Example 9-8), where we noted that Coulomb interactions are strongest between small, highly charged ions.

Small: because smaller ions, being closer together, enjoy a greater electrostatic interaction. The force between two charged particles varies inversely with the square of the distance between them.

Highly charged: because the electrostatic force is directly proportional to the product of the charges. The higher the charge, the greater the force.

(a) *NaCl or $MgCl_2$*. Mg^{2+} is both smaller and more highly charged than Na^+. The lattice energy for magnesium chloride, measured alternatively as the *enthalpy* change for the process

$$MgCl_2(s) \longrightarrow Mg^{2+}(g) + 2Cl^-(g) \qquad \Delta H° = 2524 \text{ kJ mol}^{-1},$$

is over three times greater than the corresponding value for sodium chloride:

$$NaCl(s) \longrightarrow Na^+(g) + Cl^-(g) \qquad \Delta H° = 788 \text{ kJ mol}^{-1}$$

(b) *NaCl or KCl*. Both Na^+ and K^+ are singly charged, but the sodium ion is smaller. Sodium chloride has the higher lattice enthalpy, 788 kJ mol^{-1} compared with 717.

QUESTION: Will the material with the higher lattice enthalpy be harder to dissolve?

ANSWER: Not necessarily. Magnesium chloride, for instance, is no less soluble in water than sodium chloride, despite being so much harder to take apart. Although $MgCl_2$ gets off to a bad start in our thermodynamic disassembly step (ΔH_1), it benefits from favorable changes in enthalpy later on. Breaking up is only the first step on the road to dissolution.

In the end, several interrelated factors combine to determine the solubility of a material: (1) The *total* enthalpy of solution—not just the cost to break up the solute (lattice enthalpy), but also the energetic stability derived from putting together the solution (enthalpy of solvation). (2) The total change in entropy. (3) The temperature.

EXAMPLE 15-9. Temperature and Solubility

Many materials dissolve more completely at higher temperatures, but not all. Think back now to the relationship between enthalpy, temperature, and equilibrium that we have developed over the last several chapters.

PROBLEM: Suppose that the temperature of a saturated solution is suddenly raised. Under what circumstances will the mixture (a) remain in equilibrium ($Q = K_{sp}$), (b) become unsaturated ($Q < K_{sp}$), or (c) become *supersaturated* ($Q > K_{sp}$)?

SOLUTION: Consider how Le Châtelier's principle (Chapter 12) applies to each of the three possibilities.

(a) *Equilibrium*. If the overall enthalpy of solution is zero, then the existing equilibrium is unstressed by the additional heat imposed at the higher temperature. The balance between dissolved and undissolved material remains the same.

(b) *Unsaturated*. If, instead, the dissolution is endothermic (ΔH°_{soln} is positive), then the equilibrium shifts to the right at higher temperatures. The increased thermal energy is dissipated by the conversion of reactants (undissolved material) into products (dissolved material), as if this heat were an invisible reactant in the transformation

$$A(\text{undissolved}) + \text{heat} \rightleftharpoons A(\text{dissolved})$$

A solution formerly saturated at one temperature would find itself unsaturated at a higher temperature, able to accept more solute. K_{sp}, a function of temperature, would rise correspondingly.

(c) *Supersaturated*. If the dissolution is exothermic (ΔH°_{soln} is negative), the equilibrium shifts leftward to favor the undissolved material in the reaction

$$A(\text{undissolved}) \rightleftharpoons A(\text{dissolved}) + \text{heat}$$

The solution, supersaturated at the higher temperature, must now move toward a new equilibrium with a smaller $K_{sp}(T)$ and a lower concentration of solute. Sooner or later, a precipitate will form.

And *when*, exactly, is sooner or later? We cannot say, at least not from the thermodynamic data alone. "When" is a kinetic issue (Chapter 18), a practical matter of fulfilling a thermodynamic promise, a question of actually getting the job done. When? When the dissolved particles find the time and means to reorganize themselves and crystallize from the supersaturated solution—that's when. It can take seconds; it can take weeks.

QUESTION: How do we know that the equilibrium constant for an endothermic reaction increases with temperature?

ANSWER: Rewrite the relationship

$$\Delta G^\circ = \Delta H^\circ - T\,\Delta S^\circ = -RT \ln K$$

in the form

$$K = \exp\!\left(-\frac{\Delta G^\circ}{RT}\right) = \exp\!\left(-\frac{\Delta H^\circ}{RT}\right)\exp\!\left(\frac{\Delta S^\circ}{R}\right)$$

to see the connection, noting that ΔH° is positive for an endothermic reaction. The exponent $-\Delta H^\circ/RT$, which therefore turns *negative* for such a process, grows *less* negative as T increases. Raised to a less negative power, the exponential factor for K then becomes larger.

That understood, be careful just the same. For although solubility usually does increase with temperature when ΔH°_{soln} is positive, we should also expect exceptions, surprises, and problems to add interest to our calculations. Reliance on standard enthalpies, for one, becomes questionable with sparingly soluble solutes that rarely attain standard concentrations of 1 M. But our skepticism should go even deeper than that, because sometimes a solution is just too complicated to support broad generalizations. Many factors are at work.

EXAMPLE 15-10. Enthalpy of Solution

PROBLEM: Using the formation data provided in Appendix C (Table C-16), compute ΔH°_{soln} for the dissolution of sodium chloride in water.

SOLUTION: We obtain the standard enthalpy change for the reaction

$$NaCl(s) \rightleftharpoons Na^+(aq) + Cl^-(aq)$$

by combining the heats of formation, ΔH°_f, according to Hess's law. The procedure is the same as described in Chapters 13 and 14:

$$\Delta H^\circ_{soln} = \Delta H^\circ_f[Na^+(aq)] + \Delta H^\circ_f[Cl^-(aq)] - \Delta H^\circ_f[NaCl(s)]$$

$$= -240.1 - 167.2 - (-411.2) = 3.9 \text{ kJ mol}^{-1}$$

Put the pieces together conceptually, step by step: (1) Ions are separated, but not without some large payment of enthalpy; one step. (2) Clusters of water are disturbed, again at some high cost in enthalpy; another step. (3) Finally, ions are hydrated and some large amount of enthalpy is recovered; a third step.

The net result? Almost a draw: the overall enthalpy of solution is nearly zero—slightly unfavorable, in fact, amounting to a small positive value. If enthalpy were the deciding factor, then sodium chloride would be insoluble in water. Rather it is the increase in solution *entropy* that drives forward the dissolution, overcoming the positive ΔH°_{soln} at all temperatures where

$$\Delta G^{\circ}_{soln} = \Delta H^{\circ}_{soln} - T\,\Delta S^{\circ}_{soln}$$

is negative.

QUESTION: Does NaCl become more soluble or less soluble at higher temperatures?

ANSWER: With the solution enthalpy small and positive, we expect the solubility to increase with temperature, but only slightly.
It does: from 357 g L^{-1} at 0°C to 391 g L^{-1} at 100°C.

EXAMPLE 15–11. Making Up: Enthalpy of Hydration

PROBLEM: Is Na^+ likely to have a higher or lower enthalpy of hydration than K^+ in our imagined second and third steps of dissolution ($\Delta H_2 + \Delta H_3$)? Which of the two ions should pass more easily through a semipermeable membrane?

SOLUTION: A naked sodium ion, uncoordinated to any other species, has a smaller radius than a naked potassium ion. Recall the arguments from Chapter 6: Sodium lies just above potassium in Group I; sodium contains one fewer shell than potassium; sodium is smaller.

Now with the same charge compressed into less volume, the smaller Na^+ ion produces a stronger electric field than K^+. The sodium ion draws more solvent molecules to itself as a result, enjoying a more negative enthalpy of hydration as additional weak bonds form.

And the more solvent molecules there are, the bigger is the hydrated ion *in solution*—a trend opposite to the one followed by bare species in the gas phase, where ionic radii increase going down a group. Hydrated Na^+, surrounded by more water molecules than hydrated K^+, becomes the larger of the two in solution and thus the more difficult to diffuse across a membrane. The K^+ ion passes through more easily.

EXAMPLE 15–12. All Together: Free Energy
and the Equilibrium Constant

Enthalpy and entropy come together in one number, the free energy, which ultimately determines who wins: products or reactants. With that one number we have, as always, a concise summary of the thermodynamic ups and downs.

Here is an example showing how to determine the equilibrium constant from tabulated free energies of formation.

PROBLEM: Calculate $\Delta G°$ and $K_{sp}(25°C)$ for the dissolution of $BaCO_3$ in water.

SOLUTION: From Table C-16 we extract formation data for the three species in the reaction

$$BaCO_3(s) \rightleftharpoons Ba^{2+}(aq) + CO_3^{2-}(aq)$$

The numbers are:

SUBSTANCE	$\Delta G_f°$ (kJ mol^{-1})
$Ba^{2+}(aq)$	−560.8
$CO_3^{2-}(aq)$	−527.8
$BaCO_3(s)$	−1137.6

We combine them according to a generalized Hess's law, just as described in Chapters 13 and 14, to obtain first the standard free energy of solution

$$\Delta G° = \Delta G_f°[Ba^{2+}(aq)] + \Delta G_f°[CO_3^{2-}(aq)] - \Delta G_f°[BaCO_3(s)]$$

$$= (-560.8) + (-527.8) - (-1137.6)$$

$$= 49.0 \text{ kJ mol}^{-1}$$

and then the equilibrium constant:

$$K_{sp} = \exp\left(-\frac{\Delta G°}{RT}\right) = \exp\left(-\frac{49000 \text{ J mol}^{-1}}{8.3145 \text{ J mol}^{-1} \text{ K}^{-1} \times 298 \text{ K}}\right)$$

$$= \exp(-19.78) = 2.6 \times 10^{-9}$$

The number, hypersensitive to small variations in the hard-to-measure thermodynamic data, is only approximate, but the message for barium carbonate is clear: Climbing uphill against a positive free energy of solution, $BaCO_3$ is an insoluble salt. Its equilibrium constant is considerably less than 1.

EXAMPLE 15-13. Like Dissolves Like

In the end, faced with all the thermodynamic subtleties of solutes and solvents, we resort less to exact calculation and more to qualitative argument.

PROBLEM: Which of the following substances is more soluble in water? (a) CCl_4 (carbon tetrachloride) or CH_3Cl (methyl chloride). (b) CH_3CH_2OH (ethanol) or $CH_3CH_2CH_2CH_2CH_2CH_2CH_2OH$ (1-heptanol). (c) $C_{12}H_{22}O_{11}$ (sucrose) or C_6H_6 (benzene). (d) AgCl or diamond.

SOLUTION: Ions and polar molecules mix best with water, a strongly polar solvent. The more polar the solute, the more effective is the dissolution.

Look, accordingly, for signs such as (1) polar functional groups, particularly those that create large differences in electronegativity; (2) multiple occurrences of polar bonds; (3) molecules that consist mostly of polar regions; (4) molecules with irregular shapes that allow for nonvanishing dipole moments; and (5) ions with large charges.

(a) *Carbon tetrachloride or methyl chloride.* Both are tetrahedrally bonded molecules, but the more symmetric CCl_4 is nonpolar and therefore *immiscible* with water (unable to be mixed). CH_3Cl, with its net dipole moment, interacts less obtrusively with the polar molecules of the solvent.

(b) *Ethanol or 1-heptanol.* Although each alcohol contains a polar OH group, CH_3CH_2OH is considerably more soluble than $CH_3(CH_2)_6OH$. 1-Heptanol carries its hydroxyl group at the end of a seven-carbon chain, making the OH a small polar part of a comparatively large, mostly nonpolar molecule. Ethanol, with only two carbons, presents a more concentrated target for hydrogen bonding.

(c) *Sucrose or benzene.* Eight polar OH sites make a molecule of sucrose readily soluble in water, as we know from ordinary experience. Benzene, a nonpolar molecule, is poorly soluble in water.

sucrose

benzene

(d) *Silver chloride or diamond.* Each is a strong, tightly bound solid, but AgCl is nevertheless susceptible to attack by the polar water molecules (if only slightly). Diamond, a covalently bonded network crystal, is not. Its lattice energy is too high.

EXERCISES

1. (a) What does it mean for a solution to be saturated? (b) Must a solution be saturated to attain thermal equilibrium?

2. Why is the presence of undissolved solute a telltale sign of saturation? Assume that the solution has been thoroughly mixed and has been given ample time to come to equilibrium.

3. Consider the various kinds of equilibria maintained simultaneously in a saturated solution: (a) Disorder–order equilibrium between two heterogeneous phases. (b) Chemical equilibrium between products and reactants. (c) Thermal equilibrium among all the colliding particles. (d) Osmotic equilibrium. In what ways are they different, and in what ways are they the same? Describe how each equilibrium arises from a dynamic balancing of opposing processes.

4. A supersaturated solution contains more solute than its equilibrium constant allows, yet still there is no solid precipitate. The solution exists as a single phase, poised *above* its equilibrium point, with Q (the reaction quotient) greater than K (the equilibrium constant). How can that be? Keep in mind the different demands of thermodynamics and kinetics.

5. Suppose that sugar is incompletely stirred into a cup of tea, so that some portions of the solution are sweeter than others. (a) Is the system fully at equilibrium? If not, what further changes must ensue? (b) Approximately how much time will pass before equilibrium is finally established?

6. Argue, from statistics, that concentration will inevitably be the same throughout a solution at equilibrium. Ignore any effects due to gravity.

7. Solutions A and B are separated by a membrane that allows the passage of water only, not ions. Solution A contains one mole of sodium chloride in a volume of one liter. Solution B contains three moles of sodium chloride in a volume of one liter. (a) In which direction will the water flow—from A to B or from B to A? (b) What volume of water eventually will pass from one solution to the other? (c) What will be the final concentrations of A and B?

8. Which compound in each pair is probably more soluble in water?

   (a) C_2H_6 (ethane) . . . or C_2H_5OH (ethanol)
   (b) $CH_3CH_2CH_2OH$ (propanol) . . . or $CH_3CH_2CH_2CH_2OH$ (butanol)
   (c) C_2H_4 (ethene) . . . or the amino acid serine, both drawn below:

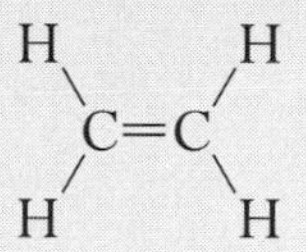

   ethene 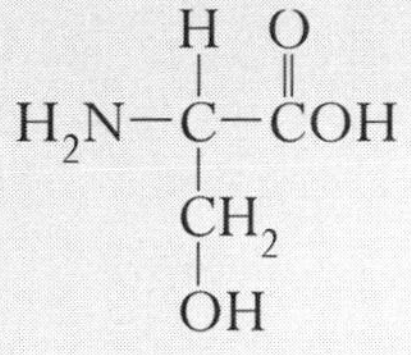   serine

   (d) Vitamin A or vitamin C:

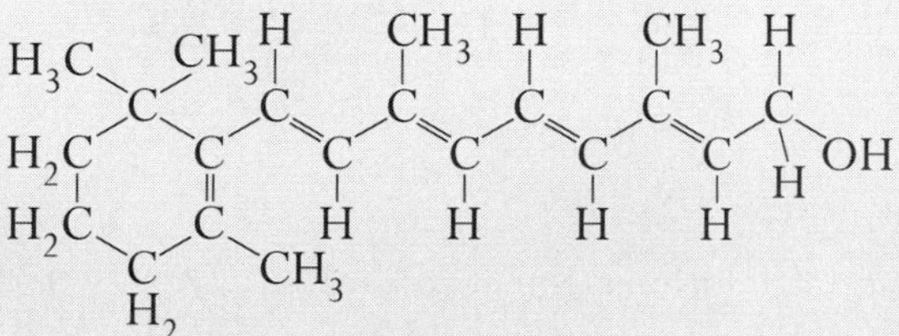

   vitamin A

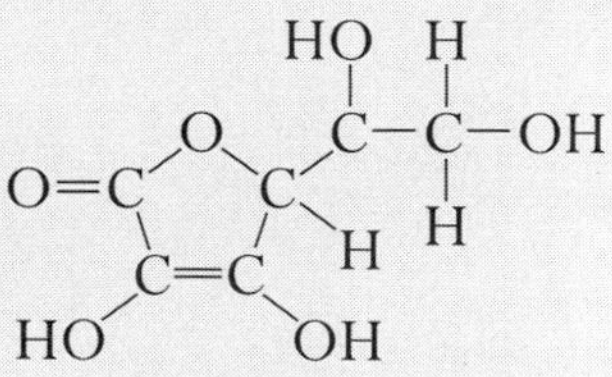

   vitamin C

9. Which compound in each pair is probably more soluble in benzene?

   (a) NaCl . . . or $C_6H_5CH_3$ (toluene), drawn below:

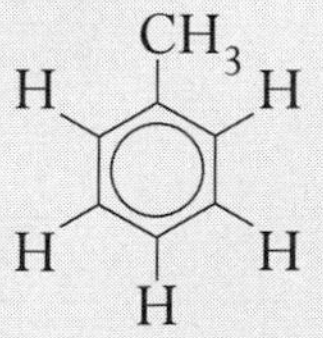

   toluene

   (b) C_2H_6 (ethane) . . . or H_2SO_4 (sulfuric acid)
   (c) $C_{12}H_{22}O_{11}$ (sucrose) . . . or $C_{10}H_8$ (naphthalene), both drawn on the following page:

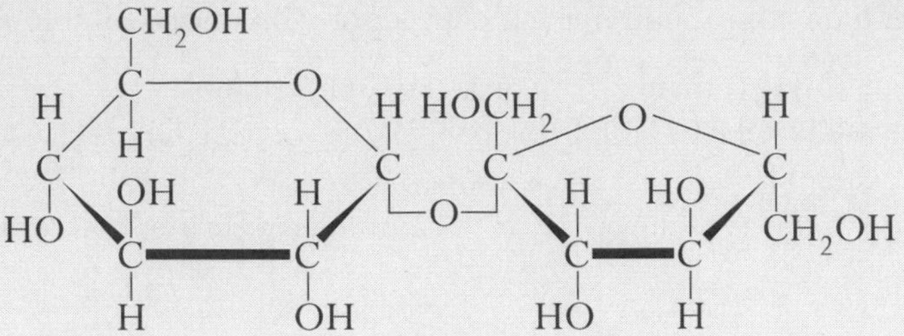

sucrose

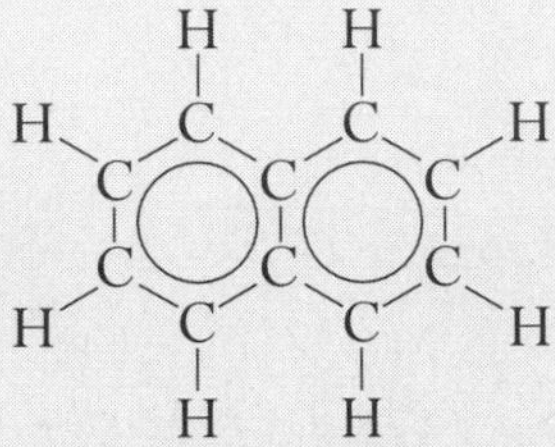

naphthalene

10. **Make some solutions.** Suggest a way to prepare 100 mL of each of these aqueous solutions:

   (a) 0.375 M Na$_2$SO$_4$
   (b) 0.150 M AgNO$_3$
   (c) 0.625 M C$_6$H$_{12}$O$_6$
   (d) 1.00 M HCl

   Review the material in Chapters 3 and 9 as needed.

11. **Make some dilutions.** Take each of the 100-mL solutions already prepared in the preceding exercise, and add sufficient water to increase the volume to 125 mL. What are the new concentrations? The original values are:

   (a) 0.375 M Na$_2$SO$_4$
   (b) 0.150 M AgNO$_3$
   (c) 0.625 M C$_6$H$_{12}$O$_6$
   (d) 1.00 M HCl

12. Suppose that additional solvent is added to 500 mL of a 2.00 M solution. What final volume is needed to realize each of the following concentrations?

   (a) 1.00 M
   (b) 0.50 M

(c) 0.25 M

(d) 0.10 M

13. How many moles of Na^+ and how many moles of SO_4^{2-} are contained in 50.0 mL of each of these Na_2SO_4 solutions?

(a) 0.750 M

(b) 1.256 M

(c) 0.105 M

(d) 3.07×10^{-3} M

(e) 4.92×10^{-5} M

14. The solubility of Na_2CO_3 in cold water (0°C) is 7.1 grams per 100 mL. (a) Calculate $[Na^+]$ and $[CO_3^{2-}]$ in a saturated solution of Na_2CO_3 at 0°C, expressing the concentrations in moles per liter. (b) Use the solubility data to estimate the corresponding equilibrium constant for the reaction

$$Na_2CO_3(s) \rightleftharpoons 2Na^+(aq) + CO_3^{2-}(aq)$$

15. The solubility of Li_2CO_3 in water is 15.4 g L^{-1} at 0°C and 7.2 g L^{-1} at 100°C. (a) Is the dissolution

$$Li_2CO_3(s) \rightleftharpoons 2Li^+(aq) + CO_3^{2-}(aq)$$

endothermic or exothermic? Answer without doing an explicit calculation. (b) Use formation data tabulated in Appendix C to calculate the enthalpy of solution at 25°C. Review the material in Chapter 13 as needed.

16. The solubility of $NaNO_3$ in water increases from approximately 900 g L^{-1} at 25°C to 1800 g L^{-1} at 100°C. (a) Is the dissolution endothermic or exothermic? (b) Is the change in entropy positive or negative? Answer both (a) and (b) using just the information given.

17. At 25°C, the solubility-product constant for the reaction

$$Hg_2Cl_2(s) \rightleftharpoons Hg_2^{2+}(aq) + 2Cl^-(aq)$$

is approximately $K_{sp} = 1.5 \times 10^{-18}$. Calculate the concentrations of Hg_2^{2+} and Cl^- in a saturated solution of mercury(I) chloride.

18. How many grams of aluminum hydroxide will saturate 50.0 mL of water at 25°C? The solubility-product constant for the dissolution

$$Al(OH)_3(s) \rightleftharpoons Al^{3+}(aq) + 3OH^-(aq)$$

is $K_{sp} = 1.9 \times 10^{-33}$.

19. A chemist measures the following equilibrium concentrations in an aqueous solution at 25°C:

$$[Ba^{2+}] = 1.05 \times 10^{-5} \ M$$

$$[SO_4^{2-}] = 1.05 \times 10^{-5} \ M$$

Calculate K_{sp} for $BaSO_4$.

20. K_{sp} for lead chloride, $PbCl_2$, is 1.6×10^{-5} at 25°C. In which of the following mixtures will solid $PbCl_2$ appear as a precipitate?

   (a) 50.0 mL of 0.025 M $Pb(NO_3)_2$ added to 50.0 mL of 0.025 M NaCl
   (b) 50.0 mL of 0.050 M $Pb(NO_3)_2$ added to 50.0 mL of 0.025 M NaCl
   (c) 50.0 mL of 0.025 M $Pb(NO_3)_2$ added to 50.0 mL of 0.050 M NaCl
   (d) 50.0 mL of 0.050 M $Pb(NO_3)_2$ added to 50.0 mL of 0.050 M NaCl
   (e) 50.0 mL of 0.0005 M $Pb(NO_3)_2$ added to 50.0 mL of 1.000 M NaCl

21. K_{sp} for calcium fluoride, CaF_2, is 3.9×10^{-11} at 25°C. (a) Calculate $[Ca^{2+}]$ and $[F^-]$ for an aqueous solution in equilibrium at that temperature. (b) Calculate the solubility in grams per liter. (c) How many moles of dissolved Ca^{2+} and F^- exist in 100 mL of a saturated solution?

22. Suppose that a saturated solution of CaF_2 (see above) loses half its solvent by evaporation, going from a volume of 0.100 L to a volume of 0.050 L. (a) Compute the reaction quotient for the out-of-equilibrium, supersaturated solution. (b) Compute the amount of CaF_2 (moles and grams) that must precipitate from the solution to restore equilibrium. (c) Compute the number of moles of Ca^{2+} and F^- that remain in solution after equilibrium is reestablished in the smaller volume.

23. Calculate the solubility, in grams per liter at 25°C, of CaF_2 in each of the aqueous solutions specified below. Consult Appendix C where necessary.

   (a) 0.050 M $AgNO_3$
   (b) 0.100 M $AgNO_3$
   (c) 0.050 M NaF
   (d) 0.100 M NaF

Explain the results by appealing to Le Châtelier's principle, specifically the common-ion effect.

24. Another example of the common-ion effect: How many grams of AgCl will dissolve in 0.100 L of 0.100 M AgNO$_3$? Compare the equilibrium concentration of Ag$^+$ in the AgCl/AgNO$_3$ mixture with the corresponding value in a solution of silver chloride alone. $K_{sp}(25°C) = 1.8 \times 10^{-10}$ for AgCl.

25. Mix together two solutions A and B (see below) to make 100.0 mL of a combined solution that contains Na$^+$, Cl$^-$, and CrO$_4^{2-}$ ions:

   A: 50.0 mL of 0.100 M NaCl

   B: 50.0 mL of 0.050 M Na$_2$CrO$_4$

   Then, drop by drop, add a highly concentrated solution of AgNO$_3$ to the mixture of A and B. For simplicity, assume that the total volume holds constant at 100.0 mL since the amount of AgNO$_3$ added is so small. (a) Which solid will precipitate first . . . AgCl ($K_{sp} = 1.8 \times 10^{-10}$) or Ag$_2CrO_4$ ($K_{sp} = 1.1 \times 10^{-12}$)? (b) After how many moles of additional AgNO$_3$ will the first solid begin to precipitate? (c) How many moles of Ag$^+$ will be present when the second solid begins to precipitate?

26. Return to the example of selective precipitation described in the preceding exercise:

   AgNO$_3$ is added to a 50:50 mixture of 0.100 M NaCl and 0.050 M Na$_2$CrO$_4$ (total volume = 100.0 mL).

   (a) Compute [Na$^+$] and [Cl$^-$] at the point where Ag$_2$CrO$_4$ just begins to precipitate. (b) How many grams of AgCl—if any—have already precipitated from solution at this stage?

27. Use the thermodynamic data in Appendix C to calculate $\Delta H°$, $\Delta S°$, and $\Delta G°$ for the dissolution of each of these salts in water at 25°C:

   (a) KCl
   (b) AgCl
   (c) Ca(OH)$_2$

   Which of the three compounds is most soluble? For which of them is the equilibrium constant greater than 1? For which of them is it less? Indicate, for each compound, whether the changes in enthalpy and entropy are favorable for dissolution.

28. Use the results of the preceding exercise to calculate $K_{sp}(25°C)$ for AgCl and $Ca(OH)_2$. Note that the values computed may differ from those tabulated in Appendix C.

29. Use the thermodynamic data in Appendix C to calculate $\Delta H°$, $\Delta S°$, and $\Delta G°$ for the dissolution of each of these salts in water at 25°C:

    (a) $MgCl_2$
    (b) $BaSO_4$
    (c) $CaSO_4$

    Which of the three compounds is most soluble? For which of them is the equilibrium constant greater than 1? For which of them is it less? Indicate, for each compound, whether the changes in enthalpy and entropy are favorable for dissolution.

30. Use the results of the preceding exercise to calculate $K_{sp}(25°C)$ for $BaSO_4$ and $CaSO_4$. Note that the values computed may differ from those tabulated in Appendix C.

31. Now, a look at another manifestation of disorder–order equilibrium: a gas dissolved in a liquid. Calculate, using Henry's law for the equilibrium between gas and liquid, the concentrations of N_2 and O_2 normally present in a solution of water open to the atmosphere. Assume that air is a mixture of 78% nitrogen, 21% oxygen, and 1% other gaseous components (by moles), and choose the Henry's law constants appropriate for aqueous solutions at 20°C. See below:

$$N_2(aq) \rightleftharpoons N_2(g) \qquad K(20°C) = 1.4 \times 10^3 \text{ atm L mol}^{-1}$$

$$O_2(aq) \rightleftharpoons O_2(g) \qquad K(20°C) = 7.2 \times 10^2 \text{ atm L mol}^{-1}$$

32. (a) Compute, using data from the previous exercise, the solubility of N_2 in aqueous solution under a partial N_2 pressure of 100 atm. (b) Do the same for oxygen under a partial O_2 pressure of 100 atm. Why do the concentrations increase?

33. Carbonated beverages are bottled with CO_2 under high pressure. Why does the mixture fizz when the container is suddenly opened to the atmosphere? Invoke Le Châtelier's principle to describe what happens.

34. Moving away, finally, from disorder–order equilibrium between solution and solute, consider yet another kind of equilibrium: osmotic equilibrium, nature's drive to spread matter uniformly in space.

(a) Calculate the osmotic pressure of a 0.10 M solution of $AgNO_3$ at 25°C. (b) What concentration of Na_2SO_4 will produce the same osmotic pressure? (c) What is the equivalent concentration of glucose $(C_6H_{12}O_6)$?

35. How many grams of glucose $(C_6H_{12}O_6)$ dissolved in 0.100 L of aqueous solution will generate an osmotic pressure of 2.00 atm at 20.0°C?

36. Say that a certain aqueous solution has an osmotic pressure of 8.4 atm at 37.0°C (body temperature). If the solution contains 7.6 g of a molecular solute dissolved in 500 mL, what is the molar mass of the solute?

16

Acids and Bases

16-1. A Wandering Ion
16-2. Conjugate Acids and Bases
        Strength and Weakness
        Stabilization of a Conjugate Base
16-3. Aqueous Equilibria
        Acidic Solutions: pK_a and pH
        Basic Solutions: pK_b and pOH
        Autoionization of Water
        K_b for a Conjugate Base
16-4. Neutralization
        Strong Acid and Strong Base
        Weak Acid and Strong Base
16-5. Weak Acids, Conjugate Bases, and Buffers
        Weak Acid: Mostly HA
        Conjugate Base: Mostly A^-
        Buffers
16-6. Titration
        REVIEW AND GUIDE TO PROBLEMS
        EXERCISES

16-1. A WANDERING ION

Transfer of a hydrogen ion—a proton, H^+—from one species to another is one of chemistry's most broadly familiar and useful transactions. The electron-hungry H^+ seems to turn up wherever one looks. It might, for a start, attack a $C=C$ double bond during the formation of an organic polymer. Or it might move on and off certain sites in a zeolite, subtly but unmistakably changing the properties of the material. At times this well-traveled particle leaves one structure and bolts itself to another; at times it acts solely as a catalytic agent, moving in and out while making no lasting attachments. It holds the key to many a reaction, and not just in the laboratory or chemical plant but in nature as well. H^+ controls life's biochemical pathways by its mere presence or absence, for a roving proton can alter the structure and function of the amino acids, proteins, nucleic acids, and all the other biological molecules.

In any number of ways, often understated, the hydrogen ion figures at the center of molecular change. It gives vinegar its bite and cabbage juice its color; it controls the chemical environment of the blood; it makes and breaks polymers; it eats away at stone. Apparently everywhere, the hydrogen cation frequently encounters a particularly favored partner—the hydroxide anion, OH^-—and the two species unite, swiftly and surely, to form water. This simplest of reactions,

$$H^+ + OH^- \longrightarrow H_2O$$

has within it the potential to raise up and also to tear down, being common both to the formation of great limestone hills and to the progressive decay of a mouthful of teeth.

We are dealing here with the chemistry of acids and bases, particularly as envisioned by Brønsted and Lowry in their conception of hydrogen donors and acceptors. Our preliminary look in Chapter 3 has revealed some of the insight gained when reactions are viewed in this yin-yang fashion, whereby two opposites merge into an entirely new form. There we saw the unifying power of the acid–base model of reactivity, brought out especially by Lewis's more general picture of an acid as an electron-pair acceptor and a base as an electron-pair donor. If chemistry, to Lewis and to all of us today, is about the giving and taking of electrons, then Brønsted and Lowry's H^+ must stand prominent among the electron takers. It is what makes an acid an acid, and its comings and goings now demand our special attention.

16-2. CONJUGATE ACIDS AND BASES

A generalized acid–base reaction (Figure 16-1) is represented by the equation

$$HA + B \rightleftharpoons A^- + BH^+$$

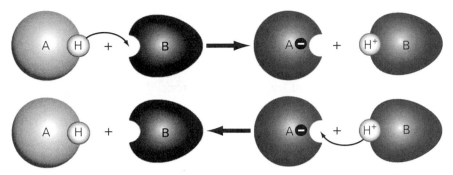

FIGURE 16-1. The acid gives, and the base takes. The deprotonated acid becomes a conjugate base. The protonated base becomes a conjugate acid.

where HA is the acid and B is the base. We take HA to be a generic (and not necessarily neutral) entity, some specific examples of which might include: (1) hydrochloric acid (HCl), where A^- is the chloride ion Cl^-; (2) acetic acid (CH_3COOH), for which A^- is the acetate ion CH_3COO^-; (3) water (H_2O), where A^- denotes the hydroxide ion OH^-; and (4) the ammonium ion (NH_4^+), where "A^-" is actually the neutral molecule NH_3. The double arrow reminds us that the transformation can go both ways; it is an equilibrium process.

It is, still more, a coordinated process—a reciprocated interaction that requires a giver and taker. Through an *ionization* or *dissociation* reaction,

$$HA \longrightarrow H^+ + A^-$$

the acid A—H *loses* its detachable proton and becomes the species A^-. The base (brandishing a pair of accessible electrons; hence B:) *gains* the H^+ and is transformed into BH^+ as

$$B\!: + H^+ \longrightarrow BH^+$$

Just like reduction and oxidation, one action implies the other; for every proton lost there is a proton gained. The hydrogen ion released by an acid can neither disappear nor avoid contact with other atoms. Something captures it instead: OH^-, H_2O, NH_3, CH_3COO^-, a $C=C$ double bond, or anything else with electrons to offer. That species—in whatever form and by whatever means (covalent bond, solvation, or some other mechanism)—acts as the base, B.

Now when an acid dissociates to form H^+ and A^-, the broken species left behind (A^-) evidently must be a base in its own right. Having come into existence by loss of a proton, A^- can reclaim the particle and rebuild an HA molecule according to the reverse reaction

$$A^- + H^+ \longrightarrow HA$$

How far this base-to-acid transformation proceeds is a question of equilibrium and free energy, but some capability is always there. What was lost can be regained, and thus A^- becomes the **conjugate base** of the acid HA.

Similarly, BH^+ is the **conjugate acid** of the base B. Having gained a proton, the conjugate acid BH^+ now can lose the H^+ and go back to being B. No attachment is really permanent. There is only the equilibrium law to govern the continual back-and-forth in any reaction, and in this instance we have a two-on-two competition involving acid/base and *conjugate* base/*conjugate* acid:

$$HA + B \rightleftharpoons A^- + BH^+$$

| acid | base | conjugate base | conjugate acid |

On one side, HA loses a proton and B gains a proton. On the other side, BH^+ loses and A^- gains.

Strength and Weakness

An acid loses a proton and becomes a base; a base acquires a proton and becomes an acid. Easy come, easy go?

Quite the opposite, for the strengths of acid and conjugate base (or base and conjugate acid) are related inversely. An acid is said to be *strong* when it loses a proton readily, and *weak* when it tends to hold on to the H^+. And what a strong acid gives up easily, it takes back only reluctantly as a conjugate base. Consider that if some strong acid HA can barely hold on to its H^+, then surely A^- cannot expect to recapture the proton with much success. The equilibrium in Figure 16-2(a) is driven to the right as the strong acid dissociates ("take my proton; you need it more than I do") to form a weak conjugate base ("keep the proton, thank you; I'm more stable now without it").

A *weak* acid, by contrast, ionizes only with some difficulty, and so presumably its conjugate base A^- must be a good scavenger of protons. The equilibrium in Figure 16-2(b) lies toward the left as the strong conjugate base sweeps up protons to regenerate the acid HA.

The same thermodynamic relationships connect a base with its conjugate acid, enabling us to summarize the four possibilities:

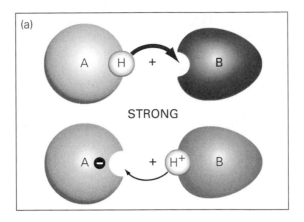

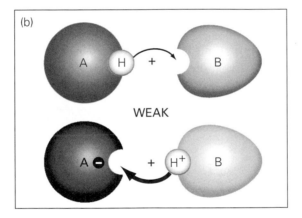

FIGURE 16-2. The equilibrium between each conjugate pair is skewed in the direction of the heavy arrow, to assert that: (a) A strong acid leaves behind a weak conjugate base. Easy to lose is hard to regain. (b) A weak acid leaves behind a strong conjugate base. Hard to lose is easy to regain.

strong acid ⟶ weak conjugate base

strong base ⟶ weak conjugate acid

weak acid ⟶ strong conjugate base

weak base ⟶ strong conjugate acid

The stronger the acid and base, then the weaker will be the conjugate base and conjugate acid . . . and the higher will be the equilibrium constant for the generalized two-on-two acid–base reaction.

Strength therefore is inextricably bound together with weakness. For an acid to be strong, its conjugate base must be weak enough to resist reprotonation while surviving as A⁻. A base is weak precisely because it is thermodynamically content to remain as it is, without benefit of reattached proton. Somehow the weak conjugate base is able to stabilize the deprotonated structure left behind after dissociation, and this stabilization of A⁻ is what makes the acid (HA) strong. Once ionized, the strong acid stays ionized.

How might it do that? How might a conjugate base become stable?

Stabilization of a Conjugate Base

Figure 16-3 reminds us that many conjugate bases are anions, acquiring their negative charge when H⁺ is removed from a neutral acid molecule. It is not a promising beginning. Charge imbalance of any kind is bad enough, but a concentrated ionic charge localized near the ruptured H—A bond is even worse. An anion stands a better chance if its charge can be dispersed throughout the structure, perhaps with the aid of some strongly electronegative atoms.

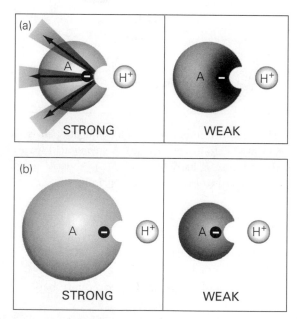

FIGURE 16-3. Its proton gone, a conjugate anion carries the negative charge abandoned by the departing H⁺. To *stay* deprotonated, A⁻ must then find a place to put these extra electrons—to spread them, where possible, throughout the structure. (a) Groups that withdraw electrons help to disperse the ionic charge and stabilize the anion. The symbolic acid on the left is presumably stronger than the one on the right. (b) A large, roomy anion (left) can dilute the additional electron density better than a smaller species.

Chlorine, for example, withdraws electrons more effectively than hydrogen, and trichloroacetic acid (CCl_3COOH) thereby becomes a stronger acid than ordinary acetic acid (CH_3COOH). A trichloroacetate ion

$$Cl-\underset{\underset{Cl}{|}}{\overset{\overset{Cl}{|}}{C}}-C\overset{O}{\underset{O}{<}}^-$$

proves better equipped to remove excess charge from the carboxylate (COO^-) group than does an acetate ion (CH_3COO^-), which otherwise has a similar geometry. Since the CCl_3 group draws electrons away from the COO^- more strongly than does CH_3, the negative charge is spread more diffusely over the trichloroacetate anion. The chlorinated species is stabilized relative to the regular acetate, and so trichloroacetic acid remains dissociated to a greater extent than acetic acid. Trichloroacetate is left as the weaker base, less prone to recapturing H^+. The reduced negative charge at the carboxylate site offers a less attractive target to the electron-poor proton.

A conjugate base can be supported in many other ways, including some already noted in our discussion of substitution reactions back in Chapter 8. Especially for organic acids, say, the presence of aromatic or conjugated π systems may provide a means to delocalize part of the charge. Delocalization is indeed one of the reasons why carboxylic acids ($R-COOH$) usually dissociate more readily than the corresponding alcohols, $R-OH$. A carboxylate anion $R-COO^-$ spreads the negative charge over (at least) a three-atom COO^- π system,

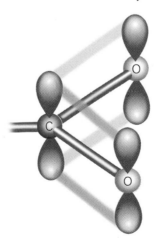

rather than leaving it concentrated on the oxygen as $R-O^-$. By stabilizing the anion, this smearing of charge allows $R-COOH$ to be a better acid than some R—OH with the same R group. A survivable, proton-resistant conjugate base makes for a strong acid.

The *size* of an anion is important to acidity as well, because larger structures generally are better able to support the extra electron pair. Other factors being equal, the anion that can simply disperse its charge over greater volume has the lower energy—even without any added benefit of delocalization. Expect, then, for acidity within a family of related compounds to increase as the conjugate bases grow larger.

But not always. Sometimes, rather than stabilizing a conjugate base, sheer bigness might leave A^- poorly adapted for solvation; and, when so, the acidity of HA in solution is impaired. Consider the disadvantage: A bulky anion, surrounded by fewer solvent molecules, is more susceptible to attack and reprotonation than a highly protected, compact structure. Hence methyl alcohol (CH_3OH), although a very, very weak acid in water, is nevertheless stronger than *tert*-butyl alcohol, with its large, hard-to-handle conjugate base:

$$CH_3OH \longrightarrow CH_3O^- \ + \ H^+$$

methyl alcohol

$$CH_3-\overset{\displaystyle CH_3}{\underset{\displaystyle CH_3}{\overset{|}{\underset{|}{C}}}}-OH \longrightarrow \left[CH_3-\overset{\displaystyle CH_3}{\underset{\displaystyle CH_3}{\overset{|}{\underset{|}{C}}}}-O\right]^- \ + \ H^+$$

tert-butyl alcohol

More of the smallish CH_3O^- species stay dissociated.

Nor should we forget one of the most obvious sources of acid strength: an easily broken H—A bond in the acid itself, polarized as $H^{\delta+}-A^{\delta-}$. Some bonds are easier to break than others, and a loosely attached proton clearly will make an acid strong from the start. Once such a proton comes off, its reattachment to A^- is hindered by the poor thermodynamics associated with that particular H—A bond.

Structural features such as these combine to determine the equilibrium between an acid and its conjugate base, together also with the entropy changes consequent to both dissociation of HA and solvation of A^-. So too will the relative strength of a *base* and its conjugate acid be controlled by similar considerations, for there is a reciprocal symmetry inherent in all acid–base phenomena. Whatever we assert for the acid and conjugate base, we imply tacitly for base and conjugate acid. There is no real difference.

Tying everything together, moreover, are the unbending constraints imposed at equilibrium. However complicated may be the interplay of acid and base, there is always the equilibrium constant (which can be measured) and the associated free energy change (which balances local entropy against local enthalpy) to summarize what happens. Characterization of that equilibrium will now become our main concern, as we turn to the important problem of acids and bases in aqueous solution.

16-3. AQUEOUS EQUILIBRIA

Much of acid–base chemistry takes place in the solution phase, where a homogeneous equilibrium is maintained among all the participants. Water, possessed of certain remarkable properties, typically is the solvent—but, far from being a passive host, this versatile molecule actively mediates the exchange of H^+.

H_2O is both acid and base. It can lose a proton to yield the hydroxide ion (OH^-); it can gain a proton to become the hydronium ion (H_3O^+). These two species, hydronium and hydroxide, are the principal acids and bases in aqueous solution, and to Arrhenius (Chapter 3) they were literally the *only* acids and bases.

Acidic Solutions: pK$_a$ and pH

Picture, first, water acting as a base to snatch a proton off the acid HA in a reaction

$$HA(aq) + H_2O(\ell) \rightleftharpoons A^-(aq) + H_3O^+(aq)$$

with equilibrium constant

$$K_a = \frac{[H_3O^+][A^-]}{[HA]}$$

There develops a balance among (1) the acid HA, (2) its conjugate base A^-, (3) H_2O acting as a proton acceptor (a *base*, whose unchanging concentration is omitted from K_a), and (4) H_3O^+, the hydronium ion, which is the acid conjugate to water. Hydronium ion then becomes *the* acid in solution, just as Arrhenius envisioned. The original acid HA has lost its proton to water, and H_3O^+ takes over the role of proton donor.

The resulting equilibrium is governed by K_a, the **acid dissociation constant** or **acid ionization constant**. A large K_a (Figure 16-4a) means

(a)   (b)

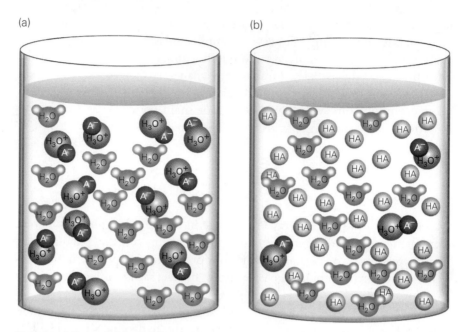

FIGURE 16-4. Acids and bases in water: winners and losers. (a) Equilibrium for
a strong acid tilts heavily in favor of the ionization products H_3O^+ and A^-, the conjugate base. Undissociated molecules of HA are scarce, and K_a is correspondingly
large. (b) For a weak acid, just the opposite: HA is the dominant species at equilibrium. K_a is small, less than 1.

a large concentration of H_3O^+ and A^- at equilibrium but not much undissociated HA. It means that HA is a *strong* acid, heavily ionized in solution,
with the liberated proton used to produce hydronium. A small K_a indicates, conversely, a relatively small concentration of H_3O^+ and A^- coexisting at equilibrium with a substantial amount of undissociated HA. Such
a species is a *weak* acid, largely unionized and therefore unable to generate
significant quantities of H_3O^+. See Figure 16-4(b).

For "large" we shall expect K_a to be greater than 1, if only to establish
a rule of thumb. Since strong acids such as hydrochloric (HCl) and sulfuric (H_2SO_4) usually are nearly 100% ionized, the exact value of K_a becomes unimportant anyway. We simply assume that every molecule of
HCl, for instance, dissolves in water to yield one hydronium ion and one
chloride ion. The equilibrium lies so far to the right that K_a is effectively
infinite, and we just treat the matter as a one-for-one stoichiometry
problem. The calculation involved is important, easy, and relatively un-
interesting. All strong acids are alike.

Weak acids, however, are shaped individually by their small equilibrium
constants ($K_a < 1$), and here the calculation is important, occasionally

complicated, and usually interesting. Much of our subsequent effort will be devoted to exploring the behavior of such systems, often using acetic acid $[K_a(25°C) = 1.76 \times 10^{-5}]$ as a model. Values of K_a for various acids, weak and strong, are compiled in Appendix C (Table C-20).

Note how the attributes "strong" and "weak" are beginning to acquire quantitative meaning. A strong acid, with a large K_a, is strong in comparison specifically with *water* in the reaction

$$HA(aq) + H_2O(\ell) \rightleftharpoons A^-(aq) + H_3O^+(aq) \qquad K_a \gg 1$$

This strong acid HA, in aqueous solution, successfully gives its proton to H_2O and is able to remain ionized as A^-. On the left of the equation are reactants HA and H_2O with some (relatively high) free energy, and on the right are products A^- and H_3O^+ with some (relatively low) free energy. The drop in free energy drives the acid–base reaction to the right, as in Figure 16-5: Strong acid goes to weak conjugate base, and strong base goes to weak conjugate acid. To call any conjugate species "weak" is, as we have seen, only to recognize it as being more stable thermodynamically—as having a lower free energy—than its parent.

The equilibrium constant for the ionization thus increases according to the strength of acid and base and the *weakness* of the conjugate base and acid. Free energy goes down in the direction strong-to-weak, and

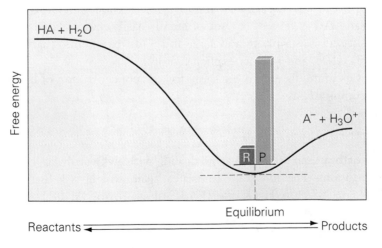

FIGURE 16-5. The weak and the strong, interpreted thermodynamically: The larger the difference in free energy, the greater is the drive toward products. Dissociation of a strong acid in water goes forward with a substantial drop in free energy, marked by a correspondingly large equilibrium constant K_a. The minimum in the curve lies to the right—toward the weak conjugate species A^- and H_3O^+.

the connection between free energy and the equilibrium constant is given at once by the master relationship from Chapter 14:

$$\Delta G° = -RT \ln K_a$$

Everything we need to know is contained herein: enthalpy, entropy, and how much of this and how much of that is present at equilibrium.

First, by using the factor 2.303 to go from a natural logarithm (base $e = 2.718$) to the base-10 form, we obtain an expression

$$\frac{\Delta G°}{2.303\ RT} = -\log K_a \equiv pK_a$$

through which the **pK_a** ("power of K_a") is defined as $-\log K_a$. The value of pK_a subsequently provides a convenient numerical measure of both the magnitude of the equilibrium constant and the associated change in standard free energy. A lower pK_a means a higher K_a. Acids with negative pK_a values are strongest of all.

For example: Acetic acid, with K_a equal to 1.76×10^{-5} at 25°C, has a pK_a of 4.75, whereas the much stronger hydrochloric acid (K_a approximately 10^7) has a pK_a of -7. It is a logarithmic scale, similar to the Richter scale used for earthquakes, so pK_a increases by one unit for every 10-fold decrease in K_a: $-\log 10^{-1} = 1$, $-\log 10^{-2} = 2$, etc. The advantage becomes clear from the wide-ranging tabulation of K_a and pK_a in Table C-20, where we see how the logarithm reduces an enormous span of K_a (and $\Delta G°$) values to a set of small, easily manipulated numbers. Better, it seems, simply to add 1, 2, and 3 (as exponents) than to multiply 10^{-1}, 10^{-2}, and 10^{-3} as explicit factors.

Let us immediately do the same for the concentration of hydronium ion, defining **pH** as

$$pH = -\log [H_3O^+]$$

Henceforth we can use pH to avoid dealing with awkwardly large and small concentrations. The log scale translates again to pH = 1 for $[H_3O^+] = 10^{-1}\ M$; pH = 7 for $[H_3O^+] = 10^{-7}\ M$; pH = 14 for $[H_3O^+] = 10^{-14}\ M$, and so on. By analogy (and for mathematical convenience only), we shall reserve the right to define p(ANYTHING) simply as $-\log$(ANYTHING).

Basic Solutions: pK_b and pOH

So much for water acting as a base. Envision, instead, H_2O in its role as an acid when it loses a proton to become the conjugate base OH^-:

$$H_2O(\ell) + B(aq) \rightleftharpoons OH^-(aq) + BH^+(aq)$$

The equilibrium constant for this reaction, designated K_b, is given by

$$K_b = \frac{[BH^+][OH^-]}{[B]}$$

and is termed the **base dissociation constant** or, equivalently, the **base ionization constant** of species B.

With this definition we now can describe the equilibrium between a base, its conjugate acid, and the hydroxide ion in exactly the same terms used for acidic equilibria. The two pictures are identical, since the fundamental symmetry between acids and bases can never be broken.

We need only interchange acid and base in the previous treatment to obtain the appropriate new description. The doctrine in brief: (1) A basic substance, dissolved in water, causes the generation of hydroxide ions. (2) OH^- subsequently becomes the primary proton acceptor in solution, taking over the role from the original base B. (3) A strong base has a pronounced tendency to accept H^+ from a water molecule, splitting the H_2O into OH^- while producing the conjugate acid BH^+. Similarly to pH, the quantity **pOH** is defined as $-\log [OH^-]$. (4) All three species (BH^+, OH^-, and B) coexist with equilibrium constant K_b and a corresponding **pK_b**. (5) The stronger the base, the greater is the concentration of OH^- and the higher is the value of K_b. The weaker the base, the lower the value of K_b. (6) Strong bases have weak conjugate acids, and weak bases have strong conjugate acids.

Analogously to a strong acid, a strong base produces nearly one mole of OH^- for every mole of B introduced into solution. Sodium hydroxide (NaOH), an ionic compound that dissociates completely into Na^+ and OH^-, will serve as our exemplary strong base, acting as a counterpart to HCl. Ammonia (NH_3), which accepts a proton from water to form the ammonium ion (NH_4^+), provides a contrasting example of a weak base. Its reaction with H_2O (called **hydrolysis**, meaning "breaking down of water") proceeds as

$$NH_3(aq) + H_2O(\ell) \rightleftharpoons NH_4^+(aq) + OH^-(aq)$$

with a K_b of 1.8×10^{-5} at 25°C (pK_b = 4.74), making the extent of this basic equilibrium comparable to that for the ionization of acetic acid.

Autoionization of Water

Water, both a giver and taker of protons, also can react with its own kind to produce hydronium and hydroxide ions in a so-called **autoionization** process. One water molecule plays acid and the other plays base, from which we have the reaction

$$H_2O(\ell) + H_2O(\ell) \rightleftharpoons H_3O^+(aq) + OH^-(aq)$$

together with its equilibrium constant

$$K_w = [H_3O^+][OH^-]$$

We designate K_w synonymously as the **autoionization constant**, **auto-dissociation constant**, or **ion-product constant** for water.

Nothing distinguishes this particular acid–base reaction from those used previously to define K_a and K_b, but the autoionization of water soon emerges as the key link between the two. Water's small ion-product constant, measured as

$$K_w(25°C) = [H_3O^+][OH^-] = 1.0 \times 10^{-14}$$

at 25°C, strictly controls the distribution of hydronium and hydroxide from whatever source these ions may arise.

Because: *How* the H_3O^+ and OH^- ions get into solution is of no importance, since the equilibrium is indifferent to all details of preparation. The equilibrium constant, $K_w(T)$, demands simply that—in the end, however it happens—the product of $[H_3O^+]$ and $[OH^-]$ must always work out to 10^{-14} at 25°C. Strong acids, weak acids; anions, cations . . . it doesn't matter. The final concentrations of hydronium and hydroxide will be governed, without exception, by K_w.

Start with pure water, absent any other acids or bases. Exactly one OH^- is produced for every H_3O^+ generated by autoionization, and hence the concentrations at 25°C must be

$$[H_3O^+] = [OH^-] = 1.0 \times 10^{-7} \ M$$

in order to satisfy the constraint

$$K_w(25°C) = [H_3O^+][OH^-] = (1.0 \times 10^{-7})(1.0 \times 10^{-7}) = 1.0 \times 10^{-14}$$

The pH is 7, rendering water a weak electrolyte with small (and equal) concentrations of hydronium and hydroxide. Pure water is pH-neutral—not acidic, not basic—because neither hydronium nor hydroxide is favored.

That condition changes when external acids and bases are introduced. Additional H_3O^+ and OH^- species are generated beyond those existing in pure water, and the pH increases or decreases accordingly. It goes below 7 in the presence of excess hydronium, corresponding to an *acidic* solution in which $[H_3O^+]$ exceeds the default concentration of $10^{-7} \ M$.

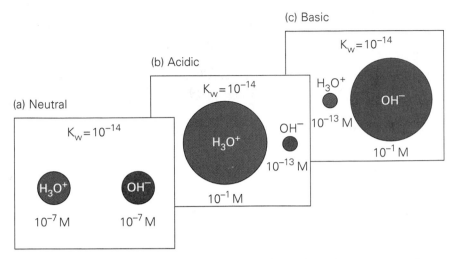

FIGURE 16-6. At equilibrium, the product of $[H_3O^+]$ and $[OH^-]$ must always equal K_w. Any increase in the population of one ion comes only at the expense of the other, as rendered here for three benchmark values of the pH at 25°C: (a) In a neutral solution (pH = 7), neither acid nor base disturbs the balance between $[H_3O^+]$ and $[OH^-]$. The ions arise solely from the self-ionization of water, and the concentrations stand equal at 10^{-7} M. (b) pH = 1. In acidic solution (generally, pH < 7), an acid *stronger than water* gives its protons to H_2O. The transfer of H^+ creates an excess of hydronium over hydroxide ions. (c) pH = 13. In basic solution (pH > 7), a *base* stronger than water takes protons from H_2O and produces a surplus of OH^- ions.

The pH of a basic solution, on the other hand, is higher than 7. Here $[OH^-]$ increases beyond 10^{-7} M while $[H_3O^+]$ falls below 10^{-7} M. But, whatever the mixture shown in Figure 16-6, the concentrations of hydronium and hydroxide always are linked by the equilibrium constant K_w. If $[H_3O^+] = 10^{-1}$ M, for example, then $[OH^-]$ must equal 10^{-13} M to maintain the proper balance at 25°C.

K_b *for a Conjugate Base*

K_w connects the concentrations of hydronium and hydroxide. Even more, it connects the equilibrium constants for an acid and its *conjugate base*. Recall that K_a for an acid is defined as

$$K_a = \frac{[H_3O^+][A^-]}{[HA]}$$

and that K_b for some arbitrary base is similarly defined as

$$K_b = \frac{[BH^+][OH^-]}{[B]}$$

Let us ask, though, that B not be any ordinary base, but rather something special: the base A^- conjugate to the acid HA in the paired reactions

(1) $HA(aq) + H_2O(\ell) \rightleftharpoons A^-(aq) + H_3O^+(aq)$ $K_1 = K_a$

(2) $H_2O(\ell) + A^-(aq) \rightleftharpoons OH^-(aq) + HA(aq)$ $K_2 = K_b$

For the basic hydrolysis (2), we recognize that A^- plays the role of B whereas HA plays the role of BH^+ in the definition of K_b above. We then multiply together K_a and K_b to find

$$K_aK_b = \frac{[H_3O^+][A^-]}{[HA]} \times \frac{[HA][OH^-]}{[A^-]} = [H_3O^+][OH^-]$$

This important result for an acid and its conjugate base,

$$K_aK_b = K_w$$

makes quantitative all the previous arguments concerning strength and weakness. The equilibrium constant K_b for the basic hydrolysis of A^- varies *inversely* with HA's dissociation constant K_a. If one is big, the other is small: strong acid $\longrightarrow$ weak conjugate base.

K_w, which determines the relationship between $[H_3O^+]$ and $[OH^-]$ at equilibrium, is the constant of proportionality between K_a and $1/K_b$. It gives us a numerical benchmark against which we can compare the acidity of HA with the basicity of A^-.

16-4. NEUTRALIZATION

When acid meets base in aqueous solution, there arises one overpowering thermodynamic tendency: to form water. The hydronium ion, with a proton to give, is the strongest acid still existing in the form HA. Hydroxide, lacking a proton, is likewise the strongest base in solution. The two come together as

$$H_3O^+(aq) + OH^-(aq) \rightleftharpoons 2H_2O(\ell)$$

and are impelled forward by an enormous equilibrium constant equal to the *reciprocal* of K_w:

$$K = \frac{1}{[H_3O^+][OH^-]} = \frac{1}{K_w} = 1.0 \times 10^{14} \qquad \text{(at } 25°C\text{)}$$

Indeed this **neutralization** of acid and base is water's autoionization re-action run in reverse, with the thermodynamic impetus going entirely to-ward the formation rather than the dissociation of water. Autoionization is, after all, a rare event with a correspondingly small equilibrium con-stant. It demands that a weak base (water) and weak acid (also water) go uphill to form a stronger conjugate acid and a stronger conjugate base (hydronium and hydroxide ions). These ions do exist, but only to the extent that the product of $[H_3O^+]$ and $[OH^-]$ is limited to 10^{-14}. Any hy-dronium and hydroxide in violation of that quota will combine to form water. The equilibrium lies overwhelmingly in favor of undissociated water.

Strong Acid and Strong Base

We have only a simple problem of stoichiometry to solve whenever a strong acid encounters a strong base—simple because everything reacts completely, sparing us any explicit calculation involving the equilibrium constant. Consider, for instance, an aqueous solution of hydrochloric acid into which a quantity of sodium hydroxide is introduced. The HCl is already dissociated fully into H_3O^+ and Cl^-, and the NaOH is similarly broken up into OH^- and Na^+. Now the neutralization begins. One hy-dronium ion meets one hydroxide ion, and the two form water. A sec-ond H_3O^+ finds a second OH^-, and they too are transformed into H_2O. And there follows a third and a fourth and then a millimole or more, and so it continues. The neutralization (Figure 16-7) proceeds until there is just enough hydronium and hydroxide left to satisfy the normal autoion-ization quota set by K_w, whereupon equilibrium is established.

What happens, meanwhile, to the conjugate species Na^+ and Cl^-, the "salt" of the neutralization? What do they do while hydronium and hy-droxide practice mutual annihilation? They do nothing. Conjugate to strong acids and bases, the Cl^- and Na^+ have little strength themselves and act simply as spectators. A chloride ion, existing as the very weak conju-gate base of a very strong acid, is in no position to recapture the proton that HCl loses so readily. In such an environment, neither Na^+ nor Cl^- can engineer any subsequent generation of hydronium or hydroxide ions.

When strong acid meets strong base, the net reaction is to form water—nothing more. One H_3O^+ picks off one OH^- until the two con-centrations are brought into equilibrium, and then everything holds steady. Any other anions and cations swim around passively while the strong acids and bases come together.

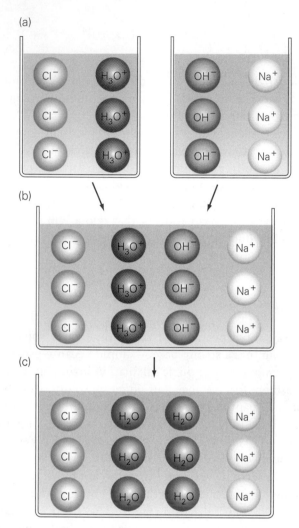

FIGURE 16-7. Mutually assured destruction: neutralization of a strong acid by a strong base. (a) Acid and base ionize completely. (b) One for one, a hydronium ion from the acid finds a hydroxide ion from the base. They meet. They merge. They become water. Each matched pair of H_3O^+ and OH^- ions is transformed into two molecules of H_2O. (c) The conjugate species, long abandoned, look on passively as spectators to the equilibrium maintained by H_2O with small concentrations of H_3O^+ and OH^- (not shown). In the end, all that remains are the hydronium and hydroxide ions produced by autodissociation—equal partners in a neutral solution at 25°C (pH = 7).

To calculate the pH after a strong-acid/strong-base neutralization, we therefore need determine only whether hydronium or hydroxide is in excess. If, say, equal concentrations of HCl and NaOH are mixed, then at equilibrium the concentrations of H_3O^+ and OH^- must also be equal because the ions are neutralized one by one. Neither species gains numerical superiority over the other, and hence the final values at 25°C are $[H_3O^+] = [OH^-] = 10^{-7} \ M$ in accordance with $K_w(T)$. The solution will be pH-neutral (pH = 7), with the hydronium and hydroxide concentrations determined solely by the autoionization of water.

Suppose, though, that there is an imbalance: maybe 0.1 mole of NaOH is added to 1 liter of 0.2 M HCl. Initially the solution contains 0.2 mole of H_3O^+ and 0.1 mole of OH^-, but then the neutralization occurs—one pair of ions at a time. The 0.1 mole of OH^- removes 0.1 mole of H_3O^+, leaving behind 0.1 mole of hydronium in a volume of 1 L. Particle by particle and mole by mole, the process unfolds as in the diagram below. Each sphere represents 0.1 mole of substance:

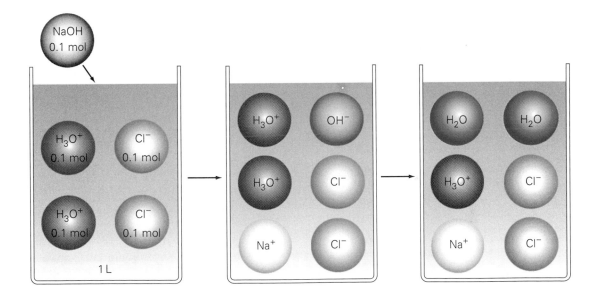

The final concentration is 0.1 M, and the pH of the resulting acidic solution is 1. Equilibrium is maintained with a hydroxide concentration of $10^{-13} \ M$, corresponding to

$$K_w = [H_3O^+][OH^-] = 10^{-1} \times 10^{-13} = 10^{-14}$$

The same final concentrations might arise in any number of ways. A 0.1 M solution of HCl, for example, would have a pH of 1 all by itself. Or we could introduce 0.2 mole of HCl into 1 L of 0.1 M NaOH, adding acid to base rather than base to acid. There is no discernible difference from mixture to mixture, and at equilibrium one knows only that $[H_3O^+] = 0.1 \ M$ and $[OH^-] = 10^{-13} \ M$. In Figure 16-8 there are

PAST   PRESENT

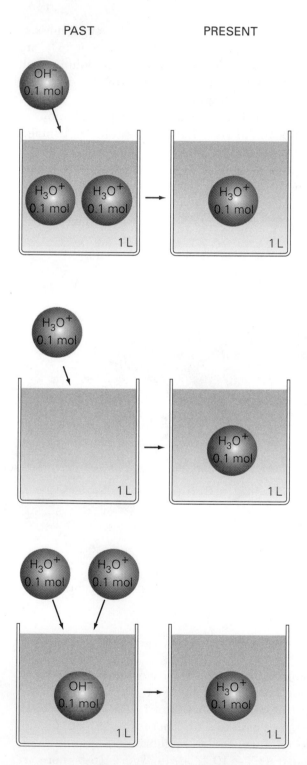

FIGURE 16-8. *How* the hydronium and hydroxide ions came to be is of no importance at equilibrium. Only the present state is relevant, not any past history of preparation. Here, although each solution takes a different route, all three systems arrive finally at the same destination: in equilibrium, with pH equal to 1. The concentrations of H_3O^+ are the same. The concentrations of OH^- are the same. The properties are the same.

merely three solutions having a pH of 1, with nothing to suggest how they got their hydronium ions.

And nothing ever will. Equilibrium always means the end of history, whether for acids, bases, or anything else. It means an unchanging present with no past or future.

Weak Acid and Strong Base

All our reactions, so far, have been between strong acids and bases, each fully dissociated in water to yield hydronium and hydroxide ions. Still, the same stoichiometric considerations apply when *weak* acid meets strong base (or weak base meets strong acid). Again the overpowering tendency is toward the formation of water, and again the huge equilibrium constant associated with that process drives the reaction to completion.

Take acetic acid as an example. Left to itself this weak acid will remain largely undissociated, ionizing just enough to satisfy its K_a of 1.76×10^{-5}. But let a substantial concentration of hydroxide be present (possibly from a strong base like NaOH), and now there will be two competing reactions: (1) the dissociation of the weak acid, and (2) the neutralization of H_3O^+ by the excess OH^-. Thus we have

$$(1) \quad CH_3COOH(aq) + H_2O(\ell) \rightleftharpoons CH_3COO^-(aq) + H_3O^+(aq)$$

$$K_a = 1.76 \times 10^{-5}$$

and

$$(2) \quad H_3O^+(aq) + OH^-(aq) \rightleftharpoons 2H_2O(\ell) \qquad \frac{1}{K_w} = 1.0 \times 10^{14}$$

which, added together, yield a net neutralization reaction

$$CH_3COOH(aq) + OH^-(aq) \rightleftharpoons CH_3COO^-(aq) + H_2O(\ell) \quad K = \frac{K_a}{K_w}$$

pushed strongly to the right.

See how it happens. As with any multistep process, the overall equilibrium constant

$$K = K_a \times \frac{1}{K_w} = \frac{[H_3O^+][CH_3COO^-]}{[CH_3COOH]} \times \frac{1}{[H_3O^+][OH^-]}$$

$$= \frac{[CH_3COO^-]}{[CH_3COOH][OH^-]}$$

$$= 1.8 \times 10^9$$

is equal to the *product* of the K for each step. Despite acetic acid's weak ionization, the combined equilibrium constant (1.8×10^9) is more than enough to drive the neutralization to completion. The tendency to produce water proves irresistible.

Thinking microscopically, as in Figure 16-9, we imagine that some hydroxide ion first finds and neutralizes a hydronium ion produced by the acetic acid. Loss of that hydronium then throws the acid dissociation reaction (K_a) suddenly out of equilibrium, and so one more molecule of CH_3COOH has to give up a proton. But since another hydroxide is still available to wipe out another hydronium, the acetic acid is forced to ionize again. It keeps ionizing and ionizing until it has no more protons to give, for there is always an OH^- present (from the strong base) waiting for another H^+. In the end, a stoichiometric quantity of acetic acid will have been neutralized by the base: one OH^- for each CH_3COOH. For every mole of strong base, one mole of weak acid is converted into its conjugate base.

Unlike the neutralization product of a strong acid like HCl, however, here the conjugate base is relatively strong. The acetate anion (CH_3COO^-) *can* snare a proton and it *can* go back to being CH_3COOH. In contrast to Cl^-, acetate is not reduced to a mere spectator; it participates actively and it will readily react with water to form OH^-. The CH_3COOH molecules are all gone—picked off one at a time by the OH^- ions—but the CH_3COO^-

FIGURE 16-9. Neutralization of a weak acid by a strong base: (a) The weak acid (HA) releases its protons sparingly. The strong base ionizes completely to produce Na^+ and OH^-. (b) The mixed system is flooded with OH^- ions, more than enough to match up against and neutralize the few H_3O^+ ions produced by the acid. For clarity, spectator cations (Na^+) are hereafter represented collectively by just one symbol. (c) Moving to restore equilibrium, the acid replaces the lost hydronium ions. Neutralization continues until not a single molecule of HA remains intact. (d) Reacting with H_2O (represented collectively), the anions A^- arrive at a new equilibrium with the hydrolysis products HA and OH^-. The final mixture is basic.

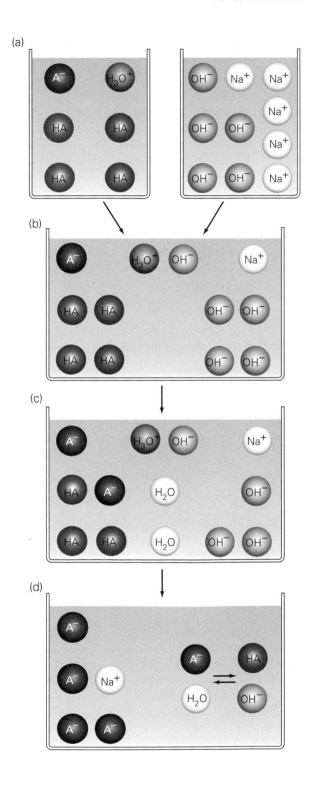

anions remain. The ensuing hydrolysis of the acetate leads to a basic solution, one where the pH is higher than 7 after neutralization.

How basic? How much OH^- is produced when CH_3COO^- reacts with H_2O? For that, straight ahead, we must look more closely into the equilibrium between an acid and its conjugate base.

16-5. WEAK ACIDS, CONJUGATE BASES, AND BUFFERS

At this point we need to understand something about the different environments available to a weak acid. At a minimum, HA might exist as:

1. HA, in equilibrium with small concentrations of H_3O^+ and A^-;
2. its conjugate base A^-, with small concentrations of OH^- and HA; or
3. HA and A^-, with substantial concentrations of both.

Three questions, but really one answer. It is only a matter of applying the mass-action law (equilibrium constant) to three specific cases, and with K_a and K_w we already have all the necessary equipment. Let us consider each mixture in turn.

Weak Acid: Mostly HA

Introduced to water with some initial concentration c, the weak acid ionizes to produce an (as yet) unknown concentration of H_3O^+ and an *equal* concentration of A^-. Let that number be x, so that the concentrations at equilibrium are

$$[HA] = c - x$$

$$[H_3O^+] = x$$

$$[A^-] = x$$

The dissociation reaction is the familiar process

$$HA(aq) + H_2O(\ell) \rightleftharpoons A^-(aq) + H_3O^+(aq)$$
$$c - x x x$$

where the equilibrium constant is represented as

$$K_a = \frac{[H_3O^+][A^-]}{[HA]} = \frac{x^2}{c - x}$$

The unknown quantity x is determined directly from this expression (which can always be solved as a quadratic equation), and thence are obtained all the concentrations at equilibrium.

Suppose that acetic acid is present initially at $c = 0.100$ M. Simpler than it seems, the equation to be solved,

$$K_a = \frac{x^2}{0.100 - x} = 1.76 \times 10^{-5}$$

lends itself to the kind of approximation we first used in Chapter 12. Since K_a is so small, x will certainly turn out to be small as well (presumably much less than the 0.100 in the denominator). Let us assume, then, that the quantity $0.100 - x$ is almost equal to 0.100, so as to obtain the even simpler equation

$$x^2 = 0.100 \cdot K_a = 1.76 \times 10^{-6}$$

It is an approximation, admittedly, but ultimately this simplification is justified because the estimate leads to $x = [H_3O^+] = [CH_3COO^-] = 0.00133$. The concentration of dissociated species is nearly 100 times smaller than the initial concentration of 0.100 M, and so the work is done. More sophisticated algebraic methods are not needed under these conditions.

The equilibrium concentration of both acetate and hydronium ion is thus 1.33×10^{-3} M, with a corresponding pH of 2.88. And although only 1.33% of the acid is dissociated, the weak acid has significantly shifted the balance of hydronium and hydroxide nonetheless. Acetic acid may be weak compared with something like hydrochloric acid, but still it is considerably stronger than H_2O. One liter of this acetic acid solution contains fully 1.33×10^{-3} mole of H_3O^+ and just 7.52×10^{-12} mole of OH^- (as demanded by K_w).

Compare that with neutral water, where the concentration of both hydronium and hydroxide is only 10^{-7} M. So great is the increase in $[H_3O^+]$—by a factor greater than 13,000—that acetic acid becomes the only real source of hydronium ion. The small amount of H_3O^+ contributed by the autoionization of water is swamped by a veritable flood of ions coming from the acid.

Conjugate Base: Mostly A^-

In this second mixture there is initially no undissociated acid in solution, but rather only the conjugate base A^-. How the A^- came to exist is, as always, unimportant. Perhaps an equivalent quantity of HA was neutralized by a strong base; perhaps somebody happened to prepare a salt solution of sodium acetate from a bottle off the shelf. Not important—our concern is only with the basic hydrolysis

$$A^-(aq) + H_2O(\ell) \rightleftharpoons HA(aq) + OH^-(aq)$$

$$c - x \qquad\qquad\qquad x \qquad\quad x$$

and how the concentrations stand at equilibrium.

For consistency with the example above, let the initial concentration of acetate be 0.100 M. The equilibrium constant is

$$K_b = \frac{[HA][OH^-]}{[A^-]} = \frac{K_w}{K_a} = \frac{1.0 \times 10^{-14}}{1.76 \times 10^{-5}} = 5.7 \times 10^{-10}$$

which is actually not as small as it first appears. Small in some absolute sense, yes, but still this K_b is a number 50,000 times larger than K_w. As the conjugate base of a weak acid, acetate is strong enough to compete successfully with water for protons. It pulls them off the H_2O molecules and thereby creates an excess of hydroxide ions relative to pure water.

The equilibrium concentrations are obtained by exactly the same procedure used for the acid dissociation, which in this instance yields the simplified equation

$$x^2 = 0.100 \cdot K_b = 5.7 \times 10^{-11}$$

Hydrolysis of the conjugate base therefore generates an OH^- concentration of 7.5×10^{-6} M (75 times more than from autoionization alone), balanced by $[H_3O^+] = 1.3 \times 10^{-9}$ M to maintain the K_w equilibrium. The pH is 8.88; it is a weakly basic solution.*

To sum up: A^- is a base. Existing alone in aqueous solution and competing only with H_2O, acetate gets its share of protons and manufactures a corresponding excess of hydroxide. No other species is present to alter the balance. Everything goes as expected.

---

*Sharp-eyed readers take note: The calculation above uses the value $K_b = (1.0 \times 10^{-14})/(1.76 \times 10^{-5}) = 5.68 \times 10^{-10}$, without initial rounding, to obtain $x = 7.54 \times 10^{-6}$. The final result is limited to two significant figures by the accuracy of $K_w = 1.0 \times 10^{-14}$.

Buffers

Having visited the two poles (HA at one extreme, A^- at the other), we move to the center: to a solution containing sizable concentrations of both weak acid and conjugate base. The mixture might have been prepared by neutralizing some fraction of the weak acid with a strong base. Alternatively it might have arisen from direct mixing of the acid and its salt (equal concentrations of acetic acid and sodium acetate, for instance). Either way will yield the same result at equilibrium.

Yet before coming to the center, we must fully appreciate the precarious balance existing at either end. There, with *only* acid or *only* conjugate base, the equilibrium is exquisitely sensitive to the addition of H_3O^+ and OH^-. Observe what happens when even a bit of hydrochloric acid is introduced to the 0.100 *M* aqueous solution of acetic acid analyzed above. Before any disturbance, the acetic acid dissociates to produce a mildly acidic solution with $[H_3O^+] = 0.00133$ *M* and pH = 2.88. Approximately 1% of the weak acid is ionized, and there is no other major source of hydronium ion. But now introduce just enough strong acid to create another 0.1 mole of H_3O^+ per liter. It might take only a drop of concentrated HCl to do so, but that drop instantly changes the complexion of the solution. The concentration of hydronium ion abruptly increases by two orders of magnitude, going from roughly 0.001 *M* all the way to 0.1 *M*:

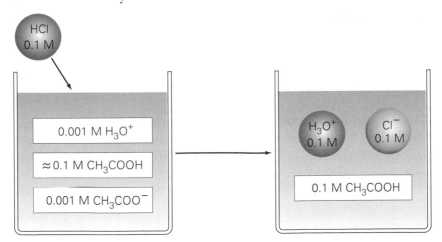

The hydrochloric acid, fully dissociated, becomes the dominant source of H_3O^+, and the delicate equilibrium between acetic acid, acetate, and hydronium ion is dramatically shifted. The pH drops to 1.

Forget the acetic acid. This new mixture is effectively an aqueous solution of a strong acid. The weak acid, forced to cope with 100 times more H_3O^+ in the mixture, has suffered a massive insult to its equilibrium and must readjust to the new conditions. There is far too much H_3O^+ in solution, and consequently most of the acetate anions are forced to take back their protons. The dissociation of acetic acid, never extensive by any standard, is suppressed even further by this sudden excess of hydronium. The ionization is reversed accordingly, driven backward as

$$CH_3COOH(aq) + H_2O(\ell) \rightleftharpoons CH_3COO^-(aq) + H_3O^+(aq)$$
$$\text{(HA)} \qquad\qquad\qquad \text{(A}^-\text{)}$$

just as Le Châtelier would predict (according to the common-ion effect, Chapter 15). Acetic acid's production of H_3O^+ decreases greatly at the new equilibrium, and the acidity of the solution is then controlled entirely by the excess hydrochloric acid.

At the other extreme (all conjugate base), a solution of pure acetate reacts similarly if excess strong base disturbs the equilibrium. The pH shoots up when NaOH is added, after which the preexisting weak base becomes irrelevant. Production of OH^- by the weak base is suppressed owing to the extra hydroxide, and the hydrolysis reaction is pushed far to the left:

$$CH_3COO^-(aq) + H_2O(\ell) \rightleftharpoons CH_3COOH(aq) + OH^-(aq)$$
$$\text{(A}^-\text{)} \qquad\qquad\qquad \text{(HA)}$$

With the acetate no longer an important source of OH^-, the solution behaves simply as a strong base in water.

Now contrast these delicate states with the more robust equilibrium attained in a **buffer solution** (Figure 16-10), where substantial amounts of both HA and A^- are found. The *joint* presence of acid and conjugate base provides an effective defense against stray hydronium and hydroxide ions. HA goes after the OH^-, and A^- goes after the H_3O^+, as we shall discover presently.

Suppose that a strong acid releases some quantity of protons (H^+) into the buffer solution. We know that a pure weak acid, having no means to counter the resulting H_3O^+ ions, would be vulnerable to even the slightest perturbation. The buffer solution, however, is guarded by the conjugate base, which captures the solvated protons and uses them to manufacture more weak acid:

$$A^-(aq) + H_3O^+(aq) \rightleftharpoons HA(aq) + H_2O(\ell)$$

The intruding protons, now made part of the weak acid, are effectively removed from solution.

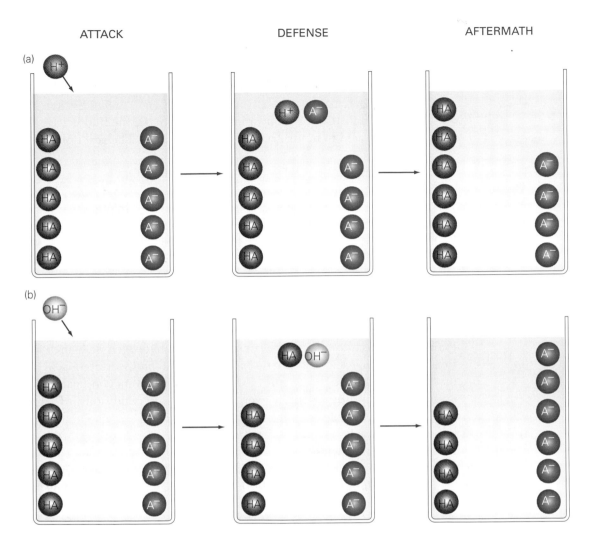

FIGURE 16-10. Mutual defense: a buffer solution. Prepared in roughly equal (and generous) concentrations, an acid and its conjugate base combine to maintain a steady pH. (a) Reserves of conjugate base neutralize extra hydronium ions attempting to invade, converting one anion (A^-) into its parent acid (HA) during each encounter. Note that hydronium ions (H_3O^+) are represented here simply as H^+, and a molecule of H_2O produced in the reaction is omitted as well. (b) Acting in analogous fashion, reserves of acid remove unwanted OH^-.

The equilibrium is strained ever so slightly by these newly created HA species, because suddenly there is too much undissociated weak acid and not enough H_3O^+. Since the reaction quotient is less than K_a, some of the newly manufactured HA must dissociate to bring the concentrations back into equilibrium. But the acid is weak to begin with (it never ionizes much anyway), and so the additional dissociation needed to reestablish equilibrium is modest. Only a handful of protons are given back, and the pH remains nearly unchanged. The buffer solution, we see, can fight off the excess acid, in stark contrast to a pure weak acid unprotected by a supply of conjugate base.

Confronted instead by an invasion of strong *base*, the buffer uses its supply of weak acid to neutralize the incoming OH^-. The reaction

$$HA(aq) + OH^-(aq) \rightleftharpoons A^-(aq) + H_2O(\ell)$$

produces additional conjugate base, A^-, followed by a minimal readjustment of the equilibrium in the manner just described. Most of the excess hydroxide ion is converted harmlessly into water, and so a large increase in pH is averted.

Buffering is a defense by the weak against the strong, and the effort succeeds precisely because the defenders are weak acids and bases working together. Their relative weakness allows them to coexist in solution, each protecting the other. Not so for a strong acid, which is completely ionized and therefore helpless against additional hydrogen or hydroxide ions. The strong acid has no reserve supply of undissociated acid to divert incoming OH^- away from existing H_3O^+, and its conjugate base is unable to capture a proton even under normal conditions. H^+ and OH^- alike can be dumped into the water, unchallenged. The buffer solution, by contrast, maintains a balanced supply of HA and A^- to deal with such contingencies.

The whole picture comes together into one compact and useful expression, the **Henderson–Hasselbalch equation**, which we obtain by recasting the equilibrium constant

$$K_a = \frac{[H_3O^+][A^-]}{[HA]}$$

in the form $-\log K_a$:

$$-\log K_a = -\log [H_3O^+] - \log \frac{[A^-]}{[HA]}$$

or

$$pH = pK_a + \log \frac{[A^-]}{[HA]}$$

Equilibrium pH is thereby connected directly to the concentrations of undissociated acid and conjugate base (salt), and both these latter quantities can be estimated from the initial stoichiometry alone. No further calculation of [HA] and [A$^-$] is necessary, provided that both acid and base are present in substantial quantity.

One species holds back the other. Were HA to dissociate, it would need to force another A$^-$ into an already large pool of the same species. But Le Châtelier's principle guards against just such additional stress, and consequently the ionization is held in check. If, in the same way, A$^-$ were to hydrolyze, then it would add another HA to a similarly large pool. Le Châtelier again says no, and a stalemate ensues: *Both* dissociation and hydrolysis are inhibited, and so [HA] and [A$^-$] stay where they are. It becomes the role of H$_3$O$^+$, manifested through the pH, to determine the balance needed to satisfy K_a.

We see quickly that pH is controlled entirely by the ratio of conjugate base to weak acid, not by the individual concentration of either species. Since only the quotient [A$^-$]/[HA] influences the pH, a buffer thus provides additional protection against changes in the solution's *volume*. If, for instance, the solution is diluted 10-fold, the unchanging ratio

$$\frac{[\text{A}^-]/10}{[\text{HA}]/10} = \frac{[\text{A}^-]}{[\text{HA}]}$$

is sufficient to hold [H$_3$O$^+$] steady. No such protection is available to a strong acid, however, where a 10-fold increase in volume is matched by a 10-fold decrease in [H$_3$O$^+$]. Its unbuffered pH, directly sensitive to volume and concentration, rapidly increases by one unit (going from 1 to 2, for example, as [H$_3$O$^+$] drops from 10^{-1} M to 10^{-2} M).

Buffers work best, too, with identical concentrations of acid and conjugate base, for which [HA]/[A$^-$] = 1 and hence

$$\text{pH} = \text{p}K_a + \log 1 = \text{p}K_a$$

Such a mixture, possessing an equally strong capacity to defend against either H$^+$ or OH$^-$, is the most pH-resistant buffer attainable. The pH departs only slightly from the central value (pH = pK_a) when small quantities of either acid or base are added.

If so, what is the pH of a 1.00-L solution containing 0.050 mole of acetic acid and 0.050 mole of sodium acetate? What happens when 0.001 mole of OH$^-$ is added?

Concerning the first question, we know at once that pH becomes identical to pK_a for any equimolar mixture of acid and conjugate base (provided that the concentrations are not excessively low). For our 50:50 acetate buffer, then, the pH is automatically equal to 4.75—the pK_a of acetic acid.

Addition of 0.001 mol OH^- brings about a decrease in [HA] from 0.050 M to 0.049 M, accompanied by a corresponding increase in $[A^-]$ to 0.051 M. There follows a modest repositioning of the equilibrium, which is approximated by the equation

$$pH = pK_a + \log \frac{[A^-]}{[HA]}$$

$$= 4.75 + \log \frac{0.051}{0.049} = 4.77$$

The buffer works: Whereas a 0.001 M solution of OH^- ordinarily has a pH of 11, the buffering action restrains it to 4.77.

How about a 10:1 buffer containing 10 times as much salt as acid? Here, with $\log([A^-]/[HA]) = 1$, the pH is centered about the value $pK_a + 1$. It is a more basic solution, entirely consistent with the higher population of the conjugate base A^-.

And a 1:10 buffer with the acid dominant? Now we have $\log([A^-]/[HA]) = -1$, and the equilibrium moves toward a more acidic solution: $pH = pK_a - 1$.

It is a good place to stop. Spanning mixtures from 10:1 to 1:10, the range

$$pH = pK_a \pm 1$$

roughly delineates the working of an effective buffer. Too little acid, and there is insufficient protection against OH^-. Too little base, and there is insufficient protection against H^+. Best of all is the midrange ([HA] = $[A^-]$, with $pH = pK_a$), where the buffer has equal reserves and where the salt-to-acid ratio varies least.

Yet no buffer is perfect, and even the most resilient of buffers can be overwhelmed by too great a shock. Forced to capture an excessive amount of stray H^+ or OH^-, any buffer will fail once its stock of A^- or HA is depleted. The system operates only within certain limits. It cannot be pushed forever.

Maintenance of a steady pH, even within these limits, is still no small achievement. Hydrogen ions insert themselves into all kinds of processes, and success or failure often depends on having just the right concentration. Acting as catalysts, movable protons can determine how fast a reaction proceeds or whether it even gets started at all. Of especial importance, of course, are metabolic systems, where an improper balance of protons might lead to malfunction, poisoning, or death. Biochemical mechanisms to regulate the concentration of hydronium ions have evolved accordingly, and a notable example is provided by the H_2CO_3/HCO_3^- buffer that maintains a pH of 7.4 in the blood.

16-6. TITRATION

The varied interplay of acids and bases—the weak and the strong, hydronium and hydroxide, the shifting equilibria—can be summarized in a single graph: a titration curve.

A chemist performs a **titration** (Figure 16-11) by carrying out a chemical reaction slowly and deliberately, pitting one reactant systematically against another. Drop by drop, millimole by millimole, a known amount of reactant B is added to an *unknown* amount of reactant A. The reaction, monitored carefully along the way, proceeds molecule by molecule until all of reactant A is consumed by some measured quantity of

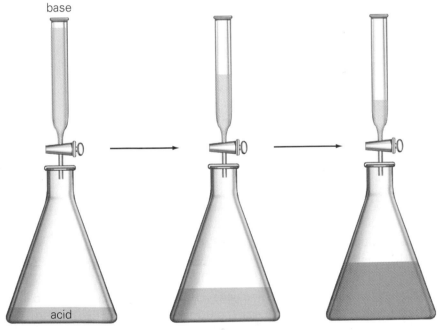

FIGURE 16-11. Titration: a carefully executed, controlled reaction between acid and base—or, more generally, between reactants A and B of any kind. In this example, a basic solution of known concentration is delivered incrementally to an acidic solution of known volume but unknown concentration. Measuring the volume of base consumed, an analyst then determines the concentration of acid. Either a pH meter or color-changing chemical indicator is used to monitor the extent of reaction.

B. An observer then knows, from the balanced equation, exactly how much of A reacted with that much of B, and thence is revealed the original number of moles and concentration of A. It is an analytical method of general significance, used not only for acid–base reactions but for reduction–oxidation and other processes as well.

Acid–base titrations can be followed in a number of ways, both by electronic instruments and by various telltale species called *indicators*. An indicator, itself a weak acid, has some measurable property (often color) that depends on its protonation state. Even something as mundane as cabbage extract can do the job, appearing red in acidic solution and blue in a basic environment. The color, which varies according to whether the acidic proton is on or off the indicator, provides a rough measure of pH. Many other organic dyes behave the same way, each displaying a characteristic color at a particular pH.

Imagine now that we have a solution of acetic acid, fitted out either with electronic pH meter or a small quantity of suitable indicator. Initially there is only the weak acid, its conjugate base, and the hydronium ions all balanced in equilibrium. This equilibrium will determine the concentration of hydronium ions and corresponding pH (which, for the example treated in Section 16-5, we know is equal to 2.88 for a 0.100 M solution of CH_3COOH).

Let that 0.100 M acidic solution be the starting point for a titration using NaOH, the results of which are shown in Figure 16-12. Into the acetic acid we inject small doses of standard base, introducing so many millimoles of OH^- with every drop. Acetic acid molecules and hydroxide ions meet one by one, and from each neutralization reaction comes one more A^- and one fewer HA species.

What are we doing? We are simply making mixtures of acetic acid and sodium acetate, gradually converting acid into conjugate base. The original acidic solution is evolving into a buffer, and its pH (described approximately by the Henderson-Hasselbalch equation) increases along with the salt-to-acid ratio, $[A^-]/[HA]$.

More and more NaOH leads to more acetate and less acid. There comes a point—at the center of the buffer region—where half the acid is neutralized and pH is therefore equal to pK_a. The titration then passes through increasingly basic buffer solutions and eventually reaches the *equivalence point*, whereupon all the acid has been finally converted into conjugate base. We know how much NaOH was used, and from that we know how many moles of CH_3COOH originally existed in the pure solution. If, for example, 0.0100 mole of NaOH was needed to neutralize a 0.100-L solution of acetic acid, then the initial concentration of the monoprotic acid must have been 0.100 M.

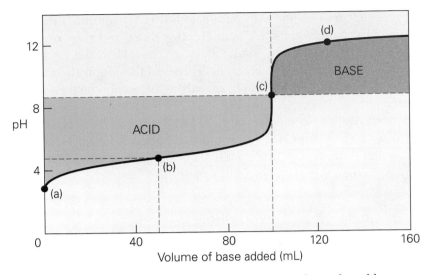

FIGURE 16-12. Titration curve for the neutralization of a weak acid by a strong base: the reaction of 100 mL of 0.100 M CH$_3$COOH with 0.100 M NaOH. (a) Beginning as pure acid, the system passes through a buffer region while becoming progressively more basic. (b) At the point where pH = pK_a, exactly half the acid has been converted into conjugate base. (c) At the *equivalence point*, neutralization is complete. Hydrolysis of the conjugate base yields a basic solution. (d) Past the equivalence point, the solution behaves like a strong base.

Thus from start to finish—from acid to buffer to conjugate base—the titration brings about merely a succession of mixtures, and this same general pattern also holds for *any* weak acid titrated by *any* strong base:

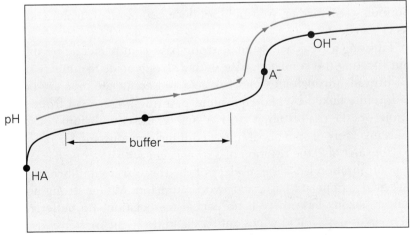

The generic curve begins with pure acid (all HA), continues through a relatively flat buffer region (A⁻ and HA together), and then rises more steeply to the equivalence point (all A⁻). The pH at the equivalence point itself is determined solely by hydrolysis of the conjugate base, after which the solution behaves simply as a strong base upon further addition of OH⁻.

Alternatively we could start with A⁻, the conjugate base. Using strong acid to convert A⁻ into HA, we would then follow an analogous titration curve—this time, though, tracing the pH *down* from the upper left (a basic solution) to the lower right (an acidic solution):

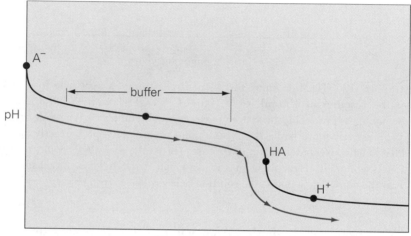

The system passes through a buffer region as A⁻ is converted into HA and eventually reaches a new equivalence point with pH less than 7: all HA. Continuing still further, we would produce a solution of strong acid as excess H⁺ is added.

Titrating a weak acid with a strong base is in principle no different from titrating the conjugate base with a strong acid. Symmetry of concept prevails throughout, for weak acid and conjugate base are but two fixed points linked by two equivalent pathways. The road from HA to A⁻ mirrors the road from A⁻ to HA, just as the equilibrium constant K_a stands inversely proportional to the equilibrium constant K_b.

Consider also the titration of a *strong* acid with a strong base, where the neutralization lacks the subtlety of the three-way equilibrium among weak acid, conjugate base, and hydronium ion. When strong acid encounters strong base there is no partial dissociation, no buffer region, and no hydrolysis. Acid wipes out base and there the matter ends, without further interplay among the conjugate species and water. The titration curve sketched in Figure 16-13, characterized by an unbuffered and

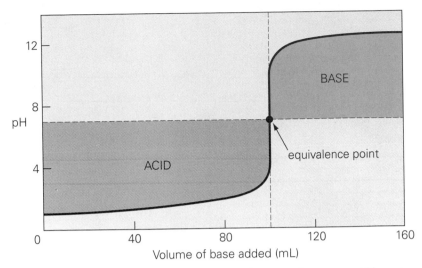

FIGURE 16-13. Titration curve for the neutralization of a strong acid by a strong base. Unbuffered, the pH rises steeply and suddenly near the equivalence point as the mixture goes from acid to base. Its value at the equivalence point is always 7 for any mixture of strong acids and bases at 25°C—as here, where 0.100 M NaOH is added to 100 mL of 0.100 M HCl.

consequently rapid change in pH, demonstrates this uncomplicated neutralization at 25°C. So uncomplicated, indeed, that one strong acid–base neutralization is just like another. The equivalence point is forever predetermined to occur at a pH of 7, leaving the salt solution neither acidic nor basic. Always it is the same, because strong acids and strong bases are all alike.

REVIEW AND GUIDE TO PROBLEMS

There are no individual actors in a chemical process; only partners, always partners: a taker for every giver, a gain for every loss, a backward for every forward. No atom, molecule, ion, or electron ever acts alone, unnoticed, without consequence to others.

When one speeds up, another slows down. If a system becomes cooler, its surroundings become hotter. What one molecule loses, another one gains. Always a pairing: solvent for solute, reductant for oxidant, acid for base.

And eventually, obeying the same laws of thermodynamics, the givers and takers all come to equilibrium: solutes and solvents (Chapter 15), reductants and oxidants (Chapter 17), and now (Chapter 16) acids and bases.

IN BRIEF: ACID–BASE EQUILIBRIA

1. GIVERS AND TAKERS AND TAKERS AND GIVERS. The article of exchange is a movable proton, a hydrogen ion (H^+), passed between Brønsted-Lowry acid and Brønsted-Lowry base. The *acid* (HA) loses a proton to become A^-, its *conjugate base*. The *base* (B) gains the proton to become BH^+, its *conjugate acid*:

$$HA + B \rightleftharpoons A^- + BH^+$$

$$\text{acid} \quad \text{base} \quad \begin{array}{c} \text{conjugate} \\ \text{base} \end{array} \quad \begin{array}{c} \text{conjugate} \\ \text{acid} \end{array}$$

The acid gives. The base takes. Then they turn the process around, with the conjugate acid (BH^+) giving the proton back to the conjugate base (A^-). On again, off again, the hydrogen ion goes back and forth until there is equilibrium. Acid and base give and take and take and give until the reverse transformation

$$A^- + BH^+ \longrightarrow HA + B$$

proceeds neither faster nor slower than the forward transformation

$$HA + B \longrightarrow A^- + BH^+$$

Stalemated, the reaction is over to any macroscopic observer. The concentrations of HA, B, A^-, and BH^+ remain unchanged thereafter.

2. WHO WINS? Ask, rather, is the equilibrium constant

$$K = \frac{[A^-][BH^+]}{[HA][B]}$$

greater than 1 or less than 1? Are there relatively more products (dissociated acid, A^-; protonated base, BH^+) or reactants (undissociated acid, HA; unprotonated base, B)? Is the standard free energy of the products less than or more than the standard free energy of the reactants? Ask, in other words, the same kind of questions we would ask about any reaction.

Such as: Is HA a strong acid? Does it give its proton to B easily and spontaneously? If so, then its conjugate base (A^-) will take back the H^+ only with considerable difficulty. Easy to lose; hard to regain. The species conjugate to a strong acid is a weak base.

And the other way: Is B a strong base? Does it take the proton from HA easily and spontaneously? If so, then its conjugate acid (BH^+) will be poorly equipped to return the H^+. Easy to gain; hard to lose. The species conjugate to a strong base is a weak acid.

Strong goes to weak. High free energy goes to low free energy. Reactants go to products.

3. STRENGTH AND WEAKNESS. A strong acid is strong precisely because it leaves behind a weak conjugate base. It leaves behind a base (A^-) that shows little tendency to recapture H^+. It leaves behind a base that is thermodynamically stable, a base that is sufficiently low in energy to stay just as it is.

Expect, accordingly, the conjugate base of a strong acid to support itself independently and resist reprotonation. One way or another, the H—A bond will be easier to break than to make; and one way or another, the conjugate base, A^-, will have a lower free energy than the undissociated acid, HA. A^- might benefit, for example, from electronegative groups that withdraw negative charge from the broken bond (as the chlorine atoms do in CCl_3COOH). Or, also possible, the structure might offer a way to disperse the excess charge throughout a system of delocalized orbitals, as in a benzene ring or conjugated chain. Or A^- might be large enough to endure the increased electron–electron repulsions it must suffer . . . or, just the opposite, small enough to attract a larger share of energy-lowering solvent molecules . . . or something else, but hardly ever will there be a simple, single, unambiguous answer. Many factors come into play.

For a base, the same thing: A strong base is strong precisely because it turns itself into a weak conjugate acid, an acid that shows little thermodynamic tendency to give back the proton it got. The BH^+ bond is easier to make than to break. The rest of the molecule is able, somehow, to pump electrons toward the attached H^+, thus reducing the deficiency of negative charge at that site.

4. WATER. Able to go both ways, water is simultaneously acid and base. The molecule pulls itself into hydronium ions (H_3O^+) and hydroxide ions (OH^-) in an *autoionization* reaction

$$H_2O(\ell) + H_2O(\ell) \rightleftharpoons OH^-(aq) + H_3O^+(aq)$$

| acid | base | conjugate base | conjugate acid |

controlled by the equilibrium constant

$$K_w(25°C) = [H_3O^+][OH^-] = 1.0 \times 10^{-14}$$

at 25°C. Equal (and small) concentrations of H_3O^+ and OH^- are produced in the pure liquid as a result:

$$[H_3O^+] = [OH^-] = 1.0 \times 10^{-7} \, M \qquad \text{(at 25°C)}$$

K_w, usually called either the *autoionization constant* or *ion-product constant* for water, then determines the fate of all other acids and bases in competition with H_2O. This one number, understand, regulates the amount of H_3O^+ and OH^- ions that can exist together in the same aqueous solution—regardless of source, regardless of history, regardless of other species. K_w demands that the product of $[H_3O^+]$ and $[OH^-]$ be fixed at equilibrium, whether the ions come from water reacting with itself or with any other acid or base.

5. ACIDS IN WATER. The acid (HA) teams with the base (H_2O) to produce a conjugate base (A^-) and conjugate acid (H_3O^+), giving us a reaction

$$HA(aq) + H_2O(\ell) \rightleftharpoons A^-(aq) + H_3O^+(aq)$$

governed by a certain *acid ionization constant* (*acid dissociation constant*), K_a:

$$K_a = \frac{[H_3O^+][A^-]}{[HA]}$$

HA is an *Arrhenius acid*: an acid that is stronger than water; an acid that, in aqueous solution, produces a concentration of H_3O^+ greater than otherwise exists in the pure liquid; an acid that can force its proton onto the H_2O molecule, which must play the base in this first give-and-take. After that, with HA converted into A^-, the hydronium ion stands alone as the only acid with a proton still to give. The species H_3O^+, the acid conjugate to H_2O, becomes the principal proton donor in aqueous solution.

This interplay of ions is measured, in part, by the quantities *pH* and *pK_a*, defined as

$$pH = -\log [H_3O^+]$$

and

$$pK_a = -\log K_a$$

The stronger the acid, the lower the pK_a.

Note, in particular, that the pH of pure water at 25°C is 7, corresponding to a neutral solution in which neither H_3O^+ nor OH^- gains an advantage:

$$[H_3O^+] = [OH^-] = 1.0 \times 10^{-7} \ M \qquad \text{neutral: } pH = 7$$

The pH of an acidic solution, by contrast, is less than 7:

$$[H_3O^+] > 1.0 \times 10^{-7} \ M \qquad \text{acidic: } pH < 7$$

There are more hydronium ions than hydroxide ions.

6. BASES IN WATER. Same idea as an acid in water, but with names and roles reversed. Here an *Arrhenius base* emerges as a stronger proton *acceptor* than H_2O, able to rob water of H^+ and liberate OH^- in the reaction

$$H_2O(\ell) + B(aq) \rightleftharpoons OH^-(aq) + BH^+(aq)$$

The *base ionization constant* (*base dissociation constant*) then follows as

$$K_b = \frac{[BH^+][OH^-]}{[B]}$$

and the related quantities *pOH* and *pK_b* are defined as

$$pOH = -\log [OH^-]$$

$$pK_b = -\log K_b$$

Bear in mind the connection between pH and pOH at 25°C,

$$pH + pOH = 14.00$$

a relationship that allows us to express pOH easily and equivalently in terms of pH.

The result: A basic solution has a pH greater than 7, corresponding to a hydroxide concentration greater than 10^{-7} M (and, since $K_w = 10^{-14}$ at 25°C, a hydronium concentration less than 10^{-7} M):

$$[OH^-] > 1.0 \times 10^{-7} \, M \qquad \text{basic: } pH > 7$$

And another result: For an acid and its conjugate base, the equilibrium constants K_a and K_b are connected by the autoionization constant for water:

$$K_a K_b = K_w$$

$$pK_a + pK_b = 14.00 \qquad \text{(at 25°C)}$$

To see why, we simply write the two reactions as

(1) $HA(aq) + H_2O(\ell) \rightleftharpoons A^-(aq) + H_3O^+(aq) \qquad K_1 = K_a$

(2) $H_2O(\ell) + A^-(aq) \rightleftharpoons OH^-(aq) + HA(aq) \qquad K_2 = K_b$

and (inserting the definitions of K_a, K_b, and K_w) we observe that

$$K_a K_b = \frac{[H_3O^+][A^-]}{[HA]} \times \frac{[HA][OH^-]}{[A^-]} = [H_3O^+][OH^-] = K_w$$

7. THEME AND VARIATIONS. Strong acids and strong bases; strong acids and weak bases; weak acids and strong bases; weak acids and weak bases; acids that release one proton, two protons, three; buffered mixtures of weak acids and conjugate bases—so many combinations, so much richness and complexity, so many variations. All, however, conform to the same rules of strength and weakness. All must satisfy the constraints imposed at equilibrium. All, in the end, are really the same. Consider

just two of the possibilities before we begin to solve problems: neutralization and buffering.

Simple and direct is a **neutralization** reaction between an Arrhenius acid and an Arrhenius base, leading immediately to the production of water:

$$H_3O^+(aq) + OH^-(aq) \longrightarrow 2H_2O(\ell) \qquad K = \frac{1}{K_w} = 1.0 \times 10^{14}$$

The reaction, supported by a large equilibrium constant, goes overwhelmingly to the right. It proceeds stoichiometrically, with the limiting reactant consumed entirely.

In a **buffer solution**, by contrast, an acid and its conjugate base work as a team to maintain a steady pH. The acid (HA) scavenges any invading OH^- ions, producing A^- and H_2O in the process. At the same time, the conjugate base (A^-) reacts with intruding H_3O^+ ions to generate HA and H_2O. If these stray hydronium and hydroxide ions are relatively few, and if [HA] and [A^-] are substantial and of comparable magnitude, then the pH stays nearly constant:

$$pH = pK_a + \log \frac{[A^-]}{[HA]}$$

The approximate expression so obtained, the **Henderson-Hasselbalch equation**, follows directly from the logarithm of the ionization constant and the definitions of pH and pK_a.

SAMPLE PROBLEMS

EXAMPLE 16-1. Relative Acidity

PROBLEM: Which is the stronger Brønsted-Lowry acid in each pair? (a) H_2O or H_3O^+. (b) H_2O or OH^-. (c) HIO or HIO_3. (d) $CH_2ClCOOH$ or $CHCl_2COOH$. (e) HCl or HF. (f) HCl or NaH. (g) H_2SO_4 or HSO_4^-.

SOLUTION: Look for the molecule or ion that will produce the most stable conjugate base upon losing H^+. Look, too, for the species in each pair that can force its proton onto the other in direct competition. That one—the giver—is the stronger acid of the two.

(a) *Water or hydronium ion.* H_3O^+ gives to H_2O, leaving behind a stable water molecule as its conjugate base. H_3O^+ is thus the stronger acid. Were H_2O to protonate H_3O^+, it would have to force an additional H^+ onto an already charged hydronium ion.

(b) *Water or hydroxide ion*. Different partner, different role. This time, H_2O is the stronger acid. It gives a proton to OH^-.

(c) *Hypoiodous acid or iodic acid*. HIO_3, with its three electronegative oxygens instead of one, is better able than HIO to spread out the charge of the conjugate base:

$$H-\overset{\overset{\ddot{O}:}{|}}{\underset{..}{O}}-\overset{..}{\underset{..}{I}}-\overset{..}{\underset{..}{O}}: \longrightarrow H^+ + \left[:\overset{..}{\underset{..}{O}}-\overset{\overset{:\ddot{O}:}{|}}{\underset{..}{I}}-\overset{..}{\underset{..}{O}}: \right]^-$$

iodic acid

$$H-\overset{..}{\underset{..}{O}}-\overset{..}{\underset{..}{I}}: \longrightarrow H^+ + \left[:\overset{..}{\underset{..}{O}}-\overset{..}{\underset{..}{I}}: \right]^-$$

hypoiodous acid

(d) *Chloroacetic acid or dichloroacetic acid*. The more electron-with-drawing groups available to stabilize the anion, the stronger is the acid. $CH_2ClCOOH$ has one chlorine. $CHCl_2COOH$ has two. Dichloroacetic acid is the stronger acid.

(e) *Hydrochloric acid or hydrofluoric acid*. An exception to the rule concerning electronegativity: Fluorine is more electronegative than chlorine, but HCl is the stronger acid. First, the H—F bond is tighter and harder to break than the H—Cl bond. Second, the F^- anion is smaller and consequently higher in potential energy than Cl^-.

(f) *Hydrogen chloride or sodium hydride*. NaH is an ionic hydride in which hydrogen exists as the anion H^-—an electron-pair *donor*, not an acceptor. NaH is a base; HCl is an acid.

(g) *Sulfuric acid or bisulfate*. The bisulfate ion, HSO_4^-, is the conjugate base of H_2SO_4, formed when the strong acid loses its first proton. For HSO_4^- now to give up the remaining proton, it must produce the doubly negative sulfate anion (SO_4^{2-}) as a conjugate base. Although a second ionization does occur (Example 16-10), the conversion of HSO_4^- into SO_4^{2-} is more difficult than the conversion of H_2SO_4 into HSO_4^-. H_2SO_4 is the stronger acid.

EXAMPLE 16-2. Hydronium Ions, Hydroxide Ions, pH

PROBLEM: $Ca(OH)_2$, a strong base, releases two moles of hydroxide ion for every mole of calcium hydroxide dissolved in water. What is the pH of a solution that contains 0.110 g $Ca(OH)_2$ in a total volume of 250. mL? Assume a temperature of 25°C.

SOLUTION: Break the problem into three steps. (1) Compute the concentration of hydroxide ion, $[OH^-]$. (2) Compute the concentration of hydronium ion, $[H_3O^+]$, knowing the relationship between $[H_3O^+]$, $[OH^-]$, and K_w. (3) Express the hydronium ion concentration as a pH value.

1. *Concentration of $[OH^-]$.* First, the number of moles; second, the number of moles per liter:

$$0.110 \text{ g } Ca(OH)_2 \times \frac{1 \text{ mol } Ca(OH)_2}{74.09 \text{ g } Ca(OH)_2} \times \frac{2 \text{ mol } OH^-}{1 \text{ mol } Ca(OH)_2}$$

$$= 0.00297 \text{ mol } OH^-$$

$$[OH^-] = \frac{0.00297 \text{ mol } OH^-}{250. \text{ mL}} \times \frac{1000 \text{ mL}}{L} = 0.0119 \ M$$

2. *Concentration of $[H_3O^+]$.* At equilibrium, the product of $[H_3O^+]$ and $[OH^-]$ is equal to K_w:

$$K_w = [H_3O^+][OH^-]$$

Thus:

$$[H_3O^+] = \frac{K_w}{[OH^-]} = \frac{1.0 \times 10^{-14}}{1.19 \times 10^{-2}} = 8.4 \times 10^{-13} \ M$$

3. *pH $= -log[H_3O^+]$.* A few keystrokes on an electronic calculator, and we shall have the answer—but first a warning and an exhortation. What if we had no calculator? What if the batteries were dead? What if we pushed the wrong buttons? How can we tell if the answer is correct?

Bad things can happen, so be prepared. Make an estimate. First, see that the number 8.4×10^{-13} is nearly equal to 10×10^{-13} or, equivalently, 10^{-12}. Next, remembering the definition of a base-10 logarithm (Appendix B), recognize that $\log 10^{-12}$ is simply -12. Finally, expecting an answer close to 12 (slightly more), use the machine to compute the exact value:

$$pH = -\log [H_3O^+] = -\log(8.4 \times 10^{-13}) = -(-12.08) = 12.08$$

Concerning precision, note that only the decimal portion of a logarithm (the *mantissa*) is significant. Since the whole-number *characteristic* (here, -12) does nothing more than fix the power of 10, a pH of 12.08 contains two significant digits rather than four.

EXAMPLE 16-3. Dilute Solution of a Weak Acid

PROBLEM: Given that $K_a(25°C)$ for acetic acid is 1.76×10^{-5}, compute the pH of a 0.200 M aqueous solution of CH_3COOH at room temperature. What is the ratio of conjugate base to acid, $[CH_3COO^-]/[CH_3COOH]$, at equilibrium?

SOLUTION: Suppose that some concentration of acetic acid (x) dissociates in water, so that at equilibrium we have

$$CH_3COOH(aq) + H_2O(\ell) \rightleftharpoons CH_3COO^-(aq) + H_3O^+(aq)$$

$$0.200 - x \qquad\qquad\qquad x \qquad\qquad\qquad x$$

$$\frac{[H_3O^+][CH_3COO^-]}{[CH_3COOH]} = \frac{x^2}{0.200 - x} = K_a = 1.76 \times 10^{-5}$$

Presuming x to be small compared with 0.200 M, we then make the approximation

$$0.200 - x \approx 0.200$$

and solve the simplified equation that follows:

$$x^2 = 0.200 \cdot K_a = 0.200(1.76 \times 10^{-5})$$

$$x = 1.88 \times 10^{-3} \ M$$

Now, does this approximate result prove small enough to justify our assumption? Check and see. The approximation led us to pretend that 0.198 (the difference between 0.200 and 0.00188) was effectively the same as 0.200. Had we solved the full quadratic equation instead, the exact solution would have been $x = 0.00187$ M rather than $x = 0.00188$ M. The error is only 0.5%.

Summary: Since x represents both $[H_3O^+]$ and $[CH_3COO^-]$ at equilibrium, we derive a pH near 2.73 and an anion-to-acid ratio near 0.0095. The calculation to three significant figures is shown below:

$$pH = -\log [H_3O^+] = -\log(0.00188) = -(-2.726) = 2.726$$

$$\frac{[CH_3COO^-]}{[CH_3COOH]} = \frac{x}{0.200 - x} = \frac{0.00188}{0.198} = 0.00949$$

Approximately 1% of the acid is dissociated at this concentration (0.00188 M out of the original 0.200 M).

At *this* concentration. Will the aqueous world look different to a weak acid under different circumstances? Read on.

EXAMPLE 16-4. A *Very* Dilute Solution of a Weak Acid

PROBLEM: How many grams of acetic acid must be dissolved in a total volume of 50.0 mL to give a pH of 4.200 at 25°C? What is the value of the ratio $[CH_3COO^-]/[CH_3COOH]$ at equilibrium? Assume, for simplicity, that all the hydronium ions come from the dissociation of the acid—not from the autoionization of water.

SOLUTION: The exercise forces us (1) to compute the hydronium concentration, $[H_3O^+]$, from the pH; and (2) to undertake an equilibrium calculation involving K_a, $[H_3O^+]$, $[CH_3COO^-]$, and $[CH_3COOH]$.

1. $pH \longrightarrow [H_3O^+]$. Since pH $= -\log [H_3O^+]$, we have

$$[H_3O^+] = 10^{-pH} = 10^{-4.200} = 6.31 \times 10^{-5} \ M$$

Is the number reasonable? Check. The antilogarithm should fall between 10^{-4} and 10^{-5}. It does.

Check also: Is $6.31 \times 10^{-5} \ M$ a large enough number to justify our neglect of water's autoionization? It is: 6.31×10^{-5} is over 600 times greater than 10^{-7}, the concentration of H_3O^+ normally present in pure water.

2. *Equilibrium.* Consider what happens. We dissolve some unknown amount of CH_3COOH in 50.0 mL to produce a solution of unknown concentration, c, in moles per liter. Some portion (x) of that concentration c then dissociates into H_3O^+ and CH_3COO^- and subsequently comes to equilibrium with the remaining acid:

$$CH_3COOH(aq) + H_2O(\ell) \rightleftharpoons CH_3COO^-(aq) + H_3O^+(aq)$$

$$ c - x x x$$

$$\frac{[H_3O^+][CH_3COO^-]}{[CH_3COOH]} = \frac{x^2}{c - x} = K_a$$

Into this equation we substitute values for both x (the concentration of H_3O^+, computed above as $6.31 \times 10^{-5} \ M$) and K_a (1.76×10^{-5}), whereupon we solve for c:

$$c = \frac{x^2 + xK_a}{K_a} = 2.89 \times 10^{-4} \ M$$

The number thus obtained, $c = 2.89 \times 10^{-4}$ M, is the concentration of acetic acid *before* dissociation, equivalent to just 8.68×10^{-4} g in a total volume of 50.0 mL:

$$\frac{2.89 \times 10^{-4} \text{ mol}}{\text{L}} \times \frac{1 \text{ L}}{1000 \text{ mL}} \times 50.0 \text{ mL} \times \frac{60.05 \text{ g}}{\text{mol}}$$

$$= 8.68 \times 10^{-4} \text{ g CH}_3\text{COOH}$$

After dissociation, the ratio of conjugate base to acid is 0.279:

$$\frac{[\text{CH}_3\text{COO}^-]}{[\text{CH}_3\text{COOH}]} = \frac{x}{c - x} = \frac{6.31 \times 10^{-5}}{(2.89 \times 10^{-4}) - (6.31 \times 10^{-5})} = 0.279$$

QUESTION: How can we speak of a "weak" acid when the extent of ionization is so high?

ANSWER: The degree of dissociation is high, but the concentrations are low. A small, overworked population of CH_3COOH molecules must release enough H^+ and CH_3COO^- ions to establish the same equilibrium ratio, K_a, demanded of all systems—and so, with scant material to begin with, the acid must dissociate more and more to meet the target. A sizable fraction of the original group is consumed to produce an even smaller concentration of H^+ and CH_3COO^- ions at equilibrium.

In a more concentrated solution (as in Example 16-3), fewer ions are needed to yield the requisite final concentrations. With more acetic acid available at the start, a smaller fraction of a larger total is enough.

QUESTION: Since the ratio of conjugate base to acid is relatively high (0.279), will the solution be an effective buffer? Will it maintain a steady pH despite any addition of H^+ and OH^-?

ANSWER: No, this mixture is no buffer solution. It exists in a fragile state of equilibrium, supported by vanishingly small reserves of CH_3COOH (0.000226 M) and CH_3COO^- (0.0000631 M). Add OH^- at a concentration of only 0.000226 M, for example, and all the undissociated CH_3COOH disappears. The solution, unbuffered, is left defenseless against further intrusion by either H^+ or OH^-.

A buffer, after all, is something that acts as a shield, a barrier, a defense. To work effectively as a buffer, a solution must be able to tap substantial reserves of acid and conjugate base. See the next example.

EXAMPLE 16-5. Calling Up the Reserves: A Buffer Solution

PROBLEM: 1.144 grams of $NaCH_3COO$ are added to 0.500 L of a 0.100 M solution of CH_3COOH at 25°C. (a) What is the approximate pH of the resulting buffer solution? Use the Henderson-Hasselbalch equation. (b) What is the pH after addition of 0.000500 mole of HCl to the buffer? (c) Similarly, what is the pH after addition of 0.000500 mole of NaOH to the original solution?

SOLUTION: Dissolved in a volume of 0.500 L, the 1.144 grams of sodium acetate produce a concentration of 0.0279 M:

$$1.144 \text{ g NaCH}_3\text{COO} \times \frac{1 \text{ mol NaCH}_3\text{COO}}{82.03 \text{ g NaCH}_3\text{COO}} \times \frac{1 \text{ mol CH}_3\text{COO}^-}{1 \text{ mol NaCH}_3\text{COO}}$$

$$= 0.01395 \text{ mol CH}_3\text{COO}^-$$

$$[CH_3COO^-] = \frac{0.01395 \text{ mol}}{0.500 \text{ L}} = 0.0279 \text{ } M$$

Together with the stated concentration of CH_3COOH,

$$[CH_3COOH] = 0.100 \text{ } M$$

and the pK_a of acetic acid,

$$pK_a = -\log(1.76 \times 10^{-5}) = 4.754$$

we have just enough information to apply the Henderson-Hasselbalch equation.

(a) *pH of the buffer solution.* Substitution of pK_a, [CH_3COOH], and [CH_3COO^-] into the Henderson-Hasselbalch equation yields an initial pH of 4.200, the same as in Example 16-4:

$$pH = pK_a + \log\frac{[CH_3COO^-]}{[CH_3COOH]} = 4.754 + \log\frac{0.0279}{0.100} = 4.200$$

The ratio of conjugate base to acid (0.279) is the same in both this solution and the one previous, but now there are sufficient quantities of A^- and HA to work together as a buffer.

(b) *Add 0.000500 mole of H^+.* Dissolved in a volume of 0.500 L, the HCl is introduced at a concentration of 0.00100 M:

$$[\text{HCl}] = \frac{0.000500 \text{ mol}}{0.500 \text{ L}} = 0.00100 \text{ } M$$

The strong acid then reacts stoichiometrically with the conjugate base in solution, lowering the concentration of CH_3COO^- and simultaneously raising the concentration of CH_3COOH. The $[A^-]/[HA]$ ratio drops from 0.279 to 0.266:

$$CH_3COO^-(aq) + H_3O^+(aq) \longrightarrow CH_3COOH(aq) + H_2O(\ell)$$

$$[CH_3COO^-] = 0.0279 \text{ } M - 0.00100 \text{ } M = 0.0269 \text{ } M$$

$$[CH_3COOH] = 0.100 \text{ } M + 0.00100 \text{ } M = 0.101 \text{ } M$$

$$\frac{[CH_3COO^-]}{[CH_3COOH]} = \frac{0.0269}{0.101} = 0.266$$

The buffered pH falls only slightly as a result, from 4.20 to just under 4.18:

$$\text{pH} = \text{p}K_a + \log \frac{[CH_3COO^-]}{[CH_3COOH]} = 4.754 + \log \frac{0.0269}{0.101} = 4.179$$

(c) *Add 0.000500 mole of OH⁻.* Again, the added concentration is 0.00100 M (0.000500 mol OH^-/0.500 L); and again, the result is a small change in the ratio of $[A^-]$ to $[HA]$. This time, though, with the re-action between the strong base (OH^-) and the weak acid (CH_3COOH), the ratio *rises* slightly as the buffer does its work:

$$CH_3COOH(aq) + OH^-(aq) \longrightarrow CH_3COO^-(aq) + H_2O(\ell)$$

The concentration of acid decreases by 0.00100 M; the concentration of conjugate base increases by 0.00100 M. The ratio of $[CH_3COO^-]$ to $[CH_3COOH]$ goes up a bit, and so does the pH—from 4.20 to nearly 4.22:

$$\text{pH} = \text{p}K_a + \log \frac{[CH_3COO^-]}{[CH_3COOH]} = 4.754 + \log \frac{0.0289}{0.0990} = 4.219$$

QUESTION: In what way is the Henderson-Hasselbalch equation merely an approximation to a more accurate treatment?

ANSWER: Assume that the concentrations of acid and conjugate base before equilibrium are $[HA]_0$ and $[A^-]_0$, respectively; and assume further that $[H_3O^+]$ is initially zero. If HA subsequently ionizes to produce x moles of H_3O^+ per liter at equilibrium, then the final concentrations are

$$[H_3O^+] = x$$

$$[A^-] = [A^-]_0 + x$$

$$[HA] = [HA]_0 - x$$

Using the equation

$$pH = pK_a + \log \frac{[A^-]_0}{[HA]_0}$$

is therefore tantamount to neglecting x as small compared with both $[HA]_0$ and $[A^-]_0$ in the full expression

$$K_a = \frac{x([A^-]_0 + x)}{[HA]_0 - x} \approx \frac{x[A^-]_0}{[HA]_0}$$

It is an approximation we make frequently, but one that still demands justification case by case. Be careful.

QUESTION: Are there other approximations buried in the calculation?

ANSWER: Yes. As usual, we neglect any production of H_3O^+ and OH^- arising from the autoionization of water. Such neglect is mostly benign, but be prepared to reexamine the entire procedure in highly dilute solutions. When acid or base becomes sufficiently scarce, autoionization can contribute substantially to the total concentration of hydronium and hydroxide.

At *high* concentrations, however, be prepared also to reconsider the assumption that the solution is ideal—a possible weak point in our argument.

EXAMPLE 16-6. Hydrolysis of a Base

Having treated a weak acid alone (Examples 16-3 and 16-4) and also a buffer solution of acid and conjugate base together (Example 16-5), we go next to the opposite end of the titration curve: a solution consisting entirely of conjugate base.

Except for a few changes in name and notation, the idea is the same. So is the arithmetic.

PROBLEM: What is the pH of a 0.200 *M* solution of $NaCH_3COO$ at 25°C?

SOLUTION: Sodium acetate forms as a soluble salt when sodium hydroxide (a strong base) neutralizes acetic acid (a weak acid):

$$CH_3COOH(aq) + NaOH(aq) \longrightarrow$$

$$Na^+(aq) + CH_3COO^-(aq) + H_2O(\ell)$$

After that, with all the acid gone and with all the hydroxide gone, there occurs yet another reaction: **hydrolysis**, the breaking apart of a water molecule. The acetate anion, a base conjugate to acetic acid, pulls a proton off water and thereby triggers a release of hydroxide ions:

$$CH_3COO^-(aq) + H_2O(\ell) \rightleftharpoons CH_3COOH(aq) + OH^-(aq)$$

		INITIAL	CHANGE		EQUILIBRIUM
$[CH_3COOH]$	=	0	+ x	=	x
$[OH^-]$	=	0	+ x	=	x
$[CH_3COO^-]$	=	0.200	− x	=	$0.200 - x$

To solve for x, recall that the base ionization constant, K_b, is inversely proportional to the corresponding acid ionization constant:

$$K_b = \frac{K_w}{K_a} = \frac{1.0 \times 10^{-14}}{1.76 \times 10^{-5}} = 5.7 \times 10^{-10}$$

$$= \frac{[CH_3COOH][OH^-]}{[CH_3COO^-]} = \frac{x^2}{0.200 - x} = 5.7 \times 10^{-10}$$

Expecting x to be much smaller than 0.200, we then solve the approximate equation

$$\frac{x^2}{0.200} = 5.7 \times 10^{-10}$$

$$x = 1.07 \times 10^{-5} = [OH^-]$$

according to our usual procedure. That done, the hydronium concentration and pH follow straightforwardly as

$$[H_3O^+] = \frac{K_w}{[OH^-]} = \frac{1.0 \times 10^{-14}}{1.07 \times 10^{-5}} = 9.4 \times 10^{-10}$$

$$pH = -\log [H_3O^+] = 9.03$$

The pH is greater than 7, corresponding to a basic solution.

Conclusion: Neutralization of a weak acid by a strong base will produce, after hydrolysis of the conjugate base, a hydroxide concentration greater than the value found in pure water. See Example 16-7.

QUESTION: What is the pH of a solution of NaCl at some arbitrary concentration?

ANSWER: 7. NaCl is the salt of a strong base (NaOH) and a strong acid (HCl). Neither Na^+ nor Cl^- has the strength to react with water in a way that can alter the pH. The concentrations of H_3O^+ and OH^- are controlled entirely by the autoionization of water.

QUESTION: Would NaCl ever make a good buffer?

ANSWER: No. What defense could either Na^+ or Cl^- possibly offer? Na^+ cannot capture OH^-; Cl^- cannot capture H^+. These ions are weak, passive spectators to the acid–base equilibria around them.

EXAMPLE 16-7. A Titration Curve

Proton by proton, molecule by molecule, mole by mole, we carry out a *titration*: a systematic neutralization of acid or base; a gradual passage, carefully monitored, from one extreme to the other; a controlled, deliberate sampling of every gradation in between.

The method: Start with one solution, usually of unknown concentration; add to it, little by little, a solution of known concentration; let the partners come together; analyze the mixture that results. Add another drop, and another, and another, until there is nothing left to react. Along the way, plot the changing pH and see all there is to see: the complete life cycle of the acid as it goes from HA to A^- (or, same thing, the life of the base as it goes from A^- to HA).

From the last few examples, we already know what to expect; but here, with some satisfaction perhaps, we can finally put everything together

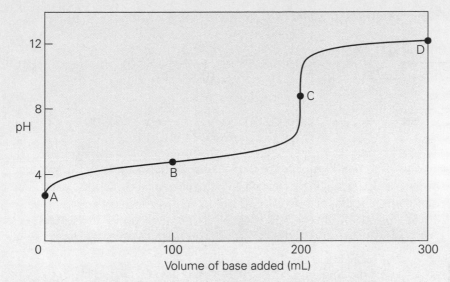

FIGURE R16-1. Titration of 50 mL of 0.200 M CH_3COOH by 0.0500 M NaOH.

into one titration curve. Later, in Example 16-8, we shall see how a titration allows us to determine the concentration of an unknown solution.

PROBLEM: To 50.0 mL of 0.200 M acetic acid at 25°C, a chemist adds a 0.0500 M solution of NaOH in the following amounts: (a) 0 mL. (b) 100.0 mL. (c) 200.0 mL. (d) 300.0 mL. Compute the pH for each mixture, and explain how the associated points fit onto the titration curve in Figure R16-1.

SOLUTION: There are two calculations to do at each stage. First, we determine the amount of acetic acid that remains after a complete, stoichiometric reaction wherein every OH^- from the strong base finds and neutralizes one molecule of the weak acid. Since the equilibrium constant is overwhelmingly large, the reaction goes all the way; no OH^- remains unattended. After that, we compute the concentrations of the remaining species when they subsequently come to a new equilibrium.

(a) *0 mL of NaOH.* The mixture is all acid at the beginning of the titration (point A), corresponding exactly to the problem already solved in Example 16-3. Aside from the small concentrations of H_3O^+ and OH^- produced by autoionization, any new hydronium ions can only come from the dissociation of CH_3COOH. To set up and solve the equilibrium equation, see Example 16-3. The resulting pH is 2.73, mildly acidic.

(b) *100.0 mL of 0.0500 M NaOH.* The molar amounts of acid and base follow directly as the product of concentration (mol L^{-1}) and volume (L):

Concentration   ×   volume   =   moles

$$\frac{0.200 \text{ mol } CH_3COOH}{L} \times 50.0 \text{ mL } \times \frac{1 \text{ L}}{1000 \text{ mL}} = 0.0100 \text{ mol } CH_3COOH$$

$$\frac{0.0500 \text{ mol } OH^-}{L} \times 100.0 \text{ mL } \times \frac{1 \text{ L}}{1000 \text{ mL}} = 0.00500 \text{ mol } OH^-$$

A total of 0.00500 mol NaOH has been added so far to the original 0.0100 mol CH_3COOH, enough to neutralize exactly half the acid. The total volume stands at 150 mL (50 mL of acid; 100 mL of added base).

In these 150 mL we now have 0.00500 mol *less* of CH_3COOH and 0.00500 mol *more* of CH_3COO^-, produced by the neutralization reaction

$$CH_3COOH(aq) + OH^-(aq) \longrightarrow CH_3COO^-(aq) + H_2O(\ell)$$

Before equilibration, the new concentrations are

$$[CH_3COOH]_0 = \frac{(0.0100 - 0.00500) \text{ mol}}{0.1500 \text{ L}} = 0.033 \ M$$

$$[CH_3COO^-]_0 = \frac{(0 + 0.00500) \text{ mol}}{0.1500 \text{ L}} = 0.033 \ M$$

and, since $[HA] = [A^-]$, the pH is equal approximately to the pK_a:

$$pH = pK_a + \log\frac{[CH_3COO^-]_0}{[CH_3COOH]_0} = 4.75 + \log 1 = 4.75$$

The titration has reached the ***half-equivalence point***: half the original acid is gone, converted into an equal amount of conjugate base. Marked as B on the titration curve, the half-equivalence point lies at the midpoint of the buffer zone—where pH varies the least.

QUESTION: The pH is "approximately" 4.75? Why not *exactly* 4.75?

ANSWER: Approximately—not exactly—because again we choose not to solve the full equilibrium expression

$$K_a = \frac{x([CH_3COO^-]_0 + x)}{[CH_3COOH]_0 - x}$$

for the reaction

$$CH_3COOH(aq) + H_2O(\ell) \rightleftharpoons CH_3COO^-(aq) + H_3O^+(aq)$$

$$[CH_3COOH]_0 - x \qquad\qquad [CH_3COO^-]_0 + x \qquad x$$

The variable x corresponds to $[H_3O^+]$, the extra hydronium concentration produced upon equilibration.

But since the initial concentrations $[CH_3COOH]_0$ and $[CH_3COO^-]_0$ are relatively high (0.033 M), we assume that this small incremental change (x) will not substantially alter the final concentrations at equilibrium. The simplified form

$$K_a = \frac{x[CH_3COO^-]_0}{[CH_3COOH]_0}$$

and the Henderson-Hasselbalch equation then follow under the assumption that

$$\frac{x([CH_3COO^-]_0 + x)}{[CH_3COOH]_0 - x} \approx \frac{x[CH_3COO^-]_0}{[CH_3COOH]_0}$$

Such is our standard practice with a buffer solution. See also Example 16-5.

(c) *200.0 mL of 0.0500 M NaOH.* By this stage, 0.0100 mole of OH^- has reacted with 0.0100 mole of CH_3COOH. Neutralization is complete; and now, in a total volume of 250.0 mL (50 mL of acid; 200 mL of added base) there is 0.0100 mole of CH_3COO^- that still must come to equilibrium with H_2O and CH_3COOH. So it does: The conjugate base undergoes hydrolysis, just as described in Example 16-6, to produce a greater-than-neutral concentration of hydroxide ions:

$$CH_3COO^-(aq) + H_2O(\ell) \rightleftharpoons CH_3COOH(aq) + OH^-(aq)$$

The equilibrium calculation is the same as in Example 16-6. Starting from an initial concentration

$$[CH_3COO^-]_0 = \frac{0.0100\ \text{mol}}{0.2500\ \text{L}} = 0.0400\ M$$

the mixture hydrolyzes to a point where

$$K_b = \frac{[CH_3COOH][OH^-]}{[CH_3COO^-]} = \frac{x^2}{0.0400 - x} = 5.7 \times 10^{-10}$$

Neglecting x compared with 0.0400, we then obtain both $[OH^-]$ and the pH at equilibrium:

$$x^2 = 0.0400(5.7 \times 10^{-10})$$

$$x = 4.77 \times 10^{-6} = [OH^-]$$

$$[H_3O^+] = \frac{K_w}{[OH^-]} = \frac{1.0 \times 10^{-14}}{4.77 \times 10^{-6}} = 2.1 \times 10^{-9}$$

$$pH = -\log [H_3O^+] = 8.68$$

It is a basic solution, just as we expect for the neutralization of a weak acid by a strong base. The equivalence point, marked C on the curve, occurs at pH = 8.68.

   (d) *300.0 mL of 0.0500 M NaOH.* Every milliliter of NaOH delivered past the equivalence point simply adds more hydroxide ions to the solution, by now an unbuffered base. Hence at point D, after the addition of 100.0 mL in excess of neutralization (see above), we have 0.00500 mol OH^- in a total volume of 350.0 mL. The pH rises to 12.15:

$$\frac{0.0500 \text{ mol OH}^-}{L} \times 0.1000 \text{ L} = 0.00500 \text{ mol OH}^-$$

$$[OH^-] = \frac{0.00500 \text{ mol OH}^-}{0.3500 \text{ L}} = 0.0143 \text{ } M$$

$$pH = -\log [H_3O^+] = -\log \frac{K_w}{[OH^-]} = -\log(7.0 \times 10^{-13}) = 12.15$$

Example 16-8.   Determination of an Unknown Concentration by Titration

PROBLEM: Compute the original concentration of the acid solution in each of the following titrations, given that the equivalence point is reached when: (a) 145.6 mL of 0.1000 M NaOH react with 100.0 mL of CH_3COOH; (b) 145.6 mL of 0.1000 M NaOH react with 100.0 mL of HCl.

SOLUTION: A known volume of base (V_b) having a known concentration (c_b) reacts stoichiometrically with a known volume of acid (V_a).

From that exchange, we want to determine the unknown concentration of the acid (c_a).

In the first titration, we have a weak acid and a strong base; in the second, a strong acid and a strong base. Either way, however, the condition for neutralization is the same: The number of moles of acid must equal the number of moles of base.

To ascertain V_a, we recognize from the dimensions of concentration and volume that

$$\frac{\text{mol}}{\text{L}} \times \text{L} = \text{mol}$$

and we impose the condition

$$\text{moles of acid} = \text{moles of base}$$

$$c_a V_a = c_b V_b$$

For the procedures as described, the numbers are identical for both titrations:

$$c_a = \frac{V_b}{V_a} c_b = \frac{145.6 \text{ mL}}{100.0 \text{ mL}} \times 0.1000 \, M = 0.1456 \, M$$

EXAMPLE 16-9. Polyprotic Acids

A giver that keeps on giving, a *polyprotic acid* has more than one proton to lose. *Diprotic* species such as sulfuric acid (H_2SO_4), carbonic acid (H_2CO_3), and ascorbic acid ($H_2C_6H_6O_6$) have two ionizable protons. *Triprotic* species such as phosphoric acid (H_3PO_4) and citric acid ($H_3C_6H_5O_7$) have three.

Now imagine, for simplicity, that a polyprotic acid sheds its protons one at a time, so that each ionization has its own equilibrium constant. Off comes the first proton, ionizing according to some value K_{a1}; and then follows the second proton, conforming to some other value, K_{a2}. In the end, though, everything comes to equilibrium and we cannot tell where and when each hydronium ion originated. All hydronium ions are created equal.

But the amounts produced during each dissociation are not. The second ionization is always harder to do than the first, and the third is still more difficult than the second. Because: The conjugate base grows progressively more negative with the loss of each H^+, holding on more tightly to the protons that remain. Each value of K_a is smaller than the one before.

PROBLEM: Carbonic acid ionizes in two stages, eventually settling into a common equilibrium in which $[H_2CO_3]$, $[H_3O^+]$, $[HCO_3^-]$, and $[CO_3^{2-}]$ are thoroughly interconnected:

$$H_2CO_3(aq) + H_2O(\ell) \rightleftharpoons HCO_3^-(aq) + H_3O^+(aq) \qquad K_{a1} = 4.3 \times 10^{-7}$$

$$HCO_3^-(aq) + H_2O(\ell) \rightleftharpoons CO_3^{2-}(aq) + H_3O^+(aq) \qquad K_{a2} = 5.6 \times 10^{-11}$$

Compute the equilibrium concentrations of all species in a solution where $[H_2CO_3]$ is initially 0.050 M.

SOLUTION: H_2CO_3, HCO_3^-, and H_3O^+ reach a quasi equilibrium when the conditions imposed by K_{a1}, the *first* ionization constant, are met:

		INITIAL		CHANGE		EQUILIBRIUM (STAGE 1)
$[HCO_3^-]$	=	0	+	x	=	x
$[H_3O^+]$	=	0	+	x	=	x
$[H_2CO_3]$	=	0.050	−	x	=	$0.050 - x$

$$K_{a1} = \frac{[H_3O^+][HCO_3^-]}{[H_2CO_3]} = \frac{x^2}{0.050 - x} = 4.3 \times 10^{-7}$$

$$x = [H_3O^+] = [HCO_3^-] = 0.00015$$

But there is more to come, because the HCO_3^- (bicarbonate) ions will ionize further to produce CO_3^{2-} (carbonate) ions and additional H_3O^+. Denoting that change in molarity by y, we now write the equilibrium concentrations symbolically as

		INITIAL		CHANGE		EQUILIBRIUM (STAGE 2)
$[CO_3^{2-}]$	=	0	+	y	=	y
$[H_3O^+]$	=	0.00015	+	y	=	$0.00015 + y$
$[HCO_3^-]$	=	0.00015	−	y	=	$0.00015 - y$

Finally, one last approximation: Since K_{a2} is almost 10,000 times smaller than K_{a1}, the additional concentration (y) should likewise be much smaller than the concentration produced by the first ionization ($x = 0.00015$ M). If so, then we can neglect y compared with 0.00015 and assert that

$$[H_3O^+] = [HCO_3^-] \approx 0.00015$$

at final equilibrium. The second ionization, we see, barely affects the concentrations of hydronium and bicarbonate; what it does, instead, is introduce a small concentration of *carbonate* ion that would otherwise not be present:

$$K_{a2} = \frac{[H_3O^+][CO_3^{2-}]}{[HCO_3^-]} \approx \frac{(0.00015)y}{0.00015} = 5.6 \times 10^{-11}$$

$$y = [CO_3^{2-}] = K_{a2} = 5.6 \times 10^{-11}$$

Collecting the results, we have an approximate (but reasonably accurate) value for the concentration of each participant:

$$[H_3O^+] \approx 0.00015 \, M \qquad (pH = 3.82)$$

$$[HCO_3^-] \approx 0.00015 \, M$$

$$[CO_3^{2-}] \approx 5.6 \times 10^{-11} \, M = K_{a2}$$

$$[H_2CO_3] \approx 0.050 \, M$$

Conclusion: For those diprotic acids in which the second ionization is much weaker than the first, the pH is determined fully by the first ionization. The second ionization constant, K_{a2}, fixes the concentration of the doubly deprotonated conjugate base (here, CO_3^{2-}).

Now contrast these findings with the special case treated next in Example 16-10: a dilute solution of sulfuric acid.

EXAMPLE 16-10. Strong Diprotic Acids

PROBLEM: H_2SO_4 behaves as a strong acid upon first ionization, dissociating completely into hydronium and bisulfate ions (HSO_4^-). The second ionization, which yields sulfate ions (SO_4^{2-}) plus additional hydronium, proceeds with an ionization constant $K_{a2} = 0.012$. Compute the pH of a 0.10 M aqueous solution of H_2SO_4.

SOLUTION: The first ionization converts effectively every H_2SO_4 molecule into one H_3O^+ ion and one HSO_4^- ion:

$$H_2SO_4(aq) + H_2O(\ell) \longrightarrow HSO_4^-(aq) + H_3O^+(aq)$$

		INITIAL		CHANGE		EQUILIBRIUM (STAGE 1)
$[HSO_4^-]$	=	0	+	0.10	=	0.10
$[H_3O^+]$	=	0	+	0.10	=	0.10
$[H_2SO_4]$	=	0.10	−	0.10	=	0

The pH at this first stage is 1.00 (since $[H_3O^+] = 0.10\ M$), but the system has not yet reached its true equilibrium. More hydronium ions are released during the second ionization.

Exactly how many? Final equilibrium arrives only when the stage-1 concentrations evolve to the point where the second equilibrium constant, $K_{a2} = 0.012$, is also satisfied:

	INITIAL	CHANGE	EQUILIBRIUM (STAGE 2)
$[SO_4^{2-}] =$	0	$+\quad y$	$=$	y
$[H_3O^+] =$	0.10	$+\quad y$	$=$	$0.10 + y$
$[HSO_4^-] =$	0.10	$-\quad y$	$=$	$0.10 - y$

$$K_{a2} = \frac{[H_3O^+][SO_4^{2-}]}{[HSO_4^-]} = \frac{(0.10 + y)y}{0.10 - y} = 0.012$$

Since K_{a2} is a relatively large number, we are forced here (for once) to solve the quadratic equation in full:

$$y^2 + 0.112y - 0.0012 = 0$$

$$y = 0.00985,\ -0.122$$

The negative solution we reject as unphysical since it demands, impossibly, that $[SO_4^{2-}]$ be less than zero. The final hydronium concentration, altered by the incremental change $y = 0.00985\ M$, is therefore 0.11 M. The pH is 0.96:

$$[H_3O^+] = 0.10 + 0.00985 = 0.11\ M$$

$$pH = -\log\ [H_3O^+] = -\log\ 0.11 = 0.96$$

Conclusion: When a second ionization is sufficiently strong, the equilibrium concentrations of both H_3O^+ and the singly deprotonated conjugate base (here, HSO_4^-) are determined by K_{a2} as well as K_{a1}.

EXERCISES

1. HCl is a strong acid. HF is not. Why not?

2. HCl is a strong acid. Is HI a strong acid as well? Why or why not?

3. Pick the stronger acid in each pair:

   (a) CH_2FCOOH, CF_3COOH
   (b) H_3BO_3, H_2SO_4
   (c) $HClO$, HIO
   (d) H_3PO_4, $H_2PO_4^-$

4. Pick the stronger acid in each pair:

   (a) CH_3COOH, CH_3CH_2OH
   (b) HSO_4^-, KOH
   (c) $HClO_2$, $HClO_4$
   (d) NH_3, NH_4^+

5. Identify the base conjugate to each acid below, and then arrange the conjugate bases in order of increasing strength:

$$CF_3COOH, \quad H_2O, \quad CH_3COOH, \quad CCl_3COOH$$

6. Identify the base conjugate to each acid below, and then arrange the conjugate bases in order of increasing strength:

$$CH_3CH_2OH, \quad HClO_4, \quad CH_3COOH, \quad H_2O$$

7. Identify the acid conjugate to each of the following bases:

   (a) NH_3
   (b) $CH_3CH_2CH_2COO^-$
   (c) HSO_4^-
   (d) Cl^-
   (e) H_2N-CH_3
   (f) CN^-

   Draw Lewis structures for both base and conjugate acid.

8. Identify the acid conjugate to each of the following bases:

   (a) HS^-

   (b) ClO^-

   (c) OH^-

   (d) $CH_3CH_2O^-$

   (e) H_2NOH

   (f) HCO_2^-

   Draw Lewis structures for both base and conjugate acid.

9. Given $[H_3O^+]$, calculate the pH:

   (a) 1 M (b) 0.1 M (c) 0.01 M
   (d) 0.001 M (e) 0.0001 M (f) 0.00001 M
   (g) 0.000001 M (h) 0.0000001 M

10. Given $[H_3O^+]$, calculate the pH:

    (a) 0.5 M (b) 0.05 M (c) 0.005 M
    (d) 0.0005 M (e) 0.00005 M (f) 0.000005 M
    (g) 0.0000005 M

11. Given $[H_3O^+]$, calculate the pH:

    (a) 10^{-8} M (b) 10^{-9} M (c) 10^{-10} M (d) 10^{-11} M
    (e) 10^{-12} M (f) 10^{-13} M (g) 10^{-14} M

12. Given $[H_3O^+]$, calculate the pH:

    (a) 5×10^{-8} M (b) 5×10^{-9} M (c) 5×10^{-10} M
    (d) 5×10^{-11} M (e) 5×10^{-12} M (f) 5×10^{-13} M
    (g) 5×10^{-14} M

13. Given $[H_3O^+]$, calculate the pH:

    (a) 0.001 M (b) 0.002 M (c) 0.003 M
    (d) 0.004 M (e) 0.005 M (f) 0.006 M
    (g) 0.007 M (h) 0.008 M (i) 0.009 M (j) 0.01 M

14. Given $[H_3O^+]$, calculate pH, $[OH^-]$, and pOH at 25°C:

    (a) 1 M (b) 1×10^{-1} M (c) 1×10^{-2} M
    (d) 1×10^{-3} M (e) 1×10^{-4} M (f) 1×10^{-5} M
    (g) 1×10^{-6} M (h) 1×10^{-7} M

15. Given $[H_3O^+]$, calculate pH, $[OH^-]$, and pOH at 25°C:

(a) 1×10^{-8} *M*   (b) 1×10^{-9} *M*   (c) 1×10^{-10} *M*
(d) 1×10^{-11} *M*   (e) 1×10^{-12} *M*   (f) 1×10^{-13} *M*
(g) 1×10^{-14} *M*

16. Using symbols rather than numbers, show that

$$pH + pOH = pK_w$$

What is the value of pK_w at 25°C?

17. Given the pH, calculate $[H_3O^+]$ and $[OH^-]$ at 25°C:

(a) 1.0   (b) 1.5   (c) 2.0   (d) 2.5   (e) 3.0
(f) 3.5   (g) 4.0   (h) 4.5   (i) 5.0   (j) 5.5
(k) 6.0   (l) 6.5   (m) 7.0

18. Given the pH, calculate $[H_3O^+]$ and $[OH^-]$ at 25°C:

(a) 7.5   (b) 8.0   (c) 8.5   (d) 9.0   (e) 9.5
(f) 10.0   (g) 10.5   (h) 11.0   (i) 11.5   (j) 12.0
(k) 12.5   (l) 13.0   (m) 13.5   (n) 14.0

19. Given the pH, calculate $[H_3O^+]$ and $[OH^-]$ at 25°C:

(a) 2.0   (b) 2.1   (c) 2.2   (d) 2.3   (e) 2.4
(f) 2.5   (g) 2.6   (h) 2.7   (i) 2.8   (j) 2.9
(k) 3.0

20. Given *K*, calculate p*K*:

(a) 1×10^{-1} (b) 5×10^{-2} (c) 1×10^{-2}
(d) 5×10^{-3} (e) 1×10^{-3} (f) 5×10^{-4}
(g) 1×10^{-4} (h) 5×10^{-5} (i) 1×10^{-5}
(j) 5×10^{-6} (k) 1×10^{-6} (l) 5×10^{-7}

21. Given *K*, calculate p*K*:

(a) 1×10^{-7} (b) 5×10^{-8} (c) 1×10^{-8}
(d) 5×10^{-9} (e) 1×10^{-9} (f) 5×10^{-10}
(g) 1×10^{-10} (h) 5×10^{-11} (i) 1×10^{-11}

22. Given p*K*, calculate *K*:

(a) 1.0   (b) 1.5   (c) 2.0   (d) 2.5   (e) 3.0
(f) 3.5   (g) 4.0   (h) 4.5   (i) 5.0   (j) 5.5
(k) 6.0   (l) 6.5   (m) 7.0

23. Given pK, calculate K:

    (a) 2.0   (b) 2.1   (c) 2.2   (d) 2.3   (e) 2.4
    (f) 2.5   (g) 2.6   (h) 2.7   (i) 2.8   (j) 2.9
    (k) 3.0

24. Compute [H_3O^+], [OH^-], and pH for each of these aqueous solutions at 25°C:

    (a) 0.345 g KOH dissolved in a total volume of 57.2 mL
    (b) 0.107 mol HCl dissolved in a total volume of 0.898 L
    (c) 300.7 g KCl dissolved in a total volume of 2.00 L
    (d) 0.000100 mol HBr dissolved in a total volume of 100.0 mL

    Indicate whether the mixture is acidic, basic, or neutral.

25. Compute [H_3O^+], [OH^-], and pH for each of the following aqueous solutions at 25°C:

    (a) 1.00 mol HNO_3 dissolved in a total volume of 1.00 L
    (b) 22.7 g $HClO_4$ dissolved in a total volume of 40.9 L
    (c) 1.0 g NaOH dissolved in a total volume of 0.100 L
    (d) 0.250 M NaBr

    Indicate whether the mixture is acidic, basic, or neutral.

26. K_a for chloroacetic acid is 1.4×10^{-3}:

    $$CH_2ClCOOH(aq) + H_2O(\ell) \rightleftharpoons CH_2ClCOO^-(aq) + H_3O^+(aq)$$

    Given an initial concentration of chloroacetic acid, calculate pH and [CH_2ClCOO^-]/[$CH_2ClCOOH$] at equilibrium:

    (a) [$CH_2ClCOOH$]$_0$ = 1.000 M
    (b) [$CH_2ClCOOH$]$_0$ = 0.500 M
    (c) [$CH_2ClCOOH$]$_0$ = 0.100 M
    (d) [$CH_2ClCOOH$]$_0$ = 0.010 M

    Assume a temperature of 25°C.

27. K_a for benzoic acid is 6.5×10^{-5}:

    $$C_6H_5COOH(aq) + H_2O(\ell) \rightleftharpoons C_6H_5COO^-(aq) + H_3O^+(aq)$$

    Given an initial concentration of benzoic acid, calculate pH and [$C_6H_5COO^-$]/[C_6H_5COOH] at equilibrium:

(a) $[C_6H_5COOH]_0 = 1.000$ M

(b) $[C_6H_5COOH]_0 = 0.500$ M

(c) $[C_6H_5COOH]_0 = 0.100$ M

(d) $[C_6H_5COOH]_0 = 0.010$ M

Assume a temperature of 25°C.

28. K_a for hydrocyanic acid is 6.2×10^{-10}:

$$HCN(aq) + H_2O(\ell) \rightleftharpoons CN^-(aq) + H_3O^+(aq)$$

Given an initial concentration of hydrocyanic acid, calculate pH and $[CN^-]/[HCN]$ at equilibrium:

(a) $[HCN]_0 = 1.000$ M

(b) $[HCN]_0 = 0.500$ M

(c) $[HCN]_0 = 0.100$ M

(d) $[HCN]_0 = 0.010$ M

Assume a temperature of 25°C.

29. (a) Review the results of the preceding three exercises and rank the acids ($CH_2ClCOOH$, C_6H_5COOH, HCN) according to strength. Which is strongest? Why? (b) How is the percent ionization affected by the initial concentration?

30. Look again at the same three reactions:

(a) $CH_2ClCOOH(aq) + H_2O(\ell) \rightleftharpoons CH_2ClCOO^-(aq) + H_3O^+(aq)$

(b) $C_6H_5COOH(aq) + H_2O(\ell) \rightleftharpoons C_6H_5COO^-(aq) + H_3O^+(aq)$

(c) $HCN(aq) + H_2O(\ell) \rightleftharpoons CN^-(aq) + H_3O^+(aq)$

In each, which base is conjugate to which acid? Rank the conjugate bases in order of increasing strength.

31. Now compute K_b, the base ionization constant, for the conjugate bases produced in the three reactions treated previously:

(a) $CH_2ClCOOH(aq) + H_2O(\ell) \rightleftharpoons CH_2ClCOO^-(aq) + H_3O^+(aq)$

(b) $C_6H_5COOH(aq) + H_2O(\ell) \rightleftharpoons C_6H_5COO^-(aq) + H_3O^+(aq)$

(c) $HCN(aq) + H_2O(\ell) \rightleftharpoons CN^-(aq) + H_3O^+(aq)$

32. Again, use data from the preceding exercises (also available in Appendix C) for this problem and the two following. . . . Given an

initial concentration of the salt $NaCH_2ClCOO$, calculate the pH of an aqueous solution in equilibrium at 25°C:

(a) $[NaCH_2ClCOO]_0 = 1.000\ M$

(b) $[NaCH_2ClCOO]_0 = 0.500\ M$

(c) $[NaCH_2ClCOO]_0 = 0.100\ M$

(d) $[NaCH_2ClCOO]_0 = 0.010\ M$

Predict, beforehand, whether the mixture is acidic, basic, or neutral.

33. Given an initial concentration of NaC_6H_5COO, calculate the pH of an aqueous solution in equilibrium at 25°C:

(a) $[NaC_6H_5COO]_0 = 1.000\ M$

(b) $[NaC_6H_5COO]_0 = 0.500\ M$

(c) $[NaC_6H_5COO]_0 = 0.100\ M$

(d) $[NaC_6H_5COO]_0 = 0.010\ M$

Predict, beforehand, whether the mixture is acidic, basic, or neutral.

34. Given an initial concentration of $NaCN$, calculate the pH of an aqueous solution in equilibrium at 25°C:

(a) $[NaCN]_0 = 1.000\ M$

(b) $[NaCN]_0 = 0.500\ M$

(c) $[NaCN]_0 = 0.100\ M$

(d) $[NaCN]_0 = 0.010\ M$

Predict, beforehand, whether the mixture is acidic, basic, or neutral.

35. Here, too, continue with the same three acid–base reactions in this exercise and the two following. . . . Given initial concentrations of $CH_2ClCOOH$ and CH_2ClCOO^-, calculate the pH of an aqueous solution in equilibrium at 25°C:

(a) $[CH_2ClCOOH]_0 = 0.500\ M,\ [CH_2ClCOO^-]_0 = 0.500\ M$

(b) $[CH_2ClCOOH]_0 = 1.000\ M,\ [CH_2ClCOO^-]_0 = 0.500\ M$

(c) $[CH_2ClCOOH]_0 = 0.500\ M,\ [CH_2ClCOO^-]_0 = 1.000\ M$

(d) $[CH_2ClCOOH]_0 = 1.000\ M,\ [CH_2ClCOO^-]_0 = 1.000\ M$

(e) $[CH_2ClCOOH]_0 = 1.000\ M,\ [CH_2ClCOO^-]_0 = 0.100\ M$

(f) $[CH_2ClCOOH]_0 = 0.100\ M,\ [CH_2ClCOO^-]_0 = 1.000\ M$

By how much does the pH of each buffer solution change upon addition of 0.005 mol hydronium or hydroxide ion to 1.00 L?

36. Given initial concentrations of C_6H_5COOH and $C_6H_5COO^-$, calculate the pH of an aqueous solution in equilibrium at $25°C$:

    (a) $[C_6H_5COOH]_0 = 0.500\ M$, $[C_6H_5COO^-]_0 = 0.500\ M$
    (b) $[C_6H_5COOH]_0 = 1.000\ M$, $[C_6H_5COO^-]_0 = 0.500\ M$
    (c) $[C_6H_5COOH]_0 = 0.500\ M$, $[C_6H_5COO^-]_0 = 1.000\ M$
    (d) $[C_6H_5COOH]_0 = 1.000\ M$, $[C_6H_5COO^-]_0 = 1.000\ M$
    (e) $[C_6H_5COOH]_0 = 1.000\ M$, $[C_6H_5COO^-]_0 = 0.100\ M$
    (f) $[C_6H_5COOH]_0 = 0.100\ M$, $[C_6H_5COO^-]_0 = 1.000\ M$

    By how much does the pH of each buffer solution change upon addition of 0.005 mol hydronium or hydroxide ion to 1.00 L?

37. Given initial concentrations of HCN and CN^-, calculate the pH of an aqueous solution in equilibrium at $25°C$:

    (a) $[HCN]_0 = 0.500\ M$, $[CN^-]_0 = 0.500\ M$
    (b) $[HCN]_0 = 1.000\ M$, $[CN^-]_0 = 0.500\ M$
    (c) $[HCN]_0 = 0.500\ M$, $[CN^-]_0 = 1.000\ M$
    (d) $[HCN]_0 = 1.000\ M$, $[CN^-]_0 = 1.000\ M$
    (e) $[HCN]_0 = 1.000\ M$, $[CN^-]_0 = 0.100\ M$
    (f) $[HCN]_0 = 0.100\ M$, $[CN^-]_0 = 1.000\ M$

    By how much does the pH of each buffer solution change upon addition of 0.005 mol hydronium or hydroxide ion to 1.00 L?

38. (a) Design a buffer to have a pH of 7.52 when the acid and its conjugate base are present in equal concentrations. Consult Appendix C as needed. (b) Show explicitly, on a graph, how the pH changes as the ratio $[A^-]/[HA^-]$ varies over the range

$$0.25 \leq \frac{[A^-]}{[HA]} \leq 4$$

    (c) Use the same acid and its conjugate base to design a buffer with a central pH of 7.71.

39. (a) Design a buffer to have a pH of 2.59 when the acid and its conjugate base are present in equal concentrations. Consult Appendix C as needed. (b) Show explicitly, on a graph, how the pH changes as the ratio $[A^-]/[HA]$ varies over the range

$$0.25 \leq \frac{[A^-]}{[HA]} \leq 4$$

(c) Use the same acid and its conjugate base to design a buffer with a central pH of 2.40.

40. A certain buffer solution contains 0.100 mole of propionic acid (CH_3CH_2COOH, $pK_a = 4.86$) and 0.100 mole of sodium propionate ($NaCH_3CH_2COO$) dissolved in a volume of 0.500 L. (a) What is the pH? (b) The amount of each component is increased by 0.010 mole. What happens to the pH? (c) The volume of the solution is doubled to 1.000 L. What happens to the pH?

41. Formic acid ($HCOOH$) has a pK_a of 3.75. If equal *masses* (grams) of $HCOOH$ and $NaHCOO$ are dissolved together, what is the pH of the resulting buffer solution?

42. The components indicated below are mixed together and then diluted to a total volume of 1.000 L. What is the pH of each solution?

(a) 0.100 L of 0.100 M HNO_3 + 0.200 L of 0.050 M NaOH
(b) 0.100 L of 0.100 M HNO_3 + 0.100 L of 0.100 M KOH
(c) 0.100 L of 0.100 M HNO_3 + 0.050 L of 0.200 M NaOH
(d) 0.100 L of 0.100 M HNO_3 + 0.050 L of 0.100 M $Ca(OH)_2$
(e) 0.100 L of 0.100 M HNO_3 + 0.100 L H_2O

43. The components indicated below are mixed together and then diluted to a total volume of 1.000 L. What is the pH of each solution?

(a) 0.100 L of 0.100 M HCl + 0.200 L of 0.100 M NaOH
(b) 0.100 L of 0.100 M HCl + 0.050 L of 0.100 M NaOH
(c) 0.100 L of 0.100 M HCl + 0.100 L of 0.100 M NaCl
(d) 0.100 L of 0.100 M HCl + 0.100 L of 0.100 M HNO_3
(e) 0.100 L of 0.100 M HCl + 0.100 L H_2O

44. The components indicated below are mixed together and then diluted to a total volume of 1.000 L. What is the pH of each solution?

(a) 0.100 L of 0.100 M HCOOH + 0.100 L of 0.100 M NaOH
(b) 0.100 L of 0.100 M HCOOH + 0.200 L of 0.050 M NaOH
(c) 0.100 L of 0.100 M HCOOH + 0.050 L of 0.100 M NaOH
(d) 0.100 L of 0.100 M HCOOH + 0.100 L of 0.050 M NaOH
(e) 0.100 L of 0.100 M HCOOH + 0.100 L of 0.100 M HCl
(f) 0.100 L of 0.100 M HCOOH + 0.100 L of 0.100 M HCOOH
(g) 0.100 L of 0.100 M HCOOH + 0.100 L H_2O

The ionization constant of formic acid ($HCOOH$) is $K_a(25°C) = 1.8 \times 10^{-4}$.

45. The components indicated below are mixed together and then diluted to a total volume of 1.000 L. What is the pH of each solution?

    (a) 0.100 L of 0.100 M NH_3 + 0.100 L of 0.100 M HCl
    (b) 0.100 L of 0.100 M NH_3 + 0.200 L of 0.050 M HCl
    (c) 0.100 L of 0.100 M NH_3 + 0.050 L of 0.100 M HCl
    (d) 0.100 L of 0.100 M NH_3 + 0.100 L of 0.050 M HCl
    (e) 0.100 L of 0.100 M NH_3 + 0.100 L of 0.100 M NaOH
    (f) 0.100 L of 0.100 M NH_3 + 0.100 L of 0.100 M NH_3
    (g) 0.100 L of 0.100 M NH_3 + 0.100 L H_2O

    The base ionization constant for ammonia is $K_b = 1.8 \times 10^{-5}$.

46. A solution of hydrochloric acid, concentration unknown, is titrated against 0.100 M NaOH. (a) If 50.0 mL of acid are neutralized by 76.2 mL of base, what was the acid's original concentration? (b) What is the pH of the neutralized solution?

47. A solution of sodium hydroxide, concentration unknown, is titrated against 0.100 M HCl. (a) If 50.0 mL of base are neutralized by 76.2 mL of acid, what was the base's original concentration? (b) What is the pH of the neutralized solution?

48. A solution of hydrofluoric acid (HF), concentration unknown, is titrated against 0.100 M NaOH. (a) If 50.0 mL of acid are neutralized by 76.2 mL of base, what was the acid's original concentration? (b) What is the pH of the neutralized solution? $K_a(25°C) = 6.8 \times 10^{-4}$ for HF.

49. A solution of ammonia, concentration unknown, is titrated against 0.100 M HCl, reacting as

    $$NH_3(aq) + HCl(aq) \rightleftharpoons NH_4^+(aq) + Cl^-(aq)$$

    (a) If 50.0 mL of base are neutralized by 76.2 mL of acid, what was the base's original concentration? (b) What is the pH of the neutralized solution? $K_b(25°C) = 1.8 \times 10^{-5}$ for NH_3.

50. Suppose that 0.250 M HCl is added incrementally to 50.0 mL of 0.100 M $C_2H_5NH_2$ (ethylamine). (a) What is the pH at the start of the titration? (b) At the half-equivalence point? (c) At the equivalence point? (d) At a volume 2.0 mL beyond the equivalence point? (e) Use these four points to sketch the titration curve (pH versus volume of acid added). The base ionization constant for $C_2H_5NH_2$ is $K_b(25°C) = 6.4 \times 10^{-4}$.

51. (a) Plot the full titration curve (pH versus volume of base added) for the addition of 0.250 M NaOH to 50.0 mL of 0.100 M CH_3CH_2COOH (propionic acid). (b) What is the pH at the start of the titration? (c) At the half-equivalence point? (d) At the equivalence point? (e) At a volume 2.0 mL beyond the equivalence point? The ionization constant for CH_3CH_2COOH is $K_a(25°C) = 1.4 \times 10^{-5}$.

52. Given an initial concentration of sulfuric acid (H_2SO_4), calculate the pH of an aqueous solution in equilibrium at 25°C:

    (a) 1.000 M (b) 0.500 M (c) 0.100 M (d) 0.010 M

    The dissociation of the first proton is effectively 100%, whereas K_{a2} for the second ionization is 0.012. In which of the four solutions is the final concentration of H_3O^+ influenced most by K_{a2}?

53. Given an initial concentration of hydrosulfuric acid (H_2S), a diprotic species, calculate the pH of an aqueous solution in equilibrium at 25°C:

    (a) 1.000 M (b) 0.500 M (c) 0.100 M (d) 0.010 M

    The successive ionization constants for H_2S are 8.9×10^{-8} and 1×10^{-19}.

54. Last words (1): (a) Is pH an intrinsic property of an acid? Is it proper simply to say that such-and-such an acid has a pH of such-and-such, without specifying anything else about the system? (b) Similarly, is pK_a an intrinsic property of an acid? Why or why not? (c) Draw an analogy between pH and pK_a for an acid (Chapter 16) and ion concentration and K_{sp} for an inorganic solute (Chapter 15).

55. Last words (2): $Ca(OH)_2$ is classified as a strong Arrhenius base even though its solubility in water is comparatively low. Explain what it means to be a strong base but a poor solute.

56. Last words (3): The autoionization constant of water, $K_w(T)$, goes from 1.0×10^{-14} at 25°C to 1.5×10^{-13} at 70°C. (a) Is the reaction

    $$2H_2O(\ell) \rightleftharpoons H_3O^+(aq) + OH^-(aq)$$

    exothermic or endothermic? (b) Is pure water acidic, basic, or neutral at 70°C? (c) How about at 0°C? (d) Guess the pH of pure water at 90°C, choosing from the following list of values: 12.8, 7.8, 7.4, 7.0, 6.6, 6.2, 1.2.

17

Chemistry and Electricity

17-1. Putting Reactions to Work
17-2. Pushing Electrons
17-3. Redox Redux
17-4. From Chemical Energy to Electrical Work
17-5. Completing the Circuit
17-6. Electrochemical Life and Death
17-7. Half-Reactions
        Standard Electrode Potentials
        Reductants and Oxidants
17-8. Charge and Mass: Balancing the Redox Equation
        Acidic Solution
        Basic Solution
17-9. Going Uphill: Electrolysis
        REVIEW AND GUIDE TO PROBLEMS
        EXERCISES

17-1. PUTTING REACTIONS TO WORK

No chemical change is without cost or consequence. No transformation goes by without bringing fresh disorder to the world, and so there is always a bill to be paid. Is a salt to dissolve? An acid to ionize? An electron to move? Whatever may happen, the universe demands its due. Increased entropy is the price of change.

Conservation of energy, maximization of entropy . . . those are the lessons of thermodynamics, already familiar to us. We know, too, that nature takes payment in the local currency of *free energy*, leaving us to

607

understand any chemical reaction in the starkest of terms, simply as

reactants ⟶ products + free energy

Just strip away the details of each process (which reactants? which products?), and discover what is common to all: Matter at a high chemical potential goes to matter at a low potential, accompanied by release of the excess free energy on the road to equilibrium. Whenever, wherever, and however matter alters its form, somewhere there is a corresponding evolution of free energy. Out comes an *energy*, denominated in units of joules per mole; out comes, in particular, the Gibbs free energy

$$\Delta G = \Delta H - T\,\Delta S$$

when a process occurs at constant temperature and pressure.

ΔG is indeed an energy, and from the application of such energy comes the ability to do work—work to move matter, work to create entropy. ΔG brings disorder to system or surroundings or both during a reaction, stirring up the particles as heat flows in and out.

Either way, the reaction does its work. Matter is made to move. Particles pick up speed; a gas expands against a piston; a matter-rearranging explosion occurs. An electron moves. Sometimes useful and sometimes wasted, work (or, equivalently, heat) is implicit in any natural process.

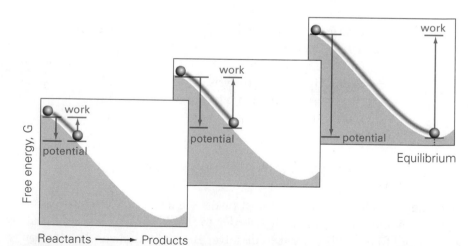

FIGURE 17-1. Like a stone rolling down a hillside, a reacting system releases energy. From top to bottom, the process converts differing amounts of *potential* energy—energy not yet realized as work—into a form that can be harnessed and used. The instantaneous difference in Gibbs free energy, ΔG, is the maximum work available from each reaction mixture under conditions of constant pressure and temperature.

With that reaffirmation, made graphic in Figure 17-1, we come now to one further manifestation of the free energy: the electrical work done during a redox reaction, whereby an electron is transferred from point to point. It is a question of some practical importance, since here is work that we can channel into an electric current and use for our own devices. Here, in the practice of **electrochemistry**, is where the hidden electrical character of matter turns into a perceptible flow of electricity; where "chemical" energy is used purposefully to push electrons along a wire and into an external circuit.

17-2. PUSHING ELECTRONS

Look to a ski slope to glimpse the essence of an electric circuit. One after another the skiers glide down the mountain, falling from high gravitational energy to low. They trade potential energy for kinetic—height for speed—and they apply that energy both to their own motion and (as heat) to the motion of their surroundings. They accelerate and decelerate. They stir up a breeze. They melt some snow. And eventually they reach the bottom, drained of gravitational energy, unable to climb up the mountain unassisted.

But here comes a lift to transport the low-potential skiers back to the top, and so the whole process can begin anew. Down they go again, releasing the potential energy acquired during the recent ascent. Soon there develops an unbroken loop of skiers climbing up and gliding down; think of it as a circuit, a gravitational circuit.

Now consider what is needed to maintain this operation. First: *a mountain*, to create a difference in gravitational potential energy. Without that difference there is no opportunity for unassisted work and no impetus for motion downhill. Second: *a skier*, to act as a mass-endowed body able to fall downward in the gravitational field. Third: *a path* for the skier (a slick layer of snow), to facilitate passage from high to low potential. Fourth: *a lift*, to bring fresh skiers back to the top of the mountain.

We, too, shall need a metaphorical mountain such as in Figure 17-2 if we intend to drive electrons around a circuit—not a mountain of gravitational potential energy, of course, but rather a mountain of electrical potential energy. Imagine that we place a charged particle in an environment where its electrical energy depends upon position in space, so as to be high at one point and low at another. The object then moves from site to site in this *field* of electrical potential (see Chapter 1 to recall how), suggesting that our mountain will best be calibrated in joules per coulomb: a **voltage**, the electrical work done to move one unit of charge. To accelerate a charge of 1 coulomb through a potential of 1 volt

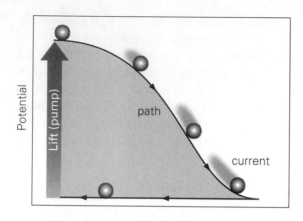

FIGURE 17-2. A symbolic circuit: Particles in a field fall from high energy to low, impelled downward by a drop in potential. At the bottom, they receive a boost back to the top. The round trips continue for as long as the lift operates.

means, for example, that the particle loses 1 joule of electrical energy while moving from point A to point B. It slides down the electrical slope in the same way that skiers come down a mountain, converting potential energy into kinetic energy and heat.

Analogous to the skiers are the electrons or ions in an electric circuit. Give these charged bodies a suitable path (a wire, say, or an electrolytic solution) and let them flow downhill from high potential energy to low. After the fall, pump them up from low energy to high and send them on their way again. Do so, over and over, and a current will run through the closed circuit without interruption.

How shall we drive it? With what kind of electrical operation might we sustain a current? With familiar devices like batteries, power supplies, and generators—each, in its own way, brings a fresh supply of charge carriers to a point of high electrical energy. Each gets the extra energy from somewhere else in the universe, taking it from sources such as falling water, solar radiation, and (our present interest) reduction–oxidation reactions.

The point is only this: To make any kind of electricity, we need (1) a difference in electrical potential between two sites, (2) a source of electrons, (3) a connecting path, and (4) a pump to recycle electrons from low to high energy. An electrochemical circuit, driven by a free energy of reaction, is no exception. Aware of these basic requirements, let us now find and assemble the components necessary to make chemical electricity.

17-3. REDOX REDUX

For electrons we shall draw upon the elementary redox reaction discussed in Chapter 3, under which one species (an *oxidant*, or *oxidizing agent*) takes electrons from something else (a *reductant*, or *reducing agent*). ***Oxidation*** denotes a loss of electrons; ***reduction***, a gain.

The terminology is logical, but potentially confusing. Consider: The reductant *loses* the electrons. The reducing agent *causes* a reduction to take place in some other species, and so the reducing agent itself undergoes oxidation (*not* reduction, despite the suggestive name):

$$\text{reductant} \longrightarrow \text{oxidized species} + \text{electrons} \qquad \text{(oxidation)}$$

$$\text{Zn} \longrightarrow \text{Zn}^{2+} + 2e^- \qquad \text{(electron loss)}$$

Its oxidation number *increases*, going from 0 to +2 in the example indicated.

The oxidant *gains* the electrons. The oxidizing agent *causes* an oxidation to take place; hence the oxidizing agent, acting as a receiver for the liberated electrons, undergoes reduction:

$$\text{oxidant} + \text{electrons} \longrightarrow \text{reduced species} \qquad \text{(reduction)}$$

$$\text{Cu}^{2+} + 2e^- \longrightarrow \text{Cu} \qquad \text{(electron gain)}$$

Its oxidation number *decreases*.

Oxidation and reduction; you can't have one without the other. For every electron lost there is an electron gained; for every oxidation there is a reduction. For every oxidant there is a reductant. Redox is—just like the transfer of H^+—a coordinated two-step process, proceeding according to the general scheme

$$\text{reductant} + \text{oxidant} \longrightarrow \text{oxidized species} + \text{reduced species}$$

$$\text{Zn} + \text{Cu}^{2+} \longrightarrow \text{Zn}^{2+} + \text{Cu}$$

set forth in Figure 17-3.

It is a reciprocal exchange, thermodynamically equivalent to the interaction between acid and base:

$$\text{acid} + \text{base} \longrightarrow \text{conjugate base} + \text{conjugate acid}$$

Just as an acid loses a proton to produce a conjugate base, a reductant loses electrons to produce a conjugate oxidant. What comes off can be reattached. The oxidized species is now an oxidant. It can gain electrons.

But just as a *strong* acid leaves behind a *weak* conjugate base, a strong reductant leaves behind a weak conjugate oxidant. What comes off easily is correspondingly hard to reattach. Once oxidized, a formerly strong reducing agent tends to stay that way: oxidized, lacking some portion of its original electron density.

And just as any acid–base equilibrium is determined by the difference

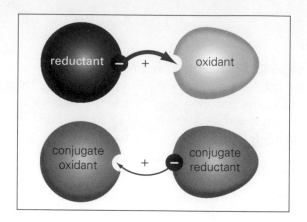

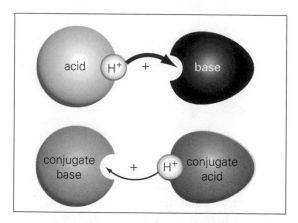

FIGURE 17-3. Reductant and oxidant; acid and base. The names are different, but the ideas are the same: Reactants high in free energy become products low in free energy. The strong giver gives to the strong taker, in the direction of the thick arrow. The strong giver, having given, becomes a weak taker. The strong taker, having taken, becomes a weak giver.

in free energy, so is a redox equilibrium. For the former, the largest drop occurs when strong acid and strong base go to weak conjugate base and weak conjugate acid. For the latter, the largest drop occurs when strong reductant and strong oxidant go to weak conjugate oxidant and weak conjugate reductant. Everything is the same. Should it be otherwise?

17-4. FROM CHEMICAL ENERGY TO ELECTRICAL WORK

Closer we come to building an electrochemical cell, having acquired both a source of electrons and a mountain of *chemical* energy—this energy coming, as always, from the difference in G between products (here the oxidized and reduced species, such as Zn^{2+} and Cu) and reactants (the reductant Zn and oxidant Cu^{2+}). The next steps are to convert chemical potential energy into electrical work and to provide a suitable conducting path for the electrons.

Start with no external path at all. Start with just a strip of metallic zinc dipping into a solution of copper sulfate, $CuSO_4$, so that the hydrated Cu^{2+} ions come into direct contact with Zn metal:

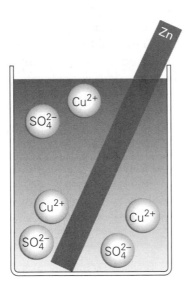

A reaction begins. Originally a rich blue, the $CuSO_4$ solution soon loses color as Cu^{2+} begins to plate out as metallic copper onto the zinc. Cu replaces Zn, and an equal number of Zn^{2+} ions are dissolved away in the process. The SO_4^{2-} ions, which suffer no change in oxidation state, play the role of spectators:

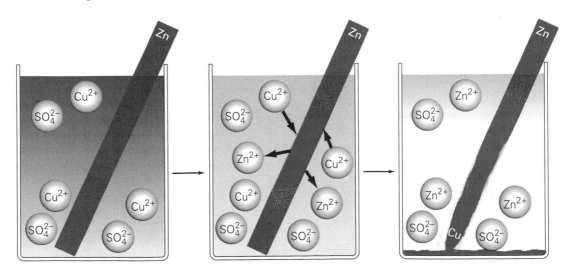

We recognize the transaction as a redox exchange of two electrons. Zn gives and Cu^{2+} takes, leaving us with the net reaction

$$\begin{aligned} Zn(s) &\longrightarrow Zn^{2+}(aq) + 2e^- \\ 2e^- + Cu^{2+}(aq) &\longrightarrow Cu(s) \\ \hline Zn(s) + Cu^{2+}(aq) &\longrightarrow Zn^{2+}(aq) + Cu(s) \end{aligned}$$

High G to low G, the redox reaction goes forward unaided. The Cu^{2+} ion, a stronger oxidant than metallic zinc, wrests away the electrons and transforms itself into solid metal. *Strong* oxidant and *strong* reductant go to weak conjugate reductant and weak conjugate oxidant, liberating free energy as they fall downhill.

Speak of "chemical" energy if you like, but remember that all chemical interactions are inherently electrical; and here, in this redox reaction, is surely a most fundamental electrical event: a transfer of electrons, a movement of charge accompanied by a change in electrical energy. It is the germ of a current.

Go to the source. Electrons have been moved, and the reactants have spent part of their potential energy to move them. Before the shift, the system stood at a certain free energy. After the shift, it stands at some other free energy. The free energy of reaction

$$\Delta G = G_{\text{after}} - G_{\text{before}}$$

thus supplies the impetus that drives two moles of electrons from one mole of Zn to one mole of Cu^{2+}; and, properly harnessed, this drop in potential energy offers us the capacity to do purposeful work outside the system. For that, we define a difference in electrical potential (an *electromotive force*, $\mathscr{E}$) to match the difference in chemical potential (ΔG per mole).

There it is, embodied in the chemical redox reaction: an electromotive force, an agency to move electrons; equivalently, a *voltage*—some number of joules sufficient to push some number of coulombs.

How many joules? The maximum work available is the free energy (ΔG) released by the electrons as they come down the slope represented in Figure 17-4, slipping from high energy to low, from strong to weak. It will be work done *by* the redox electrons on charged particles *outside* the system, presumably those in some external electrical device.

How many coulombs? Multiply the elementary charge (1.6×10^{-19} coulombs per electron) by the total number of electrons, first defining the **Faraday constant**

$$\mathscr{F} = \frac{1.60217733 \times 10^{-19} \text{ C}}{\text{electron}} \times \frac{6.0221367 \times 10^{23} \text{ electrons}}{\text{mol}}$$

$$= 96{,}485.31 \text{ C mol}^{-1}$$

as the magnitude of the charge on one mole of electrons. With n denoting the moles of electrons transferred, we then have $-n\mathscr{F}$ for the total negative charge.

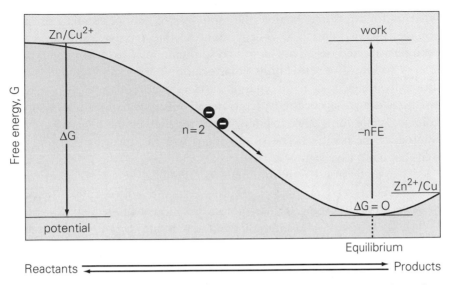

FIGURE 17-4. The goal is to use the free energy of a redox reaction to drive electrons around a circuit. As long as the reacting mixture remains out of equilibrium, its residual chemical energy (ΔG) furnishes the electrical work ($-n\mathscr{F}\mathscr{E}$) to move n moles of electrons across a voltage $\mathscr{E}$. The Faraday constant, $\mathscr{F}$, denotes the charge carried by one mole, and the value $-n\mathscr{F}\mathscr{E}$ represents the maximum amount of work available under idealized conditions—when the transformation occurs *reversibly*, inching forward in a series of infinitesimally small steps.

Work and charge, together, give us the voltage associated with a redox process,

$$\mathscr{E} = \frac{\text{work}}{\text{charge}} = -\frac{\Delta G}{n\mathscr{F}}$$

and the accompanying relationship between free energy and electrical work:

$$\Delta G = -n\mathscr{F}\mathscr{E}$$

Coulombs ($-n\mathscr{F}$) multiplied by joules per coulomb ($\mathscr{E}$) yields joules, and the result is a *change* in energy: a change in chemical energy and a matching change in electrical energy. We interpret $\mathscr{E}$, then, also as a change—as the change in voltage suffered by an electron moving from one point to another. It is the difference between values at two separate locations (such as the 12 volts *between* the + and − terminals of an automobile battery).

Note additionally that the road downhill (negative ΔG) corresponds, curiously, to an increase in voltage for the negatively charged electron, whereas for a positive particle the same loss of energy entails a drop in voltage. Inconsistent? No; energy alone is the decisive criterion, and both positive and negative charges go spontaneously from high electrical energy to low. The rest is only a matter of definition, and voltage is traditionally defined as if all charge carriers were positive. A negative charge is drawn, accordingly, from negative to positive potential (up in voltage) while going *down* in energy. It is repelled by the negative and attracted by the positive. The spontaneous work it does, *outside* the system, has a negative sign.

That understood, our mountain of chemical energy is recast as a mountain of electrical energy. With it we acquire a reservoir of work to be extracted from the redox reaction, a source from which we might build a usable electric circuit. But understand also that it is a mountain of finite height, destined never to exceed the moment-to-moment difference in free energy between a mixture and its eventual point of equilibrium. That height represents the maximum amount of work to be garnered from a spontaneous process, and then only if the free energy is harvested with 100% efficiency (which never happens).

17-5. COMPLETING THE CIRCUIT

We have, so far, only the slightest hint of an electric current in our all-in-one Zn/Cu system. True, from Zn to Cu^{2+} there is undeniably a tiny current—a current so narrowly confined, though, that we can scarcely hope to thread these electrons through a macroscopic electric circuit. The givers and takers of electrons are too close. Charge hops from one species to the other too easily, traveling only the microscopic distance between ion and atom. There is no space to interpose a wire, and hence no sustained flow of current. Undirected and unchanneled, the energy of reaction is dissipated as heat.

To produce electricity of the day-to-day sort, we must separate the electron givers from the electron takers. Suppose therefore that we spread the redox reaction out in space, so that the electrons must traverse some macroscopic distance while going from donor to receiver. Oxidation will take place at one site, and reduction at another. A charge that builds up at one pole will flow to the other through a manageable electric circuit.

Such is the construction of a **galvanic cell**, a circuit driven by the free energy of a spontaneous redox reaction. Also called *voltaic* cells, these sources of power are everywhere around us. They exist as dry cells, wet cells, fuel cells, concentration cells, and more . . . particularly in the familiar form of *batteries*, or lumped arrays of cells. The nickel–cadmium battery for electronic equipment, the lead storage battery for automotive ignition, the cardiac pacemaker, and the humble flashlight battery all come quickly to mind. Each has its own use and its own peculiarities, but each operates according to the same guiding principle: make the electrons travel some appreciable distance as they go from reductant to oxidant.

True to that principle, let us retain the Zn/Cu system and use just one basic design to expose the features common to most cells. The prototype throughout will be the simple two-compartment arrangement shown in Figure 17-5, which we shall represent by the shorthand notation

$$Zn(s)|ZnSO_4(aq)\|CuSO_4(aq)|Cu(s)$$

Here is what happens. Oxidation takes place in one compartment, at the left; reduction takes place in the other compartment, at the right. At the left is a strip of metallic zinc, dipping here into a solution of

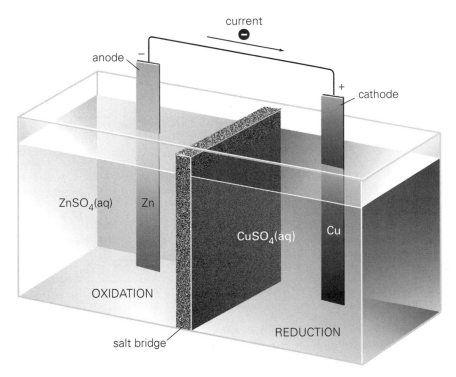

FIGURE 17-5. A simple galvanic cell, powered by the free energy of a spontaneous redox reaction: Electrons, released by oxidation at the anode, flow toward the positively charged cathode, site of reduction. A migration of ions through the connecting bridge completes the circuit.

$ZnSO_4$. The zinc metal functions as an **electrode**, a conductor able to supply mobile electrons to a circuit. So, too, does the copper electrode at the right, which dips into a solution of $CuSO_4$. Connecting the two metallic electrodes are a wire and a *salt bridge*, soon to be explained.

Oxidation occurs at one of the electrodes, traditionally called the **anode**. Reduction takes place at the other electrode, the **cathode**, yet these names remain entirely subordinate to the underlying functions. What is preserved for all cells and all materials is just this one convention: oxidation occurs at the anode, reduction at the cathode.

Names can change, after all, since redox is a competition to match the strongest oxidant with the strongest reductant. Few substances *always* undergo reduction or *always* undergo oxidation; those roles depend entirely on the nature of the competition. The same zinc electrode, say, may be an anode in a cell of one composition but a cathode in another, switching function according to the reducing power of its partner.

Paired with copper, zinc acts as a reducing agent. It undergoes oxidation, from Zn to Zn^{2+}. It loses electrons. The Zn electrode (acting as *anode*) sheds Zn^{2+} ions into the aqueous $ZnSO_4$ solution, leaving behind excess electrons on the metallic conductor:

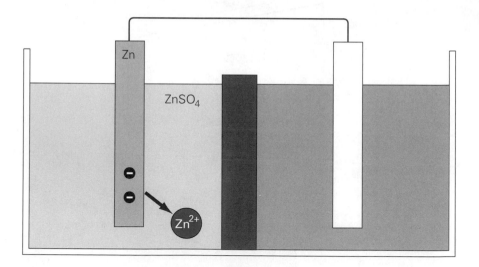

Negative charge builds there as more electrons are abandoned, making the anode a site of high electrical energy. Having hosted an oxidation reaction, the anode now harbors too many particles with the same charge. Somewhere else, perhaps, their energy might be lower.

In the other compartment, separated in space, the Cu^{2+} ion serves as oxidizing agent. It undergoes reduction, from Cu^{2+} to Cu. It gains electrons;

it gains them from the Cu electrode (the cathode), which draws upon the metal's reserves to award two electrons to each oncoming Cu^{2+}. Additional metallic copper is thereby deposited onto the electrode, and the conductor grows increasingly electron-poor:

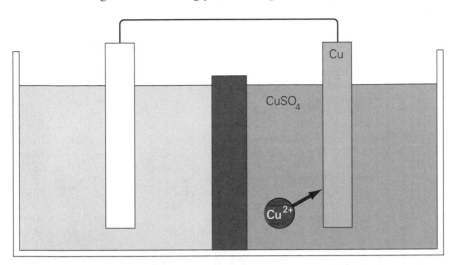

Too few electrons are distributed among too many nuclei. The cathode becomes for the electrons a site of lower electrical energy, growing less negative (more positive) than usual.

Clearly the arrangement is out of balance. It cannot persist. There is a surplus of electrons at the anode, created by the oxidation of Zn and the subsequent departure of Zn^{2+}. At the cathode there is a deficit of negative charge, drained away during the reduction of Cu^{2+} to Cu. One electrode is too negative, and the other is too positive.

The imbalance goes even further, for the aqueous solutions are threatened with excess charge as well. Originally neutral, each electrolytic solution is now being asked to accept more ions with the *same* charge—either all positive or all negative. The Zn electrode dumps extra Zn^{2+} into the $ZnSO_4$ compartment, making it more positive; the Cu electrode removes Cu^{2+} from the $CuSO_4$, leaving behind uncompensated SO_4^{2-} anions. The left-hand solution is too positive, and the right-hand solution is too negative.

Nature will not tolerate such an imbalance of charge, nor will nature allow these separated oxidation and reduction reactions to proceed uncoordinated. For if Zn is to lose its two electrons, then some other species must gain them. The orphaned electrons cannot simply be left on the anode while the Zn^{2+} cations are dropped—unescorted—into a neutral solution. Rather, Zn must donate its electrons to a receptive species like Cu^{2+} or, failing that, the metal must refrain from oxidizing.

But remember: It was *we* who opted to separate the reductant from its oxidant. *We* chose to alienate Zn from Cu^{2+}, and now *we* shall have to allow reductant and oxidant to make electrical contact. That we can do, in part, with the wire that runs from the electron-rich anode to the electron-poor cathode:

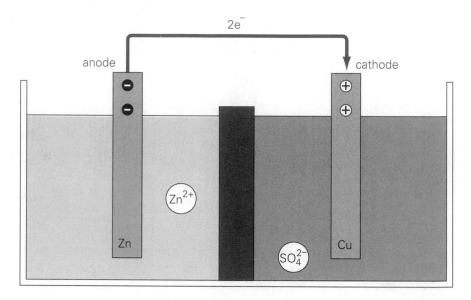

Along this conducting path the negative charge will flow from donor to acceptor, from the negative pole (whence the electron is repelled) to the positive pole (whither it is attracted). Along that path the free energy of reaction will be converted into electrical work. Along that path will develop a difference in voltage: the **cell potential**, $\mathcal{E}$, as promised in the relationship $\Delta G = -n\mathcal{F}\mathcal{E}$.

Then, to rebalance the electrolytic solutions and close the circuit, we must provide additional negative charge (anions) for the anode compartment and additional positive charge (cations) for the cathode. One alternative is to employ a **salt bridge**, which can be as simple as an inverted U-tube containing a solution of nonreactive common ions (Na_2SO_4 will do for the present example). Many designs are possible, but all aim to provide a path for charge to return from cathode to anode. A semiporous plug, for example, will allow ions to flow from the bridge into each compartment while still keeping the two electrolytes unmixed. And there, with that additional source of ions, neutrality is restored to the solutions: *anions* (SO_4^{2-}) migrate toward the *anode*, and *cations* ($2Na^+$) migrate toward the *cathode*:

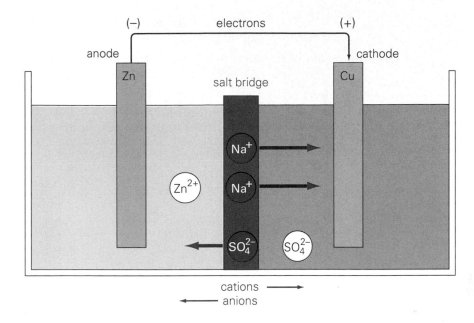

The circuit is complete. An electron from the oxidized species is deposited on the anode and drawn toward the cathode, whereupon it is attached to the species undergoing reduction. The loop is then closed by a directed ion flow within the electrolytic solutions, facilitated by the salt bridge. A current flows.

Produced by a stream of electrons, it is a regular electric current: an ordinary electric field propagating along an ordinary wire, able to do work on other charged particles. We are free to use its power to animate a radio, a computer, a television, a wristwatch, or any device similarly critical to human welfare and happiness.

17-6. ELECTROCHEMICAL LIFE AND DEATH

Thus is born a galvanic cell. It starts out away from equilibrium, with reactants higher in free energy than products, and goes downhill thereafter. The cell's working life is spent away from equilibrium—as are all lives—but less and less with each passing instant. Ultimately, having come to equilibrium, its mountain of chemical potential has been eroded down to a level plain. Nothing lasts forever.

Impelled by a fall in free energy, electrons go from anode to cathode only while such difference still endures. Fresh electrons will be released only for as long as ΔG remains negative, for only then is there a drive to form products. The difference in free energy creates the chemical energy that the galvanic cell converts into a voltage, and that difference is all there is. Take it away and the cell dies.

Indeed the mountain of chemical potential (Figure 17-6) crumbles right from the start, leaving the cell to die a bit with every redox transaction. Each oxidation of Zn to Zn^{2+} chips away at the anode and increases the concentration of Zn^{2+} in solution. Each reduction of Cu^{2+} to metallic Cu adds mass to the cathode while simultaneously diluting the $CuSO_4$ solution. Steadily the reaction quotient

$$Q = \frac{[Zn^{2+}]}{[Cu^{2+}]}$$

changes as the reaction

$$Zn(s) + Cu^{2+}(aq) \longrightarrow Zn^{2+}(aq) + Cu(s)$$

progresses toward equilibrium; and steadily the instantaneous difference in free energy,

$$\Delta G = \Delta G^\circ + RT \ln Q$$

changes along with it (as we know from Chapter 14).

The driving force, ΔG, diminishes to zero on the way to equilibrium, just as in any other spontaneous process. It becomes harder and harder to get work from the system, until finally there is nothing more to

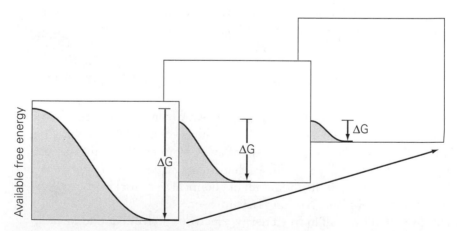

FIGURE 17-6. Proportional to the free energy available each instant, the electromotive force changes continuously as reactants turn into products. The closer the system comes to equilibrium, the smaller is the difference in chemical potential. Eventually the cell dies, drained of its driving force when ΔG falls to zero at equilibrium.

extract. Both ΔG and the corresponding cell potential

$$\mathscr{E} = -\frac{\Delta G}{n\mathscr{F}}$$

are zero when the redox reaction comes to equilibrium.

This three-way link involving ΔG (the up-to-the-moment difference in free energy between products and reactants), $\Delta G°$ (the difference under standard conditions), and Q (the reaction quotient) was our crowning thermodynamic result of Chapter 14. The same connection reappears now, upon substitution of $\mathscr{E}$ for G, as the **Nernst equation**,

$$\mathscr{E} = \mathscr{E}° - \frac{RT}{n\mathscr{F}} \ln Q$$

which similarly relates the instantaneous cell potential ($\mathscr{E}$) to the progress of reaction (Q). Given this equation, we immediately gain a way to compute the potential for a cell operating reversibly at any concentration and at any temperature.

Key to the relationship is the **standard cell potential**, $\mathscr{E}°$, derived straightforwardly from the standard change in free energy as

$$\mathscr{E}° = -\frac{\Delta G°}{n\mathscr{F}}$$

$\mathscr{E}°$ is the voltage measured when the cell operates under standard conditions (defined here so that all substances are in their normal states at 1 atm and 25°C, and all solutions are at 1 M concentration). The value of $\mathscr{E}°$ happens to be 1.10 volts for the $Zn(s)|ZnSO_4(aq, 1\ M)\|CuSO_4(aq, 1\ M)|$ $Cu(s)$ cell.

That means: The zinc–copper cell, running at 25°C and 1 M, imparts 1.10 joules of energy to *each* coulomb of charge transferred from Zn metal to Cu^{2+} ions. If 1 coulomb is moved, the maximum work done is 1.10 joules. If 2 coulombs are moved, the work doubles to 2.20 joules but the voltage (joules *per* coulomb) is still the same. It is an intensive property, dependent not on the system's total size but instead on the concentration. Large cell or small cell, $\mathscr{E}°$ remains unchanged under standard conditions.

Observe also that $\mathscr{E}°$ stands in the same relationship to the equilibrium constant, K, as does $\Delta G°$, going as

$$\mathscr{E}° = \frac{RT}{n\mathscr{F}} \ln K$$

So it must, since free energy and cell potential differ only by that one factor, $-n\mathcal{F}$, signifying the total charge. Just as ΔG goes to zero when Q goes to K, so does $\mathcal{E}$. $\mathcal{E}$ *is* ΔG. The cell voltage is nothing more than an electrical expression of the difference in free energy between products and reactants.

Yet even this simple correspondence has its consequences. The relationship between free energy and voltage allows us to measure equilibrium constants by *electrical* means, provided that we can arrange for the reaction to proceed reversibly in an electrochemical cell. Often it is a big help, especially where concentrations are excessively dilute and hence hard to measure directly (as with sparsely soluble salts). Knowing that $\mathcal{E}°$ and $\ln K$ are directly proportional at constant T, we measure the cell potential under standard conditions and then go on to predict K. Both solubility-product constants and pH values, for instance, are susceptible to such electrochemical measurement.

17-7. HALF-REACTIONS

Everything there is to know about the cell's thermodynamics is conveyed by its standard potential. One number, measured under one set of conditions, tells all—not just $\Delta G°$ and K at 25°C, but (thanks to the Nernst equation) both $\mathcal{E}$ and ΔG at any other temperature and reaction quotient as well.

$\mathcal{E}°$ emerges as an electrochemical benchmark, a way to rank one redox pair against another. A disguised equilibrium constant, the cell potential lends quantitative meaning to the terms "strength" and "weakness" as applied to electron transfer. $\mathcal{E}°$ plays the same role for electron transfer as pK plays for proton transfer. It is the electrical driving force, a concise measure of an electron's tendency to go from here to there.

The electrical work is paid for by chemical energy, and so from a negative $\Delta G°$ comes a positive $\mathcal{E}°$. The voltage drives the electrons forward, from giver to receiver. The more positive the cell potential, the more spontaneous is the forward cell reaction. More electrons go from reductant to oxidant; more product is formed; the equilibrium constant is larger. From energy comes work, and $\mathcal{E}°$ provides the connecting link between the two.

A certain difficulty remains still to be overcome, however, before the standard potential can be put to effective use. For, as matters currently stand, we appear forced to measure $\mathcal{E}°$ for every possible combination of oxidant and reductant. Each new cell seems to demand a new experiment.

Zn/Cu? Put them together, and measure the voltage. Al/Ag? Record another value. Al/Mn? Zn/Sn? Ag/Zn? Make three more measurements; and more after that for other combinations, and still more again and again. But to embark on such a road would be hopelessly inefficient, a course doomed inevitably to fail. Better, if we can, to measure something *just* for Zn alone, just for Al, just for Ag, just for Sn, just for one species at a time; and then, only later, match up oxidant with reductant. Hess's law gives us precisely that freedom to manipulate thermodynamic functions of state.

Standard Electrode Potentials

Why not, therefore, proceed by halves: by ***half-reactions***, so that the reduction is viewed separately from the oxidation. To characterize the impetus just to *give* an electron, why not define some ***standard oxidation potential***, $\mathscr{E}^\circ_{ox}$, existing all by itself? Having done so, we could then combine $\mathscr{E}^\circ_{ox}$ with the appropriate ***standard reduction potential***, $\mathscr{E}^\circ_{red}$, to obtain the total voltage as

$$\mathscr{E}^\circ = \mathscr{E}^\circ_{red} + \mathscr{E}^\circ_{ox}$$

Under this scheme, the oxidation half-reaction of Zn will contribute its proportionate share of the potential, as will the reduction of Cu^{2+}. Each will have its own voltage, and together they will sum to the total potential of the Zn/Cu cell—and, most important, each half-potential will function independently as a modular quantity, ready to contribute the same amount when paired with any other partner.

If so, then all we need is to compile a relatively short table of $\mathscr{E}^\circ_{red}$ and $\mathscr{E}^\circ_{ox}$ for various half-reactions, measured under standard conditions. The rest will be handled by mixing and matching the half-potentials.

Unconvinced? Does it seem improper to treat oxidation independently of reduction, given that the one demands the other. Half-reaction or not, a cell needs both anode and cathode. Half-reaction or not, an electron donor needs an electron acceptor. Half-reaction or not, Zn needs a place to put its two electrons.

But that place need not be Cu^{2+}. For present purposes, in fact, the reduced species can be anything at all, provided that it remains the *same* species for every other half-reaction we plan to measure. Our aim is only to find a consistent way to compare strength and weakness relative to some arbitrary zero. Let there be, accordingly, some specially designated redox partner against which all other candidates can be compared; and let that one half-reaction, chosen arbitrarily, always be implemented at one of the cell's electrodes. It is the same approach taken with acids and bases, where water sets the standard for givers and takers of protons.

What is sensible for acids and bases must, arguably, apply to reductants

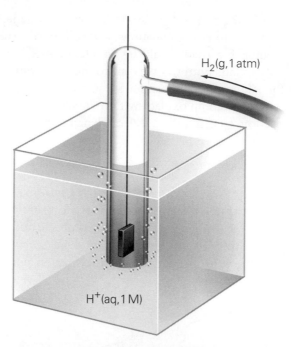

FIGURE 17-7. The standard hydrogen electrode, arbitrarily assigned a null voltage for the reduction of $2H^+$(aq, 1 M) to H_2(g, 1 atm). Hydrogen gas bubbles over a platinum electrode sensitive to changes in voltage. The platinum is electrically receptive but chemically inert to H_2 and H^+.

and oxidants too. For just as we observe whether an acid gives or takes a proton from H_2O, we can also test the strength of any reductant against H^+ and H_2 in a benchmark half-reaction

$$2H^+(\text{aq, } 1\ M) + 2e^- \longrightarrow H_2(\text{g, } 1\ \text{atm}) \qquad \mathscr{E}^\circ_{\text{red}} = 0\ \text{V}$$

With its standard reduction potential *defined* as exactly zero, this agreed-upon reaction will serve henceforth as our universal point of reference.*

   The question becomes: Will such-and-such a species *take* electrons from H_2 rather then *give* them to H^+. Pitted against our ***standard hydrogen electrode***, will it be an oxidizing agent or a reducing agent? To learn the answer, we build a test cell in which hydrogen gas bubbles over a 1 M acidic solution contained in one of the two compartments. (See Figure 17-7 for details, notably for the role of the platinum "sensing"

---

*Note that the solvated hydronium ion, H_3O^+(aq), is represented concisely throughout this chapter as H^+(aq). The half-reaction above is equivalent to an equation in which H_3O^+ ions and additional H_2O molecules appear explicitly: $2H_3O^+(\text{aq}) + 2e^- \longrightarrow H_2(g) + 2H_2O(\ell)$.

electrode.) Into the other compartment goes the half-reaction under consideration, which for the moment can be $Zn|Zn^{2+}$.

Now observe which way the electrons flow. If they go from zinc to acid, then Zn (functioning apparently as the anode) is giving electrons to H^+. Hydrogen gas should be liberated as a result.

They do. The electrons do go from Zn to H^+ in this cell, and so H^+ proves itself to be a more powerful *taker* than Zn^{2+}. Zn is oxidized to Zn^{2+} while H^+ is reduced to H_2. A positive voltage—measured as 0.76 V—pushes the charge from Zn to H^+ under standard conditions.

By agreement, we assign that whole voltage to the $Zn|Zn^{2+}$ half-cell and record the result as

$$Zn(s) \longrightarrow Zn^{2+}(aq,\ 1\ M) + 2e^- \qquad \mathscr{E}^\circ_{ox} = 0.76\ V$$

To do so is to acknowledge Zn as the stronger giver and H^+ as the stronger receiver. The products (Zn^{2+} and H_2) sit lower in free energy than the reactants (Zn and H^+), and the drop translates to an electrical potential of 0.76 volts. From a negative ΔG° thus comes a positive $\mathscr{E}^\circ_{ox}$ for the oxidation of metallic zinc, a value established by the relationship $\mathscr{E}^\circ_{ox} = -\Delta G^\circ / n\mathscr{F}$.

So defined, $\mathscr{E}^\circ_{ox}$ acquires meaning only through its connection to the standard hydrogen electrode. Alone, the number signifies nothing; it only invites the question "0.76 V compared with what?" Tied to the reduction of H^+, though, zinc's oxidation potential takes on a crisp interpretation. It says that 0.76 joules will accompany the transfer of each coulomb from Zn to H^+—and only to H^+; not to Cu^{2+}, not to MnO_4^{2-}, not to F_2, but to H^+ alone.

Immediately we know the standard reduction potential as well ($\mathscr{E}^\circ_{red}$), which differs from $\mathscr{E}^\circ_{ox}$ only in its sign. Generated by the reverse reaction

$$Zn^{2+}(aq,\ 1\ M) + 2e^- \longrightarrow Zn(s) \qquad \mathscr{E}^\circ_{red} = -0.76\ V$$

the negative voltage corresponds to an *increase* in chemical free energy, a positive ΔG°. What was a negative ΔG° in the forward direction is now positive in reverse, and a formerly positive $\mathscr{E}^\circ$ has become correspondingly negative. And always this relationship between half-cell oxidation and half-cell reduction is the same, regardless of the participants. Always one is the negative of the other; always $\mathscr{E}^\circ_{red} = -\mathscr{E}^\circ_{ox}$.

Good enough. If only the signs are different, then we may confine our attention to just one of the two potentials and further simplify the tabulation. Make it $\mathscr{E}^\circ_{red}$, for no reason other than to establish a convention; and then, one at a time, proceed to measure the voltage of an undetermined half-reaction against the standard hydrogen electrode. There are two possible outcomes:

1. If electrons flow toward the hydrogen electrode, then evidently the other half-cell is sponsoring the oxidation:

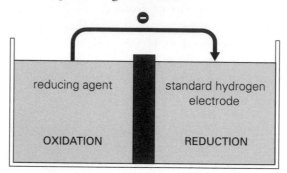

Here the species under observation undergoes reduction less readily than H^+; it *gives* electrons to H^+, in the same way that hydrochloric acid gives protons to water. Treated as an oxidation potential, the measured voltage is positive (as for zinc going from Zn to Zn^{2+}). Switched around and treated as a reduction potential (for Zn^{2+} going to Zn), the voltage recorded for $\mathscr{E}^\circ_{red}$ turns negative.

2. If, instead, electrons flow away from the hydrogen electrode, then the other species is being reduced:

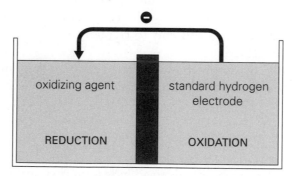

Rather than give electrons to H^+, it *takes* them from H_2. It is the more reducible entity, distinguished by a positive value for $\mathscr{E}^\circ_{red}$. Cu^{2+}, for example, is reduced by hydrogen in the test cell, going as

$$Cu^{2+}(aq,\ 1\ M) + 2e^- \longrightarrow Cu(s) \qquad \mathscr{E}^\circ_{red} = 0.34\ V$$

The Cu^{2+} ions oxidize the hydrogen molecules, deriving a profit of 0.34 joules per coulomb of charge transferred.

In the end there are three numbers (0.34 V, 0.00 V, −0.76 V), none with any absolute meaning but each recorded consistently relative to the same zero. From the resulting hierarchy

$$Cu^{2+}|Cu \qquad \mathscr{E}^{\circ}_{red} = \ \ 0.34 \ V$$

$$H^+|H_2 \qquad \mathscr{E}^{\circ}_{red} = \ \ 0.00 \ V$$

$$Zn^{2+}|Zn \qquad \mathscr{E}^{\circ}_{red} = -0.76 \ V$$

we then have the desired breakdown at last. It is a ladder with three rungs, from which the Zn/Cu cell acquires its 1.10 volts according to the following logic:

First, we know that H^+ can—if given the opportunity—take electrons from Zn, since the oxidation reaction

$$Zn(s) \longrightarrow Zn^{2+}(aq) + 2e^-$$

proceeds simultaneously with the reduction reaction

$$2H^+(aq) + 2e^- \longrightarrow H_2(g)$$

in a test cell. Passed from Zn to H^+ and incorporated into H_2, the electrons take a step *up* the ladder equal to +0.76 volts: from a reduction potential of −0.76 volts *up* to a level of 0.

Second, we know that Cu^{2+} takes electrons from H_2. The oxidation reaction

$$H_2(g) \longrightarrow 2H^+(aq) + 2e^-$$

goes forward in the test cell while the reduction reaction

$$Cu^{2+}(aq) + 2e^- \longrightarrow Cu(s)$$

proceeds with a potential of 0.34 V: a step up the ladder, again, this time from H_2 to Cu^{2+} and then to Cu.

Third, since Cu^{2+} takes electrons from H_2, and H^+ takes electrons from Zn—then, by extension, Cu^{2+} must take electrons directly from Zn as well.

The total climb up the ladder of reduction potentials becomes 1.10 V, first from zinc to hydrogen (0.76 V) and next from hydrogen to copper (0.34 V). This potential of 1.10 volts is, moreover, a true *difference* between products and reactants, a difference that exists irrespective of the arbitrary zero point. The cell potential is to be taken in the same spirit as any other difference in free energy (to which, never forget, it is intimately connected).

Such voltage stands as an immutable thermodynamic fact, possessing significance beyond the cell itself. It determines the energy needed to move electrons from reducing agent to oxidizing agent, not just in a specially constructed cell but *anywhere*. Whether through a wire or directly from atom to atom, the same number of joules is needed for the same number of coulombs. The voltage transcends the circumstances of its electrochemical measurement; it becomes something far more general, something valid wherever the redox reaction occurs. $\mathscr{E}^{\circ}_{red}$ becomes an expression of thermodynamic potential, a tendency to react or not to react.

The partitioning into standard electrode potentials now will be of especial help, because what was true for zinc and copper remains true for any other redox couple. Be it zinc–silver or fluorine–lithium or iron–oxygen, we need only consult a table of standard reduction potentials to determine who will give electrons to whom.

Reductants and Oxidants

And who does give electrons to whom? The species with the more positive reduction potential gets the electrons. It is the stronger oxidant, the stronger oxidizing agent. By taking the electrons it guarantees the largest drop in free energy.

An example: Fluorine, with a standard potential of 2.87 V for the reduction

$$F_2(g) + 2e^- \longrightarrow 2F^-(aq) \qquad \mathscr{E}^{\circ}_{red} = 2.87 \text{ V}$$

is an efficient taker of electrons, the strongest oxidizing agent commonly available. As it should be, too, considering the fluorine atom's small radius, high effective nuclear charge, and poorly screened $2s^2 2p^5$ configuration (one short of an octet)—factors that all combine to make fluorine the most electronegative of the elements. With a large ionization potential and a strong electron affinity to match, fluorine holds onto its own electrons while easily attracting negative charge from other atoms. That much we have long known, but now we see it confirmed by a large and positive reduction potential.

F_2 can take electrons from any species with a lower reduction potential, and the many contenders include much of the chemical world. Fluorine prevails even over an oxidizing agent as strong as the permanganate ion, MnO_4^-, for which $\mathscr{E}^{\circ}_{red}$ is 1.51 V in acidic solution. Under other circumstances MnO_4^- might have done the taking, since the highly positive Mn^{7+} oxidation state* usually provides a good target for electrons:

$$MnO_4^- + 8H^+ + 5e^- \longrightarrow Mn^{2+} + 4H_2O \qquad \mathscr{E}^\circ_{red} = 1.51 \text{ V}$$

Such would be permanganate's power over any lower-potential oxidant, but not so in the presence of fluorine. Confronted by the better oxidizing agent, the permanganate half-reaction turns around and becomes a supplier of electrons.

Thus arises a pecking order among electron givers and takers. F_2 is more positive than MnO_4^-; and MnO_4^- is more positive than Au^{3+}; and Au^{3+} is above $Cr_2O_7^{2-}$; and $Cr_2O_7^{2-}$ is above O_2; and O_2 is above Ag^+, and in this way the rankings start to take shape:

HALF-REACTION	$\mathscr{E}^\circ$ (V)
$F_2(g) + 2e^- \longrightarrow 2F^-(aq)$	2.87
$MnO_4^-(aq) + 8H^+(aq) + 5e^- \longrightarrow Mn^{2+}(aq) + 4H_2O(\ell)$	1.51
$Au^{3+}(aq) + 3e^- \longrightarrow Au(s)$	1.40
$Cr_2O_7^{2-}(aq) + 14H^+(aq) + 6e^- \longrightarrow 2Cr^{3+}(aq) + 7H_2O(\ell)$	1.33
$O_2(g) + 4H^+(aq) + 4e^- \longrightarrow 2H_2O(\ell)$	1.23
$Ag^+(aq) + e^- \longrightarrow Ag(s)$	0.80

Any member of this *electrochemical series* can, in principle, oxidize any other with a lower value of $\mathscr{E}^\circ_{red}$. For a more extensive tabulation, consult the data in Appendix C (Table C-21).

We see, for instance, that strong acid will react with metallic zinc to produce hydrogen gas:

$$Zn(s) + 2H^+(aq) \longrightarrow Zn^{2+}(aq) + H_2(g) \qquad \mathscr{E}^\circ = 0.76 \text{ V}$$

Zinc, with its negative reduction potential, lies below hydrogen in the table. Zn donates electrons to H^+, as do other metals similarly situated below hydrogen. Copper, by contrast, displays a positive reduction

---

*Recall (from Chapter 3) how the oxidation state is calculated for Mn: With O assigned its standard value of -2, the four oxygens contribute a total of -8. Mn therefore must be $+7$ if the ion is to have a net charge of -1.

potential; it lies above H^+, and consequently does not reduce the hydrogen ion to H_2 under standard conditions.

From the rankings we realize also that molecular oxygen is a relatively strong oxidizing agent, developing a potential of 1.23 V in the reaction

$$O_2(g) + 4H^+(aq) + 4e^- \longrightarrow 2H_2O(\ell) \qquad \mathscr{E}°_{red} = 1.23 \text{ V}$$

That number is sufficiently high to oxidize most metals, but not gold. The Au^{3+} ion, its $\mathscr{E}°_{red}$ equal to 1.40 V, tends toward pure metal instead:

$$Au^{3+}(aq) + 3e^- \longrightarrow Au(s) \qquad \mathscr{E}°_{red} = 1.40 \text{ V}$$

The progression continues downward, with the reduction potentials decreasing as the species become less electronegative. Near the bottom, as expected, are the alkali metals, which all have one electron beyond an octet. These elements are ready targets not for reduction but for oxidation: for *loss* of the electron, for a strongly positive *oxidation* potential and (conversely) a reduction potential just as strongly negative. So it happens that lithium displays a reduction potential considerably more negative than hydrogen, going as

$$Li^+(aq) + e^- \longrightarrow Li(s) \qquad \mathscr{E}°_{red} = -3.05 \text{ V}$$

Lithium metal is a giver, a reducing agent with an appropriately low ionization potential. It cedes electrons to oxidants with more positive values of $\mathscr{E}°_{red}$ (almost everything else).

17-8. CHARGE AND MASS: BALANCING THE REDOX EQUATION

With redox we have, in the most practical sense, a question of simple bookkeeping, of electrons lost and electrons gained; yet, properly valued, we also have sweeping insight into the way matter works its changes. For to understand oxidation and reduction is to make sense of a world of chemical change, both natural and of human invention, a procession seemingly without end. Change, everywhere: from the ever-present corrosion of iron to the tarnishing of silver; from the simplest electroplating process to the most ambitious of metallurgical extractions; from the violent burning of petroleum to the controlled metabolism of a cell; from the electrical transport of ions across a membrane to the beating of a heart and the twitch of a nerve. All involve redox reactions. All unfold according to the same principles. All obey the laws of thermodynamics.

Where individual redox reactions differ is in their details—important, sometimes subtle details—details such as the species that participate, the oxidation numbers that change, and the energy released or absorbed. Some processes, like the reduction of Cu^{2+} by Zn, are relatively simple; others, like the discharge of a lead storage battery,

$$Pb(s) + PbO_2(s) + 4H^+(aq) + 2SO_4^{2-}(aq) \longrightarrow 2PbSO_4(s) + 2H_2O(\ell)$$

are more complex. But to each reaction there is an equation, a *balanced* equation, and in this expression we find confirmation that matter follows the rules. The balanced equation comes not from our design, but instead from nature's demand to conserve both charge and mass throughout all chemical changes.

Our stoichiometric concern is, as always, to verify the conservation laws by writing a properly balanced equation. Without that, we have nothing; absent the same electrons and atoms left and right, there is no reaction. Everything begins with the conservation of charge and mass.

Whereas the earlier examples required no special comment, now we must proceed systematically to recognize and balance redox reactions wherever they occur (notably in acidic and basic solutions, for which simple methods of inspection often fail). The plan, built around our newfound appreciation of half-reactions, amounts to a straightforward sequence of five steps:

1. Identify reductant and oxidant.
2. Write the half-reactions.
3. Balance the atoms in each half-reaction, leaving O and H for last.
4. Balance the electrons, again separately for the half-reactions.
5. Combine the half-reactions into one complete equation.

Let us start in acidic solution, where $H^+(aq)$ and H_2O participate in the reaction along with the oxidized and reduced species.

Acidic Solution

Suppose we know only two reactants (the dichromate ion $Cr_2O_7^{2-}$, the chloride ion Cl^-), two products (Cr^{3+}, Cl_2), and the type of medium (aqueous and acidic):

$$\underline{\quad} Cr_2O_7^{2-}(aq) + \underline{\quad} Cl^-(aq) \longrightarrow \underline{\quad} Cr^{3+}(aq) + \underline{\quad} Cl_2(aq)$$

What can be done?

With a total charge of -3 on the left and $+3$ on the right, the unbalanced equation offers an inauspicious beginning. Still worse, there are two chromium atoms on the left and only one on the right; one chlorine on the left, two chlorines on the right; seven oxygens on the left, none on the right. The equation needs some work.

Ask first: Is it a redox reaction? Do the oxidation numbers change? They do. Chlorine's oxidation number increases from -1 in Cl^- to 0 in Cl_2. The chloride ion is oxidized to produce the diatomic chlorine molecule. As for chromium, it is reduced from $+6$ in $Cr_2O_7^{2-}$ ($2Cr^{6+} + 7O^{2-} \equiv Cr_2O_7^{2-}$) to $+3$ in Cr^{3+}. The dichromate ion, by siphoning electrons from the chloride ion, acts as the oxidizing agent.

Next, having discerned who gives electrons to whom, we lay out the half-reactions in raw form,

$$Cr_2O_7^{2-} \longrightarrow Cr^{3+} \qquad \text{(reduction)}$$

$$Cl^- \longrightarrow Cl_2 \qquad \text{(oxidation)}$$

and proceed to balance each one separately.

First to be considered are all the atoms except for H and O. A glance shows that the reduction equation needs two chromiums on the right, whereas the oxidation needs two chlorines on the left. We write

$$Cr_2O_7^{2-} \longrightarrow 2Cr^{3+} \qquad \text{(reduction)}$$

$$2Cl^- \longrightarrow Cl_2 \qquad \text{(oxidation)}$$

and note, after checking, that the oxidation half-reaction is now balanced with respect to atoms.

The next step is to balance the oxygens by adding H_2O molecules where warranted—one water for each missing oxygen, to be supplied by the aqueous medium. Here, with dichromate's seven unmatched oxygens, the reduction equation needs to acquire seven water molecules on the right:

$$Cr_2O_7^{2-} \longrightarrow 2Cr^{3+} + 7H_2O \qquad \text{(reduction)}$$

$$2Cl^- \longrightarrow Cl_2 \qquad \text{(oxidation)}$$

By doing that, however, we also introduce 14 new H atoms on the right, two for each molecule of H_2O added.

But this material imbalance will be corrected by the hydrogen ions in the acidic solution, and so we simply add H^+ to the appropriate side,

$$14H^+ + Cr_2O_7^{2-} \longrightarrow 2Cr^{3+} + 7H_2O \qquad \text{(reduction)}$$

$$2Cl^- \longrightarrow Cl_2 \qquad \text{(oxidation)}$$

to obtain two half-reactions with all atoms properly balanced left and right. Still needed, of course, is to conserve the charge for each half-reaction.

Count up the charges. The reactants in the reduction equation contribute a total of +12 (+14 from the $14H^+$ and −2 from the $Cr_2O_7^{2-}$), while the products bear a charge of only +6. This difference must be made up by six electrons on the left, which the dichromate ion accepts upon undergoing reduction according to the (now balanced) half-reaction

$$6e^- + 14H^+ + Cr_2O_7^{2-} \longrightarrow 2Cr^{3+} + 7H_2O \qquad \text{(reduction)}$$

The oxidation half-reaction, treated separately, takes two electrons on the right to make the charge −2 on each side:

$$2Cl^- \longrightarrow Cl_2 + 2e^- \qquad \text{(oxidation)}$$

Two electrons lost and six electrons gained? Impossible—but remember that we still must couple the oxidation to the reduction, so that precisely the number of electrons gained during reduction is lost during oxidation. The chloride reducing agent therefore will supply not two, but *six* electrons to the dichromate, forcing us to multiply left and right of the oxidation equation by a factor of 3 (= 6/2):

$$6Cl^- \longrightarrow 3Cl_2 + 6e^- \qquad \text{(oxidation)}$$

With the electrons thus matched, the oxidation properly fuels the reduction.

The two completed half-reactions then sum to a fully balanced equation,

$$6e^- + 14H^+(aq) + Cr_2O_7^{2-}(aq) \longrightarrow 2Cr^{3+}(aq) + 7H_2O(\ell) \quad \text{(reduction)}$$
$$6Cl^-(aq) \longrightarrow 3Cl_2(g) + 6e^- \qquad \text{(oxidation)}$$
$$\overline{\rule{0pt}{1em}\hspace{10cm}}$$
$$14H^+(aq) + Cr_2O_7^{2-}(aq) + 6Cl^-(aq) \longrightarrow$$
$$2Cr^{3+}(aq) + 3Cl_2(g) + 7H_2O(\ell)$$

in which six moles of electrons are transferred to each mole of dichromate ions ($n = 6$). The electrons seemingly disappear once the half-reactions are consolidated, but really they never leave the scene. Without them, there is no redox reaction. They *are* the redox reaction.

The resulting expression is a net equation from which all nonreacting (spectator) species are excluded, and our final task is to check the accounting:

atoms in: 14 hydrogen, 2 chromium, 7 oxygen, 6 chlorine
atoms out: 14 hydrogen, 2 chromium, 7 oxygen, 6 chlorine
charge in: +6
charge out: +6

Mass is in balance. Charge is in balance. The equation is complete.

Basic Solution

Redox processes in basic solution involve the hydroxide ion, OH^-, instead of H^+, but still the balancing procedure remains largely the same. First, we write the half-reactions separately and balance every atom except for H and O; and, after that, we introduce H_2O to supply O where needed. The sole difficulty arises if now we add OH^- directly to adjust the hydrogen content, since with each hydroxide comes an additional, unwanted oxygen that undoes the previous step.

To avoid such unpleasantness we shall pretend that the solution is actually acidic, thus permitting ourselves to use H^+ as a hydrogen-balancing device (just as above, where the procedure worked so smoothly). This little fiction will be a temporary measure, adopted for convenience only, following which the acidic H^+ species will be replaced by proper hydroxide ions. The switch is easily made.

As an example, take the unbalanced half-reaction

$$Cd(s) \longrightarrow Cd(OH)_2(s)$$

in which cadmium is assumed to be oxidized from Cd to Cd^{2+} in basic solution. There are two O atoms and two H atoms on the right, as yet unmatched on the left.

The equation is clearly simple enough to be balanced by inspection, but its very simplicity helps us better to understand why the systematic method always manages to add the correct number of hydroxide ions—even for equations not so simple. The first step is to add two water molecules on the left,

$$2H_2O + Cd \longrightarrow Cd(OH)_2$$

and then two H^+ ions on the right (as if the solution were acidic):

$$2H_2O + Cd \longrightarrow Cd(OH)_2 + 2H^+$$

The new ions join with the two hydrogens already there, producing an equation that is superficially mass-balanced.

But H^+ should not play such a role in a predominantly basic solution, and so we shall remove it in a chemically satisfactory way: through a pseudoneutralization, whereby one OH^- (an acceptable species) is introduced to convert every H^+ ion into an H_2O molecule (another acceptable species). The equation then becomes

$$2H_2O + Cd + 2OH^- \longrightarrow Cd(OH)_2 + \underbrace{2H^+ + 2OH^-}_{2H_2O}$$

upon addition of $2OH^-$ to both sides, simplifying to

$$Cd + 2OH^- \longrightarrow Cd(OH)_2 + 2e^-$$

after the redundant H_2O molecules are removed and two electrons are added.

This example, incidentally, describes the anode reaction of a nickel–cadmium battery, for which the complete redox transformation is

$$
\begin{array}{ll}
Cd(s) + 2OH^-(aq) \longrightarrow Cd(OH)_2(s) + 2e^- & \text{(oxidation)} \\
\underline{2e^- + 2NiO(OH)(s) + 2H_2O(\ell) \longrightarrow 2Ni(OH)_2(s) + 2OH^-} & \text{(reduction)} \\
Cd(s) + 2NiO(OH)(s) + 2H_2O(\ell) \longrightarrow Cd(OH)_2(s) + 2Ni(OH)_2(s) &
\end{array}
$$

Often the nickel–cadmium reaction is the force behind the small 9-V radio battery, long a required element for portable electronic gear. It is a *rechargeable* battery, too, a system capable of electrochemical rebirth even after reaching equilibrium.

On that optimistic-sounding note, we shall try finally to do what nature would rather not: to climb back up the mountain, to go from low G to high G.

17-9. GOING UPHILL: ELECTROLYSIS

Drained of chemical potential, a system comes to equilibrium and there it sits. It sits, unchanging, with products and reactants present in such ratios as to have equal free energies.

Left alone, the system has nowhere to go. Not down, for there is no further room to fall; not up, for to increase the free energy would (impossibly) create order in the universe. Moribund, dead as a discharged battery, a system at equilibrium has used up its potential to change.

But not so the universe itself—which, notwithstanding a few dead batteries, continues to enjoy ample opportunity for spontaneous change.

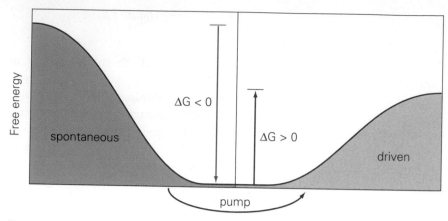

FIGURE 17-8. No free lunch: The energy to push a nonspontaneous process up the hill must come from somewhere else—from a downhill, spontaneous process that adds to the entropy of the universe. Local order is purchased at the cost of global disorder.

The universe still has a long way to go, possessing vast stores of energy to reallocate throughout.

That energy is for us to discover and use. To force a nonspontaneous process, we have only to find sufficient energy to carry low-potential matter up to a point higher in G. Viewed globally, this forcible climb uphill will be tolerated as part of a larger, disordering change somewhere else.

We pay. Nature never loses; nature always takes payment in liberated free energy. The second law allows us, for example, to reenergize all the batteries we like, provided that the requisite energy comes from the spontaneous discharge of either a new battery or any other downhill source of motive power.

We stay alive by just such a strategy (Figure 17-8), financing our order-inducing activities with energy released from food and fuel. Only in that way can a living cell create local order. The cell survives by diverting energy from thermodynamically favorable (spontaneous) processes into thermodynamically unfavorable (nonspontaneous) activities. Taking the free energy released by the combustion of carbohydrates, it does the work necessary to put together proteins and nucleic acids and to fashion these biopolymers into complex organisms. Human beings, powered by that internal energy, later burn petroleum to produce the electricity needed to create still more order in the world outside (including, among other projects, such gravity-defying activities as the transport of low-potential skiers to the top of a mountain).

And so we return to our original galvanic Zn/Cu cell in its final state: at equilibrium, dead, with no potential difference between anode and cathode. The redox reaction did its work on the way to equilibrium, losing voltage continually while products and reactants equalized their free energies. If the exhausted cell is ever to do work again, we shall have to spend some energy to rebuild its chemical mountain.

We need to run the Zn/Cu reaction in reverse, to change the stalemated ratios of products and reactants. We need to go *away* from equilibrium, to do work *on* the system and thereby recreate a difference in chemical potential.

We must, in short, apply energy to put the electrons back onto their original atoms. One species gave an electron to the other, and free energy was released; now, to reverse the reaction, the reduced species must be induced to hand back the electron. If successful, then products will disappear and reactants will reappear—along with a restored ΔG. It will be a driven, nonspontaneous process (called *electrolysis*), and it can occur only if energy is supplied from without. A live battery, operated independently, offers one way to meet that need.

The idea, illustrated in Figure 17-9, is to subject the dead cell to a voltage of reversed polarity, arranged so that electrons run not from left to right (as from the Zn electrode to the Cu) but rather are externally driven from right to left (Cu to Zn). To do so requires a driving voltage *no less* than the original electrochemical potential (in practice, sometimes

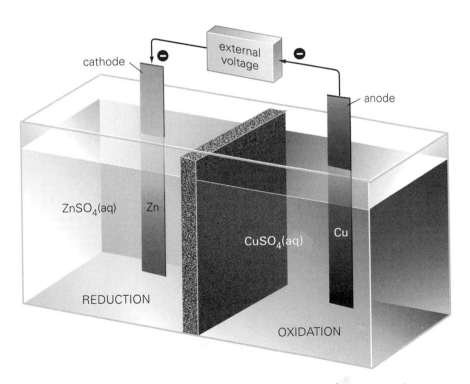

FIGURE 17-9. In an electrolytic cell, an external source of energy sustains a nonspontaneous reaction. Here the electrons are pushed uphill in energy from a copper anode to a zinc cathode, exactly the reverse of the path in a galvanic cell.

considerably more), a voltage able to overcome the system's natural tendency to seek lower free energy.

See, then, how the spontaneously operating galvanic cell compares with this new ***electrolytic cell***. In the galvanic reaction, it was the zinc electrode that functioned as the anode; it was the zinc electrode that accumulated extra electrons when zinc ions were oxidized off the metal. The copper electrode, acting as the cathode, siphoned away these internally generated electrons while it reduced Cu^{2+} ions to Cu metal. The result was a flow of current from negative electrode to positive electrode, from the zinc anode (the site of oxidation) to the copper cathode (the site of reduction). Repelled by the overly negative Zn anode, the electrons ran naturally from left to right. No outside agency intervened.

That was the natural course of reaction, arising from the drop in free energy suffered in the forward direction:

$$Zn(s) + Cu^{2+}(aq) \longrightarrow Zn^{2+}(aq) + Cu(s) \qquad \mathscr{E}^\circ = 1.10 \text{ V}$$

But now, applying our own oppositely directed voltage with magnitude greater than 1.10 V, we intervene to bring about the reverse reaction to restore standard conditions:

$$Zn^{2+}(aq) + Cu(s) \longrightarrow Zn(s) + Cu^{2+}(aq)$$

Everything changes. Oxidation becomes reduction; anode becomes cathode; left becomes right; and, most telling, voluntary becomes involuntary. Electrons are delivered from the battery *to* the zinc,

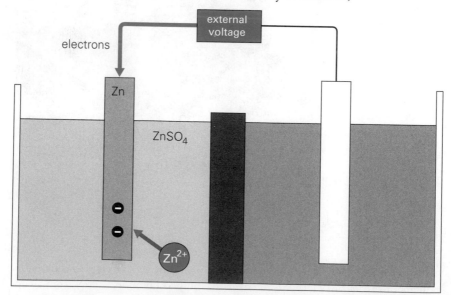

not produced *by* the zinc as in the galvanic cell. Although the zinc electrode once again grows negative, this time the electrons come from *outside* the system—from the right, from the external generator.

Suddenly the Zn^{2+} cations in solution have an opportunity for reduction. Going uphill, they plate out onto the electron-rich Zn electrode in what is now a *cathode* reaction, a gain of electrons:

$$Zn^{2+}(aq) + 2e^- \longrightarrow Zn(s)$$

Upon eventual completion of the circuit (see below), one mole of Zn will be deposited for every two moles of electrons delivered by the battery. We use this rate of delivery to define the electric **current**, taking an **ampere** as the number of coulombs passing a given point each second: $1\ A = 1\ C\ s^{-1}$.

Meanwhile, the external battery draws electrons away from the copper electrode. Playing the role of electrolytic *anode*, the copper metal hosts the oxidation reaction

$$Cu(s) \longrightarrow Cu^{2+}(aq) + 2e^-$$

and thereby grows positive as electrons are pulled off the electrode. Solution-based anions then flow toward the positive charge of this turn-about anode, properly recognized as the place of oxidation (which it is, whatever the polarity):

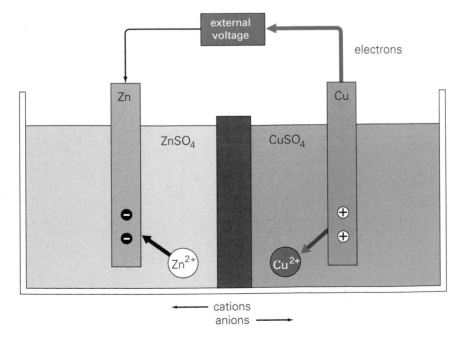

The copper electrode loses mass as the Cu^{2+} ions, robbed of electrons, fall into solution.

The electrolytic circuit, running in a sense opposite to the galvanic, is complete. Electrons are force-fed from right to left, from new anode to new cathode. Products and reactants move away from equilibrium, and the original cell is reconstituted and reenergized. It can be disconnected from the power supply and allowed, once again, to deliver current as a spontaneously operating galvanic cell.

Back in business, a recharged system proves that equilibrium need not last forever. No state is truly final if, somewhere in the world, there remains energy still to be reshuffled. To reverse a reaction, we see, is neither to do magic nor to defy nature; it is, instead, more a question of bringing the ingredients together and importing the necessary energy. If the various species stay readily available, then the recharging is made all the easier (as in the nickel–cadmium and lead storage batteries, where the redox products are deposited on the electrodes as solids). Where some of the products escape as gases, a battery may be difficult or impossible to recharge.

More than just recharged batteries, though, electrolysis gives us wide-ranging power to force changes that would not otherwise proceed. Nonspontaneous reduction of metal ions, for instance, permits the extraction of pure metals from their ores—aluminum, copper, sodium, magnesium, and more. Natural *givers* of electrons, these elements typically exist as oxidized species (Al^{3+}, Cu^{2+}, Na^+, and so on) within other substances. The metals have already given away their electrons for the sake of a lower free energy; we, if we want pure metal, must put them back. Again that is a job for electrolysis, which has grown by such applications into a huge, varied industry of great economic importance—an industry regulated strictly and impartially by the laws of thermodynamics, to which there are no exceptions and no exemptions.

REVIEW AND GUIDE TO PROBLEMS

Chemistry *and* electricity? Better to say: chemistry *is* electricity. The potential to make electricity is always there, locked within matter, latent in any reaction. Every chemical particle is an electrical particle, and every chemical reaction is an electrical event.

But what appears is mostly a fine-grained, random, elusive kind of electricity lacking duration and direction, an electricity rarely sensed. Again and again, something happens microscopically; and again and again, we see nothing macroscopically. A dipole may align or an ion may move or an electron may hop, yet who can tell? For just the briefest instant over the shortest distance, a tiny current skates from atom to atom—in this direction, in that direction, here, there, and gone.

We never suspect it. We see only a large, smoothed-out, neutral world where equal numbers of positive and negative charges huddle close together; a world in which there is no buildup of net charge, no organized flow; a macroscopic world where unseen particles move randomly and at cross purposes, sparing us from sudden shock. To bring latent, chemical electricity to that otherwise neutral world is now the challenge of *electrochemistry*: to marshal the electricity of chemistry, to direct it, to put it to work.

IN BRIEF: ELECTROCHEMISTRY

Hook together a source, a sink, a path, and a pump: a *source* of electrons or ions, a *sink* to receive them, a *path* to carry them, and a *pump* to recycle them. The resulting assembly becomes a self-driven electric circuit, powered by the free energy of a spontaneous *redox* reaction. We call the construction a *galvanic cell*; alternatively, a *voltaic cell*.

1. SOURCE AND SINK. Electrons go from giver to taker, from electron-rich to electron-poor, from have to have-not; and so, in the language of redox chemistry, we follow the negative charge as it moves from *reductant (reducing agent)* to *oxidant (oxidizing agent)*:

$$\text{reductant} + \text{oxidant} \longrightarrow \text{conjugate oxidant} + \text{conjugate reductant}$$

$$\text{Zn} + \text{Cu}^{2+} \longrightarrow \text{Zn}^{2+} + \text{Cu}$$

Down the mountain of free energy the electrons flow, spontaneously, always toward a more disordered universe—from easy giver to easy taker, from strong reactants to weak products, from high free energy to low. Like acid and base, the species conjugate to a strong reductant is a weak oxidant; and, going the other way too, the species conjugate to a strong oxidant is a weak reductant. Easy to gain is hard to lose.

Thus the source: a material prone to lose electrons, a reducing agent, an electron-rich site where negative charge builds up and then is sent away. Call it the **anode**, the place where oxidation occurs. In the example above, the anode might be a strip of zinc metal dipping into an electrolytic solution containing Zn^{2+} ions.

And the sink: a material prone to gain electrons, an oxidizing agent, an electron-poor site to which negative charge is drawn. Call it the **cathode**, the place of reduction. For the Zn/Cu reaction, a strip of copper metal dipping into a Cu^{2+} electrolyte will do.

2. PATH. Put Zn and Cu^{2+} together and, no doubt, the copper ion will grab two electrons from the metallic zinc, but not in the way we want. A direct atom-to-ion transfer of electrons is just too quick, too random, and executed over too short a distance to produce an orderly flow of electricity. For that, we shall need to make the electrons travel a far longer way, along a route *we* choose, a winding path from anode to cathode and back to anode again.

For the first part of the trip, let the negative charge flow from electron-rich anode to electron-poor cathode along (think simple) an ordinary wire. Then, with negative charge threatening to build up at the cathode, the current must find its way back to the anode. One possibility is to allow the ions in the electrolytic solution to migrate where needed: cations (positive) toward the cathode, where they can neutralize the electrons recently arrived over the wire; anions (negative) toward the anode, where they compensate for the negative charge sent away. For this return path from cathode to anode, possible to secure in many other ways, we shall use the generic name **salt bridge**.

3. PUMP. As long as the reaction remains away from equilibrium, the reactants sit collectively higher or lower in free energy than the products; and when they do, the system is always poised for change. Reactants turn into products, and products turn into reactants. They react visibly until the last difference in free energy is quenched, satisfied only when the mixture reaches just the right proportions to ensure balance.

Equilibrium has arrived, and with its onset comes an end to all macroscopic differences. The free energy, G, stands equal on both sides of the equation

$$R \longrightarrow P$$

$$G_R = G_P$$

and nothing more can be done. Nature no longer discriminates between reactants (R) and products (P). $\Delta G = 0$.

But during the rundown, however long it lasts, the reactants shed free energy continually as they grind their way irreversibly to equilibrium. And even though the cell is dying, the residual difference in free energy

$$\Delta G = G_P - G_R$$

still allows the system to do work: useful work, work that can be harnessed to push electrons around a circuit. Let any out-of-equilibrium difference in free energy, then, supply the driving force that draws the electrons down the mountain and later provides fresh ones to take their place. The stream of chemical electricity flows just as long as ΔG, ever diminishing, remains nonzero.

4. CHEMICAL ENERGY AND ELECTRICAL WORK. For every change in a body's motion, look for a force; and where electric charge is involved, look for an electric field. Look for some difference in electrical *potential energy* between anode and cathode, made manifest by a difference in **voltage** (or **cell potential**, $\mathscr{E}$) such that

$$\Delta G = -n\mathscr{F}\mathscr{E}$$

where: (1) n specifies the moles of electrons transferred from reductant to oxidant ($n = 2$ for the Zn/Cu cell). (2) $\mathscr{F}$ defines the **Faraday constant**, equal to the magnitude of charge carried by one mole of electrons ($96,485$ C mol^{-1}). The total negative charge is therefore $-n\mathscr{F}$ (moles $\times$ charge per mole). (3) $\mathscr{E}$, the cell potential expressed in volts, establishes the difference in electrical energy between anode and cathode. When a charge of one coulomb moves through a potential difference of one volt, the particle changes its energy by one joule. 1 V $= 1$ J C^{-1}.

Together, the equation says that the chemical energy of a redox reaction (ΔG, the energy "freely" available to do work) carries within it the maximum electrical work possible for a cell operating reversibly or nearly so:

$$\text{Electrical work } (\Delta G) = \text{ charge } (-n\mathscr{F}) \times \text{ potential } (\mathscr{E})$$

$$\text{Joules} \qquad = \quad \text{coulombs} \quad \times \text{joules/coulomb}$$

Not all the available energy goes to work, though, because some of it always leaks away into unintended modes. In the real world, there is heat; there is friction; there are all kinds of squeaky parts that rob the cell of its full potential. The voltage, never *more* than $-\Delta G/n\mathscr{F}$, is indeed always less, even if only slightly. We live in an irreversible, energy-dispersing, dissipative world where everything is always winding down. Blame it on the second law of thermodynamics.

5. $\mathscr{E}$ AND G. Make the connections now between chemistry and electricity. For free energy, we have the cell potential. $\mathscr{E}$ maps onto G. For the *standard* change in free energy ($\Delta G°$), we similarly have a **standard cell potential** ($\mathscr{E}°$) measured for concentrations of 1 *M* and gas pressures of 1 atm. Since $\Delta G°$ is related to K (the equilibrium constant), $\mathscr{E}°$ is related to K in the same way.

So: Whereas the quantities ΔG, $\Delta G°$, and Q (the reaction quotient) come together as

$$\Delta G = \Delta G° + RT \ln Q$$

to produce the master relationship of Chapter 14, the analogous quantities $\mathscr{E}$, $\mathscr{E}°$, and Q likewise obey the **Nernst equation** for an electrochemical cell:

$$\mathscr{E} = \mathscr{E}° - \frac{RT}{n\mathscr{F}} \ln Q$$

And further: Just as the instantaneous free energy (ΔG) diminishes to zero at equilibrium, giving us

$$\Delta G° = -RT \ln K \qquad (\Delta G = 0)$$

when Q is equal to K, so falls the instantaneous cell potential ($\mathscr{E}$) as well:

$$\mathscr{E}° = \frac{RT}{n\mathscr{F}} \ln K \qquad (\mathscr{E} = 0)$$

The potential under *standard* conditions, $\mathscr{E}°$, thus follows the ups and downs of the equilibrium constant. A high standard voltage means a high proportion of products in the final mixture.

6. HALF-REACTIONS. Recall how, in Chapter 16, we so fruitfully assigned separate equilibrium constants to acids and bases, measuring for each a strength relative to the same partner: water. Having ranked the acids by K_a and the bases by K_b, we then knew who could give a proton to whom.

For oxidizing agents and reducing agents, same thing. (1) Measure the standard potential against a common partner: $H_2|H^+$. (2) Establish thereby the ***standard reduction potential*** $(\mathscr{E}^{\circ}_{red})$ for each unknown ***half-reaction***, measuring it relative to a benchmark process

$$2H^{+}(aq, 1\ M) + 2e^{-} \longrightarrow H_{2}(g, 1\ atm) \qquad \mathscr{E}^{\circ}_{red} = 0\ V$$

in which the ***standard hydrogen electrode*** is arbitrarily assigned zero voltage. (3) Do the same, next, with the ***standard oxidation potential*** $(\mathscr{E}^{\circ}_{ox})$, but realize here that the relationship

$$\mathscr{E}^{\circ}_{ox} = -\mathscr{E}^{\circ}_{red}$$

holds between any conjugate reductant–oxidant pair. If ΔG is negative for oxidation, it turns positive for reduction. If we know the voltage in one direction, we know it in the other.

Who gives? Who takes? A species with a *more positive* half-cell potential prevails over a competitor with a less positive potential. The Cu^{2+} ion, for example, develops a standard reduction potential of 0.34 V upon reduction to copper metal; the Zn^{2+} ion, however, exhibits a potential of -0.76 V under the same conditions:

$$Cu^{2+}(1\ M) + 2e^{-} \longrightarrow Cu(s) \qquad \mathscr{E}^{\circ}_{red} = \ \ 0.34\ V$$

$$Zn^{2+}(1\ M) + 2e^{-} \longrightarrow Zn(s) \qquad \mathscr{E}^{\circ}_{red} = -0.76\ V$$

Cu^{2+} therefore *takes* the two electrons from Zn, generating in the process a potential of 1.10 V between anode and cathode:

$$
\begin{array}{ll}
Zn(s) \longrightarrow Zn^{2+}(1\ M) + 2e^{-} & \mathscr{E}^{\circ}_{ox} = 0.76\ V \\
\underline{Cu^{2+}(1\ M) + 2e^{-} \longrightarrow Cu(s)} & \underline{\mathscr{E}^{\circ}_{red} = 0.34\ V} \\
Zn(s) + Cu^{2+}(1\ M) \longrightarrow Zn^{2+}(1\ M) + Cu(s) & \mathscr{E}^{\circ} = \mathscr{E}^{\circ}_{red} + \mathscr{E}^{\circ}_{ox} = 1.10\ V
\end{array}
$$

Cu^{2+}, more easily reduced, is a stronger oxidizing agent than Zn^{2+}. Zn, more easily oxidized, is a stronger reducing agent than Cu.

7. RENEWAL AND REBIRTH: ELECTROLYSIS. We can do it the easy way, or we can do it the hard way.

Either: spontaneously—sliding down the slope of free energy, using the system to do work on its surroundings. That way, the easy way, is the path taken by a freely operating galvanic cell.

Or: nonspontaneously—climbing up the mountain instead, using the surroundings to do work on the system. This way, uphill, is the hard road traveled inside an ***electrolytic cell***, where we bring in external energy to force a process in the "wrong" direction, toward increasing free energy.

How? By finding the requisite energy somewhere else. By coupling the uphill process to a more powerful downhill process.

Why? To do something that the universe will not do on its own, whether it be to recharge a battery, plate gold onto a surface, or extract pure metal from an ore. Sometimes we want the free energy of our system to rise, not fall; and we find, remarkably, that all these uphill things can actually be done, none of them in violation of the natural order. Global entropy goes up. Global energy stays the same. We merely borrow from one account to pay the bill of another.

SAMPLE PROBLEMS

EXAMPLE 17-1. Redox in Acidic Solution: The Lead Storage Battery

PROBLEM: Complete and balance the equation

$$Pb(s) + PbO_2(s) \longrightarrow PbSO_4(s)$$

in acidic solution, using the method of half-reactions. Identify the reducing agent and oxidizing agent.

SOLUTION: We start by assigning oxidation numbers to the lead atom in each of its compounds, knowing already that oxygen exists as O^{2-} and sulfate as SO_4^{2-}. Since oxidation states in elements (Pb) and neutral compounds ($PbSO_4$, PbO_2) must all sum to zero, the assignment is straightforward:

$$Pb \equiv Pb^0 \qquad \text{(reactant)}$$

$$PbO_2 \equiv Pb^{4+} + O^{2-} + O^{2-} \qquad \text{(reactant)}$$

$$PbSO_4 \equiv Pb^{2+} + SO_4^{2-} \qquad \text{(product)}$$

One reactant, metallic lead, evidently is oxidized from Pb^0 to Pb^{2+} in the unbalanced oxidation half-reaction

$$Pb \longrightarrow PbSO_4$$

The other reactant, lead dioxide, is simultaneously reduced from Pb^{4+} to Pb^{2+} in the unbalanced reduction half-reaction

$$PbO_2 \longrightarrow PbSO_4$$

Pb(s), which undergoes oxidation, acts as the reducing agent. PbO_2, which undergoes reduction (and hence *causes* the oxidation), is the oxidizing agent.

So goes the process overall. To complete and balance the equation from here, we now need four additional ingredients: hydrogen ions (H^+), water molecules (H_2O), sulfate ions (SO_4^{2-}), and electrons (e^-). Each half-reaction is best handled separately.

1. *Oxidation (Pb $\longrightarrow$ PbSO$_4$).* The first step is to balance all atoms except O and H, which we do immediately by adding SO_4^{2-} at the left:

$$Pb + SO_4^{2-} \longrightarrow PbSO_4$$

Then, with all the atoms in place, we complete the half-cell by adding two electrons at the right to balance the charge:

$$Pb + SO_4^{2-} \longrightarrow PbSO_4 + 2e^- \qquad \text{(oxidation)}$$

2. *Reduction (PbO$_2$ $\longrightarrow$ PbSO$_4$).* We add one SO_4^{2-} ion at the left,

$$PbO_2 + SO_4^{2-} \longrightarrow PbSO_4$$

and then place two H_2O molecules at the right to balance the oxygens:

$$PbO_2 + SO_4^{2-} \longrightarrow PbSO_4 + 2H_2O$$

The four H atoms on the right now are balanced by four H^+ ions on the left, followed by two electrons to equalize the charge:

$$PbO_2 + 4H^+ + SO_4^{2-} \longrightarrow PbSO_4 + 2H_2O$$

$$PbO_2 + 4H^+ + SO_4^{2-} + 2e^- \longrightarrow PbSO_4 + 2H_2O \qquad \text{(reduction)}$$

3. *Coupling the half-reactions.* Next we multiply one or both half-reactions, as needed, to ensure that the number of electrons lost is equal to the number of electrons gained; and finally, with the oxidation properly coupled to the reduction, we add the two half-reactions to put together a complete, balanced equation:

$$Pb + SO_4^{2-} \longrightarrow PbSO_4 + 2e^- \qquad \text{(oxidation)}$$
$$PbO_2 + 4H^+ + SO_4^{2-} + 2e^- \longrightarrow PbSO_4 + 2H_2O \qquad \text{(reduction)}$$
$$\overline{Pb(s) + PbO_2(s) + 4H^+(aq) + 2SO_4^{2-}(aq) \longrightarrow}$$
$$2PbSO_4(s) + 2H_2O(\ell) \qquad \text{(redox)}$$

No multiplication is necessary for the half-reactions as written, since the oxidation and reduction are already in balance: two electrons lost, two electrons gained. $n = 2$.

EXAMPLE 17-2. Redox in Basic Solution: The Nickel–Cadmium Battery

PROBLEM: Complete and balance the equation

$$Cd(s) + NiO(OH)(s) \longrightarrow Cd(OH)_2(s) + Ni(OH)_2(s)$$

in basic solution, using the method of half-reactions. Identify the reducing agent and oxidizing agent.

SOLUTION: Remember the trick we use to balance O and H in basic solution: (1) Pretending that the solution is acidic, we add H_2O to balance the oxygens and then H^+ to balance the hydrogens. The sequence is identical to the procedure carried out in Example 17-1. (2) With all atoms in place, we add OH^- to both the left and right sides of the equation—a fictitious titration, in effect, just enough to convert all the spurious H^+ ions into H_2O molecules.

First, the oxidation numbers:

$$Cd \equiv Cd^0 \qquad \text{(reactant)}$$

$$NiO(OH) \equiv Ni^{3+} + O^{2-} + OH^- \qquad \text{(reactant)}$$

$$Cd(OH)_2 \equiv Cd^{2+} + OH^- + OH^- \qquad \text{(product)}$$

$$Ni(OH)_2 \equiv Ni^{2+} + OH^- + OH^- \qquad \text{(product)}$$

Next, the primitive half-cells:

$$Cd \longrightarrow Cd(OH)_2 \qquad \text{(oxidation)}$$

$$NiO(OH) \longrightarrow Ni(OH)_2 \qquad \text{(reduction)}$$

Cd, oxidized from Cd^0 to Cd^{2+}, is the reducing agent. NiO(OH), reduced from Ni^{3+} to Ni^{2+}, is the oxidizing agent.

1. *Oxidation [Cd* $\longrightarrow$ *Cd(OH)$_2$].* The completed half-reaction, balanced earlier in the chapter, contains OH^- and H_2O in addition to the primary reductant and conjugate oxidant:

$$Cd + 2OH^- \longrightarrow Cd(OH)_2 + 2e^-$$

2. *Reduction [NiO(OH)* $\longrightarrow$ *Ni(OH)$_2$].* The primitive half-reaction

$$NiO(OH) \longrightarrow Ni(OH)_2$$

contains nickel and oxygen already in balance. To this partial equation we then add one H^+ at the left to equalize the hydrogens,

$$NiO(OH) + H^+ \longrightarrow Ni(OH)_2$$

and, after that, one OH^- ion on both sides to remove the fictitious acid:

$$NiO(OH) + H^+ + OH^- \longrightarrow Ni(OH)_2 + OH^-$$

After matching up H^+ and OH^- to produce H_2O, we complete the half-reaction by putting one electron on the left:

$$NiO(OH) + H_2O + e^- \longrightarrow Ni(OH)_2 + OH^-$$

3. *Coupling the half-reactions.* Multiplication of the reduction half-reaction by 2 enforces conservation of charge. Two electrons are lost during oxidation; two electrons are gained during reduction:

$$Cd + 2OH^- \longrightarrow Cd(OH)_2 + 2e^- \qquad \text{(oxidation)}$$
$$\underline{2NiO(OH) + 2H_2O + 2e^- \longrightarrow 2Ni(OH)_2 + 2OH^- \qquad \text{(reduction)}}$$
$$Cd(s) + 2NiO(OH)(s) + 2H_2O(\ell) \longrightarrow$$
$$Cd(OH)_2(s) + 2Ni(OH)_2(s) \qquad \text{(redox)}$$

The hydroxide ions cancel when the half-reactions are added.

EXAMPLE 17-3. Competing for Electrons

PROBLEM: Which of the two species is the stronger oxidizing agent in each pair? (a) $Cl_2(g)$, $Br_2(\ell)$. (b) $Na^+(aq)$, $Cl_2(g)$. (c) $Ag^+(aq)$, $Zn^{2+}(aq)$. Consult the standard reduction potentials in Table C-21 before answering.

SOLUTION: Every atom, molecule, or ion competes for electrons with every other, and the winning reactions are simply those that produce the greatest drop in free energy. To rank the winners and losers, we measure their standard reduction potentials relative to hydrogen.

(a) *$Cl_2(g)$, $Br_2(\ell)$.* Look up the half-reactions

$$Cl_2(g) + 2e^- \longrightarrow 2Cl^-(aq) \qquad \mathcal{E}^\circ_{red} = 1.36 \text{ V}$$

$$Br_2(\ell) + 2e^- \longrightarrow 2Br^-(aq) \qquad \mathcal{E}^\circ_{red} = 1.07 \text{ V}$$

and compare: In a cell where Cl_2 takes electrons from H_2, there arises a potential difference of 1.36 volts between anode and cathode. But in a cell where Br_2 takes the same two electrons, the voltage is 1.07 V— smaller now, less positive. Given this difference, we know at once that chlorine is the stronger oxidizing agent.

Because: The stronger oxidizing agent is the one that *takes* the electrons, oxidizing its partner while undergoing reduction itself. For the same quantity of electrons transferred, the stronger oxidizing agent brings about the larger drop in free energy.

Here, with two moles of electrons ($n = 2$) and a positive $\mathscr{E}°$ in each reaction, the standard free energy falls by

$$\Delta G° = -n\mathscr{F}\mathscr{E}° = -2\mathscr{F}\mathscr{E}°$$

in both the chlorine and bromine reductions. Apparently it falls more, though, when chlorine takes the electrons from hydrogen, since the higher positive voltage corresponds to a more negative change in free energy.

In competition with H_2, chlorine therefore oxidizes *more* hydrogen than bromine does under exactly the same conditions. The reduction potential is more positive; the change in free energy is more negative; the equilibrium constant is larger. $Cl_2(g)$, a smaller and more electron-receptive molecule than $Br_2(\ell)$, is the stronger oxidizing agent.

(b) $Na^+(aq), Cl_2(g)$. Sodium, an alkali metal, is a giver of electrons, an atom far more likely to be oxidized than reduced. Once ionized, it tends to stay that way. The negative standard reduction potential for the process

$$Na^+(aq) + e^- \longrightarrow Na(s) \qquad \mathscr{E}°_{red} = -2.71 \text{ V}$$

thus confirms that Na^+ will consume rather than release free energy upon taking electrons from H_2. Nonspontaneous, this uphill reduction of $Na^+(aq)$ to $Na(s)$ is destined to leave more reactants than products at equilibrium. The equilibrium constant is less than 1, made that way by a positive $\Delta G°$.

Sodium *metal*, going in the other direction, is a reducing agent instead, prone to lose electrons to an oxidizing agent such as chlorine. Turn the half-cell reaction around and see:

$$Na(s) \longrightarrow Na^+(aq) + e^- \qquad \mathscr{E}°_{ox} = 2.71 \text{ V}$$

One way, the change in free energy is positive; the other way, negative. One way, a negative voltage; the other way, a positive voltage. One way, reduction; the other, oxidation.

$Cl_2(g)$ is a strong oxidizing agent, more easily reduced than $Na^+(aq)$. $Na(s)$ is a strong reducing agent, easily oxidized.

(c) *$Ag^+(aq), Zn^{2+}(aq)$.* The silver ion, $Ag^+(aq)$, with the higher (more positive) reduction potential, is the stronger oxidizing agent:

$$Ag^+(aq) + e^- \longrightarrow Ag(s) \qquad \mathscr{E}^\circ_{red} = 0.80 \text{ V}$$

$$Zn^{2+}(aq) + 2e^- \longrightarrow Zn(s) \qquad \mathscr{E}^\circ_{red} = -0.76 \text{ V}$$

Ag^+ wins the electrons more often than does Zn^{2+}.

Going the other way, of course, the tables are turned and metallic zinc here has the higher *oxidation* potential:

$$Zn(s) \longrightarrow Zn^{2+}(aq) + 2e^- \qquad \mathscr{E}^\circ_{ox} = 0.76 \text{ V}$$

$$Ag(s) \longrightarrow Ag^+(aq) + e^- \qquad \mathscr{E}^\circ_{ox} = -0.80 \text{ V}$$

Competing directly against silver, Zn therefore will give electrons to Ag^+—as, for example, in the displacement reaction

$$Zn(s) + 2Ag^+(aq) \longrightarrow Zn^{2+}(aq) + 2Ag(s)$$

We say that zinc, more easily oxidized, is more **active** than silver. Zn stands above Ag in the competitive *activity series*, where substances are arranged in order of increasing oxidation potentials.

QUESTION: Have we made a fair comparison of zinc's voltage against silver's? The half-reaction for Zn^{2+} involves two electrons, compared with only one for Ag^+.

ANSWER: The ranking is meaningful as it stands. Voltage, an intensive property, is independent of both the amount of substance and the value of *n*. The potential difference pertains to joules *per coulomb*, not joules alone and not coulombs alone. Two moles of Ag^+ produce the same voltage as one mole, although twice as many joules of free energy. See the next example.

EXAMPLE 17-4. Electrode Potential and Free Energy

PROBLEM: From the voltages given in Table C-21 and Example 17 3(c), we already know the value of $\mathscr{E}^\circ_{red}$ for the reduction of silver ion to silver metal:

$$Ag^+(aq) + e^- \longrightarrow Ag(s) \qquad \mathscr{E}^\circ_{red} = 0.80 \text{ V}$$

Use this information to compute the standard change in free energy, $\Delta G°$, both for the half-reaction as written and in the form given below:

$$2\,Ag^+(aq) + 2e^- \longrightarrow 2\,Ag(s)$$

Note that the second equation differs from the first only by a factor of 2. What is the corresponding value of $\mathscr{E}°_{red}$?

SOLUTION: Free energy and cell potential are directly proportional, related by the equation

$$\Delta G° = -n\mathscr{F}\mathscr{E}°$$

under standard conditions. Inserting the value $\mathscr{E}° = 0.80$ V from the first half-reaction, then, we substitute $n = 1$ to obtain $\Delta G°$ for the transfer of *one* mole of electrons to one mole of Ag^+ ions:

$$\Delta G° = -1 \text{ mol} \times 96{,}485 \text{ C mol}^{-1} \times 0.80 \text{ J C}^{-1} = -7.7 \times 10^4 \text{ J}$$

Now if 77,000 joules (77 kJ) accompany the formation of one mole, then exactly twice as much energy must attend the formation of two moles. Why? Because free energy is an *extensive* property, paid out molecule by molecule and mole by mole. More moles produce more joules:

$$Ag^+(aq) + e^- \longrightarrow Ag(s) \qquad \Delta G° = -77 \text{ kJ}, \quad \mathscr{E}° = 0.80 \text{ V}$$

$$2\,Ag^+(aq) + 2e^- \longrightarrow 2\,Ag(s) \qquad \Delta G° = -154 \text{ kJ}, \quad \mathscr{E}° = ?$$

But more joules do not produce more volts, because the additional joules are needed to move the additional coulombs. The ratio of joules to coulombs, $\mathscr{E}°$, stays the same when both n and $\Delta G°$ are doubled:

$$\mathscr{E}° = -\frac{\Delta G°}{n\mathscr{F}} = -\frac{(-2 \times 77{,}000 \text{ J})}{2 \text{ mol} \times 96{,}485 \text{ C/mol}} = 0.80 \text{ J C}^{-1} \equiv 0.80 \text{ V}$$

Conclusion: Free energy, an extensive property, scales directly with the coefficients of the balanced equation. Cell potential, an intensive property, remains the same however much or little material there may be. A bigger cell produces more joules, but not a bigger voltage.

EXAMPLE 17-5. Cell Potential and Free Energy

Here we combine the silver and zinc half-reactions from Examples 17-3(c) and 17-4 into a single galvanic cell, connecting a source of electrons (Zn) to a sink (Ag^+). The standard reduction potentials are repeated below for reference:

$$Ag^+(aq) + e^- \longrightarrow Ag(s) \qquad \mathscr{E}^\circ_{red} = 0.80 \text{ V}$$

$$Zn^{2+}(aq) + 2e^- \longrightarrow Zn(s) \qquad \mathscr{E}^\circ_{red} = -0.76 \text{ V}$$

PROBLEM: Calculate the standard change in free energy for the reaction

$$Zn(s) + 2Ag^+(aq) \longrightarrow Zn^{2+}(aq) + 2Ag(s)$$

What is the standard cell potential?

SOLUTION: Zn, the electron donor, is the reducing agent. Ag^+, the electron acceptor, is the oxidizing agent. We arrange, accordingly, for the oxidation half-reaction

$$Zn(s) \longrightarrow Zn^{2+}(aq) + 2e^- \qquad \mathscr{E}^\circ_{ox} = 0.76 \text{ V}$$

to occur at the anode and for the reduction half-reaction

$$Ag^+(aq) + e^- \longrightarrow Ag(s) \qquad \mathscr{E}^\circ_{red} = 0.80 \text{ V}$$

to occur at the cathode, allowing electrons to flow from giver to taker.

To match donor with acceptor, the system now must reduce two moles of Ag^+ for every mole of Zn it oxidizes. Two moles of electrons need to go from one mole of Zn to two moles of Ag^+. Nothing can be lost.

To arrive at this balanced equation for the cell, we multiply the Ag^+ half-reaction by 2 and thereby couple the oxidation to the reduction:

$$
\begin{array}{ll}
Zn(s) \longrightarrow Zn^{2+}(aq) + 2e^- & \mathscr{E}^\circ_{ox} = 0.76 \text{ V} \\
\underline{2[Ag^+(aq) + e^- \longrightarrow Ag(s)]} & \underline{\mathscr{E}^\circ_{red} = 0.80 \text{ V}} \\
Zn(s) + 2Ag^+(aq) \longrightarrow Zn^{2+}(aq) + 2Ag(s) & \Delta G^\circ = ? \; \mathscr{E}^\circ = ?
\end{array}
$$

Confident that $\mathscr{E}^\circ_{red}$ is unchanged by the multiplication (Example 17-4), we then go on to compute ΔG° and $\mathscr{E}^\circ$ for the overall redox process. The remaining steps are: (1) Determine the free energy of each half-reaction separately ($n = 2$), making use of the appropriate electrode potential. (2) Add ΔG°_{red} and ΔG°_{ox} to obtain the overall change in free energy for the cell:

$$\Delta G^\circ = \Delta G^\circ_{red} + \Delta G^\circ_{ox} = -n\mathscr{F}(\mathscr{E}^\circ_{red} + \mathscr{E}^\circ_{ox}) = -n\mathscr{F}\mathscr{E}^\circ$$

$$= -2 \text{ mol} \times 96,485 \text{ C mol}^{-1} \times (0.80 + 0.76) \text{ J C}^{-1} = -3.01 \times 10^5 \text{ J}$$

$$\mathscr{E}^\circ = \frac{\Delta G^\circ}{-n\mathscr{F}} = \mathscr{E}^\circ_{red} + \mathscr{E}^\circ_{ox} = 0.80 \text{ V} + 0.76 \text{ V} = 1.56 \text{ V}$$

The combined cell potential

$$\mathscr{E}^\circ = \mathscr{E}^\circ_{red} + \mathscr{E}^\circ_{ox}$$

thus falls out naturally during the course of solution, confirming what we already know.

EXAMPLE 17-6. Is It Spontaneous?

PROBLEM: Balance the equation

$$I^-(aq) + Al^{3+}(aq) \longrightarrow I_2(s) + Al(s)$$

and identify the oxidation and reduction half-reactions. Is the overall cell reaction spontaneous in the direction indicated? Calculate $\mathscr{E}^\circ$ and ΔG°.

SOLUTION: Al^{3+} gains electrons; I^- loses electrons. So says the equation as written. Maybe it happens; maybe it doesn't.

To decide, we balance each half-reaction separately for atoms and charge,

$$2I^-(aq) \longrightarrow I_2(s) + 2e^- \qquad \text{oxidation (anode)}$$

$$Al^{3+}(aq) + 3e^- \longrightarrow Al(s) \qquad \text{reduction (cathode)}$$

and, after that, combine them into a single process in which six electrons are lost and six electrons are gained ($n = 6$):

$$3[2I^-(aq) \longrightarrow I_2(s) + 2e^-] \qquad\qquad \mathscr{E}^\circ_{ox} = -0.54 \text{ V}$$
$$\underline{2[Al^{3+}(aq) + 3e^- \longrightarrow Al(s)]} \qquad\qquad \mathscr{E}^\circ_{red} = -1.66 \text{ V}$$
$$6I^-(aq) + 2Al^{3+}(aq) \longrightarrow 3I_2(s) + 2Al(s) \quad \mathscr{E}^\circ = \mathscr{E}^\circ_{red} + \mathscr{E}^\circ_{ox} = -2.20 \text{ V}$$

The electrode potentials, taken from Table C-21, are unaffected by the multiplication of each half-reaction.

The standard cell potential (-2.20 V) proves to be negative, signaling a positive ΔG° and hence a nonspontaneous reaction under standard conditions:

$$\Delta G^\circ = -n\mathscr{F}\mathscr{E}^\circ = -6 \text{ mol} \times 96{,}485 \text{ C mol}^{-1} \times (-2.20 \text{ J C}^{-1})$$

$$= 1.27 \times 10^6 \text{ J} = 1.27 \times 10^3 \text{ kJ}$$

To push the reaction uphill, forcing electrons from I^- onto Al^{3+}, we need therefore to pump energy continuously *into* the system. How? We find a spontaneous source of external power and hook it up to an *electrolytic* cell.

Or, rather than fight, we can simply reverse direction and take energy *out* of the system instead, running the reaction as

$$3I_2(s) + 2Al(s) \longrightarrow 6I^-(aq) + 2Al^{3+}(aq) \qquad \mathscr{E}^\circ = 2.20 \text{ V}$$

The cell potential turns positive; the free energy turns negative; and the electrons now flow downhill in a spontaneously operating *galvanic* cell. Al, the natural donor, is oxidized to Al^{3+} at the anode; I_2, the acceptor, is reduced to I^- at the cathode.

Either way, the world grows more disordered. Energy dribbles into more levels than ever before. Entropy rises.

Example 17-7. Cell Potential and the Equilibrium Constant

PROBLEM: Compute the equilibrium constant for the reaction

$$3I_2(s) + 2Al(s) \longrightarrow 6I^-(aq) + 2Al^{3+}(aq) \qquad \mathscr{E}^\circ = 2.20 \text{ V}$$

at 298.15 K (25°C).

SOLUTION: We know, from Example 17-6, that six electrons are transferred between two moles of Al and three moles of I_2. Substitution of $n = 6$ into the relationship between K, ΔG°, and $\mathscr{E}^\circ$,

$$\Delta G^\circ = -RT \ln K = -n\mathscr{F}\mathscr{E}^\circ$$

then gives:

$$\ln K = \frac{n\mathscr{F}}{RT} \mathscr{E}^\circ = 6 \text{ mol} \times \frac{96{,}485 \text{ C mol}^{-1}}{(8.3145 \text{ J K}^{-1})(298.15 \text{ K})} \times 2.20 \text{ J C}^{-1} = 514$$

$$\log K = \frac{\ln K}{2.303} = \frac{514}{2.303} = 223$$

$$K = 10^{\log K} = 10^{223}$$

It is a huge number, infinite for all purposes. The reaction goes to completion in the direction written.

QUESTION: With just a few volts, nature unleashes this process of tremendous thermodynamic power: $K = 10^{223}$, $\Delta G^\circ = -1270$ kJ. How many volts would accompany a more modest reaction, say one where $K = 10$ at 298.15 K?

ANSWER: Solve for $\mathscr{E}^\circ$ as shown below, realizing that $\log K = \log 10 = 1$:

$$\mathscr{E}^\circ = \frac{RT}{n\mathscr{F}} \ln K = 2.303 \frac{RT}{n\mathscr{F}} \log K = \frac{2.303\ RT}{n\mathscr{F}}$$

$$= \frac{0.0592\ \text{V mol}}{n}$$

The larger the value of n, the lower the potential. For a reaction that transfers one mole of electrons through a potential of 0.0592 V, the equilibrium constant, $K(25^\circ C)$, is equal to 10.

We have, as a result, a handy number for later use:

$$\frac{2.303\ RT}{\mathscr{F}} = 0.0592\ \text{V} \qquad (T = 298.15\ \text{K},\ n = 1\ \text{mol e}^-)$$

EXAMPLE 17-8. The Nernst Equation

Earlier examples showed us that the size of a cell (the amount of material it contains) has no effect on the potential. What does?

PROBLEM: Return to a Zn/Cu cell and compute the electromotive force, $\mathscr{E}$, under the following sets of conditions: (a) $[Zn^{2+}] = 1.0\ M$, $[Cu^{2+}] = 1.0\ M$. (b) $[Zn^{2+}] = 10.0\ M$, $[Cu^{2+}] = 1.0\ M$. (c) $[Zn^{2+}] = 1.0\ M$, $[Cu^{2+}] = 10.0\ M$. Assume that the cell operates reversibly at 298.15 K.

SOLUTION: The voltage depends not on how many moles of Zn^{2+} and Cu^{2+} the cell contains, but rather on how many moles per liter: the concentrations. $\mathscr{E}$ rises and falls with the instantaneous reaction quotient, Q, varying logarithmically according to the Nernst equation:

$$\mathscr{E} = \mathscr{E}^\circ - \frac{RT}{n\mathscr{F}} \ln Q$$

At 25°C (298.15 K), we have the simplified form

$$\mathcal{E} = \mathcal{E}° - \frac{0.0592}{n} \log Q \qquad (T = 298.15 \text{ K})$$

derived above in Example 17-7. Observe how the factor 2.303, implicit in the all-in-one number 0.0592, ensures the conversion of a natural logarithm into a base-10 logarithm.

(a) *Standard conditions,* $[Zn^{2+}] = [Cu^{2+}] = 1.0$ *M.* The half-cell reactions are

$$
\begin{array}{lr}
Zn(s) \longrightarrow Zn^{2+}(aq) + 2e^- & \mathcal{E}°_{ox} = 0.76 \text{ V} \\
Cu^{2+}(aq) + 2e^- \longrightarrow Cu(s) & \mathcal{E}°_{red} = 0.34 \text{ V} \\
\hline
Zn(s) + Cu^{2+}(aq) \longrightarrow Zn^{2+}(aq) + Cu(s) & \mathcal{E}° = \mathcal{E}°_{red} + \mathcal{E}°_{ox} = 1.10 \text{ V}
\end{array}
$$

and the reaction quotient is

$$Q = \frac{[Zn^{2+}]}{[Cu^{2+}]} = \frac{1.0}{1.0} = 1.0$$

The key factor in the Nernst equation, $\log Q$, thus reduces to 0, enabling the cell potential to attain its standard value:

$$\mathcal{E} = \mathcal{E}° - \frac{0.0592}{n} \log 1 = \mathcal{E}° = 1.10 \text{ V}$$

Such is the definition of $\mathcal{E}°$: the voltage that develops when all concentrations are 1.0 *M* ($Q = 1$). Any deviation from the ratio $Q = 1$ must produce a different voltage.

(b) $[Zn^{2+}] = 10.0$ *M,* $[Cu^{2+}] = 1.0$ *M.* The reaction now stands closer to equilibrium than under standard conditions, with more products and less reactants in the mix. The new reaction quotient ($Q = 10$) gives us

$$\log Q = \log \frac{[Zn^{2+}]}{[Cu^{2+}]} = \log \frac{10.0}{1.0} = 1.00$$

and (with $n = 2$) a corresponding cell voltage of

$$\mathcal{E} = \mathcal{E}° - \frac{0.0592}{n} \log Q = 1.10 \text{ V} - 0.030 \text{ V} = 1.07 \text{ V}$$

$\mathcal{E}$ falls by $RT/n\mathcal{F}$ from its standard value.

For every 10-fold increase in Q, the electromotive force decreases by 0.0592 V/n. It drops and drops until finally equilibrium sets in, whereupon $\mathscr{E} = 0$ (because $Q = K$) and all thermodynamic potential is quenched:

$$\mathscr{E} = \mathscr{E}° - \frac{0.0592}{n} \log K = 0 \qquad \text{(at equilibrium, } T = 298.15 \text{ K)}$$

QUESTION: At what ratio of concentrations does that happen?

ANSWER: First compute the equilibrium constant from the change in standard free energy, just as in Example 17-7, to obtain the value

$$K = 1.54 \times 10^{37} \qquad (T = 298.15 \text{ K})$$

Substitution of K into the Nernst equation then confirms that $\mathscr{E}$ (the Q-dependent cell potential) vanishes at equilibrium, where $Q = K$:

$$\mathscr{E} = \mathscr{E}° - \frac{0.0592}{n} \log Q$$

$$\mathscr{E}(Q = K) = 1.10 \text{ V} - \frac{0.0592 \text{ V}}{2} \log (1.54 \times 10^{37})$$

$$= 1.10 \text{ V} - 1.10 \text{ V} = 0$$

The exceedingly large value of K ensures that the reactants are entirely consumed.

(c) *$[Zn^{2+}] = 1.0$ M, $[Cu^{2+}] = 10.0$ M.* The reaction quotient is $1/10$, indicating an excess of reactants over products. The cell, driven forward by the imbalance, generates a higher voltage:

$$\mathscr{E} = \mathscr{E}° - \frac{0.0592}{2} \log \frac{1.0}{10.0} = 1.10 \text{ V} + 0.030 \text{ V} = 1.13 \text{ V}$$

The increase, dependent on log Q, amounts to 0.0592 V/n for every 10-fold decrease in Q.

EXAMPLE 17-9. Electrolysis: Where Givers Become Takers

First, some questions and answers for review. Then, a pair of numerical examples to illustrate the operation of an electrolytic cell.

QUESTION: What happens when we dip a strip of zinc into a solution of $CuSO_4$?

ANSWER: A spontaneous redox reaction occurs, driven forward by a fall in free energy. Oxidation: zinc ions drop from the metal into the solution, leaving behind two electrons for every atom oxidized. Reduction: copper ions go from the solution to the metal, finding there the electrons to take them from Cu^{2+} to Cu.

$$Zn(s) + Cu^{2+}(aq) \longrightarrow Zn^{2+}(aq) + Cu(s)$$

$$\Delta G° = -212 \text{ kJ}, \quad \mathscr{E}° = 1.10 \text{ V}$$

Metallic copper plates out onto the zinc strip.

QUESTION: And the reverse reaction? What happens when we dip a strip of copper into a solution of $ZnSO_4$?

ANSWER: Nothing. The transformation

$$Cu(s) + Zn^{2+}(aq) \longrightarrow Cu^{2+}(aq) + Zn(s)$$

$$\Delta G° = 212 \text{ kJ}, \quad \mathscr{E}° = -1.10 \text{ V}$$

is not spontaneous. Left alone, Zn^{2+} does not oxidize copper. It lacks the energy to make the climb.

QUESTION: Can we force the reverse reaction somehow?

ANSWER: Yes, if we supply the energy to an electrolytic cell.

QUESTION: How much?

ANSWER: No less, certainly, than the free energy needed for the uphill reaction ($\Delta G° = 212$ kJ)—and usually more, because some energy always trickles away during the cell's operation. We need, consequently, to apply a potential of at least -1.10 V to the zinc electrode, barely sufficient to reduce Zn^{2+} to Zn. Held at negative potential, the zinc terminal then serves as the *cathode* of the electrolytic cell. Electrons are forced—repelled by the negative polarity from the power supply—onto the zinc metal.

And, forced or not, every reduction requires a corresponding oxidation. Here the copper electrode, subjected to a positive polarity, becomes a place of oxidation, thereby reversing the role it plays in the galvanic cell. Electrons are drawn away from Cu (the electrolytic *anode*) by a positive potential from the power supply, causing Cu^{2+} ions to fall into solution.

PROBLEM: Assume that we apply a current of 5.0 A to the electrolytic cell just described. How much mass accumulates on the zinc electrode after 0.50 h of forced operation under standard conditions?

SOLUTION: A current of one ampere ($A = C\ s^{-1}$), flowing for one second, delivers a charge of one coulomb past a given point:

$$\text{Charge} = \text{current} \times \text{time}$$
$$C = C\ s^{-1} \times s$$

Knowing further that one mole of electrons carries a charge of 96,485 C (the Faraday constant, $\mathscr{F}$), we have at once a way to compute the moles of electrons (n) forced through the cell:

$$n = 5.0\ C\ s^{-1} \times 0.50\ h \times \frac{3600\ s}{h} \times \frac{1\ \text{mol}\ e^-}{96,485\ C} = 0.093\ \text{mol}\ e^-$$

Since two moles of electrons are needed to reduce one mole of Zn^{2+}, we calculate the mass of Zn deposited stoichiometrically:

$$0.093\ \text{mol}\ e^- \times \frac{1\ \text{mol}\ Zn}{2\ \text{mol}\ e^-} \times \frac{65.39\ g}{\text{mol}\ Zn} = 3.0\ g\ Zn$$

EXAMPLE 17-10. Electrolysis: Paying the Price

PROBLEM: Determine the minimum electrical work needed to carry out the operation described in Example 17-9 (5.0 A applied for 1800 s under standard conditions).

SOLUTION: Use the relationship

$$\text{Work} = -\text{charge} \times \text{voltage}$$

to obtain the work needed to move a negative charge through a difference in electrical potential. It is, note, the same expression we always employ to equate free energy with electrochemical work:

$$w = -n\mathscr{F}\mathscr{E}$$

The higher the voltage, the greater the work.

For the copper–zinc electrolytic cell, a potential difference of -1.10 V would be just enough to force two electrons from Cu to Zn^{2+} in an ideal, loss-free system. That value, combined with the total number of coulombs moved, gives us the minimum work required:

Charge $= (5.0 \text{ C s}^{-1})(1800 \text{ s}) = 9.0 \times 10^3 \text{ C}$

Work $= -\text{charge} \times \text{voltage} = (9.0 \times 10^3 \text{ C})(1.10 \text{ J C}^{-1}) = 9.9 \times 10^3 \text{ J}$

The voltage actually applied is always more, however, differing by some *overvoltage* needed to overcome the resistance present in every cell. The work scales proportionally.

EXERCISES

1. Pick the stronger oxidizing agent in each pair:

   (a) $Ba^{2+}(aq)$, $Ca^{2+}(aq)$

   (b) $Mn^{3+}(aq)$, $I_2(s)$

   (c) $Au^+(aq)$, $Hg_2^{2+}(aq)$

   (d) $Cu(s)$, $Fe^{2+}(aq)$

   (e) $Al^{3+}(aq)$, $Li^+(aq)$

   Consult the table of standard reduction potentials provided in Appendix C.

2. Pick the stronger reducing agent in each pair:

   (a) $Zn(s)$, $Ca(s)$

   (b) $I^-(aq)$, $F^-(aq)$

   (c) $Na(s)$, $Li(s)$

   (d) $Zn^{2+}(aq)$, $Ni(s)$

   (e) $Mn^{2+}(aq)$, $Ag(s)$

   Consult a table of standard reduction potentials to decide.

3. Balance each equation:

   (a) ___ $Fe^{3+}(aq)$ + ___ $I^-(aq)$ $\longrightarrow$ ___ $Fe^{2+}(aq)$ + ___ $I_2(s)$

   (b) ___ $Tl^{3+}(aq)$ + ___ $Cr^{2+}(aq)$ $\longrightarrow$ ___ $Tl^+(aq)$ + ___ $Cr^{3+}(aq)$

   (c) ___ $Br_2(\ell)$ + ___ $In^+(aq)$ $\longrightarrow$ ___ $Br^-(aq)$ + ___ $In^{3+}(aq)$

   Identify the oxidizing agents and reducing agents. How many moles of electrons are transferred from giver to receiver?

4. Complete and balance the partial redox equations below, writing each overall reaction as a sum of two half-reactions in acidic solution. Add H^+ and H_2O as needed.

   (a) ___ $MnO_4^-(aq)$ + ___ $C_2O_4^{2-}(aq)$ $\longrightarrow$

   ___ $Mn^{2+}(aq)$ + ___ $CO_2(g)$

   (b) ___ $Cu(s)$ + ___ $NO_3^-(aq)$ $\longrightarrow$ ___ $Cu^{2+}(aq)$ + ___ $NO(g)$

   (c) ___ $Ce^{4+}(aq)$ + ___ $Bi(s)$ $\longrightarrow$ ___ $Ce^{3+}(aq)$ + ___ $BiO^+(aq)$

   (d) ___ $CO(g)$ + ___ $O_2(g)$ $\longrightarrow$ ___ $CO_2(g)$

Then, from the half-reactions, identify the oxidizing agents and reducing agents. How many moles of electrons are transferred from giver to receiver?

5. Complete and balance the partial redox equations below, writing each overall reaction as a sum of two half-reactions in acidic solution. Add H^+ and H_2O as needed.

(a) ___ $I_2(s) +$ ___ $Cu^{2+}(aq) \longrightarrow$ ___ $IO_3^-(aq) +$ ___ $Cu(s)$

(b) ___ $VO_2^+(aq) +$ ___ $Ni(s) \longrightarrow$ ___ $VO^{2+}(aq) +$ ___ $Ni^{2+}(aq)$

(c) ___ $H_2S(aq) +$ ___ $NO_3^-(aq) \longrightarrow$ ___ $S(s) +$ ___ $NO(g)$

(d) ___ $NaBr(s) +$ ___ $SO_4^{2-}(aq) \longrightarrow$
___ $Br_2(\ell) +$ ___ $SO_2(g) +$ ___ $Na^+(aq)$

Then, from the half-reactions, identify the oxidizing agents and reducing agents. How many moles of electrons are transferred from giver to receiver?

6. Complete and balance the partial redox equations below, writing each overall reaction as a sum of two half-reactions in basic solution. Add OH^- and H_2O where needed.

(a) ___ $NaClO(aq) +$ ___ $NaBr(aq) \longrightarrow$
___ $NaBrO_3(aq) +$ ___ $NaCl(aq)$

(b) ___ $Ba^{2+}(aq) +$ ___ $ClO_2(aq) +$ ___ $H_2O_2(aq) \longrightarrow$
___ $Ba(ClO_2)_2(s) +$ ___ $O_2(g)$

(c) ___ $Fe(s) +$ ___ $NiO_2(s) \longrightarrow$ ___ $Fe(OH)_2(s) +$ ___ $Ni(OH)_2(s)$

(d) ___ $Ag(s) +$ ___ $O_2(g) \longrightarrow$ ___ $Ag^+(aq)$

Then, from the half-reactions, identify the oxidizing agents and reducing agents. How many moles of electrons are transferred from giver to receiver?

7. Complete and balance the partial redox equations below, writing each overall reaction as a sum of two half-reactions in basic solution. Add OH^- and H_2O where needed.

(a) ___ $Zn(s) +$ ___ $MnO_2(s) \longrightarrow$
___ $Zn(OH)_2(s) +$ ___ $Mn_2O_3(s)$

(b) ___ $PH_3(g) +$ ___ $CrO_4^{2-}(aq) \longrightarrow$
___ $P_4(s) +$ ___ $Cr(OH)_4^-(aq)$

(c) ___ $Au(s) +$ ___ $CN^-(aq) +$ ___ $O_2(g) \longrightarrow$ ___ $Au(CN)_2^-(aq)$

(d) ___ $NH_2OH(aq) +$ ___ $CO_2(g) \longrightarrow$ ___ $N_2(g) +$ ___ $CO(g)$

Then, from the half-reactions, identify the oxidizing agents and reducing agents. How many moles of electrons are transferred from giver to receiver?

8. A standard potential of 1.247 V develops in a galvanic cell containing electrodes of iron and silver:

$$Fe(s) \longrightarrow Fe^{2+}(aq) + 2e^-$$
$$\underline{2Ag^+(aq) + 2e^- \longrightarrow 2Ag(s)}$$
$$Fe(s) + 2Ag^+(aq) \longrightarrow Fe^{2+}(aq) + 2Ag(s) \qquad \mathscr{E}^\circ = 1.247 \text{ V}$$

(a) Which half-reaction occurs at the anode? Which occurs at the cathode? (b) The standard potential for the reduction of $Fe^{2+}(aq)$ to $Fe(s)$ is -0.447 V. Using just the information given, calculate $\mathscr{E}^\circ_{ox}$ and $\mathscr{E}^\circ_{red}$ for the corresponding oxidation and reduction half-reactions above.

9. A standard potential of 0.59 V develops in a galvanic cell driven by the following half-reactions:

$$3[Ni(s) + 2OH^-(aq) \longrightarrow Ni(OH)_2(s) + 2e^-]$$
$$\underline{2[CrO_4^{2-}(aq) + 4H_2O(\ell) + 3e^- \longrightarrow Cr(OH)_3(s) + 5OH^-(aq)]}$$
$$3Ni(s) + 2CrO_4^{2-}(aq) + 8H_2O(\ell) \longrightarrow$$
$$3Ni(OH)_2(s) + 2Cr(OH)_3(s) + 4OH^-(aq) \qquad \mathscr{E}^\circ = 0.59 \text{ V}$$

(a) Which half-reaction occurs at the anode? Which occurs at the cathode? (b) The standard potential for the reduction of $Ni(OH)_2(s)$ to $Ni(s)$ is -0.72 V. Using just the information given, calculate $\mathscr{E}^\circ_{ox}$ and $\mathscr{E}^\circ_{red}$ for the corresponding half-reactions above.

10. The preceding two exercises demonstrate the relationship

$$\mathscr{E}^\circ = \mathscr{E}^\circ_{red} + \mathscr{E}^\circ_{ox}$$

in an electrochemical cell. (a) Show that the connection amounts to a Hess's law summation of the free energies for the oxidation and reduction half-reactions:

$$\Delta G^\circ = \Delta G^\circ_{red} + \Delta G^\circ_{ox}$$

(b) Show further that the summation is equivalent to a multiplication of the equilibrium constants:

$$K = K_{red}K_{ox}$$

11. Press the analogy between redox and acid–base reactions. In what way is the relationship

$$\mathscr{E}^\circ = \mathscr{E}^\circ_{red} + \mathscr{E}^\circ_{ox}$$

between reductant and oxidant similar to the relationship

$$pK_w = pK_a + pK_b$$

between an acid and its conjugate base?

12. The following reaction proceeds spontaneously in a basic electrolyte:

$$HgO(s) + Zn(s) + H_2O(\ell) \longrightarrow Hg(\ell) + Zn(OH)_2(s)$$

(a) Write the overall equation as a sum of oxidation and reduction half-reactions, identifying the oxidizing agent and the reducing agent. (b) Which half-reaction occurs at the anode? Which occurs at the cathode? (c) Calculate the standard cell potential using data from Appendix C.

13. Write each of the balanced equations below as a sum of balanced oxidation and reduction half-reactions:

(a) $4Fe^{2+}(aq) + O_2(g) + 4H^+(aq) \longrightarrow 4Fe^{3+}(aq) + 2H_2O(\ell)$
(b) $2H_2(g) + O_2(g) \longrightarrow 2H_2O(\ell)$ (basic)
(c) $2H_2O(\ell) \longrightarrow 2H_2(g) + O_2(g)$ (acidic)

Identify the oxidizing agents and reducing agents. How many moles of electrons are transferred from giver to receiver?

14. Calculate the standard cell potential ($\mathcal{E}°$) and change in free energy ($\Delta G°$) for each of the reactions treated in the preceding exercise:

(a) $4Fe^{2+}(aq) + O_2(g) + 4H^+(aq) \longrightarrow 4Fe^{3+}(aq) + 2H_2O(\ell)$
(b) $2H_2(g) + O_2(g) \longrightarrow 2H_2O(\ell)$ (basic)
(c) $2H_2O(\ell) \longrightarrow 2H_2(g) + O_2(g)$ (acidic)

Which of them will go spontaneously in the direction written under standard conditions?

15. The reaction

$$2Ag(s) + Zn^{2+}(aq) + 2OH^-(aq) \longrightarrow Ag_2O(s) + Zn(s) + H_2O(\ell)$$

is not spontaneous under standard conditions, having a cell potential of -1.104 V. (a) Calculate the standard reduction potential for the half-reaction

$$Ag_2O(s) + H_2O(\ell) + 2e^- \longrightarrow 2Ag(s) + 2OH^-(aq) \qquad \mathcal{E}°_{red} = ?$$

(b) Rewrite the redox reaction so that it will proceed spontaneously. Which half-reaction now occurs at the anode? (c) Calculate the corresponding values of $\mathcal{E}°$ and $\Delta G°$ for the spontaneous process.

16. Write each of the balanced equations below as a sum of balanced oxidation and reduction half-reactions:

    (a) $F_2(g) + 2Br^-(aq) \longrightarrow 2F^-(aq) + Br_2(\ell)$

    (b) $H_2SO_3(aq) + 2Mn(s) + 4H^+(aq) \longrightarrow$
    $$S(s) + 2Mn^{2+}(aq) + 3H_2O(\ell)$$

    Identify the oxidizing agents and reducing agents. How many moles of electrons are transferred from giver to receiver?

17. Calculate the standard cell potential ($\mathscr{E}°$) and change in free energy ($\Delta G°$) for each reaction treated in the preceding exercise:

    (a) $F_2(g) + 2Br^-(aq) \longrightarrow 2F^-(aq) + Br_2(\ell)$

    (b) $H_2SO_3(aq) + 2Mn(s) + 4H^+(aq) \longrightarrow$
    $$S(s) + 2Mn^{2+}(aq) + 3H_2O(\ell)$$

    Which of them, under standard conditions, will go spontaneously in the direction written?

18. Calculate $\mathscr{E}°$, $\Delta G°$, and K (the equilibrium constant at 25°C) for the following reactions:

    (a) $2Al(s) + 3Mn^{2+}(aq) \longrightarrow 2Al^{3+}(aq) + 3Mn(s)$

    (b) $2Hg(\ell) + Zn^{2+}(aq) \longrightarrow Hg_2^{2+}(aq) + Zn(s)$

    Determine whether each process is spontaneous under standard conditions.

19. A galvanic cell contains a gold electrode and a silver electrode, with standard reduction potentials

    $$Au^{3+}(aq) + 3e^- \longrightarrow Au(s) \qquad \mathscr{E}°_{red} = 1.40 \text{ V}$$

    $$Ag^+(aq) + e^- \longrightarrow Ag(s) \qquad \mathscr{E}°_{red} = 0.80 \text{ V}$$

    (a) Write the half-reactions that occur at anode and cathode. (b) Write a balanced equation for the cell. (c) Calculate $\mathscr{E}°$, $\Delta G°$, and K (the equilibrium constant at 25°C) for the cell.

20. A galvanic cell contains a silver electrode and a copper electrode, with standard reduction potentials

    $$Ag^+(aq) + e^- \longrightarrow Ag(s) \qquad \mathscr{E}°_{red} = 0.80 \text{ V}$$

    $$Cu^{2+}(aq) + 2e^- \longrightarrow Cu(s) \qquad \mathscr{E}°_{red} = 0.34 \text{ V}$$

    (a) Write the half-reactions that occur at anode and cathode. (b) Write a balanced equation for the cell. (c) Calculate $\mathscr{E}°$, $\Delta G°$, and K for the cell at 25°C.

21. A galvanic cell contains a gold electrode and a copper electrode, with standard reduction potentials

$$Au^{3+}(aq) + 3e^- \longrightarrow Au(s) \qquad \mathscr{E}^\circ_{red} = 1.40\ V$$

$$Cu^{2+}(aq) + 2e^- \longrightarrow Cu(s) \qquad \mathscr{E}^\circ_{red} = 0.34\ V$$

(a) Write the half-reactions that occur at anode and cathode. (b) Write a balanced equation for the cell. (c) Calculate $\mathscr{E}^\circ$, ΔG°, and K for the cell at 25°C.

22. Standard potentials for the reduction of $AgCl(s)$ and $Ag^+(aq)$ are given below:

$$AgCl(s) + e^- \longrightarrow Ag(s) + Cl^-(aq) \qquad \mathscr{E}^\circ_{red} = 0.2223\ V$$

$$Ag^+(aq) + e^- \longrightarrow Ag(s) \qquad \mathscr{E}^\circ_{red} = 0.7996\ V$$

(a) Use them to calculate the solubility-product constant for silver chloride at 25°C:

$$AgCl(s) \longrightarrow Ag^+(aq) + Cl^-(aq) \qquad K_{sp} = [Ag^+][Cl^-] = ?$$

(b) Is the equilibrium constant greater than 1 or less than 1? Is the equivalent cell potential positive or negative? Is the reaction spontaneous?

23. Use a suitable combination of reduction potentials to calculate K_{sp} for silver chromate at 25°C:

$$Ag_2CrO_4(s) \longrightarrow 2Ag^+(aq) + CrO_4^{2-}(aq)$$

$$K_{sp} = [Ag^+]^2[CrO_4^{2-}] = ?$$

24. Which ionic compound—AgI or AgCl—is less soluble in water at 25°C? Use just the information below to decide:

$$AgCl(s) + e^- \longrightarrow Ag(s) + Cl^-(aq) \qquad \mathscr{E}^\circ_{red} = 0.2223\ V$$

$$AgI(s) + e^- \longrightarrow Ag(s) + I^-(aq) \qquad \mathscr{E}^\circ_{red} = -0.1522\ V$$

Calculate the ratio of solubility-product constants.

25. Reminders: (a) What voltage does a galvanic cell produce when its reaction quotient (Q) is equal to the equilibrium constant (K)? (b) What voltage does a cell produce when all dissolved species have a concentration of 1 M and all gaseous species have a pressure of 1 atm?

26. Write the reaction

$$Cu(s) + 2Ag^+(aq) \longrightarrow Cu^{2+}(aq) + 2Ag(s)$$

as a sum of oxidation and reduction half-reactions, and first calculate $\mathscr{E}°$ for the cell. Then determine the cell voltage at 25°C under the following conditions:

(a) $[Ag^+] = 1.000\ M$ $[Cu^{2+}] = 1.000\ M$
(b) $[Ag^+] = 0.500\ M$ $[Cu^{2+}] = 1.000\ M$
(c) $[Ag^+] = 1.000\ M$ $[Cu^{2+}] = 0.500\ M$
(d) $[Ag^+] = 0.500\ M$ $[Cu^{2+}] = 0.500\ M$

27. Continue with the reaction

$$Cu(s) + 2Ag^+(aq) \longrightarrow Cu^{2+}(aq) + 2Ag(s)$$

at 25°C. (a) Compute $\Delta G°$ and K. (b) If $[Cu^{2+}]$ is 1.000 M, what concentration of Ag^+ will produce zero voltage in the cell? (c) If $[Cu^{2+}]$ is 5.000 M, what concentration of Ag^+ will likewise give a null voltage? (d) If, instead, the concentration of Ag^+ is 1.000 M, what concentration of Cu^{2+} will produce zero voltage? Comment on the size of the numbers involved.

28. One last time, consider the reaction

$$Cu(s) + 2Ag^+(aq) \longrightarrow Cu^{2+}(aq) + 2Ag(s)$$

at 25°C. Suppose that $[Cu^{2+}]$ is 1.000 M. (a) What concentration of Ag^+ will produce a voltage of 0.5000 V? (b) What concentration of Ag^+ will produce 0.4600 V? (c) 0.4200 V?

29. A galvanic cell driven by the reaction

$$2Fe^{3+}(aq) + H_2(g) \longrightarrow 2Fe^{2+}(aq) + 2H^+(aq)$$

operates at 25°C. Calculate the cell potential under each set of conditions below:

	$[Fe^{3+}]$	P_{H_2}	$[Fe^{2+}]$	pH
(a)	1.00 M	1.000 atm	1.00 M	0.000
(b)	1.00 M	1.000 atm	1.00 M	1.000
(c)	1.00 M	1.000 atm	1.00 M	2.000
(d)	1.00 M	1.000 atm	1.00 M	3.000
(e)	1.00 M	1.000 atm	1.00 M	4.000
(f)	1.00 M	1.000 atm	1.00 M	5.000
(g)	1.00 M	1.000 atm	1.00 M	6.000
(h)	1.00 M	1.000 atm	1.00 M	7.000

30. The same reaction:

$$2Fe^{3+}(aq) + H_2(g) \longrightarrow 2Fe^{2+}(aq) + 2H^+(aq)$$

The same temperature: 25°C. New conditions:

	$[Fe^{3+}]$	P_{H_2}	$[Fe^{2+}]$	pH
(a)	1.00 M	0.250 atm	1.00 M	0.000
(b)	1.00 M	0.500 atm	1.00 M	0.000
(c)	1.00 M	0.750 atm	1.00 M	0.000
(d)	1.00 M	1.000 atm	1.00 M	0.000

Calculate the cell potential at each of the four different pressures.

31. Again, consider a voltaic cell in which the reaction

$$2Fe^{3+}(aq) + H_2(g) \longrightarrow 2Fe^{2+}(aq) + 2H^+(aq)$$

goes forward at 25°C, and assume the following conditions:

1. Except for $[Fe^{3+}]$, all aqueous concentrations stand at 1.00 M.
2. The partial pressure of hydrogen is 1.00 atm.
3. The (nonstandard) cell potential, $\mathscr{E}$, is 0.800 V.

(a) What is the concentration of Fe^{3+}? (b) How far from equilibrium is the reaction? Compute $\Delta G - \Delta G°$. (c) Is the out-of-equilibrium system overweighted toward products or toward reactants?

32. An ordinary galvanic cell runs down and dies when its reactants and products come to equilibrium, but a *fuel cell* lives on: It survives, persistently out of equilibrium, on a steady resupply of reactants from the outside. Thus replenished, a fuel cell will produce a steady electric current as long as fresh reactants are available. Question: By how much will the voltage of a fuel cell driven by the reaction

$$CH_4(g) + 2O_2(g) \longrightarrow CO_2(g) + 2H_2O(g)$$

change when all the partial pressures are simultaneously doubled? Tripled? Quadrupled?

33. A Zn/Cu cell, reacting as

$$Zn(s) + Cu^{2+}(aq) \longrightarrow Zn^{2+}(aq) + Cu(s)$$

at 25°C, produces a voltage of 1.10 V under each of the following sets of conditions:

$$[Zn^{2+}] \quad [Cu^{2+}]$$

0.50 *M*   0.50 *M*
1.00 *M*   1.00 *M*
2.00 *M*   2.00 *M*

Show how.

34. Finally, having come down the hill, we climb back up to consider an *electrolytic cell*: the nonspontaneous counterpart of a galvanic cell. Start with the problem posed by H_2O, which can be either forcibly reduced to H_2 or forcibly oxidized to O_2:

$$2H_2O(\ell) + 2e^- \longrightarrow H_2(g) + 2OH^-(aq) \qquad \mathscr{E}^\circ_{red} = -0.83 \text{ V}$$

$$2H_2O(\ell) \longrightarrow 4H^+(aq) + O_2(g) + 4e^- \qquad \mathscr{E}^\circ_{ox} = -1.23 \text{ V}$$

Now answer: In aqueous solution at 25°C, is the electrolytic reaction

$$2Na^+(aq) + 2Cl^-(aq) \longrightarrow 2Na(s) + Cl_2(g)$$

likely to proceed as written? What product will appear at the cathode?

35. How about, instead, a reaction of Na^+ and Cl^- in *molten* (liquefied) sodium chloride?

$$2Na^+(\ell) + 2Cl^-(\ell) \longrightarrow 2Na(\ell) + Cl_2(g)$$

Here, no H_2O molecules compete for electrons with the Na^+ and Cl^- ions. (a) Write the anode and cathode reactions appropriate for an electrolytic cell made from molten NaCl. (b) Which electrode (anode or cathode) must be held at positive potential?

36. Suppose that an electrolytic cell is designed to reduce molten sodium chloride to metallic sodium:

$$2Na^+(\ell) + 2Cl^-(\ell) \longrightarrow 2Na(\ell) + Cl_2(g)$$

For how long, under standard conditions, must a current of 10.0 A be applied to obtain 100.0 grams of Na?

37. During the electrolysis of molten NaCl, how many liters of chlorine gas are produced per gram of sodium ion reduced? Assume that the gas is captured at 25°C and 1 atm.

38. Will the electrolysis of Ni^{2+} in aqueous solution produce metallic nickel? In other words, can the desired reduction

$$Ni^{2+}(aq) + 2e^- \longrightarrow Ni(s)$$

successfully compete against the reduction of H_2O? Consult the table of standard electrode potentials in Appendix C.

39. How many grams of each metal will plate out after electrolysis for 2.00 hours by a current of 35.0 A?

    (a) Cu from aqueous Cu^{2+}
    (b) Ni from aqueous Ni^{2+}
    (c) Fe from aqueous Fe^{2+}

40. An electrolytic current of 3.00 A is applied to a solution in which $[Ag^+] = 0.200 \ M$. (a) How long will it take to remove all the silver ions from 0.200 L of solution? (b) What will be the concentration of Ag^+ after 100 s of electrolysis? (c) 200 s? (d) 300 s?

18

Kinetics—The Course
of Chemical Reactions

18-1. Of Potentiality and Actuality

18-2. Prerequisites

18-3. The Road to Equilibrium: A Macroscopic View
Rate of Reaction
Rates and Concentrations
Concentrations and Time
Rates and Temperature

18-4. Crisis
Collision Theory
Thermodynamics of the Transition State

18-5. Control: Thermodynamics, Kinetics, and Catalysis

18-6. Mechanism: Step by Step
Elementary Rate Laws
Rate-Determining Step
Pre-equilibria
At Equilibrium: A Detailed Balance
REVIEW AND GUIDE TO PROBLEMS
EXERCISES

18-1. OF POTENTIALITY AND ACTUALITY

To be able is not necessarily to do, for there is a difference between what *can* happen and what *does* happen. It is the difference between latent potential and realized potential, the difference between mere possibility and completed action. It is precisely the difference between thermodynamics (the energetics of reaction; what can happen) and **chemical kinetics** (the rates and mechanisms of reaction; what does happen).

643

Thermodynamics, looking just at the start and finish of a process, is all about possibility and potential—not action. Thermodynamics asks only if, in the end, free energy has increased or decreased, and from this sparse information answers either *yes* or *no*:

No, the reaction is disallowed because $\Delta G°$ is positive. Were it to occur spontaneously, global entropy would be forced to decrease. The universe would, most improbably, become more ordered.

Or . . . yes, the reaction is legal. The products have lower free energy than the reactants, and so the transformation *can* go forward. If it does (and we are not told either when or how), then eventually there will be an equilibrium. Concerning this state we learn only the proportions of the final mixture, taking note of the equilibrium constant, $K(T)$, at each temperature. About the events prior to equilibrium we know nothing, since $\Delta G°$ and $K(T)$ stay the same regardless of the path taken.

Nowhere is there any mention of atoms or molecules or bonds, nor does the *time* appear in any equation of thermodynamics: not in the equilibrium expression, not in the free energy, not in the enthalpy, not in the entropy. A spontaneous reaction may be fast (like the dissociation of hydrochloric acid in water), slow (like the rusting of iron), or imperceptible (like the natural conversion of graphite into diamond). A large, negative $\Delta G°$ does not make a spontaneous process fast, and a small (but still negative) $\Delta G°$ does not necessarily make it slow.

Indeed $\Delta G°$ has nothing to do with speed and nothing to do with time; thermodynamics asserts itself, instead, at the *end* of time, where a large drop in free energy guarantees that products will predominate at equilibrium. Timeless and tolerant, thermodynamics sets bounds for energy and entropy but remains unaffected by the nuts and bolts of a reaction. From the state functions of thermodynamics comes a driving force, but not a speed; a tendency to react, but not a mechanism; a direction, but not a map. Thermodynamics is all about potential, not action.

Yet chemistry must ultimately be about action, the kind of action embodied in the words "reactants *go to* products" and symbolized by the arrow in the equation

$$\text{reactants} \longrightarrow \text{products}$$

And action we shall get, despite our former preoccupation with start and finish to the exclusion of process. We have, until now, stressed stoichiometry (what goes in and what comes out), structure (atoms and molecules, quantum mechanics), and equilibrium (thermodynamics, energy, entropy). We have stressed the left and right sides of the equation separately (reactants and products), but not so much the arrow itself; not the intermediate species that come and go along the way; not the intimate encounters between molecule and molecule that turn possibility into fact.

Let that be our new task: to understand, in the language of chemical kinetics, how the potential to react becomes the reality of reaction.

18-2. PREREQUISITES

Expect no grand new laws to arise from kinetics—nothing as sweeping as the conservation of energy, the maximization of global entropy, the conservation of momentum, the conservation of mass, the conservation of charge, the laws of electromagnetic and gravitational force, the Schrödinger equation, the Heisenberg uncertainty principle, the Pauli exclusion principle, the laws of symmetry. This small body of principles forms much of nature's constitution, and all other phenomena are shaped according to its dictates. Chemical kinetics offers no exception. The same "big" laws govern all reactions, and so the new features must be found in the smaller details.

Such details are by no means lacking. Molecules can come together in seemingly innumerable ways, enough to make the field of reaction resemble, at times, a free-for-all. From the bare ingredients of charge and mass (electrons and nuclei) arise more than 10 million known compounds, delightfully varied and complex, each with its own structural peculiarities. Every reaction brings together some different combination; every reaction has peculiarities of its own.

But chemical kinetics really is no free-for-all, of course, because certain basic requirements must always be satisfied. Start with the obvious, right away in Figure 18-1: Bonds need to be broken. Like omelettes, new molecules take shape only after the starting materials are destroyed. Since to make one molecule is to break another, most reactions begin when something is torn away from something else. Perhaps a covalent bond is ripped apart, or an ion is pried loose from a solvent cluster, or an electron is removed from an atom or molecule; some tie, whatever it may be, often comes undone right at the outset.

Destruction requires energy. The typical reaction therefore incurs a start-up cost, an investment to be recovered when new bonds are formed and energy is released. The payoff comes not until later, however, and some agency first must provide sufficient *activation energy* to start the process. Nothing happens without that impetus in the beginning, no matter how much thermodynamic *potential* there may be to liberate free energy in the end.

Although energy sometimes enters from without the system (principally as heat or light), ultimately it is the molecules that do the work. The activation energy is drawn from the thermal energy of the molecules themselves, associated chiefly with their energy of motion—translational, vibrational, rotational. Molecules, simply put, smash into each other. They collide and exchange energy.

Collisions are the lifeblood of a chemical reaction. A reaction's speed increases when there are *more* collisions per second, more *energetic* collisions, and more *properly oriented* collisions. No collision, no reaction. Too weak a collision, no reaction. A misdirected, poorly aimed collision, no reaction.

Take these prerequisites one at a time, beginning with the frequency of collisions. We expect, first, that proportionally more collisions will develop among a crowded, dense population of molecules such as one finds at higher concentrations and (for gases) at higher pressures. Rates should rise accordingly when reactant concentrations are increased, and typically they do. Soon we shall investigate the experimentally observed relationships between concentration and speed.

Second requirement: more energetic collisions. Faster-moving molecules collide more violently, and here is where temperature has its

FIGURE 18-1. First a climb *up* the hill, and then a slide down: Molecules break bonds to make bonds. They collide at different angles, at different rates, with different energies, and with different results. Some collisions are reactive; some are not.

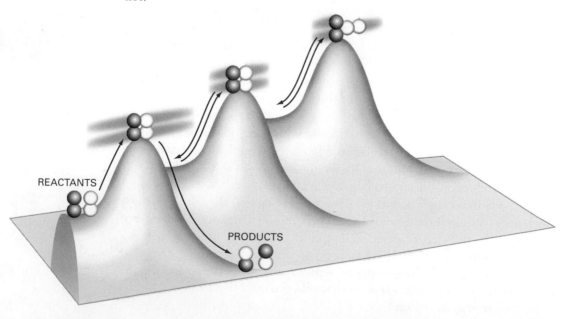

effect. The higher the temperature, the faster the molecules; the higher the temperature, the more rapid should be the reaction. Again our hunch will be confirmed experimentally. Most reactions do proceed faster at higher temperatures.

Third requirement: Molecules must collide in just the right way. Species A must present itself properly to species B, so as to bring electrons into a position where they can interact discriminatingly. If not, the energy is wasted. Consequently we expect the *state* of matter to influence the rate and mechanism of reaction, since spatial relations clearly depend on the intimate details of aggregation. A particular reaction might be slow in the gas phase, for example, where collisions are brief, haphazard, and hence unlikely to bring the molecules properly together. The same reaction in solution, though, may be accelerated owing to the more prolonged contacts between reactants. The interacting molecules, traveling together for longer intervals, might have more time to accumulate the necessary energy and to position themselves properly for reaction.

Better still, for some reactions, might be a solid surface on which one species could be oriented optimally to receive the impact of another. Available *area* is important here too, since more collisions can occur amidst the greater working space of a finely divided support—a broken-up powder, say, compared with a pellet. And a catalyst (like H^+ or a zeolite or a platinum surface or an enzyme) often will make a crucial difference to the course of reaction, as will solute–solvent interactions and many other environmental factors.

At various times we have touched on such issues pertaining to reaction kinetics, but now we return for a closer, more systematic examination. Our analysis begins with observation of the changes that accompany a chemical system on its way to equilibrium.

18-3. THE ROAD TO EQUILIBRIUM: A MACROSCOPIC VIEW

Away from equilibrium, a reaction *moves*. Things change. Molecules change. Concentrations change. Macroscopic properties change. It might be the pH of an acid; the voltage of a galvanic cell; spectroscopic frequencies; color, maybe; maybe the state of matter itself. There is, until equilibrium smothers all variation, a pervading sense of *time*. Time is change.

Here are visible signs that we can measure and compare against a clock. The study of kinetics begins with an awareness of just what is changing, how fast, and under what conditions.

First comes the most elementary question of all: rate.

Rate of Reaction

Suppose we bring together two reactants, A and B, and allow them to form products C and D in some generic reaction

$$A + B \longrightarrow C + D$$

Suppose further (to keep the numbers simple) that concentrations [A] and [B] are both 1.00 mol L^{-1} at the start of reaction, with one mole of each reactant dispersed over a volume of one liter.

From this point forward, the molecules react and we observe. [A] and [B] diminish over time while [C] and [D] grow from zero to their equilibrium values. For every mole of A that disappears, a mole of B disappears as well. For every mole of A or B that disappears, one mole each of C and D eventually appear as products.

Imagine that we make the following observations, spaced at intervals of one second: (1) After 1 s, the vessel contains 0.99 mole of A and 0.01 mole of C in its volume of 1.00 L. (2) After 2 s, the amounts are 0.98 mole of A and 0.02 mole of C. (3) After 3 s, 0.97 and 0.03. What do we say?

We say that the concentration of each species is changing at an average rate of 0.01 mole per liter (M) per second (s). The reaction is consuming reactants and forming products at a rate (or speed) of 0.01 M s^{-1}. During the time from $t = 0$ to $t = 1$ s, [A] and [B] drop from 1.00 M to 0.99 M. [C] and [D], meanwhile, rise from 0.00 M to 0.01 M. In the next second, the respective changes are from 0.99 M to 0.98 M and from 0.01 M to 0.02 M. In the third interval, the concentrations go from 0.98 M and 0.02 M to 0.97 M and 0.03 M.

So far, the rate appears to be constant; but what if now, between $t = 3$ s and $t = 4$ s, the concentrations change from 0.97 to 0.95 and from 0.03 to 0.05? What do we say then? We say, unperturbed, that the speed in this most recent one-second period is 0.02 M s^{-1}. It changed. Why not? Just as a vehicle can accelerate or decelerate, so can a reaction—but, for both, we should be clear about the time during which the speed is recorded.

Thus is defined, in Figure 18-2(a), the **average rate of reaction** over an interval Δt,

$$\text{Rate of reaction} = \frac{\text{change in concentration}}{\text{change in time}} = \frac{\Delta[\cdots]}{\Delta t}$$

in complete analogy with our earlier definition of velocity as $\Delta(\text{distance})/\Delta t$ (Chapter 1). If we agree, moreover, to monitor the concentrations only at an exceedingly short interval ($\Delta t \longrightarrow 0$), then the speed so obtained will be

(a)

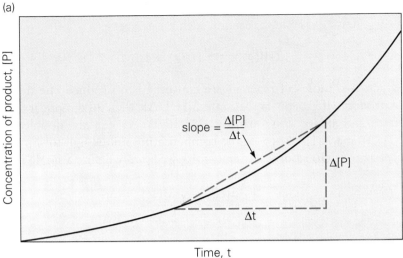

(b)

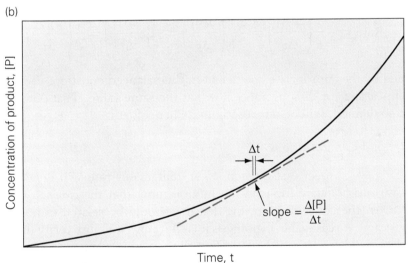

FIGURE 18-2. Rate of reaction, as measured by the progressive appearance of a certain product P; rate = $\Delta[P]/\Delta t$. (a) For points spaced widely in time, the slope of the connecting line (equal to $\Delta[P]/\Delta t$) gives the average rate. (b) Where Δt is small, the line tangent to the curve marks the instantaneous rate. Interpret it this way: The concentration goes from [P] to [P] + $\Delta[P]$ during the instant Δt.

the *instantaneous rate* set forth in Figure 18-2(b): a measure of how fast the reaction is progressing at a particular instant.

That understood, we need finally to extend the definition to include equations of the sort

$$a\text{A} + b\text{B} \longrightarrow c\text{C} + d\text{D}$$

where each stoichiometric coefficient may be different. Select, as a sufficiently general example, a reaction such as

$$2\,HI(g) \longrightarrow H_2(g) + I_2(g)$$

in which two moles of reactant are consumed to produce one mole of each product. If H_2 appears at a rate $\Delta[H_2]/\Delta t$, then HI disappears twice as fast; and hence the quantity $-\Delta[HI]/\Delta t$ is exactly double both $\Delta[H_2]/\Delta t$ and $\Delta[I_2]/\Delta t$ (note the compensating minus sign to show that [HI] *decreases*). To resolve any ambiguity we simply define a single rate as

$$\text{Rate} = -\frac{1}{2}\left(\frac{\Delta[HI]}{\Delta t}\right) = \frac{\Delta[H_2]}{\Delta t} = \frac{\Delta[I_2]}{\Delta t}$$

which generalizes to

$$\text{Rate} = -\frac{1}{a}\left(\frac{\Delta[A]}{\Delta t}\right) = -\frac{1}{b}\left(\frac{\Delta[B]}{\Delta t}\right) = \frac{1}{c}\left(\frac{\Delta[C]}{\Delta t}\right) = \frac{1}{d}\left(\frac{\Delta[D]}{\Delta t}\right)$$

Scaled by the appropriate coefficient in the balanced equation, each rate of appearance or disappearance now has the same value. That one number, so defined, will be our reaction rate henceforth.

Rates and Concentrations

Already we suspect that a reaction's instantaneous rate will depend on the various concentrations in the mix, arguing that the *closeness* of the particles (the concentration, the density) should be related to how *fast* they react. A reasonable hypothesis it is, but now we must verify this expectation experimentally and then put the results into quantitative form.

Let us do so by recording the **initial rate** of reaction, making measurements just after the reactants come together (at $t = 0$) but before the products appear in appreciable number. Such simplification is welcome and appropriate, because the early kinetics will be dominated by the concentrations of reactants alone. And, although the rate will change as products begin to form, this first impression of a reaction's speed is revealing nonetheless. Arising from the pure clash of reactant against reactant, it tells us something directly about the forward-going process; it provides a macroscopic summary of the microscopic scene, a report not yet complicated by the competing reverse transformation.

Returning to the general $a A + b B$ reaction, we proceed to vary the concentration of each reactant and measure the initial rate. First to be examined is the rate's dependence upon $[A]_0$ (concentration at $t = 0$), for which one set of hypothetical results might be:

$[A]_0$	INITIAL RATE
1.00 M	0.01 M s^{-1}
2.00 M	0.02 M s^{-1}
3.00 M	0.03 M s^{-1}
4.00 M	0.04 M s^{-1}

When $[A]_0$ is doubled, the rate doubles. When $[A]_0$ is tripled, the rate triples. When $[A]_0$ is quadrupled, the rate quadruples. And? Evidently the speed is directly proportional to [A], provided that [B] is held constant. We say that such a reaction is *first order* with respect to A, meaning that its rate varies as the first power of [A]: rate goes as $[A]^1$.

Next we apply the same treatment to [B], this time obtaining the following data:

$[B]_0$	INITIAL RATE
1.00 M	0.01 M s^{-1}
2.00 M	0.04 M s^{-1}
3.00 M	0.09 M s^{-1}
4.00 M	0.16 M s^{-1}

With $[A]_0$ fixed, the reaction clearly exhibits a *second-order* dependence upon the initial concentration of B. The rate is proportional to $[B]^2$. It increases fourfold when $[B]_0$ is doubled, ninefold when $[B]_0$ is tripled, and sixteenfold when $[B]_0$ is quadrupled.

Put both results together now, and out comes the consolidated *rate equation* (or *rate law*) for the forward reaction in its early stages:

$$\text{Rate} = k[A][B]^2$$

Overall our imaginary process is *third order* (first order for A plus second order for B), with the *rate constant* k serving as a constant of proportionality. An important indicator of a reaction's intrinsic speed, k and its origins will figure prominently in our subsequent analysis. At present, however, take the rate constant at face value: as an empirically determined proportionality constant between rate and concentration (and not, incidentally, to be confused with the Boltzmann constant k_B).

The units of k? Whatever it takes to yield a rate with the proper dimensions of M s^{-1}—in this instance,

$$k = \frac{\text{rate}}{[A][B]^2} \sim \frac{M\ \text{s}^{-1}}{M\ M^2} = M^{-2}\ \text{s}^{-1}$$

yet *this* instance (a third-order rate law) is only one of many possible outcomes. Molecules have other options as well.

There are, in fact, third-order reactions of this kind, such as

$$2NO(g) + 2H_2(g) \longrightarrow N_2(g) + 2H_2O(g) \qquad \text{rate} = k[H_2][NO]^2$$

but there are also pure first-order reactions such as

$$2N_2O_5(g) \longrightarrow 4NO_2(g) + O_2(g) \qquad \text{rate} = k[N_2O_5]$$

and pure second-order reactions such as

$$2CH_3(g) \longrightarrow C_2H_6(g) \qquad \text{rate} = k[CH_3]^2$$

and even reactions with fractional exponents such as

$$H_2(g) + Br_2(g) \longrightarrow 2HBr(g) \qquad \text{rate} = k[H_2][Br]^{1/2}$$

Look closely, too: There is no clear connection between the rate exponents and the coefficients of the balanced equation. Typically one has nothing to do with the other, because the kinetic expressions depend on what actually happens *during* the reaction. It is not enough to say that H_2 and Br_2 "go to" $2HBr$; rather we must uncover the step-by-step mechanism that unfolds over time, and only then will the rate equation make sense microscopically.

Does one molecule of H_2 collide with one molecule of Br_2, thereby giving birth to two molecules of HBr in one step? Or is there a sequence of elementary reactions? Are intermediate species generated along the way? Do free radicals like H· or Br· play a role? Therein lies the microscopic genesis of the macroscopic rate equation, but to each question our answer at the moment is: we don't know. The balanced equation—an expression of stoichiometry alone, of conservation of mass—tells us nothing about what happens during the time symbolized by the arrow.

Later we shall delve more closely into the microscopic events, but here let us pause to collect our observations into one all-purpose rate equation,

$$\text{Rate} = k[A]^n[B]^m \cdots$$

while remembering these general features: (1) Any species (product, reactant, catalyst, intermediate, solvent) may appear in the expression. (2) The exponents have no essential connection to the stoichiometric coefficients. (3) The overall order of reaction is given by the sum of the exponents. (4) Any of the exponents may prove to be zero, signifying that the rate is independent of the associated concentration. (5) The exponents need not be integers. (6) The rate equation need not even follow this simple form; and, when it does not, the numerical order of reaction is left undefined.

The formulation appears to be loose and open to all kinds of special cases, but bear in mind that kinetics is not the free-for-all it seems. Seeking simplicity, we specialize now to pure first-order and second-order processes, two very common (and very important) examples of kinetic phenomena in general. These prototypical reactions will become especially significant when later we turn to the microscopic world of colliding molecules.

Concentrations and Time

Reaction rate (molarity per second) is to concentration as linear velocity (miles per hour) is to distance. One quantity tells us how much a certain concentration changes from one time to the next; the other, how much a distance similarly changes over time.

The problem is familiar, almost automatic. Riding a train at a constant speed of 60 miles per hour, a traveler knows exactly *where* the train will be and *when*: 10 miles down the track after 10 minutes, 20 miles after 20 minutes, 60 miles at the end of an hour. The rider converts "rate = distance/time" into "distance = rate × time."

Even if the speed changes from moment to moment (60 miles per hour for the first minute, 58 mph for the second minute, 63 mph for the next half-hour, ...), the idea remains the same. The traveler breaks the trip into small bits over which the speed is nearly constant, computes "distance = rate × time" for each such bit, and finally adds together all the bits to get the total distance. It is precisely the task for which integral calculus was developed.

The same scheme applies to concentration versus time, and so (omitting the calculus-based derivation) we go on to quote the results for simple rate equations. Assume that we know the rate constant and order of reaction, and ask: What will be the concentrations at each instant? How much of each species will be present as time unfolds?

Begin with a *first-order* reaction (forward direction only), where a reactant disappears at the rate $k[A]$. Turned around, or *inverted*, the rate equation

$$\text{Rate} = -\frac{\Delta[A]}{\Delta t} = k[A]$$

becomes

$$\ln [A]_t = -kt + \ln [A]_0$$

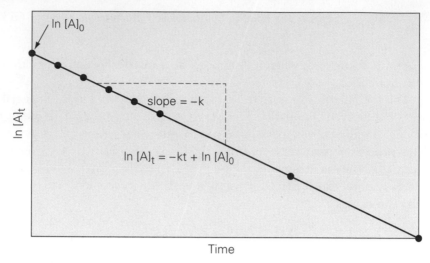

FIGURE 18-3. A logarithmic plot tracking the depletion of species A during a first-order reaction. The initial value, $\ln [A]_0$, appears as the vertical intercept of a line with slope $-k$. The number k is the first-order rate constant; rate $= k[A]$.

Starting from $[A]_0$, the concentration decreases steadily with time. The value of $[A]_t$, expressed as a natural logarithm, goes from $\ln [A]_0$ (at $t = 0$) to $\ln [A]_0 - k$ (at $t = 1$ s) to $\ln [A]_0 - 2k$ (at $t = 2$ s), and so forth, for a rate constant stated in s^{-1}. The points correspond to the equation of a straight line,

$$y = mx + b$$

with slope (m) equal to $-k$ and y-intercept (b) equal to $\ln [A]_0$. The steeper the line shown in Figure 18-3, the faster the consumption of A.

Aware of that dependence, an experimenter will plot reactant concentration logarithmically versus time for a reaction of undetermined kinetics. If the points fall on a straight line, the graph provides evidence for a first-order process and (from the measured slope) the rate constant as well.

Upon rearrangement to the form

$$\ln \frac{[A]_t}{[A]_0} = \ln [A]_t - \ln [A]_0 = -kt$$

the first-order equation quickly supplies a convenient single number to characterize the decay: the **half-life**, $t_{1/2}$, over which the concentration decreases to *half* its original value. Since $[A]_t$ is equal to $\frac{1}{2}[A]_0$ when $t = t_{1/2}$, we make the appropriate substitutions and obtain

$$t_{1/2} = -\frac{1}{k} \ln \frac{\frac{1}{2}[A]_0}{[A]_0} = -\frac{\ln \frac{1}{2}}{k} = \frac{\ln 2}{k} = \frac{0.6931}{k}$$

for the half-life of a first-order reaction. Note that $t_{1/2}$ is uniquely determined by the rate constant alone, independently of the initial concentration.

Observe: After one half-life, $[A]_0$ is cut in half. After two half-lives, the amount remaining is $\frac{1}{4}[A]_0$ (half of half). After three half-lives, it falls to $\frac{1}{8}[A]_0$ (half again). Smaller by half each time, the pattern is common to all supposed first-order rate processes—radioactive decay, births or deaths in a population, and many others—not just unilaterally disappearing molecules.

Observe further: First-order kinetics means that the rate *at each instant* is directly proportional to the amount of material currently present. Thus a highly concentrated set of molecules begins to decay faster than a more dilute system, just as a larger population of wolves initially dies off more rapidly than a smaller group. The mechanics of the two processes are vastly different, clearly, but not the kinetics. Plotted on a sheet of semilog graph paper, decomposing molecules and soon-to-be-decomposing wolves suffer the same sequence of changes.

We have only to express the log quantities directly as exponentials,

$$[A]_t = [A]_0 \exp(-kt)$$

to appreciate the interrelationship of time, population, and rate in a first-order reaction (Figure 18-4). At the outset, when $[A]_t$ is at a maximum, so also is the rate of decay: rate$(t = 0) = k[A]_0$. The concentration falls most quickly during that time, going from $[A]_0$ to $\frac{1}{2}[A]_0$ over the first half-life.

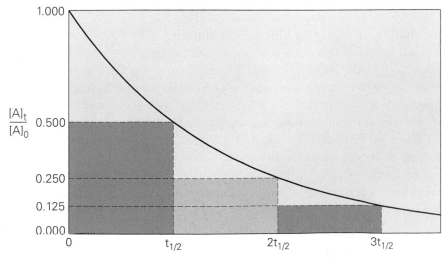

FIGURE 18-4. Exponential behavior is the mark of all first-order processes. For first-order decay, the measured quantity decreases by half during each successive half-life: $t_{1/2} = (\ln 2)/k$.

Fully half the material disappears, but then (with less of A remaining to decay) the instantaneous rate slows accordingly. The velocity, $k[A]_t$, is already cut to $\frac{1}{2}k[A]_0$ at $t = t_{1/2}$, after which succeeding half-lives see further slowdowns from rates of $\frac{1}{2}k[A]_0$ to $\frac{1}{4}k[A]_0$ and then to $\frac{1}{8}k[A]_0$ and less. Whereas the same current *fraction* (half of what remains) steadily decays over each interval, the numbers involved get smaller and smaller—and always according to the same schedule, too, whatever the material might be. All first-order processes are alike.

Turning briefly now to a *second-order* process of the type

$$\text{Rate} = k[A]^2$$

we find that

$$\frac{1}{[A]_t} = kt + \frac{1}{[A]_0}$$

Again the equation describes a straight line ($y = mx + b$), although here instead of ln $[A]_t$ the *reciprocal* concentration, $1/[A]_t$, is plotted against the time (Figure 18-5). The line's slope is equal in magnitude, as before, to the rate constant k, but the second-order half-life introduces a new aspect to the decay. The half-life comes to depend on the initial concentration as well as k, going as

$$t_{1/2} = \frac{1}{k[A]_0}$$

and thereby changing the kinetics significantly.

No longer is there a universal pattern for the decay. The higher the value of $[A]_0$, the shorter the half-life and the quicker the reaction. More reactant (present at the beginning) is consumed correspondingly faster in a pure second-order process, where now the rate scales with the square of the concentration. This extra sensitivity to the amount of material is reflected by the half-life, which is different for every mixture. Not all second-order reactions are alike.

More complicated, still, is a second-order process that goes as $k[A][B]$, being first order in each of two reactants. Here the course of reaction depends not just on one initial concentration, but on $[A]_0$ and $[B]_0$ together.

Rates and Temperature

Kinetic phenomena are so varied that "typical" becomes a dangerous word, but certainly one generalization has survived for over a century:

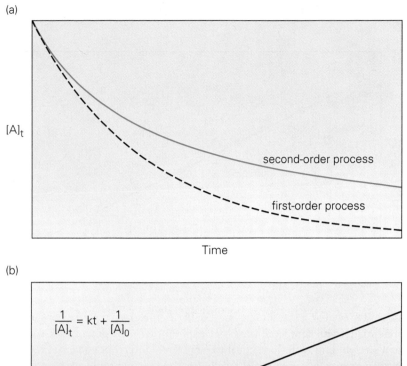

FIGURE 18-5. Consumption of species A during a second-order process governed by the rate law $k[A]^2$. (a) Plotted against time, the concentration $[A]_t$ (solid curve) clearly deviates from an exponential profile (broken curve). (b) Instead, a plot of the *reciprocal* concentration $1/[A]_t$ yields a straight line with slope k and intercept $1/[A]_0$. The half-life, in turn, depends on the initial concentration $[A]_0$.

Most reactions proceed faster at higher temperatures. A rate constant usually does grow larger as T increases, often going exponentially as

$$k = A \exp\left(-\frac{E_a}{RT}\right)$$

or (cast in logarithmic form) as

$$\ln k = -\frac{E_a}{RT} + \ln A$$

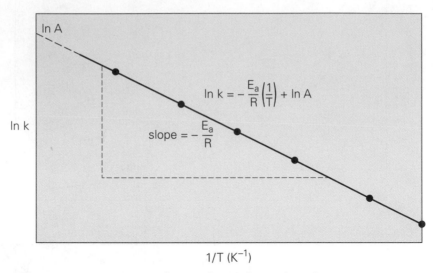

FIGURE 18-6. For a system obeying the Arrhenius law, the rate constant varies exponentially with temperature: $k = A \exp(-E_a/RT)$. A plot of $\ln k$ versus $1/T$ gives a straight line with slope equal to $-E_a/R$ and intercept equal to $\ln A$. E_a, the activation energy, poses an initial hurdle for the reacting molecules—a minimum energy needed to start the process.

Both expressions are equivalent representations of the **Arrhenius equation** (1889), a relationship characterized by the two parameters E_a and A. The quantity E_a has dimensions of molar energy, and the **pre-exponential factor** A (not to be mistaken for generic reactant A) has the same units as k.

To use the Arrhenius equation, we measure the rate constant at a set of temperatures and plot $\ln k$ against $1/T$. If the points fall on a straight line, as they do in Figure 18-6, the slope is identified as $-E_a/R$ and the y-intercept as $\ln A$. But there is more to it than just numbers, because suddenly, with the Arrhenius law, a window opens onto the microscopic world of reaction.

For we recognize the form $\exp(-E_a/RT)$ as a *Boltzmann factor*, the same expression first encountered in the kinetic theory of gases (Chapter 10). E_a, appearing in the exponent $-E_a/RT$, is our sought-after **activation energy**, and the exponential factor $\exp(-E_a/RT)$ tells us how likely a particle is to have at least this minimum amount. Recall, in particular, how both kinetic energy and speed are distributed in a gas, and note again (from Figure 18-7) how more and more particles move faster as the temperature increases. Always there is this dimensionless number, $\exp(-E_a/RT)$, which governs the populations. Large when T is large and small when T is small, it tracks the fraction of particles with energy equal to or greater than E_a.

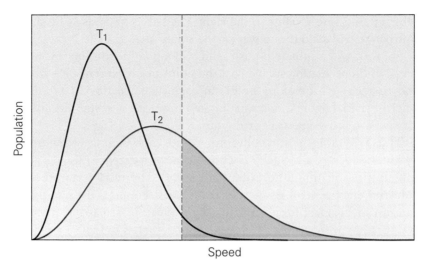

FIGURE 18-7. Speeds are allocated to molecules in an ideal gas according to the Maxwell-Boltzmann distribution: the higher the temperature, the faster the average speed. As temperature rises, more molecules acquire enough kinetic energy to hurdle the barrier (dashed line) and begin to react. The curves are drawn so that $T_2 > T_1$.

An energy equal to or greater than some "activation" energy . . . the description sounds like a threshold for a bond-ripping, molecule-breaking, reaction-inducing collision. We have been anticipating just such a requirement, and now our initial suspicions are beginning to acquire experimental support. Let us use this Arrhenius equation, an empirical fact, as the first step toward building a microscopic understanding of those critical moments when chemistry actually happens.

18-4. CRISIS

A ball climbing a hill, a pendulum set in motion, a weight bouncing on a spring, a molecule in collision: four images, each grossly different yet all fundamentally the same.

At rest in a valley, the ball is initially in equilibrium; it moves neither up nor down until impelled by an outside force. But suddenly, given just such a push, the ball starts up the hill—rapidly at first, and then more slowly as it fights against gravity. Upward it climbs, slower and slower, steadily converting kinetic energy into gravitational potential energy.

If the kinetic energy runs out too soon, the ball stops momentarily and immediately rolls back down the slope. Suppose, though, the ball does have enough energy to reach the summit, just barely, upon which it comes to rest in a precarious state. Sitting unsteadily with the old valley to the left and a new valley to the right, the ball is in crisis. The slightest disturbance will send it one way or the other.

So, too, will a sufficient push send a swinging pendulum up and over its arc, or an oscillating spring past the point of no return, or a vibrating bond past *its* own breaking point. In each instance there is a climb to higher potential energy, culminating in a moment of instability followed by a decisive move one way or the other.

Thus two molecules in collision: Alone, each is endowed with a certain kinetic energy and each (as a condition of existence) jiggles about a configuration of minimum potential energy. The nuclei vibrate to and fro in the electric field of the electrons, as if connected by a spring of quantum mechanical force. Initially the molecules are apart, in equilibrium, resting in the valley of the reactants depicted in Figure 18-8.

Then, by accident, they approach and collide in what promises initially to be an energy-raising, endothermic encounter. How could it *not* be endothermic, at least at first, given that each structure is already optimized and thermodynamically perfect in its own way? Yet into this perfection comes one molecule with its cloud of electrons smashing into another, bringing negative against negative and positive against positive; here comes a sudden, distorting impact that forces the molecules out of their configurations of minimum energy.

We view the crash along the symbolic "reaction coordinate" shown in the energy profile, where (for simplicity of concept) all events are referred to some generalized direction. Compressed ever closer, the colliding molecules begin to mount a hill of electrical potential energy as the encounter progresses. Like a ball climbing a slope of gravitational potential, the molecules fuel their ascent by spending kinetic energy. They *slow down*, drawing upon reserves of heat to force electrons into positions of higher potential energy.

Eventually a crisis comes. The molecules either bounce apart and recover from the crash, shaken but unbroken, or else they pause uncertainly at the top of the hill. Teetering at this point of maximum potential energy, they can roll back to reactants or on to products. It takes only a little push in one direction or the other, and either alternative will lead to a lower potential energy.

To go over the top is to go a bit too far. Pushed to the limit, a bond acquires enough extra energy to vibrate out of control. The nuclei and their tissue of electrons vibrate back and forth until finally it is too much; the spring is stretched too far, and the elasticity is gone. The molecule breaks, and something new is formed in its place. The new valley of potential energy may be lower than before, making the net reaction exothermic (Figure 18-8a); or it may be higher than previously, and hence endothermic (Figure 18-8b).

(a)

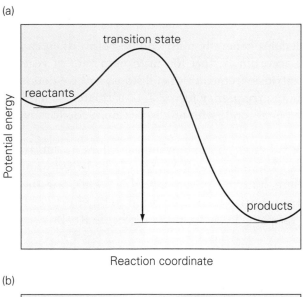

(b)

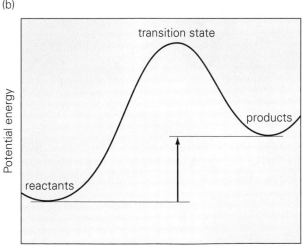

FIGURE 18-8. Colliding molecules climb to higher and higher potential energy as they approach, coming eventually to a point of crisis: an untenable *transition state*, from which they must either go forward to products or back to reactants. (a) The products of an exothermic reaction lie in a valley of enthalpy deeper than the reactants. (b) The products of an endothermic reaction lie in a higher valley. Either way, there is still a hill to climb in between.

Collision Theory

From the scene of this accident comes a first model of chemical reactivity, useful primarily for not-too-complex species in the gas phase. Called **collision theory**, the model gathers up the ideas discussed above and fashions them plausibly into the Arrhenius equation. Its arguments are as follows:

1. Molecules generally must collide before they can react. We acknowledge, accordingly, that the maximum rate of reaction can never exceed the intrinsic frequency of collisions, and we denote that underlying rate by some *frequency factor Z*. *Z* is treated strictly as the number of collisions per second, with no distinction made for energy and orientation.

2. Still, we argue, a random impact offers no substitute for a well-aimed attack. There are specific targets (a hydroxyl group here, a halogen atom there, a methyl group, a double bond), and the partners must be oriented appropriately for a collision to be effective. If, for example, a Br^- ion is to spring loose from the molecule R—Br,

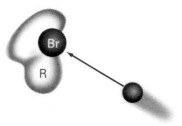

then the incoming species had best attack somewhere near the bromide. Since a lucky strike becomes less likely as the rest of the molecule (R) grows larger, the chances of a reactive collision diminish in bulky molecules of complicated shape:

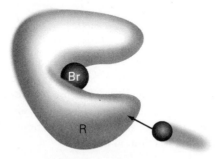

Reasoning that only some of the encounters will be geometrically acceptable, we then define a *steric factor P*, with value between 0 and 1. Different for every reaction, *P* denotes the fraction of collisions in which the molecules are aligned suitably for reaction.

3. Climbing a hill of potential energy, our colliding molecules create

a structure called an ***activated complex*** as they reach the summit. At the very top, separated from the original reactants by an activation energy E_a, the activated complex assumes a short-lived configuration called the ***transition state*** (see again, Figure 18-8). Formation of the transition state provokes the anticipated crisis, after which either products will emerge or reactants will be regenerated.

4. To propel the reactants over the top, the collision must supply energy no less than E_a. Hence we arrive at the Boltzmann factor, $\exp(-E_a/RT)$, which winnows out any molecule unable to contribute the requisite energy of activation.

5. Putting together, finally, the frequency, steric, and energy factors, we obtain the expression

$$\text{Rate of reactive collisions} = ZP \exp\left(-\frac{E_a}{RT}\right)$$

or

$$\text{Rate of reactive collisions} = A \exp\left(-\frac{E_a}{RT}\right) [A][B]$$

once P and Z are consolidated into a pre-exponential factor (A) multiplied by the concentrations of colliding molecules A and B. The higher the concentrations, the greater the chance of a collision.

The result is the Arrhenius equation for k, as it appears in the rate law $k[A][B]$. It says that the rate constant is determined both by the number of properly oriented collisions per second (through A) and by the Boltzmann energy factor, $\exp(-E_a/RT)$. Of these two contributions, however, the energetic requirement (Figure 18-9) proves to be particularly variable from reaction to reaction, and also more subject to experimental control. A low activation energy and a high temperature usually make for high speed.

Because: The lower the barrier, the more molecules there are that can make the climb; $\exp(-E_a/RT)$ increases as E_a decreases. And, varying exponentially, the dependence is especially sensitive even to seemingly modest changes, as we can quickly verify by straightforward evaluation. Start with an activation energy of, say, 10 kJ mol^{-1}, for which $\exp(-E_a/RT)$ has the value 0.04 at a temperature of 373 K. That number drops 25-fold, to 0.0016, when E_a is doubled to 20 kJ mol^{-1}, and it falls all the way to 0.0000025 when E_a is doubled again to 40 kJ mol^{-1}. Whereas 40,000 collisions per million would be sufficiently "hot" to hurdle a 10 kJ mol^{-1} barrier at 373 K, only 2.5 per million would succeed at $E_a = 40$ kJ mol^{-1}. It is a big effect.

At higher *temperatures*, too, the Boltzmann factor increases along with

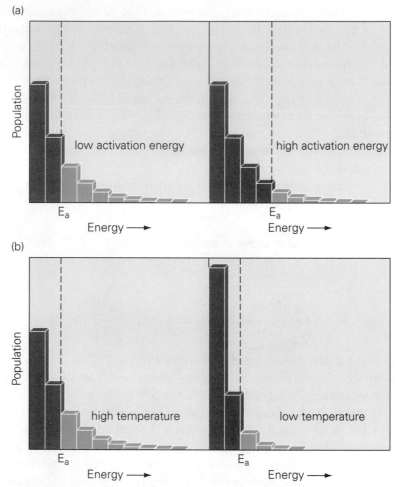

FIGURE 18-9. Kinetic energies are partitioned according to the Boltzmann distribution. (a) When the requisite activation energy is low, more molecules (those to the right of the dashed line) are eligible to react. The reaction goes faster. (b) High temperature, too, allows more molecules to meet the threshold for activation. Rates increase with temperature.

the average kinetic energy of the molecules. To attack the 10 kJ mol^{-1} barrier at 473 K rather than 373 K, for instance, is to double $\exp(-E_a/RT)$ from 0.04 to 0.08. Going the other way, the rate at 273 K would be cut fully to 30% of the value observed at 373 K.

So here we have, in broad outline, the collision theory of reaction rates. It is a simplified model—misleading in some ways, wrong in other ways, disproved by numberless exceptions—although, not to be forgotten, also a description of tremendous value. Much about it is right.

Much also remains unexplained. Appropriate only for gases (and really just the simplest of gases), collision theory leaves us adrift in a world of solution chemistry, enzymatic catalysis, and complicated molecular structure. Nor can the steric factor P, so important for predicting the correct rate, be computed with any reliability. In most instances we can only assign its value retrospectively, by choosing a number to make the theoretical rate agree with the experimental data. P then begins to resemble something disturbingly like a fudge factor, more a confession of ignorance than a satisfying microscopic explanation.

Yet there is an even more serious flaw built directly into the original assumptions. For, naturally enough, the model tends to emphasize the *act* of collision over the *result* of collision: the transition state. It is the transition state that brings the reaction to a climax; it is the transition state that serves as the true gateway to products. Here, clearly, is where we must look next.

Thermodynamics of the Transition State

A fleeting, unstable structure, the transition state serves as a way station between reactants and products. It is neither one nor the other, but rather an inchoate blend of both at once. Having some old bonds and some new bonds, the transition state can go either way.

It is the last stop before products. If the reactants are ever to realize their potential, they will have to pass through that transition state $X^{\ddagger}$ perched atop the hill. They are forced to take an indirect route, like passengers flying from New York to Nairobi through London; and, like those travelers, the reacting molecules must wait for their connection. Arrival at the final destination (products) depends on how fast they can get out of London.

This commonplace idea finds molecular expression in **transition-state theory**, a model that equates the overall rate of a process with the rate of transit through $X^{\ddagger}$. It proposes that the rate depends on how *many* transition states are present (the concentration, $[X^{\ddagger}]$) and on how *fast* they decompose into products. These two factors translate into an equation first order in $[X^{\ddagger}]$,

$$\text{Rate} = v^{\ddagger}[X^{\ddagger}]$$

where $v^{\ddagger}$ denotes the frequency at which the transition states break apart into products (how many times per second). The values of $v^{\ddagger}$ and $[X^{\ddagger}]$ thus determine the rate of the entire process, which we represent symbolically as

$$\text{reactants} \rightleftharpoons X^{\ddagger} \longrightarrow \text{products}$$

There may be one reactant, two reactants, three, four, or more. They may acquire the requisite activation energy through collisions, through photons of light, through interactions with a solvent, or through some other agency. Such is not our concern at the moment; what we do recognize, more fundamentally, is that these reactants (call them A and B) may enter into a pseudoequilibrium with the transition state:

$$A + B \rightleftharpoons X^{\ddagger} \longrightarrow \text{products}$$

It is a funny kind of equilibrium, not quite proper, because sometimes the transition state goes off by itself without ever returning to reactants. But suppose that the pseudoequilibrium

$$A + B \rightleftharpoons X^{\ddagger}$$

is established well *before* the subsequent decay of $X^{\ddagger}$ into products. Then, notwithstanding that one oddity, we see the characteristic features of equilibrium emerge. First, the process does go both ways. Reactants A and B do form $X^{\ddagger}$, and $X^{\ddagger}$ does turn around to regenerate A and B. Second, the rates of the forward and reverse reactions do eventually become equal, at which point the concentrations of all species cease to change.

Under such conditions we may associate an equilibrium constant

$$K^{\ddagger} = \frac{[X^{\ddagger}]}{[A][B]}$$

with formation of the transition state and, furthermore, a corresponding standard *free energy of activation*, *enthalpy of activation*, and *entropy of activation*:

$$\Delta G^{\ddagger} = \Delta H^{\ddagger} - T\,\Delta S^{\ddagger} = -RT \ln K^{\ddagger}$$

Given these parameters, we will be able (keep reading) to express $[X^{\ddagger}]$ in terms of [A], [B], $K^{\ddagger}$, $\Delta G^{\ddagger}$, $\Delta H^{\ddagger}$, and $\Delta S^{\ddagger}$—because the two equations above, remember, define the relationship between standard free energy and the equilibrium constant.

Along with $K^{\ddagger}$ and $\Delta G^{\ddagger}$ also will come an apparent *thermodynamic* influence on the rate of reaction, curious at first since both free energy and equilibrium should stand independent of time. And indeed they do (they always do), so really there is no surprise at all. Equilibrium thermodynamics remains as timeless as ever.

For who knows *when* A, B, and $X^{\ddagger}$ will reach their version of equilibrium? $K^{\ddagger}$ and $\Delta G^{\ddagger}$ tell us not when, but merely *how much* to expect when that state is attained. The lower the free energy of activation, the more transition states exist at equilibrium. The more we have of these ready-to-react complexes, the faster the products can form. We assume only that pseudoequilibrium will come soon enough to allow the transition states symbolized in Figure 18-10 to build up to some steady concentration, $[X^{\ddagger}]$.

If so, then from the definition of $K^{\ddagger}$ we write

$$[X^{\ddagger}] = K^{\ddagger}[A][B]$$

and also

$$K^{\ddagger} = \exp\left(-\frac{\Delta G^{\ddagger}}{RT}\right) = \exp\left(-\frac{\Delta H^{\ddagger}}{RT}\right)\exp\left(\frac{\Delta S^{\ddagger}}{R}\right)$$

Inserted into the original rate expression, $[X^{\ddagger}]$ and $K^{\ddagger}$ now yield a measurable rate law involving reactants A and B. The result is

$$\text{Rate} = \nu^{\ddagger}[X^{\ddagger}] = \nu^{\ddagger}K^{\ddagger}[A][B] = k[A][B]$$

with the rate constant given by

$$k = \nu^{\ddagger}K^{\ddagger} = \nu^{\ddagger}\exp\left(\frac{\Delta S^{\ddagger}}{R}\right)\exp\left(-\frac{\Delta H^{\ddagger}}{RT}\right)$$

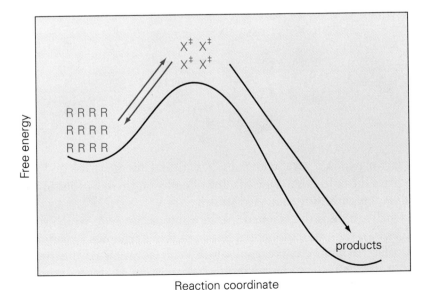

FIGURE 18-10. For as long as reactants and their transition state can maintain a condition of pseudoequilibrium, the concentration of $X^{\ddagger}$ will hold steady.

What do we have?

We have an expression that looks remarkably similar to the Arrhenius equation. It contains, first, an exponential term involving an enthalpy of activation, $\Delta H^{\ddagger}$, which differs from the exponentiated *energy* of activation, E_a, only by a constant factor. The exact value of that factor (derivable from the definition $H = E + PV$ in Chapter 13) is of little concern here, though, and we shall simply assert that $\exp(-\Delta H^{\ddagger}/RT)$ is nearly equivalent to $\exp(-E_a/RT)$. It is effectively the same Boltzmann factor found in the Arrhenius equation.

Next there is an exponential term involving the *entropy* of activation, $\exp(\Delta S^{\ddagger}/R)$, which tells us something about how the transition state is organized. We expect that the change in entropy will be negative in many instances, consistent with the creation of a *more ordered* structure, $X^{\ddagger}$, from the previously unassociated reactants. We expect, too, that the magnitude of $\Delta S^{\ddagger}$ will depend on the geometric details of the encounter—so that maybe this transition state,

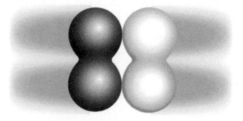

has an entropy of activation different from this one:

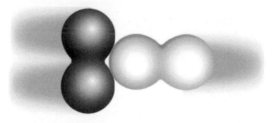

The factor $\exp(\Delta S^{\ddagger}/R)$, with value less than 1 for negative $\Delta S^{\ddagger}$, therefore plays the role of the steric factor of collision theory—but in a more satisfying, natural way than previously.

Finally there is the frequency of decomposition, $\nu^{\ddagger}$, which similarly merges into the pre-exponential factor of the Arrhenius equation. This particular frequency corresponds to a fatal vibration of the transition state, a stretch too far. The complex $X^{\ddagger}$, existing as a pseudomolecule, will have a set of allowed vibrations with discrete frequencies; and, every so often, thermal energy will excite a lethal mode and blow the unstable structure apart. Products are born from such accidents.

18-5. CONTROL: THERMODYNAMICS, KINETICS, AND CATALYSIS

Recognize, then, the two controlling destinies of any chemical reaction, plain to see in Figure 18-11: thermodynamics (ΔG° between products and reactants) and kinetics ($\Delta G^{\ddagger}$ between transition state and reactants).

(a)

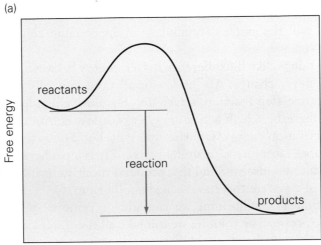

(b)

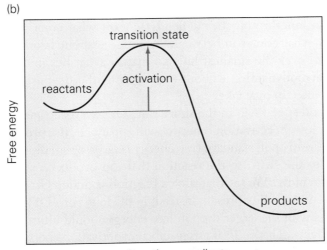

FIGURE 18-11. (a) Thermodynamics: how much. The composition at equilibrium depends only on the difference in free energy between reactants and products, the free energy of *reaction*. No change in reaction path can ever alter this fixed relationship under standard conditions. (b) Kinetics: how fast. The rate of reaction varies with the path taken, rising and falling with the hill (or hills) separating the valleys. The controlling quantity is the free energy of *activation*.

About the first, there is little to do but wait for equilibrium to come; about the second, there is room for some tinkering. Here is why:

If we want an abundance of products at equilibrium, our only recourse is to make them *well*. Products must be strong and stable, with substantially lower free energy than the reactants. The larger the standard difference in G between products and reactants, the larger will be the equilibrium constant.

If, on the other hand, we want the products fast, then we must make them *cheaply*. The reaction demands a low free energy of activation, for only then will the system accumulate an appreciable concentration of transition states.

Some things, like the difference in free energy between products and reactants, never change. $\Delta G°$ is set for all time, not subject to appeal. Let it be a one-step reaction, a two-step reaction, a hundred-step reaction. Let there be a high activation energy, a low activation energy, or even no activation energy. Do what you will, but $\Delta G°$ is fixed at a given temperature—absolutely, unalterably fixed. It is a property inherent in the molecules themselves, and that property alone determines the long-term fate of a system that has reached equilibrium. We can apply stress to a reaction at equilibrium; we can add or remove species; we can change the pressure or volume; we can take all these steps and more, yet none of them will change the underlying difference in free energy. That *net* difference depends solely on the start and finish of the transformation, regardless of what comes in between.

Not so kinetics; kinetics depends *only* on what comes in between. Kinetics *is* what comes in between. Kinetics is about taking the easiest, fastest road over the smallest hill, no matter what lies in the valley beyond. Thermodynamics is about settling into the deepest valley, however long the trip may take.

Ease and speed win in the short run. Of two reactions, the process with the lower activation energy will produce its products faster. Whether the drop in standard free energy is large or small, the advantage goes first to the swift: to the reaction that can produce transition states easily and rapidly. We say that such a reaction is under **kinetic control**.

Depth and stability prevail, instead, in the long run. Of two reactions, the process with the larger drop in free energy should ultimately produce more products. That is, *if*: If given sufficient activation energy and time to reach equilibrium (perhaps microseconds, perhaps millennia), all the products that *can* be formed *will* be formed. All the eligible reactants eventually climb the hill, if given the chance, and any initial kinetic advantage is forgotten at equilibrium. Such a process, where forward and reverse reactions face off in perpetual stalemate, is under **thermodynamic control**—a sound bet for the long term. So long as the system permits free traffic in both directions, thermodynamics will win . . . someday.

But what good is thermodynamics if the rate of reaction is too fast or too slow? We want our products and we want them on our own timetable. The good products of chemical reaction (like breakdown of metabolic waste, or digestion of food), we want fast; the bad ones (like corrosion) can wait. Enter, therefore, the various **catalysts** and **inhibitors**, agents able to speed up or slow down the course of reaction.

Catalysts and inhibitors alter the kinetics by tampering with the transition state. A catalyst lowers the energy of activation. An inhibitor raises it. Each intervenes during that critical stage when reactants are actually "going" to products.

They are go-betweens, emerging untouched when the reaction is over. These are species that participate in the transformation without becoming products themselves. They change only the kinetic pathway of reaction, only E_a or $\Delta G^{\ddagger}$—not the thermodynamics of reaction, not ΔG° and not the equilibrium constant. We get what we get in the same proportions as before, but we get it either faster (with a catalyst) or slower (with an inhibitor). See Figure 18-12.

Let nature or the chemist do something to make a transition state easier to put together. Create an environment with favorable nonbonded interactions to stabilize the activated complex. Arrange for the positively charged portions of a catalyst to coordinate closely with the negatively charged portions of the transition state. Arrange for the activated complex to be pinned down, and hold the structure together long enough for it to ripen and mature into products. Or, even better, enable the reactants to form an entirely new transition state with a lower energy of activation, a structure possible only with the active assistance of some external agent. Do any of these things, and the reaction will proceed faster; it will be speeded by a catalyst. Do the opposite—make it harder to produce a transition state, say by blocking the active site of an enzyme—and the reaction will proceed slower. It will be hobbled by an inhibitor.

Thus in the *uncatalyzed* decomposition of formic acid into carbon monoxide and water,

$$HCOOH \longrightarrow CO + H_2O$$

a hydrogen is perhaps forced to migrate, unassisted, from the carbon to an oxygen. The resulting transition state,

transition state

(a)

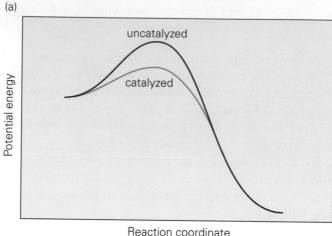

Reaction coordinate

(b)

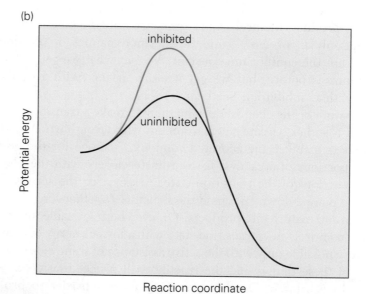

Reaction coordinate

FIGURE 18-12. Resetting the bar: (a) A catalyst lowers the energy of activation and increases the rate. (b) An inhibitor raises the barrier and decreases the rate. Each agent alters in some way the makeup and function of the transition state.

is an awkwardly shaped, hard-to-form structure, and the energy of activation is correspondingly high. But when strong acid is introduced, as in the reaction

transition state

the external hydrogen ion acts as a catalyst. H^+ makes possible the new, easier-to-form transition state drawn above, and the reaction speeds up considerably.

The hydrogen ion goes in at the beginning and comes out at the end, making no net contribution to the thermodynamics. Still, H^+ is no spectator here. It is an active catalytic agent, a matchmaker that helps to fashion a thermodynamically more favorable transition state and then departs (after passing through the intermediate form HCO^+, not shown). It opens up an alternative *pathway* for the reaction.

In previous chapters we have encountered other catalysts too, among them various metals, zeolites, and enzymes. Metals such as palladium or platinum, recall, are good at *adsorbing* molecules onto their surfaces, upon which the reactants are assembled and then eased over a lower activation barrier. With their bonds loosened by reactant–metal interactions, these surface-bound species can be primed for reaction and held in place for effective collision.

The catalytically crucial process of adsorption (binding to the exterior) is made easier by the unsatisfied valence usually existing at a surface, where the internal network of metallic bonds is suddenly terminated. Atoms at the surface, after all, would have made more bonds had not the metal come to an abrupt end. They had (and still have) the capacity to combine, and now they can do so. The new bonds come from the reactants, temporary visitors who use the surface to build an easier-to-attain transition state.

Recall as well, from Chapter 15, the shape-selective catalysis that goes on in certain zeolitic cavities, and the lock-and-key configuration of an enzyme's active site. Here it is *molecular recognition* that proves decisive; the specific, complementary matching of the catalyst's structure to the structure of the transition state serves to lower the barrier. And, still more, molecular recognition proves key to life itself: both to metabolism, made possible only by enzymatic catalysis; and also to immunity, which depends upon the ability of antibodies to recognize and bind to specific foreign proteins.

Might an antibody's ability to recognize—to bind selectively—be combined with a catalyst's facility to speed a reaction? Recently, yes. Using the techniques of recombinant DNA, researchers have induced certain organisms to produce *monoclonal antibodies*: antibodies that conform to a specific structure chosen by the experimenter. If that structure is designed to match the transition state of a particular reaction, then the result may be a custom-tailored catalyst, an artificially manufactured *catalytic antibody*.

18-6. MECHANISM: STEP BY STEP

A few molecules smash together and emerge as something else. Molecules A and B become molecules C and D, passing through an unstable transition state $X^{\ddagger}$ along the way. From one valley to another they go, forced to scale one hill of activation energy. Reactants are broken apart and rearranged into products.

That, so far, is our image of a single chemical event: a *one*-step process that combines an endothermic tearing-apart with an exothermic coming-together. We call such a reaction an ***elementary step***, and it is a good, serviceable way to picture what happens during a simple encounter.

To be truly elementary, the encounter must go both ways. If A and B can produce C and D in one step, then so must C and D be able to regenerate A and B in one step:

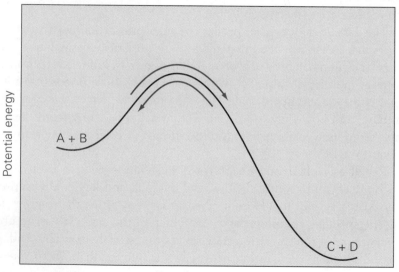

Reaction coordinate

The rate of travel need not be the same in each direction, but the connecting road must be open nonetheless. For here sit two valleys separated by one hill, and the hill can be climbed in either direction. Any elementary step consequently must be a reversible transformation, one that ultimately promises thermodynamic equilibrium between the valleys. Instantaneous rates will vary over time, and eventually the forward and reverse reactions will go at the same speed. After that, the concentrations of the species no longer change.

Now, to this elementary picture add a measure of chemical reality: Not all reactions are executed in a single step. Allow, accordingly, for not just one hill, but (as in Figure 18-13) for an entire range of hills and

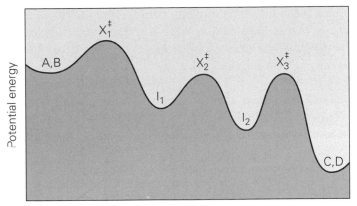

FIGURE 18-13. Between reactants and products there may be more than one hill—and more than one valley as well. Lodged in each of the intervening valleys is a reactive *intermediate*, a stable though short-lived compound produced along the way. The peaks, representing unstable structures, are the transition states.

valleys between reactants and products, consistent with a sequence of elementary steps (a **reaction mechanism**). No longer will we have A and B always going directly to C and D; instead the complete reaction might take a circuitous route like the one below:

Step 1:	$A + B \rightleftharpoons I_1$
Step 2:	$I_1 \rightleftharpoons I_2$
Step 3:	$I_2 \rightleftharpoons C + D$
Net:	$A + B \rightleftharpoons C + D$

Here I_1 and I_2 are sets of **intermediate species**, temporary by-products of reaction that come and go before the final products, C and D, appear. Added together, the individual steps yield the net reaction with no intermediates on either side of the equation.

Although often short-lived and hard to find, an intermediate is thermodynamically stable. Unlike transition states, these structures rest in valleys rather than perch on hilltops. Each interior valley represents one set of intermediates, and each surrounding peak represents a transition state that overlooks the valley.

Elementary Rate Laws

Expect simplicity. Everywhere there is simplicity to be found, if only one looks hard enough. Always there is some small set of building blocks and some simple, repetitive pattern.

For chemical reactions, simplicity comes from the elementary steps. All processes, even those with the most intricate mechanisms, string together the stingiest of elementary processes, one at a time. Look closely, then, and discover simplicity:

Each elementary reaction involves, first, a bare minimum of interacting species, with no more than one, two, or (less likely) three particles taking part in any single encounter. The combinations shown in Figure 18-14 are few and easy to classify according to their **molecularity**, the number of participating molecules. There are **unimolecular** reactions, in which *one* entity is transformed; **bimolecular** reactions, in which two species clash; and **termolecular** reactions in which three molecules collide at once, a coincidence rarely observed.

But simplicity comes not from these small numbers alone. More

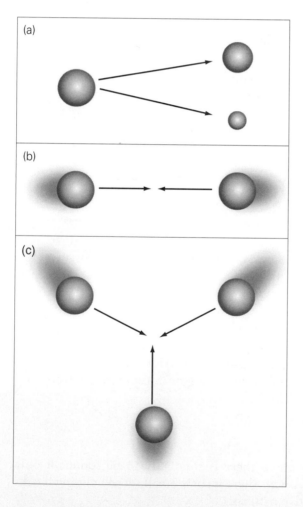

FIGURE 18-14.
Elementary reactions and molecularity.
(a) Unimolecular decomposition: rate = $k[A]$.
(b) Bimolecular collision: rate = $k[A][B]$.
(c) Termolecular collision: rate = $k[A][B][C]$.
Each elementary step is reversible.

telling is the simplicity that comes from the predictable kinetics of an elementary step. For here is the one instance (the only instance) where we can use stoichiometry to predict the rate law: when the reaction is elementary, proceeding in a single step. Only then.

Only then—just for an elementary, one-step, potentially *reversible* reaction—do the exponents of a rate equation coincide with the coefficients of the balanced equation. Thus a *uni*molecular step, signaling either a rearrangement or decomposition of a single species,

$$A \longrightarrow A' \qquad \text{rate} = k[A]$$

$$A \longrightarrow B + C \qquad \text{rate} = k[A]$$

always proceeds at a rate proportional to the concentration of that one species. Whether the reaction is the isomerization of methyl isonitrile,

$$H_3C-N\equiv C \longrightarrow H_3C-C\equiv N$$

or the formation of a carbocation from *tert*-butyl bromide,

$$
\begin{array}{ccc}
\underset{\displaystyle \underset{CH_3}{|}}{\overset{\displaystyle \overset{CH_3}{|}}{CH_3-C-Br}}
& \longrightarrow &
\underset{\displaystyle \underset{CH_3}{|}}{\overset{\displaystyle \overset{CH_3}{|}}{CH_3-C^+}} + Br^-
\end{array}
$$

the rate law takes this same predictable, first-order form: rate $= k[A]$.

Understand, though, that first-order kinetics does not mean that the species A acts entirely alone (as if, unaided, it could break its own bonds and change spontaneously into something else). The step here is unimolecular, yes, but A needs prior assistance before the final transformation can go forward. Even a unimolecular reactant first must acquire additional energy from some other species M, so that actually we have an initial *activation* step

$$A + M \rightleftharpoons A^* + M$$

which only then is followed by a decay:

$$A^* \longrightarrow A^{\ddagger} \longrightarrow \text{products}$$

M might be any molecule (including another A) that collides with a reactant A, boosts the energy of A, and then moves away. Or, for the example involving *tert*-butyl bromide, M might represent the molecules of a polar solvent, which surround the alkyl halide and help stretch the C—Br bond just to the point of breaking. Either way, the reactant A becomes the energized species A^*; and this A^* then goes off on its own, passing through a transition state $A^{\ddagger}$ on the way to final products. The concentration [M] is absent from the unimolecular rate law.

A true *bimolecular* reaction, by comparison, displays second-order kinetics,

$$A + A \longrightarrow \text{products} \qquad \text{rate} = k[A]^2$$

$$A + B \longrightarrow \text{products} \qquad \text{rate} = k[A][B]$$

just as we pictured for a one-on-one collision in the models above. Such is the rate law characteristic of the S_N2 pathway for nucleophilic substitution (Chapter 8), which proceeds as an elementary reaction. Hydroxide ion, for instance, will collide with methyl bromide, and (all in one uninterrupted step, from reactants to transition state to products) will displace the bromide to produce methanol:

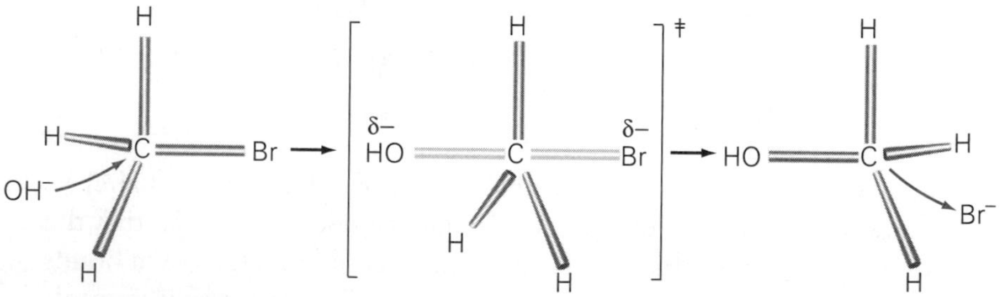

The OH^-, attacking from behind the Br, injects extra electrons into an otherwise satisfied molecule of CH_3Br. A trigonal bipyramidal transition state is the consequence. With the carbon coordinated to five other atoms (one too many), the unstable structure contains a half-broken C—Br bond and a half-formed C—OH bond. Eventually the bromide pops loose and the bimolecular reaction is over, having occurred at a rate equal to $k[OH^-][CH_3Br]$.

Rate-Determining Step

We recognize, in a multistep reaction (Figure 18-15), an operation similar to an assembly line: a sequence of individual tasks, each linked to the

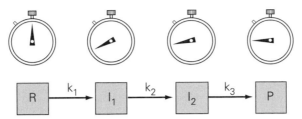

FIGURE 18-15. Building a product, step by step: The intermediate produced at one stage serves as the reactant for another, passed along as if on an assembly line. The overall reaction proceeds no faster than its slowest step (here the formation of I_1).

one before and the one after. Material manufactured in step 1 is used in step 2; material manufactured in step 2 is used in step 3; material manufactured in step 3 is used in step 4; and on it goes, with every stage fed by the one just completed. One step after another the reaction unfolds, each task in its own time.

Now suppose, for clarity, that there are only two elementary reactions, and let them occur in successive intervals of 0.05 s apiece. The entire two-step reaction then will be over by 0.1 s, and by that we shall set our clock: 0.1 s for one complete reaction, 0.2 s for two, . . . , one full second for ten. It is an uncomplicated result, needing little elaboration, and the overall rate clearly is ten per second (10 s^{-1}). By adding together the time spent at each step, we arrive straightforwardly at the time required for the whole process.

Even more straightforward, however, is a process in which one of the steps takes far more time than all the others. How much time will elapse, then, if step 1 requires 1 s while step 2 requires only 3×10^{-8} s—or, to scale the difference up to human-size numbers, one *year* and one second? Surely the extra second is meaningless on a scale of one year, and so we acknowledge the entire process to be nearly over after that slow first step is completed. And if the one-second reaction were to occur before the one-year reaction? We still must wait a year. The effective rate is determined by the time consumed during the slowest step, whenever it may occur. A multistep reaction will invest most of its time building the transition state with the highest activation energy.

The slowest elementary reaction thus becomes the **rate-determining step**, the bottleneck holding up the assembly line. This one process proves key to the kinetics overall, because from the rate-determining molecularity comes the rate law. If, say, the first step is rate-determining *and* unimolecular, then the net transformation will be first order, as for an S_N1 reaction:

$$1. \quad CH_3-\underset{\underset{CH_3}{|}}{\overset{\overset{CH_3}{|}}{C}}-Br \quad \xrightarrow[\text{(slow)}]{k_1} \quad CH_3-\underset{\underset{CH_3}{|}}{\overset{\overset{CH_3}{|}}{C}}^+ \quad + \quad Br^-$$

$$2. \quad CH_3-\underset{\underset{CH_3}{|}}{\overset{\overset{CH_3}{|}}{C}}^+ \quad + \quad OH^- \quad \xrightarrow[\text{(fast)}]{k_2} \quad CH_3-\underset{\underset{CH_3}{|}}{\overset{\overset{CH_3}{|}}{C}}-OH$$

$$\text{Net:} \quad (CH_3)_3CBr \quad + \quad OH^- \quad \longrightarrow \quad (CH_3)_3COH \quad + \quad Br^-$$

$$\text{rate} = k_1\,[(CH_3)_3CBr]$$

Production of a carbocation during the first step determines the overall rate. The hard, slow part of the proposed sequence is the manufacture of this high-energy species, after which the OH^- attacks rapidly to complete the reaction.

If true, then we should be able to accelerate the substitution by lowering the barrier for the first step alone, where such improvement will be felt directly. Any action designed to ease formation of the carbocation (such as using a strongly polar solvent to stabilize the charge) should help—and it does, lending credence to the S_N1 mechanism.

But be clear about what we have. The suggested sequence is just one of several plausible microscopic scenarios, not the only scheme conceivable. Yet this mechanism does agree with the experimentally observed rate law (if not, we would reject it out of hand) and, moreover, we have been able to verify some of its implications indirectly. So we find with all mechanisms: They are *proposals*, meaningful only when confirmed by experiment.

Pre-equilibria

Suppose next that the initial step is not rate-determining, leaving us unable to impose the kinetics of the first elementary reaction on the entire process. Such are the circumstances governing the reaction

$$2NO(g) + O_2(g) \longrightarrow 2NO_2(g)$$

which proceeds at a rate equal to $k[NO]^2[O_2]$. Ruling out the unlikely event of a ternary collision (which would also yield the correct rate), we consider the following mechanism instead:

STEP 1. In the first of two steps, two molecules of NO react to produce the intermediate species N_2O_2. We assume that this reversible reaction

$$1. \ NO + NO \underset{k_{-1}}{\overset{k_1}{\rightleftharpoons}} N_2O_2 \quad \text{(fast)}$$

progresses rapidly to equilibrium, whereupon the forward and reverse rates become equal and the concentrations of all species remain constant. We write, accordingly, the condition for *dynamic* equilibrium as

$$\begin{array}{cc} \text{forward} & \text{reverse} \end{array}$$
$$k_1[NO]^2 = k_{-1}[N_2O_2]$$

and thus obtain the mass-action expression for the equilibrium constant (K_1) in terms of the *rate* constants k_1 and k_{-1}:

$$K_1 = \frac{k_1}{k_{-1}} = \frac{[N_2O_2]}{[NO]^2}$$

From this **pre-equilibrium**, rapidly established, there arises a sensibly constant concentration of the intermediate N_2O_2, with

$$[N_2O_2] = K_1[NO]^2$$

The N_2O_2 is now processed in the rate-determining second step.
STEP 2. N_2O_2 reacts slowly with O_2, going as

$$2. \ N_2O_2 + O_2 \xrightarrow{k_2} 2NO_2 \quad \text{(slow)}$$

with a bimolecular rate law:

$$\text{Rate} = k_2[N_2O_2][O_2]$$

And, having just determined the pre-equilibrium concentration of the intermediate, we substitute $[N_2O_2] = K_1[NO]^2$ to obtain the correct *third*-order rate law:

$$\text{Rate} = k_2 K_1[NO]^2[O_2] = k_2 \frac{k_1}{k_{-1}}[NO]^2[O_2]$$

That done, we see that the observed rate constant

$$k = k_2 \frac{k_1}{k_{-1}}$$

is expressed in terms of the elementary rate constants k_1, k_{-1}, and k_2.

At Equilibrium: A Detailed Balance

In the end there is equilibrium. Products and reactants coexist in their predestined thermodynamic proportions, and every forward-going reaction is balanced by a reverse transformation proceeding at the same speed. All concentrations remain constant. The reaction is over.

Equilibrium represents final victory for thermodynamics, which enforces the equilibrium constant and thereby realizes the reaction's chemical potential. What *could* happen evidently *has* happened.

For kinetics, the end of time signals stalemate. Microscopic equilibrium brings about not a cessation of reaction, but a perfectly balanced tug of war instead. Amid the back-and-forth of reactants becoming products and products becoming reactants, one party no longer gains on the other. Rather for every elementary reaction

$$A + B \underset{k_{-1}}{\overset{k_1}{\rightleftharpoons}} C + D$$

the rate forward becomes equal to the rate backward,

$$k_1[A][B] = k_{-1}[C][D]$$

and so the equilibrium constant, K_1, follows directly as

$$K_1 = \frac{k_1}{k_{-1}} = \frac{[C][D]}{[A][B]}$$

The expression is the same form we noted, in passing, while studying reactions with a fast initial step, but here it looms larger. This simple ratio of rate constants, k_1/k_{-1}, serves to close the loop between thermodynamics (K_1) and kinetics (k_1 and k_{-1}). With it we have, for the first time, a link between the thermodynamic picture of equilibrium (static, macroscopic, finished) and the kinetic picture of equilibrium (dynamic, microscopic, ongoing). From start to finish, we can now see how kinetics and thermodynamics come finally into harmony.

Consider: In the beginning, with so much A and B, the thermodynamic tendency is to consume the high-potential reactants. The kinetic drive, meanwhile, responds to the same high concentrations by running the forward reaction at a correspondingly fast rate, $k_1[A][B]$. Reactants begin to disappear and products begin to form.

Inevitably the thermodynamic impetus ($\Delta G = \Delta G^\circ + RT \ln Q$) diminishes as products appear and the reaction quotient moves toward equilibrium. So, too, does the *rate* of production vary as the different concentrations grow and shrink. The instantaneous rates depend, we know, on how *much* material is present at any time.

Forward and reverse reactions adjust their speeds continually until both products and reactants eventually find themselves at the same chemical potential. That so, there are just enough molecules of each to guarantee equal free energy for all. Those proportions are precisely where the opposing rates become equal as well, and thus kinetics ends where equilibrium begins. The kinetic competition, like the thermodynamic drive to equalize G between reactants and products, reaches stalemate when $k_1[A][B]$ matches $k_{-1}[C][D]$.

There is our link. The equilibrium constant for an elementary reaction is determined by the ratio of the forward and reverse rate constants. When k_1 is much greater than k_{-1}, the products will dominate at equilibrium, as if the double arrow were skewed to the right:

$$A + B \; \underset{k_{-1}}{\overset{k_1}{\rightleftharpoons}} \; C + D$$

If, instead, the reverse rate constant dominates, then the reaction goes to the left.

Everything comes together finally at system-wide equilibrium, where the ***principle of detailed balance*** (or ***microscopic reversibility***) holds sway. The rule: To maintain a proper equilibrium, every elementary reaction must be separately in balance, with each *equilibrium* constant, K_i, equal to the corresponding ratio of *rate* constants, k_i/k_{-i}. Hence in a two-step reaction

$$1. \quad A + B \; \underset{k_{-1}}{\overset{k_1}{\rightleftharpoons}} \; E$$

$$2. \quad\quad E \; \underset{k_{-2}}{\overset{k_2}{\rightleftharpoons}} \; C + D$$

the overall equilibrium constant becomes the product of K_1 and K_2:

$$K = K_1 K_2 = \frac{k_1}{k_{-1}} \frac{k_2}{k_{-2}} = \frac{[E]}{[A][B]} \frac{[C][D]}{[E]} = \frac{[C][D]}{[A][B]}$$

What is true for two elementary reactions is true, moreover, for any other sequence; and so this fundamental connection between big K and little k carries through from step to step, from intermediate to intermediate, and ultimately to the final products. All the forward reactions are counterbalanced by the reverse reactions; all concentrations hold steady; the system is at rest.

REVIEW AND GUIDE TO PROBLEMS

Picture a kind of equilibrium: pebbles in a valley, mountains all around, other valleys beyond, some high, some low. Timeless. Unchanging. At rest.

Then the earth starts to shake and pebbles begin to fly—some to the north, some to the south, some over one mountain, some over another, some into a new valley, others trapped in the old. Suddenly the sky is raining pebbles, and we wonder when and how it all will end.

Of this much we are sure: the dust will settle. Pulled by gravity, the pebbles will sink eventually to the lowest possible levels. Shake them long enough; shake them hard enough; shake them often enough, and the outcome is always the same. Fast or slow, easy or hard, this way or that—when the dust settles, the pebbles come to rest in the deepest valleys. The lower a valley, the more dust it contains. Those are the laws of thermodynamics, and the law must be obeyed.

Must. But when? By what means? By what path? Be specific. Ask: How violently does the ground shake? Is there a wind blowing? Do the pebbles have enough energy to clear the peaks? Is the northern range taller than the southern range? How steep is the climb? These are the practical, workmanlike questions of reaction **kinetics**. These are questions of rates and mechanism, questions of means and opportunity rather than thermodynamic motive. These are the questions to be answered now as *atoms* fly like pebbles from one valley of free energy to another, coming to rest who knows where and when.

Concerning thermodynamics, then, we ask and answer: A deep valley? Many pebbles. A large drop in free energy? Many molecules. A larger drop in free energy? Still more molecules. Expect to see a new equilibrium, skewed even further toward the products.

With that, some of the wondering is over—but only the thermodynamic wondering, not the kinetic. Thermodynamics looks to the ends, not the means; thermodynamics sees just the timeless state of equilibrium, not the events before or after. *When* the products form, *where* they form, *how* they form, and all such questions have nothing to do with thermodynamics, which is strictly nature's long-term, statistically driven, dead-certain strategy for the dispersal of energy. Thermodynamics asks simply if the entropy of the world is greater at the end than it was at the beginning. Yes? Then there will be more products than reactants. No? Then there won't.

For kinetics, though, the questions are pointedly those of a tactician or engineer, not those of a patient strategist waiting for the dust to settle. A large activation energy? A long wait. How long, exactly? Better see what particular molecules you have in play, how they interact, how they collide, and what bruises they suffer in the process. The answers are different for every transformation; and, unlike thermodynamic imperatives, there is no immutable number like $\Delta G°$ to fix the course of reaction.

Realize, after all, that saying (thermodynamics) is one thing, but doing (kinetics) is another; and now the reactants must indeed do something in a very practical, very specific way. There are choices to make, different for each structure. The atoms must take some definite route over those mountains of energy, going from the valley of reactants through (maybe) several intermediate valleys before coming at last to the valley of products. Along the way, there are peaks of energy to be scaled. There are temporary structures to build. There are steps to be taken and mechanisms to follow, all the while against a background of random collision and accidental encounter. Reactions take time. Reactions take energy. Reactions take luck.

And so, with the nitty-gritty of kinetics, we move finally from the realm of the possible to the realm of the probable; we progress from promise to fulfillment, from potentiality to actuality.

IN BRIEF: REACTION MECHANISM AND KINETICS

1. START-UP COSTS: ACTIVATION. Businessmen spend money to make money; and so too, in a way, does nature. Molecules spend energy to get a reaction started. They knock together, searching blindly for stronger combinations. They break bonds to make bonds. Profits come later, but at first there are only costs.

First things first. Before the thermodynamic promise of a lower free energy can be fulfilled, the reactants must climb out of a local equilibrium. They must go up before they can go down. They have a hill to climb, and at the top of that hill—rising above the valley floor by some *activation energy* (E_a)—is an unstable structure, a structure not-quite-products and not-quite-reactants: a *transition state*, an *activated complex* in its critical configuration. Once formed, the transition state can go either way, back to reactants or on to products in another valley.

2. CUTTING COSTS: CATALYSIS. Sometimes there is an easy way, a route that goes over a lower hill or through a pass in the thermodynamic mountains: a catalytic path. A *catalyst* allows the reactants to build a transition state with less activation energy than otherwise, thereby enabling more molecules to start down the road to products. It might be a surface on which certain key bonds are stretched and nearly broken. It might be a porous zeolite in which molecules are confined

within cavities and primed for reaction. It might be a biological enzyme, able to bind a reacting molecule onto a perfectly matching active site. Any of these it might be, and more; there are catalysts and catalysts.

3. PAYING THE BILL: THERMAL ENERGY. High or low, the activation barrier stands between reactants and products; and, catalyzed or not, no reaction can proceed unless the reactants first make that jump. Their prime resource to do so is *thermal energy*, the energy that comes naturally with temperature and works its way into various forms: (1) Translational kinetic energy ($\frac{3}{2}nRT$ for a monatomic gas), which can bring a shattering, bond-breaking, smashing kind of momentum to a molecular collision. (2) Electronic energy, which gives an atom, molecule, or ion its chemical hooks, its capacity to combine. (3) Rotational energy, which allows a molecule to whirl around like a top. (4) Vibrational energy, which can force bonds to bend and eventually break. The higher the temperature, the more thermal energy goes into the system and all its different modes.

Now consider: How many randomly jittering, flittering, whirling molecules can hurdle a barrier of height E_a at some temperature T? Thinking statistically, we should expect more to succeed when both E_a is low (the jump is easier) and the temperature is high (the molecules have more energy). The exact number depends on the statistical Boltzmann factor described in Chapters 10 and 14, here called the **Arrhenius factor**: $\exp(-E_a/RT)$. It tells us, in general, that reactions run faster when T goes up and E_a goes down, the rate roughly proportional to $\exp(-E_a/RT)$ for a simple process.

4. ECONOMIES OF SCALE: CONCENTRATION AND RATE. Most reactions also speed up at higher concentrations of reactants, in the same way that accidents increase on a crowded highway. A dense, tightly packed system promises more molecular accidents than a dilute environment.

To measure the effect, we define the rate of an arbitrary reaction

$$a\text{A} + b\text{B} \longrightarrow c\text{C} + d\text{D}$$

as

$$\text{Rate} = -\frac{1}{a}\left(\frac{\Delta[\text{A}]}{\Delta t}\right) = -\frac{1}{b}\left(\frac{\Delta[\text{B}]}{\Delta t}\right) = \frac{1}{c}\left(\frac{\Delta[\text{C}]}{\Delta t}\right) = \frac{1}{d}\left(\frac{\Delta[\text{D}]}{\Delta t}\right)$$

and go on to study as many systems as possible. Doing so, we find that not all reactions conform to a common pattern, but nevertheless many processes do obey a **rate law** of the form

$$\text{Rate} = k[\text{A}]^n[\text{B}]^m$$

in which the overall speed of reaction is proportional to various powers of the concentrations. The proportionality constant k, the **rate constant**, is itself often proportional to the Arrhenius factor, varying with temperature as $\exp(-E_a/RT)$.

Especially common are first-order and second-order processes—*first order*, where the rate of reaction is proportional to the concentration of a single reactant,

$$\text{Rate} = k[A]$$

and *second order*, where the rate now depends on the product of two concentrations:

$$\text{Rate} = k[A]^2 \qquad \text{or} \qquad \text{Rate} = k[A][B]$$

Both first-order and second-order reactions exhibit characteristic **half-lives**, fixed times during which the concentrations decrease by half.

5. ELEMENTARY REACTIONS. Looking at the stoichiometry of a chemical equation, we are ignorant of its rate law except in one special instance: if the transformation is an **elementary reaction**, completed in a single, reversible, molecule-to-molecule encounter. When so, the order of reaction is simply the number of particles that collide (the **molecularity**). If, instead, the reaction proceeds as a series of elementary steps, the overall rate depends more intricately on the entire sequence.

For any *one* of these elementary steps, however, we have direct **unimolecular**, **bimolecular**, or (rarely) **termolecular** engagements, which are first order, second order, and third order in the respective concentrations:

Unimolecular: $\qquad\qquad\qquad A \longrightarrow$ products $\qquad$ rate $= k[A]$

Bimolecular: $\qquad\qquad A + B \longrightarrow$ products $\qquad$ rate $= k[A][B]$

Termolecular: $\qquad A + B + C \longrightarrow$ products $\qquad$ rate $= k[A][B][C]$

6. COLLISION THEORY. A simple doctrine for bimolecular reactions in the gas phase: (1) Since molecules must collide before they can react, the overall reaction rate should vary with the frequency of collisions. Denote that number by a *frequency factor*, Z. (2) Since, further, molecules must collide with the proper orientations, the rate should also depend on a certain **steric factor** (from the Greek, meaning "spatial"). Let P specify the fraction of collisions in which the molecules are suitably

aligned. (3) Finally, since molecules must have enough energy to hurdle the activation barrier E_a, the Arrhenius factor joins with Z and P to fix the net rate for on-target, energy-rich, reactive collisions:

$$\text{Rate} = ZP \exp\left(-\frac{E_a}{RT}\right)$$

The combined expression, with ZP rewritten as a single factor A, is identical to the Arrhenius equation

$$k = A \exp\left(-\frac{E_a}{RT}\right)$$

for the rate constant. The pre-exponential factor, A, gives the number of properly oriented collisions per second. The Arrhenius factor, $\exp(-E_a/RT)$, selects only those collisions energetic enough to clear the barrier.

7. MECHANISM: PRE-EQUILIBRIUM AND RATE-DETERMINING STEP. Faced with reactions that unfold in more than one stage, our best kinetic clue is the **rate-determining step**—the slowest elementary reaction in the sequence. For if we know, say, that the second step of some process is much slower than the first,

$$1. \quad A + B \underset{k_{-1}}{\overset{k_1}{\rightleftharpoons}} X \qquad \text{(fast)}$$

$$2. \quad X \xrightarrow{k_2} \text{products} \qquad \text{(slow)}$$

then the system-wide rate will be limited by the concentration of X at the bottleneck:

$$\text{Rate} = k_2[\text{X}]$$

Sometimes, too, X (either an **intermediate** species or a transition state) will reach a steady concentration in a rapidly established **pre-equilibrium** with the reactants. When it does, we can calculate that concentration thermodynamically and insert the value into the rate expression.

This way: Knowing that dynamic equilibrium sets in only when the rates forward and backward become equal, we uncover a thermodynamic–kinetic connection between the equilibrium constant (K) and the forward and reverse rate constants (k_+, k_-) for any two-way process. In step 1 of our sample reaction, for instance, the equilibrium between A, B, and X develops as

$$\text{forward} \quad \text{reverse}$$

$$k_1[\text{A}][\text{B}] = k_{-1}[\text{X}]$$

$$K_1 = \frac{[\text{X}]}{[\text{A}][\text{B}]} = \frac{k_1}{k_{-1}}$$

so that $[\text{X}] = K_1[\text{A}][\text{B}]$. If step 1 then reaches equilibrium before step 2, the overall rate becomes

$$\text{Rate} = k_2[\text{X}] = k_1 K_1[\text{A}][\text{B}] = \frac{k_2 k_1}{k_{-1}}[\text{A}][\text{B}]$$

and we have an answer. Each reaction, of course, has its own mechanism.

For specific demonstrations, see Examples 18-10 through 18-12. Take note, meanwhile, of the ***principle of detailed balance***, which connects the equilibrium constant with the rate constant in so satisfying a way, saying: Thermodynamic equilibrium ensues when kinetic processes reach a dynamic stalemate, with rates equal in both directions. The system, as a whole, reaches equilibrium only when all its elementary steps are individually in balance.

SAMPLE PROBLEMS

EXAMPLE 18-1. Stoichiometry and Rate

PROBLEM: Suppose that nitric oxide (NO) disappears at an initial rate of 6.0×10^{-5} M s^{-1} when reacting with hydrogen at 1000 K:

$$2\,\text{NO}(g) + 2\,\text{H}_2(g) \longrightarrow \text{N}_2(g) + 2\,\text{H}_2\text{O}(g)$$

At what rate does the concentration of hydrogen change under the same conditions? At what rates do the products appear?

SOLUTION: Unless we know the mechanism of reaction, the stoichiometry of the balanced equation tells us nothing about the rate. Except this: Appearances and disappearances are interconnected through the stoichiometric ratios. For every two molecules of NO consumed, two molecules of H_2 are destroyed as well. For every two molecules of NO and two molecules of H_2 that react, eventually one molecule of N_2 and two molecules of H_2O are produced. If products follow reactants without undue delay (our simplifying assumption), then we can express the rate of reaction consistently as

$$\text{Rate} = -\frac{1}{2}\left(\frac{\Delta[\text{NO}]}{\Delta t}\right) = -\frac{1}{2}\left(\frac{\Delta[\text{H}_2]}{\Delta t}\right) = \frac{\Delta[\text{N}_2]}{\Delta t} = \frac{1}{2}\left(\frac{\Delta[\text{H}_2\text{O}]}{\Delta t}\right)$$

and so determine the signed rate of change for each species:

NO $-6.0 \times 10^{-5}\ M\ s^{-1}$ Given.
H_2 $-6.0 \times 10^{-5}\ M\ s^{-1}$ H_2 disappears at the same rate as NO.
N_2 $+3.0 \times 10^{-5}\ M\ s^{-1}$ N_2 appears at half the rate that NO disappears.
H_2O $+6.0 \times 10^{-5}\ M\ s^{-1}$ H_2O appears at the same rate that NO disappears.

Products come; reactants go. One sign is positive; the other, negative.

EXAMPLE 18-2. Rate, Time, Concentration

PROBLEM: Assume that the rates specified in Example 18-1 are maintained, on the average, while the reaction

$$2\text{NO}(g) + 2\text{H}_2(g) \longrightarrow \text{N}_2(g) + 2\text{H}_2\text{O}(g)$$

progresses for several seconds. If the initial concentration of hydrogen is 0.00100 M, to what value does it fall after 3.0 s?

SOLUTION: We assume, perhaps naively, that H_2 disappears at a constant rate of $6.0 \times 10^{-5}\ M\ s^{-1}$ (Example 18-1). If so, then $[H_2]$ drops by 6.0×10^{-5} mol L^{-1} during every second that the reaction runs.

Now just as a constant velocity (change in distance per change in time) multiplied by time yields distance traveled, for a chemical reaction we have the equivalent relationship linking rate (change in concentration per change in time), time, and concentration:

$$[\text{H}_2]_t = [\text{H}_2]_0 + \frac{\Delta[\text{H}_2]}{\Delta t}\,\Delta t$$

$$[\text{H}_2]_t = 0.00100\ M - (6.0 \times 10^{-5}\ M\ s^{-1})(3.0\ s) = 0.00082\ M$$

Note that hydrogen's rate of change, $\Delta[H_2]/\Delta t$, is negative. H_2 disappears as time passes.

QUESTION: Turn the problem around and ask: Given that $[H_2]$ falls from 0.00100 *M* to 0.00082 *M* over the course of 3.0 s, what is the average rate of reaction during this period?

ANSWER: Again, the same three quantities (rate, time, concentration); and again, the same relationship:

$$\text{Rate} = \frac{\text{change in concentration}}{\text{change in time}} = \frac{0.00082\ M - 0.00100\ M}{3.0\ \text{s}}$$

$$= -6.0 \times 10^{-5}\ M\ \text{s}^{-1}$$

EXAMPLE 18-3. Rate Law

We continue with the reaction of nitric oxide and hydrogen,

$$2NO(g) + 2H_2(g) \longrightarrow N_2(g) + 2H_2O(g)$$

this time analyzing the relationship between initial concentration and initial rate.

PROBLEM: Extract the rate law and rate constant from the following experimental data at 1000 K:

TRIAL	$[NO]_0$	$[H_2]_0$	INITIAL RATE	RATIO
1	$1.0 \times 10^{-3}\ M$	$1.0 \times 10^{-3}\ M$	$6.0 \times 10^{-5}\ M\ \text{s}^{-1}$	1.0
2	$1.0 \times 10^{-3}\ M$	$2.0 \times 10^{-3}\ M$	$1.2 \times 10^{-4}\ M\ \text{s}^{-1}$	2.0
3	$1.0 \times 10^{-3}\ M$	$3.0 \times 10^{-3}\ M$	$1.8 \times 10^{-4}\ M\ \text{s}^{-1}$	3.0
4	$2.0 \times 10^{-3}\ M$	$1.0 \times 10^{-3}\ M$	$2.4 \times 10^{-4}\ M\ \text{s}^{-1}$	4.0
5	$2.0 \times 10^{-3}\ M$	$2.0 \times 10^{-3}\ M$	$4.8 \times 10^{-4}\ M\ \text{s}^{-1}$	8.0

The normalized rates in the last column are ratios calculated relative to the rate measured in trial 1.

SOLUTION: Are the numbers consistent with a rate expression of the type

$$\text{Rate} = k[NO]^n[H_2]^m$$

in which each reactant has a certain *order*? Look and see.

When $[H_2]_0$ doubles from 0.0010 M to 0.0020 M (trial 2 versus trial 1), the initial rate doubles as well. We suspect, accordingly, that the rate varies linearly with $[H_2]$, and we draw further support from the comparison of trial 3 with trial 1. There the rate triples when $[H_2]_0$ increases threefold from 0.0010 M to 0.0030 M, suggesting again that $m = 1$ in the general rate law above.

Turn next to a comparison of trial 4 with trial 1, where $[NO]_0$ doubles while $[H_2]_0$ remains constant. Since the rate quadruples (going as 2^2, so that $n = 2$), we seem to have a second-order dependence on $[NO]$.

Trial 5 versus trial 1, finally, confirms the other results: With both $[NO]_0$ and $[H_2]_0$ doubled, the rate increases eightfold—a factor of 2 due to H_2, a factor of 4 due to NO. Putting everything together, we thus have a rate law

$$\text{Rate} = k[NO]^2[H_2]$$

that is first order in H_2, second order in NO, and third order overall.

We have, as well, just enough information now to compute the rate constant k by substituting the experimental data back into the equation just derived. For trial 1, say, a quick calculation shows that $k = 6.0 \times 10^4 \ M^{-2} \ s^{-1}$:

$$k = \frac{\text{rate}}{[NO]^2[H_2]} = \frac{6.0 \times 10^{-5} \ M \ s^{-1}}{(1.0 \times 10^{-3} \ M)^2(1.0 \times 10^{-3} \ M)} = 6.0 \times 10^4 \ M^{-2} \ s^{-1}$$

EXAMPLE 18-4. First-Order Kinetics

PROBLEM: A chemist monitoring the decomposition of dinitrogen pentoxide,

$$2N_2O_5(g) \longrightarrow 4NO_2(g) + O_2(g)$$

measures the initial reaction rate versus initial concentration at 343 K:

TRIAL	$[N_2O_5]_0$	INITIAL RATE
1	0.200 M	0.00136 $M \ s^{-1}$
2	0.400 M	0.00270 $M \ s^{-1}$
3	0.600 M	0.00412 $M \ s^{-1}$
4	0.800 M	0.00545 $M \ s^{-1}$
5	1.000 M	0.00682 $M \ s^{-1}$

Deduce the rate law and rate constant from the information given, and then consider how the concentrations change with time under different sets of conditions: (a) If $[N_2O_5]$ is initially 0.200 M (as in trial 1), what concentration remains after the reaction proceeds for 100 s? 200 s? (b) Repeat for an initial concentration of 1.000 M, as in trial 5.

SOLUTION: We see right away that the initial speed of reaction grows linearly with initial concentration, consistent with a first-order rate law proportional to $[N_2O_5]$:

$$\text{Rate} = k[N_2O_5]$$

The rate doubles, triples, and quadruples in direct proportion to the concentration of dinitrogen pentoxide at each instant. The more there is, the faster it disappears.

Taking $[N_2O_5]_0$ as a starting point, we can now determine k for each experiment by the same method as in Example 18-3. Analysis of trial 4, for example, produces an apparent k equal to 0.00681 s^{-1}:

$$k = \frac{\text{rate}}{[N_2O_5]_0} = \frac{0.00545 \ M \ \text{s}^{-1}}{0.800 \ M} = 0.00681 \ \text{s}^{-1}$$

A better way, however, is to treat everything all at once—on a single graph, where all the data are considered together. To do so, we plot rate versus $[N_2O_5]_0$ and ask: Do the points fall on a straight line passing through the origin? If yes, then we have a first-order reaction and the slope of the line gives us the first-order rate constant. The straight line in Figure R18-1 thus confirms that the decomposition of N_2O_5 is indeed first order and that k, the slope, has the value 0.00681 s^{-1}.

Having learned the magnitude of k, we know everything there is to know about the fortunes of $[N_2O_5]_t$ at the stated temperature. Remember, especially, that for any first-order decay (chemical, biological, nuclear), the amount of material decreases exponentially with a **half-life**, $t_{1/2}$, that depends solely on this single rate constant k and nothing else—not the kind of material, not the concentration, not the mass, not the volume; *just* k, equal here to 0.00681 s^{-1}:

$$t_{1/2} = \frac{\ln 2}{k} = \frac{0.693}{0.00681 \ \text{s}^{-1}} = 102 \ \text{s}$$

So: After 102 s, one half-life, exactly half the original concentration of

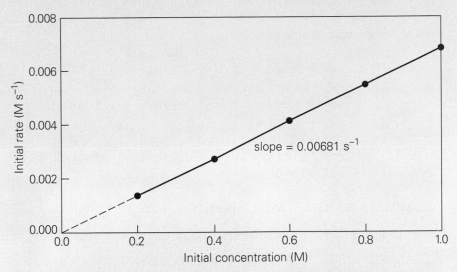

FIGURE R18-1. Initial rate versus initial concentration for the decomposition of N_2O_5. The straight line extrapolated to the origin is consistent with a first-order reaction.

N_2O_5 will remain; after 204 s (two half-lives), one-quarter; after 306 s, one-eighth; after 408 s, one-sixteenth, and so on. And, moreover, the concentration at any time in general is specified by the exponential equation

$$[N_2O_5]_t = [N_2O_5]_0 \exp(-kt)$$

(a) $[N_2O_5]_0 = 0.200\ M;\ t = 100\ s,\ 200\ s.$ Since an interval of 100 s is slightly less than one half-life, we expect to have slightly more than one-half the starting material still present: a little over 0.100 M. After 200 s, nearly two half-lives, the concentration should be roughly one-fourth the initial value: just over 0.050 M, half again what it was before.

$$[N_2O_5]_{t=100\,s} = [N_2O_5]_0 \exp(-kt) = 0.200\ M \exp(-0.00681\ s^{-1} \times 100\ s)$$
$$= 0.101\ M$$

$$[N_2O_5]_{t=200\,s} = [N_2O_5]_0 \exp(-kt) = 0.200\ M \exp(-0.00681\ s^{-1} \times 200\ s)$$
$$= 0.0512\ M$$

A decay of $0.98t_{1/2}$ (100 s) scales the concentration by a factor of 0.506, whereas a decay of $1.96t_{1/2}$ (200 s) brings it down by a factor of 0.256.

(b) $[N_2O_5]_0 = 1.000\ M;\ t = 100\ s,\ 200\ s$. Determined solely by k, the scaling factors are the same:

$$\exp(-kt) = \exp(-0.00681\ s^{-1} \times 100\ s) = \exp(-0.681) = 0.506$$

$$\exp(-kt) = \exp(-0.00681\ s^{-1} \times 200\ s) = \exp(-1.362) = 0.256$$

Starting out with $1.000\ M\ N_2O_5$, we are left with $0.506\ M$ after 100 s and $0.256\ M$ after 200 s. As expected: all first-order processes adhere to a universal schedule, dictated entirely by the rate constant.

EXAMPLE 18-5. First-Order Kinetics, Again

Same reaction as before, but with a slightly different emphasis.

PROBLEM: Use the data below to determine k for the first-order decomposition of N_2O_5:

t	$[N_2O_5]_t$	t	$[N_2O_5]_t$
0 s	1.000 M	160 s	0.336 M
40 s	0.762 M	200 s	0.256 M
80 s	0.580 M	400 s	0.066 M
120 s	0.442 M	600 s	0.017 M

SOLUTION: Expressed logarithmically, the first-order exponential equation

$$[N_2O_5]_t = [N_2O_5]_0 \exp(-kt)$$

reduces to the linear form

$$\ln\ [N_2O_5]_t = -kt + \ln\ [N_2O_5]_0$$

Plotting $\ln\ [N_2O_5]_t$ against t, as in Figure R18-2, we obtain a straight line with slope equal to $-k$ and intercept equal to $\ln\ [N_2O_5]_0$.

The slope is best computed from two well-separated points on the line, preferably not actual data. Any suitable pair (such as points 1 and 2 in Figure R18-2) will do:

$$\text{Slope} = \frac{\ln[N_2O_5]_{t_2} - \ln[N_2O_5]_{t_1}}{t_2 - t_1} = \frac{-2.043 - (-0.681)}{300\ s - 100\ s} = -0.00681\ s^{-1}$$

$$k = -\text{slope} = 0.00681\ s^{-1}$$

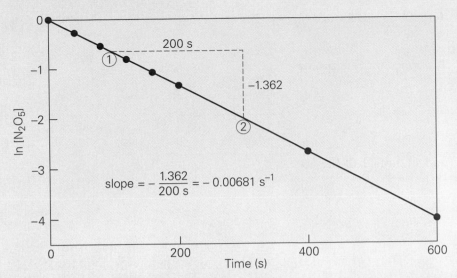

FIGURE R18-2. A logarithmic plot of concentration versus time for the decomposition of N_2O_5. The first-order rate constant is determined from the slope of the line.

QUESTION: How long must we wait for the concentration of N_2O_5 to fall to, say, 11.0% of its initial value?

ANSWER: Rearrange the equation

$$\ln [N_2O_5]_t = -kt + \ln [N_2O_5]_0$$

to solve for t when $[N_2O_5]_t/[N_2O_5]_0 = 0.110$. For that, we recall from Appendix B that $\ln([N_2O_5]_t/[N_2O_5]_0) = \ln [N_2O_5]_t - \ln [N_2O_5]_0$:

$$t = -\frac{1}{k} \ln\left(\frac{[N_2O_5]_t}{[N_2O_5]_0}\right) = -\frac{\ln 0.110}{0.00681 \text{ s}^{-1}} = 324 \text{ s}$$

The result is 324 seconds.

Not just N_2O_5, note, but *all* first-order systems with this particular rate constant ($k = 0.00681$ s^{-1}) fall to 11.0% after 324 s of decay ($t = 3.18t_{1/2}$), regardless of initial concentration. Since k is independent of $[A]_0$, a highly concentrated system decays no faster or slower than a dilute system.

A second-order process follows a different course. See Example 18-6.

EXAMPLE 18-6. Second-Order Kinetics

Reminder: Corresponding to the rate law

$$\text{Rate} = k[A]^2$$

is a concentration profile

$$\frac{1}{[A]_t} = kt + \frac{1}{[A]_0}$$

and an initial half-life

$$t_{1/2} = \frac{1}{k[A]_0}$$

A plot of $1/[A]_t$ versus time yields a straight line with slope equal to k and intercept equal to $1/[A]_0$.

PROBLEM: Decide from the graphs in Figure R18-3 whether the process represented is first order or second order. (a) Starting from an initial concentration of 1.00 M, how much of species A remains after 15 s of decay? (b) Repeat the calculation for an initial concentration of 0.50 M.

SOLUTION: The points shown in Figure R18-3(a), a plot of ln $[A]_t$ against t, deviate from a straight line. The reaction is not first order.

It is, rather, a second-order process—clear from the straight line produced when $1/[A]_t$ is graphed as a function of time (Figure R18-3b). The slope, computed in the diagram, gives the rate constant: $k = 0.067$ M^{-1} s^{-1}.

(a) $[A]_0 = 1.00$ M. Rewrite the equation

$$\frac{1}{[A]_t} = kt + \frac{1}{[A]_0}$$

in the more convenient form

$$[A]_t = \frac{[A]_0}{kt[A]_0 + 1}$$

and insert the values $k = 0.067$ M^{-1} s^{-1}, $t = 15$ s, and $[A]_0 = 1.00$ M:

$$[A]_{t=15s} = \frac{1.00\ M}{(0.067\ M^{-1}\ s^{-1})(15\ s)(1.00\ M) + 1} = 0.50\ M$$

Exactly one-half the original concentration remains after 15 s of decay,

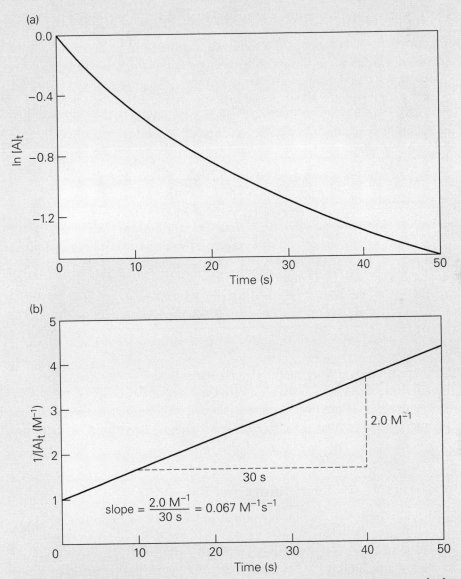

FIGURE R18-3. Concentration versus time for a hypothetical reaction in which A is the sole reactant. The data are presented in two ways. (a) Logarithmic: $\ln[A]_t$ versus t. The nonlinear graph rules out first-order kinetics. (b) Reciprocal: $1/[A]_t$ versus t. The straight line is consistent with a second-order rate law.

an interval corresponding to the first half-life:

$$t_{1/2} = \frac{1}{k[A]_0} = \frac{1}{(0.067 \; M^{-1} \; s^{-1})(1.00 \; M)} = 15 \; s$$

QUESTION: Is it really a true half-life?

ANSWER: No, not in the sense we use the term to describe a first-order decay. Here, in the second-order reaction, the value of $t_{1/2}$ varies with the starting concentration. It changes—it grows longer—as the decay progresses and the material becomes more dilute. See below.

(b) $[A]_0 = 0.50\ M$. With the concentration cut in two, the half-life doubles in sympathy: from 15 seconds to 30 seconds.

$$t_{1/2} = \frac{1}{k[A]_0} = \frac{1}{(0.067\ M^{-1}\ s^{-1})(0.50\ M)} = 30\ s$$

The rate constant stays the same, but the more dilute system (0.50 M) falls off more slowly. Whereas a concentration of 1.00 M is halved over an interval of 15 s, a 0.50 M system drops to only two-thirds of its original value during the same time:

$$[A]_t = \frac{[A]_0}{kt[A]_0 + 1} = \frac{0.50\ M}{(0.067\ M^{-1}\ s^{-1})(15\ s)(0.50\ M) + 1} = 0.33\ M$$

Less stuff, but it stays around more persistently.

Just think: A weakly concentrated poison, if decaying in a second-order process, might pose a more sustained environmental threat than a strongly concentrated system. The less there is, the longer it takes to react.

EXAMPLE 18-7. Temperature and Rate

Molecules move faster at higher temperatures. They vibrate faster. They rotate faster. They have more energy. They are more likely to hurdle an activation barrier and begin to react.

How much more?

PROBLEM: Suppose that the activation energy for a certain process is 30.0 kJ mol^{-1}. If the rate constant is $1.0 \times 10^{-4}\ M^{-1}\ s^{-1}$ at 300°C, what will it be at 400°C? Assume here that k obeys the Arrhenius equation

$$k = A\ \exp\left(-\frac{E_a}{RT}\right)$$

and that A (the frequency factor) stays the same at the two temperatures.

SOLUTION: Compute the Arrhenius factors

$$f = \exp\left(-\frac{E_a}{RT}\right)$$

at each temperature, being careful to (1) express T in kelvins, not degrees Celsius; (2) express E_a in joules, not kilojoules, to match the units of R; (3) check that the exponent E_a/RT is properly dimensionless:

$$f_1 = \exp\left(-\frac{E_a}{RT_1}\right) = \exp\left[-\frac{3.00 \times 10^4 \text{ J mol}^{-1}}{(8.3145 \text{ J mol}^{-1} \text{ K}^{-1})(573 \text{ K})}\right]$$
$$= \exp(-6.30) = 1.84 \times 10^{-3}$$

$$f_2 = \exp\left(-\frac{E_a}{RT_2}\right) = \exp\left[-\frac{3.00 \times 10^4 \text{ J mol}^{-1}}{(8.3145 \text{ J mol}^{-1} \text{ K}^{-1})(673 \text{ K})}\right]$$
$$= \exp(-5.36) = 4.69 \times 10^{-3}$$

Given that

$$k_1 = A \exp(-E_a/RT_1) = Af_1$$

$$k_2 = A \exp(-E_a/RT_2) = Af_2$$

we then have

$$k_2 = k_1 \frac{f_2}{f_1} = (1.0 \times 10^{-4} \ M^{-1} \ s^{-1}) \times \frac{4.69 \times 10^{-3}}{1.84 \times 10^{-3}} = 2.5 \times 10^{-4} \ M^{-1} \ s^{-1}$$

The barrier stays in place, but more molecules are able to climb it at the higher temperature. The rate constant scales according to the ratio f_2/f_1, increasing by a factor of 2.5.

QUESTION: What if the net reaction is exothermic, so that the equilibrium constant diminishes with increasing temperature?

ANSWER: Overall exothermicity or endothermicity has nothing to do with the kinetics of activation. Exothermic means only that the products have less enthalpy than the reactants, making the release of excess heat inevitable in the long term:

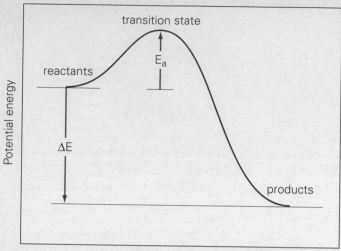

Reaction coordinate

That difference in energy, shown as ΔE, is a fixed thermodynamic property of state. It never changes.

The activation energy, marked as E_a on the diagram, looms instead as a short-term kinetic obstacle standing in the way of thermodynamic destiny. Nothing happens until the reactants muster sufficient energy to overcome the barrier. The more they can do so, the faster the reaction *starts*.

It starts, and eventually it ends. It ends where all reversible reactions, fast or slow, must end if we wait long enough: in equilibrium, with reactants and products in the fixed proportions demanded by the equilibrium constant. And, yes, the equilibrium constant does grow smaller when an exothermic reaction operates at a higher temperature; and so we find, in one of nature's little ironies, that the reactants work harder and faster to produce less and less.

QUESTION: What if we lower the barrier? Will the reactants produce more products if given an easier way to do so?

ANSWER: Again, the activation energy (kinetics) has nothing to do with the eventual equilibrium (thermodynamics). The thermodynamic outcome is forever predetermined by the standard difference in free energy between the reactants and products—not between the reactants and a transition state, not between the reactants and an intermediate, not between anything else but the reactants at the beginning and the products *at the end*.

So what does a catalyst do? It lowers the price of admission. By granting to the reactants a lower E_a, a catalyst allows more of them to enter the thermodynamic sweepstakes. Catalysis hastens the end, but does not *change* the end. The same number of molecules will eventually get there, fast or slow, easy or hard:

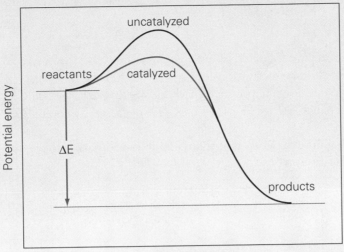

Reaction coordinate

Does a catalyst in any way alter the net change in free energy? No. The free energy of formation is built into the very structure of a molecule, and no tinkering with an activation barrier can ever change that. A catalyst offers the system a new transition state, not a new set of products.

EXAMPLE 18-8. Activation Barrier: A Lopsided Hill

PROBLEM: Continue with the hypothetical reaction of Example 18-7, for which we stipulated that $E_a = 30.0$ kJ mol^{-1} in the forward direction. If we specify further that the overall heat of reaction is -75.0 kJ mol^{-1}, what activation energy is required to initiate the reconversion of products into reactants? Assume that $\Delta E \approx \Delta H$.

SOLUTION: Going forward, the exothermic reaction

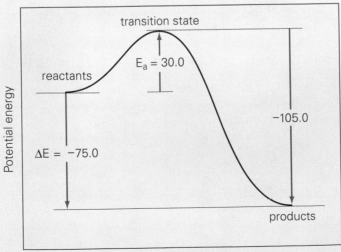

Reaction coordinate

takes one step up followed by a bigger step down: (1) From reactants up to the transition state (activation energy $= E_a = 30.0$ kJ mol^{-1}). (2) From the transition state down to products (payback $= -E_a + \Delta E = -105.0$ kJ mol^{-1}). The net change, $\Delta E = -75.0$ kJ mol^{-1}, is negative: a drop of 75 kilojoules per mole.

Going in reverse, so that the reaction coordinate is flipped left-right, the same reaction

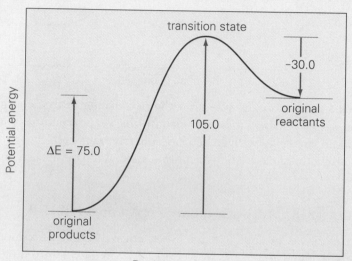

now has to climb the larger hill first: (1) From the original products up to the transition state (activation energy $= 105.0$ kJ mol^{-1}). (2) From the transition state down to the original reactants (payback $= -30.0$ kJ mol^{-1}). The net change, $\Delta E = 75.0$ kJ mol^{-1}, is positive.

Conclusion: If the forward reaction is exothermic, then the reverse reaction is endothermic and its activation energy is higher.

EXAMPLE 18-9. Activation Barrier: The Arrhenius Equation

Given the value of a rate constant at two temperatures, can we determine the activation energy? Yes, if the Arrhenius law applies.

Do it this way: Write the exponential equation

$$k = A \, \exp\left(-\frac{E_a}{RT}\right)$$

in logarithmic form first as

$$\ln k = -\frac{E_a}{RT} + \ln A$$

and then

$$\ln A = \ln k + \frac{E_a}{RT}$$

Next equate the expression for $\ln A$ at (k_1, T_1) with the expression at (k_2, T_2),

$$\ln k_1 + \frac{E_a}{RT_1} = \ln k_2 + \frac{E_a}{RT_2}$$

and finally rearrange into the form

$$\ln\left(\frac{k_2}{k_1}\right) = -\frac{E_a}{R}\left(\frac{1}{T_2} - \frac{1}{T_1}\right)$$

Note, again from Appendix B, that $\ln(k_2/k_1)$ is equal to $\ln k_2 - \ln k_1$.

PROBLEM: In Example 18-4, we found that N_2O_5 decomposes with a first-order rate constant equal to 6.8×10^{-3} s^{-1} at 343 K. Compute the Arrhenius activation energy, E_a, given that $k = 3.5 \times 10^{-5}$ s^{-1} at $T = 298$ K.

SOLUTION: Substitute the values

$$k_1 = 3.5 \times 10^{-5} \text{ s}^{-1} \qquad T_1 = 298 \text{ K}$$

$$k_2 = 6.8 \times 10^{-3} \text{ s}^{-1} \qquad T_2 = 343 \text{ K}$$

into the equation above and solve for E_a:

$$\ln\left(\frac{k_2}{k_1}\right) = -\frac{E_a}{R}\left(\frac{1}{T_2} - \frac{1}{T_1}\right)$$

$$\ln\frac{6.8 \times 10^{-3} \text{ s}^{-1}}{3.5 \times 10^{-5} \text{ s}^{-1}} = \frac{-E_a}{8.3145 \text{ J mol}^{-1} \text{ K}^{-1}}\left(\frac{1}{343 \text{ K}} - \frac{1}{298 \text{ K}}\right)$$

$$E_a = \frac{(8.3145)(5.27)}{4.40 \times 10^{-4}} \text{ J mol}^{-1} = 1.0 \times 10^5 \text{ J mol}^{-1}$$

The activation energy is approximately 100 kJ mol^{-1}.

QUESTION: Can we solve the same problem graphically?

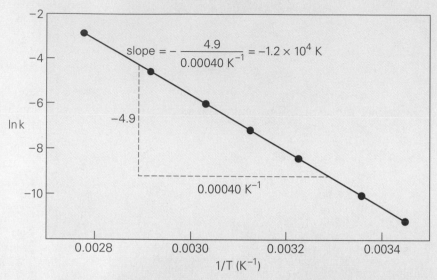

FIGURE R18-4. An Arrhenius plot: $\ln k$ versus $1/T$. Analysis of the line yields both the activation energy and the pre-exponential factor in the Arrhenius form of the rate constant. The idealized data shown are for the decomposition of N_2O_5.

ANSWER: Yes—and better, too, if we have a good set of rate constants measured over a range of temperatures. The procedure, illustrated in Figure R18-4 for the N_2O_5 reaction, includes three steps: (1) Plot $\ln k$ versus $1/T$. (2) If the points fall on a straight line, equate the slope of that line to $-E_a/R$ and solve: $E_a = -\text{slope} \cdot R$. (3) If desired, extrapolate to locate the vertical intercept, $\ln A$, at the point where $1/T = 0$ (not shown). The intercept then supplies the frequency factor in the Arrhenius equation:

$$\ln k = \underbrace{\left(-\frac{E_a}{R}\right)\left(\frac{1}{T}\right)}_{\text{slope}} + \underbrace{\ln A}_{\text{intercept}}$$

EXAMPLE 18-10. Molecularity and Mechanism

PROBLEM: The rate of the reaction

$$H_2(g) + Br_2(g) \longrightarrow 2\,HBr(g)$$

goes as

$$\text{Rate} = k[H_2][Br_2]^{1/2}$$

at early stages of the process, when the concentration of HBr is still low. Does the reaction take place in one step?

SOLUTION: No. How could it? If two molecules of HBr were actually produced by the collision of one H_2 with one Br_2, the balanced equation would tell us not just the stoichiometry but the mechanism as well. The single-step reaction

$$H_2 + Br_2 \longrightarrow 2\,HBr$$

would then be an *elementary reaction* with rate law given by

$$Rate = k[H_2][Br_2]$$

Since the observed rate depends, instead, on the square root of $[Br_2]$, the reaction evidently follows a multistep mechanism.

Try again.

EXAMPLE 18-11. Evaluating a Mechanism

PROBLEM: Does the following three-step mechanism plausibly describe the hydrogen–bromine reaction presented in Example 18-10, for which the rate varies as $k[H_2][Br_2]^{1/2}$?

1. $Br_2 + M \underset{k_{-1}}{\overset{k_1}{\rightleftharpoons}} Br + Br + M$ 　　(fast, pre-equilibrium)

2. $Br + H_2 \overset{k_2}{\longrightarrow} HBr + H$ 　　(slow)

3. $H + Br_2 \overset{k_3}{\longrightarrow} HBr + Br$ 　　(fast)

M represents some unspecified molecule that splits Br_2 into the free radicals $Br\cdot$ and $Br\cdot$, without M itself suffering any lasting chemical change owing to the collision.

SOLUTION: Since each step of a mechanism describes an elementary process, we already know the form of its rate law. The order of reaction is equal to the molecularity, and the powers in the rate law are equal to the coefficients in the elementary equation. A unimolecular elementary reaction is first order; a bimolecular elementary reaction is second order; a termolecular elementary reaction is third order.

In the first step, for example, the rate varies with concentration as

$$Rate_1 = k_1[Br_2][M]$$

in the forward direction and

$$\text{Rate}_{-1} = k_{-1}[\text{Br}]^2[\text{M}]$$

in the reverse. This first step, presumed to be fast, then establishes a nearly steady concentration of Br radicals as soon as Br and Br_2 come to their own separate quasi equilibrium. At that point, with the forward and reverse rates equal,

$$k_1[\text{Br}_2][\text{M}] = k_{-1}[\text{Br}]^2[\text{M}]$$

the concentrations of Br_2 and Br are controlled by the equilibrium constant K_1 (uppercase K):

$$K_1 = \frac{k_1}{k_{-1}} = \frac{[\text{Br}]^2}{[\text{Br}_2]}$$

The effect is to replace Br_2 as a reactant by the free radical Br, which now enters the reaction with an initial concentration set by the pre-equilibrium:

$$[\text{Br}] = K_1^{1/2}\,[\text{Br}_2]^{1/2}$$

This steady-state population of Br then feeds step 2 of the sequence (the slow step), which obeys the elementary rate law

$$\text{Rate}_2 = k_2[\text{Br}][\text{H}_2] = k_2\,K_1^{1/2}[\text{H}_2][\text{Br}_2]^{1/2}$$

and which, being slow, determines the kinetics for the reaction as a whole. Step 3, the fast reaction of H with Br_2, follows rapidly and makes no contribution to the overall rate. The rate set by the slow step alone agrees with the known concentration dependence.

But, remember, a word of caution: The mechanism is plausible, although not proven. To prove it, we need to study the reaction microscopically and detect the intermediate species. Our analysis on paper shows only that this particular sequence *can* produce the correct products with the correct rate law. Does it really? Design some experiments and see.

EXAMPLE 18-12. Dynamic Equilibrium

Balanced concentrations, balanced rates, balanced free energies—equilibrium, nothing more to do, a good point to stop.

PROBLEM: At 1000 K, the equilibrium constant for the elementary reaction

$$C_2H_6(g) \rightleftharpoons 2CH_3(g)$$

is $K_c = 1.30 \times 10^{-13}$ M, and the rate constant in the forward direction is $k_+ = 1.57 \times 10^{-3}$ s^{-1}.* (a) Write the rate laws for both the forward and reverse reactions. (b) What is the value of k_-, the rate constant in the reverse direction?

SOLUTION: The key word is *elementary*, an elementary reaction: a single, reversible step, with the order of reaction equal to the molecularity in each direction.

(a) *Rate laws*. Going forward, the elementary dissociation of *one* ethane molecule into two methyl radicals is a first-order process:

$$\text{Rate forward} = k_+[C_2H_6]$$

Going backward, the elementary recombination of *two* methyl radicals into one ethane molecule is a second-order process:

$$\text{Rate backward} = k_-[CH_3]^2$$

The higher the concentration of CH_3, the faster the recombination. The lower the concentration of C_2H_6, the slower the dissociation. Higher and lower, faster and slower; and, sooner or later, there is equilibrium.

(b) *Reverse rate constant*. At equilibrium, the rate forward is equal to the rate backward, giving us

$$k_+[C_2H_6] = k_-[CH_3]^2$$

or, equivalently, the mass-action expression for K_c:

$$K_c = \frac{[CH_3]^2}{[C_2H_6]} = \frac{k_+}{k_-} = 1.30 \times 10^{-13} \ M$$

---

*For dimensional harmony with k_+ and k_- (see part b), we assign explicit units to K_c in this problem—a departure from our usual practice, but an approach fully consistent with the understanding reached in Example 12-4. Recall that each concentration [X] in K_c is properly taken as the dimensionless *activity* [X]/c°, where c° is equal to 1 mol L^{-1} (1 M) and where the system is assumed to be ideal. If we express K_c directly in terms of molarity, omitting the factors of c°, then the ratio $[CH_3]^2/[C_2H_6]$ has units of M. The magnitude of K_c remains the same either way.

The equilibrium constant thus fixes the ratio of k_+ to k_-,

$$k_- = \frac{k_+}{K_c} = \frac{1.57 \times 10^{-3} \text{ s}^{-1}}{1.30 \times 10^{-13} \text{ M}} = 1.21 \times 10^{10} \text{ M}^{-1} \text{ s}^{-1}$$

and so, at equilibrium, just enough CH_3 recombines with a second-order rate constant

$$k_- = 1.21 \times 10^{10} \text{ M}^{-1} \text{ s}^{-1}$$

to balance the breakup of C_2H_6 with a first-order rate constant

$$k_+ = 1.57 \times 10^{-3} \text{ s}^{-1}$$

The end: the competition is stalemated, and the concentrations stay the same thereafter. The final thermodynamic ratio, predetermined long ago by the standard difference in free energy between CH_3 and C_2H_6, is enforced kinetically in the dynamic equilibrium that ensues. All else is forgotten.

EXERCISES

1. The accompanying diagram tracks a system's potential energy during the course of a certain reaction:

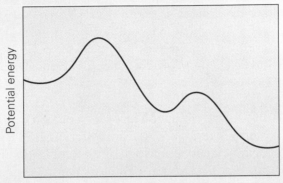

   Identify the following points on the curve: (a) Reactants. (b) Products. (c) Transition states. (d) Intermediates. What is the difference between a transition state and an intermediate?

2. Mark the following differences in energy on the diagram used just above: (a) ΔE_r, the net change in energy during the reaction. (b) E_a, the activation energy for the slowest step.

3. $\Delta G°$, the standard change in free energy, is -100 kJ mol^{-1} for reaction 1 and -1000 kJ mol^{-1} for reaction 2. Assume that molecules collide at the same rate and with the same effect in both processes. (a) Which reaction produces the greater proportion of products at equilibrium? (b) Which reaction occurs faster? Can you tell?

4. $\Delta G°$ for reaction 1 is -100 kJ mol^{-1}. $\Delta G°$ for reaction 2 is -1000 kJ mol^{-1}. E_a, the activation energy, is 10 kJ mol^{-1} for reaction 1 and 50 kJ mol^{-1} for reaction 2. (a) Which reaction produces the greater proportion of products at equilibrium? (b) Which reaction occurs faster? Assume that collision rates and efficiencies are the same for both.

5. $\Delta E°$ for reaction 1, an endothermic process, is 50 kJ mol^{-1}. Its activation energy is 55 kJ mol^{-1}. $\Delta E°$ for reaction 2, an exothermic process, is -50 kJ mol^{-1}. The activation energy is 50 kJ mol^{-1}.

(a) Draw a diagram that shows the change in energy during the course of each reaction. (b) Which process runs faster at a given temperature? Again, for simplicity, assume that reactive collisions occur at the same rate. (c) Which reaction would yield more product at equilibrium if the temperature were higher?

6. One more: ΔE° for reaction 1, an endothermic process, is 50 kJ mol^{-1}. Its activation energy is 55 kJ mol^{-1}. ΔE° for reaction 2, an exothermic process, is -100 kJ mol^{-1}. The activation energy is 60 kJ mol^{-1}. (a) Draw the energy profile for each reaction. (b) Which process runs faster at a given temperature? (c) Which process would yield more product at equilibrium if the temperature were higher?

7. Suppose that the rate constant k for some reaction has the value 1 at 0°C. Recalculate k at 10°C, 20°C, 30°C, and 100°C as predicted by the Arrhenius equation for each of these activation energies:

(a) $E_a = 1$ kJ mol^{-1}
(b) $E_a = 10$ kJ mol^{-1}
(c) $E_a = 50$ kJ mol^{-1}
(d) $E_a = 100$ kJ mol^{-1}

8. The diagram below shows the effect of a catalyst on a reaction:

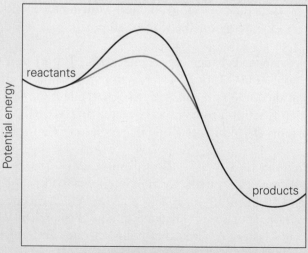

(a) Which of the two curves represents the catalyzed reaction? (b) Which of the two reactions produces the greater yield of products at equilibrium?

9. Estimate the amount of time needed for the following reactions:

   (a) Water evaporates.
   (b) Graphite becomes diamond.
   (c) Sodium chloride dissolves in water.

10. Calculate $\Delta G°$ and $K(25°C)$, the equilibrium constant at $25°C$, for each reaction mentioned in the exercise above:

   (a) $H_2O(\ell) \longrightarrow H_2O(g)$
   (b) $C(s, graphite) \longrightarrow C(s, diamond)$
   (c) $NaCl(s) \longrightarrow Na^+(aq) + Cl^-(aq)$

   Do the results help to explain the previous exercise in any way?

11. A certain reaction

$$A + 2B + C \longrightarrow D + 3E$$

   is first order in A, first order in B, and second order in C. (a) Write the rate law in the form

$$Rate = k[A]^n[B]^m \cdots$$

   (b) What are the proper units for k? (c) Calculate k, given that the observed rate is $1\ M\ s^{-1}$ when $[A] = [B] = [C] = 1\ M$. (d) Calculate the rate when $[A] = [B] = [C] = 2\ M$. (e) Does the process occur in a single, elementary step?

12. Suppose that each of these arbitrary reactions proceeds in a single step:

   (a) $A + B \longrightarrow 2C + D + E$
   (b) $A \longrightarrow 3B + C$
   (c) $2A \longrightarrow B$
   (d) $A + B + C \longrightarrow D$
   (e) $A + 2B \longrightarrow C$

   Write the corresponding rate laws in the form

$$Rate = k[A]^n[B]^m \cdots$$

13. Suppose that each rate law describes an elementary reaction:

   (a) $Rate = k[A]$
   (b) $Rate = k[A]^2$
   (c) $Rate = k[A][B]$
   (d) $Rate = k[A]^2[B]$
   (e) $Rate = k[A][B][C]$

Reconstruct the left-hand side of the balanced equation, showing in what stoichiometric ratios the reactants combine. Which of the five elementary steps are least likely to occur?

14. Yes, no, or maybe? The reaction

$$2NO_2(g) + F_2(g) \longrightarrow 2NO_2F(g)$$

is governed by the rate law

$$\text{Rate} = k[NO_2]^2[F_2]$$

Careful—remember the difference between stoichiometry and mechanism. What is it?

15. True or false? The three reactions below, similar in reactant stoichiometry, must all share the same third-order rate law: Rate = $[A]^2[B]$.

$$2NO(g) + O_2(g) \longrightarrow 2NO_2(g)$$

$$2NO(g) + Cl_2(g) \longrightarrow 2NOCl(g)$$

$$2NO_2(g) + F_2(g) \longrightarrow 2NO_2F(g)$$

16. Compute the average rate (per second) of each process:

    (a) A sprinter runs 100 meters in 10.1 seconds.
    (b) 1634 leaves fall from a tree in two weeks.
    (c) 2,993,000 babies were born last year.
    (d) The concentration of $NOCl(g)$ rises from 0 to 0.40 M in two minutes.

    Do these numbers enable us, say, to predict with certainty the number of leaves that will fall between midnight and one second after midnight? What further information is required for such a prediction?

17. Suppose that chlorine gas, reacting as

$$2NO(g) + Cl_2(g) \longrightarrow 2NOCl(g)$$

at 263 K, disappears at an initial rate of 0.0030 M s^{-1} when both $[NO]_0$ and $[Cl_2]_0$ have the value 0.10 M. (a) At what rate does NO disappear? (b) At what rate does NOCl appear? (c) If only NO and Cl_2 are present at the start, approximately what concentration of NOCl will develop during the first 5.0 s of reaction?

18. The rate law for the disappearance of Cl_2 in the reaction

$$2NO(g) + Cl_2(g) \longrightarrow 2NOCl(g)$$

has been found experimentally to be

$$\text{Rate} = k[NO]^2[Cl_2]$$

Estimate the initial rate for each set of concentrations below, given that k has the value 3.0 M^{-2} s^{-1} at 263 K:

	$[NO]_0$	$[Cl_2]_0$
(a)	0.050 M	0.050 M
(b)	0.050 M	0.100 M
(c)	0.100 M	0.050 M
(d)	0.100 M	0.100 M

19. Extract the rate law and rate constant for the reaction

$$2NO(g) + O_2(g) \longrightarrow 2NO_2(g)$$

from the following experimental data:

TRIAL	$[NO]_0$	$[O_2]_0$	$-\dfrac{\Delta[NO]}{\Delta t}$ (INITIAL RATE)
1	0.0100 M	0.0100 M	7.11×10^{-3} M s^{-1}
2	0.0100 M	0.0200 M	1.42×10^{-2} M s^{-1}
3	0.0100 M	0.0300 M	2.13×10^{-2} M s^{-1}
4	0.0200 M	0.0100 M	2.84×10^{-2} M s^{-1}
5	0.0300 M	0.0100 M	6.40×10^{-2} M s^{-1}

20. Using results from the previous exercise, estimate an initial rate $(-\Delta[NO]/\Delta t)$ for the reaction

$$2NO(g) + O_2(g) \longrightarrow 2NO_2(g)$$

under each set of initial concentrations:

TRIAL	$[NO]_0$	$[O_2]_0$
1	0.0050 M	0.0050 M
2	0.0050 M	0.0100 M
3	0.0200 M	0.0300 M
4	0.0300 M	0.0300 M

Assume that the implicit temperature is the same as in the example above.

21. Show, using the data below, that the reaction

$$SO_2Cl_2(g) \longrightarrow SO_2(g) + Cl_2(g)$$

is first order:

TRIAL	$[SO_2Cl_2]_0$	$-\dfrac{\Delta[SO_2Cl_2]}{\Delta t}$ (INITIAL RATE)
1	0.100 M	2.21×10^{-6} $M\,s^{-1}$
2	0.200 M	4.39×10^{-6} $M\,s^{-1}$
3	0.300 M	6.60×10^{-6} $M\,s^{-1}$
4	0.400 M	8.80×10^{-6} $M\,s^{-1}$

Then: (a) Calculate the first-order rate constant and half-life. (b) Starting with $[SO_2Cl_2]_0 = 0.0150$ M, calculate the concentration that remains after 3.00, 6.00, 9.00, 12.00, 15.00, and 18.00 hours of reaction.

22. The following kinetic data were measured for the decomposition of glucose in aqueous solution:

TIME	$[C_6H_{12}O_6]_t$
0 min	0.200 M
100 min	0.195 M
200 min	0.190 M
300 min	0.184 M
500 min	0.175 M
1000 min	0.152 M
2000 min	0.116 M

(a) Show, graphically, that the reaction is first order rather than second order. (b) Calculate the rate constant and half-life. (c) Predict the concentration of glucose after 24.0 hours of reaction.

23. The decomposition of chloroethane at high temperature is first order, with $k = 1.5 \times 10^{-4}$ s^{-1} at 720 K:

$$CH_3CH_2Cl(g) \longrightarrow H_2C{=}CH_2(g) + HCl(g)$$

(a) If $[CH_3CH_2Cl]_0$ is 0.100 M at the start of reaction, what is the initial rate? (b) What concentration of chloroethane will remain after 100 s, 1000 s, and 10,000 s?

24. The formation of methyl radicals from ethane,

$$C_2H_6(g) \longrightarrow 2\,CH_3(g)$$

obeys a first-order rate law with $k = 5.36 \times 10^{-4}$ s^{-1} at 700°C. (a) Compute $t_{1/2}$. (b) How much time is required to cut $[C_2H_6]$ from 1.000 M to 0.680 M? (c) From 2.000 M to 1.360 M? (d) From 2.730 M to 1.856 M?

25. The half-life for the decomposition of $N_2O_5(g)$ into $NO_2(g)$ and $O_2(g)$, a first-order process, is 22.7 minutes at 45°C. Calculate the *rate* of reaction,

$$\text{Rate} = k[N_2O_5]$$

at $t = 0$, 10.0, 20.0, and 30.0 minutes, beginning with an initial concentration of 1.00 M. Assume the initial rate law to be valid throughout.

26. In a first-order reaction at 595 K, the concentration of $FClO_2$ falls from 0.0600 M to 0.0354 M over a period of 1127 s. (a) Calculate k and $t_{1/2}$. (b) How long will it then take for $[FClO_2]$ to decrease from 0.0354 M to 0.0200 M?

27. Are all first-order reactions unimolecular? Do all reactions of the form

$$A \longrightarrow \text{products}$$

proceed in a single, elementary step?

28. Here is a second-order reaction, the gas-phase recombination of iodine radicals to form molecular iodine:

$$2I(g) \longrightarrow I_2(g)$$

The second-order rate constant is $7.0 \times 10^9 \ M^{-1} \ s^{-1}$ at 25°C, and the initial concentration of radicals, $[I]_0$, is $1.0 \times 10^{-3} \ M$. (a) Calculate the first half-life. (b) Calculate $[I]$ after 1.0×10^{-8} s, 5.0×10^{-8} s, 1.0×10^{-7} s, 5.0×10^{-7} s, and 1.0×10^{-6} s of reaction. What percentage of the original concentration of I remains after 1.0×10^{-6} s? (c) Calculate the concentration of I_2 after 1.0×10^{-6} s. (d) Is it reasonable to suppose that the reaction goes far to the right, so that all the I atoms eventually become I_2 molecules? Use the thermodynamic data in Appendix C to estimate the equilibrium constant.

29. Same reaction:

$$2I(g) \longrightarrow I_2(g)$$

Same second-order rate constant: $7.0 \times 10^9 \ M^{-1} \ s^{-1}$. Different initial concentration: $[I]_0 = 1.0 \times 10^{-4} \ M$. (a) Calculate the first half-life. (b) Calculate $[I]$ after 1.0×10^{-8} s, 5.0×10^{-8} s, 1.0×10^{-7} s, 5.0×10^{-7} s, and 1.0×10^{-6} s of reaction. What percentage of the original concentration of I remains after 1.0×10^{-6} s? (c) Calculate the concentration of I_2 after 1.0×10^{-6} s. (d) Does the reaction proceed faster or slower than in the preceding exercise? Is there any change in the equilibrium constant?

30. Another second-order reaction, the gas-phase decomposition of nitrosyl bromide:

$$2NOBr(g) \longrightarrow 2NO(g) + Br_2(g)$$

The second-order rate constant is $0.80\ M^{-1}\ s^{-1}$ at $10°C$, and the initial concentration of NOBr is $0.200\ M$. (a) Calculate the first half-life. (b) Plot [NOBr] from $t = 0$ to $t = 30$ s. (c) Plot $1/[NOBr]$ from $t = 0$ to $t = 30$ s. (d) Calculate the rate of reaction at $t = 0$ s, 15 s, and 30 s.

31. Repeat the preceding exercise, but this time assume that the initial concentration of NOBr is $0.100\ M$.

32. One more second-order reaction, the decomposition of nitrogen dioxide:

$$2NO_2(g) \longrightarrow 2NO(g) + O_2(g)$$

The rate constant is $0.54\ M^{-1}\ s^{-1}$ at $300°C$. (a) How much time is required to cut the initial concentration of NO_2 from $0.1850\ M$ to $0.0925\ M$? (b) Plot $[NO_2]_t/[NO_2]_0$ from $t = 0$ to $t = 4t_{1/2}$, where $t_{1/2}$ is the first half-life. Does the concentration regularly decrease by $\frac{1}{2}$ over each interval $t_{1/2}$?

33. Imagine, wrongly, the reaction

$$2NO_2(g) \longrightarrow 2NO(g) + O_2(g)$$

to be *first* order, but with a rate constant of the same magnitude as given above: $0.54\ s^{-1}$. (a) How much time is required to cut the initial concentration of NO_2 from $0.1850\ M$ to $0.0925\ M$? (b) Plot $[NO_2]_t/[NO_2]_0$ from $t = 0$ to $t = 4t_{1/2}$. (c) Does the concentration regularly decrease by $\frac{1}{2}$ over each interval $t_{1/2}$? Compare the first-order kinetic profile with the curve obtained in the previous exercise.

34. The rate constant k_1 for reaction 1, a first-order process, is $1000\ s^{-1}$:

$$\text{Rate} = k_1[A] \qquad k_1 = 1000\ s^{-1}$$

The rate constant k_2 for reaction 2, a second-order process, is $1000\ M^{-1}\ s^{-1}$:

$$\text{Rate} = k_2[B]^2 \qquad k_2 = 1000\ M^{-1}\ s^{-1}$$

What initial concentration $[B]_0$ will give reaction 2 the same initial rate as reaction 1 for each choice of $[A]_0$ specified below?

(a) $[A]_0 = 1\ M$
(b) $[A]_0 = 2\ M$
(c) $[A]_0 = 3\ M$
(d) $[A]_0 = 4\ M$

Compute $t_{1/2}$ for reactions 1 and 2 under each set of conditions.

35. Continue with the same two hypothetical processes:

$$\text{Reaction 1:} \quad \text{rate} = k_1[A] \quad\quad k_1 = 1000\ s^{-1}$$

$$\text{Reaction 2:} \quad \text{rate} = k_2[B]^2 \quad\quad k_2 = 1000\ M^{-1}\ s^{-1}$$

Carrying forward the results of the preceding exercise, plot

$$[A]\ \text{versus}\ t \quad\quad \text{and} \quad\quad [B]\ \text{versus}\ t$$

for each of the four cases previously considered:

(a) Initial rate $= 1000\ M\ s^{-1}$
(b) Initial rate $= 2000\ M\ s^{-1}$
(c) Initial rate $= 3000\ M\ s^{-1}$
(d) Initial rate $= 4000\ M\ s^{-1}$

36. Assume the Arrhenius equation to be valid, and start to consider the dependence of rate upon temperature. Example: The rate of a certain reaction doubles when the temperature goes from 10°C to 20°C. (a) Compute the activation energy, E_a. (b) Compute the ratio of the rate constant at 16°C to the rate constant at 10°C.

37. Continue: The rate constant for a certain first-order reaction goes from $100\ s^{-1}$ at 298 K to $600\ s^{-1}$ at 350 K. (a) Compute E_a and A, the pre-exponential factor in the Arrhenius equation for the rate constant. (b) Compute the rate constant at 320 K.

38. Rate constants for the second-order decomposition of acetaldehyde (CH_3CHO) are given below:

$T\ (°C)$	$k\ (M^{-1}\ s^{-1})$
550	1.12
600	5.21
650	20.57
700	70.50

Use them to determine the parameters E_a and A in the Arrhenius equation.

39. Rate constants for the isomerization of acetonitrile,

$$CH_3NC(g) \longrightarrow CH_3CN(g)$$

are given below:

T (K)	k (s^{-1})
470	5.79×10^{-5}
480	1.36×10^{-4}
490	3.10×10^{-4}
500	6.81×10^{-4}
510	1.45×10^{-3}
520	3.01×10^{-3}
530	6.05×10^{-3}

Use them to determine the parameters E_a and A in the Arrhenius equation.

40. Arrhenius parameters for the first-order decomposition of N_2O_5 were found to be

$$E_a = 103 \text{ kJ mol}^{-1}$$
$$A = 4.94 \times 10^{13} \text{ s}^{-1}$$

(a) Calculate $k(T)$, the rate constant, at $T = 300$ K, 325 K, 350 K, 375 K, and 400 K. (b) For each temperature, calculate the initial rate of reaction when $[N_2O_5]_0$ is 0.0100 M. (c) Calculate initial rate versus temperature, again, when $[N_2O_5]_0$ is 0.0200 M.

41. Study the Arrhenius parameters for two first-order reactions:

	E_a	A
1.	100 kJ mol^{-1}	10^{13} s^{-1}
2.	110 kJ mol^{-1}	10^{14} s^{-1}

(a) Calculate the rate constants k_1 and k_2 at 300 K and 1000 K. (b) Which reaction is faster at each temperature? Why? (c) At what temperature do the two rate constants become equal?

42. The reaction

$$C_2H_5Br(aq) + OH^-(aq) \longrightarrow C_2H_5OH(aq) + Br^-(aq)$$

follows second-order kinetics, with Arrhenius parameters

$$E_a = 89.5 \text{ kJ mol}^{-1}$$

$$A = 4.30 \times 10^{11} \ M^{-1} \ s^{-1}$$

Recognize the process as an S_N2 nucleophilic substitution, described earlier in Chapter 8. (a) Calculate $k(T)$, the rate constant, at $T = 10°C$, $20°C$, $30°C$, $40°C$, and $50°C$. (b) For each temperature, calculate the initial rate of reaction when both $[C_2H_5Br]_0$ and $[OH^-]_0$ are $0.100 \ M$. (c) Calculate the initial rate versus temperature when each initial concentration is doubled to $0.200 \ M$.

43. Arrhenius parameters for three S_N2 reactions are given below:

REACTANTS	E_a	A
1. $C_2H_5I + C_2H_5O^-$	86.6 kJ mol^{-1}	$1.49 \times 10^{11} \ M^{-1} \ s^{-1}$
2. $CH_3I + C_2H_5O^-$	81.6 kJ mol^{-1}	$2.42 \times 10^{11} \ M^{-1} \ s^{-1}$
3. $C_2H_5Br + OH^-$	89.5 kJ mol^{-1}	$4.30 \times 10^{11} \ M^{-1} \ s^{-1}$

Assume that the initial concentration of each reactant is $0.10 \ M$. (a) Calculate the rate constants for reactions 1, 2, and 3 at $40°C$. (b) Which reaction begins the fastest? (c) From just the data given, can we tell which reaction will yield the greatest proportion of products at equilibrium? If not, what information do we need?

44. Consider some elementary reaction

$$A + B \longrightarrow C + D$$

for which the Arrhenius parameters are

$$E_a = 100.0 \text{ kJ mol}^{-1}$$

$$A = 1.00 \times 10^{10} \ M^{-1} \ s^{-1}$$

in the forward direction. At equilibrium, the concentrations are

$$[A] = 1.00 \ M \quad [B] = 2.00 \ M \quad [C] = 5.00 \ M \quad [D] = 4.00 \ M$$

when the temperature is 700 K. (a) Calculate the equilibrium constant. (b) Calculate the forward rate constant, $k_+(T)$, at 700 K. (c) Calculate the rate constant, $k_-(T)$, for the reverse reaction

$$C + D \longrightarrow A + B$$

at 700 K.

45. Repeat the calculations made in the preceding exercise, this time using a higher temperature: 800 K. The equilibrium concentrations are now

$$[A] = 0.10 \ M \quad [B] = 0.50 \ M \quad [C] = 10.00 \ M \quad [D] = 15.00 \ M$$

(a) Calculate the equilibrium constant. (b) Calculate the forward rate constant, $k_+(T)$, at 800 K. (c) Calculate $k_-(T)$, the rate constant for the reverse reaction

$$C + D \longrightarrow A + B$$

at 800 K. (d) Is the reaction endothermic or exothermic?

46. At 500 K, the rate constant for the reaction

$$H_2(g) + I_2(g) \rightleftharpoons 2HI(g)$$

is $k_+ = 4.3 \times 10^{-7} \ M^{-1} \ s^{-1}$ in the forward direction. Assume, as well, that the standard free energy of reaction ($\Delta G°$) is -20.2 kJ at the stated temperature. (a) Calculate K, the equilibrium constant. (b) Suppose that the reaction occurs as written, in a single, elementary step. If so, what is the value of the rate constant k_- in the reverse direction?

47. Here are two possible mechanisms for the reaction

$$H_2(g) + I_2(g) \longrightarrow 2HI(g)$$

The first, a bimolecular elementary collision of H_2 and I_2. The second, a two-step pathway involving iodine atoms as shown below:

$$I_2 + M \underset{k_{-1}}{\overset{k_1}{\rightleftharpoons}} I + I + M \qquad \text{(rapid, equilibrium)}$$

$$H_2 + I + I \overset{k_2}{\longrightarrow} 2HI \qquad \text{(slow)}$$

Each mechanism produces the same overall rate law. What is it?

48. The mechanism of a certain reaction

$$A \longrightarrow B$$

proceeds in two steps, with individual rate constants as shown:

$$
\begin{array}{lll}
\text{Step 1} & \text{(forward)} & k_1 = 10 \\
\text{Step 1} & \text{(reverse)} & k_{-1} = 1 \\
\text{Step 2} & \text{(forward)} & k_2 = 3 \\
\text{Step 2} & \text{(reverse)} & k_{-2} = 30
\end{array}
$$

At system-wide equilibrium, [A] is 1 M. What is the concentration of B?

49. A proposed mechanism for the reaction

$$2NO_2(g) + F_2(g) \longrightarrow 2NO_2F(g)$$

proceeds in two steps:

$$NO_2 + F_2 \xrightarrow{\;k_1\;} NO_2F + F \qquad \text{(slow)}$$

$$F + NO_2 \xrightarrow{\;k_2\;} NO_2F \qquad \text{(fast)}$$

(a) Which step is rate-determining? (b) What overall rate law will this mechanism produce?

50. The rate law for the reaction

$$I^-(aq) + OCl^-(aq) \longrightarrow Cl^-(aq) + OI^-(aq)$$

in basic solution is

$$\text{Rate} = k\frac{[I^-][OCl^-]}{[OH^-]}$$

(a) Show how the overall rate law may arise from the following mechanism:

$$OCl^-(aq) + H_2O(\ell) \underset{k_{-1}}{\overset{k_1}{\rightleftharpoons}} HOCl(aq) + OH^-(aq) \qquad \text{(fast)}$$

$$I^-(aq) + HOCl(aq) \xrightarrow{\;k_2\;} HOI(aq) + Cl^-(aq) \qquad \text{(slow)}$$

$$OH^-(aq) + HOI(aq) \xrightarrow{\;k_3\;} H_2O(\ell) + OI^-(aq) \qquad \text{(fast)}$$

(b) Express the empirical rate constant, k, in terms of the elementary rate constants k_1, k_{-1}, k_2, etc.

51. A proposed mechanism for the reaction

$$2O_3(g) \longrightarrow 3O_2(g)$$

proceeds as follows:

$$O_3 \underset{k_{-1}}{\overset{k_1}{\rightleftharpoons}} O_2 + O \qquad \text{(rapid, equilibrium)}$$

$$O_3 + O \overset{k_2}{\longrightarrow} 2O_2 \qquad \text{(slow)}$$

(a) What overall rate law is predicted? (b) Express the empirical rate constant, k, in terms of the elementary rate constants k_1, k_{-1}, and k_2.

19

Chemistry Coordinated—
The Transition Metals
and Their Complexes

19-1. Coordination Complexes
Structure and Coordination
The Metal
The Ligands: Electrons and Isomers
19-2. Bonding
Crystal Field Theory
Beyond Crystal Field Theory
19-3. Thermodynamics
Water: Ligand and Solvent
Aqueous Equilibria
Chelation
19-4. Kinetics
REVIEW AND GUIDE TO PROBLEMS
EXERCISES

Charge and mass, the wave function, bonding, energy and entropy, equilibrium—some of the intangible, conceptual elements we use to understand matter and its transformations. Such has been our sustained effort to this point.

Before long, however, the abstract must merge with the concrete. From the abstractions of quantum mechanics, thermodynamics, and kinetics must come the very real, very tangible facts of the material world: the literal elements of chemistry, the atoms of the periodic table and the compounds they form.

The chemistry of the transition metals is a good place to start.

19-1. COORDINATION COMPLEXES

Spread out in Figures 19-1 and 19-2 are diagrams of various **coordination complexes**, each a grouping of electron-rich **ligands** around a transition-metal atom or ion. A mere glimpse of what nature has to offer, these drawings should be just enough to spark our interest and provoke some questions. The complexes all have odd-looking formulas, unusual structures, intriguing and important properties. There is an anticancer drug

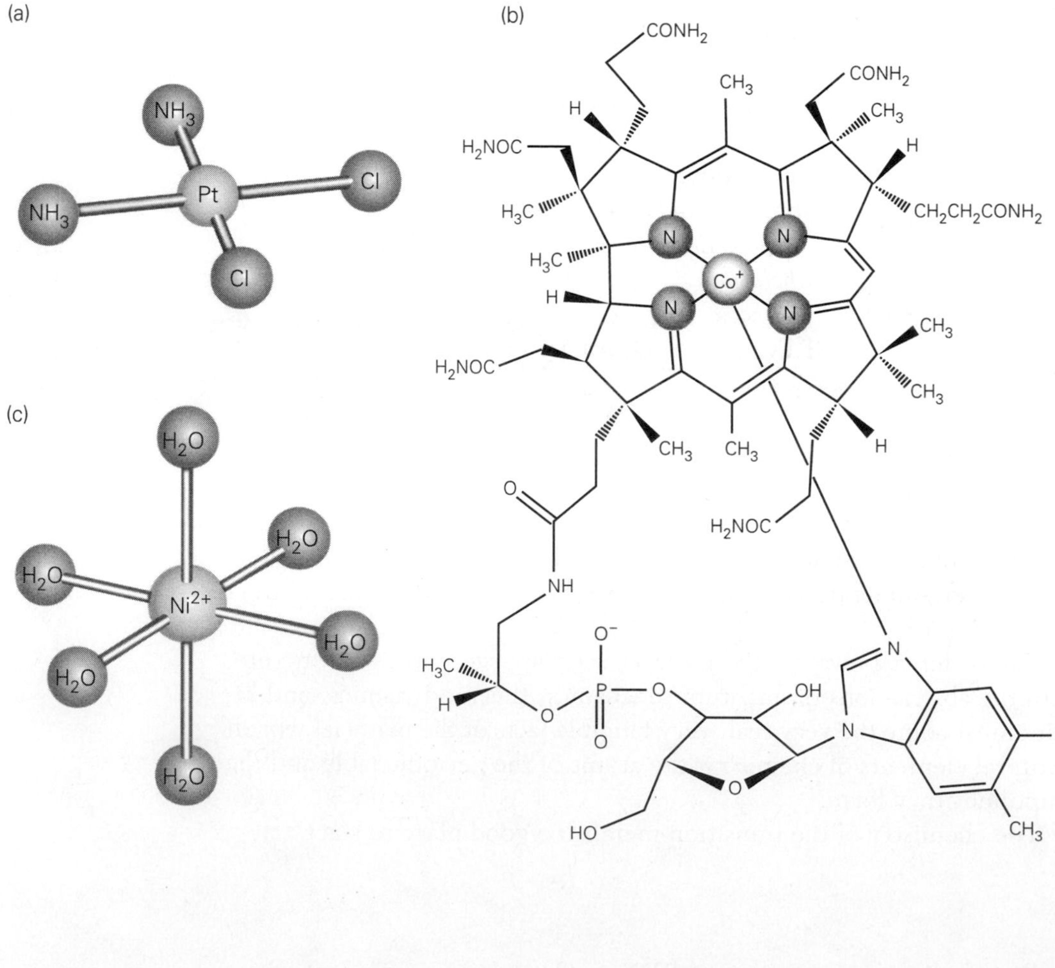

FIGURE 19-1. Selected coordination complexes. (a) The drug *cisplatin*, *cis*-[Pt(NH$_3$)$_2$Cl$_2$], a common anticancer agent. Two molecules of ammonia and two chloride ions lie in a plane containing the Pt^{2+} ion. (b) *Vitamin B$_{12}$*. At the heart of the molecule sits a cobalt ion coordinated to four nitrogen atoms, again forming a square plane. Each unmarked vertex is occupied by a carbon atom and its associated hydrogens. (c) [Ni(H$_2$O)$_6$]$^{2+}$, a six-coordinate ion with Ni^{2+} at the center. Each water molecule occupies one of the six vertices of an octahedron.

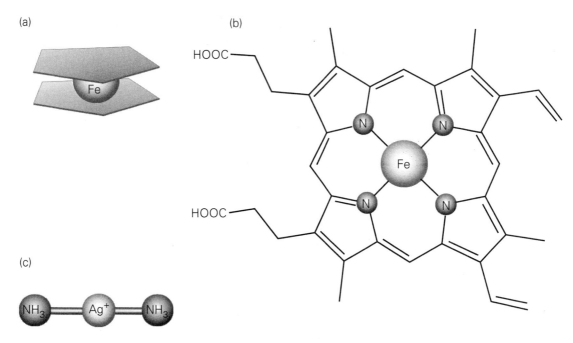

FIGURE 19-2. More coordination complexes. (a) The sandwich-like *ferrocene* molecule, Fe(C_5H_5)$_2$, holds an iron atom between two pentagonal hydrocarbon rings (C_5H_5). (b) *Heme*, a critical component of hemoglobin and certain other proteins. Four nitrogen atoms form a square planar complex about the iron. (c) [Ag(NH_3)$_2$]$^+$, a linear ion built from Ag$^+$ and two molecules of ammonia.

among them, a vitamin, a piece of the protein hemoglobin, a molecule that looks like a sandwich.

There are, looking ahead, complexes with two and three ligands, and with four, five, six, and greater. Some are paramagnetic; others are diamagnetic. Particularly conspicuous are the deep colors of many complexes, covering the visible spectrum from red to green to violet. Some systems occur naturally; others we synthesize in the laboratory.

There are complexes, self-contained and complete, in which up to six ions or molecules are coordinated to the metal—small structures, typically, such as [Ag(CN)$_2$]$^-$, [PtCl$_4$]$^{2-}$, and [Fe(CN)$_6$]$^{4-}$. Other complexes embed themselves in the larger molecules essential to life, including chlorophyll (with nitrogen coordinated to magnesium), hemoglobin (built around iron), and the cytochromes (also around iron). And even with a limited selection, we can appreciate the importance of the transition metals and their compounds. These elements, present in the tiniest amounts, make an organism come to life. Without them, nothing remains but an inanimate mass of carbon, hydrogen, oxygen, and nitrogen.

Large or small, each coordination complex is a product of chemistry's defining interaction: the giving and taking of electrons. The metal, upon losing part of its electron density (as metals tend to do), becomes a Lewis acid, an acceptor of electrons. The ligands, with electrons to give, are Lewis bases able to supply what the metal lacks. There is a giving and there is a taking, and a coordination complex is the result.

Every complex has its own chemical identity. It holds together as a group, maintaining a regular structure and stoichiometry. We recognize that integrity—that oneness—by using square brackets in a formula such as $[Co(NH_3)_4Cl_2]^+$, not as a symbol for molarity but rather to suggest that this **complex ion** is a unified, intact species. The independent entity $[Co(NH_3)_4Cl_2]^+$ is a legitimate molecular ion, able to form neutral **coordination compounds** with a suitable number of oppositely charged *counterions* (as with Cl^-, to make $[Co(NH_3)_4Cl_2]Cl$).

A complex has, like any compound, a new and unique set of properties, different from its parts. It will be colorless or colored depending on what ligands are coordinated to what metal in what oxidation state. It will be magnetic or nonmagnetic depending, again, on what ligands are coordinated to what metal in what oxidation state. So, too, is the stability of a complex (free energy of formation) similarly determined, as is the *speed* of formation (the kinetic question).

We wonder, first, about what kind of structure we have. Is it a covalent molecule, like benzene? An ionic compound, like sodium chloride? Something in between? Or is it a loose confederation like the clusters in a liquid? The answers hinge mostly on the organization of the electrons, since all other properties necessarily follow from the bonding pattern (and, for that, read on).

Here is a sketch of what is to come. A complex will offer more than a casual, liquidlike association between ligands and metal, even if something less than the homogeneity of a fully delocalized molecule. Often a coordination complex stands closer to an ionic compound like NaCl, appearing as a collection of still-recognizable parts set inside a new system with new properties. Ligands complexed to a metal may retain a significant portion of their original electron density, and they can be liberated as intact ions or molecules. Just as NaCl will release Na^+ and Cl^- ions in aqueous solution, so can a complex of, say, $[Ni(CO)_4]$ be taken apart to yield the nickel atom and four molecules of carbon monoxide. Contrast that hierarchy of organization with the delocalized environment of a benzene molecule, where we no longer can distinguish six carbon *atoms* and six hydrogen *atoms*. In benzene there is only a shared community of nuclei and electrons, distributed uniformly and blended thoroughly into a molecule.

Structure and Coordination

First we need the facts. What *is*. We must ascertain, by experiment, the stoichiometry and geometry of a coordination complex. First will come the atoms, the oxidation states, the bond distances, the bond angles. Later will come the models, our attempts to explain why the structure is what it is.

Metal and ligands together—the complex proper, ML_n—define the so-called ***coordination sphere***, which need not be spherical at all. The geometry of the structure is determined, instead, by the number of bound ligands, the ***coordination number*** n. There are two-coordinate arrangements

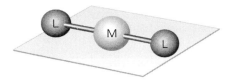

such as $[H_3N-Ag-NH_3]^+$, usually linear. There are three-coordinate complexes, which incorporate the metal either at the center of a triangle

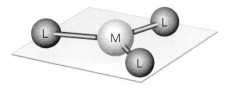

or at the apex of a pyramid:

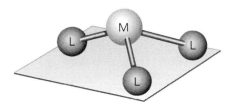

There are five-coordinate complexes, existing mostly in the form of trigonal bipyramids (below left) and square pyramids (below right),

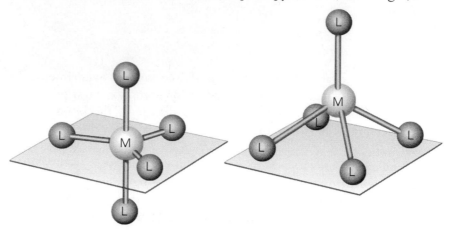

and there are even structures with seven, eight, nine ligands, and more. But most common are the six-coordinate octahedral complexes,

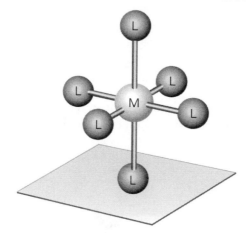

followed by the four-coordinate systems, which typically are either tetrahedral (left) or square planar (right):

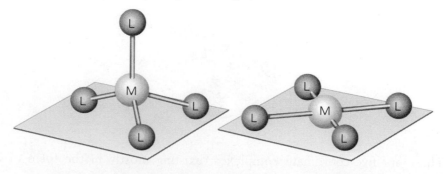

Every metal ion has its own preferred coordination numbers, and these numbers will vary according to the oxidation state and the particular kind of ligand. Some metals accept ligands in only one geometric configuration, whereas others can support two or more different arrangements—yet, even then, there are severe restrictions imposed on a complex. Stoichiometry, shape, bond distances, and bond angles remain fixed properties of the structure. Coordination is no haphazard event, likely to change from moment to moment.

The Fe^{2+} ion, for example, binds six CN^- ions to produce the complex $[Fe(CN)_6]^{4-}$, always with a consistent stoichiometry and structure. Expect not to see $[Fe(CN)_5]^{3-}$ today and $[Fe(CN)_7]^{5-}$ tomorrow, but $[Fe(CN)_6]^{4-}$ today and $[Fe(CN)_6]^{4-}$ every day. Nor will these six ligands settle into some arbitrary configuration, but always they will appear at the vertices of an octahedron, every time in the same way. There is a geometry to the process, a regularity. The predictable construction of the complex is symptomatic of directed chemical bonding, not the random jostling of six uncoordinated species around a common center.

The Metal

At the center is the metal, key to the complex. It is the metal that pulls the structure together, drawing inward the ligands with their available electrons. The geometry and stability of the complex depend on the strength of that attraction.

A strong positive charge on the metal will help. A strong positive charge, especially if compressed into a small volume, pulls in nearby negative particles most effectively. If we knew nothing else—if we were to suppose that all the interactions are purely electrostatic—then our first guess would be this: The smaller the metal ion and the larger its charge, the stronger is the attraction to the ligand.

We know better, of course, because we know (from Chapter 6) that the metal is not simply a pinprick of charge, but rather a community of negative particles about a positive nucleus: a quantum mechanical layering of electrons from the inside out. Whatever guise the coordination will take, it will have to adapt to the configuration of electrons already present.

Our informed instinct, then, is to take the existing electronic structure as a rough gauge of how the metal will respond to an electron-bearing ligand. How many valence electrons are already there? Where are they? Is there room for more?

Big questions, but we start small: with the oxidation number of the metal. Analysis begins with the assignment of oxidation numbers to all atoms in a complex, leaving the metal for last. Count the charges in $[Ag(NH_3)_2]^+$, for instance, and see why silver evidently must take the form Ag^+. To make that assessment, we note simply that (1) each NH_3, a neutral molecule, has a charge of 0, and (2) the net charge on the complex is +1. One Ag^+ ion together with two NH_3 molecules therefore will produce a monopositive complex ion $[Ag(NH_3)_2]^+$.

From Ag^+ we go on to examples of higher oxidation numbers, beginning with $[Pt(Cl)_4]^{2-}$. Here the same kind of arithmetic shows that the platinum exists in the dipositive state Pt^{2+}, also written as platinum(II). With chlorine assigned its usual oxidation number of -1, the various contributions add up to an overall charge of -2. The complex now supports four ligands, typical of *this* metal in *this* oxidation state surrounded by *these* Lewis bases. Other species will behave differently.

Other species do behave differently. Some elements will assume different oxidation states in different materials, acquiring thereby a special versatility. So it is with iron, which exists as Fe^{2+} in the neutral compound $[Fe(CN)_6]K_4$ yet as something else entirely—Fe^{3+}, iron(III)—in the deceptively similar compound $[Fe(CN)_6]K_3$. Any similarity exists only in the formula, though, and by that we shall not be misled. Each of these two complexes adopts a six-coordinate, octahedral arrangement, but the electronic distribution about the central metal changes dramatically with the total chemical environment. Iron(II) is not the same as iron(III).

A shift of just one electron is all that distinguishes iron(II) from iron(III), but what a difference that makes. It means the considerable difference between a species with five d electrons and a species with six d electrons. Think what one electron more or less might bring: a change in stability, a change in magnetism, a change in color, perhaps a change in reactivity.

We take off the electrons now, one at a time, and see how these oxidation states translate into patterns of filled and empty orbitals. Relative to a neutral atom, Fe^0, with configuration

$$[Ar]3d^64s^2$$

the iron is oxidized to $[Ar]3d^6$ in potassium hexacyanoferrate(II) and then to $[Ar]3d^5$ in potassium hexacyanoferrate(III). Recognize, in that sequence, the Aufbau model of Chapter 6 run in reverse, with the $4s$ electrons ionized before the $3d$. The result is a d^6 configuration for iron(II) and a d^5 configuration for iron(III).

Writing these ionic configurations, we are reminded again of the quantum mechanical basis for electronic receptivity: empty orbitals, homes for visiting electrons. There are limits, remember, on where an electron may go, on the energies it may acquire, on the angular momentum it may have, on the spin it may take. For an electron to be accepted in a chemical bond, it must have a properly prepared home; and that home, subject to all the warnings and fine print of Chapters 6 and 7, will be an empty orbital. An empty orbital: a volume of space, an energy, an angular momentum, a spin; a place to go.

And it will be the *d* functions in particular, unique to the transition metals, that give these complexes their special properties. Other metals also can attract ligands and form complexes, but only the transition atoms and ions have available *d* orbitals. For a partial listing of the manifold oxidation states possible, see Appendix C (Table C-10).

The Ligands: Electrons and Isomers

To *ligate* is to *bind*, and it is through the electrical interaction (as always) that ligands are bound into the complex. They enter the coordination sphere of the metal, offering some portion of their charge density to stabilize the electron-receptive species. Complementary to the empty orbitals of the metal, the ligands come with filled orbitals. They bring pairs of electrons poised to play the role of Lewis base.

The ammonia molecule, $:NH_3$, carrying a lone pair on the nitrogen, serves as a typical ligand:

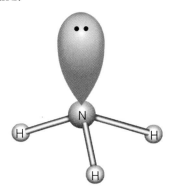

Housed in an orbital concentrated near the N nucleus, these two electrons provide a prominent, accessible source of negative charge for an electron-deficient metal ion such as Co^{3+}. Such a species will act as a **monodentate** ligand, having just this one "tooth" to bite its way into the metal's sphere of influence. Each ammonia molecule approaches with its electron-rich nitrogen directed inward, and the resulting Lewis acid–base interaction creates a complex:

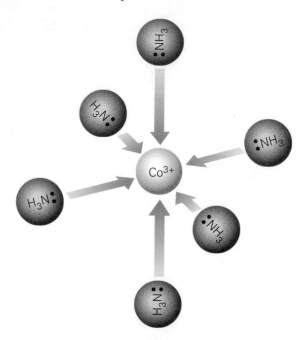

The structure thus begins with an outright gift of electrons, termed a ***dative*** (or ***coordinate covalent***) ***bond***. One party brings electrons, and the other party brings empty orbitals. After that initial donation, however, ligands and metal are bound into one complex, and "who gave what" is no longer important. More important is what the complex does with those electrons and orbitals, as we shall soon see.

Other common ligands (see also Table A-3 in Appendix A) include the halide ions (F^-, Cl^-, Br^-, I^-), the cyanide ion (CN^-), the hydroxide ion (OH^-), the water molecule (H_2O), and the carbon monoxide molecule (CO). Any of these anions or polar molecules will make a coordinate covalent bond by donating the equivalent of two electrons. Each offers some exposed portion of its electron cloud, a filled orbital that can be aimed at the metal.

Examine the possibilities. With chemically distinct ligands we should have, just as with organic molecules (Chapter 8), the potential to make ***isomers***: complexes with the same atoms in different spatial arrangements. Allow, say, three ammonia molecules and three chloride ions to compete for places around an octahedral cobalt(III) ion. Two possible ***geometric isomers*** arise, each a reconfiguration of the same ligands and the same bonds about the central atom:

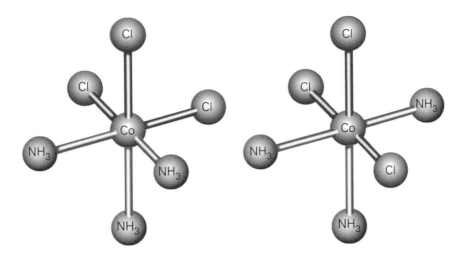

Or, given four ammonias and just two chlorides, we have another pair of geometric isomers,

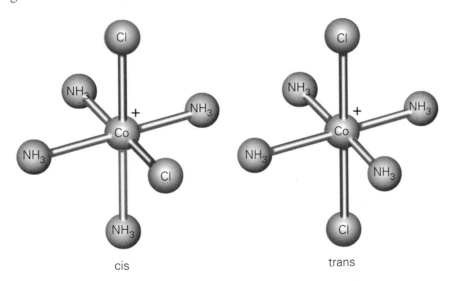

cis   trans

distinguished here by the positioning of the two chlorides around the metal: same side (cis), or across (trans). The choices are similar to those offered around a C=C double bond, as in

H   H   and   H   Cl
 \   /   \   /
  C=C   C=C
 /   \   /   \
Cl   Cl   Cl   H

cis   trans

and, not surprisingly, this same sort of arrangement is available in square planar *complexes* as well:

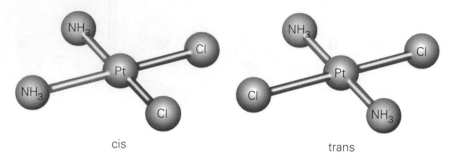

cis   trans

Small differences are important. The cis form of this complex, *cis-platin*, interferes with cell division and is used to fight cancer. The trans form, by contrast, has no known pharmaceutical use. It is a different compound, with its own arrangement of nuclei and electrons, as different from the cis isomer as one brother might be from another. So we should expect, aware by now of the subtlety of matter. Small differences amount to big changes.

Ask another question. What if just one ligand were to be replaced in a regular octahedral or square planar complex, with the remaining ligands all the same? How many isomers would there be then? . . . One; here there can be only one, no matter how we view the structure. Since no vertex looks different from any other, the odd ligand finds itself in an identical environment wherever it lands. This, for an octahedron:

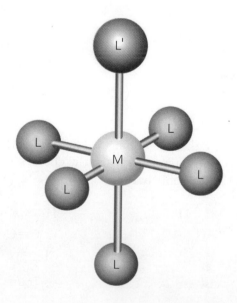

This, for a square:

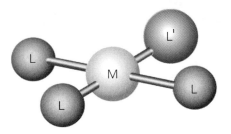

Symmetry demands no less.

But suppose that a ligand offers two or more possible points of attachment, having additional pockets of accessible electron density. A good example is the nitrite ion, NO_2^-, which can coordinate either through electrons on the nitrogen (as the nitro form, NO_2^-) or through electrons on one of the oxygens (as the nitrito form, ONO^-). It is an *ambidentate* ligand, equipped with two teeth but allowed only one bite; and from such ambiguity comes the opportunity for further *linkage isomerism*:

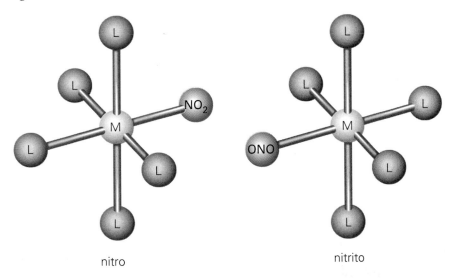

nitro   nitrito

Once again, too, small-scale differences can lead to large-scale consequences, typified by compounds such as $[Co(NH_3)_5NO_2]^{2+}$. The nitro form of the complex is yellow whereas the nitrito structure is red.

There is more. There are true *bidentate* ligands such as ethylenediamine (abbreviated en),

$$H_2\ddot{N}-CH_2-CH_2-\ddot{N}H_2$$

with two teeth and two bites; as well as wraparound, **polydentate** ligands such as the ethylenediaminetetraacetate anion (EDTA) and the *six* accessible sites it brings:

$$
\begin{array}{c}
:\!\ddot{O}\!: \qquad\qquad\qquad\qquad :\!\ddot{O}\!: \\
\| \qquad\qquad\qquad\qquad \| \\
\overline{\ddot{O}}\!-\!\overset{\displaystyle}{C}CH_2 \qquad\qquad CH_2\overset{\displaystyle}{C}\!-\!\ddot{\overline{O}} \\
\diagdown \ddot{N}CH_2CH_2\ddot{N} \diagup \\
\overline{\ddot{O}}\!-\!\overset{\displaystyle}{C}CH_2 \diagup \qquad\qquad \diagdown CH_2\overset{\displaystyle}{C}\!-\!\ddot{\overline{O}} \\
\| \qquad\qquad\qquad\qquad \| \\
:\!\ddot{O}\!: \qquad\qquad\qquad\qquad :\!\ddot{O}\!:
\end{array}
$$

The extra teeth allow a single ligand to tie up two or more sites on the metal simultaneously, and so (for the bidentate species) we observe fully coordinated octahedral complexes formed by just three ligands:

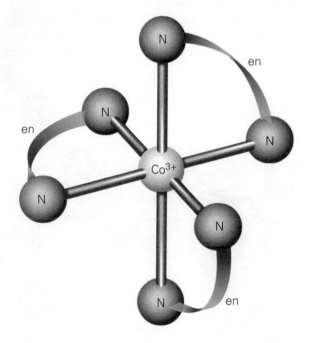

Imagining the cobalt ion in $[Co(en)_3]^{3+}$ to be seized by three pairs of pincers, we call the resulting structure a **chelate** complex from the Greek word for claw.

Particularly with chelates, there is the further possibility of creating *optical isomers*—nonsuperimposable mirror images, *enantiomers*, analogous to the chiral configurations of tetrahedral carbon (Chapter 8). One enantiomer rotates polarized light clockwise; the other, counterclockwise. Observe, then, the two mirror-image isomers produced from the octahedral complex of ethylenediamine shown just above:

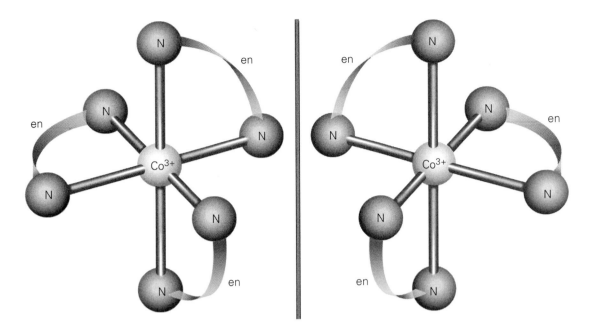

Chelates bring still more surprises. Notably voracious is the afore-mentioned EDTA, which all by itself can take a six-coordinate metal out of circulation. One EDTA anion, wrapped around the cation, fills the coordination sphere and inhibits further reaction by the metal:

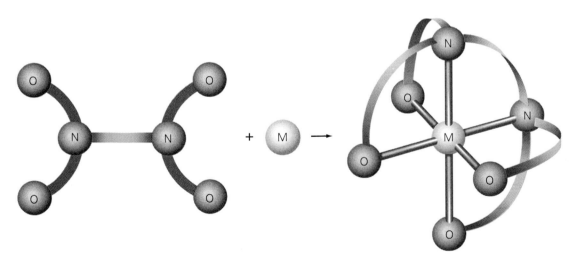

This chelating agent is famous to consumers as a common preservative in salad dressing, which it preserves and protects by capturing metallic species that would otherwise act to catalyze undesirable reactions.

Beyond even the triumph of salad dressing, though, chelating ligands offer a general way to sequester and render harmless certain species in solution. The effect, reminiscent of micellation (Chapter 15), finds medical application as a way to remove poisonous heavy metals from an organism. Polydentate ligands play a crucial biochemical role as well, serving to bind transition metals into critical sites on proteins, chloroplasts, vitamins, and other molecules. In many instances it is a *porphine-like* group

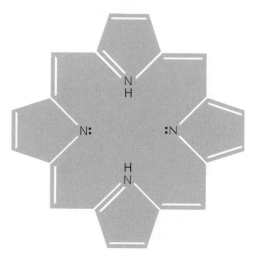

that does the work, presenting four nitrogen atoms to a metal. Two H^+ ions leave the nitrogens during coordination.

19-2. BONDING

No model of bonding is without its compromises, because for no molecule—not even tiny H_2—are we able to solve the Schrödinger equation exactly. Always we do the best we can, making whatever simplifications are reasonable for the problem at hand. The idea is never to insist upon some impossibly complete description of nature, but rather to develop working models consistent with the known laws. Compromises of that sort are well worth making.

Here is one worthy expedient: crystal field theory.

Crystal Field Theory

For transition-metal complexes, the simplest assumption is to pretend that the ligands are mere point charges or dipoles, lacking any shape and structure of their own. Positioned about the coordination sphere, these tightly drawn, pointlike ligands would not actually donate electrons to the metal. They would, instead, keep their electrons while subjecting the metal (and *its* electrons) to an electric field. Such is the view of **crystal field theory**, where we picture the complex almost as an ionic compound, a creation of the electrostatic force alone.

Admittedly it is a drastic step to ignore all electron density between ligand and metal, but it is just the kind of compromise we need to get started. Consider the consequences:

For the metal, the world has changed. No longer is one direction the same as any other; no longer are the surroundings spherically symmetric. Suddenly there are neighbors to contend with. There are neighbors bearing negative charges, neighbors that produce an electric field, neighbors that affect the energy of the metallic *d* electrons. No longer do all five *d* orbitals have the same energy, as once they did in isolation. There are markers in space.

Most important is the distribution of charge: where the ligands lie. A line is different from a triangle. A triangle is different from a tetrahedron. A tetrahedron is different from an octahedron. An octahedron is different from a sphere, and, above all, a sphere is different from everything else.

A sphere is, in some ways, like no distribution at all. A sphere does nothing to change an atom's perspective, nothing to distinguish one orbital from another. Spatial discrimination becomes possible only when the ligands occupy discrete, lumpy positions around the metal. No markers could exist if the ligands' electrons were spread uniformly over the coordination sphere, for then it would still be a world without direction (as pictured in the first two panels of Figure 19-3). Everywhere surrounding the metal would be the same negative charge: the same at the north pole, the same at the south pole, the same at the equator, everywhere the same.

Viewed by the metal, a spherical landscape of charge preserves the spherically symmetric world of an *atom*; and in that isolated state, recall, one *d* orbital is no different from another. Bathed in a hypothetical sphere of negative charge, all the *d* functions must feel the same extra repulsion; and all, in consequence, must rise to the same higher level in energy:

spherical field

— — — — —

no field

— — — — —

A sphere is a sphere. It has no direction.

Imagine, however, that the smooth surface of a sphere turns into the pointy vertices of an octahedron. Let there occur, by some means, a

(a)   (b)   (c)

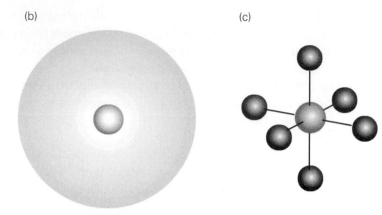

FIGURE 19-3. Space, direction, and symmetry. (a) Seen from the perspective of a free atom, the view is the same all around: empty. No marker exists to define a unique direction. (b) For an atom set within an unbroken sphere of negative charge, the view is still the same: indistinguishable left and right, up and down, here and there. The electrostatic energy is everywhere identical. (c) Finally, something different: When the spherical shell of charge coalesces into distinct points, space acquires a newfound directionality. The potential energy varies from site to site.

regrouping of the surrounding charge, so as to place one ligand along each of six perpendicular axes $(x, -x, y, -y, z, -z)$, equidistant from the center. Do that, and we have a coordination complex. The sphere of Figure 19-3(b) is now the octahedron of Figure 19-3(c).

Even then, much remains the same. Clumped around the metal—at the same radius as before—is the same total amount of negative charge, and hence the electrons on the metal will suffer the same increase in their total energy. The electrostatic energy depends only on the charge and the distance. Neither has changed.

And so, despite the rearrangement, the total energy is indeed conserved, although there is still one critical qualification. Here, in the nonspherical environment of Figure 19-4, different orbitals are affected differently. To any electron on the metal, clearly, the intrusion of negative charge brings higher energy, but in the octahedron there is a newfound sense of direction. Some orbitals go up more than others.

Since two of the d functions (d_{z^2} and $d_{x^2-y^2}$) have lobes pointing directly at the ligands, they move to higher energy than before. For them, the neighborhood has become more repulsive. The lobes of the d_{xy}, d_{xz}, and d_{yz} orbitals, by contrast, all fall *between* the ligands and therefore these three orbitals are lowered in energy relative to a spherical field. Compared with the d_{z^2} and $d_{x^2-y^2}$ sites, the new electrostatic neighborhood is less repulsive.

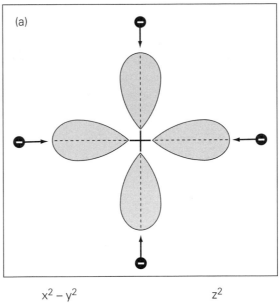

x²–y² z²

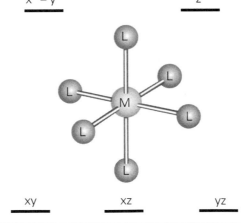

xy xz yz

FIGURE 19-4. No longer indistinguishable, the five d orbitals fare differently in an octahedral crystal field, shown here as a projection in two dimensions: (a) Lobes of the z^2 and $x^2 - y^2$ orbitals face the ligands directly. Electrons in these territories come closer to a ligand's negative charge and suffer increased repulsion as a result. They rise to higher energy, both by the same amount. (b) The xy, xz, and yz orbitals lie staggered between the ligands and move, as a group, to lower energy relative to z^2 and $x^2 - y^2$. See Figure 5-12 for the shapes and orientations of all five orbitals.

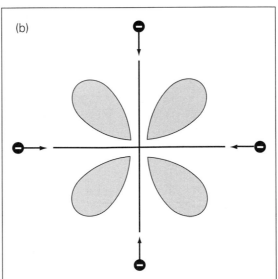

Thus develops a difference in energy between the group of three and the group of two, a splitting we shall call Δ. Disturbed by an octahedral field, the d orbitals move, first, to higher energy overall owing to the increased repulsion and, second, split into two sets called t_{2g} and e_g:

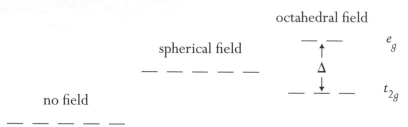

The former z^2 and $x^2 - y^2$ functions, now the upper levels, become the e_g orbitals, while the former xy, xz, and yz functions become the t_{2g}. Each of the three t_{2g} levels is depressed by 0.4Δ (for a total lowering of 1.2Δ), and each of the two e_g levels is elevated by 0.6Δ (for a total raising of 1.2Δ). The result is a wash: no change in total energy among all five levels, just as we predicted.

Yet for *one* electron it is not a wash. For one electron, the splitting offers an opportunity to acquire a lower energy. For one electron, the octahedral field allows the complex a certain stability. A d^1 configuration in a spherical environment becomes, instead, a t_{2g}^1 configuration in the octahedral complex, its energy lowered to -0.4Δ: the **crystal field stabilization energy (CFSE)**. With this one unpaired electron, moreover, the complex also becomes paramagnetic.

Nor is the octahedral field a wash for two electrons, which, according to the rules for filling orbitals (Chapter 6), file into the lower levels with parallel spins and enjoy a combined stabilization energy of -0.8Δ:

$$\left.\begin{array}{c} - \ - \end{array}\right\} 0.6\Delta \quad e_g$$
$$\underline{\uparrow \ \uparrow \ \_} \ \Big\} \ 0.4\Delta \quad t_{2g}$$

Again the ground state is paramagnetic, this time housing two unpaired electrons.

So continues the buildup for three electrons, where the t_{2g}^3 configuration

$$\left.\begin{array}{c} - \ - \end{array}\right\} 0.6\Delta \quad e_g$$
$$\underline{\uparrow \ \uparrow \ \uparrow} \ \Big\} \ 0.4\Delta \quad t_{2g}$$

yields a *CFSE* of -1.2Δ (and three unpaired electrons); but what after that? What comes after these first three configurations, which we have declared unambiguously to be t_{2g}^1, t_{2g}^2, and t_{2g}^3?

What happens next will be decided by the value of Δ, case by case, and herein lies the key to understanding magnetism in transition-metal complexes: The separation Δ, combined with the normal building-up sequence, determines the number of unpaired electrons and hence the extent of paramagnetism.

Ask, then, how many half-filled orbitals will remain in a formerly d^4 configuration, upcoming in our series. Will the configuration be **high spin** with four unpaired electrons,

$$\underline{\uparrow} \quad \underline{} \qquad e_g$$
$$\underline{\uparrow} \quad \underline{\uparrow} \quad \underline{\uparrow} \qquad t_{2g}$$

$$CFSE = -0.6\Delta \quad \text{(high spin)}$$

or will it be **low spin** with only two?

$$\underline{} \quad \underline{} \qquad e_g$$
$$\underline{\uparrow\downarrow} \quad \underline{\uparrow} \quad \underline{\uparrow} \qquad t_{2g}$$

$$CFSE = -1.6\Delta \quad \text{(low spin)}$$

The outcome rests with the size of the splitting, Δ, compared with the *pairing* energy required for double occupancy. To pair or not to pair is a recurring question during the building-up process, and the choice is always the same. Either: Pay the energetic premium needed to force two electrons into the same orbital, where they share the same space and consequently suffer greater repulsions. Or: Send the electron to a nominally higher orbital, where it can remain unpaired. The system, for the sake of greater stability, will accept a small increase in orbital energy to avoid an even larger increase in pairing energy.

A *low*-spin complex develops when groups of perturbed d orbitals are widely split, making it advantageous to take the penalty associated with a spin pair. A high-spin complex, by comparison, forms when the separation is small enough to encourage pair avoidance. Correspondingly more unpaired electrons are supported under the high-spin arrangement, and such a system displays a stronger paramagnetism. The various possibilities are summarized in Figure 19-5.

The splitting determines the *color* of a complex as well, since Δ typically falls in the visible spectrum. Photons with energy

$$E = h\nu \approx \Delta$$

are *absorbed* by the system, so that their energy goes to promote a d

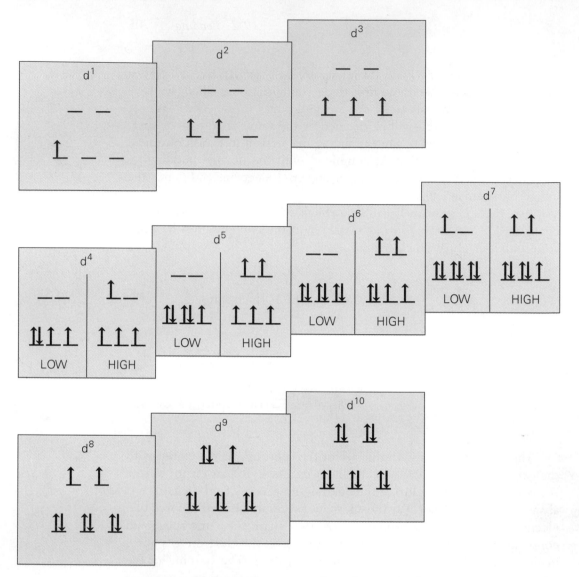

FIGURE 19-5. Spin and paramagnetism: electron configurations in an octahedral field, d^1 through d^{10}. High-spin and low-spin alternatives exist for systems with four, five, six, and seven d electrons.

electron to a higher level. Or, for some complexes, an excited electron might be transferred from a ligand to one of the unoccupied d orbitals on the metal, a phenomenon known as *charge transfer*. Whatever the mechanism, the color we see is determined by what photons go in and what photons come out.

When any portion of the original white light is captured by a complex, the particular frequency corresponding to the energy Δ fails to emerge. That one portion of the spectrum has been absorbed by an

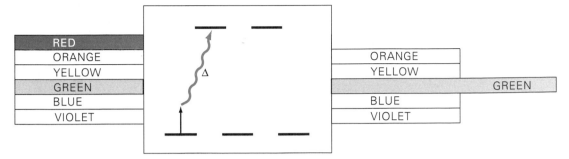

FIGURE 19-6. Color: what you get is what you see—after all the other frequencies are lost in transit. Photons able to excite a *d* electron are absorbed by the complex and fail to emerge. The wavelengths that pass through combine to produce the complementary color.

excitable *d* electron, and what we see (Figure 19-6) is the absorbed frequency's **complementary color**—its opposite number on the artist's color wheel. If red has been absorbed, then we see green transmitted. If blue has been absorbed, we see orange. If yellow has been absorbed, we see violet.

A simple idea, but it works. Crystal field theory, for all its drastic assumptions, accounts in many ways for the color and magnetism of the various complexes. Nor are the benefits limited just to octahedral complexes, because the same arguments apply equally well to other geometries. The predicted splittings and groupings are different in each instance, as they should be, but still the basic effect remains. The *d* orbitals, placed in a nonspherical electric field, respond to space differently. Some regions are more repulsive, and some are less repulsive. Some orbitals move higher, and some move lower.

Under *tetrahedral* coordination, for example, the (former) $d_{x^2-y^2}$ and d_{z^2} orbitals find themselves lower in energy than the d_{xy}, d_{xz}, and d_{yz}. To understand why, we inspect the position of the ligands relative to the five original *d* functions, just as we did for the octahedral structure. But now circumstances have changed; now, in the tetrahedron, the ligands approach from four vertices of a cube, as shown in Figure 19-7. The *xy*, *xz*, and *yz* orbitals are situated closer to those repulsive positions than either the $x^2 - y^2$ or z^2, and therefore the group of three rises higher than the group of two. The arrangement is reversed compared with an octahedral splitting, and (with two fewer ligands) it is a weaker interaction as well, marked by a smaller Δ. Such complexes tend to be high spin.

For all geometries, then, color and magnetism are determined by the pattern of perturbed *d* orbitals and by the extent of the crystal field

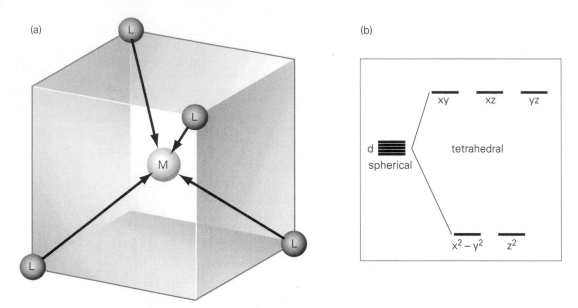

FIGURE 19-7. *d* electrons in a tetrahedral crystal field. (a) Four ligands approach the metal from opposite corners of a cube. The exact geometric relationships are not pictured, but the negative charges draw closer to the set of three *d* orbitals (xy, xz, yz) than to the set of two. (b) The xy, xz, and yz orbitals go up, and the z^2 and $x^2 - y^2$ orbitals go down. The tetrahedral splitting is smaller compared with an octahedral field, where contacts with the ligands are more direct and differences are more pronounced.

splitting—although not all ligands are equally effective. *Weak-field ligands* (such as the halide ions) produce smaller splittings and hence more high-spin complexes than *strong-field ligands* (such as carbon monoxide and the cyanide ion). A partial ranking of ligand strength, from weak to strong, is provided by the following **spectrochemical series**, a useful tabulation:

$$I^- < Br^- < Cl^- < F^- < OH^- < H_2O < NCS^- < NH_3 < en < CO, CN^-$$

weak field   intermediate field   strong field

The series enables us to guess whether a complex will be paramagnetic or diamagnetic, colored or colorless, closer to the red or closer to the blue.

We run into a conceptual stumbling block, though, because crystal field theory cannot explain why the series takes the form it does. For that, a broader model would recognize that ligands do have shape and structure; that they do have electrons to share with the metal; that they are species with orbitals of their own, not merely points in space.

Beyond Crystal Field Theory

Can we do better? We have developed, so far, this crude but surprisingly effective model based entirely on ionic interactions: crystal field theory. A simple electrostatic treatment, almost naive, it manages to predict the crucial splitting of the *d* orbitals; and this it does in the presumed absence of any covalent bonding. In return for that one assumption, we get a plausible explanation both for the color of a complex and for the number of unpaired electrons.

Now such benefits are considerable, but surely the underlying assumption is unrealistic; for surely a coordination complex is, despite many extraordinary features, something very ordinary. Isn't it just a molecule, a quantum mechanical gathering of electrons and nuclei? It is. It is a united structure, a *molecule* with a *molecular* wave function. Better we should consider, without prejudice, a set of molecular orbitals delocalized over the entire structure, similar to those of the benzene molecule (Chapter 7). Let the Schrödinger equation tell the story, unencumbered by any prior notion of "ligand" or "metal" or "ion" or "bond." Let the electrons do what they have to do.

Take that approach—form true molecular orbitals—and you will find, in the end, that many of the predictions of crystal field theory are vindicated in a more refined treatment. There is, most important, a splitting of the *d* orbitals, bringing with it the same implications for color and magnetism. Also predicted, under many circumstances, are distinctly ionlike concentrations of electron density local to the metal and to each ligand. Yet unlike a purely ionic description, the molecular orbitals show further how ligand and metal *mix* their orbitals to make the complex. Computation of electronic density in the molecular orbitals (Figure 19-8) shows the electrons clustered near the metal, near the ligands, and at points in between.

Where do we stand? At one extreme, there is a simple yet successful picture based on ionic bonding: the crystal field theory. At the other extreme lies the more elaborate, more accurate model of unrestricted covalent bonding: **ligand field theory**. We wonder, finally, if some intermediate description might exist as well, a covalent treatment less elaborate than full-dress molecular orbitals.

There is. The middle ground is occupied by the **valence bond approach** of Chapter 7, in which localized molecular orbitals are formed between pairs of nuclei. Each two-site orbital describes a single covalent bond, a sharing of two electrons by two nuclei.

Compared with crystal field theory, a valence bond description offers something new—not something necessarily better or worse, just different. To its credit is a more satisfying description of the ligand–metal bonding, bolstered by a generally correct accounting for the unpaired

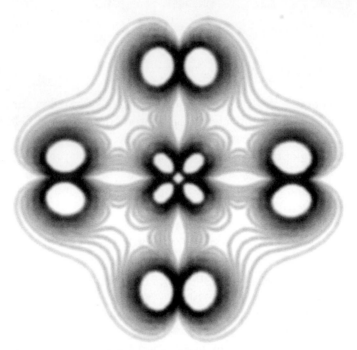

FIGURE 19-8. Contours of electron density for a selected molecular orbital in an octahedral complex, taken in cross section to show four of the six ligands. Recognize the signature shapes of the *d* and *p* orbitals, with the metal at the center and the ligands acting collectively. The extent of delocalization varies from system to system, but rarely does a ligand act entirely as a point charge. Some of its electrons are shared with the metal.

electrons. Valence bond theory provides, as portrayed in Figure 19-9, a more realistic distribution of electrons between donor and acceptor; and these covalent bonds, once conceived, allow us to picture the movement of charge density from one species to the other.

To its detriment, unfortunately, is an inability to explain the color of a complex, because the localized bonds include no provision for excited states. Nor can the valence bond model predict, in advance, how many electrons will be unpaired. Any judgment on paramagnetism can be made only after the fact, *after* the magnetic properties have already been measured. So, yes, there are good points and bad points, and there are compromises to be made with all the models; that much we have already conceded. The only sensible course is to know and beware, so as to exploit the strengths and avoid the weaknesses of each description.

Concerning valence bonds, our remaining inquiry will be brief and directed only at geometry, a clear strength of the model. We want to describe the suggested pair bonds between ligand and metal, and we must determine whether the shape of the complex is consistent with the properties of the orbitals. Our question becomes: What hybrid orbitals, localized on the metal, will conform to the observed geometries?

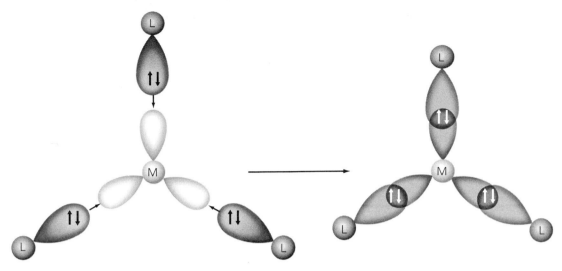

FIGURE 19-9. Ligand and metal: a valence bond approach. The sharing of electrons between donor and acceptor is modeled as a covalent bond localized between two sites.

Addressed here is the same problem already posed in Chapter 7, with an added twist: the metal's d functions. Into the hybridized mixture now will go these higher orbitals, which offer geometric possibilities beyond the previous assortment of lines (sp hybrids), triangles (sp^2), and tetrahedra (sp^3). To make, for example, the six equivalent orbitals needed to build an octahedral complex, the metal might combine two $3d$ functions, one $4s$ function, and the three $4p$ functions. Out will come six new combinations (d^2sp^3 hybrids) pointing toward the vertices of an octahedron, available to combine with orbitals on the ligands:

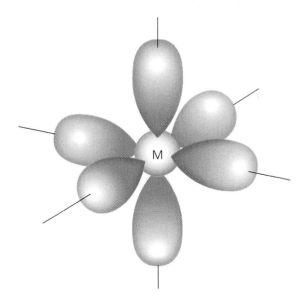

Combine they do (in the valence bond picture), and from this mixing we get six bonding orbitals—not the only six possible, but six acceptable combinations nonetheless. Each connects one d^2sp^3 hybrid on the metal with the appropriate orbital on a ligand. Each is filled with the two electrons contributed by the ligand, and each establishes a localized covalent bond between ligand and metal. Together, the six bonds produce the octahedral framework of the complex:

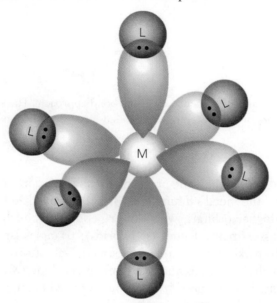

It could not have happened without the *d* orbitals. The strong directionality of these functions is needed to sustain the elaborate structure of the octahedron, which draws support primarily from the $d_{x^2-y^2}$ and d_{z^2} components. With far-reaching lobes pointing in the six directions, these two orbitals provide the right-angled scaffolding needed to build an octahedron. A different geometry would require a different set of hybrids— four dsp^2 functions, say, to make the square planar arrangement shown below:

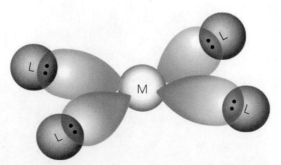

So far, good; the valence bond model is consistent with the shape of the complex. But still there remains the question of paramagnetism, since no unpaired electrons can be accommodated in the pair bonds assigned to the framework. Each of those orbitals is filled. Each of those orbitals contains the two electrons originally housed on one of the ligands. Unpaired electrons can reside, however, in the *unhybridized d* orbitals remaining on the metal, in just those orbitals that do not form bonds with the ligands. There, according to both valence bond and crystal field theory, is where the magnetic properties must originate.

19-3. THERMODYNAMICS

Going beyond the bare facts of existence, we turn next to the quality of that existence: the strength and stability of a complex. Our focus shifts from design to fate, from "how is it put together?" to "how likely is it to survive?" We look ahead to the onset of equilibrium, where enthalpy and entropy will determine which structures live and which structures die.

Consider some of the contenders.

Water: Ligand and Solvent

Water, once again, plays host. For many of our coordination complexes, aqueous solution provides the environment in which structures come together, break up, act as catalysts, or change into other forms themselves. Whatever reactions may unfold, they unfold in the presence of water.

Simultaneously acid and base, the H_2O molecule is both an aggressive solvent and a potential reactant. We already know, from Chapters 15 and 16, what it can do. The molecule, small and polar, is able to swarm around cations, anions, and polar molecules of all sizes; and, by its intrusive presence, this strongly polar entity will influence the behavior of nonpolar neighbors as well. Few species remain unaffected by water.

Certainly the transition metals are not immune, because the H_2O molecule is a potent Lewis base and intermediate-field ligand in its own right. More than just a solvent, water will compete with other ligands to form legitimate *complexes*—not solvent clusters of variable size (like a hydrated Li^+ cation), but regular coordination complexes with fixed stoichiometry.

Six H_2O molecules bind with Fe^{3+}, for example, to produce the acid-soluble ion $[Fe(H_2O)_6]^{3+}$, which undergoes further hydrolysis to yield $[Fe(H_2O)_5(OH)]^{2+}$, an iron(II) complex:

$$[Fe(H_2O)_6]^{3+} + H_2O \rightleftharpoons [Fe(H_2O)_5(OH)]^{2+} + H_3O^+$$

Yet water's business is not finished even then, for now there is a new target: the exterior of the complex itself. To other water molecules, a complex such as $[Fe(H_2O)_5(OH)]^{2+}$ is simply a new cation to be hydrated. The bonded complex, once formed, becomes just another electrical body around which the solvent molecules can congregate.

So they do, forming hydrated clusters of variable size around the already coordinated ion. The complex ion, a structure with predictable stoichiometry, thus is surrounded by *other* water molecules not bound into any fixed pattern. This secondary association produces a solvation of the ordinary kind: the looser, shifting kind of association characteristic of noncovalent interactions in a liquid.

We recognize this additional aggregation by writing a revised equation,

$$[Fe(H_2O)_6]^{3+}(aq) + H_2O(\ell) \rightleftharpoons [Fe(H_2O)_5(OH)]^{2+}(aq) + H_3O^+(aq)$$

in which the symbol *aq* has its usual meaning. Each species so indicated is surrounded by some unspecified number of H_2O molecules, held there by the usual forms of noncovalent interaction.

Aqueous Equilibria

Much is going on. Water molecules are competing with other ligands for a share of the transition metal. Solvent molecules are swarming around each complex as well, piling up layer upon hydrated layer *outside* the primary coordination sphere. Complexes stand in various stages of completion. Some have all their ligands attached; others are yet unfinished. Ligands are coming and ligands are going. Such is the scene as the system comes to equilibrium.

It is a complicated, busy equilibrium in which all these different processes must be balanced simultaneously. Suppose, for instance, we consider the formation of $[Ag(NH_3)_2]^+$ in aqueous solution. There will be—both at once—equilibria involving one $Ag^+(aq)$, *one* $NH_3(aq)$, and the half-formed complex $[Ag(NH_3)]^+(aq)$; plus equilibria involving $[Ag(NH_3)]^+(aq)$, $NH_3(aq)$, and the full complex $[Ag(NH_3)_2]^+(aq)$. One process is connected with the other, each determined by its own equilibrium constant.

Here, for simplicity, we shall skip to the overall equilibrium constant

$$K_f = \frac{[Ag(NH_3)_2^+]}{[Ag^+][NH_3]^2}$$

coming from the net equation

$$Ag^+(aq) + 2NH_3(aq) \rightleftharpoons [Ag(NH_3)_2]^+(aq)$$

This ***formation constant***, K_f, summarizes the thermodynamics in one number. It encapsulates, for all the dissolved species, the change in both enthalpy and entropy. Like all equilibrium constants, K_f parallels the change in free energy; and for $[Ag(NH_3)_2]^+(aq)$, specifically, the value tells us where the complex stands relative to the $Ag^+(aq)$ and $NH_3(aq)$ species alone. Everything is included: metal, ligands, complex, solvent.

K_f for $[Ag(NH_3)_2]^+$ is a big number (larger than 10^7), as are many of the other formation constants given for reference in Appendix C (Table C-22). Such complexes offer formidable thermodynamic stability. The formation reactions, once started, continue until scarcely any uncombined reactants remain.

Note the echo of acid–base phenomena in those words. The nearly 100% *association* of the $[Ag(NH_3)_2]^+$ complex is analogous to the nearly 100% *dissociation* of a strong acid or base. Merely the names and symbols have changed, not the ideas. Similar to acid–base phenomena, too, is the way in which a strongly associated complex will pull an otherwise insoluble species into solution—just as a strong base will completely neutralize a weak acid.

Observe the effect of ammonia on the solubility of silver chloride. With K_{sp} for the reaction

$$AgCl(s) \rightleftharpoons Ag^+(aq) + Cl^-(aq)$$

barely above 10^{-10}, we would ordinarily expect a scant concentration of dissolved AgCl; but now, in the presence of NH_3, new competition arises for the Ag^+ ions. Impelled by an enormous equilibrium constant, any dissolved silver ion will be snatched by two ammonia molecules to form the complex $[Ag(NH_3)_2]^+$:

$$Ag^+(aq) + 2NH_3(aq) \rightleftharpoons [Ag(NH_3)_2]^+(aq) \qquad (K_f = 1.7 \times 10^7)$$

For every silver ion picked off, a new one is released from the solid; and so, one by one, the combined reaction

$$AgCl(s) + 2NH_3(aq) \rightleftharpoons [Ag(NH_3)_2]^+(aq) + Cl^-(aq)$$

is driven to the right. When equilibrium sets in, the solution contains a far greater amount of dissolved AgCl than would be possible in pure water.

The effect is another demonstration of Le Châtelier's principle, no different from the way a strong base (NaOH) pulls apart a weak acid (CH_3COOH):

$$CH_3COOH(aq) + NaOH(aq) \rightleftharpoons Na^+(aq) + CH_3COO^-(aq) + H_2O(\ell)$$

For every ionized H^+ picked off by the ubiquitous OH^-, a new one will be released by CH_3COOH. The weak dissociation of acetic acid, spurred by the unremitting removal of H^+ (to form H_2O), is shifted to the right by the stronger base.

Chelation

Although coordination complexes are often favored by large formation constants, there is one special structure—the chelate—that stands above the rest. A chelated complex typically enjoys remarkable thermodynamic stability compared with its monodentate alternatives. The differences can be impressive.

Formation of the monodentate $[Ni(NH_3)_6]^{2+}$ complex, for example, proceeds with an equilibrium constant on the order of 10^8, a number large by any standard. Yet that value pales beside the number associated with the chelated complex $[Ni(en)_3]^{2+}$, for which K_f is roughly 10^{18}. We go merely from the structure

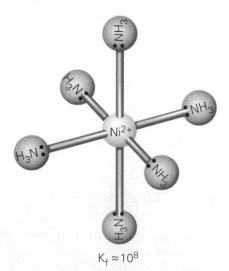

$$K_f \approx 10^8$$

to the not-so-different structure

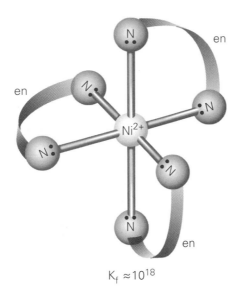

$$K_f \approx 10^{18}$$

but still we enhance the equilibrium constant by nearly a factor of 10^{10}. Why?

It is not a question of enthalpy, not a question of stronger bonds between ligand and metal. The bonds seem hardly to change. In the one complex, six ammonia molecules are attached via six $:N$ sites. In the other, three ethylenediamine molecules are similarly attached via six $:N$ sites. There is, for each, the same number and kind of bonds between $:N$ and Ni^{2+}. The differences in enthalpy are minor.

The chelation effect derives, instead, from a favorable change in entropy—an *increase* in entropy—registered despite the attachment of the once-free ligands to the metal. True, the actual coming-together of ligands and metal does bring order to the system, not disorder. True, but the final attachment of ligands is only part of the whole change. There are goings as well as comings.

Since the dissolved complexes are not fashioned from naked metal ions, we have always to consider the water molecules. They are ligands by default. In aqueous solution, an incoming ligand first must displace one of these ever-present water molecules; then, only then, does it gain a place in the coordination sphere. A new ligand inevitably loses its freedom, to be sure, but an old ligand (H_2O) is released simultaneously into solution. The net change in entropy depends on both sets of events.

Compare the alternatives. To form the chelated complex, three bidentate ligands (the ethylenediamine molecules) displace six monodentate water molecules. Three molecules are captured while six are released, and thus the solution becomes more disordered overall. To form the monodentate complex, however, six ammonia molecules displace six water molecules. The change in entropy is smaller than before, and the drop in free energy is reduced correspondingly.

19-4. KINETICS

Coordination complexes live and die by the exchange of ligands. They substitute one ligand for another, stumbling blindly down the mountain of free energy; and they stumble their way, eventually, into the most stable structure, arriving there purely by happenstance. The larger the drop in free energy, the more products exist at equilibrium. Such is the way of thermodynamics.

Complexes have ample opportunity to stumble toward greater stability. To exist, a complex must defend itself against any other potential ligands simultaneously present—beginning with the water molecule, a constant competitor. Ligands come and ligands go. Complexes come and go.

According to thermodynamics, a structure like $[Co(NH_3)_6]^{3+}$ is destined to *go*. One mole of this ammonia-based complex stands hundreds of kilojoules higher in free energy than its hydrated counterpart, $[Co(H_2O)_6]^{3+}$, which ought to be obtainable by reaction in acid:

$$[Co(NH_3)_6]^{3+} + 6\,H_3O^+ \rightleftharpoons [Co(H_2O)_6]^{3+} + 6\,NH_4^+$$

Corresponding to the large ΔG° is an equilibrium constant variously reported as between 10^{20} and 10^{64}—at either extreme, a number little different from infinity. The thermodynamic drive forward is overwhelming.

Nevertheless, the reactants are not overwhelmed. Exposed to acid, the ammonia complex will defend itself for weeks and months against the water molecules. This $[Co(NH_3)_6]^{3+}$ ion, a high-energy structure, persists despite the more stable alternative. It is an **inert** complex, slow to react.

The reaction that does not happen (or, rather, occurs so very slowly) illustrates the difference between thermodynamics and kinetics, our principal result from Chapter 18. One drive has nothing to do with the other. The thermodynamic tendency to form *strong* products is unrelated to the kinetic tendency to form *any* products as fast as possible. "Good" and "cheap" need not be incompatible requirements, but neither are they linked in any way.

So there sits the $[Co(NH_3)_6]^{3+}$ ion, trapped, unable temporarily to realize its thermodynamic potential. Disadvantaged thermodynamically but also kinetically inert, the complex does not teeter precipitously at the top of a mountain of free energy. It rests, instead, in a high-altitude valley, confined there by a hill of activation energy. To roll down the mountain, the complex first must be jarred up and over the hill of activation energy. It must pass through a transition state. It must go up before it can go down.

For the Co^{3+} ion, a d^6 species, the price of activation becomes too high. Octahedral complexes of cobalt(III), including the $[Co(NH_3)_6]^{3+}$ structure, resist change because they already enjoy the special stability of a filled set of t_{2g} orbitals:

$$
\left. \begin{array}{c} \text{—} \quad \text{—} \end{array} \right\} 0.6\Delta \quad {}^{e_g}
$$

$$
\cdots\cdots\cdots\cdots\cdots
$$

$$
\left. \begin{array}{c} \uparrow\downarrow \;\; \uparrow\downarrow \;\; \uparrow\downarrow \end{array} \right\} 0.4\Delta \quad {}_{t_{2g}} \qquad CFSE = -2.4\Delta
$$

Here, with the *CFSE* optimized at -2.4Δ, any disturbance of the environment will add significantly to the energy. To tamper with this t_{2g}^6 configuration or crystal field in any way—no matter what the eventual outcome might be—is therefore to begin climbing a large hill of energy, a kinetic difficulty. But, difficult or not, such tampering is unavoidable when ligands try to come and go. Departing groups must take away their stabilizing negative charges; incoming groups must bring them.

There arises, then, a temporary conflict between thermodynamics and kinetics, with the transition state caught unfavorably in the middle. Few transition states are built; and so, for the moment, the reaction is blocked by this steep kinetic barrier. The d^6 complex is inert, its fate to be decided eventually by thermodynamics. When that might happen, we do not know.

So, too, do octahedral d^3 complexes (Cr^{3+}) similarly resist alteration in the crystal field of their *half*-filled t_{2g}^3 levels, again a preferred configuration:

$$
\left. \begin{array}{c} \text{—} \quad \text{—} \end{array} \right\} 0.6\Delta \quad {}^{e_g}
$$

$$
\cdots\cdots\cdots\cdots\cdots
$$

$$
\left. \begin{array}{c} \uparrow \;\; \uparrow \;\; \uparrow \end{array} \right\} 0.4\Delta \quad {}_{t_{2g}} \qquad CFSE = -1.2\Delta
$$

Octahedral structures with other populations, however, lack the high activation barriers posed by t_{2g}^3 and t_{2g}^6 distributions, and most of these *labile* complexes do exchange their ligands rapidly. Indeed some react so readily that they will trade good for bad (thermodynamically stable for thermodynamically unstable) simply because it is *easy* to do so: the activation barrier is low.

And some labile complexes just as readily exchange bad for good, easily overcoming a small barrier on the way to more stable products. One drive has nothing to do with the other.

REVIEW AND GUIDE TO PROBLEMS

Not knowing any better, one might suppose that all atoms are alike—that a copper atom, say, is just a heavier version of a titanium atom, which itself is simply an overgrown form of carbon or boron or lithium or hydrogen. Why not? The electrons are the same. The interactions are the same. The architecture is the same.

Look at it this way: Banish everything that allows an atom to distinguish up from down. Remove all magnetic fields, all electric fields, all other atoms, all other molecules, and let the atom stand alone. Take away all markers in space, and see what an atom really is: a sphere; a sphere with no ups, no downs, no hooks, no handles. Hydrogen, carbon, copper? Except for size and weight, all the same.

What a surprise, then, to discover such a thing as *valence*; to realize that, despite the symmetry common to all, each element does have its own quirks and special affinities. What a surprise to learn that hydrogen forms only one bond; carbon, four; copper, as many as six. What a surprise to find, buried within the roundness of an atom, a hidden capacity to reach out in certain preferred directions—to make bonds, molecules, and complexes, and to do so in a way different for every element.

Be surprised, yes; mystified, no. Know the reasons why. Know that the electrons occupy a hierarchy of quantized states. Know that the entropic dispersal of energy favors some structures and excludes others. Know that a reaction can run slow or fast, climbing over an activation barrier the way a driver eases over a speed bump. Know what is always true. Know what is sometimes true. Know the elements and what they do.

What we attempted earlier for carbon (in Chapter 8), we do now for the transition metals and their complexes, finally able to enjoy a connected grasp of bonding, thermodynamics, and kinetics.

IN BRIEF: COORDINATION COMPLEXES

1. ELECTRONS AND ENVIRONMENT. Take an electron from copper, replace it with an electron from carbon, and ask: Where is the difference? Answer: no difference. We see the same atom of copper as before. Electrons are interchangeable.

But atomic *environments* are not. The charge of the nucleus, the number and distribution of the electrons, the balance of attraction and repulsion, penetration and shielding—these are the features that, taken together, make copper different from carbon; these are the features that turn a generic atom into a distinctive chemical element.

For carbon and similar ***representative elements***, the environment is sufficient to support only s and p electrons. In copper and the other ***transition metals***, however, the atom has room for up to 10 d electrons. It is a world with new possibilities.

2. d ELECTRONS. A free electron says to a transition-metal atom or ion, "If I were to join your group, where would I go and how much energy would I have?" The nucleus and bound electrons reply, "Pick any of the d_{xy}, d_{yz}, d_{xz}, d_{z^2}, and $d_{x^2-y^2}$ orbitals you like. The various territories contain two angular nodes and (usually) four main lobes pointing away from the nucleus, but all five orbitals have the same energy and are fully equivalent in the isolated atom. So long as we, the atom, stay away from other atoms and electromagnetic fields, you may occupy any of these regions subject to the usual rules. The labels xy, yz, xz, z^2, and $x^2 - y^2$ are meaningless out here in the open.

"Expect changes, though, whenever we join together with additional atoms or groups of atoms. Some of you d electrons will find yourselves closer to other electrons and suffer increased repulsions as a result. Others will feel it less. In a coordination complex, for instance, your location will determine your energy, angular momentum, and other properties."

3. COORDINATION COMPLEXES. Bring together a transition metal and a ***ligand*** (an atom or group of atoms with electrons to give). The ligand, a Lewis base, has the electrons. The metal, a Lewis acid, has a place to put them. Base meets acid; giver meets taker; electrons go from ligand to metal, and a ***coordinate covalent bond*** (or ***dative bond***) is thus formed. Allow two or more ligands to attach themselves by these Lewis acid–base interactions, and a ***coordination complex*** takes shape.

The shape depends on the number and kind of ligands. When six ligands fill the ***coordination sphere*** around the central metal, they stake out the six corners of an octahedron. With four ligands, the complex usually is tetrahedral or square planar. Five, a trigonal bipyramid or a square pyramid. Three, a triangle or pyramid. Two, a straight line.

Results vary. The ***coordination number*** differs for different metals and different ligands; and, adding further richness, the ligands will shuffle themselves around to create ***isomers***: geometric isomers (such as cis-trans), linkage isomers (where the same ligand attaches itself in different ways), optical isomers (nonsuperimposable mirror images). Some groups, called ***monodentate ligands***, offer only one pair of electrons. Others, the ***polydentate ligands***, have more than one lone pair to give (two in a ***bidentate ligand***; more in general).

4. CRYSTAL FIELD THEORY. Consider a proposition, simple but not unreasonable: Suppose that the ligands, Lewis bases, keep their electrons intact as they bathe the metal atom with an electric field. The complex then takes on the character of an ionic compound, bound together

not by covalent bonds but by electrostatic interactions between ligands and metal. The partners say: What's mine is mine; what's yours is yours. Whatever electrons the electron-rich ligands had before, they have still. Whatever electrons the electron-poor metal had before, it has now; but now, sitting in this "crystal field" imposed by the ligands, the metal is different. No longer is it an atom, alone in a directionless space. Now it is part of a molecule, and now its five *d* orbitals are no longer equivalent.

For *d* electrons positioned closest to a ligand, energy goes up the most. Nearby electrons bring more repulsion. For *d* electrons farther from a ligand, the effect is less. Some orbitals rise above the mean; some orbitals fall below; the average energy remains the same. Thus **crystal field theory**.

Whereas in a free atom all five *d* orbitals share the same energy, in a complex the levels break apart into different classes. Under octahedral geometry, for example, two of the *d* orbitals move up and three move down (by $+0.6\Delta$ and -0.4Δ, respectively, where the *crystal field splitting energy*, Δ, is an attribute of the individual complex):

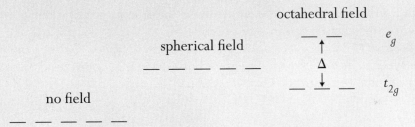

The former z^2 and $x^2 - y^2$ orbitals, pointing directly at the ligands, become the e_g orbitals in the crystal field. The former xy, xz, and yz orbitals, staggered between the ligands, fall to lower energy and become the t_{2g} orbitals.

Optical and magnetic properties change as well. If Δ falls within the range of visible light, the complex is colored. If Δ is such that some electrons stay unpaired, the complex is paramagnetic. If Δ exceeds the spin-pairing energy, the complex is **low spin** (the electrons pair up in the lower levels). If not, the complex is **high spin** (the unpaired electrons spread out over the split levels).

Certain attached groups (**weak-field ligands**) produce small splittings; others (**strong-field ligands**) produce large splittings. The pecking order is summarized in the **spectrochemical series**:

$$I^- < Br^- < Cl^- < F^- < OH^- < H_2O < NCS^- < NH_3 < en < CO, CN^-$$

weak field   intermediate field   strong field

5. COVENANT MODELS OF BONDING. *Valence bond theory*: Here we concede that electrons are indeed *shared* between ligands and metal, but we imagine that they hook up through localized pair bonds only. What develops is a picture of hybrid orbitals, put together as various mixtures of s, p, and d functions. Common examples include d^2sp^3 (octahedral) and dsp^2 (square planar) combinations.

Ligand field theory: The motto is, let the electrons fall where they may. We treat the complex as a whole, allowing the Schrödinger equation to set molecular orbitals for the entire structure; and only then, with minds open, do we ask: Are electrons concentrated mostly between the central metal and each ligand separately? Yes? Then the bonds are localized. Does the electron distribution resemble the hybrid orbitals of valence bond theory? Does it make sense to speak of a dsp^2 or d^2sp^3 or sp^3 configuration? Or, if not, are the electrons shared between ligand and ligand and thereby delocalized throughout the complex?

6. EQUILIBRIUM. Water, the natural medium for most complexes, plays host to a richly textured equilibrium involving metal, ligands, and the complex in various stages of completion. The water molecule itself, an aggressive Lewis base, competes for the transition metal along with whatever other ligands may be present.

Every complex has its own metal, its own ligands, its own bonds, and hence its own thermodynamics; but usually the long-term outlook is favorable. *Formation constants*, defined as

$$K_f = \frac{[ML_n]}{[M][L]^n}$$

for the reaction

$$M + nL \rightleftharpoons ML_n,$$

tend to be large. Often it is the change in entropy, more than the change in enthalpy, that decides the outcome—particularly for the **chelation effect**, where the binding of a polydentate ligand simultaneously displaces an even greater number of H_2O molecules from the metal. The aqueous host becomes more disordered, and the free energy falls. Thus encouraged, the complex has a reason to be.

7. KINETICS. Some complexes are *inert*, slow to react. Others are *labile*, able to exchange ligands swiftly and easily. The difference lies in the activation energies that come between reactant and product, determined mostly by the structure of the unstable transition state (not the stable complex on either side of the reaction arrow). Look for such features as filled or half-filled configurations, for instance, to identify exceptionally stable arrangements.

SAMPLE PROBLEMS

We begin with an architectural issue (isomerism) and then move on to questions of bonding.

EXAMPLE 19-1. Geometric Isomers

PROBLEM: Shown below is the trans isomer of the complex ion $[Co(NH_3)_4Cl_2]^+$:

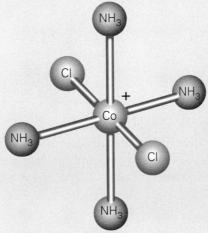

Draw the cis isomer.

SOLUTION: The chlorines in the trans isomer lie *across* the central cobalt atom, situated on opposite sides of a plane containing the four ammonias. The Cl—Co—Cl bond angle is 180°, a straight line.

In the cis isomer, the angle is 90°. The two Cl atoms lie on the *same side* of the cobalt:

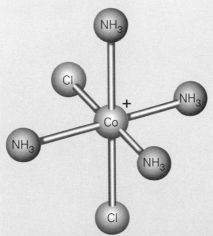

QUESTION: Are any other structural isomers possible for $[Co(NH_3)_4Cl_2]^+$? How, for example, should we classify the configuration shown in this drawing?

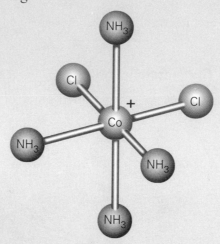

ANSWER: Look closely: Except for a 90° rotation about one of the Co—Cl bonds, the structure is identical to the cis isomer we just drew. No unique cis-trans isomers exist beyond the two shown above.

Might there be an *optical* isomer? See the next example.

EXAMPLE 19-2. Optical Isomers

PROBLEM: Does either the cis or trans form of $[Co(NH_3)_4Cl_2]^+$ have an optical isomer?

SOLUTION: To be ***chiral*** is to possess the property of handedness, meaning to have a nonsuperimposable mirror image (like left and right hands). For the two octahedra shown in Example 19-1, we therefore draw the corresponding mirror images and compare them with the original structures.

First, the trans isomer:

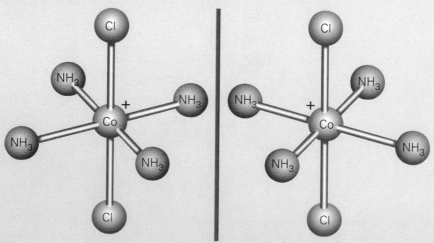

The object and its image are clearly identical, so *trans*-$[Co(NH_3)_4Cl_2]^+$ is not optically active.

And neither is the cis isomer:

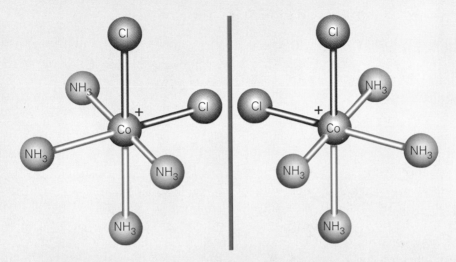

A simple rotation about the vertical axis brings the two structures into alignment.

No optical isomers are possible for $[Co(NH_3)_4Cl_2]^+$.

QUESTION: What change of environment, if any, might make such a complex optically active?

ANSWER: The lower symmetry afforded by a bidentate ligand. See what happens, in particular, to the cis structure when we use two bidentate ethylenediamine ligands in place of the four monodentate ammonia molecules:

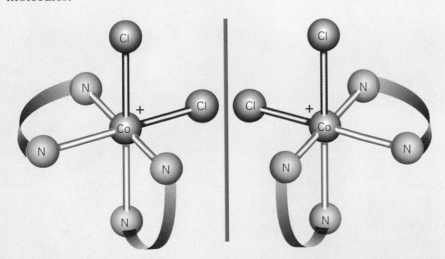

Try as we might, no single rotation or series of rotations can superimpose these two mirror images. The complex is optically active.

EXAMPLE 19-3. Oxidation States

PROBLEM: Determine the oxidation number for the central atom in each complex: (a) $[Co(NH_3)_4Cl_2]^+$. (b) $[Co(NH_3)_6]^{3+}$. (c) $[Co(OH)_4]^{2-}$. (d) $[Cr(H_2O)_6]^{2+}$. (e) $[Cd(NH_3)_4](NO_3)_2$.

SOLUTION: We sum the oxidation numbers for the ligands and compare the total with the net charge on the complex, taking the difference as the metal's oxidation state. The usual rules from Chapter 3 apply.

(a) *$[Co(NH_3)_4Cl_2]^+$*. Each Cl contributes -1 to make a total of -2, whereas each NH_3 molecule carries an oxidation number of 0. The cobalt(III) ion, Co^{3+}, brings the structure's net charge to $+1$.

(b) *$[Co(NH_3)_6]^{3+}$*. Another example of cobalt(III). The cobalt atom, oxidized to Co^{3+}, bears the full charge of the complex. The ammonia molecules contribute zero.

(c) *$[Co(OH)_4]^{2-}$*. Each of the four hydroxide ions, OH^-, has a charge of -1. Cobalt exists as cobalt(II), Co^{2+}, so that the net ionic charge is -2:

$$Co^{2+} + 4OH^- \longrightarrow Co(OH)_4^{2-}$$

(d) *$[Cr(H_2O)_6]^{2+}$*. Chromium, oxidized to Cr^{2+}, is responsible for the net charge of $+2$ in the complex as a whole. Each water molecule is assigned an oxidation state of 0.

(e) *$[Cd(NH_3)_4](NO_3)_2$*. The four-coordinate complex ion (charge $+2$) is counterbalanced by two nitrate ions (NO_3^-):

$$[Cd(NH_3)_4]^{2+} + 2NO_3^- \longrightarrow [Cd(NH_3)_4](NO_3)_2$$

Cadmium inside the complex interacts with the four ammonia ligands as Cd^{2+}.

EXAMPLE 19-4. Counting *d* Electrons

PROBLEM: How many *d* electrons are associated with each of the following oxidation states? (a) Co^{2+}. (b) Co^{3+}. (c) Cr^{3+}. (d) Cu^+. (e) Ti^{2+}.

SOLUTION: The standard building-up sequence (Chapter 6) calls for atoms to lose electrons first from the $4s$ orbital and then from the $3d$. But note, by way of exception, that a half-shell $3d^5 4s^1$ configuration such as

↑_ ↑_ ↑_ ↑_ ↑_ ↑_

    3d   4s

incurs less electron–electron repulsion and thus is more stable than a $3d^4 4s^2$ configuration housing the same six electrons:

↑_ ↑_ ↑_ ↑_ __ ↑↓

    3d   4s

Similarly, a $3d^{10} 4s^1$ configuration (stabilized by a filled d subshell) is favored over $3d^9 4s^2$.

(a) Co^{2+}. Neutral cobalt (atomic number = 27) supports nine electrons past its 18-electron, argonlike core configuration. Losing two $4s$ electrons, Co goes from $[Ar]3d^7 4s^2$ as an atom to $[Ar]3d^7$ in the oxidized state Co^{2+}.

(b) Co^{3+}. Starting from Co^{2+}, the next electron to be lost comes from a $3d$ orbital. Cobalt(III) has the configuration $[Ar]3d^6$.

(c) Cr^{3+}. Neutral chromium, with six valence electrons ($Z = 24$), qualifies for the aforementioned exception granted to half-filled d shells. Cr is oxidized from $[Ar]3d^5 4s^1$ in Cr^0 to $[Ar]3d^3$ in Cr^{3+}.

(d) Cu^+. Copper's 11 valence electrons ($Z = 29$) produce the configuration $[Ar]3d^{10} 4s^1$ in a neutral atom. A Cu^+ ion, formed by removal of the $4s$ electron, leaves the filled d subshell intact: $[Ar]3d^{10}$.

(e) Ti^{2+}. Neutral Ti, with configuration $[Ar]3d^2 4s^2$, becomes Ti^{2+} after losing the two $4s$ electrons. The ionic configuration is $[Ar]3d^2$.

EXAMPLE 19-5. Crystal Field Theory: Magnetism

Next we use crystal field theory to predict the electron configuration of a complex, bearing in mind the ordering of the spectrochemical series:

$$I^- < Br^- < Cl^- < F^- < OH^- < H_2O < NCS^- < NH_3 < en < CO, CN^-$$

  weak field   intermediate field   strong field

PROBLEM: Compare the two complexes $[CoF_6]^{3-}$ and $[Co(CN)_6]^{3-}$. Which of the two is likely to be paramagnetic?

SOLUTION: For each, the coordination number is 6; and for each, we expect an octahedral complex built around Co^{3+}, a d^6 ion (see Example 19-4b). Six electrons occupy a set of crystal field orbitals split into three degenerate t_{2g} levels and two degenerate e_g levels:

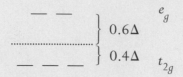

There are two possible arrangements. (1) Low spin: If Δ is large compared with the spin-pairing energy, the six electrons pair up in the t_{2g} orbitals to produce a diamagnetic complex (no unpaired electrons):

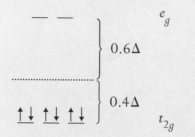

(2) High spin: If Δ is small, single electrons occupy first the t_{2g} levels and then the e_g levels before returning to the t_{2g}. With four unpaired electrons, the resulting high-spin complex is paramagnetic:

Which is which? Consult the spectrochemical series. CN^-, a strong-field ligand, produces a low-spin, diamagnetic complex. F^-, a weak-field ligand, produces a high-spin, paramagnetic complex. $[CoF_6]^{3-}$ is paramagnetic.

EXAMPLE 19-6. Valence Bond Theory

Can we account for magnetic properties using a valence bond approach? Yes—the interpretation differs from crystal field theory, but the number of unpaired electrons does not.

PROBLEM: Describe the bonding in (a) $[CoF_6]^{3-}$ and (b) $[Co(CN)_6]^{3-}$ as if the interactions involved hybrid orbitals. How many electrons are left unpaired in each complex?

SOLUTION: Knowing that the geometry is octahedral, we envision six *empty* hybrid orbitals coming from one s, three p, and two d functions on the metal. Each ligand, a Lewis base with an electron pair to give, then can place its two electrons into one of the vacant hybrids to form a covalent sigma bond.

(a) $[CoF_6]^{3-}$. Standing alone, a Co^{3+} ion distributes its six d electrons throughout all five $3d$ orbitals, leaving four electrons unpaired:

Co^{3+} [Ar]   ↑↓ ↑ ↑ ↑ ↑   __ __ __ __   __ __ __ __ __
                      $3d$ $4s$ $4p$ $4d$

With the $3d$ subshell thus occupied, the ion now must reach into the $n = 4$ shell to offer vacant d space to the ligands.

Let it do so. Let the Co^{3+} put together six hybrid sp^3d^2 orbitals by mixing one $4s$, three $4p$, and two $4d$ functions in six different ways, pointing them toward the six vertices of an octahedron.

Here they are, vacant and ready to accept donations from each of six F^- ions:

Co^{3+} [Ar]   ↑↓ ↑ ↑ ↑ ↑   __ __ __ __ __ __   __ __ __
                      $3d$ sp^3d^2 $4d$

And here they are after these dative bonds are made, each sp^3d^2 orbital now filled with a bonding pair of electrons from a ligand:

Co^{3+} [Ar]   ↑↓ ↑ ↑ ↑ ↑   ↑↓ ↑↓ ↑↓ ↑↓ ↑↓ ↑↓   __ __ __
in $3d$ sp^3d^2 $4d$
$[CoF_6]^{3-}$

Remaining behind on the metal ion, then, are four unpaired electrons spread over five d orbitals—just as we described them in Example 19-5, although using different language.

Compare: The crystal field approach sees the d orbitals as modified from their native state. It presents them as split into a set of three and a set of two, no longer degenerate in the presence of the ligands. It tells us that the complex contains four unpaired electrons, and it suggests that a t_{2g} electron might absorb energy and move to an e_g orbital. So far, so good.

Except for this one deficiency: Crystal field theory takes away our usual picture of bonding. The model pretends, against reality, that the ligands keep their electrons entirely to themselves.

Valence bond theory, with its hybrid orbitals, gives us back the covalent bonds and registers the correct number of unpaired electrons as well—although, again, not without a certain deficiency. Using valence bond theory, we cannot readily predict the absorption spectrum. With crystal field theory, we can (Example 19-10). No model is perfect.

(b) *[Co(CN)$_6$]$^{3-}$*. Assume, from Example 19-5, that the *d* electrons are paired so as to make a low-spin Co^{3+} ion:

Co^{3+} [Ar] ↑↓ ↑↓ ↑↓ __ __   __ __ __ __   __ __ __ __ __

                                3*d*   4*s*   4*p*   4*d*

If so, then the two vacant 3*d* orbitals can mix with the 4*s* and 4*p* to produce six d^2sp^3 hybrid combinations:

Co^{3+} [Ar] ↑↓ ↑↓ ↑↓   __ __ __ __ __ __   __ __ __ __ __

                                3*d* d^2sp^3 4*d*

The 4*d* orbitals, higher in energy, do not participate.

The bonded complex thus is diamagnetic after final assembly, with six electrons paired up in cobalt's three unhybridized *d* orbitals:

Co^{3+} [Ar] ↑↓ ↑↓ ↑↓   ↑↓ ↑↓ ↑↓ ↑↓ ↑↓ ↑↓   __ __ __ __ __
in
[Co(CN)$_6$]$^{3-}$ 3*d* d^2sp^3 4*d*

Each CN$^-$ forms a sigma bond with one of the d^2sp^3 hybrid orbitals localized on the Co^{3+}.

Note, incidentally, the difference in terminology between d^2sp^3 and sp^3d^2 functions. One uses 4*d* orbitals, the other 3*d*.

QUESTION: Doesn't our presumed low-spin configuration for Cr^{3+} ([Ar] ↑↓ ↑↓ ↑↓ __ __) violate Hund's rule, which states that electrons preferentially fill a subshell with their spins unpaired?

ANSWER: Yes, but Hund's rule is a rule, not a commandment, and rules can be broken.

To do so, a system must provide the additional energy needed to support two electrons in the same orbital. In a low-spin complex, an especially strong set of ligand–metal bonds apparently compensates for any higher electron–electron repulsion suffered by the metal.

QUESTION: Even so, how can we say that the 3*d* rather than the 4*d* orbitals control the hybridization? The 4*d* orbitals are empty as well.

ANSWER: True, but the 4*d* orbitals have a higher energy than the 3*d*. A set of d^2sp^3 hybrids made from the lower-lying 3*d* levels describes more aptly the strong ligand–metal bonds in [Co(CN)$_6$]$^{3-}$.

EXAMPLE 19-7. Crystal Field Theory:
The d^8 Octahedral Complex

PROBLEM: Use crystal field theory to predict a configuration for the octahedral complex $[Ni(H_2O)_6]^{2+}$.

SOLUTION: The eight d electrons of Ni^{2+} can be distributed in only one way:

$$\underline{\uparrow} \quad \underline{\uparrow} \qquad\qquad\qquad e_g$$
$$\left.\rule{0pt}{20pt}\right\} \; 0.6\Delta$$
$$\underline{\uparrow\downarrow} \; \underline{\uparrow\downarrow} \; \underline{\uparrow\downarrow} \Big\} \; 0.4\Delta \qquad t_{2g}$$

Result: a paramagnetic complex, with two unpaired electrons.

There is no difference between high-spin and low-spin configurations in an octahedral d^8 complex. Both are $t_{2g}^6 e_g^2$. Only in d^4, d^5, d^6, and d^7 systems does such a possibility exist under octahedral geometry.

EXAMPLE 19-8. Crystal Field Theory:
Four-Coordinate d^8 Complexes

PROBLEM: Ni^{2+} also forms four-coordinate structures, both tetrahedral (high spin) and square planar (low spin). Use crystal field theory to predict the electron distribution in each.

SOLUTION: A tetrahedral arrangement of ligands will push the former $d_{x^2-y^2}$ and d_{z^2} orbitals below the former d_{xy}, d_{xz}, and d_{yz} orbitals:

$$\underline{\quad} \; \underline{\quad} \; \underline{\quad} \qquad xy, \, xz, \, yz$$
$$\underline{\quad} \; \underline{\quad} \qquad x^2 - y^2, \, z^2$$

The splitting, smaller than for a comparable octahedral structure, usually produces a high-spin, paramagnetic complex. For d^8, as in Ni^{2+}, the distribution leaves two electrons unpaired:

$$\underline{\uparrow\downarrow} \; \underline{\uparrow} \; \underline{\uparrow} \qquad xy, \, xz, \, yz$$
$$\underline{\uparrow\downarrow} \; \underline{\uparrow\downarrow} \qquad x^2 - y^2, \, z^2$$

Turn next to a square planar complex (something new), and imagine that we first pull two ligands off an octahedron—one from the z-axis above the central plane, and one from the z-axis below:

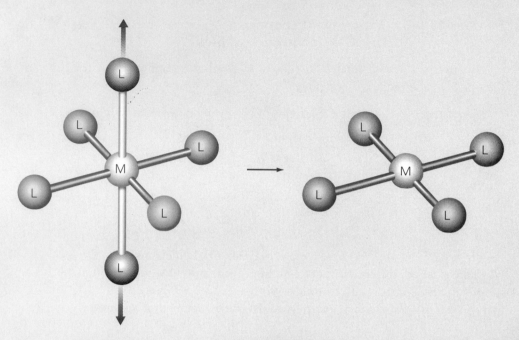

The removal leaves a grouping of four ligands all in a plane, each at the corner of a square with the metal atom at the center. Some of the *d* orbitals go up, and some of the *d* orbitals go down.

Which ones? Those orbitals near the *z*-axis (z^2, *yz*, *xz*), relieved of two repulsive ligands, enjoy a lower energy in the square planar complex. The $x^2 - y^2$ and *xy*, burdened with proportionally more repulsion, move higher:

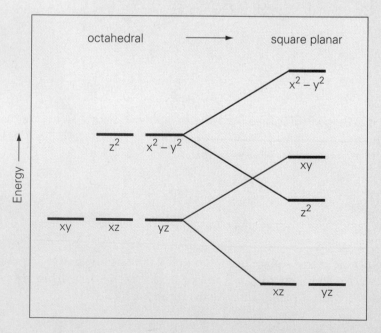

The low-spin d^8 configuration is then diamagnetic in a square planar complex:

$$\underline{} \quad x^2 - y^2$$

$$\underline{\uparrow\downarrow} \quad xy$$

$$\underline{\uparrow\downarrow} \quad z^2$$

$$\underline{\uparrow\downarrow}\ \underline{\uparrow\downarrow} \quad xz,\, yz$$

EXAMPLE 19-9. Valence Bond Theory: Four-Coordinate d^8 Complexes

PROBLEM: Use a valence bond approach to describe the systems treated in the preceding example: (a) A high-spin, tetrahedral d^8 complex. (b) A low-spin, square planar d^8 complex.

SOLUTION: We follow the same procedure as in Example 19-6, this time employing sp^3 hybrid orbitals for a tetrahedron and dsp^2 hybrids for a square plane.

(a) *Tetrahedron: high spin.* Start with a high-spin d^8 configuration for the free ion,

[Ar] $\underline{\uparrow\downarrow}\,\underline{\uparrow\downarrow}\,\underline{\uparrow\downarrow}\,\underline{\uparrow}\,\underline{\uparrow}$ $\underline{}$ $\underline{}\,\underline{}\,\underline{}$ $\underline{}\,\underline{}\,\underline{}\,\underline{}\,\underline{}$

       $3d$ $4s$ $4p$ $4d$

and use the vacant $4s$ and $4p$ orbitals to produce the four sp^3 hybrid combinations:

[Ar] $\underline{\uparrow\downarrow}\,\underline{\uparrow\downarrow}\,\underline{\uparrow\downarrow}\,\underline{\uparrow}\,\underline{\uparrow}$ $\underline{}\,\underline{}\,\underline{}\,\underline{}$ $\underline{}\,\underline{}\,\underline{}\,\underline{}\,\underline{}$

       $3d$ sp^3 $4d$

The four ligand–metal bonds thus formed leave the complex with two unpaired d electrons:

[Ar] $\underline{\uparrow\downarrow}\,\underline{\uparrow\downarrow}\,\underline{\uparrow\downarrow}\,\underline{\uparrow}\,\underline{\uparrow}$ $\underline{\uparrow\downarrow}\,\underline{\uparrow\downarrow}\,\underline{\uparrow\downarrow}\,\underline{\uparrow\downarrow}$ $\underline{}\,\underline{}\,\underline{}\,\underline{}\,\underline{}$

       $3d$ sp^3 $4d$

(b) *Square plane: low spin.* The low-spin free ion

[Ar] $\underline{\uparrow\downarrow}\,\underline{\uparrow\downarrow}\,\underline{\uparrow\downarrow}\,\underline{\uparrow\downarrow}\,\underline{}$ $\underline{}$ $\underline{}\,\underline{}\,\underline{}$ $\underline{}\,\underline{}\,\underline{}\,\underline{}\,\underline{}$

       $3d$ $4s$ $4p$ $4d$

contributes its empty $3d$ orbital to a set of four dsp^2 hybrid functions,

[Ar]   ↑↓ ↑↓ ↑↓ ↑↓   _ _ _ _   _   _ _ _ _ _ _
          $3d$ dsp^2 $4p$ $4d$

and thereby produces a diamagnetic complex:

[Ar]   ↑↓ ↑↓ ↑↓ ↑↓   ↑↓ ↑↓ ↑↓ ↑↓   _   _ _ _ _ _ _
          $3d$ dsp^2 $4p$ $4d$

EXAMPLE 19-10. Crystal Field Theory: Absorption and Color

Let crystal field theory close the chapter, showing us the way to predict the absorption spectrum and color of an octahedral complex.

PROBLEM: Both $[Cr(H_2O)_6]^{3+}$ and $[Cr(NH_3)_6]^{3+}$ are octahedral d^3 complexes built around Cr^{3+}. (a) The ammonia complex absorbs light most strongly at a wavelength near 460 nm. What color do we perceive for the structure? (b) Which of the two complexes absorbs light at the shorter wavelength?

SOLUTION: An octahedral complex will promote an electron from a t_{2g} level to an e_g level if hit with a photon of matching energy. The larger the splitting, the more energy goes into the transition.

For the chromatic progression of the visible spectrum, see Figure 4-4(b). Note that photon energies increase in the order of *red*, *orange*, *yellow*, *green*, *blue*, *indigo*, *violet*—ROYGBIV.

(a) *Color.* White light, containing all colors, enters the $[Cr(NH_3)_6]^{3+}$ complex and exits with something missing: photons with a wavelength near 460 nm, their energy taken up by chromium's newly excited e_g electrons. Emerging with everything *except* these violet wavelengths (400 nm to 500 nm), the beam we see (white minus violet) therefore has a color complementary to violet: yellow.

(b) *Splitting and absorption.* NH_3, positioned above H_2O in the spectrochemical series, produces the greater crystal field splitting. $[Cr(NH_3)_6]^{3+}$ absorbs light at the higher energy, hence the higher frequency and shorter wavelength.

EXERCISES

1. Draw structures for the following complexes, indicating both the coordination number and geometry around the metal:

   (a) $[Co(NH_3)_6]^{3+}$
   (b) $[Co(CN)_6]^{3-}$
   (c) $[Ni(en)_3]^{2+}$
   (d) $[NiCl_4]^{2-}$, a tetrahedral complex
   (e) $[Ni(CN)_4]^{2-}$, a square planar complex

2. Draw structures for the following complexes, indicating both the coordination number and geometry around the metal:

   (a) $[Ni(en)Cl_2]$, a square planar complex
   (b) $[Ti(H_2O)_6]^{3+}$
   (c) $[Ag(NH_3)_2]^{+}$
   (d) $[FeCl_4]^{2-}$, a tetrahedral complex
   (e) $[Fe(CO)_5]$, a trigonal bipyramid

3. How many geometric and optical isomers are possible for the tetrahedral complex $[FeCl_4]^{2-}$?

4. $[Pt(NH_3)_2Cl_2]$ forms a square planar complex. Draw the structures of its two geometric isomers (cis and trans).

5. Is square planar $[Pt(NH_3)_2Cl_2]$ a chiral structure in any of its forms? Does it have optical isomers?

6. Treat the problem of a square planar complex in general: How many geometric and optical isomers are possible for square planar complexes with formulas MX_4, MX_3Y, MX_2Y_2, MX_2YZ, and MWXYZ? Draw them.

7. Likewise, for a tetrahedral complex: Draw all the geometric and optical isomers formed by tetrahedral complexes with formulas MX_4, MX_3Y, MX_2Y_2, MX_2YZ, and MWXYZ.

8. Consider now an octahedral structure, the complex ion $[Co(en)_2Cl_2]^{+}$:
   (a) Draw the cis and trans forms. (b) Show that the trans form has

no optical isomers. (c) Draw the two enantiomers possible for the cis form. Remember that ethylenediamine ($NH_2CH_2CH_2NH_2$) is a bidentate ligand.

9. Show that an octahedral complex with the formula MX_6 or MX_5Y exists in only one form, with no distinguishable geometric isomers or optical isomers. Assume that neither X nor Y is an ambidentate ligand.

10. Suppose that an octahedral complex has the formula MX_2Y_4. How many geometric isomers and optical isomers are possible? Draw them.

11. Similar: Draw the geometric isomers and optical isomers belonging to an octahedral complex with formula MX_3Y_3.

12. The thiocyanate ion, SCN^-, is an ambidentate ligand, able to coordinate either as M—SCN or M—NCS. How many linkage isomers, geometric isomers, and optical isomers are possible for the complex $[Co(NH_3)_5NCS]^{2+}$? Draw them.

13. Suppose that X is an ambidentate ligand. Count the various isomers possible for an octahedral complex MX_6.

14. Determine the metal's oxidation number in each complex:

   (a) $[Co(NH_3)_6]^{3+}$
   (b) $[Co(CN)_6]^{3-}$
   (c) $[Ni(en)_3]^{2+}$
   (d) $[NiCl_4]^{2-}$
   (e) $[Ni(CN)_4]^{2-}$

15. Determine the metal's oxidation number in each complex:

   (a) $[Ni(en)Cl_2]$
   (b) $[Ti(H_2O)_6]^{3+}$
   (c) $[Ag(NH_3)_2]^+$
   (d) $[FeCl_4]^{2-}$
   (e) $[Fe(CO)_5]$

16. Determine the metal's coordination number and oxidation number in each coordination compound:

(a) $[Cr(H_2O)_6]Cl_3$

(b) $[Cr(H_2O)_5Cl]Cl_2 \cdot H_2O$

(c) $[Cr(H_2O)_4Cl_2]Cl \cdot 2H_2O$

(d) $K_4[Fe(CN)_6]$

(e) $[Co(NH_3)_4(H_2O)CN]Cl_2$

17. Count the d electrons:

   (a) $[Co(NH_3)_6]^{3+}$

   (b) $[Co(CN)_6]^{3-}$

   (c) $[Ni(en)_3]^{2+}$

   (d) $[NiCl_4]^{2-}$

   (e) $[Ni(CN)_4]^{2-}$

18. Count the d electrons:

   (a) $[Ni(en)Cl_2]$

   (b) $[Ti(H_2O)_6]^{3+}$

   (c) $[Ag(NH_3)_2]^+$

   (d) $[FeCl_4]^{2-}$

   (e) $[Fe(CO)_5]$

19. Count the d electrons:

   (a) $[Cr(H_2O)_6]Cl_3$

   (b) $[Cr(H_2O)_5Cl]Cl_2 \cdot H_2O$

   (c) $[Cr(H_2O)_4Cl_2]Cl \cdot 2H_2O$

   (d) $K_4[Fe(CN)_6]$

   (e) $[Co(NH_3)_4(H_2O)CN]Cl_2$

20. Write the electron configuration and count the d electrons for each oxidation state listed below. The species all belong to the first-series transition metals:

   (a) Sc^{3+}

   (b) Ti^{2+}, Ti^{3+}, Ti^{4+}

   (c) V^{2+}, V^{3+}, V^{4+}, V^{5+}

   (d) Cr^{2+}, Cr^{3+}, Cr^{4+}, Cr^{5+}, Cr^{6+}

   (e) Mn^{2+}, Mn^{3+}, Mn^{4+}, Mn^{5+}, Mn^{6+}, Mn^{7+}

21. Write the electron configuration and count the *d* electrons for each oxidation state listed below. The species all belong to the first-series transition metals:

    (a) Fe^{2+}, Fe^{3+}, Fe^{4+}, Fe^{6+}
    (b) Co^{2+}, Co^{3+}
    (c) Ni^{2+}, Ni^{3+}, Ni^{4+}
    (d) Cu^+, Cu^{2+}, Cu^{3+}
    (e) Zn^{2+}

22. Write the electron configuration and count the *d* electrons for each oxidation state listed below. The species all belong to the second-series transition metals:

    (a) Y^{3+}
    (b) Zr^{3+}, Zr^{4+}
    (c) Nb^{4+}, Nb^{5+}
    (d) Mo^{2+}, Mo^{3+}, Mo^{4+}, Mo^{5+}, Mo^{6+}
    (e) Tc^{2+}, Tc^{3+}, Tc^{4+}, Tc^{5+}, Tc^{6+}, Tc^{7+}

23. Write the electron configuration and count the *d* electrons for each oxidation state listed below. The species all belong to the second-series transition metals:

    (a) Ru^{2+}, Ru^{3+}, Ru^{4+}, Ru^{8+}
    (b) Rh^{3+}, Rh^{4+}, Rh^{6+}
    (c) Pd^{2+}, Pd^{4+}
    (d) Ag^+, Ag^{2+}, Ag^{3+}
    (e) Cd^{2+}

24. Which process is probably more favorable thermodynamically for the free ion?

    $$Mn^{3+} + e^- \longrightarrow Mn^{2+} \qquad \text{(reduction of } Mn^{3+} \text{ to } Mn^{2+}\text{)}$$

    or

    $$Mn^{2+} \longrightarrow Mn^{3+} + e^- \qquad \text{(oxidation of } Mn^{2+} \text{ to } Mn^{3+}\text{)}$$

    To decide, consider the electron configurations of product and reactant.

25. Similar: Which process should have the higher reduction potential?

$$Mn^{3+} + e^- \longrightarrow Mn^{2+}$$

or

$$Fe^{3+} + e^- \longrightarrow Fe^{2+}$$

Why?

26. Write down, according to crystal field theory, the electron configurations possible in an octahedral complex where the metal ion exists as:

(a) d^1 (b) d^2 (c) d^3 (d) d^4 (e) d^5

Draw an energy-level diagram for each configuration, and include both high-spin and low-spin alternatives where applicable.

27. For each configuration determined in the exercise above, give the following information: (1) The stabilization energy in units of Δ, the energy gap in an octahedral crystal field. (2) The number of unpaired electrons.

28. Write down, according to crystal field theory, the electron configurations possible in an octahedral complex where the metal ion exists as:

(a) d^6 (b) d^7 (c) d^8 (d) d^9 (e) d^{10}

Draw an energy-level diagram for each configuration, and include both high-spin and low-spin alternatives where applicable.

29. For each configuration determined in the exercise above, give the following information: (1) The stabilization energy in units of Δ, the energy gap in an octahedral crystal field. (2) The number of unpaired electrons.

30. Around which oxidation state is an octahedral complex more likely to be built . . . Cr^{2+} or Cr^{3+}?

31. Pick the ligand more likely to produce a high-spin octahedral complex:

(a) CO or F^-

(b) CN^- or OH^-

(c) en or H_2O

(d) NH_3 or I^-

32. One of these two complexes is high spin: $[Fe(CN)_6]^{3-}$ or $[Fe(NCS)_6]^{3-}$. (a) Which is it? (b) Use crystal field theory to predict the electron configuration for each system. How many electrons are unpaired?

33. Use a valence bond model to account for the magnetic properties of $[Fe(CN)_6]^{3-}$ and $[Fe(NCS)_6]^{3-}$, mentioned above. How many electrons are left unpaired in each complex?

34. One of these two complexes is high spin: $[Mn(CN)_6]^{3-}$ or $[MnCl_6]^{3-}$. (a) Which is it? (b) Use crystal field theory to predict the electron configuration for each system. How many electrons are unpaired?

35. Use a valence bond model to account for the magnetic properties of $[Mn(CN)_6]^{3-}$ and $[MnCl_6]^{3-}$, mentioned above. How many electrons are left unpaired in each complex?

36. Which metal ion should produce the more stable octahedral complex?

    (a) V^{2+} or V^{3+}
    (b) Fe^{2+} (high spin) or Fe^{2+} (low spin)

37. Use crystal field theory to show the electron configuration arising in a tetrahedral complex where the metal ion exists as:

    (a) d^1 (b) d^2 (c) d^3 (d) d^4 (e) d^5

    Draw an energy-level diagram for each system, and assume that all the configurations are high spin.

38. For each configuration determined in the exercise above, give the following information: (1) The stabilization energy in units of Δ_t, the energy gap in a tetrahedral crystal field. Hint: Remember that the *net* splitting, averaged over all five levels, is zero. (2) The number of unpaired electrons.

39. Use crystal field theory to show the electron configuration arising in a tetrahedral complex where the metal ion exists as:

    (a) d^6 (b) d^7 (c) d^8 (d) d^9 (e) d^{10}

    Draw an energy-level diagram for each system, and assume that all the configurations are high spin.

40. For each configuration determined in the exercise above, give the following information: (1) The stabilization energy in units of Δ_t, the energy gap in a tetrahedral crystal field. Hint: Remember that the *net* splitting, averaged over all five levels, is zero. (2) The number of unpaired electrons.

41. $[NiCl_4]^{2-}$ is tetrahedral. $[Ni(CN)_4]^{2-}$ is square planar. (a) Use crystal field theory to predict the electron configuration in each complex. (b) Which of the two structures is paramagnetic?

42. Use a valence bond approach to describe the bonding in the two systems noted above:

    (a) $[NiCl_4]^{2-}$, tetrahedral
    (b) $[Ni(CN)_4]^{2-}$, square planar

43. Group these six colors into three complementary pairs: red, orange, yellow, green, blue, violet.

44. The crystal field splitting in $[Cr(CN)_6]^{3-}$ is 5.21×10^{-19} J. (a) At what wavelength will the $[Cr(CN)_6]^{3-}$ ion absorb light most strongly? (b) What color will the structure take on?

45. Which color is $[CrCl_6]^{3-}$ most likely to have . . . green, yellow, or red? Make a guess based on the position of chloride in the spectrochemical series.

46. Which complex absorbs light at the shorter wavelength . . . $[Co(NH_3)_5Cl]^{2+}$ or $[Co(NH_3)_5NCS]^{2+}$?

47. Solutions containing the complex ion $[Cu(H_2O)_4]^{2+}$ are bright blue, but solutions containing the complex ion $[Cu(CN)_2]^-$ are colorless. Why?

48. A 0.1 *M* solution of $[CoCl_4]^{2-}$ absorbs light at a wavelength of 690 nm. (a) Will the wavelength of absorption increase, decrease, or remain the same if the concentration of $[CoCl_4]^{2-}$ is increased? (b) Will the intensity of absorption increase, decrease, or remain the same if the concentration of $[CoCl_4]^{2-}$ is increased? (c) Draw an analogy to the photoelectric effect.

20

Spectroscopy and Analysis

20-1. The Interaction of Light and Matter
Spectroscopic Vision
Probing Matter with Light
Spectroscopy: An Appreciation
20-2. Manipulations in Space and Time
Mass Spectrometry
Chromatography
REVIEW AND GUIDE TO PROBLEMS
EXERCISES

There is more to matter than meets the eye. Unseen is the quiet order of a crystal, the slippery disarray of a liquid, the unrelenting chaos of a gas. Unseen are the molecules, the atoms, the electrons. Unseen are the nuclei, the protons, the neutrons. It might be a motto for chemistry, an exhortation to dig deeper. Dig deeper, because there is more to matter than meets the eye.

What does meet the eye? Not matter directly, but matter represented by the medium of light: the electromagnetic field, photons. The photon serves as matter's messenger, the agency by which one charged particle announces itself to another. Light transmitted by matter is first focused onto the retina, where specially receptive molecules absorb the photons. A chemical change is induced; charges are rearranged; electrical impulses are triggered; signals are carried through the optic nerve to the brain; and, finally, the brain registers the subjective sensation of sight. The picture developed by the brain is too coarse to reveal an individual molecule, but still there is a microscopic message plain to see: *color*, a measure of the electronic energy levels.

All the senses are electrical. Touch? A finger presses against a wall, and the wall presses back; neither moves the other. The resistance arises from the repulsive clash of electrons forced too close together, a clash mandated by the Pauli exclusion principle. Pressure on the skin triggers electrical impulses in a nerve; the impulses travel to the brain; the brain, again, registers the subjective sensation of touch. Hearing? It starts with the impact of gas molecules on the eardrum, and the rest follows as before. Taste? Smell? Each comes from the binding of guest molecules onto specially prepared host sites on the sensory organ. The binding is a consequence of molecular recognition. Binding is chemical. Chemical is electrical.

Wherever there is an electron, an atom, a molecule, a crystal, a liquid, a gas, a phase transition . . . wherever there is *chemistry*, there is an electromagnetic influence. Every chemical event outside the nucleus arises from an electromagnetic interaction between charged particles—not from gravity, not from the strong nuclear force, not from the weak nuclear force. Every normal chemical transaction is brokered in some way by the electromagnetic field: by the agency of photons; by *light*, understood broadly to include the entire electromagnetic spectrum. Chemistry, at its most fundamental level, operates entirely in the realm of the electromagnetic.

If we want to dig deeper into the molecular structure of matter, our only tool is the electromagnetic interaction. We shall use it.

But we do need better eyes.

20-1. THE INTERACTION OF LIGHT AND MATTER

Whether by eye or instrument, to *see* matter is to illuminate it. There is a special, profound relationship between light and matter, never more apparent than when matter is bathed in light.

From Chapter 4, we already know how light and matter blur together in form and function. Light conveys the electromagnetic force from point to point, from instant to instant, tugging incessantly on the bits of charge that constitute matter. Light arises from charge in motion; light sets charge in motion. Light would be lost without matter, and matter would be lost without light.

One demands the other, for to bring light to matter is also to bring matter to light. Light moves electrons, and electrons move to make light. Charged particles are both the source and target of electromagnetic fields.

The source of light: Flip a switch, and there is light. An electromagnetic field sweeps through space as the wave of force depicted in

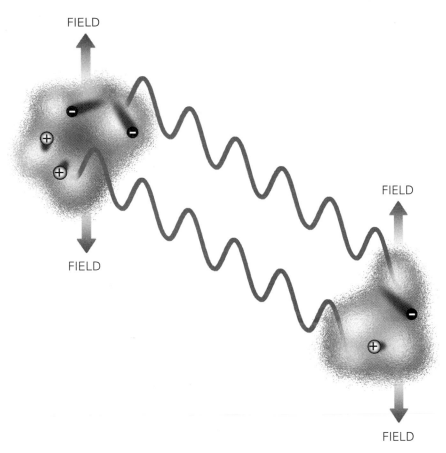

FIELD

FIELD

FIELD

FIELD

FIELD

FIGURE 20-1. The means to move matter, and thus the means to *probe* matter: light, the electromagnetic field. From the motion of charged particles comes a wave of electromagnetic force, carrying energy from sender to receiver. Wrapped up in it is the essence of chemistry: the interplay of electromagnetic fields at the atomic level.

Figure 20-1, a source of energy, an ability to do work. It carries a stream of photons, each with quantized wavelength, frequency, energy, and momentum. But the light—this intangible, immaterial force—had to come from somewhere; it had to come from something material. It was born, inside the lightbulb, as the product of charge in motion.

Before there was light, there had to be matter in motion. Charged particles, moving over *there*, first generated the electromagnetic fields that reach us, later, over *here*. Matter **radiated** the light. Radio transmitters, microwave ovens, heating coils, lightbulbs, X-ray tubes—all contain moving electrons.

The target of light: Waiting at the receiving end is a charged particle. It can absorb energy and emit energy, always going between one quantum state and another. An electron moves in response to the electromagnetic force, making new light of its own. So does a proton. So does a whole nucleus. All charged particles take light and make light.

Spectroscopic Vision

Light comes from somewhere (the sun? a fluorescent tube?) and shines on something (a blade of grass? a beaker of aqueous $CuSO_4$ solution?). Photons of different energies pass through the material. Some photons, unable to bridge the gap between any pair of quantized energy levels, are transmitted straight through. Others, having precisely the right energy, are captured by the particles. A suitably matched photon thus may be **absorbed** by a system, whereby it supplies the energy needed to pump a transition from one quantum state up to another. Or, in the related process of **emission**, a photon may be released while an excited structure falls to a state of lower energy.

Either way, something happens. Something happens to the light, and something happens to the matter. The light coming out is not the same as the light going in. Matter intervenes. The matter changes the light.

So it happens that white light is partially absorbed by the magnesium complex in chlorophyll,

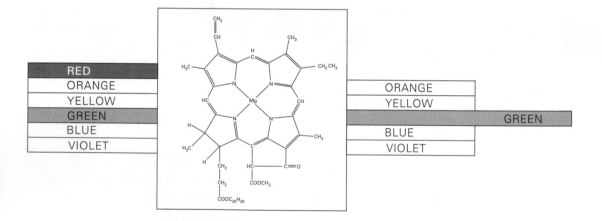

leaving the grass to transmit a complementary spectrum of wavelengths deemed "green" by the brain. Later, perhaps, the same brain interprets the event in another way: as a promotion of d electrons between two quantized energy levels. Call the result "green" or call it "evidence for an electronic transition," but both interpretations seek to describe the same phenomenon. Each attempts to reconstruct what happened when light met matter.

Eye and brain, together, make a natural **spectroscope**, an instrument able to sort the various wavelengths of light emerging from an object. Sensitive to different wavelengths, the brain sees different colors. It registers blue for the copper solution, red for blood, black for the absence of all color, and white for all the colors together. It distinguishes between photons of different energies, although only over a relatively narrow range extending from approximately 400 nm (violet) to 700 nm (red): the *visible* portion of the electromagnetic spectrum.

We need stronger, more discerning eyes to look deeply into matter, but even the most sophisticated spectroscopic instruments follow this overall plan of eye and brain. Step 1: Light goes in. Step 2: Light does its work on the matter. Step 3: Light comes out. We observe what happens during steps 1 and 3,

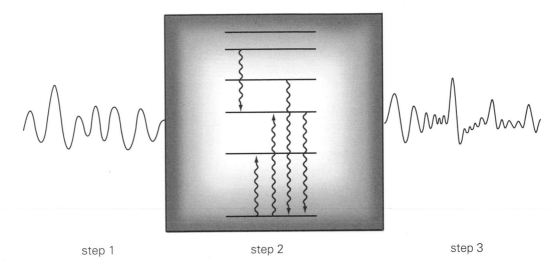

step 1   step 2   step 3

and then we interpret what must have happened during step 2. For that final interpretation, the brain emerges as the most important component.

Probing Matter with Light

What light does to matter it does specifically and discretely, one photon at a time. Each packet of light is matched to a difference between two energy levels in the molecule (or atom, or fragment, or ion, or radical, or any other microscopic entity). Photons are absorbed and emitted according to the basic rules we discovered in Chapter 5.

When electromagnetic radiation is *absorbed*, for example, the photon's energy

$$E = h\nu$$

goes to promote the system from an initial level ϵ_i, low in energy, to a higher final level, ϵ_f. The boost is positive, a step *up* the quantum ladder shown in Figure 20-2:

$$\Delta\epsilon = \epsilon_f - \epsilon_i = h\nu \qquad (\epsilon_f > \epsilon_i)$$

(a)

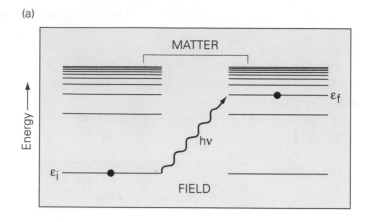

(b)

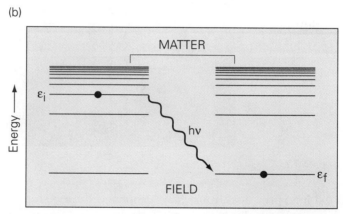

FIGURE 20-2. Energy passes between matter and the electromagnetic field, but only in packets of regulated size: photons. For a swap to occur, the energy of the photon, *hν*, must match the gap between a pair of levels in the material. Energy is conserved. (a) Absorption. (b) Emission.

Energy is transferred from the electromagnetic field to the material particles. Nothing is lost and nothing is gained. Energy is conserved.

When electromagnetic radiation is *emitted*, a photon is released. The system drops from a higher energy (ϵ_i) to a lower energy (ϵ_f). It is a step *down*, a demotion, and now the change

$$\Delta\epsilon = \epsilon_f - \epsilon_i = h\nu \qquad (\epsilon_f < \epsilon_i)$$

is negative. The final state, ϵ_f, is lower than the initial state. Energy is delivered from the matter to the electromagnetic field.

Take note: During absorption, light makes a charge move; during emission, a moving charge makes light. But nothing is lost and nothing is gained during either transaction. Energy is conserved. Momentum is conserved.

Those are the rules, never yet violated. Now, reminded of the law, let us see exactly what light can do to matter. We shall take a brief tour from one end of the electromagnetic spectrum to the other, from low energy to high, recalling the relationship between wavelength (λ), frequency (ν), and speed ($c = 3.00 \times 10^8$ m s^{-1}) set forth below and in Figure 20-3:

$$\lambda\nu = c$$

The longer the wavelength, the lower the frequency and the lower the energy. The shorter the wavelength, the higher the frequency and the higher the energy. Long wavelengths carry low energy; short wavelengths carry high energy.

Our exploration will cover five regions: radiofrequency waves, microwaves, infrared radiation, visible and ultraviolet radiation, and X rays.

1. RADIOFREQUENCY WAVES. Here are the longest wavelengths, far exceeding the dimensions of any atom or molecule. With λ ranging from a few meters to infinity, radiofrequency (rf) waves oscillate at correspondingly low frequencies: from hundreds of megahertz (1 MHz = 10^6 s^{-1}) down to tens of kilohertz (1 kHz = 10^3 s^{-1}) in most applications.

FIGURE 20-3. Electromagnetic waves oscillate in space and time, rising and falling as they move from point to point. Although wavelength and frequency vary across the electromagnetic spectrum, the speed in a vacuum is fixed.

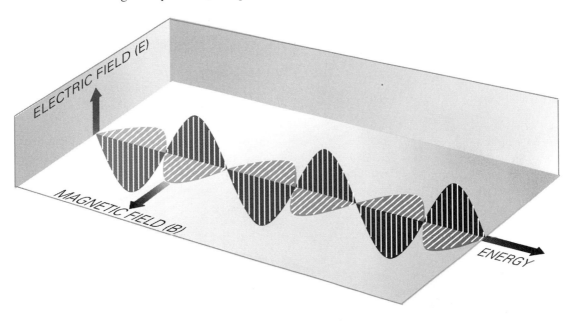

As a typical example, consider a frequency of 100 MHz (station 100 on the FM radio band). Over a wavelength of 3 m—an infinite expanse for a molecule—the field of force rises from zero to its crest, falls from crest to trough, and finally rises back to zero:

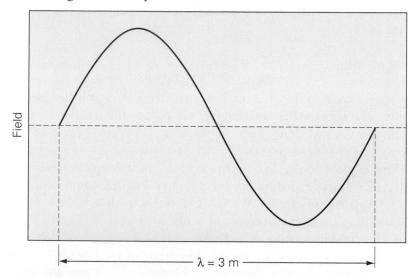

The pattern repeats itself every three meters. Crest to crest and trough to trough, the spacing is approximately 10 feet.

Superimposed upon this pattern, too, is an additional disturbance in time, characterized by the frequency of oscillation v. The field rises, falls, and rises again 100 million times per second at every point in space. A particle hit by the radiation thus suffers an electromagnetic push-pull, albeit one delivered comparatively slowly (!) and with relatively low energy:

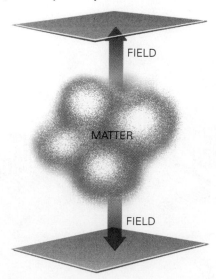

What happens? To the electrons, nothing. The electrons are unmoved; their orbitals are separated by more energy than a radio wave can deliver. Nor can the radio wave move the nuclei from point to point, as if to alter the geometry of a molecule. Bond distances are untouched. Bond angles are untouched. The structure vibrates and rotates as before.

Matter is not immune to radiofrequency irradiation, though, for there can yet be a change in the **nuclear spin**—the *nuclear* spin, a quantity analogous to the spin of an electron (Chapter 5). Within a nucleus there are protons and neutrons, and each of these subnuclear particles is endowed (like the electron) with a special kind of angular momentum: a spin, so called. Like the electron as well, the proton and neutron have spin quantum numbers equal to $\frac{1}{2}$.

Each proton and *each* neutron has a spin angular momentum, although not necessarily the nucleus as a whole. For ^1H, yes. The nucleus of ordinary hydrogen has only that one proton, and so it retains a net spin. But for all other nuclei, the outcome depends on the exact distribution of protons and neutrons. Just as an "up" electron (↑) will cancel the spin of some other "down" electron (↓) in an atom or molecule, so also will protons and neutrons combine their individual spins within a nucleus. Results vary. The dominant ^{12}C isotope of carbon (99% abundant) has no net nuclear spin, but the minority ^{13}C isotope (1%) does emerge with a resultant spin of $\frac{1}{2}$. ^1H, ^2H, ^3H, ^{13}C, ^{15}N, ^{31}P—all these nuclei, and many more, possess a nuclear spin equal to $\frac{1}{2}$.

Now remember: The combination of spin and charge creates a *magnetic moment*, an effect analogous to a microscopic current loop. Similar to an unpaired electron, a nucleus with spin-$\frac{1}{2}$ becomes a tiny magnet. Placed into an external magnetic field, this nuclear magnet then behaves like a quantum mechanical compass needle. It can align either up or down with the field, selecting either of the states ↑ or ↓:

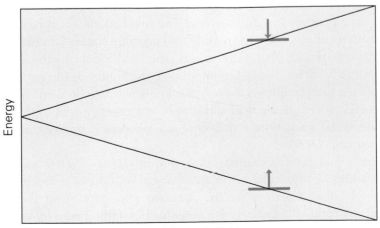

Magnetic field

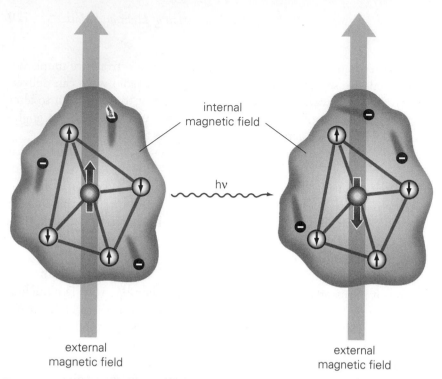

internal
magnetic field

hv

external
magnetic field

external
magnetic field

FIGURE 20-4. Nuclear magnetic resonance: atom by atom. Sensitive to the slightest differences in local field, a nuclear magnet broadcasts its position as it flips between spin states. The frequency of the NMR signal depends critically on the local magnetic field present at each site.

Suppose that ↑ is the state of lower energy, the ground state. There is another level as well (↓), but it lies higher in energy; and the larger the external field, the larger is the gap between ↑ and ↓. To flip the nuclear spin requires a corresponding push, a push of sufficient energy to match the exact difference between up and down. If this magnet really were a compass needle, we would use the simple push of a finger to select any position between north and south. For a *nuclear* magnet, however, we must use a photon; and, furthermore, we must accept the restrictions of quantum mechanics: There are only two possible states, ↑ and ↓.

Expressed as a photon frequency, the field-induced gap between ↑ and ↓ usually varies from a few megahertz to hundreds of megahertz—perfect for radiofrequency waves. We know what to do next. We place the nuclei in a magnetic field and supply just enough rf energy to reorient the spins. Each time a spin flips, we observe a **nuclear magnetic resonance (NMR)**.

And? And now, with that flip of a spin (Figure 20-4), we can see deep within a molecule. From this strange notion of a nuclear spin, seemingly divorced from reality, comes a vivid picture of the microscopic environment. For here is a neighborhood filled with *internal* magnetic fields, different at each chemically distinct site.

A magnetic field, after all, arises from charges in motion; and always there are plenty of charges moving in the molecular world. Small fields, emanating from other nuclei and also from nearby electrons, are present everywhere. These internal magnetic fields supplement the external magnetic field we impose by design.

Indeed it is the internal, locally varying field that determines the exact gap between ↑ and ↓ at each site. Every nucleus responds to its own local field, and so every nucleus has its own resonant energy. By measuring this energy, we learn the strength and direction of the local magnetic field; and, from there, we infer what the surrounding neighborhood of electrons and nuclei must look like. The nuclear spin serves as a kind of spy, a passive observer of the local scene.

We shall not attempt to interpret, peak by peak, the NMR spectra shown as illustrations in Figure 20-5; to do so would take us far beyond

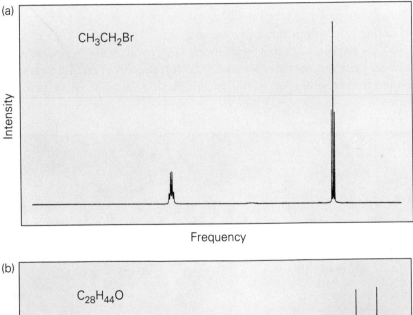

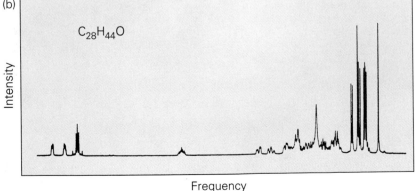

FIGURE 20-5. ^{1}H NMR spectra, simple and complex: (a) Ethyl bromide, CH_3CH_2Br. (b) Ergosterol, $C_{28}H_{44}O$. Each signal or group of signals arises from a hydrogen atom in a distinct environment within the molecule.

our present purposes. But do realize what information must be contained in these signals: the structure of the molecule, revealed one atom at a time. NMR spectroscopy gives us the bonding pattern. It tells us that this carbon has one hydrogen attached, whereas some other carbon has two hydrogens and still another has three hydrogens. It tells us that a certain molecule contains one double bond and one triple bond. It tells us that a structure exists as the cis isomer, not the trans. It can even give us a visual image of a brain (Figure 20-6), as sharp and clear as an X-ray picture.

Altogether, we reap a surprisingly abundant harvest from these weak radiofrequency waves. Our tour now proceeds to its next stop: the microwave region.

2. MICROWAVES. With wavelengths ranging from centimeters to millimeters, microwave dimensions still dwarf any atomic diameter. Frequencies run from a few gigahertz (1 GHz = 10^9 s^{-1}) to tens of gigahertz. Picture, say, a force field repeating itself every foot or so, rising and falling one billion times per second.

More energetic than rf radiation, microwaves are responsible for ***electron paramagnetic resonance (EPR)***: the flip of an electron's spin, the electronic counterpart of nuclear magnetic resonance. Basic principles

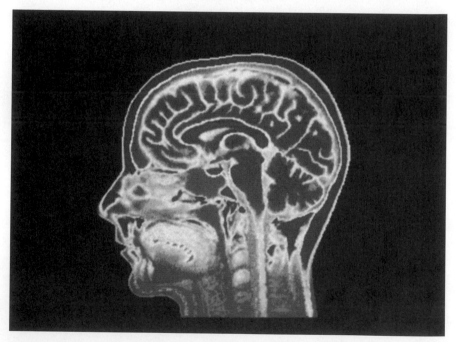

FIGURE 20-6. Nuclear magnetic resonance imaging, used widely for medical diagnosis since the 1980s.

are the same for the two forms of magnetic resonance, except that an electron is more strongly magnetic than any nucleus. Placed in an external magnetic field, an unpaired electron needs proportionately more energy to make a transition.

EPR requires that there be *unpaired* electrons, each housed in a half-filled orbital so as to create a net electronic spin. Look for transition-metal complexes (Chapter 19), free radicals (Chapter 3), rare-earth ions (which contain *f* electrons; Chapter 6), various odd-electron structures, and certain electronically excited systems to be eligible.

Like the NMR signal, an EPR spectrum portrays the molecular environment from the perspective of a sensitive but unobtrusive reporter: a spin magnetic moment. Analogous to the nuclear spins, electrons in different sites demand different amounts of energy to flip from up to down or down to up. Small variations in the local magnetic field are sensed by the unpaired electrons, and the information is coded into the resonant frequencies recorded in a spectrum. EPR allows us, for instance, to follow the reactions of free radicals, including those species present only briefly as intermediates. It supplies structural information as well, showing us where the unpaired electrons exist and also which nuclei lie nearby; and it provides a way to measure low concentrations of paramagnetic ions in biological systems.

Microwave radiation touches more than electron spin, though, for these same photons also induce changes in molecular *rotation*. Rotation, like any other form of internal motion, is quantized in a molecule, and only certain movements are allowed. A molecule is no freely spinning top, allowed to rotate about any axis at any speed; rather, in the close quarters of a molecule, there arise rotational quantum numbers and rotational energies. Axis, energy, and angular momentum are restricted to certain discrete choices, just as a bound electron is constrained by a set of orbital quantum numbers.

Microwaves carrying the proper energy will excite a molecule from one rotational energy level to another. Such excitation occurs, among other places, in the microwave oven, where H_2O molecules acquire additional energy by jumping to higher rotational levels. The water molecules rotate faster, and the meat gets hotter.

Again, the elucidation of molecular structure is the prize we extract from a rotational spectrum. To know how a molecule rotates is to know something about its construction: bond lengths, bond angles, how the masses are distributed. Working backward from the spectrum, we deduce precisely what kind of structure must be responsible for the measured signals.

3. INFRARED RADIATION. Photon energies continue to increase as we pass through the infrared (ir) region, where wavelengths run from

1 millimeter to just below 1 micrometer (1 μm $= 10^{-6}$ m). At the short end, near 1 micrometer, the waves finally start to enter the realm of the microscopic—not as small as atomic diameters (10^{-10} m), but close to the dimensions of very large polymers. The radiation is noticeably more compressed, energetic, and rapid in its oscillation than, say, rf waves. Frequencies go from approximately 3×10^{11} s^{-1} up to 3×10^{14} s^{-1}.

Up and down, push and pull, 300 trillion times per second . . . Every second, the electric force accompanying a micrometer-sized ir wave delivers 300 trillion push-pull cycles to any charged particle it encounters. A millimeter-sized wave does the same, only slower: 300 billion times per second, a thousandfold decrease. How does a molecule respond to such treatment?

It vibrates. Absorbed by a molecule, the energy of an infrared photon is invested in the gross motion of the nuclei. Infrared radiation pushes and pulls the nuclei of a molecule. The bonds stretch and compress, bend and wag, spurred to higher energies under the impetus of the light. The molecule, excited from one quantum state to another, vibrates faster and more energetically than before:

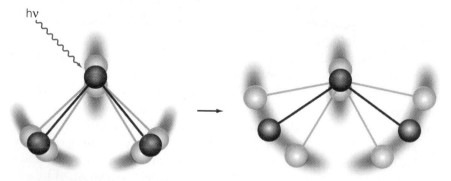

To higher energies and to new modes of vibration it goes. But remember also that a molecule—any molecule—will vibrate all by itself, with or without an infrared wave to drive the motion. Thermal energy and the electric forces inherent in the structure provide drive enough, as first we discovered for simple H_2 (Chapter 7). There is an *internal* tug of war between positive and negative particles, never ending.

Unceasing vibration is indeed the normal condition for these delicately balanced structures, which hover about a point of minimum energy without ever coming to rest. Two nuclei are pulled together by their mutual attraction for the same electron, and so they move closer. Yet not *too* close; too close, and they are pushed apart by their growing repulsion for each other. Too far apart, once again, and they come back together; and so it continues: too far, too close, too far, too close. The nuclei overshoot and undershoot their equilibrium positions like masses on a spring, back and forth, always vibrating.

They vibrate; and, as it happens, they usually vibrate in the infrared range of frequencies (between 10^{11} and 10^{14} cycles per second). Back and forth goes a nucleus, and back and forth goes the electromagnetic force—each at the same rate. The oscillating light is in tune with the oscillating matter, and therefore this external driving force, synchronized with the internal motion, acts like a series of well-timed, rhythmical pushes delivered to a child on a swing. Just as the child picks up energy, so does the vibrating molecule pick up energy.

The effect is called **resonance**, one of the great phenomena of nature and the driving force behind all spectroscopic transitions. To be in rhythm is to transfer energy most efficiently, whether on the playground or in a molecule. NMR, EPR, rotational transitions, vibrational transitions, electronic transitions—each derives from the phenomenon of resonance. Resonance provides the connection between energy and frequency, tacitly expressed in the relationship $\Delta\epsilon = h\nu$.

We appreciate, then, this general feature of resonance, not just for vibrations but for all spectroscopic transitions: that radiofrequency waves can be tuned to match the energies of nuclear spins; that microwaves can be tuned both to EPR frequencies and to rotational energies; that infrared radiation can be tuned to vibrational transitions, and so on. Light, with its back-and-forth field of force, comes into resonance with the natural frequencies of matter. Energy is absorbed, and spectroscopic transitions take place.

So much for generalities. What can we say specifically about these vibrating bonds, which remind us of pendulums and springs? Are the vibrations of nuclei really like masses on a spring?

In some ways, they are. We stretch a spring, feel the pull in the opposite direction, and then let go. To and fro moves the vibrating mass, overshooting and undershooting its equilibrium point on each pass. The stiffer the spring, the stronger the force. The stronger the force, the faster the vibration. The lighter the mass, the faster the vibration as well.

Like masses on a spring? The masses are the nuclei. The spring is the elastic tissue of electrons. Strong bonds, like stiff springs, generate high-frequency vibrations; small masses, too, generate high-frequency vibrations. A carbon–hydrogen bond, for example, vibrates faster than its carbon–deuterium counterpart. Such is the way of a spring.

Yet this molecular spring is no ordinary, macroscopic spring. Yes, there is a characteristic frequency of vibration, ν_0; and yes, this frequency does increase as the bond becomes stiffer and as the nuclear masses become smaller. But, no, the possible energies of vibration do not assume arbitrary and continuous values, as they do for an ordinary stretched spring. Since the microscopic oscillations are quantized, a molecular spring can vibrate only with certain discrete energies. The quantum mechanical energies are restricted by a quantum number v such that only the values

$$\epsilon_{vib} = (v + \tfrac{1}{2})h\nu_0 \qquad (v = 0, 1, 2, \ldots)$$

are permitted. Gaps appear in the pattern. Just as there are distinct electronic energy levels and distinct rotational levels and distinct spin levels, there are distinct vibrational levels as well (with energies $\tfrac{1}{2}h\nu_0$, $\tfrac{3}{2}h\nu_0$, ...). Infrared vibrational spectroscopy provides a map of them.

Matched to the difference between two suitable levels, an infrared photon carries the molecule from one allowed state of vibration to another. Measuring these differences in energy, we then discover something about the strength and elasticity of the vibrating bonds. We learn, furthermore, how to recognize the various bonds from their infrared signatures, as if the vibrational spectrum (Figure 20-7) were a molecular fingerprint.

This too we learn: that there is a certain portability to the chemical bond; that a C—H or C=C or O—H bond vibrates at nearly the same frequency in whatever molecule it resides; and that, consequently, a vibrating molecule seems to be roughly the sum of its vibrating *bonds*, pair by pair. Here is clear evidence supporting chemistry's all-important notion of pair bonds within molecules, a partial justification for the valence bond model of Chapter 7.

Let us note, finally, one further characteristic of microscopic vibration,

FIGURE 20-7. Infrared spectrum of 1-hexanol. Even in different molecules, the vibrational frequency of a particular functional group usually falls somewhere within a narrow range.

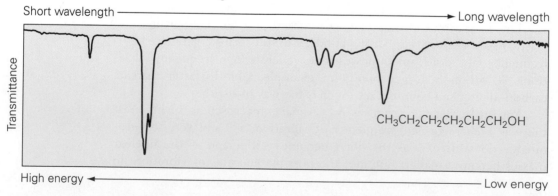

a subtle but important peculiarity: The quantized expression

$$\epsilon_{vib} = (v + \tfrac{1}{2})hv_0$$

can never be zero. Even in the lowest state, $v = 0$, there is still a finite energy $\tfrac{1}{2}hv_0$, called the **zero-point energy**. There is always some motion, and this unquenchable motion is unavoidable in the quantum world.

Zero-point energy ensures that the vibrating nuclei never come to rest, not even at 0 K. To do so would violate the Heisenberg uncertainty principle, for then they would disclose their exact positions and exact momenta simultaneously. There would be no indeterminacy, no quantum interference, no probability amplitude.

So, does *all* motion cease at absolute zero? No. The quantum world is too jittery ever to come to a complete halt. Does the third law of thermodynamics, which assigns zero entropy to a perfect crystal at 0 K, ever come fully into force? Can we even hope to reach absolute zero?

4. VISIBLE AND ULTRAVIOLET RADIATION. Infrared radiation (with photon energies *below the red*) gives way to ordinary red light at wavelengths between 800 nm and 700 nm (1 nm = 10^{-9} m = 10^{-3} μm = 10 Å), the onset of the visible region. Here begins the perception of color, and here begins the excitation of the most weakly bound *electrons*: the valence electrons, those responsible largely for chemical bonding.

Screened by the core, the valence electrons are attracted by an attenuated and distant nuclear charge. They occupy orbitals far from the nucleus—highest in energy—and as a result, they are the electrons easiest to disturb. These outermost electrons are the first to be ionized and, under milder excitation, the first to be promoted to higher (but still confined) levels of energy:

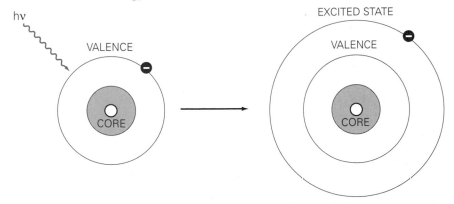

There is always room to go higher. In a transition-metal complex, an electron might be promoted from a t_{2g} state to an e_g state. In a conjugated organic molecule, the jump might be between an occupied bonding orbital and an unoccupied antibonding orbital. In a rare-earth compound, the transition might involve *f* orbitals. There is always a frontier, an open territory beyond the valence region.

Some valence transitions are stimulated by red light, whereas others require orange, yellow, green, blue, violet, or even shorter wavelengths. There is no sharp cutoff between visible and invisible, though, and these same valence excitations continue *beyond the violet* into the ultraviolet (uv) region, which runs from approximately 380 nm to 10 nm. Visibility is not the issue. Color is not the issue. *Energy* is. Systems that interact with ultraviolet light may be colorless, but the mechanism of excitation remains the same: the promotion of valence electrons.

Absorbing a properly tuned photon, an excited atom or molecule will move to a state of higher electronic energy. A valence electron goes from an occupied orbital to an unoccupied orbital, and afterward the structure is no longer what it once was. Size and shape may be different. Bond strengths may be different. Polarity may be different. Magnetic properties may be different. Color may be different. The distribution of electrons has been altered, and there are consequences—consequences including, possibly, the complete removal of the valence electron and the creation of an ion, as below:

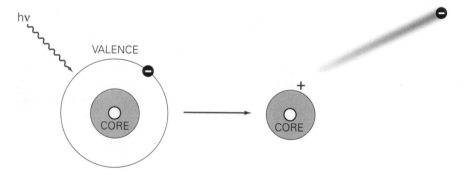

We call this latter consequence **photoionization**, an ionization attributable to the energy of a photon. It is an ionization by light, the same *photoelectric effect* described in Chapter 4: A photon is absorbed, and an electron breaks loose. For most molecules, the requisite ionization energy can be provided by ultraviolet radiation at wavelengths less than 200 nm.

5. X Rays. What is true for the valence is true, in general, for the core as well, but with one important qualification: Core electrons are harder to dislodge. Bound tightly around the nucleus, they bear the brunt of a large, poorly shielded positive charge. They sit low in energy, far below the first vacant orbital. They are not easily moved.

To promote a core electron requires the considerable energy of an X-ray photon, for which wavelengths range from 10 nm down to atomic-sized dimensions of 0.1 nm (1 Å) and below. Lesser energies are insufficient to rip a core electron away from the fierce grip of a nearly bare nucleus.

Wielding X rays, however, we are able to pierce the interior in increasingly clever ways (some of which are sketched in Figure 20-8). We can, for a start, eject an electron from the core and measure both its final kinetic energy and original orbital energy. This ejected *photoelectron* then leaves behind a vacancy in a low-lying orbital, giving us opportunity for further exploration. One possibility: Into the newly vacant orbital

FIGURE 20-8. Strategies for probing electrons in the core. (a) *X-ray photoelectron spectroscopy*. The incoming photon ejects an electron from a low-lying orbital, giving the freed particle a certain kinetic energy. (b) *X-ray fluorescence*. First, the X-ray photon ($h\nu$) expels an electron and leaves a vacancy in one of the orbitals. Another electron, falling to lower energy, then fills the hole while a second photon ($h\nu'$) is emitted. (c) *Auger spectroscopy*. Similar, but now a second *electron* is emitted instead of a photon. The energy to expel the additional electron comes from the energy released when the first vacancy is filled.

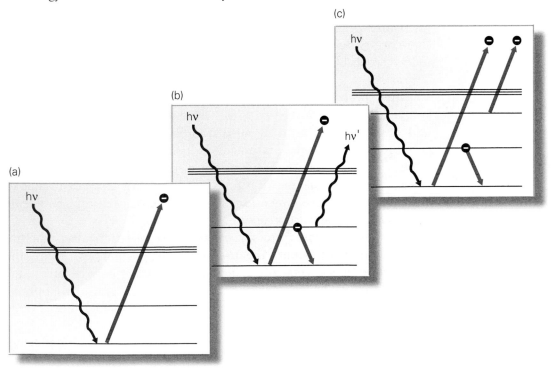

may fall an electron from a higher level, accompanied by the emission of another X-ray photon. Another possibility: Rather than appearing as emitted light, the energy of the falling electron is used to eject a *second* photoelectron from a different core orbital. Either way, we study the energy evolved as the perturbed system relaxes, and, in the process, we map out the innermost orbitals of the structure.

Alternatively, we may use the data to analyze the elemental composition of our material. These core electrons, understand, associate more with their original nuclei than with the molecule as a whole. They do not contribute to the bonding; they are not shared; they are not delocalized. Confined near a nucleus, a core electron resides in an environment little different from an uncombined *atom*; and so the electron's energy, measured by X-ray spectroscopy, hardly varies from molecule to molecule. The energy of an inner orbital therefore serves as an atomic fingerprint, a distinctive signature for each atom.

Spectroscopy: An Appreciation

Eyes opened, we now see the microscopic world in a new light; we see a world illuminated by the entire electromagnetic spectrum, a world of new possibilities. Every atom, every molecule, every particle in that world has options. Nothing is fixed; everything can change.

Before spectroscopy, we might describe a water molecule as follows: There are two atoms of hydrogen and one atom of oxygen, covalently bonded, probably arranged in the form H—O—H with a molar mass of 18 grams.

After spectroscopy, we say: The H_2O molecule contains two hydrogen nuclei and one oxygen nucleus, together with 10 electrons distributed over a defined set of molecular orbitals. There are no unpaired electrons, but the hydrogen nuclei do possess nuclear magnetic moments. Further: Each O—H bond has, in the ground state, an average length of 0.958 Å, and the H—O—H bond angle has an average value of 104.5°. Both of these dimensions fluctuate constantly as the molecule vibrates about its average structure.

But there is far more to say, because this one particular structure (the ground state) is only the first of many possibilities. Above the ground state rises a seemingly endless progression of *excited* states, each a variation on the original structure. Spectroscopy reveals these options to us, and spectroscopy tells us precisely what the molecule needs to go up the ladder: measured, quantized doses of energy.

Up to where? To new energies, to new forms of the same molecule.

The nuclear magnets can flip. The molecule can rotate faster, vibrate faster, move in new ways. The electrons can acquire energy and move into different regions. Making spectroscopic measurements, we count all the possibilities and thus uncover the pattern of energy levels. Each energy level, measured spectroscopically, represents just one of the many rotating, vibrating, electronically excited versions of the structure. How else, absent spectroscopy, could we have known that?

And spectroscopy does more than simply excite a system to a higher level. Excited states, although rarer than the ground state, already exist. Even without the spectroscopist's laser beam or microwave apparatus, atoms and molecules promote *themselves* all the time. They have, as we recall from Chapter 10, a standard allotment of *thermal energy*, randomly apportioned according to the temperature. They jump around. Some systems, endowed with sufficient thermal energy, exist naturally in an excited state; others, less energetic, cannot rise above the ground state. The lower the temperature, the greater will be the number of molecules confined to the ground state. It is a game of chance, with the odds decided by the Boltzmann distribution as described in Chapter 14.

But nature's random game is already over for the system. For to conform to the Boltzmann distribution—to have a temperature—is to be in thermal equilibrium: to be macroscopically unchanging, dead. Nothing seems to happen. Masked by the dynamic equilibrium is any overt sign of ongoing change, any jumping from state to state, any expression of the frequencies hidden within. Viewed macroscopically, a system in thermal equilibrium is like a bell that does not ring.

Enter spectroscopy (Figure 20-9), which does what nature will not. Spectroscopy shakes the system out of equilibrium. It rings the bell, and we listen to the sounds produced. Hearing the different pitches and durations, we recreate a picture of the bell: size, shape, strength, composition.

Spectroscopy replaces the randomness of equilibrium with a determined, systematic stimulation of the system. Wavelength by wavelength, we probe the material for its resonances. Level by level, we identify the quantum states and measure their energies. Energy by energy, we ascribe physical attributes to the states.

Structure and identity follow. A spectrum may be just a set of lines recorded on a sheet of paper, but realize what these lines mean. Each is an electromagnetic message from an electron, nucleus, atom, or molecule; each is one piece of a puzzle. The challenge is to put together the structure from its spectroscopic signals—or, failing that, to identify the unknown material based on its unique spectral fingerprint.

It is a puzzle we have learned to solve; it is a puzzle we *had* to solve. In the past, spectroscopy illuminated for us the quantized structure of

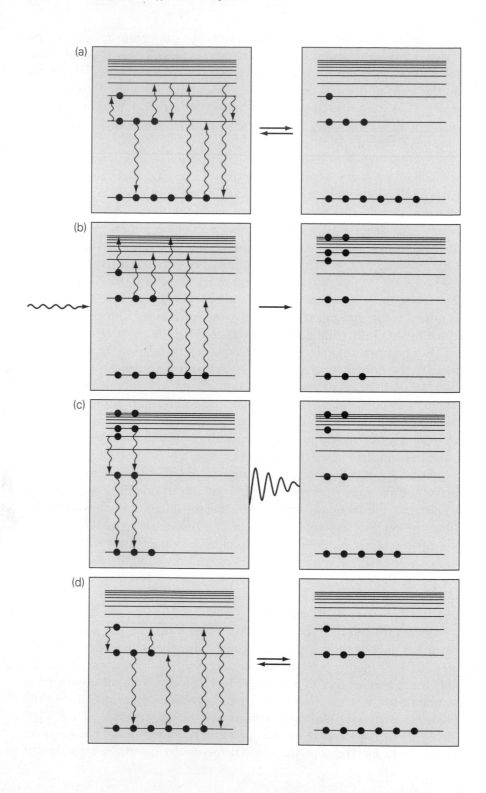

Figure 20-9 (opposite). Spectroscopy: excitation and response. (a) Left alone, a system comes to thermal equilibrium and there it remains. Its particles hop from level to level, but to no net effect. The Boltzmann distribution holds throughout (shown schematically as six particles, three particles, and one particle in the lowest three states). (b) Supplying energy from the outside, a spectroscopist excites the system and disturbs the equilibrium. The populations on the right (after excitation) differ from those on the left. (c) The system begins to return to equilibrium, resonating at its natural frequencies while the particles fall back toward the original thermal distribution. (d) Relaxation is complete, and equilibrium is restored.

atoms and molecules. In the present, spectroscopy assists in everything from routine analysis to the development of new materials and processes. In the future, expect more of the same.

20-2. MANIPULATIONS IN SPACE AND TIME

There is yet one more way for a molecule to express its energy. In addition to nuclear spin flips (at radiofrequencies), in addition to electronic spin flips (at microwave frequencies), in addition to rotations (also microwave), in addition to vibrations (infrared), in addition to valence electronic excitation (visible, ultraviolet), in addition to core electronic excitation (X ray) . . . in addition to all these *degrees of freedom*, there is still one more option available to a molecule. It can, simply, *move*.

Like a billiard ball, a molecule can move from point A to point B; like a billiard ball, a molecule can undergo **translational motion**. Set free, a molecule will travel north, south, east, west, up, down. The entire body will move, independent of its internal goings-on, in any of three perpendicular directions: x, y, z on the coordinate grid.

The larger the distance covered, the more a moving molecule begins to resemble a billiard ball. Gaps between energy levels shrink to nothing when a particle gains its freedom; and so the little ball just moves ahead, seemingly in any direction and at any speed. Unconfined, the molecule suffers no quantum restrictions on its translational energy ($\frac{1}{2}mv^2$). We can change that energy at will, employing the appropriate force to redirect the motion.

Macroscopic rules apply. To a stationary charge q, we administer an electric field E. To a charge moving at some velocity v, we administer a magnetic field B. The stationary charge receives an electric force

$$F = qE,$$

and the moving charge receives a magnetic force

$$F = qvB \sin \theta$$

that varies with θ, the angle between v and B:

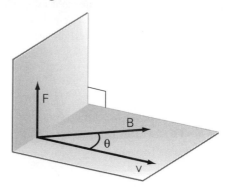

The magnetic force is perpendicular both to the direction of motion and to the magnetic field. The electric force, simpler mathematically, lies parallel to the corresponding electric field.

These fields E and B, independent of time ($v = 0$), will be our tools to move ions and molecules.

We begin with the magnetic force.

Mass Spectrometry

Return now to the basic laws of force and motion (Chapter 1), and recall: Nothing starts to move unless pushed. A body changes its velocity only when subjected to a force. No force, no acceleration.

Left alone, a particle moving in a straight line at constant speed will simply keep on going—its speed unchanged, its direction unchanged. Apply a force, however, and the object will *accelerate*. It will change velocity at the rate

$$a = \frac{F}{m}$$

where m denotes the mass, F denotes the force, and a denotes the acceleration. The speed can change. The direction can change. Both can change.

There are only three variables to consider:

1. *The mass of the object.* Heavier masses are harder to move; lighter masses are easier to move. A lighter body develops a proportionally greater acceleration under the same force.

2. *The magnitude of the force.* A strong push induces a big acceleration; a weak push induces a small acceleration.

3. *The direction of the force.* A force delivered along the line of motion keeps the particle moving in the same direction; a force delivered *perpendicular* to the line of motion deflects the particle into a curved path.

Just imagine trying to throw a ball parallel to the ground. Will the moving body trace out a straight line? Never. The force of gravity, acting at right angles to the initial motion, forces the ball to follow a curved trajectory. Gravity pulls the mass down while it tries to move forward.

Now suppose that this ball is really a molecular ion, some unknown species M^{q+} with a charge equal to $+q$. Suppose further, as in Figure 20-10, that the ion is traveling in a straight line—left to right—until suddenly

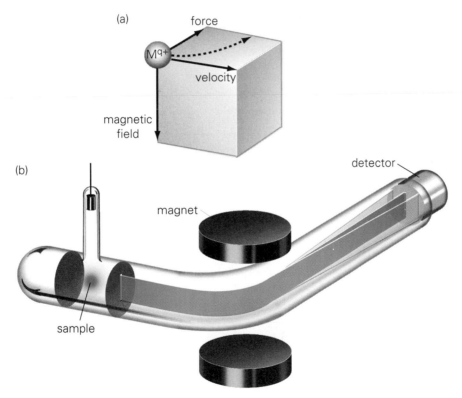

FIGURE 20-10. Mass spectrometry. (a) Application of a magnetic force causes a moving charge to deflect from its linear path. The extent of the deflection depends on the ratio of charge to mass. (b) A mass spectrometer uses the same principle to measure the mass of an ion. A beam of gas molecules is first ionized and then accelerated to high speed by an electric field. After passing through a magnetic field, ions of different masses follow different paths and are detected separately.

we disturb it. To do so, we subject the moving charge to a magnetic field, B, so as to deflect the particle with a perpendicular force

$$F = qvB$$

when v and B are at right angles. Knocked off course, the ion changes path. It starts to move along a curve, coming to rest only after hitting some backstop placed in the instrument.

Our instrument is called a ***mass spectrometer***, and our experiment is about to give us the *mass* of the unknown molecular ion. The mass spectrometer sends a beam of unknown structures through a magnetic field, sorts the mixture according to molecular mass (including all possible isotopes), and presents the results as a *mass spectrum*. For a pair of examples, see Figure 20-11.

Unsure whether some material contains $C_{12}H_{26}$ (mass number = 170) or $C_{12}H_{24}$ (mass number = 168)? Run a sample through the mass

FIGURE 20-11. (a) Mass spectrum of dodecane, $C_{12}H_{26}$. (b) Mass spectrum of 1-dodecene, $C_{12}H_{24}$. The various peaks correspond to fragments produced during the violent initial ionization of the molecule. Each spectrum is different.

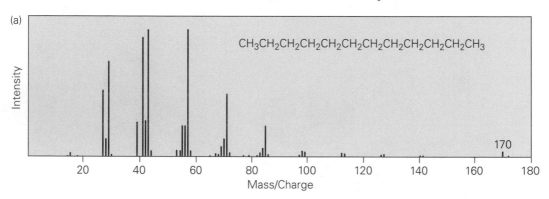

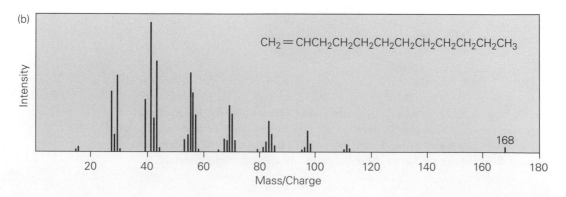

spectrometer and study the spectrum. The question will be decided by the patterns of measured mass. It is a technique of extraordinary power, able to distinguish between masses that differ by as little as one part in several thousand.

The operation proceeds as follows. A mass spectrometer (1) takes a collection of molecules, (2) ionizes them, (3) forms the ions into a directed beam, (4) shoots the beam through a magnetic field, and (5) monitors the arrival of the beam at a detector. From that measurement, we learn the mass.

Here is what we know. Having installed the magnetic field ourselves, we know both the magnitude and direction of B. Having set the ion on its original path, we also know the value of v. Understanding, moreover, that electric charge is quantized, we know that q must be a small integer: 1, 2, 3 units, probably not higher.

We know, all told, the magnitude and direction of the force ($F = qvB$) to within some multiple of q. And, after monitoring the curved trajectory from beginning to end, we know the acceleration (a) as well.

Put all this knowledge together, and the result is

$$a = \frac{F}{m} = \frac{q}{m}\, vB$$

The classical acceleration is determined, as always, by force and mass together; and here, specifically, the acceleration (known) depends on the original velocity (known), the magnetic field (known), and the ratio of mass to charge (unknown). Unknown until *now*, that is—because now, after measuring all the other variables, we can solve for the value of m/q.

Since q typically is equal to +1, this determination of m/q usually gives us the mass of the particle directly. Even if not directly, though, an analyst soon learns to recognize patterns involving a number of tricky complications: higher ionic charges, combinations of isotopes, and broken fragments of molecules caused by excessively violent ionization. The reward is a clean, highly accurate measurement of the molecular mass, often the critical link in a chemical puzzle.

Chromatography

The object shooting through a mass spectrometer is, under the prescribed conditions, a shapeless, featureless, formless *particle*. It has a mass, and it has a charge. Anything else is irrelevant.

The object passing through a chromatographic apparatus is, by contrast, a true *molecule*. It has a mass, certainly, but much more. It has a size. It has a shape. It has a distribution of charge. It has a polarity.

This assembly of electrons and nuclei is no undifferentiated particle;

it is an individual molecule, having characteristics unique to a particular compound. It interacts—in its own special way—with atoms, with ions, with other molecules. It is attracted toward some species and repelled by others. It dissolves in some materials, not others. It reacts selectively, depending on the partner: sometimes fast, sometimes slow, sometimes completely, sometimes partially, sometimes not at all. It has energetic preferences. It discriminates.

Each interaction is different, yet each interaction is really the same. The same: because every molecule-to-molecule contact derives, ultimately, from the electromagnetic attraction and repulsion of charged particles. But different too: because every interaction has a slightly different strength, a slightly different mechanism, a slightly different geometry.

These *differences* we can exploit deliberately and systematically, using the various techniques of **chromatography** to analyze a mixture. Chromatography enables us to address molecules as individuals, manipulate their interactions, and finally separate them according to chemical type.

Chromatographic separation is best pictured as a competition between two solvents for the same solute. Start with a solvent A and a solvent B, each with an affinity for some species X. Impose no specific requirements on A or B (for the moment), except this one: X must be able to dissolve in both A and B, but to differing extents. Thus equipped, we proceed to stage a molecular race over a macroscopic distance.

The racecourse has two tracks, depicted conceptually in Figure 20-12. Over track A (the **stationary phase**) is distributed solvent A, its material packed in place from end to end. The stationary phase might be a porous solid packed in a tube, a liquid adsorbed onto a porous solid, a sheet of moistened paper, or any of a number of possibilities; but, viewed *macroscopically*,

FIGURE 20-12. Chromatography: separation of a mixture. Two solvents compete for the same solute as it passes through a chromatographic column. One of the solvents, the *mobile phase* (B), moves rapidly compared with the other solvent, the *stationary phase* (A). Since the solute interacts preferentially with the solvent offering a larger equilibrium constant and hence a deeper equilibrium, different kinds of molecules (X, Y, Z) linger to different extents in each phase. They emerge from the column one after the other, separated in space and time.

the stationary phase always displays one salient feature: it appears not to move. The stationary phase seems to stand immobile, like a line of parked cars.

On track B, we arrange for solvent B to flow continuously over the same stretch laid out for track A. Gas or liquid, the molecules of B are carried along with the macroscopic flow. They form a **mobile phase**, running directly over the stationary phase. Solvent A stands. Solvent B runs. Each will be given a chance to capture the solute molecule X.

We introduce the solute at the start of the course and await the outcome. If X dissolves preferentially in the mobile phase, then the solute is carried swiftly to the end of the track. If, however, the stationary phase is the preferred host, then passage is delayed. Caught up in the slower solvent, the solute exits at a later time. Its motion is retarded.

X will prefer A to B (or B to A) for all the usual thermodynamic considerations: lower enthalpy, higher entropy, lower free energy. The outcome depends on the detailed structures of solute and solvent, encapsulated in Chapter 15's adage of "like dissolves like." Either A or B might be closer in polarity to X. Like dissolves like. Either A or B might be better matched to X in size or shape or charge distribution. Like dissolves like. Or, also possible, one of the solvents might capture X by engaging it in a chemical reaction (acid–base, say), so as to convert X into Y. Whatever the mechanism, there will be some difference that distinguishes the A–X interactions from the B–X interactions; and that difference will give one solvent a thermodynamic advantage over the other.

The competition is described compactly by the attempted equilibrium between the two phases,

$$X(B) \rightleftharpoons X(A)$$

for which we define the equilibrium constant (here called the **partition coefficient**) as

$$K = \frac{[X(A)]}{[X(B)]}$$

The greater the affinity of X for A, the larger the concentration of X in the stationary phase. A large partition coefficient means a slower passage for the solute, whereas a strong affinity of X for B ensures a fast trip through the mobile phase.

So much for one solute. Now add X and Y together; and then X, Y, Z; and then W, X, Y, Z; and U, V, W, X, Y, Z; or as many as you like. Each component is a unique, individual molecule with its own set of interactions. Each component responds differently to the two solvent phases.

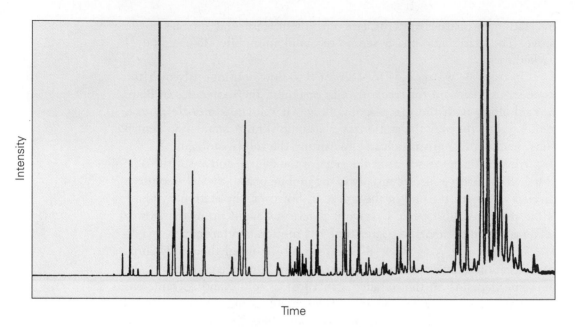

FIGURE 20-13. Chromatographic separation of a sample of gasoline into its individual hydrocarbon components. Each signal arises from a different species of molecule, present in a concentration proportional to the intensity of the corresponding peak.

Each component has its own partition coefficient. Each component emerges *in its own time*. We give the mixture access to both phases, and we collect the individual compounds (X, Y, Z) at the end. That done, the mixture is separated. Its components are identified.

Chromatography succeeds on a fine scale, well suited for sensitive analysis. Trace amounts of substances can be separated and collected chromatographically, as little as one part in 10^9 by mass and sometimes even finer. Mixtures containing scores of components can be analyzed, as in Figure 20-13. Unknown compounds can be identified by their chromatographic profiles.

And there are dozens of ways to do it: gas–liquid chromatography, gas–solid chromatography, thin-layer chromatography, liquid chromatography, gel permeation chromatography, paper chromatography, ion-exchange chromatography, adsorption chromatography, partition chromatography, column chromatography, and many more. These assorted methods employ different phases, different kinds of selectivity, and different methods of detection, but all derive from a common principle. All take advantage of the unique electrical properties of each molecule's charge distribution.

So they must. What else is there?

REVIEW AND GUIDE TO PROBLEMS

How else, if not with the electromagnetic field, shall we probe deeply into the molecular structure of matter? There is no other way. Without the electromagnetic interaction—without *light*—matter disappears. Without light, there is nothing to see.

Through spectroscopy, light becomes a tool; and what eye and brain do naturally, spectroscopist and chemist now do by design: peer into the atoms and molecules, spy out the resonances buried within, make visible the invisible.

IN BRIEF: SPECTROSCOPY—
THE INTERACTION OF LIGHT AND MATTER

1. THE TARGET: MATTER. Atoms and molecules, electrons and nuclei, particles with mass and charge, particles that can move, particles that can absorb and emit energy—there, in sum, is the whole substance of the universe; and there, no less, is our whole aim: to make that hidden substance clear . . . to know how the particles are put together . . . to know how near they come, how far they go . . . to know which ways they move and how much energy they need.

To do so, we put matter to the test. We strike it like a bell and listen to it ring. We smash it to bits and sort through the debris. We excite the particles; we await their response.

How? With the electromagnetic field. There is no other way to provoke an electrical object into action.

2. THE PROBE: THE ELECTROMAGNETIC FIELD, the means by which energy goes into and comes out of a charged particle. As waves spanning the **electromagnetic spectrum**, the force moves through space and interacts with any electron, nucleus, atom, or molecule in its path. Packaged in discrete bits, **photons**, the energy is delivered in quantized doses

$$E = h\nu = \frac{hc}{\lambda}$$

where the frequency (ν) and wavelength (λ) are linked to the speed of light (c):

$$\lambda\nu = c$$

The higher the frequency, the higher the energy. The higher the frequency, the shorter the wavelength.

3. THE RANGE: THE ELECTROMAGNETIC SPECTRUM. Low energy to high, the oscillations run from zero frequency (a constant field) up through radiofrequency waves, microwaves, infrared radiation, visible light, ultraviolet radiation, and X rays, all the way to gamma rays at the very highest frequencies and shortest wavelengths.

Each bullet hits a specific mark. A steady, *zero*-frequency field, simplest of all, takes hold of the overall charge of an ion or the dipole moment of a polar molecule, giving the body a push forward in space—a "translational" motion, we say, with energy so finely spaced as to be effectively unquantized. *Radiofrequency* waves, oscillating at frequencies up to several hundred megahertz, stimulate the magnetic moments arising from the spin of a nucleus, changing the direction with a small shot of energy. *Microwaves*, more energetic, flip the spin of an unpaired electron and trigger the rotation of a molecule as well. *Infrared*, the vibration of a molecule. *Visible* and *ultraviolet*, the state of a valence electron. *X rays*, the energy of a core electron. *Gamma* radiation, the protons and neutrons within a nucleus.

4. HITTING A MARK: TRANSITION. When a photon's energy ($E = h\nu$) matches the spacing between two levels of a molecule, the electromagnetic field induces a spectroscopic ***transition*** from ϵ_i to ϵ_f:

$$\Delta E = h\nu = \epsilon_f - \epsilon_i$$

Energy may be absorbed, so that ϵ_f is greater than ϵ_i; or it may be emitted, so that ϵ_f is less than ϵ_i. In the one instance, the molecule takes a photon from the electromagnetic field and rises to higher energy. In the other, the molecule gives a photon *to* the electromagnetic field and falls to lower energy.

Either way, the molecule's energy levels stand waiting to be probed, quantum by quantum, frequency by frequency, either by ***absorption spectroscopy*** or ***emission spectroscopy***. The choice is ours. Measuring how much light disappears into an excited system, we map the levels by absorption spectroscopy. Measuring how much light comes out, we map the same levels by emission spectroscopy.

SAMPLE PROBLEMS

The first step is to deliver light with energy $h\nu$ comparable to the allowed transitions of the matter, $\epsilon_f - \epsilon_i$. Examples 20-1 through 20-4 give some idea of the numbers involved, for transitions ranging from the flip of a nuclear spin (radiofrequency) to the ejection of a core electron (X ray).

Example 20-1. Radiofrequency Excitation: Nuclear Magnetic Resonance

NMR, nuclear magnetic resonance; a wealth of chemical information to be gleaned, nearly too good to be true. The energies are low, and the corresponding radiation (radiofrequency) is easy to produce and easy to control. A spectroscopist's dream, it seems—almost too easy, except for this one complication: a dearth of targets. Too many of the particles are already excited.

PROBLEM: Placed into a magnetic field, a hydrogen nucleus has access to two states: spin-up (↑) and spin-down (↓). The difference in energy thus created,

$$\Delta \epsilon = \epsilon_2 - \epsilon_1 = \gamma_H B_0$$

increases directly with the strength of the external magnetic field, B_0, measured in gauss (G):

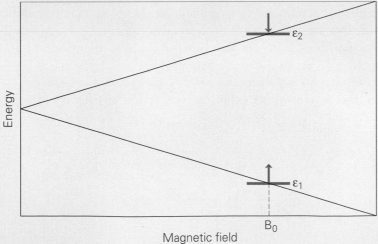

Completing the relationship is a proportionality constant, γ_H, which has the value 2.82×10^{-30} J G^{-1}. (a) Allow a sample containing hydrogen nuclei to equilibrate in the earth's magnetic field, approximately 1 G. Then compute $\Delta \epsilon$ and the Boltzmann population ratios at 4.2 K (the temperature of liquid helium), 77 K (liquid nitrogen), and 273 K (liquid water). (b) Do the same for magnetic fields of 10^4 G and 10^5 G.

SOLUTION: We have two spin states for each nucleus, a ground state (↑) with energy ϵ_1 and an excited state (↓) with energy ϵ_2; and we want to stimulate a transition from ↑ to ↓:

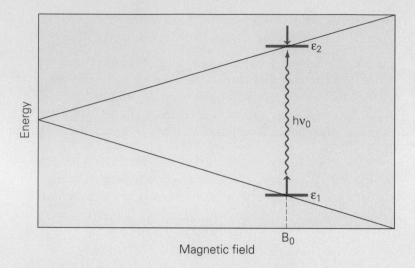

What do we need?

We need, first of all, a nucleus that *can* be excited, a nucleus not *already* excited, a nuclear magnet still able to soak up energy. For that, we need a nuclear spin in the ground state, its energy ϵ_1 at a minimum.

Ask, then, more sharply: At thermal equilibrium, absent our direct intervention, how many nuclear spins (N_1) are in the ground state and how many (N_2) are in the excited state?

To answer, we recall nature's recipe for thermal equilibrium in any system: the Boltzmann distribution, elaborated in Chapters 10 and 14. Defined as

$$f_{12} = \frac{N_1}{N_2} = \frac{\exp\left(-\dfrac{\epsilon_1}{k_B T}\right)}{\exp\left(-\dfrac{\epsilon_2}{k_B T}\right)} = \exp\left(-\frac{\epsilon_1 - \epsilon_2}{k_B T}\right) = \exp\left(\frac{\Delta\epsilon}{k_B T}\right)$$

for a stated difference in energy ($\Delta\epsilon = \epsilon_2 - \epsilon_1$), the Boltzmann ratio fixes the ratio of N_1 to N_2 at constant temperature. The higher the value of f_{12}, the more systems (N_1) are in the ground state and, consequently, the more spectroscopic targets we have.*

*Notational alert: Since the energy of the excited state, ϵ_2, is greater than the energy of the ground state, the value of f_{12} is always greater than 1. Were we to define $\Delta\epsilon$ as $\epsilon_1 - \epsilon_2$, then $\Delta\epsilon$ would be negative and f_{12} would be written as $\exp(-\Delta\epsilon/k_B T)$. Instead, we take $\Delta\epsilon$ as equal to $\epsilon_2 - \epsilon_1$ and write the exponential simply as $\exp(\Delta\epsilon/k_B T)$. Both definitions yield the same value for f_{12}. The presence or absence of the minus sign has no physical significance.

Start with the smallest magnetic field given, 1 G. Measured *per particle*, the difference in energy between the excited state ($\downarrow$) and ground state ($\uparrow$) is

$$\Delta\epsilon = \gamma_H B_0 = (2.82 \times 10^{-30} \text{ J G}^{-1})(1.00 \text{ G}) = 2.82 \times 10^{-30} \text{ J}$$

and, equivalently, the difference in energy *per mole* (ΔE) is

$$\Delta E = N_0 \, \Delta\epsilon = (6.02 \times 10^{23} \text{ mol}^{-1})(2.82 \times 10^{-30} \text{ J})$$
$$= 1.70 \times 10^{-6} \text{ J mol}^{-1}$$

Either one will do, provided that we correctly use either Boltzmann's constant ($k_B = 1.38 \times 10^{-23}$ J K^{-1}) or the universal gas constant ($R = N_0 k_B = 8.3145$ J mol^{-1} K^{-1}) when computing the ratio N_1/N_2:

$$f_{12}(B_0, \, T) = \frac{N_1}{N_2} = \exp\left(\frac{\Delta\epsilon}{k_B T}\right) = \exp\left(\frac{\Delta E}{RT}\right)$$

k_B goes with $\Delta\epsilon$, and R goes with ΔE.*

At 4.2 K, the lowest temperature, we find:

$$f_{12}(B_0, \, T) = \exp\left(\frac{\Delta E}{RT}\right)$$

$$f_{12}(1 \text{ G}, \, 4.2 \text{ K}) = \exp\left[\frac{1.70 \times 10^{-6} \text{ J mol}^{-1}}{(8.3145 \text{ J mol}^{-1} \text{ K}^{-1})(4.2 \text{ K})}\right]$$

$$= \exp(4.87 \times 10^{-8}) = 1.0000000487 \approx 1$$

The exponent ($\Delta E/RT$) is only 4.87×10^{-8}, nearly zero, and the resulting Boltzmann ratio differs scarcely from 1.

QUESTION: What inference should we draw?

ANSWER: See it this way: For every 10,000,000,487 hydrogen nuclei with spin up ($\uparrow$), there are fully 10,000,000,000 with spin down ($\downarrow$)—a difference of less than one in 20 million, a mere speck, but from this tiny excess must come the whole NMR spectrum. These few low-energy particles are the only ones that can, on balance, absorb the radio

---

*Hence the symbol E will do double duty for the remainder of this chapter, its role always made clear in context: either as a photon energy ($E = h\nu$) or as a *molar* material energy in combination with RT.

frequency energy we supply and undergo a transition from ↑ to ↓. At equilibrium in this small field ($B_0 = 1$ G), nearly half of the nuclei already have enough thermal energy to exist naturally in the higher state.

If, for example, 10,000,000,000 nuclear spins absorb energy and go from ↑ to ↓, then another 10,000,000,000 nuclear spins (already excited) simultaneously emit energy and go from ↓ to ↑. Except for the extra 487 ground-state nuclei, there would be neither absorption nor emission. Only from the slight excess of ↑ over ↓ can we stimulate any net absorption under the conditions described.

That one odd nucleus in 20 million is our unlikely target.

QUESTION: What can be done to improve the odds?

ANSWER: We can either change the temperature or change the magnetic field. On the one hand, any increase in temperature will push more nuclear spins into the excited state and consequently make the ratio even closer to 1. Increasing the magnetic field, however, will *increase* the value of ΔE and thus skew the Boltzmann factor in favor of the ground state. Results are given in the table below:

	B_0 (G)	T (K)	ΔE (J mol^{-1})	$\Delta E/RT$	$\exp(\Delta E/RT)$
(a)	1	4.2	1.70×10^{-6}	4.87×10^{-8}	1.0000000487
	1	77	1.70×10^{-6}	2.66×10^{-9}	1.0000000027
	1	273	1.70×10^{-6}	7.49×10^{-10}	1.0000000007
(b)	10^4	4.2	1.70×10^{-2}	4.87×10^{-4}	1.000487
	10^4	77	1.70×10^{-2}	2.66×10^{-5}	1.000027
	10^4	273	1.70×10^{-2}	7.49×10^{-6}	1.000007
	10^5	4.2	1.70×10^{-1}	4.87×10^{-3}	1.00488
	10^5	77	1.70×10^{-1}	2.66×10^{-4}	1.00027
	10^5	273	1.70×10^{-1}	7.49×10^{-5}	1.00007

Conclusion: Even at low temperatures and high magnetic fields, there are nearly equal numbers of ↑ nuclei and ↓ nuclei. At odds of approximately one in a hundred thousand under normal conditions, nuclear magnetic resonances are still hard to find.

Nevertheless, they *can* be found; they *can* be induced; and, benefitting from advanced techniques and instrumentation, today's NMR spectroscopist overcomes any natural handicaps with ease. The rewards are great.

For just the barest hint of those rewards, see Example 20-8.

EXAMPLE 20-2. Microwave Radiation and Rotational Energies

Microwave energies usually correspond to differences between the rotational states of a structure. From a microwave spectrum, a spectroscopist deduces the bond lengths and bond angles of a polar molecule.

PROBLEM: Microwave ovens use 0.122-m radiation to excite H_2O molecules to higher rotational energies. Calculate the energy and Boltzmann population ratio for a transition at 273 K.

SOLUTION: A single photon ($\lambda = 0.122$ m) delivers an energy of

$$E = h\nu = \frac{hc}{\lambda} = \frac{(6.63 \times 10^{-34} \text{ J s})(3.00 \times 10^8 \text{ m s}^{-1})}{0.122 \text{ m}} = 1.63 \times 10^{-24} \text{ J}$$

while oscillating at a frequency of

$$\nu = \frac{c}{\lambda} = \frac{3.00 \times 10^8 \text{ m s}^{-1}}{0.122 \text{ m}} = 2.46 \times 10^9 \text{ s}^{-1}$$

The ground-state-to-excited-state ratio, although more favorable than the values we found for NMR transitions at the same temperature (Example 20-1), is still barely greater than 1:

$$f_{12} = \exp\left(\frac{\Delta\epsilon}{k_BT}\right) = \exp\left[\frac{1.63 \times 10^{-24} \text{ J}}{(1.38 \times 10^{-23} \text{ J K}^{-1})(273 \text{ K})}\right]$$

$$= \exp(4.33 \times 10^{-4}) = 1.000433 \approx 1$$

EXAMPLE 20-3. Infrared Transitions: Energy, Wavenumbers, Vibrations

Spectroscopists often express photon energies in **wavenumbers** ($\overline{\nu}$), defining a *reciprocal wavelength*

$$\overline{\nu} = \frac{1}{\lambda}$$

to be directly proportional to E:

$$E = h\nu = \frac{hc}{\lambda} = hc\overline{\nu}$$

They do so for convenience, to replace an inverse proportionality ($E = hc/\lambda$) with a direct proportionality ($E = hc\bar{\nu}$).

The new quantity tells us how many complete cycles fit into a given unit of length—the centimeter, by long practice. If, say, λ is 0.1 cm for a certain wave, then $1/\lambda$ is 10 cm^{-1}: There are 10 wavelengths per centimeter, each extending over a distance of 0.1 cm. If λ is 0.01 cm, then $\bar{\nu}$ is 100 cm^{-1}: 100 wavelengths per centimeter. If λ is 0.001 cm, then $\bar{\nu}$ is 1000 cm^{-1}; and so forth, $\bar{\nu}$ being simply an inverse wavelength. The units are cm^{-1}, reciprocal centimeters.

PROBLEM: Stimulated by infrared radiation, the vibrations of most organic molecules fall between 500 cm^{-1} and 4000 cm^{-1}. (a) Compute the photon energy corresponding to infrared light at these two endpoints. (b) Compute the Boltzmann ratio at 273 K for states separated by 4000 cm^{-1}.

SOLUTION: Except for the wavenumber, defined above, there is little new here to consider. Be careful, though, to use $c = 3.00 \times 10^{10}$ *centimeters* per second (not meters per second) in the equation $E = hc\bar{\nu}$.

(a) *Energy*. We start with the energy of a transition at 500 cm^{-1} ($\lambda = 0.002$ cm $= 0.00002$ m, $\nu = 1.5 \times 10^{13}$ s^{-1}), computing a value first for a single photon

$$E = hc\bar{\nu} = (6.63 \times 10^{-34} \text{ J s})(3.00 \times 10^{10} \text{ cm s}^{-1})(500 \text{ cm}^{-1})$$
$$= 9.95 \times 10^{-21} \text{ J}$$

and then converting the result into kilojoules per mole:

$$N_0 E = (6.02 \times 10^{23} \text{ mol}^{-1})(9.95 \times 10^{-21} \text{ J}) \times \frac{1 \text{ kJ}}{1000 \text{ J}} = 5.99 \text{ kJ mol}^{-1}$$

Call it 6 kJ mol^{-1}, the energy delivered by infrared light at 500 cm^{-1}.

Moving next to 4000 cm^{-1}, we quickly come to appreciate the convenience of the wavenumber as a unit of energy. Since E and $\bar{\nu}$ are directly proportional, the energy at 4000 cm^{-1} is simply eight times the value at 500 cm^{-1}:

$$6 \text{ kJ mol}^{-1} \times \frac{4000 \text{ cm}^{-1}}{500 \text{ cm}^{-1}} = 48 \text{ kJ mol}^{-1}$$

(b) *Boltzmann ratio*. Vibrational levels evidently are separated by far more energy than either nuclear spin levels (approximately

0.00002 kJ mol^{-1} to 0.0002 kJ mol^{-1} in typical fields, Example 20-1) or rotational states (0.001 kJ mol^{-1} to 1 kJ mol^{-1}, Example 20-2). The Boltzmann ratio of ground-state-to-excited-state population, evaluated at 273 K, shoots up to greater than 10^9 as the gap between the ground and excited states widens to 4000 cm^{-1} (48 kJ mol^{-1}):

$$f_{12} = \exp\left(\frac{\Delta E}{RT}\right) = \exp\left[\frac{48 \times 10^3 \text{ J mol}^{-1}}{(8.3145 \text{ J mol}^{-1} \text{ K}^{-1})(273 \text{ K})}\right]$$

$$= \exp 21.15 = 1.5 \times 10^9$$

For every system already excited thermally, there are over 1 billion still in the ground state.

Meaning: With ΔE ranging up to 50 kilojoules per mole, most of the molecules remain trapped in the ground state, unable to muster enough thermal energy to promote themselves higher. They need our help, and they have only one way to go: up.

Plentiful targets, these ground-state systems respond all the more to any infrared radiation supplied from without. They soak up the energy and invest it in faster, more energetic vibrations of the chemical bonds.

QUESTION: Can they soak up enough energy to shatter a covalent bond?

ANSWER: If the linkage is particularly weak, yes; but with bond energies of several hundred kilojoules per mole, most molecules stay intact under infrared excitation. Like young trees in a windstorm, the bonds stretch, bend, and wag but usually they do not break. For a demonstration, see Example 20-7.

EXAMPLE 20-4. Electronic Excitations: Visible, Ultraviolet, X Ray

Infrared energies are typically too low to promote an electron to a higher level. To ionize a valence electron, to break a bond, to expel a core electron, or even just to bring color to a molecule is to do real violence to the structure. We need photons with more energy.

PROBLEM: Compute photon energies (kJ mol^{-1}) over each of the following portions of the electromagnetic spectrum: (a) Visible: 7000 Å through 4000 Å. (b) Ultraviolet: 400 nm through 10 nm. (c) X ray: 10 nm through 0.01 nm.

SOLUTION: We use the relationship $E = hc/\lambda$ once again, careful to convert both angstroms and nanometers into meters:

$$1 \text{ nm} = 10^{-9} \text{ m} \qquad\qquad 10^9 \text{ nm} = 1 \text{ m}$$

$$1 \text{ Å} = 10^{-10} \text{ m} \qquad\qquad 10^{10} \text{ Å} = 1 \text{ m}$$

$$10 \text{ Å} = 1 \text{ nm}$$

One last calculation of this sort should suffice:

$$7000 \text{ Å} \times \frac{10^{-10} \text{ m}}{\text{Å}} = 7 \times 10^{-7} \text{ m}$$

$$E = h\nu = \frac{hc}{\lambda} = \frac{(6.63 \times 10^{-34} \text{ J s})(3.00 \times 10^8 \text{ m s}^{-1})}{7.00 \times 10^{-7} \text{ m}} = 2.84 \times 10^{-19} \text{ J}$$

$$N_0 E = (6.02 \times 10^{23} \text{ mol}^{-1})(2.84 \times 10^{-19} \text{ J}) \times \frac{1 \text{ kJ}}{1000 \text{ J}} \approx 170 \text{ kJ mol}^{-1}$$

The full set of values is tabulated below:

	RANGE	WAVELENGTH (nm)	ENERGY (kJ mol^{-1})
(a)	visible	700	170
	visible	400	300
(b)	ultraviolet	400	300
	ultraviolet	10	12,000
(c)	X ray	10	12,000
	X ray	0.01	12,000,000

EXAMPLE 20-5. X-Ray Photoelectron Spectroscopy: Core Energies

What next? We have at our disposal the entire electromagnetic spectrum, the tools of a surgeon, a different probe for each degree of freedom. What shall we do with them? Shall we tickle the nuclear spins with a radiofrequency pulse? Flip the spin of an unpaired electron with a burst of microwaves? Make a molecule tumble? Or, aiming higher, shall we pump infrared energy into the vibrating bonds, or use visible and ultraviolet light to excite the valence electrons, or use X rays to expel electrons from the core?

We can do it all—map out the allowed energies, deduce structure,

identify unknown substances, detect impurities, analyze mixtures, solve
crimes . . . all by listening to the conversation between light and matter.
    The measurement of orbital energies provides a ready first example.

    PROBLEM: Take note that ionization energies for the 1s electrons of
the second-row atoms are:

Li	Be	B	C	N	O	F
4820	10,600	18,300	27,000	38,600	51,100	66,600 kJ mol^{-1}

Now suppose that a certain substance is bombarded by X rays having a
wavelength of 0.989 nm:

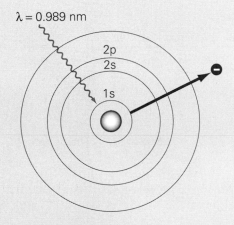

If photoelectrons with kinetic energies of 94,000 kJ mol^{-1} and 69,900 kJ
mol^{-1} are ejected from the material, which of the elements listed above
must be present in the sample?

    SOLUTION: Track the energy. Each quantum of radiation delivers a
jolt

$$E = h\nu = \frac{hc}{\lambda}$$

to the electrons, part of which pays for the ionization (I) while the rest
goes into the kinetic energy of the ejected photoelectron ($\frac{1}{2}mv^2$):

$$E = I + \tfrac{1}{2}mv^2$$

Energy is conserved.
    Given both E and $\frac{1}{2}mv^2$, we then compute the ionization energy (I)
needed to tear the electron loose from its 1s orbital—just enough,

understand, to park the electron outside the atom, not to impart any kinetic energy beyond:

$$I = E - \tfrac{1}{2}mv^2$$

That done, we compare the measured ionization energies with the standard values above to learn which elements are responsible.

For the photoelectron with kinetic energy equal to 94,000 kJ mol^{-1}, explicit calculation yields an ionization energy of 27,000 kJ mol^{-1}, appropriate for a carbon atom:

$$E = \frac{hc}{\lambda}$$

$$= \frac{(6.63 \times 10^{-34} \text{ J s})(3.00 \times 10^8 \text{ m s}^{-1})}{0.989 \times 10^{-9} \text{ m}} \times \frac{1 \text{ kJ}}{1000 \text{ J}} \times \frac{6.02 \times 10^{23}}{\text{mol}}$$

$$= 1.21 \times 10^5 \text{ kJ mol}^{-1}$$

$$I = E - \frac{1}{2}mv^2 = (1.21 - 0.94) \times 10^5 \text{ kJ mol}^{-1} = 2.7 \times 10^4 \text{ kJ mol}^{-1}$$

Similar treatment for the second photoelectron (kinetic energy = 69,900 kJ mol^{-1}) reveals an ionization energy of 51,100 kJ mol^{-1}, attributable to oxygen.

EXAMPLE 20-6. Picking Up the Pieces: Mass Spectrometry

Imagine some object, completely unknown, that we can neither touch, see, nor weigh—unless, perversely, we smash it to pieces and put it together again like a jigsaw puzzle.

Not unlike mass spectrometry, perhaps, where first we shatter a molecule, identity unknown, and then pass the ionized rubble through a magnetic field. Bending to the force, the ions sort themselves according to mass; while we, picking up the pieces, mentally reassemble the original structure from its fragments, better able now to see the molecule as a whole.

PROBLEM: Included in the mass spectrum of CH_3CH_2OH (Figure R20-1) are peaks attributable to masses of 15, 29, 31, and 46. From what fragments do these signals originate?

SOLUTION: Stripped of an electron, ethanol produces a parent ion of mass 46 ($CH_3CH_2OH^+$)—which, although it does appear in the

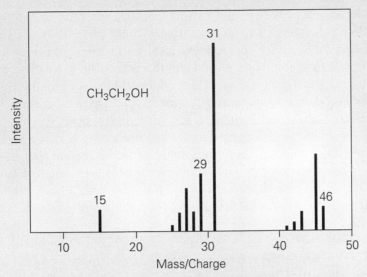

FIGURE R20-1. Mass spectrum of ethanol.

spectrum, is not the strongest signal. More intense is the peak from the fragmented ion $^+CH_2OH$ (mass = 31), which arises when the C—C bond cleaves as follows:

$$CH_3-CH_2OH \longrightarrow CH_3 + {}^+CH_2OH + e^-$$

The violence of the ionization tears the molecule in two, leaving behind the $^+CH_2OH$ ion and a neutral CH_3 radical. A weaker signal at mass 15, indicating a CH_3^+ ion, also develops when the division of electrons goes (less often) the other way:

$$CH_3-CH_2OH \longrightarrow CH_3^+ + CH_2OH + e^-$$

Cleavage of the C—OH bond to form $CH_3CH_2^+$, another option, similarly creates an ion with mass 29:

$$CH_3CH_2-OH \longrightarrow CH_3CH_2^+ + OH + e^-$$

There are other fragments and other peaks too, but already we sense what a mass spectrum has to say. It says, foremost, that we are witnessing *chemistry*, not the random destruction of a molecule. It says that the fragmentation, although violent, is not haphazard. It says that bonds cleave selectively; that the pieces make sense chemically; that charge appears where it can be best supported; that there is a pattern, a structure, a meaning.

Why a weak peak here and a strong peak there? Why no peak at this mass? Why this fragment and not that? Think why: Maybe a certain bond is unstable either thermodynamically or kinetically; maybe it is more fragile than some other bond and therefore able to be broken with comparatively little extra energy. A weak bond breaks all the more readily, leaving behind its signature fragment. Or maybe a certain ion is particularly stable, well suited to accommodate a positive charge. Or perhaps one ion can be formed easily by eliminating a stable molecule such as H_2O or H_2, just as another might profit by rearranging its functional groups to produce a new isomer. If the reasons sound familiar, it is only because they *are* familiar; they are the same reasons we cite to explain the reactions of acids and bases, reductants and oxidants, solutes and solvents, and any other chemical process.

Complicated? Yes. Even for a molecule as simple as ethanol, there are several peaks in the spectrum; and for larger, more complex structures the thicket of lines grows denser and denser. Many atoms exist, moreover, as isotopes, adding still more signals to the mass spectrum. Mass spectrometry, like other spectroscopies, is a puzzle, and puzzles sometimes are complicated.

But puzzles also can be solved, even if only partially; and to solve a mass spectrum is to take a great step in the analysis of an unknown molecule. Nor is complexity always a bad thing. Some spectra are indeed so complex as to be unique—like fingerprints, no two alike, instantly recognizable.

At best, we deduce the full structure. At worst, we learn the mass. Often, the reward falls somewhere in between: identification of the key functional groups, atoms, and structural features.

EXAMPLE 20-7. Vibrations: Infrared Absorption Spectroscopy

Light goes in and light comes out, but not every wavelength gets through. Some of the radiant energy, resonant with the matter, disappears into the molecules. It has a place to go.

For light, the choice is one or the other: either into the translational, rotational, vibrational, electronic, or magnetic energy of atom or molecule, *or not*; either into a spectroscopic transition, or out the back door.

If a wavelength emerges with intensity undiminished, it does so after finding no matching pair of energy levels within. When part of the light disappears en route, however, we witness instead a transfer of energy from light to matter, an absorption. The transition shows up in the absorption spectrum (Figure R20-2) as a dimming of the light passing through the sample, sometimes represented as a dip at the appropriate frequency.

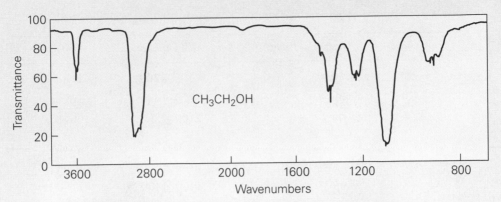

FIGURE R20-2. Infrared spectrum of ethanol.

Absorbing infrared radiation, a molecule drives itself into more agitated states of vibrations; and from the absorption spectrum we learn what kind of structure we have—its functional groups, its bonds, its stiffness, its elasticity.

QUESTION: How so?

ANSWER: Each functional group vibrates at a certain characteristic frequency, distinguishable from atoms in other environments. Bonds vary in strength and elasticity in the same way that springs made from different materials vary in springiness. Single bonds differ from double bonds. Double bonds differ from triple bonds. Carbon–hydrogen bonds differ from carbon–nitrogen bonds, and carbon–nitrogen bonds differ from carbon–oxygen bonds. Carbonyl groups vibrate at one frequency, ethers at another, alcohols at another. Every group is slightly different. Every group has its own energy of vibration.

Some of the most common infrared transitions are collected in Table R20-1, with energies quoted in wavenumbers (cm^{-1}) according to the definition given in Example 20-3.

QUESTION: What are *stretches* and *bends*?

ANSWER: Different modes of vibration. When a bond *stretches*, the two nuclei move back and forth along their interconnecting axis:

The distance alternately grows and shrinks, as, for example, when H_2 vibrates. The vibration is a simple springlike oscillation, the only mode available to a diatomic molecule.

When, by contrast, a polyatomic molecule such as H_2O vibrates, the

TABLE R20-1. Selected Infrared Vibrational Frequencies

Bond	Functional Group	Range of Absorption (cm^{-1})	Mode
C—C	alkane	700–1200	bend
C=C	alkene	1620–1680	stretch
C≡C	alkyne	2100–2260	stretch
C—H	alkane	2850–3000	stretch
C—H	alkane (CH_2, CH_3)	1370–1380	bend (1)
C—H	alkane (CH_2, CH_3)	1450–1470	bend (2)
C—H	alkene	3000–3150	stretch
C—H	alkene	700–1000	bend
C—H	alkyne	3300–3350	stretch
C—H	aldehyde	2700–2725	stretch
C—H	aldehyde	2820–2900	stretch
C=O	amide	1630–1700	stretch
C=O	carboxylic acid	1700–1730	stretch
C=O	ketone	1700–1730	stretch
C=O	aldehyde	1720–1740	stretch
C=O	ester	1730–1750	stretch
C=O	anhydride	1750–1820	stretch
C—O	carboxylic acid	1050–1350	stretch
C—O	alcohol	1050–1200	stretch
C—O	ester	1050–1350	stretch
C—O	ether	1070–1150	stretch
C—I	halide	200–500	stretch
C—Br	halide	500–680	stretch
C—Cl	halide	750–850	stretch
C—F	halide	1000–1350	stretch
C≡N	nitrile	2240–2260	stretch
O—H	alcohol (H-bonded)	3300–3550	stretch
O—H	alcohol (free)	3580–3650	stretch
O—H	carboxylic acid	2500–3300	stretch

bonds also can *bend* back and forth in various scissoring and rocking motions:

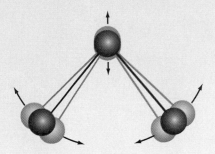

The bond angle expands and contracts periodically.

PROBLEM: The infrared spectrum of CH_3CH_2OH is shown in Figure R20-2 on p. R20.15. Identify the major bands.

SOLUTION: Better to ask: Which functional groups are present in the molecule, and (just as important) which are *absent*?

Present: (1) A telltale signal near 3600 cm^{-1}, strongly suggesting an alcoholic OH group. (2) A band extending from approximately 2800 cm^{-1} to 3000 cm^{-1}, arising from both the CH_2 and CH_3 groups in ethanol. (3) A tangle of signals between 800 cm^{-1} and 1500 cm^{-1}, assignable to the low-frequency bending vibrations of the molecule. The most complex portion of any infrared spectrum, this region stamps each molecule with a unique vibrational fingerprint.

Absent: everything else. Nothing close to 1700 cm^{-1}, where the C=O group resonates; hence no ketone, aldehyde, carboxylic acid, ester, amide, or anhydride groups. Nothing near 2250 cm^{-1}, where we would find a C≡N vibration. No C≡C signal (expected between 2100 and 2260 cm^{-1}); no C=C signal (1620 cm^{-1} to 1680 cm^{-1}); apparently nothing but CH_3, CH_2, and OH. So it is: CH_3CH_2OH.

A simple molecule, a simple spectrum; but enough to show how infrared spectroscopy can help us piece together a structure from its bonds and functional groups.

QUESTION: C=C double bonds stretch at energies ranging from 1620 cm^{-1} to 1680 cm^{-1}; C≡C triple bonds, from 2100 cm^{-1} to 2260 cm^{-1}. How might we account for the higher frequency of the triple bond?

ANSWER: Think of each bond as a spring, imagining the nuclei to suffer some *restoring force*,

$$F = -k \, \Delta x,$$

when displaced from equilibrium by a distance Δx:

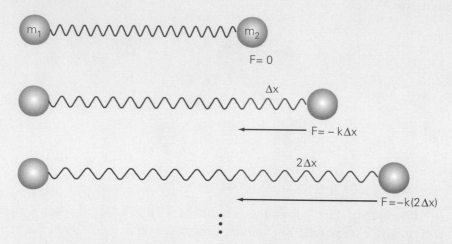

Let the masses of the nuclei be m_1 and m_2.

Now take the equation apart: (1) The direction of the force is opposite to the deformation, a condition made explicit by the negative sign. Stretch the spring (Δx is positive); it pulls back (F is negative). Compress the spring (Δx is negative); it pushes out (F is positive). (2) The magnitude of F is proportional to Δx. A big stretch demands a strong restoration; a small stretch, a weak restoration. (3) The force constant, k, gives the spring its overall stiffness. A stiff spring (a strong bond) has a large force constant and strongly resists deformation. A loose spring (a weak bond) has a small force constant.

Note further that the frequency of vibration

$$\nu = \frac{1}{2\pi} \sqrt{\frac{k}{\mu}}$$

depends on both the force constant (k) and the *reduced mass* (μ) of the two connected atoms, defined below:

$$\mu = \frac{m_1 m_2}{m_1 + m_2}$$

Draw from these results two rules: (1) The stiffer the bond, the larger the force constant and the higher the frequency. (2) The smaller the reduced mass, the higher the frequency as well.

So, thinking *crudely*, we expect a C≡C triple bond to be roughly $\frac{3}{2}$ times stiffer than a C=C double bond. Going from C=C to C≡C, we then expect the force constant to go from k to $3k/2$, and the frequency to go from ν to $\sqrt{\frac{3}{2}}\,\nu$, an increase of some 20% to 25%. Crudely enough, it does.

QUESTION: The C—H bonds in CH_3 and CH_2 groups stretch at frequencies near 2900 cm^{-1}. How would the vibrations change if hydrogen were replaced by deuterium?

ANSWER: The vibrational frequency varies inversely with the square root of the reduced mass. Since a carbon–deuterium bond has almost twice the reduced mass of carbon–hydrogen, a C—D stretch should appear lower in energy by about 25% to 30%:

$$\mu_{C-D} = \frac{m_C m_D}{m_C + m_D} = \frac{12 \times 2}{12 + 2} = 1.714 \text{ atomic mass units}$$

$$\mu_{C-H} = \frac{m_C m_H}{m_C + m_H} = \frac{12 \times 1}{12 + 1} = 0.923 \text{ atomic mass units}$$

$$\nu_{C-D} = \sqrt{\frac{\mu_H}{\mu_D}} \; \nu_{C-H} = \sqrt{\frac{0.923}{1.714}} \; (2900 \text{ cm}^{-1}) \approx 2130 \text{ cm}^{-1}$$

EXAMPLE 20-8. Nuclear Spies: NMR Spectroscopy

They sit passively, affected by their environment but barely able to influence it themselves. They are the nuclear spins, tiny magnets, their energies sensitive to the smallest magnetic fields.

Like a compass needle, a nuclear spin acquires its magnetic energy from a surrounding field; but unlike a compass needle, a nuclear spin cannot point any which way without restriction. For a spin-$\frac{1}{2}$ nucleus such as ^1H, the choices are only two: up or down, parallel to the field or against it. The two states are separated by an energy $\Delta\epsilon$, slightly different for every functional group:

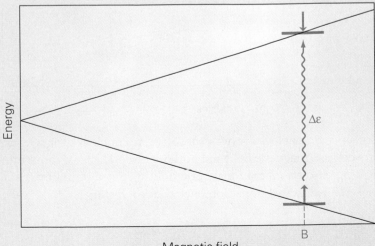

To measure that energy is to learn the strength of the prevailing magnetic field B—and from that, the NMR spectroscopist infers the arrangement of atoms and electrons around the nucleus. A clandestine witness, the nuclear magnet observes and reports on its environment from deep within the molecule. It testifies as to where the electrons are, how the atoms are connected, how the molecule might be moving, and how the molecule might be reacting.

The message is in code: a coded description of the local magnetic field at each nucleus.

QUESTION: How do we receive the message and break the code?

ANSWER: We place the material in a large magnetic field, B_0, which establishes a baseline transition energy $\Delta\epsilon_0$ for each nucleus (Example 20-1):

$$\Delta\epsilon_0 = \gamma_H B_0$$

B_0 thereby gives us a point of reference, meaning that if the nucleus lay bare, bereft of all electrons, isolated from all other atoms—completely barren and alone—this energy, $\Delta\epsilon_0$, would accurately reflect the nearby magnetic field:

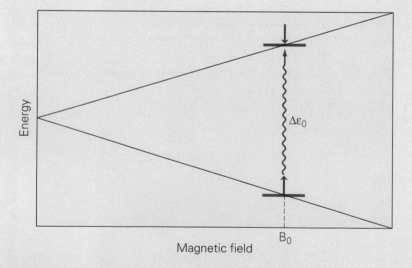

But since no nucleus is ever so desolate, no nucleus ever sees just the spectroscopist's magnetic field and nothing else. There is, more interesting than B_0, also an *internal* magnetic field B_{int}, generated by the atoms themselves. The actual transition energy at each site,

$$\Delta\epsilon = \gamma_H(B_0 + B_{int})$$

thus depends not only on B_0 (which we know) but also on B_{int} (which we want to know):

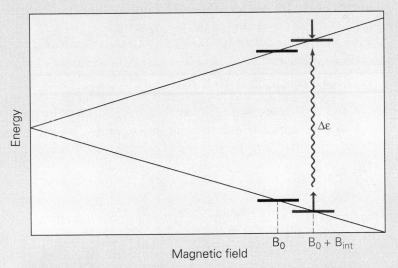

QUESTION: From where does the internal magnetic field arise?

ANSWER: Look outward from a nucleus and see the world as a nuclear magnet would. A small island in a sea of electrons and other nuclei, it responds to the magnetic fields of all charged particles moving nearby.

There are, first, electrons swarming around the site, which in the presence of B_0 produce an internal field analogous to the magnetism generated by wires in an electric circuit:

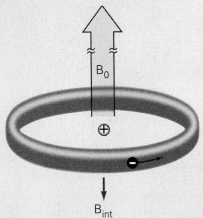

A nucleus encased within a dense electron cloud, affected strongly by such currents, therefore sees an external magnetic field altered by its own internal field; the more electrons there are, the greater is the opposition. A nucleus with a sparse electron cloud, affected less, sees a total magnetic field equal more nearly to B_0.

Different atoms differ in their ability to attract electrons. Bonded to an oxygen atom, for example, a hydrogen atom gives away more electrons than it does when bonded to carbon. Less of a presence around the nucleus, the electrons closest to hydrogen in an OH group produce less of an internal field. The hydrogen in an OH group resonates at an energy closer to ϵ_0 than, say, the hydrogen in a CH_3 group.

These electron-generated fields give rise to the **chemical shift**, B_{cs}, the primary local interaction at each site.

QUESTION: Is that all?

ANSWER: No. There is also a local field manufactured by other *nuclei* in the vicinity, most important for atoms separated by one, two, or three bonds. The effect produced is called the **spin–spin coupling** (or **J coupling**).

Suppose that we have two hydrogen nuclei, H_A and H_B, close enough to interact via the spin–spin coupling. Nucleus H_A subjects nucleus H_B to some local field $J/2$, and H_B subjects H_A to the same field in turn. How does each see the other?

With double vision: H_A presents not one but two local fields to H_B—one component ($J/2$) corresponding to H_A in its ↑ state, the other ($-J/2$) corresponding to H_A in its ↓ state. H_B responds both to a field

$$B = B_0 + B_{cs} + \frac{J}{2}$$

and to a field

$$B = B_0 + B_{cs} - \frac{J}{2}$$

and consequently its NMR spectrum splits into two peaks, a *doublet*. H_A, likewise responding to a doubled local field ($\pm J/2$), produces two signals as well:

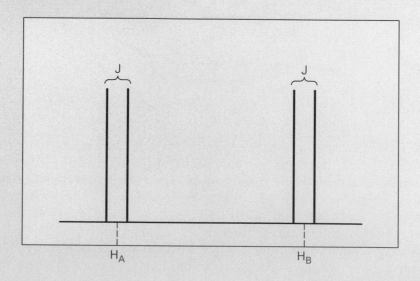

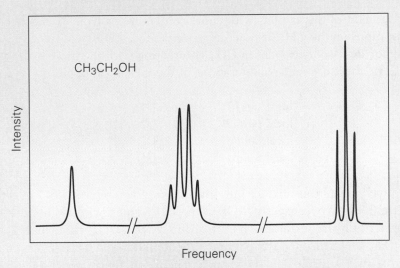

CH₃CH₂OH

FIGURE R20-3. ¹H NMR spectrum of ethanol, CH₃CH₂OH (expanded view).

PROBLEM: The ¹H NMR spectrum of CH₃CH₂OH is shown in Figure R20-3. Account for each peak.

SOLUTION: There are six hydrogen nuclei in ethanol, but only three distinguishable environments: the CH₃ group, which contains three hydrogens; the CH₂ group, which contains two; and the OH group, which contains one. The three ¹H nuclei on the methyl group all share the same magnetic field, as do the two ¹H nuclei on the methylene group, as does the one hydrogen on the hydroxyl.

First, the chemical shifts: a distinct value for each of the three equivalent groups. The three hydrogen atoms in CH₃, suffering the least withdrawal of electrons, appear at one extreme of the spectrum. They are shielded the most from the external magnetic field by their own electrons. The two hydrogen atoms in CH₂, shielded less, are shifted differently, their electrons ceded more to the electronegative oxygen in H—O—CH₂—CH₃. The hydroxyl hydrogen appears at the other end of the spectrum (although its position will vary in solutions of different concentration).

Next, the *J* couplings. The three CH₃ hydrogens interact as a unified group with the two CH₂ hydrogens, seeing *three* configurations of the CH₂ spins. Count them:

local field 1	Both CH₂ spins are up.	↑↑
local field 2	One is up; one is down.	↑↓ or ↓↑
local field 3	Both are down.	↓↓

The CH₃ signal, reflecting these three variations of the CH₂ local field, splits accordingly into three peaks in the ratio 1:2:1. The central component of this triplet, with equal contributions from ↑↓ and ↓↑ configurations on the CH_2, is twice as strong as the other two.

For the two hydrogens in CH_2, interacting with the three hydrogens in CH_3, there are four configurations of the CH_3 local field:

local field 1	↑↑↑		
local field 2	↑↑↓	↑↓↑	↓↑↑
local field 3	↑↓↓	↓↑↓	↓↓↑
local field 4	↓↓↓		

The CH_2 signal thus divides into four peaks, forming a 1:3:3:1 quartet. Its total intensity is exactly $\frac{2}{3}$ that of the CH_3 triplet, consistent with the number of hydrogen atoms in each group: two in CH_2, three in CH_3.

The OH hydrogen stands alone, uncoupled to the CH_2 group. It *can* couple, but it rarely gets the chance. The alcoholic proton usually is on the move, passing rapidly from one molecule to another, never staying in one place long enough to interact fully with its neighbors. The OH signal appears as a singlet, unsplit, its intensity corresponding to one hydrogen.

QUESTION: Can we obtain an NMR spectrum from the carbon nuclei in ethanol?

ANSWER: Yes; but only from some of the carbons, approximately one in a hundred. The ^{13}C nucleus, with an odd number of protons and neutrons (6p + 7n), possesses a nuclear spin. The ^{12}C isotope, with six protons and six neutrons, does not.

Although each proton and neutron is individually magnetic, no net magnetism arises when a nucleus contains even numbers of each particle. Paired in nuclear *shells* (Chapter 21), the nuclear magnetic moments cancel one another just the way electronic spins do in filled orbitals.

Oxygen, incidentally, also exists as both a magnetic and a nonmagnetic isotope. ^{16}O (99.76% abundant) has eight protons, eight neutrons, and hence no magnetic moment. The rare ^{17}O species (0.04%), with nine neutrons, does have a spin and does produce an NMR spectrum.

QUESTION: Can we obtain an electron paramagnetic resonance (EPR) spectrum from ethanol?

ANSWER: No. Since there are no unpaired electrons, there is no net electronic spin. In a transition-metal complex, yes. In a free radical, yes. In a closed-shell molecule like ethanol, no.

EXERCISES

1. Nuclei that contain an even number of both protons and neutrons have no magnetic moment. Which of these nuclei, then, are *not* paramagnetic?

   ^{1}H, ^{2}H, ^{3}H, ^{4}He, ^{10}B, ^{11}B, ^{12}C, ^{13}C, ^{14}N, ^{15}N, ^{16}O, ^{17}O

2. A single hydrogen nucleus—a proton—can present its local magnetic field in two ways: spin up (↑) or spin down (↓). A pair of spin-$\frac{1}{2}$ particles can do so in four ways: ↑↑, ↑↓, ↓↑, ↓↓. How many such configurations are possible for groups of three and four ^{1}H nuclei?

3. How many distinct lines appear in the proton NMR spectrum of each molecule?

   (a) CH_4 (b) C_2H_6 (c) C_2H_4 (d) C_2H_2 (e) C_6H_6

4. Do hydrogens H_a and H_b have the same chemical shift?

$$\begin{array}{ccc} Br & & H_b \\ & C=C & \\ H_a & & CCl_3 \end{array}$$

5. How many lines appear in the proton NMR spectrum?

$$\begin{array}{ccc} Br & & H \\ & C=C & \\ H & & CCl_3 \end{array}$$

   Include signals arising from both the chemical shift and J couplings.

6. How many lines appear in the ^{1}H NMR spectra of the two molecules below?

$$\begin{array}{ccccccc} Br & & H & \qquad\qquad & H & & H \\ & C=C & & & & C=C & \\ H & & CCl_3 & \qquad\qquad & Br & & CCl_3 \end{array}$$

   Are the spectra identical?

7. Which of these species can produce an EPR (electron paramagnetic resonance) spectrum? H, H_2, H^+, H^-

8. Again: Which can produce an EPR spectrum? CH_3, CH_4, Cl_2, Cl, Cl^-, NH_3

9. Rank these nitrogen bonds in order of increasing vibrational frequency: $N—N$, $N=N$, $N\equiv N$. Which will absorb infrared photons at the longest wavelength?

10. A carbonyl functional group, $C=O$, absorbs infrared photons with energies near 1750 cm^{-1}. For the $C—O$ single bond, however, the energy absorbed is considerably lower, falling in a range centered at approximately 1150 cm^{-1}. (a) Convert these two photon energies from wavenumbers to kilojoules per mole. (b) Compare the electromagnetic energies to the bond energies of $C—O$ (360 $kJ\ mol^{-1}$) and $C=O$ (740 $kJ\ mol^{-1}$). Doing so, show that the excited bonds bend and stretch but do not break.

11. Pretend that a vibrating bond is like a vibrating spring. Using typical vibrational frequencies for $C—O$ (1150 cm^{-1}) and $C=O$ (1750 cm^{-1}), estimate the ratio of force constants for the two bonds.

12. Given the bond dissociation enthalpies listed below, estimate the longest wavelength able to break apart the molecules H_2, N_2, and O_2:

BOND ENTHALPIES ($kJ\ mol^{-1}$)

$H—H$	436	$N—N$	163	$O—O$	146
		$N=N$	409	$O=O$	497
		$N\equiv N$	946		

In what portion of the electromagnetic spectrum does the light fall?

13. Vibrational energies are quantized according to the formula

$$\epsilon_{vib} = (v + \tfrac{1}{2})\, hv_0 \qquad (v = 0, 1, 2, 3, \ldots \infty)$$

where v_0 is the fundamental frequency and v is a quantum number. For a molecule of HCl, v_0 has the value 8.95×10^{13} Hz. (a) Calculate the vibrational energy in the ground state. (b) Calculate the energies of the first, second, and third vibrational states above the ground state.

14. Continue with the vibrational frequencies of HCl: (a) Calculate the wavelength and energy corresponding to a transition between $v = 0$ and $v = 1$. (b) Calculate the corresponding Boltzmann ratio at 298 K.

For every molecule in the $v = 1$ state, how many other molecules are in the ground state?

15. More HCl: (a) Calculate the wavelength and energy corresponding to a transition between $v = 1$ and $v = 2$. Compare the result to the energy similarly computed for the $v = 0 \longrightarrow v = 1$ transition. (b) Calculate the wavelength and energy corresponding to a transition between $v = 2$ and $v = 3$.

16. Consider now the excitation of a molecule's valence electrons, in particular the delocalized π electrons in ethylene ($H_2C{=}CH_2$) and 1,3-butadiene ($H_2C{=}CH{-}CH{=}CH_2$). Think, crudely, of these π orbitals as standing waves on a string—ethylene being a "string" of two carbons, and 1,3-butadiene a string of four. (a) Sketch the first two molecular orbitals for each molecule, recalling the description of standing waves in Chapters 4, 5, and 7. (b) Which of the two molecules will absorb light with the shorter wavelength?

17. Same argument: Conjugated organic molecules such as

$$H_2C{=}CH{-}CH{=}CH{-}CH{=}CH{-}CH{=}CH_2$$

go from colorless to colored as the $-C{=}C{-}C{=}C{-}C{=}C{-}$ chains grow longer. Why?

18. The ionization energy for a $1s$ electron in atomic nitrogen is 38,600 kJ mol^{-1}. (a) Calculate the wavelength, frequency, and energy of a matching photon. (b) Calculate the speed of a $1s$ electron ejected from a nitrogen atom by an X-ray photon with a wavelength of 0.989 nm.

19. Electronic energies are quantized, true—yet, even so, a bound electron will apparently accept an overpayment from a photon. A photon with energy *greater* than the ionization energy is, we know, quite able to knock an electron out from the core. Explain why. In particular, explain why the so-called overpayment is no violation of the quantization rules.

20. NMR and EPR frequencies are exquisitely sensitive to slight variations in chemical environment, as are transitions between rotational, vibrational, and valence electronic levels. By contrast, core electrons seem almost to ignore their wider surroundings. A nitrogen $1s$ electron, for example, has nearly the same ionization energy whether in N_2 or NO_2 or glycine or even a strand of DNA. Why?

21. The mass spectrum of titanium contains five signals:

Mass	Intensity
45.952629	0.1084
46.951764	0.0989
47.947947	1.0000
48.947871	0.0745
49.944792	0.0732

Calculate the average molar mass of titanium.

21

Worlds Within Worlds— The Nucleus and Beyond

21-1. Unfinished Business

21-2. Beginnings

21-3. Relativity and the Meaning of $E = mc^2$
Space and Time
Energy and Mass
Matter and Antimatter

21-4. Nuclear Structure: Binding Energy and the Strong Force

21-5. Nuclear Reactions
Kinetics of Radioactive Decay
Mothers and Daughters
Building a Chemical Universe: Fusion
Splitting Up: Fission
Latter-Day Alchemy: Transmutation

21-6. Fields and Particles
Interactions: The Medium and the Message
Mesons and the Strong Force
Quarks

21-7. Epilogue
REVIEW AND GUIDE TO PROBLEMS
EXERCISES

What do we want from our science except this: to know how things are put together and how they work. No detail, however small, should escape scrutiny.

For the chemist, the job is mostly to observe the grain of matter at the level of atoms and molecules—on a scale where spacings become as small as 10^{-10} meters, where intervals become as short as 10^{-15} seconds, where molar energies range from fractions of kilojoules to several hundred. Starting from the smooth and regular world of macroscopic objects, the road to chemical understanding soon takes us through a discontinuous, lumpy microworld of electrons and nuclei. It takes us from the gravitational interaction to the electromagnetic, from a world of mass to a world of charge.

A miniaturized, quantized, electromagnetic world of charge: atoms and molecules. In that realm we see how electrons and nuclei weave themselves into the cloth of ordinary matter, appearing in everything from hydrogen atoms to stars and planets, single cells, dogs and cats. For the chemist, it is a universe of atoms and molecules interacting on a small scale.

There is an economy to this chemical universe, evident in nature's persistent use of the same few components and tools. Count among them (1) a single negative particle, the electron; (2) several dozen elemental nuclei from hydrogen through uranium, positively charged; (3) one basic force, the electromagnetic interaction; and (4) a few rules of engagement governing motion and energy.

Momentum is conserved. Charge is conserved. Energy is conserved. Entropy is maximized. Electrons and nuclei, interacting electromagnetically, do so within the probabilistic laws of quantum mechanics: the Schrödinger equation, the Pauli exclusion principle, the Heisenberg uncertainty principle.

We know the particles of chemistry. We know the forces. We know the laws. We know the equations. We think, perhaps, that we are finished. Never mind how difficult the equations are to solve, because (in principle) we know everything. Chemistry, the study of matter and its transformations, is now a solved puzzle—provided, of course, that microscopic matter is confined to the electromagnetic world, the world of charge.

But what if there are other worlds?

21-1. UNFINISHED BUSINESS

What about the nucleus? How can *that* unlikely piece of matter survive in a world ruled only by gravity and electromagnetism? With nothing to hold together its fantastically dense structure, a nucleus should blow into pieces. Its two major particles, proton and neutron (jointly called **nucleons**), lack the mass to adhere by gravitational attraction. Nearly 2000 times more massive than the electron, the nucleons still can muster only a pitiably small gravitational force in the face of an enormous, unrelieved electrical repulsion. For here, after all, are positive particles—the protons—crammed one against the other at distances of only 10^{-15} m, a spacing 1 / 100,000 the diameter of the smallest neutral atom. Yet nuclei do manage to hold together, despite the repulsion; and, more surprising, they hold together far more strongly than any atom or molecule.

Should they? No negative particles exist within the nucleus to cement the protons together electrically. There are, instead, *uncharged* particles (the neutrons), which suffer neither electrical attraction nor repulsion and which themselves display no visible means of support. In a world of charge and mass alone, the neutrons should simply float free of the nucleus. There is nothing to hold them. Uncharged, they are electromagnetically immune. Low in mass, they are gravitationally irrelevant.

Why, too, are there only dozens of elements and not thousands or millions? Why are certain nuclei rare and certain nuclei abundant? Why do elements exist as different isotopes, so as to give each variant nucleus (a *nuclide*) the same complement of protons but a different number of neutrons? Why are some isotopes more plentiful than others? Why do some nuclei occasionally spew out radioactive particles? Why, also, if a neutron is a neutral particle, does it exhibit a spin magnetic moment?

Why . . . because, inside the nucleus, there must be forces *other* than the irrelevant gravity and the overmatched electromagnetism. There must be other forms of matter, different from the electron. There must be particles with endowments other than mass and charge, subject to forces other than gravity and electromagnetism. There must be particles smaller and more fundamental than either proton or neutron. There must be a substructure within each nucleon itself, a world within a world.

Too long have we treated the nucleus as a piece of dead weight, serving merely as a sturdy anchor for atom or molecule. Granted, its positive charge supplies the electrical attraction needed to organize the electrons. Granted, its dense mass gives heft to the atom and skeletal structure to the molecule. Granted, its magnetic moment gives us (through NMR spectroscopy, Chapter 20) a useful probe of the nearby chemical environment. But surely we are not finished, not yet; not with the nucleus and not with matter.

There is more to matter than mass and charge, and there is more to the nucleus than proton and neutron.

21-2. BEGINNINGS

The story begins with three discoveries, each made at roughly the same time. One was an accident. Another was the result of deliberate action. The third seemed to be irrelevant.

1. Becquerel's 1896 discovery of *radioactivity* was an accident—an

unexpected, unintended fogging of a photographic plate upon exposure to certain uranium salts. Soon after, Marie and Pierre Curie isolated two new radioactive elements (polonium and radium), and thus was the door to the nucleus opened inadvertently.

Nobody knew, in 1896, how atoms were constructed. Still less did one suspect that these exposed plates provided evidence of a spontaneous *nuclear* reaction, the decay of a particular uranium isotope. What they did recognize, though, was that all these substances were emitting some kind of *radiation*; and so they undertook to measure the charge, mass, and energy of the particles produced. Here, briefly, is what they found, interpreted in the light of our later understanding.

They found **alpha particles**, observed to have a charge of +2 and a mass number of 4. We recognize them now as the bare nuclei of helium atoms, each containing two protons and two neutrons:

$$\alpha \text{ particle:} \qquad {}^{4}_{2}\text{He} \quad \overset{\longleftarrow A \text{ (mass number: protons + neutrons)}}{\longleftarrow Z \text{ (atomic number: protons)}}$$

They found **beta particles**, eventually identified as ordinary electrons. These we represent by the symbol

$$\beta^{-} \text{ particle:} \qquad {}^{0}_{-1}\text{e}$$

to indicate the electron's charge of −1 and its negligible mass compared with either proton or neutron.

They found **gamma rays** (γ) as well: purely electromagnetic radiation, ordinary photons. Yet these are photons packing energies higher even than X rays, short-wavelength radiation more energetic than any spectroscopic emission originating outside the nucleus. Somewhere inside the nucleus, then, there must be a charged particle falling precipitously from one energy level to another—falling far enough, apparently, to produce these highly energetic photons.

Such are the $\alpha\beta\gamma$'s of nuclear decay, the first signs that nuclei were not really fundamental particles; the first signs that nuclei themselves were made of particles; the first signs that nuclei could undergo transformation. There are other forms of radiation and other phenomena still to unfold, but here is where we begin.

We learn, right away, that some nuclei are unstable. Some nuclei break spontaneously into smaller pieces. There can be chemistry within a nucleus.

2. Next came a frontal assault on the atom, executed dramatically by Rutherford in his scattering experiments circa 1911 (Chapter 4). He shot a beam of α particles at a sheet of thin gold foil, allowed the positively charged masses to bounce off the neutral atoms, and collected the ricochets.

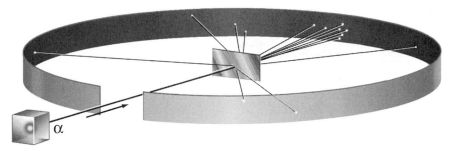

FIGURE 21-1. Rutherford's beam experiment, carried out by Geiger and Marsden early in the 20th century: A beam of α particles, shot through a thin layer of gold, remains largely undeflected after interacting with the atoms. Occasionally an α particle bounces straight back, repelled by a small but dense kernel in the atom it encounters—the nucleus. The experiment led to both the discovery of the nucleus and a new way of probing atoms and subatomic particles.

Rutherford's conclusion: The mass of an atom is concentrated within a dense core of positive charge, surrounded by a diffuse cloud of electrons. He took aim, fired, and discovered something small and hard. He discovered the nucleus.

Rutherford's barrage (Figure 21-1) was the beginning of a long line of particle-beam experiments, still continuing today. Firing projectiles of ever higher energy, physicists try to break the nucleus and its nucleons into small bits. They select a target, smash it, and pick up the pieces. From the debris, they attempt to discover the most basic components used in the original construction (as if to blow up a house and collect the individual bricks).

There is considerable debris produced in a high-energy particle accelerator, more little bits than one cares to imagine. Most, however, are not the ultimate, irreducible bricks of matter; most are composite particles themselves, two and three bricks clumped together at once.

But we are getting ahead of our story, for the most important initial discovery is still to come: the constancy of the speed of light. Thence follows Einstein's theory of *special relativity*, by which we learn (almost incidentally) that mass is a form of energy. Nothing in the nuclear world makes sense without it.

3. Relativity has humble origins, far from the exotic world of subatomic particles. A railroad station will do for starters.

Picture a train moving at constant velocity, traveling in a straight line at, say, 30 meters per second (67 miles per hour). An observer, standing on the ground, notes distance and time in the way indicated in Figure 21-2(a). The train passes a marker that reads 30 m while simultaneously a clock reads 1 s. The train passes another marker at 60 m, and the clock reads 2 s; it passes a third marker at 90 m, and the clock reads 3 s; and

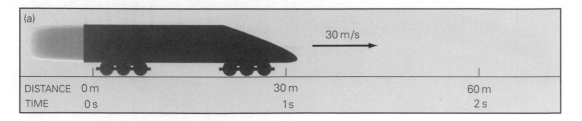

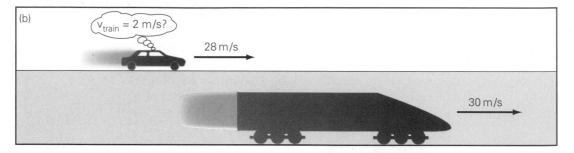

FIGURE 21-2. Frames of reference. (a) A ground-based observer measures velocity by recording time and distance as the train passes. (b) Perception of the train's motion changes when measured from a moving vehicle. Chasing the train, the observer measures a velocity relative to her own; she sees only the *difference* in speed as the train moves ahead. (c) When both train and observer move at the same speed, each views the other as standing still. The relative velocity is zero.

on forever. Standing still, our observer (Jane) measures a velocity of 30 m s^{-1}.

Then, suddenly wishing to catch the train, Jane takes off in pursuit. She hops into a car and begins to close the distance. Looking out from the moving car (Figures 21-2b and 21-2c), Jane now sees the train advancing at a rate slower than the speed she first measured at the station. When she reaches a steady 28 m s^{-1} in the car, for example, the train (*she* says) is gaining ground at only 2 m s^{-1}. Looking out from the windshield, she sees the faster vehicle advancing by just 2 meters each second—not the

30 meters that a stationary observer claims. Jane's perception of the train's motion is influenced by her own motion; she measures a velocity *relative* to her own: 30 minus 28. It is the train's velocity on the ground minus her own velocity on the ground.

Eventually she catches up to the train, riding alongside at exactly 30 m s^{-1}; but be careful: that 30 m s^{-1} is the speed as perceived by somebody anchored to the ground, not riding in the car. To Jane, the train at her side is going nowhere. To Jane, the train is standing still. Seeing no such motion, Jane decides that she, too, is standing still relative to the train. Nothing is changing. The relative velocity between train and car—30 minus 30—is zero. She can board the train as easily as if it were parked at the station.

Good. Jane has caught up with and boarded the train, and we shall move on to a new point of view as well. Suppose next that some other observer, George, tries to pursue not a train but a light ray. He sets out, quixotically, to pursue the light coming off his own face, hoping to overtake his own image before it reaches a distant mirror. Assume further that the electromagnetic radiation is propagating in a vacuum, where we know the light's velocity to be constant:

$$c = 3.00 \times 10^8 \text{ m s}^{-1}$$

The speed is 10,000,000 times faster than Jane's train, but George is willing to try nonetheless. He has a fast rocket at his disposal.

Standing still, George first notes the progress of the electromagnetic wave as it advances steadily by 3.00×10^8 meters every second. Then he takes off, pushes the rocket to full throttle, and soon reaches his final speed: 2.00×10^8 m s^{-1}, measured relative to the earth's surface. A stationary observer on the ground (Elroy) claims that George is whizzing by the earthly landmarks at the rate of 200,000,000 meters per second.

Question: What velocity of light does George measure? How fast is his rocket moving relative to the light? Think back to Jane's low-speed experience, and make a similar prediction here. What should we say?

The answer is obvious, we claim, since there seems only to be a change in wording. George's rocket in Figure 21-3 plays the role of Jane's car in Figure 21-2, and George's light wave plays the role of Jane's train. George is in hot pursuit ($v = 2c/3$) of a light beam that advances at velocity c once it leaves his forehead. He should observe, presumably, that (relative to the rocket) the light advances only at the speed $c/3$, namely 300,000,000 *minus* 200,000,000.

Wrong. Completely and utterly wrong. Completely, utterly, and quite unexpectedly wrong, no less, for our argument was a good one. We appealed to common sense, the accumulated experience of Jane's

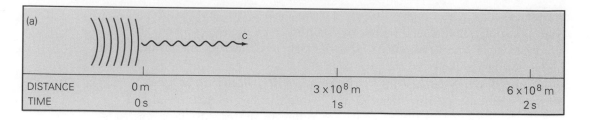

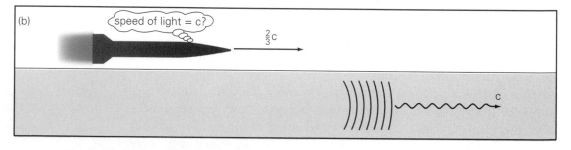

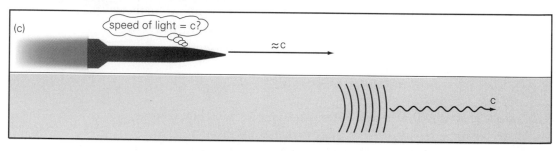

FIGURE 21-3. Always chasing, but never catching: a rocket sets out to overtake a light wave. (a) Anchored to the ground, an observer first measures the speed of the advancing light: $c = 3 \times 10^8$ m s^{-1}. (b) Pursuing at high speed (two-thirds the velocity of light), he expects to gain ground. He expects to measure a *smaller* velocity relative to his own, but he is wrong; the light, even when viewed from a moving reference frame, runs away at the same speed as before: $c = 3 \times 10^8$ m s^{-1}. (c) Traveling almost at the velocity of light itself, the pilot still comes no closer. The beam continues to travel at the same apparent speed: $c = 3 \times 10^8$ m s^{-1}. Unlike a train fixed to a track, an electromagnetic wave is anchored to nothing. A pulse of light enjoys no uniquely privileged frame of reference.

everyday life; but it was a step too far. Since everyday life proceeds at a pace far slower than the speed of light, what does all our previous experience matter? Once again, we shall have to revise our view of the world. Just as we did for things very small (quantum mechanics), we shall now do for things very fast (relativity).

Here is the correct answer, revealed not by speculation but by experiment. In unaccelerated motion or at rest, George always measures the same velocity of light: *c*. So does Elroy: also *c*.

George, in fact, will never catch up to his light wave. However fast he travels, he will always measure the same velocity of light; and both George and Elroy must come to the following understanding:

> "To an observer moving at constant velocity, the speed of light in a vacuum never changes. Electromagnetic waves propagate only at the velocity $c = 3.00 \times 10^8$ m s^{-1}. Every observer in uniform motion will always record the same value, independent of his own speed."

Such was the conclusion drawn from an experiment first reported by Michelson and Morley in 1887 and verified repeatedly over the next century.

Electromagnetic waves are different from trains, different from sound waves, different from water waves, different from everything we know. There is no special frame of reference for an electromagnetic wave; there is no fixed, ground-based track along which it must propagate. Light propagates on *nothing*. It is supported by nothing. It makes its own way through empty space, requiring no material medium to pass it along.

Not so a train. A train moves at 30 m s^{-1} only when viewed from a unique perspective: along a track, permanently anchored to the earth. Move yourself relative to that track (as Jane did), and the train appears to go faster or slower depending on your own motion.

For a light wave, though, relative to what track do you plan to move? Vacuum is vacuum. In what special environment should George expect to catch up to his light wave? *No* special environment. Once the light leaves George's face, there is no material "conveyor belt" to carry it forward in Elroy's ground-based world. Vacuum is vacuum. There is no unique state of motion from which to observe light.

Sound waves, yes. Sound waves are propagated by the compression and decompression of molecules in the air, and here we do have an unambiguous frame of reference to measure the speed of sound: motionless air. Sound will travel at its standard speed only when the air is still. Let there be a wind of 30 m s^{-1}, however, and now the vibrations will travel that much faster. The moving air will carry the sound wave along with it.

Light waves, no. In vacuum, the electromagnetic disturbance is anchored to no fixed points in space. Attached to nothing, light has no medium to carry it along in a strong wind. Elroy therefore cannot see George's light swept forward at a velocity higher than *c*, for there is no medium available to capture and carry it. Neither George nor Elroy enjoys a preferred vantage point. All uniformly moving observers record the same velocity.

Hard to believe, maybe, but from this fact of life comes Einstein's celebrated formula equating mass with energy: $E = mc^2$, gateway to the nucleus. In the next section, we shall see why.

But to do so, we shall first need a new map of the world, a new understanding of the most basic concepts of space and time.

21-3. RELATIVITY AND THE MEANING OF $E = mc^2$

Jane now in her train moving at a constant velocity of 30 m s^{-1} ($v = c/10^7$), George in his rocket moving at a constant velocity of 2.00×10^8 m s^{-1} ($v = 2c/3$), Elroy standing on the ground ($v = 0$)—each has something in common. Each is subject to Newton's first law, the principle of inertia as laid down in Chapter 1:

> "An object at rest tends to remain at rest, and an object moving in a straight line at constant speed will continue that same motion unless acted on by an outside force."

The cup of coffee on Jane's tray stays in its place as long as the train moves smoothly. So does the cup in George's rocket, and so does the one on Elroy's desk.

Each observer claims, furthermore, to be standing still while everybody else is moving. Jane points to her stationary cup; George to his; Elroy to his. Without looking out the window, none of them even has any sensation of motion.

Elroy says that *Jane* is moving to the right, and Jane says that Elroy is moving to the left. George says that Elroy is in motion, and Elroy says that George is in motion. Each makes a valid claim, and so we can only say that there is no state of absolute rest. There is no absolutely stationary state compared with which one can declare, "I have zero velocity."

What about earth, you ask? Doesn't Elroy have his feet on the ground? Yes, but the earth itself is moving and Elroy has no way to tell. Elroy's earth is no more stationary than Jane's train and George's rocket. Each, moving uniformly, serves as an equivalent **inertial reference frame**. One perspective is not any better than the other.

If so, then our laws of physics must take the same form in every such equivalent frame. If energy is conserved on the ground, then energy must be conserved in the rocket. If there is some relationship linking mass, force, and acceleration in the rocket, then the same relationship must hold in the train as well. If light travels at velocity c in the train, it must travel at the same velocity in the rocket. For if there were to be any discernible difference, then either Jane, George, or Elroy could claim—could claim absolutely—to be moving or not moving; and that, we assert, is impossible.

Here, then, is Einstein's **principle of relativity**: The laws of nature are the same in all inertial reference frames. Combined with the constancy of c, this profoundly simple idea forces us now to take a new look at space and time, energy and mass.

Space and Time

Jane, George, and Elroy must cast their laws in a way consistent with the principle of relativity. They must come up with the same equations, the same conservation laws, the same view of how the world works.

But first they must reconcile their individual ideas of where and when, of space and time. Jane, measuring distances relative to the back of the train ($x = 0$), claims that her seat in the center is positioned at $x = 25$ m. Elroy, seeing the car's rear end pass the 500-meter line, says that (at this instant) Jane is located at $x' = 500 + 25$ m. She sees herself 25 m from the end of the train, always in the same position, never changing. He sees her, right now, 500 m down the track *plus* 25 m into the train; and each second after that, he sees the train move another 30 m down the track. Each second, Elroy adds 30 m to his reckoning of x'. Jane, meanwhile, always reports the same position: $x = 25$ m.

So far, it seems, they face no particular problem concerning x and x'. They have only to compare notes, after which they can devise some suitable way to convert from train coordinates to ground coordinates.

Next, Jane and Elroy want to compare their perception of time—to see, for example, whether they can agree if two events take place simultaneously. To do so, Jane stands at the midpoint of the train shown in Figure 21-4(a) and fires two pulses of light, perfectly synchronized: one light pulse to the left and one to the right. Each pulse travels until it hits a wall at either end.

Jane, stationary in the railway car, notes that both pulses must travel the same distance to their respective bulkheads. They do, and (moving at the same speed, c) they do so in the same time. She reports, accordingly, that the two signals arrive simultaneously, and Elroy agrees. From his perspective, he also concludes that the leftward and rightward beams hit their targets at the same time. Elroy and the *slow*-moving Jane come, therefore, to an understanding. They agree to use the appropriate formula to reconcile their respective positions, x and x', but not their measurements of time, which they believe to be the same (a mistake, by the way).

Now it is George's turn to trigger two light pulses inside his frame of reference, a very fast rocket. He stands at the midpoint of the cabin in Figure 21-4(b), fires left and right simultaneously, and observes (just like Jane) that the two signals also arrive simultaneously at their destinations.

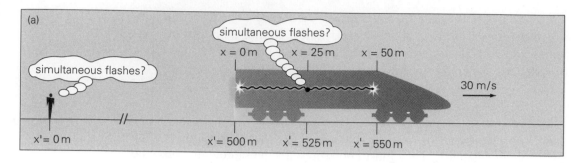

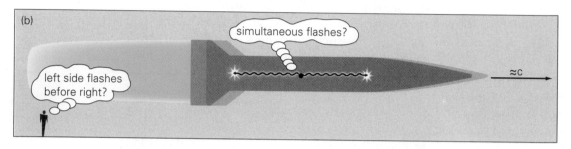

FIGURE 21-4. Time, like space, is relative; its perception is in the eye of the be-holder. Observers moving at different speeds will disagree on the elapsed time be-tween two events. (a) At low velocity, the discrepancies are present but masked—as here, where both ground-based and train-based observers claim that the two pulses arrive simultaneously at opposite walls. They deceive themselves, unable to resolve the slight separation in time. The beams of light move too fast for them to notice. (b) At velocities approaching c, the true state of affairs becomes clear: The ground-based observer sees the rocket's nose as running away from one pulse and the rocket's tail as running toward the other. He sees the left-going pulse hit its bulkhead *before* the right-going pulse. Inside the rocket, an observer no less privileged claims that the two signals arrive at the same time.

No surprise: The walls are fixed in place. The distance to the left is the same as the distance to the right. The two pulses travel the same dis-tance in the same time.

This time, though, Elroy disagrees. Elroy (still standing on the ground) claims that the leftward pulse hits the left wall before the right-ward pulse arrives at the right wall. Elroy says, "Our positions are differ-ent and our *times* are different as well. We must devise a formula to convert not just our three spatial coordinates (x, y, z), but also the time (t). We must consider all four *space-time* coordinates together in a single package (x, y, z, t)."

Elroy is right. He remembers that George's light pulses do not, in any sense, belong preferentially to George's rocket frame. As if disembodied, the light propagates in empty space at velocity c relative to any observer. To Elroy, the left side of the rocket now appears to be gaining on the left-going light ray while, up front, the right side of the rocket seems to be running away. Moving *toward* the light at an extraordinarily high speed, the left wall of the rocket meets the pulse first. Moving *away* from the light at the same high speed, the right wall is intercepted at a later time (according to Elroy).

Jane was moving too, but moving very slowly compared with the speed of light—so slowly, evidently, that Elroy failed to note that the leftward pulse actually arrived first by a vanishingly small margin. But it did.

Conclusion: Perceptions of both space and time are inextricably bound up with one's state of motion. No longer can we speak of x, y, z separately from t. There is no absolute space and there is no absolute time. Space is relative. Time is relative. The principle of relativity, combined with the constancy of c, demands that all our laws and equations must be convertible from one inertial reference frame (x, y, z, t) into another (x', y', z', t').

Energy and Mass

We have some revising to do, clearly, for all our laws of motion were derived from observations of a world in slow motion. Take momentum, for instance, which in Chapter 1 we defined as

$$p = mv$$

for understandable reasons. We observed that the total momentum, so stated, remains constant amidst all other changes in a closed system. Here, surely, was no arbitrary definition chosen on a whim; rather, the quantity mv stood out as something special, something enduring, something conserved. The conservation of momentum had meaning. It stemmed from the symmetry of space itself, noted (in passing) in Chapter 13.

Asserting that this special quantity mv is always conserved, we were able to make prediction after prediction concerning the motion of particles. Not once did we fail. Conservation of mv was a pillar of our laws.

Then George took off in his rocket, and the whole system broke down. Upon converting our space-time coordinates (x, y, z, t) from one frame into the other, we now find that the old momentum, mv, is no longer special. Mass times velocity is *not* conserved when one views an encounter in a high-speed frame of reference. It is a troubling observation, seeming to imply that conservation of momentum is a low-speed phenomenon only.

Not so fast. Proper relativistic treatment shows that a new quantity

$$p = \frac{mv}{\sqrt{1 - \dfrac{v^2}{c^2}}}$$

is indeed conserved in all uniformly moving reference frames. Good for speeds high and low, this new expression thus becomes our all-purpose, relativistic definition of momentum. There is no need to abandon the conservation law.

And look: When v is substantially less than c, the relativistic momentum reduces straightaway to $p = mv$. Even for a particle traveling at $c/10$ (fast!), we scarcely notice the difference between Newton's momentum (mv) and Einstein's momentum ($1.005\ mv$). "Fast" blends seamlessly into "slow" according to relativity, just as "small" blends seamlessly into "large" according to quantum mechanics.

And just as Planck's constant (h; Chapter 4) sets limits in the quantum world through the uncertainty principle, so does the value of c establish a *speed* limit throughout the universe: 3.00×10^8 m s^{-1}, the speed of light in vacuum. No particle with mass can be accelerated up to or beyond a velocity equal to c.

There is not enough energy in the world to do so. To accelerate means to change the momentum, and to change the momentum requires application of a force and, ultimately, energy. But we see, from the formula above, that p becomes infinite when $v = c$ (making the denominator equal to zero). A particle's momentum therefore grows without limit as its velocity approaches the speed of light. Nowhere in the universe is there sufficient energy to reach that threshold.

We come, then, to the central question of energy, looking to define a quantity that (1) assumes the same mathematical form for all freely moving observers, and (2) is *conserved* in all equivalent reference frames. Newton's old expression for kinetic energy, $\frac{1}{2}mv^2$, fails on both counts once v becomes sufficiently large. George's understanding of $\frac{1}{2}mv^2$ differs from Elroy's.

To reconcile their observations, George and Elroy soon discover that they must cast the energy as

$$E = \frac{mc^2}{\sqrt{1 - \dfrac{v^2}{c^2}}} = \sqrt{(mc^2)^2 + (pc)^2}$$

Not by coincidence does this expression so plainly connect E with the relativistic definition of p, because it happens that momentum and energy are related to each other in precisely the same way as space and time. They have very much in common.

1. There are three directions in space (x, y, z) and three directions of the momentum (p_x, p_y, p_z). Position is connected to momentum.

2. One number suffices for the time (t) and one number suffices for the energy (E). Energy is connected to time.

3. The three spatial coordinates (x, y, z) must be treated on equal footing with the time coordinate (t). Always we speak of space and time together, combining x, y, z, and t into a unified set of four numbers. So must we do with energy and momentum as well, uniting the three components of p with the single value of E. Energy is conserved; momentum is conserved; and p_x, p_y, p_z, and E, together, blend into momentum-energy, just as x, y, z, and t blend into space-time.

We are now ready to draw out the equivalence of energy and mass from the formula $E = mc^2/\sqrt{1 - v^2/c^2}$. One can show mathematically (but not here) that this relativistic expression for E is approximately equal to

$$E \approx mc^2 + \tfrac{1}{2}mv^2 + \cdots$$

where the omitted terms $(\cdots)$ become important only at very high velocities.

Now examine the two quantities on the right. The second term, $\tfrac{1}{2}mv^2$, we have long known to be the Newtonian kinetic energy of a particle of mass m moving at velocity v. The first term is new. It has the proper dimensions of energy (kg m² s⁻², mass times the square of a velocity), but what does it mean?

It means that there is energy inherent in just *standing still*, because we are left with

$$E = mc^2$$

even when $v = 0$. Absent all motion, there is still this **rest energy** mc^2. It is the energy that a particle possesses just by virtue of its mass.

What if a particle has zero mass, like a photon? If so, then it has zero rest energy and moves at the speed of light. Only a massless particle can attain this limiting speed, whereby its energy (simplifying to $E = pc$) becomes proportional to its momentum, consistent with de Broglie's $p = h/\lambda$. By contrast, any massive particle has a nonzero rest energy.

Note that we say nothing about gravitational energy, electrical energy, nuclear energy, or any other specific form of energy. We remain silent on the kind of particles, their arrangement, and the nature of mass itself. We say only that energy can be stored *as* mass and extracted *from* mass; and we treat "mass" (recall Chapter 1) just as a body's resistance to acceleration. The conclusion that mass is a form of energy then follows inevitably from the principle of relativity and the constancy of *c*.

Our old chemical law, the conservation of mass, therefore cannot be strictly correct. Not mass, but total *energy* is conserved. The sum of the rest energy (which depends on mass) and the relativistic kinetic energy (which depends on velocity) is forever constant when particles in systems and surroundings interact. Mass is but a form of energy, subject to ups and downs in the same way that gravitational or electrical potential energy varies with position.

The rest energy, mc^2, changes whenever the energy of a system increases or decreases, and accompanying this transformation is a corresponding change in rest *mass*:

$$\Delta E = (\Delta m)c^2$$

When a system loses energy to its surroundings (as in an exothermic reaction), it loses mass. It becomes easier to push around. Less force is needed to produce the same acceleration. When a system gains energy, it also gains mass. More force must be applied to effect a given acceleration.

How much mass? Work it out . . . For a typical chemical change of 100 kJ mol^{-1}, the loss or gain is negligible: 10^{-12} kilograms per mole, an infinitesimal 10^{-36} kilograms per particle. For purely chemical changes, clearly, we need not worry about the slight alteration in mass that accompanies the exchange of energy.

Not so for nuclear reactions. There the energies involved are far higher, and there the gains or losses of rest mass are significant. Just how significant will soon become apparent.

Matter and Antimatter

We have still to make the Schrödinger equation comply with the principle of relativity. As it now stands, George and Elroy will offer conflicting physical descriptions of a fast electron.

For a not-so-fast particle, the differences are minor, but we do have a problem with, say, the innermost core electrons of a heavy atom. Held sway by a highly charged, unshielded nucleus, these particles can acquire velocities close to the speed of light. It is a hard problem to solve.

More important here, however, are the consequences that follow for *any* electron once we have a proper quantum mechanics. Dirac, applying the principle of relativity specifically to the one-electron atom, was the first to try, and from his work came these two results:

1. The electron spin (Chapter 5) appears automatically once space and time are put on equal footing. Spin, otherwise inexplicable, thus becomes part of the natural fabric of a four-dimensional world.

2. Along with the electron (charge $= -e$) there must also exist, according to Dirac, a **positron**, a particle equal in mass but with a positive charge ($+e$). The positron is to be the electron's evil twin (a positive electron), normally unseen in our low-energy world. But if we muster sufficient energy, we should be able (so said Dirac) to make a positron "materialize" from the vacuum under certain conditions. If a positron then should encounter an electron, the particle and its "antiparticle" would **annihilate** each other. All their mass would be converted into energy, and the matter would disappear in a burst of gamma rays.

Were these predictions not borne out, they would be worthy of the most creative science fiction. But they have indeed been confirmed experimentally, notably four years *after* Dirac derived his relativistic quantum mechanics. Positrons and other antiparticles are now observed routinely in high-energy processes and in some forms of radioactive decay.

So there is matter, and there is **antimatter**: same mass, same spin, opposite charge, opposite magnetic moment. Every particle of matter has its deadly twin. Proton, antiproton. Neutron, antineutron. Neutrino (still to come), antineutrino. Many more. Meson, antimeson. Quark, antiquark.

21-4. NUCLEAR STRUCTURE: BINDING ENERGY AND THE STRONG FORCE

Define, first, a handy measure of mass, the **atomic mass unit (u)**, taken to be exactly one-twelfth the mass of a single $^{12}_{6}C$ atom. Since one *mole* of $^{12}_{6}C$ is similarly defined to contain Avogadro's number of atoms in 12.0000 g, we then have an exact relationship between grams and atomic mass units:

$$1 \text{ g} = 6.022137 \times 10^{23} \text{ u}$$

$$1 \text{ u} = 1.660540 \times 10^{-27} \text{ kg}$$

Masses for an assortment of isotopes are collected in Appendix C (Tables C-11 and C-12).

Using these values, we can compute the mass (and hence the mass-energy) lost when protons and neutrons come together from afar to make a nucleus. Helium's principal isotope, $_2^4$He, will provide our first example: two protons and two neutrons, the whole process summarized schematically in Figure 21-5.

A proton ($_1^1$p) has a mass of 1.00727647 u, whereas a neutron ($_0^1$n) has a slightly larger mass of 1.00866490 u. Kept apart, two protons and two neutrons have a combined mass of 4.03188274 u:

$$2 \times 1.00727647 \text{ u} + 2 \times 1.00866490 \text{ u} = 4.03188274 \text{ u}$$

A neutral atom of $_2^4$He, however, has a total mass of only 4.00260324 u, from which we must still subtract two electronic masses (each = 0.00054857990 u) to reveal the mass of the $_2^4$He nucleus alone. The result is 4.00150608 u for the bare helium nucleus, manifestly a loss of 0.03037666 atomic mass units during formation.

Something is missing, and we know where it went. Mass has been converted into energy during the formation of the nucleus, leaving the

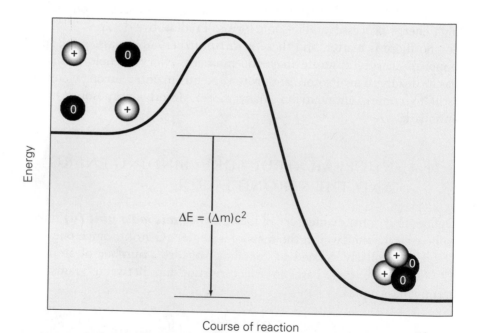

Course of reaction

FIGURE 21-5. When mass becomes energy: nuclear binding. Two protons and two neutrons, uncombined, lie at higher energy than a united nucleus of $_2^4$He. Overcoming a formidable activation barrier, they join together exothermically and release the statutory binding energy prescribed by Einstein's relativistic formula: $\Delta E = (\Delta m)c^2$. The intact nucleus is less massive and less energetic than its separated constituents.

assembled structure more stable than its individual components. The whole (the bound nucleus) is less than the sum of its parts (the isolated nucleons): less in mass, less in energy.

The missing matter, or **mass defect** (Δm), translates into a nuclear **binding energy** (ΔE) according to the Einstein formula:

$$\Delta E = (\Delta m)c^2$$

$$= \left(-0.03037666 \text{ u} \times \frac{1.66054 \times 10^{-27} \text{ kg}}{\text{u}}\right) \times (2.99792 \times 10^8 \text{ m/s})^2$$

$$= -4.53346 \times 10^{-12} \text{ J/nucleus}$$

At nearly 3 billion kilojoules per mole, the binding energy far exceeds any molecular heat of formation; but such an enormous, unearthly ΔE is just the amount we need to tear a helium nucleus apart. Here, plain to see, is an extraordinarily stable piece of matter, sure to withstand the comparatively small quantities of energy exchanged during ordinary chemical reactions.

Aside from a negligibly small change in entropy, this mass-become-energy marks the thermodynamic difference between product (nucleus) and reactants (nucleons). Consider it the free energy of nuclear formation—arising, as it does, not from the interaction of electrons and nuclei, but from the interaction of protons and neutrons.

Attracted neither by the electromagnetic force nor by gravity, the nucleons are bound instead by the so-called **strong interaction**, a force operating only within the close confines of a nucleus. The new interaction had better be strong, too, if it is to overcome the enormous electrostatic repulsion suffered by protons at close range. Realize just how powerful that repulsion is: a macro-sized 231 N of force (some 50 pounds) exerted by two tiny protons over a distance of only 1×10^{-15} m. Were there no countervailing force, the nucleus could not exist.

Nature answers with her aptly named "strong" force, an interaction fundamentally different from any we have seen so far. Unaffected by either mass or charge, the strong force ignores the electron entirely but treats proton and neutron as equals. Proton–proton, proton–neutron, and neutron–neutron interactions are all the same. Differences in charge and mass are irrelevant here.

Effective only inside the nucleus, the strong force is powerful but short-range. Short and sharp, it is (1) *very* repulsive at distances just under 10^{-15} m (1 femtometer, abbreviated fm); (2) *very* attractive at distances between approximately 1 fm and 2 fm; and (3) practically zero at

distances beyond 2 fm. It traps a pair of nucleons within a very deep,
very narrow well of the sort sketched in Figure 21-6.

Since both neutron and proton have radii of roughly 1×10^{-15} m,
two nucleons easily escape the strong force unless they are nearly touching. Let them come within range, though, and abruptly they are caught:
attracted by a force that outweighs the electrostatic repulsion hundred-
fold. And there they stay, confined on either side by a steep repulsive
wall, unable to squeeze closer, unable to move apart.

Yet the strong interaction also grants a curious freedom to particles
deep inside a nucleus. A balance of forces develops there, since each interior nucleon is pulled equally in all directions by its immediate neighbors (and just by these nearest neighbors, because all others are too far
away). The outcome? Nobody wins. For every tug this way, there is an
equal tug that way. Only at the outer surface does a nucleon feel a net
pull inward, for only at the surface is there no next layer to pull the particle away.

Provided, then, that a nucleon stays within a pack of other nucleons,

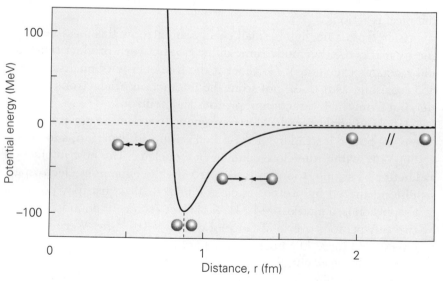

FIGURE 21-6. Potential energy for two nucleons interacting via the strong force,
plotted (roughly) versus distance. Very strong and very sharp, the force operates
with different effectiveness in three ranges. (1) Separated by more than 2×10^{-15} m
(2 fm), the particles are immune. (2) Squeezed much closer than 1 fm, they repel
explosively. (3) At distances in between, with the nucleons nearly touching, the trap
springs shut: a ferociously strong attraction, millions of times greater than ordinary
chemical interactions, more than enough to overcome any electrical repulsion. The
MeV (million electron volt), a standard unit in nuclear physics, is equivalent to
9.65×10^7 kJ mol^{-1}.

it is free to migrate throughout the nucleus. Away from the surface, there is no energetic advantage or disadvantage to be had at any point. The net potential energy remains unchanged—flat—until a nucleon floats to the surface, whereupon the particle is abruptly restrained. It is restrained by the strongest force that nature can muster.

Protons and neutrons therefore move as if, on the average, they are confined in a flat-bottomed well of potential energy (Figure 21-7). The rules are: Roam freely within the interior of the nucleus, but maintain center-to-center separations of between 1×10^{-15} m and 2×10^{-15} m; and on the surface, hold tight. The arrangement is similar to the forces

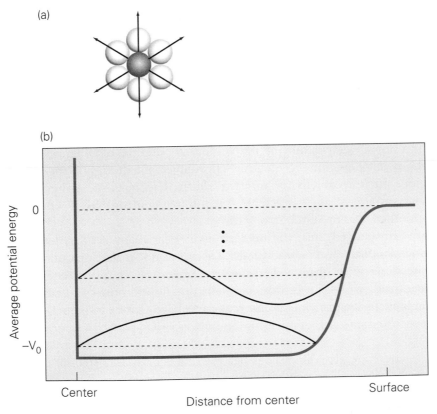

FIGURE 21-7. Potential energy within a many-nucleon nucleus. (a) Like a drop of liquid, particles inside the nucleus suffer forces different from those on the surface. Responding only to its immediate neighbors, an interior nucleon is pulled equally in all directions—and, as a result, is pulled nowhere at all. It sloshes around unrestrained in a pack of other nucleons. Only at the surface is there a net force from a neighboring nucleon. (b) The result is a flat-bottomed well of energy, measured from the center of the nucleus outward. Supported within the walls of the well, like standing waves, are nuclear wave functions (*shells*) analogous to the orbitals in atoms and molecules.

holding together a drop of liquid, where an analogous *surface tension* causes the fluid to bead up. So for the nucleus as well.

The big question now is whether our quantum mechanics, conceived originally for electromagnetic interactions, will also hold for strong nuclear interactions. Let us suppose that it does, subject to later proof by experiment, and let us then try to determine the energy levels of a nucleus. Here is one such approach, the **nuclear shell model**, patterned after the orbital method developed for the electronic states of atoms and molecules:

We devise, first, an appropriate Schrödinger equation, consistent with the known laws of quantum mechanics but suitably modified for the nuclear environment.

Next, we replace the sharp, spikelike interaction suffered in *each* one-on-one encounter (Figure 21-6) with the smooth, flat-bottomed well that emerges when all the interactions are averaged together (Figure 21-7). Our inspiration comes from the treatment already applied to atoms (Chapter 6), where we similarly smoothed out the electron–electron potential energy.

The result then was a set of electronic *orbitals*, each a wave function for a single electron moving in the blurred potential of all its neighbors. The result now, for the nuclear Schrödinger equation, is a set of analogous nuclear orbitals, or **nuclear shells**. Each shell describes a wave function for a single nucleon (proton or neutron) as it endures a smoothed version of the strong force.

Constructed thus, the nuclear shells bear many similarities to the more familiar electronic orbitals. There are nodes and quantum numbers. Energy is quantized. Angular momentum is quantized. Spin is quantized, with both proton and neutron allowed only two states. The Pauli exclusion principle is enforced as well, so that no two nucleons can have the same set of quantum numbers. Spin pairing occurs where needed.

There are also differences between nuclear and electronic orbitals, particularly in their angular momenta, but the similarities are still more compelling. Above all, this: The nuclear shell model, by assigning the protons and neutrons to individual compartments, establishes a hierarchy of orbitals. Arranged in order of increasing energy, the nuclear orbitals are subsequently filled in accord with the exclusion principle.

Between certain shells, moreover, there are large gaps in energy—gaps similar to those that appear in the noble-gas atoms. And just as closed-shell configurations lend great stability to atoms, so they do for nuclei as well. There are nuclear **magic numbers** (2, 8, 20, 28, 50, 82, 126 protons or neutrons) associated with particularly stable nucleonic configurations, arising when all shells below some large gap are filled. The magic numbers, so called, are the nuclear equivalents of such electronic configurations as $1s^2$ (for He, with 2 electrons), $1s^2 2s^2 2p^6$ (for Ne, with 10 electrons), $1s^2 2s^2 2p^6 3s^2 3p^6$ (for Ar, with 18 electrons), and others.

$^4_2$He, with two protons and two neutrons, enjoys the extra stability of a closed shell. So does $^{16}_8$O, with eight protons and eight neutrons (the next magic number); and so does $^{100}_{50}$Ca, with 50 protons and 50 neutrons (another magic number). Even-integer configurations of protons or neutrons, in general, promote stability by minimizing the number of half-filled shells. We seem to be on the right track, a good sign that quantum mechanics indeed does govern the nucleus.

The shell model cannot explain everything, but it does allow us to predict the relative stability of various nuclei. Let us try next to explain, using experimental data as well, why some nuclei are radioactive while others are not, and why unstable nuclei emit the particular particles they do.

21-5. NUCLEAR REACTIONS

Nuclei submit to thermodynamic laws as surely as do molecules, atoms, waterfalls, or any other system. They stumble toward greater stability. They are compelled, like the directors of a publicly traded company, to take the best offer: the lowest energy. If a system can gain stability by changing its structure or breaking into pieces, so it must. Inevitably, the change will occur if allowed the proper conditions; maybe not now, but eventually.

For most nuclides, there is room for improvement. Some nuclei have too many protons; some have too few. Some have too many neutrons, others too few. Some atoms become more stable when their nucleus captures an electron from the outside; some become more stable when their nucleus emits an electron. In all, close to 90% of the more than 2200 known species are unstable. They spontaneously emit particles and electromagnetic radiation in a drive toward lower energy. They are radioactive.

The stable nuclei, by comparison, have no better alternatives. Indeed they *are* the better alternatives; they are the final destinations reached by the radioactive nuclides. Stable isotopes have just the right numbers of neutrons and protons to fill their shells efficiently and, furthermore, to dilute the proton–proton electrical repulsions (which, bear in mind, are always working to burst open the nucleus).

Held in check by the strong force, the electromagnetic repulsions are overpowered but not eliminated. They never disappear. They continue to affect every proton too, not just the nearest neighbors; and

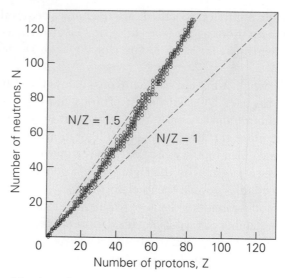

FIGURE 21-8. Number of neutrons versus protons for the stable nuclei, a minority of all nuclides. The heavier the nucleus, the more neutrons are required for stability.

sometimes, if there are not enough neutrons, the electrical repulsions will prevail over the strong attractions. The neutron-to-proton ratio (N/Z) determines whether a particular nucleus is stable or unstable.

From a scatter plot of N versus Z (Figure 21-8), we discover the *belt of stability*: a band containing the neutron-to-proton ratios for the nonradioactive nuclei. Note how the less massive species ($Z < 20$) tend toward equal numbers of protons and neutrons ($N/Z = 1$), whereas the heavier nuclei demand ever more neutrons to ensure stability. The requisite ratio increases to 1.5 up through $^{209}_{83}\text{Bi}$, after which all succeeding nuclei are radioactive. Decay they must; it is a thermodynamic imperative, a foregone conclusion.

Kinetics of Radioactive Decay

Sooner or later, one of these radioactive nuclei will fulfill its energetic destiny. It will decay spontaneously into a **daughter nucleus** and one or more smaller particles (see below), forming a set of products with less combined mass than the parent. The daughter's birth thus releases energy to the surroundings, packaged variously as the kinetic energy of the emitted particles and often as the radiant energy of gamma photons. The process is called radioactive decay.

Which nucleus will decay? We cannot say. When will it decay? Impossible to predict. Each reaction is a random quantum mechanical event, a roll of the statistical dice. So—not when, not which—but on this much we can count: *Some* nucleus will react, and then another, and another, and another after that; always at random, but always with a known set of odds.

After a certain time $t_{1/2}$, fully one-half of the original nuclei will be gone, replaced by their daughters. If there were 1000 potentially radioactive systems at the start, then 500 will remain after this first *half-life*; and 250 after the second half-life, and 125 after the third, and so on until none is left. It is an example of first-order kinetics, for which we use the same procedures laid down in Chapter 18. The characteristic exponential decay is shown again in Figure 21-9.

Each radioactive nuclide has its own half-life, some long and some short. Naturally occurring $^{238}_{92}U$, an alpha emitter with a half-life of 4.5 billion years, clearly will be with us for a long time. The synthetic radioisotope $^{131}_{53}I$, by comparison, takes just eight days to go through a half-life. After two months (approximately eight half-lives), less than 1% of the original sample remains radioactive.

During iodine-131's radioactive existence, it emits beta particles: high-speed electrons, energetic enough to damage body tissue. Taken up largely by the thyroid gland, a sufficiently large dose of $^{131}_{53}I$ consequently will destroy parts of that organ (a circumstance to be avoided by any healthy, reasoning person). Administered to a patient with an overactive thyroid, however, this radioactive isotope acts not as an assassin but as a surgeon: It goes straight to the designated target, destroys just enough

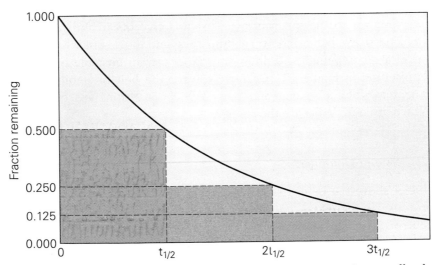

FIGURE 21-9. Unstable isotopes decay in a variety of ways, but usually the process follows first-order kinetics. The number of intact nuclei falls exponentially with time, decreasing by half with the passage of every half-life.

tissue to solve the problem, and then obligingly disappears over the course of two or three months.

Radioactivity evidently has two faces, benevolent and malevolent at the same time. Unwanted radiation damage can be deadly; controlled use of radioactivity, as with iodine-131, can be a lifesaver. And iodine-131 is but one example of how physicians use radioactivity to treat localized problems. Many other applications for radioisotopes have been developed as well, including diagnostic roles as **radiotracers**: small amounts sent to different parts of the body as local reporters. Using the proper isotopes in proper quantity, a diagnostician can monitor (yet not kill) the thyroid's activity, or measure blood flow through the coronary arteries, or, say, detect liver tumors, to cite just a few possibilities.

These *radiochemical* techniques work because the radioactive disintegrations are easily detected. Think, for example, of the phosphor-coated "glow in the dark" wristwatch, now banned, where radium nuclei pepper a thin layer of ZnS with α particles. Excited by the high-energy impacts, the electrons in ZnS emit visible light as they relax. The same principle finds application in the modern **scintillation counter**, an analytical instrument used to measure decay rates.

In chemical research, too, radiotracers such as $^{14}_{6}C$ allow one to monitor the ins and outs of selected atoms during a reaction or, in a procedure called **radiocarbon dating**, to determine the age of a material by estimating the number of half-lives undergone by the carbon-14. Having the same electron distribution as nonradioactive carbon, the $^{14}_{6}C$ isotope generally displays the same chemical properties (except for its vibrational frequencies, which depend on nuclear mass).

Now, on to the mechanisms of radioactive decay and let us ask: Alpha particles, beta particles, gamma rays, and more, but what kinds of nuclei are compelled to disintegrate? How can an unstable nucleus, as represented in Figure 21-10, move toward the belt of stability? What daughters does it spawn, and why? Which quantities are conserved? Which are not? What are the rules?

Mothers and Daughters

Nuclei above the belt of stability have too many neutrons and too few protons. To rectify the imbalance, a proton-poor nucleus can convert a neutron into a proton by casting out a beta particle ($\beta^- \equiv {}^{0}_{-1}e$):

$$^{1}_{0}n \longrightarrow {}^{1}_{1}P + {}^{0}_{-1}e + \tilde{\nu}$$

neutron   proton   electron   antineutrino

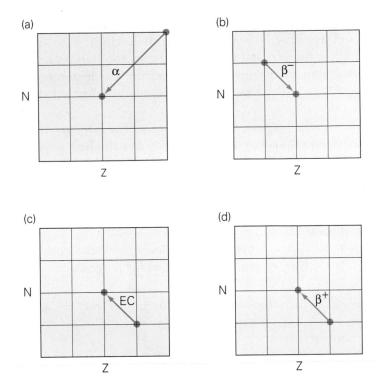

FIGURE 21-10. Made unstable by either too many neutrons or too few, a radioactive nucleus moves toward stability. The nuclide at the head of an arrow is stable; the nuclide at the tail is not. (a) Alpha emission takes a nucleus from upper right to lower left on a plot of N versus Z: two fewer neutrons, two fewer protons. Each grid line corresponds to one nucleon. (b) Beta emission brings the structure down and to the right: one more proton, one fewer neutron. (c) For electron capture the shift is from lower right to upper left: one more neutron, one fewer proton. (d) For positron emission, similar: one more neutron, one fewer proton.

The neutron spits out a high-speed electron (the β^- particle), together with a piece of antimatter: a presumably massless, chargeless, momentum-carrying, energy-carrying *antineutrino* ($\bar{\nu}$). Left behind is a proton. Energy is conserved. Momentum is conserved. Charge is conserved.

And, additionally, there is a new kind of bookkeeping to consider: the conservation of particle *number* according to particle *class*. Some particles are classified as leptons (the "light ones"), and some are classified as baryons (the "heavy ones"). Leptons transform into other leptons, and baryons transform into other baryons. According to our present experience—and our present experience may be incomplete—the two classes seem not to mix.

Which is which? The **leptons** include all those particles not susceptible

to the strong force, among them the electron, the neutrino, and their respective antileptons (the positron and antineutrino). Each lepton is assigned a *lepton number*, and the sum of the lepton numbers remains the
same throughout any change. For an electron, the lepton number is 1;
for a positron, −1. For a neutrino, the lepton number is 1; for an antineutrino, −1.

Applied to the baryons, a similar scheme: The **baryons** include those
particles that do respond to the strong force and also have a spin quantum
number equal to an odd multiple of $\frac{1}{2}$. The proton, the neutron, and their
respective antibaryons are the best known examples. Each baryon is assigned
a *baryon number*, and the sum of the baryon numbers is conserved. For a proton, the baryon number is 1; for an antiproton, −1. For a neutron, the
baryon number is 1; for an antineutron, −1.

With these assignments, we see at once how the conservation of
both lepton and baryon number is enforced for β decay:

Particle:	neutron	$\longrightarrow$	proton	+	electron	+	antineutrino
Lepton number:	0	=	0	+	1	−	1
Baryon number:	1	=	1	+	0	+	0

The lepton number is 0 on the left and 0 on the right; electron and antineutrino appear together, thus preserving the particle number within
the lepton family. Within the baryon family, the baryon number remains
fixed at 1 throughout. The number of protons and neutrons (the mass
number) remains constant. One neutron becomes one proton.

Now this decay of a neutron is a consequence of the **weak interaction**—something new, another fundamental force in nature. Distinct
from the strong interaction, which binds together protons and neutrons,
the weak force mediates, instead, the interconversion of protons and
neutrons. A neutron becomes a proton by emitting a β^- particle (an
electron) plus an antineutrino, and a proton inside a nucleus becomes a
neutron by emitting a β^+ particle (an antielectron, ^0_1e, Dirac's positron)
plus a neutrino:

$$^1_1P \longrightarrow {^1_0n} + {^0_1e} + \nu$$

$$\text{proton} \qquad \text{neutron} \quad \text{positron} \quad \text{neutrino}$$

In each case one baryon transforms itself into another baryon, accompanied by the appearance of a lepton and antilepton.

Shall we say that the baryon contains within itself a lepton and antilepton, ready to spring loose during a β decay? Emphatically not: A proton is not composed of a neutron, positron, and neutrino stuck together, nor is a neutron built from a proton, electron, and antineutrino. β decay does not break a nucleon into its most basic building blocks; rather, it rearranges those building blocks and throws off some dust in the process. Concerning these more fundamental units of matter, the *quarks*, we shall inquire briefly toward the end of the chapter.

Meanwhile, understand that β decay allows a proton-deficient nucleus such as $^{14}_{6}C$ to become $^{14}_{7}N$:

$$^{14}_{6}C \longrightarrow {}^{14}_{7}N + {}^{0}_{-1}e + \tilde{\nu}$$

Originally containing 6 protons and 8 neutrons, the nucleus now has 7 protons and 7 neutrons. Z increases by 1. N decreases by 1. A neutron becomes a proton; the atomic number changes; one element (carbon) becomes another (nitrogen). The criterion for spontaneity is a net decrease in mass, including contributions from the "chemical" electrons bound outside the nucleus of each whole atom.

Understand also that a proton-rich nucleus such as $^{19}_{10}Ne$ can emit a positron (β$^+$) and so become $^{19}_{9}F$:

$$^{19}_{10}Ne \longrightarrow {}^{19}_{9}F + {}^{0}_{1}e + \nu$$

The original neon nucleus (10 protons and 9 neutrons) is now a fluorine nucleus (9 protons and 10 neutrons). Z decreases by 1. N increases by 1. A proton turns into a neutron, and the result, again, is that one element changes into another.

In the related process of **electron capture (EC)**, a nucleus will seize an electron from *outside* the nucleus and similarly convert a proton into a neutron:

$$\underset{\text{proton}}{^{1}_{1}p} \quad + \quad \underset{\text{electron}}{^{0}_{-1}e} \quad \longrightarrow \quad \underset{\text{neutron}}{^{1}_{0}n} \quad + \quad \underset{\text{neutrino}}{\nu}$$

Certain heavy atoms, which otherwise cannot lower their energy sufficiently by positron emission, are particularly prone to electron capture—as, for example, the $^{201}_{80}Hg$ isotope of mercury:

$$^{201}_{80}Hg + {}^{0}_{-1}e \longrightarrow {}^{201}_{79}Au + \nu$$

The net result is to convert an atom of $^{201}_{80}Hg$ (including its 80 electrons) into a complete atom of $^{201}_{79}Au$, with the attendant release of a massless neutrino. Total mass decreases during the process, and hence the reaction is spontaneous.

For the medieval alchemist it would have been a dream fulfilled, the transmutation of a base metal into gold. He could not have known, though, that an element's identity comes only from the protons in its nucleus, not the electrons outside. Only a nuclear reaction can transform one element into another.

Heavy elements are also likely to disgorge alpha particles ($\alpha \equiv {}^{4}_{2}He$) and thus decrease both Z and N by 2, taking a route *not* controlled by the weak interaction. The conversion of radium into radon offers a ready example:

$$ {}^{226}_{88}Ra \longrightarrow {}^{222}_{86}Rn + {}^{4}_{2}He $$

Whereas β emission (a weak interaction) preserves the mass number of a nuclide, α emission does not. The daughter nucleus, with two fewer protons and two fewer neutrons, has a mass number

$$ A = Z + N $$

smaller by 4. The α particle carries away the difference.

Proton emission, another mode of decay, similarly reduces both atomic number and mass number, this time by only 1. The daughter inherits $Z - 1$ protons plus all the original neutrons. **Neutron emission**, also possible, preserves the atomic number while eliminating one unit of mass from the parent. Neither pathway involves the weak interaction.

Finally, consider the nonmaterial emissions called ***gamma rays***. Here we have highly energetic electromagnetic radiation, generated when an excited nucleus relaxes to a state of lower energy. Released during the rearrangement of the nuclear shells, gamma photons accompany most of the emissions described above. They carry away part of the energy.

Building a Chemical Universe: Fusion

In the beginning was energy, and from energy came matter. Simple matter: protons and neutrons, electrons, neutrinos, some positrons. Born with enormous kinetic energy, the matter soon began to cool. It cooled just enough to collect into clumps here and there; into clumps still very hot and still very dense; into sparsely scattered clumps that, even today, punctuate the vast emptiness of the universe. We call them stars.

They are chemical factories, the stars, the fires in which are forged all atoms heavier than hydrogen. Only here, where temperatures run into the millions, can one hydrogen nucleus overcome the electrical repulsion of another. Only in the stars, nowhere else in nature, can two unbound protons get close enough for the strong interaction to take hold.

Blocking that proton–proton contact is a kinetic activation barrier taller than any we have seen—more than 100 million kilojoules per mole—but if such a formidable obstacle can be overcome, then eventually there is still greater energy to be released. Given the right conditions, even the most kinetically disadvantaged partners can realize their thermodynamic potential. Somehow, nuclei more massive than hydrogen can be built.

The process is thought to begin when two highly energetic protons *fuse* into a deuterium nucleus, letting fly a positron and neutrino:

$$\,^1_1\text{H} + \,^1_1\text{H} \longrightarrow \,^2_1\text{H} + \,^0_1\text{e} + \nu$$

Thus is "heavy hydrogen" made by a process of **nuclear fusion**.

Smashing into another proton, the newborn deuteron reacts further to produce helium-3 and an energetic gamma photon. This second fusion,

$$\,^2_1\text{H} + \,^1_1\text{H} \longrightarrow \,^3_2\text{He} + \gamma$$

is followed next by a third reaction in which two helium-3 nuclei combine to form helium-4, the dominant isotope. At the same time, two hydrogen nuclei are recycled for continued use:

$$\,^3_2\text{He} + \,^3_2\text{He} \longrightarrow \,^4_2\text{He} + 2\,^1_1\text{H}$$

From hydrogen to helium, mass is lost and energy is released—a great deal of energy, despite the huge amounts invested to scale the barriers at each step.

It is the energy of the sun, where 4 million tons of matter are converted into 100 billion billion kilowatt-hours of energy every second. It is (through a different fusion process) the energy of the thermonuclear bomb, a woeful attempt to unleash the destructive potential of nuclear fusion. It is also the hoped-for energy of a *controlled* fusion reaction, an attempt to harness the power of mass-energy. It is the way elements are built.

So begins, in the stars, the long climb from hydrogen to uranium, with the heavier nuclei produced first by fusion of lighter particles. Some products (such as helium-5 and lithium-5) are unstable and disappear with half-lives as short as 10^{-21} s; we never see them. Of the stable nuclei, certain isotopes (helium-4) are more stable than others (helium-3); and these favored nuclei, lower in energy, are the ones that prevail in the end.

Nothing new here: Thermodynamics is thermodynamics, and good thermodynamics is why 99.999863% of all helium nuclei contain two protons and two neutrons. Helium-4, with its closed shell and double magic number, is low in energy and profoundly stable.

This process of stellar **nucleosynthesis**, not fully understood, unfolds as an intricate sequence of dependent steps: (1) hydrogen to helium-4, (2) helium-4 to beryllium-8, (3) beryllium-8 to carbon-12, . . . , and upward to heavier and heavier nuclei. Gaps in the pattern are created and filled along the way so that, over billions of years, the elements gradually build up in their natural proportions. Then the star, collapsing finally under its own weight, might die in the shattering explosion of a **supernova** and generate even heavier elements in the process. Atoms fly like shrapnel from the exploding star, raining down throughout the universe and giving us our chemical legacy.

Splitting Up: Fission

The heavier the nucleus, the stronger the nucleus . . . but only up to a point. See why: We take as our measure of stability the **binding energy per nucleon**, which for ^4_2He is computed as

$$E_A = -\frac{\Delta E}{A} = \frac{4.53346 \times 10^{-12} \text{ J}}{4 \text{ nucleons}} = 1.13337 \times 10^{-12} \text{ J/nucleon}$$

Doing so, we apportion the total binding energy, ΔE, among the two protons and two neutrons; and this per-nucleon equivalent, E_A, then becomes a standard for comparison. The higher the binding per nucleon, the more stable the nucleus.

Figure 21-11 shows us, then, that the nucleons are bound more tightly in beryllium-9 than in lithium-7; and more tightly in boron-11 than in beryllium-9; and still more tightly in carbon-12. Helium-4 is an unusually stable exception to the rule, but the general trend is *up*. Light nuclei *gain* stability by fusing into heavier nuclei until, however, the mass numbers grow to between 50 and 60. Near $^{56}_{26}\text{Fe}$, the curve peaks and then begins slowly to tail downward; fusion is no longer a growth industry after that.

The nuclei are becoming too heavy, and the proton–proton repulsions are starting to become troublesome. Remember: Every proton continues to repel every other proton, whereas the strong force remains strictly an interaction between near neighbors. The balance grows more precarious as the nuclear mass increases, and so any small change (a new shape, an extra particle, an internal redistribution) may shift the advantage toward the repulsions.

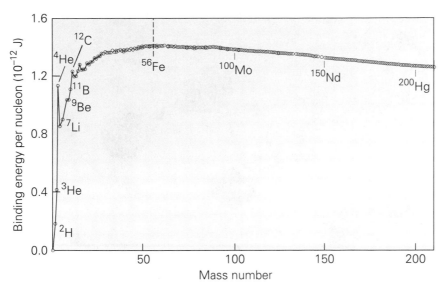

FIGURE 21-11. Diminishing returns: the battle between strong attraction and electromagnetic repulsion. The average binding energy per nucleon reaches a peak for mass numbers between 50 and 60, beyond which the growth of a nucleus becomes less profitable thermodynamically. Light nuclei shed energy by merging together (fusion), whereas heavy nuclei benefit by splitting apart (fission).

Do something to disturb that balance. Change the shape of the nucleus, for example, by injecting an extra neutron. It can happen, too, because neutrons are stealthy particles. Having no charge, they slip through the outer electromagnetic barrier and are sucked into the nucleus by the strong attraction. Let that occur, and a precariously balanced heavy nucleus might suddenly explode into two smaller nuclei of roughly equal mass (but still with less combined mass than the parent). The lost mass then shows up as energy; again, a great deal of energy. The process is called **nuclear fission**.

The fission of $^{235}_{92}U$ provided the energy for the first atomic bomb. Able to absorb a slow-moving neutron, uranium-235 subsequently splits into two lighter nuclei, which, even after the release of energy, may be unstable themselves and so behave as radioactive *fallout*. Fission is a complicated reaction, with many available pathways and dozens of possible daughter nuclides. Some of the known pairs include $^{91}_{36}Kr$ and $^{142}_{56}Ba$, $^{97}_{40}Zr$ and $^{137}_{52}Te$, $^{94}_{36}Kr$ and $^{139}_{56}Ba$, $^{72}_{30}Zn$ and $^{162}_{62}Sm$, $^{80}_{38}Sr$ and $^{153}_{54}Xe$. A typical process, illustrated in cartoonlike fashion, is pictured in Figure 21-12.

There is one more product as well, a product critical to sustain the reaction: neutrons. Accompanying each fission are two or three neutrons, which can provide the fuel for a self-propagating chain reaction.

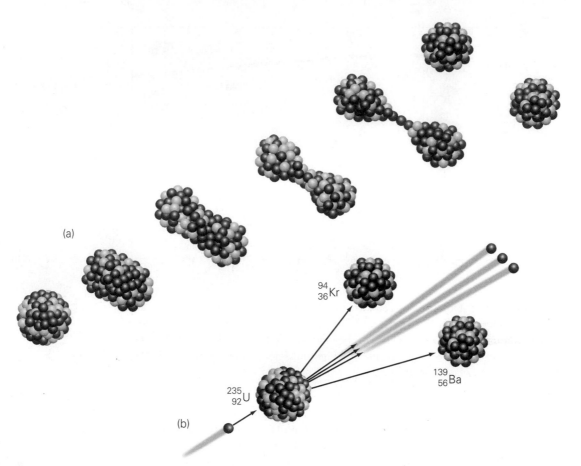

FIGURE 21-12. Sometimes, the electromagnetic force wins. (a) Asked to hold together more and more protons, the strong force—strong, but severely limited in range—eventually fails. Ordinary electrical repulsions, present everywhere, overwhelm the nearest-neighbor strong force and make the nucleus ripe for violent disintegration by fission. (b) Destroying a fragile balance, the addition of a 144th neutron forces the $^{235}_{92}$U nucleus to undergo fission. The overweight structure splits into two lighter nuclei, releasing (in this example) three unbound neutrons and abundant energy as well. Numerous other products are possible besides the ones shown here.

Each liberated neutron may induce another fission reaction, which itself produces two or three neutrons; and each of *these* neutrons may find its way into yet another uranium-235 nucleus; and so on, more and more, until possibly the chain reaction goes out of control—but only if there are enough fissionable nuclei. Too small a sample, and the neutrons are lost before they can do further damage.

There must be, instead, some ***critical mass*** of $^{235}_{92}$U for the chain reaction to be sustained; and, take note, this threshold mass must hold together for sufficient time while the reaction proceeds. A controlled, moderated chain reaction can power an electrical generator; an uncontrolled, unmoderated, *supercritical* chain reaction produces a bomb.

Latter-Day Alchemy: Transmutation

Adding a neutron (or some other particle) to a nucleus does not always trigger a fission reaction. Sometimes the result is a new element, a nucleus not otherwise found in nature.

So it was in 1940, when researchers created a 93rd element (neptunium) by injecting a neutron into $^{238}_{92}$U:

$$^{238}_{92}\text{U} + {}^{1}_{0}\text{n} \longrightarrow {}^{239}_{92}\text{U} \longrightarrow {}^{239}_{93}\text{Np} + {}^{0}_{-1}\text{e} + \tilde{\nu}$$

The intermediate product, $^{239}_{92}$U (half-life = 23.5 minutes), decays by beta emission into this first **transuranic element**, $^{239}_{93}$Np, also a beta emitter (half-life = 2.4 days). From the natural decay of neptunium-239 then comes fissionable plutonium-239 ($^{239}_{94}$Pu, half-life = 24,000 years), and, by various particle bombardments, even more. There follows the rest of the *actinide* series, which concludes with lawrencium (Lr, $Z = 103$); and, beyond that, some nine additional elements discovered to date.

Just past lawrencium, for instance, the 104th element takes its place in the periodic table directly below hafnium (Hf). Its electronic configuration is $[\text{Rn}]7s^2 5f^{14} 6d^2$. The 118th element, expected to have the noble-gas configuration $[\text{Rn}]7s^2 5f^{14} 6d^{10} 7p^6$, should fall below radon in the table.

The chemistry is difficult, because all the transuranic nuclei are radioactive. Some have exceedingly short half-lives, and some can only be produced in infinitesimal quantities. But still there is new *chemistry* beyond uranium, for here are the first man-made elements—new *atoms, not new molecules*. Until the 20th century, atom making was a task reserved for the stars alone.

21-6. FIELDS AND PARTICLES

Matter interacts with matter because it has mass. We say that a gravitational force is transmitted by a gravitational field.

Charged matter interacts with charged matter because it has, well, charge. We say that an electromagnetic force is transmitted by an electromagnetic field.

So protons and neutrons interact under the strong and weak forces because they have . . . ? Because they have, precisely, *what*? Have we neglected something fundamental concerning the nucleus?

And do we really know what it means for matter to have charge and mass? Are our definitions the product of circular reasoning, little more than an attempt to mask a profound ignorance; or do they hold some deeper meaning? What really are the fundamental forces that shape the world?

Provocative, disturbing questions all—and still no fully satisfactory answers; but, even so, there are partial solutions and glimpses into what might be. The questions go straight to the core of matter and energy, and the still-evolving answers have a simple beauty of their own.

Let us say good-bye to chemistry, then, while we dig just a little deeper into nucleus and nucleon, if only to knock on one of nature's innermost doors. For only by looking *inside* the proton and neutron can we hope to tie up the loose ends; and only then, with broadened perspective, can we better appreciate the electron's role in ordinary chemistry.

This final journey begins outside the nucleus, in the electromagnetic world.

Interactions: The Medium and the Message

Two boys tossing a ball back and forth are *interacting*. They are occupied with each other; they have something in common. The ball is a kind of bond, a link.

Whether the link suggested in Figure 21-13 is attractive or repulsive depends on how the boys throw the ball. A hard, straight pitch will drive the catcher back a step. The one boy will recoil, having absorbed the momentum of the ball; and then, returning the pitch in the same way, he will drive his friend back a step or two as well. Pitch and catch (recoil), pitch and catch (recoil), the game drives them apart. The interaction is repulsive.

But if their tosses are just a bit short (or if they throw a boomerang), then each boy must take a step *in* to make a catch. The game brings them together now. The interaction is attractive. It all depends on how they throw the ball.

The *range* of the interaction depends on the nature of the ball. A tennis ball will travel farther than a steel shot, enabling the two parties to communicate over a longer distance. There is also a time scale to the interaction; it takes time for the ball to travel from pitcher to catcher. Perhaps, too, the ball spins clockwise or counterclockwise, or turns end over end like a football—all these characteristics should affect the quality of the interaction.

Now make the connections:

FIGURE 21-13. Playing catch: exchange of a particle. (a) Sharing the ball, thrower and catcher are forced to interact; but they do so repulsively, moving farther away with each toss. The continual transfer of momentum pushes them apart. (b) With a different kind of toss, the interaction now becomes attractive. Sender and receiver move closer together.

1. The boys are charge-bearing particles of matter.
2. The ball is a photon.
3. The game of catch is the electromagnetic interaction.

Because: To be charged means, precisely, to be able to emit and absorb a photon. To be charged is forever to pitch and catch a *massless*, *chargeless* ball moving at the speed of light. Only a charged particle has this capability, and only two charged particles can play catch with a photon.

Because: To emit a photon means to *give* one additional quantum of energy ($h\nu$) and momentum (h/λ) to the electromagnetic field. To absorb a photon means to *take* a quantum of energy and momentum from the field.

Because: To interact electromagnetically means to exchange a photon. One particle throws out a quantum; the quantum propagates through space; the quantum is caught by another charged particle. If the particles have like charges, they play a repulsive game of catch. If the particles have unlike charges, they play an attractive game.

The electron thus becomes a giver and taker of photons, constantly throwing out and taking in lumps of energy and momentum. Any charged particle sprays out massless billiard balls at the speed of light, and only *other* charged particles have the facility to catch them. It is a long-range game, played from one end of the universe to another; and we derive from it the electromagnetic force.

Stripped down, chemistry then becomes a game of electromagnetic catch: photons tossed between electrons, photons tossed between electrons and nuclei, photons tossed between nuclei. Underlying every chemical interaction is this simple means of communication. Atomic energy levels, molecular energy levels, covalent bonding, ionic bonding, noncovalent intermolecular forces—each is a manifestation of the electromagnetic force. Each, at the most elementary level, emerges as an exchange of photons. A complicated exchange, yes, but still basically that: a game of catch.

We would go mad trying to break down every chemical transaction into individual photons and electrons, and why should we? As chemists, we want to shape atoms and molecules to our own ends, not be distracted by the most minute features of these already minute systems. We want, thinking practically, not to lose sight of the proverbial forest for the trees.

But, as it happens, the game of catch *is* the forest. The exchange of a photon by charge-bearing particles becomes a prototype for all of nature's interactions, including those responsible for nuclear binding and decay.

Back into the nucleus we go.

Mesons and the Strong Force

Protons and neutrons play their own game of catch, although not with a photon. Tossed back and forth between nucleons is, instead, a *massive* particle generically called a **meson**; and it is this exchange of mesons that creates the strong nuclear force. The mesons, with integral spin quantum numbers (0, 1, 2, . . .), form a class distinct from leptons (spin $= \frac{1}{2}$) and baryons (spin $= \frac{1}{2}, \frac{3}{2}, \frac{5}{2}, \ldots$).

To interact via the strong force is to emit and absorb a suitable meson, in many cases a positive, negative, or neutral *pion*. Predicted in 1935 and discovered in 1947, these pions (250 to 300 times more massive than an electron) are too heavy to throw a long way. So we have, again, a game of pitch and catch just like the electromagnetic force, but here the game is played only at the shortest distances. The meson interaction is effective only at short range, over nucleon–nucleon distances barely greater than 10^{-15} m.

Is nucleon–meson–nucleon exchange our strong force? Yes . . . but. Yes, neutrons and protons do draw together by sharing a meson. But, no, the meson is not the end of the story. There are dozens of different mesons, and many of these particles also contribute to nucleon–nucleon binding. Nor is a meson a truly irreducible, indivisible, *fundamental* brick of matter. It, too, has an internal structure, and its internal components are held together themselves by strong forces.

We have still to dig deeper.

Quarks

We are led to a new building block of matter, an even tinier bit that appears (we think) to be a true brick, a genuinely elementary particle. It is called the **quark**.

Like the electron, a quark seems to have no internal structure. Like the electron, a quark has mass; it is subject to gravitation. Like the electron, a quark has charge; it is subject to the electromagnetic force. Unlike the electron, however, a quark has two other endowments: the whimsically named properties of *flavor* and **color**.

Flavor: which is affected, in particular, by the *weak* interaction.

Color: which gives rise to the *strong* interaction.

They seem to be silly names, flavor and color, except when we realize that the sober word *charge* conveys scarcely more meaning. Take the example of the electromagnetic interaction, then, and draw the analogies:

1. FLAVOR: A MENU OF UP, DOWN, STRANGE, CHARMED, TOP, AND BOTTOM QUARKS. Just as a charged piece of matter will emit and absorb photons, so will a "flavored" quark emit and absorb special messengers of its own. Called the $W^\pm$ and Z^0 particles, these are some of the (very) heavy balls exchanged when two quarks play catch. The interaction transforms neutrons into protons and thereby plays a part in beta decay. It is a short-range effect.

Of the six possible flavors, we need mention just two at this stage: *up* (*u*) and *down* (*d*). Each of the up and down flavors bears an electric charge as well. The *u* quark carries a charge of $+2e/3$, whereas the *d* quark has a charge of $-e/3$ ($e = 1.6 \times 10^{-19}$ C). Each quark has a baryon number of $\frac{1}{3}$, and each antiquark has a baryon number of $-\frac{1}{3}$.

Nature mixes the different flavors into the more familiar bits of matter shown in Figure 21-14. She glues together a quark (baryon number $= \frac{1}{3}$) and an *anti*quark (baryon number $= -\frac{1}{3}$), and we have a meson (baryon number $= 0$). From two *u* quarks and one *d* quark (*uud*) we obtain the monopositive electric charge of a proton,

$$uud = \tfrac{2}{3} + \tfrac{2}{3} - \tfrac{1}{3} = 1 \qquad \text{(proton)},$$

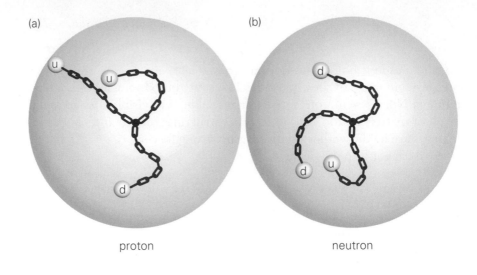

(a)

u
u
d

proton

(b)

d
d
u

neutron

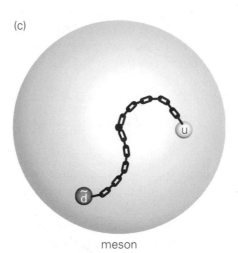

(c)

u
d̃

meson

FIGURE 21-14. A nucleon, unlike an electron, is built from smaller components: quarks, bound together in twos and threes to form a composite particle. The quarks move about freely until, at some very short distance, they are reined in by the strong *color* force symbolized by the chains. (a) A proton contains two *u* quarks and one *d*, giving it a net charge of +1. (b) A neutron: one *u* and two *d*. (c) A meson: a quark (here, *u*) and an antiquark (here, *d̃*). The example shown is just one of many.

and from one *u* quark and two *d* quarks (*udd*) we get the uncharged neutron:

$$udd = \tfrac{2}{3} - \tfrac{1}{3} - \tfrac{1}{3} = 0 \qquad \text{(neutron)}$$

Together, the three quarks produce a baryon number of 1 for both proton and neutron.

Beta decay? A down quark changes to an up quark, and thus a neutron becomes a proton. The flavor changes, but the number of quarks remains the same. The quark number (and, with it, the baryon number) is always conserved.

Two quarks per meson, three quarks per proton, three quarks per neutron; and now, finally, on to the endowment of color.

2. COLOR: A PALETTE OF RED, GREEN, AND BLUE. Just as a charged piece of matter will emit and absorb photons, so will a "colored" quark emit and absorb special messengers too; but these messengers are different from the photon and different from the $W^{\pm}$ and Z^0 particles. They are the **gluons**, the exchange particles believed to produce the strong *color* interaction between quarks themselves. Here is an interaction more fundamental than the exchange of a meson between quark-built *nucleons*, and it is from this color force that the mesonic strong force ultimately derives.

Red, green, and blue—the colors of quarks, albeit colors in name only. There are "red" *u* quarks, green *u* quarks, blue *u* quarks, red *d* quarks, green *d* quarks, . . . manifold possibilities, rich and complicated. Nevertheless, the basic effect is analogous to the interaction between electric charges: quarks of various colors attract and repel one another.

It is a peculiar kind of force, sometimes likened to the shackles that bind the prisoners in a chain gang. Provided that the bound quarks stay reasonably close together, they remain free to slosh around inside a nucleon. Let them try to break out, though, and the powerful color force yanks them back. No single quark ever escapes.

Instead, three quarks always stay confined inside proton and neutron: one red, one green, one blue. The three colors, equally represented, make the nucleon *color-neutral* in the same way that an atom or molecule has no net electric charge. Proton and neutron are colorless overall; they have no net color charge. The three quarks inside the nucleon flicker continually from red to green to blue as they exchange gluons, colored themselves.

Molecules are similarly *electric*-charge-neutral, yet still they manage to interact electromagnetically. They exchange photons, messengers of the electromagnetic field. One molecule momentarily distorts the electron cloud of another, whereupon there develops the fleeting electrical interaction recalled in Figure 21-15(a): a noncovalent intermolecular force, a *secondary* force rooted squarely in the electromagnetic world (Chapter 9). We have no new fundamental force here, but rather a complicated mosaic of electron–photon–electron and electron–photon–quark exchanges. The electromagnetic force remains fundamentally at the heart of it all.

(a)

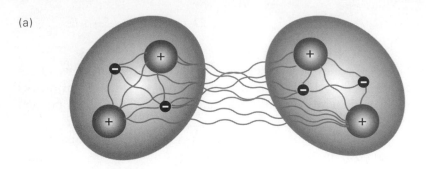

(b)

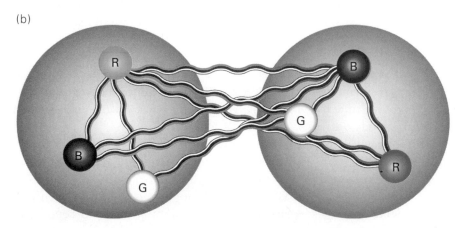

FIGURE 21-15. Composite particles; composite interactions. (a) Two molecules, each uncharged, interact electromagnetically despite their overall neutrality. Acting separately, every charged particle in the mix exchanges photons with every other and is individually attracted or repelled. A simpler net force (such as between induced dipoles) emerges from the web of fundamental transactions. (b) For two nucleons, each an "uncolored" blend of red, green, and blue quarks, the effect is the same: a tangled network of strong interactions between individually colored quarks, mediated by gluons (colored as well). The result is a composite strong attraction between the two color-neutral nucleons, as if carried by a meson.

The same goes for nucleons and the strong force. Protons and neutrons are color-neutral, yet still they manage to interact by the color force. Not directly, but indirectly: Momentary distortions produce indirect color interactions between proton and neutron, which add up to the toss of a meson. The color force remains fundamentally at the heart of it all.

End of story?

21-7. EPILOGUE

The ending might never be written, but we are beginning to glimpse a universe far simpler, far more unified than ever imagined.

It is a universe in which just two fundamental classes of particles, quarks and leptons (especially the electron), combine to build most of what we call ordinary matter: the proton (from three quarks), the neutron (from three quarks), the atom (from electrons, protons, and neutrons), the molecule (from atoms).

It is a universe ruled by no more than four fundamental forces, each associated with a particular endowment of matter: (1) gravity, the attraction of mass for mass; (2) electromagnetism, the attraction and repulsion of electric charge; (3) the weak force, by which quarks change their flavor; (4) the strong force, arising from the color of a quark. Particles of matter interact by exchanging their messenger particles, throwing up a field in space and time. For gravity, the graviton (as yet undetected) is thought to carry the message. For electromagnetic interactions, it is the photon. For the weak force, the $W^{\pm}$ and Z^0 particles. For the strong force, the gluons.

It is a universe where the four forces also display signs of a deep commonality, signs that maybe we have just one fundamental force, not four. The electromagnetic, weak, and strong interactions already seem to be variant expressions of a single, primeval agency under which all the messenger particles are equivalent. Gravity remains apart, at least for now.

It is a universe in which energy is exchanged according to the rules of thermodynamics, and particles move according to the rules of quantum mechanics. There is no absolute space; there is no absolute time. A few conservation laws are rigorously enforced. Entropy grows and grows.

For the chemist, it is a universe mostly of charge. All the action unfolds in the electromagnetic world, where the building blocks are known and the rules are set. The chemist tries first to understand what nature does with the electrons, protons, neutrons, atoms, molecules, ions, and states of matter already present; and then, having observed carefully, sets out to make improvements.

Such as: building molecules that can recognize and interact with each other; molecules that fit together like hand in glove; molecules that can fight disease, remediate environmental problems, accelerate and inhibit selected reactions.

Such as: devising tailored catalysts to alter the speed and pathways of chemical processes; catalysts to provide needed products swiftly and cheaply; catalysts to act selectively amidst a welter of competing reactions; catalysts as diverse as zeolites, enzymes, surfaces, and metallic clusters.

Such as: developing synthetic materials with unique, molecular-based properties; materials strong yet lightweight; materials with unusual electrical and magnetic characteristics; materials like polymers, ceramics, semiconductors, superconductors, crystals, and nanocrystals. That, and more.

Much remains to be done.

REVIEW AND GUIDE TO PROBLEMS

Nature tolerates no waste, no extra parts, no useless adornment. Everything fits together. Everything has a place.

Especially the nucleus—which, deep inside the atom, far from the action at the edges, plays a role in chemistry as important as any electron. True, a nucleus stays put; true, a nucleus rarely changes; true, a nucleus is like a brick; but try, against nature, to imagine a nuclear-free chemistry: it fails. Without nuclei to rein in the electrons, there would be no atoms. There would be no molecules. There would be no reactions.

Yet for all that, a nucleus lives in a world of its own—not the chemical, electromagnetic world of the photon, but the even smaller, less understood world of the strong force and weak force, the quark and the gluon.

IN BRIEF: NUCLEAR STRUCTURE AND REACTIONS

1. SMALL AND SMALLER. Atoms are small, just a few angstroms across (10^{-10} m). Nuclei are smaller, with radii as little as 10^{-15} m. The ratio of nuclear to atomic dimensions goes down to one part per hundred thousand, the difference between an inch and a mile.

It is, for the positively charged protons, an almost impossible crowding. The electromagnetic repulsion, barely lightened by a sprinkling of neutrons, swells to macroscopic proportions.

How then, we ask, can a nucleus exist, and how can it be so tenaciously strong? With only the merest gravitational force to hold neutrons and protons together, the particles should fly away unconnected. There must be something else. There is: the strong force.

2. NUCLEAR FORCES. Nature needs a strong, different kind of force to confine the **nucleons** (protons and neutrons) within a nucleus, a force strong enough to overcome the repulsion between positive particles and different enough to allow neutral particles to interact at all. And so there arises, just right, the aptly named nuclear **strong force**, an interaction fundamentally different from both gravity and electromagnetism, an interaction dependent on neither mass nor charge.

Nearly zero at distances greater than the usual nucleon–nucleon separation, the strong force becomes overwhelmingly attractive at just under 1×10^{-15} m. Less than that, it grows impenetrably repulsive.

The result: protons and neutrons stay together like marbles in a bag, free to move about inside but unable either to squeeze closer or to break free.

3. OUTSIDE/INSIDE. Outside the nucleus, electrons and nuclei group themselves into atoms and molecules, solids, liquids, and gases—complex assemblies, each outwardly different yet each sewn together with the same thread: the electromagnetic interaction. Outside, electrically charged particles exchange energy and momentum by dipping into the electromagnetic field. The photon, the quantum of that field, serves as the common currency.

Chemistry, then, stripped bare: an exchange of photons outside the nucleus, whether between elementary particles (electrons) or composite structures built from them (atoms and molecules). Some transactions involve a single photon, as when two electrons interact. Other exchanges are indescribably complex, as when all the electrons and all the nuclei of one molecule interact with all the electrons and all the nuclei of another. But photons there are, in every chemical event, be it the repulsion of two electrons, the maintenance of a covalent bond, or the flickering of a London interaction.

Inside the nucleus, a parallel universe, the electromagnetic interaction gives way to the strong force. **Quarks** play the role of elementary particles here, coming together to form nucleons just as electrons and whole nuclei come together to form atoms or molecules. Whereas the electromagnetic interaction arises from the *electric* charge of both electron and quark, the strong interaction applies only to the special **color** attribute of the different quarks.

Not the photon, then, but another messenger—the **gluon**—binds quarks into protons, neutrons, and diverse **mesons**. Protons and nucleons, composite structures themselves, subsequently form larger associations just as atoms and molecules do in the world outside. They weave a web of tangled interactions, each one born of a fundamental force, all of them blended together for a simpler effect overall. Between intact atoms and molecules, the elementary electromagnetic transactions blur into dipole–dipole forces, hydrogen bonds, dispersion forces, and the various other noncovalent interactions. Between neutrons and protons, the strong force manifests itself as an exchange of a meson.

Inside and outside. Two worlds or one? In our minds, we order them in the same way: according to structure (quantum mechanics), energy (thermodynamics), and reactivity (kinetics and mechanism).

4. NUCLEAR QUANTUM MECHANICS. Schrödinger, Heisenberg, Bohr, Dirac, and the other coinventors of quantum mechanics knew nothing, at first, about the nucleus, but still they produced a theoretical framework good both inside and outside. So good, indeed, that if we know the particles involved . . . if we know the interactions . . . if we know the potential energy, then we can construct a Schrödinger-like wave equation for any system, including the protons and neutrons within a nucleus. Whether such an equation might be soluble is another matter,

but we can always try. We can always make approximations.

Take, for example, the **nuclear shell model**, which gives us the nucleonic equivalent of electronic orbitals: individual wave functions for each proton or neutron (a *shell*), wherein the nucleon is imagined to move in a smoothed-out sea of interactions produced by all its neighbors. Mimicking the electronic Aufbau process *outside* a nucleus, the protons and neutrons fill the nuclear shells in sequence, pairing their spins in deference to the Pauli exclusion principle. The configuration of lowest energy corresponds to the ground state. Transitions to states of higher energy are induced by gamma photons.

Especially stable is a nucleus that contains a **magic number** of protons or neutrons (2, 8, 20, 28, 50, 82, 126 nucleons of either type), an arrangement similar to the filled electronic shells of the noble gases (2, 10, 18, 36, 54, 86 electrons).

5. NUCLEAR THERMODYNAMICS. Different nuclei have different energy levels, different configurations, different proportions of protons and neutrons. Some systems stick together more tightly than others. Some are prone to change; others not.

A useful benchmark is the **binding energy**, taken as the difference in energy between (1) the protons and neutrons separated to infinity and (2) the nucleons bound into a nucleus. The greater the binding energy per nucleon, the more stable the nucleus. Since differences in entropy are minimal among nuclei, the binding energy effectively becomes a nuclear free energy of formation.

The tightness of binding is manifested as a **mass defect**, a "missing" mass, a mass (Δm) converted into an energy

$$\Delta E = (\Delta m)c^2$$

when the nucleons mesh together. This equivalence of mass and energy, made famous by the equation $E = mc^2$, we understand now to be an inevitable consequence of Einstein's **theory of relativity**: the demand that all physical laws look the same in any nonaccelerated frame of reference (any **inertial reference frame**). We combine this simple, reasonable expectation with the observed constancy of c, the speed of light, and we have the mass-energy equation. It tells us that there is energy not just in motion (the kinetic energy, $\frac{1}{2}mv^2$), but also in the *mass* of any body at rest (the rest energy, mc^2).

6. NUCLEAR KINETICS. Molecules high in energy eventually transform themselves into molecules low in energy. Nuclei with small binding energies per nucleon eventually transform themselves into nuclei with large binding energies per nucleon. Wishing to know when they do and how they do, we turn once more to questions of kinetics and mechanism.

For a nucleus, most decay processes are **first order**. The rate of decay (called the **activity**, A) is therefore proportional to the number of unreacted nuclei (N) present at each instant,

$$A = -\frac{\Delta N}{\Delta t} = kN$$

so that the changing activity, A_t, decreases exponentially from the starting value A_0:

$$A_t = A_0 \exp(-kt)$$

Here, note, we define the activity explicitly for the first time, specifying the **becquerel (Bq)** as the SI unit of radioactive decay: one disintegration per second. The first-order rate constant, k, plays the same role in nuclear kinetics that it does in chemical kinetics.

The half-life

$$t_{1/2} = \frac{\ln 2}{k}$$

then measures the time needed for the activity to decrease by exactly half, or, equivalently, for half the population of nuclei to disintegrate. From either k or $t_{1/2}$, moreover, we know the number of nuclei responsible for a given activity ($A = kN$):

$$N = \frac{A}{k} = \frac{A t_{1/2}}{\ln 2}$$

For further details, review the discussion of first-order chemical kinetics in Chapter 18. The mathematics is the same.*

7. NUCLEAR REACTIONS. Common mechanisms of decay include **alpha emission** ($\alpha \equiv {}_2^4\mathrm{He}$, a helium nucleus), **beta emission** ($\beta^- \equiv {}_{-1}^{0}e$, an electron), **gamma emission** ($\gamma \equiv h\nu$, a high-energy photon), **positron emission** ($\beta^+ \equiv {}_1^0 e$, a positron, the electron's **antiparticle**), **proton emission** (${}_1^1\mathrm{p}$), **neutron emission** (${}_0^1\mathrm{n}$), and **electron capture**. Both electron capture and the emission of electrons or positrons involve the interconversion of protons and neutrons, a process governed by the **weak interaction**.

---

*In this context, the symbols for activity (A) and number of nuclei (N) should not be confused with the same symbols used elsewhere for mass number (A) and neutron number (N). The notation, although conventional, is not without ambiguity.

Recall, finally, the difference between fission and fusion. *Fission*: the breakup of a nucleus into two lighter nuclei, usually of roughly equal mass. *Fusion*: the joining of two or more light nuclei to produce a heavier nucleus.

SAMPLE PROBLEMS

EXAMPLE 21-1. Counting Nucleons

PROBLEM: How many protons and neutrons are contained in each of the following nuclei? (a) Uranium-238. (b) Thorium-234. (c) Lead-206.

SOLUTION: Every element has a unique atomic number (Z, the number of protons), which we read directly from the periodic table:

uranium	U	$Z = 92$
thorium	Th	$Z = 90$
lead	Pb	$Z = 82$

To this value are added N neutrons, giving us the mass number

$$A = Z + N$$

for each isotope $_Z^A X$ shown below:

		Z	N	A
(a)	$_{92}^{238}U$	92	146	238
(b)	$_{90}^{234}Th$	90	144	234
(c)	$_{82}^{206}Pb$	82	124	206

EXAMPLE 21-2. Mass and Energy

PROBLEM: Taking note of the atomic masses given in Appendix C (Tables C-11 and C-12), compute the binding energy per nucleon for (a) $_{92}^{238}U$ and (b) $_{82}^{206}Pb$.

SOLUTION: To determine the binding energy, we compare the mass of the nucleus to the total mass of the nucleons taken separately. The difference,

$$\Delta m = \text{mass of nucleus} - \text{mass of nucleons}$$

is then proportional to the nuclear formation energy

$$\Delta E = (\Delta m)c^2$$

Where mass is lost (Δm is negative), the energy of the system decreases (ΔE is negative).

The nucleonic masses may be converted from atomic mass units into kilograms if we wish to express E in joules ($J = kg\ m^2\ s^{-2}$):

$$1\ u = 1.660540 \times 10^{-27}\ kg$$

The speed of light, c, has its fixed value of $2.99792458 \times 10^8\ m\ s^{-1}$.

(a) *Uranium-238.* Beginning with 92 separated protons and 146 separated neutrons, we first compute an unbound mass of 239.9345106 u:

Mass of nucleons = mass of protons + mass of neutrons

$$= 92 \times 1.00727647\ u + 146 \times 1.00866490\ u$$

$$= 92.6694352\ u + 147.2650754\ u = 239.9345106\ u$$

The mass of the bound nucleus, by contrast, is only 238.000315 u, as we find by subtracting the mass of 92 electrons from the atomic mass tabulated in Table C-12:

Mass of nucleus = atomic mass − mass of electrons

$$= 238.050784\ u - 92 \times 0.0005485799\ u = 238.000315\ u$$

The mass defect

$$\Delta m = 238.000315\ u - 239.9345106\ u = -1.934196\ u$$

is −1.934196 u, and the change in energy for one nucleus is therefore

$$\Delta E = (\Delta m)c^2$$

$$= \left(-1.934196\ u \times \frac{1.660540 \times 10^{-27}\ kg}{u}\right) \times (2.99792458 \times 10^8\ m\ s^{-1})^2$$

$$= -2.886631 \times 10^{-10}\ J$$

or $-1.738 \times 10^{11}\ kJ\ mol^{-1}$. Expressed per nucleon, the binding energy is

$$E_A = -\frac{\Delta E}{A} = \frac{2.886631 \times 10^{-10}\ J}{238\ nucleons} = 1.212870 \times 10^{-12}\ J/nucleon$$

Our convention is to state the binding energy per nucleon, E_A, as a positive number: $E_A = -\Delta E / A$.

(b) *Lead-206.* A similar calculation for $^{206}_{82}\text{Pb}$ yields $\Delta m = -1.741662$ u and a corresponding binding energy of 1.261791×10^{-12} J/nucleon:

Bound mass:	$205.974440 \text{ u} - 82e^- = 205.929456 \text{ u}$
Unbound mass:	$82\,\text{p} + 124\,\text{n} = 207.671118 \text{ u}$

$$\Delta m = -1.741662 \text{ u}$$

$$\Delta E = (\Delta m)c^2 = -2.599289 \times 10^{-10} \text{ J}$$

$$E_A = 1.261791 \times 10^{-12} \text{ J/nucleon}$$

Take note: Lead-206, with its magic number of protons (82) is a stronger, more stable nucleus than uranium-238. Nucleon for nucleon, the binding energy is higher.

Uranium-238, in common with all elements above bismuth ($Z = 83$), is radioactive. It has too many neutrons.

Example 21-3. Alpha and Beta Decay

Must uranium-238 decay promptly into a more stable nucleus such as lead-206? No. Favorable thermodynamics does not necessarily mean favorable kinetics.

Does uranium-238 eventually become lead-206? Yes. See below.

Problem: Uranium-238 undergoes a series of 14 radioactive decays before finally settling into a stable lead-206 configuration:

$$^{238}_{92}\text{U} \longrightarrow {}^{234}_{90}\text{Th} \longrightarrow {}^{234}_{91}\text{Pa} \longrightarrow {}^{234}_{92}\text{U} \longrightarrow {}^{230}_{90}\text{Th} \longrightarrow$$

$$^{226}_{88}\text{Ra} \longrightarrow {}^{222}_{86}\text{Rn} \longrightarrow {}^{218}_{84}\text{Po} \longrightarrow {}^{214}_{82}\text{Pb} \longrightarrow {}^{214}_{83}\text{Bi} \longrightarrow$$

$$^{214}_{84}\text{Po} \longrightarrow {}^{210}_{82}\text{Pb} \longrightarrow {}^{210}_{83}\text{Bi} \longrightarrow {}^{210}_{84}\text{Po} \longrightarrow {}^{206}_{82}\text{Pb}$$

Identify the particle emitted at each step.

Solution: There are only two modes of decay here, alpha and beta. We have, first, reactions where the atomic number (Z) decreases by 2 and the mass number (A) decreases by 4, as in the initial step from $^{238}_{92}\text{U}$ to $^{234}_{90}\text{Th}$. An α particle is ejected from the parent nucleus, hence:

$$^{238}_{92}\text{U} \longrightarrow {}^{234}_{90}\text{Th} + {}^{4}_{2}\text{He}$$

In all, 238 units of mass number appear on the left; 238 units of mass

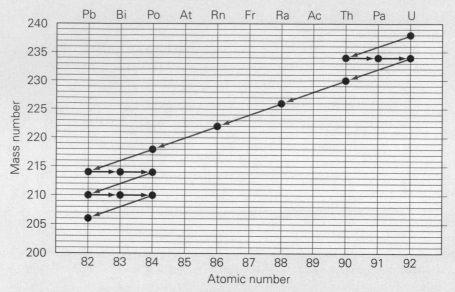

FIGURE R21-1. Radioactive series initiated by the disintegration of uranium-238 into thorium-234. The series terminates with lead-206, a stable isotope. Diagonal arrows represent α decays; horizontal arrows represent β decays.

number appear on the right. Similarly, 92 units of positive charge appear on the left; 92 units of positive charge appear on the right.

Second, there are reactions where Z increases by 1 and A remains the same, as in the next step of the series:

$$^{234}_{90}\text{Th} \longrightarrow {}^{234}_{91}\text{Pa} + {}^{0}_{-1}\text{e} + \tilde{\nu}$$

A β^- particle and antineutrino are emitted.

The complete **radioactive series**, diagrammed in Figure R21-1, proceeds as follows:

$$^{238}_{92}\text{U} \xrightarrow{\alpha} {}^{234}_{90}\text{Th} \xrightarrow{\beta} {}^{234}_{91}\text{Pa} \xrightarrow{\beta} {}^{234}_{92}\text{U} \xrightarrow{\alpha} {}^{230}_{90}\text{Th} \xrightarrow{\alpha}$$

$$^{226}_{88}\text{Ra} \xrightarrow{\alpha} {}^{222}_{86}\text{Rn} \xrightarrow{\alpha} {}^{218}_{84}\text{Po} \xrightarrow{\alpha} {}^{214}_{82}\text{Pb} \xrightarrow{\beta} {}^{214}_{83}\text{Bi} \xrightarrow{\beta}$$

$$^{214}_{84}\text{Po} \xrightarrow{\alpha} {}^{210}_{82}\text{Pb} \xrightarrow{\beta} {}^{210}_{83}\text{Bi} \xrightarrow{\beta} {}^{210}_{84}\text{Po} \xrightarrow{\alpha} {}^{206}_{82}\text{Pb}$$

EXAMPLE 21-4. Other Modes of Decay

Before addressing the kinetics of the uranium-238 series, we pause here and in Example 21-5 to recall some of the options available to other radioactive nuclei: positron emission, electron capture, neutron emission, proton emission, to cite just a few possibilities.

We shall, in the process, also reshape our view of stoichiometry and the conservation laws, looking more critically at the broader world already introduced in Examples 21-2 and 21-3: a world where mass can disappear and where electrons, protons, and neutrons all can come and go.

PROBLEM: Write a balanced equation for each of the transformations proposed below.

(a) $^{36}_{17}Cl$ captures an electron to become $^{36}_{16}S$.

SOLUTION: Z decreases by 1 unit while A stays the same. A neutrino (ν, not to be confused with the symbol for photon frequency) appears as a by-product:

$$^{36}_{17}Cl + {}^{0}_{-1}e \longrightarrow {}^{36}_{16}S + \nu$$

QUESTION: Why a neutrino? Why not an antineutrino?

ANSWER: *Why* is perhaps the wrong question. We do not know *why* nature demands that the total number of leptons in the universe must remain constant. We know only that when one lepton (an electron, say) disappears, some other lepton (not necessarily another electron) must arise simultaneously to take its place. Lepton number is conserved.

Do the bookkeeping: At the left, an electron enters the equation as a reactant. The total lepton number is 1, and there it must remain. But the right-hand side of the equation tells us that the "captured" electron is gone—truly gone, too, gone in a sense that we never contemplate for a chemical reaction. The electron that undergoes a weak nuclear interaction is deleted from the universe's stock of leptons, not simply pasted into some other structure as in a chemical transformation. Whereas on the left we have a proton inside the nucleus and an electron outside, on the right we have a *neutron* inside the nucleus and a neutrino flying out into space. And the neutron, remember, is made from three inseparable quarks, not from a detachable proton and electron. The electron is gone.

To balance the books, a suitable lepton (not an *anti*lepton) must take the place of the lost electron. Here it is the neutrino, which contributes a lepton number of 1 to the right-hand side of the equation. The leptons are in balance. The laws of nature are obeyed.

QUESTION: Yes, but what happens to our notion of stoichiometry—the idea that the same number of protons, neutrons, and electrons must be present at the start and finish of every reaction?

ANSWER: For transformations taking place outside the nucleus—for ordinary chemical reactions—we observe again and again that changes in mass are slight, usually too slight even to detect. So small are the shifts in energy that a typical chemical reaction produces no discernible difference in mass. Moreover, no weak nuclear interaction exists to interconvert electrons, protons, neutrons, and neutrinos in a chemical process. To say, under such circumstances, that "matter cannot be created or destroyed" is then to make only a trivially small mistake, a mistake worth making for the tremendous simplification it brings. Because for chemistry outside the nucleus, everything indeed does seem simple: There are no transformations of leptons and of baryons, no neutrinos, no antineutrinos, no positrons. There are only immutable nuclei and electrons, and they shuffle about with unchanging mass and in unchanging number. It is by these almost-but-not-quite-true ideas of mass conservation and the indestructibility of matter that we justify the laws of chemical stoichiometry.

Inside the nucleus, or inside a high-energy particle accelerator, or inside the sun, or inside the primeval material of the early universe . . . there, under more drastic conditions, the "chemistry" proceeds according to the more general laws. Not the conservation of mass, but the conservation of energy and mass *together* as prescribed by Einstein's formula $E = mc^2$. Not the indestructibility of protons, neutrons, and electrons, but the conservation of baryon number and lepton number. Not the electromagnetic force alone, but the weak force and the strong force as well.

Such is the world at high energy. It forces upon us a new kind of stoichiometry, a stoichiometry still being discovered.

(b) ^8_5B emits a positron to become ^8_4Be.

SOLUTION: Again, the atomic number decreases by 1 unit while the mass number remains unchanged:

$$^8_5B \longrightarrow {}^8_4Be + {}^0_1e + \nu$$

QUESTION: The mass *number* is the same on both sides of the equation, but is the total mass before and after the reaction also unchanged?

ANSWER: No. In a nuclear reaction, a change in mass plays the same role that a change in energy does for a chemical reaction. Mass is energy.

As reactants, we have 5 protons and 3 neutrons bound into a nucleus of ^8_5B. The mass number, equivalent to the baryon number, is 8.

As products, we have 4 protons and 4 neutrons bound into a nucleus of $^8_4$Be. The baryon number is still 8. Appearing as well are a positron (lepton number $= -1$) and a neutrino (lepton number $= 1$) to preserve the lepton number at its original value of 0.

Baryons, tallied by the mass number, are in balance, and leptons are in balance too. What does not balance, however, is the mass: The combined mass of $^8_4$Be plus a positron plus a neutrino is less than the mass of $^8_5$B alone, and the difference is invested in the energies of the ejected positron and neutrino.

In a spontaneous chemical reaction, we say that a system releases "free" energy as reactants go to products—energy free to do work on the surroundings. In a nuclear reaction, we speak analogously of the energy carried away by such agents as high-speed electrons, positrons, neutrinos, antineutrinos, and gamma photons. They fly off with energy distributed in various proportions, always conserving total energy and momentum as they do.

(c) $^9_4$Be absorbs an α particle and releases a neutron.

SOLUTION: From the incomplete equation

$$^4_2\text{He} + {}^9_4\text{Be} \longrightarrow {}^A_Z\text{X} + {}^1_0\text{n}$$

we see immediately that species X must have a mass number of 12:

$$4 + 9 = A + 1$$

$$A = 12$$

Knowing also that neutron emission preserves the number of protons, we realize that X has a positive charge (atomic number) of 6:

$$2 + 4 = Z + 0$$

$$Z = 6$$

The nucleus produced is therefore $^{12}_6$C, the abundant isotope of carbon:

$$^4_2\text{He} + {}^9_4\text{Be} \longrightarrow {}^{12}_6\text{C} + {}^1_0\text{n}$$

QUESTION: Does the process involve the weak interaction?

ANSWER: No. There are 6 protons and 7 neutrons on the left of the equation and exactly the same distribution on the right: 6 protons, 7 neutrons. No neutron is converted into a proton; no proton is converted into a neutron.

Leptons, the signature of weak nuclear interactions, are nowhere to be found. There are no electrons, no positrons, no neutrinos, no antineutrinos. The nucleons are unaffected by any form of β decay such as electron emission, positron emission, or electron capture.

(d) $^{14}_{7}N$ absorbs a neutron and is subsequently transformed into $^{14}_{6}C$.

SOLUTION: The process as described,

$$^{14}_{7}N + ^{1}_{0}n \longrightarrow ^{14}_{6}C$$

is unbalanced. We need to add one proton on the right to conserve mass number and charge:

$$^{14}_{7}N + ^{1}_{0}n \longrightarrow ^{14}_{6}C + ^{1}_{1}p$$

QUESTION: Must the neutron be accelerated to high energy?

ANSWER: No, the uncharged neutron suffers no electromagnetic repulsion from the positive nucleus. It slips in with low energy.

EXAMPLE 21-5. Uphill or Downhill: Positron Emission
Versus Electron Capture

With changes in the Gibbs free energy as a yardstick, we confidently predict whether or not a chemical reaction can proceed spontaneously. If $\Delta G°$ is negative, yes. If $\Delta G°$ is positive, no.

For a nuclear reaction, the equivalent measure is the difference in mass between products and reactants. If Δm is negative for a certain mechanism of decay, the transformation can occur. If Δm is positive, the way is blocked.

PROBLEM: $^{231}_{92}U$, a radioactive atom with a half-life of 4.2 days, produces $^{231}_{91}Pa$ as a daughter nucleus. The atomic mass of uranium-231 is 231.036270 u, and the atomic mass of protactinium-231 is 231.035880 u. By what mode of decay does the process go forward?

SOLUTION: There are two possible routes: positron emission or electron capture. Either way, the atomic number of $^{231}_{92}U$ will decrease by 1 while the mass number remains constant at 231. One proton becomes one neutron, and the same daughter is produced whether the decay goes by positron emission,

$$^{231}_{92}U \longrightarrow ^{231}_{91}Pa + ^{0}_{1}e + \nu \qquad (\beta^{+})$$

or by electron capture:

$$^{231}_{92}\text{U} + {}^{0}_{-1}\text{e} \longrightarrow {}^{231}_{91}\text{Pa} + \nu \qquad \text{(EC)}$$

To distinguish between the two routes, we compute the overall changes in mass.

First, the positron emission (β^+): For the mass of reactant, we take one complete atom of $^{231}_{92}\text{U}$ with 92 electrons outside the nucleus. For the products, we take the mass of a complete atom of $^{231}_{91}\text{Pa}$ (including its 91 electrons) plus the mass of a positron (equal to $m_e = 0.0005485799$ u, the mass of an electron) *plus* the mass of one more electron (originally the 92nd electron outside the uranium nucleus). The neutrino, with its rest mass assumed to be zero, does not enter into the calculation:

$$\Delta m(\beta^+) = m(\text{Pa-231}) + 2m_e - m(\text{U-231})$$

$$= 231.035880 \text{ u} + 2 \times 0.0005485799 \text{ u} - 231.036270 \text{ u}$$

$$= 0.000707 \text{ u}$$

The change in mass for positron emission is therefore equal to $+0.0007$ u, signifying a net increase and hence an uphill climb in energy. Positron emission is unfavorable here, even though the daughter atom has a slightly lower mass than the parent. The difference is too small—less than twice the mass of two electrons—to overcome the additional mass contributed by the positron produced in the decay.

Nevertheless, the same end (an atom of protactinium-231) is attainable by a different means: electron capture. Here we have one complete atom of Pa-231 as the only product with mass, and we have one complete atom of U-231 as the sole reactant. Don't be misled: The mass of the electron shown explicitly on the left-hand side of the equation

$$^{231}_{92}\text{U} + {}^{0}_{-1}\text{e} \longrightarrow {}^{231}_{91}\text{Pa} + \nu \qquad \text{(EC)}$$

is already included in the atomic mass of neutral U-231. We write the electron separately only to show the net transformation in the clearest way possible.

That done, we compute the change in mass for electron capture as

$$\Delta m(\text{EC}) = m(\text{Pa-231}) - m(\text{U-231})$$

$$= 231.035880 \text{ u} - 231.036270 \text{ u}$$

$$= -0.000390 \text{ u}$$

and this time we find a net decrease. Beta decay by electron capture (but not by positron emission) is energetically possible for U-231. The transformation goes forward spontaneously with a half-life of 4.2 days.

QUESTION: Is the half-life related in any way to the change in mass?

ANSWER: No, the half-life is a kinetic parameter. The change in mass is an energetic parameter, analogous to the free energy in a chemical reaction. A large drop in mass does not guarantee a short half-life, nor does a small drop in mass necessitate a long one. For an exploration of kinetic issues, see the next two examples.

EXAMPLE 21-6. Kinetics: Half-Life

Knowing simply that a reaction *can* happen is not enough. We need to know *whether* it happens; we need to know *when* it happens.

The radioactive series introduced in Example 21-3 will serve again as a prototype.

PROBLEM: Half-lives for each of the nuclides involved in the decay of $^{238}_{92}$U are listed below:

STEP	ISOTOPE	HALF-LIFE	STEP	ISOTOPE	HALF-LIFE
1	U-238	4.5×10^9 y	8	Po-218	3.0 min
2	Th-234	24.1 d	9	Pb-214	27 min
3	Pa-234	6.7 h	10	Bi-214	19.9 min
4	U-234	2.5×10^5 y	11	Po-214	1.6×10^{-4} s
5	Th-230	7.5×10^4 y	12	Pb-210	22.6 y
6	Ra-226	1600 y	13	Bi-210	5.0 d
7	Rn-222	3.82 d	14	Po-210	138.4 d

Taking one mole of uranium-238 to start, estimate the composition of the sample after 4.5 billion years of radioactive decay (approximately the age of the earth).

SOLUTION: Think of the process as a 14-step reaction in which the first step,

$$^{238}_{92}U \longrightarrow {}^{234}_{90}Th + {}^4_2He \qquad t_{1/2} = 4.5 \times 10^9 \text{ years}$$

is rate-limiting, slower than all the others combined. Doing so, we soon realize that after 4.5 billion years (one half-life) exactly half the original sample of uranium-238 will be converted into lead-206. The partially decayed material must contain one-half mole $^{238}_{92}U$ and one-half mole $^{206}_{82}Pb$. Nothing else.

See what happens: The process begins gradually, with the sporadic disintegration of $^{238}_{92}U$. Little by little, the uranium-238 nuclei randomly emit alpha particles—so slowly, though, that half the sample remains unchanged even after 4.5 billion years. Every now and then (we don't know when), one of the original uranium-238 atoms turns into thorium-234; and then, much later, another one suddenly ticks away; then another after that, and another, and another; very slowly, tick-tock, a clock ticking imperceptibly since the beginnings of the earth.

And while the clock yet ticks, it is time for the newly formed atoms of $^{234}_{90}Th$ to disappear. Sporadically too, but much more often than $^{238}_{92}U$, a thorium-234 nucleus turns into protactinium-234; again and again it happens until all the thorium-234 is gone . . . and so the decays continue, one at a time, nucleus after nucleus, all the way down to lead-206, where the series finally stops.

What do we find after a few billion years? Nothing but the end product, lead-206, together with any uranium-238 still intact. Relative to that excruciatingly slow first step with its half-life of 4.5 billion years, all subsequent decays proceed rapidly—none having a half-life greater than 250,000 years, the blink of an eye on our billion-year time scale. The $^{234}_{90}Th$ daughters disappear with a half-life of only 24 days; the $^{234}_{91}Pa$ granddaughters decay in a matter of hours; even the longer-lived $^{234}_{92}U$ great-granddaughters ($t_{1/2} = 2.5 \times 10^5$ years) are gone after a few million years. Over billions of years, no intermediate product builds up. The second hand and minute hand sweep around the dial while the hour hand barely moves.

QUESTION: What would we find after only 1 billion years?

ANSWER: A billion years, give or take, is still a long time compared with either 24 days, 6.7 hours, 250,000 years, 75,000 years, 1600 years, 3.8 days, 3 minutes, 27 minutes, 20 minutes, 22.6 years, 5 days, or 138 days. The sample contains only $^{238}_{92}U$ and $^{206}_{82}Pb$.

To compute the amount of $^{238}_{92}U$ remaining, we follow the usual procedure for any first-order process (Chapter 18). The first step is to determine the rate constant from the stated half-life:

$$k = \frac{\ln 2}{t_{1/2}}$$

Next, we insert k into the exponential expression for the activity at any time t:

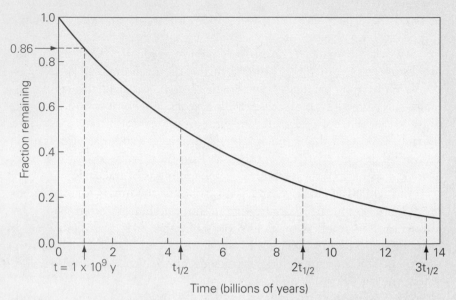

FIGURE R21-2. First-order decay of uranium-238. The half-life is 4.5 billion years.

$$A_t = A_0 \exp(-kt) = A_0 \exp\left[-(\ln 2)\left(\frac{t}{t_{1/2}}\right)\right]$$

$$\frac{A_t}{A_0} = \exp\left[-0.693 \, \frac{1.0 \times 10^9 \text{ y}}{4.5 \times 10^9 \text{ y}}\right] = \exp(-0.154)$$

$$= 0.86$$

Approximately 14% of the uranium-238 disintegrates over a period of 1 billion years. See Figure R21-2.

EXAMPLE 21-7. Turning Back the Clock: Radioactive Dating

PROBLEM: A certain rock contains 0.805 g of uranium-238 and 0.150 g of lead-206. No other isotopes of lead are present. How long ago did the uranium first begin to decay?

SOLUTION: Lead exists naturally as a mixture of four isotopes: lead-204 (1.4%), lead-206 (24.1%), lead-207 (22.1%), lead-208 (52.4%). Since the rock contains only the one species, lead-206, we suppose that every atom of $^{206}_{82}$Pb must have come expressly from the decay of uranium-238. Had the lead arisen from any other process, the rock today would show substantial amounts of the other isotopes.

The present-day 0.150 g of lead-206, then, once must have been 0.173 g of radioactive uranium-238, each new atom of $^{206}_{82}Pb$ having come from an old atom of $^{238}_{92}U$:

$$0.150 \text{ g Pb-206} \times \frac{238.0508 \text{ g U-238}}{205.9744 \text{ g Pb-206}} = 0.173 \text{ g U-238}$$

What began as 0.978 g of $^{238}_{92}U$ (0.805 g + 0.173 g) has apparently been whittled down to 0.805 g.

For uranium-238, the decay clock ticks with a half-life of 4.5×10^9 years. To wind it back, we insert the present population (N_t), the past population (N_0), and the rate constant [$k = (\ln 2)/t_{1/2}$] into the decay equation

$$N_t = N_0 \exp(-kt)$$

and compute the time elapsed from the equivalent logarithmic expression:

$$-kt = \ln N_t - \ln N_0 = \ln\left(\frac{N_t}{N_0}\right)$$

For N_t/N_0 we use the ratio of the two masses, thus:

$$t = -\frac{1}{k} \ln\left(\frac{N_t}{N_0}\right) = -\frac{t_{1/2}}{\ln 2} \ln\left(\frac{N_t}{N_0}\right)$$

$$= -\frac{4.5 \times 10^9 \text{ y}}{0.693} \ln\left(\frac{0.805}{0.978}\right) = 1.3 \times 10^9 \text{ y}$$

The procedure is the same as in Example 21-6.

QUESTION: What is the remaining *activity* of the sample after these 1.3 billion years?

ANSWER: Remember that activity is defined as

$$A = kN$$

where N is the number of radioactive nuclei and

$$k = \frac{\ln 2}{t_{1/2}} = \left(\frac{0.693}{4.5 \times 10^9 \text{ y}}\right) \times \frac{1 \text{ y}}{365.25 \text{ d}} \times \frac{1 \text{ d}}{86,400 \text{ s}} = 4.8(8) \times 10^{-18} \text{ s}^{-1}$$

is the first-order rate constant implicit above. Given the mass of the sample in grams, we then use Avogadro's number to compute N and A:

$$A = kN = (4.88 \times 10^{-18} \text{ s}^{-1}) \times \left(0.805 \text{ g U-238} \times \frac{6.02 \times 10^{23} \text{ nuclei}}{238.0508 \text{ g U-238}} \right)$$

$$= 9.9 \times 10^3 \text{ s}^{-1} \equiv 9.9 \times 10^3 \text{ Bq}$$

QUESTION: What was the activity 1.3 billion years before, when the rock originally contained 0.978 g of uranium-238 and first began to decay?

ANSWER: The same calculation, repeated for 0.978 g, yields an activity of 1.2×10^4 Bq. The ratio A_t / A_0 is equal to the ratio N_t / N_0, as expected. We defined it that way.

QUESTION: Work the problem in the other direction now, and ask: What mass of uranium-238 would generate an activity of, say, 12,400 Bq?

ANSWER: First, solve for N. Then, convert N into its equivalent mass in grams:

$$N = \frac{A}{k} = \left(\frac{12,400 \text{ nuclei s}^{-1}}{4.9 \times 10^{-18} \text{ s}^{-1}} \right) \times \frac{238.0508 \text{ g U-238}}{6.02 \times 10^{23} \text{ nuclei}} = 1.0 \text{ g}$$

The activity produced *per gram* of radioactive material is called its **specific activity**, reminiscent of the usage "specific heat" to denote the heat capacity of a substance per gram.

EXAMPLE 21-8. Antimatter and Annihilation

Mass is energy and energy is mass; and where mass disappears, energy appears in its stead. It is one of the great certainties in a randomly driven universe: energy, including mass-energy, is conserved.

At high velocities, at low velocities . . . for molecules, for atoms, for nuclei, for electrons, for quarks, for matter, for antimatter . . . all the same. Energy is conserved.

$E = mc^2$. Let Einstein have the last word.

PROBLEM: What wavelength of electromagnetic radiation is generated by the annihilation of a proton and antiproton?

SOLUTION: When a particle meets its antiparticle, their total mass is converted into energy—usually carried away by two gamma rays. Here, proton and antiproton each have mass

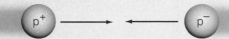

FIGURE R21-3. Conversion of mass into energy: the annihilation of a proton and antiproton.

$$m_p = 1.00728 \text{ u} \times \frac{1.66054 \times 10^{-27} \text{ kg}}{\text{u}} = 1.67263 \times 10^{-27} \text{ kg}$$

and hence a combined value of $2m_p$. Upon annihilation (Figure R21-3), the mass-energy

$$\Delta E = (\Delta m)c^2 = (2m_p)c^2$$

is split equally between two gamma rays, each carrying the amount

$$E = \frac{\Delta E}{2} = m_p c^2 = h\nu = \frac{hc}{\lambda}$$

Solving for λ, we then find that the wavelength is comparable to the diameter of a proton, approximately 10^{-15} m:

$$\lambda = \frac{h}{m_p c} = \frac{6.63 \times 10^{-34} \text{ J s}}{(1.67 \times 10^{-27} \text{ kg})(3.00 \times 10^8 \text{ m s}^{-1})} = 1.32 \times 10^{-15} \text{ m}$$

QUESTION: Express the photon's energy in million electron volts (MeV).

ANSWER: We define an electron volt as the energy acquired by a single electron ($e = 1.602177 \times 10^{-19}$ coulomb) moving through a potential difference of one volt (1 V = 1 joule per coulomb):

$$1 \text{ eV} = (1.602177 \times 10^{-19} \text{ C})(1 \text{ J C}^{-1}) = 1.602177 \times 10^{-19} \text{ J}$$

The larger unit

$$1 \text{ MeV} = 10^6 \text{ eV} = 1.602177 \times 10^{-13} \text{ J}$$

is particularly suited for the measurement of nuclear energies.

Each photon, carrying away the rest energy of a single proton, thus has an energy of 938.273 MeV (1.50328×10^{-10} J):

$$E = m_p c^2 = (1.67263 \times 10^{-27} \text{ kg})(2.99792 \times 10^8 \text{ m s}^{-1})^2$$

$$= 1.50328 \times 10^{-10} \text{ J}$$

$$= 1.50328 \times 10^{-10} \text{ J} \times \frac{1 \text{ MeV}}{1.602177 \times 10^{-13} \text{ J}} = 938.273 \text{ MeV}$$

QUESTION: Compute the rest energy of an electron ($m_e = 9.10939 \times 10^{-31}$ kg).

ANSWER: 0.511 MeV. The ratio of the electron's mass to the proton's mass is 1:1836. Annihilation of an electron and positron releases 1.022 MeV, corresponding to $2m_e$. Again, mass is converted into energy.

QUESTION: Energy is conserved, granted, but can two particles simply "disappear" without violating some other conservation principle?

ANSWER: Pairs of particles appear and disappear during annihilation, β decay (Examples 21-4 and 21-5), and many other processes, always in full compliance with the laws of nature. For electron and positron, the law calls for the total lepton number to be conserved. The sum of all the leptons (such as electrons and neutrinos) *minus* all the antileptons (such as positrons and antineutrinos) must stay the same.

It does. Before the annihilation, we have one electron (lepton number = 1) together with one antielectron (lepton number = −1) to give us a total lepton number of 0. After the annihilation, we still have a lepton number of 0: no electron, no positron.

QUESTION: What about the simultaneous disappearance of a proton and antiproton, our original problem? Are the particle numbers properly conserved?

ANSWER: Yes, the *baryon number* is maintained at 0: one proton (baryon number = 1) plus one antiproton (baryon number = −1) before annihilation; no proton and no antiproton after annihilation.

QUESTION: A proton, put together with two up quarks and one down quark (*uud*), has a baryon number of 1. From what combination of quarks does an antiproton derive?

ANSWER: Not from quarks, but from *antiquarks*—two up antiquarks and one down antiquark ($\tilde{u}\,\tilde{u}\,\tilde{d}$). Since each antiquark has a baryon number of $-\frac{1}{3}$, the antiproton as a whole has a baryon number of −1.

The electric charge, opposite in sign to a proton, is −1 as well. Each of the two up antiquarks contributes $-\frac{2}{3}$, and the down antiquark contributes $+\frac{1}{3}$.

Exercises

1. How many protons and neutrons are contained in each nucleus?

   (a) ^{10}B, ^{11}B, ^{12}B
   (b) ^{12}C, ^{13}C, ^{14}C
   (c) ^{24}Mg, ^{25}Mg, ^{26}Mg
   (d) ^{90}Zr, ^{91}Zr, ^{92}Zr

2. Given the number of protons and neutrons, write the corresponding nuclear symbol in the form ^A_ZX:

   (a) 86p, 123n
   (b) 87p, 122n
   (c) 88p, 121n
   (d) 89p, 120n

   Reminder: Z is the atomic number; A is the mass number.

3. Elemental chlorine is normally a mixture of two isotopes:

   $$X_{35} \cdot {}^{35}Cl + X_{37} \cdot {}^{37}Cl$$

   The atomic mass of chlorine-35, including the electrons outside the nucleus, is 34.968852 u, whereas the mass of chlorine-37 is 36.965903 u. The average atomic mass of chlorine is the value given in the periodic table: 35.453 u. Calculate the mole fractions X_{35} and X_{37}.

4. Elemental magnesium (average atomic mass = 24.305 u) is normally a mixture of three isotopes:

   $$X_{24} \cdot {}^{24}Mg + X_{25} \cdot {}^{25}Mg + X_{26} \cdot {}^{26}Mg$$

   The dominant isotope, magnesium-24, makes up 78.99% of the mixture and has an atomic mass of 23.985042 u. The other two stable isotopes have atomic masses of 24.985837 u (^{25}Mg) and 25.982593 u (^{26}Mg). Calculate the mole fractions X_{24}, X_{25}, and X_{26}.

5. Stable carbon is a mixture of ^{12}C (atomic mass = 12.000000 u) and ^{13}C (atomic mass = 13.003355 u). Carbon-12, the dominant isotope, has a natural abundance of 98.90%. Calculate the average atomic mass of carbon, and compare the value to the mass given in the periodic table.

6. A deuterium nucleus (^{2}H) contains one proton (mass = 1.00727647 u) and one neutron (mass = 1.00866490 u), and thus the mass of the nucleons taken separately is 2.01594137 u. The mass of an intact deuterium atom, however, is only 2.0140 u, including the single electron outside the nucleus. (a) Why is the mass of the bound nucleus lower than the masses of the individual particles? (b) Compute the total binding energy of the ^{2}H nucleus. (c) Compute the binding energy per nucleon.

7. Calculate the mass defect, total binding energy, and binding energy per nucleon for each of the following nuclei:

NUCLEUS	ATOMIC MASS (u)
(a) hydrogen-1	1.007825
(b) hydrogen-2	2.0140
(c) hydrogen-3	3.01605

Take note: (1) The atomic masses, here and everywhere, include the electrons outside the nucleus. (2) The calculation for deuterium is also undertaken in the preceding exercise. (3) ^{1}H enjoys a special status. Explain why.

8. Calculate the mass defect, total binding energy, and binding energy per nucleon for each of the following nuclei:

NUCLEUS	ATOMIC MASS (u)
(a) carbon-12	12.000000
(b) carbon-13	13.003355
(c) carbon-14	14.003241

9. Calculate the mass defect, total binding energy, and binding energy per nucleon for each of the following nuclei:

NUCLEUS	ATOMIC MASS (u)
(a) uranium-235	235.043924
(b) uranium-238	238.050784
(c) uranium-239	239.054289
(d) uranium-240	240.056587

10. The results of the preceding exercises show that some nuclei are bound more tightly than others. But ask: Does a small binding energy

necessarily doom a nucleus to early and rapid decay? Distinguish between thermodynamic stability and kinetic stability.

11. Nucleus A has a half-life of 1 second. Nucleus B has a half-life of 1 day. Can we use these numbers to predict which of the two nuclei is more tightly bound?

12. According to the shell model, a nucleus that contains a "magic" number of protons or neutrons will acquire special thermodynamic stability—even more so when the numbers of both protons *and* neutrons are magic. Pick out the magic and extra-magic nuclei from the following list:

$$^{16}O, \quad ^{17}O, \quad ^{16}F, \quad ^{19}F, \quad ^{40}Ca, \quad ^{40}Sc, \quad ^{208}Pb$$

13. Predict, on the basis of magic numbers, which isotope in each pair should be more stable:

(a) ^{3}He, ^{4}He
(b) ^{206}Pb, ^{208}Pb
(c) ^{208}Bi, ^{209}Bi

14. Identify the daughter nucleus produced when each mother isotope undergoes alpha decay:

(a) polonium-193
(b) actinium-222
(c) radium-222
(d) lawrencium-253

15. Write a balanced nuclear equation for each of the processes described above.

16. Identify the mother nucleus that spawns each of these daughters by alpha decay:

(a) thorium-218
(b) actinium-218
(c) radium-213
(d) curium-236

17. Write a balanced nuclear equation for each of the processes described above.

18. Identify the daughter nucleus produced when each mother isotope undergoes beta decay:

    (a) barium-142
    (b) cesium-137
    (c) carbon-14
    (d) boron-12

19. Write a balanced nuclear equation for each of the processes described above, introducing neutrinos and antineutrinos where appropriate.

20. Identify the mother nucleus that spawns each of these daughters by beta decay:

    (a) erbium-168
    (b) californium-251
    (c) oxygen-16
    (d) chlorine-35

21. Write a balanced nuclear equation for each of the processes described above, introducing neutrinos and antineutrinos where appropriate.

22. Thorium-232 undergoes a series of radioactive decays before settling into a stable lead-208 configuration:

$$^{232}_{90}\text{Th} \longrightarrow {}^{228}_{88}\text{Ra} \longrightarrow {}^{228}_{89}\text{Ac} \longrightarrow {}^{228}_{90}\text{Th} \longrightarrow$$
$$^{224}_{88}\text{Ra} \longrightarrow {}^{220}_{86}\text{Rn} \longrightarrow {}^{216}_{84}\text{Po} \longrightarrow {}^{212}_{82}\text{Pb} \longrightarrow$$
$$^{212}_{83}\text{Bi} \longrightarrow {}^{212}_{84}\text{Po} \longrightarrow {}^{208}_{82}\text{Pb}$$

    Identify the particle emitted at each step.

23. Uranium-235 undergoes a series of α and β decays terminating at lead-207. How many α particles and how many β particles are emitted during the course of reaction? Hint: First determine the number of α particles.

24. Give the daughter nucleus produced in each of the following reactions:

(a) Arsenic-73 captures an electron.
(b) Fluorine-15 emits a proton.
(c) Helium-5 emits a neutron.
(d) Aluminum-24 emits a positron.
(e) Vanadium-49 captures an electron.

25. Write a balanced nuclear equation for each of the processes described above, introducing neutrinos and antineutrinos where appropriate.

26. Give the daughter nucleus produced in each of the following reactions:

(a) Titanium-45 captures an electron.
(b) Titanium-45 emits a positron.
(c) Neon-16 emits two protons.
(d) Helium-7 emits a neutron.
(e) Lanthanum-135 captures an electron.

27. Write a balanced nuclear equation for each of the processes described above, introducing neutrinos and antineutrinos where appropriate.

28. Identify the mother nucleus responsible for each of the following reactions:

(a) Neodymium-143 is produced by electron capture.
(b) Tellurium-108 is produced by proton emission.
(c) Indium-103 is produced by positron emission.
(d) Helium-8 is produced by neutron emission.
(e) Potassium-35 is produced by positron emission.

29. Write a balanced nuclear equation for each of the processes described above, introducing neutrinos and antineutrinos where appropriate.

30. Complete the equation, and then summarize the reaction in words:

(a) $^{14}_{7}N + ^{4}_{2}He \longrightarrow \underline{} + ^{1}_{1}H$
(b) $^{59}_{27}Co + \underline{} \longrightarrow ^{60}_{27}Co$
(c) $^{14}_{6}C \longrightarrow ^{14}_{7}N + \underline{} + \underline{}$
(d) $^{235}_{92}U + ^{1}_{0}n \longrightarrow ^{137}_{52}Te + ^{97}_{40}Zr + \underline{}$
(e) $\underline{} + \alpha \longrightarrow ^{242}_{96}Cm + ^{1}_{0}n$

31. Complete the equation, and then summarize the reaction in words:

    (a) $^{16}_{8}O + \underline{\quad} \longrightarrow ^{13}_{7}N + ^{4}_{2}He$

    (b) $\underline{\quad} + ^{1}_{0}n \longrightarrow ^{24}_{11}Na + ^{4}_{2}He$

    (c) $^{58}_{26}Fe + \underline{\quad} \longrightarrow ^{59}_{26}Fe$

    (d) $\underline{\quad} + ^{1}_{1}H \longrightarrow ^{4}_{2}He + ^{0}_{1}e + \underline{\quad}$

    (e) $^{235}_{92}U + ^{1}_{0}n \longrightarrow \underline{\quad} + ^{91}_{36}Kr + 3^{1}_{0}n$

32. Each of the following transformations is impossible as written:

    (a) $^{1}_{0}n + ^{1}_{0}n \longrightarrow ^{1}_{1}p + ^{1}_{-1}p$

    (b) $^{238}_{92}U \longrightarrow ^{233}_{90}Th + ^{4}_{2}He + ^{1}_{1}p$

    (c) $^{59}_{26}Fe \longrightarrow ^{59}_{27}Co + ^{0}_{-1}e + \nu$

    (d) $^{0}_{-1}e + ^{0}_{1}e \longrightarrow \nu$

    State the reasons why.

33. Which transformations are spontaneous?

    (a) $^{234}_{94}Pu$ (atomic mass = 234.043299 u) produces $^{234}_{93}Np$ (234.042888 u) by positron emission.

    (b) $^{234}_{94}Pu$ (234.043299 u) produces $^{234}_{93}Np$ (234.042888 u) by electron capture.

    (c) $^{12}_{6}C$ (12.00000 u) produces $^{11}_{6}C$ (11.01143 u) by neutron emission.

    (d) $^{10}_{4}Be$ (10.013534 u) produces $^{10}_{5}B$ (10.012937 u) by electron emission.

    The masses are for the neutral atoms.

34. Which transformations are spontaneous?

    (a) $^{98}_{47}Ag$ (97.921560 u) produces $^{98}_{46}Pd$ (97.912722 u) by positron emission.

    (b) $^{98}_{47}Ag$ (97.921560 u) produces $^{98}_{46}Pd$ (97.912722 u) by electron capture.

    (c) $^{98}_{47}Ag$ (97.921560 u) produces $^{98}_{48}Cd$ (97.92711 u) by electron emission.

    (d) $^{98}_{47}Ag$ (97.921560 u) produces $^{94}_{45}Rh$ (93.921670 u) by alpha emission.

    The masses are for the neutral atoms. Helium-4 has an atomic mass of 4.002603 u, and an electron has a mass of 0.005485799 u.

35. The half-life of ^{14}C is 5730 years. (a) What fraction of a carbon-14 sample is undecayed after 2500, 5000, 10,000, and 20,000 years? (b) After what time will only 10% of the original nuclei remain intact?

36. Plutonium-234 has a half-life of 8.8 hours. (a) Starting from a population of 1 million ^{234}Pu atoms, how many of the nuclei will be untouched after 2.0, 4.0, 8.0, 16.0, and 32.0 hours? (b) If 976,432 plutonium-234 nuclei are still present from the original sample, how much time has elapsed?

37. Recall the definition of *activity*, A, as used to express the rate of radioactive decay for a first-order process:

$$A = -\frac{\Delta N}{\Delta t} = kN = \text{rate}$$

The activity, thus defined, is proportional to the number of unreacted nuclei (N) present at each instant. The rate constant k then determines the variation of activity with time:

$$A_t = A_0 \exp(-kt)$$

Question: What curve is obtained from a plot of $\ln A_t$ versus t? How may the half-life be extracted from such a plot?

38. Below is a sequential set of activities measured for the beta decay of ^{232}Ac (atomic mass = 232.042130 u):

TIME (s)	A (Bq)	TIME (s)	A (Bq)
0	1000	180	354
45	771	225	273
90	595	270	210
135	459	315	162

(a) Calculate the first-order rate constant and half-life. (b) What mass of actinium-232, in grams, is responsible for the initial activity of 1000 Bq? (c) Calculate the expected activity and corresponding mass after 360 seconds of decay.

39. Tritium (^3H, atomic mass = 3.01605 u) has a half-life of 12.3 years. (a) What mass of tritium, in grams, will generate an activity of 500 Bq? (b) Starting from an initial activity of 500 Bq, calculate the remaining activity and mass of tritium after 6, 12, 18, and 24 years of beta decay.

40. Iodine-131, a radioisotope used in the treatment of thyroid disease, has a half-life of 8.04 days. How much time must elapse before the activity falls to 1% of its original value?

41. Phosphorus-32, a beta emitter with atomic mass = 31.973907 u, has a half-life of 14.28 days. (a) To which of the following isotopes does ^{32}P decay?

	ATOMIC MASS (u)	HALF-LIFE
^{31}Si	30.975362	2.62 h
^{32}Si	31.974148	160 y
^{33}Si	32.977920	6.1 s
^{31}P	30.973762	stable
^{33}P	32.971725	25.3 d
^{31}S	30.979554	2.56 s
^{32}S	31.972070	stable
^{33}S	32.971456	stable

(b) Calculate the percent composition of mother and daughter (by mass) after a sample of pure phosphorus-32 has decayed for 21.00 days.

42. The half-life of uranium-238 is 4.5×10^9 years. Determine the age of a rock that contains 0.726 g of uranium-238 (atomic mass = 238.050784 u) and 0.101 g of lead-206 (atomic mass = 205.974440 u), with no other isotopes of lead present in the sample.

43. The stable isotopes of lead occur naturally in the following proportions:

$$^{208}\text{Pb} \ (207.976627 \text{ u}) \quad 52.4\%$$
$$^{207}\text{Pb} \ (206.975872 \text{ u}) \quad 22.1\%$$
$$^{206}\text{Pb} \ (205.974440 \text{ u}) \quad 24.1\%$$
$$^{204}\text{Pb} \ (203.973020 \text{ u}) \quad 1.4\%$$

Now suppose that a rock contains 0.9050 g of uranium-238, together with the isotopic mixture of lead tabulated below:

$$^{208}\text{Pb} \quad 0.2620 \text{ g}$$
$$^{207}\text{Pb} \quad 0.1105 \text{ g}$$
$$^{206}\text{Pb} \quad 0.2205 \text{ g}$$
$$^{204}\text{Pb} \quad 0.0070 \text{ g}$$

Estimate the age of the rock.

44. Final thoughts (1): What combination of quarks (including anti-quarks) will give rise to each of the following particles?

    (a) a positive pion, π^+

    (b) a neutral pion, π^0

    (c) a negative pion, π^-

    (d) an antineutron, $\tilde{n}$

45. Final thoughts (2): How small is a nucleus, and exactly how much mass is crammed into it? (a) Experiments show that the radius of a nucleus goes roughly as

$$R = (1.2 \times 10^{-15} \text{ m}) \cdot A^{1/3}$$

for mass number A. Use this relationship to gauge the radius of the hydrogen-1 nucleus, the carbon-12 nucleus, and the uranium-238 nucleus. (b) Estimate the density of each nucleus (in g cm^{-3}), assuming the shape to be a perfect sphere. (c) How many tons of nuclear material, packed at these densities, would fill a volume of 1 cubic centimeter? Relevant conversion factors are:

$$1 \text{ cm}^3 = 10^{-6} \text{ m}^3$$

$$1 \text{ ton} = 2000 \text{ lb}$$

$$1 \text{ kg} = 2.2 \text{ lb}$$

46. Final thoughts (3): The previous exercise demonstrates that the average *volume per nucleon* is roughly constant for all nuclei. (a) Prove it, using the empirical relationship between radius (R) and mass number (A):

$$R = (1.2 \times 10^{-15} \text{ m}) \cdot A^{1/3}$$

(b) How many nucleons would occupy 1 cubic meter? What, approximately, is the average volume per nucleon? (c) Given the nature of the strong force, why is this similarity in packing to be expected?

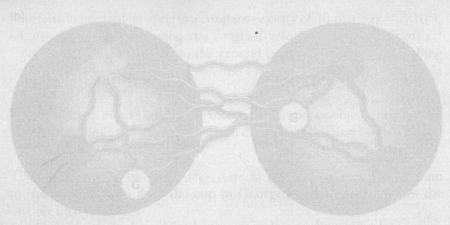

FIGURE 21-15. Composite particles; composite interactions. (a) Two molecules, each uncharged, interact electromagnetically despite their overall neutrality. Acting separately, every charged particle in one exchanges photons with every other and is individually attracted or repelled. A simpler net force (such as between spaced dipoles) emerges from this web, of fundamental interactions. (b) For two nucleons, each one uncolored (blend of red, green, and blue) quarks, the effect is the same: a tangled network of strong interactions between individually colored quarks, mediated by gluons (colored as well). The result is a composite interaction, between the two color-neutral nucleons, as if carried by a meson.

APPENDIX A

Nomenclature and Vocabulary

A-1. Naming the Atoms

A-2. Prefixes and Suffixes
   Number of Atoms
   Charge and Oxidation State
   Arrangement

A-3. Combining the Elements: Inorganic Systems
   Ionic Compounds
   Binary Molecular Compounds
   Inorganic Acids
   Coordination Complexes

A-4. Root and Branch: Organic Molecules
   Alkanes
   Alkenes and Alkynes
   Alcohols and Esters

A-5. Last Words: Toward a Chemical Vocabulary
   EXERCISES

> What's in a name? That which we call a rose
> By any other name would smell as sweet.
>
> —*Romeo and Juliet* II, ii

Yes it would; but still we must call it something, for without names there are only grunts and gestures. A *rose*? Maybe not; maybe *shoshannah* or *hoa hòng* or *waridi* instead—but every flower, every rock, every animal, every person, every emotion . . . everything in the world, visible or not, needs a name, and it is our prerogative and power to bestow those names however we like.

We make them up. Sometimes by design, more often on a whim, someone conjures up a name and then, as if by magic, the made-up name becomes real. It becomes the thing itself. To name a thing is to own it.

How else, if not with words, shall we give substance to a world we can only imagine? How else, except by their names, shall we distinguish tulip from rose, love from hate, particle from wave? How else shall we organize a few dozen elements into millions of compounds? We need names. We need words. We need a language.

Better to say, we need a working translation of nature's language into a speech of our own: simple at the core, like the universe itself; powerful enough, though, to describe a world deceivingly complex on the surface.

We have it—a chemical and scientific language strung together partly with words, partly with numbers; a language expressed oddly at times, more often sensibly and systematically; a language laid out, by necessity, in a way befitting the design and structure of matter. Atoms combine only in certain ways, and so must the words and names chosen to represent them.

A-1. NAMING THE ATOMS

For those elements known since antiquity, we usually inherit the names given long ago, come down to us in the various languages:

COPPER (Cu): originally from the Latin *cuprum* and *Cyprium*, named for the island of Cyprus; translated as *copper* in English, *Kupfer* in German, *cuivre* in French. Its two-letter symbol is Cu, short for the Latin form.

GOLD (Au): descended, unchanged, from Old English *gold*; and, with modest change, from Latin *aurum* to French *or*. The symbol, again taken from the Latin, is Au.

IRON (Fe): from *iren* in Old English to *iron* in Modern English; from *ferrum* in Latin to *fer* in French, whence the symbol Fe.

LEAD (Pb): *lead*, also unchanged from the Old English; and *plumbum*, Latin, from which comes "plumbing" and "plumb line" and the symbol Pb.

SILVER (Ag): from *siolfur* in Old English to *silver* in Modern English; from *argentum* in Latin to *argent* in French.

SULFUR (S): *sulphur*, Latin; *sulvere*, Sanskrit.

TIN (Sn): *tin*, Old English; *stannum*, Latin.

ZINC (Zn): *Zink*, of obscure Germanic origin.

For some, a borrowed or resurrected name brings to light a substance previously known only in compounds, not as an element alone; elements such as:

ALUMINUM (Al): from the oxide *alum* (formula $K_2SO_4 \cdot Al_2(SO_4)_3 \cdot 24 H_2O$), used by the Greeks and Romans as a dye and a medicinal astringent.

ANTIMONY (Sb): roots uncertain, but perhaps a combination of the Greek words *anti* and *monos* ("not found alone"). The symbol Sb derives from the mineral stibnite (Sb_2S_3), the principal ore containing antimony.

ARSENIC (As): denoting *arsenikon*, male, from the Greek. The ancients believed that metals possessed masculine and feminine properties.

BORON (B): from Latin *borax* (formula $Na_2B_4O_7 \cdot 10 H_2O$) and from Persian *burag* and Arabic *buraq*.

CADMIUM (Cd): from *kadmeia* in Greek and *cadmia* in Latin, the ancient names for the mineral calamine ($ZnCO_3$); also *Cadmus*, founder of Thebes.

CALCIUM (Ca): from the Latin *calx*, lime.

CARBON (C): from the French *carbone*, taken from the Latin *carbo* (charcoal, coal burning, burnt wood).

For others, unearthed within the last few centuries, a new name may evoke some special property or some hallmark of the element's discovery, as:

ACTINIUM (Ac): *aktis*, ray, beam (Greek).

ARGON (Ar): *argos*, inactive (Greek).

ASTATINE (At): *astatos*, unstable (Greek).

BARIUM (Ba): *barys*, heavy (Greek).

BERYLLIUM (Be): *beryllos*, beryl (Greek), a mineral with formula $Be_3Al_2Si_6O_{18}$.

BISMUTH (Bi): *Wissmuth*, white mass (*Weisse Masse*, German).

BROMINE (Br): *bromos*, stench (Greek).

CERIUM (Ce): *Ceres* (Greek goddess), for an asteroid discovered two days before.

CESIUM (Cs): *caesius*, sky blue (Latin).

CHLORINE (Cl): *chloros*, greenish yellow (Greek).

CHROMIUM (Cr): *chroma*, color (Greek).

DYSPROSIUM (Dy): *dysprositos*, hard to get at (Greek).

FLUORINE (F): *fluere*, flow or flux (Latin, later French).

GADOLINIUM (Gd): *gadolinite*, a mineral named for Johann Gadolin.

HELIUM (He): *helios*, sun (Greek); first found on the sun, 1868.

HYDROGEN (H): *hydro*, water + *genes*, forming; thus "water forming" (Greek).

INDIUM (In): *indigo*, for a prominent line in the atomic spectrum of the metal.

IODINE (I): *iodes*, violet (Greek).

IRIDIUM (Ir): *iris*, rainbow (Latin), for the multicolored salts it forms.

KRYPTON (Kr): *kryptos*, hidden (Greek).

LANTHANUM (La): *lanthanein*, to lie hidden (Greek).

LITHIUM (Li): *lithos*, stone (Greek).

MANGANESE (Mn): *magnes*, magnet (Greek).

MOLYBDENUM (Mo): *molybdos*, lead (Greek).

NEODYMIUM (Nd): *neos*, new + *didymos*, twin (Greek).

NEON (Ne): *neos*, new (Greek).

NITROGEN (N): *nitron*, soda + *genes*, forming; thus "soda forming" (Greek).

OSMIUM (Os): *osme*, a smell (Greek).

OXYGEN (O): *oxys*, sharp (keen, acid) + *genes*, forming; thus "acid forming" (Greek).

PHOSPHORUS (P): *phosphoros*, light-bearing (Greek).

PLATINUM (Pt): *platina*, silverlike (Spanish).

POTASSIUM (K): *potash* (English); *kalium* (Latin); *qali* (Arabic).

PRASEODYMIUM (Pr): *prasios*, green + *didymos*, twin (Greek).

PROTACTINIUM (Pa): *protos*, first + *aktis*, ray (Greek).

RADIUM (Ra), RADON (Rn): *radius*, ray (Latin).

RHODIUM (Rh): *rhodon*, rose (Greek).

RUBIDIUM (Rb): *rubidus*, deepest red (Latin).

SAMARIUM (Sm): *samarskite*, a mineral.

SILICON (Si): *silex*, flint (Latin).

SODIUM (Na): *soda* (English); *natrium* (Latin).

TECHNETIUM (Tc): *technetos*, artificial (Greek).

TELLURIUM (Te): *tellus*, earth (Latin).

THALLIUM (Tl): *thallos*, green shoot (Greek).

TUNGSTEN (W): *tung sten*, heavy stone (Swedish); *wolframite*, a mineral.

XENON (Xe): *xenos*, stranger (Greek).

ZIRCONIUM (Zr): *zargun*, like gold (Persian).

Or, also common, a newly invented name may offer a fanciful or mythological characterization, often shared by a planet or other heavenly body:

COBALT (Co): *Kobold* (German) and *kobalos* (Greek), goblin or evil spirit.

MERCURY (Hg): *Mercury* (Greco-Roman god); *hydrargyrum*, liquid silver (Latin).

NEPTUNIUM (Np): *Neptune* (Greco-Roman god), name of a planet.

NICKEL (Ni): *Kupfernickel*, Old Nick's copper (Satan's copper, German).

NIOBIUM (Nb): *Niobe* (figure from Greek mythology).

PALLADIUM (Pd): *Pallas* (Greek goddess), name of an asteroid.

PLUTONIUM (Pu): *Pluto* (Greco-Roman god), name of a planet.

PROMETHIUM (Pm): *Prometheus* (Greco-Roman god).

SELENIUM (Se): *selene*, moon (Greek).

TANTALUM (Ta): *Tantalus* (from Greek mythology, father of Niobe).

THORIUM (Th): *Thor* (Scandinavian god).

TITANIUM (Ti): *Titans* (Greco-Roman gods).

URANIUM (U): *Ouranos* (Greco-Roman god), name of a planet.

VANADIUM (V): *Vanadis* (Scandinavian goddess).

A name may denote a place:

AMERICIUM (Am): *America*.

BERKELIUM (Bk): University of California at *Berkeley*.

CALIFORNIUM (Cf): *California*.

ERBIUM (Er), TERBIUM (Tb), YTTERBIUM (Yb), YTTRIUM (Y): *Ytterby* (Swedish town).

EUROPIUM (Eu): *Europe*.

FRANCIUM (Fr): *France*.

GALLIUM (Ga): *Gallia* (France, Latin; also, a pun on the name of *Lecoq*).

GERMANIUM (Ge): *Germania* (Germany, Latin).

HAFNIUM (Hf): *Hafnia* (Copenhagen, Latin).

HOLMIUM (Ho): *Holmia* (Stockholm, Latin).

LUTETIUM (Lu): *Lutetia* (Paris, Latin).

MAGNESIUM (Mg): *Magnesia*.

POLONIUM (Po): *Poland*.

RHENIUM (Re): *Rhenus* (Rhine, Latin).

RUTHENIUM (Ru): *Ruthenia* (Russia, Latin).

SCANDIUM (Sc): *Scandia* (Scandinavia, Latin).

STRONTIUM (Sr): *Strontian* (Scottish town).

THULIUM (Tm): *Thule* (early name for Scandinavia).

Or a person:

CURIUM (Cm): Pierre and Marie *Curie*.

EINSTEINIUM (Es): Albert *Einstein*.

FERMIUM (Fm): Enrico *Fermi*.

LAWRENCIUM (Lr): Ernest *Lawrence.*

MENDELEVIUM (Md): Dmitri *Mendeleev.*

NOBELIUM (No): Alfred *Nobel.*

But for whatever the reason, there is always a name; and to the discoverer goes the almost unconditional right to choose it—whether to honor a person or place, to commemorate the circumstances of discovery, to describe color, form, or some other property of the element, or even to make a personal statement. They are *names*, after all, and all names are made up by the namer. We take them and use them in the spirit offered.

A-2. PREFIXES AND SUFFIXES

The names of the elements stand in place of the atoms; and so the names, like the atoms, must combine according to a strict pattern, not willy-nilly. To comply with nature's own rules, our system of **nomenclature** (names and naming) needs the tools to recognize differences in structure and oxidation state. A set of prefixes and suffixes, aptly used, bears much of the burden.

Number of Atoms

First there is the matter of valence and stoichiometry, of number, of proportion; the question of how to convey, from the names alone, the compositional differences in a series of compounds such as N_2O, NO, NO_2, NO_3, N_2O_3, N_2O_4, and N_2O_5.

We want, for that, a device to indicate how *many* atoms come together, a device well provided by a set of numerical prefixes sufficient to cover the possible whole-number combinations. Taken from the Greek, the first 10 are:

mono: one	*hexa*: six
di: two	*hepta*: seven
tri: three	*octa*: eight
tetra: four	*nona*: nine
penta: five	*deca*: ten

They allow us to distinguish nitrogen *di*oxide (NO_2) from nitrogen *tri*-oxide (NO_3), carbon *mon*oxide (CO) from carbon *di*oxide (CO_2), *penta*ne (C_5H_{12}) from *hexa*ne (C_6H_{14}). In the sections that follow, we shall see them applied systematically to both organic and inorganic compounds.

Charge and Oxidation State

Next come questions pertaining to oxidation state, questions bearing on the *form* assumed by the atom in a compound. We ask: Does the atom take electrons and grow more negative, or does it give electrons and grow more positive? Does it become either a full-fledged anion or cation? Can it exist in more than one oxidation state?

For monatomic (one-atom) cations, the rule is simple: Call the element an *ion* rather than an atom, keeping the name otherwise the same:

Na is the sodium *atom*. Na^+ is the sodium *ion*.

Mg is the magnesium *atom*. Mg^{2+} is the magnesium *ion*.

Al is the aluminum *atom*. Al^{3+} is the aluminum *ion*.

In the corresponding symbol, a superscript integer followed by a plus sign denotes the number of electrons lost.

For elements that support two or more positive oxidation states (Fe^{2+} and Fe^{3+}, say), a Roman numeral helps separate the possibilities:

Fe^{2+} is the *iron(II)* ion.

Fe^{3+} is the *iron(III)* ion.

An alternative method, still popular, uses a pair of suffixes instead: *ic* for the ion with the higher charge, *ous* for the lower. Here we would contrast the *ferrous* ion (Fe^{2+}) with the *ferric* ion (Fe^{3+}); the *mercurous*, or *mercury(I)* ion (Hg_2^{2+}), with the *mercuric*, or *mercury(II)* ion (Hg^{2+}); and the *cuprous*, or *copper(I)* ion (Cu^+) with the *cupric*, or *copper(II)* ion (Cu^{2+}):

Fe^{2+} is the *ferrous* ion. Fe^{3+} is the *ferric* ion.

Hg_2^{2+} is the *mercurous* ion. Hg^{2+} is the *mercuric* ion.

Cu^+ is the *cuprous* ion. Cu^{2+} is the *cupric* ion.

With suffixes, too, we assign the negative ions—the anions—their names. For monatomic species, the suffix *ide* attaches to the root name of the atom to give, for example, the *hydride* ion (H^-), the *chloride* ion (Cl^-), the *oxide* ion (O^{2-}), the *sulfide* ion (S^{2-}), the *nitride* ion (N^{3-}), and the *phosphide* ion (P^{3-}).

Polyatomic anions, if unique, also take the *ide* suffix, as we find with the *hydroxide* ion (OH^-), the *cyanide* ion (CN^-), the *peroxide* ion (O_2^{2-}), and the *azide* ion (N_3^-). More often, however, the same element will combine in different proportions to produce a series of anions distinguished by different prefixes and suffixes—such as the set of two oxyanions (oxygen-containing polyatomic anions) formed with nitrogen,

$$NO_2^-\qquad \text{nit}rite\ \text{ion}$$
$$NO_3^-\qquad \text{nit}rate\ \text{ion}$$

and the set of four oxyanions formed with chlorine:

$$ClO^-\qquad hypo\text{chlor}ite\ \text{ion}$$
$$ClO_2^-\qquad \text{chlor}ite\ \text{ion}$$
$$ClO_3^-\qquad \text{chlor}ate\ \text{ion}$$
$$ClO_4^-\qquad per\text{chlor}ate\ \text{ion}$$

Look at the names and discover the rules:

1. The primary suffixes are *ite* and *ate*. Where just two oxyanions exist (NO_2^-, NO_3^-), we use *ite* for the ion with fewer oxygens and *ate* for the ion with more. NO_2^-, containing two oxygens, becomes nit*rite*; NO_3^-, with three, becomes nit*rate*. SO_3^{2-}, sulf*ite*; SO_4^{2-}, sulf*ate*.

2. In a series of three or four related anions, the prefixes *hypo* and *per* provide additional comparison. The oxyanion with one oxygen *fewer* than the "ite" form takes the prefix *hypo* (from the Greek, meaning "under"). The oxyanion with one oxygen *more* than the "ate" form, being the most oxygen-rich of the set, is labeled with the prefix *per* (from the Latin, meaning "thorough"). Thus *hypo*chlorite (ClO^-) contains one oxygen to chlorite's two (ClO_2^-), whereas *per*chlorate (ClO_4^-) contains four oxygens to chlorate's three (ClO_3^-). *Hypo* means one *fewer*; *per* means one *more*.

Another possibility: An anion captures a hydrogen cation, H^+, thereby reducing its charge by one unit, as when the carbonate ion (CO_3^{2-}) becomes the *hydrogen* carbonate ion (HCO_3^-), or when the sulfate ion (SO_4^{2-}) becomes the *hydrogen* sulfate ion (HSO_4^-). We show the acquisition of H^+ simply by including the word *hydrogen* or, in another notation, by using the prefix *bi*. Hence the names *bicarbonate* ion and *hydrogen carbonate* ion both refer to the same HCO_3^- species; and, similarly, the names *bisulfate* ion and *hydrogen sulfate* ion each designate HSO_4^-.

Or: The anion binds two H^+ ions, not one, and the union is then marked by the prefix *dihydrogen*. The phosphate ion, for instance, first goes from PO_4^{3-} to HPO_4^{2-} to become the hydrogen phosphate ion, and subsequently from HPO_4^{2-} to $H_2PO_4^-$ to become the *di*hydrogen phosphate ion. The charge drops from -3 to -1 with the attachment of the two hydrogen ions.

Are we making up names? Of course. There is nothing special about *ite* or *ate*, *ic* or *ous*, *hypo* or *per*, or any other arbitrary combination of sounds and symbols.

But are we making up names *arbitrarily*? No. Our suffixes and prefixes prove themselves worthy not by chance, but precisely because chemistry itself is not arbitrary: because atoms combine only in certain fixed ways; because, really, there is no other choice.

Arrangement

Ask now, *where* are the atoms? What spatial arrangements do they make? How should the name of one isomer be distinguished from another?

Again, a well-chosen set of prefixes and suffixes works to capture the most important structural variations: cis and trans; D and L; ortho, meta, and para; *n* (normal chain); others, too, which we can call upon as needed.

Example: The geometric isomers

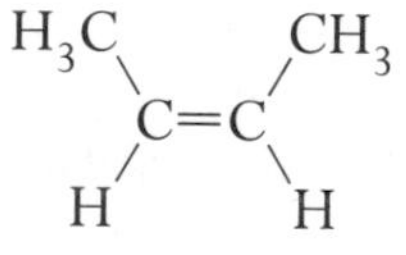

cis-2-butene 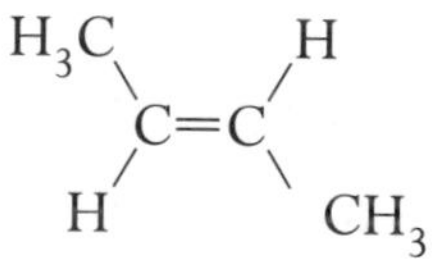   trans-2-butene

are differentiated as *cis* (the two CH_3 substituents lie on the *same* side of the double bond) and *trans* (the two methyl groups lie *across* the double bond, on opposite sides).

Example: The optical isomers

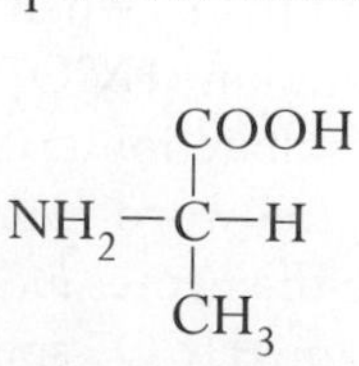

L-alanine 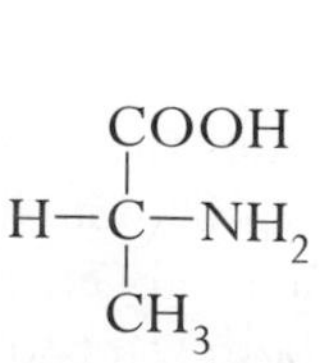   D-alanine

are differentiated as L and D according to their arrangement of four different groups around a chiral carbon.

Example: The structural isomers

o-dichlorobenzene   *m*-dichlorobenzene   *p*-dichlorobenzene

are differentiated as *ortho* (*o*), *meta* (*m*), and *para* (*p*) according to the placement of the two Cl atoms around the hexagonal ring. Each unmarked vertex represents a carbon atom, its attached hydrogen omitted for clarity.

A-3. COMBINING THE ELEMENTS: INORGANIC SYSTEMS

Names of common inorganic species, assigned according to the conventions discussed above, are provided in Table A-1 for reference. What remains now is to put them together and to give each structure a name.

TABLE A-1. Selected Inorganic Species

Class	Name	Formula	Alternative Name
anion	acetate (organic)	$C_2H_3O_2^-$	
	arsenate	AsO_4^{3-}	
	azide	N_3^-	
	borate	BO_3^{3-}	
	bromide	Br^-	
	carbonate	CO_3^{2-}	
	chlorate	ClO_3^-	
	chloride	Cl^-	
	chlorite	ClO_2^-	

Continued on next page

TABLE A-1. Selected Inorganic Species (*Continued*)

Class	Name	Formula	Alternative Name
anion	chromate	CrO_4^{2-}	
	cyanide	CN^-	
	dichromate	$Cr_2O_7^{2-}$	
	dihydrogen phosphate	$H_2PO_4^-$	
	fluoride	F^-	
	hydride	H^-	
	hydrogen carbonate	HCO_3^-	bicarbonate
	hydrogen sulfate	HSO_4^-	bisulfate
	hydrogen sulfite	HSO_3^-	bisulfite
	hydroxide	OH^-	
	hypochlorite	ClO^-	
	iodate	IO_3^-	
	iodide	I^-	
	nitrate	NO_3^-	
	nitride	N^{3-}	
	nitrite	NO_2^-	
	oxalate	$C_2O_4^{2-}$	
	oxide	O^{2-}	
	perchlorate	ClO_4^-	
	permanganate	MnO_4^-	
	peroxide	O_2^{2-}	
	phosphate	PO_4^{3-}	
	phosphide	P^{3-}	
	phosphite	PO_3^{3-}	
	sulfate	SO_4^{2-}	
	sulfide	S^{2-}	
	sulfite	SO_3^{2-}	
	thiocyanate	SCN^-	
	thiosulfate	$S_2O_3^{2-}$	
cation	aluminum	Al^{3+}	
	ammonium	NH_4^+	

TABLE A-1. Selected Inorganic Species (*Continued*)

Class	Name	Formula	Alternative Name
cation	barium	Ba^{2+}	
	cadmium	Cd^{2+}	
	calcium	Ca^{2+}	
	cesium	Cs^{+}	
	chromium(III)	Cr^{3+}	chromic
	cobalt(II)	Co^{2+}	cobaltous
	copper(I)	Cu^{+}	cuprous
	copper(II)	Cu^{2+}	cupric
	hydrogen	H^{+}	
	hydronium	H_3O^{+}	
	iron(II)	Fe^{2+}	ferrous
	iron(III)	Fe^{3+}	ferric
	lead(II)	Pb^{2+}	plumbous
	lithium	Li^{+}	
	magnesium	Mg^{2+}	
	manganese(II)	Mn^{2+}	manganous
	mercury(I)	Hg_2^{2+}	mercurous
	mercury(II)	Hg^{2+}	mercuric
	phosphonium	PH_4^{+}	
	potassium	K^{+}	
	silver	Ag^{+}	
	sodium	Na^{+}	
	tin(II)	Sn^{2+}	stannous
	uranyl	UO_2^{2+}	
	vanadyl	VO^{2+}	
	zinc	Zn^{2+}	
molecule	ammonia	NH_3	
	carbon dioxide	CO_2	
	carbon monoxide	CO	
	hydrazine	N_2H_4	
	hydrogen peroxide	H_2O_2	
	water	H_2O	

Ionic Compounds

Composed of two oppositely charged species, a simple ionic compound has both a first name and a last name. The first name: the cation, often a metal. The second name: the anion, usually a nonmetal.

Where the oxidation state or ionic charge of a metal is unambiguous, we omit the Roman numeral, as below:

$NaCl$	sodium chloride
Na_2SO_4	sodium sulfate
$NaC_2H_3O_2$	sodium acetate (from the acetate anion, CH_3COO^-)
$CaCl_2$	calcium chloride
CaO	calcium oxide
Mg_3N_2	magnesium nitride
Al_2O_3	aluminum oxide
NH_4OH	ammonium hydroxide (from the ammonium cation, NH_4^+)

Otherwise, we either label the cation explicitly with a Roman numeral

$FeBr_2$	iron(II) bromide	[iron existing as Fe^{2+}, balanced by $2\,Br^-$]
$FeBr_3$	iron(III) bromide	[iron existing as Fe^{3+}, balanced by $3\,Br^-$]

or use the *ous-ic* terminology:

$FeBr_2$	ferrous bromide	[iron(II), the lower oxidation state]
$FeBr_3$	ferric bromide	[iron(III), the higher oxidation state]

Binary Molecular Compounds

Two nonmetallic elements, neither able to capture the other's electron completely, bond covalently in various proportions to form **binary** molecular compounds. The less electronegative element (like the cation in an ionic compound) is named first, followed by the more electronegative element carrying the suffix *ide* (as if it were an anion). The standard numerical prefixes then indicate the number of atoms of each type, although *mono* is omitted for the first element and is usually optional for the second.

Some examples:

HCl	hydrogen chloride
CO	carbon monoxide
CO_2	carbon dioxide

SF_6 sulfur hexafluoride
P_4S_{10} tetraphosphorus decasulfide
N_2O dinitrogen (mon)oxide
NO nitrogen (mon)oxide
NO_2 nitrogen dioxide
NO_3 nitrogen trioxide
N_2O_3 dinitrogen trioxide
N_2O_4 dinitrogen tetroxide
N_2O_5 dinitrogen pentoxide

Note that prefixes ending in *a* or *o* are often contracted when followed by a vowel, as in

P_4O_6 tetraphosphorus hexoxide (not hex*a*oxide)
Cl_2O_7 dichlorine heptoxide (not hept*a*oxide)
CO carbon monoxide (not mon*o*oxide)

Note also that some common names still survive, familiar from long usage even if assigned unsystematically. Water (H_2O), ammonia (NH_3), nitrous oxide (N_2O), nitric oxide (NO), and hydrazine (N_2H_4) are among the best known.

Inorganic Acids

Dissolved in water, an Arrhenius acid

$$H_nX(aq) \longrightarrow nH^+(aq) + X^{n-}(aq)$$

dissociates into *n* hydrogen ions (H^+) and a counterbalancing X^{n-} anion. A *monoprotic* acid releases one H^+ ($n = 1$); a *polyprotic* acid releases two or more.

The acid takes its name from the anion. Wherever X^{n-} normally carries the suffix *ide*, its ending in the acid is first modified to *ic*:

Cl^- chloride $\longrightarrow$ chloric

The prefix *hydro* then marks the presence of the hydrogen, so that we have

HCl hydrochloric acid

In acids containing sulfur, however, the form *sulfuric* is used rather than the expected *sulfic*; this slight exception yields names such as

$$S^{2-} \qquad \text{sulfide ion}$$
$$H_2S \qquad \text{hydrosulfuric acid}$$

Where, next, the acid releases a polyatomic oxyanion such as ClO_2^- or ClO_3^-, both the suffixes *ous* and *ic* are used: *ic* for any anion originally ending in *ate* (the oxygen-rich species); *ous* for an anion originally ending in *ite* (the oxygen-poor species). The chlor*ate* anion, ClO_3^-, therefore produces chlor*ic* acid; whereas the chor*ite* anion, ClO_2^-, produces chlor-*ous* acid:

$HClO$ hypochlor*ous* acid (from the hypochlor*ite* anion, ClO^-)
$HClO_2$ chlor*ous* acid (from the chlor*ite* anion, ClO_2^-)
$HClO_3$ chlor*ic* acid (from the chlor*ate* anion, ClO_3^-)
$HClO_4$ perchlor*ic* acid (from the perchlor*ate* anion, ClO_4^-)

Except for the change of suffix, the only other modification is the addition of the term *acid*.

Other examples include

HNO_2 nitrous acid (from the nitrite anion, NO_2^-)
HNO_3 nitric acid (from the nitrate anion, NO_3^-)

and

H_2SO_3 sulfurous acid (from the sulfite anion, SO_3^{2-})
H_2SO_4 sulfuric acid (from the sulfate anion, SO_4^{2-})

and

H_3PO_3 phosphorous acid (from the phosphite anion, PO_3^{3-}; but see below)
H_3PO_4 phosphoric acid (from the phosphate anion, PO_4^{3-})

Similar to sulfur, the combining form for phosphorus oxyanions is modified slightly to *phosphor* from the expected *phosph*. Be sure to distinguish also the spelling phosphor*us* (for the element) from the acidic form phosphor*ous*.

Phosphorous acid, incidentally, is actually a diprotic acid, not a triprotic acid as its empirical formula implies. One of the three hydrogens attaches itself directly to the phosphorus, producing the structure

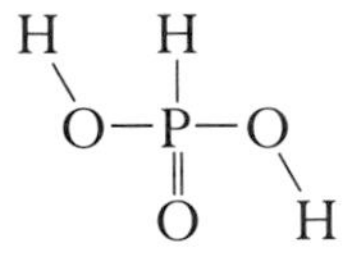

$$
\begin{array}{ccc}
\text{H} & & \text{H} \\
\diagdown & & | \\
\text{O} & \!\!-\text{P}-\!\! & \text{O} \\
& \| & \diagdown \\
& \text{O} & \text{H}
\end{array}
$$

The PO_3^{3-} ion is not released upon dissociation of H_3PO_3.
See Table A-2 for a summary of common inorganic acids.

TABLE A-2. Selected Inorganic Acids

Acid	Name	Anion[a]	Name
HF	hydrofluoric acid	F^-	fluoride ion
HCl	hydrochloric acid	Cl^-	chloride ion
HBr	hydrobromic acid	Br^-	bromide ion
HI	hydr(o)iodic acid	I^-	iodide ion
HN_3	hydr(o)azoic acid	N_3^-	azide ion
HCN	hydrocyanic acid	CN^-	cyanide ion
H_2S	hydrosulfuric acid	S^{2-}	sulfide ion
HClO	hypochlorous acid	ClO^-	hypochlorite ion
$HClO_2$	chlorous acid	ClO_2^-	chlorite ion
$HClO_3$	chloric acid	ClO_3^-	chlorate ion
$HClO_4$	perchloric acid	ClO_4^-	perchlorate ion
H_2CO_3	carbonic acid	CO_3^{2-}	carbonate ion
HNO_2	nitrous acid	NO_2^-	nitrite ion
HNO_3	nitric acid	NO_3^-	nitrate ion
H_3PO_4	phosphoric acid	PO_4^{2-}	phosphate ion
H_2SeO_3	selenous acid	SeO_3^{2-}	selenite ion
H_2SeO_4	selenic acid	SeO_4^{2-}	selenate ion
H_2SO_3	sulfurous acid	SO_3^{2-}	sulfite ion
H_2SO_4	sulfuric acid	SO_4^{2-}	sulfate ion

[a]After release of all ionizable protons.

Coordination Complexes

The name of a complex ion or compound should convey at least three features of the structure: (1) The central metal and its oxidation state, along with the overall electrical character of the complex (positive, negative, or neutral). (2) The number and kind of ligands coordinated to the metal. (3) The counterbalancing ions, if any, that exist outside the coordination sphere.

Easiest, perhaps, is to take a pair of names and see how they conform to a systematic nomenclature. First the names, then the rules.

The names:

COMPOUND 1: $[Cr(NH_3)_6](NO_3)_3$. The formula describes a salt containing one $[Cr(NH_3)_6]^{3+}$ cation for every three NO_3^- anions. Six NH_3 ligands are coordinated around each Cr^{3+} ion. The name is

hexaamminechromium(III) nitrate

Break it down: *hexa* (six) + *ammine* (with two *m*'s: a special designation for the ammonia ligand, NH_3) + *chromium(III)* (the Cr^{3+} ion) + *nitrate* (the NO_3^- ion).

COMPOUND 2: $NH_4[Cr(NH_3)_2(NCS)_4]$. Here we have a salt with NH_4^+ and $[Cr(NH_3)_2(NCS)_4]^-$ ions in a one-to-one ratio. Its name is

ammonium diamminetetraisothiocyanatochromate(III)

Again, break it down: *ammonium* (NH_4^+) + *di* (two) + *ammine* (NH_3) + *tetra* (four) + *isothiocyanato* (NCS^-) + *chromate(III)* (Cr^{3+}).

The rules:

1. *The cation is specified first, consistent with the convention of positive before negative.*

    In compound 1, the complex ion $[Cr(NH_3)_6]^{3+}$ (the cation) is positive, and the counterbalancing NO_3^- ion (the anion) is negative. The complex ion, designated as *hexaamminechromium(III)*, comes first; the *nitrate* ion follows: *hexaamminechromium(III) nitrate*.

    In compound 2, the ammonium ion outside the complex is positive and hence precedes the name of the anionic complex ion in the full name *ammonium diamminetetraisothiocyanatochromate(III)*.

2. *The ligands are given in alphabetical order, listed before the metal*

and modified by the Greek numerical prefixes to indicate number. Prefixes are ignored for purposes of alphabetization.

In compound 1, the six NH_3 ligands appear as *hexaammine* in the name *hexaamminechromium(III) nitrate*. There is no contraction between *hexa* and *ammine*.

In compound 2, the two NH_3 ligands (*ammine*) are alphabetized before the four NCS^- ligands (*isothiocyanate*) to produce the form *diamminetetraisothiocyanato*. For the meaning of the suffix *o*, see the next rule.

3. *Neutral ligands are assigned their usual names, whereas anionic ligands end with the letter* o.

In compound 2, the neutral ammonia molecule has its special name *ammine*. The anionic NCS^- species changes its ending from *ate* in *isothiocyanate* to *ato* in *isothiocyanato*.

Other special names for molecules include *aqua* for water (H_2O) and *carbonyl* for carbon monoxide (CO). See Table A-3 for more.

TABLE A-3. Common Ligands

Ligand	Formula	Name in Complex
ammonia	NH_3	ammine
bromide ion	Br^-	bromo
carbonate ion	CO_3^{2-}	carbonato
carbon monoxide	CO	carbonyl
chloride ion	Cl^-	chloro
cyanide ion	CN^-	cyano
ethylenediamine	$NH_2CH_2CH_2NH_2$	ethylenediamine (en)
hydroxide ion	OH^-	hydroxo
nitrite ion	NO_2^-	nitro (bound as $:NO_2^-$)
nitrite ion	ONO^-	nitrito (bound as $:ONO^-$)
oxalate ion	$C_2O_4^{2-}$	oxalato
oxide ion	O^{2-}	oxo
pyridine	C_5H_5N	pyridine
sulfate ion	SO_4^{2-}	sulfato
thiocyanate ion	SCN^-	thiocyanato (bound as $:SCN^-$)
thiocyanate ion	NCS^-	isothiocyanato (bound as $:NCS^-$)
water	H_2O	aqua

4. *The metal is designated simply by name, and its oxidation state is indicated by a Roman numeral in the usual way.*

In compound 1, which contains a complex *cation*, the Cr^{3+} oxidation state appears as *chromium(III)*. In compound 2, the Cr^{3+} state in an overall *anionic* complex is named *chromate(III)*. For the distinction between cationic and anionic complex ions, see the next rule.

5. *A metal in an anionic complex takes the suffix* ate.

The name of the ion goes from chrom*ium* in the positively charged $[Cr(NH_3)_6]^{3+}$ complex to chrom*ate* in the negatively charged $[Cr(NH_3)_2(NCS)_4]^-$ complex.

Finally, a few more examples to bring out some of the finer points:

COMPOUND 3: $K_4[Fe(CN)_6]$, which contains the anionic $[Fe(CN)_6]^{4-}$ complex, is called *potassium hexacyanoferrate(II)*. The suffix *ate*, used in accordance with rule 5, applies not to the English form (which would become "ironate") but to the Latin root instead: ferrate. The same treatment is given copper (cuprate), gold (aurate), lead (plumbate), silver (argentate), and tin (stannate).

Note, otherwise, how the name conforms to the rules exactly as stated: (1) The cation (*potassium*) comes first. (2) A Greek prefix (*hexa*) indicates the number of ligands (six). (3) The cyan*ide* anion, CN^-, becomes the cyan*o* ligand in the complex. (4) A Roman numeral (II) designates the oxidation state. (5) The suffix *ate* shows that the metal exists inside an *anionic* complex.

COMPOUND 4: $[CoCl_2(en)_2]NO_3$, containing the bidentate ligand *ethylenediamine* (en: one *m* in *amine*, the NH_2 group). Since the name ethylenediamine already contains the Greek prefix *di*, we use the alternative form *bis* instead. For clarity, the *di*-containing name is enclosed within parentheses, resulting in the full name

dichloro*bis*(ethylenediamine)cobalt(III) nitrate

The alternative prefixes are *bis* (two times), *tris* (three times), *tetrakis* (four times), *pentakis* (five times), *hexakis* (six times), and so forth.

COMPOUND 5: $[Al(OH)_4]^-$, the *tetrahydroxoaluminate ion*. Since aluminum has only the one oxidation state Al^{3+}, we omit the Roman numeral (III) as superfluous.

COMPOUND 6: $[Fe(CO)_5]$, *pentacarbonyliron(0)*. The zero oxidation state is indicated explicitly. Also, observe that the English form *iron* is retained for the *neutral* complex in accord with rules 4 and 5.

COMPOUND 7: $K_3[Fe(CN)_6]$. The common name *potassium ferri-cyanide*, given long before the development of systematic nomenclature, still enjoys occasional use. Similar exceptions crop up for compounds of all types.

A-4. ROOT AND BRANCH: ORGANIC MOLECULES

From the Greek words for wine and wood (*methy*, *hyle*), chemists of the 19th century coined the word *methylene* (suggesting "wood spirits") and thence *methane*: CH_4, the simplest hydrocarbon.

From the Greek word *aither*, meaning "sky" (or, in another sense, a burning or glowing) came *ether* and *ethyl* and also *ethane*: C_2H_6, the next simplest hydrocarbon.

From Greek *pion* (fat) came *propionic acid* and *propane*, a hydrocarbon with three carbons: C_3H_8.

From the Latin *butyrum* (butter) came *butyric acid*, so named for its odor of rancid butter; and from butyric acid came the name of the parent hydrocarbon, *butane*: C_4H_{10}.

And so on. Unsystematically, like the elements, the organic molecules got their names for obscure reasons at first—sometimes to recall a telling property, sometimes with less cause than that. But such assignment of common names soon must break down, doomed by carbon's sheer versatility. There are too many molecules. Faced with millions upon millions of organic structures, we need a system.

We need, for each molecule, a name that sketches out both the hydrocarbon skeleton and the location of all its functional groups. Ask: Are there shoots and branches, or is there just a single, continuous chain? Are there closed rings in the structure? Are there groups other than hydrogen attached to the framework? What? Where? How many? Double bonds? Triple bonds? How shall we put everything into just a few words?

Try one and see:

$$CH_3-\overset{\overset{\displaystyle CH_3}{|}}{\underset{\underset{\displaystyle H}{|}}{C}}-CH_2-CH_2-CH_3$$

Think of the molecule as a five-unit hydrocarbon chain

$$CH_3-\overset{|}{\underset{\underset{\displaystyle H}{|}}{C}}-CH_2-CH_2-CH_3$$

to which we attach a $-CH_3$ group at the second position:

$$CH_3-\underset{\underset{H}{|}}{\overset{\overset{CH_3}{|}}{C}}-CH_2-CH_2-CH_3$$

$$1 \qquad 2 \quad 3 \qquad 4 \qquad 5$$

The name so far: "CH_3 bonded to C_5H_{11}, second carbon from the end"— accurate, but wordy.

To shorten such descriptions, we coin names first for *unbroken* hydrocarbon chains on the model of meth*ane*, eth*ane*, prop*ane*, but*ane*, using the suffix *ane* to mean "single bonds only" (alkane). After four carbons, the common names give way to the numerical Greek prefixes *penta*, *hexa*, *hepta*, and up. Hence:

methane	CH_4
ethane	CH_3CH_3
propane	$CH_3CH_2CH_3$
butane	$CH_3CH_2CH_2CH_3$
pentane	$CH_3CH_2CH_2CH_2CH_3$
hexane	$CH_3CH_2CH_2CH_2CH_2CH_3$
heptane	$CH_3CH_2CH_2CH_2CH_2CH_2CH_3$
octane	$CH_3CH_2CH_2CH_2CH_2CH_2CH_2CH_3$
nonane	$CH_3CH_2CH_2CH_2CH_2CH_2CH_2CH_2CH_3$
decane	$CH_3CH_2CH_2CH_2CH_2CH_2CH_2CH_2CH_2CH_3$

Next we decide that any alkane (CH_4, say) *minus* one hydrogen ($-CH_3$) shall be called an *alkyl* group (the suffix *ane* becomes *yl*), and so we have:

methyl	$-CH_3$
ethyl	$-CH_2CH_3$
propyl	$-CH_2CH_2CH_3$
butyl	$-CH_2CH_2CH_2CH_3$
pentyl	$-CH_2CH_2CH_2CH_2CH_3$
hexyl	$-CH_2CH_2CH_2CH_2CH_2CH_3$
heptyl	$-CH_2CH_2CH_2CH_2CH_2CH_2CH_3$
octyl	$-CH_2CH_2CH_2CH_2CH_2CH_2CH_2CH_3$
nonyl	$-CH_2CH_2CH_2CH_2CH_2CH_2CH_2CH_2CH_3$
decyl	$-CH_2CH_2CH_2CH_2CH_2CH_2CH_2CH_2CH_2CH_3$

Returning, then, to our molecule

2-methylpentane

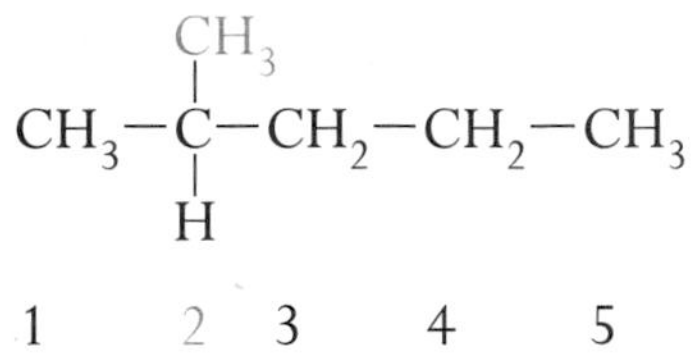

$$CH_3-\underset{\underset{\displaystyle H}{|}}{\overset{\overset{\displaystyle CH_3}{|}}{C}}-CH_2-CH_2-CH_3$$

1   2   3   4   5

we put together at last a concise, informative name: 2-methylpentane. A single $-CH_3$ group is attached to the second carbon of a five-carbon pentane chain.

And if there are two methyl groups attached at the same position, as in

2,2-dimethylpentane

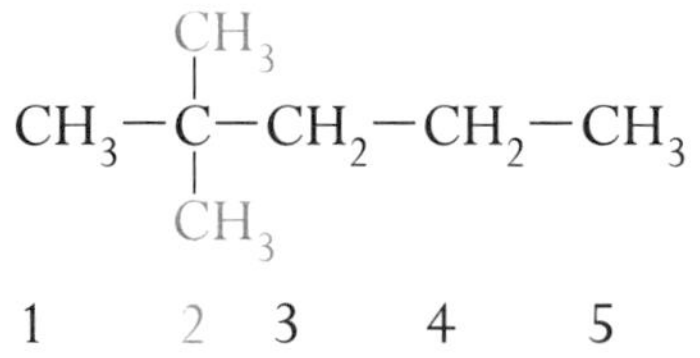

$$CH_3-\underset{\underset{\displaystyle CH_3}{|}}{\overset{\overset{\displaystyle CH_3}{|}}{C}}-CH_2-CH_2-CH_3$$

1   2   3   4   5

then we use the Greek prefix *di* to say so: 2,2-dimethylpentane.

Or if two methyl groups are arranged this way,

2,3-dimethylpentane

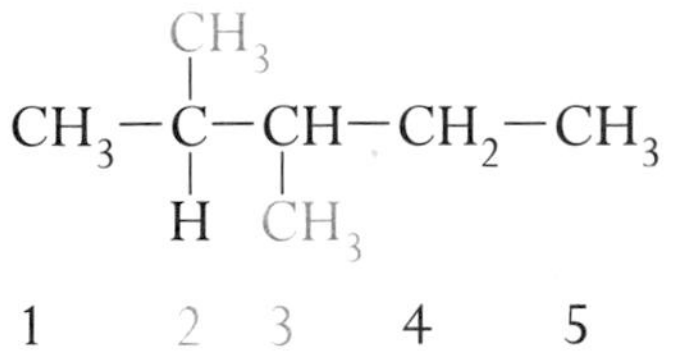

$$CH_3-\underset{\underset{\displaystyle H}{|}}{\overset{\overset{\displaystyle CH_3}{|}}{C}}-\underset{\underset{\displaystyle CH_3}{|}}{CH}-CH_2-CH_3$$

1   2   3   4   5

then the name is 2,3-dimethylpentane.

Or if this way,

2,3,4-trimethylpentane

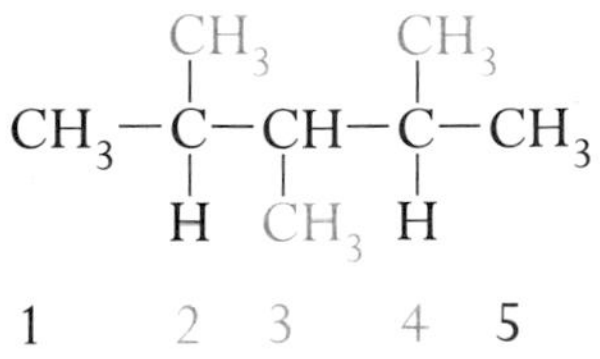

$$CH_3-\underset{\underset{\displaystyle H}{|}}{\overset{\overset{\displaystyle CH_3}{|}}{C}}-\underset{\underset{\displaystyle CH_3}{|}}{CH}-\underset{\underset{\displaystyle H}{|}}{\overset{\overset{\displaystyle CH_3}{|}}{C}}-CH_3$$

1   2   3   4   5

then we call it 2,3,4-trimethylpentane.

From there, the possibilities multiply. The same numbering system permits double and triple bonds to be assigned to precise locations, with the cis-trans notation available to specify geometric isomers. Functional groups such as −Cl (*chloro*), CH_3O− (*methoxy*), −OH (*hydroxyl* or *alcohol*), and others can be described. Rings too.

Alkanes

The rules are:

1. *Identify the longest unbroken chain, and name the molecule as the corresponding derivative*: something-something-butane for a substituted four-carbon chain, something-something-pentane for a five-carbon chain.

2. *Number the carbons in the parent chain, proceeding consecutively from one end to the other. Where there is a choice, pick the alternative that assigns the lowest numbers to the substituted positions.* Not 4-methylpentane, for example, but 2-methylpentane:

2-methylpentane (yes)

"4"-methylpentane (no)

$$CH_3-\underset{\underset{H}{|}}{\overset{\overset{CH_3}{|}}{C}}-CH_2-CH_2-CH_3 \qquad CH_3-\underset{\underset{H}{|}}{\overset{\overset{CH_3}{|}}{C}}-CH_2-CH_2-CH_3$$

1   2 3   4   5   5   4 3   2   1

3. *For each distinct alkyl group, use the appropriate Greek prefix (di, tri, . . .) to indicate the number of times it appears. List the alkyl groups in alphabetical order ("ethyl" before "methyl"), ignoring any attached prefix.*

4. *Label each alkyl branch (methyl, ethyl, . . .) by its position on the main chain, repeating the number if two or more groups are attached to the same carbon:* 2,2-dimethylpentane. Note, additionally, the names of the *branched* isopropyl and *tertiary*-butyl (*tert*-butyl) groups among the first several:

$$CH_3-\underset{\underset{H}{|}}{\overset{|}{C}}-CH_3 \qquad\qquad CH_3-\underset{\underset{CH_3}{|}}{\overset{|}{C}}-CH_3$$

isopropyl

tert-butyl

Alkenes and Alkynes

From eth*ene* (CH_2=CH_2; common name, ethylene) we extract the ending *ene* to mean "C=C double bond." Past ethene, the smallest of the
alkenes, there is first propene (CH_3—CH=CH_2, propylene), then
butene (butylene), pentene, hexene, and more.

Take, as an example, the molecule

4-ethyl-*trans*-2-heptene

$$CH_3CH_2CH_2-\overset{\overset{\displaystyle CH_2CH_3}{|}}{CH}\quad\overset{}{\underset{\overset{C=C}{\diagup\ \ \diagdown}}{}}\quad\begin{matrix}H\\\\CH_3\end{matrix}$$

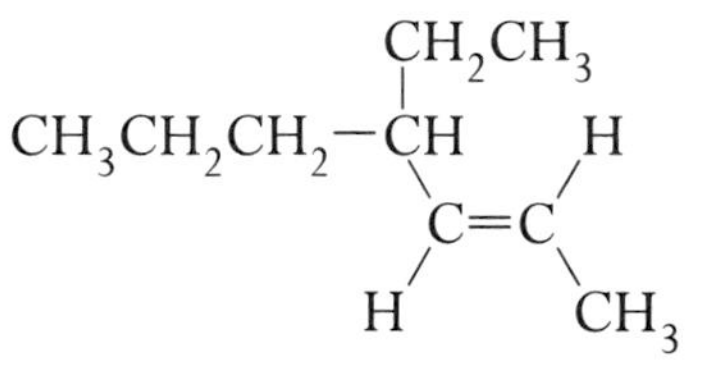

for which we derive the name 4-ethyl-*trans*-2-heptene by following a
procedure similar to the steps taken for the alkanes:

1. *Identify the longest unbroken chain containing the double bond, and
   change the* ane *suffix of the original alkane to* ene. *For a chain of
   seven carbons, the parent alkene is* heptene:

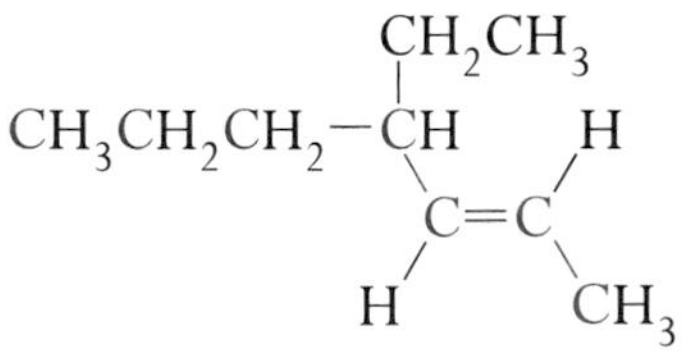

$$CH_3CH_2CH_2-\overset{\overset{\displaystyle CH_2CH_3}{|}}{CH}\quad H$$
$$C=C$$
$$H\qquad CH_3$$

2. *Number the chain so that the double bond appears closer to carbon 1,
   and label the* ene *position accordingly.*

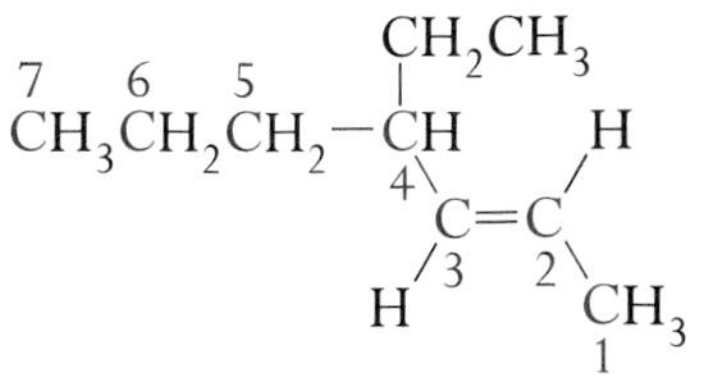

$$\overset{7}{C}H_3\overset{6}{C}H_2\overset{5}{C}H_2-\overset{\overset{\displaystyle CH_2CH_3}{|}}{\underset{4}{C}H}\quad H$$
$$\underset{3}{C}=\underset{2}{C}$$
$$H\qquad \underset{1}{C}H_3$$

   The name so far: 2-heptene, a onetime heptane molecule with a
   double bond now between carbons 2 and 3.

3. *Specify a geometric isomer, if appropriate.* Here the molecule displays

the *trans* configuration, showing the two alkyl groups on opposite sides of the double bond: *trans*-2-heptene.

4. *Identify and number the alkyl substituents in standard fashion, employing numerical prefixes where necessary*: 4-ethyl-*trans*-2-heptene.

For structures with other features, we have additional rules:

5. *Molecules containing more than one double bond* (polyenes) *are labeled with Greek prefixes* (diene, triene, . . .) according to the model

$$CH_2=C=CH-CH_3 \qquad 1,2\text{-butadiene}$$
$$CH_2=CH-CH=CH_2 \qquad 1,3\text{-butadiene}$$
$$CH_2=CH-CH_2-CH=CH_2 \qquad 1,4\text{-pentadiene}$$

Cis and trans configurations are specified in the usual way, as are alkyl substituents.

6. *Alkynes are similarly named, with the ending* yne *used to indicate a triple bond*. The series runs as ethyne (HC≡CH, acetylene), propyne, butyne, and so forth:

$$HC≡CH \qquad \text{ethyne}$$
$$HC≡C-CH_3 \qquad \text{propyne}$$
$$HC≡C-CH_2-CH_3 \qquad 1\text{-butyne}$$
$$CH_3-C≡C-CH_3 \qquad 2\text{-butyne}$$

Alcohols and Esters

Of the many other functional groups, we conclude with just two—alcohols and esters—and even then, only briefly.

To name an organic alcohol ($R-OH$), modify the *ane* suffix of the alkane to incorporate the *ol* suffix of the alcohol:

CH_4 (methane) becomes CH_3OH (methanol or methyl alcohol).

CH_3CH_3 (ethane) becomes CH_3CH_2OH (ethanol or ethyl alcohol).

$CH_3CH_2CH_3$ (propane) becomes $CH_3CH_2CH_2OH$ (1-propanol) or $CH_3CHOHCH_3$ (2-propanol).

An ester ($R-COO-R'$), formed by the condensation of an alcohol ($R'-OH$) and a carboxylic acid ($R-COOH$), then takes its first name from the alcohol and its second name from the acid (with the suffix *ate*):

$$CH_3-\overset{\displaystyle O}{\overset{\|}{C}}-OH \; + \; HO-CH_2CH_3 \; \overset{H_2O}{\longrightarrow} \; CH_3-\overset{\displaystyle O}{\overset{\|}{C}}-O-CH_2CH_3$$

acetic acid   ethyl alcohol   ethyl acetate

A-5. LAST WORDS: TOWARD A CHEMICAL VOCABULARY

The names of the compounds, like nouns in any language, are not enough. A name says only that something *is*, not what it does and how it does it. For that, we need verbs. We need adverbs. We need adjectives. We need a big language to describe a small world.

More than big, though, we need a new, artificial, *planned* language to depict a world alien to the senses—a foreign language, a language that sinks in only by repeated usage, by habit, by practice so persistent that the unfamiliar finally becomes familiar.

Unfamiliar and big it surely is. Borrowed mostly from ancient Greek and Latin, the thousand words defined in the Glossary (Appendix D) present only a fraction of the arcane, polysyllabic language of matter and energy. Even then, the list exceeds the vocabulary most people use in ordinary conversation.

But why quibble? We learn the language of chemistry as we would any other speech, just as a child does: by hearing it, repeating it, using it; and in the end, without conscious effort, committing it to memory. Yet here, far better, we have a language designed expressly for one goal alone, and the deliberate effort to be reasonable shines through. There is a plan and a structure to the words. The scientific vernacular suffers far less from the usual hodgepodge of old misunderstandings, mistakes, and inconsistencies that confound all other languages. It makes sense, usually.

See the logic in a word like *hydrolysis*, coined knowingly from the Greek words *hydor* ("water") and *lysis* ("a breaking down, loosening"): the reaction of a substance with water.

See next how *hydrolysis* is joined by *photolysis*, the cleavage of a substance by light (*photos*) . . . and by *electrolysis*, describing a reaction driven by an electric current . . . and by *catalysis*, and *analysis*, and all the other *lysis* words that mean "to break down or dissolve."

See how the *photo* in photolysis then brings its "light" to *photocurrent* (where *current* comes from the Latin *currere*, "to run") . . . and also to *photograph* (where *graph* is Greek for "writing") . . . and *photosynthesis* (to "put together" with light, Greek *synthesis*).

And there, not so strange after all, lies the key to the language: scholarly awareness of its interchangeable roots, prefixes, and suffixes, taken mostly from the Greek and Latin, each readily understood in combination. Stripped to the roots, the vocabulary is smaller than it seems.

As an aid to understanding, consult either the Glossary or a comprehensive dictionary for the etymological details, as in:

ETYMOLOGY: the history or derivation of a word. Introduced into English in approximately 1350 to 1400 A.D. from the Latin *etymologia*, originally from the Greek: *etymos* (true, real, actual; hence a *true* meaning) + *logos* (word, reason).

EXERCISES

1. Write empirical formulas for the following compounds: (a) Barium bromide. (b) Zinc carbonate. (c) Sodium phosphate. (d) Potassium sulfate. (e) Calcium hydride. (f) Lithium hydroxide.

2. Write empirical formulas for the following compounds: (a) Iron(II) sulfide. (b) Iron(III) sulfide. (c) Iron(II) sulfate. (d) Iron(III) sulfate. (e) Iron(II) sulfite.

3. Write empirical formulas for the following compounds: (a) Beryllium chloride. (b) Beryllium nitride. (c) Lead iodide. (d) Lead iodate. (e) Cadmium phosphate. (f) Lithium nitrate.

4. Write empirical formulas for the following compounds: (a) Cobalt(III) oxide. (b) Cobalt(II) fluoride. (c) Cobalt(III) fluoride. (d) Sodium chromate. (e) Magnesium acetate.

5. Name the following compounds: (a) CdSe. (b) $Ca(ClO_3)_2$. (c) $Ca(ClO_4)_2$. (d) Pr_2S_3. (e) $Pr_2(SO_4)_3$.

6. Name the following compounds: (a) $Zn(CN)_2$. (b) $Zn(IO_3)_2$. (c) ZnTe. (d) $RbMnO_4$. (e) $Bi(C_2H_3O_2)_2$. (f) NH_4N_3.

7. Name the following compounds: (a) $Al(OH)_3$. (b) AlN. (c) NH_4Br. (d) FeF_2. (e) FeF_3. (f) Cr_2O_3.

8. Name the following compounds: (a) $OsSO_3$. (b) Tl_2SO_4. (c) $TlHSO_4$. (d) Tl_2S. (e) Tl_2S_3. (f) KSCN.

9. Write empirical or molecular formulas: (a) Uranium hexafluoride. (b) Dichlorine monoxide. (c) Sodium hypochlorite. (d) Xenon trioxide. (e) Osmium tetrasulfide.

10. Write empirical or molecular formulas: (a) Mercurous chloride. (b) Mercuric chloride. (c) Cuprous chloride. (d) Cupric chloride. (e) Ferrous bromide. (f) Ferric bromide.

11. Name the following compounds: (a) AsF_3. (b) AsF_5. (c) Si_2Br_6. (d) Cu_2O. (e) CuO.

12. Name the following compounds: (a) $NaNO_2$. (b) $NaNO_3$. (c) $NaClO$. (d) $NaClO_2$. (e) Na_2CO_3. (f) $NaHCO_3$.

13. Write formulas for the following acids: (a) Phosphoric acid. (b) Phosphorous acid. (c) Hypobromous acid. (d) Hydrochloric acid. (e) Hydrocyanic acid.

14. Write formulas for the following acids: (a) Carbonic acid. (b) Hydrofluoric acid. (c) Hydrobromic acid. (d) Chloric acid. (e) Perchloric acid.

15. Name the following acids: (a) HNO_2. (b) HNO_3. (c) H_2S. (d) H_2SO_3. (e) H_2SO_4.

16. Name the following acids: (a) HI. (b) HIO. (c) HIO_2 (a hypothetical substance). (d) HIO_3. (e) HIO_4. (f) H_3AsO_4.

17. Write formulas for the following coordination compounds:

    (a) potassium hexacyanoferrate(II)
    (b) potassium hexacyanoferrate(III)
    (c) tris(ethylenediamine)platinum(IV) bromide
    (d) triaquabromoplatinum(II) chloride

18. Write formulas for the following coordination compounds and complex ions:

    (a) hexaamminechromium(III) nitrate
    (b) aquapentachlororuthenate(III) ion
    (c) tetracyanonickelate(II) ion
    (d) pentaamminesulfatocobalt(III) chloride

19. Name the following coordination compounds:

    (a) $Na[Ru(H_2O)_2(C_2O_4)_2]$
    (b) $[Co(NH_3)_4(H_2O)CN]Cl_2$
    (c) $[Ni(CO)_6]Br_2$
    (d) $K_2[CuCl_4]$

20. Name the following complex ions:

    (a) $[Co(NH_3)_6]^{3+}$
    (b) $[Co(NH_3)_4Cl_2]^+$
    (c) $[FeCl_4]^-$
    (d) $[Pt(en)_3]^{4+}$

21. Write the structural formula for each organic compound:

    (a) 3-bromo-2-methyloctane
    (b) 3-bromo-2-methyl-4-octyne
    (c) *cis*-4-octene
    (d) *trans*-4-octene

22. Write the structural formula for each organic compound:

    (a) 1,1,1-trichloroethane
    (b) 1,1-dichloroethene
    (c) dichloroethyne
    (d) 2,3,5,5,6-pentamethyldecane

23. Write the structural formula for each organic compound:

    (a) tetraiodomethane
    (b) *meta*-dibromobenzene
    (c) *para*-dibromobenzene
    (d) 3-hexanol

24. Name the following alkanes:

$$
\text{(a)} \quad \underset{}{CH_3-CH_2-CH_2-CH_2-\overset{\overset{\displaystyle CH_3}{|}}{CH}-CH_2-CH_2-CH_3}
$$

$$
\text{(b)} \quad CH_3-\overset{\overset{\displaystyle CH_3}{|}}{CH}-CH_2-CH_2-\overset{\overset{\displaystyle CH_3}{|}}{CH}-CH_2-CH_2-CH_3
$$

$$
\text{(c)} \quad CH_3-\overset{\overset{\displaystyle CH_3}{|}}{CH}-CH_2-CH_2-\overset{\overset{\displaystyle Cl}{|}}{CH}-CH_2-\overset{\overset{\displaystyle CH_3}{|}}{\underset{\underset{\displaystyle CH_3}{|}}{C}}-CH_3
$$

25. Name the following alkanes and alkenes:

(a) 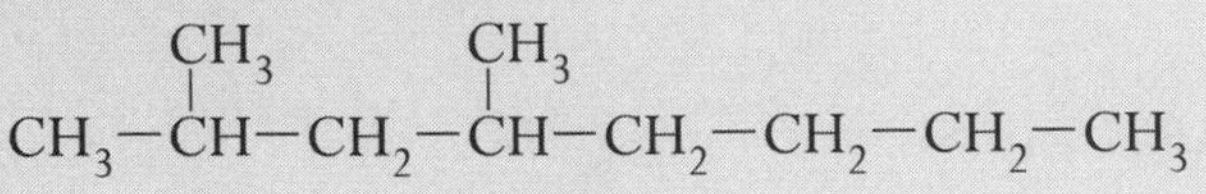 $CH_3-\overset{\overset{\displaystyle CH_3}{|}}{CH}-CH_2-\overset{\overset{\displaystyle CH_3}{|}}{CH}-CH_2-CH_2-CH_2-CH_3$

(b) 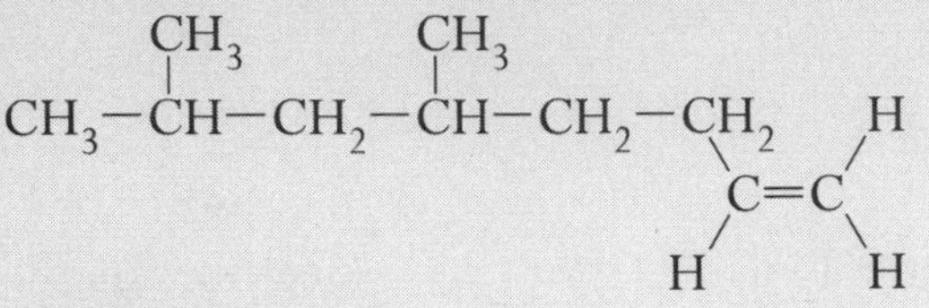 $CH_3-\overset{\overset{\displaystyle CH_3}{|}}{CH}-CH_2-\overset{\overset{\displaystyle CH_3}{|}}{CH}-CH_2-CH_2$ $\underset{\underset{\displaystyle H}{}}{}C=C\underset{\underset{\displaystyle H}{}}{\overset{\overset{\displaystyle H}{}}{}}$

(c) $CH_3-\overset{\overset{\displaystyle CH_3}{|}}{CH}-CH_2-\overset{\overset{\displaystyle CH_3}{|}}{CH}-CH_2-CH_2$ $\underset{\underset{\displaystyle H}{}}{}C=C\underset{\underset{\displaystyle CH_3}{}}{\overset{\overset{\displaystyle H}{}}{}}$

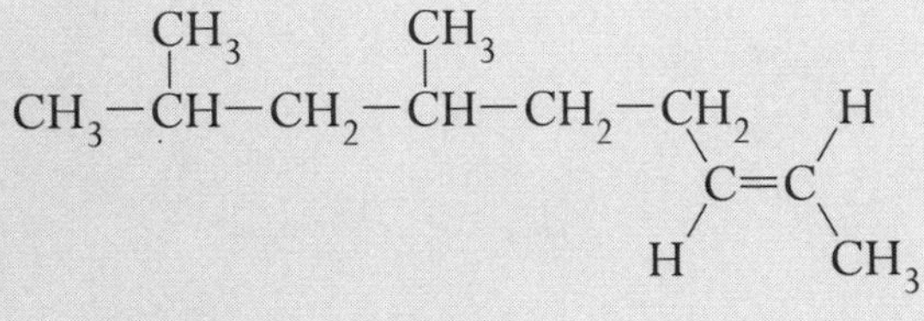

(d) 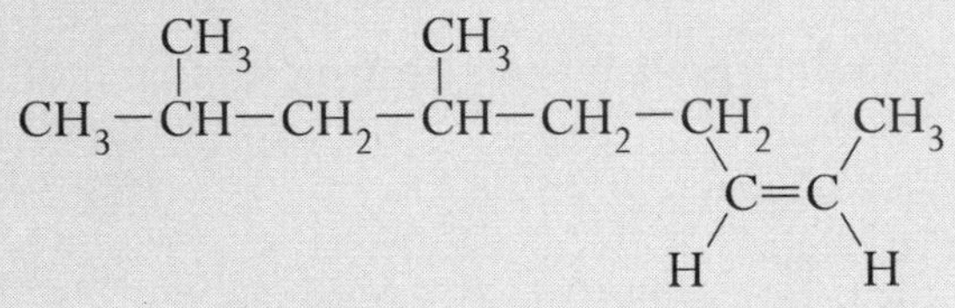 $CH_3-\overset{\overset{\displaystyle CH_3}{|}}{CH}-CH_2-\overset{\overset{\displaystyle CH_3}{|}}{CH}-CH_2-CH_2$ $\underset{\underset{\displaystyle H}{}}{}C=C\underset{\underset{\displaystyle H}{}}{\overset{\overset{\displaystyle CH_3}{}}{}}$

26. Name the following alcohols:

(a) $CH_3CH_2CH_2CH_2OH$

(b) $CH_3CHOHCH_2CH_3$

(c) CH_3OH

(d) 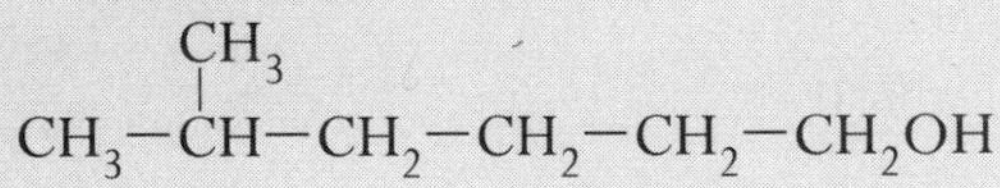 $CH_3-\overset{\overset{\displaystyle CH_3}{|}}{CH}-CH_2-CH_2-CH_2-CH_2OH$

(e) 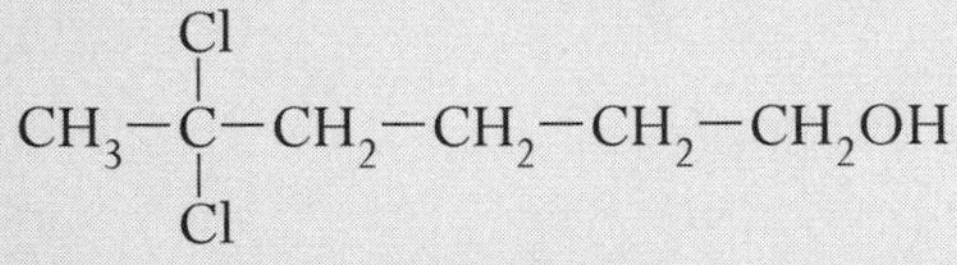 $CH_3-\overset{\overset{\displaystyle Cl}{|}}{\underset{\underset{\displaystyle Cl}{|}}{C}}-CH_2-CH_2-CH_2-CH_2OH$

APPENDIX B

Pertinent Mathematics

B-1. Powers and Logarithms
Scientific Notation
Common Logarithms: Definition and Properties
Evaluation of Common Logarithms
Natural Logarithms
B-2. Functions and Graphs
B-3. Quadratic Equations
EXERCISES

Better to solve an equation such as

$$6x^2 - x - 2 = 0$$

than to ask:

> "What number, which, multiplied by itself to give a second number, which, added to itself six times to give a third number which, diminished by the original number to give a fourth number which, diminished by two, is equal to zero?"

Where words fail, mathematics speaks.

It speaks in a language of sparse vocabulary and grammar, a language all the more powerful for its underlying simplicity. Simple, yes—because much of mathematics is only arithmetic in disguise; and, simpler still, because all of arithmetic is just *addition* in disguise. From $1 + 1 = 2$, a modest beginning, comes everything else: subtraction (addition in reverse), multiplication (a shorthand addition), division (a shorthand subtraction), powers, logarithms, functions, equations, calculus . . . numerical words made into quantitative thoughts, strung together step by step. With $1 + 1 = 2$, we take the first step toward grasping the mathematical precision of the universe.

B-1. POWERS AND LOGARITHMS

Begin with addition. First, combine some quantity b with itself to double the amount:

$$b + b = 2b$$

Then do it three times,

$$b + b + b = 3b$$

four times,

$$b + b + b + b = 4b$$

and, in general, n times,

$$b + \cdots + b = nb$$

to discover the meaning of *multiplication*: repeated addition, $b + b + b + b$. To multiply b by n ("n times b") is to add the quantity b to itself n times. Multiplication is a shortcut for addition.

One shortcut then leads to another, as we find upon adding b to itself b times to create the *square* of b:

$$\overset{b \text{ times}}{b + \cdots + b} = b \times b = b^2$$

We say, by way of definition, that b denotes the **base** and 2 denotes the **exponent** (or **power**) of the number b^2; but our understanding of b-squared soon will blossom into far more than just a new notation.

As a start, realize that the term *square* derives from a persuasive geometric analogy, an analogy best appreciated when the problem is made concrete. To do so we set b equal to 10,

$$10 + 10 + 10 + 10 + 10 + 10 + 10 + 10 + 10 + 10 = 10 \times 10$$

and watch how a line in one dimension becomes a square in two dimensions:

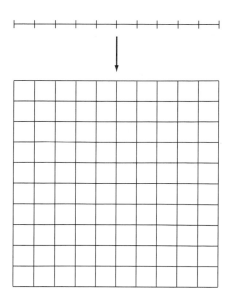

The line, 10 units in a row, expands into a two-dimensional grid of 100 squares, each extending 1 unit on a side. Our notation

$$10^2 = 10^1 \times 10^1 = 100$$

then reminds us that 100 is the product of two factors of 10. *One* power of 10 (i.e., 10^1) means one factor of 10; *two* powers of 10 (i.e., 10^2) means two factors.

Three powers? Keep going. Attach another factor, and the square turns into a cube

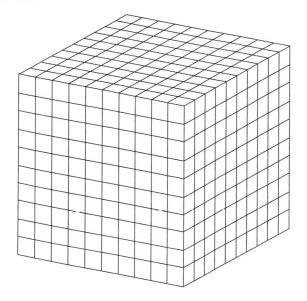

as each of the 100 squares (1 × 1) grows vertically into a line of 10 cubes (1 × 1 × 1):

$$10^3 = 10 \times 10 \times 10 = 1000$$

A pattern emerges: A point becomes a line; one power of 10. A line becomes a square; two powers. A square becomes a cube; three powers. Each power of 10 corresponds to one factor in the product.

   Beyond three powers, of course, a cube has nowhere to grow in a concrete three-dimensional world, but the expression

$$10^n = 10 \times 10 \times \overset{n \text{ times}}{\cdots} \times 10$$

retains its arithmetic meaning nevertheless. The exponent, n, specifies the number of factors in the product; and so for four powers, five, and more, we simply continue the geometric progression one factor at a time:

$$10^1 = 10 = 10$$

$$10^2 = 10 \times 10 = 100$$

$$10^3 = 10 \times 10 \times 10 = 1000$$

$$10^4 = 10 \times 10 \times 10 \times 10 = 10{,}000$$

$$10^5 = 10 \times 10 \times 10 \times 10 \times 10 = 100{,}000$$

And look: For every product, the combined exponent is equal to the sum of the exponents from all the factors. We find, for example, that

$$10^1 \times 10^1 \times 10^1 = 10^{(1+1+1)} = 10^3$$

just as

$$10^1 \times 10^2 = 10^{(1+2)} = 10^3$$

just as

$$10 \times 10{,}000 = 10^1 \times 10^4 = 10^{(1+4)} = 10^5 = 100{,}000$$

and just as

$$10^n \times 10^m = 10^{(n+m)}$$

When powers of 10 are multiplied, the exponents *add*.

Not only 10, moreover, but any other base obeys the same rule,

$$b^n \times b^m = b^{(n+m)}$$

giving us such direct combinations as (for $b = 2$):

$$2^3 = 2 \times 2 \times 2 = 8$$

$$2^2 = 2 \times 2 = 4$$

$$2^3 \times 2^2 = 2^5 = 32$$

The rule applies also to a nonintegral base (such as $b = 2.7$),

$$2.7 \times 2.7^2 = 2.7^3 = 2.7 \times 2.7 \times 2.7 = 19.683$$

although not to a mixture of different bases (such as $b_1 = 2, b_2 = 3$):

$$2^3 \times 3^2 = 8 \times 9 = 72$$

The pattern for multiplication holds only when the base factors are identical. When they are, the exponents add.

And when bases are divided? Again, look and see:

$$\frac{100{,}000}{1000} = \frac{10^5}{10^3} = \frac{10 \times \cancel{10} \times \cancel{10} \times \cancel{10} \times 10}{\cancel{10} \times \cancel{10} \times \cancel{10}} = 10^2 = 100$$

Each factor on the bottom cancels one factor above to leave only the difference between the two,

$$\frac{10^5}{10^3} = 10^{(5-3)} = 10^2$$

and thence follows the general rule: Exponents *subtract* when identical bases are divided:

$$\frac{10^n}{10^m} = 10^{(n-m)}$$

$$\frac{b^n}{b^m} = b^{(n-m)}$$

The power of the denominator is taken away from the power of the numerator.

We discover, by the same operation, the meaning of a *zero* exponent, which appears whenever a number is divided by itself—an action for which there can be only one outcome: 1.

$$\frac{b^n}{b^n} = b^{(n-n)} = b^0 = 1$$

Any number raised to the zero power therefore has the value *one*, always the same, regardless of the base. 2^0? Answer: 1. 10^0? Answer: 1. 1000^0? Answer: 1. $(-435{,}324{,}553)^0$? Answer: 1.

We discern, too, the meaning of a *negative* exponent, which arises whenever the number b^0 is divided by the number b^n:

$$b^{-n} = \frac{b^0}{b^n} = \frac{1}{b^n}$$

The minus sign thus is a signal to form the reciprocal of a quantity, just as we do in our representations of certain units:

$$g\ mol^{-1} = \frac{g}{mol}$$

Conclusion: Any expression b^n multiplied by its reciprocal, b^{-n}, gives b^0, *one*.

Put together these positive, negative, and zero exponents, and our earlier table expands now to include all integral powers of 10:

$$10^{-2} = \frac{1}{10^2} = 0.01$$

$$10^{-1} = \frac{1}{10^1} = 0.1$$

$$10^0 = 1$$

$$10^1 = 10$$

$$10^2 = 100$$

But still there is more to do, for we have yet to consider the meaning of *nonintegral* powers in expressions such as $10^{1/2}$ or $3^{1/3}$ or $2.7^{-0.368}$. How shall we account for them?

In the same way as before: by enforcing the rule that exponents add when bases are multiplied. If, say, $10^{1/2}$ is multiplied by $10^{1/2}$, for consistency we demand that

$$10^{1/2} \times 10^{1/2} = 10^{(1/2+1/2)} = 10^1$$

$10^{1/2}$ is therefore the square root of 10—that number which, multiplied by itself, yields 10.

By the same reasoning, $3^{1/3}$ is the cube root of 3:

$$3^{1/3} \times 3^{1/3} \times 3^{1/3} = 3^{(1/3+1/3+1/3)} = 3^1$$

and $b^{1/n}$ is the *n*th root of *b*:

$$\overbrace{b^{1/n} \times b^{1/n} \times \cdots \times b^{1/n}}^{n \text{ times}} = b^{(\overbrace{1/n+1/n+\cdots+1/n}^{n \text{ times}})} = b^1$$

So it goes for all fractional exponents, positive and negative.

Observe, as well, what happens in expressions of the kind

$$(10^{1/2})^2 = (10^{1/2})(10^{1/2}) = 10^{(1/2+1/2)} = 10^{(2\times1/2)} = 10^1$$

$$(3^{1/3})^3 = 3^{(3\times1/3)} = 3^1$$

$$(2^2)^3 = 2^6 = 64$$

where an exponentiated base (b^n) is itself raised to a power (m). The combined exponent is equal to the product of the two powers:

$$(b^n)^m = b^{(n\times m)}$$

Such are the facts of exponents. Next, see what advantages they offer.

Scientific Notation

Without exponents, even the simplest chemical calculation would be awash in zeros—cluttered equally with small numbers like

$e = 0.0000000000000000001602$ C (the charge on an electron or proton)

and large ones like

$$N_0 = 602{,}200{,}000{,}000{,}000{,}000{,}000{,}000 \text{ (number of particles per mole)}$$

That, after all, is the way of the microworld: tiny particles with tiny masses and tiny charges, but present always in huge numbers.

Our recourse is to use an exponent, which collects all the zeros into a tidy power of 10. We can, for instance, compute the Faraday constant,

$$\mathcal{F} = N_0 e \qquad \text{(charge per mole of electrons)}$$

by multiplying the values N_0 and e directly:

$$\mathcal{F} = 602{,}200{,}000{,}000{,}000{,}000{,}000{,}000 \text{ mol}^{-1}$$
$$\times\ 0.0000000000000000001602 \text{ C}$$

Far easier, however, would be to represent the one number as

$$N_0 = 6.022 \times 100{,}000{,}000{,}000{,}000{,}000{,}000{,}000 = 6.022 \times 10^{23} \text{ mol}^{-1}$$

and the other as

$$e = 1.602 \times 0.0000000000000000001 = 1.602 \times 10^{-19} \text{ C}$$

so that we have only a straightforward combination of exponents:

$$N_0 e = (6.022 \times 10^{23} \text{ mol}^{-1})(1.602 \times 10^{-19} \text{ C})$$

$$= (6.022 \times 1.602)(10^{23} \times 10^{-19}) \text{ C mol}^{-1}$$

$$= 9.647 \times 10^{(23-19)} \text{ C mol}^{-1}$$

$$= 9.647 \times 10^{4} \text{ C mol}^{-1} \equiv 96{,}470 \text{ C mol}^{-1}$$

To do so is to use **scientific notation**, a method whereby we represent numbers as powers of 10, a great convenience.

The rules governing scientific notation follow directly from the properties of exponents, noted above. For any quantity, large or small, we write the value as a signed decimal portion times the appropriate power of 10. The decimal portion should lie between 1 and 10 and should contain all the number's significant figures—and no more. The exponent will then be a positive integer (≥ 1), provided that the absolute value of the number is greater than 10.

Here are two examples:

1. $12{,}345 = 1.2345 \times 10{,}000 = 1.2345 \times 10^4$
2. $-567 = -5.67 \times 100 \quad = -5.67 \times 10^2$

In each, the exponent (*n*) tells us to move the decimal point *n* places to the right, thus:

1. $1.2345 \times 10^4 \equiv 1.2345 \equiv 12{,}345$

2. $-5.67 \times 10^2 \equiv -5.67 \equiv -567$

For numbers with absolute values less than 1, we use a *negative* exponent instead (always an integer) and move the decimal to the *left*. Again, two examples:

1. $0.0012345 \quad = 1.2345 \times 0.001 = 1.2345 \times 10^{-3}$
    $1.2345 \times 10^{-3} \equiv \quad 001.2345 \quad \equiv 0.0012345$

2. $-0.0000567 = -5.67 \times 0.00001 = -5.67 \times 10^{-5}$
    $-5.67 \times 10^{-5} \equiv \quad -00005.67 \quad \equiv -0.0000567$

With each number expressed in the form

$$D \times 10^n$$

we can now handle the decimal portions (*D*) separately from the powers of 10 while applying exactly the same operations to each. The decimal portions are added, subtracted, multiplied, divided, and raised to powers as need be. Likewise, in identical fashion, the factors 10^n receive a matching treatment.

For that, we use the procedures already developed for exponents in general. Where factors are multiplied, exponents are added. Where factors are divided, exponents are subtracted. Where factors are raised to a power, exponents are multiplied.

One caution, though: When we add or subtract numbers by hand (forgoing calculator or computer), the powers of 10 must match. If not—as in a sum such as

$$(1.2345 \times 10^4) + (-5.67 \times 10^2)$$

where one power (10^4) combines with another (10^2)—we shift the decimal points as required:

$$(1.2345 \times 10^4) - (5.67 \times 10^2)$$

$$= (123.45 \times 10^2) - (5.67 \times 10^2)$$

$$= 117.78 \times 10^2 = 1.1778 \times 10^4$$

Or (same thing):

$$(1.2345 \times 10^4) - (5.67 \times 10^2)$$

$$= (1.2345 \times 10^4) - (0.0567 \times 10^4) = 1.1778 \times 10^4$$

The usual rules for significant figures apply.

Common Logarithms: Definition and Properties

We lose nothing (and gain much) by going from this,

$$1,000,000,000,000,000 \times 10,000,000,000$$

$$= 10,000,000,000,000,000,000,000,000$$

to this:

$$10^{15} \times 10^{10} = 10^{(15+10)} = 10^{25}$$

The arithmetic is the same, but the zeros are gone.

The next step, even more economical, is to strip away the bases (the tens) and pose the problem as

$$15 + 10 = 25$$

so that we consider only the exponents. The bases, which never change, tell us nothing new. Everything of importance is in the exponents themselves.

Let us define, accordingly, the base-10 logarithm (the *common* logarithm) of any number A through the equation

$$A = 10^{\log A}$$

and agree: log A is the *exponent*, or **logarithm**, needed to convert the base (10) into the specified number A, the so-called **antilogarithm**. Given A, we determine log A (usually with the help of a calculator or published table). Given log A, we determine the antilogarithm, A, by reversing the procedure; we *raise* 10 to the power log A to compute A.

A common logarithm, so defined, thus has all the properties of an exponent. The values *add* when the underlying numbers are multiplied:

$$\log AB = \log A + \log B$$

They *subtract* when the numbers are divided:

$$\log\left(\frac{A}{B}\right) = \log A - \log B$$

They *scale* when the numbers are raised to a power:

$$\log(A^n) = n \log A$$

Whatever we do with exponents, we do with logarithms.

A negative number, note, has no logarithm in the real number system, because no real exponent can raise the base 10 to a negative value. Zero, on the other hand, does possess a logarithm ($-\infty$) in the sense of a limiting quantity:

$$10^{-\infty} \rightarrow 0 \qquad \log 0 \rightarrow -\infty$$

Evaluation of Common Logarithms

If A happens to be an exact power of 10, then log A is simply the corresponding exponent, an integer:

A	$10^{\log A}$	$\log A$
0.01	10^{-2}	-2
0.1	10^{-1}	-1
1	10^{0}	0
10	10^{1}	1
100	10^{2}	2

For all points in between, log A is a decimal number.

Try one: the logarithm of 3. See first, from the table above, that the value of log 3 lies between log 1 (value = 0) and log 10 (value = 1)— somewhere near 0.5, presumably, since 3 is "almost" the square root of 10 (note that $10^{1/2} = 3.16$). If 3 were exactly the square root of 10, then its logarithm would be 0.5. The actual value is 0.4771, a little less.

Try another: the logarithm of 30, which falls between log 10 (value = 1) and log 100 (value = 2). Here we write the number as

$$A = 3 \times 10^1$$

and apply the rule for multiplication: The exponent (logarithm) of a product is equal to the sum of the exponents (logarithms) from each factor. The result is

$$\log 30 = \log(3 \times 10^1) = \log 3 + \log 10^1 = 0.4771 + 1 = 1.4771$$

The logarithm of 300? Answer: 2.4771, following precisely the same reasoning. The logarithm of 3000? Answer: 3.4771. The logarithm of 3×10^{9999}? Answer: 9999.4771.

Go the other way now, and compute

$$\log 0.3 = \log \frac{3}{10} = \log 3 - \log 10 = 0.4771 - 1 = -0.5229$$

using the rule for division: The exponent (logarithm) of a quotient is equal to the difference of the exponents (logarithms) from each factor.

Do it again for log 0.03,

$$\log 0.03 = \log \frac{3}{100} = \log 3 - \log 100 = 0.4771 - 2 = -1.5229$$

and also for log 0.003,

$$\log 0.003 = \log \frac{3}{1000} = \log 3 - \log 1000 = 0.4771 - 3 = -2.5229$$

and then consider what we have:

A	$\log A$	A	$\log A$
0.001	−3.0000	1	0.0000
0.003	−2.5229	3	0.4771
0.01	−2.0000	10	1.0000
0.03	−1.5229	30	1.4771
0.1	−1.0000	100	2.0000
0.3	−0.5229	300	2.4771
1	0.0000	1000	3.0000

All at once, the pieces start to fall into place.

See why. Since the decimal portions of the logarithms are identical between one power of 10 and the next, we need only tabulate values over the range $1 \leq A < 10$. For most applications, too, logarithms of just the first few prime numbers

$$\log 2 = 0.3010$$

$$\log 3 = 0.4771$$

$$\log 5 = 0.6990$$

will take us a long way, appearing over and over again as factors in larger numbers. Evaluation of log 540,000, for example, unfolds directly as

$$\log 540{,}000 = \log(2 \times 3^3 \times 10^4)$$

$$= \log 2 + \log(3^3) + \log(10^4)$$

$$= \log 2 + 3 \log 3 + \log(10^4) = 5.7324$$

with no further assistance from computer, calculator, or table. The result is consistent with a rough guess of 5.7, the value corresponding to log 500,000.

Natural Logarithms

Take, finally, a broader view of a logarithm, expanded to include bases other than 10:

$$A = b^{\log_b A}$$

In this more explicit notation, $\log_b A$ is the exponent needed to transform the base b into the number A. When $b = 10$, the exponent reduces to the common logarithm, $\log_{10} A \equiv \log A$.

Of particular interest is the ***natural logarithm***

$$\log_e A \equiv \ln A$$

defined for the special base

$$e = 2.718281828459045 \ldots$$

Interpret it simply to indicate

$$A = e^{\ln A}$$

or, equivalently, the same operation represented by the symbol *exp*:

$$A = \exp(\ln A)$$

Meaning: Whatever power is attached to the base e becomes, without question, the natural logarithm of the exponentiated quantity, whether the exponent be an algebraic expression or actual number. We know at once, for example, that

$$\text{if } A = \exp(-kt), \qquad \text{then } \ln A = -kt$$

$$\text{if } A = \exp(x^2), \qquad \text{then } \ln A = x^2$$

$$\text{if } A = \exp(-2.34), \quad \text{then } \ln A = -2.34$$

and also that

$$\text{if } \ln A = -kt, \qquad \text{then } A = \exp(-kt)$$

$$\text{if } \ln A = x^2, \qquad \text{then } A = \exp(x^2)$$

$$\text{if } \ln A = -2.34, \quad \text{then } A = \exp(-2.34)$$

See an e; pick off a natural logarithm. One goes with the other.

Good enough, then, but why use this peculiar "e" of all numbers? Again, for convenience: The base e, which arises explicitly in integral calculus, acts in some ways as a more natural base than 10. Nevertheless, the mathematical properties are the same as for the common, base-10 logarithm:

1. $\ln AB = \ln A + \ln B$

2. $\ln\left(\dfrac{A}{B}\right) = \ln A - \ln B$

3. $\ln(A^n) = n \ln A$

Whatever the base, an exponent is an exponent.

Recognize, in particular, that natural and common logarithms obey the relationship

$$\ln A = 2.303 \log A$$

since

$$\exp 2.303 = 10 \qquad \text{and} \qquad \ln 10 = 2.303$$

Key values of the natural logarithm are listed in the table below:

A		$\ln A$
$\exp(-3) =$	0.0498	-3
$\exp(-2) =$	0.1353	-2
$\exp(-1) =$	0.3679	-1
$\exp 0 =$	1.0000	0
$\exp 1 =$	2.7183	1
$\exp 2 =$	7.3891	2
$\exp 3 =$	20.0855	3

B-2. FUNCTIONS AND GRAPHS

Let some quantity y depend on another quantity x, and imagine that we have the power to control x at will—as, for instance, when we alter the temperature of a gas to effect a corresponding change in the energy. The *controlling* quantity (x, the temperature) is the **independent variable**, and the *responding* quantity y (the energy that results) is the **dependent variable**. For every change made in the independent variable x, there is a change provoked in the dependent variable y. We say that y is a **function** of x; or, in plain language, that y *depends* on x.

For example: $y = 3x + 2$. When x is equal to 0 (our choice), y is equal to 2 (the response). When x is equal to 1, y becomes equal to 5. When x is 2, y is 8; and so on, and so on, and so on.

Represented on a graph as **ordered pairs** of points x and y (Figure B-1), the function

$$y = 3x + 2$$

then traces out a straight line with just two defining characteristics: (1) The line rises 3 units in y for every unit of x it advances horizontally. (2) The function intersects the y-axis when $x = 0$ and $y = 2$.

From that, we construct a generalized **linear equation** of the form

$$y = mx + b$$

in which the parameter m (value = 3) tells us the **slope** of the line and the parameter b (value = 2) tells us the **y-intercept**. It is an equation of the **first degree**, containing powers no higher than x^1.

The molar energy of an ideal gas,

$$E = \left(\frac{3R}{2}\right)T$$

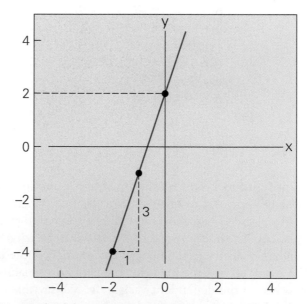

FIGURE B-1. Graph of the equation $y = 3x + 2$: a line with slope equal to 3 and vertical intercept equal to 2. The independent variable is x; the dependent variable is y.

is a case in point, a linear function where E (the dependent variable) goes as the first power of the temperature. A plot of E versus T falls on a straight line with slope $3R/2$ and intercept zero.

Another example: The equation

$$°F = \tfrac{9}{5}(°C) + 32$$

describing the conversion from degrees Celsius (independent variable) to degrees Fahrenheit (dependent variable) also traces out a straight line. The slope is $\tfrac{9}{5}$, and the vertical intercept is 32. Turned around, the conversion

$$°C = \tfrac{5}{9}(°F - 32)$$

from $°F$ (independent variable) to $°C$ (dependent variable) yields a line with slope equal to $\tfrac{5}{9}$ and vertical intercept equal to $-160/9$.

And another: Any exponential equation of the form

$$p = A \exp mx$$

turns into the linear equation

$$\ln p = mx + \ln A$$

when expressed logarithmically. A plot of $\ln p$ versus x produces a line with slope equal to m and intercept equal to $\ln A$. Thus from the exponential dependence of a rate constant on temperature,

$$k = A \exp\left(-\frac{E_a}{RT}\right)$$

we derive the linear Arrhenius equation:

$$\ln k = -\frac{E_a}{R}\left(\frac{1}{T}\right) + \ln A$$

The independent variable is $1/T$. The dependent variable is $\ln k$. The slope is $-E_a/R$. The intercept is $\ln A$.

Beyond a simple linear dependence, there are functions containing terms other than just x to the first power: quadratic equations (x^2), cubic equations (x^3), quartic equations (x^4), and higher; equations that contain $\sin x$ or $\cos x$, logarithms, exponentials, and other functions. Figure B-2

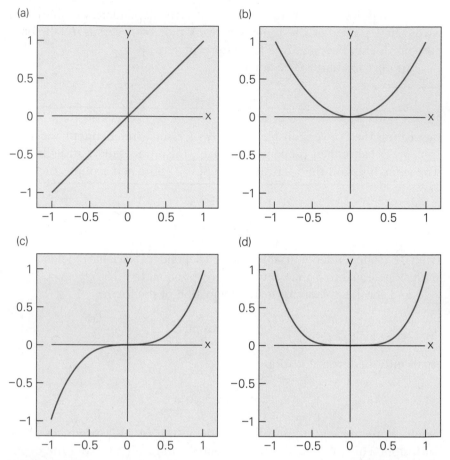

FIGURE B-2. Functions of various degrees. (a) Linear: $y = x$. (b) Quadratic: $y = x^2$.
(c) Cubic: $y = x^3$. (d) Quartic: $y = x^4$.

provides a few examples, among which is a parabolic curve derived from the quadratic function

$$y = ax^2 + bx + c$$

For further analysis of such equations, which appear frequently in equilibrium calculations, see Section B-3.

B-3. QUADRATIC EQUATIONS

An equation of the form

$$ax^2 + bx + c = 0$$

is said to be **quadratic** in x, limited to powers no higher than the second degree (x^2). The name derives from the Latin *quadrare* ("to make square"), descriptive of the x-squared term that defines the class.

Hence the equation

$$6x^2 - x - 2 = 0 \qquad a = 6, b = -1, c = -2$$

is quadratic, as is the equation

$$x^2 - x = 0 \qquad a = 1, b = -1, c = 0$$

as is the equation

$$x^2 - 1 = 0 \qquad a = 1, b = 0, c = -1$$

Each contains a term involving x^2 (nothing higher), and each is satisfied by *two* values of x. Since the equation contains a squared quantity, the solution must reflect, in some way, a square root—and every number, remember, has two square roots, one positive and one negative.

Take, for example, the expression

$$x^2 - 1 = 0$$

and rearrange it to produce the equivalent form

$$x^2 = 1$$

Put into words, the equation simply asks: What numbers, when squared, are equal to 1? Answer: 1 and −1, the two possible square roots of 1.

See what happens also with the equation

$$x^2 - x = 0$$

which presents itself more suggestively as the product of two factors:

$$x(x - 1) = 0$$

If either the first factor (x) or the second factor $(x - 1)$ is zero, then so must be the entire expression. The two solutions are $x = 0$ and $x = 1$.

Similar, but only slightly more complicated, is the equation

$$6x^2 - x - 2 = 0$$

which again factors into two linear terms:

$$(3x - 2)(2x + 1) = 0$$

The two solutions are $x = \frac{2}{3}$ and $x = -\frac{1}{2}$:

$$3x - 2 = 0 \qquad\qquad 2x + 1 = 0$$

$$3x = 2 \qquad\qquad 2x = -1$$

$$x = \frac{2}{3} \qquad\qquad x = -\frac{1}{2}$$

All our problems are solved, it seems, if we can split a quadratic expression cleanly into two linear factors—but what if we cannot? How shall we handle an equation that permits no quick and easy solution? What we need, clearly, is a general method to solve *any* problem of the kind

$$ax^2 + bx + c = 0$$

regardless of the particular values of a, b, and c.

Here it is, the **quadratic formula**:

$$x = \frac{-b \pm \sqrt{b^2 - 4ac}}{2a}$$

and here is its message: A square in the equation demands a square root, positive and negative, somewhere in the solution. One solution, or *root*, corresponds to the positive square root (call it x_+); the other solution, x_-, corresponds to the negative square root:

$$x_+ = \frac{-b + \sqrt{b^2 - 4ac}}{2a} \qquad x_- = \frac{-b - \sqrt{b^2 - 4ac}}{2a}$$

If the *discriminant* ($b^2 - 4ac$) is positive, both roots are real. If the discriminant is negative, the roots are complex conjugates. If the discriminant is zero, the two roots are identical (or *degenerate*). Those are the only three possibilities.

The formula itself, cumbersome but effective, we obtain by the algebraic trick of *completing the square*. The derivation goes as follows:

1. Divide all terms by a:

$$x^2 + \frac{b}{a}x + \frac{c}{a} = 0$$

2. Subtract c/a from both sides:

$$x^2 + \frac{b}{a}x = -\frac{c}{a}$$

3. Add $b^2/4a^2$ to both sides:

$$x^2 + \frac{b}{a}x + \frac{b^2}{4a^2} = \frac{b^2}{4a^2} - \frac{c}{a}$$

4. Write the left-hand side as a perfect square:

$$\left(x + \frac{b}{2a}\right)^2 = \frac{b^2}{4a^2} - \frac{c}{a}$$

5. Reduce the right-hand side to a single fraction:

$$\left(x + \frac{b}{2a}\right)^2 = \frac{b^2 - 4ac}{4a^2}$$

6. Take the square root of both sides:

$$\pm\left(x + \frac{b}{2a}\right) = \frac{\sqrt{b^2 - 4ac}}{2a}$$

7. Rearrange and solve for x:

$$x = \frac{-b \pm \sqrt{b^2 - 4ac}}{2a}$$

The resulting formula works without fail, provided that a is unequal to zero—which means only that the equation is indeed quadratic. Applied, say, to the last problem solved above, the quadratic formula confirms right away that the two roots are $\frac{2}{3}$ and $-\frac{1}{2}$:

$$6x^2 - x - 2 = 0 \qquad a = 6, \ b = -1, \ c = -2$$

$$x = \frac{-b \pm \sqrt{b^2 - 4ac}}{2a}$$

$$= \frac{-(-1) \pm \sqrt{(-1)^2 - 4(6)(-2)}}{2(6)} = \frac{1 \pm \sqrt{49}}{12} = \frac{1 \pm 7}{12}$$

$$x_+ = \frac{1 + 7}{12} = \frac{2}{3}$$

$$x_- = \frac{1 - 7}{12} = -\frac{1}{2}$$

To check, we substitute x_+ and x_- into the original equation and show that the three terms sum to zero:

$$6x^2 - x - 2 = 0$$

$$x_+: \qquad 6(\tfrac{2}{3})^2 - \tfrac{2}{3} - 2 = \tfrac{8}{3} - \tfrac{2}{3} - 2 = 0$$

$$x_-: \qquad 6(-\tfrac{1}{2})^2 + \tfrac{1}{2} - 2 = \tfrac{3}{2} + \tfrac{1}{2} - 2 = 0$$

The roots are correct.

Foolproof? Almost, but beware of a potential ambiguity whenever a quadratic equation is used to describe a natural process: There are two mathematical solutions, but usually only one value is reasonable. To pick the physically valid solution and discard the other, we need to use proper intuition. A root that leads, for instance, to a negative concentration is inadmissible. We drop it.

EXERCISES

1. Evaluate each expression as a decimal number:

   (a) 3.7×10^3
   (b) -0.005×10^{-4}
   (c) $7.36 \times 10^2 \times 10^{-2}$
   (d) $3.28 \div (8.02 \times 10^{-4})$

   Use a calculator only where necessary.

2. Same. Evaluate each expression as a decimal number, using a calculator only where necessary:

   (a) $10^{3.4}$ (b) $10^{-3.4}$ (c) $2^3 \times 10^3$ (d) $(3^2)^3$

3. Once more:

   (a) $e^{-1.65}$ (b) $\exp(-1.65)$ (c) $\exp 1.65$ (d) $\dfrac{1}{\exp 1.65}$

4. Simplify or expand where possible:

   (a) $\log(a^3 b^2)$
   (b) $\ln(a^3 b^2)$
   (c) $\log(10^{1.17})$
   (d) $\ln(10^{1.17})$
   (e) $\ln(e^{1.17})$

5. Simplify or expand where possible:

   (a) $\log\left(\dfrac{3a^4}{2b^{1/2}}\right)$

   (b) $\ln\left(\dfrac{3a^4}{2b^{1/2}}\right)$

   (c) $\log abcdef$
   (d) $\log(a + b + c)$

6. Evaluate without using a calculator:

   (a) log 8
   (b) log 80
   (c) log 0.125
   (d) ln 0.5

   Note that log 2 = 0.301030 and ln 2 = 0.693147.

7. Evaluate without using a calculator:

   (a) $\log(6.00 \times 10^{-7})$
   (b) $\log(1.50 \times 10^{1/2})$
   (c) log 1.8
   (d) log 0.0027

   Note that log 2 = 0.301030 and log 3 = 0.477121.

8. Evaluate without using a calculator:

   (a) $\ln(e^{\ln 3})$
   (b) ln 0.0027
   (c) $\ln(e^2)$
   (d) ln 6

   Note that ln 2 = 0.693147, ln 3 = 1.098612, and ln 10 = 2.302585.

9. Evaluate using a calculator:

   (a) $\log(3.31345 \times 10^{-0.3451})$
   (b) log 0.000053321
   (c) log 432,134.9
   (d) $\log(12.345 \times 10^{52})$

10. Evaluate using a calculator:

    (a) ln 0.1534
    (b) ln 12.34
    (c) $\ln(8.08 \times 10^{-8.08})$
    (d) $\ln(6.02 \times 10^{23})$

11. (a) What number has the common logarithm −7? (b) What number has the common logarithm 2.06?

12. (a) What number has the natural logarithm 1? (b) What number has the natural logarithm −1?

13. Suppose that the slope of a straight line is −7.2 and its vertical intercept is 17.6. What value does the dependent variable assume when the independent variable has the value 1.2?

14. Determine the slope and vertical intercept of a line described by the function

$$y = 3x - 7 + 2x - 1$$

15. Solve for x:

   (a) $3.45x - 12.3 = 2.5$
   (b) $\pi - 7 - 8x = 1$
   (c) $x + 2 = 3x + 6$
   (d) $\ln[(e^x)^3] + x = 4$

16. Graph the following equations over the range $-3 \le x \le 3$:

   (a) $y = x^{-1}$
   (b) $y = x$
   (c) $y = x^2$
   (d) $y = x^4$

   How do the functions behave as x approaches $\pm\infty$?

17. Graph the equation

$$y = 4x^4 + 3x^3 + 2x^2 + 1$$

over the range $-2 \le x \le 2$. Use the curve to estimate the value of y at the point $x = 1.67$.

18. Graph the equation

$$y = (x + 1.5)(x - 1.5)$$

over the range $-2 \le x \le 2$. (a) Show that the equation is quadratic, and estimate the values of its two roots by marking the points on the curve where $y = 0$. (b) Compare the graphical results to the exact values.

19. Factor each of the following quadratic equations and solve for x:

   (a) $x^2 + 5x + 6 = 0$
   (b) $x^2 + 3x - 4 = 0$
   (c) $x^2 - x + 0.25 = 0$
   (d) $x^2 - 0.25 = 0$
   (e) $2x^2 - 3x - 2 = 0$

20. Solve, once again, the same equations as before, but this time use the quadratic formula:

    (a) $x^2 + 5x + 6 = 0$
    (b) $x^2 + 3x - 4 = 0$
    (c) $x^2 - x + 0.25 = 0$
    (d) $x^2 - 0.25 = 0$
    (e) $2x^2 - 3x - 2 = 0$

21. Solve each equation using the quadratic formula:

    (a) $3x^2 - 7x + 3 = 0$
    (b) $4x^2 + x - 6 = 0$
    (c) $x^2 - 4.234x - 1.92 = 0$
    (d) $-2x^2 + 11x - 1 = 0$
    (e) $x^2 + x - 2 = 0$

22. Solve using the quadratic formula:

$$\frac{x^2}{0.1 - x} = 100$$

23. The End (something big): $2^{1,398,269} - 1$, the 35th Mersenne Prime, the largest prime number to be found as of late 1996. (a) Use scientific notation to express the 35th Mersenne Prime as an approximate power of 10. (b) Assume that 60 zeros can be written on a single line and 60 lines can be written on a page. How many pages are needed to write all the digits in the number?

APPENDIX C

Data

C-1. Dimensions, Constants, and Symbols
C-2. Elements
C-3. Molecules and Interactions
C-4. Equilibrium

The following tables of data should provide a modest experimental basis—plain, simple numbers—to underpin the abstract principles developed throughout the text. Use the numbers to add detail, to solve problems, to challenge assertions, to make predictions, and to raise questions and doubts.

Unsatisfied with the graph of ionization energies in Chapter 6? Turn to Table C-8 and inspect the data used to plot the points.

Need to check the electron configuration of tungsten? The atomic radius of beryllium? The element with 72 protons? Table C-8, again.

Is praseodymium a solid, liquid, or gas under standard conditions? Is silver more dense than tin? At what temperature does sodium melt? Consult Table C-9, Miscellaneous Physical Properties of the Elements.

For oxidation states and configurations of the transition metals, see Table C-10. For masses and binding energies of selected isotopes, see Tables C-11 and C-12. For average bond lengths and dissociation energies, see Table C-13. And so on.

The collection, then, in short:

Dimensions, Constants, and Symbols

TABLE C-1. SI Base Units
TABLE C-2. SI Derived Units
TABLE C-3. SI Prefixes

TABLE C-4. Special Units and Conversion Factors
TABLE C-5. Physical Constants
TABLE C-6. Greek Alphabet
TABLE C-7. Mathematical and Physical Symbols

Information provided here is used first in Chapter 1 and repeatedly thereafter.

Elements

TABLE C-8. Ground Configurations, Atomic Radii, and Ionization Energies of the Elements   (Chapter 6 and beyond)
TABLE C-9. Miscellaneous Physical Properties of the Elements (includes standard state, density, melting point, and boiling point; useful everywhere, but especially in Chapters 2, 3, 9, 10, and 11)
TABLE C-10. Valence Configurations and Oxidation States of the Transition Metals   (Chapter 19)
TABLE C-11. A Selection of Stable Isotopes   (natural abundance, atomic number, neutron number, mass number, atomic mass, binding energy per nucleon; Chapters 2 and 21)
TABLE C-12. A Selection of Radioactive Isotopes   (similar, but with half-life and mode of decay as well; Chapters 2 and 21)

Molecules and Interactions

TABLE C-13. Length and Strength of Covalent Bonds (Chapters 7, 8, and more)
TABLE C-14. Critical Temperatures (T_c) and van der Waals Parameters (a, b) of Real Gases   (Chapters 9 through 11)
TABLE C-15. Properties of Liquid Water Between $0°C$ and $100°C$ (density, vapor pressure, heat capacity, enthalpy of vaporization, ion-product constant; handy throughout—notably in Chapter 9 and Chapters 11 through 15)
TABLE C-16. Thermodynamic Properties at $25°C$ (molar mass, formation data, and heat capacity for over 200 substances to supplement Chapters 13 through 17)

Equilibrium

TABLE C-17. Solubility of Ionic Compounds in Water
            (a qualitative guide to various inorganic
            compounds; Chapter 15)

TABLE C-18. Approximate Solubilities of Selected Inorganic
            Salts   (Chapters 3 and 15)

TABLE C-19. Solubility-Product Constants at 25°C   (Chapter 15)

TABLE C-20. Ionization Constants for Acids at 25°C   (Chapter 16)

TABLE C-21. Standard Reduction Potentials and Free Energies
            at 25°C   (Chapter 17)

TABLE C-22. Formation Constants of Complexes at 25°C
            (Chapter 19)

Conversion factors and fundamental constants conform to the values recommended in 1986 by the Committee on Data for Science and Technology. For a brief summary and further references, see the August 1995 issue of *Physics Today*. A companion article in the same number provides a sketch of SI units accepted by international convention.

Much of the experimental data has been culled from the 75th edition of the *CRC Handbook of Chemistry and Physics* (David R. Lide, editor; CRC Press, 1994) and reorganized and expanded here into a new compilation to suit the present work. References to original sources are given in the *Handbook*; and, of these, the *NBS Tables of Chemical Thermodynamics* (1982) provides most of the raw data selected in Table C-16.

Finally, take heed: Be tolerant of the small discrepancies, errors, and variations that inevitably plague such tabulations. No measurement is perfect, and some measurements are less perfect than others. Expect, for example, to find different values of solubilities and solubility-product constants given in different sources. Equilibrium constants, in general, and thermodynamic data of all sorts vary throughout the literature—sometimes widely, sometimes not—but measurements of very small or very large values are always hard to make. Different methods, different conditions, and different instruments all combine to produce a spread in results.

The lessons of Chapter 1 remain true: Experimentation is not philosophy. Numbers are not ideologies. Digits are significant for a reason.

C-1. DIMENSIONS, CONSTANTS, AND SYMBOLS

TABLE C-1. SI Base Units
TABLE C-2. SI Derived Units
TABLE C-3. SI Prefixes
TABLE C-4. Special Units and Conversion Factors
TABLE C-5. Physical Constants
TABLE C-6. Greek Alphabet
TABLE C-7. Mathematical and Physical Symbols

TABLE C-1. SI Base Units

SI Base Quantity	*Unit*	*Symbol*
length	meter	m
mass	kilogram	kg
time	second	s
amount of substance	mole	mol
temperature	kelvin	K
electric current	ampere	A
luminous intensity	candela	cd

TABLE C-2. SI Derived Units[a]

SI Derived Quantity	*Unit*	*Symbol*	*Dimensions*
acceleration	meter per second per second	—	$m\ s^{-2}$
electric charge	coulomb	C	$A\ s$
electric field	volt per meter[b]	—	$V\ m^{-1}$
electric potential	volt	V	$J\ C^{-1}$
force	newton	N	$kg\ m\ s^{-2}$
frequency	hertz	Hz	s^{-1}
momentum (impulse)	newton second	—	$kg\ m\ s^{-1}$
power	watt	W	$J\ s^{-1}$
pressure	pascal	Pa	$N\ m^{-2}$
radioactivity	becquerel	Bq	s^{-1}
speed or velocity	meter per second	—	$m\ s^{-1}$
work, energy, or heat	joule (newton meter)	J	$kg\ m^2\ s^{-2}$

[a]A partial listing.
[b]Equivalent to newton per coulomb ($N\ C^{-1}$).

TABLE C-3. SI Prefixes

Prefix	Symbol	Multiplier	Prefix	Symbol	Multiplier
deci	d	10^{-1}	deka	da	10^{1}
centi	c	10^{-2}	hecto	h	10^{2}
milli	m	10^{-3}	kilo	k	10^{3}
micro	μ	10^{-6}	mega	M	10^{6}
nano	n	10^{-9}	giga	G	10^{9}
pico	p	10^{-12}	tera	T	10^{12}
femto	f	10^{-15}	peta	P	10^{15}
atto	a	10^{-18}	exa	E	10^{18}

TABLE C-4. Special Units and Conversion Factors

Quantity	Unit	Symbol	Conversion
energy	electron volt	eV	$1 \text{ eV} = 1.60217733 \times 10^{-19}$ J
force	pound	lb	$1 \text{ lb} = 0.453592$ kg (earth)
heat	calorie	cal	$1 \text{ cal} = 4.184$ J
length	angstrom	Å	$1 \text{ Å} = 10^{-8} \text{ cm} = 10^{-10}$ m
length	inch	in	$1 \text{ in} = 2.54$ cm
mass	atomic mass unit	u	$1 \text{ u} = 1.6605402 \times 10^{-27}$ kg
pressure	atmosphere	atm	$1 \text{ atm} = 1.01325 \times 10^{5}$ Pa
pressure	torr	torr	$1 \text{ torr} = 1 \text{ atm}/760$
temperature	Celsius scale	°C	$°C = K - 273.15$
temperature	Fahrenheit scale	°F	$°F = \frac{9}{5}(°C) + 32$
time	minute	min	$1 \text{ min} = 60$ s
time	hour	h	$1 \text{ h} = 60 \text{ min} = 3600$ s
time	day	d	$1 \text{ d} = 24 \text{ h} = 86{,}400$ s
time	year	y	$1 \text{ y} = 365.25$ d
volume	liter	L	$1 \text{ L} = 1 \text{ dm}^3 = 10^{-3} \text{ m}^3$
volume	cubic centimeter	cm^3, cc	$1 \text{ cm}^3 = 1 \text{ mL} = 10^{-3}$ L

TABLE C-5. Physical Constants

Quantity	Symbol	Value
acceleration due to gravity (earth)	g	9.80665 m s^{-2} (exact)
Avogadro's number	N_0	6.0221367×10^{23} mol^{-1}
Bohr radius	a_0	$0.529177249 \times 10^{-10}$ m
Boltzmann's constant	k_B	1.380658×10^{-23} J K^{-1}
electronic charge-to-mass ratio	$-e/m_e$	$1.75881962 \times 10^{11}$ C kg^{-1}
elementary charge	e	$1.60217733 \times 10^{-19}$ C
Faraday constant	$\mathcal{F}$	9.6485309×10^4 C mol^{-1}
gravitational constant (universal)	G	6.67259×10^{-11} m^3 kg^{-1} s^{-2}
mass of an electron	m_e	$9.1093897 \times 10^{-31}$ kg
mass of a neutron	m_n	$1.6749286 \times 10^{-27}$ kg
mass of a proton	m_p	$1.6726231 \times 10^{-27}$ kg
molar volume of ideal gas at STP	V_m	22.41410 L mol^{-1}
permittivity of vacuum	ϵ_0	$8.854187817 \times 10^{-12}$ C^2 N^{-1} m^{-2}
Planck's constant	h	$6.6260755 \times 10^{-34}$ J s
Rydberg constant	R_∞	$2.1798741 \times 10^{-18}$ J
speed of light in vacuum	c	2.99792458×10^8 m s^{-1} (exact)
universal gas constant	R	8.314510 J mol^{-1} K^{-1}
		0.08205783 atm L mol^{-1} K^{-1}

TABLE C-6. Greek Alphabet

Symbol		Name	Symbol		Name	Symbol		Name
α	A	alpha	ι	I	iota	ρ	P	rho
β	B	beta	κ	K	kappa	σ	Σ	sigma
γ	Γ	gamma	λ	Λ	lambda	τ	T	tau
δ	Δ	delta	μ	M	mu	υ	Υ	upsilon
ε	E	epsilon	ν	N	nu	φ	Φ	phi
ζ	Z	zeta	ξ	Ξ	xi	χ	X	chi
η	H	eta	ο	O	omicron	ψ	Ψ	psi
θ	Θ	theta	π	Π	pi	ω	Ω	omega

TABLE C-7. Mathematical and Physical Symbols

Symbol	Name	Meaning
~	comparison	approximately; on the order of
≈	comparison	approximately; on the order of
≡	comparison	equivalent to; defined as
>	comparison	greater than
≥	comparison	greater than or equal to
≳	comparison	greater than or approximately equal to
<	comparison	less than
≤	comparison	less than or equal to
≫	comparison	much greater than
≪	comparison	much less than
Δ	delta	finite change in quantity: $\Delta x = x_2 - x_1$
∞	infinity	infinity
∫	integral sign	integration
λ	lambda	wavelength
log	logarithm (common)	$\log 10^x = x$
ln	logarithm (natural)	$\ln e^x = \ln \exp(x) = x$; $e = 2.718281828459045 \ldots$
ν	nu	frequency; neutrino
ϕ	phi	general-purpose angle or wave function
π	pi	3.14159265 (approximately)
′	prime	distinguishing mark
∝	proportionality	$y \propto x$ implies $y = \text{constant} \cdot x$
ψ	psi	wave function
Σ	sigma (large)	summation
$\sqrt{}$	square root	$\sqrt{x} = x^{1/2}$
θ	theta	general-purpose angle

C-2. ELEMENTS

TABLE C-8. Ground Configurations, Atomic Radii, and
Ionization Energies of the Elements

TABLE C-9. Miscellaneous Physical Properties of the Elements

TABLE C-10. Valence Configurations and Oxidation States
of the Transition Metals

TABLE C-11. A Selection of Stable Isotopes

TABLE C-12. A Selection of Radioactive Isotopes

TABLE C-8. Ground Configurations, Atomic Radii,
and Ionization Energies of the Elements

Element	Symbol	Atomic Number Z	Ground-State Configuration	Atomic Radius (Å)	Ionization Energy (kJ mol^{-1})
hydrogen	H	1	$1s^1$	0.37	1312.0
helium	He	2	$1s^2$	0.32	2372.3
lithium	Li	3	$[He]2s^1$	1.52	520.2
beryllium	Be	4	$[He]2s^2$	1.13	899.5
boron	B	5	$[He]2s^22p^1$	0.88	800.6
carbon	C	6	$[He]2s^22p^2$	0.77	1086.5
nitrogen	N	7	$[He]2s^22p^3$	0.70	1402.3
oxygen	O	8	$[He]2s^22p^4$	0.66	1313.9
fluorine	F	9	$[He]2s^22p^5$	0.64	1681.0
neon	Ne	10	$[He]2s^22p^6$	0.69	2080.7
sodium	Na	11	$[Ne]3s^1$	1.86	495.8
magnesium	Mg	12	$[Ne]3s^2$	1.60	737.7
aluminum	Al	13	$[Ne]3s^23p^1$	1.43	577.5
silicon	Si	14	$[Ne]3s^23p^2$	1.17	786.5
phosphorus	P	15	$[Ne]3s^23p^3$	1.10	1011.8
sulfur	S	16	$[Ne]3s^23p^4$	1.04	999.6
chlorine	Cl	17	$[Ne]3s^23p^5$	0.99	1251.2
argon	Ar	18	$[Ne]3s^23p^6$	0.97	1520.6
potassium	K	19	$[Ar]4s^1$	2.27	418.8
calcium	Ca	20	$[Ar]4s^2$	1.97	589.8
scandium	Sc	21	$[Ar]4s^23d^1$	1.61	633.1
titanium	Ti	22	$[Ar]4s^23d^2$	1.45	658.8
vanadium	V	23	$[Ar]4s^23d^3$	1.31	650.9
chromium	Cr	24	$[Ar]4s^13d^5$	1.25	652.9
manganese	Mn	25	$[Ar]4s^23d^5$	1.37	717.3
iron	Fe	26	$[Ar]4s^23d^6$	1.24	762.5
cobalt	Co	27	$[Ar]4s^23d^7$	1.25	760.4
nickel	Ni	28	$[Ar]4s^23d^8$	1.25	737.1
copper	Cu	29	$[Ar]4s^13d^{10}$	1.28	745.5
zinc	Zn	30	$[Ar]4s^23d^{10}$	1.34	906.4

TABLE C-8. Ground Configurations, Atomic Radii,
and Ionization Energies of the Elements (*Continued*)

Element	Symbol	Atomic Number Z	Ground-State Configuration	Atomic Radius (Å)	Ionization Energy (kJ mol^{-1})
gallium	Ga	31	$[Ar]4s^2 3d^{10} 4p^1$	1.22	578.8
germanium	Ge	32	$[Ar]4s^2 3d^{10} 4p^2$	1.22	762.2
arsenic	As	33	$[Ar]4s^2 3d^{10} 4p^3$	1.21	947.0
selenium	Se	34	$[Ar]4s^2 3d^{10} 4p^4$	1.17	941.0
bromine	Br	35	$[Ar]4s^2 3d^{10} 4p^5$	1.14	1139.9
krypton	Kr	36	$[Ar]4s^2 3d^{10} 4p^6$	1.10	1350.8
rubidium	Rb	37	$[Kr]5s^1$	2.47	403.0
strontium	Sr	38	$[Kr]5s^2$	2.15	549.5
yttrium	Y	39	$[Kr]5s^2 4d^1$	1.78	599.8
zirconium	Zr	40	$[Kr]5s^2 4d^2$	1.59	640.1
niobium	Nb	41	$[Kr]5s^1 4d^4$	1.43	652.1
molybdenum	Mo	42	$[Kr]5s^1 4d^5$	1.36	684.3
technetium	Tc	43	$[Kr]5s^2 4d^5$	1.35	702.4
ruthenium	Ru	44	$[Kr]5s^1 4d^7$	1.32	710.2
rhodium	Rh	45	$[Kr]5s^1 4d^8$	1.34	719.7
palladium	Pd	46	$[Kr]4d^{10}$	1.38	804.4
silver	Ag	47	$[Kr]5s^1 4d^{10}$	1.44	731.0
cadmium	Cd	48	$[Kr]5s^2 4d^{10}$	1.49	867.8
indium	In	49	$[Kr]5s^2 4d^{10} 5p^1$	1.63	558.3
tin	Sn	50	$[Kr]5s^2 4d^{10} 5p^2$	1.40	708.6
antimony	Sb	51	$[Kr]5s^2 4d^{10} 5p^3$	1.41	833.6
tellurium	Te	52	$[Kr]5s^2 4d^{10} 5p^4$	1.43	869.3
iodine	I	53	$[Kr]5s^2 4d^{10} 5p^5$	1.33	1008.4
xenon	Xe	54	$[Kr]5s^2 4d^{10} 5p^6$	1.30	1170.4
cesium	Cs	55	$[Xe]6s^1$	2.65	375.7
barium	Ba	56	$[Xe]6s^2$	2.17	502.9
lanthanum	La	57	$[Xe]6s^2 5d^1$	1.87	538.1
cerium	Ce	58	$[Xe]6s^2 4f^1 5d^1$	1.82	534.4
praseodymium	Pr	59	$[Xe]6s^2 4f^3$	1.82	527.2
neodymium	Nd	60	$[Xe]6s^2 4f^4$	1.81	533.1

Continued on next page

TABLE C-8. Ground Configurations, Atomic Radii,
and Ionization Energies of the Elements (*Continued*)

Element	Symbol	Atomic Number Z	Ground-State Configuration	Atomic Radius (Å)	Ionization Energy (kJ mol^{-1})
promethium	Pm	61	$[Xe]6s^2 4f^5$	1.81	535.5
samarium	Sm	62	$[Xe]6s^2 4f^6$	1.80	544.5
europium	Eu	63	$[Xe]6s^2 4f^7$	2.00	547.1
gadolinium	Gd	64	$[Xe]6s^2 4f^7 5d^1$	1.79	593.4
terbium	Tb	65	$[Xe]6s^2 4f^9$	1.76	565.8
dysprosium	Dy	66	$[Xe]6s^2 4f^{10}$	1.75	573.0
holmium	Ho	67	$[Xe]6s^2 4f^{11}$	1.74	581.0
erbium	Er	68	$[Xe]6s^2 4f^{12}$	1.73	589.3
thulium	Tm	69	$[Xe]6s^2 4f^{13}$	1.72	596.7
ytterbium	Yb	70	$[Xe]6s^2 4f^{14}$	1.94	603.4
lutetium	Lu	71	$[Xe]6s^2 4f^{14} 5d^1$	1.72	523.5
hafnium	Hf	72	$[Xe]6s^2 4f^{14} 5d^2$	1.56	658.5
tantalum	Ta	73	$[Xe]6s^2 4f^{14} 5d^3$	1.43	761.3
tungsten	W	74	$[Xe]6s^2 4f^{14} 5d^4$	1.37	770.0
rhenium	Re	75	$[Xe]6s^2 4f^{14} 5d^5$	1.34	760.3
osmium	Os	76	$[Xe]6s^2 4f^{14} 5d^6$	1.34	839.4
iridium	Ir	77	$[Xe]6s^2 4f^{14} 5d^7$	1.36	878.0
platinum	Pt	78	$[Xe]6s^1 4f^{14} 5d^9$	1.37	868.4
gold	Au	79	$[Xe]6s^1 4f^{14} 5d^{10}$	1.44	890.1
mercury	Hg	80	$[Xe]6s^2 4f^{14} 5d^{10}$	1.50	1007.1
thallium	Tl	81	$[Xe]6s^2 4f^{14} 5d^{10} 6p^1$	1.70	589.4
lead	Pb	82	$[Xe]6s^2 4f^{14} 5d^{10} 6p^2$	1.75	715.6
bismuth	Bi	83	$[Xe]6s^2 4f^{14} 5d^{10} 6p^3$	1.55	703.3
polonium	Po	84	$[Xe]6s^2 4f^{14} 5d^{10} 6p^4$	1.67	812.1
astatine	At	85	$[Xe]6s^2 4f^{14} 5d^{10} 6p^5$	1.40	924.6
radon	Rn	86	$[Xe]6s^2 4f^{14} 5d^{10} 6p^6$	1.45	1037.1
francium	Fr	87	$[Rn]7s^1$	2.7	380
radium	Ra	88	$[Rn]7s^2$	2.23	509.3
actinium	Ac	89	$[Rn]7s^2 6d^1$	1.88	499
thorium	Th	90	$[Rn]7s^2 6d^2$	1.80	587

TABLE C-8. Ground Configurations, Atomic Radii,
and Ionization Energies of the Elements (*Continued*)

Element	Symbol	Atomic Number Z	Ground-State Configuration	Atomic Radius (Å)	Ionization Energy (kJ mol^{-1})
protactinium	Pa	91	$[\text{Rn}]7s^2 5f^2 6d^1$	1.61	568
uranium	U	92	$[\text{Rn}]7s^2 5f^3 6d^1$	1.38	587
neptunium	Np	93	$[\text{Rn}]7s^2 5f^4 6d^1$	1.30	597
plutonium	Pu	94	$[\text{Rn}]7s^2 5f^6$	1.51	585
americium	Am	95	$[\text{Rn}]7s^2 5f^7$	1.84	578
curium	Cm	96	$[\text{Rn}]7s^2 5f^7 6d^1$	—	581
berkelium	Bk	97	$[\text{Rn}]7s^2 5f^9$	—	601
californium	Cf	98	$[\text{Rn}]7s^2 5f^{10}$	—	608
einsteinium	Es	99	$[\text{Rn}]7s^2 5f^{11}$	—	619
fermium	Fm	100	$[\text{Rn}]7s^2 5f^{12}$	—	627
mendelevium	Md	101	$[\text{Rn}]7s^2 5f^{13}$	—	635
nobelium	No	102	$[\text{Rn}]7s^2 5f^{14}$	—	642
lawrencium	Lr	103	$[\text{Rn}]7s^2 5f^{14} 6d^1$	—	—
rutherfordium	Rf	104	$[\text{Rn}]7s^2 5f^{14} 6d^2$	—	—
dubnium	Db	105	$[\text{Rn}]7s^2 5f^{14} 6d^3$	—	—
seaborgium	Sg	106	$[\text{Rn}]7s^2 5f^{14} 6d^4$	—	—
bohrium	Bh	107	$[\text{Rn}]7s^2 5f^{14} 6d^5$	—	—
hassium	Hs	108	$[\text{Rn}]7s^2 5f^{14} 6d^6$	—	—
meitnerium	Mt	109	$[\text{Rn}]7s^2 5f^{14} 6d^7$	—	—
element 110	[110]	110	$[\text{Rn}]7s^2 5f^{14} 6d^8$	—	—
element 111	[111]	111	$[\text{Rn}]7s^2 5f^{14} 6d^9$	—	—
element 112	[112]	112	$[\text{Rn}]7s^2 5f^{14} 6d^{10}$	—	—

TABLE C-9. Miscellaneous Physical Properties of the Elements[a]

Element	Symbol	Atomic Number	Standard State[b,c]	Density[d] (g mL^{-1})	Melting Point (°C)	Boiling Point (°C)
hydrogen	H	1	gas	0.000090	−259.14	−252.87
helium	He	2	gas	0.000179	<−272.2	−268.93
lithium	Li	3	solid	0.534	180.5	1347
beryllium	Be	4	solid	1.848	1283	2484
boron	B	5	solid	2.34	2300	3650
carbon	C	6	solid (gr)	1.9–2.3	≈3350	sublimes
nitrogen	N	7	gas	0.00125	−210.00	−195.8
oxygen	O	8	gas	0.00143	−218.8	−182.95
fluorine	F	9	gas	0.00170	−219.62	−188.12
neon	Ne	10	gas	0.00090	−248.59	−246.08
sodium	Na	11	solid	0.971	97.72	883
magnesium	Mg	12	solid	1.738	650	1090
aluminum	Al	13	solid	2.6989	660.32	2467
silicon	Si	14	solid	2.33	1414	2355
phosphorus	P	15	solid (wh)	1.82	44.15	280
sulfur	S	16	solid (rh)	2.07	115.21	444.60
chlorine	Cl	17	gas	0.00321	−101.5	−34.04
argon	Ar	18	gas	0.00178	−189.3	−185.9
potassium	K	19	solid	0.862	63.28	759
calcium	Ca	20	solid	1.55	842	1484
scandium	Sc	21	solid	2.989	1541	2830
titanium	Ti	22	solid	4.54	1668	3287
vanadium	V	23	solid	6.11	1910	3407
chromium	Cr	24	solid	7.19	1857	2671
manganese	Mn	25	solid	7.3	1246	1962
iron	Fe	26	solid	7.874	1538	2750
cobalt	Co	27	solid	8.9	1495	2870
nickel	Ni	28	solid	8.902	1455	2730

[a]For relative atomic masses and an alphabetical listing of the elements, see the flyleaf at the front of this volume.
[b]Normal state at 25°C and 1 atm.
[c]Allotropes: gr = graphite, gy = gray, rh = rhombic, wh = white.
[d]Liquids and solids at 25°C and 1 atm; gases at 0°C and 1 atm (STP).

TABLE C-9. Miscellaneous Physical Properties of the Elements (*Continued*)

Element	Symbol	Atomic Number	Standard State[b,c]	Density[d] (g mL^{-1})	Melting Point (°C)	Boiling Point (°C)
copper	Cu	29	solid	8.96	1084.6	2562
zinc	Zn	30	solid	7.133	419.53	907
gallium	Ga	31	solid	5.904	29.76	2403
germanium	Ge	32	solid	5.323	938.25	2833
arsenic	As	33	solid (gy)	5.727	614	sublimes
selenium	Se	34	solid (gy)	4.79	221	685
bromine	Br	35	liquid	3.12	−7.2	58.78
krypton	Kr	36	gas	0.00373	−157.36	−153.22
rubidium	Rb	37	solid	1.532	39.31	688
strontium	Sr	38	solid	2.54	777	1382
yttrium	Y	39	solid	4.469	1526	3336
zirconium	Zr	40	solid	6.506	1855	4409
niobium	Nb	41	solid	8.57	2477	4744
molybdenum	Mo	42	solid	10.22	2623	4639
technetium	Tc	43	solid	11.50	2157	4538
ruthenium	Ru	44	solid	12.41	2334	3900
rhodium	Rh	45	solid	12.41	1964	3695
palladium	Pd	46	solid	12.02	1555	2963
silver	Ag	47	solid	10.50	961.78	2212
cadmium	Cd	48	solid	8.65	321.07	767
indium	In	49	solid	7.31	156.60	2072
tin	Sn	50	solid (wh)	7.31	231.9	2270
antimony	Sb	51	solid	6.691	630.63	1750
tellurium	Te	52	solid	6.24	449.5	998
iodine	I	53	solid	4.93	113.7	184.4
xenon	Xe	54	gas	0.00589	−111.75	−108.0
cesium	Cs	55	solid	1.873	28.44	671
barium	Ba	56	solid	3.5	727	1640
lanthanum	La	57	solid	6.145	920	3455
cerium	Ce	58	solid	6.770	799	3424
praseodymium	Pr	59	solid	6.773	931	3510

Continued on next page

TABLE C-9. Miscellaneous Physical Properties of the Elements (*Continued*)

Element	Symbol	Atomic Number	Standard State[b,c]	Density[d] (g mL^{-1})	Melting Point (°C)	Boiling Point (°C)
neodymium	Nd	60	solid	7.008	1016	3066
promethium	Pm	61	solid	7.264	1042	≈3000
samarium	Sm	62	solid	7.520	1072	1790
europium	Eu	63	solid	5.244	822	1596
gadolinium	Gd	64	solid	7.901	1314	3264
terbium	Tb	65	solid	8.230	1359	3221
dysprosium	Dy	66	solid	8.551	1411	2561
holmium	Ho	67	solid	8.795	1472	2694
erbium	Er	68	solid	9.066	1529	2862
thulium	Tm	69	solid	9.321	1545	1946
ytterbium	Yb	70	solid	6.966	824	1194
lutetium	Lu	71	solid	9.841	1663	3393
hafnium	Hf	72	solid	13.31	2233	4603
tantalum	Ta	73	solid	16.654	3017	5458
tungsten	W	74	solid	19.3	3422	5660
rhenium	Re	75	solid	21.02	3186	5596
osmium	Os	76	solid	22.57	3033	5012
iridium	Ir	77	solid	22.42	2446	4130
platinum	Pt	78	solid	21.45	1768.4	3825
gold	Au	79	solid	19.3	1064.18	2856
mercury	Hg	80	liquid	13.546	−38.83	356.73
thallium	Tl	81	solid	11.85	304	1473
lead	Pb	82	solid	11.35	327.46	1749
bismuth	Bi	83	solid	9.747	271.4	1564
polonium	Po	84	solid	9.32	254	962
astatine	At	85	solid	unknown	302	337
radon	Rn	86	gas	0.00973	−71	−61.7
francium	Fr	87	solid	unknown	27	677

[b]Normal state at 25°C and 1 atm.
[c]Allotropes: gr = graphite, gy = gray, rh = rhombic, wh = white.
[d]Liquids and solids at 25°C and 1 atm; gases at 0°C and 1 atm (STP).

TABLE C-9. Miscellaneous Physical Properties of the Elements (*Continued*)

Element	Symbol	Atomic Number	Standard State[b, c]	Density[d] (g mL^{-1})	Melting Point (°C)	Boiling Point (°C)
radium	Ra	88	solid	5	700	1737
actinium	Ac	89	solid	10.07	1051	≈3200
thorium	Th	90	solid	11.72	1750	4788
protactinium	Pa	91	solid	15.37	1572	unknown
uranium	U	92	solid	18.95	1132	3818

TABLE C-10. Valence Configurations and Oxidation States of the Transition Metals

Series	Element	Symbol	Atomic Number	Valence	Oxidation State[a]							
					+1	+2	+3	+4	+5	+6	+7	+8
1st	scandium	Sc	21	$4s^2 3d^1$			x					
	titanium	Ti	22	$4s^2 3d^2$		x	x	x				
	vanadium	V	23	$4s^2 3d^3$		x	x	x	x			
	chromium	Cr	24	$4s^1 3d^5$		x	x	x	x	x		
	manganese	Mn	25	$4s^2 3d^5$		x	x	x	x	x	x	
	iron	Fe	26	$4s^2 3d^6$		x	x			x		
	cobalt	Co	27	$4s^2 3d^7$		x	x					
	nickel	Ni	28	$4s^2 3d^8$		x	x	x				
	copper	Cu	29	$4s^1 3d^{10}$	x	x	x					
	zinc	Zn	30	$4s^2 3d^{10}$		x						
2nd	yttrium	Y	39	$5s^2 4d^1$			x					
	zirconium	Zr	40	$5s^2 4d^2$			x	x				
	niobium	Nb	41	$5s^1 4d^4$				x	x			
	molybdenum	Mo	42	$5s^1 4d^5$		x	x	x	x	x		
	technetium	Tc	43	$5s^2 4d^5$		x	x	x	x	x	x	
	ruthenium	Ru	44	$5s^1 4d^7$		x	x	x				x
	rhodium	Rh	45	$5s^1 4d^8$		x	x		x			
	palladium	Pd	46	$4d^{10}$		x		x				
	silver	Ag	47	$5s^1 4d^{10}$	x	x						
	cadmium	Cd	48	$5s^2 4d^{10}$		x						
3rd	lanthanum	La	57	$6s^2 5d^1$			x					
	hafnium	Hf	72	$6s^2 4f^{14} 5d^2$			x	x				
	tantalum	Ta	73	$6s^2 4f^{14} 5d^3$				x	x			
	tungsten	W	74	$6s^2 4f^{14} 5d^4$				x	x	x		
	rhenium	Re	75	$6s^2 4f^{14} 5d^5$		x	x	x	x	x	x	
	osmium	Os	76	$6s^2 4f^{14} 5d^6$		x	x	x				x
	iridium	Ir	77	$6s^2 4f^{14} 5d^7$			x	x		x		
	platinum	Pt	78	$6s^1 4f^{14} 5d^9$		x		x		x		
	gold	Au	79	$6s^1 4f^{14} 5d^{10}$	x		x					
	mercury	Hg	80	$6s^2 4f^{14} 5d^{10}$	x	x						

[a]A partial listing. The most common oxidation numbers are marked x; others, less frequent, are marked x.

TABLE C-11. A Selection of Stable Isotopes[a]

Isotope ^AX	Natural Abundance (%)	Atomic Number Z	Neutron Number N	Mass Number A	Atomic Mass (u)	Binding Energy per Nucleon (MeV)
^1H	99.985	1	0	1	1.007825	—
^2H	0.015	1	1	2	2.014000	1.160
^3He	0.000137	2	1	3	3.016030	2.572
^4He	99.999863	2	2	4	4.002603	7.075
^6Li	7.5	3	3	6	6.015121	5.333
^7Li	92.5	3	4	7	7.016003	5.606
^9Be	100.0	4	5	9	9.012182	6.463
^{10}B	19.9	5	5	10	10.012937	6.475
^{11}B	80.1	5	6	11	11.009305	6.928
^{12}C	98.90	6	6	12	12.000000	7.680
^{13}C	1.10	6	7	13	13.003355	7.470
^{14}N	99.634	7	7	14	14.003074	7.476
^{15}N	0.366	7	8	15	15.000108	7.699
^{16}O	99.762	8	8	16	15.994915	7.976
^{17}O	0.038	8	9	17	16.999131	7.751
^{18}O	0.200	8	10	18	17.999160	7.767
^{19}F	100.0	9	10	19	18.998403	7.779
^{20}Ne	90.48	10	10	20	19.992435	8.032
^{21}Ne	0.27	10	11	21	20.993843	7.972
^{22}Ne	9.25	10	12	22	21.991383	8.081
^{23}Na	100.0	11	12	23	22.989770	8.112
^{24}Mg	78.99	12	12	24	23.985042	8.261
^{25}Mg	10.00	12	13	25	24.985837	8.223
^{26}Mg	11.01	12	14	26	25.982593	8.334
^{27}Al	100.0	13	14	27	26.981538	8.331
^{28}Si	92.23	14	14	28	27.976927	8.448
^{29}Si	4.67	14	15	29	28.976495	8.449
^{30}Si	3.10	14	16	30	29.973770	8.521

[a]Selection is complete through cobalt-59. Where natural abundances do not add to 100%, the differences are made up by radioactive isotopes with exceedingly long half-lives: potassium-40 (0.0117%, $t_{1/2} = 1.3 \times 10^9$ y); vanadium-50 (0.250%, $t_{1/2} > 1.4 \times 10^{17}$ y).

Continued on next page

TABLE C-11.   A Selection of Stable Isotopes (*Continued*)

Isotope ^AX	Natural Abundance (%)	Atomic Number Z	Neutron Number N	Mass Number A	Atomic Mass (u)	Binding Energy per Nucleon (MeV)
^{31}P	100.0	15	16	31	30.973761	8.481
^{32}S	95.02	16	16	32	31.972070	8.493
^{33}S	0.75	16	17	33	32.971456	8.498
^{34}S	4.21	16	18	34	33.967866	8.584
^{36}S	0.02	16	20	36	35.967080	8.575
^{35}Cl	75.77	17	18	35	34.968852	8.520
^{37}Cl	24.23	17	20	37	36.965903	8.570
^{36}Ar	0.337	18	18	36	35.967545	8.520
^{38}Ar	0.063	18	20	38	37.962732	8.614
^{40}Ar	99.600	18	22	40	39.962384	8.595
^{39}K	93.258	19	20	39	38.963707	8.557
^{41}K	6.730	19	22	41	40.961825	8.576
^{40}Ca	96.941	20	20	40	39.962591	8.551
^{42}Ca	0.647	20	22	42	41.958618	8.617
^{43}Ca	0.135	20	23	43	42.958766	8.601
^{44}Ca	2.086	20	24	44	43.955480	8.658
^{46}Ca	0.004	20	26	46	45.953689	8.669
^{48}Ca	0.187	20	28	48	47.952533	8.666
^{45}Sc	100.0	21	24	45	44.955910	8.619
^{46}Ti	8.0	22	24	46	45.952629	8.656
^{47}Ti	7.3	22	25	47	46.951764	8.661
^{48}Ti	73.8	22	26	48	47.947947	8.723
^{49}Ti	5.5	22	27	49	48.947871	8.711
^{50}Ti	5.4	22	28	50	49.944792	8.756
^{51}V	99.750	23	28	51	50.943962	8.742
^{50}Cr	4.345	24	26	50	49.946046	8.701
^{52}Cr	83.789	24	28	52	51.940509	8.776
^{53}Cr	9.501	24	29	53	52.940651	8.760
^{54}Cr	2.365	24	30	54	53.938882	8.778
^{55}Mn	100.0	25	30	55	54.938049	8.765
^{54}Fe	5.9	26	28	54	53.939612	8.736

TABLE C-11. A Selection of Stable Isotopes (*Continued*)

Isotope ^{A}X	Natural Abundance (%)	Atomic Number Z	Neutron Number N	Mass Number A	Atomic Mass (u)	Binding Energy per Nucleon (MeV)
^{56}Fe	91.72	26	30	56	55.934939	8.790
^{57}Fe	2.1	26	31	57	56.935396	8.770
^{58}Fe	0.28	26	32	58	57.933277	8.792
^{59}Co	100.0	27	32	59	58.933200	8.768
^{204}Pb	1.4	82	122	204	203.973020	7.880
^{206}Pb	24.1	82	124	206	205.974440	7.875
^{207}Pb	22.1	82	125	207	206.975872	7.870
^{208}Pb	52.4	82	126	208	207.976627	7.868
^{209}Bi	100.0	83	126	209	208.980380	7.848

TABLE C-12. A Selection of Radioactive Isotopes

Isotope $^A X$	Decay Mode[a]	Half-Life $t_{1/2}$	Atomic Number Z	Neutron Number N	Mass Number A	Atomic Mass (u)	Binding Energy per Nucleon (MeV)
^{3}H	β^-	12.3 y	1	2	3	3.01605	2.827
^{8}Be	α	$\approx 7 \times 10^{-17}$ s	4	4	8	8.005305	7.062
^{14}C	β^-	5.7×10^3 y	6	8	14	14.003241	7.520
^{22}Na	β^+	2.6 y	11	11	22	21.994434	7.916
^{24}Na	β^-	15.0 h	11	13	24	23.990961	8.064
^{32}P	β^-	14.3 d	15	17	32	31.973907	8.464
^{35}S	β^-	87.2 d	16	19	35	34.969031	8.538
^{59}Fe	β^-	44.5 d	26	33	59	58.934877	8.755
^{60}Co	β^-	5.3 y	27	33	60	59.933819	8.747
^{90}Sr	β^-	29.1 y	38	52	90	89.907738	8.696
^{99}Tc	β^-	2.1×10^5 y	43	56	99	98.906524	8.611
^{109}Cd	EC	462 d	48	61	109	108.904953	8.539
^{125}I	EC	59.4 d	53	72	125	124.904620	8.450
^{131}I	β^-	8.04 d	53	78	131	130.906114	8.422
^{137}Cs	β^-	30.3 y	55	82	137	136.907073	8.389
^{222}Rn	α	3.82 d	86	136	222	222.017570	7.695
^{226}Ra	α	1600 y	88	138	226	226.025402	7.662
^{232}Th	α	1.4×10^{10} y	90	142	232	232.038054	7.615
^{235}U	α	7.0×10^8 y	92	143	235	235.043924	7.591
^{238}U	α	4.5×10^9 y	92	146	238	238.050784	7.570
^{239}Pu	α	2.4×10^4 y	94	145	239	239.052157	7.560

[a]Modes of decay include alpha emission (α), beta emission (β^-), positron emission (β^+), electron capture (EC).

C-3. MOLECULES AND INTERACTIONS

TABLE C-13. Length and Strength of Covalent Bonds
TABLE C-14. Critical Temperatures (T_c) and van der Waals
Parameters (a, b) of Real Gases
TABLE C-15. Properties of Liquid Water Between 0°C
and 100°C
TABLE C-16. Thermodynamic Properties at 25°C

TABLE C-13. Length and Strength of Covalent Bonds

Atom	Bond	Bond Length (Average) (Å)	Dissociation Enthalpy (Average) (kJ mol^{-1})
H	H—H	0.74	436
	H—F	0.92	565
	H—Cl	1.27	431
	H—Br	1.41	366
	H—I	1.61	299
C	C—C	1.54	348
	C=C	1.34	612
	C≡C	1.20	838
	C—H	1.09	413
	C—N	1.43	305
	C=N	1.38	613
	C≡N	1.16	890
	C—O	1.43	360
	C=O	1.23	743
	C≡O	1.13	1076
	C—F	1.27	484
	C—Cl	1.65	338
	C—Br	1.79	276
	C—I	2.15	238
N	N—N	1.47	163
	N=N	1.24	409
	N≡N	1.10	946
	N—H	1.01	388
	N—O	1.36	157
	N=O	1.22	630
O	O—O	1.48	146
	O=O	1.21	497
	O—H	0.96	463
F	F—F	1.41	155
Cl	Cl—Cl	1.99	242
Br	Br—Br	2.28	193
I	I—I	2.67	151

TABLE C-14.   Critical Temperatures (T_c) and van der Waals
Parameters (a, b) of Real Gases

Gas^a	Molar Mass (g mol^{-1})	T_c (K)	a (atm L^2 mol^{-2})	b (L mol^{-1})
H_2O	18.015	647.14	5.537	0.0305
Br_2	159.808	588	9.75	0.0591
CCl_3F	137.367	471.2	14.68	0.1111
Cl_2	70.906	416.9	6.343	0.0542
CO_2	44.010	304.14	3.658	0.0429
Kr	83.80	209.41	2.325	0.0396
CH_4	16.043	190.53	2.300	0.0430
O_2	31.999	154.59	1.382	0.0319
Ar	39.948	150.87	1.355	0.0320
F_2	37.997	144.13	1.171	0.0290
CO	28.010	132.91	1.472	0.0395
N_2	28.013	126.21	1.370	0.0387
H_2	2.016	32.97	0.245	0.0265
He	4.003	5.19	0.035	0.0238

[a]Listed in descending order of critical temperature.

TABLE C-15. Properties of Liquid Water Between 0°C and 100°C

Temperature (°C)	Density (g mL^{-1})	Vapor Pressure (torr)	Molar Heat Capacity (c_P) (J mol^{-1} K^{-1})	Enthalpy of Vaporization (kJ mol^{-1})	Ion-Product Constant[a] K_w
0	0.99984	4.6	75.98	45.0	1.2×10^{-15}
4	1.00000	6.1	—	—	1.9×10^{-15}
10	0.99970	9.2	75.52	—	3.0×10^{-15}
20	0.99821	17.5	75.34	—	6.9×10^{-15}
25	0.99705	23.8	75.29	44.0	1.0×10^{-14}
30	0.99565	31.8	75.28	—	1.5×10^{-14}
40	0.99222	55.4	75.28	43.4	2.9×10^{-14}
50	0.98803	92.6	75.31	—	5.3×10^{-14}
60	0.98320	149.5	75.38	42.5	9.2×10^{-14}
70	0.97778	233.8	75.48	—	1.5×10^{-13}
80	0.97182	355.3	75.60	41.6	2.4×10^{-13}
90	0.96535	525.9	75.75	—	3.7×10^{-13}
100	0.95840	760.0	75.95	40.7	5.4×10^{-13}

[a]Also called autoionization constant or autodissociation constant.

TABLE C-16. Thermodynamic Properties at 25°C

Substance[a,b]	$\mathcal{M}$ (g mol^{-1})	c_P (J mol^{-1} K^{-1})	ΔH_f° (kJ mol^{-1})	S° (J mol^{-1} K^{-1})	ΔG_f° (kJ mol^{-1})
		Elements and Monatomic Ions			
Ag$^+$(aq)	107.868	21.8	105.6	72.7	77.1
Ag(g)	107.868	20.8	284.9	173.0	246.0
Ag(s)	107.868	25.4	0.0	42.6	0.0
Al^{3+}(aq)	26.982	—	−531	−321.7	−485
Al(g)	26.982	21.4	330.0	164.6	289.4
Al(s)	26.982	24.4	0.0	28.3	0.0
Ar(g)	39.948	20.8	0.0	154.8	0.0
Au(g)	196.967	20.8	366.1	180.5	326.3
Au(s)	196.967	25.4	0.0	47.4	0.0
B(g)	10.811	20.8	565.0	153.4	521.0
B(s)	10.811	11.1	0.0	5.9	0.0
Ba^{2+}(aq)	137.327	—	−537.6	9.6	−560.8
Ba(g)	137.327	20.8	180.0	170.2	146.0
Ba(s)	137.327	28.1	0.0	62.8	0.0
Be(g)	9.012	20.8	324.0	136.3	286.6
Be(s)	9.012	16.4	0.0	9.5	0.0
Br$^-$(aq)	79.904	−141.8	−121.6	82.4	−104.0
Br(g)	79.904	20.8	111.9	175.0	82.4
Br$_2$(g)	159.808	36.0	30.9	245.5	3.1
Br$_2$(ℓ)	159.808	75.7	0.0	152.2	0.0
C(g)	12.011	20.8	716.7	158.1	671.3
C(s, diamond)	12.011	6.1	1.9	2.4	2.9
C(s, graphite)	12.011	8.5	0.0	5.7	0.0
Ca^{2+}(aq)	40.078	—	−542.8	−53.1	−553.6
Ca(g)	40.078	20.8	177.8	154.9	144.0
Ca(s)	40.078	25.9	0.0	41.6	0.0

[a]Substances are arranged alphabetically by chemical formula within each class: (1) elements and monatomic ions; (2) polyatomic ions; (3) inorganic compounds (including CO and CO$_2$); (4) organic molecules (hydrocarbon-based).

[b]Symbols denote molar mass ($\mathcal{M}$), molar heat capacity at constant pressure (c_P), standard enthalpy of formation (ΔH_f°), standard third-law entropy (S°), and standard Gibbs free energy of formation (ΔG_f°). Entropies in aqueous solution are referred to $S^\circ[\text{H}^+(\text{aq})] = 0$, not to absolute zero.

Continued on next page

TABLE C-16.   Thermodynamic Properties at 25°C (*Continued*)

Substance[a,b]	$\mathcal{M}$ (g mol^{-1})	c_P (J mol^{-1} K^{-1})	ΔH_f° (kJ mol^{-1})	S° (J mol^{-1} K^{-1})	ΔG_f° (kJ mol^{-1})
Cl^-(aq)	35.453	−136.4	−167.2	56.5	−131.2
Cl(g)	35.453	21.8	121.3	165.2	105.3
Cl_2(g)	70.906	33.9	0.0	223.0	0.0
Co^{2+}(aq)	58.933	—	−58.2	−113	−54.4
Co^{3+}(aq)	58.933	—	92	−305	134
Co(g)	58.933	23.0	424.7	179.5	380.3
Co(s)	58.933	24.8	0.0	30.0	0.0
Cr(g)	51.996	20.8	396.6	174.5	351.8
Cr(s)	51.996	23.4	0.0	23.8	0.0
Cs^+(aq)	132.905	−10.5	−258.3	133.1	−292.0
Cs(g)	132.905	20.8	76.5	175.6	49.6
Cs(s)	132.905	32.2	0.0	85.2	0.0
Cu^+(aq)	63.546	—	71.7	40.6	50.0
Cu^{2+}(aq)	63.546	—	64.8	−99.6	65.5
Cu(g)	63.546	20.8	337.4	166.4	297.7
Cu(s)	63.546	24.4	0.0	33.2	0.0
F^-(aq)	18.998	−106.7	−332.6	−13.8	−278.8
F(g)	18.998	22.7	79.4	158.8	62.3
F_2(g)	37.997	31.3	0.0	202.8	0.0
Fe^{2+}(aq)	55.845	—	−89.1	−137.7	−78.9
Fe^{3+}(aq)	55.845	—	−48.5	−315.9	−4.7
Fe(g)	55.845	25.7	416.3	180.5	370.7
Fe(s)	55.845	25.1	0.0	27.3	0.0
H^+(aq)	1.008	0.0	0.0	0.0	0.0
H(g)	1.008	20.8	218.0	114.7	203.3
H_2(g)	2.016	28.8	0.0	130.6	0.0
He(g)	4.003	20.8	0.0	126.2	0.0

[a]Substances are arranged alphabetically by chemical formula within each class: (1) elements and monatomic ions; (2) polyatomic ions; (3) inorganic compounds (including CO and CO_2); (4) organic molecules (hydrocarbon-based).

[b]Symbols denote molar mass ($\mathcal{M}$), molar heat capacity at constant pressure (c_P), standard enthalpy of formation (ΔH_f°), and standard third-law entropy (S°), and standard Gibbs free energy of formation (ΔG_f°). Entropies in aqueous solution are referred to $S^\circ[H^+(aq)] = 0$, not to absolute zero.

TABLE C-16. Thermodynamic Properties at 25°C (*Continued*)

Substance[a,b]	$\mathcal{M}$ (g mol^{-1})	c_P (J mol^{-1} K^{-1})	ΔH_f° (kJ mol^{-1})	S° (J mol^{-1} K^{-1})	ΔG_f° (kJ mol^{-1})
Hg_2^{2+}(aq)	401.18	—	172.4	84.5	153.5
Hg^{2+}(aq)	401.18	—	171.1	−32.2	164.4
Hg(g)	200.59	20.8	61.4	175.0	31.8
Hg(ℓ)	200.59	28.0	0.0	75.9	0.0
I$^-$(aq)	126.905	−142.3	−55.2	111.3	−51.6
I(g)	126.905	20.8	106.8	180.8	70.2
I_2(g)	253.809	36.9	62.4	260.7	19.3
I_2(s)	253.809	54.4	0.0	116.1	0.0
K$^+$(aq)	39.098	21.8	−252.4	102.5	−283.3
K(g)	39.098	20.8	89.0	160.3	60.5
K(s)	39.098	29.6	0.0	64.7	0.0
Li$^+$(aq)	6.941	68.6	−278.5	13.4	−293.3
Li(g)	6.941	20.8	159.3	138.8	126.6
Li$^+$(g)	6.941	20.8	685.7	133.0	648.5
Li(s)	6.941	24.8	0.0	29.1	0.0
Mg^{2+}(aq)	24.305	—	−466.9	−138.1	−454.8
Mg(g)	24.305	20.8	147.1	148.6	112.5
Mg(s)	24.305	24.9	0.0	32.7	0.0
Mn^{2+}(aq)	54.938	50	−220.8	−73.6	−228.1
Mn(g)	54.938	20.8	280.7	173.7	238.5
Mn(s)	54.938	26.3	0.0	32.0	0.0
N(g)	14.007	20.8	472.7	153.3	455.5
N_2(g)	28.013	29.1	0.0	191.5	0.0
Na$^+$(aq)	22.990	46.4	−240.1	59.0	−261.9
Na(g)	22.990	20.8	107.5	153.7	77.0
Na$^+$(g)	22.990	20.8	609.3	148.0	574.3
Na(s)	22.990	28.2	0.0	51.3	0.0
Ne(g)	20.180	20.8	0.0	146.3	0.0
Ni^{2+}(aq)	58.693	—	−54.0	−128.9	−45.6
Ni(g)	58.693	23.4	429.7	182.2	384.5
Ni(s)	58.693	26.1	0.0	29.9	0.0
O(g)	15.999	21.9	249.2	161.1	231.7

Continued on next page

TABLE C-16. Thermodynamic Properties at 25°C (*Continued*)

Substance[a,b]	$\mathcal{M}$ (g mol^{-1})	c_P (J mol^{-1} K^{-1})	ΔH_f° (kJ mol^{-1})	S° (J mol^{-1} K^{-1})	ΔG_f° (kJ mol^{-1})
$O_2(g)$	31.999	29.4	0.0	205.0	0.0
$P(g)$	30.974	20.8	314.6	163.1	278.3
$P_4(s, \text{red})$	123.895	21.2	−17.6	22.8	−12.1
$P_4(s, \text{white})$	123.895	23.8	0.0	41.1	0.0
$Pb^{2+}(aq)$	207.2	—	−1.7	10.5	−24.4
$Pb(g)$	207.2	20.8	195.2	162.2	175.4
$Pb(s)$	207.2	26.4	0.0	64.8	0.0
$Rb^+(aq)$	85.468	—	−251.2	121.5	−284.0
$Rb(g)$	85.468	20.8	80.9	170.1	53.1
$Rb(s)$	85.468	31.1	0.0	76.8	0.0
$S(g)$	32.066	23.7	277.2	167.8	236.7
$S(s, \text{rhombic})$	32.066	22.6	0.0	32.1	0.0
$Sc(g)$	44.956	22.1	377.8	174.8	336.0
$Sc(s)$	44.956	25.5	0.0	34.6	0.0
$Si(g)$	28.086	22.3	450.0	168.0	405.5
$Si(s)$	28.086	20.0	0.0	18.8	0.0
$Sn(g)$	118.710	21.3	301.2	168.5	266.2
$Sn(s, \text{gray})$	118.710	25.8	−2.1	44.1	0.1
$Sn(s, \text{white})$	118.710	27.0	0.0	51.2	0.0
$Sr^{2+}(aq)$	87.62	—	−545.8	−32.6	−559.5
$Sr(g)$	87.62	20.8	164.4	164.6	130.9
$Sr(s)$	87.62	26.4	0.0	52.3	0.0
$Ti(g)$	47.867	24.4	473.0	180.3	428.4
$Ti(s)$	47.867	25.0	0.0	30.7	0.0
$V(g)$	50.942	26.0	514.2	182.2	468.5
$V(s)$	50.942	24.9	0.0	28.9	0.0
$Zn^{2+}(aq)$	65.39	46	−153.9	−112.1	−147.1

[a]Substances are arranged alphabetically by chemical formula within each class: (1) elements and monatomic ions; (2) polyatomic ions; (3) inorganic compounds (including CO and CO_2); (4) organic molecules (hydrocarbon-based).

[b]Symbols denote molar mass ($\mathcal{M}$), molar heat capacity at constant pressure (c_P), standard enthalpy of formation (ΔH_f°), standard third-law entropy (S°), and standard Gibbs free energy of formation (ΔG_f°). Entropies in aqueous solution are referred to $S^\circ[H^+(aq)] = 0$, not to absolute zero.

TABLE C-16. Thermodynamic Properties at 25°C (*Continued*)

Substance[a,b]	$\mathcal{M}$ (g mol^{-1})	c_P (J mol^{-1} K^{-1})	ΔH_f° (kJ mol^{-1})	S° (J mol^{-1} K^{-1})	ΔG_f° (kJ mol^{-1})
Zn(g)	65.39	20.8	130.4	161.0	94.8
Zn(s)	65.39	25.4	0.0	41.6	0.0
Polyatomic Ions					
CH_3COO^-(aq)	59.045	−6.3	−486.0	86.6	−369.3
CO_3^{2-}(aq)	60.009	—	−677.1	−56.9	−527.8
$C_2O_4^{2-}$(aq)	88.020	—	−825.1	45.6	−673.9
CrO_4^{2-}(aq)	115.994	—	−881.2	50.2	−727.8
$Cr_2O_7^{2-}$(aq)	215.988	—	−1490.3	261.9	−1301.1
$HCOO^-$(aq)	45.018	−87.9	−425.6	92	−351.0
HCO_3^-(aq)	61.017	—	−692.0	91.2	−586.8
HSO_4^-(aq)	97.072	−84	−887.3	131.8	−755.9
MnO_4^-(aq)	118.936	−82.0	−541.4	191.2	−447.2
NH_4^+(aq)	18.038	79.9	−132.5	113.4	−79.3
NO_3^-(aq)	62.005	−86.6	−205.0	146.4	−108.7
OH^-(aq)	17.007	−148.5	−230.0	−10.8	−157.2
PO_4^{3-}(aq)	94.971	—	−1277.4	−222	−1018.7
SO_4^{2-}(aq)	96.064	−293	−909.3	20.1	−744.5
Inorganic Compounds					
AgCl(s)	143.321	50.8	−127.1	96.2	−109.8
AgI(s)	234.773	56.8	−61.8	115.5	−66.2
$AgNO_3$(s)	169.873	93.1	−124.4	140.9	−33.4
Al_2O_3(s)	101.961	79.0	−1675.7	50.9	−1582.3
$BaCO_3$(s)	197.34	85.4	−1216.3	112.1	−1137.6
$BaSO_4$(s)	233.39	101.8	−1473.2	132.2	−1362.2
$CaCO_3$(s)	100.087	81.9	−1206.9	92.9	−1128.8
$CaCl_2$(s)	110.984	72.9	−795.4	108.4	−748.8
CaF_2(s)	78.075	67.0	−1228.0	68.5	−1175.6
CaO(s)	56.077	42.0	−634.9	38.1	−603.3
$Ca(OH)_2$(s)	74.093	87.5	−985.2	83.4	−897.5
$CaSO_4$(s)	136.142	99.7	−1434.5	106.5	−1322.0
CO(g)	28.010	29.1	−110.5	197.7	−137.2

Continued on next page

TABLE C-16. Thermodynamic Properties at 25°C (*Continued*)

Substance[a,b]	$\mathcal{M}$ (g mol^{-1})	c_p (J mol^{-1} K^{-1})	ΔH_f° (kJ mol^{-1})	S° (J mol^{-1} K^{-1})	ΔG_f° (kJ mol^{-1})
$CO_2(g)$	44.010	37.1	−393.5	213.6	−394.4
$CsCl(s)$	168.358	52.5	−443.0	101.2	−414.6
$CuSO_4(s)$	159.610	100.0	−771.4	109.2	−662.2
$FeCl_2(s)$	126.750	76.7	−341.8	118.0	−302.3
$FeCl_3(s)$	162.203	96.7	−399.5	142.3	−334.0
$FeO(s)$	71.844	—	−271.9	60.8	−255.2
$Fe_2O_3(s)$	159.688	103.9	−824.2	87.4	−742.2
$HBr(g)$	80.912	29.1	−36.3	198.7	−53.4
$HCl(g)$	36.461	29.1	−92.3	186.9	−95.3
$HF(g)$	20.006	29.1	−273.3	173.8	−275.4
$HI(g)$	127.912	29.2	26.5	206.6	1.7
$HNO_3(g)$	63.013	53.4	−135.1	266.4	−74.7
$HNO_3(\ell)$	63.013	109.9	−174.1	155.6	−80.7
$HgCl_2(s)$	271.50	—	−224.3	146.0	−178.6
$Hg_2Cl_2(s)$	472.09	102	−265.4	191.6	−210.7
$H_2O(g)$	18.015	35.6	−241.8	188.7	−228.6
$H_2O(\ell)$	18.015	75.3	−285.8	70.0	−237.2
$H_2O_2(g)$	34.015	43.1	−136.3	232.7	−105.6
$H_2O_2(\ell)$	34.015	89.1	−187.8	109.6	−120.4
$H_2SO_4(\ell)$	98.079	138.9	−814.0	156.9	−690.0
$KBr(s)$	119.002	52.3	−393.8	95.9	−380.7
$KCl(s)$	74.551	51.3	−436.5	82.6	−408.5
$LiBr(s)$	86.845	—	−351.2	74.3	−342.0
$LiCl(s)$	42.394	48.0	−408.6	59.3	−384.4
$Li_2CO_3(s)$	73.891	99.1	−1215.9	90.4	−1132.1
$MgCl_2(s)$	95.211	71.4	−641.3	89.6	−591.8
$Mg(OH)_2(s)$	58.320	77.0	−924.5	63.2	−833.5

[a]Substances are arranged alphabetically by chemical formula within each class: (1) elements and monatomic ions; (2) polyatomic ions; (3) inorganic compounds (including CO and CO_2); (4) organic molecules (hydrocarbon-based).

[b]Symbols denote molar mass ($\mathcal{M}$), molar heat capacity at constant pressure (c_p), standard enthalpy of formation (ΔH_f°), standard third-law entropy (S°), and standard Gibbs free energy of formation (ΔG_f°). Entropies in aqueous solution are referred to $S^\circ[H^+(aq)] = 0$, not to absolute zero.

TABLE C-16. Thermodynamic Properties at 25°C (*Continued*)

Substance[a,b]	$\mathcal{M}$ (g mol^{-1})	c_P (J mol^{-1} K^{-1})	ΔH_f° (kJ mol^{-1})	S° (J mol^{-1} K^{-1})	ΔG_f° (kJ mol^{-1})
$MgSO_4(s)$	120.369	96.5	−1284.9	91.6	−1170.6
$MnO_2(s)$	86.937	54.1	−520.0	53.1	−465.1
$NaBr(s)$	102.894	51.4	−361.1	86.8	−349.0
$NaCH_3COO(s)$	82.034	79.9	−708.8	123.0	−607.2
$NaCl(s)$	58.443	50.5	−411.2	72.1	−384.2
$Na_2CO_3(s)$	105.989	112.3	−1130.7	135.0	−1044.4
$NaHCO_3(s)$	84.007	87.6	−950.8	101.7	−851.0
$NaNO_3(s)$	84.995	92.9	−467.9	116.5	−367.0
$NaOH(s)$	39.997	59.5	−425.6	64.5	−379.5
$Na_2SO_4(s)$	142.043	128.2	−1387.1	149.6	−1270.2
$NF_3(g)$	71.002	53.4	−132.1	260.8	−90.6
$NH_3(g)$	17.031	35.1	−46.1	192.3	−16.5
$NH_4NO_3(s)$	80.043	139.3	−365.6	151.1	−183.9
$NiCl_2(s)$	129.60	71.7	−305.3	97.7	−259.0
$NiO(s)$	74.69	44.3	−239.7	38.0	−211.7
$NO(g)$	30.006	29.8	90.3	210.7	86.6
$NO_2(g)$	46.006	37.2	33.2	240.0	51.3
$N_2O(g)$	44.013	38.5	82.1	219.9	104.2
$N_2O_4(g)$	92.011	77.3	9.2	304.2	97.8
$NOCl(g)$	65.459	44.7	51.7	261.7	66.1
$O_3(g)$	47.998	39.2	142.7	238.8	163.2
$PbCl_2(s)$	278.1	—	−359.4	136.0	−314.1
$PbSO_4(s)$	303.3	103.2	−920.0	148.5	−813.0
$SO_2(g)$	64.065	39.9	−296.8	248.2	−300.1
$SO_3(g)$	80.064	50.7	−395.7	256.8	−371.1
$ZnCl_2(s)$	136.30	71.3	−415.1	111.5	−369.4
$ZnSO_4(s)$	161.45	99.2	−982.8	110.5	−871.5
Organic Molecules					
$CCl_4(g)$	153.823	83.3	−102.9	309.7	−60.6
$CCl_4(\ell)$	153.823	131.8	−135.4	216.4	−65.3
$CH_4(g)$	16.043	35.3	−74.8	186.2	−50.8

Continued on next page

TABLE C-16. Thermodynamic Properties at 25°C (*Continued*)

Substance[a,b]	$\mathcal{M}$ (g mol^{-1})	c_P (J mol^{-1} K^{-1})	ΔH_f° (kJ mol^{-1})	S° (J mol^{-1} K^{-1})	ΔG_f° (kJ mol^{-1})
$CH_3COOH(g)$	60.053	66.5	−432.8	282.5	−374.5
$CH_3COOH(\ell)$	60.053	123.3	−484.5	159.8	−389.9
$CH_3OH(g)$	32.042	43.9	−200.7	239.7	−162.0
$CH_3OH(\ell)$	32.042	81.6	−238.7	126.8	−166.4
$C_2H_2(g)$	26.038	43.9	226.7	200.8	209.2
$C_2H_4(g)$	28.054	43.6	52.3	219.5	68.1
$C_2H_6(g)$	30.070	52.6	−84.7	229.5	−32.9
$C_2H_5OH(g)$	46.069	65.4	−235.1	282.6	−168.6
$C_2H_5OH(\ell)$	46.069	111.5	−277.7	160.7	−174.9
$C_3H_8(g)$	44.097	73.5	−103.9	269.9	−23.5
$n\text{-}C_4H_{10}(g)^c$	58.123	97.5	−124.7	310.0	−15.7
$n\text{-}C_4H_{10}(\ell)^c$	58.123	140.9	−147.6	231.0	−15.0
$C_6H_6(g)$	78.114	81.6	82.9	269.2	129.7
$C_6H_6(\ell)$	78.114	136.3	49.0	172.8	124.5
$C_6H_{12}O_6(s)$	180.158	218.8	−1274.4	212.1	−910.1
$n\text{-}C_8H_{18}(\ell)^c$	114.231	254.6	−249.9	361.1	6.4
$C_{12}H_{22}O_{11}(s)$	342.300	425.5	−2221.7	360.2	−1543.8
$HCOOH(\ell)$	46.026	99.0	−424.7	129.0	−361.4

[a]Substances are arranged alphabetically by chemical formula within each class: (1) elements and monatomic ions; (2) polyatomic ions; (3) inorganic compounds (including CO and CO_2); (4) organic molecules (hydrocarbon-based).

[b]Symbols denote molar mass ($\mathcal{M}$), molar heat capacity at constant pressure (c_P), standard enthalpy of formation (ΔH_f°), standard third-law entropy (S°), and standard Gibbs free energy of formation (ΔG_f°). Entropies in aqueous solution are referred to $S^\circ[H^+(aq)] = 0$, not to absolute zero.

[c]The symbol n denotes the "normal" unbranched alkane.

C-4. EQUILIBRIUM

TABLE C-17. Solubility of Ionic Compounds in Water
TABLE C-18. Approximate Solubilities of Selected Inorganic Salts
TABLE C-19. Solubility-Product Constants at 25°C
TABLE C-20. Ionization Constants for Acids at 25°C
TABLE C-21. Standard Reduction Potentials and Free Energies at 25°C
TABLE C-22. Formation Constants of Complexes at 25°C

TABLE C-17.   Solubility of Ionic Compounds in Water[a]

	Br^-	CH_3COO^-	CO_3^{2-}	Cl^-	F^-	I^-	NO_3^-	OH^-	PO_4^{3-}	S^{2-}	SO_4^{2-}
Ag^+	i	s	i	i	s	i	s	i	i	i	ss
Ba^{2+}	s	s	i	s	ss	s	s	s	i	d	i
Ca^{2+}	s	s	i	s	i	s	s	ss	i	ss	ss
Cs^+	s	s	s	s	s	s	s	s	—	s	s
Hg_2^{2+}	i	s	i	i	d	i	s	—	i	i	ss
Hg^{2+}	ss	s	i	s	d	i	s	i	i	i	d
Li^+	s	s	s	s	ss	s	s	s	ss	s	s
K^+	s	s	s	s	s	s	s	s	s	s	s
Mg^{2+}	s	s	i	s	i	s	s	i	i	d	s
NH_4^+	s	s	s	s	s	s	s	s	s	s	s
Na^+	s	s	s	s	s	s	s	s	s	s	s
Pb^{2+}	ss	s	i	ss	ss	i	s	ss	i	i	i
Rb^+	s	s	s	s	s	s	s	s	—	s	s
Sr^{2+}	s	s	i	s	ss	s	s	ss	i	i	i

[a]Compounds of cation and anion are understood to form in proper stoichiometric ratios.

s: soluble (solubility greater than 10 g L^{-1} at room temperature)
ss: slightly soluble (solubility between 0.1 g L^{-1} and 10 g L^{-1})
i: insoluble (solubility less than 0.1 g L^{-1})
d: decomposes

TABLE C-18. Approximate Solubilities of Selected Inorganic Salts

Compound	Molar Mass (g mol^{-1})	Grams Dissolved per Liter H$_2$O	
		0°C	100°C
AgF	126.87	870	2030
AgNO$_3$	169.87	1220	9500
CaCl$_2$	110.99	570	1590
Ca(NO$_3$)$_2$	164.09	1210[a]	3760
Ce$_2$(SO$_4$)$_3$	568.42	200	20
KCl	74.55	270	570
KClO$_4$	138.55	10	20
K$_2$Cr$_2$O$_7$	294.18	50	1000
KNO$_3$	101.10	130	2470
NaCH$_3$COO	82.03	1190	1700
NaCl	58.44	360	390
NaNO$_3$	84.99	750	1800
Pb(NO$_3$)$_2$	331.21	380	1270

[a]$T = 18$°C

TABLE C-19. Solubility-Product Constants at 25°C

Cation	Anion	Heterogeneous Equilibrium[a]	K_{sp}
aluminum	hydroxide	$Al(OH)_3 \rightleftharpoons Al^{3+} + 3OH^-$	1.9×10^{-33}
	phosphate	$AlPO_4 \rightleftharpoons Al^{3+} + PO_4^{3-}$	9.8×10^{-21}
barium	carbonate	$BaCO_3 \rightleftharpoons Ba^{2+} + CO_3^{2-}$	2.6×10^{-9}
	fluoride	$BaF_2 \rightleftharpoons Ba^{2+} + 2F^-$	1.0×10^{-6}
	sulfate	$BaSO_4 \rightleftharpoons Ba^{2+} + SO_4^{2-}$	1.1×10^{-10}
calcium	carbonate	$CaCO_3 \rightleftharpoons Ca^{2+} + CO_3^{2-}$	5.0×10^{-9}
	fluoride	$CaF_2 \rightleftharpoons Ca^{2+} + 2F^-$	3.9×10^{-11}
	hydroxide	$Ca(OH)_2 \rightleftharpoons Ca^{2+} + 2OH^-$	4.7×10^{-6}
	phosphate	$Ca_3(PO_4)_2 \rightleftharpoons 3Ca^{2+} + 2PO_4^{3-}$	2.1×10^{-33}
	sulfate	$CaSO_4 \rightleftharpoons Ca^{2+} + SO_4^{2-}$	7.1×10^{-5}
copper(I)	bromide	$CuBr \rightleftharpoons Cu^+ + Br^-$	6.3×10^{-9}
	chloride	$CuCl \rightleftharpoons Cu^+ + Cl^-$	1.0×10^{-6}
	iodide	$CuI \rightleftharpoons Cu^+ + I^-$	1.3×10^{-12}
copper(II)	phosphate	$Cu_3(PO_4)_2 \rightleftharpoons 3Cu^{2+} + 2PO_4^{3-}$	1.4×10^{-37}
iron(II)	carbonate	$FeCO_3 \rightleftharpoons Fe^{2+} + CO_3^{2-}$	3.1×10^{-11}
	fluoride	$FeF_2 \rightleftharpoons Fe^{2+} + 2F^-$	2.4×10^{-6}
	hydroxide	$Fe(OH)_2 \rightleftharpoons Fe^{2+} + 2OH^-$	4.9×10^{-17}
lead	bromide	$PbBr_2 \rightleftharpoons Pb^{2+} + 2Br^-$	6.6×10^{-6}
	carbonate	$PbCO_3 \rightleftharpoons Pb^{2+} + CO_3^{2-}$	1.5×10^{-13}
	chloride	$PbCl_2 \rightleftharpoons Pb^{2+} + 2Cl^-$	1.6×10^{-5}
	fluoride	$PbF_2 \rightleftharpoons Pb^{2+} + 2F^-$	7.1×10^{-7}
	iodide	$PbI_2 \rightleftharpoons Pb^{2+} + 2I^-$	8.5×10^{-9}
	sulfate	$PbSO_4 \rightleftharpoons Pb^{2+} + 2SO_4^{2-}$	1.8×10^{-8}
lithium	carbonate	$Li_2CO_3 \rightleftharpoons 2Li^+ + CO_3^{2-}$	8.2×10^{-4}
magnesium	carbonate	$MgCO_3 \rightleftharpoons Mg^{2+} + CO_3^{2-}$	6.8×10^{-6}
	fluoride	$MgF_2 \rightleftharpoons Mg^{2+} + 2F^-$	6.5×10^{-9}
	hydroxide	$Mg(OH)_2 \rightleftharpoons Mg^{2+} + 2OH^-$	5.6×10^{-12}
manganese(II)	carbonate	$MnCO_3 \rightleftharpoons Mn^{2+} + CO_3^{2-}$	2.2×10^{-11}
	hydroxide	$Mn(OH)_2 \rightleftharpoons Mn^{2+} + 2OH^-$	2.1×10^{-13}
mercury(I)	bromide	$Hg_2Br_2 \rightleftharpoons Hg_2^{2+} + 2Br^-$	6.4×10^{-23}

[a]Equilibrium is between solid phase and aqueous solution.

TABLE C-19. Solubility-Product Constants at 25°C (*Continued*)

Cation	Anion	Heterogeneous Equilibrium[a]	K_{sp}
mercury(I)	carbonate	$Hg_2CO_3 \rightleftharpoons Hg_2^{2+} + CO_3^{2-}$	3.7×10^{-17}
	chloride	$Hg_2Cl_2 \rightleftharpoons Hg_2^{2+} + 2Cl^-$	1.5×10^{-18}
	iodide	$Hg_2I_2 \rightleftharpoons Hg_2^{2+} + 2I^-$	5.3×10^{-29}
	sulfate	$Hg_2SO_4 \rightleftharpoons Hg_2^{2+} + SO_4^{2-}$	8.0×10^{-7}
mercury(II)	hydroxide	$Hg(OH)_2 \rightleftharpoons Hg^{2+} + 2OH^-$	3.1×10^{-26}
	iodide	$HgI_2 \rightleftharpoons Hg^{2+} + 2I^-$	2.8×10^{-29}
silver	bromide	$AgBr \rightleftharpoons Ag^+ + Br^-$	5.4×10^{-13}
	carbonate	$Ag_2CO_3 \rightleftharpoons 2Ag^+ + CO_3^{2-}$	8.5×10^{-12}
	chloride	$AgCl \rightleftharpoons Ag^+ + Cl^-$	1.8×10^{-10}
	chromate	$Ag_2CrO_4 \rightleftharpoons 2Ag^+ + CrO_4^{2-}$	1.1×10^{-12}
	hydroxide	$AgOH \rightleftharpoons Ag^+ + OH^-$	1.8×10^{-8}
	iodide	$AgI \rightleftharpoons Ag^+ + I^-$	8.3×10^{-17}
	phosphate	$Ag_3PO_4 \rightleftharpoons 3Ag^+ + PO_4^{3-}$	8.9×10^{-17}
	sulfate	$Ag_2SO_4 \rightleftharpoons 2Ag^+ + SO_4^{2-}$	1.2×10^{-5}
strontium	carbonate	$SrCO_3 \rightleftharpoons Sr^{2+} + CO_3^{2-}$	5.6×10^{-10}
	fluoride	$SrF_2 \rightleftharpoons Sr^{2+} + 2F^-$	4.3×10^{-9}
	sulfate	$SrSO_4 \rightleftharpoons Sr^{2+} + SO_4^{2-}$	3.4×10^{-7}
zinc	carbonate	$ZnCO_3 \rightleftharpoons Zn^{2+} + CO_3^{2-}$	1.2×10^{-10}

TABLE C-20. Ionization Constants for Acids at 25°C

Acid	Stage	Aqueous Equilibrium	K_a	pK_a
acetic	1	$CH_3COOH \rightleftharpoons H^+ + CH_3COO^-$	1.76×10^{-5}	4.75
ammonium ion	1	$NH_4^+ \rightleftharpoons H^+ + NH_3$	5.7×10^{-10}	9.25
arsenic	1	$H_3AsO_4 \rightleftharpoons H^+ + H_2AsO_4^-$	5.5×10^{-3}	2.26
	2	$H_2AsO_4^- \rightleftharpoons H^+ + HAsO_4^{2-}$	1.7×10^{-7}	6.77
	3	$HAsO_4^{2-} \rightleftharpoons H^+ + AsO_4^{3-}$	5.1×10^{-12}	11.29
benzoic	1	$C_6H_5COOH \rightleftharpoons H^+ + C_6H_5COO^-$	6.46×10^{-5}	4.19
boric	1	$H_3BO_3 \rightleftharpoons H^+ + H_2BO_3^-$	5.4×10^{-10}	9.27
	2	$H_2BO_3^- \rightleftharpoons H^+ + HBO_3^{2-}$	$< 10^{-14}$	> 14
bromoacetic	1	$CH_2BrCOOH \rightleftharpoons H^+ + CH_2BrCOO^-$	2.0×10^{-3}	2.70
butanoic	1	$C_3H_7COOH \rightleftharpoons H^+ + C_3H_7COO^-$	1.5×10^{-5}	4.82
carbonic	1	$H_2CO_3 \rightleftharpoons H^+ + HCO_3^-$	4.3×10^{-7}	6.37
	2	$HCO_3^- \rightleftharpoons H^+ + CO_3^{2-}$	5.6×10^{-11}	10.25
chloric	1	$HClO_3 \rightleftharpoons H^+ + ClO_3^-$	$\gg 1$ (strong)	< 0
chlorous	1	$HClO_2 \rightleftharpoons H^+ + ClO_2^-$	1.1×10^{-2}	1.96
chloroacetic	1	$CH_2ClCOOH \rightleftharpoons H^+ + CH_2ClCOO^-$	1.4×10^{-3}	2.85
dichloroacetic	1	$CHCl_2COOH \rightleftharpoons H^+ + CHCl_2COO^-$	5.5×10^{-2}	1.26
ethanol	1	$CH_3CH_2OH \rightleftharpoons H^+ + CH_3CH_2O^-$	1.3×10^{-16}	15.9
fluoroacetic	1	$CH_2FCOOH \rightleftharpoons H^+ + CH_2FCOO^-$	2.6×10^{-3}	2.59
formic	1	$HCOOH \rightleftharpoons H^+ + HCOO^-$	1.77×10^{-4}	3.75
germanic	1	$H_2GeO_3 \rightleftharpoons H^+ + HGeO_3^-$	9.8×10^{-10}	9.01
	2	$HGeO_3^- \rightleftharpoons H^+ + GeO_3^{2-}$	5×10^{-13}	12.3
hydr(o)azoic	1	$HN_3 \rightleftharpoons H^+ + N_3^-$	1.9×10^{-5}	4.72
hydr(o)iodic	1	$HI \rightleftharpoons H^+ + I^-$	$\gg 1$ (strong)	< 0
hydrobromic	1	$HBr \rightleftharpoons H^+ + Br^-$	$\gg 1$ (strong)	< 0
hydrochloric	1	$HCl \rightleftharpoons H^+ + Cl^-$	$\gg 1$ (strong)	< 0
hydrocyanic	1	$HCN \rightleftharpoons H^+ + CN^-$	6.2×10^{-10}	9.21
hydrofluoric	1	$HF \rightleftharpoons H^+ + F^-$	6.8×10^{-4}	3.17
hydronium ion	1	$H_3O^+ \rightleftharpoons H^+ + H_2O$	1.0	0.00
hydrosulfuric	1	$H_2S \rightleftharpoons H^+ + HS^-$	8.9×10^{-8}	7.05
	2	$HS^- \rightleftharpoons H^+ + S^{2-}$	$\approx 10^{-19}$	≈ 19
hypobromous	1	$HBrO \rightleftharpoons H^+ + BrO^-$	2.8×10^{-9}	8.55
hypochlorous	1	$HClO \rightleftharpoons H^+ + ClO^-$	3.0×10^{-8}	7.52

TABLE C-20. Ionization Constants for Acids at 25°C (*Continued*)

Acid	Stage	Aqueous Equilibrium	K_a	pK_a
hypoiodous	1	$HIO \rightleftharpoons H^+ + IO^-$	3×10^{-11}	10.5
iodic	1	$HIO_3 \rightleftharpoons H^+ + IO_3^-$	1.7×10^{-1}	0.77
iodoacetic	1	$CH_2ICOOH \rightleftharpoons H^+ + CH_2ICOO^-$	7.6×10^{-4}	3.12
lactic	1	$CH_3CHOHCOOH \rightleftharpoons H^+ + CH_3CHOHCOO^-$	1.4×10^{-4}	3.85
malonic	1	$HOOCCH_2COOH \rightleftharpoons H^+ + HOOCCH_2COO^-$	1.5×10^{-3}	2.82
	2	$HOOCCH_2COO^- \rightleftharpoons H^+ + {}^-OOCCH_2COO^-$	2.0×10^{-6}	5.70
nitric	1	$HNO_3 \rightleftharpoons H^+ + NO_3^-$	$\gg 1$ (strong)	<0
oxalic	1	$HOOCCOOH \rightleftharpoons H^+ + HOOCCOO^-$	5.9×10^{-2}	1.23
	2	$HOOCCOO^- \rightleftharpoons H^+ + {}^-OOCCOO^-$	6.4×10^{-5}	4.19
perchloric	1	$HClO_4 \rightleftharpoons H^+ + ClO_4^-$	$\gg 1$ (strong)	<0
periodic	1	$HIO_4 \rightleftharpoons H^+ + IO_4^-$	2.3×10^{-2}	1.64
phenol	1	$C_6H_5OH \rightleftharpoons H^+ + C_6H_5O^-$	1.3×10^{-10}	9.89
phosphoric	1	$H_3PO_4 \rightleftharpoons H^+ + H_2PO_4^-$	7.52×10^{-3}	2.12
	2	$H_2PO_4^- \rightleftharpoons H^+ + HPO_4^{2-}$	6.2×10^{-8}	7.21
	3	$HPO_4^{2-} \rightleftharpoons H^+ + PO_4^{3-}$	2.2×10^{-13}	12.66
propionic	1	$CH_3CH_2COOH \rightleftharpoons H^+ + CH_3CH_2COO^-$	1.4×10^{-5}	4.86
sulfuric	1	$H_2SO_4 \rightleftharpoons H^+ + HSO_4^-$	$\gg 1$ (strong)	<0
	2	$HSO_4^- \rightleftharpoons H^+ + SO_4^{2-}$	1.2×10^{-2}	1.92
sulfurous	1	$H_2SO_3 \rightleftharpoons H^+ + HSO_3^-$	1.4×10^{-2}	1.85
	2	$HSO_3^- \rightleftharpoons H^+ + SO_3^{2-}$	6.3×10^{-8}	7.20
thiocyanic	1	$HSCN \rightleftharpoons H^+ + SCN^-$	$\gg 1$ (strong)	<0
trichloroacetic	1	$CCl_3COOH \rightleftharpoons H^+ + CCl_3COO^-$	2.3×10^{-1}	0.64
trifluoroacetic	1	$CF_3COOH \rightleftharpoons H^+ + CF_3COO^-$	5.9×10^{-1}	0.23
water	1	$H_2O \rightleftharpoons H^+ + OH^-$	1.0×10^{-14}	14.00

TABLE C-21. Standard Reduction Potentials and Free Energies at 25°C

Half-Reaction[a]	$\Delta G°$ (kJ)[b]	n	$\mathscr{E}°_{red}$ (V)[c]
$F_2 + 2e^- \longrightarrow 2F^-$	-553.1	2	2.866
$H_2N_2O_2 + 2H^+ + 2e^- \longrightarrow N_2 + 2H_2O$	-511	2	2.65
$O(g) + 2H^+ + 2e^- \longrightarrow H_2O$	-467.2	2	2.421
$Cu^{3+} + e^- \longrightarrow Cu^{2+}$	-230	1	2.4
$XeO_3 + 6H^+ + 6e^- \longrightarrow Xe + 3H_2O$	-1220	6	2.10
$O_3 + 2H^+ + 2e^- \longrightarrow O_2 + H_2O$	-400.6	2	2.076
$OH + e^- \longrightarrow OH^-$	-195	1	2.02
$Co^{3+} + e^- \longrightarrow Co^{2+}$	-185	1	1.92
$H_2O_2 + 2H^+ + 2e^- \longrightarrow 2H_2O$	-342.7	2	1.776
$N_2O + 2H^+ + 2e^- \longrightarrow N_2 + H_2O$	-340.8	2	1.766
$Au^+ + e^- \longrightarrow Au$	-163.3	1	1.692
$PbO_2 + SO_4^{2-} + 4H^+ + 2e^- \longrightarrow PbSO_4 + 2H_2O$	-326.37	2	1.6913
$MnO_4^- + 4H^+ + 3e^- \longrightarrow MnO_2 + 2H_2O$	-486.0	3	1.679
$NiO_2 + 4H^+ + 2e^- \longrightarrow Ni^{2+} + 2H_2O$	-323.8	2	1.678
$Mn^{3+} + e^- \longrightarrow Mn^{2+}$	-148.8	1	1.542
$MnO_4^- + 8H^+ + 5e^- \longrightarrow Mn^{2+} + 4H_2O$	-727.0	5	1.507
$ClO_3^- + 6H^+ + 5e^- \longrightarrow \frac{1}{2}Cl_2 + 3H_2O$	-709	5	1.47
$PbO_2 + 4H^+ + 2e^- \longrightarrow Pb^{2+} + 2H_2O$	-280.8	2	1.455
$Au^{3+} + 3e^- \longrightarrow Au$	-405	3	1.40
$Cl_2 + 2e^- \longrightarrow 2Cl^-$	-262.11	2	1.3583
$Cr_2O_7^{2-} + 14H^+ + 6e^- \longrightarrow 2Cr^{3+} + 7H_2O$	$-770.$	6	1.33
$O_2 + 4H^+ + 4e^- \longrightarrow 2H_2O$	-474.3	4	1.229
$IO_3^- + 6H^+ + 5e^- \longrightarrow \frac{1}{2}I_2(s) + 3H_2O$	-576.5	5	1.195
$IO_3^- + 6H^+ + 6e^- \longrightarrow I^- + 3H_2O$	-628.1	6	1.085
$Br_2(\ell) + 2e^- \longrightarrow 2Br^-$	-205.7	2	1.066
$Hg^{2+} + 2e^- \longrightarrow Hg$	-164	2	0.851
$Ag^+ + e^- \longrightarrow Ag$	-77.15	1	0.7996
$Hg_2^{2+} + 2e^- \longrightarrow 2Hg$	-153.9	2	0.7973

[a]For species in their normal states at 25°C and 1 atm. Ions are in aqueous solution.
[b]Gibbs free energy of reaction, computed from the relationship $\Delta G° = -n\mathscr{F}\mathscr{E}°$. Values tabulated here are independent of those given in Table C-16.
[c]In descending electrochemical order.

TABLE C-21. Standard Reduction Potentials and Free Energies at 25°C (*Continued*)

Half-Reaction[a]	$\Delta G°$ (kJ)[b]	n	$\mathscr{E}°_{red}$ (V)[c]
$Fe^{3+} + e^- \longrightarrow Fe^{2+}$	−74.3	1	0.770
$I_2(s) + 2e^- \longrightarrow 2I^-$	−103.3	2	0.5355
$H_2SO_3 + 4H^+ + 4e^- \longrightarrow S + 3H_2O$	−173	4	0.449
$Ag_2CrO_4 + 2e^- \longrightarrow 2Ag + CrO_4^{2-}$	−86.26	2	0.4470
$O_2 + 2H_2O + 4e^- \longrightarrow 4OH^-$	−155	4	0.401
$Ag_2O + H_2O + 2e^- \longrightarrow 2Ag + 2OH^-$	−66.0	2	0.342
$Cu^{2+} + 2e^- \longrightarrow Cu$	−65.98	2	0.3419
$AgCl + e^- \longrightarrow Ag + Cl^-$	−21.45	1	0.2223
$HgO + H_2O + 2e^- \longrightarrow Hg + 2OH^-$	−18.9	2	0.0977
$2H^+ + 2e^- \longrightarrow H_2$	0.0	2	0.0000
$CrO_4^{2-} + 4H_2O + 3e^- \longrightarrow Cr(OH)_3 + 5OH^-$	38	3	−0.13
$AgI + e^- \longrightarrow Ag + I^-$	14.69	1	−0.1522
$Ni^{2+} + 2e^- \longrightarrow Ni$	49.6	2	−0.257
$Fe^{2+} + 2e^- \longrightarrow Fe$	86.3	2	−0.447
$Ni(OH)_2 + 2e^- \longrightarrow Ni + 2OH^-$	140	2	−0.72
$Zn^{2+} + 2e^- \longrightarrow Zn$	147.0	2	−0.7618
$2H_2O + 2e^- \longrightarrow H_2 + 2OH^-$	159.7	2	−0.8277
$Mn^{2+} + 2e^- \longrightarrow Mn$	228.7	2	−1.185
$Zn(OH)_2 + 2e^- \longrightarrow Zn + 2OH^-$	241.0	2	−1.249
$Al^{3+} + 3e^- \longrightarrow Al$	481.1	3	−1.662
$Na^+ + e^- \longrightarrow Na$	261	1	−2.71
$Ca^{2+} + 2e^- \longrightarrow Ca$	553.4	2	−2.868
$Ba^{2+} + 2e^- \longrightarrow Ba$	561.9	2	−2.912
$Li^+ + e^- \longrightarrow Li$	294	1	−3.05

TABLE C-22.　Formation Constants of Complexes at 25°C

Complex Ion	Aqueous Equilibrium	K_f
$[Ag(NH_3)_2]^+$	$Ag^+ + 2NH_3 \rightleftharpoons Ag(NH_3)_2^+$	1.7×10^7
$[AgCl_2]^-$	$Ag^+ + 2Cl^- \rightleftharpoons AgCl_2^-$	2.5×10^5
$[Ag(CN)_2]^-$	$Ag^+ + 2CN^- \rightleftharpoons Ag(CN)_2^-$	1.0×10^{21}
$[Al(OH)_4]^-$	$Al^{3+} + 4OH^- \rightleftharpoons Al(OH)_4^-$	7.7×10^{33}
$[Au(CN)_2]^-$	$Au^+ + 2CN^- \rightleftharpoons Au(CN)_2^-$	2.0×10^{38}
$[Co(NH_3)_6]^{2+}$	$Co^{2+} + 6NH_3 \rightleftharpoons Co(NH_3)_6^{2+}$	7.7×10^4
$[Co(NH_3)_6]^{3+}$	$Co^{3+} + 6NH_3 \rightleftharpoons Co(NH_3)_6^{3+}$	5.0×10^{31}
$[HgCl_4]^{2-}$	$Hg^{2+} + 4Cl^- \rightleftharpoons HgCl_4^{2-}$	1.2×10^{15}
$[Ni(NH_3)_6]^{2+}$	$Ni^{2+} + 6NH_3 \rightleftharpoons Ni(NH_3)_6^{2+}$	5.5×10^8
$[PbCl_4]^{2-}$	$Pb^{2+} + 4Cl^- \rightleftharpoons PbCl_4^{2-}$	2.5×10^1
$[Zn(NH_3)_4]^{2+}$	$Zn^{2+} + 4NH_3 \rightleftharpoons Zn(NH_3)_4^{2+}$	2.9×10^9
$[Zn(OH)_4]^{2-}$	$Zn^{2+} + 4OH^- \rightleftharpoons Zn(OH)_4^{2-}$	2.8×10^{15}

APPENDIX D

Glossary

Abbreviations:
Ar: Arabic
Ger: German
Gk: Greek
L: Latin

absolute temperature [L *absolutus* free, unrestricted] The Kelvin scale for temperature, calibrated so that the hypothetical lowest value is absolute zero.

absolute zero The lowest temperature on the Kelvin scale (0 K = −273.15°C), impossible for any system to reach exactly.

absorption [L *absorbere* to suck in, to swallow] (1) In spectroscopy: the transfer of energy from light to matter. Contrasted with *emission*. (2) In general: the taking up of one quantity by another.

absorption spectrum The intensity of electromagnetic radiation absorbed by a material, recorded as a function of wavelength, frequency, or energy. Contrasted with *emission spectrum*.

acceleration [L *accelerare* to quicken] The rate at which a body changes its speed, direction, or both. Symbol: *a*.

acceptor [L *acceptare* to take] Something that takes or receives, such as an electron-pair acceptor (a Lewis acid) or a proton acceptor (a Brønsted-Lowry base). Contrasted with *donor*.

accuracy [L *cura* care] The extent to which a measured value agrees with an accepted value, distinct from *precision*.

acid [L *acidus* sour] Stated thus, without qualification: usually a Brønsted-Lowry acid (a proton donor). Contrasted with *base*. See *Arrhenius acid*; *Brønsted-Lowry acid*; *Lewis acid*.

acid dissociation constant, acid ionization constant   Equilibrium constant for the breakup of an acid (HA) into a hydrogen ion (H^+) and a conjugate base (A^-):

$$K_a = \frac{[H^+][A^-]}{[HA]}$$

In aqueous solution, a hydronium ion (H_3O^+) is produced when H^+ is transferred to H_2O.

actinide series   The 14 radioactive elements that follow actinium (Ac, $Z = 89$), from thorium (Th, $Z = 90$) through lawrencium (Lr, $Z = 103$). Most of the actinides have partially filled $5f$ orbitals.

activated complex   [L *agere* (*actus*) to do] A way station on the road from reactants to products: the transient structure near a maximum in potential energy. After assembling the reactants into a critical configuration, the unstable system will either go on to form products or fall back to regenerate reactants. Often used interchangeably with *transition state*.

activation barrier, activation energy   The difference in potential energy between the reactants and a transition state: a hurdle that must be overcome before products are formed. An activation barrier represents the minimum energy needed to initiate a reaction.

active site   For an enzyme: the specific location where a substrate is bound and where catalysis occurs.

activity   (1) For a radioactive nucleus: the rate of disintegration. (2) In thermodynamics of ideal systems: the dimensionless ratio of a concentration [X] or pressure P_X to some standard value, typically $c^\circ = 1\ M$ or $P^\circ = 1$ atm. These ratios $[X]/c^\circ$ and P_X/P° can then be corrected for nonideal behavior. (3) Of a metal: pertaining to the relative ease of oxidation.

addition reaction   A reaction in which two new chemical groups are added to the two atoms of a multiple bond. Example: $H_2 + CH_2{=}CH_2 \rightarrow CH_3{-}CH_3$.

adsorption   [L *ad* to, toward + *sorbere* to suck in, to swallow] Binding of a chemical species to a surface.

affinity   [L *affinitas* connection by marriage] A selective attraction or preference of one thing for another.

alchemy   [Ar *al* the + Gk *kemeia* transmutation] A blend of chemistry and philosophy practiced especially in the Middle Ages, seeking to discover (1) a way to transmute base metals (such as lead) into gold, (2) a universal solvent, and (3) an elixir of life.

alcohol   An organic compound with the general formula $R{-}OH$, R being the hydrocarbon portion of the molecule. Examples: CH_3OH, methanol; CH_3CH_2OH, ethanol.

aldehyde An organic compound with the general formula $R-\overset{\overset{\displaystyle O}{\|}}{C}-H$, R being the hydrocarbon portion of the molecule. Example: CH_3CHO, acetaldehyde.

alkali metal [Ar *al-qali* saltwort ashes] A Group I element (Li, Na, K, . . .), distinguished by the valence configuration ns^1. Hydrogen ($1s^1$), although part of Group I, is classified as a *nonmetal*.

alkaline earth metal A Group II element (Be, Mg, Ca, . . .), distinguished by the valence configuration ns^2.

alkane A saturated hydrocarbon with the general formula C_nH_{2n+2}, hence a compound of carbon and hydrogen having single bonds throughout. Examples: CH_4, methane; C_2H_6, ethane.

alkene A hydrocarbon containing one or more C=C double bonds. Where there is only one double bond, the general formula is C_nH_{2n}. Example: CH_2=CH_2, ethene (ethylene).

alkyl group A radical hydrocarbon fragment, derived from an alkane by loss of a single hydrogen atom. Symbol: $R-$. Examples: $-CH_3$, methyl; $-CH_2CH_3$, ethyl.

alkyl halide An organic compound with the general formula $R-X$, formed by substitution of a halogen atom (X = F, Cl, Br, I) for hydrogen at one position on an alkane. Example: CH_3Cl, methyl chloride.

alkyne A hydrocarbon containing one or more C≡C triple bonds. Where there is only one triple bond, the general formula is C_nH_{2n-2}. Example: CH≡CH, ethyne (acetylene).

allotropes [Gk *allos* other + *tropos* turn, turning] Distinct chemical forms of a single element, differing in the arrangement and linkage of the atoms. Examples: Diamond, graphite, and buckminsterfullerene are allotropes of carbon.

alloy [L *alligare* to bind up] An intimate mixture of two or more metals. Example: Bronze is an alloy of copper and tin.

alpha particle A helium nucleus ($^4_2He^{2+}$), emitted by certain radioactive nuclei. Symbol: α.

ambidentate ligand [L *ambi* both + *dens* tooth] A ligand offering a choice between two points of attachment: two "teeth" but one "bite." When bound, an ambidentate ligand occupies only one position in the coordination sphere.

amide An organic compound containing the functional group $-\overset{\overset{\displaystyle O}{\|}}{C}-NR_2$, formed by reaction of a carboxylic acid and amine. Example: CH_3CONH_2, acetamide.

amine A compound with the general formula NR_3, derived from ammonia (NH_3) by replacement of one or more hydrogen atoms. Replacement of one hydrogen atom creates a *primary* amine; two, a *secondary* amine; three, a *tertiary* amine. Examples: $CH_3CH_2NH_2$, primary; $(CH_3CH_2)_2NH$, secondary; $(CH_3CH_2)_3N$, tertiary.

amino acid   A carboxylic acid with the general formula

$$H_2N-\underset{\underset{R}{|}}{\overset{\overset{H}{|}}{C}}-\overset{\overset{O}{\|}}{C}-OH$$

containing an amino group ($-NH_2$). In an α-amino acid, the amino group is attached at the position adjacent to the carboxylic group as shown. Example: glycine (R = H).

amorphous   [Gk *amorphos* shapeless] Lacking definite shape or form.

amorphous solid   A solid lacking long-range order in the arrangement of its particles; often called a *glass*.

ampere   [André Marie *Ampère*, 1775–1836, French physicist] The base SI unit of electric current, equivalent to 1 coulomb per second. The standard ampere is defined by reference to the force exerted on a specific configuration of electrical conductors. Symbol: $A = C\ s^{-1}$.

amphipathic   [Gk *amphi* both + *pathos* suffering, sensation] Having two different affinities, such as a hydrophobic region and a hydrophilic region in the same molecule.

amphoteric   [Gk *amphoteros* both] Able to act as either an acid or a base.

amplitude   [L *amplitudo* breadth, size] The maximum displacement attained during an oscillation.

analytical chemistry   [Gk *analyein* to loosen up, to take apart] The branch of chemistry concerned with quantitative and qualitative measurement of composition and structure.

angstrom   [Anders Jonas *Ångström*, 1814–1874, Swedish astronomer and physicist] A non-SI unit of length, equal to 10^{-10} meter. Symbol: Å.

angular momentum   The rotational counterpart of linear momentum. For a classical particle of mass m moving in a circle of radius r at linear velocity v, the angular momentum has magnitude mvr. Its direction is perpendicular to the plane of motion.

angular momentum quantum number   A quantum number pertaining to the shape of an orbital for a one-electron atom. Symbol: ℓ. Allowed values range from $\ell = 0$ to $\ell = n - 1$, where n is the principal quantum number. Also called *azimuthal quantum number*.

angular node   A region in space where the angular portion of a wave function changes sign, manifested as a surface of zero electron density. Contrasted with *radial node*.

anhydride   A compound formed by removal of water from a structure. An organic acid anhydride, produced by a condensation reaction between two carboxylic acids, contains the functional group

$$-\overset{\overset{\displaystyle O}{\|}}{C}-O-\overset{\overset{\displaystyle O}{\|}}{C}-\ .$$

anhydrous   [Gk *anydros* waterless] With all water removed.

anion   [Gk *ana* up + *ion* going] A negatively charged ion, formed from either an atom or a group of atoms. The anion is understood to be the species that moves toward (goes "up" to) the anode in an electrochemical cell. Contrasted with *cation*.

anisotropic   [Gk *a* without + *iso* equal + *tropos* turn] Having different properties along different axes. Contrasted with *isotropic*.

annihilation   [L *nihil* nothing] A process whereby a particle and its antiparticle meet and convert all their mass into energy. No mass remains.

anode   [Gk *ana* up + *hodos* way → *anodos* a way up] The electrode, or terminal, at which oxidation occurs. Contrasted with *cathode*.

antibody   [Gk *anti* against, opposite] Any of various protein molecules able to bind selectively with a foreign protein, or *antigen*. Antibodies are employed by the immune system as a defense against viruses and bacteria.

antibonding orbital   A molecular orbital formed by out-of-phase combination of atomic orbitals and positioned higher in energy than the corresponding in-phase, bonding combination. Electrons occupying an antibonding orbital are concentrated away from the region between the nuclei.

antielectron   A *positron*: an elementary particle with the same mass and spin as an electron, but having a positive charge equal in magnitude to the electron's negative charge. See also *antimatter*.

antigen   [Gk *anti* against, opposite + *genes* born, produced] A substance that stimulates the production of antibodies and subsequently binds to them.

antimatter   Matter composed of *antiparticles* such as positrons, antiprotons, antineutrons, antimesons, antineutrinos, and antiquarks. A particle and its antiparticle have the same mass and spin but opposite charge, magnetic moment, lepton number, baryon number, and other internal attributes.

antineutrino   The uncharged antiparticle of a neutrino, emitted in certain kinds of radioactive decay.

antiproton   An elementary particle with the same mass and spin of a proton, but having a negative charge equal in magnitude to the proton's positive charge.

aqueous   [L *aqua* water] Watery, containing water—as in an *aqueous solution*, where water is the solvent.

area   [L *area* level or open space] Extent of space in two dimensions; the space occupied by a two-dimensional surface.

aromatic compound   A planar organic molecule containing one or more benzene rings or similar cyclic structures, characterized by a delocalized π system with $4n + 2$ electrons ($n = 0, 1, 2, \ldots$). Many such compounds have characteristic odors. Examples: C_6H_6, benzene; $C_{10}H_8$, naphthalene.

Arrhenius acid   [Svante *Arrhenius*, 1859–1927, Swedish physicist and chemist: Nobel prize for chemistry, 1903] A compound which, in aqueous solution, releases hydrogen cations (H^+) in sufficient quantity to increase the concentration of H_3O^+ beyond its value in pure water. Examples: CH_3COOH, acetic acid; H_2SO_4, sulfuric acid.

Arrhenius base   A compound which, in aqueous solution, releases hydroxide ions (OH^-) in sufficient quantity to increase the concentration of OH^- beyond its value in pure water. Example: NaOH, sodium hydroxide.

Arrhenius equation, Arrhenius law   A mathematical relationship between the rate constant for a reaction (k) and the absolute temperature (T), stated in either exponential or logarithmic form:

$$k = A \, \exp\left(-\frac{E_a}{RT}\right)$$

$$\ln k = -\frac{E_a}{RT} + \ln A$$

E_a is the activation energy. A is the pre-exponential factor, also called the frequency factor.

association   [L *socius* partner, ally] Noncovalent interactions between atoms, molecules, and ions, producing weakly bonded aggregates.

atmosphere   [Gk *atmos* vapor, smoke + *sphaira* ball, globe] A non-SI unit of pressure, equivalent to 760 torr and 101.325 kilopascals (the normal air pressure at sea level, approximately 14.7 pounds per square inch). Symbol: atm.

atom   [Gk *atomos* undivided, uncut] The smallest sample of an element that still retains the chemical properties of that element. An atom, containing a positively charged nucleus amidst a number of negatively charged electrons, is electrically neutral overall.

atomic mass   (1) The absolute mass of a particular isotope of an element, expressed in a specific set of units as either an individual or molar quantity. (2) The average mass of all the isotopes of an element. See also *molar mass*; *relative atomic mass*.

atomic mass unit   A unit of mass equal to $\frac{1}{12}$ the mass of a single atom of carbon-12, the isotope containing six protons and six neutrons in its nucleus. Symbol: u (amu) $= 1.66 \times 10^{-27}$ kg.

atomic number   The number of protons in a nucleus, matched in a neutral atom by an equal number of electrons outside. Symbol: Z.

atomic orbital   A wave function for a single electron in an atom. For a many-electron system, an atomic orbital is patterned after the functions obtained by exact solution of the Schrödinger equation for the hydrogen atom. Each electron, considered separately, is envisioned to move through a smoothed-out, average field produced by all its neighbors.

atomic radius   A measure of the size of an atom, equal to half the center-to-center distance between neighboring atoms in a diatomic molecule or homonuclear crystal of a particular element.

atomic weight   See *atomic mass*; *molar mass*.

attraction   [L *attrahere* (*attractus*) to be drawn to]   To be drawn by a force emanating from a certain object or position, so as to move toward that position.

Aufbau principle   [Ger *Aufbau* building, construction]   The *building-up* sequence for atoms and molecules, whereby electrons are imagined to occupy rungs on a ladder of orbitals. Orbitals are populated in order of increasing energy, with no more than two electrons (with opposite spins) in each.

autodissociation, autoionization   [Gk *autos* self]   For water: the formation of H_3O^+ and OH^- ions by transfer of a proton from one H_2O to another.

autodissociation constant, autoionization constant   For water: the equilibrium constant for the autoionization reaction. Symbol: K_w.

$$H_2O(\ell) + H_2O(\ell) \rightleftharpoons H_3O^+(aq) + OH^-(aq)$$

$$K_w(25°C) = [H_3O^+][OH^-] = 1.0 \times 10^{-14}$$

Also called *ion-product constant*.

Avogadro's law   [Count Amadeo *Avogadro*, 1776–1856, Italian physicist and chemist]   The principle that equal volumes of all gases at the same temperature and pressure contain the same number of particles—valid, in practice, for gases in the limit of ideal behavior. At 273 K and 1 atm, the molar volume of an ideal gas is 22.4 L.

Avogadro's number   The number of carbon-12 atoms in exactly 12 grams of the pure isotope, generalized to mean the number of objects (of any kind) in 1 mole. Symbol: $N_0 = 6.02 \times 10^{23}$ mol^{-1}.

azimuthal quantum number   A quantum number pertaining to the shape of an orbital for a one-electron atom. Symbol: ℓ. Allowed values range from $\ell = 0$ to $\ell = n - 1$, where n is the principal quantum number. Also called *angular momentum quantum number*.

Balmer series   [J. J. *Balmer*, 1825–1898, Swiss physicist] A series of lines in the visible emission spectrum of hydrogen, corresponding to transitions terminating at the level $n = 2$.

barometer   [Gk *baros* weight + *metron* measure] An instrument for measuring atmospheric pressure.

baryon   [Gk *barys* heavy] A particle with half-integral spin that responds to the strong force. Examples: the proton, the neutron, and their antiparticles. See also *lepton*; *meson*.

baryon number   A conserved quantum number: 1 for baryons, −1 for antibaryons.

base   Stated thus, without qualification: usually a Brønsted-Lowry base (a proton acceptor). Contrasted with *acid*. See *Arrhenius base*; *Brønsted-Lowry base*; *Lewis base*.

base dissociation constant, base ionization constant   Equilibrium constant for the protonation of a base (B) and the resulting production of a conjugate acid (BH^+). In aqueous solution, the hydroxide ion (OH^-) is formed when a molecule of H_2O transfers H^+ to the base:

$$K_b = \frac{[BH^+][OH^-]}{[B]}$$

battery   An electrochemical cell or a combination of two or more such cells, arranged to do electrical work.

becquerel   [Antoine Henri *Becquerel*, 1852–1908, French physicist: Nobel prize for physics, 1903] A derived SI unit of radioactive decay, equivalent to 1 disintegration per second. Symbol: $Bq = 1 \ s^{-1}$.

belt of stability   The set of neutron-to-proton ratios within which all nonradioactive nuclei exist. Also called *zone of stability*, *band of stability*, *peninsula of stability*, etc.

beta particle   A high-energy electron emitted by certain radioactive nuclei. Symbol: β^- or $_{-1}^{0}e$.

bidentate ligand   [L *bis* twice + *dens* tooth] A ligand that attaches in two places, thus occupying two positions in the coordination sphere.

bilayer   A structure built from two molecular layers, such as the *phospholipid bilayer* found in many cell membranes.

bimolecular reaction   An elementary reaction between two species.

binary   [L *bini* two each, by twos] (1) Of an ionic or molecular compound: containing two elements. (2) Of an alloy or other mixture: containing two principal components. (3) In general: indicating or involving *two*.

binding energy   (1) For a nucleus: the energy, equal to the mass defect, required to decompose the structure into its constituent protons and neutrons. (2) In general: the energy needed to decompose any system (atom, molecule, complex) into its constituent particles.

biochemistry   [Gk *bios* life] The study of chemical transformations in living matter.

biopolymer   Any macromolecule manufactured by a living system, such as a protein or nucleic acid.

body-centered cubic cell   A unit cell that contains a lattice point at the center of a cube, in addition to a point at each of the eight vertices. Compare with *face-centered cubic cell*; *primitive cubic cell*.

Bohr atom   [Niels *Bohr*, 1885–1962, Danish physicist: Nobel Prize for physics, 1922] A model of the hydrogen atom in which the electron is constrained to a well-defined circular orbit around the nucleus. Part of the "old" quantum theory, the Bohr model was superseded by the quantum mechanics of Schrödinger and Heisenberg.

Bohr radius   The radius of the lowest orbital in Bohr's theory of the hydrogen atom, later proven equivalent to the most probable radial distance for a $1s$ electron. Symbol: $a_0 = 0.529$ Å.

boiling   A rapid transition from a liquid to a gas, accompanied by a decrease in order and an increase in enthalpy.

boiling point, boiling temperature   The temperature at which a liquid boils (called the *normal boiling temperature* when measured at 1 atm). The vapor pressure at the boiling point becomes equal to atmospheric pressure, and turbulent gas bubbles begin to form in the liquid—a process termed *ebullition*.

boiling-point elevation   The increase in a solvent's boiling temperature brought about by the presence of a solute; a colligative property.

Boltzmann distribution, Boltzmann factor   [Ludwig *Boltzmann*, 1844–1906, Austrian physicist] The population of energy levels characteristic of a system in thermal equilibrium. For each energy level ϵ_i, the exponential function

$$f(\epsilon_i) = \frac{1}{Z} \exp\left(-\frac{\epsilon_i}{k_B T}\right)$$

defines the Boltzmann factor. Z, called the *partition function*, is a normalization constant (not to be confused with the atomic number). T is the absolute temperature, and k_B is Boltzmann's constant. At thermal equilibrium, energy levels ϵ_i and ϵ_j are occupied according to the ratio $f(\epsilon_i)/f(\epsilon_j)$. See also *Maxwell-Boltzmann distribution*.

Boltzmann's constant   A physical constant equal to the ratio R/N_0, where R is the universal gas constant and N_0 is Avogadro's number. Boltzmann's constant figures prominently in many microscopic equations involving the energy. Symbol: $k_B = 1.38 \times 10^{-23}$ J K^{-1}.

bomb calorimeter   An instrument that measures the heat produced when a substance undergoes combustion at constant volume.

bond   Any linkage by which particles are held together in a stable arrangement, as in a covalent bond, ionic bond, noncovalent (weak) bond, or the like.

bond dipole moment   A local electric dipole moment in a molecule, considered as arising from the separation of charge in a particular chemical bond. See *electric dipole, electric dipole moment*.

bond dissociation energy   The energy or enthalpy required to separate 1 mole of bonded atoms from their equilibrium distance to infinity. Also called simply *bond energy* or *bond enthalpy*.

bond distance   The average separation between two bonded atoms as they vibrate about a position of minimum potential energy.

bond energy   See *bond dissociation energy*.

bonding orbital   A molecular orbital formed by in-phase combination of atomic orbitals and positioned lower in energy than the wave function for the separated atoms. Electrons occupying a bonding orbital are concentrated in the region between the nuclei.

bonding pair   In the Lewis model of chemical bonding: a pair of electrons shared between two atoms.

bond order   A measure of the strength of a covalent bond, designed roughly to correlate with single, double, and triple bonds in the Lewis model: bond order $= \frac{1}{2} \times$ (number of bonding electrons − number of antibonding electrons).

Boyle's law   [Robert *Boyle*, 1627–1691, English chemist and physicist] For a fixed amount of gas at constant temperature, the variables pressure and volume are inversely proportional: $PV = $ constant.

Brønsted-Lowry acid   [Johannes *Brønsted,* 1879–1947, Danish chemist; Thomas *Lowry*, 1874–1936, English chemist] A species able to donate a hydrogen ion (H^+, a proton). Examples: HNO_3, CH_3COOH, NH_4^+.

Brønsted-Lowry base   A species able to accept a hydrogen ion (H^+, a proton). Examples: NO_3^-, CH_3COO^-, NH_3.

Brownian motion   [Robert *Brown*, 1773–1858, English botanist] The random movement of particles suspended in a liquid or gas (dust motes, for example), caused by the continual but otherwise unseen bombardment of the particles by the molecules of the solvent.

buckminsterfullerene   [R. Buckminster *Fuller*, 1895–1983, U.S. architect and engineer] A third allotrope of carbon, discovered in 1985: C_{60}, a network of 60 carbon atoms organized into a truncated icosahedral structure resembling a soccer ball. See also *diamond; graphite*.

buffer solution   A solution able to maintain an approximately constant pH upon the addition of small quantities of acid or base.

building-up principle   See *Aufbau principle*.

bulk   Pertaining to the whole assembly as distinct from a small portion. A macroscopic crystal is a bulk material; a surface or cluster is not.

calorie   [L *calor* heat] A non-SI unit for heat, equal to exactly 4.1840 joules. The calorie was formerly defined as the amount of heat needed to raise the temperature of 1 gram of water by 1°C. Symbol: cal.

calorimeter   [L *calor* heat + Gk *metron* measure] An instrument that measures the absorption or evolution of heat. *Calorimetry* is the measurement of heat flow during a process.

carbocation   An organic ion in which a carbon atom is trivalent and partially positive. Formerly called *carbonium ion*.

carbohydrate   An organic compound with the general formula $C_n(H_2O)_m$. Example: $C_{12}H_{22}O_{11}$, sucrose.

carbonyl group   The fragment $-\overset{\overset{\displaystyle O}{\|}}{C}-$, found especially in aldehydes and ketones.

carboxyl group   The fragment $-\overset{\overset{\displaystyle O}{\|}}{C}-OH$, found in carboxylic acids.

carboxylic acid   An organic compound with the general formula $R-COOH$, containing at least one carboxyl group. Example: CH_3COOH, acetic acid.

catalyst   [Gk *katalyein* to dissolve] A substance that accelerates the rate of a reaction without suffering any permanent change itself. A *homogeneous* catalyst works in the same phase as the reactants; a *heterogeneous* catalyst works in a different phase. Contrasted with *inhibitor*.

catalytic antibody   An antibody that binds to a structure similar to the transition state of a particular reaction, so as to function as a catalyst.

cathode   [Gk *kata* down + *hodos* way → *kathodos* a way down] The electrode, or terminal, at which reduction occurs. Contrasted with *anode*.

cation   [Gk *kata* down + *ion* going] A positively charged ion, formed from either an atom or group of atoms. The cation is understood to be the species that moves toward (goes "down" to) the cathode in an electrochemical cell. Contrasted with *anion*.

cell potential   [L *potens* power] The electrical driving force that moves electrons between the terminals of an electrochemical cell, measured in volts. A cell potential is related to the change in free energy by the equation $\Delta G = -n\mathscr{F}\mathscr{E}$. Also called the *electromotive force*. Symbol: $\mathscr{E}$.

Celsius scale   [Anders *Celsius*, 1701–1744, Swedish astronomer] A temperature scale calibrated so that water freezes at $0°C$ and boils at $100°C$. Formerly called *centigrade*. Symbol: $°C$ (degree Celsius).

centimeter   [L *centum* hundredth; used as a numerical prefix (*centi*) to mean 10^{-2}] One hundredth of a meter (2.54 cm = 1 in). Symbol: cm.

ceramic   [Gk *keramikos* potter's clay] Made of clay or similar materials, usually containing various silicates, oxides, and carbides.

chain reaction   A self-sustaining sequence of processes in which products from one step are used to initiate the transformation in the next step.

chalcogen   [Gk *chalkos* copper + *genes* born, produced] A Group VI element (O, S, Se, . . .), distinguished by the valence configuration ns^2np^4.

chaos   [Gk *chaos* vast chasm, void] (1) In general: a condition of utter confusion and disorder, as in a gas. (2) "Deterministic" chaos: mathematical behavior whereby an apparently random sequence of events is found to conform to a predictable pattern.

charge, electric   A fundamental property of matter, source of the electromagnetic interaction. There are two kinds of charge: positive and negative. A particle or structure with no net electric charge is said to be *neutral*.

charge transfer   In a complex: movement of an electron from a ligand to an unoccupied *d* orbital on the metal.

Charles's law   [J. A. C. *Charles*, 1746–1823, French physicist] For a fixed amount of gas (number of moles) at constant pressure, the variables volume and temperature are directly proportional: $V/T =$ constant.

chelating agent   [Gk *chele* claw] A polydentate ligand that attaches itself to a central metal atom through two or more donor sites.

chelation effect   An increase in entropy that often results when a polydentate ligand replaces two or more monodentate ligands in a complex.

chemical bond   See *bond*.

chemical compound   See *compound*.

chemical equation   An accounting of the reactants and products in a chemical reaction, showing the formula and combining ratio for each participant.

chemical equilibrium   A persistent, self-sustaining condition in which the concentrations of reactants and products remain constant during a chemical reaction. Interpreted thermodynamically, chemical equilibrium sets in when the combined free energy of the products becomes equal to the combined free energy of the reactants. Interpreted kinetically, the equilibrium reflects a dynamic balance between forward and reverse reactions proceeding at the same rate.

chemical formula   A representation of the elemental composition of a substance (ionic, molecular, or any other form), written as a sequence of chemical symbols with subscripts to indicate the proportion of each element in the compound. A *molecular formula* is a chemical formula that pertains specifically to a molecule. Compare with *empirical formula*.

chemical kinetics   [Gk *kinetikos* moving] Study of the rates and mechanisms of chemical reactions. Also called simply *kinetics*.

chemical potential   [L *potens* power] A measure of the tendency for matter to flow from one point to another, related to the free energy per mole of substance.

chemical property   An attribute of a substance made manifest when the individual molecules or ions undergo a change in structure or composition.

chemical reaction   A transformation of one or more substances into other substances, usually involving a rearrangement of atoms.

chemical shift   In NMR spectroscopy: an internal magnetic field arising from the interaction of an external magnetic field with the electronic currents around a nucleus. The chemical shift causes a nuclear magnetic signal to be displaced from a standard reference point in the spectrum. Often called *chemical shielding*. See also *nuclear magnetic resonance spectroscopy (NMR)*; *spin–spin coupling*.

chemical symbol   An abbreviation for the name of an element, consisting typically of one or two letters. Examples: C, carbon; He, helium.

chemistry   [Gk *kemeia* transmutation] The study of matter, its structure, composition, and transformation. Chemistry is concerned chiefly with the interactions of electrons and atomic nuclei.

chiral   [Gk *cheir* hand] Having the property of handedness. Of a molecule: not superimposable on its mirror image. See also *enantiomers*; *optical isomers*.

chromatography   [Gk *chroma* color + *graphein* to write] Separation of a mixture into its components, accomplished by differential adsorption onto various media.

cis   [L *cis* on this side] Describing a geometric isomer in which two identical groups appear on the same side—as around a C=C double bond, or relative to the metal atom in a complex. Contrasted with *trans*.

cis-trans isomerization   Conversion between cis and trans geometric isomers.

classical mechanics   Study of the motion and energy of bodies large enough to obey Newton's laws. Prodded by a force, a classical particle moves in a predictable path. Contrasted with *quantum mechanics*. Also called *Newtonian mechanics*, *Newtonian physics*.

close packing   An arrangement whereby atoms occupy the smallest volume with the least empty space.

cluster   A small or intermediate-sized aggregate of atoms or molecules, characterized by properties different from either the isolated particles or the bulk material.

codon   A "coding" sequence of three nitrogen bases that corresponds to a particular amino acid in the synthesis of a protein by DNA and RNA.

coherent light   [L *cohaerere* to hold together] Electromagnetic radiation in which the waves sustain a fixed phase relationship over long distances and times.

colligative property   [L *colligare* to bind] A property that depends solely on the concentration of particles, e.g., osmotic pressure, vapor-pressure lowering, freezing-point depression, boiling-point elevation.

collision   [L *collidere* to collide] A meeting of particles that results in an exchange of energy and momentum.

collision theory   A model for bimolecular reactions in the gas phase, built on the assumption that molecules must collide with sufficient energy and proper orientation in order to react.

colloid   [Gk *kolla* glue] A substance in which relatively large (but still microscopic) particles are dispersed in a solid, liquid, or gaseous medium. Also called a *colloidal suspension* or a *colloidal dispersion*.

color   (1) The brain's subjective interpretation of the frequency of electromagnetic radiation. (2) A fundamental property of quarks, responsible for the strong force.

combining capacity   The number of bonds an atom is able to form. Also called *valence*.

combustion   [L *ustus* burned] The burning of a substance, usually in combination with oxygen to produce heat and (sometimes) a flame.

common–ion effect   Repositioning of an ionic equilibrium caused by addition of a participating ion from a source other than the nominal reactants and products.

complementary colors   [L *complere* to fill up] Two colors that, when added together, will produce white light. Example: red and green.

complex   [L *complector* (*complexus*) to embrace, encompass, include] (1) A structure held together by coordinate covalent bonds, usually involving a metal atom and two or more ligands. Also called *coordination complex*. See also *complex ion*; *coordination compound*. (2) In general: any freestanding structure in which the interacting species maintain a weak association by means of noncovalent bonds, thereby retaining much of their individual chemical identities. Examples: enzyme–substrate complex, host–guest complex, solvent–solute complex.

complex ion   An ion in which two or more ligands, functioning as Lewis bases, attach themselves to a metal atom or ion by coordinate covalent bonds.

compound   [L *componere* to put together] A substance formed from two or more elements in definite proportions.

compressibility   For a gas: the ratio of the fractional change in volume ($\Delta V/V$) to the pressure applied (ΔP). Symbol: $\gamma = -(1/V)(\Delta V/\Delta P)$.

compression   [L *comprimere* (*compressus*) to squeeze together] The squeezing of a gas into a smaller volume.

Compton effect   [Arthur Holly *Compton*, 1892–1962, U.S. physicist: Nobel prize for physics, 1927] The increase in wavelength suffered by a photon that has been scattered by an electron.

concentration   [L *centrum* center] The amount of solute present in a stated amount of solution, often given as *molarity* (mol solute/L solution) or *molality* (mol solute/kg solvent).

condensation   [L *densare* to thicken] (1) A phase transition from gas to liquid, accompanied by an increase in order and a decrease in enthalpy. (2) A reaction in which the joining of two organic molecules is accomplished by the elimination of a molecule of H_2O. Example: $R-COOH + HO-R \rightarrow R-COO-R + H_2O$.

condensed phase   A solid, liquid, liquid crystal, or other state of matter in which intermolecular interactions determine the structure and properties. A gas is not a condensed phase.

conductor   [L *ducere* to lead] A substance that allows the passage of heat or electricity. Contrasted with *insulator*.

configuration   [L *figura* figure, shape] (1) An arrangement of atoms that cannot be changed without the breaking of bonds. Contrasted with *conformation*. (2) The distribution of electrons among the orbitals of an atom or molecule.

conformation   [L *forma*, form, shape] An arrangement of atoms that can be altered by rotation about a bond, usually a C—C single bond.

conformational isomers, conformers   Molecules that differ only by the rotation of atoms about a single bond.

conjugate acid   [L *conjugare* to yoke together] The Brønsted-Lowry acid formed upon protonation of a Brønsted-Lowry base. Example: NH_4^+ is the conjugate acid of NH_3. Together, the two constitute a *conjugate acid–base pair*.

conjugate base   The Brønsted-Lowry base created when a Brønsted-Lowry acid loses its proton. Example: CH_3COO^- is the conjugate base of CH_3COOH.

conjugation   The alternation of carbon single and double bonds in a Lewis structure.

conservation of angular momentum   [L *conservare* to keep, to preserve] A physical law: In a system not subjected to any outside rotational forces (torques), the angular momentum is constant in both magnitude and direction.

conservation of charge   A physical law: The total electric charge stays the same during any transformation.

conservation of energy   A physical law: The total energy of an isolated system is constant. Energy may be converted into other forms, but may not be created or destroyed.

conservation of mass and energy   A physical law deriving from the equivalence of mass and energy: In a system closed to the outside, the total amount of all energy, including mass-energy ($E = mc^2$), is constant. In chemical reactions (where mass is not appreciably converted into energy), the law reduces to the more restrictive *conservation of mass*: Total mass is constant. Matter may not be created or destroyed.

conservation of momentum, conservation of linear momentum   A physical law: In a system not subjected to any outside forces, the linear momentum is constant in both magnitude and direction.

constructive interference   [L *construere* (*constructus*) to build + *inter* between + *ferire* to strike] In-phase, crest-to-crest combination of two waves (ψ_A and ψ_B) with the same frequency, producing a combined amplitude ($\psi_A + \psi_B$) and intensity ($\psi_A + \psi_B)^2$. See also *destructive interference*; *interference*.

coordinate covalent bond   [L *cum* (*co*) with + *ordinare* to arrange] A linkage in which one atom supplies both electrons in a bonding pair. Also called *dative bond*.

coordinates   [L *cum* (*co*) with + *ordinare* to arrange] Numbers used to specify the position of an object in space. Three coordinates are needed to locate a point in three-dimensional space. Examples: (1) Cartesian coordinates: x, y, z. (2) Spherical coordinates: r (radius), θ (polar angle, latitude), ϕ (azimuth, longitude).

coordination complex   See *complex*.

coordination compound   A neutral compound containing a complex ion in electrical balance with one or more counterions.

coordination number   (1) In general: the number of linkages between an atom or molecule and its nearest neighbors, whether by covalent bonds (as in a molecule), dative bonds (as in a coordination complex), noncovalent bonds (as in a solute–solvent complex), ionic bonds (as in an ionic solid), or some other mechanism. (2) In a transition-metal complex, specifically: the number of ligands coordinated to a central metal atom or ion.

coordination sphere   In a complex or similar structure: the central species and the groups around it.

core   The inner, fully occupied electronic shells of an atom or molecule.

core electron   An electron that occupies an orbital in the filled core of an atom or molecule; an inner-shell electron.

correspondence principle   A guiding principle of quantum mechanics, enunciated by Niels Bohr: The laws and equations of quantum mechanics reduce to those of classical mechanics under conditions where Planck's constant may be regarded as negligible (typically where de Broglie wavelengths are small and system dimensions are large).

coulomb   [Charles Augustin de *Coulomb*, 1736–1806, French physicist] A derived SI unit of electrical quantity, equal to the amount of charge transported in 1 second by a current of 1 ampere. The charge on an electron is -1.60×10^{-19} coulomb. Symbol: C.

coulomb force   See *Coulomb's law*; *electrostatic force*.

Coulomb's law   A law of electrostatic interaction: The force between two point charges is (1) directed along the line between the two particles, (2) repulsive between like charges and attractive between unlike charges, (3) directly proportional to the product of the charges, and (4) inversely proportional to the square of the distance between them.

counterion   [L *contra* against, opposite; pronounced "counter-ion"] An ion that interacts with other ions of opposite charge to ensure electrical neutrality in a solution, ionic solid, coordination compound, or similar material.

covalent bond   [L *cum* (*co*) with + *valens* strong] A linkage arising from the sharing of two electrons by two atoms.

covalent crystal   A solid in which the atoms are joined by covalent bonds throughout the entire three-dimensional structure. Also called *network crystal*; *network solid*. Example: diamond.

crest   The highest point on a wave.

critical   [Gk *krisis* (*kritikos*) decision, judgment] A state in which a system undergoes a change.

critical mass   In nuclear fission: the amount of fissionable material that will produce sufficient neutrons to perpetuate a chain reaction.

critical point   The conditions under which a liquid has the same density, pressure, and temperature as the corresponding gas, realized when a system is at its critical temperature and pressure. The critical point terminates the liquid–gas line on a phase diagram.

critical pressure   The lowest pressure sufficient to liquefy a gas at its critical temperature. Symbol: P_c.

critical temperature   The highest temperature at which a gas can be liquefied, coinciding with the termination point of the liquid–vapor line in a phase diagram. Symbol: T_c.

cross-linking   The formation of bonds between individual chains of a polymer.

crown ether   A cyclic ether formed in the shape of a crown, with a central cavity able to accommodate species of a certain size. The structure contains $-CH_2CH_2-$ groups alternating with oxygen atoms.

crystal   [Gk *krystallos* clear ice] A solid with an internal structure characterized by translational symmetry and long-range order. Also called *crystalline solid*.

crystal field stabilization energy (*CFSE*)   Lowering of the d electrons' energy in a complex relative to that of a free atom in a spherical field, modeled according to crystal field theory.

crystal field theory   A model of the electrical interactions in a complex, derived from an idealized conception of the ligands as negative point charges. The d orbitals on the metal atom are split in the presence of a field generated by the ligands (the crystal field).

crystal lattice   See *lattice*.

crystalline solid   See *crystal*.

cubic centimeter   A unit of volume enclosing a cube 1 cm on a side, thus equivalent to one-thousandth of a liter (1 milliliter, mL). Symbol: cm^3, cc.

cubic close packing   [Gk *kybos* cube, die] A close-packing arrangement with the following characteristics: (1) The second layer of spheres (B) sits in the depressions of the first layer (A). (2) The third layer of spheres (C) is offset slightly from the first. (3) The fourth layer is identical to the first, whereafter the pattern repeats indefinitely as ABCABC . . . .

current   [L *currere* to run] The rate at which electric charge is transported past a certain point, measured in amperes (coulombs per second).

cycle   [Gk *kyklos* circle, wheel] A sequence of different states that concludes with the reappearance of the initial state, as in the oscillation of a wave.

cycloalkane   An alkane hydrocarbon with the generic formula C_nH_{2n}, formed when the carbon atoms are connected in a closed ring. Example: C_4H_8, cyclobutane.

Dalton's law of partial pressures   [John *Dalton*, 1766–1844, English chemist and physicist] A principle pertaining to mixtures of ideal gases: The total pressure is equal to the sum of the partial pressures of all the components. See *partial pressure*.

dative bond   [L *datus* given] A linkage in which one atom supplies both electrons in a bonding pair. Also called *coordinate covalent bond*.

daughter   (1) In nuclear reactions: a nucleus generated by the radioactive decay of another isotope. (2) In biology: a structure generated by replication or division (such as daughter cell, daughter DNA, etc.).

d block   The transition metals: elements in which the d orbitals receive their electrons. The d block is laid out as 10 columns in the center of the periodic table.

de Broglie wave   [Louis Victor *de Broglie*, 1892–1987, French physicist: Nobel prize for physics, 1929] A hypothetical "matter wave" proposed to explain the diffraction of particles by crystals, later understood to be a solution to Schrödinger's equation (thus a probability amplitude). For a particle with mass *m* and velocity *v*, the *de Broglie wavelength* is given by the relationship

$$\lambda = \frac{h}{mv}$$

where *h* is Planck's constant.

debye [Peter *Debye*, 1884–1966, Dutch physicist: Nobel prize for chemistry, 1936] A unit for the electric dipole moment, equivalent to 3.336×10^{-30} C m. Symbol: D. The electric dipole moment between an electron and proton (charge = 1.60×10^{-19} C) separated by one Bohr radius (0.529×10^{-10} m) is 2.54 D.

decomposition [L *de* from + *compositus* put together → taken apart] A chemical change in which one compound breaks up into two or more new species.

degenerate [L *de* from + *genus* family, race → from a common ancestor] Having the same value of energy, frequency, or some other quantity.

degree of freedom Each of the independent variables by which the energy of an atom or molecule is determined, understood usually as independent modes of motion (translation, vibration, rotation).

dehydration [L *de* from + Gk *hydor* water → water (taken) from] Removal of water.

delocalization The presence of an electron over extended portions of a molecule. Contrasted with *localization*.

delta (Δ) [Uppercase form of the fourth letter of the Greek alphabet; lowercase form is δ] (1) In mathematical expressions, the symbol Δ means "change in" a quantity. Example: ΔE = "change in E" = $E_{\text{final}} - E_{\text{initial}}$. (2) In chemical equations, Δ over the arrow denotes the addition of heat.

density [L *densus* thick] (1) For bulk matter, the ratio of mass (*m*) to volume (*V*): m/V. (2) In quantum mechanics: the point-by-point probability of locating an electron throughout space (proportional to the square of the wave function, ψ^2). Also called *electron density*; *probability density*.

deposition [L *de* down + *ponere* (*positus*) to put → to put down] Direct condensation of a gas to a solid phase; the reverse of sublimation.

destructive interference [L *destruere* (*destructus*) to destroy + *inter* between + *ferire* to strike] Out-of-phase, crest-to-trough combination of two waves (ψ_A and ψ_B) with the same frequency, producing a combined amplitude ($\psi_A - \psi_B$) and intensity ($\psi_A - \psi_B$)^2. See also *constructive interference*; *interference*.

detailed balance, principle of   A description of microscopic equilibrium: Every elementary reaction comes to equilibrium separately, with forward and reverse transformations proceeding at equal rates. Also called *principle of microscopic reversibility*.

deterministic   [L *determinare* to fix the limits of] Traceable to a direct cause. Newton's equations of motion are deterministic because they attribute a unique path to a particle under the influence of a well-defined force.

deuterium   [Gk *deuteros* second] An isotope of hydrogen containing one proton and one neutron in its nucleus. Also called *heavy hydrogen*.

dextrorotatory   [L *dexter* right] Of an optical isomer: tending to rotate plane-polarized light in the clockwise direction, as judged by an observer looking into the oncoming beam. Contrasted with *levorotatory*.

diamagnetic   [Gk *dia* opposed] Describing a substance with no unpaired electrons. Placed in an external magnetic field, a diamagnetic material produces its own magnetic field in the opposite direction. Contrasted with *paramagnetic*.

diamond   [L *adamas* hard metal] An allotrope of carbon, formed as an extremely pure and hard covalent crystal. The diamond crystal is built from a network of tetrahedrally bonded sp^3 carbon atoms. See also *buckminsterfullerene*; *graphite*.

diatomic   [Gk *di* two, twice] Of a molecule, ion, or similar structure: having two atoms, as distinct from *monatomic* and *polyatomic*.

diffraction   [L *diffractus* broken up] Interference by waves as they recombine after spreading past an obstacle. The effect is manifested in a pattern of alternating strong and weak bands.

diffusion   [L *diffusus* spread, poured forth] The spreading and dispersal of a substance, resulting in intermingling of the particles. See also *effusion*.

dimension   [L *dimensus* measured out] (1) A property of space denoting extent in a particular direction. Space has three dimensions. (2) A specific amount of some quantity; a unit.

dimensional analysis   A general approach to computation in which the units of the relevant quantities are used to derive the proper relationships.

dimer   [Gk *di* two, twice + *meros* part, portion] A structure composed of two identical substructures (monomers). Example: the dimeric molecule N_2O_4, formed by the union of two NO_2 monomers.

dipeptide   [Gk *di* two, twice + *peptos* digested] Two amino acids connected by a single peptide bond. The peptide bond is an amide linkage created by a condensation reaction between an amino group (NH_2) on one amino acid and a carboxylic acid group (COOH) on the other. See also *polypeptide*; *protein*.

dipole   [Gk *di* two, twice + *polos* axis, pole] A molecule having a nonzero electric dipole moment; a polar molecule. See also *electric dipole*, *electric dipole moment*; *magnetic moment*.

dipole–dipole interaction   The force or potential energy between two dipoles, thus a source of noncovalent interaction for polar molecules. The positive end of one dipole is attracted to the negative end of the other, and the potential energy is inversely proportional to the cube of the distance between the two.

dipole moment   See *electric dipole*, *electric dipole moment*.

diprotic acid   [Gk *di* two, twice] An acid with two ionizable protons. Compare with *monoprotic acid*; *triprotic acid*; *polyprotic acid*.

direct proportionality   [Gk *pro* before + L *portio* share, part] A relationship between two numbers, *A* and *B*, such that

$$A = kB$$

where *k* is a constant.

discrete   [L *discretus* separated] Separated or detached from others; distinct.

disorder   [L *dis* apart, asunder + *ordinare* to arrange] Lack of regularity; randomness.

disorder–order transition   A change from a state of relative disorder to one of relative order, as during the condensation of a gas.

dispersion interaction   The force or potential energy between two induced dipoles, usually the only source of noncovalent interaction for atoms and nonpolar molecules. The potential energy is inversely proportional to the sixth power of the separation between the dipole moments. Also called *London interaction*, *London dispersion interaction*.

disproportionation   [L *dis* apart, asunder + Gk *pro* before + L *portio* share, part] The simultaneous oxidation and reduction of a species. Example: $2NO_2 + 2H_2O \rightarrow H_3O^+ + NO_3^- + HNO_2$.

dissociation   [L *dissociatus* divided, severed] Separation of a structure into two pieces, often by the breaking of a bond.

dissolution   [L *dis* apart, asunder + *solutus* loosened] The dispersal of one substance into another, leading to intimate mixing of the molecules; the reverse of precipitation.

distillation   [L *de* down, from + *stillare* to drip] Separation of the components in a mixture by selective vaporization and subsequent condensation.

distribution   [L *distributus* divided up] A tabulation of numbers in which each value is paired with the fraction of a total population having that particular value. Example: the Maxwell-Boltzmann distribution, which records the percentage of molecules traveling at each speed.

DNA   [*deoxyribonucleic acid*] An extremely large polymer that carries genetic information for living organisms. The structure, a double helix formed from two strands of nucleotides, contains the following principal components: (1) deoxyribose, a sugar; (2) a phosphate group; (3) the four nitrogen bases adenine, thymine, cytosine, and guanine. See also *nucleotide*; *RNA*.

donor   [L *donare* to give] Something that donates or gives, as an electron-pair donor (a Lewis base) or a proton donor (a Brønsted-Lowry acid). Contrasted with *acceptor*.

d **orbital**   [*d* = *diffuse*, used originally to describe a spectroscopic line] A one-electron wave function with angular momentum quantum number $\ell = 2$, inclusive of five magnetic sublevels: $m_\ell = -2, -1, 0, 1, 2$.

double bond   A linkage arising from the sharing of two electron pairs by two atoms, often existing as one σ bond and one π bond.

dsp^2 **hybrid orbitals**   A set of one-electron wave functions constructed from one *d* function, one *s* function, and two *p* functions on the same atom. The four dsp^2 orbitals point to the vertices of a square.

dsp^3 **hybrid orbitals**   A set of one-electron wave functions constructed from one *d* function, one *s* function, and three *p* functions on the same atom. The five dsp^3 orbitals point to the vertices of a trigonal bipyramid.

d^2sp^3 **hybrid orbitals**   A set of one-electron wave functions constructed from two *d* functions, one *s* function, and three *p* functions on the same atom. The six d^2sp^3 orbitals point to the vertices of an octahedron.

dynamic   [Gk *dynamis* force, power] Pertaining to power or force, especially as related to motion; vigorously active, forceful, energetic.

dynamic equilibrium   [Gk *dynamis* force, power + L *aequus* equal + *libra* balance, weight] An exact balance between opposing forces or reactions, maintained from within so that a system undergoes no net change. The presence of a constant vapor pressure, for example, is proof of a dynamic equilibrium between condensation and vaporization.

dynamics   The study of forces, motion, and equilibrium.

eclipsed conformation   [Gk *ekleipein* to leave out, to fail to appear] For groups attached on each side of a single bond, an arrangement in which the atoms in each group are directly aligned across the axis of rotation. Contrasted with *staggered conformation*.

effective nuclear charge   In a many-electron atom, the net positive charge that acts on an outer electron—usually smaller than the full nuclear charge owing to shielding by the inner electrons. Symbol: Z_{eff}.

effusion   [L *ex* (*ef*) out + *fusus* poured] The escape of a gas through an opening. See also *diffusion*.

elastic collision   [Gk *elastos* ductile, beaten]   A collision in which the total kinetic energy of the particles is the same before and after. No energy is diverted into heat. Contrasted with *inelastic collision*.

electrical potential, electrical potential energy   [Gk *elektron* amber, an alloy of gold and silver + L *potens* power]   The work needed to move a small charge from a given reference point in an electric field.

electric charge   See *charge, electric*.

electric dipole, electric dipole moment   [Gk *di* two, twice + *polos* axis, pole]   A pair of point charges Q and $-Q$, equal in magnitude but opposite in sign, separated by a distance d. The electric dipole moment, μ, is a vector quantity with magnitude

$$\mu = Qd$$

and direction along the axis from positive to negative.

electric field   The force exerted on a small charged particle, expressed as quantity per unit charge.

electrochemical cell   A device that transmits an electric current between a reducing agent and an oxidizing agent, arranged typically as two electrodes in contact with an electrolytic solution. In a *galvanic cell* (also called a *voltaic cell*), the free energy of a spontaneous reaction produces an electric current able to do work. In an *electrolytic cell*, the energy of an electric current is used to drive a nonspontaneous reaction. See also *battery*.

electrochemistry   Study of the relationship between chemical reactions and the flow of electricity.

electrode   [Gk *hodos* way, road]   A conductor, often metallic, through which an electric current enters or leaves an electrochemical cell.

electrolysis   [Gk *lysis* a breaking down, loosening]   The driving of a nonspontaneous reaction by energy delivered from an electric current.

electrolyte   [Gk *lyein* to loosen, to release]   A substance that produces ions when dissolved. An *electrolytic solution*, populated by equal numbers of positive and negative charges, is a conductor of electricity. Contrasted with *nonelectrolyte*.

electrolytic cell   An electrochemical cell in which the energy of an electric current is used to drive a nonspontaneous reaction.

electromagnetic field   [Gk *elektron* amber, an alloy of gold and silver + *Magnes lithos* stone of Magnesia (a city in Turkey)]   The joint electric and magnetic field produced by a charge in motion.

electromagnetic force   The force imparted to a small charged particle in an electromagnetic field.

electromagnetic radiation   The energy emitted by an electromagnetic wave.

electromagnetic spectrum   The full range of electromagnetic radiation produced by electromagnetic waves, classified according to frequency and wavelength. In order of increasing energy, the electromagnetic spectrum includes radiofrequency, microwave, infrared, visible, ultraviolet, X-ray, and gamma radiation.

electromagnetic wave   An oscillating disturbance of the electromagnetic field that varies in space and time. An electromagnetic wave is produced by the motion of charged particles.

electromotive force   [L *movere* (*motus*) to move] The electrical driving force that develops in an electrochemical reaction, equivalent to the change in free energy expressed in volts. Symbols: $\mathscr{E}$, emf.

electron   [Gk *elektron* amber, an alloy of gold and silver] An elementary particle found outside the atomic nucleus, less massive than a proton by a factor of 1836. The negative charge of an electron ($-e = -1.60 \times 10^{-19}$ C) has the same magnitude as the positive charge carried by a proton.

electron affinity   [L *affinitas* connection by marriage] The change in energy produced when an electron attaches itself to an atom or ion.

electron capture   A process of radioactive decay whereby an electron is absorbed into a nucleus.

electron configuration   The distribution of electrons among the orbitals of an atom or molecule.

electron density   In quantum mechanics: the point-by-point probability of locating an electron throughout space (proportional to the square of the wave function, ψ^2). Also called *probability density*.

electron diffraction   The wavelike diffraction of an electron beam through a regularly spaced array of atoms in a crystal. See also *diffraction*; *interference*; *neutron diffraction*; *X-ray diffraction*.

electronegativity   A relative measure of the tendency of a bonded atom to draw electrons to itself, estimated in various ways.

electron microscope   A microscope capable of great magnification and very high resolution, made possible by the relatively small de Broglie wavelength of an electron compared with that of visible light.

electron paramagnetic resonance (EPR)   A form of spectroscopy that measures the energy absorbed or emitted when an electron undergoes a change in spin.

electron spin   An angular momentum intrinsic to all electrons, not arising from any orbital motion about the nucleus but producing a magnetic moment nevertheless. For a single electron, the spin angular momentum quantum number ($s = \frac{1}{2}$) is analogous to the orbital angular momentum quantum number ℓ. Corresponding to $s = \frac{1}{2}$ are two magnetic sublevels: $m_s = \pm\frac{1}{2}$, spin-*up* and spin-*down*.

electron volt A unit of energy, equivalent to the energy gained by a single electron as it accelerates through an electric field of 1 volt. Symbol: $eV = 1.60 \times 10^{-19}$ J.

electrophile [Gk *philos* dear, beloved] A species that is attracted to electron-rich sites. Example: H^+.

electrophilic substitution The replacement of one electrophile by another at an electron-rich site.

electropositive Of a bonded atom in a compound: suggesting the relative tendency to surrender electrons to another atom. Contrasted with *electronegative*.

electrostatic force [Gk *statikos* standing] The electric force acting on a stationary charged particle in an electric field. Also called *Coulomb force*.

element [L *elementum* first principle, rudiment; originally denoting one of the four Greek elements] A substance containing only atoms with the same atomic number.

elemental analysis Various techniques that measure the composition of a substance according to type of atom.

elementary reaction A reversible process occurring in a single step. Multistep reactions proceed as sequences of elementary reactions.

elimination reaction [L *eliminare* to turn out of doors] A reaction in which groups attached to two adjacent atoms depart from the molecule, leaving behind a multiple bond.

emission [L *emittere* (*emissus*) to send out] The release of electromagnetic energy by a material system, accompanied by a transition to a quantum state of lower energy. Contrasted with *absorption*.

emission spectrum The intensity of electromagnetic radiation emitted by a material, recorded as a function of wavelength, frequency, or energy. Contrasted with *absorption spectrum*.

empirical formula [Gk *empeirikos* experienced] The simplest chemical formula for a substance, showing relative numbers of the atoms but not necessarily the molecular proportions. Compare with *molecular formula*.

enantiomers [Gk *enantios* opposite] Mirror-image structures that cannot be superimposed; optical isomers. See also *chiral*; *optical isomers*.

endothermic reaction [Gk *endon* within + *therme* heat] A chemical change accompanied by absorption of heat from the surroundings. Contrasted with *exothermic reaction*.

energy [Gk *energeia* activity] The ability to do work. See also *kinetic energy*; *potential energy*.

energy level A quantized value of energy and its associated quantum state.

enthalpy [Gk *enthalpein* to warm within] An extensive thermodynamic property of state, equivalent to the heat exchanged under constant pressure. Denoted by the letter H, the enthalpy is expressed in terms of the internal energy (E), pressure (P), and volume (V) as

$$H = E + PV$$

enthalpy of activation   The difference in enthalpy between a transition state and a set of reactants. Similar: entropy of activation, free energy of activation.

enthalpy of binding   The difference in enthalpy between any complex (including an enzyme–substrate complex) and its unbound constituents. Similar: entropy of binding, free energy of binding.

enthalpy of formation   The difference in enthalpy between a compound and its constituent elements in their most stable states. See also *standard enthalpy change*.

enthalpy of fusion, enthalpy of melting   The difference in enthalpy between a substance in its liquid and solid forms; equivalently, the heat that must be added at constant pressure and temperature to melt a solid. Similar: entropy of melting, free energy of melting.

enthalpy of reaction   The difference in enthalpy between the products and reactants in a chemical transformation; equivalently, the heat absorbed or released during a process at constant pressure. Similar: entropy of reaction, free energy of reaction. See also *standard enthalpy change*.

enthalpy of solution   The enthalpy absorbed or released when a substance is dissolved. Similar: entropy of solution, free energy of solution.

enthalpy of vaporization   The difference in enthalpy between a substance in its liquid and gaseous forms; equivalently, the heat that must be added at constant pressure and temperature to vaporize a liquid. Similar: entropy of vaporization, free energy of vaporization.

entropy   [Gk *endon* within + *tropos* turn, turning → a turning within, suggesting the direction of time] (1) In thermodynamics: a function of state, S, defined such that

$$\Delta S = \frac{q_{rev}}{T}$$

where q_{rev} is the heat transferred during a hypothetically reversible process at temperature T. (2) In statistical mechanics: a measure of microscopic disorder defined as $S = k_B \ln W$, where k_B is Boltzmann's constant and W is the number of microstates. The two definitions are equivalent. Symbol: S.

entropy of reaction   The difference in entropy between the products and reactants in a chemical transformation. Similar: enthalpy of reaction, free energy of reaction. For specific kinds of reactions (for example, *entropy of solution*), see corresponding enthalpy entry (*enthalpy of solution*). See also *standard entropy change*.

enzyme   [Gk *enzymos* leavened] A catalyst for a biochemical reaction, usually a protein molecule containing a highly specific active site.

enzyme–substrate complex   The assembly formed when a reactant is bound to the active site of an enzyme by noncovalent bonds.

equation of state   A mathematical relationship applicable to the state variables of a system at equilibrium, such as the ideal gas equation: $PV = nRT$.

equilibrium   [L *aequus* equal + *libra* balance, weight] The balanced condition that results when two opposing forces act with equal effectiveness. See also *chemical equilibrium*; *thermal equilibrium*.

equilibrium constant   A mathematical relationship governing the concentrations of reactants and products in a state of chemical equilibrium. The form of the expression is given by the law of mass action. See also *mass action, law of*.

equilibrium distance   The separation between atoms in a bond that corresponds to the lowest potential energy.

equipartition of energy   A physical law: For a system in thermal equilibrium, the same amount of energy is allocated to each degree of freedom.

equivalence point   [L *aequus* equal + *valentia* strength] In a titration, the stage when just enough material has been added to react completely with the substance being titrated.

error   See *random error*; *systematic error*.

ester   An organic compound containing the functional group $-\overset{\displaystyle O}{\overset{\|}{C}}-O-$, formed by reaction of a carboxylic acid and alcohol. Example: CH_3COOCH_3, methyl acetate.

ether   An organic compound with the generic formula $R-O-R'$. Example: $CH_3CH_2OCH_2CH_3$, diethyl ether.

evaporation   [L *evaporare* to disperse in vapor] A change in state from liquid to vapor, sometimes carrying the connotation of complete vaporization.

exchange reaction   A chemical transformation of the form

$$AB + CD \rightarrow AC + BD$$

in which groups of atoms switch places. Also called *double displacement reaction*, *double replacement reaction*, *metathesis reaction*.

excited state   [L *excitare* to arouse] A quantum state with energy higher than the ground state—thus any condition other than the ground state.

exclusion principle   [L *excludere (exclusus)* to shut out] A physical law, enunciated by Wolfgang Pauli: No two electrons in an atom may have the same four quantum numbers: spatial (n, ℓ, m_ℓ) and spin (m_s). As a consequence, an orbital may contain no more than two electrons, one with spin up and the other with spin down.

exothermic reaction   [Gk *exo* outside + *therme* heat] A chemical change accompanied by release of heat to the surroundings. Contrasted with *endothermic reaction*.

experiment   [L *experiri* to try, to test] A controlled test undertaken either to discover something unknown or to examine a hypothesis.

extensive property   [L *extendere (extensus)* to stretch out] A property of a substance that varies with the amount of material. Examples: mass, volume, internal energy, enthalpy, entropy. Contrasted with *intensive property*.

face-centered cubic cell   A cubically shaped unit cell that contains a lattice point at the center of each of its six faces, in addition to a point at each of the eight vertices. Compare with *body-centered cubic cell*; *primitive cubic cell*.

factor-label method   A method of dimensional analysis in which unit factors are used to convert one quantity into another. Also called *unit-factor method*.

Fahrenheit scale   [Gabriel *Fahrenheit*, 1686–1736, German physicist] A temperature scale calibrated so that water freezes at 32°F and boils at 212°F. Symbol: °F (degree Fahrenheit).

Faraday constant   [Michael *Faraday*, 1791–1867, English physicist and chemist] The charge carried by 1 mole of electrons, equal to 96,485 coulombs. Symbol: $\mathscr{F}$.

fat   An ester formed by glycerol and fatty acids.

fatty acid   A long-chained carboxylic acid.

f block   The lanthanides and actinides: elements in which the f orbitals receive their electrons. The f block is laid out as two 14-atom rows at the bottom of the periodic table.

ferromagnetism   [L *ferrum* iron] The ability of certain substances to exhibit magnetism even in the absence of an external magnetic field.

field   An influence due to some agency that pervades space and time, measured as the force exerted on a small particle at each point. Examples: gravitational field, electromagnetic field.

first harmonic   See *fundamental frequency*.

first law of thermodynamics   A physical law, equivalent to the conservation of energy: The change in internal energy for a system (ΔE) is equal to the sum of the heat transferred (q) and the work done (w):

$$\Delta E = q + w$$

If the system has no connection to its surroundings, the internal energy is constant.

first-order reaction   A process for which the rate is directly proportional to the concentration of a single substance: rate = $k[A]$.

fission   [L *fissio* a splitting, dividing] The splitting of an atomic nucleus into two lighter nuclei, usually of roughly equal mass. An amount of energy equal to the mass defect is released during the process.

flavor   A label with six values used to describe different kinds of quark: up, down, strange, charmed, bottom, top.

fluctuation   [L *fluctuare* to flow, to undulate] A change from one condition to another, often used to denote a sudden, random, or unpredictable event.

fluid   [L *fluidus* flowing] A substance able to flow and change its shape easily, such as a liquid or gas.

***f* orbital**   [$f = fundamental$, used originally to describe a spectroscopic line] A one-electron wave function with angular momentum quantum number $\ell = 3$, inclusive of seven magnetic sublevels: $m_\ell = -3, -2, -1, 0, 1, 2, 3$.

force   [L *fortis* strong] An agency that changes the velocity of a massive object; a push or a pull. Symbol: F.

formal charge   The hypothetical atomic charge that would result if all electrons in chemical bonds were shared equally between the atoms.

formation   The process of making or assembling, as in the formation of a complex or the formation of a compound from its elements.

formation constant   The equilibrium constant associated with the formation of a complex. Symbol: K_f.

formula   See *chemical formula*.

formula weight   The combined weight (properly, the combined *mass*) of all the atoms in a chemical formula, useful especially when applied to the empirical formula of a nonmolecular compound. For a molecule, the formula weight is identical to the molecular weight.

free energy   The maximum energy available to a system to do work, related to the combined change in entropy of system and surroundings. The *Gibbs free energy*, G, is defined for processes that occur at constant temperature and pressure.

free energy of reaction   The difference in free energy between the products and reactants in a chemical transformation. Similar: enthalpy of reaction, entropy of reaction. For specific kinds of reactions (for example, *free energy of solution*), see corresponding enthalpy entry (*enthalpy of solution*). See also *standard free energy change*.

free radical   A structure containing one or more unpaired electrons.

freezing   A phase transition from a liquid to a solid, accompanied by an increase in order and a decrease in enthalpy.

freezing point, freezing temperature   The temperature at which a liquid freezes (called the *normal freezing temperature* when measured at 1 atm).

freezing-point depression   The decrease in a solvent's freezing temperature brought about by the presence of a solute; a colligative property.

frequency   [L *frequentia* crowd] The number of complete cycles that a wave or other oscillation executes in 1 second. Symbol: ν.

frequency factor   (1) The factor A in the Arrhenius expression for a rate constant:

$$k = A \, \exp\!\left(-\frac{E_a}{RT}\right)$$

The magnitude of A depends on the frequency and efficiency of reaction-inducing molecular collisions. Also called *pre-exponential factor*. (2) In the collision theory of bimolecular reactions: the total number of collisions per second.

friction   [L *frictio* a rubbing] A dissipative force that opposes the motion of a body, generated by the resistance of atoms in the material being traversed. A body slowed by friction loses energy in the form of heat.

fullerene   [R. Buckminster *Fuller*, 1895–1983, U.S. architect and engineer] One of a class of compounds in which clusters of carbon atoms form a closed cage. The prototype is buckminsterfullerene, C_{60}.

functional group   An atom or assembly of atoms that can be incorporated, intact, into an organic molecule. Once attached, a functional group imparts a characteristic set of chemical properties to the site. The local chemical identity of such a group is largely preserved even in molecules with substantially different structures. Examples: OH, hydroxyl; COOH, carboxylic acid.

fundamental forces   [L *fundare* to found, lay the foundation] Four primitive agencies by which matter can interact: gravity, the electromagnetic force, the weak nuclear force, the strong nuclear force. Of the four, electromagnetic interactions are decisive for all chemical processes outside the nucleus.

fundamental frequency   The lowest frequency at which a system can oscillate; the ground state. Also called *first harmonic, fundamental mode*.

fusion   [L *fusio* a pouring out, melting] (1) Melting. (2) A nuclear reaction in which two or more light nuclei join to produce nuclei of heavier atoms.

galvanic cell   [Luigi *Galvani*, 1737–1798, Italian physiologist] See *electrochemical cell*.

gamma radiation   The most energetic radiation in the electromagnetic spectrum, produced either by an annihilation or by a nuclear transition. Wavelengths are typically less than 0.1 Å.

gas   [Gk *chaos* chasm, atmosphere] A fluid state of matter lacking both long-range and short-range order. Highly compressible, a gas has no definite shape and adapts itself to whatever vessel it fills. See also *vapor*.

geometric isomers   Compounds in which atoms joined in the same sequence are arranged differently in space, as in: (1) the cis-trans arrangements of two substituents around a C=C double bond; (2) the placement of ligands around a metal ion.

Gibbs free energy   [Josiah Willard *Gibbs*, 1839–1903, U.S. physicist] The maximum energy available to a system to do work at constant temperature and pressure, related to the combined change in entropy of system and surroundings. The Gibbs function is defined as

$$G = H - TS$$

where H is the system's enthalpy, S is the system's entropy, and T is the absolute temperature. At constant temperature and pressure, the change in free energy is given by

$$\Delta G = \Delta H - T\Delta S$$

glass   An amorphous solid produced by fusion, often containing a mixture of SiO_2, CaO, and other oxides. Although solid, the material lacks long-range order.

globular protein   [L *globus* a globe, sphere] A protein, usually water-soluble, having a roughly spherical three-dimensional structure. Globular proteins serve a number of metabolic and immunological functions. Example: myoglobin.

gluon   [*glue*] One of a family of massless particles thought to transmit the strong force—properly, the *color* force—between quarks.

Graham's law   [Thomas *Graham*, 1805–1869, Scottish chemist] A law concerning the effusion of a gas: The rate of effusion varies inversely with the square root of the particle mass.

gram   [Gk *gramma* something drawn, a small weight] A unit of mass, equal to one-thousandth of a kilogram. Symbol: g.

graphite   [Gk *graphein* to write, to draw] The "lead" in a pencil. Graphite is an allotrope of carbon that exists as a soft, black, lustrous solid. Consisting of planar sheets of sp^2 hybridized carbon stacked internally, the material is a lubricant and conductor of electricity. See also *buckminsterfullerene*; *diamond*.

gravitational constant   The constant of proportionality in the equation that states the law of gravity. Symbol: $G = 6.67 \times 10^{-11}$ N m^2 kg^{-2}. See also *gravity*; *standard acceleration of gravity*.

gravitational field   The attractive influence of mass upon mass, envisioned to spread through space. At each point in a gravitational field, there exists a corresponding force on a massive body.

gravitational potential, gravitational potential energy   [L *potens* power] The work needed to move a mass from a certain reference point in a gravitational field. At a height h above the surface of the earth, the potential energy for a mass m is equal to mgh, where g is the acceleration due to terrestrial gravity (9.81 m s^{-2}).

gravity   [L *gravitas* heavy] A fundamental force by which massive bodies interact. The gravitational force, always attractive, is directly proportional to the product of the two masses (m_1, m_2) and inversely proportional to the square of the distance r between them:

$$F = G\frac{m_1 m_2}{r^2}$$

G is the universal gravitational constant.

ground state   The state of lowest energy for a system.

group   (1) A vertical column of the periodic table, containing elements with analogous valence configurations. Example: Group I (the alkali metals plus hydrogen) includes all the atoms with ns^1 configurations. (2) See *functional group*.

guest   A structure that binds to a complementary *host* by means of noncovalent interactions, often highly specific.

guest–host chemistry   See *host–guest chemistry*.

Haber-Bosch process   [Fritz *Haber*, 1868–1934, German chemist: Nobel prize for chemistry, 1918; Karl *Bosch*, 1874–1940, German chemist: Nobel prize for chemistry, 1931] A catalytic synthesis of ammonia from nitrogen and hydrogen, carried out at high temperature and high pressure.

half-life   The time needed for a diminishing quantity to drop to half its initial amount—as, for example, the concentration of a reactant or the number of radioactive nuclei in a sample. Symbol: $t_{1/2}$.

half-reaction   A single reduction or oxidation reaction, pictured in isolation to highlight the gain or loss of electrons. Example: Ag$^+$(aq) + e$^-$ $\longrightarrow$ Ag(s).

halogen   [Gk *hals* salt + *genes* born, produced] A Group VII element (F, Cl, Br, . . .), distinguished by the valence configuration ns^2np^5.

harmonic frequency   [Gk *harmonikos* musical, suitable] A mode of oscillation such that the frequency is an integral multiple of the fundamental frequency.

harmonic motion   Periodic motion in which the path is symmetric about a point of equilibrium. Example: the oscillations of an idealized spring or pendulum.

heat   A flow of thermal energy from a body at higher temperature to a body at lower temperature. The energy is invested in the microscopic motion of the atoms and molecules.

heat capacity   An extensive thermodynamic property: the amount of heat needed to increase the temperature of a material by 1 K (equivalently, 1°C). See also *molar heat capacity*; *specific heat*, *specific heat capacity*.

Heisenberg uncertainty principle   [Werner *Heisenberg*, 1901–1976, German physicist: Nobel prize for physics, 1932] See *uncertainty principle*.

helix   [Gk *helix* anything twisted] A spiral.

Henderson-Hasselbalch equation   [Lawrence Joseph *Henderson*, 1878–1942, American biochemist; Karl *Hasselbalch*, Danish biochemist, 1874–1962] A relationship connecting the pH, pK_a, and concentrations of acid and conjugate base in a buffer solution:

$$HA(aq) \rightleftharpoons H^+(aq) + A^-(aq)$$

$$pH = pK_a + \log\frac{[A^-]}{[HA]}$$

Henry's law   [William *Henry*, 1775–1836, English chemist] The dissolved concentration of a gas is proportional to the pressure of that gas over the solution: $P_A = K\,[A(aq)]$, where K is called the *Henry's law constant*.

hertz   [Heinrich *Hertz*, 1857–1894, German physicist] The SI unit of frequency, equal to 1 cycle per second. Symbol: Hz.

Hess's law   [Germain Henri *Hess*, 1802–1850, Russian chemist born in Germany] A principle, originally stated for enthalpy, that holds for all functions of state: For a process that can be divided into a sequence of component processes 1, 2, 3, . . . , the overall change in any state function X is given by

$$\Delta X = \Delta X_1 + \Delta X_2 + \Delta X_3 + \cdots$$

heterogeneous   [Gk *heteros* the other of two, different + *genes* born, produced] Made from different parts, as a patchwork. A system containing different phases is said to be heterogeneous, as is a mixture in which the different components are segregated into distinct regions. Contrasted with *homogeneous*.

heterogeneous catalysis   A catalyzed reaction in which the catalyst exists in a phase different from that of the reactants.

heterogeneous equilibrium   An equilibrium between substances that exist in different phases.

heteronuclear   Containing elements of different kinds. A heteronuclear diatomic molecule, such as HCl, is formed from two distinguishable atoms.

heuristic   [Gk *heuriskein* to discover, to find out] Serving to point out or awaken interest, especially to encourage further investigation—sometimes, an approach based upon trial and error.

hexagon   [Gk *hex* six + *gonia* angle] A two-dimensional figure containing six sides and six angles.

hexagonal close packing   A close-packing arrangement with the following characteristics: (1) The second layer of spheres (B) sits in the depressions of the first layer (A). (2) The third layer is identical to the first, whereafter the pattern repeats indefinitely as ABAB . . . .

high-spin complex   A complex with the largest possible number of unpaired *d* electrons.

homogeneous   [Gk *homos* one and the same + *genes* born, produced] Either made from only one kind of part, or else blended together uniformly. In a homogeneous mixture (a solution), all components are mixed intimately and uniformly. Contrasted with *heterogeneous*.

homogeneous catalysis   A catalyzed reaction in which the catalyst and reactants all exist in the same phase.

homogeneous equilibrium   An equilibrium between substances that exist in the same phase.

homonuclear   Containing only one kind of element. A homonuclear diatomic molecule, such as H_2, is formed from two indistinguishable atoms.

hormone   [Gk *horman* to set in motion, to excite, to stimulate] A biological molecule that exerts an influence on specifically receptive cells, tissues, or organs.

host   A receptive structure that binds to a complementary *guest* by means of noncovalent interactions, often highly specific.

host–guest chemistry   Study of the structure, thermodynamics, and kinetics of complexes formed by the binding of complementary species. See also *guest*; *host*; *molecular recognition*.

Hund's rule   [Friedrich *Hund*, 1896–1997, German physicist and spectroscopist] An empirical observation relating to electron configurations: Within a set of degenerate orbitals, the configuration of lowest energy will contain the largest number of parallel electron spins.

hybrid orbital [L *hibrida* a crossbred animal] A combination of atomic orbitals on the same atom, resulting in a blended wave function with new directional properties. Example: An sp^3 hybrid orbital is formed by combining one *s* function with three *p* functions.

hydration [Gk *hydor* water] (1) In general: the addition of H_2O to a compound. (2) For solutes or complexes: the coordination of water molecules around a central species.

hydration shell A layer of water molecules surrounding a solute.

hydride ion The hydrogen anion, H^-.

hydrocarbon A compound consisting entirely of carbon and hydrogen atoms.

hydrogen bond A relatively strong noncovalent interaction, mediated by a bonded hydrogen atom. When attached to a suitably electronegative partner (N, O, or F), the partially positive hydrogen attracts a lone pair on a similarly polar bond elsewhere.

hydrolysis [Gk *hydor* water + *lysis* a breaking down, loosening] Reaction of a substance with water.

hydronium ion The H^+ ion in aqueous solution, customarily represented as H_3O^+.

hydrophilic [Gk *hydor* water + *philia* friendship, affinity, love] Having an affinity for water; attracted to water.

hydrophobic [Gk *hydor* water + *phobos* fear] Repelled by water.

hydroxide ion The OH^- ion.

hypertonic solution See *isotonic solutions*.

hypothesis [Gk *hypo* under + *thesis* a setting down] A proposed explanation for a set of observations.

hypotonic solution See *isotonic solutions*.

icosahedron [Gk *eikosi* twenty + *hedra* seat, face of a geometric form] A three-dimensional figure with 20 faces and 12 vertices.

ideal gas A gas that satisfies the equation of state $PV = nRT$, where P is the pressure, V is the volume, T is the absolute temperature, n is the number of moles, and R is a constant. The hypothetically ideal gas—an abstraction existing in the mind—is imagined to consist entirely of volumeless, noninteracting particles in random motion. All gases approach ideality at sufficiently low density.

ideal solution A solution that satisfies Raoult's law: The vapor pressure of the solvent or any other volatile component is reduced in proportion to its mole fraction. In the hypothetically ideal solution, there are no differences between solute–solvent, solute–solute, and solvent–solvent interactions. All solutions approach ideality at sufficiently low solute concentrations.

idiosyncratic [Gk *idios* one's own, personal + *syn* with + *krasis* a blending] Describing a characteristic or property peculiar to an individual.

immiscible   [L *in* (*im*) not + *miscere* to mix, mingle] Mutually insoluble; unable to be mixed.

inchoate   [L *inchoare* (*inchoatus*) to hitch up] Incompletely or imperfectly formed; incipient, recently begun.

incoherent light   [L *in* not + *cohaerere* to hold together] Electromagnetic radiation in which the waves have random phases.

indeterminacy principle   See *uncertainty principle*.

indicator   [L *indicare* (*indicatus*) to point, make known] A substance that changes color or some other property at a point of transition. For an acid–base indicator, the transition is from acid to conjugate base. For a redox indicator, the transition is from reductant to conjugate oxidant.

induced dipole   See *instantaneous dipole*.

inelastic collision   [L *in* not + Gk *elastos* ductile, beaten] A collision in which the total kinetic energy of the particles is not the same before and after. Part of the energy is diverted either into heat or into some other mode. Contrasted with *elastic collision*.

inert   [L *iners* unskillful] Having little or no tendency to react; inactive. Contrasted with *labile*.

inert complex   A complex for which the rate of ligand exchange is very low.

inertia   The tendency for an object to retain its present state of motion unless acted upon by a force.

inertial reference frame   A reference frame in which an object not subject to a force will either remain at rest or else move in a straight line at constant speed. See also *reference frame*.

infrared radiation   [L *infra* below → below the red] Electromagnetic radiation with a wavelength longer than visible red light and shorter than microwaves. The energy of an infrared photon is comparable to the vibrational energies of most chemical bonds.

infrared spectroscopy   Measurement of the intensity of infrared radiation absorbed or emitted by a molecule, corresponding generally to excitation of molecular vibrations.

inhibitor   [L *inhibere* (*inhibitus*) to restrain] A species that slows or stops a reaction. Contrasted with *catalyst*.

initial rate   [L *inire* (*initus*) to go into, begin] The rate of a reaction as the process just begins, with products mostly unformed.

inorganic chemistry   [L *in* not + Gk *organon* bodily organ] The branch of chemistry concerned with the structure, properties, and reactions of all compounds not derived from hydrocarbons.

instantaneous dipole   [L *instare* (*instans*) to be present, to be urgent] A transient dipole moment arising from a fluctuation in the electron distribution of an atom, molecule, or ion. An *induced* dipole in one species may be created by the momentary skewing of the electron cloud in another.

instantaneous rate   In kinetics: the change in concentration per unit time, $\Delta[A]/\Delta t$, when measured over an infinitesimally short interval ($\Delta t \rightarrow 0$). The *average rate* corresponds to the change measured over a comparatively long interval.

insulator   [L *insulatus* made into an island] A substance that does not allow the passage of heat or electricity. Contrasted with *conductor*.

intensity   [L *intendere* (*intensus*) to stretch toward, to aim at] The energy transported by a wave across a unit area per unit time, proportional to the square of the amplitude.

intensive property   A property of a substance that is independent of the amount of material. Examples: density (mass per unit volume), temperature (average energy per particle). Contrasted with *extensive property*.

interaction   [L *inter* between + *actus* a doing] The effect that one particle has on another, expressed either as a force or as potential energy.

interatomic potential   The variation of the interaction energy between two atoms as a function of their separation. Short for *interatomic potential energy*.

interface   The boundary between two different phases.

interference   [L *inter* between + *ferire* to strike] Combination of two waves with the same frequency, leading to either reinforcement or cancellation of the joint intensity. See also *constructive interference*; *destructive interference*.

intermediate   [L *inter* between + *medius* middle, in the middle] A species that comes and goes during the course of a reaction, produced in one step and consumed in another.

intermediate-field ligand   A ligand that imposes a crystal field splitting of moderate extent, falling in the center of the spectrochemical series. Examples: NCS^-, H_2O.

intermolecular   Pertaining to a relationship between two different molecules.

intermolecular interaction   The influence that one molecule exerts on another through various quantum mechanical and electromagnetic effects, expressed either as a force or as potential energy.

intermolecular potential   The variation of the interaction energy between two molecules as a function of either their separation or their orientation or both. Short for *intermolecular potential energy*.

internal energy   A system's total energy (E), governed by the first law of thermodynamics: $\Delta E = q + w$, where q is the heat transferred between system and surroundings and w is the work done.

intramolecular   [L *intra* within] Pertaining to a relationship within a single molecule.

invariance   [L *in* not + *variare* to change, vary] The constancy of a particular property despite changes in other characteristics of a system.

inverse proportionality   [L *invertere* (*inversus*) to turn upside down or inside out] A relationship between two numbers, A and B, such that $AB = k$ where k is a constant.

inverse square law   A relationship where one quantity is inversely proportional to the square of another. Example: Coulomb's law, which states that the electrostatic force between two charged particles separated by a distance r varies as $1/r^2$.

ion   [Gk *ion* going] An electrically charged atom (monatomic ion) or group of atoms (polyatomic ion). A *cation*, deficient in electrons, carries a positive charge. An *anion*, with a surplus of electrons, carries a negative charge.

ion–dipole interaction   The force or potential energy between an ion and a polar molecule.

ionic bond   A net electrostatic attraction between positive and negative ions. In an ionic crystal, every ion interacts with every other ion.

ionic crystal, ionic solid   A compound composed of positive and negative ions, held together by a balance of attractive and repulsive electrostatic forces (ionic bonds).

ion–ion interaction   The Coulomb force or potential energy between two ions; an electrostatic interaction.

ionization   (1) In general: the formation of an ion, as by the removal or addition of an electron. (2) For acids of the form HA: the loss of a proton to yield the conjugate base A^-. (3) For bases of the form B: the gain of a proton to yield the conjugate acid BH^+.

ionization constant   See *acid dissociation constant*, *acid ionization constant*; *base dissociation constant*, *base ionization constant*.

ionization energy, ionization potential   The energy needed to strip away a ground-state electron from a species in the gas phase.

ion pairing   The electrical association of an anion and cation, responsible for deviations from ideality in solution.

ion-product constant   See *autodissociation constant*, *autoionization constant*.

irreversible process   [L *in* (*ir*) not + *revertere* (*reversus*) to turn back] A transformation that cannot be accomplished through a series of small steps in which the system remains infinitesimally close to equilibrium.

isobaric   [Gk *isos* equal + *baros* weight] At constant pressure. An *isobar* is a line that connects points of equal pressure on a phase diagram.

isoelectronic species   Structures having the same numbers of valence electrons. Example: F^- and Ne.

isomers   [Gk *isomeres* having equal parts] Molecules with the same number and kinds of atoms arranged in different ways.

isothermal   [Gk *isos* equal + *therme* heat] At constant temperature. An *isotherm* is a line that connects points of equal temperature on a phase diagram or *P-V* plot.

isothermal compression or expansion   A decrease (compression) or increase (expansion) in volume accomplished at constant temperature.

isotonic solutions   [Gk *isotonos* having equal accent or tone] Solutions having the same osmotic pressure and thus not susceptible to osmosis. A solution with a lower osmotic pressure is *hypotonic*, whereas one with higher osmotic pressure is *hypertonic*.

isotopes   [Gk *isos* equal + *topos* place] Atoms of a single element that differ in the number of neutrons in their nuclei. Isotopes have the same atomic number but different masses.

isotropic   [Gk *isos* equal + *tropos* turn] Having equal properties along different axes—invariant to direction, angle, or orientation. Contrasted with *anisotropic*.

J coupling   See *spin–spin coupling*.

joule   [James Prescott *Joule*, 1818–1889, English physicist] The SI unit of energy, defined as the work done when a force of 1 newton is applied over a distance of 1 meter. Symbol: $J = N\ m = kg\ m^2\ s^{-2}$.

kelvin   [William Thomson, 1st Baron *Kelvin*, 1824–1907, English physicist] The base SI unit of temperature, defined to be $1/273.16$ times the triple point of water. The kelvin is the interval on the absolute scale of temperature, equal in magnitude to the Celsius degree. Symbol: K.

Kelvin scale   See *absolute temperature*.

ketone   An organic compound with the general formula $R-\overset{\displaystyle O}{\overset{\|}{C}}-R'$, containing at least one carbonyl group. Example: CH_3COCH_3, acetone.

kilocalorie   [Gk *chilioi* a thousand; used as a numerical prefix (*kilo*) to mean 10^3 + L *calor* heat] A non-SI unit of heat, equal to 1000 calories. One kilocalorie is equivalent to 4.1840 kilojoules. Symbol: kcal.

kilogram   [Gk *chilioi* a thousand + *gramma* something drawn, a small weight] The base SI unit of mass, defined by reference to an international prototype (a platinum-iridium cylinder) maintained in Sèvres, France. A *mass* of 1 kilogram, equal to 1000 grams, has a terrestrial *weight* of 9.8 newtons (2.2 pounds). Symbol: kg.

kilohertz   A derived SI unit of frequency, equal to 1000 hertz (1000 cycles per second). Symbol: kHz.

kilojoule   A derived SI unit of energy, equal to 1000 joules. Symbol: kJ.

kilometer   A multiple of the base SI unit of length, equal to 1000 meters (roughly 0.62 mile). Symbol: km.

kinetic control   [Gk *kinein* to move; *kinetos* movable] Describing a reaction in which the dominant products are those that can be formed fastest. Contrasted with *thermodynamic control.*

kinetic energy   Energy associated with the motion of a body. An object with mass m and velocity v has a kinetic energy equal to $\frac{1}{2}mv^2$. Contrasted with *potential energy*, the energy associated with a body's position in a field.

kinetics   Study of the rates and mechanisms of chemical reactions.

kinetic theory   A statistical mechanical model that describes an ideal gas as a collection of volumeless, noninteracting particles in random motion.

labile   [L *labi* to slip] Prone to change; short-lived. Contrasted with *inert.*

labile complex   A complex that undergoes rapid exchange of ligands.

lanthanide contraction   A gradual diminution of atomic radii caused by the poor shielding capability of $4f$ orbitals, observed across the lanthanide series and affecting the elements that immediately follow.

lanthanide series   [Gk *lanthanein* to escape notice] The 14 elements that follow lanthanum (La, $Z = 57$), from cerium (Ce, $Z = 58$) through lutetium (Lu, $Z = 71$). Most of the lanthanides have partially filled $4f$ orbitals. Also called *rare-earth series.*

laser   [*lightwave amplification by stimulated emission of radiation*] A device that produces an intense beam of coherent light, usually over a narrow range of frequency.

latent   [L *latere* (*latens*) to lie hidden] Present but invisible; not realized or actualized.

lattice   A regular arrangement of points in space, derived from a repeating unit cell and possessing one or more elements of symmetry.

lattice energy   The difference in energy between the crystalline and gaseous phases of a substance, particularly an ionic solid.

Le Châtelier's principle   [H. L. *Le Châtelier*, 1850–1936, French chemist] An observation concerning dynamic equilibria: When disturbed suddenly from equilibrium by an external stress, a system will restore its equilibrium by moving in a direction that alleviates the stress. If, for example, products accumulate in excess of the mass-action ratio, the process will shift toward the formation of reactants until Q (the reaction quotient) becomes equal to K (the equilibrium constant).

length   One-dimensional extent in space.

lepton   [Gk *leptos* small, light] A spin-$\frac{1}{2}$ elementary particle subject to gravity and the weak nuclear force, but not to the strong nuclear force. Electrically charged leptons (such as electrons) respond to the electromagnetic force as well. Examples of leptons: electrons, neutrinos, and their antiparticles. Others include various forms of the muon and the tau, not treated in this text. See also *baryon; meson.*

lepton number   A conserved quantum number: 1 for leptons, -1 for antileptons.

levorotatory   [L *laevus* left] Of an optical isomer: tending to rotate plane-polarized light in the counterclockwise direction, as judged by an observer looking into the oncoming beam. Contrasted with *dextrorotatory*.

Lewis acid   [Gilbert Newton *Lewis*, 1875–1946, U.S. chemist] A species that can accept a pair of electrons. Examples: H^+, BF_3.

Lewis base   A species that can donate a pair of electrons. Examples: OH^-, NH_3.

Lewis structure   A simplified depiction of the valence electrons in an atom, molecule, or ion. Also called a *Lewis dot diagram*, the Lewis structure is drawn as an arrangement of dots around chemical symbols representing the various nuclei. A covalent pair bond is usually shown as a dash.

ligand   [L *ligare* to bind, tie] An atom or group of atoms that bonds to the central species in a complex.

ligand field theory   A molecular orbital model of bonding in a complex.

light   Electromagnetic radiation, especially those wavelengths in the visible portion of the electromagnetic spectrum.

limiting reactant, limiting reagent   The species that is entirely consumed when a reaction goes to completion.

linear combination of atomic orbitals (LCAO)   Construction of approximate molecular orbitals by addition and subtraction of atomic orbitals in various proportions.

linear dependence   A functional variation of the form $y = mx + b$, where some quantity y (the dependent variable) depends on the first power of another quantity x (the independent variable). The points (x, y) lie on a straight line with constant slope m and y-intercept b.

line spectrum   A spectrum in which signals appear only for certain discrete values of energy. See *spectrum*.

linkage isomers   Complexes that differ only in what part of the ligand or ligands attaches to the central species. The positions of the ligands in the coordination sphere are otherwise the same. Example: The nitrite ion, NO_2^-, can bond to a metal either through the nitrogen's lone pair (as the nitro form, $:NO_2^-$) or through electrons on one of the oxygens (as the nitrito form, $:ONO^-$).

lipid   [Gk *lipos* fat] An organic compound with a greasy feel, soluble in alcohol and ether but not water; a fat.

lipid bilayer   See *phospholipid bilayer*.

liquid   [L *liquere* (*liquidus*) to be liquid] A fluid state of matter lacking long-range order but possessing short-range order. Relatively incompressible, a liquid has definite volume although no definite shape. It adapts itself to whatever vessel it fills.

liquid crystal   A state of matter possessing, on average, some of the rotational symmetry of a crystal but lacking the translational order of a rigid lattice. Although the molecules flow like a liquid, they preserve their rotational orientations while so doing. Examples of liquid crystals: nematic, smectic, discotic, cholesteric, hexatic, columnar.

liter   [Gk *litra* pound] A derived SI unit of volume, equivalent to a cube 10 centimeters on a side. One kilogram of pure water at 4°C occupies a volume of 1 liter. Symbol: L.

localization   [L *locus* place] The presence of high electron density in a restricted region of a structure, often near one or two nuclei. Contrasted with *delocalization*.

local order   Synonymous with *short-range order*: regularity or organization that persists only over a limited portion of space. Example: clusters in a liquid. The molecules in each cluster are arranged according to some pattern, but there is no correlation between the locations of one cluster and another.

logarithm   [Gk *logismos* a counting, calculation + *arithmein* to count, reckon] The exponent to which a specified base number must be raised to produce a given value. See also *natural logarithm*.

London interaction, London dispersion interaction   [Fritz *London*, 1900–1954, U.S. physicist born in Germany] The force or potential energy between two induced dipoles, a source of noncovalent interaction most important for nonpolar molecules. The potential energy is proportional to the inverse sixth power of the separation between the dipole moments.

lone pair   A localized pair of valence electrons, not shared with any other atom.

long-range order   Regularity or organization that persists over an extended region of space, as in a crystalline lattice.

low-spin complex   A complex with the least possible number of unpaired d electrons.

luminescence   [L *lumin* light, window] The emission of light from a system, caused by a transition to a lower energy level.

Lyman series   [Theodore *Lyman*, 1874–1954, U.S. physicist] A series of lines in the ultraviolet emission spectrum of hydrogen, corresponding to transitions terminating at the level $n = 1$.

macromolecule   [Gk *makros* long, large] A very large molecule, formed from hundreds or thousands of atoms; often, a polymer. Examples: proteins, nucleic acids.

macroscopic   [Gk *makros* long, large + *skopein* to look at] Large; visible to the unaided eye.

macrostate   The complete set of macroscopic properties needed to characterize a system. Example: The variables pressure (P), volume

(V), temperature (T), and amount (n) describe the macrostate of an ideal gas. If any three of the values are known, then the equation of state fixes the value of the fourth.

macroworld   Those aspects of nature that involve large objects and large transfers of energy.

magic numbers   Combinations of protons and neutrons found in exceptionally stable nuclei, analogous to octets and closed shells for electrons.

magnetic field   [Gk *Magnes lithos* stone of Magnesia (a city in Turkey)] The force field produced by a charge in motion; the component of the electromagnetic field that arises from a moving charge.

magnetic force   The force on a small charged particle moving in a magnetic field, considered separately from the electrostatic force.

magnetic moment   A quantity associated with electronic orbital angular momentum and spin angular momentum, nuclear spin angular momentum, current loops, and certain other phenomena. The magnitude and direction of a magnetic moment determine its potential energy in a magnetic field.

magnetic quantum number   A quantum number pertaining to the orientation of a one-electron orbital in a magnetic field; in particular, the component of orbital angular momentum along the field. Symbol: m_ℓ. For a given value of the angular momentum quantum number ℓ, there are $2\ell + 1$ values for m_ℓ: $0, \pm 1, \pm 2, \ldots, \pm \ell$.

main-group elements   The elements in the *s* and *p* blocks of the periodic table, also known as the *representative elements*.

manometer   [Gk *manos* loose, sparse + *metron* measure] An instrument for measuring the pressure of a gas.

mass   [L *massa* mass; Gk *massein* to knead] The amount of matter in a sample, defined (for example) through Newton's second law:

$$m = \frac{F}{a}$$

A body's mass (m) determines its acceleration (a) when subjected to a force (F).

mass action, law of   A statement concerning the composition of a chemical system at equilibrium: For any reaction

$$aA + bB \rightleftharpoons cC + dD$$

in a state of equilibrium, the concentrations stand in the relationship

$$K(T) = \frac{[C]^c[D]^d}{[A]^a[B]^b}$$

This mass-action ratio $K(T)$, better known as the equilibrium constant, has a fixed value at a given temperature.

mass defect   The difference between the total mass of an atomic nucleus and the sum of the masses of the constituent protons and neutrons in isolation. The missing mass, related to energy through Einstein's formula $E = mc^2$, provides the binding energy for the nucleus.

mass number   The combined number of protons and neutrons in a nucleus. Symbol: A.

mass spectrometry   A technique for measuring the distribution of atomic and molecular mass in a system. The method involves analysis of the path of a charged particle in a magnetic field.

matter   [L *materia* material, substance, timber] The tangible stuff of the universe: anything that has mass and occupies space.

Maxwell–Boltzmann distribution   [James Clerk *Maxwell*, 1831–1879, Scottish physicist; Ludwig *Boltzmann*, 1844–1906, Austrian physicist] The apportionment of speeds among the particles of an ideal gas, involving the Boltzmann factor and a factor proportional to the square of the velocity. See also *Boltzmann distribution*.

mean   [L *medianus* in the middle] An average value, intermediate between two extremes.

mean free path   The average distance traveled by a particle between collisions.

mean-square speed   The average value of the quantity v^2, computed as

$$<v^2> = \frac{1}{N}(v_1^2 + v_2^2 + \cdots + v_N^2)$$

for a system of N particles. The symbol v_i denotes the speed of particle i.

measurement   [L *metiri* (*mensus*) to measure, to mete out] Determination of dimensions, extent, amount, and other properties, usually by comparison with a standard quantity.

mechanics   [Gk *mechane* machine, contrivance] Study of the motion of objects in relation to force and energy.

mechanism   The sequence of elementary steps by which a reaction proceeds.

megahertz   [Gk *megas* large, great; used as a numerical prefix (*mega*) to mean 10^6, 1 million] A derived SI unit of frequency, equal to 10^6 hertz (10^6 cycles per second). Symbol: MHz.

melting   A phase transition from a solid to a liquid, accompanied by a decrease in order and an increase in enthalpy.

melting point, melting temperature   The temperature at which a solid melts (called the *normal melting temperature* when measured at 1 atm).

membrane   [L *membrana* a thin skin, parchment] The thin surface that encloses the contents of a cell, consisting of a phospholipid bilayer and various proteins.

meson   [Gk *mesos* middle, in the middle] One of various elementary particles carrying the strong force between nucleons. Mesons, possessing integral values of the spin angular momentum quantum number, are neither baryons nor leptons.

metal   [Gk *metallon* mine, quarry, metal] An atom that tends to lose electrons. Bulk metals are usually lustrous (shiny), ductile (easily drawn into wires), and malleable (easily hammered). They conduct heat and electricity. Contrasted with *nonmetal*. Examples: Na, sodium; Ag, silver; Au, gold.

metallic crystal   A crystal in which the bonding electrons are delocalized throughout the entire lattice, shared indistinguishably among a fixed array of cations.

metalloid   See *semimetal*.

metallurgy   Study of the structure and properties of metals, including techniques for extraction, purification, and the formation of alloys.

metastable state   [Gk *meta* after, beyond] A system resting in a *local* mininum of energy, thus lying above a level of greater stability. A metastable state will return to its original configuration following a *small* disturbance. Pushed with sufficient energy, however, it will undergo a transition to the lower level. Compare with *stable state*; *unstable state*.

metathesis reaction   [Gk *meta* after, along with, beyond + *thesis* a setting down, putting down] An exchange of ions between two species, as in the reaction

$$AB + CD \rightarrow AC + BD$$

Example: $AgNO_3$ and NaCl react in solution to produce AgCl(s) and $NaNO_3$(aq). Also called *double displacement reaction*, *double replacement reaction*.

meter   [Gk *metron* measure] The base SI unit of length (39.37 inches), approximately equal to one 10-millionth of the distance from the Equator to the North Pole when measured along a meridian. From 1960 to 1983, the meter was defined to be 1,650,763.73 wavelengths of the orange-red emission of krypton-86 under a set of standard conditions. Since 1983, one meter has been defined as $1/299{,}792{,}458$ of the distance traveled by light in 1 second through a vacuum.

micelle   [L *mica* crumb, grain] An aggregated particle held together by noncovalent interactions between oriented molecules, often involving the coordination of hydrophobic and hydrophilic structures.

microscopic   [Gk *mikros* small + *skopein* to look at] Small; invisible to the unaided eye.

microscopic reversibility, principle of   For a system in overall equilibrium, every elementary reaction stands individually in dynamic equilibrium. The forward and reverse rates are equal at each step. Also called *principle of detailed balance*.

microstate   A microscopic arrangement of energy and position consistent with a given set of macroscopic properties.

microwave radiation   Electromagnetic radiation with a wavelength shorter than radiofrequency waves and longer than infrared radiation. The energy of a microwave photon is comparable both to the rotational energies of most molecules and to the magnetic energies of unpaired electronic spins.

microwave spectroscopy   Measurement of the intensity of microwave radiation absorbed or emitted by a molecule, corresponding generally to molecular rotations or electronic spin flips.

microworld   Those aspects of nature that involve small objects and small transfers of energy; atoms and molecules, quantum mechanics, statistics.

milliliter   [L *mille* thousand; used as a numerical prefix (*milli*) to mean 10^{-3}, one-thousandth] A derived SI unit of volume, equal to 10^{-3} liter—thus equivalent to 1 *cubic centimeter* (cm^3, cc), a cube 1 centimeter on a side. Symbol: mL.

mirror image   The image of an object as it appears reflected in a mirror, with left and right reversed. See also *enantiomer*.

miscible   [L *miscere* to mix, mingle] Mutually soluble; able to be mixed.

mixture   A combination of chemically distinct substances in which the different components interact noncovalently, thus permitting separation by physical properties such as boiling point, polarity, mass, and so forth.

mobile phase   [L *mobilitare* to set in motion] The solvent that moves over the column in a chromatographic apparatus. Contrasted with *stationary phase*.

molal boiling-point-elevation constant   The number K_b in the expression for boiling-point elevation:

$$\Delta T_b = K_b m$$

ΔT_b denotes the increase in boiling point for a solution of molality m. K_b is a solvent-specific constant that determines the magnitude of the colligative effect. For water, $K_b = 0.51°C\ m^{-1}$.

molal freezing-point-depression constant   The number K_f in the expression for freezing-point depression:

$$\Delta T_f = K_f m$$

ΔT_f denotes the decrease in freezing point for a solution of molality m. K_f is a solvent-specific constant that determines the magnitude of the colligative effect. For water, $K_f = 1.86°C \ m^{-1}$.

molality　A measure of a solution's concentration: the number of moles of solute dissolved per kilogram of solvent. Symbol: $m = mol \ kg^{-1}$. See also *molarity*.

molar　(1) Pertaining to quantity per mole of substance. (2) Pertaining to a solution in which 1 mole of solute is dissolved per liter of solution.

molar heat capacity　Heat capacity per mole of substance. Symbols: c_p, molar heat capacity at constant pressure; c_V, molar heat capacity at constant volume. Contrasted with *specific heat*. See also *heat capacity*.

molarity　A measure of a solution's concentration: the number of moles of solute dissolved per liter of solution (combined volume of solvent and solute). Symbol: $M = mol \ L^{-1}$. See also *molality*.

molar mass　The mass of 1 mole of substance measured in grams, identical numerically to the particle mass expressed in atomic mass units. Formerly called *atomic weight*, *formula weight*, or *molecular weight*, as appropriate. Symbol: $\mathcal{M}$.

mole　[L *moles* mass] The base SI unit for amount of substance, equal to the number of atoms in exactly 12 grams of carbon-12. Symbol: mol.

molecular crystal　A crystalline array of individual atoms or molecules held together by noncovalent interactions.

molecular formula　A representation of the elemental composition of a molecule, written as a sequence of chemical symbols with subscripts to indicate the actual number of atoms in the structure. Compare with *chemical formula*; *empirical formula*.

molecularity　The number of reacting species in an elementary reaction. See also *unimolecular reaction*; *bimolecular reaction*; *termolecular reaction*.

molecular mass　The absolute mass of a particular molecule, expressed in a specific set of units as either an individual or molar quantity. See also *atomic mass*; *molar mass*; *relative atomic mass*; *relative molecular mass*.

molecular orbital　A wave function for a single electron in a molecule, patterned after the functions obtained by exact solution of the Schrödinger equation for the hydrogen atom. A molecular orbital is usually approximated as a linear combination of atomic orbitals.

molecular orbital theory　A model of bonding in which the electrons arc assigned to molecular orbitals, usually extending over two or more atoms. The orbitals are filled according to the Aufbau principle.

molecular recognition　Selective binding of one structure to another, with preference determined on the basis of size, shape, polarity, and various other characteristics.

molecular sieve   A material having internal pores of molecular dimension, thus able to discriminate between microscopic particles on the basis of size.

molecular weight   See *molar mass*.

molecule   [L *moles* mass + *culus* a diminutive] The smallest sample of a compound that still retains the chemical properties of that compound. The structure exists as an electrically neutral assembly of atoms held together by covalent bonds.

mole fraction   A measure of the proportion of a certain component i in a mixture, defined as

$$X_i = \frac{n_i}{n_1 + n_2 + \cdots + n_N}$$

The value of n_i, the number of moles of i, is specified in the numerator. The total number of moles in the N-component mixture, equal to the sum $n_1 + n_2 + \cdots + n_N$, appears in the denominator.

momentum   [L *momentum* motion, cause of motion] A vector quantity (p) associated with the motion of an object, its magnitude equal to the product of mass and velocity:

$$p = mv$$

The direction of the momentum is parallel to the velocity. Plural: momenta.

monatomic   [Gk *monos* alone, only] Having only one atom, as distinct from *diatomic* and *polyatomic*.

monoclonal antibody   [Gk *monos* alone, only + *klon* a slip, twig] An antibody produced by genetic engineering, designed to be more abundant than a natural antibody and able to bind specifically to a designated site on an antigen.

monodentate ligand   [Gk *monos* alone, only + L *dens* tooth] A ligand offering just one point of attachment. Example: NH_3, which bonds to a metal only through the lone pair on the nitrogen atom.

monomer   [Gk *monos* alone, only + *meros* part, portion] A repeating unit of a polymer, usually a small molecule able to form chainlike structures.

monoprotic acid   An acid with only one ionizable proton. Compare with *diprotic acid*; *triprotic acid*; *polyprotic acid*.

monotonic   [Gk *monos* alone, only + *tonos* a stretching, tone, mode] Pertaining to a function that steadily increases or decreases.

most probable radius   The distance from the nucleus ($r = r_{MP}$) for which the radial probability distribution,

$$P = 4\pi r^2 \psi^2(r)$$

is maximum. $\psi(r)$ denotes an orbital that depends solely on the electron's distance from the nucleus, not on its angular coordinates. The electron is found most frequently anywhere on the surface of a sphere with radius $r = r_{MP}$.

nanocrystal [Gk *nanos* dwarf] A small cluster with crystalline symmetry, ranging from a few atoms, molecules, or ions to many hundreds or thousands.

nanometer [Gk *nanos* dwarf; used as a numerical prefix (*nano*) to indicate 10^{-9}, one-billionth] A submultiple of the base SI unit of length, equal to 10^{-9} meter.

natural logarithm A logarithm that uses $e = 2.718281828459045$ for a base. The natural logarithm of a number A, symbolized $\ln A$, is the exponent needed to satisfy the expression

$$A = e^{\ln A}$$

By contrast, the base-10 logarithm ($\log A$) is defined as

$$A = 10^{\log A}$$

The two systems are related by the factor $\ln 10 = 2.303$.

Nernst equation [Walther *Nernst*, 1864–1941, German physicist and chemist: Nobel prize for chemistry, 1920] For an electrochemical cell, the relationship between the cell potential ($\mathscr{E}$), the standard cell potential ($\mathscr{E}°$), the reaction quotient (Q), and the number of electrons transferred (n):

$$\mathscr{E} = \mathscr{E}° - \frac{RT}{n\mathscr{F}} \ln Q$$

R is the universal gas constant. T is the absolute temperature. $\mathscr{F}$ is the Faraday constant.

net ionic equation A chemical equation describing a reaction of ions exclusive of all spectator ions. See also *spectator ion*.

network crystal, network solid See *covalent crystal*.

neutral [L *neuter* neither of two] (1) Of a structureless particle: having no electric charge. (2) Of a collection of particles: having an equal number of positive and negative electric charges.

neutralization Reaction of an acid and a base, resulting in the formation of a salt (and, in aqueous solution, water as well).

neutrino ["the little neutral one," coined by Enrico Fermi] One of a class of uncharged leptons, massless or nearly so.

neutron   An elementary particle, one of the primary constituents of an atomic nucleus. Slightly more massive than a proton, the neutron is electrically neutral and has a spin of $\frac{1}{2}$.

neutron diffraction   Scattering and interference of a beam of neutrons through a crystal, used to ascertain the spacings between atoms. Neutron diffraction is analogous to X-ray diffraction, but better able to resolve the positions of lightweight atoms (such as hydrogen) owing to the neutron's shorter de Broglie wavelength.

neutron emission   An avenue of radioactive decay: the ejection of a neutron from an atomic nucleus.

Newman projection   [Melvin S. *Newman*, 1908–1993, U. S. chemist] A drawing designed to show the conformation of an organic molecule, rendered as a view down the axis of a carbon–carbon bond.

newton   [Isaac *Newton*, 1642–1727, English natural philosopher and mathematician] The SI unit of force, derived from Newton's second law as the force needed to accelerate a 1-kilogram mass ($m = 1$ kg) by 1 meter per second per second (1 m s^{-2}):

$$F = ma$$

Symbol: N = kg m s^{-2}.

Newtonian mechanics, Newtonian physics   Study of the motion and energy of bodies large enough to obey Newton's laws; classical mechanics. Contrasted with *quantum mechanics*.

Newton's laws   The classical laws of motion: (1) A body at rest remains at rest unless acted on by an external force. A body moving in a straight line at constant speed continues that motion unless acted on by an external force. (2) A force (F) acting on a mass (m) produces an acceleration (a) such that $F = ma$. (3) For every force that acts on a body, that same body produces a reciprocating force equal in magnitude and opposite in direction.

noble gas   [L *nobilis* notable, of high rank] A Group VIII element (He, Ne, Ar, . . .), distinguished from other atoms by its filled valence shell. Most noble gases are unreactive, except under special conditions. Formerly called *inert gas*.

node   [L *nodus* knot] A point, line, or surface of a standing wave where the disturbance is zero. For an orbital, a node corresponds to a point or region of vanishing electron probability.

nonbonding orbital   A molecular orbital formed by atomic orbitals that do not overlap, leading to neither enhancement nor diminution of the electron density between the nuclei.

noncovalent interactions   Any of various interactions that do not involve the direct sharing of electrons between two nuclei, sometimes referred to as nonbonded interactions. Also called *van der Waals interactions*, especially when applied to neutral, closed-shell structures. Examples: dipole–dipole interaction, London dispersion interaction.

nonelectrolyte A substance that does not produce ions when dissolved. A nonelectrolytic solution, populated by neutral species, does not conduct electricity. Contrasted with *electrolyte*.

nonmetal An atom that tends to gain electrons. Bulk nonmetals are usually not lustrous, not ductile, and not malleable. They tend to be poor conductors of heat and electricity. Contrasted with *metal*. Examples: N, nitrogen; O, oxygen; S, sulfur.

nonpolar Describing a vanishing dipole moment caused by a symmetric distribution of charge—as in nonpolar molecules, nonpolar bonds, and so on. Contrasted with *polar*.

nonvolatile Contrasted with *volatile*.

nuclear fission See *fission*.

nuclear fusion See *fusion (2)*.

nuclear magnetic moment The magnetic moment that arises in a nucleus endowed with nonzero spin angular momentum.

nuclear magnetic resonance spectroscopy (NMR) Measurement of the intensity of radiofrequency radiation absorbed or emitted by a nucleus in a magnetic field, corresponding generally to nuclear spin flips.

nuclear shell model A model of nuclear binding that uses one-particle wave functions to represent the nucleons. The approach is analogous to the orbital model commonly used for molecules and many-electron atoms.

nuclear spin An angular momentum intrinsic to certain nuclei, not arising from any overt rotational motion but producing a magnetic moment nevertheless. The quantum numbers corresponding to nuclear spin, I and m_I, are analogous to the quantum numbers s and m_s for electron spin. For a single proton (the ^{1}H nucleus), where $I = \frac{1}{2}$, there are two magnetic sublevels: $m_I = \frac{1}{2}$ (spin-up) and $m_I = -\frac{1}{2}$ (spin-down).

nucleic acid A biopolymer such as DNA or RNA, composed of condensed nucleotides and carrying genetic information.

nucleon A proton or neutron inside a nucleus.

nucleophile [Gk *philos* dear, beloved] A species that is attracted to electron-poor sites (where the charge distribution tends to be partially positive, like a nucleus). Example: OH^-.

nucleophilic substitution The replacement of one nucleophile by another at an electron-poor site. See also S_N1 *mechanism*; S_N2 *mechanism*.

nucleosynthesis [Gk *syn* with + *tithenai* (*thesis*) to put, place] The formation of atomic nuclei by nuclear reactions, presently occurring in the interior of stars and believed to have occurred in the early life of the universe.

nucleotide   One of various monomers that can be linked together to form nucleic acids. A nucleotide is constructed from (1) a nitrogen base (adenine, thymine, cytosine, guanine, or uracil), (2) a molecule of phosphoric acid (H_3PO_4), and (3) a sugar (deoxyribose or ribose). See also *DNA*; *RNA*.

nucleus   [L *nucleus* kernel, little nut] The very small, very dense, electrically positive particle found at the center of an atom, containing most of the atom's mass.

nuclide   The nucleus of a specified isotope.

octahedron   [Gk *okto* eight + *hedra* seat, face of a geometric form] A three-dimensional figure with eight faces and six vertices.

octet   A completed valence shell of eight electrons, particularly for an element in the *p* block.

optical isomers   [Gk *optos* seen, visible] Nonsuperimposable mirror-image structures; enantiomers.

orbital   [L *orbita* wheel, track, course, circuit] A wave function for a single particle (for example, an electron), usually viewed within a many-particle system. See also *atomic orbital*; *molecular orbital*.

order   [L *ordinare* to arrange] (1) Regularity; absence of randomness. (2) The sum of the exponents on an algebraic term. Example: The rate expression

$$Rate = k[A]^2$$

is second order in the concentration [A].

organic chemistry   [Gk *organon* bodily organ] The branch of chemistry concerned with the structure, properties, and reactions of compounds that contain carbon. Organic molecules figure prominently in the architecture and function of living organisms.

oscillation   [L *oscillum* a swing] A regular motion between two extremes; a vibration.

osmosis   [Gk *osmos* thrust, push] A colligative property: the movement of solvent between solutions of different concentrations, as through a semipermeable membrane.

osmotic pressure   The pressure needed to suppress an osmotic flow. Symbol: Π.

overlap   The extent to which two wave functions (orbitals) occupy a common region of space; a measure of the strength of bonding.

oxidant, oxidizer   See *oxidizing agent*.

oxidation   (1) Loss of electrons by a species in a chemical reaction; complementary to reduction. (2) Reaction with oxygen. (3) In organic chemistry, especially: addition of oxygen or removal of hydrogen.

oxidation number, oxidation state   A hypothetical atomic charge that results when all bonding electrons are assigned to the more electronegative atom in each pair bond.

oxidation–reduction reaction   A reaction in which electrons are transferred between species. Synonymous with *reduction–oxidation* or *redox* reaction.

oxide   A compound in which oxygen has an oxidation state of -2. The *oxide ion* has the form O^{2-}. Compare with *peroxide*; *superoxide*.

oxidizing agent   The species that gains electrons (is reduced) in an oxidation–reduction reaction. An oxidizing agent is the agent that causes some other species to undergo oxidation. Contrasted with *reducing agent*.

oxyacid   An inorganic acid that contains oxygen. The structure typically includes two or more OH groups bonded to a central atom, often with additional O atoms attached as well. Also called *oxoacid*.

pair bond   The covalent bond as envisioned by G. N. Lewis: two electrons shared between two nuclei.

parallelepiped   [Gk *parallelepipedon* body with parallel surfaces] A six-sided figure with three pairs of parallel faces, all parallelograms.

paramagnetic   [Gk *para* beside, alongside] Describing a substance that contains unpaired electrons. A paramagnetic material is drawn into an external magnetic field. Contrasted with *diamagnetic*.

partial pressure   In a mixture of gases, the pressure that a single component would exert if it occupied the vessel by itself.

particle   [L *pars* part, portion; *particula* small part] (1) Used informally, a small bit or fragment: an electron, a nucleus, an atom, a molecule, and so forth. (2) In more precise usage: a body in which the internal structure and internal motions can be ignored. Example: the volumeless, noninteracting particles of an ideal gas.

particulate   Composed of distinct, separated particles.

partition coefficient   The equilibrium constant that determines the concentrations of solute in two immiscible phases.

pascal   [Blaise *Pascal*, 1623–1662, French philosopher and mathematician] The SI unit of pressure, equal to 1 newton per square meter. Symbol: $Pa = N\ m^{-2}$ (1 atm = 1.01325×10^5 Pa).

Pauli exclusion principle   [Wolfgang *Pauli*, 1900–1958, Austrian physicist: Nobel prize for physics, 1945] See *exclusion principle*.

p block   Elements in which the p orbitals receive their electrons. The p block is laid out as six columns on the right-hand side of the periodic table.

penetration   [L *penetrare* to put into] The extent to which an electron spends time in the core (inner shells) of an atom, close to the nucleus. See also *shielding*.

peptide bond   [Gk *peptos* digested] The amide linkage created when H_2O is eliminated between amino acids during a condensation reaction.

percent yield   See *yield*.

period   [Gk *periodos* circuit, way around] (1) A horizontal row of the periodic table, presenting a series of atoms in which valence subshells are filled sequentially. Each period opens with an element having a configuration ns^1. Example: The second period begins with lithium ($[He]2s^1$) and ends with neon ($[He]2s^2 2p^6$). (2) The time during which a wave executes one complete cycle of oscillation.

periodicity   The tendency to repeat at regular intervals.

periodic law   For chemistry, an important organizing principle: The properties of the elements are periodic functions of atomic number. They recur at regular intervals as protons and electrons are added to each atom.

periodic table   An arrangement of the chemical elements by atomic number, presented so as to display their recurring properties. The table consists of horizontal rows (*periods*) and vertical columns (*groups*).

peroxide   [L *per* through, thoroughly, utterly] A compound in which oxygen has an oxidation state of -1. The *peroxide ion* has the form O_2^{2-}. Compare with *oxide*; *superoxide*.

perturbation   [L *perturbare* to throw into confusion] A disturbance or disordering.

petroleum   [Gk *petra* rock + *oleum* oil] A complex mixture of hydrocarbons, occurring naturally as crude oil.

pH   [*p* power + *H*$^+$ hydronium ion] A measure of the concentration of hydronium ion in solution, defined as the negative base-10 logarithm:

$$pH = -\log [H_3O^+]$$

Compare with *pOH*.

phase   [Gk *phasis* appearance] (1) A distinct, homogeneous component within a heterogeneous system, physically separated from another component by a phase boundary or *interface*. (2) A state of matter, such as solid, liquid, or gas. (3) A particular point of advancement in a cycle, expressed as a fraction of a completed oscillation.

phase diagram   A graphical summary of the phases present in a system under different combinations of temperature and pressure. Single phases are represented by open areas in the diagram. Phase boundaries, corresponding to conditions where two states of matter coexist in equilibrium, are represented by lines. See also *critical point*; *triple point*.

phase transition   A transformation from one state of matter to another, as from solid to liquid (melting) or from liquid to solid (freezing). See also *disorder–order transition*.

phospholipid bilayer A bilayer formed by hydrophobic lipid molecules and hydrophilic phosphate groups, oriented in water with the phosphate groups pointing outward and the lipids pointing inward. The phospholipid bilayer is a standard architectural feature of cell membranes.

photocurrent [Gk *photos* light + L *currere* to run] A current of electricity produced by the *photoelectric effect*.

photoelectric effect The expulsion of an electron from a metallic surface after the absorption of a photon.

photoelectron An electron ejected from a system by means of a photon.

photoionization An ionization induced by electromagnetic radiation.

photolysis [Gk *photos* light + *lysis* a breaking down, a loosening] The decomposition of a molecule or breaking of a bond as induced by electromagnetic radiation.

photon [Gk *photos* light] A quantized excitation of the electromagnetic field. Characterized by a wavelength and frequency, a photon acts as a neutral, massless particle carrying momentum, energy, and spin.

photosynthesis [Gk *photos* light + *syn* with + *tithenai* (*thesis*) to put, place] A complicated, vital process that occurs in green plants: the formation of carbohydrates and other organic molecules from carbon dioxide, water, and inorganic salts. The reactions are assisted by chlorophyll and other pigments, with sunlight providing the requisite energy (hence "synthesis with light").

physical chemistry [Gk *physike* science of nature] The branch of chemistry concerned generally with the structure, energy, and transformation of matter, placing special emphasis on the principles of quantum mechanics, thermodynamics, statistical mechanics, and kinetics.

physical property An attribute that can be studied without changing the molecular structure or composition of a substance.

pi antibonding orbital A molecular orbital formed by the out-of-phase overlap of two atomic orbitals oriented side to side, typically p or d functions. Symbol: π^*.

pi bond Usually a bond formed by the overlap of p orbitals oriented side to side, with the upright lobes standing perpendicular to the line connecting the nuclei. The electron density is concentrated above and below a nodal plane that includes the nuclei. d orbitals form π bonds as well. Contrasted with *sigma bond*.

pi bonding orbital A molecular orbital formed by the in-phase overlap of two atomic orbitals oriented side to side, typically p or d functions. Symbol: π.

pion A particular kind of meson, thought to participate in the strong nuclear interaction.

pK_a   [*p* power + K_a acid ionization constant] A measure of the magnitude of an acid ionization constant, defined as a negative logarithm:

$$pK_a = -\log K_a$$

Similarly for *pK_b*, the negative logarithm of a base ionization constant K_b.

planar   Flat, level; two-dimensional; lying in a single plane. Example: Benzene (C_6H_6) is a planar molecule.

Planck's constant   [Max *Planck*, 1858–1947, German physicist: Nobel prize for physics, 1918] The fundamental unit of quantum mechanics, a measure of the "smallness" of a quantum. Planck's constant, *h*, is the proportionality constant between energy and frequency:

$$E = h\nu$$

Symbol: $h = 6.63 \times 10^{-34}$ J s.

plane-polarized light   Electromagnetic radiation in which the electric field rises and falls in a single plane. The magnetic field vibrates in a second plane rotated by 90°. Also called *polarized light*; *linearly polarized light*.

pOH   [*p* power + *OH^-* hydroxide ion] A measure of the concentration of hydroxide ion in solution, defined as the negative base-10 logarithm:

$$pOH = -\log [OH^-]$$

Compare with *pH*.

polar   [Gk *polos* pivot, axis, pole] Describing a nonzero dipole moment caused by an asymmetric distribution of charge—as in polar molecules, polar bonds, and so on. Contrasted with *nonpolar*.

polarity, polarization   The appearance of two opposite tendencies, such as the separation of charge to create an electric dipole moment.

polarized light   See *plane-polarized light*.

polyatomic   [Gk *polys* much, many] Of a molecule, ion, or similar structure: having many atoms, as distinct from *monatomic* and *diatomic*.

polydentate ligand   [Gk *polys* much, many + L *dens* tooth] A ligand offering two or more points of attachment.

polymer   [Gk *polys* much, many + *meros* part, portion] A large molecule formed by the joining together (polymerization) of smaller molecules or units called *monomers*. See also *macromolecule*.

polypeptide   [Gk *polys* much, many + *peptos* digested] A condensation polymer consisting of amino acids joined by peptide bonds, usually understood to have a molecular mass of less than 10,000. See also *protein*.

polyprotic acid   An acid that releases more than one proton (H^+) when dissolved in water. The term embraces both *diprotic* and *triprotic* acids, which have two and three ionizable protons, respectively. Compare with *monoprotic acid*.

p orbital   [$p = principal$, used originally to describe a spectroscopic line] A one-electron wave function with angular momentum quantum number $\ell = 1$, inclusive of three magnetic sublevels: $m_\ell = -1$, 0, 1.

porous   [Gk *poros* passage] Having small openings; admitting passage. Zeolites are porous materials.

position   [L *ponere* (*positus*) to put, place] Location; situation in space. See also *coordinates*.

positional disorder   Randomness in location; a multiplicity of spatial arrangements.

positron   An *antielectron*: an elementary particle with the same mass and spin as an electron, but having a positive charge equal in magnitude to the electron's negative charge. Symbol: β^+ or ^0_1e.

potential   [L *potens* power] Possible; latent; capable of actualization. (1) Synonymous with *potential energy*. (2) In general: a measure of an imbalance in some quantity at two positions in space. Example: A system's temperature serves as a *potential* for the transfer of heat. Heat flows from a body at high temperature to a body at low temperature.

potential difference, electrical   The work needed to move a charge between two points in an electric field, measured in volts (joules per coulomb): $1 \text{ V} = 1 \text{ J C}^{-1}$.

potential energy   The energy associated with the position of a body in space or, more precisely, in a field. Contrasted with *kinetic energy*, the energy associated with the motion of a body.

potential energy surface   The variation of potential energy as a function of a system's coordinates. A full potential surface for a molecule would display the potential energy for every possible combination of bond distances and bond angles.

potential well   The range of structures existing at and near the point of lowest potential energy for a system.

precipitation   [L *praecipitare* to cast down headlong] Formation of an insoluble solid during a solution-phase reaction.

precision   [L *praecidere* (*praecisus*) to cut off, to cut short] The extent to which measured values agree among themselves, distinct from *accuracy*. The more precise the measurements, the lower is the random error; the values are clustered tightly about their mean.

pre-equilibrium   [L *pre* before] A rapidly attained equilibrium for an elementary step of a reaction, resulting in a steady concentration of one or more intermediates. Example: A pre-equilibrium may exist in the two-step reaction

$$(1) \quad A + B \rightleftharpoons C$$

$$(2) \qquad C \rightarrow D$$

if the first step comes to equilibrium much faster than the second step:

$$K_1 = \frac{[C]}{[A][B]}$$

pre-exponential factor   The factor A in the Arrhenius expression for a rate constant:

$$k = A \exp\left(-\frac{E_a}{RT}\right)$$

The magnitude of A depends on the frequency and efficiency of reaction-inducing molecular collisions. Also called *frequency factor*.

preorganized   Describing an already-formed, preexisting structure that has the ability to recognize and bind with a designated substrate.

pressure   [L *pressare* to press] Force (F) exerted per unit area (A). Symbol: $P = F/A$.

primary structure   [L *prima* first] The sequence of amino acids in a protein, specified strictly in order of connection without reference to three-dimensional arrangement.

primitive cubic cell   A cubically shaped unit cell that contains a lattice point at each of its eight vertices. Compare with *body-centered cubic cell*; *face-centered cubic cell*.

principal quantum number   [L *principalis* first, chief] In a one-electron atom: the quantum number associated with the radial coordinate. The principal quantum number controls the energy and size of the orbital. Symbol: n.

probability   [L *probabilis* likely, capable of standing a test] The likelihood that an event will occur, expressed as the quotient A/T: the number of actual occurrences (A) divided by the total number of events possible (T).

probability amplitude   The value of a wave function, $\psi(x, y, z)$, at each point in space (x, y, z).

probability density   The square of a wave function, $\psi^2(x, y, z)$, at each point in space (x, y, z); also called *electron density*. The probability density provides the point-by-point probability of locating an electron.

product   [L *producere* to lead forward] A species formed as a result of chemical reaction.

protein   [Gk *proteios* primary] A biopolymer consisting of one or more long polypeptide chains, often folded into an intricate three-dimensional structure.

proton   [Gk *protos* first] An elementary particle, one of the primary constituents of an atomic nucleus. More massive than an electron by a factor of 1836, a proton is positively charged and has a spin of $\frac{1}{2}$. The positive charge ($e = 1.60 \times 10^{-19}$ C) has the same magnitude as the negative charge carried by an electron.

protonation   Attachment of a proton, as in the transfer of H^+ from a Brønsted-Lowry acid to a Brønsted-Lowry base.

proton emission   An avenue of radioactive decay: the ejection of a proton from an atomic nucleus.

pyramid   [Gk *pyramis* pyramid] A three-dimensional figure with triangular sides and a polygonal base. The sloping sides meet at a single point.

quadratic equation   [L *quadrare* (*quadratus*) to make square] An equation having the general form $ax^2 + bx + c = 0$, where a, b, and c are constants. See also *linear dependence*; *order (2)*.

qualitative [L *qualis* of what sort?] Concerned with the attributes or properties of a system in general, independent of their numerical values. Contrasted with *quantitative*.

quantitative   [L *quantus* how much?] Concerned with numerical values and amounts. Contrasted with *qualitative*.

quantum   [L *quantus* how much?] A discrete packet of some quantity, such as energy, linear momentum, or angular momentum. A quantum of energy is related to a radiant frequency (ν) by Planck's equation

$$E = h\nu$$

The symbol h represents Planck's constant. Plural: quanta.

quantum mechanics   Study of the motion and energy of bodies too small to obey Newton's laws of classical mechanics. Governed by Heisenberg's uncertainty principle, Pauli's exclusion principle, and Schrödinger's equation, a quantum mechanical particle exchanges energy in discrete packets called *quanta*.

quantum number   A number (generally integral or half-integral) associated with a particular quantum state. The quantum number helps both to label the state and to determine the value of the corresponding property. Example: In a one-electron atom, the energy of an orbital is proportional to n^2 (the square of the principal quantum number n).

quantum state   An allowed condition of a physical system, character-ized typically by a wave function and associated quantum numbers.

quark   [From James Joyce's *Finnegans Wake*, coined by U.S. physicist Murray Gell-Mann] A fundamental particle having a charge of $-\frac{1}{3}$ or $\frac{2}{3}$, occurring only in groups of two or three. Three quarks combine to make a proton or neutron; one quark and one antiquark combine to make a meson. Quarks are subject to all four fundamental forces: the strong interaction, the weak interaction, electromagnetism, gravity.

quasicrystal   [L *quasi* as if, as though] A solid material characterized by a "forbidden" rotational symmetry such as fivefold. A quasicrystal possesses long-range order but lacks translational symmetry.

radial   [L *radius* spoke, rod] Pertaining only to distance from a central point, as along the radius of a circle or sphere. A radial wave function depends solely on the electron's distance from the nucleus, not on its angular orientation.

radial node   A region in space where the radial portion of a wave function changes sign, manifested as a surface of zero electron density. Contrasted with *angular node*.

radial probability distribution   The likelihood (P) of finding an electron at some distance r from the nucleus, expressed as

$$P = 4\pi r^2 \psi^2(r)$$

for a wave function $\psi(r)$. The radial probability is constant over the entire surface of a sphere with radius r.

radiation   The emission of energy from a particle or wave. Example: electromagnetic radiation.

radical   [L *radix* root] (1) A species with one or more unpaired elec-trons, usually highly reactive and susceptible to chain reaction. (2) Two or more atoms in a recognizable arrangement, used in a sense similar to *functional group*. (3) In mathematics: a root.

radioactivity   The emission of energy or particles from an atomic nucleus. Examples of radioactive emissions include alpha particles, beta particles, and gamma rays. Also called *radioactive decay*.

radiocarbon dating   A method of determining the age of a sample by analysis of the decay pattern of carbon-14.

radiofrequency radiation   The least energetic radiation in the electromagnetic spectrum, corresponding usually to transitions be-tween nuclear spin states. Wavelengths are typically greater than 1 meter, with frequencies ranging up to several hundred megahertz.

radioisotope   An isotope that undergoes radioactive decay. Many ra-dioisotopes are produced artificially for medical or scientific applications.

radiotracer   A radioisotope used specifically to illuminate the path of an element through a system.

random   Pertaining to a set of equally probable outcomes, with no tendency toward one or the other.

random coil   An unfolded protein existing as a completely disordered, fluctuating chain of amino acids.

random error   An unpredictable fluctuation, sometimes above a median value and sometimes below.

Raoult's law   [Francois *Raoult*, 1830–1901, French chemist and physicist] A relationship that describes the vapor pressure above an ideal solution: The vapor pressure (P_i) of component *i* is proportional to its mole fraction (X_i) in the mixture:

$$P_i = X_i P_i^\circ$$

P_i° is the vapor pressure exerted by component *i* as a pure liquid.

rare-earth series   See *lanthanide series*.

rarefied   [L *rarus* loose, wide apart, thin, infrequent] Sparsely distributed; not dense; widely scattered.

rate   The change in a quantity per unit time. Examples: reaction rate = concentration/time; speed = distance/time.

rate constant   The proportionality constant that appears in a rate law, connecting the rate of reaction with the concentrations of the reactants. Symbol: k.

rate-determining step   The slowest, most time-consuming elementary step in a complex reaction. The rate-determining step establishes the minimum time needed for the entire reaction to occur.

rate equation, rate law   A mathematical relationship between the instantaneous rate of a reaction and the concentrations of the species involved. The expression is often (but not always) in the form

$$\text{Rate} = k[A]^n[B]^m \cdots$$

for reactants A, B, ...

reactant   A species that participates in a chemical reaction, specifically a substance present at the outset. Reactants are indicated to the left of the arrow in a chemical equation.

reaction   [L *re* back, again + *agere* (*actus*) to do] A chemical transformation, brought about by the mutual effect of one substance or agent upon another.

reaction coordinate   A generalized variable used to indicate the progress of a reaction, particularly along an axis of a potential energy curve or surface.

reaction mechanism   The sequence of elementary steps by which a reaction proceeds.

reaction quotient   The mass-action ratio of product concentrations to reactant concentrations, evaluated at an arbitrary point during a reaction. If the process has reached equilibrium, then the reaction quotient is equal to the equilibrium constant. Symbol: Q.

reaction rate   The change in concentration of a reacting species per unit time. The *instantaneous* reaction rate measures the change over an infinitesimally small interval of time. The *average* rate measures the change over a finite interval.

reagent   A substance that can be made to undergo a reaction, often for a very specific purpose such as detection or analysis of another species.

rearrangement reaction   An intramolecular change involving the movement of a specified fragment from one position to another.

receptor   [L *recipere* (*receptus*) to receive] A structure that binds to a complementary species by means of noncovalent interactions, often highly specific. Similar to a *host* structure, but usually with strong biological connotations: hormone receptors, antibodies, etc.

redox   See *reduction–oxidation reaction*.

reducing agent   The species that loses electrons (is oxidized) in a reduction–oxidation reaction. A reducing agent is the agent that causes some other species to undergo reduction. Contrasted with *oxidizing agent*.

reductant, reducer   See *reducing agent*.

reduction   (1) Gain of electrons by a species in a chemical reaction; complementary to oxidation. (2) Removal of oxygen or addition of hydrogen.

reduction–oxidation reaction   A reaction in which electrons are transferred between species. Synonymous with *oxidation–reduction* or *redox* reaction.

redux   [L *reducere* to bring back, lead back] Brought back again.

reference frame   [L *referre* to carry back, report] An observational framework for characterizing an event in space and time. A frame of reference combines a coordinate system for measuring position with a set of clocks for measuring time. See also *inertial reference frame*.

relative atomic mass   [L *relatus* carried back] The mass of a particular isotope expressed as a dimensionless ratio against $\frac{1}{12}$ the mass of carbon-12. An atom of carbon-12 is assigned a mass of exactly 12.

relative molecular mass   The mass of a particular molecule expressed as a dimensionless ratio against $\frac{1}{12}$ the mass of carbon-12.

relativity, theory of   [Albert *Einstein*, 1879–1955, German physicist: Nobel prize for physics, 1921] An interpretation of space and time due to Einstein, showing that all motion must be considered relative to an observer's reference frame. The theory of relativity demonstrates that neither position nor time has any absolute meaning, since observers in different states of motion will measure different values for each. See also *special relativity*.

representative elements   The elements in the *s* and *p* blocks of the periodic table, also known as the *main-group elements*.

repulsion   [L *repellere* (*repulsus*) to drive back] To be pushed back by a force emanating from an object or position, so as to move away from that position.

resonance   [L *resonare* to resound, to echo] (1) Reinforcement and enhancement of a system's vibrational motion, caused by application of an external stimulus oscillating at or near the natural frequency of the affected system. (2) The merging of two or more Lewis structures into a single representation of an atom or molecule (a *resonance hybrid*). Whereas each of the individual resonance structures, taken separately, is inadequate to describe the bonding, the blended form usually suggests a more accurate picture involving delocalized electrons.

rest energy   The energy (E) inherent in the mass (m) of a body observed to be at rest in an inertial reference frame, a consequence of Einstein's theory of relativity:

$$E = mc^2$$

The symbol c denotes the speed of light in a vacuum: $c = 3.00 \times 10^8$ m s^{-1}.

rest mass   The mass of a body observed to be at rest in an inertial reference frame. Rest mass is a property intrinsic to the object.

reversible process   [L *revertere* (*reversus*) to turn back] A transformation that proceeds in a series of infinitesimally small steps, enabling the system to remain in equilibrium at each stage and, if so directed, to return to its previous state. Also called *quasi-static process*.

RNA   [*ribonucleic acid*] A biopolymer: any of various nucleic acids carrying genetic information transcribed from DNA; used by cells to manufacture proteins. Considerably smaller than DNA, ribonucleic acids consist of a single strand of nucleotides containing (1) the sugar ribose, not deoxyribose; and (2) the nitrogen base uracil, not thymine. See also *nucleotide*; *DNA*.

root-mean-square speed   The square root of the mean-square speed, computed as

$$v_{rms} = <v^2>^{1/2} = \left[\frac{1}{N}(v_1^2 + v_2^2 + \cdots + v_N^2) \right]^{1/2}$$

for a system of N particles. The symbol v_i denotes the speed of particle i.

rotation, rotational motion   [L *rotare* to move in a circle, to roll] A turning around an axis or central point; revolution.

rotational symmetry   Invariance to rotation. A system possesses rotational symmetry if its appearance and properties remain unchanged after the structure is turned through a specified angle.

Rydberg constant   [Robert *Rydberg*, 1854–1919, Swedish physicist] The constant that appears in the expression for the quantized energies (E_n) of a one-electron atom (nuclear charge $= Z$):

$$E_n = -R_\infty \frac{Z^2}{n^2}$$

Symbol: $R_\infty = 2.18 \times 10^{-18}$ J.

salt   The product, other than water, of a neutralization reaction between an acid and a base. The term is conventionally applied to an ionic compound produced by neutralization of an Arrhenius acid by an Arrhenius base, but the concept extends to Brønsted-Lowry and Lewis acid–base reactions as well.

salt bridge   Any of various constructions that can secure a conducting path in an electrochemical cell. The salt bridge completes the electric circuit and allows current to flow between anode and cathode.

saturated hydrocarbon   [L *saturare* to fill] A hydrocarbon in which there are no double or triple bonds. Each carbon makes four single bonds, leaving the valence fully satisfied, or *saturated*, at each site. Examples: CH_4, methane; CH_3CH_3, ethane; $CH_3CH_2CH_3$, propane. Contrasted with *unsaturated hydrocarbon*.

saturated solution   A solution in which equilibrium is maintained between dissolved and undissolved solute, thereby preventing the dissolution of additional solute. Compare with *supersaturated solution*; *unsaturated solution*.

***s* block**   The alkali metals and alkaline earth metals, plus H and He: elements in which the *s* orbitals receive their electrons. The *s* block is laid out as two columns on the left-hand side of the periodic table.

scalar   [L *scalaris* represented as a point on a scale or ladder] A quantity possessing only magnitude and not direction. Examples: speed, temperature. Contrasted with *vector*.

scanning tunneling microscope (STM)   A device capable of exceedingly high magnification and resolution, greater than that of an electron microscope. The scanning tunneling microscope produces an image of an atomic surface by detecting electrical signals with a moving needle.

scattering   Deflection and dispersal of a wave or particle beam as it collides with particles in a material.

Schrödinger equation   [Erwin *Schrödinger*, 1887–1961, German physicist: Nobel prize for physics, 1933] The quantum mechanical wave equation, applicable to particles traveling at velocities far slower than the speed of light. Given the masses and potential energy of the particles, the Schrödinger equation determines the allowed wave functions, energies, and quantum numbers.

scientific method   [L *scientia* knowledge] A systematic approach to the investigation of nature, rooted in empirical observation, hypothesis, experiment, and theory.

scintillation counter   [L *scintillare* to send out sparks, to flash] An apparatus for detecting and measuring radioactivity. The device registers the flash of light produced when certain atoms are struck by an energetic particle or photon.

screening   Synonymous with *shielding*: the apparent reduction of the nuclear attraction as perceived by a valence electron. The strength of attraction is diminished by the repulsion suffered from inner-shell, core electrons. See also *penetration*.

second   [L *secundus* following, next, second] The base SI unit of time, defined as the interval required to complete 9,192,631,770 cycles of the electromagnetic radiation produced by a certain emission of the cesium atom.

secondary structure   The way in which a protein is wound into a coil, often in the form of an α helix. Secondary structure is enforced by noncovalent interactions between nearby amino acids on the polypeptide chain.

second law of thermodynamics   A physical law, complementary to the first law of thermodynamics (conservation of energy): The universe moves inexorably toward a state of greater disorder, deriving increased global entropy from every spontaneous change. All isolated systems run down and come to equilibrium eventually, having attained a condition of maximum entropy.

second-order reaction   A process for which the rate law is of the second degree, taking the form $k[A]^2$ or $k[A][B]$. The exponents on the concentrations sum to 2.

selective precipitation   Separation of different solutes by forcing them out of solution sequentially, typically by exploitation of differing equilibrium constants.

semiconductor   [L *semi* half, partially] A material with electrical conductivity falling between that of an insulator and a conductor. The conductivity increases with temperature as electrons are excited to a higher state. Examples: Si, silicon; Ge, germanium.

semimetal   A material with properties intermediate between those of a metal and a nonmetal; also called a *metalloid*. Example: Sb, the element antimony. Although antimony does exhibit the characteristic luster of a metal, it is nevertheless a poor conductor of heat and electricity.

semipermeable membrane   [L *semi* half, partially + *permeare* to pass through] A membrane that permits only selective passage of certain ions and molecules.

shape selectivity   A form of molecular recognition dependent on the shape of a species, as in the separation of the para isomer from a mixture of dichlorobenzene molecules.

shell   (1) For atoms: the complete set of electron orbitals associated with a given principal quantum number. Example: The $n = 3$ shell contains one $3s$ orbital, three $3p$ orbitals, and five $3d$ orbitals. (2) In models of nuclear binding: a *nuclear shell*, a one-particle wave function used to represent a nucleon moving in a smoothed-out field of interactions. The one-nucleon states are constructed by analogy to electron orbitals.

shielding   Synonymous with *screening*: the apparent reduction of the nuclear attraction as perceived by a valence electron. The strength of attraction is diminished by the repulsion suffered from inner-shell, core electrons. See also *penetration*.

short-range order   Synonymous with *local order*: regularity or organization that persists only over a limited region of space, as with the clusters found in a liquid. The molecules within each cluster are arranged according to some pattern, but there is no correlation between the location of one cluster and another.

SI   [*Système International d'Unités*] The International System of Units, widely used for scientific applications; informally, the *metric system*. The fundamental SI units for length, mass, time, electric current, temperature, amount of substance, and luminous intensity are, respectively: meter (m), kilogram (kg), second (s), ampere (A), kelvin (K), mole (mol), and candela (cd). All other units are derived from these seven base quantities.

sigma (σ, Σ)   [eighteenth letter of the Greek alphabet] (1) In mathematical expressions, uppercase sigma (Σ) means "take the sum." Example: $\Sigma\, n_i = n_1 + n_2 + \cdots$ (2) Lowercase sigma (σ) is used frequently to denote a sigma orbital.

sigma antibonding orbital   A molecular orbital formed by the out-of-phase overlap of two atomic orbitals oriented head to head, appearing cylindrically symmetric around the internuclear vector. Symbol: σ^*.

sigma bond   A bond formed by the overlap of two atomic orbitals oriented directly along the internuclear vector. The electron density is cylindrically symmetric around the bond axis. Contrasted with *pi bond*.

sigma bonding orbital   A molecular orbital formed by the in-phase overlap of two atomic orbitals oriented head to head, appearing cylindrically symmetric around the internuclear vector. Symbol: σ.

significant digits, significant figures   [L *significare* to make a sign, to indicate] The meaningful digits reported for a measured quantity, inclusive of the first uncertain digit. All subsequent digits are considered not significant.

single bond   A linkage arising from the sharing of one electron pair by two atoms.

S_N1 mechanism   [*Substitution, Nucleophilic, 1* (unimolecular)] A two-step pathway for nucleophilic substitution: First, unimolecular loss of the group X^- from R—X creates a carbocation, R^+. Second, attack by the group Y^- produces the substituted compound R—Y.

S_N2 mechanism   [*Substitution, Nucleophilic, 2* (bimolecular)] A one-step pathway for nucleophilic substitution: The attacking group, Y^-, displaces the leaving group, X^-, from R—X to produce the substituted molecule R—Y. A five-membered transition state (pentavalent carbon) is formed during the concerted reaction.

solid   [L *solidus* dense, solid] A dense, tightly bound state of matter usually possessing both short-range and long-range order. Difficult to compress, a solid has a definite shape as well as definite volume.

solubility   [L *solvere* (*solutus*) to loosen, to release] The concentration of solute in a saturated solution—thus the maximum amount of solute that can be dissolved in a specified volume of solvent at a specified temperature.

solubility-product constant   The equilibrium constant for a dissolution–precipitation reaction, usually expressed as a product of ion concentrations. Symbol: K_{sp}. Example: $K_{sp} = [Ag^+][Cl^-]$ for the reaction $AgCl(s) \rightleftharpoons Ag^+(aq) + Cl^-(aq)$.

solute   A substance that is uniformly dispersed throughout a solvent, thus forming a solution. The solute, considered to be the dissolved species, is usually less abundant than the solvent.

solute–solute interactions   The force or potential energy between solute particles in a solution.

solute–solvent complex   A weakly bound structure produced by the coordination of solvent particles around a solute particle; the product of *solvation*. Synonymous with *solvent–solute complex*. See also *complex (2)*; *hydration*.

solute–solvent interactions   The force or potential energy between solute and solvent particles in a solution.

solution   A homogeneous phase produced by dispersal of a solute throughout a solvent, intermingled thoroughly at the molecular level.

solvation   The coordination of solvent particles around a solute, enforced by noncovalent interactions. See also *hydration*.

solvation sphere   A layer of solvent particles surrounding a solute.

solvent   The host medium of a solution, into which the solute particles are dissolved.

solvent–solute complex   See *solute–solvent complex*.

solvent–solvent interactions   The force or potential energy between solvent particles in a solution.

***s* orbital**   [*s* = *sharp*, used originally to describe a spectroscopic line] A spherically symmetric one-electron wave function with angular momentum quantum number $\ell = 0$, supporting the single magnetic sublevel $m_\ell = 0$.

space-time   A four-dimensional coordinate system, defined with reference to a given inertial frame: Three spatial coordinates (x, y, z) and one temporal coordinate (t) are used to specify an event in space-time.

special relativity   Einstein's theory of relativity, restricted to an inertial reference frame (see *relativity, theory of*). Special relativity is distinguished from *general relativity*, Einstein's theory of gravity as viewed in an accelerated reference frame.

specific heat, specific heat capacity   Heat capacity per gram of substance, contrasted with *molar heat capacity*. The specific heat is the amount needed to increase the temperature of a material by 1 K (equivalently, 1°C). Symbol: c_s.

spectator ion   [L *spectare* to look at] An ion that does not participate directly in a solution-based reaction. Unchanged throughout the process, spectator ions are omitted from the net ionic equation.

spectrochemical series   A ranking of ligands according to the strength of the crystal field they impose on a metal atom. Strong-field ligands produce the largest *d*-orbital splittings in a complex; weak-field ligands, the smallest.

spectroscopy   [L *spectrum* appearance + Gk *skopein* to look at] Measurement of the absorption and emission of electromagnetic radiation by matter, indicative of a system's energy levels. Various spectroscopic techniques apply to different portions of the electromagnetic spectrum: nuclear magnetic resonance spectroscopy, microwave spectroscopy, infrared spectroscopy, etc. General terms for spectroscopic instruments include *spectrograph*, *spectrometer*, and *spectroscope*.

spectrum   [L *spectrum* appearance] The intensity of electromagnetic radiation absorbed or emitted by a material, recorded as a function of wavelength, frequency, or energy. Plural: spectra.

speed   A scalar quantity equal to the magnitude of velocity: change in distance per unit time, independent of direction.

spherical symmetry   Invariance to angular orientation; equivalence in all directions. Example: The electric field produced by a point charge is spherically symmetric, a function of radial distance alone.

sp **hybrid orbitals**   A set of one-electron wave functions constructed from one *s* function and one *p* function on the same atom. The two *sp* orbitals, spaced 180 degrees apart, point in opposite directions.

sp^2 **hybrid orbitals**   A set of one-electron wave functions constructed from one *s* function and two *p* functions on the same atom. The three *sp*^{2} orbitals, spaced 120 degrees apart, point toward the vertices of an equilateral triangle.

sp^3 **hybrid orbitals**   A set of one-electron wave functions constructed from one *s* function and three *p* functions on the same atom. The four *sp*^{3} orbitals, spaced 109.5 degrees apart, point toward the vertices of a regular tetrahedron.

sp^3*d*^{2} **hybrid orbitals**   A set of one-electron wave functions constructed from one *s* function, three *p* functions, and two *d* functions on the same atom. The six *sp*^{3}*d*^{2} orbitals, spaced 90 degrees apart, point toward the vertices of a regular octahedron.

spin, spin angular momentum   An angular momentum intrinsic to electrons, protons, neutrons, photons, and certain other particles (including nuclei). Although unrelated to any overt rotation of the body, the spin angular momentum generates a corresponding magnetic moment for a charged particle.

spin quantum numbers   A particle endowed with spin is described by two quantum numbers, s and m_s, analogous to an electron's orbital angular momentum quantum numbers ℓ and m_ℓ. For an electron, $s = \frac{1}{2}$. For other particles, s may take the values $0, \frac{1}{2}, 1, \frac{3}{2}, 2, \frac{3}{2}, \ldots$. (1) The spin angular momentum quantum number, s, is defined such that the total spin angular momentum has magnitude $\sqrt{s(s+1)}$. (2) The spin magnetic quantum number, m_s, is defined such that the spin angular momentum along *one* specified axis has the value m_s, restricted to integers or half-integers from the set $s, s-1, s-2, \ldots, -s$. All magnitudes are expressed as multiples of $h/2\pi$.

spin–spin coupling   An interaction between nuclei endowed with nuclear spin, manifested in a nuclear magnetic resonance (NMR) spectrum as a splitting of the affected signals into patterns of two or more lines. Often called *J coupling*. See also *chemical shift*; *nuclear magnetic resonance spectroscopy (NMR)*.

spontaneous process   [L *sponte* of one's free will, voluntarily] A transformation that adds to the entropy of the universe, producing a broader dispersal of global energy. A spontaneous reaction, impelled forward by a decrease in a system's free energy, can proceed on its own absent

any external driving force. Spontaneity, a thermodynamic consideration, is unrelated to the *rate* of change.

square planar geometry An arrangement of five atoms in a plane,
configured with each of four atoms at one vertex of a square and the
fifth atom at the center. Removal of two diametrically opposed ligands
from an octahedral complex will produce a square planar complex.

square pyramid   A pyramid with a square base.

square pyramidal geometry   An arrangement of five atoms to
form a square pyramid: One atom appears at the apex and each of the
other four occupies one vertex of the base. In a square pyramidal
molecule or complex, a central (sixth) atom is usually positioned at
the midpoint of the pyramid.

stable state   A state resting in a *global* minimum of energy, thus lying
lower than any other level. A stable state will return to its original
configuration following a small disturbance. Compare with *metastable
state*; *unstable state*.

staggered conformation   For groups attached on each side of a single bond, an arrangement in which the atoms in each group are offset
across the axis of rotation. Contrasted with *eclipsed conformation*.

standard acceleration of gravity   The rate (g) at which a mass is
accelerated toward the center of the earth, expressed in the relationship between force (F, equivalent to *weight*) and mass (m): $F = mg$.
Symbol: $g = 9.8$ m s^{-2}.

standard cell potential, standard emf   The voltage that exists
across an electrochemical cell at 25°C when all solution concentrations are 1 M and all gas pressures are 1 atm. Symbol: $\mathscr{E}°$.

standard conditions   Circumstances under which all participants in
a reaction exist in their standard states. See also *standard state*.

standard enthalpy change   A difference in enthalpy measured under
standard conditions: standard enthalpy of formation, standard enthalpy
of reaction, standard enthalpy of activation, etc. Symbol: $\Delta H°$.

standard entropy change   A difference in entropy measured under
standard conditions: standard entropy of formation, standard entropy
of reaction, etc. The standard entropy of formation, $S°$, is reported
relative to a perfectly ordered crystal at 0 K. For all other changes,
the symbol is $\Delta S°$.

standard free energy change   A difference in free energy measured under standard conditions: standard free energy of formation,
standard free energy of reaction, etc. Symbol: $\Delta G°$.

standard hydrogen electrode   A half-reaction used to establish a
point of zero voltage in an electrochemical cell. The standard reduction potential of the reaction $2H^+(aq, 1\ M) + 2e^- \longrightarrow H_2(g, 1\ atm)$ is
arbitrarily assigned a value of 0 V.

standard oxidation potential   The voltage measured under standard conditions for an oxidation half-reaction coupled to a standard hydrogen electrode. Symbol: $\mathscr{E}^{\circ}_{ox}$.

standard reduction potential   The voltage measured under standard conditions for a reduction half-reaction coupled to a standard hydrogen electrode. Symbol: $\mathscr{E}^{\circ}_{red}$.

standard state   The most stable state of a material as it normally exists under a specific set of standard reference conditions: pressure = 1 atm, concentration = 1 M (for solutions), and a stated temperature (usually 25°C, but not necessarily). Compare with *standard temperature and pressure*.

standard temperature and pressure (STP)   Reference conditions for measurement of gas properties: temperature = 0°C (273 K), pressure = 1 atm. Compare with *standard state*.

standing wave   A mode of vibration that develops in a confined space, giving rise to a pattern of stationary nodes. The disturbance oscillates up and down in the same place, unable to move forward as would a *traveling wave*.

state   [L *status* condition, standing] The condition of a system. A thermodynamic state is defined by the values of all macroscopic properties subject to independent variation. A quantum mechanical state is specified by the wave function and associated quantum numbers. See also *state of matter*.

state function, state property   A property that depends only on the current state of a system, thus determined entirely by the values of its state variables and independent of history and preparation. Examples: internal energy, enthalpy, entropy. Counterexamples: work, heat.

state of matter   An aggregation of atoms, molecules, or ions shaped by interparticle forces and often macroscopic in extent; a phase. Solid, liquid, and gas are the most common forms, but there are also intermediate states such as clusters and liquid crystals.

state variables   The external constraints on a system that suffice to fix its macroscopic properties. Examples: pressure, volume, temperature.

stationary phase   The solvent that remains fixed relative to the column in a chromatographic apparatus. Contrasted with *mobile phase*.

stationary state   A state with properties that do not change with time.

statistical mechanics, statistical thermodynamics   The explanation of macroscopic mechanical and thermodynamic properties by microscopic models. All such methods exploit the probabilistic constraints governing large numbers of particles.

steady state   A condition under which a system's macroscopic properties do not vary with time, although not as in a true dynamic equilibrium arising from an internal balance of forces. Unlike a self-sustaining equilibrium state, a steady state must be maintained actively by some outside source or sink. Material is usually added or withdrawn to maintain a constant level.

stereogenic   [Gk *stereos* solid, three-dimensional + *genes* born, produced] Of an atom: able to present two nonequivalent spatial configurations of its attached groups.

steric factor   [Gk *stereos* solid, three-dimensional] In the collision theory of bimolecular reactions: the fraction of collisions in which the molecules are aligned suitably for reaction.

stoichiometry   [Gk *stoicheion* element, one of a row + *metron* measure] The quantitative relationships governing the amounts of material in chemical combinations. Fixed, predictable stoichiometric ratios are evidence of the particulate structure of matter.

STP   See *standard temperature and pressure*.

strong acid, strong base   An acid or base that ionizes completely in aqueous solution. Examples: HCl, hydrochloric acid; NaOH, sodium hydroxide.

strong electrolyte   Any substance that ionizes completely in solution. Examples: strong acids and bases; soluble salts such as NaCl.

strong-field ligand   A ligand that imposes a crystal-field splitting of relatively large extent, appearing toward the right in the spectrochemical series. Examples: CO, CN^-.

strong nuclear force   The short-range, very powerful interaction that binds together the nucleons in a nucleus. Independent of mass or charge, the strong force is a manifestation of the *color* interaction between quarks.

structural formula   [L *struere (structus)* to heap up, build] A chemical formula that shows the bonding pattern and sometimes the geometry of a molecule.

structural isomers   Molecules built from the same set of atoms, arranged in different bonding patterns to produce distinguishable structures. Unlike conformational isomers, structural isomers differ in the order and sequence of their bonds. See also *geometric isomers*; *optical isomers*.

subcritical   [L *sub* below] (1) Below the critical point of, for example, a vapor–liquid transition. (2) Below the critical mass in a nuclear fission reaction.

sublimation   [L *sublimare* to elevate] Direct transformation from a solid phase to a gas; the reverse of deposition.

subnuclear particle   A component particle of a nucleus. Examples: proton, neutron, meson.

subshell   The complete set of electron orbitals associated with a given angular momentum quantum number. Example: The $\ell = 2$ subshell contains five *d* orbitals with magnetic quantum numbers $m_\ell = -2, -1,$ 0, 1, 2.

substance   [L *substantia* essence, substance (that which stands under)] A pure sample of matter with definite atomic composition and characteristic properties.

substitution reaction   [L *substituere* to put in place of] A reaction in which one group in a molecule is replaced by another group.

substrate   [L *sub* below + *stratum* a cover] A substance that binds to an enzyme; in general, a supporting structure of any kind.

superconductor   [L *super* above, exceeding] A material that offers zero resistance to the flow of electricity.

supercritical   (1) Above the critical point of, for example, a vapor–liquid transition. (2) Above the critical mass in a nuclear fission reaction.

supercritical fluid   The phase of matter that exists above a system's critical point.

supernova   [L *super* above, exceeding + *novus* new] The catastrophic explosion of a star, accompanied by the release of heavy elements accumulated during prior nucleosynthesis.

superoxide   A compound in which oxygen has an oxidation state of $-\frac{1}{2}$. The *superoxide ion* has the form O_2^-. Compare with *oxide*; *peroxide*.

supersaturated solution   A solution in which the concentration of solute exceeds the equilibrium value. Compare with *saturated solution*; *unsaturated solution*.

surface tension   A force, attributable to intermolecular attractions, that constrains a liquid to minimize its surface area.

surroundings   The portion of the universe that can interact with or affect a system.

symmetry   [Gk *syn* with, together + *metron* measure ⟶ *symmetria* commensurateness, having the same measure] A property that remains the same despite some transformation of an object or its environment; invariance. Example: A square, which is unchanged whether rotated by 90°, 180°, 270°, or 360°, has fourfold rotational symmetry.

symmetry operation   A transformation—often a rotation, translation, or reflection—that leaves a system apparently unchanged. Symmetry operations apply both to physical objects (such as crystals) and to abstract concepts (such as mathematical equations). See also *symmetry*.

system   [Gk *systema* an organized whole] Any part of the universe chosen for investigation, treated as distinct from its surroundings. Together, system and surroundings include all portions of the universe relevant to a particular experiment.

systematic error   A persistent, nonrandom discrepancy that does not average to zero over repeated trials.

tangible   [L *tangere* to touch] Able to be felt or touched; substantial; real. Opposite to *intangible*.

taxonomy   [Gk *taxis* arranged, put in order + *nomia* law, management, distribution] A systematic scheme for classification and definition.

temperature   [L *temperare* to divide judiciously, to be moderate, restrained, proportionate] (1) Interpreted macroscopically: a measure of the tendency for thermal energy to migrate from one body to another—thus a potential that governs the transfer of heat. Heat flows from a body at higher temperature to a body at lower temperature until thermal equilibrium is established at an intermediate temperature. (2) Interpreted microscopically: a parameter that, through the Boltzmann distribution, determines the allocation of particles over the energy levels available to a system.

termolecular reaction   [L *ter* (*tres*) three] An elementary reaction involving three species.

tertiary structure   [L *tertius* third] The three-dimensional shape or folding pattern of a protein, enforced by noncovalent interactions between peptides on distant parts of the polypeptide chain.

tetrahedral geometry   An arrangement of four atoms to form a tetrahedron, configured with one atom at each vertex. In a tetrahedral molecule or complex, a central (fifth) atom is usually positioned at the midpoint of the tetrahedron.

tetrahedron   [Gk *tettares* four + *hedra* seat, face of a geometric form] A three-dimensional figure with four faces and four vertices; a triangular (trigonal) pyramid.

theoretical yield   See *yield*.

theory   [Gk *theoria* a viewing, contemplation] A model developed to explain a body of prior observations and to predict new phenomena.

thermal energy   [Gk *therme* heat] Energy associated with a temperature and distributed among the internal modes of an atom, molecule, or other particle. The energy of a system in thermal equilibrium is partitioned among various electronic, vibrational, rotational, and translational motions according to the Boltzmann distribution.

thermal equilibrium   (1) Interpreted macroscopically: a persistent condition in which temperature remains uniform throughout a system, with no net flow of heat between any two points. (2) Interpreted microscopically: a persistent condition in which the distribution of internal energy conforms to a Boltzmann distribution at a single temperature.

thermochemistry   A subdiscipline of thermodynamics: the measurement and analysis of heat (enthalpy) absorbed or released by chemical processes.

thermodynamic control   Describing a reaction in which the dominant products are those that are most stable. Contrasted with *kinetic control*.

thermodynamics   [Gk *therme* heat + *dynamis* force, power] Study of the interconversion, transfer, and dispersal of energy among its various forms, particularly the relationship between mechanical work (macroscopic force and motion) and nonmechanical heat (microscopic motion).

thermometer   [Gk *therme* heat + *metron* measure] An instrument for measuring temperature.

third law of thermodynamics   A physical law, supplementary to the second law of thermodynamics (maximization of entropy): The entropy of a perfect crystal at 0 K is exactly zero.

time   A measure of the sequence of events, giving quantitative meaning to the notion of before and after.

titration   [L *titulus* title $\longrightarrow$ qualification, fineness of composition] A controlled, monitored reaction between a solution of known concentration (the *titrant*) and a solution of known volume but unknown concentration (the *analyte*). The analyte's concentration is determined from the stoichiometry of the reaction and the volume of titrant needed to carry it out.

torr   [Evangelista *Torricelli*, 1608–1647, Italian physicist and mathematician] A non-SI unit of pressure, equivalent to 1/760 atm. One torr corresponds to the pressure needed to raise a column of mercury by 1 millimeter at 0°C.

trans   [L across, beyond] Describing a geometric isomer in which two identical groups appear on opposite sides—as around a C=C double bond, or relative to the metal atom in a complex. Contrasted with *cis*.

transition   [L *transire* (*transitus*) to go across] A change from one state to another.

transition elements, transition metals   Atoms with partially occupied *d* orbitals, arranged in the center of the periodic table between main groups II and III.

transition state   A way station on the road from reactants to products: the transient structure corresponding to maximum potential energy. After assembling the reactants into this unstable structure, the system will either go on to form products or fall back to regenerate reactants. Often called *activated complex*.

transition-state theory   A kinetic model of chemical reactions, emphasizing the role played by the transition state.

translation, translational motion   [L *transferre* (*translatus*) to carry across] Displacement from one point in space to another; the overall motion of a body in a specified direction, distinct from rotational motion.

translational symmetry   Invariance to translation. A system possesses translational symmetry if its appearance and properties remain unchanged after the structure is displaced through a specified distance in a specified direction.

transmutation   [L *transmutare* to shift, to go across] Transformation of one nuclide into another, usually with a change in atomic number—hence a change from one *element* to another.

transuranic elements, transuranium elements   [L *trans* across, beyond] Atoms with atomic numbers greater than 92, appearing after uranium in the periodic table. All the transuranic elements are radioactive and can only be produced by nuclear reactions.

triatomic   [L *tres, tria;* Gk *treis, tria* three] Of a molecule, ion, or similar structure: having three atoms, as distinct from *monatomic* and *diatomic*.

triclinic   [*tri* + Gk *klinein* to lean, to slope] A unit cell with three unequal axes that intersect at oblique angles (thus *sloping,* not perpendicular).

trigonal bipyramid   [*tri* + Gk *gonia* angle + L *bis* twice] A three-dimensional figure with six sides and five vertices, constructed by joining two trigonal pyramids base to base.

trigonal bipyramidal geometry   An arrangement of five atoms to form a trigonal bipyramid: Each of two atoms occupies one of the apexes, and each of the other three atoms occupies one vertex of the triangular central plane. In a trigonal bipyramidal molecule or complex, a central (sixth) atom is usually positioned at the midpoint of the common triangular plane.

trigonal planar geometry   An arrangement of four atoms in a plane, configured with each of three atoms at one vertex of a triangle and the fourth atom at the center.

trigonal pyramid   A pyramid with a triangular base; synonymous with *tetrahedron*.

trimer   [Gk *treis, tria* three + *meros* part, portion] A structure composed of three identical substructures (monomers).

triple bond   A linkage arising from the sharing of three electron pairs by two atoms, often existing as one σ bond and two π bonds.

triple point   The temperature and pressure at which three phases (solid, liquid, gas) exist together in dynamic equilibrium, appearing on a phase diagram as the intersection of three phase boundaries.

triprotic acid   An acid with three ionizable protons. Compare with *monoprotic acid; diprotic acid; polyprotic acid.*

tritium   A radioactive isotope of hydrogen containing one proton and two neutrons in its nucleus.

trough   The lowest point on a wave.

ultraviolet radiation [L *ultra* on the far side of, beyond] Electromagnetic radiation with a wavelength longer than X rays and shorter than visible violet light. The energy of an ultraviolet photon is comparable to the energy of the valence electrons in many molecules.

uncertainty   Doubt concerning the value of a quantity, arising from random error.

uncertainty principle   A fundamental principle of quantum mechanics, enunciated by Heisenberg: The position and momentum of a particle cannot be measured simultaneously to unlimited accuracy. The uncertainties in position (Δx) and momentum (Δp) are related inversely as

$$\Delta x \, \Delta p \geq \frac{h}{4\pi}$$

where h is Planck's constant. The more precisely one quantity is known, the less certain is the other. A similar relationship holds for uncertainty in energy (ΔE) and time (Δt). Also called *indeterminacy principle*.

unimolecular reaction   [L *unus* one] An elementary reaction undergone by a single species.

unionized   [pronounced "un-ionized"] Not dissociated into anions and cations; electrically neutral and intact.

unit   [L *unus* one] A specific amount of some quantity; a dimension. Examples: 1 kilogram, 1 meter, 1 second.

unit cell   The smallest repeating block in a crystal lattice. Every unit cell in a lattice is related to every other unit cell by translational symmetry.

unit-factor method   See *factor-label method*.

universal   Applicable to or characteristic of the whole; pertinent to everything.

universal gas constant   The constant, R, that appears in the equation of state for an ideal gas ($PV = nRT$). Symbol: $R = 8.3145$ J mol^{-1} K^{-1} $= 0.082058$ atm L mol^{-1} K^{-1}.

universe   [L *universus* all, entire, turned into one] (1) In general, everything: the world, the cosmos, and all phenomena known or unknown. (2) In thermodynamics, the totality of system and surroundings.

unsaturated hydrocarbon   [*un* not + L *saturare* to fill] A hydrocarbon containing double or triple bonds, thus able to bond with additional atoms at the site of each multiple bond. Examples: $CH_2{=}CH_2$, ethylene; $CH{\equiv}CH$, acetylene. Contrasted with *saturated hydrocarbon*.

unsaturated solution   A solution in which the concentration of dissolved solute is insufficient to establish equilibrium with an undissolved solid phase. Compare with *saturated solution*; *supersaturated solution*.

unstable state   A system resting precariously at a local maximum of energy, thus lying above a level of greater stability. An unstable state is likely to undergo a downward transition upon suffering even a minimal push. Compare with *metastable state*; *stable state*.

valence   [L *valentia* strength] (1) The combining capacity of an atom, indicating the number of bonds that can be formed. (2) Pertaining to the outer, partially occupied shell of an atom or molecule.

valence bond theory   A model of chemical bonding that postulates the existence of localized pair bonds. According to the theory, a valence bond is treated as an in-phase combination of atomic orbitals localized around each nucleus. The overlap creates a two-atom bonding orbital, occupied by two electrons with opposite spins.

valence electron   An electron that occupies an orbital in the unfilled valence of an atom or molecule; an outer-shell electron.

valence shell   See *valence (2)*.

valence-shell electron-pair repulsion (VSEPR)   An electrostatic description of chemical bonding, developed to predict the shape of a molecule that contains a central atom. VSEPR asserts that the repulsion between electron pairs is minimized in the most stable configuration.

van der Waals equation   [J. D. *van der Waals*, 1837–1923, Dutch physicist] A proposed equation of state for a real gas, containing parameters to account for excluded particle volume and intermolecular interactions.

van der Waals forces, van der Waals interactions   Intermolecular interactions, particularly those between neutral, closed-shell structures. Examples: dipole–dipole forces, London dispersion forces.

van't Hoff factor   [Jacobus Hendricus *van't Hoff*, 1852–1911, Dutch chemist: Nobel prize for chemistry, 1901] A correction factor that expresses the extent to which the colligative properties of a real solution deviate from ideality. The van't Hoff factor, often called the *i factor*, provides a measure of ion pairing in an electrolytic solution.

vapor   [L *vapor* steam] A gas that exists at a subcritical temperature, able to engage in an equilibrium with the corresponding liquid phase.

vaporization   A transition from a liquid to a vapor, accompanied by a decrease in order and an increase in enthalpy.

vapor pressure   The constant pressure exerted by a vapor in dynamic equilibrium with a condensed phase.

vapor-pressure lowering   The decrease in a solvent's vapor pressure brought about by the presence of a solute; a colligative property.

vector   [L *vehere* (*vectus*) to carry] A quantity possessing both magnitude and direction. Examples: velocity, momentum, acceleration, force, dipole moment. Contrasted with *scalar*.

velocity   [L *velocitas* speed] A vector quantity that indicates the speed and direction of a moving object—sometimes used, incorrectly, as a synonym for speed alone. Symbol: v.

vibration, vibrational motion   [L *vibrare* to move to and fro] The periodic displacement and return of a moving object; a regular movement back and forth; an oscillation.

visible radiation   [L *videre* (*visus*) to see] Electromagnetic radiation with a wavelength longer than ultraviolet and shorter than infrared radiation, visible to humans and many other animals. The energy of a visible photon is comparable to the energy of the valence electrons in typical molecules.

volatile   [L *volare* (*volatus*) to fly] Tending to vaporize easily and rapidly.

volt   [Count Alessandro *Volta*, 1745–1827, Italian physicist] The SI unit of electrical potential difference, equivalent to the work needed to move 1 coulomb between two points in an electric field. Symbol: $V = J\ C^{-1}$.

voltage   Electrical potential difference expressed in volts.

voltaic cell   An electrochemical cell in which the free energy of a spontaneous reaction produces an electric current able to do work. Also called *galvanic cell*.

volume   [L *volumen* roll (of sheets)] The amount of space an object occupies in three dimensions; capacity. Symbol: V.

volumetric   Pertaining to the measurement of volume.

wave   A periodic disturbance in space and time, either fixed in place (a standing wave) or moving (a traveling wave).

wave equation   A mathematical relationship that describes the propagation of a wave—in quantum mechanics, the *Schrödinger equation*.

wave function   A solution to a wave equation. In quantum mechanics, the wave function ψ serves as a probability amplitude. The probability of finding a particle at some particular point at some particular time derives from the square of the amplitude, expressed as ψ^2.

wavelength   The distance between equivalent points on consecutive cycles of a wave—thus the spatial separation between successive crests or troughs.

wavenumber   A spectroscopic quantity proportional to energy, equivalent to inverse wavelength. Wavenumbers are often measured in units of *reciprocal centimeters* (cm^{-1}). Symbol: $\bar{\nu}$.

wave–particle duality   A principle of quantum mechanics: The characteristics of particles and waves are intertwined and fundamentally unified. Matter can suffer interference like a wave, and radiation can deliver energy and momentum like a particle.

weak acid, weak base   An acid or base that ionizes incompletely in aqueous solution, typically with an ionization constant less than 1. Examples: CH_3COOH, acetic acid (a weak acid); NH_3, ammonia (a weak base).

weak electrolyte   Any substance that ionizes incompletely in solution. Examples: weak acids and bases.

weak-field ligand   A ligand that imposes a crystal field splitting of relatively small extent, appearing toward the left in the spectrochemical series. Examples: I^-, Br^-, Cl^-, F^-.

weak nuclear force   The interaction that governs beta decay in a nucleus. Common manifestations include normal beta emission (of electrons), positron emission, and electron capture.

weight   The force (W) exerted by gravity on an object, proportional to its mass (m). Near the earth's surface, the relationship between weight and mass is

$$W = mg$$

where g represents the standard acceleration due to gravity.

work   The application of force over a distance, matched by an equivalent change in energy.

work function   The threshold energy needed to expel an electron from a substance by means of the photoelectric effect.

X ray   Electromagnetic radiation with a wavelength shorter than ultraviolet radiation and longer than gamma rays. The energy of an X-ray photon is comparable to the energy of an inner-shell electron.

X-ray diffraction   Scattering and interference of a beam of X rays through a crystal, used to ascertain the spacings between atoms. See also *diffraction*.

yield   The *theoretical yield* of a reaction is the maximum amount of product that can be obtained, realizable only if there are no side reactions and if all product is recovered without loss. The *percent yield* is the ratio of actual yield to theoretical yield, expressed as a percentage.

zeolite   [Gk *zein* to boil] Any one of various crystalline minerals built largely from aluminum, silicon, and oxygen atoms. Zeolitic structures often contain networks of channels, cavities, and pores that allow the materials to function as specialized catalysts or molecular sieves. Some zeolites occur naturally; others are synthesized in the laboratory.

zero-point energy   The quantized energy of vibration that persists in the ground vibrational state, even at absolute zero. Zero-point energy is equal to $\frac{1}{2}h\nu_0$, where ν_0 is the fundamental frequency.

zwitterion   [Ger *Zwitter* hybrid, hermaphrodite] A dipolar ion: an ion that carries both a positive and negative charge, separated in space. The structure as a whole is electrically neutral. Example: an amino acid with an ionized COO^- and a protonated NH_3^+ group.

ANSWERS TO SELECTED EXERCISES

Note: Answers are subject to variation in the last significant digit. Round-off of both physical constants and intermediate values in a calculation may produce small discrepancies in the final result.

Chapter 1 **1.** (a) 8.64×10^4 s (d) 555 cm
2. (a) 0.155 in^2 (d) 1000 cm^3 **3.** (a) 0.017
(c) 144.77 **4.** (a) g cm s^{-2} (b) 10^5 dynes
6. Choices (a), (b), (c), and (e) are dimensionally incorrect. **8.** Choice (c) is dimensionally and physically correct. **11.** Only choice (c) describes unaccelerated motion. **15.** $v = 44.3$ m s^{-1}, $p = 44.3$ kg m s^{-1} **16.** (a) Potential energy decreases by 98.1 J for every 10 m that the mass falls. (b) Kinetic energy increases by 98.1 J for every 10 m that the mass falls. **19.** (a) 5.00 kg m s^{-1} (b) 2.50 kg m s^{-1} **21.** (a) 2.50×10^2 J (b) 31.3 J
22. particle A for 0.01 s: (a) 1×10^3 N (b) 1×10^2 N **26.** weightlifter A: 1.2×10^3 J; weightlifter B: 1.5×10^3 J **28.** -1.99×10^{20} N
30. 5.67×10^2 N **31.** 1.88×10^7 m s^{-1}

Chapter 2 **5.** (a) ^{9}Be $\equiv$ 4p, 5n; ^{10}Be $\equiv$ 4p, 6n
(c) ^{14}N $\equiv$ 7p, 7n; ^{15}N $\equiv$ 7p, 8n **9.** (a) Li $\equiv$ 3p, 3e$^-$; Li$^+$ $\equiv$ 3p, 2e$^-$; He $\equiv$ 2p, 2e$^-$ (c) Cs $\equiv$ 55p, 55e$^-$; Cs$^+$ $\equiv$ 55p, 54e$^-$; Xe $\equiv$ 54p, 54e$^-$
11. (a) O$^-$ $\equiv$ 7 valence e$^-$, Cl$^-$ $\equiv$ 8 valence e$^-$, Ar $\equiv$ 8 valence e$^-$ **13.** (a) metal (b) nonmetal (c) nonmetal (d) metal (e) metal **15.** (a) Cs and Cl will form an ionic compound. (c) N and O will form various covalent molecules. **16.** (a) RbF (c) Al$_2$O$_3$ **17.** (a) CaCl$_2$ $\longrightarrow$ Ca^{2+} + 2 Cl$^-$
(c) CsCl $\longrightarrow$ Cs$^+$ + Cl$^-$ **18.** (a) Cl$_2$O is a mo-

lecular compound. (c) Na$_2$CrO$_4$ is an ionic compound formed from Na$^+$ and CrO$_4^{2-}$ ions. The covalently bonded chromate species, CrO$_4^{2-}$, is a molecular ion. **20.** NaCl $\equiv$ 60.663% Cl (50% by mole); KCl $\equiv$ 47.555% Cl (50% by mole); RbCl $\equiv$ 29.319% Cl (50% by mole) **22.** SnBr$_2$
24. 11.5 g **26.** (a) NH$_3$ $\equiv$ 10e$^-$, 17.0305 g mol^{-1}
(c) HI $\equiv$ 54e$^-$, 127.9124 g mol^{-1} **27.** CH$_4$ $\equiv$ 74.869% C, 25.131% H; C$_2$H$_6$ $\equiv$ 79.888% C, 20.112% H **28.** 1.00 g CH$_4$ $\equiv$ 6.23×10^{-2} mol, 3.75×10^{22} molecules; 1.00 g C$_2$H$_6$ $\equiv$ 3.33×10^{-2} mol, 2.00×10^{22} molecules **29.** 1.732 mol CH$_4$ $\equiv$ 27.79 g; 1.732 mol C$_2$H$_6$ $\equiv$ 52.08 g
30. CH$_2$O **33.** (a) tetrahedral (b) trigonal planar
35.

$$\overset{\displaystyle \ddot{\underset{\displaystyle |}{Cl}}:}{H-C-H} \quad \text{(approximately tetrahedral)}$$
$$\underset{\displaystyle H}{}$$

36.

$$:\ddot{Cl}-\underset{|}{P}-\ddot{Cl}: \quad \begin{array}{l}\text{(trigonal pyramid, with} \\ \text{electron pairs tetrahedral} \\ \text{around P)}\end{array}$$

41. (a) N$_2$ + 3 H$_2$ $\longrightarrow$ 2 NH$_3$ (c) CH$_4$ + 2 O$_2$ $\longrightarrow$ CO$_2$ + 2 H$_2$O **43.** (a) $n = 2$, $m = 6$ **44.** 1.70 g NCl$_3$ and 1.54 g HCl are produced; 0.760 g NH$_3$ remains. Cl$_2$, the limiting reactant, is completely consumed. **46.** CH$_4$ + H$_2$O $\longrightarrow$ CO + 3 H$_2$ (a) 92.70 g H$_2$O react. (b) 144.1 g CO and 31.12 g H$_2$ are produced.

Chapter 3   1. (a) +1 (c) +5   **3.** (a) +1 (c) +4 (e) +3   **5.** PO_4^{3-} **7.** (a) oxidation numbers: $N \equiv +5$, $O \equiv -2$; formal charges: $N \equiv +1$, O (double bond) $\equiv 0$, O (single bond) $\equiv -1$ **9.** (a) $HNO_3 + NaOH \longrightarrow H_2O + NaNO_3$ (c) $H_3O^+ + OH^- \longrightarrow 2H_2O$ **11.** (a) acid $\equiv Fe^{3+}$; base $\equiv CN^-$ (c) acid $\equiv H_3O^+$; base $\equiv OH^-$. Fe^{3+} acts only as a Lewis acid.   **13.** (a) oxidizing agent $\equiv MnO_2$, reducing agent $\equiv Zn$ (b) oxidizing agent $\equiv N_2$, reducing agent $\equiv H_2$ **15.** (a) reduction: $2e^- + F_2(g) \longrightarrow 2F^-(aq)$ (c) oxidation: $H_2(g) + 2OH^-(aq) \longrightarrow 2H_2O(\ell) + 2e^-$ **17.** (a) oxidation: $Zn \longrightarrow Zn^{2+} + 2e^-$; reduction: $2e^- + 2H^+ \longrightarrow H_2$; moles of electrons transferred: $n = 2$ (c) oxidation: $2Mn \longrightarrow 2Mn^{2+} + 4e^-$; reduction: $4e^- + H_2SO_3 + 4H^+ \longrightarrow S + 3H_2O$; moles of electrons transferred: $n = 4$ **18.** (a) OH, H (c) Br, CH_3 **21.** (a) Mg^{2+}, SO_4^{2-} (c) Ca^{2+}, OH^- (e) NH_4^+, OH^- **23.** (a) acid–base   **27.** (a) 25.56 g $ZnCl_2$ (b) 1.129×10^{23} Zn^{2+} ions (c) 2.258×10^{23} Cl^- ions   **29.** (a) 8.70 g $Zn(OH)_2$ (b) The spectator ions Na^+ and Cl^-, together with excess Zn^{2+}, stay dissolved. (c) OH^- is the limiting reactant.

Chapter 4   1. period of 0.1 s $\equiv$ frequency of 10 s^{-1} ($1 \text{ s}^{-1} = 1$ Hz)   **3.** (a) 2.0 s^{-1} (b) 0.50 s   **5.** frequency $= 50 \text{ s}^{-1}$, period $= 0.02$ s   **10.** 5.0×10^{-7} m   **11.** (a) 2.9×10^{-3} m   **13.** (a) 2.998×10^8 m   **15.** (a) $3.94 \times 10^{18} \text{ s}^{-1}$ (X ray) (c) $2.61 \times 10^6 \text{ s}^{-1}$ (radiofrequency) (e) $1.38 \times 10^{15} \text{ s}^{-1}$ (ultraviolet)   **17.** (a) 1.99×10^{-25} J/photon $\equiv 1.20 \times 10^{-4}$ kJ mol^{-1} (radiofrequency) (c) 1.99×10^{-19} J/photon $\equiv 1.20 \times 10^2$ kJ mol^{-1} (infrared) (e) 1.99×10^{-13} J/photon $\equiv 1.20 \times 10^8$ kJ mol^{-1} (gamma radiation)   **19.** (a) 2.2×10^5 J (b) 5.4×10^{23} photons   **21.** 5.79×10^{-7} m $\equiv$ 579 nm $\equiv$ 5790 Å   **23.** (a) $E = 1.1 \times 10^{-20}$ J (rounded down from 1.149×10^{-20} J); $p = 1.4 \times 10^{-25}$ kg m s^{-1}; $v = 1.6 \times 10^5$ m s^{-1} (b) 1.10×10^{19} photons (rounded down from 1.10499×10^{19})   **25.** (b) 2.2×10^{-9} m   **27.** $p = 6.63 \times 10^{-24}$ kg m s^{-1}, $v = 3.97 \times 10^3$ m s^{-1} **28.** $v = 5.93 \times 10^7$ m s^{-1}, $\lambda = 1.23 \times 10^{-11}$ m   **30.** (a) 9.5×10^{-37} m   **32.** (a) $\Delta p > 7 \times 10^{-31}$ kg m s^{-1}, $\Delta v > 7 \times 10^{-1}$ m s^{-1} (d) $\Delta p > 7 \times 10^{-28}$ kg m s^{-1}, $\Delta v > 7 \times 10^2$ m s^{-1}

Chapter 5   1. There is one quantum number, the principal quantum number: $n = 1, 2, 3, \ldots \infty$. Number of nodes: $n - 1$ **3.** (a) $n = 1, 2, 3, \ldots \infty$ (b) frequencies: v ($n = 1$), $2v$ ($n = 2$), $3v$ ($n = 3$), $4v$ ($n = 4$)   **5.** (a) 25% (c) 14%   **10.** (a) $n = 3$, $\ell = 1$ (b) $n = 2$, $\ell = 0$ **11.** (a) $n = 3$, $\ell = 2$ (e) $n = 4$, $\ell = 3$ **12.** (a) $3d$ (d) $1s$ **13.** (a) $2s$ (c) $3p$ **15.** (a) 0 (c) 0, ± 1 **16.** (a) 0, ± 1, ± 2 (e) 0, ± 1, ± 2, ± 3 **18.** (a) 1 ($E_{3p} = E_{3s}$) (b) 3 (c) 6   **20.** (a) forbidden (would have $\ell > n$) (b) allowed (c) allowed (d) forbidden (would have $\ell = n$) (e) allowed (f) allowed   **21.** (a) $3d$ (c) $4f$ **22.** (a) $2s$ (c) $3p$ **23.** (a) forbidden (would have $m_s = 0$) (b) forbidden (would have $m_\ell > \ell$) (c) forbidden (would have $\ell = n$) (d) forbidden (would have $|m_\ell| > \ell$)   **25.** $4s$: 3 radial nodes, 0 angular nodes; $4p$: 2 radial nodes, 1 angular node; $4d$: 1 radial node, 2 angular nodes; $4f$: 0 radial nodes, 3 angular nodes   **27.** (a) $\frac{3}{4}R_\infty = 1.63 \times 10^{-18}$ J (rounded down from 1.6349×10^{-18} J) (b) no (c) no   **29.** absorption or emission between $n = 2$ and $n = 4$ **30.** (a) absorption: $E = 1.63 \times 10^{-18}$ J, $\lambda = 1.22 \times 10^{-7}$ m (c) absorption: $E = 2.42 \times 10^{-19}$ J, $\lambda = 8.20 \times 10^{-7}$ m   **31.** (a) absorption: $E = 6.54 \times 10^{-18}$ J, $\lambda = 3.04 \times 10^{-8}$ m (c) absorption: $E = 9.69 \times 10^{-19}$ J, $\lambda = 2.05 \times 10^{-7}$ m   **33.** (a) -6.98×10^{-19} J (c) -6.98×10^{-19} J   **35.** $E = -1.06 \times 10^{-19}$ J, -1.55×10^{-19} J, -1.82×10^{-19} J, -1.98×10^{-19} J; $\lambda = 1.87 \times 10^{-6}$ m, 1.28×10^{-6} m, 1.09×10^{-6} m, 1.00×10^{-6} m (infrared)   **37.** $E = -4.24 \times 10^{-19}$ J (4 times greater than for H); $\lambda = 4.69 \times 10^{-7}$ m (4 times shorter than for H)   **38.** $n_f = 5$ **41.** H atom: $\lambda = 9.11 \times 10^{-8}$ m. He^+ ion: $\lambda = 2.28 \times 10^{-8}$ m. Li^{2+} ion: $\lambda = 1.01 \times 10^{-8}$ m.   **43.** $Z = 5$ **45.** The atom contains more than one electron.

Chapter 6　**5.** $2 \leq Z_{\text{eff}}(1s) \leq 3$; $1 \leq Z_{\text{eff}}(2s) \leq 3$
7. Be, with a larger value of Z_{eff}, has a higher ionization energy than Li.　**9.** (a) The configuration is forbidden under all circumstances. No more than two electrons can occupy an atomic orbital. (b) The configuration is forbidden under all circumstances. Two electrons in the same orbital must have opposite spins.　**11.** Choice (b), in which the three p electrons are distributed over the three sublevels with parallel spins, is the most likely ground-state configuration (Hund's rule). The other two configurations are less likely, although not absolutely forbidden by the Pauli principle.　**13.** ↑__ ↑__ ↑__ ↑__ __ __ __ (paramagnetic, containing four unpaired electrons) **15.** ↑__ ↑__ ↑__ ↑__ __ (paramagnetic, four unpaired electrons)　**17.** The $4s^1 3d^5$ configuration, with a half-filled d subshell, has the lower energy. **19.** Cr ≡ [Ar]$4s^1 3d^5$; Cu ≡ [Ar]$4s^1 3d^{10}$　**21.** See Table C-8 and Figure 6-10.　**23.** Sg ≡ [Rn]$7s^2 5f^{14} 6d^4$ (d block)　**25.** (a) All the ions are isoelectronic with neon: Ne ≡ [He]$2s^2 2p^6$ (b) All the ions are isoelectronic with argon: Ar ≡ [Ne]$3s^2 3p^6$ **27.** (a) P (b) Kr (c) In　**29.** (a) Al$^+$ (b) Ca^{2+} (c) Br$^-$　**31.** (a) $3p$ would be occupied before $2p$. (b) Lowest occupied orbital would not be $1s$. (c) Three electrons would occupy $2s$, a violation of the Pauli exclusion principle.　**32.** $R(\text{H}) = 0.3707$ Å, $R(\text{F}) = 0.7060$ Å, $R(\text{HF}) \approx 1.08$ Å **34.** (a) $R(\text{Li}) < R(\text{Rb}) < R(\text{Cs})$ (c) $R(\text{F}) < R(\text{Cl}) < R(\text{Ba})$　**35.** (a) $R(\text{O}) < R(\text{O}^-) < R(\text{O}^{2-})$ (b) $R(\text{Be}^{2+}) < R(\text{Li}^+) < R(\text{Li})$　**36.** (a) $EA(\text{Li}^-) < EA(\text{Li}) < EA(\text{F})$ (c) $EA(\text{Ca}) < EA(\text{S}) < EA(\text{Cl})$ **37.** (a) $I_2(\text{Li})$

Chapter 7　**7.** (a) cylindrical symmetry and a zero dipole moment (b) No. All homonuclear diatomic molecules are nonpolar as a result of "left-right" symmetry.　**10.** (a) cylindrical symmetry and a nonzero dipole moment (b) The difference in electronegativity determines the size of the dipole moment.　**13.** (a) H$_2^+$ is a one-electron molecule, free from electron–electron repulsion. (b) The environment is similar in some ways to a hydrogen atom, but with cylindrical rather than spherical symmetry.　**15.** (a) bonding electrons: 1; antibonding electrons: 0 (b) bond order: $\frac{1}{2}$ (c) paramagnetic (one unpaired electron)　**17.** (a) B$_2$ ≡ $(\sigma_{2s})^2 (\sigma_{2s}^*)^2 (\pi_{2p})^2$ (b) paramagnetic (two unpaired electrons) (c) bond order: 1　**19.** B$_2^+$ ≡ $(\sigma_{2s})^2 (\sigma_{2s}^*)^2 (\pi_{2p})^1$; bond order: $\frac{1}{2}$; paramagnetic (one unpaired electron). B$_2$ ≡ $(\sigma_{2s})^2 (\sigma_{2s}^*)^2 (\pi_{2p})^2$; bond order: 1; paramagnetic (two unpaired electrons). B$_2^-$ ≡ $(\sigma_{2s})^2 (\sigma_{2s}^*)^2 (\pi_{2p})^3$; bond order: $\frac{3}{2}$; paramagnetic (one unpaired electron). B$_2^-$ has the highest dissociation energy and the shortest bond length. B$_2^+$ has the lowest dissociation energy and the longest bond length.　**22.** (a) Be$_2$: diamagnetic (b) Li$_2$: diamagnetic (c) N$_2$: diamagnetic. N$_2$ has the highest dissociation energy and the shortest bond length. Be$_2$ has the lowest dissociation energy and the longest bond length.　**23.** (a) Polarity increases in the order NO < PN < BeN. (b) Be$^{\delta+}$N$^{\delta-}$, P$^{\delta+}$N$^{\delta-}$, N$^{\delta+}$O$^{\delta-}$　**24.** (a) $\delta(\text{HF}) = 0.415$, $\delta(\text{HCl}) = 0.181$, $\delta(\text{HBr}) = 0.122$, $\delta(\text{HI}) = 0.0580$ (b) The halogen is partially negative and the hydrogen is partially positive in each molecule.　**27.** (a) VSEPR predicts a linear structure for F—Be—F. (b) Be$^{\delta+}$F$^{\delta-}$ (c) The linear molecule is nonpolar.　**29.** (a) VSEPR predicts a tetrahedral arrangement of the electron pairs and a trigonal pyramidal arrangement of the atoms:

$$\text{H}-\overset{\overset{\displaystyle ..}{|}}{\underset{\underset{\displaystyle \text{H}}{|}}{\text{N}}}-\text{H}$$

(b) Each bond is polarized as N$^{\delta-}$—H$^{\delta+}$. (c) The molecule has a net dipole moment along the axis of the pyramid.　**31.** Characteristics of bonding in NH$_3$ include: (1) sp^3 hybridization around N; (2) σ bonds between N(sp^3) and H($1s$); (3) a lone pair in one of the four sp^3 orbitals.　**33.** (a) VSEPR predicts a trigonal arrangement of electron pairs and oxygens around sulfur. Three equivalent resonance structures contain one double bond and two single bonds each:

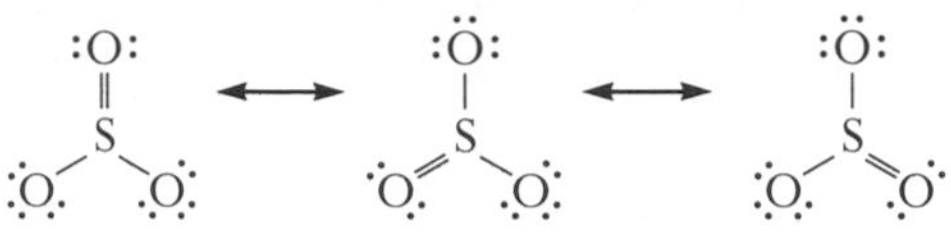

(b) Each bond is polarized as $S^{\delta+}-O^{\delta-}$. (c) The net dipole moment is zero.   **35.** (a) Four pairs of valence electrons are disposed tetrahedrally around sulfur, similar to the arrangement in NH_3. Since only three of the four sites are occupied by atoms, the SO_3^{2-} ion is trigonal pyramidal: S at the apex, one O at each vertex of the base. (b) Sulfur, hybridized as sp^3, forms single bonds with each of the three oxygens. The fourth sp^3 orbital houses a lone pair.

37. (a)

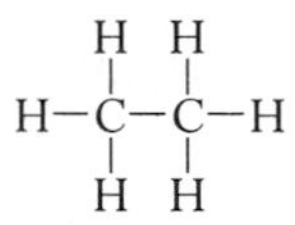

(tetrahedral around each carbon, sp^3 hybridization, single bonds)

(b) Each carbon–hydrogen bond is polarized slightly as $C^{\delta-}-H^{\delta+}$. (c) The net dipole moment for the molecule is zero.

39.

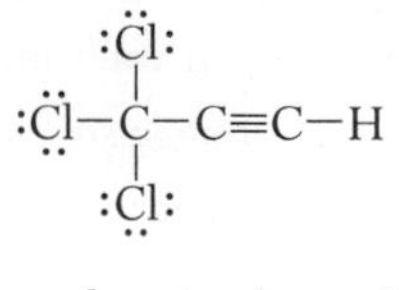

The carbon in the $-CCl_3$ group is hybridized as sp^3, consistent with four σ bonds arranged tetrahedrally. The carbons in the C≡C triple bond (one σ bond plus two π bonds) are hybridized as sp, consistent with a 180° angle.   **41.** Nitrogen lacks the d orbitals it would need to support sp^3d hybridization.   **43.** (a) SbF_5, isoelectronic with PF_5, forms a trigonal bipyramid. (b) One $5s$ function, three $5p$ functions, and one $5d$ function combine to form five sp^3d hybrid atomic orbitals on antimony.

Chapter 8   1. (a) $CH_3-C≡C-CH_3$ (C_4H_6, unsaturated) (b) $CH_3-CH_2-CH_2-CH_3$ (C_4H_{10},

saturated) (c) $CH_3-CH=CH-CH_3$ (C_4H_8, unsaturated)   **3.** $C_{13}H_{22}$:

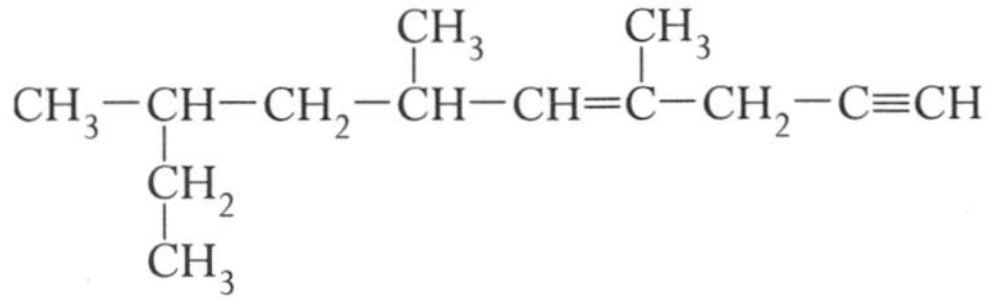

C—C bonds: sp^3 hybridization, 109.5° angles. C=C bonds: sp^2 hybridization, 120° angles. C≡C bonds: sp hybridization, 180° angles.   **5.** (a) There is one π bond, not delocalized over three sites or more:

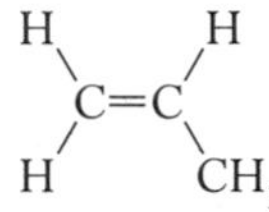

(b) There is one π bond, not delocalized over three sites or more:

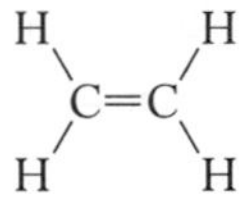

(c) There are no π bonds: $CH_3-CH_2-CH_2-CH_3$
(d) There is one π bond, not delocalized over three sites or more:

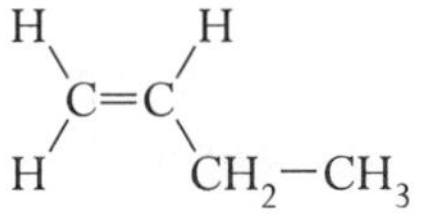

(e) There are three π bonds, delocalized over six carbons:

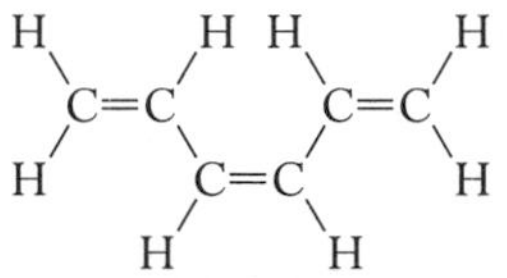

(one of two geometric isomers)

7. Hydrogens are omitted for simplicity:

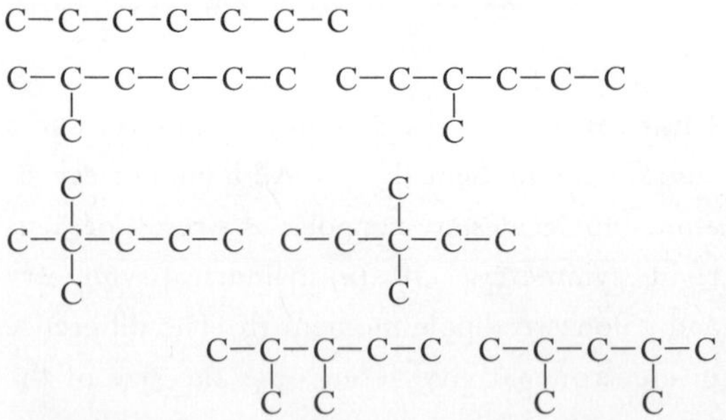

10. carbon: 84.12%; hydrogen: 15.88%

11. $CH_3-CH_2-CH_2-CH_2-CH_2-CH_2-CH_2-CH_2-CH_2-CH_3$
($C_{10}H_{22}$)   **13.** Choices (d) and (e) each contain one stereogenic carbon.

15. The cis form has a net dipole moment:

cis   trans

17.

cis   trans

19. Choices (b), (c), and (e) may exist as conformers.   **20.** functional groups: alcohol, alkene, carboxylic acid, ether, halide   **21.** functional groups: aldehyde, alkene, ester, halide, ketone   **23.** Molecule II has an anhydride in place of a carboxylic acid.

24. An anhydride is the product of a condensation reaction between two carboxylic acids:

Molecule I   Molecule X

Molecule II   Molecule Y

Other reactions are possible as well—for example, the condensation of *acyl chloride*,

with Molecule I to produce Molecule II plus the by-product Y, HCl.

27. Each process is a condensation reaction. (a) Molecule X: $HO-CH_2CH_2CCl_3$. Molecule Y: H_2O.

(b) Molecule X: Molecule Y: H_2O.
(c) Molecule X: Molecule Y: H_2O.

Molecule X could also be, for example, acyl chloride, which would then react to produce Molecule II and HCl (Molecule Y).

29. (a) CH_3CH_3 (1 mol H_2 added)
(c) $CH_3CH_2CH_2CH_2CH_2CH_3$ (1 mol H_2 added)
31. (a) elimination of H_2 (also, oxidation of alcohol to aldehyde) (c) nucleophilic substitution

Chapter 9   9. X is a gas. Y is a liquid.   **11.** X has the higher boiling point.   **12.** (a) Low temperature and high pressure favor formation of a solid. (b) High temperature and low pressure favor formation of a gas.   **14.** 2.8023 g cm^{-3} **16.** Br   **17.** (a) 2 (b) 0.97002 g cm^{-3} **19.** (a) 10.27 mL mol^{-1}, 5.862×10^{19} atoms mm^{-3} (c) 21.3 mL mol^{-1}, 2.82×10^{19} atoms mm^{-3} **21.** Br$_2$ and I$_2$ exist as condensed phases; Cl$_2$ is a gas. (a) 3.214×10^{-3} g mL^{-1} (b) 3.12 g mL^{-1} (c) 4.93 g mL^{-1} **23.** 1.11 *M*   **25.** *X*(solute) = 0.0177, *X*(solvent) = 0.9823   **27.** (a) 0.364 *M* (b) $X(Li^+) = 0.0129$, $X(SO_4^{2-}) = 0.00644$, $X(H_2O) = 0.981$ **29.** 0.333 *M*   **31.** (a) ionic bonds (b) metallic bonds (c) London dispersion interactions (d) hydrogen bonds   **34.** nylon 66   **35.** He < Ne < CO$_2$ < H$_2$O < Ag   **36.** (a) hydrogen bonds (b) London dispersion interactions (c) London dispersion interactions (d) dipole–dipole interactions (e) hydrogen bonds   **37.** CH$_4$ (dispersion) < CF$_4$ (dispersion) < CH$_3$OH (hydrogen bonds)
38. (a) $CH_3CH_2CH_2CH_2CH_2CH_2CH_2CH_2CH_2CH_3$
(b) $CH_3CH_2CH_2OH$ (c) $CH_3CH_2CH_2CH_2CH_3$
39. (a) C_2H_4 (b) C_2H_6 (c) NH_3

Chapter 10   1. (a) $P_A > P_B$ (b) $P_A = 9.81$ Pa, $P_B = 0.981$ Pa   **4.** 7.96 N   **5.** 14.7 lb in^{-2} **6.** 44.7 lb in^{-2} **8.** 13.6 g cm^{-3} **9.** 33.9 ft   **10.** (a) 620. torr (c) 94.97 kPa (e) 0.075589 atm   **11.** (a) 0.1 L

(c) 10 L   **13.** $P_2 = 2.516 P_1$ **15.** (a) 273.2 K
(c) 42.60°C   **17.** 0.817 L   **19.** 2.07 L   **21.** (a) 6.02 ×
10^{23} molecules CO_2 (c) 6.02 × 10^{23} molecules
O_2 **23.** (a) 6.02 × 10^{23} molecules CO (c) 3.01 ×
10^{23} molecules CO   **24.** 0.0821 atm L mol^{-1} K^{-1}
25. (a) 62.4 torr L mol^{-1} K^{-1} **27.** $P_2 = 1.47$ atm
29. 2.97 atm   **31.** 433 K (rounded down from
433.48 K)   **33.** 0.116 mol L^{-1} **35.** O_2
37. (a) $E_k(150\ K) = 1.87$ kJ mol^{-1}, $E_k(300\ K) =$
3.74 kJ mol^{-1}, $E_k(450\ K) = 5.61$ kJ mol^{-1}
(c) $E_k(150\ K) = 1.87$ kJ mol^{-1}, $E_k(300\ K) =$
3.74 kJ mol^{-1}, $E_k(450\ K) = 5.61$ kJ mol^{-1}
39. (a) $\epsilon_k(150\ K) = 3.11 \times 10^{-21}$ J, $\epsilon_k(300\ K) =$
6.21 × 10^{-21} J, $\epsilon_k(450\ K) = 9.32 \times 10^{-21}$ J
(c) $\epsilon_k(150\ K) = 3.11 \times 10^{-21}$ J, $\epsilon_k(300\ K) =$
6.21 × 10^{-21} J, $\epsilon_k(450\ K) = 9.32 \times 10^{-21}$ J
41. (a) $E_k(150\ K) = 0.935$ kJ, $E_k(300\ K) = 1.87$ kJ,
$E_k(450\ K) = 2.81$ kJ (c) $E_k(150\ K) = 3.74$ kJ,
$E_k(300\ K) = 7.48$ kJ, $E_k(450\ K) = 11.2$ kJ
43. (a) $v_{rms}(150\ K) = 306$ m s^{-1}, $v_{rms}(300\ K) =$
433 m s^{-1}, $v_{rms}(450\ K) = 530.$ m s^{-1} (b) $v_{rms}(Ar)/$
$v_{rms}(H_2) = 0.225$ **45.** (a) same (c) $v_{rms}(Ar) >$
$v_{rms}(Xe)$ **46.** (a) H_2 effuses approximately 40%
faster. (c) N_2 effuses at a slightly faster rate
(≈3.5%).   **47.** (a) $RT \approx 436$ kJ mol^{-1} at $T =$
52,400 K; $\frac{3}{2}RT \approx 436$ kJ mol^{-1} at $T = 35,000$ K
(hot!) (b) no   **49.** (a) 500 m s^{-1} (c) 1500 m s^{-1}
(e) 2000 m s^{-1} **50.** (a) 44,100 per million
(b) 1,077,000 per million (c) 44,400 per million
(d) 55 per million

Chapter 11   7. (a) $P_{vdW}(Kr) = 23.3$ atm, $P_{ideal} =$
24.6 atm. $P_{vdW}(Ar) = 24.1$ atm, $P_{ideal} = 24.6$
atm. $P_{vdW}(H_2) = 25.0$ atm, $P_{ideal} = 24.6$ atm.
11. (a) CH_3F (b) CH_4 (c) $C_{18}H_{38}$ (d) Xe (e) H_2O
13. (a) supercritical fluid (c) supercritical fluid
19. (a) 12.8 torr (b) 9.94 mL $H_2O(\ell)$ **21.** (a) no
(b) 9.97 torr   **23.** (a) 779 L   **26.** (a) 12.7 torr
(c) 12.3 torr   **27.** (a) 1.77% (c) 1.77%   **29.** (a) 62.2
g mol^{-1} (b) $C_2H_6O_2$ **31.** (a) $T_f = -1.76$°C, $T_b =$

100.48°C (c) $T_f = -5.08$°C, $T_b = 101.4$°C
32. (a) 0.010 m $Pb(NO_3)_2$ (c) 0.040 m $C_2H_6O_2$
33. (a) −6.2°C (b) 20.7 g $C_2H_6O_2$ **35.** (a) 2.69 mol
(b) larger   **37.** (a) 1.5 m KCl (b) 1.0 m $C_{12}H_{22}O_{11}$

Chapter 12   4. $K(T)$, by itself, determines only
the mass-action ratio at equilibrium. To compute
the final concentrations, we need the values of the
initial concentrations.   **5.** no   **7.** (a) no (b) no
(c) no (d) yes (e) no   **9.** (a) The equilibrium
moves to the right to yield more products; $K > 1$.
(c) The equilibrium moves to the right to yield
more products; $K = 1$.   **11.** no   **13.** (a) heterogeneous (b) homogeneous (c) heterogeneous
(d) homogeneous (e) heterogeneous (f) hetero-
geneous   **15.** (a) $K_c/K_p = 5.13 \times 10^3$ (c) $K_c/K_p =$
71.6   **21.** $K_2 = 1.91$, $K_3 = 3.65$ **23.** K =
$$\frac{1}{[H_3O^+][OH^-]} = 1.0 \times 10^{14}$$ **25.** 1.8×10^{-5}
26. 1.41×10^{-3} **27.** $CH_2ClCOOH$ **29.** (a) $[PCl_5]_{eq}$
$= 0.80$ M, $[PCl_3]_{eq} = [Cl_2]_{eq} = 1.20$ M (b) $K_c =$
1.8 (c) $K_p = 77$ **30.** (a) $[H_3O^+] = [Cl^-] \approx 10.$ M,
$[HCl] \approx 0$ (b) $[H_3O^+] = [Cl^-] \approx 1.00$ M, $[HCl] \approx$
0 (c) $[H_3O^+] = [Cl^-] \approx 0.100$ M, $[HCl] \approx 0$
31. (a) ≈1.00 mol H_3O^+, ≈1.00 mol Cl^- (b) ≈1.00
mol H_3O^+, ≈1.00 mol Cl^- (c) ≈1.00 mol H_3O^+,
≈1.00 mol Cl^- **32.** (a) $[HN_3] = (10. - 0.014)$ M
≈ 10. M, $[H_3O^+] = [N_3^-] = 0.014$ M (b) $[HN_3] =$
$(1.00 - 0.0044)$ M ≈ 1.00 M, $[H_3O^+] = [N_3^-] =$
0.0044 M (c) $[HN_3] = (0.100 - 0.0014)$ M
≈ 0.099 M, $[H_3O^+] = [N_3^-] = 0.0014$ M
33. (a) 0.0014 mol H_3O^+, 0.0014 mol N_3^-
(b) 0.0044 mol H_3O^+, 0.0044 mol N_3^- (c) 0.014
mol H_3O^+, 0.014 mol N_3^- **36.** $[Hg_2^{2+}] = 7.2 \times$
10^{-7} M, $[Cl^-] = 1.4 \times 10^{-6}$ M   **37.** (a) 1.3 ×
10^{-5} mol Ag^+, 1.3 × 10^{-5} mol Cl^- (b) 1.3 × 10^{-8}
mol Ag^+, 1.3 × 10^{-8} mol Cl^- (c) 1.3 × 10^{-2} mol
Ag^+, 1.3 × 10^{-2} mol Cl^- **39.** (a) $Q = K_p/2 =$
1.6 × 10^{-4}. The reaction moves to the right.
(b) $P_{CO} = 1.3 \times 10^{-3}$ atm. $P_{CO_2} = P_{H_2} = 1.9994$ atm
≈ 2.0 atm   **40.** (a) $K_c = K_p = 322$ (b) $K_c = K_p$,
independent of volume   **41.** (a) $K_c = K_p = 322$

(b) The equilibrium constant is independent of the composition of the mixture.   **42.** (a) $Q < K$; the mixture is not in equilibrium. (b) The reaction proceeds to the right.   **44.** (a) Changes in volume have no effect on the reaction. (c) The reaction proceeds to the left; K decreases. (e) The reaction proceeds to the right; K is unchanged. (g) The reaction proceeds to the left; K is unchanged.   **45.** (a) Processes (1) and (3) are driven out of equilibrium. Process (2) is unperturbed. (b) Process (1) moves to the right. Process (3) moves to the left.   **46.** (a) steady state (b) equilibrium (c) steady state (d) steady state

Chapter 13   1. (a) intensive (b) intensive (c) extensive (d) extensive (e) intensive (f) intensive (g) intensive   **3.** (a) 2.69×10^{19} atoms (b) 2.69×10^{16} atoms (c) 2.69×10^7 atoms (d) ≈ 27 atoms   **5.** Functions of state: (a), (b), (f). Functions of path: (c), (d), (e).   **7.** $T = 1000$ K, $P = 100$ atm, $E = 12.5$ kJ   **9.** (a) Heat flows from the surroundings to the system (the gas). (b) $T = 273$ K, $P = 1.58$ atm, $V = 14.2$ L, $n = 1$ mol (c) 1.25 kJ (d) faster (e) $v_{rms}(173\text{ K}) = 1.04 \times 10^3$ m s^{-1}, $v_{rms}(273\text{ K}) = 1.30 \times 10^3$ m s^{-1} (f) $v_{rms}(173\text{ K}) = 4.62 \times 10^2$ m s^{-1}, $v_{rms}(273\text{ K}) = 5.81 \times 10^2$ m s^{-1} **10.** (a) Total thermal energy is greater in the Atlantic Ocean at 33°F. Average thermal energy per molecule is greater in the swimming pool at 80°F. (b) Total thermal energy is greater in 1 mL H_2O at 100°C. Average thermal energy per molecule is greater in 1 mL H_2O at 100°C. (c) Total thermal energy is greater in 1 L H_2O at 10°C. Average thermal energy per molecule is the same for both systems. (d) Total thermal energy is greater in the universe at 5 K. Average thermal energy per molecule is greater in 1 L H_2O at 1000 K.   **11.** (a) Heat flows from the swimming pool at 80°F to the ocean at 33°F. (b) Heat flows from H_2O at 100°C to H_2O at 0°C. (c) There is no net flow of heat. (d) Heat flows from H_2O at 1000 K to the universe at 5 K.   **13.** (a) molar

volume at 100°C (liquid) = 0.018797 L mol^{-1}; molar volume at 100°C (vapor) = 30.620 L mol^{-1} (b) $\Delta E = 37.6$ kJ mol^{-1} (c) $\Delta E = -37.6$ kJ mol^{-1} (d) ΔH **15.** $V(g) - V(\ell) \gg V(\ell) - V(s)$ **17.** (a) 9.870×10^{-3} L (c) 7.501×10^{-7} L   **19.** (a) The system does work on the surroundings. (b) $w = -100$ J (c) The work of expansion, by itself, lowers the internal energy of the system. (d) Heat flows from the surroundings into the system. (e) The flow of heat, by itself, raises the internal energy of the system by 100 J. (f) System: $\Delta E = 0$, $\Delta H = 100$ J. Surroundings: $\Delta E = 0$, $\Delta H = -100$ J.   **21.** (a) $\Delta E = 312$ J (b) $w = -208$ J (c) $\Delta V = 2.05$ L (d) $\Delta H = 520$ J (e) endothermic   **23.** (a) $\Delta E = 0$ (b) irreversible   **25.** (a) X   **26.** (a) C_3H_8 (c) Cl_2 **27.** (a) $q_V = 6.18$ kJ (b) $q_P = 10.3$ kJ   **28.** (a) $\Delta T = 0.0235$ K (b) $\Delta T = 0.235$ K   **30.** (a) 2.55 J g^{-1} K^{-1} (b) 105 g   **32.** (a) 89.0 kJ (endothermic) (c) -44.5 kJ (exothermic) (e) -92.2 kJ (exothermic)   **33.** (a) -296.8 kJ (exothermic) (c) 74.8 kJ (endothermic) (e) -285.4 kJ (exothermic)   **35.** 82.1 kJ mol^{-1} **36.** (a) -546.6 kJ (c) 395.7 kJ   **37.** (a) -13.7 kJ (c) -4.94 kJ   **38.** (a) $C_3H_8(g) + 5O_2(g) \longrightarrow 3CO_2(g) + 4H_2O(\ell)$ $\Delta H° = -2219.8$ kJ (b) -50.3 kJ   **40.** (a) -2222 kJ mol^{-1} (b) -6.491 kJ g^{-1} **41.** (a) 3.9 kJ (b) 21.8°C

Chapter 14   3. (a) spontaneous (b) nonspontaneous (c) nonspontaneous (d) spontaneous (e) spontaneous (f) nonspontaneous
4. (a) (1, 1) (1, 2) (1, 3) (1, 4) (1, 5) (1, 6)
(2, 1) (2, 2) (2, 3) (2, 4) (2, 5) (2, 6)
(3, 1) (3, 2) (3, 3) (3, 4) (3, 5) (3, 6)
(4, 1) (4, 2) (4, 3) (4, 4) (4, 5) (4, 6)
(5, 1) (5, 2) (5, 3) (5, 4) (5, 5) (5, 6)
(6, 1) (6, 2) (6, 3) (6, 4) (6, 5) (6, 6)
(b) The most likely macrostate (sum = 7) is represented by 6 compatible microstates out of the total 36: (1, 6) (6, 1) (2, 5) (5, 2) (3, 4) (4, 3). Each of the least likely macrostates (sum = 2 and sum = 12) is represented by only 1 microstate: (1, 1) for

sum = 2, and (6, 6) for sum = 12.   **5.** (a) $6^4 = 1296$ (b) Most likely macrostate: sum = 14. Least likely macrostates: sum = 4 (1, 1, 1, 1), and sum = 24 (6, 6, 6, 6). Each of the least likely macrostates is represented by only 1 microstate. **17.** (a) exothermic (b) higher (c) higher   **19.** (a) exothermic (b) higher (c) The reverse reaction does occur, but with lower probability than the forward reaction.   **20.** (a) No, a certain amount of time is needed to establish thermal equilibrium. (b) The particles exchange energy by means of collisions. (c) 12.2 K   **21.** (a) No, the gas is not in equilibrium with its surroundings at STP. (b) Energy flows from the surroundings at 273 K to the system at 12.2 K. The exchange of energy, brought about by collisions between particles, occurs at the boundary between system and surroundings. (c) No. A temperature, defined by the Boltzmann distribution, applies only to a system in thermal equilibrium. (d) 273 K   **22.** (a) 200 mL H_2O (c) 1 mol Ar at 1 atm   **23.** (a) C_8H_{18} (c) the drink   **25.** (a) -163.1 J K^{-1} (c) -242.6 J K^{-1} **27.** Each substance is a liquid at 25°C and 1 atm. (a) 8.6 kJ mol^{-1} (c) 4.7 kJ mol^{-1} (e) 4.4 kJ mol^{-1} **29.** (a) 6.3 kJ mol^{-1} (The gas has the higher Gibbs free energy.) (b) ≈-0.2 kJ mol^{-1} (assuming that $\Delta H°$ and $\Delta S°$ are independent of temperature). The theoretical difference in free energy is 0 for two phases in equilibrium.   **31.** $\approx$93 J mol^{-1} K^{-1} **33.** (a) -8.9 kJ mol^{-1} (c) -3.6 kJ mol^{-1} **35.** (a) per mol $C_2H_4(g)$: $\Delta H° = -137.0$ kJ, $\Delta S° = -120.6$ J K^{-1}, $\Delta G° = -101.0$ kJ (c) per mol $C_4H_{10}(\ell)$: $\Delta H° = -2855.4$ kJ, $\Delta S° = -359.1$ J K^{-1}, $\Delta G° = -2748.6$ kJ **36.** (a) $T < 1339$ K (c) $T > 0$ K   **37.** (a) $T < 780$ K (c) $T > 1124$ K   **39.** (a) 949 K (b) 888 K (c) 1019 K   **41.** (a) $K(25°C) \approx 1 \times 10^{-23}$, $Q = 1$. The system is not in equilibrium: $Q \gg K$. (b) 1124 K **43.** 27 kJ mol^{-1} **46.** Hint: Consider the possible effect of enthalpy on spontaneity.

Chapter 15   5. (a) The solution is not in equilibrium as long as its concentration is not uni-

form. (b) Equilibrium is reached within seconds. **7.** (a) Water will flow from A to B until [A] = [B]. (b) 0.5 L (c) [A] = [B] = 2 M **8.** (a) C_2H_5OH (c) serine   **9.** (a) $C_6H_5CH_3$ (c) $C_{10}H_8$ **10.** (a) Dissolve 5.33 g Na_2SO_4 in a total volume of 100 mL. (c) Dissolve 11.3 g $C_6H_{12}O_6$ in a total volume of 100 mL.   **11.** (a) 0.300 M (c) 0.500 M **13.** (a) 0.0750 mol Na^+, 0.0375 mol SO_4^{2-} (c) 0.0105 mol Na^+, 0.00525 mol SO_4^{2-} **14.** (a) $[Na^+] =$ 1.34 M (2 significant figures $\longrightarrow$ 1.3 M). $[CO_3^{2-}] =$ 0.67 M. (b) $K_{sp} = 1.2$ **15.** (a) exothermic (b) $\Delta H° = -18.2$ kJ mol^{-1} Li_2CO_3 **17.** $[Hg_2^{2+}] =$ 7.2×10^{-7} M. $[Cl^-] = 1.44 \times 10^{-6}$ M (2 significant figures $\longrightarrow$ 1.4×10^{-6} M).   **19.** $K_{sp} = 1.10 \times 10^{-10}$ **20.** (a) no precipitate: $Q = 1.95 \times 10^{-6} <$ K_{sp} (c) no precipitate: $Q = 7.81 \times 10^{-6} < K_{sp}$ **21.** (a) $[Ca^{2+}] = 2.14 \times 10^{-4}$ M (2 significant figures $\longrightarrow$ 2.1×10^{-4} M). $[F^-] = 4.27 \times 10^{-4}$ M (2 significant figures $\longrightarrow$ 4.3×10^{-4} M). (b) 0.017 g CaF_2 L^{-1} (c) 2.1×10^{-5} mol Ca^{2+}; 4.3×10^{-5} mol F^- **23.** (a) 0.017 g CaF_2 L^{-1} (c) 1.2×10^{-6} g CaF_2 L^{-1} **25.** (a) AgCl (b) 3.6×10^{-10} mol Ag^+ (c) 6.6×10^{-7} mol Ag^+ **27.** KCl is the most soluble. $K > 1$ if $\Delta G° < 0$. Favorable: $\Delta H° <$ 0 and $\Delta S° > 0$. (a) $\Delta H° = 16.9$ kJ mol^{-1}, $\Delta S° = 76.4$ J mol^{-1} K^{-1}, $\Delta G° = -6.0$ kJ mol^{-1} (b) $\Delta H° = 65.5$ kJ mol^{-1}, $\Delta S° = 33.0$ J mol^{-1} K^{-1}, $\Delta G° = 55.7$ kJ mol^{-1} (c) $\Delta H° = -17.6$ kJ mol^{-1}, $\Delta S° = -158.1$ J mol^{-1} K^{-1}, $\Delta G° = 29.5$ kJ mol^{-1} **28.** K_{sp}(AgCl, 25°C) $\approx 1.7 \times 10^{-10}$, K_{sp}(Ca(OH)$_2$, 25°C) $\approx 6.8 \times 10^{-6}$ **31.** $[N_2(aq)] = 5.6 \times 10^{-4}$ M, $[O_2(aq)] =$ 2.9×10^{-4} M **34.** (a) 4.9 atm (b) 0.067 M (c) 0.20 M **35.** 1.50 g

Chapter 16   3. (a) CF_3COOH (b) H_2SO_4 (c) $HClO$ (d) H_3PO_4 **5.** $CF_3COO^- < CCl_3COO^- <$ $CH_3COO^- < OH^-$ **7.** (a) conjugate acid: NH_4^+ (b) conjugate acid: $CH_3CH_2CH_2COOH$ (c) conjugate acid: H_2SO_4 (d) conjugate acid: HCl (e) conjugate acid: $^+H_3N-CH_3$ (f) conjugate acid: HCN   **9.** (a) 0.0 (c) 2.0 (e) 4.0   **11.** (a) 8

(c) 10 (e) 12   **13.** (a) 3.0 (e) 2.3 (j) 2.0
15. (a) pH = 8.0, $[OH^-] = 1 \times 10^{-6}$ M, pOH = 6.0
(c) pH = 10.0, $[OH^-] = 1 \times 10^{-4}$ M, pOH = 4.0
(e) pH = 12.0, $[OH^-] = 1 \times 10^{-2}$ M, pOH = 2.0
17. (a) $[H_3O^+] = 1 \times 10^{-1}$ M, $[OH^-] = 1 \times 10^{-13}$ M
(c) $[H_3O^+] = 1 \times 10^{-2}$ M, $[OH^-] = 1 \times 10^{-12}$ M
(e) $[H_3O^+] = 1 \times 10^{-3}$ M, $[OH^-] = 1 \times 10^{-11}$ M
19. (a) $[H_3O^+] = 1 \times 10^{-2}$ M, $[OH^-] = 1 \times 10^{-12}$ M
(d) $[H_3O^+] = 5 \times 10^{-3}$ M, $[OH^-] = 2 \times 10^{-12}$ M
(k) $[H_3O^+] = 1 \times 10^{-3}$ M, $[OH^-] = 1 \times 10^{-11}$ M
21. (a) 7.0 (c) 8.0 (e) 9.0   **23.** (a) 1×10^{-2} (d) 5×10^{-3} (k) 1×10^{-3} **24.** (a) $[H_3O^+] = 9.3 \times 10^{-14}$ M,
$[OH^-] = 1.08 \times 10^{-1}$ M, pH = 13.03 (basic)
(c) $[H_3O^+] = 1.0 \times 10^{-7}$ M, $[OH^-] = 1.0 \times 10^{-7}$ M,
pH = 7.00 (neutral)   **25.** (a) $[H_3O^+] = 1.00$ M,
$[OH^-] = 1.0 \times 10^{-14}$ M, pH = 0.000 (acidic)
(c) $[H_3O^+] = 4.0 \times 10^{-14}$ M, $[OH^-] = 0.25$ M,
pH = 13.40 (basic)   **26.** (a) pH = 1.44; $[A^-]/[HA] =$
0.038 (c) pH = 1.95; $[A^-]/[HA] = 0.13$ **27.** (a) pH =
2.10; $[A^-]/[HA] = 0.0081$ (c) pH = 2.60; $[A^-]/[HA] =$
0.026   **29.** (a) $HCN < C_6H_5COOH < CH_2ClCOOH$
(b) $[A^-]/[HA]$ at equilibrium increases as $[A]_0$
decreases.   **31.** (a) 7.1×10^{-12} (b) 1.5×10^{-10}
(c) 1.6×10^{-5} **32.** (a) 8.43 (c) 7.93   **33.** (a) 9.09
(c) 8.59   **35.** (a) pH = 2.85, ΔpH = ±0.009
(c) pH = 3.15, ΔpH = ±0.007 (e) pH = 1.85,
ΔpH = ±0.024   **38.** (a) Mix equal (and substantial) concentrations of HClO and ClO^-.
(b) $6.92 \leq$ pH ≤ 8.12 (c) $[ClO^-]_0/[HClO]_0 = 1.55$
39. (a) Mix equal (and substantial) concentrations
of CH_2FCOOH and CH_2FCOO^-. (b) $1.99 \leq$ pH $\leq$
3.19 (c) $[CH_2FCOO^-]_0/[CH_2FCOOH]_0 = 0.65$
41. 3.58   **42.** (a) 7.00 (c) 7.00   **43.** (a) 12.00
(c) 2.000   **44.** (a) $\approx$7.87 (if autoionization of H_2O
is neglected) (c) 3.74 (e) 2.000 (g) 2.90   **45.** (a) 5.63
(c) 9.26 (e) 12.00 (g) 10.62   **47.** (a) 0.152 M
(b) 7.00   **49.** (a) 0.152 M (b) 5.24   **51.** (a) half-
equivalence point: V_b = 10.0 mL NaOH; equivalence point: V_b = 20.0 mL NaOH (b) 2.93 (c) 4.85
(d) 8.85 (e) 11.84   **52.** (a) −0.0051 (c) 0.96
53. (a) 3.53 (c) 4.03   **56.** (a) endothermic (b) neutral (c) neutral (d) 6.2

Chapter 17   1. (a) Ca^{2+} (c) Au^+ **2.** (a) Ca (c) Li

3. (a) $2Fe^{3+} + 2I^- \longrightarrow 2Fe^{2+} + I_2$ (n = 2 mol e^-)
oxidizing agent: Fe^{3+}; reducing agent: I^-

5. (a)
$$I_2 + 6H_2O \longrightarrow 2IO_3^- + 12H^+ + 10e^-$$
$$\text{(oxidation)}$$
$$\underline{5Cu^{2+} + 10e^- \longrightarrow 5Cu \text{ (reduction)}}$$
$$I_2 + 5Cu^{2+} + 6H_2O \longrightarrow 2IO_3^- + 5Cu + 12H^+$$
$$(n = 10 \text{ mol } e^-)$$
oxidizing agent: Cu^{2+}; reducing agent: I_2

(b)
$$Ni \longrightarrow Ni^{2+} + 2e^- \text{ (oxidation)}$$
$$\underline{2VO_2^+ + 4H^+ + 2e^- \longrightarrow 2VO^{2+} + 2H_2O}$$
$$\text{(reduction)}$$
$$Ni + 2VO_2^+ + 4H^+ \longrightarrow Ni^{2+} + 2VO^{2+} + 2H_2O$$
$$(n = 2 \text{ mol } e^-)$$
oxidizing agent: VO_2^+; reducing agent: Ni

(c)
$$3H_2S \longrightarrow 3S + 6H^+ + 6e^-$$
$$\text{(oxidation)}$$
$$\underline{2NO_3^- + 8H^+ + 6e^- \longrightarrow 2NO + 4H_2O}$$
$$\text{(reduction)}$$
$$3H_2S + 2NO_3^- + 2H^+ \longrightarrow 3S + 2NO + 4H_2O$$
$$(n = 6 \text{ mol } e^-)$$
oxidizing agent: NO_3^-; reducing agent: H_2S

(d)
$$2NaBr(s) \longrightarrow Br_2 + 2Na^+ + 2e^-$$
$$\text{(oxidation)}$$
$$\underline{SO_4^{2-} + 4H^+ + 2e^- \longrightarrow SO_2 + 2H_2O}$$
$$\text{(reduction)}$$
$$2NaBr(s) + SO_4^{2-} + 4H^+ \longrightarrow Br_2 + SO_2 + 2Na^+$$
$$+ 2H_2O \ (n = 2 \text{ mol } e^-)$$
oxidizing agent: SO_4^{2-}; reducing agent: NaBr(s)

7. (a)
$$Zn + 2OH^- \longrightarrow Zn(OH)_2 + 2e^-$$
$$\text{(oxidation)}$$
$$\underline{2MnO_2 + H_2O + 2e^- \longrightarrow Mn_2O_3 + 2OH^-}$$
$$\text{(reduction)}$$
$$Zn + 2MnO_2 + H_2O \longrightarrow Zn(OH)_2 + Mn_2O_3$$
$$(n = 2 \text{ mol } e^-)$$
oxidizing agent: MnO_2; reducing agent: Zn

(b)
$$4PH_3 + 12OH^- \longrightarrow P_4 + 12H_2O$$
$$+ 12e^- \text{ (oxidation)}$$
$$4CrO_4^{2-} + 16H_2O + 12e^- \longrightarrow 4Cr(OH)_4^-$$
$$+ 16OH^- \text{ (reduction)}$$
$$\overline{4PH_3 + 4CrO_4^{2-} + 4H_2O \longrightarrow P_4 + 4Cr(OH)_4^-}$$
$$+ 4OH^- \ (n = 12 \text{ mol } e^-)$$
oxidizing agent: CrO_4^{2-}; reducing agent: PH_3

(c)
$$4Au + 8CN^- \longrightarrow 4Au(CN)_2^- + 4e^-$$
$$\text{(oxidation)}$$
$$O_2 + 2H_2O + 4e^- \longrightarrow 4OH^-$$
$$\text{(reduction)}$$
$$\overline{4Au + 8CN^- + O_2 + 2H_2O \longrightarrow 4Au(CN)_2^-}$$
$$+ 4OH^- \ (n = 4 \text{ mol } e^-)$$
oxidizing agent: O_2; reducing agent: Au

(d) $2NH_2OH + 2OH^- \longrightarrow N_2 + 4H_2O + 2e^-$
$$\text{(oxidation)}$$
$$CO_2 + H_2O + 2e^- \longrightarrow CO + 2OH^-$$
$$\text{(reduction)}$$
$$\overline{2NH_2OH + CO_2 \longrightarrow N_2 + CO + 3H_2O}$$
$$(n = 2 \text{ mol } e^-)$$
oxidizing agent: CO_2; reducing agent: NH_2OH

9. (a) anode: $3Ni + 6OH^- \longrightarrow 3Ni(OH)_2(s)$
$$+ 6e^- \text{ (oxidation)}$$
cathode: $2CrO_4^{2-} + 8H_2O + 6e^- \longrightarrow$
$$2Cr(OH)_3(s) + 10OH^- \text{ (reduction)}$$
(b) $\mathscr{E}_{ox}^\circ = 0.72$ V, $\mathscr{E}_{red}^\circ = -0.13$ V

12. (a) $Zn + 2OH^- \longrightarrow Zn(OH)_2(s) + 2e^-$
$$\text{(oxidation)}$$
$$HgO + H_2O + 2e^- \longrightarrow Hg + 2OH^-$$
$$\text{(reduction)}$$
$$\overline{Zn + HgO + H_2O \longrightarrow Zn(OH)_2(s) + Hg}$$
$$(n = 2 \text{ mol } e^-)$$
oxidizing agent: HgO; reducing agent: Zn
(b) anode: oxidation of Zn; cathode: reduction of HgO (c) $\mathscr{E}^\circ = 1.347$ V

13. (a) $4Fe^{2+} \longrightarrow 4Fe^{3+} + 4e^-$
$$\text{(oxidation)}$$
$$O_2 + 4H^+ + 4e^- \longrightarrow 2H_2O \text{ (reduction)}$$
$$\overline{4Fe^{2+} + O_2 + 4H^+ \longrightarrow 4Fe^{3+} + 2H_2O}$$
$$(n = 4 \text{ mol } e^-)$$
oxidizing agent: O_2; reducing agent: Fe^{2+}

14. (a) $\mathscr{E}^\circ = 0.459$ V, $\Delta G^\circ = -177$ kJ (spontaneous)

15. (a) $\mathscr{E}_{red}^\circ = 0.342$ V
(b) $$Zn \longrightarrow Zn^{2+} + 2e^-$$
$$\text{(oxidation, anode)}$$
$$Ag_2O + H_2O + 2e^- \longrightarrow 2Ag + 2OH^-$$
$$\text{(reduction, cathode)}$$
$$\overline{Zn + Ag_2O + H_2O \longrightarrow Zn^{2+} + 2Ag + 2OH^-}$$
$$(n = 2 \text{ mol } e^-)$$
(c) $\mathscr{E}^\circ = 1.104$ V, $\Delta G^\circ = -213.0$ kJ

16. (a) $2Br^- \longrightarrow Br_2 + 2e^-$ (oxidation, anode)
$$F_2 + 2e^- \longrightarrow 2F^- \quad \text{(reduction, cathode)}$$
$$\overline{2Br^- + F_2 \longrightarrow Br_2 + 2F^-} \ (n = 2 \text{ mol } e^-)$$
oxidizing agent: F_2; reducing agent: Br^-

17. (a) $\mathscr{E}^\circ = 1.800$ V, $\Delta G^\circ = -347.3$ kJ (spontaneous)   **18.** (a) $\mathscr{E}^\circ = 0.477$ V, $\Delta G^\circ = -276$ kJ (spontaneous), $K \approx \exp 111 = 2 \times 10^{48}$ (∞)

19. (a) $3[Ag \longrightarrow Ag^+ + e^-]$ $\mathscr{E}_{ox}^\circ = -0.80$ V
$$Au^{3+} + 3e^- \longrightarrow Au \quad \mathscr{E}_{red}^\circ = 1.40 \text{ V}$$
(b) $3Ag + Au^{3+} \longrightarrow 3Ag^+ + Au$ (c) $\mathscr{E}^\circ = 0.60$ V, $\Delta G^\circ = -1.7 \times 10^2$ kJ, $K \approx \exp 70 = 3 \times 10^{30}$ (∞)
22. (a) 1.74×10^{-10} (b) $K < 1$, $\mathscr{E}^\circ < 0$ (nonspontaneous)   **23.** 1.20×10^{-12} **25.** (a) 0 V (b) $\mathscr{E}^\circ$
26. (a) 0.4577 V (c) 0.4666 V   **27.** (a) $\Delta G^\circ = -88.32$ kJ, $K = 2.97 \times 10^{15}$ (c) $[Ag^+] = 4.10 \times 10^{-8}$ M **29.** (a) 0.770 V (c) 0.888 V (e) 1.007 V (g) 1.125 V   **30.** (a) 0.752 V (c) 0.766 V
31. (a) 3.2 M (b) -5.8 kJ (c) The process, overweighted toward reactants, will proceed to the right.   **33.** $Q = 1$ for each set of concentrations; hence $\mathscr{E} = \mathscr{E}^\circ$.   **34.** $H_2(g)$ will be produced instead of Na(s).

35. (a) $2Cl^-(\ell) \longrightarrow Cl_2(g) + 2e^-$
$$\text{(oxidation, anode)}$$
$$2[Na^+(\ell) + e^- \longrightarrow Na(\ell)]$$
$$\text{(reduction, cathode)}$$
$$\overline{2Na^+(\ell) + 2Cl^-(\ell) \longrightarrow 2Na(\ell) + Cl_2(g)}$$
(b) The anode is held at positive potential.   **36.** 11.7 h
37. 0.5321 L   **39.** (a) 83.0 g   **40.** (a) 21.4 min (c) 0.169 M

Chapter 18   **3.** (a) reaction 2 (b) insufficient information   **4.** (a) reaction 2 (b) reaction 1   **5.** (b) reaction 2 (c) reaction 1   **6.** (b) reaction 1 (c) reaction 1   **7.** (a) $k(10°C) = 1.016$, $k(20°C) = 1.030$, $k(30°C) = 1.045$, $k(100°C) = 1.125$ (c) $k(10°C) = 2.176$, $k(20°C) = 4.491$, $k(30°C) = 8.835$, $k(100°C) = 365.0$ **9.** (a) seconds, minutes (b) geological time scale (c) seconds   **10.** (a) $\Delta G° = 8.6$ kJ mol^{-1}, $K = 0.031$. Fixed thermodynamic differences provide no insight into the kinetics of a process.   **11.** (a) Rate $= k[A][B][C]^2$ (b) M^{-3} s^{-1} (c) $k = 1\ M^{-3}$ s^{-1} (d) Rate $= 16\ M$ s^{-1} (e) no   **12.** (a) Rate $= k[A][B]$ (c) Rate $= k[A]^2$ **13.** (a) A $\longrightarrow$ products (c) A + B $\longrightarrow$ products   **14.** Maybe, maybe not—only an experimental determination can tell.   **15.** False—a rate law depends on the microscopic mechanism of reaction, not the macroscopic stoichiometry represented by the balanced equation.   **16.** (a) 9.90 m s^{-1} (c) 0.09491 babies per second (based on 1 y = 365 d)   **17.** (a) $-0.0060\ M$ s^{-1} (b) $0.0060\ M$ s^{-1} (c) $0.030\ M$ **18.** (a) $3.8 \times 10^{-4}\ M$ s^{-1} (c) $1.5 \times 10^{-3}\ M$ s^{-1} **19.** Rate $= k[NO]^2[O_2]$; $k = 7.10 \times 10^3\ M^{-2}$ s^{-1} (average of 5 trials)   **20.** Trial 1: $8.9 \times 10^{-4}\ M$ s^{-1}. Trial 3: $8.52 \times 10^{-2}\ M$ s^{-1}.   **21.** (a) $k = 2.20 \times 10^{-5}$ s^{-1}, $t_{1/2} = 3.15 \times 10^4$ s (b) $[SO_2Cl_2]_{t=3h} = 1.18 \times 10^{-2}\ M$, $[SO_2Cl_2]_{t=9h} = 7.35 \times 10^{-3}\ M$, $[SO_2Cl_2]_{t=15h} = 4.57 \times 10^{-3}\ M$ **22.** (b) $k = 2.73 \times 10^{-4}$ min^{-1}, $t_{1/2} = 2.54 \times 10^3$ min (c) $0.135\ M$ **23.** (a) $1.5 \times 10^{-5}\ M$ s^{-1} (b) $[CH_3CH_2Cl]_{t=100\ s} = 9.9 \times 10^{-2}\ M$, $[CH_3CH_2Cl]_{t=1000\ s} = 8.6 \times 10^{-2}\ M$, $[CH_3CH_2Cl]_{t=10,000s} = 2.2 \times 10^{-2}\ M$ **25.** Rate(0) $= 3.05 \times 10^{-2}\ M$ min^{-1}. Rate(10.0 min) $= 2.25 \times 10^{-2}\ M$ min^{-1}. Rate(20.0 min) $= 1.66 \times 10^{-2}\ M$ min^{-1}. Rate(30.0 min) $= 1.22 \times 10^{-2}\ M$ min^{-1}.   **27.** No. Species other than A may be involved in the mechanism, and the reaction may require more than one step.   **28.** (a) 1.4×10^{-7} s (b) Define $\Delta t = 1.0 \times 10^{-8}$ s: $[I]_{t=\Delta t} = 9.3 \times 10^{-4}\ M$, $[I]_{t=10\Delta t} = 5.9 \times 10^{-4}\ M$, $[I]_{t=100\Delta t} = 1.3 \times 10^{-4}\ M$

(c) $4.4 \times 10^{-4}\ M$ (d) $K(298\ K) \approx 1.7 \times 10^{21}$ **30.** (a) 6.3 s (rounded up from 6.25 s) (d) Rate(0) $= 3.2 \times 10^{-2}\ M$ s^{-1}. Rate(15 s) $= 2.8 \times 10^{-3}\ M$ s^{-1}. Rate(30 s) $= 9.5 \times 10^{-4}\ M$ s^{-1}.   **32.** (a) 10. s   **33.** (a) 1.3 s   **34.** (a) $[B]_0 = 1\ M$. Reaction 1: $t_{1/2} = 6.93 \times 10^{-4}$ s. Reaction 2: $t_{1/2} = 1.00 \times 10^{-3}$ s. (c) $[B]_0 = 1.732\ M$. Reaction 1: $t_{1/2} = 6.93 \times 10^{-4}$ s. Reaction 2: $t_{1/2} = 5.77 \times 10^{-4}$ s.   **36.** (a) 47.8 kJ mol^{-1} (b) 1.52   **37.** (a) $E_a = 29.9$ kJ mol^{-1}, $A = 1.73 \times 10^7$ s^{-1} (b) 229 s^{-1} **38.** $E_a = 184$ kJ mol^{-1}, $A = 5.23 \times 10^{11}\ M^{-1}$ s^{-1} **39.** $E_a = 161$ kJ mol^{-1} (160.5 kJ mol^{-1} before roundoff), $A = 4.00 \times 10^{13}$ s^{-1} **41.** (a) $k_1(300\ K) = 3.88 \times 10^{-5}$ s^{-1}, $k_1(1000\ K) = 5.98 \times 10^7$ s^{-1}. $k_2(300\ K) = 7.04 \times 10^{-6}$ s^{-1}, $k_2(1000\ K) = 1.80 \times 10^8$ s^{-1}. (b) Reaction 1 is faster at 300 K. Reaction 2 is faster at 1000 K. (c) 522 K   **43.** (a) $k_1 = 5.35 \times 10^{-4}\ M^{-1}$ s^{-1}, $k_2 = 5.93 \times 10^{-3}\ M^{-1}$ s^{-1}, $k_3 = 5.07 \times 10^{-4}\ M^{-1}$ s^{-1} (b) reaction 2 (c) There is insufficient information. Either K (the equilibrium constant), $\Delta G°$, or the rate constant for the reverse reaction is needed.   **44.** (a) $K = 10.0$ (b) $k_+ = 345\ M^{-1}$ s^{-1} (c) $k_- = 34.5\ M^{-1}$ s^{-1} **45.** (d) endothermic   **46.** (a) $K = 129$ (b) $k_- = 3.3 \times 10^{-9}\ M^{-1}$ s^{-1} **47.** Rate $= k[H_2][I_2]$ **48.** 1 M **49.** (a) step 1 (b) Rate $= k_1[NO_2][F_2]$ **50.** (b) $k = k_2k_1/k_{-1}$ **51.** (a) Rate $= k[O_3]^2/[O_2]$ (b) $k = k_2k_1/k_{-1}$

Chapter 19   **1.** (a) octahedral; coordination number = 6 (b) octahedral; coordination number = 6 (c) octahedral; coordination number = 6 (d) tetrahedral; coordination number = 4 (e) square planar; coordination number = 4   **3.** one form only: no geometric or optical isomers   **5.** no optical isomers   **10.** geometric isomers: two (cis and trans); optical isomers: none   **11.** geometric isomers: two; optical isomers: none   **12.** linkage isomers: two; geometric isomers: none; optical isomers: none   **13.** linkage isomers: MX_6, MX_5Y, MX_4Y_2, MX_3Y_3, MX_2Y_4, MXY_5, MY_6;

geometric isomers: two apiece for MX_4Y_2, MX_3Y_3, and MX_2Y_4; optical isomers: none. The symbol Y represents the ambidentate ligand in its alternative bonding configuration.   **14.** (a) +3 (c) +2   **15.** (a) +2 (c) +1   **16.** (a) coordination number: 6; oxidation number: +3 (c) coordination number: 6; oxidation number: +3   **17.** (a) d^6 (c) d^8 **18.** (b) d^1 (c) d^{10} **19.** (a) d^3 (d) d^6 **20.** (a) $Sc^{3+} \equiv$ [Ar] (0 d electrons) (c) $V^{2+} \equiv$ [Ar]$3d^3$ (3 d electrons); $V^{3+} \equiv$ [Ar]$3d^2$ (2 d electrons); $V^{4+} \equiv$ [Ar]$3d^1$ (1 d electron); $V^{5+} \equiv$ [Ar] (0 d electrons)   **21.** (a) $Fe^{2+} \equiv$ [Ar]$3d^6$ (6 d electrons); $Fe^{3+} \equiv$ [Ar]$3d^5$ (5 d electrons); $Fe^{4+} \equiv$ [Ar]$3d^4$ (4 d electrons); $Fe^{6+} \equiv$ [Ar]$3d^2$ (2 d electrons) (c) $Ni^{2+} \equiv$ [Ar]$3d^8$ (8 d electrons); $Ni^{3+} \equiv$ [Ar]$3d^7$ (7 d electrons); $Ni^{4+} \equiv$ [Ar]$3d^6$ (6 d electrons)   **22.** (a) $Y^{3+} \equiv$ [Kr] (0 d electrons) (c) $Nb^{4+} \equiv$ [Kr]$4d^1$ (1 d electron); $Nb^{5+} \equiv$ [Kr] (0 d electrons)   **23.** (a) $Ru^{2+} \equiv$ [Kr]$4d^6$ (6 d electrons); $Ru^{3+} \equiv$ [Kr]$4d^5$ (5 d electrons); $Ru^{4+} \equiv$ [Kr]$4d^4$ (4 d electrons); $Ru^{8+} \equiv$ [Kr] (0 d electrons) (c) $Pd^{2+} \equiv$ [Kr]$4d^8$ (8 d electrons); $Pd^{4+} \equiv$ [Kr]$4d^6$ (6 d electrons)   **24.** $Mn^{3+} +$ $e^- \longrightarrow Mn^{2+}$. Mn^{2+} has a d^5 configuration. **25.** $Mn^{3+} + e^- \longrightarrow Mn^{2+}$. Mn^{2+} has a d^5 configuration.   **26.** (a) $(t_{2g})^1$ (c) $(t_{2g})^3$ (e) high spin: $(t_{2g})^3(e_g)^2$; low spin: $(t_{2g})^5$ **27.** (a) $CFSE = -0.4\Delta$ (1 unpaired electron) (c) $CFSE = -1.2\Delta$ (3 unpaired electrons) (e) high spin: $CFSE = 0$ (5 unpaired electrons); low spin: $CFSE = -2.0\Delta$ (1 unpaired electron)   **28.** (a) high spin: $(t_{2g})^4(e_g)^2$; low spin: $(t_{2g})^6$ (c) $(t_{2g})^6(e_g)^2$ **29.** (a) high spin: $CFSE = -0.4\Delta$ (4 unpaired electrons); low spin: $CFSE = -2.4\Delta$ (0 unpaired electrons) (c) $CFSE = -1.2\Delta$ (2 unpaired electrons)   **30.** Cr^{3+} **31.** (a) F^- (c) H_2O **32.** (a) $[Fe(NCS)_6]^{3-}$ (b) $[Fe(NCS)_6]^{3-} \equiv$ $(t_{2g})^3(e_g)^2$ (5 unpaired electrons); $[Fe(CN)_6]^{3-} \equiv$ $(t_{2g})^5$ (1 unpaired electron)   **33.** $[Fe(NCS)_6]^{3-} \equiv$ 6 sp^3d^2 hybrid orbitals $+ 3d^5$ (5 unpaired electrons); $[Fe(CN)_6]^{3-} \equiv$ 6 d^2sp^3 hybrid orbitals $+ 3d^5$ (1 unpaired electron)   **36.** (a) V^{2+} **37.** Label the z^2 and $x^2 - y^2$ orbitals in a tetrahedral

complex as e, and label the xy, xz, and yz orbitals as t_2 to write the configurations: (a) $(e)^1$ (c) $(e)^2(t_2)^1$ **38.** (a) $CFSE = -0.6\Delta_t$ (1 unpaired electron) (c) $CFSE = -0.8\Delta_t$ (3 unpaired electrons)   **39.** (a) $(e)^3(t_2)^3$ (c) $(e)^4(t_2)^4$ **40.** (a) $CFSE = -0.6\Delta_t$ (4 unpaired electrons) (c) $CFSE = -0.8\Delta_t$ (2 unpaired electrons)   **41.** (b) $[NiCl_4]^{2-}$ **43.** red and green; orange and blue; yellow and violet   **44.** (a) 381 nm (b) yellow   **45.** green   **47.** Cu^+ has no vacant d orbitals.

Chapter 20   1. ^{4}He, ^{12}C, and ^{16}O have no nuclear spin.   **2.** 3 nuclei: 8 configurations; 4 nuclei: 16 configurations   **3.** (a) 1 (c) 1   **4.** no   **5.** 4   **6.** Each spectrum contains 4 lines. The spectra are not identical.   **7.** H   **8.** CH_3, Cl   **9.** N—N < N=N < N≡N (N—N absorbs at the longest wavelength.)   **10.** (a) 1750 cm^{-1} = 20.93 kJ mol^{-1}, 1150 cm^{-1} = 13.76 kJ mol^{-1} (b) $h\nu$ (infrared) < covalent bond energy   **11.** $k_{C-O} \approx$ $0.4 k_{C=O}$ **12.** All three wavelengths correspond to ultraviolet radiation. H—H: $\lambda = 2.74 \times 10^{-7}$ m (274 nm, 2740 Å). N≡N: $\lambda = 1.26 \times 10^{-7}$ m (126 nm, 1260 Å). O=O: $\lambda = 2.41 \times 10^{-7}$ m (241 nm, 2410 Å)   **13.** (a) $\epsilon_0 = 2.97 \times 10^{-20}$ J (b) $\epsilon_1 = 8.90 \times 10^{-20}$ J, $\epsilon_2 = 1.48 \times 10^{-19}$ J, $\epsilon_3 =$ 2.08×10^{-19} J   **14.** (a) $\lambda_{0,1} = 3.35 \times 10^{-6}$ m (3.35 μm), $\epsilon_{0,1} = 5.93 \times 10^{-20}$ J (b) $f_{0,1} = 1.82 \times$ 10^6 (≈2 million molecules in ground state for every molecule in the first excited state)   **15.** (a) $\lambda_{1,2} = 3.35 \times 10^{-6}$ m (3.35 μm), $\epsilon_{1,2} =$ 5.93×10^{-20} J (b) $\lambda_{2,3} = 3.35 \times 10^{-6}$ m (3.35 μm), $\epsilon_{2,3} = 5.93 \times 10^{-20}$ J   **16.** (b) $H_2C=CH_2$ **18.** (a) $\lambda = 3.10 \times 10^{-9}$ m (3.10 nm), $\nu = 9.67 \times$ 10^{16} s^{-1}, $\epsilon = 6.41 \times 10^{-17}$ J (b) 1.73×10^7 m s^{-1} **19.** The kinetic energy of a free electron is quasi-continuous (not quantized).   **20.** Core electrons are effectively removed from the chemical influence of the valence region. The environment near an atomic nucleus remains largely unaltered from molecule to molecule.   **21.** 47.88 g mol^{-1}

Chapter 21 **1.** (a) $^{10}B \equiv 5$ p, 5 n; $^{11}B \equiv 5$ p, 6 n; $^{12}B \equiv 5$ p, 7 n (c) $^{24}Mg \equiv 12$ p, 12 n; $^{25}Mg \equiv 12$ p, 13 n; $^{26}Mg \equiv 12$ p, 14 n **2.** (a) $^{209}_{86}Rn$ (c) $^{209}_{88}Ra$ **3.** $X_{35} = 0.758$, $X_{37} = 0.242$ **5.** 12.01 **6.** (b) $\Delta E = -3.716 \times 10^{-13}$ J/nucleus (-2.238×10^8 kJ mol^{-1}) (c) $E_A = 1.858 \times 10^{-13}$ J/nucleon **7.** (c) $\Delta m = -0.009105$ u; $\Delta E = -1.359 \times 10^{-12}$ J/nucleus; $E_A = 4.529 \times 10^{-13}$ J/nucleon **8.** (a) $\Delta m = -0.098940$ u; $\Delta E = -1.477 \times 10^{-11}$ J/nucleus; $E_A = 1.230 \times 10^{-12}$ J/nucleon **9.** (a) $\Delta m = -1.915061$ u; $\Delta E = -2.858 \times 10^{-10}$ J/nucleus; $E_A = 1.216 \times 10^{-12}$ J/nucleon (c) $\Delta m = -1.939356$ u; $\Delta E = -2.894 \times 10^{-10}$ J/nucleus; $E_A = 1.211 \times 10^{-12}$ J/nucleon **11.** no **13.** (a) ^4He (b) ^{208}Pb **14.** (a) ^{189}Pb (c) ^{218}Rn **15.** (a) $^{193}_{84}Po \longrightarrow {}^{189}_{82}Pb + {}^4_2He$ (c) $^{222}_{88}Ra \longrightarrow {}^{218}_{86}Rn + {}^4_2He$ **16.** (a) ^{222}U (c) ^{217}Th **17.** (a) $^{222}_{92}U \longrightarrow {}^{218}_{90}Th + {}^4_2He$ (c) $^{217}_{90}Th \longrightarrow {}^{213}_{88}Ra + {}^4_2He$ **18.** (a) ^{142}La (c) ^{14}N **19.** (a) $^{142}_{56}Ba \longrightarrow {}^{142}_{57}La + {}^0_{-1}e + \tilde{\nu}$ (c) $^{14}_6C \longrightarrow {}^{14}_7N + {}^0_{-1}e + \tilde{\nu}$ **20.** (a) ^{168}Ho (c) ^{16}N **21.** (a) $^{168}_{67}Ho \longrightarrow {}^{168}_{68}Er + {}^0_{-1}e + \tilde{\nu}$ (c) $^{16}_7N \longrightarrow {}^{16}_8O + {}^0_{-1}e + \tilde{\nu}$ **23.** $^{235}_{92}U \longrightarrow {}^{207}_{82}Pb + 7{}^4_2He + 4{}^0_{-1}e + 4\tilde{\nu}$ **24.** (a) ^{73}Ge (b) ^{14}O (c) ^4He (d) ^{24}Mg (e) ^{49}Ti **25.** (a) $^{73}_{33}As + {}^0_{-1}e \longrightarrow {}^{73}_{32}Ge + \nu$ (b) $^{15}_9F \longrightarrow {}^{14}_8O + {}^1_1p$ (c) $^5_2He \longrightarrow {}^4_2He + {}^1_0n$ (d) $^{24}_{13}Al \longrightarrow {}^{24}_{12}Mg + {}^0_1e + \nu$ (e) $^{49}_{23}V + {}^0_{-1}e \longrightarrow {}^{49}_{22}Ti + \nu$ **28.** (a) ^{143}Pm (b) ^{109}I (c) ^{103}Sn (d) ^9He (e) ^{35}Ca **29.** (a) $^{143}_{61}Pm + {}^0_{-1}e \longrightarrow {}^{143}_{60}Nd + \nu$ (b) $^{109}_{53}I \longrightarrow {}^{108}_{52}Te + {}^1_1p$ (c) $^{103}_{50}Sn \longrightarrow {}^{103}_{49}In + {}^0_1e + \nu$ (d) $^9_2He \longrightarrow {}^8_2He + {}^1_0n$ (e) $^{35}_{20}Ca \longrightarrow {}^{35}_{19}K + {}^0_1e + \nu$ **30.** (a) $^{14}_7N + {}^4_2He \longrightarrow {}^{17}_8O + {}^1_1H$ (b) $^{59}_{27}Co + {}^1_0n \longrightarrow {}^{60}_{27}Co$ (c) $^{14}_6C \longrightarrow {}^{14}_7N + {}^0_{-1}e + \tilde{\nu}$ (d) $^{235}_{92}U + {}^1_0n \longrightarrow {}^{137}_{52}Te + {}^{97}_{40}Zr + 2{}^1_0n$ (e) $^{239}_{94}Pu + \alpha \longrightarrow {}^{242}_{96}Cm + {}^1_0n$ **31.** (a) $^{16}_8O + {}^1_1p \longrightarrow {}^{13}_7N + {}^4_2He$ (b) $^{27}_{13}Al + {}^1_0n \longrightarrow {}^{24}_{11}Na + {}^4_2He$ (c) $^{58}_{26}Fe + {}^1_0n \longrightarrow {}^{59}_{26}Fe$ (d) $^3_2He + {}^1_1H \longrightarrow {}^4_2He + {}^0_1e + \nu$ (e) $^{235}_{92}U + {}^1_0n \longrightarrow {}^{142}_{56}Ba + {}^{91}_{36}Kr + 3{}^1_0n$ **32.** (a) violates conservation of baryon number (b) violates conservation of charge (c) violates conservation of lepton number (d) violates conservation of lepton number **33.** (a) nonspontaneous (b) spontaneous (c) nonspontaneous (d) spontaneous **35.** (a) $N_{t=2500y}/N_0 = 0.739$, $N_{t=10,000y}/N_0 = 0.298$ (b) 1.90×10^4 y **37.** straight line; slope $= -k$; $t_{1/2} = (\ln 2)/k$ **38.** (a) $k = 0.00578$ s^{-1}, $t_{1/2} = 120$ s (b) 6.67×10^{-17} g (c) 125 Bq, $m = 8.34 \times 10^{-18}$ g **39.** (a) 1.40×10^{-12} g (b) $A_{t=6y} = 357$ Bq, $m = 1.00 \times 10^{-12}$ g; $A_{t=12y} = 254$ Bq, $m = 7.12 \times 10^{-13}$ g; $A_{t=18y} = 181$ Bq, $m = 5.08 \times 10^{-13}$ g; $A_{t=24y} = 129$ Bq, $m = 3.62 \times 10^{-13}$ g **41.** (a) ^{32}S (b) 36.08% ^{32}P and 63.92% ^{32}S by mass **42.** 9.7×10^8 y **44.** (a) $u\tilde{d}$ (b) $u\tilde{u}$ or $d\tilde{d}$ (c) $\tilde{u}d$ (d) $\tilde{u}\tilde{d}\tilde{d}$ **45.** (a) $R(^1H) = 1.2 \times 10^{-15}$ m, $R(^{12}C) = 2.7 \times 10^{-15}$ m, $R(^{238}U) = 7.4 \times 10^{-15}$ m (b) The three nuclei all have the same density: 2.3×10^{14} g cm^{-3} (c) 2.5×10^8 tons

Appendix A **1.** (a) $BaBr_2$ (b) $ZnCO_3$ (c) Na_3PO_4 (d) K_2SO_4 (e) CaH_2 (f) $LiOH$ **2.** (a) FeS (b) Fe_2S_3 (c) $FeSO_4$ (d) $Fe_2(SO_4)_3$ (e) $FeSO_3$ **3.** (a) $BeCl_2$ (b) Be_3N_2 (c) PbI_2 (d) $Pb(IO_3)_2$ (e) $Cd_3(PO_4)_2$ (f) $LiNO_3$ **4.** (a) Co_2O_3 (b) CoF_2 (c) CoF_3 (d) Na_2CrO_4 (e) $Mg(C_2H_3O_2)_2$ **5.** (a) cadmium selenide (b) calcium chlorate (c) calcium perchlorate (d) praseodymium(III) sulfide (e) praseodymium(III) sulfate **6.** (a) zinc cyanide (b) zinc iodate (c) zinc telluride (d) rubidium permanganate (e) bismuth(II) acetate (f) ammonium azide **7.** (a) aluminum hydroxide (b) aluminum nitride (c) ammonium bromide (d) iron(II) fluoride (e) iron(III) fluoride (f) chromium(III) oxide **8.** (a) osmium(II) sulfite (b) thallium(I) sulfate (c) thallium(I) hydrogen sulfate or thallium(I) bisulfate (d) thallium(I) sulfide (e) thallium(III) sulfide (f) potassium thiocyanate **9.** (a) UF_6 (b) Cl_2O (c) $NaClO$ (d) XeO_3 (e) OsS_4 **10.** (a) Hg_2Cl_2 (b) $HgCl_2$ (c) $CuCl$ (d) $CuCl_2$ (e) $FeBr_2$ (f) $FeBr_3$ **11.** (a) arsenic trifluoride (b) arsenic pentafluoride (c) disilicon hexabromide (d) copper(I) oxide or cuprous oxide

(e) copper(II) oxide or cupric oxide
12. (a) sodium nitrite (b) sodium nitrate (c) sodium hypochlorite (d) sodium chlorite (e) sodium carbonate (f) sodium hydrogen carbonate or sodium bicarbonate   **13.** (a) H_3PO_4 (b) H_3PO_3 (c) HBrO (d) HCl (e) HCN   **14.** (a) H_2CO_3 (b) HF (c) HBr (d) $HClO_3$ (e) $HClO_4$ **15.** (a) nitrous acid (b) nitric acid (c) hydrosulfuric acid (d) sulfurous acid (e) sulfuric acid   **16.** (a) hydroiodic acid (b) hypoiodous acid (c) iodous acid (d) iodic acid (e) periodic acid (f) arsenic acid   **17.** (a) $K_4[Fe(CN)_6]$ (b) $K_3[Fe(CN)_6]$ (c) $[Pt(en)_3]Br_4$ (d) $[Pt(H_2O)_3Br]Cl$ **18.** (a) $[Cr(NH_3)_6](NO_3)_3$ (b) $[Ru(H_2O)Cl_5]^{2-}$ (c) $[Ni(CN)_4]^{2-}$ (d) $[Co(NH_3)_5(SO_4)]Cl$ **19.** (a) sodium diaquadioxalatoruthenate(III) (b) tetraammineaquacyanocobalt(III) chloride (c) hexacarbonylnickel(II) bromide (d) potassium tetrachlorocuprate(II)   **20.** (a) hexaamminecobalt(III) ion (b) tetraamminedichlorocobalt(III) ion (c) tetrachloroferrate(III) ion (d) tris(ethylenediamine)platinum(IV) ion

21. (a)

(b)

(c)   (d)

22. (a)   (b)

(c) $Cl-C{\equiv}C-Cl$
(d)

23. (a)

(b) See analogous structure in Example 8-2. (c) See analogous structure in Example 8-2.
(d)

24. (a) 4-methyloctane (b) 2,5-dimethyloctane (c) 4-chloro-2,2,7-trimethyloctane   **25.** (a) 2,4-dimethyloctane   (b)   5,7-dimethyl-1-octene (c) *trans*-6,8-dimethyl-2-nonene (d) *cis*-6, 8-dimethyl-2-nonene   **26.**   (a)   1-butanol (b) 2-butanol (c) methanol (d) 5-methyl-1-hexanol (e) 5,5-dichloro-1-hexanol

Appendix B   *Note*: Calculated values are treated as exact numbers in the following exercises.   **1.** (a) 3700 (b) −0.0000005 (c) 7.36 (d) 4089.775561   **2.** (a) 2511.886432 (b) 0.0003981071706 (c) 8000 (d) 729 **3.** (a) 0.1920499086 (b) 0.1920499086 (c) 5.206979827 (d) 0.1920499086   **4.** (a) $3 \log a + 2 \log b$ (b) $3 \ln a + 2 \ln b$ (c) 1.17 (d) $1.17 \ln 10$ (e) 1.17   **5.** (a) $\log 3 + 4 \log a - \log 2 - \frac{1}{2} \log b$ (b) $\ln 3 + 4 \ln a - \ln 2 - \frac{1}{2} \ln b$ (c) $\log a + \log b + \log c + \log d + \log e + \log f$ (d) $\log(a + b + c)$ **6.** (a) $3 \log 2 = 0.903090$ (b) $1 + 3 \log 2 = 1.903090$ (c) $-3 \log 2 = -0.903090$ (d) $-\ln 2 = -0.693147$ **7.** (a) $\log 2 + \log 3 - 7 = -6.221849$ (b) $\log 3 - \log 2 + \frac{1}{2} = 0.676091$ (c) $2 \log 3 + \log 2 - 1 = 0.255272$ (d) $3 \log 3 - 4 = -2.568637$ **8.** (a) $\ln 3 = 1.098612$ (b) $3 \ln 3 - 4 \ln 10 = -5.914504$ (c) $2 \ln e = 2$ (d) $\ln 2 + \ln 3 = 1.791759$ **9.** (a) 0.175180 (b) −4.273102 (c) 5.635619 (d) 53.091491   **10.** (a) −1.874706 (b) 2.512846 (c) −16.515496 (d) 54.754544 **11.** (a) 10^{-7} (b) 114.8153621   **12.** (a) $e = 2.718281828459045$ (b) $e^{-1} = 0.3678794412$ **13.** 8.96   **14.** $m = 5$, $b = -8$ **15.** (a) $x = 4.289855$ (b) $x = (\pi - 8)/8 \approx -0.607301$ (c) $x = -2$

(d) $x = 1$ **16.** (a) hyperbola (goes to 0 as x approaches $\pm\infty$) (b) straight line (goes to ∞ as x approaches ∞; goes to $-\infty$ as x approaches $-\infty$) (c) parabola (goes to ∞ as x approaches $\pm\infty$) (d) quartic (goes to ∞ as x approaches $\pm\infty$) **17.** $y(x = 1.67) \approx 51.66$ **18.** (b) $x = \pm1.5$ **19.** (a) $x = -2, -3$ (b) $x = -4, 1$ (c) $x = \frac{1}{2}$ (double root) (d) $x = \pm\frac{1}{2}$ (e) $x = -\frac{1}{2}, 2$ **20.** (a) $x = -2, -3$ (b) $x = 1, -4$ (c) $x = \frac{1}{2}$ (double root) (d) $x = \pm\frac{1}{2}$ (e) $x = 2, -\frac{1}{2}$

21. (a) $x = \dfrac{7 \pm \sqrt{13}}{6} \approx 1.767592, 0.565741$

(b) $x = \dfrac{-1 \pm \sqrt{97}}{8} \approx 1.106107, -1.356107$

(c) $x = \dfrac{4.234 \pm \sqrt{25.60676}}{2} \approx 4.647156, -0.413156$

(d) $x = \dfrac{-11 \pm \sqrt{113}}{-4} \approx 0.092464, 5.407536$

(e) $x = \dfrac{-1 \pm \sqrt{9}}{2} = 1, -2$

22. $x = \dfrac{-100 \pm \sqrt{10{,}040}}{2} \approx 0.09990, -100.09990$

23. (a) $M_{35} \approx 8.148918982 \times 10^{420{,}920}$ (b) 116.9 pages

CREDITS

Introduction: Figure 1 © 1998 Kenneth Eward/BioGrafx.

Figure 2-1 courtesy of Robert J. Hamers, Department of Chemistry, University of Wisconsin–Madison.

Figure 2-2 © 1950 Rosalind Franklin, Science Source/Photo Researchers.

Figure 3-13 courtesy of Ch. Baerlocher and L. B. McCusker, Atlas of Zeolite Structure Types on the Web (www.kristall.ethz.ch/IZA-SC/Atlas/AtlasHome.html).

Figure 4-8 courtesy of John Kendrew, Medical Research Center of Molecular Biology.

Figure 4-18 © 1993 E. R. Degginger, Science Source/Photo Researchers.

Figure 6-5 (curves $3s$ and $3p$ only) adapted by permission of Oxford University Press from P. W. Atkins, *Quanta: A Handbook of Concepts,* 2nd edition, Oxford University Press, 1991, Figure P.10, p. 270.

Figure 7-20 adapted by permission of Oxford University Press from P. W. Atkins, *Quanta: A Handbook of Concepts,* 2nd edition, Oxford University Press, 1991, Figure P.18(a), p. 286.

Figure 9-16 courtesy of Frank W. Gayle, National Institute of Standards and Technology (*J. Metals* v. 40, n. 5, 52–53, 1988).

Figure 9-17(a) courtesy of Kenji Hiraga, Institute for Materials Research, Tohoku University.

Figure 9-17(b) Penrose tile pattern created in ceramic tile and image supplied by Saxe-Patterson, Inc.; Patent Pending Pentaplex Ltd. (No. 4,133,152).

Figure 9-23 adapted from 1995 *Yearbook of Science and the Future,* Figure 47, p. 261 (as excerpted from the 1994 *Macropaedia* article "Matter: Its Properties, States, Varieties, and Behaviour"); © 1994 Encyclopaedia Britannica, Inc.

Figure 9-31 and uncaptioned diagrams in Section 9-9 drawn from sketches kindly provided by Edward T. Samulski, Department of Chemistry, University of North Carolina at Chapel Hill.

Figure 15-15(a) adapted from Lubert Stryer, *Biochemistry,* 3rd edition, Freeman, 1988, Figure 9.31, p. 217. Adapted from D. M. Blow and T. A. Steitz, "X-Ray Diffraction Studies of Enzymes," *Ann. Rev. of Biochemistry* 39:79; © 1970.

Figure 15-15(b) structural formula adapted from Lubert Stryer, *Biochemistry,* 3rd edition, Freeman, 1988, Figure 9.30, p. 217.

Figure 15-16 drawn from a sketch kindly provided by Maitland Jones, Jr., Department of Chemistry, Princeton University.

Figure 15-17 adapted from an image courtesy of Clarence D. Chang, Mobil Research Laboratory.

Figure 20-6 © 1991 Mehau Kulyk, Science Photo Library/Photo Researchers.

Figure 20-7 spectrum reprinted with permission of Aldrich Chemical Company, Inc.

Figure R20-2 spectrum reprinted with permission of Aldrich Chemical Company, Inc.

Appendix C: Raw data selected principally from David R. Lide, ed., *CRC Handbook of Chemistry and Physics,* 75th edition, CRC Press, 1994. Primary sources for the data in Table C-16 are D. D. Wagman *et al.,* "The NBS Tables of Chemical Thermodynamic Properties," *J. Phys. Chem. Ref. Data* v. 11, Suppl. 2, 1982; and M.W. Chase *et al.,* "JANAF Thermochemical Tables, Third Edition," *J. Phys. Chem. Ref. Data* v. 14, Suppl. 1, 1985.

Flyleaf (atomic weights): Commission on Atomic Weights and Isotopic Abundances, International Union of Pure and Applied Chemistry, *Pure and Applied Chemistry* v. 70, 237–257, 1998. © 1998 IUPAC

Epigraph quoted from Max Planck, *A Survey of Physical Theory,* translated by R. Jones and D. H. Williams, Dover (1993 reprint), p. 56.

INDEX

Doubly enumerated pages prefixed with "R" and a chapter number (e.g., R1.1) refer to the REVIEW AND GUIDE TO PROBLEMS for a particular chapter.

Italicized page numbers refer exclusively to an illustration, with no relevant accompanying text.

The symbol *n* indicates a footnote.

The symbol *t* indicates a table.

Data tables in Appendix C (pages A59 through A102) and glossary entries in Appendix D (pages A103 through A182) are indexed selectively, not exhaustively.

absolute temperature (Kelvin scale), 363–64, A103
absolute zero, 364, 383–84, 514, 737, A103
absorption (spectroscopic), 129, R4.4, 168–70, R5.4, 725–26, R20.2, A103
    color and, 705–7, R19.16, 724
    infrared, R20.15
    population difference and, R20.6, R20.8–9
acceleration, 18, 19, R1.3, A103
    due to gravity, 24, R1.5, A65*t*
    mass independence of, R1.8
    in mass spectrometer, 744–47
acceptor, A103
accuracy, 38, A103
acetic acid, A98*t*
    acidity, relative to halogenated forms, R8.12, 575
    in Brønsted-Lowry acid–base reactions, R3.13–14
    equilibrium constants, R12.6, R12.12–13
    ester formation, R8.11–12
    as prototypical weak acid, 579–605, R16.9–21
    synthesis from acetyl chloride, R8.12–13
acetylene (ethyne)
    formation and combustion, enthalpies of, 486–87
    Lewis structure and oxidation state, R2.12, R3.6
    as prototypical alkyne, 263
    valence bond model, 246–47, 248, 252–53, 274–76, R8.2
    VSEPR model, R2.12
acid, A103. *See also* acid–base interactions.
    Arrhenius, 83–86, R3.5, 577–605, R16.3–4, A15, A108

    selected examples, R3.13–14
    Brønsted-Lowry, 82–83, 84, R3.5, R16.6, A112
      selected examples, R3.13–14, 570
    diprotic, 86, R16.21–24, A123
    inorganic, A15–17, A17*t*
    ionization (dissociation), 83–84, 571, R16.3–4, A140
    ionization (dissociation) constant, R12.6, 577–78, R16.3–4, A98*t*–99*t*, A104
    Lewis, 64, 79–82, 84, A143
      affinity for pi electrons, 252, 277–78, 280, 283–84, R8.4
      analogy to oxidizing agent, 92, R3.4–5
      selected examples, R3.13–14, 570, 713–16
      as transition metal, 81, 688, 693–94, 713–16, R19.2
      as vacant orbital, R7.10–12
    monoprotic, A15, A150
    nomenclature, inorganic, A15–17
    organic (carboxylic acid), 284, R8.11, R8.12, R8.13
    oxyacid, R3.15, A155
    pH, 580, R16.4, R16.7–8, A156
    pK_a, 577, R16.4, A158
    polyprotic, 86, R16.21–24, A15, A159
    strong and weak, 85, 572–80, R16.2–3, R16.6–7, A174, A181
    triprotic, 86, R16.21, A178
acid–base interactions, 79–86, R3.4–5, 569–605, R16.1–34
    Arrhenius type, 83–86, R3.5, 577–605, R16.3–4, A108
      selected examples, R3.13–14

acid–base interactions (*continued*)
  autoionization (autodissociation) of water,
      581–83, R16.3, R16.5, R16.14, A109
  Brønsted-Lowry type, 82–83, 84, R3.5, R16.6,
      A112
    selected examples, R3.13–14, 570
  in buffer solution, 595–601, 602–4, R16.6,
      R16.12–14, R16.16, A113
  catalysis, examples of, 562, 570, 647, 671–73
  conjugate relationships, 570–77, 583–84,
      R16.1–3, R16.5, A117
  equilibria, 572–74, 577–605, R16.1–5,
      R16.7–24
  hydrolysis, 581, 584, 590–92, 594, 596, 599,
      604, R16.14–16, A137
  ionization (dissociation), 83–84, 571, R16.3–5,
      A140
  Lewis type, 64, 79–82, 84, A143
    analogy to redox, 92, R3.4–5
    as involving pi electrons, 252, 277–78, 280,
        283–84, R8.4
    as involving transition metals, 81, 688,
        693–94, 713–16, R19.2
    as involving vacant orbitals, R7.10–12
    selected examples, R3.13–14, 570, 713–16
  neutralization, R3.4, A151
    of Arrhenius acid and base, 84–85, R3.13,
        R16.6
    of Brønsted-Lowry acid and base, 83
    of Lewis acid and base, 80, 82
    precipitation as analog of, R3.14
    solvation as analog of, 93
    strong acid and strong base, 584–89
    weak acid and strong base, R12.13–14,
        589–92
  pH and pK_a, 577–80, R16.4, R16.7–8, A156,
      A158
  pOH and pK_b, 580–81, R16.4–5, A158
  redox, analogy to, 92, R3.5, 571, 611–12,
      R17.1–2, R17.4–5, R17.24–25
  salt, R3.4, A166
    of Arrhenius acid and base, 85, R13.13
    basic hydrolysis of, 594, R16.15–16
    of Brønsted-Lowry acid and base, 83
    buffering capacity of, 595–601
    of Lewis acid and base, 80–81, R3.4
    precipitate as analog of, R3.14
    solute–solvent complex as analog of, 93

  of strong acid and strong base, 585, 605
  strength and weakness of, 85, 572–80,
      R16.2–3, R16.6–7, A174, A181
  titration, 601–5, R16.16–21, A177
  water, as acid and base, 550, 577–84, R16.3, 713
acid halide (acyl halide), 278t, R8.12
acid ionization (dissociation) constant, R12.6,
    577–78, R16.3–4, A98t–99t, A104
actinide elements, 168, 199–200, 202, 785, A104
activated complex, 662–63, R18.2, A104. *See also*
    kinetics, transition state.
activation energy. *See under* kinetics.
active site, 556–58, 673, A104
activity, A104
  radioactive nucleus, R21.4, R21.17–18
  thermodynamic, 449n, R12.10
acyl chloride, hydrolysis of, R8.12
acyl halide (acid halide), 278t, R8.12
addition reaction, 283, 293, *294*, R8.5, A104
  hydrogenation, of ethylene, *100–1*, R8.5
  hydrohalogenation, R8.13
  polymerization, of ethylene, 344
adenine, 289–92, 343
adsorption, 673, A104
affinity, A104. *See also* electron affinity.
air, 300, 352
alchemy, 780, 785, A104
alcohols, 278t, 279, R8.11, A104
  nomenclature, A26–27
  oxidation to aldehyde, R8.6, R8.14
aldehydes, 278t, 283, A105
  as oxidation product, R8.6, R8.14
alkali metals, A105
  combining ratios, *58–59*, 59
  as *s*-block elements, 200–1
  valence configurations, 52, *53*, R2.3, R6.4
alkaline earth metals, A105
  electron affinity, lack of, 213
  as *s*-block elements, 200–1
  valence configurations, 52–54, R2.3
alkanes, 262, R8.2, A105
  bonding and isomerism, 263–71, R8.2–3,
      R8.3–4, R8.6–7
  nomenclature, A21–24
  reactions of, 276–77
alkenes, 263, R8.2, A105
  geometric isomerism, 276
  as Lewis bases, 277–78, R8.4

nomenclature, A25–26
structure and bonding, 271–74
alkyl group, A22, A24, A105
alkyl halide, 278t, 280, R8.13, A105
alkynes, 263, R8.2, A105
    as Lewis bases, 277–78
    nomenclature, A25–26
    structure and bonding, 274–76
allotropes, 324, A105
alloy, 432, A105
allyl cation, R7.16–18
alpha particle. *See under* radioactive decay.
aluminum
    atomic radius, compared with calcium, R6.9–10
    electron configuration, 194
    Lewis structure, 52
ambidentate ligand, 697, A105
amide, 278t, R8.10, 340, A105
amine, 278t, A105
amino acids, 287t, A106
    chiral (stereogenic) centers in, 286, R8.9–10
    DNA coding for, 291–92
    peptide formation (condensation), 288, R8.5
    structure, 286–87
amino group, 285, A106
ammonia
    as Brønsted-Lowry base, 83, R3.13–14
    Haber-Bosch synthesis of, 456–58, A134
    hydrolysis of, 581
    as Lewis base, 64–65, 80, 693–94, 716–17
    as prototypical weak base, R12.7–8
    VSEPR model of, 64–65
ammonium
    ammonium chloride, as neutralization product, 83
    ammonium hydroxide, as electrolyte, R3.15
    ammonium ion, as conjugate acid, R3.13–14, 571, A98t
    ammonium nitrate, as electrolyte, R3.15
amorphous solid, 311, *313*, 338, 345–47, R9.4–5, A106
ampere, R17.20, A106
amphipathic species, 552–55, 560, A106
amphoteric species, R3.13, A106
amplitude, wave, 105, R4.2, A106
analytical chemistry, A106
analytical methods. *See specific kinds, such as* chromatography; mass spectrometry; spectroscopy.

angstrom, 107, A106
angular momentum, 158, 470, A106
    orbital, of electron, 151, 158, 163–64, 168, 173, R5.2, R5.4
    spin, of electron, 173–74, R5.4–5
    spin, of neutron, 174, 753
    spin, of nucleus, 174, 729, R20.2, R20.3, R20.19, R20.24
angular momentum quantum number, 151, 154, 163, 165, 174–75, R5.2, A106
angular node. *See under* node.
anhydride, 278t, 285, R8.13, A106
anhydrous, A107
anion, 56, R2.3, A107
anisotropic, A107
annihilation, 767, R21.18–20, A107
anode, A107
    in electrolytic cell, 640–41, R17.19
    in galvanic cell, 618, R17.2
antibaryon, 778
antibody, 673, A107
antibonding orbital, 223, 229–33, R7.2, A107
antielectron (positron). *See* positron.
antigen, 529, A107
antilepton, 778–79
antimatter (antiparticle), 767, R21.18–20, A107. *See also specific kinds.*
antineutrino, in beta decay, 776–79, R21.9–10, A107
antiproton, R21.18–20, A107
antiquark
    as component of antibaryon, R21.20
    as component of meson, 789, *790*
aqueous equilibrium. *See under specific kinds, such as* acid–base interactions; solution.
area, 15, A107
argon
    boiling temperature, compared with helium, R9.15
    clusters of, 332–33
    critical temperature, 406, R11.8–9, A83t
    electron configuration, 194
    Lewis structure, 52, *53*, 57
aromatic molecules, 272–74, R8.2, A108
Arrhenius acids and bases. *See under* acid; acid–base interactions; base.
Arrhenius equation (or law). *See under* kinetics.
association, A108

atmosphere (unit), 353, A108
atom, A108
  as building block, 6–7, 46–54
  building-up (Aufbau) principle, 191–200, *201*, R6.3, R6.7
  dimensions and scale, 43, 47, R2.2, R4.5–7, 751–52, R21.1
  evidence for existence, 43–46, 120–23
  Lewis model, 50–54
  mass and size of, 47, 50, 207–10, R6.9–10
  Millikan experim.. (*e* and *m*), 120, *121*
  quantum mechanical model
    many-electron atom, 177–215, R6.1–5
    one-electron atom, 148–68, R5.1–5
  Rutherford experiment, 120–23, 754–55
  shells, 51–54, R2.3, 152, *153*, R5.2, R5.5–6, 192–200, R6.4–5
  substructure, 6–7, 46–54, R2.2–3, R2.6–7, 119–24
  Thomson experiment (*e/m*), 120, *121*
atomic mass, A77*t*–79*t*, A80*t*, A108. *See also* molar mass; relative mass.
atomic mass unit, 67, 767, A108
atomic nucleus. *See* nucleus.
atomic number, 7, 48, R2.2, A109
atomic orbitals, A109
  hybrid, 247–55, R7.3, 271–76, 709–13, R19.4, A137
    dsp^2, 712, R19.15–16, A124
    dsp^3, R7.12–13, R7.15, A124
    d^2sp^3, R7.12–15, 711–12, R19.12, A124
    sp, 252–53, A171
    sp^2, 251–52, 254–55, R7.10–12, R7.16–17, A171
    sp^3, 248–51, R7.9–10, R19.15, A171
    sp^3d^2, R19.10–11, A171
  in many-electron atom, 178–99, R6.1–5
  molecular orbital, as linear combination of, 227, R7.2–3
  in one-electron atom, 148–68, R5.1–4
atomic radius, 207–8, A68*t*–71*t*, A109
  lanthanide contraction, 209–10, A142
  periodic trends, 208–9, R6.9–10
atomic spectroscopy, 129–30, R4.4–5, 168–72, R5.4
atomic weight. *See* atomic mass. *See also* molar mass; relative mass.

Aufbau (building-up) principle, 191, R6.3, R6.7, A109
  atoms, 191–200, *201*
  complexes, 704–6
  diatomic molecules, 223–29, 233–36
Auger spectroscopy, *739*
autoionization (autodissociation), of water, 581–83, R16.3, R16.5, R16.14, A109
autoionization constant, water (ion-product constant), 582, R16.3, A84*t*, A109
average rate, 648, *649*. *See also* kinetics, rate of reaction.
Avogadro's law, 365–66, R10.2, A109
Avogadro's number, 68, R2.2, A65*t*, A109
azimuthal quantum number, 151, 154, 163, 165, 174–75, R5.2, A109

Balmer series, 170–72, A110
barometer, 354–55, A110
baryon number, conservation of, 777–78, R21.9–11, R21.20, A110
baryons, 777–79, 788, A110
base, A110. *See also* acid–base interactions.
  Arrhenius, 83–86, R3.5, 577–605, R16.4–5, A108
    selected examples, R3.13–14
  Brønsted-Lowry, 82–83, 84, R3.5, R16.6, A112
    selected examples, R3.13–14, 570
  hydrolysis of, 581, 584, 590–92, 594, 596, 599, 604, R16.14–16, A137
  ionization, 83–84, 571, R16.4–5, A140
  ionization constant, 577–78, R16.3, A98*t*–99*t*, A110
  Lewis, 64, 79–82, 84, A143
    affinity for transition metals, 81, 688, 693–94, 713–16, R19.2
    affinity for vacant orbitals, R7.10–12
    analogy to reducing agent, 92, R3.4–5
    as donor of pi electrons, 252, 277–78, 280, 283–84, R8.4
    selected examples, R3.13–14, 570, 713–16
  pK_b, 581, R16.4–5, A158
  pOH, 581, R16.4–5, A158
base ionization (dissociation) constant, 581, R16.4, A110
battery, 616–17, 637, R17.6–9, A110
becquerel, R21.4, A110
belt of (nuclear) stability, 774, A110–11

benzene
    bonding in, 254–58, 272–74
    disubstituted isomers, R8.8, 561–63
    insolubility in water, 551–52, R15.18
beryllium
    atomic radius, compared with nitrogen, R6.9
    diamagnetism of, R6.7
    diatomic molecule, instability of, 229
    effective nuclear charge, 204, *205*
    electron affinity, lack of, 213, *214*, R6.11
    electron configuration, 193
    Lewis structure, 52, *53*
beta particle. *See under* radioactive decay.
bidentate ligand, 697–99, R19.2, A110
    optical isomerism and, 698–99, R19.7–8
bilayer, A110. *See also* lipid bilayer.
bimolecular reaction, 676, 678, R18.4, A110
binary molecular compounds, A14–15, A110
binding
    enzyme–substrate, 342, 556, *557*, 558–59
    nuclear, 767–74, 782, R21.3, R21.5–7,
        A77*t*–79*t*, A80*t*, A110–11
    selective, 558–63, 564–67, 722
    thermodynamics of, 564–67
biochemistry, 2, 262, 285–92, 339–44, A111. *See
    also* organic chemistry.
biopolymers, 285–92, 339–44, A111
bisulfate ion
    acidity, relative to sulfuric acid, R16.7, A99*t*
    aqueous equilibrium, R16.23–24
    nomenclature, A9, A12*t*
body-centered cubic cell, R9.12–13, A111
Bohr atom, A111
Bohr radius, 161, A111
boiling, R9.14–15, 411, A111
    selective, 9, 270–71, 299
boiling point, 7, 9, A72*t*–75*t*, A111
    elevation of, 418, R11.4, A111
    noncovalent interactions and, 269–*70*,
        R9.14–15, 411–12
Boltzmann distribution, 389–91, A111
    Arrhenius equation and, 658, 663–64, R18.3
    as most likely macrostate, 497–501
    spectroscopy and, R20.3–9
    thermal equilibrium and, 522–26, 741
Boltzmann's constant, 379–80, R10.5, A65*t*,
    A111
bomb calorimeter, 476, A112

bond. *See* chemical bonding.
bond dipole moment, 62, R2.4, A112
bond dissociation energy or enthalpy, 229, 243,
    A82*t*, A112
bond distance (bond length), 219, 242, 245, R7.7,
    A82*t*, A112
bonding. *See* chemical bonding.
bonding orbital, 223, 229–33, R7.2, A112
bond order, 228, 243, R7.4, R7.5–7, A112
boron
    atomic radius, compared with bromine, R6.9
    as diatomic molecule, 233, *234*
    effective nuclear charge, 204–5
    electron affinity, compared with beryllium,
        R6.11
    electron configuration, 193
    ionization energy, anomaly in, 212
    Lewis structure, 52
boron trifluoride
    as Lewis acid, 65, 80, R7.10–12
    structure, 63–64
Boyle's law, 355–62, R10.2, R10.6–7, A112
    kinetic theory of, 370–74, R10.4
    undersea diving and, 361–62
Brackett series, R5.15
bromine
    atomic radius, compared with boron, R6.9–10
    selected reactions, R3.11–12, R12.20,
        R18.22–24
Brønsted-Lowry acids and bases. *See under* acid;
    acid–base interactions; base.
Brownian motion, 120, A112
buckminsterfullerene, 333–34, A112
buffer solution, 595–601, 602–4, R16.6,
    R16.12–14, R16.16, A113
building-up principle. *See* Aufbau principle.
butane, structural isomerism in, 269
1-butene
    rearrangement into 2-butene, R8.14–15
2-butene
    cis-trans isomerism in, R8.8–9
    synthesis from 1-butene, R8.14–15

calcium
    atomic radius, comparative, R16.9–10
    electron configuration, 195–96
    Lewis structure, *53*

calcium chloride, stoichiometry of, R2.7–8
calorie, 465–66, A113
calorimetry, R13.17–18, A113
  bomb calorimeter (constant volume), 476, A112
  "coffee cup" calorimeter (constant pressure), 478, *479*
  sample problems, R13.17–20
carbocation, 281–83, R8.6, 677, 680, A113
carbohydrate, A113
carbon. *See also* organic chemistry.
  atomic radius, compared with silicon, R6.9–10
  buckminsterfullerene, 333–34, A112
  carbon-12, as standard for atomic mass, 67
  carbon-13, *49*, R20.24
  diamond, 322–24, R9.6–7, R14.9, R15.19, A122
  as diatomic molecule, 234
  effective nuclear charge, 205
  electron configuration, 193
  graphite, *323*, 324, R9.6–7, R14.9, A133
  ionization energy, compared with nitrogen, R6.10–11
  Lewis structure, 52
  oxidation states, R3.6–7
  paramagnetism of, R6.7–8
  uniqueness of, 261–62
carbon dioxide
  Lewis structure and oxidation state, 59, R3.6
  phase diagram, *427*, 428
  selected reactions, R12.9, R12.20
carbonic acid, A98*t*
  as biological buffer, 600–1
  as diprotic acid, R16.21–23
carbon monoxide
  dipole–dipole interactions of, R9.14
  Lewis structure and oxidation state, 59, R3.6
  selected reactions, R12.9, R12.20
carbon tetrachloride
  aqueous solubility, compared with methyl chloride, R15.18
carbonyl group, 283, A113
carboxylic acid, 278*t*, 284, R8.11, R8.12, R8.13, 576–77, A113
catalysis, 77, 98–102, 671–73, R18.2–3, R18.18–19, A113
  acidic, 562, 570, 647, 671–73
  catalytic antibodies and, 673, A113
  clusters, possible role in, 331

enzymatic, 342, 556, *557*, 562
heterogeneous, 100–1, A136
homogeneous, 100, A136
solvent, role in, 647
surface, role in, 100–1, 647
transition state and, 100, 556, 671–73, R18.2–3
zeolitic, 101, 647
catalytic antibody, 673, A101
catalytic converter, 101
cathode, A113
  in electrolytic cell, 640–41, R17.19
  in galvanic cell, 618, R17.2
cation, 56, R2.3, A113
cell potential, 620, R17.3, A113
Celsius scale, 32, A114
centi, prefix, A64*t*
centimeter, 15, A114
ceramic, A114
CFSE. See crystal field stabilization energy.
chain reaction, 97, 783–85, A114
chalcogen, 54, R2.3, A114
chaos and chaotic processes, 445–46, A114
charge, electric. *See* electric charge.
charge transfer, coordination complex, 706, A114
Charles's law, 362–65, R10.2, R10.7–8, A114
chelate complex, 698–99
chelation effect, 716–18, R19.4, A114
chemical bonding, 8, 54–60, R2.4, 736. *See also*
    quantum mechanics, of molecule; molecular orbitals.
  combination, laws of, 45–46, 59–60, 69–70, R2.1–2, 120, R19.1
  coordinate covalent (dative) bond, 81, 694, R19.2, R19.11, A118
  covalent bond, 8, 56, R2.4, 220, 225, 227–28, 236, 239, 241–44, R7.1, R7.8–9, A119.
    See main entry covalent bond *for detailed breakdown into subentries.*
  double bond, 58, R2.4, 236, 243, 247, 252, 271–74, R8.4, A124
  ionic bond, 8–9, 56–57, R2.4, 239–40, *241*, R7.8–9, A140
  Lewis model, 55–56, 225, 228, 235–36, 242, 256, R7.4
  pi bond, 230–31, R7.3–4, 271–76, A157
  quantum mechanical model, 218–59, R7.1–18
  resonance, 256–57, A165
  sigma bond, 223, 229, R7.3, 271–76, A168

single bond, 55–56, R2.4, 243, 247, 248, A169
triple bond, 58, R2.4, 235, 243, 253, 271, 274–76, A178
chemical combination. *See under* chemical bonding; chemical reaction.
chemical energy, 609, 612–16. *See also* free energy, Gibbs.
chemical equation, 66, A114
    balancing, R2.12–14, R3.10–12
    redox, R3.10–12, 632–37, R17.6–9
chemical equilibrium, 435–37, 446–49, R12.3–4, 526, A114. *See also* equilibrium.
chemical formula, A115
    empirical, 72, R2.8–9, R2.12–13, A127
    molecular, 73, R2.8–9, R2.12–13, A149
chemical interactions, 8–9, 23
chemical kinetics. *See* kinetics.
chemical periodicity. *See under* periodicity.
chemical potential, 519, 527–28, A115
    in electrochemical cell, 621–23
    nonuniform concentrations and, 540–43, R15.4
chemical property, A115
chemical reaction. *See also specific kinds, such as* acid–base interactions; redox reactions.
    chirality and, 558–59
    classification of, 79–98, R3.12–14
    combination, laws of, 45–46, 59–60, 69–70, R2.1–2, 120
    endothermic, 439–41, 450–51, 483–87, 660, *661*
    exothermic, 441, 451, 483–87, 660, *661*
    general nature of, 75–79, R3.1–2, R3.8, R11.1–2, R20.13–14
    organic, 280–85, 287–88, 293–95, R8.5–6, R8.12–15. *See also* organic chemistry.
    spontaneity, 507–18, 612–16, 637–38, R17.1–2
chemical shift (chemical shielding), in NMR, R20.22, R20.23, A115
chemical symbol, A115
chemistry, general nature of, 1–12, 13–15, R1.1–2, R2.1–2, 75–79, R3.1–2, 203, R19.1, 722, *723*, R20.13–14, A115
chirality, 265–66, 286, R8.3, 558–59, 698, R19.6, A115
chloric acid, R3.15, A16, A98*t*
chlorine
    atomic radius, compared with silicon, R6.9–10
    as chloride ion, 56–57, R2.7–8

electron affinity, comparative, 215, R6.11
electron configuration, 194–95
    Lewis structure and electronegativity, 56–58
chloroacetic acid, relative acidity of, R16.7, A98*t*
chlorofluorocarbons (CFCs), 98
chlorous acid, R3.15, A16, A98*t*
chromatography, 747–50, A115
chromium, as anomaly in Aufbau sequence, 197–98
circuit, electric, 609–10, 612–16
cis isomer, 276, R8.3, 696, R19.5, A10, A115
cisplatin, *686*, 696
cis-trans isomerism, 276, R8.3, R8.8–9, R8.14–15, 695–96, R19.5–7, A10, A115
classical mechanics, 5–6, 16–21, R1.2–3, 104, 123, R4.1, 145–46, A115
    sample problems, R1.6–11
close packing, 328–29, A116
cluster, 330–34, R9.5, A116
codon, 292, A116
coenzyme, R8.14
coherent light, 433, A116
colligative properties, 415–20, R11.4–5, A116
    boiling-point elevation, 418, R11.4, A111
    deviations from ideality, 417–18, R11.14–16
    freezing-point depression, 418–20, R11.4, R11.14–16, A132
    molar mass and, R11.16, R15.11–12
    osmotic pressure, 542–43, R15.4–5, R15.11–12, A154
    vapor-pressure lowering, 413–16, R11.4, R11.12–14, A180
collisions, 20–21, A116
    elastic, R1.5–9, R1.10–11, A125
    ideal gas, role in, 370–74, 378–79, 384–85
    inelastic, R1.9–10, A138
    kinetic activation and, 646–47, 659–61
collision theory (in kinetics), 661–65, R18.4–5, A116
colloid, A116
color
    complementary color, 707, R19.16, A116
    in quark theory, 789, 791–92, R21.2, A116
    visible light and, 705–7, R19.16, 721, 724
combining capacity, 60, R2.2, A116
combustion, 92, 276–77, R8.14, 487, 517–18, A116

common-ion effect, 537, R15.8–9, 596, 599, A116

complementary colors, 707, R19.16, A116

complex. *See* coordination complex.

complex ion. *See* coordination complex.

compound, 8, A14–15, A117

compressibility, 308–9, 359–60, 408, 423, A117

compression or expansion, isothermal. *See* isothermal compression or expansion.

Compton effect, R4.9–10, A117

concentration, A117
  molality, 418, R11.12–14, A149
  molarity, R3.15–16, 301, R9.8–9, A149
  mole fraction, 301, 369, R10.11–12, 414, A150

condensation (organic reaction), 284–85, 288, 293, *294*, R8.5, R8.10–11, R8.13, 340, A117

condensation (phase transition), 10–11, 364, 407–9, 412, 413–14, R11.1–4, R11.8, 436, A117
  exothermic nature of, 451–52, 485
  sample problems, R11.12, R11.16–19

condensed phase. *See under* phase.

conductor, 324, 328, 433, A117

configuration, A117. *See also under* electron.

conformational isomerism (conformers), 266–68, R8.4, A117

conjugate acids and bases, 570–77, 583–84, R16.1–3, R16.5, A117

conjugation, R7.18, 274–76, 284, A117

conservation laws
  of angular momentum, 470, A117
  of baryon number, 777–78, R21.9–11, R21.20, A110
  of charge, 87, A118
    in beta decay, 777
    in redox reactions, 90, R3.10, 633, 635, R17.9
  of energy, 30–34, R1.2, A118
    as first law of thermodynamics, 461–69, R13.3–4, A130–31
    neutrino and, 777, A151
    selected examples, 39–40, 117–18, 127–28, 726, R20.11
    in theory of relativity, 763–66, R21.10
    time-reversal symmetry and, 469–70
  of energy and mass, 766, R21.10, A118

of lepton number, 777–78, R21.9–11, R21.20, A143

of mass, 67, R2.1, R2.16, 633, 766, R21.10

of momentum, 21, R1.2, A118
    neutrino and, 777, A151
    selected examples, R1.10–11, 370–72, 726
    spatial symmetry and, 470
    in theory of relativity, 763–66
  relativistic formulation, 763–66, R21.10
  sample problems, R1.6–11, R1.14–15
  symmetry and, 469–70, 763

constructive interference, *116*, 117–18, R4.3, 221, R7.2, A118

coordinate covalent bond (dative bond), 81, 694, R19.2, R19.11, A118

coordinates, *150*, A118

coordination complex, 81, 685–719, R19.1–16, A116
  aqueous equilibria, 714–16, R19.4
  chelate complex, 698–99
  chelation effect, 716–18, R19.4, A114
  color, 687, 705–7, R19.16
  crystal field theory, 700–9, R19.2–3, R19.9–10, R19.11, R19.13–15, R19.16, A120
  examples of, 81, 686–87
  formation constant, 715, R19.4, A102*t*, A131
  geometry, 689–91
  isomerism, A140
    geometric, 694–96, R19.5–7, A133
    linkage, 697, A143
    optical, 698–99, R19.6–8, A154
  kinetics, 718–19, R19.4
  labile versus inert, 718–19, R19.4, A138, A142
  ligand, 81, 686, 689, 691, 693–94, 697–98, 708, R19.2, R19.3, A19*t*, A143
  ligand field theory, 709, *710*, R19.4, A143
  metal
    d electrons, 692–93, R19.2, R19.8–9
    oxidation state, 691–92, R19.8, A76*t*
  nomenclature, A18–21
  sample problems, R19.5–16
  valence bond theory, 709–13, R19.4, R19.10–12, R19.15–16, A180

coordination compound, 688, A18–21, A118. *See also* coordination complex.

coordination number, 689, 691, R19.2, A118

coordination sphere, 689, R19.2, A118

copper, as anomaly in Aufbau sequence, 197–98
core electrons, A118
    as internal noble-gas configuration, 194, 213
    in Lewis model, 51, *53*, R2.3
    relationship to valence, 198, 211, 228
    screening effect of, 203–7, 737–38
    X rays and, 111, 211, 738–40, R20.2, R20.9–12
correspondence principle, Bohr, 136, 147, A118–19
corrosion, 92
coulomb (unit), 24, R1.4, A119
Coulomb force. *See* electrostatic force.
Coulomb's law, 24–25, A119
counterion, 688, A119
covalent bond, A119
    in Lewis model (electron-pair bond), 8, 56, R2.4
    as localized valence bond, 246–53, R7.9–15
    as molecular building block, 241–44
    in molecular orbital theory
        delocalized, 253–59, R7.16–18
        heteronuclear diatomic, 236–39
        homonuclear diatomic, 219–36
    as partially ionic, 239–40, R7.8–9
    as quantum mechanical concept, 220, 225,
        R7.1–4
covalent crystal (network crystal, network solid),
    322–24, R9.5, A119
cracking, petroleum, 277
crest, 106, R4.2, R4.3, A119
critical mass, 784–85, A119
critical opalescence, 423
critical phenomena, 422–24
critical point, 422–23, R11.3, A119
critical pressure, 422–23, R11.3, A119
critical temperature, 405–6, 412, R11.3, A83*t*,
    A119
    noncovalent interactions and, R11.8–9
cross-linking, 339, 9.21–22, A119
crown ether (18-crown-6), 559–60, 565–66,
    R15.2, A119
crude oil (unrefined petroleum), 102, 270–71,
    277, 300
crystal (crystalline solid), 311–12, R9.4, A119
crystal field stabilization energy (*CFSE*), 704–5,
    A120
crystal field theory, 700–9, R19.2–3, R19.9–10,
    R19.11, R19.13–15, R19.16, A120
    octahedral field, 701–6, R19.3, R19.9–10,
        R19.13, R19.16

square planar field, R19.13–16
    tetrahedral field, 707, *708*, R19.13, R19.15
crystal lattice. *See* lattice.
cubic centimeter, *15*, 16, A120
cubic close packing, 328–29, A116
cubic unit cell, 316–19, R9.11–13
current, electric, R17.20, A120
cycle, wave, 106, A120
cycloalkanes, 270, R8.2, A120
cylindrical symmetry, 218, 225, *237*, 239
cytosine, 289–92, 343

Dalton's law of partial pressures, 368–69, R10.2,
    R10.11–12, R11.4, A120
dative bond. *See* coordinate covalent bond.
daughter nucleus, 774, A120
d block, 202, R6.4, A120
de Broglie wavelength, 133–36, 141, R4.4,
    R4.11, 147, 765, A120
debye, R7.7, A121
decomposition reaction, 437, R12.6–7, A121
degeneracy, 182–83, 190–91, 193, R6.2, R6.3,
    A121
degrees of freedom, 218, 380, R13.7, 743, A121
    entropy and, R14.9
    heat capacity and, R13.14
    kinetic activation and, R18.3
dehydration, A121
dehydrohalogenation reaction, R8.13
delocalization, 253–58, R7.3, R7.16–18, 272–76,
    A121
density, A121
    material, 7–8, 299, R9.6–8, R9.11–12,
        R10.10, A72*t*–75*t*, A84*t*
    quantum mechanical. *See* probability density.
deoxyribonucleic acid. *See* DNA.
dependent variable, A47
deposition, 425, A121
destructive interference, *116*, 117–18, R4.3, 221,
    R7.2, A121
detailed balance, principle of, 682–84, R18.6,
    A122
determinism, 5, A122
deuterium, 48, R20.19, A122
dew, R11.12, R11.18–19
dextrorotatory enantiomer, 266, A122
diamagnetism, selected examples, A122
    atoms, 192, 193, R6.7–8

diamagnetism, selected examples (*continued*)
  coordination complexes, R19.9–16
  in crystal field theory, 704–8, R19.3
  molecules, R7.6, R7.18
  in valence bond theory, 713
diamond, 322–24, R9.6–7, R14.9, R15.19, A122
diatomic molecule, 60, *61*, R2.3, 219–39,
      R20.15, A122
dichloroacetic acid, relative acidity of, R16.7, A98*t*
dichloromethane, oxidation state, R3.6
diffraction, 113, *116*, R4.3, R4.8–9, A122
  of particles (e.g., electron), 134–35, R4.12–13,
      A126
  X-ray, 45, *46*, 118–19, R4.9, A182
diffusion, 378–79, A122
dimension, 34, A122
dimensional analysis, 35, A122
dimer, 406, A122
dimerization reaction, 437, 456, 517
dinitrogen pentoxide, unimolecular decomposition
      of, R18.9–13
dipeptide, 288, A122
dipole–dipole interaction, 305, R9.3, 396, A123
dipole moment, electric, 60–62, *237*, 238,
      R7.7–9, 304, 396, R20.2, A125
diprotic acid, 86, R16.21–24, A123
Dirac equation, 175–76, 767
disorder. *See under* entropy.
disorder–order transition, A123
  examples and analogies, 428–33
  paramagnet–ferromagnet, 429–30
  solution–solid, 531–35, R15.2–4
  vapor–liquid, 11–12, 409, 531–33
dispersion interaction. *See* noncovalent interac-
      tions, London dispersion.
disproportionation reaction, 92, A123
dissociation reaction, 84, 437, 456, 483–85, 571,
      R16.3–4. *See also* acid–base interactions.
dissolution reaction, 93–94, R3.5, R12.7,
      530–31, 546–48, R15.1–2, A123
distributions, 490–501, R14.6–8, A123. *See also*
      statistics and statistical mechanics.
D-L convention, 286, 558–59, A10
DNA (deoxyribonucleic acid), A124
  primary structure, 289–92, 340, A160
  secondary and tertiary structure, 2–3, 342–44,
      A167, A176
  X-ray diffraction pattern, 45, *46*

dodecane, mass spectrum of, 746
dodecene, mass spectrum of, 746
donor, A124
d orbital, 166–68, R19.2, A124
dot diagram. *See* Lewis structure.
double bond, 58, R2.4, 236, 243, 247, 252,
      271–74, R8.4, A124
double helix. *See* DNA.
dsp^2 hybrid orbital, 712, R19.4, R19.15–16, A124
dsp^3 hybrid orbital, R7.12–13, R7.15, A124
d^2sp^3 hybrid orbital, R7.12–15, 711–12, R19.4,
      R19.12, A124
dynamic equilibrium. *See under* equilibrium.
dyne, R1.17

eclipsed conformation, 267, *268*, A124
EDTA (ethylenediaminetetraacetate anion), 698
effective nuclear charge, 184, R6.3, A124
  periodic trends, 203–7, R6.4
effusion, 378–79, A124
eka-aluminum, 203
elastic collision. *See under* collisions.
electrical potential energy. *See under* potential
      energy.
electrical work. *See under* work.
electric charge, 4, 24–25, R1.4, A114
  conservation of, 87, A118
    in beta decay, 777
    in redox reactions, 90, R3.10, 633, 635,
      R17.9
  as source of electromagnetic field, 27, 105,
      722–23, 785–88, R21.2
electric circuit, 609–10, 612–16
electric current, R17.20, A120
electric dipole moment. *See* dipole moment,
      electric.
electric field, 27–28, R1.5, A125
electric force. *See* electromagnetic force; electro-
      static force.
electrochemical cell, A125. *See also* electrolytic
      cell; galvanic cell.
electrochemistry, 607–42, R17.1–21, A125. *See
      also* redox reactions.
  cell potential, 620, R17.3, A113
  chemical energy and, 609, 612–16
  electrolysis, 637–42, R17.5, R17.18–21, A125
  electrolytic cell, 639–42, R17.5, R17.19, A125
  Faraday constant, 614, R17.3, A65*t*, A130

galvanic cell (voltaic cell), 616–21, R17.1–2, A132
  free energy and, 621–24, R17.3–4, R17.11–18
  half-reactions, 624–32, R17.4–5, A134
  Nernst equation, 623–24, R17.4, R17.16–18, A151
  reductants and oxidants, strength of, 610–12, 625–32, R17.9–11
  sample problems, R17.6–21
  standard cell potential, 623–24, R17.4, A172
    equilibrium constant and, R17.15–16
    free energy and, 623–24, R17.4, R17.11–15
    spontaneity and, R17.14–15
  standard electrode potential (oxidation, reduction), 625–30, A173
  standard hydrogen electrode, 625–30, R17.5, A172
electrode, 618, A125
electrolysis, 637–42, R17.5, R17.18–21, A125
electrolyte, 93, R3.15, A125
electrolytic cell, 639–42, R17.5, R17.19, A125
electrolytic solution, 93, R3.15
electromagnetic field, A125. *See also* electromagnetic force; electromagnetic radiation.
  particulate nature of (photons), 124–29, R4.3–4, R4.9–11, 723–26, R20.1–2
  wavelike nature of, 105–11, 118, R4.2, 727–40
electromagnetic force (or interaction), 22–23, 24–25, 105, A125
  chemistry and, 23, R3.1, 301–3, R9.1–2, 722–23
electromagnetic radiation (or waves), 105–11, R4.2, 727–40, A125, A126
  interaction with matter, 44–45, 124–30, R4.3, R4.5, 169–70, R5.4, 721–43, R20.1–2
  particulate nature of (photons), 124–29, R4.3–4, R4.9–11, 723–26, R20.1–2
  propagation of, in vacuum, 107, 759, 762–63
  sample problems, R4.5–11
  spectrum, 45, 110–11, R4.3, 725, 727–40, R20.1–2, A126
  speed of light, 107, R4.2–3, 727, A65t
    constancy of, 757–59, 761, 763, R21.3
    time and distance, examples of, R4.7–8
  wave–particle duality, 137, R4.4–5, A181
electromagnetic spectrum, 45, 110–11, R4.3, 725, 727–40, R20.1–2, A126

electromotive force, 614, A126. *See* cell potential.
electron, 6–7, 46–47, R2.2, A65t, A126
  configuration, A126
    in atoms, 192–200, R6.3, R6.8–9, A68t–71t
    in diatomic molecules, 225–29, 233–39, R7.5–7
  density. *See* probability density.
  diffraction, 134–35, A126
  Millikan experiment (*e* and *m*), 120, *121*
  spin, 173–76, R5.4–5, 181–82, 767, A126, A171
  Thomson experiment (*e/m*), 120, *121*
electron affinity, 89, 213–15, A126
  periodic trends, 214–15, R6.11
electron capture. *See under* radioactive decay.
electronegativity, 58, R2.4, 244–45, A126
electron–electron repulsion, 178–79, 186–91, 197, R6.1–2, R6.6
electron microscope, 134, A126
electron-pair bond, 8, 56, R2.4. *See also* covalent bond.
electron paramagnetic resonance (EPR), 732–33, R20.24, A126
electron volt, R1.14, R21.19
electrophile, 284, A127
electrophilic substitution, 295, A127
electropositive species, 58, A127
electrostatic force (Coulomb force), 24–25, R1.4, 154, 743, A127
  Coulomb's law, 24–25, R1.4, A119
  sample problems, R1.11–14
elementary reaction, 674–78, R18.4, A127
elements, 7, 46–54, A127
  building blocks, role as, 46–54
  electronic properties, A68t–71t
  isotopes, A77t–79t, A80t
  names, origin of, A2–7
  physical properties, A72t–75t
elimination reaction, 293, *294*, R8.5, R8.13, A127
emission (spectroscopic), 129, R4.5, 168–72, R5.4, 724, 726, R20.2, A127
  Balmer series, 170–72, A110
  Lyman series, 170, *171*, R5.9, A144
  Paschen series, 170, *171*
  Pfund and Brackett series, R5.15
empirical formula, 72, R2.9, R2.12–13, A127
enantiomers, 265–66, R8.3, 558–59, 698, A127

endothermic process, 439–41, 483–87, A127
  activation energy of, 660, *661*, R18.19–20
  temperature and spontaneity, 512–13, 516–17
  vaporization, 450–51, 485, *486*, A180
energetic disorder, 497–501
energy, 4, 12, 28–32, 33–34, R1.1–5, A127
  conservation of, 30–34, R1.2, A118
    as first law of thermodynamics, 461–69,
      R13.3–4, A130–31
    neutrino and, 777, A151
    selected examples, 39–40, 117–18, 127–28,
      726, R20.11
    in theory of relativity, 763–66, R21.10
    time-reversal symmetry and, 469–70
  kinetic, 30–31, 39, R1.4, 462–64, A142
  of one-electron atom, 153–56, 169–72, R5.3–4,
    R5.7–9
  potential, 28–31, 39, R1.5–10, R1.14,
    462–64, A159
  quantization of, 126–27, 129–30, R4.3, R20.1–2
  relativistic, 763–66
  sample problems, R1.6–11, R13.7–12
  as state function, 468, 474, *475*, R13.4–5, A173
  thermal, 380–84, R10.4, 464, R13.7, A176
energy level. *See* energy, quantization of.
enthalpy, A127–28
  of activation, 666–68, A128
  of binding (host–guest), 564–67, A128
  chelation and, 716–17, R19.4
  of combustion, 487
  endothermicity and exothermicity, 483–87,
    R13.5
  as extensive property, 477, 483, R13.21, A130
  of formation, standard, 479–80, A85t–92t,
    A128, A172
  of freezing, 506
  Hess's law, 481–83, R13.5–6, R13.20–22, 544,
    A135
  of hydration, R15.16
  internal energy, relation to, 477–78,
    R13.15–17, A139
  of melting (fusion), 506
  of reaction, standard, 480–83, R13.5–6, A128,
    A172
  sample problems, R13.7–22
  of solution, 543–46, R15.1–2, R15.13–16,
    A128
  as state function, 477–79, R13.5–6, A173

thermochemistry, role in, 483–87, A176
  of vaporization, 485, 516, R14.18–19, A84t,
    A128
entropy, A128
  of activation, 666–68
  of binding (host–guest), 564–67, A128
  chelation and, 716–18, R19.4
  disorder and, 12, 76–77, 409, 430, 437–42,
    491–501, 507–9
    energetic disorder, 497–501
    positional disorder, 491–96, A159
  as extensive property, 501–2, R14.9, A130
  of freezing, 506
  global entropy, as driving force, 12, 76–77,
    409, 437–42, 495, 509–11, R14.4–5. *See
    also* free energy, Gibbs.
  of melting, 506
  of phase transition, 505–7, R14.18–19
  of reaction, A128–29
  sample problems, R14.9–15
  of solution, 543–46, R15.1–2, R15.15–16
  standard (third-law) entropy, 514–16, R14.5,
    A85t–92t, A172
  as state function, 474–75, 505–7, 514, A173
  statistical interpretation of, 501–5, R14.3
  thermodynamic interpretation of, 474–75,
    502–5, R14.3–4
  third law of thermodynamics, 514, R14.3, 737,
    A177
  of vaporization, R14.18–19
enzyme, 102, R8.14, A129
  substrate, binding of, 342, 556, *557*, 558–59,
    A129, A175
EPR. *See* electron paramagnetic resonance.
equation, chemical. *See* chemical equation.
equation of state, A129
  equilibrium constant, as analogy, 446–47,
    R12.3–4
  ideal gas, 366–69, R10.2, R10.8–11, 400, A137
  real gas (van der Waals equation), 402–4,
    R11.2–3, R11.5–8, A83t, A180
equilibrium, 435–60, R12.1–22, A129
  acid–base, 572–74, 577–605, R16.1–5,
    R16.7–24
  approaches to
    chaotic, 445–46, A114
    monotonic, 443, *444*, A150
    oscillatory, 443–45

chemical, 435–37, 446–49, R12.3–4, 526, A114
complex formation, 714–16
detailed balance, principle of, 682–84, R18.6, A122
drive to, 437–42, 447–49, 511–12
dynamic equilibrium, R12.1–3, A124
    chemical equilibrium, as example, 435–37, R12.3, A114
    detailed balance, principle of, 682–84, R18.6, A122
    dissolution–precipitation, as example, 531–33, R15.3–4
    evaporation–condensation, as example, 11–12, 409–10, R11.3–4
    opposing rates, constraint on, 683–84, R18.24–26
    osmotic equilibrium, as example, 540–43, R15.4
    thermal equilibrium, as example, R12.2–3, 523–26, A176
equilibrium constant, 446–52, 460, A129. *See also main entry* equilibrium constant.
in gas phase, 452–56
heterogeneous, 437, A136
homogeneous, 437, A136
Le Châtelier's principle, 454–58, R12.5, A142
    common-ion effect and, 537, R15.8–9, 596, 599, A116
    composition, changes in, 457–58, R12.18–19
    pressure, changes in, 452–55, R12.20–21
    temperature, changes in, 456–57, R12.21–22, 512–13, 517, R15.13–15
liquid–vapor, 11–12, 409–11, R11.9–12, 436
local nature of, 438–39
macroscopic properties of, 435–36, 441, R12.2–3, 531, 533, 540–41, R15.4
mass action, law of, 446, R12.3–4, A145–46
microscopic reversibility of, 682–84, R18.24–26, A148
osmotic, R15.4
pseudoequilibrium, 666–67
reaction quotient, 446–49, 452–53, R12.4–5, 520, A164
sample problems, 458–60, R12.5–22
solute–solution, 413–20, 531–40, 543–46, R15.2–4, R15.5–10
statistical interpretation, 493–94, 507, 522–26

stresses to, 452–58, R12.5, R12.18–22, R15.13–15. *See also main entry* stressed equilibria.
thermal, R12.3, 466, 522–26, 540, R15.4, 741–43, A176
time independence of, 435–36, R12.2–3, 540, R15.4
vapor pressure, 410–11, 449–52, A180
equilibrium constant, A129
    algebraic manipulation of, R12.8–13
    equation of state, analogy to, 446–47, R12.3–4
    free energy and, 519–22, R14.5–6, R14.16–17
        in acid–base reaction, 577, 579–80, R16.1–2
        in complex formation, 714–16
        in dissolution–precipitation reaction, 543–46, R15.2–3, R15.16–17
        in electrochemical cell, 621–24, R17.3–4, R17.11–18
        in host–guest binding, 564–67
    K_c versus K_p, 460, R12.9–11
    mass action, law of, 446, R12.3–4, A145–46
    pressure independence of, 453
    reverse reaction, form for, 452, R12.8–9
    sample problems, R12.5–18
    temperature dependence of, 450–52, 456–57, R15.14–15
    vapor pressure as, 410–11, 449–52
equilibrium distance, 396, A129
equipartition theorem, 380, R10.4–5, A129
equivalence point, 602, A129
erg, R1.17
ergosterol, NMR spectrum, 731
error. *See* experimental error.
ester, 278*t*, 285, R8.11, A129
    nomenclature, A26–27
ethane
    as addition product, 100–1, R8.4–5
    boiling temperature, compared with ethanol, R9.15
    conformational isomerism in, 266–67
    elimination reaction, R8.5
    entropy, compared with methane, R14.9
    formation and combustion, enthalpies of, 486–87
    geometric structure, 266
    Lewis structure and oxidation state, R3.6
    radical production, R18.24–26
ethanol (ethyl alcohol)
    boiling temperature, compared with ethane, R9.15

ethanol (ethyl alcohol) (*continued*)
    infrared spectrum, *R20.15*, R20.17
    mass spectrum, R20.12–13
    NMR spectrum, R20.23–24
    solubility in water, compared with heptanol,
        R15.18
ether, 278*t*, 559–60, 566, A129
ethyl bromide, NMR spectrum, 731
ethylene (ethene)
    addition reaction, 100–1, R8.4–5
    cis-trans isomerism in, 276, R8.3
    as elimination product, R8.5
    formation and combustion, enthalpies of, 486–87
    Lewis structure and oxidation state, R3.6
    polymerization of, 344–49
    as prototypical alkene, 263
    valence bond model, 246–47, 251–52, 271–72,
        R8.1–2
ethylenediaminetetraacetate anion (EDTA), 698
evaporation, 11, 409–10, 413–14, R11.3–4, 436,
        449–52, A129
    sample problems, R11.9–11
exchange reaction, 95, A129
excited state, 168–72, A129. *See also* spectroscopy.
exclusion principle, Pauli, 181–82, R6.1–2, A155
    chemistry and, 191, R6.3, R7.2, 301, 722
    sample problems, R6.6
exothermic process, 441, 451, 483–87, A130
    activation energy of, 660, *661*, R18.19–20
    condensation, 451–52, 485, A117
    temperature and spontaneity, 512–13, 517–18,
        R14.15–16
expansion, isothermal. *See* isothermal compression
        or expansion.
experiment (scientific method), 14, A130
experimental error
    random, 38, A163
    systematic, 38, 39, A175
exponent, A34–39
extensive property, 7, 468, *469*, A130
    examples of
        enthalpy, 477, 483, R13.21
        entropy, 501–2, R14.9
        free energy, R17.11–12
        internal energy, R13.12–13

face-centered cubic cell, R9.11–13, A130
factor-label method, 35–36, A130

Fahrenheit scale, 32, A130
fallout, radioactive, 783
Faraday constant, 614, R17.3, A65*t*, A130
fat, 554, A130
fatty acid, 552, 553, A130
f block, 202, R6.4, A130
femto, prefix, A64*t*
femtometer, 107, A64*t*
    as nuclear dimension, 769–70
ferrocene, *687*
ferromagnetism, 429–30, A130
field, general concept, 27, R1.5, 785–88, A130
first law of thermodynamics, 468–69, *470*,
        R13.3–5, R13.7–12, A130–31
first-order reaction, 651, R18.4, 774–75, R21.4,
        A131
    concentration versus time, 653–56
    half-life, 654–56, R18.4, 774–75, R21.4, A134
    sample problems, R18.9–13, R21.14–18
    unimolecular process, as prototype, 677
fission. *See* nuclear fission.
flavor, in quark theory, 789–91, A131
fluid, 423, A131
fluorine
    as diatomic system
        anion and cation, bonding in, R7.5–7
        neutral species, bonding in, *234*, R7.5–7
        as oxidizing agent, 630–31
    effective nuclear charge, 205
    electron affinity, 213–15
    electron configuration, 193
    ionic radius, comparative, R6.10
f orbital, 168, A131
force, 4, 18–19, 20, 21–25, R1.2–3, A131. *See
        also specific kind of force.*
formal charge, R3.8–10, A131
formation constant, of complex, 715, R19.4,
        A102*t*, A131
formula weight, A131
free energy, Gibbs, 509–13, 528, R14.4–6,
        A131, A133
    of activation, 666–68
    available work and, 607–9, 614–16, R17.3–4
    as driving force, 12, 76–77, 441–42, 509–11,
        519–22, R14.17–18, 621–24
    equilibrium constant and, 519–22, R14.5–6,
        R14.16–17
        in acid–base reaction, 577, 579–80, R16.1–2

in complex formation, 714–16
in dissolution–precipitation reaction, 543–46, R15.2–3, R15.16–17
in electrochemical cell, 621–24, R17.3–4, R17.11–18
in host–guest binding, 564–67
as extensive property, R17.11–12, A130
of formation, standard, 513–14, R14.5, A85t–92t
global entropy and, 509–11, R14.4–5
of reaction, 516–18, A131
of redox reaction (and electrochemical cell), 610–16, 621–24, R17.3–4, R17.11–18
of solution, 545–46, R15.1–2, R15.16–17
as state function, 514, A173
temperature and, 512–13, R14.15–16
free radical. *See* radical.
freezing, 409, 418–20, 425, R11.4, A131
  enthalpy of, 506
  entropy of, 506
  sample problems, R11.14–16, R11.16–18
  water, expansion during, 426
freezing-point depression, 418–20, R11.4, R11.14–16, A132
frequency, 107, R4.2, A132
frequency factor, 662, R18.4, A132
friction and dissipative forces, 39, 464
frost, R11.18–19
fuel cell, R17.29
fullerenes, 333, A132
function, mathematical definition, A47
functional groups. *See under* organic chemistry.
function of state. *See* state function.
fundamental forces, 21–25, R1.3–4, A132
fundamental frequency, R5.10, A132
fusion (melting). *See* melting.
fusion (nuclear). *See* nuclear fusion.

galvanic cell (voltaic cell), 616–21, R17.1–2, A132
  free energy and, 621–24, R17.3–4, R17.11–18
gamma radiation. *See under* radioactive decay.
gas. *See* ideal gas; real gas; states of matter, gas.
gasoline, 8, 101, 277, *750*
genetic code, 291–92
geometric isomerism, A133
  alkenes, 276, R8.3, R8.8–9, R8.15, A10
  coordination complexes, 694–96, R19.5–7
Gibbs free energy. *See* free energy, Gibbs.

giga, prefix, A64t
gigahertz, 107
glass, A133. *See also* amorphous solid.
globular protein, 342, A133
glucose
  combustion of, 487
  as nonelectrolyte, R3.15
gluons, 791, R21.2, A133
Gouy balance, R6.8
Graham's law, 379–80, A133
gram, 20, A133
graphite, *323*, 324, R9.6–7, R14.9, A133
gravitational constant, 23, A65t, A134
gravitational field, 29–30, R1.5, A134
gravitational potential energy. *See under* potential energy.
gravitational work. *See under* work.
gravity, 22, 23–25, 29–30, R1.4, R1.5, A134
ground state, 156, 169, R5.3, 191, R6.3, 740, R20.3–6, A134
group (column in periodic table), 52, R2.3, R6.4, A134
guanine, 289–92, 343
guest–host chemistry. *See* host–guest chemistry.

Haber-Bosch process, 456–58, A134
half-equivalence point, R16.18
half-life, 654–56, A134
  first-order process, 654–56, R18.9–13
    radioactive decay, 774–75, R21.4, R21.14–18, A80t
  second-order process, 656, R18.13–16
half-reactions, 624–32, R17.4–5, A134
halide. *See* alkyl halide.
halogen, 54, R2.3, R6.4, A134
harmonic frequency, R5.10, A135
HDPE. *See* high-density polyethylene.
heat, 31–32, 39, 464–67, R13.3, A135
  calorimetry, 476–78, 479, R13.17–20, A113
  at constant pressure, 477–78, R13.5
  at constant volume, 475–77
  enthalpy and, 477–78, A127–28
  macroscopic interpretation of, 464–67, R13.3
  mechanical equivalent of, 465–66
  microscopic interpretation of, 31–32, 39, 464
  path dependence of, 474, *475*
heat capacity, 476–78, R13.3–4, A135
  of ideal gas, R13.13–15

heat capacity (*continued*)
  molar, at constant pressure, 478, 483, A84*t*, A85*t*–92*t*, A149
  molar, at constant volume, 476–77, A149
  sample problems, R13.13–15, R13.17–20
  specific, 477, A170
heavy hydrogen (deuterium), 48, A122
Heisenberg uncertainty principle. *See* uncertainty principle, Heisenberg.
helium
  boiling temperature, compared with argon, R9.15
  critical temperature, comparative, 406
  diamagnetism of, R6.7
  diatomic molecule, nonexistence of, 224–25
  electron configuration, 187–88, 192
  electron–electron repulsion in, 178–79, R6.1, R6.3
  helium-4, nuclear binding of, 768–69
  helium ion, diatomic, R7.5
  helium ion, monatomic, R5.9
  inert nature of, 51, 192, 202
  ionization energy, R6.5–6
  isotopes, 48, *49*
  Lewis structure, 51
  orbital approximation, 178–81
helix, 289, 340–44, 432, A135
heme, *687*
Henderson-Hasselbalch equation, 598, R16.6, A135
  approximate nature of, R16.14, R16.18–19
  use of, R16.12–13, R16.17–18
Henry's law, 534–35, R15.12, A135
heptanol, comparative solubility in water, R15.18
hertz, 107, R4.2, A135
Hess's law, 481–83, R13.5–6, R13.20–22, 514, 544, A135
heterogeneous, A135–36
  catalysis, 100–1, A136
  equilibrium, 437, A136
  material, 299–300, A135–36
heteronuclear, A136
  bond, R2.4
  diatomic molecule, 60, *61*, 236–39, R7.7
hexagonal close packing, 328–29, A136
hexane, structural isomers, 270, R8.6–8
hexanol, infrared spectrum, 736
high-density polyethylene (HDPE), 345–47

high-spin complex, 705, *706*, R19.3, R19.10, R19.13, R19.15, A136
homogeneous, A136
  catalysis, 100, A136
  equilibrium, 437, A136
  material, 299–301, A136
homonuclear, A136
  bond, R2.4
  diatomic molecule, 60, *61*, 219–36
host–guest chemistry, 529, 555–67, A136
  crown ether (18-crown-6), 559–60, 565–66, R15.2, A119
  enzyme–substrate interactions, 342, *556*, *557*, 558–59, A129, A175
  molecular recognition, 558–63, 722, A149
  preorganization, 556, 565–67, R15.2, A160
Hund's rule, 191, R6.3, R6.6, A136
  in atoms, 192, 193, 197, R6.7
  in complexes, R19.12
  in diatomic molecules, 233, R7.2, R7.6
hybrid atomic orbital. *See under* atomic orbitals.
hydration, 86, 95–96, 531, 549, 714, A137
hydride ion, 91, 293–94, R8.6, R16.7, A137
hydrocarbons, R2.8, 262–76, R8.2, A137
hydrochloric acid, A98*t*
  acidity, compared with hydrofluoric acid, R16.7
  boiling temperature, comparative, R9.14–15
  nomenclature, A15
  as prototypical strong acid, 84–85, 585–89
hydrofluoric acid, R16.7, A98*t*
hydrogen
  as diatomic molecule
    bonding, 55–56, 219–25
    critical temperature, comparative, R11.8–9
    electronic structure, atomic, 46–48, 148–68, 192
    emission spectrum, atomic, *130*, 168–72, R5.9
    selected reactions, 100–1, 452–54, R12.9, R12.20–21, R18.6–7, R18.22–24
hydrogen bonding, 305–6, R9.3–4, A137
  in aqueous solution, 548, R15.18
  in DNA, 342–44
  in enzyme–substrate complex, 558
  in protein structure, 340–42, 554–55
  in water, 306, 326, *327*, R9.14, 426, *428*
hydrogen chloride, 56. *See also* hydrochloric acid.
hydrogen fluoride

boiling temperature, compared with hydrogen chloride, R9.14–15

as hydrofluoric acid, R16.7, A98*t*

as prototypical heteronuclear diatomic molecule, 238–39

hydrogen iodide, formation equilibrium, R12.14–16, R12.18–19

hydrogen sulfate ion. *See* bisulfate ion.

hydrohalogenation reaction, R8.13

hydrolysis, 581, 584, 590–92, 594, 596, 599, 604, R16.14–16, A137

hydronium ion, 84, 86, 577, R16.3, R16.7–8, A137

hydrophilic and hydrophobic effects, 550–55, 560, A137

hydroxide ion, 82, 83, 577, R16.3, R16.7–8, A137

hypertonic solutions, R15.11, A141

hypochlorous acid, R3.15, A16, A98*t*

hypoiodous acid, relative strength, R16.7, A99*t*

hypothesis (scientific method), 14, A137

hypotonic solutions, R15.11, A141

ice, 326, *327*, R9.14, 426–*28*

icosahedron, 332–33, A137

ideal gas, 351–91, R10.1–15, A137
  assumptions, 351–52, R10.2–3, 399–400, 404–5, R11.4
  Avogadro's law, 365–66, R10.2, A109
  Boyle's law, 355–62, R10.2, R10.6–7, A112
  Charles's law, 362–65, R10.2, R10.7–8, A114
  compressibility, 308–9, 359–60, 408, A117
  Dalton's law of partial pressures, 368–69, R10.2, R10.11–12, R11.4, A120
  energy, 375–76, 379–84, R10.4, R10.12–13
  equation of state, 366–69, R10.2, R10.8–11, 400
  interparticle potential, lack of, 399–400
  isotherms, 356–58, 359, 399–400, 405, A141
  kinetic theory of, 369–91, R10.4–5, A142
  Maxwell-Boltzmann distribution, 386–89, R10.5, R10.14–15, A146
  molar volume at STP, 367, R10.9, R10.10–11, A65*t*
  pressure, 353, R10.1–2
    measurement of, 353–55
    statistical interpretation of, 370–74

root-mean-square speed, 376–78, R10.4, R10.13–15, A165

sample problems, R10.6–15

temperature, 363–65, 374–78, R10.1–2, R10.7–8

ideal solution, 412–20, R11.4–5, A137
  assumptions, 412–13, 415–16, R11.4
  colligative properties, 415–20, R11.4–5, R11.12–14, A116
  deviations, 417–18, R11.14–16
  Raoult's law, 415, R11.4, A163
  vapor pressure, 413–16, R11.4, R11.12–14, A180

immiscibile liquids, R15.18, A138

incoherent light, 433, A138.

independent variable, A47

indeterminacy principle. *See* uncertainty principle, Heisenberg.

indicator, 602, A138

induced dipole–dipole interaction, 306–8, 396–97

inelastic collision. *See under* collisions.

inert complex. *See* coordination complex, labile versus inert.

inertia, 19, R1.2, 760, A138

inertial reference frame, 760, 763, R21.3, A138

infrared
  radiation, *110*, 111, 733–34, R20.2, R20.7–9, A138
  spectroscopy, 736, R20.14–19, A138

inhibitor, 671, *672*, A138

initial rate, 650, A138. *See also* kinetics, rate of reaction.

inorganic chemistry, A138

instantaneous rate, 648–49, A139. *See also* kinetics, rate of reaction.

insulator, 324, A139

intensity, wave, 110, 115–19, A139

intensive property, 7–8, 468, *469*, R13.12–13, R13.23, A139

interatomic potential, 394, A139

interface, 421, A139

interference, A139
  in quantum mechanics, 134–42, R4.4, 221, R7.2
  as wave phenomenon, 111–19, R4.3, R4.8–9

intermediate-field ligand, 708, R19.3, A139

intermediate species, 77, 674–75, R18.5, A139

intermolecular interactions. *See* intermolecular potential; noncovalent interactions.

intermolecular potential, 394–96, 399–400, 405–6, R11.2, A139
internal energy, in thermodynamics, 467–69, A139
International System of Units. *See* SI units.
inverse-square law, 25, R1.4, A140
iodic acid, relative strength, R16.7, A99*t*
iodine, selected reactions, R3.11–12, R12.14, R12.20
iodous acid, relative strength, R16.7
ion, 7, 56–57, R2.3, A140
  nomenclature, A8–10, A11*t*–13*t*, A18–21
  size relative to atom, 211, 213, R6.10
  structure, examples of, R2.6–7
ion–dipole interaction, 304–5, 326, R9.3, A140
ionic bond, 8–9, 56–57, R2.4, 239–40, *241*, R7.8–9, A140
ionic radius
  in gas phase, 211, 213, R6.10
  in solution, 550, R15.16
ionic solid (or crystal), 8–9, 57, 73, R2.7–8, 240–41, 324–26, R9.5, R9.14–16, A14, A140
ion–ion interaction, 303–4, R9.2–3, A140
ionization, A140
  electron transfer, 87
  photoionization, 738, R20.10–12
  proton transfer, 83–84, 571, R16.3–5
ionization energy (ionization potential), 89, R5.7–8, A68*t*–71*t*, A140
  periodic trends, 210–12, R6.5–6, R6.10–11
ion pairing, *417*, 418, R11.15, A140
ion-product constant, water. *See* autoionization constant.
irreversible process, 471–72, A140
isobar, 425–26, A140
isobutane, 269
isoelectronic structures, 57, 59, R2.6–7, A140
isomerism, 263–71, 276, 698–99, A10–11, A140
  in benzene, disubstituted forms, R8.8, 561, A11
  conformational, 266–68, R8.4, A117
  D-L convention, 286, 558–59, A10
  geometric, 276, R8.3, R8.8–9, R8.15, 694–96, R19.5–7, A10, A133
  linkage, 697, A143
  optical, 264–66, 286, R8.3, R8.9–10, 558–59, 698–99, R19.6–8, A10, A154

structural, 268–71, R8.2–3, R8.6–8, 561, A11, A174
isomerization reaction, 277, 295, R8.6, R8.14–15, 677
isopentane, 269
isotherm, 356–58, 359, 399–400, 405–7, 450, A141
isothermal compression or expansion, A141
  Boyle's law and, 355–60, R10.6–7, R10.12–13, 470–71, A112
  constant energy of, R10.12–13, 470–71
  heat and work during, 470–74
  as irreversible process, 471–72, A140
  phase transition and, 407–8, 424–25, A156
  as reversible process, 472–74, A165
isotonic solutions, 543, R15.11, A141
isotope, 7, 48, *49*, R2.2–3, A141, A77*t*–80*t*
  sample problems, R2.6–7, R21.5

J coupling (spin–spin coupling), R20.22, R20.23–24, A171
joule (unit), 26, R1.4, A141
Joule experiment, 465–66

kelvin (unit), 32, 364, A141
ketone, 278*t*, 283, A141
Kevlar, R9.20–21
kilo, prefix, A64*t*
kilocalorie, A141
kilogram, 20, A141
kilohertz, 107, A141
kilojoule, A141
kilometer, 106–7, A141
kinetic control, 669–71, A142
kinetic energy, 30–31, 39, R1.4, 462–64, A142
  of ideal gas, 375–76, R10.4, R10.12–13
kinetics (chemical kinetics), 643–84, R18.1–26, A142
  activation energy, A104
    in Arrhenius equation, 658, A108
    in collision theory, 663, A116
    qualitative aspects, 99, 280, 645–47, R18.2
    sample problems, R18.16–22
    in transition-state theory, 665–68, A177
    in unimolecular reaction, 676–78, A179
  Arrhenius equation (or law), 656–59, A108
    Boltzmann factor and, 658, 663–64, R18.3, A111

collision theory and, 661–65, R18.4–5, A116
   sample problems, R18.16–17, R18.20–22
   transition-state theory and, 665–68, A177
bimolecular reaction, 676, 678, R18.4, A110
collisions, activation by, 646–47, 659–61
collision theory, 661–65, R18.4–5, A116
concentrations, time course of, 653–56, *657*
contrasted with thermodynamics, 12, 77–79,
      R3.1–2, 443, 456, 518, R15.14, 643–45,
      669–73, 682–84, R18.1–2, R18.17–19,
      718–19, 781, R21.7
detailed balance, principle of, 682–84, R18.6,
      A122
elementary reaction, 674–78, R18.4, A127
general nature of, 12, 72, 77, R3.1–2, 645–47
kinetic control, 669–71, A142
mechanism of reaction, A146
   bimolecular, 678, A110
   elementary steps, 674–75, R18.4, A127
   pre-equilibrium, 680–82, R18.5–6, A159–60
   rate-determining step, 678–80, R18.5, A163
   sample problems, R18.22–24
   unimolecular, 676–78, A179
molecularity of reaction, 676, R18.4,
      R18.22–23, A149
order of reaction, 650–53, R18.3–4, A154
pre-equilibrium, 680–82, R18.5–6, A159–60
rate constant, 651, R18.4, A163
rate-determining step, 678–80, R18.5, A163
rate equation (or law), 651–53, R18.3–4,
      R18.8–9, A163
rate of reaction, 648–50, A163
   average rate, 648, *649*
   concentration dependence of, 650–53,
      R18.3–4, R18.6–8
   initial rate, 650, A138
   instantaneous rate, 648–49, A139
sample problems, R18.6–22
termolecular reaction, 676, R18.4, A176
transition state (activated complex), 99, *661*,
      662–63, R18.2, A177
   catalysis and, 100, *556*, 671–73, R18.2–3
   d electron configuration and, 718–19, R19.4
   enzyme active site and, *556*
transition-state theory, 665–68, A177
unimolecular reaction, 676–78, R18.4, A179
kinetic theory (of ideal gas), 369–91, R10.4–5,
      A142

Boyle's law, 370–74, R10.4, A112
energy, 375–76, 380–84, R10.4
Maxwell-Boltzmann distribution, 386–89,
      R10.5, R10.14–15, A146
pressure, 370–74
root-mean-square speed, 376–78, R10.4,
      R10.13–15, A165
statistical assumptions, 369–70, R10.3–4
temperature, 374–78, A176
krypton
   critical temperature, comparative, R11.8–9
   electron configuration, 198
   krypton-fluorine compounds, existence of, 202
   Lewis structure, 52

labile complex. *See* coordination complex, labile
      versus inert.
lanthanide contraction, 209–10, A142
lanthanide elements, 168, 199, 202, A142
laser, 433, A142
laser cooling, 383
lattice, 316–19, R9.4, A142
lattice energy or enthalpy, R9.15–16, 544,
      R15.12–13, A142
law. *See under specific law.*
LCAO. *See* linear combination of atomic orbitals.
lead chloride, as prototype of sparingly soluble salt,
      R15.5–9
lead storage battery, R17.6–8
Le Châtelier's principle, 454–58, R12.5, A142
   common-ion effect and, 537, R15.8–9, 596,
      599, A116
   composition, changes in, 457–58, R12.18–19
   pressure, changes in, 452–55, R12.20–21
   temperature, changes in, 456–57, R12.21–22,
      512–13, 517, R15.13–15
length, 15, A142
lepton number, conservation of, 777–78,
      R21.9–11, R21.20, A143
leptons, 777–79, 788, R21.12, A142
levorotatory enantiomer, 266, A143
Lewis acids and bases. *See under* acid; acid–base in-
      teractions; base.
Lewis structure (Lewis dot diagram), 55–56,
      R2.9–12, R2.13, A143
ligand, 81, 686, 693–94, A143
   ambidentate, 697, A105
   bidentate, 697–99, R19.2, A110

ligand (*continued*)
    chelating, 698–99, A114
    intermediate-field, 708, R19.3, A139
    monodentate, 693, R19.2, A150
    polydentate, 698–99, R19.2, A158
    strong-field, 708, R19.3, A174
    weak-field, 708, R19.3, A181
ligand field theory, 709, *710*, R19.4, A143
light. *See* electromagnetic radiation.
"like dissolves like" rule, 549, 552, 558,
    R15.18–19
limiting reactant, 69, R2.15–16, A143
linear combination of atomic orbitals (LCAO),
    227, R7.3, A143
linear equation, A48–49
linear geometry
    in coordination complexes, 689, R19.2
    in valence bond model, 246–47, 248, 252–53,
        274–76, R8.2
    in VSEPR model, *66*, R2.5, R2.12
line spectrum, 129–30, 168–72, A143
linkage isomerism, 697, A143
lipid, A143
lipid bilayer, 553–54, A143
liquid, 309–11, R9.4–5, R11.1–2, A143
liquid crystal, 334–36, *337*, R9.5, 430–31, A144
liter, *15*, 16, A144
lithium
    as diatomic molecule, 227–29, *234*
    effective nuclear charge, 204–5, 206–7
    electron configuration, 193
    Lewis structure, 51–52, *53*
    radius, compared with cation, R6.10
    as reducing agent, 632
lithium hydride, 91
local order, 310, 311, R9.4, 426, A144
lock-and-key model, *556*, 673
logarithm, A144
    common, A42–45
    estimation of, R16.8, R16.10, A45
    natural, A45–47, A151
    significant figures and, R16.8
London (dispersion) interaction, 306–8, R9.4,
    398, A144
    in molecular crystals, 326
    in nonpolar solvents, 548
lone pair, 59, R7.10, A144

long-range order, 311–12, 319–22, R9.4, 426,
    A144
low-density polyethylene, 347–49
low-spin complex, 705, *706*, R19.3, R19.10,
    R19.13–15, A144
luminescent reaction (luminescence), 72, A144
Lyman series, 170, *171*, R5.9, A144

macromolecule. *See* polymer.
macroscopic world, microscopic interpretation of,
    1–12, R1.1–2, 67, 126, 297–98, 319,
    R9.6, 351–91, R10.1–15, 428, 435–36,
    464–65, 489–90, 522–26, R14.1–6
macrostate. *See under* statistics and statistical me-
    chanics.
magic numbers, 772–73, R21.3, R21.7, A145
magnesium
    electron configuration, 194
    ionic radius, compared with sodium cation,
        R6.10
    Lewis structure, 52, *53*
magnesium chloride
    lattice energy, comparative, R15.12–13
magnesium oxide
    lattice energy, comparative, R9.15–16
magnetic field, 105, 109, A145
magnetic force, 22, 105, 744, A145
magnetic moment, dipole, *172, 174*, 729, R20.2,
    R20.24, A145
magnetic quantum number, 152, 154, 165, 175,
    R5.2–3, A145
magnetism, 172–73
    in crystal field theory, 704–8, R19.3
    paramagnet–ferromagnet transition, 429–30
    selected examples, R6.7–8, 233–36, R7.5–6,
        R19.9–13
    in valence bond theory, 713
main-group elements. *See* representative elements.
manometer, 355, A145
many-electron atom. *See under* quantum mechanics.
mass, 4, 19–20, 24, A145
    acceleration and, 4, 19, R1.3, 744, 747
    conservation of, 67, R2.1, R2.16, 633, 766,
        R21.10
    critical mass, in fission reaction, 784–85
    mass and energy, conservation of, 765–66,
        R21.10
    reduced mass, R20.18, R20.19

relativistic mass, 765–66

mass action, law of, 446, R12.3–4, A145–46

mass defect, nuclear binding and, 767–69, R21.3, A146

massless particle, 765

mass number, 7, 47, R2.2, A146

mass spectrometry, 744–47, R20.12–14, A146

matter, 4, 298–99, A146

  light and, 44–45, 124–30, R4.3, R4.5, 169–70, R5.4, 721–43, R20.1–2

  particulate nature of, 1, 43–46, 66–67, R2.1–2, 77–78, R4.2

  states of. *See* states of matter.

  wavelike nature of, 133–36, R4.4

  wave–particle duality, 137, R4.4–5, A181

Maxwell-Boltzmann distribution, 386–89, R10.5, R10.14–15, A146

mean free path, 378, A146

mean-square speed, 374, A146

measurement, 33–41, A146

  space and time, 15–16

mechanics, classical. *See* classical mechanics.

mechanics, quantum. *See* quantum mechanics.

mechanism of reaction. *See under* kinetics.

mega, prefix, A64*t*

megahertz, 107

melting, 409, 418–20, 425–28, A146

  enthalpy of, 506

  entropy of, 506

melting point, 7, A72*t*–75*t*, A146

  of cluster, 332

membrane, 542, 554, A147

mercury, clusters of, 332

meson, 788–89, A147

meta isomer, R8.8, 561, A11

metal, R2.3, A147

  transition metal, 54, R2.3, A177. *See also main entry* transition metals.

metallic solid (or crystal), 326–29, R9.5, A147

metalloid (semimetal), 54, 262, 329, A167

metastable state, A147

metathesis reaction, 95, A147

meter, 15, A147

methane

  boiling temperature, compared with water, R9.15

  entropy, compared with ethane, R14.9

formation and combustion, enthalpies of, 486–87

Lewis structure and oxidation state, R2.8–10, R3.6

as prototypical alkane, 262, 263–66

valence bond model, 246–50, R8.1

VSEPR model, R2.10

methanol (methyl alcohol)

  acidity of, compared with *tert*-butyl alcohol, 576

methyl chloride

  aqueous solubility, compared with carbon tetrachloride, R15.18

micelle, 552, 553, A147

micro, prefix, A64*t*

micrometer (micron), 107

microscopic reversibility. *See* detailed balance, principle of.

microscopic world. *See* macroscopic world, microscopic interpretation of.

microstate. *See under* statistics and statistical mechanics.

microwave

  oven, 733, R20.7

  radiation, *110*, 111, 732, R20.2, R20.7, A148

  spectroscopy, 733, A148

milli, prefix, A64*t*

Millikan experiment, 120, *121*

milliliter, *15*, 16, A148

millimeter, 107

million electron volt, R21.19

miscible liquids, A148

mixture, 299–300, A148

  of gases (Dalton's law), 368–69, R10.2, R10.11–12, R11.4

mobile phase, *748*, 749, A148

molal boiling-point-elevation constant, 418, R11.4–5, A148

molal freezing-point-depression constant, 418, R11.4–5, R11.14, A148

molality, 418, R11.12–14, A149

molar heat capacity. *See under* heat capacity.

molarity, R3.15–16, 301, R9.8–9, A149

molar mass, 68, R2.5, A149

  colligative properties and, R11.16, R15.11–12, A116

molar volume, R9.6–8, R13.15

  of ideal gas at STP, 367, R10.9, R10.10–11, A65*t*

mole, 68, R2.2, R2.5–6, A149
molecular crystal, 326, R9.5, A149
molecular formula, 73, R2.9, R2.12–13, A149
molecularity of reaction, 676, R18.4, R18.22–23, A149
molecular mass, A149. *See also* molar mass; relative mass.
molecular orbitals, 219–40, 253–59, R7.1–4, A149
  in allylic systems, R7.16–18
  bonding and antibonding, 223, 229–33, R7.2, A107, A112
  delocalized, 253–58, R7.3, R7.16–18, 272–76, A121
  in heteronuclear diatomic molecules, 236–39
  in homonuclear diatomic molecules, 219–36, R7.5–7
  in hydrogen, 221–23
  nonbonding, 235, R7.2, R7.18, A152
  overlap, *222*, 223, 229, 230, R7.3, A154
  pi, 230–33, R7.3–4, A157
  sigma, 223, 229–30, *232*, 233, R7.3, A168–69
molecular recognition, *557*, 558–63, 673, 722, A149
molecular sieve, 101, A150
molecular weight. *See* molar mass; molecular mass; relative mass.
molecules, 8, 46, 55–56, R2.4, A150
  heteronuclear diatomic, 60, *61*, 236–39, R7.7–9, A122
  homonuclear diatomic, 60, *61*, 219–36, A122
  polyatomic, *60–66*, R2.3, 242, 246, R20.15, R20.17, A158
  quantum mechanical description, 217–59, R7.1–4, 700–13, R19.2–4. *See also under* quantum mechanics.
  triatomic, 60–62, A178
mole fraction, 301, 369, R10.11–12, 414, A150
momentum, 20–21, R1.3, A150
  angular versus linear, 158, A106, A150
  conservation of, 21, R1.2, A118
    neutrino and, 777, A151
    selected examples, R1.10–11, 370–72, 726
    spatial symmetry and, 470
    in theory of relativity, 763–66
  force and, 20, R1.10–11, 370–72
  of photon, 127
  quantization of, 127, R4.3, R4.9

relativistic, 763–65
monatomic ion, R2.3, A150
monoclonal antibody, 673, A150
monodentate ligand, 693, R19.2, A150
monomer, 336, R9.5, 431, A150
monoprotic acid, A15, A150
monotonic process, 443, *444*, A150
most probable radius (one-electron atom), *160*, 161, A150–51
motion, 16–18, 18–21, R1.1–5. *See also* Newton's laws of motion.
  sample problems, R1.6–11, R1.14–15
MRI. *See* nuclear magnetic resonance imaging.
myoglobin, *118*, 342

NAD (nicotinamide adenine dinucleotide), R8.14
nano, prefix, A64t
nanocrystal, 332, R9.5, A151. *See also* cluster.
nanometer, 107, A151
natural gas, R2.8
natural logarithm, A45–47, A151
neon
  effective nuclear charge, 205
  electron affinity, lack of, 213
  electron configuration, 193
  inert nature of, 52, 202
  Lewis structure, 52, *53*
neopentane, 269, *270*, 398, *399*
Nernst equation, 623–24, R17.4, R17.16–18, A151
net ionic equation, 95, R3.14, A151
network crystal (network solid). *See* covalent crystal.
neutralization. *See under* acid–base interactions.
neutrino, 777–78, R21.9–10, A151
neutron, 6–7, 46, R2.2, A65t, A152
  beta decay, 776–79
  dimensions and scale, 770
  emission, 780, R21.4, R21.11, A152
  as fission product, 783–84
  quark structure, 789–90, R21.9
  spin, 174, 753, A171
Newman projection, 266–67, A152
newton (unit), 20, R1.4, A152
Newtonian mechanics. *See* classical mechanics; Newton's laws of motion.
Newton's laws of motion, 5–6, 18–19, R1.2–3, R1.7–8, 371–72, 760, A152. *See also* classical mechanics.

nickel–cadmium battery, 637, R17.8–9
nicotinamide adenine dinucleotide (NAD), R8.14
nitric acid, 85, 92, A16, A99*t*
nitric oxide
    dipole moment and ionic character, R7.7–9
    formal charge and oxidation state, R3.9–10
    Lewis structure, R2.12
    selected reactions, R3.12–13, R18.6–7
nitrogen
    atomic radius, compared with beryllium, R6.9
    as diatomic molecule, 58–59, 234–36
    effective nuclear charge, 205
    electron affinity, lack of, 214
    electron configuration, 193
    ionization energy, comparative, 212, R6.10–11
    liquid form, R9.14
    selected reactions, R3.12–13, 452–54,
        R12.20–21, R18.6–7
nitrosonium ion, Lewis structure, R2.12
nitrous acid
    ionization, R12.16–18
    nomenclature, A16, A17*t*
    as product of disproportionation reaction, 92
NMR. *See* nuclear magnetic resonance.
noble gas, A152
    in Aufbau sequence, 192, 193, 194, 198, 199
    stability of, 51, 52, R2.2, 201, 213
    periodic table, position in, 54, R2.3, 201–2
node, A152
    in hydrogenic orbitals
        angular node, 163–64, 166–68, R5.2,
            R5.6–7, R19.2, A106
        radial node, 157–58, *160*, 161, *162*, R5.6–7,
            A162
    in molecular orbitals, 223, 229–31, 257,
        R7.17–18
    in standing waves, 132, R4.4, 148, *149*
nomenclature
    acids, A15–17
    binary molecular compounds, A14–15
    coordination compounds, A18–21
    elements, A2–7
    inorganic species, A11–13
    ionic compounds, A14
    isomers, A10–11
    organic molecules, A21–27
    prefixes and suffixes, A7–11
nonbonding orbital, 235, R7.2, R7.18, A152

noncovalent (nonbonded) interactions, 9–10, 23,
        289, 301–8, R9.1–6, 393–*99*, 529,
        R15.1, A152–53
    dipole–dipole, 305, R9.3, 396, A123
    hydrogen bonding, 305–6, R9.3–4, A137
        in aqueous solution, 548, R15.18
        in DNA, 342–44
        in enzyme–substrate complex, 558
        in protein structure, 340–42, 554–55
        in water, 306, 326, *327*, R9.14, 426, *428*
    ion–dipole, 304–5, 326, R9.3, A140
    ion–ion, 303–4, R9.2–3, A140
    London (dispersion), 306–8, R9.4, 398, A144
        in molecular crystals, 326
        in nonpolar solvents, 548
    sample problems, R9.13–16, R11.8–9,
        R15.12–13
nonelectrolyte, R3.15, A153
nonequilibrium system, 540–41, R15.4, R15.6–8,
        R15.14. *See also* stressed equilibria.
    electrochemical cell, 621–24, R17.2–3
    reaction quotient and, 447–49, 452–53,
        R12.4–5, 520, R14.17–18, 622–24
    statistics of, 493–94, 525, R14.9–11
    steady state, R12.33
nonmetal, 54, R2.3, A153
nonpolar bond or molecule, 60, *61*, R2.4, *237*,
        238, A153
nuclear fission, 782–84, R21.5, A131
nuclear fusion, 48, 780–82, R21.5, A132
nuclear magnetic moment, 729, 740, R20.2,
        R20.24, A153
nuclear magnetic resonance (NMR), 730–32,
        R20.19–24, A153
    Boltzmann distribution and, R20.3–6
    chemical shift (chemical shielding), R20.22,
        R20.23, A115
    imaging, in medicine, 732
    internal magnetic field, 730–32, R20.20–22
    spin–spin (*J*) coupling, R20.22, R20.23–24,
        A171
nuclear magnetic resonance imaging (MRI), 732
nuclear shell model, R20.24, *771*–73, R21.2–3,
        A153
nuclear spin, 174, 729, R20.2, R20.3, R20.19,
        R20.24, A153
nucleic acid, 290, R8.5, 339–40, A153. *See also*
        DNA; RNA.

nucleon, 752, A153
quark structure of, 789–91, R21.2, R21.9,
R21.20
nucleophile, 280, A153
nucleophilic substitution, 280–83, A153, A169
kinetics of, 678–80
selected examples, R8.6, R8.12–13
nucleosynthesis, stellar, 48, 781–82, A153
nucleotide, 289–90, 340, A154
nucleus, 6–7, 46–50, R2.2, 751–93, R21.1–20,
A154
binding energy, 767–69, 782, R21.3, A110–11
sample problems, R21.5–7
data, A77*t*–79*t*, A80*t*
dimensions and scale, 47, 752–53, 769–71,
R21.1
electrostatic repulsion in, 23, R1.4, 50*n*,
752–53, 769, 773–74, 782, *784*, R21.1
fission, 782–84, R21.5, A131
fusion, 48, 780–82, R21.5, A132
isotopes, 7, 48, *49*, R2.2–3, A141
sample problems, R2.6–7, R21.5
radioactive decay, 753–54, 774–80, R21.8–18,
A162
Rutherford experiment, 120–23, 754–55
shell model, R20.24, *771*–73, R21.2–3, A153
spin, 174, 729, R20.2, R20.3, R20.19, R20.24,
A153
stability of, 774
strong force in, 23, R1.4, R1.14, 50*n*, 769–72,
788–92, R21.1, A174
transmutation of, 200, 785, A178
weak force in, 23, R1.4, 778–79, R21.4,
R21.10, A182
nuclide, 753, A154

observation (scientific method), 14
octahedral geometry, A154
in coordination complexes, 690, R19.2
in crystal field theory, 701–6, R19.3,
R19.9–10, R19.13, R19.16
isomerism in, 694–99, R19.5–8
in valence bond model, R7.12–14, 710–12,
R19.10–12
in VSEPR model, 65, *66*
octane
boiling temperature, compared with propane,
R9.14–15

conformational isomers, 267–68
octet, 52, 56, R2.3, 193, 195, R6.11, A154
octet-deficient species, 64, R7.10–12
one-electron atom. *See under* quantum mechanics.
optical isomerism, A154
chiral carbon, 264–66, R8.3, R8.9–10
coordination complexes, 698–99, R19.6–8
D-L convention, 286, 558–59, A10
orbital approximation, 178–81, R6.2
orbitals, A154. *See also* atomic orbitals; molecular
orbitals; quantum mechanics; quantum
numbers.
in many-electron atom
assumptions, 178–81, R6.2
Aufbau principle, 191, *201*, R6.3, R6.7,
A109
electron configurations, 192–200, R6.3,
R6.8–9, A126
relationship to one-electron atom, 179–81,
R6.2
in molecule
atomic orbitals, mixing of, 219–22, 225–27,
R7.3, A143
Aufbau principle, 223–29, 233–36, A109
heteronuclear diatomic, 236–39
homonuclear diatomic, 219–36, R7.5–7
in one-electron atom, 150–68, R5.1–4, 179–81
angular momentum, 151, 158, 163–64,
166–68, 173, R5.2, R5.4
d orbital, 166–68, R19.2, A124
energy levels, 153–56, R5.3–4, R5.7–9,
A127
f orbital, 168, A131
nodes, 157–64, 166–68, R5.2, R5.6–7,
A152
p orbital, 163–66, A159
quantum numbers, 150–52, *153*, A161
s orbital, 156–62, A170
order
long-range, 311–12, 319–22, R9.4, 426, A144
of reaction, 650–53, R18.3–4, A154
rotational, 335–36
short-range (local), 310, 311, R9.4, 426, A168
translational, 335–36
organic chemistry, 261–95, R8.1–15, A154
alkanes, 262, R8.2, A105
bonding and isomerism, 263–71, R8.2–3,
R8.3–4, R8.6–7

nomenclature, A21–24
reactions of, 276–77
alkenes, 263, R8.2, A105
  geometric isomerism, 276
  as Lewis bases, 277–78, R8.4
  nomenclature, A25–26
  structure and bonding, 271–74
alkynes, 263, R8.2, A105
  as Lewis bases, 277–78
  nomenclature, A25–26
  structure and bonding, 274–76
allylic systems, R7.16–18
amino acids, 287*t*, A106
  chiral (stereogenic) centers in, 286, R8.9–10
  DNA coding for, 291–92
  peptide formation (condensation), 288, R8.5
  structure, 286–87
aromatic molecules, 272–74, R8.2, A108
biopolymers, 285–92, 339–44, A111
carbocation, 281–83, R8.6, 677, 680, A113
carbon, uniqueness of, 261–62
conjugation, R7.18, 274–76, 284, A117
cycloalkanes, 270, R8.2, A120
DNA, 2–3, 289–92, 340, 342–44, A124
functional groups, 262, 276–85, 278*t*, R8.4–5, A132
  acid halide (acyl halide), 278*t*, R8.12
  alcohol, 278*t*, 279, R8.6, R8.11, R8.14, A26–27, A104
  aldehyde, 278*t*, 283, R8.6, R8.14, A105
  alkyl halide (halide), 278*t*, 280, R8.13, A105
  amide, 278*t*, R8.10, 340, A105
  amine, 278*t*, A105
  amino group, 285, A106
  anhydride, 278*t*, 285, R8.13, A106
  carbonyl group, 283, A113
  carboxylic acid, 278*t*, 284, R8.11, R8.12, R8.13, 576–77, A113
  ester, 278*t*, 285, R8.11, A26–27, A129
  ketone, 278*t*, 283, A141
  pi bond (alkene or alkyne), 277–78, A157
hydrocarbons, R2.8, 262–76, R8.2, A137
isomerism, 263–71, 276, A10–11, A140
  conformational, 266–68, R8.4, A117
  D-L convention, 286, 558–59, A10
  geometric, 276, R8.3, R8.8–9, R8.15, A10, A133

optical, 264–66, 286, R8.3, R8.9–10, 558–59, A10, A154
  structural, 268–71, R8.2–3, R8.6–8, 561, A11, A174
nomenclature, A21–27
nucleotide, 289–90, 340, A154
peptide bond, 287, R8.5, 340, A156
protein, 288, *289*, R8.5, 340–42, A161
reactions, *294*
  addition, *100–1*, 283, 293, R8.5, R8.13, 344, A104
  combustion, 92, 276–77, R8.14, 487, 517–18, A116
  condensation, 284–85, 288, 293, R8.5, R8.10–11, R8.13, 340, A117
  dehydrohalogenation, R8.13
  elimination, 293, R8.5, R8.13, A127
  hydrohalogenation, R8.13
  isomerization, 277, 295, R8.6, R8.14–15, 677
  rearrangement, 293–95, R8.6, R8.14–15, A164
  redox (reduction–oxidation), 295, R8.6, R8.14, A164
  substitution, nucleophilic, 280–83, 295, R8.6, R8.12–13, 678–80, A153, A169, A175
  RNA, 289–92, 340, A165
  sample problems, R8.6–15
ortho isomer, R8.8, 561, A11
oscillatory process, 443–45
osmosis, 542–43, A154
osmotic equilibrium, R15.4
osmotic pressure, 542–43, R15.4–5, R15.11–12, A154
overlap, *222*, 223, 229, 230, R7.3, A154
overvoltage, R17.21
oxidant. *See* oxidizing agent.
oxidation, 87, R3.4, 610–12, A154. *See also* redox reactions.
oxidation number (oxidation state), 90–92, R3.2–3, A155
  computation of, 90, R3.6–8, R17.6–8
  formal charge and, R3.8–10
  nomenclature, A8–10
  redox reactions and, 90, 92, R3.5, 611
  of transition metals in complexes, 691–92, R19.8, A76*t*
oxide, A155

oxidizer. *See* oxidizing agent.
oxidizing agent, 89, R3.4, R17.1–2, A155
    identification of, R3.12–14, R17.6–9
    relative strength, 610–12, 625–32, R17.9–11
oxyacid, R3.15, A155
oxyanion, A9, A15
oxygen
    as diatomic molecule
        bonding, 58, 235–36
        paramagnetism, 235–36
    ionization energy, comparative, 212, R6.10–11
    oxidation state of +2, as exceptional, R3.7
    peroxides and superoxides, R3.7, A156, A175
    radical, in smog reaction, 97
    selected reactions, 92, R3.12–13
ozone layer, depletion of, 98, R3.12–13

pair bond, 8, 56, R2.4. *See also* covalent bond.
para isomer, R8.8, 561, A11
paramagnetism, selected examples, A155
    in atoms, 192, 193, R6.7–8
    in coordination complexes, R19.9–13
    in crystal field theory, 704–8, R19.3
    in disorder–order transition, 429–30
    in molecules, 233–36, R7.5–6
    in valence bond theory, 713
partial pressure, A155
    Dalton's law, 368–69, R10.2, R10.11–12,
        R11.4
    in equilibrium mixture, R11.18–19, R12.20–21
particle, 16, 124, 127, R4.2, R4.4, A155
partition coefficient, 749, A155
pascal, 353, A155
Paschen series, 170, *171*
Pauli exclusion principle. *See* exclusion principle,
        Pauli.
p block, 201–2, R6.4, A155
penetration, 158, 182–86, 195–96, R6.2–3,
        A155
Penrose tiling, *321*
pentane, structural isomers, *269–70, 399*
peptide bond, 287, R8.5, 340, A156
percent yield, R2.16, A182
perchloric acid, R3.15, A16, A99*t*
period, A156
    row in periodic table, 52, R2.3, R6.4
    wave, 107
periodicity, A156

chemical (periodic law), 52–54, 194–95,
        200–15, R6.4–5, R6.8–11, A156
    of lattice, 314, 319–20, 332
periodic law and periodic properties. *See* periodic-
        ity, chemical.
periodic table, 52–54, R2.3, R6.4, A156
    organization into blocks, 200–2, R6.4–5
permanganate ion, R3.7–8, 631
permittivity constant, 24, A65*t*
peroxide, R3.7, A156
petroleum, 102, 270–71, 277, 300, A156
Pfund series, R5.15
pH, 580, R16.4, R16.7–8, A156
    electrochemical measurement of, 624
phase, A156
    condensed phase, 298
    of oscillation (wave), 107–8, R4.3
    on phase diagram, 421
    state of matter, 10, 299–301, R9.1
phase diagram, 420–28, A156
    sample problems, R11.16–19
phase equilibrium
    liquid–vapor, 11–12, 409–11, R11.9–12, 436
    on phase diagram, 421–22, R11.17–19, 450
    reversibility of, 505–6
    solute–solution, 413–20, 531–40, 543–46,
        R15.2–4, R15.5–10
phase transition, 10–11, 393, 407–9, R11.1–4,
        A156
    enthalpy change, 485, *486*, 505–6
    entropy change, 505–7, R14.18–19
    on phase diagram, 424–28, R11.17–19
phospholipid molecule, 553, A157
phosphoric acid, 86, A16, A99*t*
phosphorous acid, A16–17
phosphorus
    combustion of, 72–73
    electron configuration, 194
    phosphorus(V) oxide, empirical formula, 72–73
    phosphorus pentachloride, structure, R7.12–15
photoelectric effect, 124–29, R4.3, R4.10–11,
        738, A157
photoelectron, 125, R4.10–11, A157
photoelectron spectroscopy, 739–40, R20.10–12
photoionization, 738, A157
photolysis, 97, A157
photon, 124–29, R4.3–4, R20.1–2, A157
    as massless particle, 765

as messenger particle, 721–22, 786–88, 791, R21.2

selected examples, R4.9–11, 723–26

physical chemistry, A157

physical interactions. *See* noncovalent interactions.

pi bond, 230–31, R7.3–4, 271–76, A157

pico, prefix, A64*t*

picometer, 107

pion, 788, A157

pi orbital, 230–33, R7.3–4, A157

pK_a, 577, R16.4, A158

pK_b, 581, R16.4–5, A158

Planck's constant, 126, R4.3, 764, A65*t*, A158

plane-polarized light, 109, 266, A158

pOH, 581, R16.4–5, A158

polar bond or molecule, 60, *61*, R2.4, *237*, 238–39, 396, R20.2, A158

polarizability, electric, 308, 398

polarization, A158
    of wave, 109

polarized light (plane-polarized light), 109, 266, A158

polyatomic molecule, 60–*66*, R2.3, 242, 246, R20.15, R20.17, A158

polydentate ligand, 698–99, R19.2, A158

polyethylene, 344–49

polymer (macromolecule), 285–92, 336–49, R9.5, A158
    biopolymer, 285–92, 339–44, A111
    disorder–order analogy, 431
    synthetic, 344–49

polymerization
    addition reaction, 344, A104
    radical reaction, 97–98

polypeptide, 288, 340, A158

polyprotic acid, 86, R16.21–24, A15, A159

p orbital, 163–66, A159

porphine group, 700

positional disorder, 491–96, A159

positron (antielectron), 767, 778, A159
    emission, 779, R21.4, R21.10–14

potassium, and Aufbau sequence, 196

potassium bromide
    lattice energy, comparative, R9.15–16

potassium chloride
    lattice energy, comparative, R15.12–13

potassium sulfate, colligative properties, R11.14–16

potential, generalized concept, A159

chemical potential, 519, 527–28, 607–9, A115
    in electrochemical cell, 621–23
    nonuniform concentrations and, 540–43, R15.4
    thermal potential, 466–67

potential difference, electrical, A159. *See also* voltage.

potential energy, 28–31, R1.5, 462–64, A159
    electrical, 28–29, R1.14, 609–10, A125
    gravitational, 29–31, 39, R1.5–10, 463, A134

potential energy surface, 245–46, A159

power (exponent), A34–39

precipitation, 94–96, R3.5, 533, A159, A167
    selected examples, R3.14, R3.16, 437, R12.7
    selective precipitation, 537–40, 564, R15.9–10
    temperature and, R15.14

precision, 38, A159

pre-equilibrium, 680–82, R18.5–6, A159–60

pre-exponential factor, 658, 668, A160

preorganization, 556, 560, 565–67, R15.2, A160

pressure, 353–55, 370–74, R10.1–2, A160

primary structure (protein), 288, *289*, 340, A160

primitive cubic cell, R9.12–13, A160

principal quantum number, 151, 154–56, 165, R5.2, 207, A160

principle. *See under specific principle*.

probability amplitude, 141, R4.5, 156, 220, A160

probability density (or distribution), 141–42, 145, 147, 150, 157–61, *162*, 163–64, R5.1, A160

propane
    boiling temperature, compared with propanol, R9.14–15

protein, 288, *289*, R8.5, 340–42, A161
    denaturation, 432
    peptide bond, 287, R8.5, 340, A156
    polypeptide, 288, 340, A158
    primary structure, 288, *289*, 340, A160
    secondary structure, 288, *289*, 340, A167
    tertiary structure, 288, *289*, 340–41, A176

proton, 6–7, 46, R2.2, A65*t*, A161
    beta decay, 778–80
    dimensions and scale, 47, 752–53, 770
    emission, 780, R21.4, R21.12, A161
    quark structure, 789–90, R21.20
    spin, 174, A171

pseudoequilibrium, 666–67

quadratic equation, A49, A50–54, A161

quantum, 126, R4.4, A161

quantum mechanics, 104, 136–44, R4.5, 145–76, 177–215, A161. *See also* atomic orbitals; molecular orbitals; orbitals; quantum numbers.

chemistry and, 6, 51, 54–56, 77–79, 146, R6.3, R7.2, 301, R9.2, 722

evidence for

classical atom, instability of, 123, R4.1

Compton effect, R4.9–10, A117

electron diffraction, 134–35, A126

photoelectric effect, 124–29, R4.3, R4.10–11, 738, A157

photons, 124–29, R4.3–4, R20.1–2, A157

quantization of energy, 126–27, 129–30, R4.3, R20.1–2

quantization of momentum, 127, R4.3, R4.9

spectroscopy, 129–30, R4.4–5, 168–72, R5.4, A170

spin, 173–76, R5.4–5, 181–82, 767, A171

of many-electron atom, 177–215, R6.1–5

Aufbau (building-up) principle, 191, *201*, R6.3, R6.7, A109

electron configurations, 192–200, R6.3, R6.8–9, A126

electron–electron repulsion, 178–79, 186–91, 197, R6.1–2, R6.6

exclusion principle, Pauli, 181–82, R6.1–2, A155

orbital approximation, 178–81, R6.2

penetration and shielding, 158, 182–86, 195–96, R6.2–3, R6.4, R6.5–6, A155, A168

periodic properties, 194–95, 200–15, R6.4–5, R6.8–11

sample problems, R6.5–11

of molecule, 217–59, R7.1–4, 700–13, R19.2–4

Aufbau principle, 223–29, 233–36, A109

delocalization, 253–58, R7.3, R7.16–18, 272–76, A121

heteronuclear diatomic, 236–39

homonuclear diatomic, 219–36

hybridization, 247–53, 254–55, R7.9–15, A137

molecular orbitals, 219–40, 253–58, A149

sample problems, R7.5–18

valence bond theory, 246–53, 254, R7.3, A180. *See also main entry* valence bond theory.

of nucleus, R20.24, *771–73*, R21.2–3, A153

of one-electron atom (e.g., hydrogen), 148–72

angular momentum, 151, 158, 163–64, 166–68, 173, R5.2, R5.4, A106

energy levels, 153–56, R5.3–4, R5.7–9, A127

nodes, 157–64, 166–68, R5.2, R5.6–7, A152

orbitals, 150–68, R5.1–4, 179–81, A154

sample problems, R5.5–9

shells and subshells, 152, *153*, R5.2, R5.5–6, A168, A174

quantum numbers, 148–52, A161

relativistic formulation, 175–76, 767

theory and principles

correspondence principle, Bohr, 136, 147, A118

de Broglie wave, 133–36, 141, R4.4, R4.11, 147, A120

exclusion principle, Pauli, 181–82, R6.1–2, A155

interference, of particles, 134–42, R4.4, 221, R7.2

photons, 124–29, R4.3–4, R20.1–2, A157

probability, 141–42, R4.5, 145–46, 150, R5.1, A160

quantization, 124–30, R4.4–5, A161

Schrödinger equation, 147, R5.1, 178, R6.1–2, A167

uncertainty principle, Heisenberg, 6, 142–44, R4.5, R4.12–13, A179

wave function, 141, R4.5, 145, R5.1, 220, A181

wave–particle duality, 137, R4.4–5, A181

quantum numbers, A161

for many-electron atom

orbital approximation, 178–81, R6.2

principal quantum number, 207

for one-electron atom, 148–52

angular momentum (azimuthal), 151, 154, 163, 165, 174–75, R5.2, A106

magnetic, 152, 154, 165, 175, R5.2–3, A145

principal, 151, 154–56, 165, R5.2, A160

sample problems, R5.5–7

spin angular momentum, 174–75, R5.5, A171

spin magnetic, 174–75, R5.5, A171

for vibrating string, 132, 148, *149*

quark, 789–92, R21.2, R21.9, R21.20, A162

quasicrystal, 319–22, A162

radar, 142

radial node. *See under* node.

radial probability distribution, 159–61, 185–86, A162

radical (free radical), 96–97, 277, A131, A162

chain reaction, 97, A114

in EPR, 733, R20.24

polymerization reaction, 97–98

production, kinetics of, R18.24–26

recombination reaction, 96–97, R3.6

radioactive decay, 753–54, 774–80, R21.8–18, A162

alpha particle (alpha radiation), 754, R21.4, A105

absorption, R21.11

emission, 780, R21.7–8

in Rutherford experiment, 120–23, 754–55

becquerel, unit of, R21.4, A110

beta particle (beta radiation), 754, R21.4, A110

modes of decay, 776–79, R21.7–11

electron capture, 779–80, R21.4, R21.9, A126

compared with positron emission, 779–80, R21.12–14

fallout, 783

first-order kinetics of, 774–75, R21.4

gamma radiation, *110*, 111, 754, R21.4, A133

from nuclear excited state, 780

as product of annihilation, 767, R21.18–20

half-life, 774–75, R21.4, R21.14–18, A134

neutron emission, 780, R21.4, R21.11, A152

positron emission, 779, R21.4, R21.10–14, A159

proton emission, 780, R21.4, R21.12, A161

radioactive dating, 776, R21.16–18

radioactive series, R21.7–8, R21.14–16

sample problems, R21.8–18

radioactive series, R21.7–8, R21.14–16

radiocarbon dating, 776, A162

radiofrequency radiation. *See* radio waves.

radiotracer, 776, A163

radio waves (radiofrequency radiation), *110*, 111, 727–28, R20.2, R20.3–6, A162

radius. *See specific kind, such as* atomic radius.

random error, 38, A163

Raoult's law, 415, R11.4, A163

rare-earth elements. *See* lanthanide elements.

rate constant, 651, R18.4, A163

rate-determining step, 678–80, R18.5, A163

rate equation (or law), 651–53, R18.3–4, R18.8–9, A163

rate of reaction. *See under* kinetics.

reaction mechanism. *See* kinetics, mechanism of reaction.

reaction order, 650–53, R18.3–4, A154

reaction quotient, 446–49, 452–53, R12.4–5, 520, A164

reaction rate. *See* kinetics, rate of reaction.

reactions, organic. *See under* organic chemistry.

real gas, 393

condensation, 364, 407–9, 412, 413–14, R11.1–4, R11.8, A117

critical temperature, 405–6, 412, R11.3, R11.8–9, A83*t*, A119

intermolecular potential, 394–96, 400, *401*, 405–6, A139

isotherms, 405–7, A141

van der Waals equation, 402–4, R11.2–3, R11.5–8, A83*t*, A180

rearrangement reaction, 293–95, R8.6, R8.14–15, A164

redox reactions (reduction and oxidation), 86–92, R3.4, 610–12, A164

acid–base reactions, analogy to, 92, R3.5, 571, 611–12, R17.1–2, R17.4–5, R17.24–25

balancing, R3.10–12, 632–37, R17.6–9

conjugate relationships, 611–12, A117

half-reactions, 624–32, R17.4–5, A134

organic, 295, R8.6, R8.14

oxidation number and, 90, 92, R3.5, 611

reductants and oxidants, relative strength, 610–12, 625–32, R17.9–11

reduced mass, R20.18, R20.19

reducer. *See* reducing agent.

reducing agent, 89, R3.4, R17.1–2, A164

identification of, R3.12–14, R17.6–9

relative strength, 610–12, 625–32, R17.9–11

reductant. *See* reducing agent.

reduction, 88, R3.4, 610–12, A164. *See also* redox reactions.

relative mass

atomic, 67–68, R2.5, A164

relative mass (*continued*)
  molecular, 67–68, R2.5, A164
relative motion, perception of, 755–59
relativity, theory of (special), 755–67, R21.3, A164–65
  antimatter and, 766–67, A107
  energy, relativistic, 763–66
  inertial reference frame, 760, 763, R21.3, A138
  momentum, relativistic, 763–65
  principle of, 761, 763
  quantum mechanics and, 175–76, 767
  rest energy and rest mass, 765–66, R21.3, R21.18–20, A165
  simultaneity, perception of, 761–63
  space-time coordinates, 761–63, 765, A170
  speed of light, constancy of, 757–59, 761, 763, R21.3
representative elements (main-group elements), R2.3, 202, R19.2, A165
resonance, A165
  in chemical bonding, 256–57, 273
  in oscillating system, 735
rest energy and rest mass, 765–66, R21.3, R21.18–20, A165
reversible process, 472–74, A165
RNA (ribonucleic acid), 289–92, 340, A165
root-mean-square speed, 376–78, R10.4, R10.13–15, A165
rotational motion, *218*, 733, R20.7, A166
rotational order, 335–36
rotational symmetry, 314–16, 319–22, A166
rubber, R9.21–22
rusting, of iron, 92
Rutherford experiment, 120–23, 754–55
Rydberg constant, 155, R5.4, A65*t*, A166

salt. *See under* acid–base interactions.
salt bridge, 620–21, R17.2
saturated hydrocarbon, 262, R8.4, A166
saturated solution, 531, R15.3, R15.5–6, A166
s block, 200, R6.4, A166
scandium, and Aufbau sequence, 196–97
scanning tunneling microscope (STM), 43–44, A166
Schrödinger equation, 147, R5.1, A167
  approximations of, 178, R6.1–2
  chemistry and, 301, R9.2
  selected applications, 254, 772

scientific method, 14, A167
scientific notation, A40–42
scintillation counter, 776, A167
screening. *See* shielding.
second, 16, A167
secondary structure (protein), 288, *289*, 340, A167
second law of thermodynamics, 507–9, R14.1–3, A167
second-order reaction, 651, R18.4, A167
  concentration versus time, 656, *657*, R18.13–16
  half-life, 656, R18.13–16
selective binding, 558–63, 564–67
selective precipitation, 537–40, 564, R15.9–10, A167
semiconductor, 330, A167
semimetal (metalloid), 54, 262, 329–30, A167
semipermeable membrane, 542, R15.4, A168
shape selectivity, 561–63, A168
shell, A168
  electronic
    in Lewis model, 51–54, R2.3
    in quantum mechanics, 152, *153*, R5.2, R5.5–6, 192–200, R6.4–5
  nuclear, R20.24, *771*–73, R21.2–3
shielding (screening), 182–86, 195–96, R6.2–3, R6.4, R6.5–6, A168
short-range order, 310, 311, R9.4, 426, A168
sigma bond, 223, 229, R7.3, 271–76, A168
sigma orbital, 223, 229–30, *232*, 233, R7.3, A168–69
significant figures, 38, 39–41, A169
  logarithms and, R16.8
silicon
  atomic radius, compared with carbon, R6.9–10
  electron configuration, 194
  selected reactions, R12.6–7
silver chloride
  as prototype of insoluble salt, 535–40
  as prototypical precipitate, 94–96
  solubility, compared with diamond, R15.19
simultaneity, perception of, 761–63
single bond, 55–56, R2.4, 243, 247, 248, A169
SI units, 16, A63*t*–64*t*, A168
smog, 97
S_N1 reaction, 282–83, 679–80, A169
S_N2 reaction, 281–82, 678, A169
sodium
  atom and cation, magnetic properties, R6.7–8

electron configuration, 194
ionic radius, comparative, R6.10
Lewis structure, 52, *53*
sodium acetate
   basic hydrolysis of, R16.14—16
   in prototypical buffer solution, 594—600,
      R16.13—14
sodium chloride
   colligative properties, R11.14—16, R15.11
   crystal structure and density, R9.11—12
   enthalpy of solution, R15.15—16
   lattice energy, comparative, R9.15—16,
      R15.12—13
   oxidation number and formal charge, R3.10
   as prototypical ionic compound, 8—9, 56—57,
      324—26
   as prototypical neutral salt, 84—85, 585, R16.14
   as prototypical solute and strong electrolyte,
      93—94
   strength of ionic interactions in, 324—26, R9.15
sodium hydride, as Lewis base, R16.7
sodium hydroxide, as prototypical Arrhenius base,
   84
solid. *See under* states of matter.
solubility, 535, 546—50, R15.5, R15.7,
      R15.18—19, A94*t*, A95*t*, A169
   as basis for chromatography, 749
   complex formation and, 715—16
   electrochemical measurement of, 624
   hydrophilic and hydrophobic effects, 550—55,
      560, A137
   immiscibility, R15.18, A138
   structure and, 546—50, R15.18—19
   temperature and, R15.13—16
solubility-product constant, R12.6, 535—37,
      R15.3, R15.5—6, A96*t*—97*t*, A169
solute, 93, *94*, R3.14—16, 300—1, 529, 530—31,
      R15.1—3, A169
solute—solvent complex, 93, *94*, 95, A169
solution, 93—96, R3.14—16, 300—1, 530—55,
      R15.1—19, A169
   of acid or base, 83—86, 577—605, R16.3—24
   concentration of, R3.15—16, 301, R9.8—9,
      A117
   disorder—order equilibrium in, 531—35, R15.2—4
   electrolytic, 93
   enthalpy and entropy of, 543—46, R15.1—2,
      R15.13—16, A128

free energy of, 545—46, R15.1—2, R15.16—17
ideal solution, theoretical model of, 412—20,
   R11.4, A137
molarity of, R3.15—16, 301, R9.8—9, A149
saturation of, 531, R15.3, R15.5—6, A166
stoichiometry in, R3.15—16
supersaturation of, R15.14, A175
temperature and solubility, R15.13—16
solvation, 86, 530, 714, A170
solvent, 93, *94*, 300—1, 529, 530—31, R15.2—3,
   A170
s orbital, 156—62, A170
sound waves, propagation of, 759
space-time, 761—63, 765, A170
special relativity. *See* relativity, theory of.
specific activity, R21.18
specific heat capacity, 477, R13.19, A170
spectator ion, 95, A170
spectrochemical series, 708, R19.3, R19.9—10,
   R19.16, A170
spectrograph, 168, *169*
spectroscopy, 721—43, R20.1—28, A170
   absorption, 169—70, R5.4, 724—26, R20.2,
      R20.9, A103
   atomic, 129—30, R4.4—5, 168—72, R5.4
   Auger, *739*
   electron paramagnetic resonance (EPR),
      732—33, R20.24, A126
   emission, 168—72, R5.4, R5.9, 724, 726,
      R20.2, A127
   infrared, 736, R20.14—19, A138
   microwave, 733, A148
   nuclear magnetic resonance (NMR), 730—32,
      R20.3—6, R20.19—24, A153
   sample problems, R20.2—24
   visible, 737—38
   X-ray fluorescence, *739*
   X-ray photoelectron, *739*, R20.10—12
spectrum, 129, 168—72, A170. *See also* spectroscopy.
speed, 17—18, A170
   of gas particles
      Maxwell-Boltzmann distribution, 386—89,
         R10.5, R10.14—15, A146
      root-mean-square, 376—78, R10.4,
         R10.13—15, A165
   of light, 107, R4.2—3, 727, A65*t*
      constancy of, 757—59, 761, 763, R21.3

speed (*continued*)
  of light (*continued*)
    as limiting value, 764–65
    time and distance, examples of, R4.7–8
  of wave, 107, R4.2
spherical symmetry, 153, 157, 165–66, 168, 218, 239, 701–2, A171
sp hybrid orbital, 252–53, R7.3, A171
sp^2 hybrid orbital, 251–52, 254–55, R7.3, R7.10–12, R17.16–17, A171
sp^3 hybrid orbital, 248–51, R7.3, R7.9–10, R19.15, A171
sp^3*d*^{2} hybrid orbital, R19.10–11, A171
spin (spin angular momentum), A171
  of electron, 173–76, R5.4–5, 181–82, 767, A126
  of nucleus, 174, 729, R20.2, R20.3, R20.19, R20.24, A153
  pairing energy, 187–91, R6.6
spin angular momentum quantum number, 174–75, R5.5, A171
spin magnetic quantum number, 174–75, R5.5, A171
spin–spin (*J*) coupling, R20.22, R20.23–24, A171
spontaneity, 507, A171
  free energy and, 509–13, R14.4–6, R14.15–17, 528. *For specific examples, see also* equilibrium constant, free energy and.
  of nuclear reaction, 780, R21.12–14
  second law of thermodynamics and, 507–9, A167
  system enthalpy and entropy, effect on, 512–13, 516–18, R14.15–16, 545–46, 548–50, 564–67
  temperature and, 512–13, 516–18, R14.15–16
spring, restoring force, R20.17–18
square planar geometry, A172
  cis-trans isomerism in, 695–96
  in coordination complexes, 690, R19.2
  in crystal field theory, R19.13–15
  in valence bond model, 712, R19.4
square pyramidal geometry, 690, R19.2, A172
stable state, 439, A172
staggered conformation, 267, *268*, R8.4, A172
standard cell potential, 623–24, R17.4, A172
  equilibrium constant and, R17.15–16
  free energy and, 623–24, R17.4, R17.11–15
  spontaneity and, R17.14–15

standard conditions, 479, R13.5, A172
standard electrode potential, 625–30
standard enthalpy of formation, 479–80, A85*t*–92*t*, A128, A172
standard enthalpy of reaction, 480–83, R13.5–6, A128, A172
standard entropy, 514–16, R14.5, A85*t*–92*t*, A172
standard free energy of formation, 513–14, R14.5, A85*t*–92*t*, A172
standard free energy of reaction, 516–18, A85*t*–92*t*, A172
standard hydrogen electrode, 625–30, R17.5, A172
standard oxidation potential, 625–30, R17.5, A173
standard reduction potential, 625–30, R17.5, A100*t*–1*t*, A173
standard state, A72*t*–75*t*, A173
standard temperature and pressure (STP), 367, A173
standard thermodynamic properties, A85*t*–92*t*
standing waves, 130–33, R4.4, 148, *149*, A173
state function, 467–68, 480–81, R13.4–12, 513–14, 625, A173
states of matter, 10–11, 298, 308–12, R9.1–6, A173
  gas, 10, 308–9, R9.4–8, R11.1, A133
    ideal, 351–69, R10.1–3, R10.6–12, 399–400, *401*, A137
    kinetic theory of, 369–91, R10.3–5, R10.12–15, A142
    real, 393, 400–8, R11.2–3, R11.5–9
  liquid, 309–11, R9.4–5, R11.1–2, A143
  nontraditional states
    cluster, 330–34, R9.5, A116
    liquid crystal, 334–36, *337*, R9.5, 430–31, A144
    polymer, 285–92, 336–49, R9.5, 431, A158
  solid, 310–12, R9.4–5, A169
    amorphous (glass), 311, *313*, 338, 345–47, R9.4–5, A106, A133
    covalent (network), 322–24, R9.5, A119
    crystalline, 311–12, R9.4, A119
    ionic, 8–9, 57, 73, R2.7–8, 240–41, 324–26, R9.5, R9.14–16, A14, A140
    metallic, 326–29, R9.5, A147
    molecular, 326, R9.5, A149
    quasicrystal, 319–22, A162

state variables, 353, R10.8, 467–68, 477–78, R13.6–15, A173
stationary phase, 748–49, A173
stationary state, 151, 153, 155, 169, A173
statistics and statistical mechanics, A173
    assumptions, 78–79, R3.1, 369–70, R10.3–4, 489–90
    distributions, 490–501, R14.6–8, A123
    entropy, 501–5, A128
    equilibrium, as most likely macrostate, 493–94, 507, 522–26
    heat, mechanical conception of, 464
    kinetic theory, 370–80, 384–91, R10.4–5, A142
    large systems, 369–70, 493, 524–25, R14.9–11
    macrostates and microstates, 441, 489–94, 497–501, R14.1–3, R14.6–8, A144–45, A148
    thermodynamics, relation to, 467
steady state, nonequilibrium, R12.33, A173
stereogenic atom, 265–66, R8.9–10, A174
steric factor, 662, 664, 668, R18.4–5, A174
Stern-Gerlach experiment, *174*
STM. *See* scanning tunneling microscope.
stoichiometry, 66–73, R2.2–5, A174
    sample problems, R2.5–16
    of solutions, R3.15–16
STP. *See* standard temperature and pressure.
stressed equilibria, 452–58
    common-ion effect, 537, R15.8–9, 596, 599, A116
    composition, changes in, 457–58, R12.18–19
    Le Châtelier's principle, 454–58, R12.5, A142
    pressure, changes in, 452–55, R12.20–21
    sample problems, R12.18–22
    temperature, changes in, 456–57, R12.21–22, 512–13, 517, R15.13–15
strong acids and bases. *See* acid–base interactions, strength and weakness of.
strong electrolyte, 93, A174
strong-field ligand, 708, R19.3, A174
strong (nuclear) force, A174
    magnitude and range, 769–72
    meson theory of, 788–89
    qualitative description of, 23, R1.4, R1.14, 50*n*, R21.1
    quark theory of, 791–92, R21.2
structural isomerism, 268–71, R8.2–3, R8.6–8, 561, A11, A174

sublimation, 425, R11.17–19, A174
subshell, 152, *153*, R5.2, R5.5–6, A174
substance, 299, A175
substitution reaction, *294*, 295, A175
    electrophilic, 295
    nucleophilic, 280–83, 295, R8.6, R8.12–13, 678–80, A153, A169, A175
substrate, binding to enzyme. *See under* enzyme.
sucrose
    colligative properties, R11.14–16, R15.11–12
    as prototypical molecular solute, 530, R15.18
sulfur
    electron affinity, compared with chlorine, R6.11
    ionization energy, compared with oxygen, R6.10–11
    sulfur dioxide, redox reaction, R3.12
    sulfur hexafluoride, bonding in, R7.12–14
    sulfur tetrafluoride, bonding in, R2.10–12
sulfuric acid, 85, R16.21, A99*t*
    as diprotic acid, 86, R16.7, R16.23–24
    nomenclature, A16
sulfurous acid, A16, A99*t*
sun
    distance from earth, R4.7–8
    mass-energy of, 781
superconductor, 433, A175
supercritical fission reaction, 784–85
supercritical fluid, 423, R11.3, A175
supernova, 782, A175
superoxide, R3.7, A175
supersaturated solution, R15.14, A175
surface tension, 771–72, A175
surroundings, in thermodynamics, 467, *468*, R13.3, A175
symmetry, 182, 189, 222–23, 236, 256, 265, 312–22, R9.4, 697, A175
    conservation and, 469–70, 763
    cylindrical, 218, 225, *237*, 239
    lattice, 316–19
    rotational, 314–16, 319–22, A166
    spherical, 153, 157, 165–66, 168, 218, 239, 701–2, A171
    translational, 313–14, 319–22, R9.9–11, A177
system, in thermodynamics, 467, *468*, R13.3, A175
systematic error, 38, 39, A175

television waves, 111
temperature, 32, 362–63, A176

temperature (*continued*)

absolute (Kelvin scale), 363–64, A103

Boltzmann distribution and, 389–91, 522–26, R20.3–6, A111

Celsius scale, 32, A114

critical temperature, 405–6, 412, A119

energy and, 374–75

Fahrenheit scale, 32, A130

solubility and, R15.13–15

speed and, 376–78, 384–89

as thermal potential, 466–67

volume and, 363–65, R10.2, R10.7–8

termolecular reaction, 676, R18.4, A176

tert-butyl alcohol, poor acidity of, 576

tertiary structure (protein), 288, *289*, 340–41, A176

tetrahedron and tetrahedral geometry, 263–64, A176

in coordination complexes, 690, R19.2

in crystal field theory, 707, *708*, R19.13, R19.15

optical isomerism in, 264–66, R8.3, R8.9–10

in valence bond model, 246–50, R7.9–10, R8.1

in VSEPR model, 62, 65, *66*

theoretical yield, R2.16, A182

theory (scientific method), 14, A176

thermal energy, 380–84, R10.4, 464, R13.7, A176

thermal equilibrium, R12.3, 466, 522–26, 540, R15.4, 741–43, A176

thermal potential, 466–67

thermochemistry, 480–87, R13.17–22, A176

thermodynamic control, 669–71, A176

thermodynamics, 461–87, R13.1–22, 489–526, R14.1–19, A177. *See also related main entries, such as* enthalpy; entropy; extensive property; free energy.

calorimetry, 476–78, *479*, R13.17–20, A113

contrasted with kinetics, 12, 77–79, R3.1–2, 443, 456, 518, R15.14, 643–45, 669–73, 682–84, R18.1–2, R18.17–19, 718–19, 781, R21.7

enthalpy, 475–80, A127–28

entropy, 474–75, 501–7, R14.3–4, A128–29

extensive and intensive properties, 7–8, 468, *469*, R13.12–13, R13.21, R13.23, A130, A139

first law, 468–69, *470*, R13.3–5, R13.7–12, A130–31

free energy, Gibbs, 509–14, 516–22, R14.4–6, A131, A133

general nature of, 12, 72, 76–77, 79, R3.2, 527–29, 572–74, 749

Hess's law, 481–83, R13.5–6, R13.20–22, 544, A135

internal energy, 467–69, A139

second law, 507–9, R14.1–3, A167

state functions, 467–68, 480–81, R13.4–12, 513–14, A173

statistical (microscopic) interpretation, 489–505, 522–26

system and surroundings, 467, *468*, R13.3, A175

thermochemistry, 480–87, R13.17–22, A176

thermodynamic control, 669–71, A176

third law, 514–15, R14.3, 737, A177

of transition state, 665–68

work and heat, 461–80

thermometer, 37, 362–63, A177

thermonuclear bomb, 23, 781

third law of thermodynamics, 514–15, R14.3, 737, A177

third-order reaction, 651, R18.8–9

Thomson experiment, 120, *121*

threonine, stereogenic centers, R8.9–10

thymine, 289–92, 343

titration, 601–5, R16.16–21, A177

torr, 353, A177

trans isomer, 276, R8.3, 696, R19.5, A10, A177

transition metals (transition elements), 54, R2.3, 168, 202, 686, R19.2, A177

in complexes, 81, 685–719, R19.1–16

electron configurations of, 197–98, R6.9, A76*t*

oxidation states of, A76*t*

transition state. *See under* kinetics.

transition-state theory, 665–68, A177

translational motion, *218*, 743, R20.2, A177

translational order, 335–36

translational symmetry, 313–14, 319–22, R9.9–11, A177

transmutation, nuclear, 200, 785, A178

transuranic elements, 785, A178

transverse wave, 109

triatomic molecule, 60–62, R2.3, A178

triclinic unit cell, 317, A178

trifluoroacetic acid, R8.12, A99t
trigonal bipyramidal geometry, A178
   in coordination complexes, 690, R19.2
   in valence bond model, R7.12–15
   in VSEPR model, 65, *66*, R2.11–12
trigonal planar geometry, A178
   cis-trans isomerism in, 276, R8.3, R8.14–15
   in coordination complexes, 689, R19.2
   in valence bond model, 246–47, 251–52,
      271–74, *275*, R8.2
   in VSEPR model, 63–64, *66*, R7.10–12
trigonal pyramidal geometry, A178
   in coordination complexes, 689, R19.2
   in VSEPR model, 64, *66*
trimer, 406, A178
tripeptide, 288
triple bond, 58, R2.4, 235, 243, 253, 271,
    274–76, A178
triple point, *420*, 422, R11.3, R11.17–19, A178
triprotic acid, 86, R16.21, A178
tritium, 48, A178
trough, 106, R4.2, R4.3, A178

ultraviolet radiation, *110*, 111, 738, R20.2,
    R20.9–10, A178
uncertainty, instrumental, 37, A179
uncertainty (indeterminacy) principle, Heisenberg,
    6, 142–44, R4.5, A179
   chemistry and, 146, 301
   sample problems, R4.12–13
   zero-point energy and, 737
unimolecular reaction, 676–78, R18.4, A179
unit cell, 316–19, R9.4, A179
   cubic, 316–19, R9.11–13
unit-factor method, 35–36
units, 34–36, A63t–64t, A179
   conversion of, 35–36, 72, A64t
   SI, 16, A63t–64t
universal gas constant, 366, R10.2, A65t, A179
universe, in thermodynamics, 467, *468*, A179
unsaturated hydrocarbon, 263, R8.4, A179
unsaturated solution, 538–40, R15.14, A179
unstable state, 439, A179
uracil, 289

valence bond theory, A180
   bonds-in-molecules, assumption of, 246–48,
     254, R7.3, 709–13, R19.4, 736

hybridization schemes
   dsp^2, 712, A124
   dsp^3, R7.12–13, R7.15, A124
   d^2sp^3, R7.12–15, 711–12, A124
   sp, 252–53, A171
   sp^2, 251–52, A171
   sp^3, 248–51, A171
   sp^3d^2, R19.10–11, A171
selected examples
   coordination complexes, R19.10–12,
     R19.15–16
   hydrocarbons, 254–55, R7.16–17, 271–76
   inorganic molecules, R7.9–15
valence electrons (valence shell), 52, *53*, R2.3, 111,
    196–98, 737–38, R20.2, R20.9, A180
   configurations, periodicity of, 52–54, 194–95,
     200–15, R6.4–5, R6.8–11
   in metals, 327–28
valence-shell electron-pair repulsion model
    (VSEPR), 62–*66*, A180
   assumptions, 62, R2.5, 246
   molecular shapes
     bent, 62–*63*, R7.9
     linear, *66*, R2.12
     octahedral, 65, *66*, R7.13–14
     seesaw, R2.10–12
     tetrahedral, 65, *66*, 250, 263
     trigonal bipyramidal, 65, *66*, R7.15
     trigonal planar, 63–64, *66*
     trigonal pyramidal, 64–65
van der Waals equation, 402–4, R11.2–3, A83t,
    A180
   sample problems, R11.5–8
van der Waals forces, A180
van't Hoff factor, A180
vapor, 410, A180
vaporization, 450–51, 485, *486*, A180
vapor pressure, A180
   as equilibrium constant, 410–11, 449–52
   lowering of, 413–16, R11.4, A180
   of pure liquid, 410–12, R11.3
   sample problems, R11.9–14
   of solution, 412–18, R11.4
vector, A180
velocity, 17–18, A180
vibration, modes of, A180
   bend, R20.17
   frequencies, R20.16t

vibration, modes of (*continued*)
  isotope effect, R20.19
  quantum mechanical, 218, 219, 734–37, R20.2, R20.9, R20.15–19
  spring, analogy to, 735, R20.17–18
  standing wave, 130–33, R4.4, 148, *149*, A173
  stretch, R20.15
  zero-point, 737, A182
visible
  radiation, *110*, 111, 705–7, R19.16, 725, 737–38, R20.2, R20.9–10, A181
  spectroscopy, 737–38
vitamins, various, R15.21, *686*
volt, R1.14, R17.3
voltage, 29, R1.14, 462, 609–10, A181
  of redox reaction, 614–16, R17.3–4, R17.11–14
voltaic cell. *See* galvanic cell.
volume, 15–16, A181
VSEPR. *See* valence-shell electron-pair repulsion model.
vulcanized rubber, R9.21–22

water
  as acid and base, 550, 577–84, R16.3, 713
  autoionization, 581–83, R16.3, R16.5, R16.14, A109
  formation, 69–71
  freezing, expansion upon, 426–28
  ice, hydrogen bonding in, 326, *327*, R9.14, 426, *428*
  as Lewis base, 81, 694, 718
  physical properties, A84*t*
  as prototypical covalent molecule, 8, 60–63
  solvent, role as, 550–55, 713–14
  valence bond model, R7.9–10
  VSEPR model, 60–63
wave, A181
  diffraction and interference, 111–19, A122, A139
  electromagnetic, 105–10, R4.2, 727–40, R20.1–2, A126
  general characteristics, 105–10, R4.2–3
  sample problems, R4.5–11
  standing, 130–33, R4.4, 148, *149*, A173
  transverse, 109
  wave equation, 146–47, A181
wave equation, A181
  general, 146–47, A181

Schrödinger, 147, R5.1, 178, R6.1–2, A167
wave function, 141, R4.5, 145, R5.1, 220, A181
wavelength, 106, R4.2, A181
  de Broglie, 133–36, 141, R4.4, R4.11, 147, 765, A120–21
wavenumber, R20.7–8, A181
wave–particle duality, 137, R4.4–5, A181
weak acids and bases. *See* acid–base interactions, strength and weakness of.
weak electrolyte, 93, A181
weak-field ligand, 708, R19.3, A181
weak (nuclear) force, 23, R1.4, 778–79, R21.4, R21.10, A182
weight, 4, 24, A182
work, 4, 25–32, R1.4, 461–62, R13.1–3, A182
  electrical, 26–28, R1.14, 462
    from chemical energy, 609, 612–16, R17.3–4
    forced, in electrolytic cell, R17.20–21
  free energy and, 607–8, 612–16, R17.3–4
  gravitational, 30, R1.5, 462
  heat and, 464–66
  path dependence of, 471–74, *475*, R13.10–12
  $P\Delta V$ (pressure × volume), 366, 471–74, 475–76, 477, *478*, R13.2–3, R13.7–10, R13.15–17
work function, 125, A182
$W^{\pm}$ particles, 789, 791

X ray, A182
  diffraction, 45, *46*, 118–19, R4.9, A182
  fluorescence, *739*
  photoelectron spectroscopy, *739*, R20.10–12
  radiation, 45, *110*, 111, 127, 738–40, R20.9–10
xylene, isomers of, 561–63

yield, R2.16, A182

zeolites, 101, *102*, 561–63, 673
zero-point energy, 737, A182
zinc
  electron configuration, 197–98
  selected reactions, R3.13
  zinc ion
    as Lewis acid, R3.13
    as reducing agent, R3.14
zinc–copper cell, as prototype, 612–23, 638–42
Z^0 particle, 789, 791, A182
zwitterion, 286, A182

SI Base Quantities		Symbol
length	meter	m
mass	kilogram	kg
time	second	s
amount of substance	mole	mol
temperature	kelvin	K
electric current	ampere	A
luminous intensity	candela	cd

SI Derived Units

SI Derived Quantity	Unit	Symbol	Dimensions
acceleration	meter per second per second	—	$m\ s^{-2}$
Celsius temperature	degree Celsius	°C	K
electric charge	coulomb	C	A s
electric field	volt per meter	—	$V\ m^{-1}$ or $N\ C^{-1}$
electric potential	volt	V	$J\ C^{-1}$
force	newton	N	$kg\ m\ s^{-2}$
frequency	hertz	Hz	s^{-1}
momentum (impulse)	newton second	—	$kg\ m\ s^{-1}$
power	watt	W	$J\ s^{-1}$
pressure	pascal	Pa	$N\ m^{-2}$
radioactivity	becquerel	Bq	s^{-1}
speed or velocity	meter per second	—	$m\ s^{-1}$
work, energy, or heat	joule (newton meter)	J	$kg\ m^2\ s^{-2}$

SI Prefixes

Prefix	Symbol	Multiplier	Prefix	Symbol	Multiplier
deci	d	10^{-1}	deka	da	10^{1}
centi	c	10^{-2}	hecto	h	10^{2}
milli	m	10^{-3}	kilo	k	10^{3}
micro	μ	10^{-6}	mega	M	10^{6}
nano	n	10^{-9}	giga	G	10^{9}
pico	p	10^{-12}	tera	T	10^{12}
femto	f	10^{-15}	peta	P	10^{15}
atto	a	10^{-18}	exa	E	10^{18}